Plant Parasitic Nematodes in Subtropical and Tropical Agriculture

2nd Edition

Edited by

Michel Luc

IRD, Paris, France

Richard A. Sikora

University of Bonn, Bonn, Germany

and

John Bridge

CABI Bioscience, Egham, UK

CABI Publishing

CABI is a trading name of CAB International

CABI Head Office
Nosworthy Way
Wallingford
Oxfordshire OX10 8DE
UK

Tel: +44 (0)1491 832111
Fax: +44 (0)1491 833508
Email: cabi@cabi.org
Web site: www.cabi.org

CABI North American Office
875 Massachusetts Avenue
7th Floor
Cambridge, MA 02139
USA

Tel: +1 617 395 4056
Fax: +1 617 354 6875
Email: cabi-nao@cabi.org

A catalogue record for this book is available from the British Library, London, UK

Library of Congress Cataloging-in-Publication data
Plant parasitic nematodes in subtropical and tropical agriculture / edited by Michel Luc, Richard A. Sikora, John Bridge.—2nd ed.
p.cm.
Includes bibliographical references and index.
ISBN 0-85199-727-9 (alk. paper)
1. Plant nematodes—Tropics. I. Luc, Michel. II. Sikora, Richard A. III. Bridge, John. IV. Title.

SB998.N4P582 2005
632.6′257′0913—dc22

2004024550

ISBN-13: 978-0-85199-727-8
ISBN-10: 0-85199-727-9

First published 2005
Reprinted 2010

Printed and bound in the UK by MPG Books Ltd, Bodmin

Contents

The colour plate section can be found following p. 10

Contributors

Walter J. Apt, Department of Plant and Environmental Protection Sciences, University of Hawaii, 3190 Maile Way, Honolulu, HI 96822, USA

Kenneth R. Barker, Department of Plant Pathology, North Carolina State University, Raleigh, NC 27695-7616, USA; E-mail: krbpp@unity.ncsu.edu

John Bridge, CABI Bioscience, Bakeham Lane, Egham, Surrey TW20 9TY, UK; E-mail: J.Bridge@cabi.org

Patrice Cadet, Institut de Recherche pour le Développement (IRD), 213 Rue La Fayette, 75480 Paris, Cedex 10, France. Present address: South African Sugar Association Experiment Station, Private Bag X02, Mount Edgecombe 4300, South Africa; E-mail: cadet@sugar.org.za

Vicente P. Campos, Departamento de Fitopatologia, Universidade Federal de Lavras, Caixa Postal 37, 37200-000 Lavras – MG, Brazil; E-mail: nema@ufla.br; dfp@ufla.br; vpcampos@ufla.br

Rui Gomes Carneiro, IAPAR – Instituto Agronômico do Paraná, Rodovia Celso Garcia Cid, km 375, Caixa Postal 481, 86001-970 Londrina, PR, Brazil; E-mail: rucar@iapar.br

E.P. Caswell-Chen, Department of Nematology, University of California, Davis, CA 95616, USA; E-mail: epcaswell@ucdavis.edu

Danny L. Coyne, International Institute of Tropical Agriculture (IITA), Ibadan, Nigeria, c\o Lambourn & Co., Carolyn House, 26 Dingwall Road, Croydon CR9 3EE, UK; E-mail: d.coyne@cgiar.org

Dirk De Waele, Katholieke Universiteit Leuven, Laboratory of Tropical Crop Improvement, Kasteelpark Arenberg 13, 3001 Leuven, Belgium; E-mail: dirk.dewaele@agr.kuleuven.ac.be

Don W. Dickson, Entomology and Nematology Department, Building 970, Natural Area Drive, PO Box 110620, University of Florida, Gainesville, FL 32611-0620, USA; E-mail: dwd@ufl.edu

Larry W. Duncan, University of Florida, Citrus Research and Education Center, 700 Experiment Station Road, Lake Alfred, FL 33850, USA; E-mail: lwdn@lal.ufl.edu

S.J. Eapen, Division of Crop Protection, Indian Institute of Spices Research, Calicut-673 012, India

Fahiem E. El-Borai, Plant Protection Department, Faculty of Agriculture, El Zagazig University, Egypt

Emilio Fernández, Instituto de Investigaciones de Sanidad Vegetal, Calle 110 #516, Entre BY5F Playa Miramar, Havana, Cuba; E-mail: efernandez@inisav.cu

Roger Fogain, CARBAP (Centre Africain de Recherche sur Bananiers et Plantains), PO Box 832, Douala, Cameroon; E-mail: rfogain@yahoo.fr

Robin M. Giblin-Davis, Professor and Associate Center Director, Fort Lauderdale Research and Education Center, University of Florida/IFAS, 3205 College Avenue, Davie, FL 33314, USA; E-mail: giblin@ufl.edu

Nalini C. Gnanapragasam, 78/3 Temple Road, Hatton, Sri Lanka; E-mail: nalinic@sltnet.lk

Simon R. Gowen, School of Agriculture, Policy and Development, University of Reading, Reading RG6 2AT, UK; E-mail: aasgowen@reading.ac.uk

Nicola Greco, CNR, Istituto per la Protezione delle Piante, Sezione di Bari, Via G. Amendola, 165/A, 70126 Bari, Italy; E-mail: nemang02@ area.ba.cnr.it

Reginald Griffith, Coconut Research, Ministry of Food Production, Marine Exploitation, Central Experiment Station, Centeno, Via Arima PO, Trinidad, West Indies; E-mail: reginaldg@tstt.net.tt

Johannes Hallmann, Federal Biological Research Center for Agriculture and Forestry, Nematologie, Toppheideweg 88, D-48161 Münster, Germany; E-mail: j.hallmann@bba.de

David J. Hooper, 23 Lentune Way, Lymington, Hampshire SO41 3PE, UK

David J. Hunt, CABI Bioscience, Bakeham Lane, Egham, Surrey TW20 9TY, UK; E-mail: D.hunt@cabi.org

Charles S. Johnson, Southern Piedmont AREC, Virginia Polytechnic Institute and State University, 2375 Darvills Road, Blackstone, VA 23824, USA; E-mail: spcdis@vt.edu

P.K. Koshy, Central Plantation Crops Research Institute, Regional Station, Krishnapuram-690533, Kayangulam, Kerala, India; E-mail: cpcrikgm@vsnl.com

Charles K. Kwoseh, Department of Crop Science, University of Science and Technology, Kumasi, Ghana; E-mail: ckwoseh@hotmail.com

Michel Luc, Ex Institute de Recherche pour le Dévelopment (IRD), 6 rue Boutard, 92200 Neuilly-sur-Seine, Paris, France

Rosa H. Manzanilla-López, Plant Nematode Interactions Unit, Rothamsted Research, Harpenden, Herts AL5 2JQ, UK; E-mail: rosa.manzanilla-lopez@bbsrc.ac.uk

Peter Marczok, Bayer Crop Science, Research Insecticides, Agriculture Centre, D-40789 Monheim, Germany

Alexander H. McDonald, ARC-Grain Crops Institute, Private Bag X1251, Potchefstroom 2520, Republic of South Africa; E-mail: alex@igg2.agric.za

Keerthi M. Mohotti, Tea Research Institute of Sri Lanka, Talawakelle 22100, Sri Lanka; E-mail: mohottik@yahoo.com

Julie M. Nicol, International Wheat and Maize Improvement Center (CIMMYT), Wheat Program, PK 39, Emek, 06511 Ankara, Turkey; E-mail: jnicol@cgiar.org

Björn Niere, Biologische Bundesanstalt für Land- und Forstwirtschaft, Toppheideweg 88, D-48161 Münster, Germany

Rakesh Pandey, Central Institute of Medicinal and Aromatic Plants (CIMAP-CSIR), PO CIMAP, Lucknow (U.P.)-226 015, India; E-mail: rakeshpandey66@hotmail.com; pandey66@yahoo.com; nematolcimap@rediffmail.com

Deliang Peng, Institute of Plant Protection of Chinese Academy of Agricultural Sciences, Beijing 100094, China; E-mail: dlpeng@ippcaas.cn

Richard A. Plowright, 29 Huntstile, Goathurst, Bridgwater, Somerset TA5 2DQ, UK; E-mail: RPlowright@SoilAssociation.org

Patrick Quénéhervé, Pôle de Recherche Agronomique de la Martinique (PRAM), Laboratoire de Nématologie Tropicale, IRD, BP 8006, 97259 Fort-de-France, Martinique; E-mail: queneherve@ird-mq.fr

Onaur Ruano, IAPAR – Instituto Agronômico do Paraná, Rodovia Celso Garcia Cid, km 375, Caixa Postal 481, 86001-970 Londrina, PR, Brazil; E-mail: ruano@pr.gov.br; onaur.ruano@uol.com.br

Jean-Louis Sarah, CIRAD/UMR BGPI, TA 41/K, 34398 Montpellier, Cedex 5, France; E-mail: sarah@cirad.fr

Maria L. Scurrah, Department of Nematology and Entomology, International Potato Centre, PO Box 5969, Lima, Peru; E-mail: m.scurrah@cgiar.org

Richard A. Sikora, University of Bonn, Soil Ecosystem Phytopathology and Nematology, Nussallee 9, D-53115 Bonn, Germany; E-mail: rsikora@uni-bonn.de; rsikora@gmx.de

João Flávo Veloso Silva, Research – Embrapa Soybean, Phytopathology and Nematology, PO Box 231, 86001-970, Londrina, PR, Brazil; E-mail: veloso@cnpso.embrapa.br

Brent S. Sipes, University of Hawaii, Plant and Environmental Protection Sciences, 3190 Maile Way, Honolulu, HI 96822, USA; E-mail: sipes@hawaii.edu

V.K. Sosamma, Central Plantation Crops Research Institute, Regional Station, Krishnapuram-690533, Kayangulam, Kerala, India; E-mail: sosammavk@hotmail.com

Vaughan W. Spaull, South African Sugarcane Research Institute, Private Bag X02, Mount Edgecombe 4300, South Africa; E-mail: spaull@sugar.org.za

James L. Starr, Department of Plant Pathology & Microbiology, Texas A&M University, College Station, TX 77843-2132, USA; E-mail: j-starr@tamu.edu

Sergei Subbotin, Institute of Parasitology of the Russian Academy of Sciences, Leninskii Prospect 33, Moscow 117071, Russia

Luc Villain, CIRAD-CP, Boulevard de la Lironde, TA 800/PS3, 34398 Montpellier, Cedex 5, France; E-mail: luc.villain@cirad.fr; lvillain@anacafe.org

Jennifer Way, Tobacco Research Board, PO Box 1909, Harare, Zimbabwe; E-mail: teganza@zol.co.zw

About the Editors

Michel Luc Michel Luc has spent his career at ORSTOM (now IRD or Institut de Recherche pour le Développement): first as a plant pathologist in ORSTOM Centre of Adiopodoumé, Côte d'Ivoire. He then turned to nematology and established, at the same place, the first nematology laboratory in West Africa. He conducted extensive nematode surveys in Côte d'Ivoire and other countries, such as Madagascar, where little or no information was available on nematodes. His 18 year career in the Côte d'Ivoire ended with him being Director of the Centre, the most important of ORSTOM, for 6 years. He then established a nematology laboratory in Dakar, Senegal, devoted to subsahelian areas, where he worked for a 5 year period. In both these laboratories, he developed teams of researchers and technicians and promoted research programmes. From 1975 he was based in the Paris Muséum working on taxonomy of plant parasitic nematodes. He was the founder and inspiration behind the *Revue de Nématologie* (renamed *Fundamental and Applied Nematology* and now *Nematology* after fusion with *Nematologica*). He is Doctor *honoris causa* of the University of Neuchâtel, Switzerland and Honorary Member of the Society of Nematologists (USA). Notwithstanding his official retirement in 1992, he is still active in nematology, namely as a member of the Editorial Board of *Nematology*.

Richard A. Sikora Richard Sikora has headed Nematology and Soil-Ecosytem Phytopathology at the Institut für Pflanzenkrankheiten of the University of Bonn, Germany since 1971. He received his BSc and MSc degrees in zoology and botany at Eastern Illinois University in 1966 and 1967, where he specialized in field ecology, bacteriology and helminth physiology. His research with helminths led to the development of a bioassay using Tubificid worms as indicators of heavy metal pollution in freshwater streams. In 1967, he began research on complex disease inter-relationships at the Department of Plant Pathology at the University of Illinois in Urbana, completing his PhD in 1970. This was followed by a year at the G.B. Pant Agricultural University in India where he was introducted to biological control and the use of organic amendments for nematode management. He has worked extensively in tropical and subtropical countries of Africa, the Middle East, India, South and South-east Asia, the Pacific and Central and South America. Most of his work was conducted for the German Gesellschaft für Technische Zusammenarbeit (GTZ), but also USAID, FAO and for a number of CGIAR and associated international research centres. His experience covers problems associated with food legumes, vegetable crops and, more recently, banana and plantain, where his research interests include complex disease inter-relationships, integrated nematode management and biological control. Richard has helped to train 65 PhD and 55 MSc students in his capacity as University Professor, with students representing 21 countries of the tropics and subtropics. He has published over 200 research papers, three books and 30 book chapters, as well as co-edited the proceeding of four meetings dealing with nematology and multitrophic interactions in the rhizosphere. He was made a Fellow of the Society of Nematology, received the Van den Brande Award for Science from the University of Ghent, Beligum and was recently honoured with the Award of Merit by the University of Illinois for his contributions to international agricultural research and education. He was also granted the German Industry Award for his work with nematicides

John Bridge John Bridge graduated in Botany from the University of Hull, UK and, in 1966, after lecturing in biology at a teacher's training college, he took an MSc in Plant Pathology at McGill University, Canada where he was introduced to nematodes. He returned to the UK to begin a PhD in nematology at Imperial College, London University, which he completed in 1970, and was then recruited by the UK Overseas Development Administration (ODA) to be their Tropical Plant Nematology Scientific Liaison Officer. He has worked continuously on tropical nematode problems since that date. His first assignment as Tropical Nematologist was a 2 year appointment to the University of Ibadan, Nigeria, after which he was based in the UK taking on numerous advisory visits to many different countries each year providing advice on the management of nematodes and conducting nematode surveys in most of the countries that he visited. Initially based at Imperial College where he was a lecturer on the MSc Nematology course and a supervisor of both MSc and PhD nematology students, he moved to Rothamsted Experimental

Station in 1979 where he continued supervision of students. In 1983, he joined CAB International as their Tropical Plant Nematology Advisor and established a Tropical Plant Nematology Unit at what became the CABI International Institute of Parasitology (IIP) in St Albans, UK. He took on the role of Deputy Director until IIP was amalgamated with the other CABI Institutes to become CABI Bioscience based at Egham, UK, in 1998, where he has continued as their Tropical Plant Nematology Advisor until the present.

His work on nematodes of a very wide range of tropical crops has taken him to most continents and many countries in the tropical world in Africa, South and Central America, the Caribbean, South Asia, South-east Asia, the Middle East and the Pacific.

Foreword

Published in 1990, the first edition of *Plant Parasitic Nematodes in Subtropical and Tropical Agriculture* is now out of print. The continuous demand for the book led the Editors and CABI Publishing to consider a second edition, and thus this present work was produced.

It was decided not to simply reprint the first edition but to completely update and revise the book. It is largely based on the first edition, but some changes have occurred. We have deliberately brought in many new authors – reflecting the turnover among subtropical and tropical nematologists. The number of authors has increased from 32 to 48, with the majority of chapters being written by more than two authors, again with a very wide span of experience and working environments.

However, the book remains conceived as a truly practical book for use by agriculturists, researchers, teachers, students, extension workers and also administrators. This new edition again covers the major, economically important crops of the subtropics and tropics and their main nematode parasites. The aim was not simply to produce an encyclopaedia of nematode associations with crops but to concentrate on those nematode species that have been shown to cause yield loss.

Although in this second edition the arrangement of each chapter remains broadly the same, the text has been completely updated and revised taking into consideration the new observations, records and results published since 1990. New figures have also been introduced and there are new colour photographs and an increase in the numbers of colour plates. Some of the chapters are modified from the original. The 'Methods' chapter now has a section on molecular techniques. The 'Root and Tuber Crops' chapter has been split into two separate chapters, 'Solanum and Sweet Potatoes' and 'Tropical Root and Tuber Crops' because of the great increase in the literature on nematodes of these crops and to introduce new authors. Also the 'Coffee, Cocoa and Tea' chapter has been split into 'Coffee and Cocoa' and 'Tea' chapters, again to reflect the different types of cultivation and climatic demands of the crops and to bring in new authors. New crops such as 'Medicinal Plants' have been introduced in the chapter that includes 'Spices'. The last chapter, Chapter 22, is entirely new and deals with 'An Overview of Integrated Nematode Management Technologies' and replaces the chapter 'Effects of Tropical Climates on the Distribution and Host–Parasite Relationship of Plant Parasitic Nematodes' that can be found in the first edition. This change provides the reader with more practical data concerning the various elements for efficient management of plant parasitic nematodes – a management rendered more and more difficult due to the reduction in the availability of nematicides.

We are extremely grateful for the full cooperation given by the authors who now know the amount of work that goes into a publication of this nature. The multi-author format was again used and authors were chosen on the basis of their practical expertise, research work and their understanding of different regions of the world, as well as their experience with different crops and different types of agriculture. Twenty different nationalities are represented.

Conceived in this way, we hope that this new edition will again be a truly useful and practical book for anyone dealing with plant parasitic nematodes and working in subtropical and tropical agriculture. We wish you success in your work to improve crop yields.

The Editors

Preface to 1st Edition

The science of plant nematology developed dramatically from 1950 to the present day. Progress was founded, in part, on the availability of excellent texts on plant parasitic nematodes. This text, focusing on those nematodes affecting crop plants grown in tropical and subtropical regions of the world, is the first volume addressing tropical nematology to be published in more than 20 years.

Drs Richard A. Sikora, Michel Luc and John Bridge conceived the idea for this book at the 1986 ESN meeting in Antibes, France, and the proposal gained further momentum when Peter Gooch of CAB International offered his support for publication. At the first editorial meeting in Bonn, Germany, 12–14 January 1987, the overall goals, chapter outlines and general style of the book were formulated. Additional editorial meetings were held in Paris and St Albans and a workshop for authors of the chapters was conducted in August, 1988, at the German Physic Centre in Bad Honnef.

A unique feature of this treatise is the collaboration of two or more authors in the writing of each chapter. The authors, deliberately chosen from different geographic areas, were selected on the basis of their having worked, often for many years, on particular crop/nematode combinations, for their hands-on experience, and for their understanding of the interactions among hosts, parasites, and the environment. This approach brings diversity, experience and knowledge to the discussions of each major crop and its associated nematode pests.

A noteworthy aspect of this volume is that the authors have taken into account the various ecological differences between the tropical and temperate regions of the world and have shown how and why different approaches to nematode management are necessary. Although losses due to nematodes can be great in almost any region of the world, they are especially severe in the tropical and subtropical regions which comprise most of the developing world and where severe shortages of food and fibre are prevalent.

Tropical and subtropical agriculture differs from that of temperate regions and growers must consider the many ecological differences when they decide on approaches to nematode management. Environmental factors affecting nematode development, reproduction, survival and ability to suppress crop production include temperature, rainfall, soil types, patterns of wet and dry seasons, local vegetation and sometimes the absence of distinct seasons in the tropics.

In the tropical and subtropical regions there are more weed hosts for many nematode species. In general, tropical and subtropical soils have lower organic matter and nutrient levels. There usually are more botanical plants per unit area in the tropics than in temperate regions and cultural practices vary greatly. The target nematode genera and species will also vary, although several important genera are common to both tropical and temperate regions.

In this volume, the authors have delineated those nematode problems which have the greatest economic impact on the particular crops grown in the tropical and subtropical regions. With this information, knowledgeable administrators can facilitate allocation of their available resources to the development and employment of management tactics most appropriate for those nematodes which are judged to be most serious.

The opening chapters constitute a theoretical and practical initiation to nematology. These chapters on morphology, methods, and techniques for determining the impact of nematodes on crop growth are augmented by indexes and a section of high quality colour plates showing symptoms of damage. Altogether they comprise an invaluable handbook which can be used even by scientists with little practical experience of nematodes.

The editors, authors and publisher are to be commended for producing this valuable and timely volume on nematode problems in the tropics. They are providing an authoritative resource book for agriculturists and all plant nematologists, especially for those working in tropical regions, where sustainable agriculture is the goal. While there are many constraints to economic production of food and fibre crops in most developing countries, this volume will greatly enhance the ability of scientists whose responsibility it is to minimize the damage caused by plant nematodes.

J.N. Sasser
Professor Emeritus
Department of Plant Pathology
North Carolina State University
Raleigh, NC 27695-7616, USA

Acknowledgements

We are pleased to acknowledge the financial support given to us in the preparation of this book by the following organizations:

Bayer Crop Science AG, Leverkusen, Germany

Syngenta Crop Protection, Stein, Switzerland

BASF AG, Limburgerhof, Germany

Technical Centre for Agricultural and Rural Cooperation (CTA), Wageningen, The Netherlands

We also very much wish to thank the many nematologists, both past and present, who we have had the good fortune to meet and exchange views and experiences with that have greatly helped in producing this second edition.

Sincere thanks goes out to all the scientists, students and farmers that we have had the pleasure to work with and to visit in many different countries around the world, who have provided us with much of our information and insights into the importance of tropical nematodes and their management.

Many thanks also go to the staff of CAB International, especially Tim Hardwick, Jenny Dunhill and Tracy Ehrlich, for their support and patience during the long and complex process of compiling the final version.

Dedication

This book is dedicated to our long-suffering wives who have supported us throughout our careers in nematology and without whom we could not have completed our work.

Mariette, Ingrid and Monica

1 Reflections on Nematology in Subtropical and Tropical Agriculture

Michel Luc,[1] John Bridge[2] and Richard A. Sikora[3]

[1] *Ex Nematologist ORSTOM, 6 rue Boutard, 92200 Neuilly-sur-Seine, Paris, France;* [2] *Tropical Plant Nematology Advisor, CABI Bioscience, Bakeham Lane, Egham, Surrey TW20 9TY, UK;* [3] *Soil-Ecosystem Phytopathology and Nematology, Institut für Pflanzenkrankheiten der Rhein, Friedrich-Wilhelms-Universität, Nussallee 9, D-53115 Bonn, Germany*

If the birth of nematology in temperate areas can be dated to 1743 with the observations by Needham of the wheat seed gall nematode or 'ear cockle eelworm', nematology in the tropics was initiated at a much later date.

The first tropical nematodes were described from Oceania during the late 19th and early 20th centuries. Cobb (1891) reported finding nearly 30 species in banana soil and plant tissues from Fiji; among them, he described (Cobb, 1893) several new species, such as *Radopholus similis* and *Helicotylenchus multicinctus*, now well known, even though their names have changed from the original descriptions. Species now known as *Meloidogyne javanica* and *Hirschmanniella oryzae* were identified at an early date from Java, Indonesia, by Treub (1885) and by van Breda de Haan (1902), respectively. Few records are available for this period from other parts of the tropics, a notable exception being the description of the genus *Meloidogyne* and its type species *M. exigua* on coffee trees in Brazil by Göldi (1889, 1892); following an earlier report from Jobert (1880), he made an extensive study of the nematode problem in coffee plantations.

In the following four or five decades, nearly all descriptions of tropical nematode species were done in laboratories in temperate countries, particularly in the USA by Cobb, Steiner and Thorne, in England by T. Goodey and J.B. Goodey and in The Netherlands by Schuurmans Stekhoven. Observations and experiments based on field work were rare in countries outside the temperate regions until the 1950s. Two other exceptions were first, the study of red ring disease of coconuts in the Caribbean by Nowell (1919, 1920) who established that a nematode was the cause of the disease and instigated further work in the area; and, secondly, some outstanding field work by Butler (1913, 1919) in East Bengal (Bangladesh) who identified 'ufra disease' of rice and described its causal organism, *Ditylenchus angustus*. One other finding in the early part of the 20th century, which was to have a profound effect on nematology, was the discovery in 1935 of a serious nematode parasite in the pineapple fields of Hawaii, later to be described by Linford and Oliveira (1940) as *Rotylenchulus reniformis*. This led, in the early 1940s, to the discovery of the first effective nematicidal soil fumigant, D-D (1,2-dichloropropane, 1,3-dichloropropene), from work done at the Pineapple Research Institute, Hawaii. Notwithstanding these and other evident successes, the amount of nematological work in the tropics was very meagre in the first half of the last century. For example, when the first nematology laboratory was established in West Africa (by ORSTOM in the Côte d'Ivoire) in 1955, there were only nine published references relating to plant parasitic nematodes found in the whole of West Africa and Zaire.

With the strong support of the nematologists working in the UK, a thrust was made to develop the field of nematology in many of the Commonwealth countries during the 20th century. The first laboratories were established in India and Kenya, with a great deal of our initial information on nematodes of the tropics and subtropics gained in these countries.

Nematology laboratories have now been established in many, but by no means all, subtropical and tropical countries, especially in Africa, South America and India. Up to 1983, approximately 278 scientists working on nematodes in the tropics were recorded (Thomason *et al.*, 1983) not including those in India or Pakistan, nor those in the semi-arid regions. We would estimate that there are now at least 400 scientists working and/or teaching full- or part-time on the nematode problems and in the areas to which the present book is devoted. Most editions of all the nematological journals now contain a number of articles dealing with nematodes or nematological problems from outside the temperate regions, and some journals such as *Nematologia Mediterranea*, *Nematropica*, *Indian Journal of Nematology*, *International Journal of Nematology*, *Nematologia Brasileira* and *Pakistan Journal of Nematology* deal almost exclusively with such work.

Nematology laboratories established in the second half of the last century in the tropical regions had to look afresh at nematode problems. Often they needed to determine initially which problems existed by basic survey work, and accurately identify which nematodes were present (determination, systematics), followed by establishing which nematodes are harmful or economically important by pathogenicity tests and field trials, and finally deciding on which treatments or methods are appropriate for management of the nematodes. It has been, and continues to be, a long and difficult task and, if many problems are now rather well known, few of them have been fully solved. This is not surprising when we consider that over a large part of the past century, approximately 100 nematologists worked in temperate countries on the problems caused by the potato, soybean and sugarbeet cyst nematodes; satisfactory results were only attained towards the end of the century, with the bias on plant resistance and integrated control.

It is, therefore, safe to predict that the future for subtropical and tropical nematology will be long and full of complex and economically important problems especially with regards to subsistence agriculture. Of utmost importance to nematology in the future will be access to centres with competence in systematics. Due to the present trend of down-sizing in all fields of agricultural research, and thereby the loss of many diagnostic laboratories, qualified taxonomic identification will be a problem in many countries. This will be important especially in quarantine where decisions on nematodes detected in samples, in particular species and race designations, need to be made almost spontaneously. In the future, it may be necessary to develop 'virtual-centres of excellence' in diagnostics for use by nematologists working in the tropics to support nematology in the field of species identification. We have been referring to nematology in 'temperate' compared with 'subtropical and tropical' regions. It is appropriate here to raise the obvious questions of whether there are fundamental differences or whether they differ only in degrees because of the different species of nematodes and types of crop present. The fundamental differences have been discussed in detail by Noe and Sikora (1990) in the first edition of this book. Climate definitely affects nematode distribution on a geographical scale since most nematode life processes have thermic optima that determine the ideal geographic ranges of nematodes. Presumably, there are southern and northern hemisphere bands of appropriate temperatures for each nematode species, that would be contiguous and would meet at the equator for true tropical species. We can state with some certainty and without too many dissenting voices that nearly all the major problems that can be caused directly by nematodes have been detected in temperate countries. However, even here, forgotten problems can reappear all of a sudden as rotation sequences are altered or new cultivars introduced, as has been seen with new outbreaks of the potato cyst nematode and sugarbeet stem nematode *Ditylenchus dipsaci*. A problem new to a particular country could arise through the introduction and subsequent spread of a known nematode parasite from another temperate country. It is, therefore, the case in temperate countries that surveys are designed to determine the distribution of known nematodes causing known damage. In contrast, in the subtropical and tropical areas, new problems are being, and have yet to be, discovered involving new nematode species and even genera, or species not previously recorded as harmful to a crop. Examples are the 'legume Voltaic chlorosis' of leguminous crops, discovered in Burkina Faso, associated with a new species, *Aphasmatylenchus straturatus*, and a genus not previously known to be a harmful parasite (Germani and Luc, 1982); 'miti miti' disease of taro (*Colocasia esculenta*) in the Pacific caused by a new species; *Hirschmanniella miticausa* (Bridge *et al.*, 1983); and, in the semi-arid areas, the new cyst species *Heterodera ciceri* causing damage to chickpeas and lentils (Greco *et al.*, 1984; Vovlas *et al.*, 1985); *Meloidogyne mayaguensis* (Rammah and Hirschmann, 1988) now widespread on many crops; *Achlysiella*, a new genus and potentially damaging pest of sugarcane (Hunt *et al.*, 1989); *Radopholus citri* very pathogenic on citrus in Indonesia (Machon and Bridge, 1996); *M. paranaensis* (Carneiro *et al.*, 1996) now a devastating pest on coffee in Brazil; and most recently *Radopholus duriophilus* found widely distributed on durian in Vietnam associated with decline and death of trees in many durian nursery gardens (Nguyen *et al.*, 2003). There is little doubt that many more new nematodes and their associated problems will be found in the tropics.

The lack of trained nematologists has often meant a lack of awareness of the importance of nematology in the development of quarantine guidelines. This has led to the movement of both tropical and temperate plant parasitic species into new uninfested areas. Good examples in the past are the dissemination of the banana burrowing and root lesion nematodes (*Radopholus similis*, *Pratylenchus* spp.) and of the citrus slow decline nematode (*Tylenchulus semipenetrans*) to nearly all areas where these crops are grown, as well as the movement of *Globodera rostochiensis* into the high altitude tropical growing areas of the Philippines (Sikora, 1982).

The spread of known economically important plant parasitic nematodes has occurred in the recent past and is still occurring today, e.g. the spread of *Heterodera glycines* to South America, *Globodera pallida* to Europe, and *Bursaphelenchus cocophilus* to Central and South America.

The detection, description and recognition of possible new species of nematodes is highly relevant to both practical nematology and quarantine departments around the world. The lack of trained nematologists will lead to the spread of such plant parasites as *Radopholus similis*, *Pratylenchus coffeae*, *P. goodeyi*, *Meloidogyne chitwoodi*, *M. graminicola*, *M. mayaguensis*, *M. floridensis*, *Globodera pallida*, *Heterodera glycines*, *Ditylenchus dipsaci* and *Bursaphelenchus cocophilus*, to mention but a few. The ever increasing movement of food in the form of dried seed and fresh produce ensures future spread and underscores the need for trained nematologists in quarantine. The use of distribution maps to track important species and to make decisions on designation of new species for quarantine exclusion as presented in Chapter 22 of this 2nd edition gives support to the future need for a geophytonematological approach to monitoring the distribution of new and important species for quarantine use. The development of races within species will also make work in quarantine difficult.

There is a greater diversity of nematode genera and species in subtropical and tropical countries than in temperate ones. As many of these nematodes are new taxa, it is evident that there is a great deal of work for nematode taxonomists in the tropics. This indeed is happening, but a big disadvantage of concentrating on the taxonomic aspect is that the surveys are designed to collect nematodes and not to determine the problems caused by nematodes. This is often the only possible means of establishing new nematology laboratories with limited staff and financial means. The danger is that such laboratories can limit their activities to systematics and so become production lines for new species and genera, to the exclusion of determining the importance of the nematode being described.

Knowing which nematode genera and species occur is the necessary first step, but establishing the pathogenicity of the nematodes involved in subtropical and tropical agriculture has to be made a main priority. Many nematodes are now recognized as serious or potentially serious pests of tropical crops, as detailed in the following chapters, but information on the actual yield losses caused by the nematodes in different situations and on different crops is still sadly lacking for a large proportion of these nematodes. This knowledge is essential to provide agricultural scientists, extension officers and administrators with the information needed to recommend practical and economic means of controlling the harmful nematodes in the face of all the other constraints on crop production. The chapters in this book contain pertinent information on nematodes of the most widely grown crops in subtropical and tropical agriculture, but there are still gaps in our knowledge. The chapters show the extent of damage that can be caused by nematodes, which is recognized by the nematologists concerned but generally not by other agriculturists. This crop damage by nematodes invariably remains hidden by the many other limiting factors operating in subtropical and tropical agriculture, especially the presence of multiple biotic and abiotic stress factors operating simultaneously on the crop. Nematodes have rarely been considered or recognized as major limiting factors until all other constraints on yield increase have been removed (Bridge, 1978).

The practical problems of determining nematode pathogenicity in the tropics can often be far more difficult than in temperate countries (Noe and Sikora, 1990). Problems such as maintaining controlled conditions in glasshouses or screen houses with air-conditioning or cooling tanks because of the excessive heat can be a daunting and expensive task. The stories behind failure of field experiments are legendary in the tropical countries, with everything from lizards to elephants and from hurricanes to volcanoes doing their utmost to frustrate the attempts of nematologists to obtain accurate and replicated results. Isolated, irrigated field trials during the dry season tend to result in every hungry pest and predator for some distance around descending in droves on the plots with thanks to the irate research worker. It does mean that nematologists in the tropical countries have to be more resourceful and patient than their counterparts in the temperate countries.

There are more intrinsic differences between temperate and tropical areas based mainly on the wide diversity of nematodes, crops and agricultural systems.

The range and severity of parasitism on all living organisms, humans, animals and plants, is greater in the subtropical and tropical countries. Plant parasitic nematodes generally have shorter life cycles resulting in a more rapid population explosion than in temperate areas. For example, in temperate areas, *Heterodera* spp. generally produce one or two generations per year, whereas *H. oryzae*, in West Africa, produces one generation every 25 days (Merny, 1966). The life cycle of the northern root knot nematode *M. hapla* compared with tropical/subtropical species such as *M. incognita* and *M. javanica* is similar, and one tropical species, *M. graminicola,* has a life cycle of less than 20 days. More often than not a crop is attacked by a number of damaging nematodes. In temperate areas, there are also 'secondary species', but most often there is only one main nematode parasite of a crop which is easily recognizable and upon which control efforts can be focused. This is not the case for many tropical crops where a number of species of several different genera may be major parasites of a crop. For instance, sugarcane can be damaged by 10–20 different species of genera such as *Meloidogyne*, *Heterodera*, *Achlysiella*, *Pratylenchus*, *Xiphinema* and *Paratrichodorus*. The component species of a nematode population do differ from country to country, making predictions of damage that much more difficult. Such types of multispecies populations have a number of consequences concerning control of the nematodes. First, they can seriously hinder the establishment of an effective crop rotation as the host status of each crop will differ depending on the nematode species present. We have an example of such a phenomenon in the Côte d'Ivoire where *Crotalaria* was recommended as an intercrop to control *Meloidogyne* spp. on pineapple. The intercrop produced an effective control of the root knot nematodes but increased the populations of *Pratylenchus brachyurus* to levels which were at least as harmful to the crop as *Meloidogyne* spp. A second consequence is that multispecies populations increase the complexity of the search for crop resistance to nematodes; targeting one nematode species for resistance is normally not sufficient. The lesson of breeding for resistance to one species of nematode should have been learned with the emergence of the potato cyst nematode *Globodera pallida* following extensive planting of *G. rostochiensis*-resistant cultivars. The recent detection of a new species, *M. floridensis*, a new and aggressive species of root knot, that was detected because it was not parasitized by the obligate bacterial parasite *Pasteuria penetrans*, should also be mentioned. Strong differences in the level of aggressiveness between populations of *Radopholus similis* attacking banana will also affect future integrated pest management strategies. The most fundamental facts of subtropical and tropical agriculture that differ from the temperate regions and markedly affect the study and control of plant nematodes are the crops grown, the cultural practices and the farming systems. Commercial, plantation crops are a common feature of subtropical and tropical agriculture, but by far the largest proportion of cultivated land in most of the tropical countries is farmed by farmers with smallholdings, using traditional cropping practices. The crops grown cover a very wide range of grain, root and vegetable food crops, also many different cash and utility crops. Mono-cropping is practised, but multiple or intercropping is more common. Much of the traditional agriculture in the tropics is based on the reproduction of crops by vegetative propagation, in contrast to the dependence upon seed-reproduced plants in the temperate countries. This can increase the dissemination of nematodes in plant tissues. The outstanding feature of traditional agriculture, and one that makes life difficult for nematologists, is the complexity of the methods involved (Bridge, 1996). In contrast, modern farming in temperate countries is comparatively simple and the study and control of the nematodes is also, in comparison, relatively straightforward. The many different farming systems operating in the tropics fall into four main categories: (i) shifting cultivation; (ii) fallow farming; (iii) permanent upland cultivation; and (iv) systems with arable irrigation (Ruthenberg, 1983). In some of these farming systems, nematodes are less likely to be causing damage; in others, the cultivation practices will greatly increase the risk of nematodes causing serious yield losses (Bridge, 1987).

The nematode management methods that theoretically can be employed in subtropical and tropical countries differ little from those used in temperate countries, but in practice they are more difficult to implement and need to be considerably modified in many circumstances. There will be obvious differences in the methods to manage nematodes in developed countries compared with developing countries and in large, modern farms or plantations compared with small rural farms with more traditional cultivation systems.

Chemical soil treatment is recognized as an essential means of controlling nematodes on a number of cash crops in the tropics. In many instances, these crops cannot be grown economically without the use of nematicides. The use of nematicides and pesticides to control nematodes is of limited or no importance in developing countries on most field crops, especially at the subsistence level. Nematicide usage in many countries and by small-scale growers in the past has been strongly limited by their high price. The choice and availability of many nematicides was limited years ago due to the banning from most of the world markets of the fumigants D-D, ethylene dibromide (EDB) and dibromochloropropane (DBCP). More importantly, the recent global movement to ban the highly effective and broad spectrum fumigant methyl bromide by the year 2005, because of its side effects on atmospheric ozone, has had a major impact on how many horticultural crops are and will be grown in the near future. Some of the more easily applied granular, non-volatile nematicides are effective and are used extensively on a number of crops. They have disadvantages in being expensive and extremely toxic to man and animals when used improperly. Their availability is often curtailed because of their solubility and threat to groundwater as well as long waiting periods between use and marketing of some crops. The future of nematicides for the control of nematodes will depend on the formulation of new compounds that are effective and environmentally safe. The development of other application technology, for example treatment by seed coating or chemicals applied through drip-irrigation systems as well as development of systemic nematicides that move basipetally, is urgently needed.

The modification of existing agricultural practices in order to manage nematode populations is one of the most acceptable alternatives to chemical control for both the small- and large-scale farmers in the tropics. Crop rotation can vary from non-existent, where there is continuous cultivation of a susceptible crop or crops often planted sequentially in 1 year, through what can be termed random rotation, to a relatively sophisticated form of rotation. However, most of the rotation schemes in operation have been designed to prevent disease outbreaks or increase available nutrients, and are not always compatible with nematode control. With an understanding of the nematodes involved and the accepted cropping systems, modifications can be made to produce effective control by rotation of crops. Many other cultural methods, apart from rotation, can be used and are outlined in the following chapters and summarized in Chapter 22.

Resistant cultivars can produce the most dramatic increases in the yields of many crops and appear to hold the solution to most nematode problems, particularly with the work on gene transfer. Unfortunately, this solution is more apparent than real, as it is now clear that such cultivars mainly show resistance to only a limited number of nematode genera. These nematodes tend to belong to the groups of parasites, such as the Heteroderidae, which have a highly developed host–parasite relationship where cell modification occurs and is required for successful reproduction of the nematodes (Luc and Reversat, 1985). Many of the major subtropical and tropical plant parasitic nematodes belong to the group of migratory endoparasites which cause cell destruction without modifying the host tissues, e.g. species in the genera *Radopholus*, *Pratylenchus*, *Hirschmanniella*, *Scutellonema*, *Helicotylenchus* and *Hoplolaimus*. With the exception of one banana cultivar resistant to *R. similis*, no true resistance has been found for this large group of nematodes. Even when the possibility does exist, for nematodes such as *Heterodera*, *Meloidogyne* and *Rotylenchulus*, such research nevertheless remains aleatory and very costly: many years and several millions of US dollars were necessary to obtain a cultivar of soybean resistant to *Heterodera glycines*. A major limiting factor affecting the effectiveness of newly introduced resistant cultivars is the selection of pathotypes or races that are able to break down the resistance. The existence of resistance-breaking pathotypes is a major problem in breeding programmes in temperate crops. Similar complications must be expected when resistant cultivars are bred for tropical crops. Another difficulty which applies more to subtropical and tropical countries is in the practical introduction of these resistant cultivars. Where resistant cultivars are available and suited to the conditions prevailing in a country, many other factors have to be taken into account before their successful introduction. For instance, subsistence farmers are not aware that the *Mi* gene in tomato breaks down at high temperatures or that nematode-resistant tissue culture banana plantlets are still susceptible to damage in the seedling stage. There will again be a marked contrast in what can be achieved with the big producer compared with the rural farmer, but consideration has to be given to local needs. A good illustration of this difficulty was when dwarf rice cultivars were introduced to prevent lodging (Mydral, 1974): people in South-east Asia were deprived of their normal source of rice straw for animal feed, bedding and thatching material. The recent development of transgenic plants with resistance to insects, and the detection of genes in the plant that are responsible for giant cell formation as well as genes in plants needed for protein synthesis by the nematodes may lead to new forms of resistance. The importance of this technology to small and large growers, to the different nematode groups and crops, although highly publicized, will take years to have an impact as well as trickling down to the subsistence growers. The cost of developing transgenic crops is enormous and the time it will take from detection to market will outlive this edition of the book.

Because of economic constraints, research in nematode management in the tropics often focuses on low-input methods involving crop rotations, multicropping, adjustment of planting and harvest dates, use of various soil amendments and mulches, trap and antagonistic crops, fallow, flooding, etc. Emphasis on these forms of control strategies by agricultural scientists working in the tropics and subtropics reflects increased awareness of the need for nematode management systems that rely less on the use of nematicides. However, new management tools have been developed that have widened the integrated pest management tool-box, including: solarization, biological control, trap cropping, resistant rootstocks, biofumigation, molecular kits for root knot identification, remote sensing and precision farming, nematicide formulation and application technology.

We have outlined some of the differences and difficulties facing nematology in the tropics, but wish to emphasize that none of the problems is insurmountable with the appropriate effort, expertise and backing. You will see, reading through the chapters, that a great deal of new knowledge on the importance of nematodes as plant parasites and, more relevantly, the successes in their management has been accumulated by nematologists since the printing of the first edition. For example, a literature search of CABI abstracts for plant parasitic nematodes and vegetables yielded over 2800 citations for the period between 1990 and 2003.

However, nematology in the tropics is underfunded and there is a shortage of nematologists to work on the problems. Sasser and Freckman (1987) estimated that less than 0.2% of the crop value lost to nematodes worldwide is used to fund nematological research to combat these losses, which probably exceed US$100 billion annually. In our opinion, support has dropped from this level due to the overall reduction in emphasis on funding for agricultural research worldwide. Furthermore, the percentage funding for nematological research in the tropics is considerably less than it is in most of the temperate countries, which makes the amount infinitesimal. With few exceptions, the efforts and resources directed towards research on plant parasitic nematodes within the International Agricultural Research Centres (IARCs) have been and remain much less than even a conservative assessment of their significance as crop pests would merit (Sharma *et al.*, 1997). Examination of the Senior Scientific Staff in the IARCs over a 20-year period showed that numbers of nematologists remained unchanged at a bare minimum even though there was an increase in other disciplines (Sharma *et al.*, 1997) (Fig. 1.1).

However, the need for such research in subtropical and tropical agriculture is greater than in temperate agriculture. Many temperate countries are suffering the embarrassment of massive surpluses in food production which are not transferable. In contrast, the majority of countries in the tropics have shortfalls in the production of most crops. An increase is needed in food crops, to improve the nutritional level of the populations, and in export cash crops, to obtain essential foreign currency. Solving nematode problems can play an important part in improving crop yields to the benefit of commercial and subsistence farms, the consumers and governments.

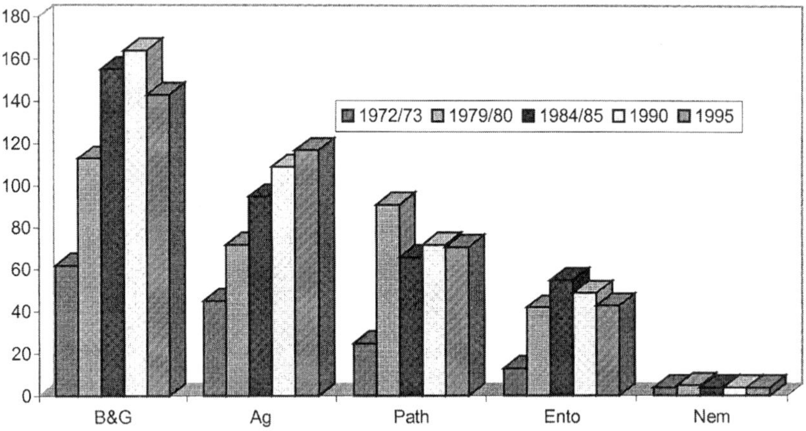

Fig. 1.1. Numbers of senior staff by specialism in seven International Agricultural Research Centres (CIAT, CIMMYT, CIP, ICARDA, ICRISAT, IITA and IRRI). B&G, breeders and geneticists; Ag, agronomists; Path, pathologists; Ento, entomologists; Nem, nematologists. (Courtesy of Nigel S. Price published in Sharma *et al.*, 1997.)

References

Bridge, J. (1978) Agricultural aspects – comments and discussion 1. In: Taylor, A.E. and Muller, R. (eds) *The Relevance of Parasitology to Human Welfare Today*. Blackwell Scientific Publications, Oxford, pp. 111–117.

Bridge, J. (1987) Control strategies in subsistence agriculture. In: Brown, R.H. and Kerry, B.R. (eds) *Principles and Practice of Nematode Control in Crops*. Academic Press, Melbourne, Australia, pp. 389–420.

Bridge, J. (1996) Nematode management in sustainable and subsistence agriculture. *Annual Review of Phytopathology* 34, 201–225.

Bridge, J., Mortimer, J.J. and Jackson, G.V. (1983) *Hirschmanniella miticausa* n.sp. (Nematoda: Pratylenchidae) and its pathogenicity on taro (*Colocasia esculenta*). *Revue de Nématologie* 6, 285–290.

Butler, E.J. (1913) *Diseases of Rice*. Bulletin of the Agriculture Research Institute, Pusa, India, 34.

Butler, E.J. (1919) The rice worm (*Tylenchus angustus*) and its control. *Memoirs of the Department of Agriculture of India* 10, 1–37.

Carneiro, R.M.D.G., Carneiro, R.G., Abrantes, I.M.O., Santos, M.S.N.A. and Almeida, M.R.A. (1996) *Meloidogyne paranaensis* n. sp. (Nemata: Meloidogynidae), a root knot nematode parasitizing coffee in Brazil. *Journal of Nematology* 28, 177–189.

Cobb, N.A. (1891) Diseased banana plants. *Agricultural Gazette of New South Wales* 2, 622–624.

Cobb, N.A. (1893) Nematodes, mostly Australian and Fijian. *McLeay Memorial Volume, Linnean Society of New South Wales*, 252–308.

Germani, G. and Luc, M. (1982) Etudes sur la chlorose voltaïque des légumineuses due au Nématode *Aphasmatylenchus straturatus* Germani. 1 & 2. *Revue de Nématologie* 5, 139–146 and 161–168.

Göldi, E.A. (1889) Der Kaffeenematode Brasiliens (*Meloidogyne exigua* G). *Zoologische Jahrbücher, Abteilung Systematik* 4, 261–267.

Göldi, E.A. (1892) Relatório sobre a moléstia do cafeeiro na província do Rio de Janeiro. *Archivos Museo Nacional de Rio de Janeiro* 8 (1887), 1–121.

Greco, N., Di Vito, M., Reddy, M.V. and Saxena, M.C. (1984) A preliminary report of a survey of plant-parasitic nematodes of leguminous plants in Syria. *Nematologia Mediterranea* 12, 87–93.

Hunt, D.J., Bridge, J. and Machon, J.E. (1989) On *Achlysiella*, a new genus of obese Pratylenchidae (Nematoda: Tylenchoidea). *Revue de Nématologie* 12, 401–407.

Jobert, C. (1880) [Maladie du caféier au Brésil.] *Comptes Rendus des Séances de la Société de Biologie*, 6 ème série 5, 360–361.

Linford, M.B. and Oliveira, J.M. (1940) *Rotylenchulus reniformis*, nov.gen., nov.sp., a nematode parasite of roots. *Proceedings of the Helminthological Society of Washington* 7, 35–42.

Luc, M. and Reversat, G. (1985) Possibilités et limites des solutions génétiques aux affections provoques par les nématodes sur les cultures tropicales. *Comptes Rendus des Séances de l'Académie d'Agriculture de France* 71, 781–791.

Machon, J.E. and Bridge, J. (1996) *Radopholus citri* n.sp. (Tylenchida: Pratylenchidae) and its pathogenicity on citrus. *Fundamental and Applied Nematology* 19, 127–133.

Merny, G. (1966) Biologie d'*Heterodera oryzae* Luc & Berdon, 1961. 11. Rôle des massess d'oeufs dans la dynamique des populations et la conservation de l'espèce. *Annales des Epiphyties* 17, 445–449.

Mydral, G. (1974) The transfer of technology to underdeveloped countries. *Scientific American* 231, 173–182.

Nguyen, C.N., Subbotin, S.A., Madani, M., Trinh, P.Q. and Moens, M. (2003) *Radopholus duriophilus* sp.n. (Nematoda: Pratylenchidae) from Western Highland of Vietnam. *Nematology* 5, 549–558.

Noe, J.P. and Sikora, R.A. (1990) Effects of tropical climates on the distribution and host–parasite relationship of plant parasitic nematodes. In: Luc, M., Sikora, R.A. and Bridge, J. (eds) *Plant Parasitic Nematodes in Subtropical and Tropical Agriculture*. CAB International, Wallingford, UK, pp. 583–597.

Nowell, W. (1919) The red ring or 'root' disease of coconut palms. *West Indies Bulletin* 17, 189–202.

Nowell, W. (1920) The red ring disease of coconut palms. Infection experiments. *West Indies Bulletin* 18, 74–76.

Rammah, A. and Hirschmann, H. (1988) *Meloidogyne mayaguensis* n.sp. (Meloidogynidae), a root-knot nematode from Puerto Rico. *Journal of Nematology* 20, 58–69.

Ruthenberg, H. (1983) *Farming Systems in the Tropics*, 3rd edn. Oxford University Press, London.

Sasser, J.N. and Freckman, D.W. (1987) A world perspective on nematology: the role of the society. In: Veech, J.A. and Dickson, D.W. (eds) *Vistas on Nematology*. Society of Nematologists Inc., Hyattsville, Maryland, pp. 7–14.

Sharma, S.B., Price, N.S. and Bridge, J. (1997) The past, present and future of plant nematology in International Research Centres. *Nematological Abstracts* 66 (3), 119–142.

Sikora, R.A. (1982) *Globodera rostochiensis* on potato in the Philippines. *Zeitschrift für Pflanzenkrankheiten und Pflanzenschutz* 89, 532–533.

Thomason, I.J. (1987) Challenges facing nematology: environmental risks with nematicides and the need for new approaches. In: Veech, L.A. and Dickson, D.W. (eds) *Vistas on Nematology*. Society of Nematologists Inc., Hyattsville, Maryland, pp. 469–476.

Thomason, L.J., Freckman, D.W. and Luc, M. (1983) Perspectives in nematode control. *Revue de Nématologie* 6, 315–323.

Treub, M. (1885) Onderzoekingen over sereh-zieh suikerriet gedaan in's Lands Plantentuin te Buitenzorg. *Mededeelingen's Lands Plantentuin (Buitenzorg)*, 11–39.

van Breda de Haan, J. (1902) Een aaltjes-ziekte der rijst 'omo mentek' of 'omo-bambang'. Voorlooping rapport. *Medeleelingen's Lands Plantentuin (Buitenzorg)* 53, 1–65.

Vovlas, N., Greco, N. and Di Vito, M. (1985) *Heterodera ciceri* sp. n. (Nernatoda: Heteroderidae) on *Cicer arietinum* from Northern Syria. *Nematologia Mediterranea* 13, 239–252.

Plant Parasitic Nematodes in Subtropical and Tropical Agriculture

2nd Edition

CTA

partageons les connaissances au profit des communautés rurales

sharing knowledge, improving rural livelihoods

The Technical Centre for Agricultural and Rural Cooperation (CTA) was established in 1983 under the Lomé Convention between the ACP (African, Caribbean and Pacific) Group of States and the European Union Member States. Since 2000, it has operated within the framework of the ACP-EC Cotonou Agreement.

CTA's tasks are to develop and provide services that improve access to information for agricultural and rural development, and to strengthen the capacity of ACP countries to produce, acquire, exchange and utilize information in this area. CTA's programmes are designed to: provide a wide range of information products and services and enhance awareness of relevant information sources; promote the integrated use of appropriate communication channels and intensify contacts and information exchange (particularly intra-ACP); and develop ACP capacity to generate and manage agricultural information and to formulate ICM strategies, including those relevant to science and technology. CTA's work incorporates new developments in methodologies and cross-cutting issues such as gender and social capital.

CTA
Postbus 380
6700 AJ Wageningen
The Netherlands

Website: www.cta.int

Plate 1. **(A)** White patches on rice leaf base caused by *Ditylenchus angustus* (Photo: J. Bridge). **(B)** Twisted and distorted panicles of rice caused by *Ditylenchus angustus* (Photo: J. Bridge). **(C)** Ufra disease. Brown patch of dead and dying rice (left) caused by *Ditylenchus angustus* (Photo: R.A. Plowright). **(D)** White tip symptoms on rice infested with *Aphelenchoides besseyi* (Photo: J. Bridge). **(E)** Characteristic hooked root tip galls caused by *Meloidogyne graminicola* (Photo: J. Bridge). **(F)** Stained females and eggs of *Meloidogyne graminicola* within rice root (Photo: J. Bridge).

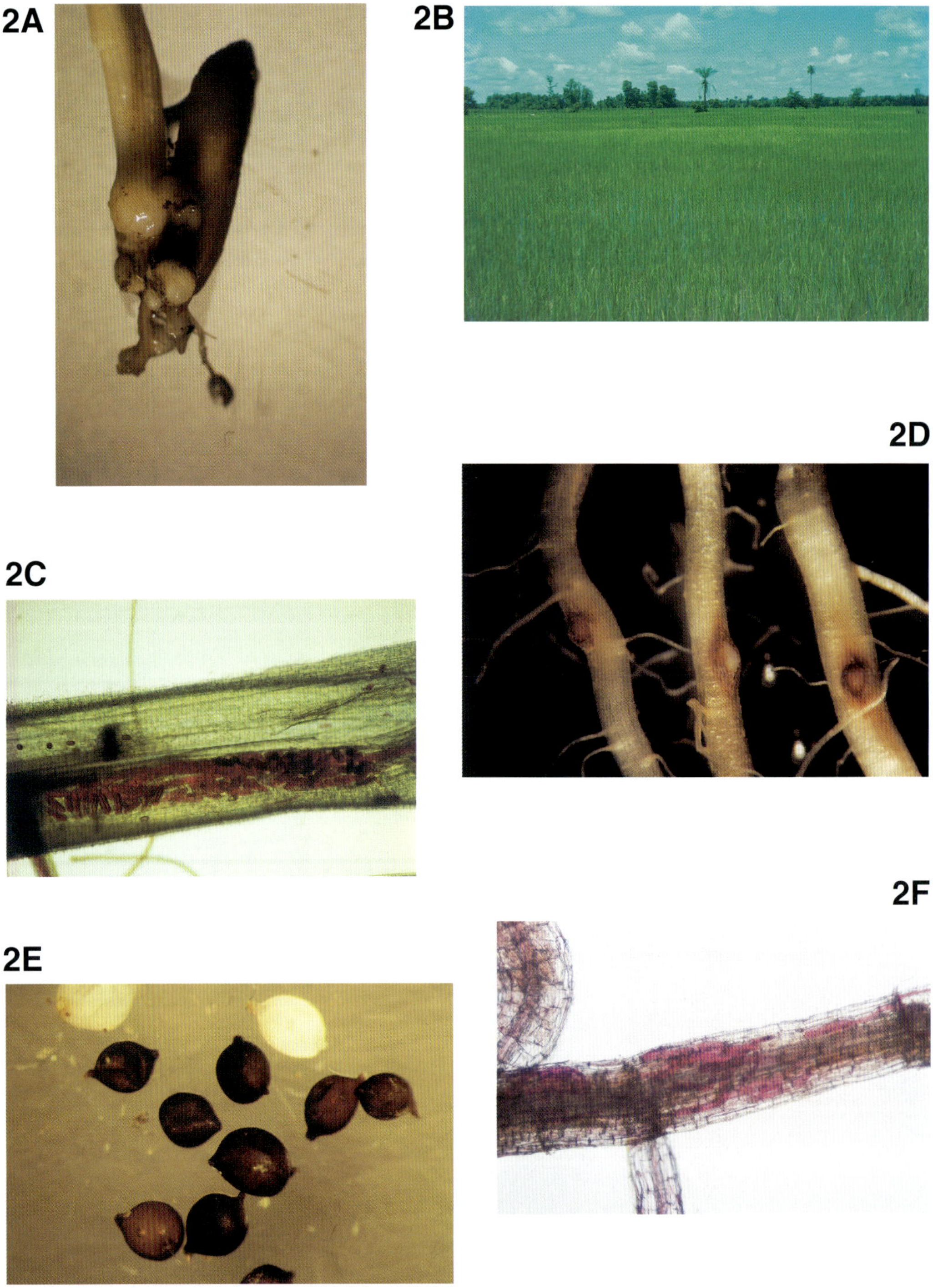

Plate 2. **(A)** Newly germinated rice seedling severely galled by *Meloidogyne graminicola* (Photo: R.A. Plowright). **(B)** Yellow patch of plants infested with *Hirschmanniella* spp. in swamp rice in the Gambia (Photo: J. Bridge). **(C)** Stained *Hirschmanniella oryzae* female and eggs endoparasitic in rice root (Photo: J. Bridge). **(D)** *Heterodera oryzicola* cysts and white female emerging from rice root (Photo: R.A. Plowright). **(E)** *Heterodera sacchari* cysts and white females (Photo: J. Bridge). **(F)** Stained *Pratylenchus zeae* endoparasitic in rice root (Photo: J. Bridge).

3A 3B 3C 3D 3E 3F

Plate 3. (A) Symptoms of cereal cyst nematode, *Heterodera avenae*, on wheat roots, showing a bushy appearance (Photo: R. Rivoal). **(B)** Uneven patchy growth of a wheat crop in a field infested with *Heterodera avenae* (Photo: R.A. Sikora). **(C)** Symptoms of root lesion nematode, *Pratylenchus thornei*, on susceptible wheat, showing extensive lesions, cortical degradation and reduction in both seminal and lateral root systems with increasing nematode density from top to bottom under natural field infestation (Photo: J.M. Nicol). **(D)** Different stages of *Anguina tritici* infection of wheat in India along with symptoms of 'yellow ear-rot disease' caused by the interaction of the nematode with *Corynebacterium michiganese*. Healthy ears on far right and far left (Photo: R.A. Sikora). **(E)** Close-up of stem nematode, *Ditylenchus dipsaci*, damage on susceptible oats indicating severe dwarfing, twisting of leaves, and an abnormal number of tillers giving the plant a bushy stunted appearance (Photo: S. Taylor, SARDI, Australia). **(F)** *Xiphinema* root-tip galling of maize (Photo: B.J. Jacobsen and R.A. Sikora).

4A

4B

4C

4D

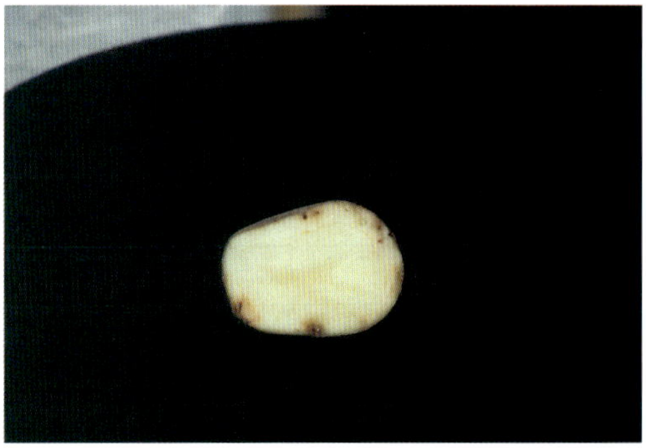

4E

4F

Plate 4. **(A)** Yellowing and stunting of potato plant infested with *Globodera rostochiensis* (Photo: J. Bridge). **(B)** Cysts of *Globodera rostochiensis* on root of potato (Photo: BBA Münster). **(C)** Swellings on surface of tubers caused by *Meloidogyne incognita* (Photo: J. Bridge). **(D)** Section through potato tuber showing females and necrotic spots below surface caused by *Meloidogyne incognita* (Photo: J. Bridge). **(E)** Rounded bead-like galls on roots of potato cv. Waych'a caused by *Nacobbus aberrans* in Bolivia (Photo: J. Bridge). **(F)** Internal necrosis of sweet potato tubers around females of *Meloidogyne incognita* in Papua New Guinea (Photo: J. Bridge).

Plate 5. (A) Galled cassava roots infested with *Meloidogyne incognita* (Photo: J. Bridge). **(B)** Galled cassava root system infested with *Meloidogyne incognita* (Photo: D. Coyne). **(C)** Section though cassava root gall showing females of *Meloidogyne incognita* surrounded by necrotic spots below the surface (Photo: J. Bridge). **(D)** Dry rot disease of yam (*Dioscorea rotundata*) tubers caused by *Scutellonema bradys* and secondary infection of wet rot (light brown) caused by fungi or bacteria (Photo: C.K. Kwoseh). **(E)** Yam tubers (*Dioscorea rotundata*) with dry rot disease caused by *Scutellonema bradys* showing cracking and flaking off of epidermis in Nigeria (Photo: J. Bridge). **(F)** Dry rot disease of yam (*Dioscorea rotundata*) tuber caused by *Pratylenchus coffeae* in Papua New Guinea (Photo: J. Bridge).

6A

6B

6C

6D

6E

6F

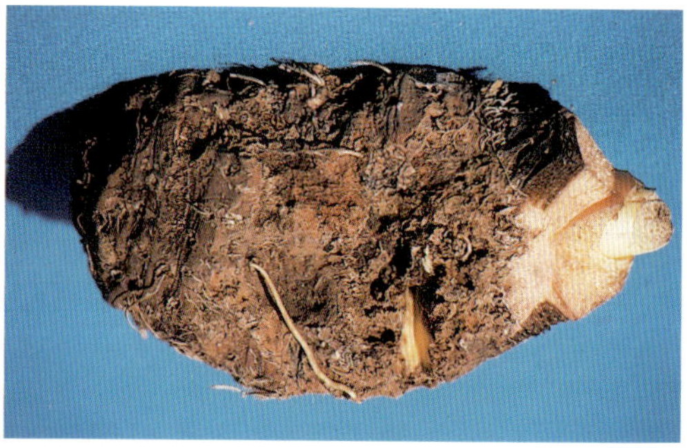

Plate 6. (A) All stages of *Scutellonema bradys* endoparasitic in dry rot tissues of yam tuber (Photo: J. Bridge). **(B)** Taro (*Colocasia esculenta*) growing in Uganda (Photo: J. Bridge). **(C)** Early stages of miti-miti disease caused by *Hirschmanniella miticausa* in taro corm showing reddening of tissues in longitudinal section of corm (Photo: J. Bridge). **(D)** Red miti-miti diseased tissues caused by *Hirshmanniella miticausa* in longitudinal section of taro corm plus secondary rot (Photo: J. Bridge). **(E)** Surface of swamp taro (*Cyrtosperma chammissonis*) corm removed to expose lesions caused by *Radopholus similis* in Yap, South Pacific (Photo: J. Bridge from material collected by G.V.H. Jackson). **(F)** Swamp taro (*Cyrtosperma chamissonis*) corm damaged by *Radopholus similis* in Yap, South Pacific (Photo: J. Bridge from material collected by G.V.H. Jackson).

7A

7B

7D

7C

7E

7F

Plate 7. **(A)** Darkened, reddened stems on broad bean, *Vicia faba*, infested with *Ditylenchus dipsaci* 'Giant Race' in Syria showing reduced tillering (Photo: R.A. Sikora). **(B)** Dark black spots on the seeds of *Vicia faba* infested with *Ditylenchus dipsaci* 'Giant Race' (Photo: R.A. Sikora). **(C)** Broad bean crop showing a patch of stunted plants in a field infested with *Heterodera goettingiana* (Photo: N. Greco). **(D)** *Meloidogyne artiellia*: chickpea roots with large eggsacs of the nematode protruding from roots resembling cysts and a noticeable lack of root galling (Photo: M. De Vito). **(E)** *Heterodera ciceri*: white, lemon-shaped females on the roots of chickpea (Photo: N. Greco). **(F)** Roots of chickpea exhibiting necrotic lesions caused by a lesion nematode *Pratylenchus* sp. (Photo: N. Greco).

8A

8B

8C

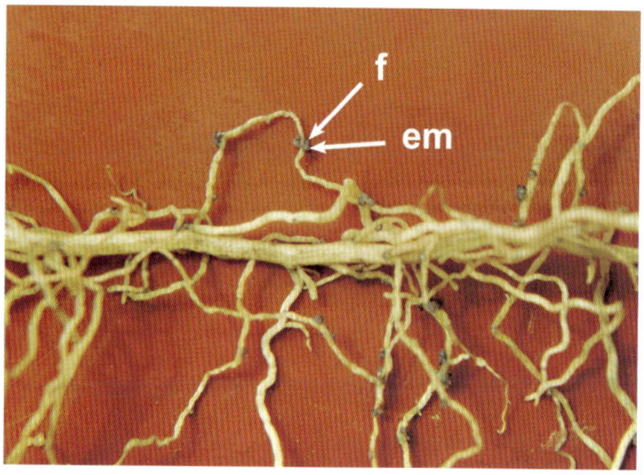

8D

8E

8F

Plate 8. **(A)** *Meloidogyne incognita* galls on cowpea in Nigeria (Photo: J. Bridge). **(B)** *Meloidogyne incognita:* galling and root rotting of haricot bean roots in the Philippines due to the interaction between nematodes and soil fungi (Photo: R.A. Sikora). **(C)** Roots of pigeonpea showing egg masses of *Rotylenchulus reniformis* (Photo: S.B. Sharma). **(D)** Root of a pigeonpea showing severe galling by *Meloidogyne javanica* (Photo: S.B. Sharma). **(E)** Soybean plants exhibiting chlorosis and early senescence caused by *Heterodera glycines* in North Carolina, USA (Photo: D. Schmitt). **(F)** Growth differences between soybean cultivars Clark-63 (susceptible, left) and Custer (resistant, right) to soybean cyst nematode, *Heterodera glycines* (Photo: R.A. Sikora).

9A

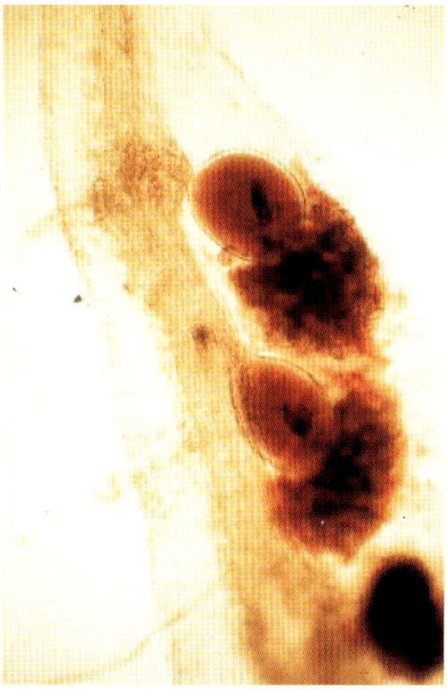

9B

9C

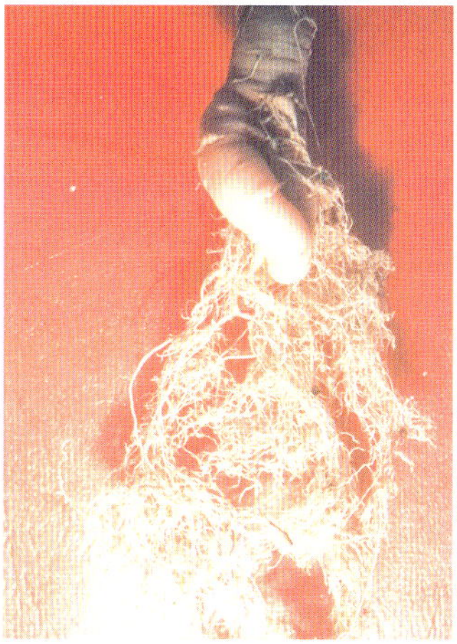

9D

9F

9E

Plate 9. (A) Adult females of a species of *Meloidogyne* inside the root of carrot with protruding egg masses (Photo: D.P.P. Taylor and R.A. Sikora). **(B)** Typical firm large galls produced by species of *Meloidogyne* on most vegetables crops grown in the tropics and subtropics, here *M. incognita* on beet in Bahrain (Photo: R.A. Sikora). **(C)** Root knot, *Meloidogyne hapla*, induced 'bearded root' with deformed tap root of carrot (Photo: R.A. Sikora). **(D)** Severely galled tomato root system, *Meloidogyne incognita*, with secondary root rot symptoms caused by soil-borne fungi (Photo: R.A. Sikora). **(E)** Yellowing and death of tomato infested with *Meloidogyne incognita*, being intercropped with papaya in Yemen (Photo: R.A. Sikora). **(F)** Tomato in plastic greenhouse exhibiting chlorosis, wilting and leaf necrosis due to concomitant infection by *Meloidogyne incognita* and *Fusarium oxysporum* in Crete (Photo: R.A. Sikora).

10A

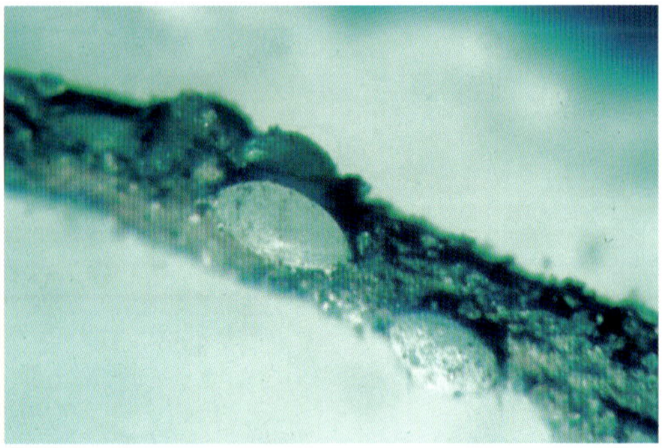

10B

10C

10D

10E

10F

Plate 10. (A) Bead-like galls produced by *Nacobbus aberrans* on the roots of tomato (Photo: J. Bridge). **(B)** A typical white female of the sugarbeet cyst nematode *Heterodera schachtii* on the surface of a root (Photo: R.A. Sikora). **(C)** Deformed garlic bulbs in a field infested with *Ditylenchus dipsaci* (Photo: J.L. Starr). **(D)** Reduced root biomass and root necrosis of maize due to damage by a species of *Pratylenchus* with (right) and without (left) nematicide treatment (Photo: B.J. Jacobsen). **(E)** 'Stubby-root' symptoms caused by the feeding of *Paratrichodorus minor* on maize (Photo: D.W. Dickson, in SON Slide Set 1). **(F)** Arrested root growth, root-tip galling and deformed carrots caused by *Longidorus* spp. in Israel (Photo: R.A. Sikora).

11A

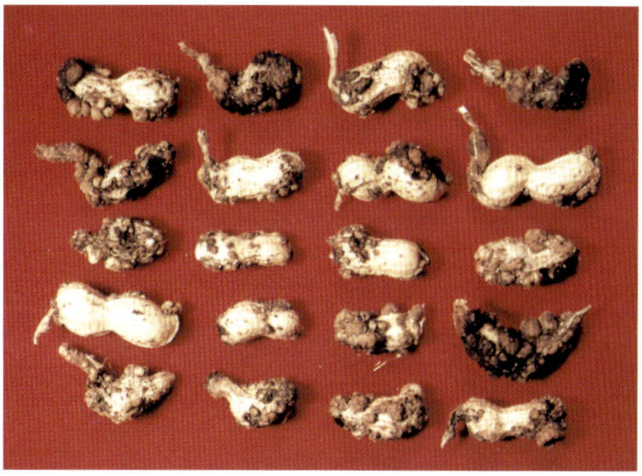

11B

11C

11D

11F

11E

Plate 11. (A) *Meloidogyne arenaria*: portion of roots with galling and matting (top), and uninfected root portion with nitrogen-fixing nodules (Photo: D.W. Dickson). **(B)** *Meloidogyne arenaria*: pods and a short portion of pegs with light to heavy galling (Photo: D.W. Dickson). **(C)** *Meloidogyne arenaria*: peanut (groundnut) field in Florida, USA, treated with 1,3-D (right) and untreated (left) (Photo: D.W. Dickson). **(D)** *Pratylenchus brachyurus*: lesions on pods (Photo: D.W. Dickson). **(E)** *Aphelenchoides arachidis*: brown and wrinkled infested seed (top), uninfested healthy seed (bottom) (Photo: J. Bridge). **(F)** *Ditylenchus africanus*: infected pod (right) and uninfected pod (left) (Photo: D. De Waele).

12A

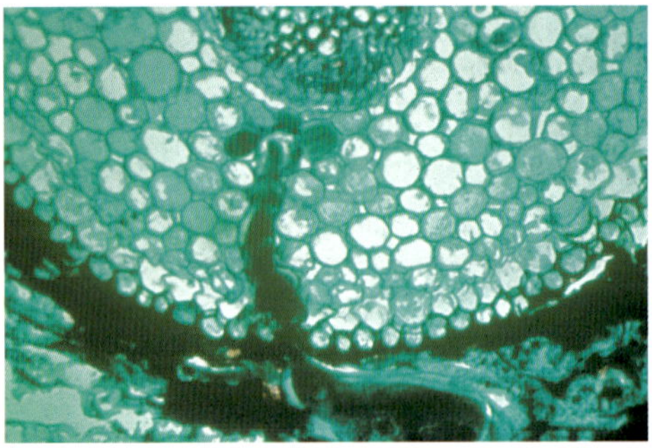

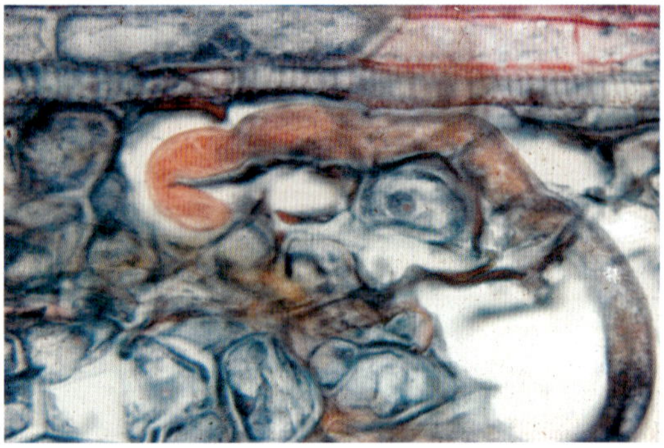

12B

12C

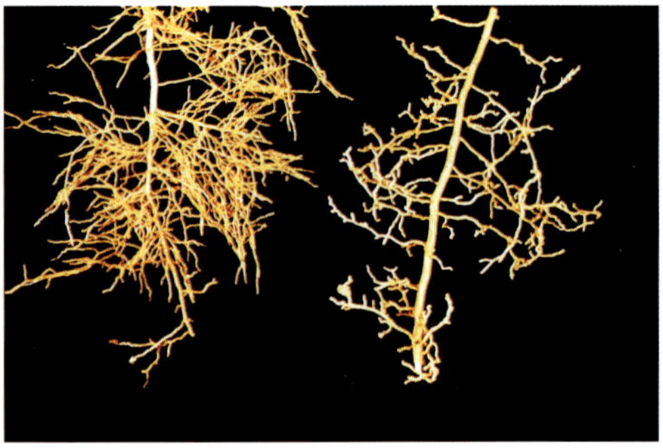

12D

12F

12E

Plate 12. Nematode parasites of economic importance in citrus. **(A)** Cross section of a feeder root showing extension of the *T. semipenetrans* female's body into the root cortex and densely stained nurse cells surrounding the head (Photo: R. Inserra). **(B)** Cavity created in fibrous root cortical tissue by *Radopholus similis* (note that the nematode does not penetrate the stellar tissues) (Photo: J. O'Bannon). **(C)** Valencia orange trees on rough lemon rootstock in various stages of decline (note the large numbers of replanted trees) due to infection by the lesion nematode, *Pratylenchus coffeae*. **(D)** Stubby root tips and reduced fibrous root system due to feeding by the sting nematode, *Belonolaimus longicaudatus*. (E and F) Effect of sting nematode on young trees. **(E)** Eight-year-old citrus tree on Swingle citrumelo rootstock (1.6-m height) that was planted in an area of the orchard heavily infested with sting nematodes. **(F)** Same age and variety tree (2.2-m height) planted in an uninfested area of the same orchard.

13A

13B

13C

13D

13E

13F

Plate 13. **(A)** *Meloidogyne* sp.: galling of guava roots, Niger (Photo: R.A. Sikora). **(B)** *Meloidogyne* sp.: severely infested guava exhibiting dieback symptoms, Niger (Photo: R.A. Sikora). **(C)** Simultaneous infestations of papaya and tomato intercropped (Photo: P. Baujard). **(D)** Galling of roots of papaya caused by *Meloidogyne incognita* from Bahia, Brazil (Photo: R. Ritzinger). **(E)** Galling of roots of Acerola (*Malpighia glabra*) caused by *Meloidogyne javanica* from Bahia, Brazil (Photo: R. Ritzinger). **(F)** Dieback and decline of Acerola caused by *Meloidogyne javanica* in Cruz das Almas, Bahia, Brazil (Photo: R. Ritzinger).

14A

14B

14C

14D

14E

14F

Plate 14. **(A)** Coconut tree infested with the red ring nematode, *Bursaphelenchus cocophilus* (Photo: K. Gerber). **(B)** Cross section of coconut stem showing red ring symptoms caused by *Bursaphelenchus cocophilus* (Photo: K. Gerber). **(C)** Longitudinal section of old coconut stem showing diffuse reddened tissues caused by *Bursaphelenchus cocophilus* becoming one solid block (Photo: J. Bridge). **(D)** Brownish ring of diseased tissue caused by *Bursaphelenchus cocophilus* in cross section of oil palm stem (Photo: H. Gerber). **(E)** Drying out and browning of leaves of oil palm associated with red ring disease caused by *Bursaphelenchus cocophilus* (Photo: K. Gerber). **(F)** Roots of arecanut palm showing lesions, blackening and rotting due to *Radopholus similis* (Photo: V.K. Sosamma).

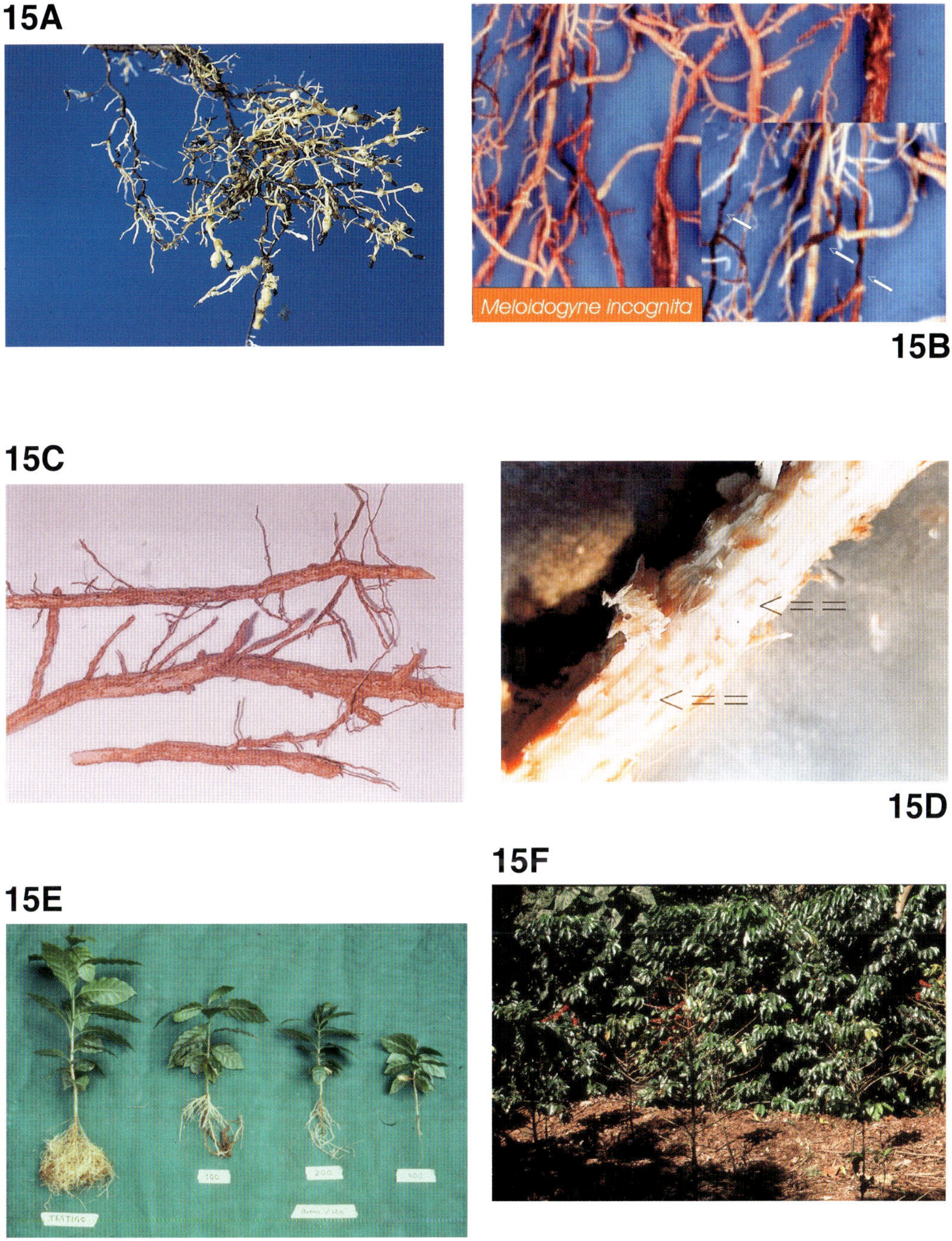

Plate 15. (A) *Meloidogyne exigua* galls on coffee roots (Photo: J. Bridge). **(B)** Segments of coffee roots infested with *Meloidogyne incognita* showing brown lesions and dark rings (Photo: V.C. Campos). **(C)** Peeling and cracking of older coffee roots where females of *Meloidogyne coffeicola* are developing (Photo: V.C. Campos). **(D)** Dissected root showing location of *Meloidogyne coffeicola* females (arrowed) (Photo: V.C. Campos). **(E)** Eight-month-old *Coffea arabica* plants infested with *Pratylenchus* sp. from Guatemala (species under description). From left to right: control plant (without nematode inoculation) and plants infested with 100, 200 and 400 nematodes (Photo: L. Villain). **(F)** *Coffea arabica* cv. Caturra infested by *Pratylenchus* sp. in Guatemala: non-grafted plants in the foreground and grafted on to *Coffea canephora* in the second plane, planted at the same time and both without chemical treatment.

16A

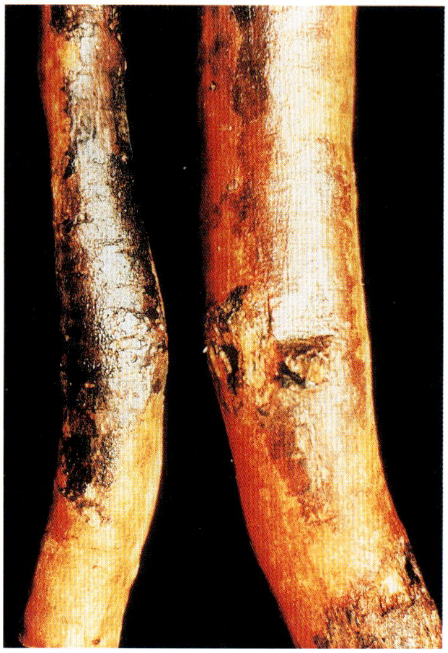

16B

16C

16D

Plate 16. **(A)** A declining patch of tea infested with *Pratylenchus loosi* showing typical symptoms of early flowering and fruiting (Photo: N.C. Gnanapragasam). **(B)** Large storage roots of tea displaying necrotic patches caused by *Pratylenchus loosi* (Photo: N.C. Gnanapragasam). **(C)** Typical galling of mature tea roots caused by *Meloidogyne brevicauda* (Photo: N.C. Gnanapragasam). **(D)** Susceptible tea clone (TRI 2025) damaged by *Radopholus similis* (right group) compared with uninfested plants of similar age (left group) (Photo: N.C. Gnanapragasam).

17A

17B

17C

17D

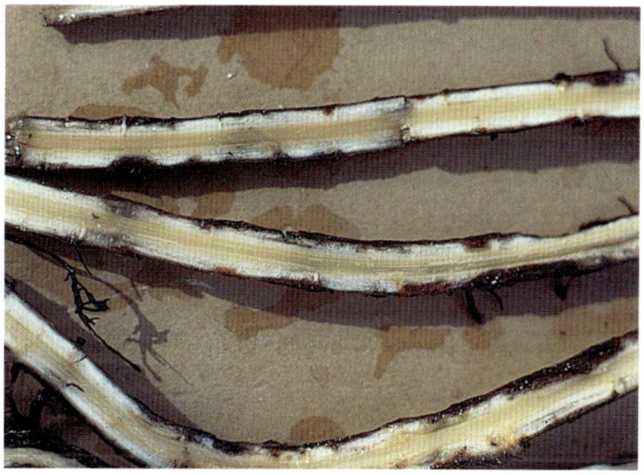

17E

Plate 17. (A) Toppling and uprooting of banana plants due to *Radopholus similis* (Photo: S.R. Gowen). **(B)** Lesions in banana roots caused by *Radopholus similis* (Photo: B. Pembroke). **(C)** Poor growth and toppling of cooking bananas infested with *Pratylenchus goodeyi* (Photo: J. Bridge). **(D)** Necrosis of outer cortex of banana roots caused by *Helicotylenchus multicinctus* (Photo: S.R. Gowen). **(E)** Root galling of banana caused by *Meloidogyne* sp. (Photo: B. Pembroke).

18A

18B

18C

18D

18E

Plate 18. (A) Shoot roots of sugarcane showing composite symptoms of nematode community damage. **(B)** Sett roots of cane damaged by nematodes. **(C)** Increased vegetative growth and cover in sugarcane treated with the nematicides aldicarb (foreground) and DD (middle distance) compared with untreated cane (centre). **(D)** Intercropping sugarcane with sweet potatoes in Barbados. **(E)** Effect of nematodes on the sustainability of sugarcane. In a field trial in South Africa, over a period of two annual crops, the susceptible cultivar N24 died (centre), whereas plots of this cultivar treated with nematicide were still producing economic yields (left).

19A

19B

19C

19D

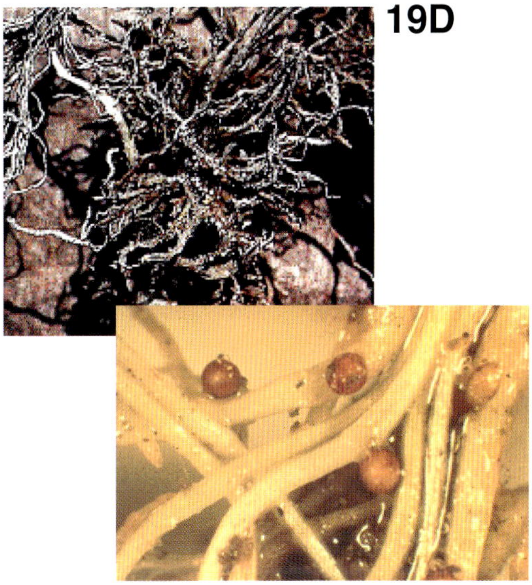

19E

19F

Plate 19. **(A)** Galling and root rot of burley tobacco caused by *Meloidogyne* spp. and associated soil microorganisms (Photo: C. Johnson). **(B)** Stunting of flue-cured tobacco by *Pratylenchus coffeae* in South Carolina, USA (Photo: S.A. Lewis, Clemson University). **(C)** Brown root rot on flue-cured tobacco. Necrotic roots on a young flue-cured tobacco plant (left); close-up on discrete necrosis of small feeder roots (right) (Photo: C. Johnson). **(D)** Cysts of *Globodera tabacum solanacearum* on roots of flue-cured tobacco. White females on the roots of a tobacco transplant (top); brown cysts on tobacco roots (bottom) (Photo: C. Johnson). **(E)** Plant mortality in flue-cured tobacco caused by a *Globodera tabacum solanacearum*–*Fusarium* nematode–disease complex (Photo: C. Johnson). **(F)** Plant mortality in Spain from a nematode–disease complex involving *Meloidogyne arenaria*, *Globodera tabacum* and *Fusarium oxysporum* (Photo: E.A. Wernsman, North Carolina State University).

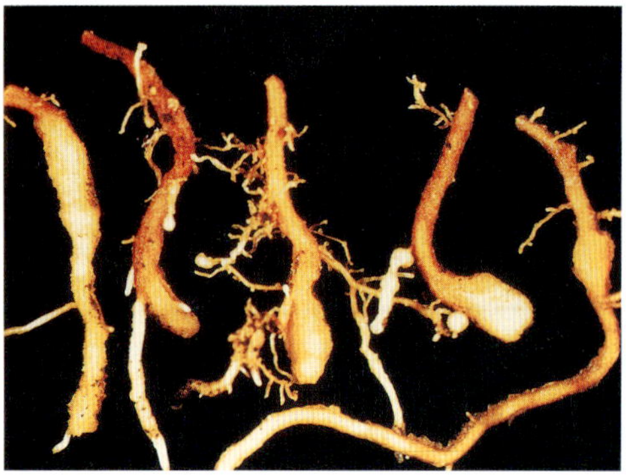

20A

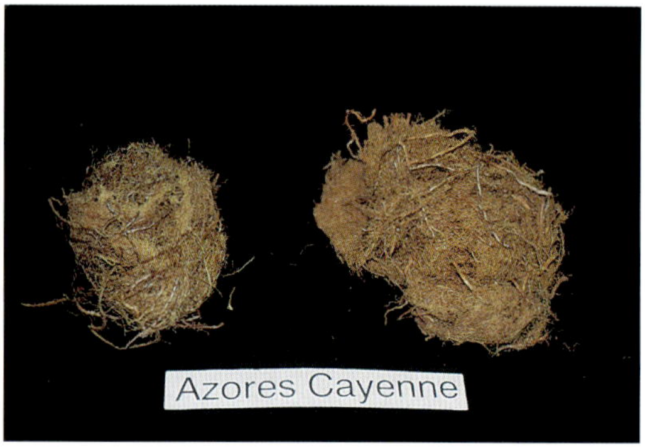

20B

20C

20D

20F

20E

Plate 20. **(A)** Symptoms of root knot nematode (*Meloidogyne* sp.) infection on pineapple roots. **(B)** Symptoms of reniform nematode (*Rotylenchulus reniformis*) infection on pineapple roots. **(C)** Symptoms of lesion nematode infection on pineapple roots. **(D)** A field showing symptoms of severe nematode damage. **(E)** Pineapple rooting pattern of plants treated for nematode control. **(F)** Pineapple rooting pattern of plants damaged by nematodes.

21A

21B

21C

21D

21E

21F

Plate 21. (A) Speckled leaf symptom of cotton due to *Meloidogyne incognita* (Photo: O. Ruano). **(B)** Moderate root galling of cotton caused by *Meloidogyne incognita* (Photo: J.L. Starr). **(C)** Symptoms of *Fusarium* wilt–root knot nematode complex of cotton (Photo: J.L. Starr). **(D)** Brown egg masses of *Rotylenchulus reniformis* on cotton roots (Photo: O. Ruano). **(E)** Stunting of cotton due to *Hoplolaimus columbus* (Photo: S.A. Lewis). **(F)** Severe root stunting of cotton due to *Belonolaimus longicaudatus* (Photo: W.T. Crowe).

22A

22B

22D

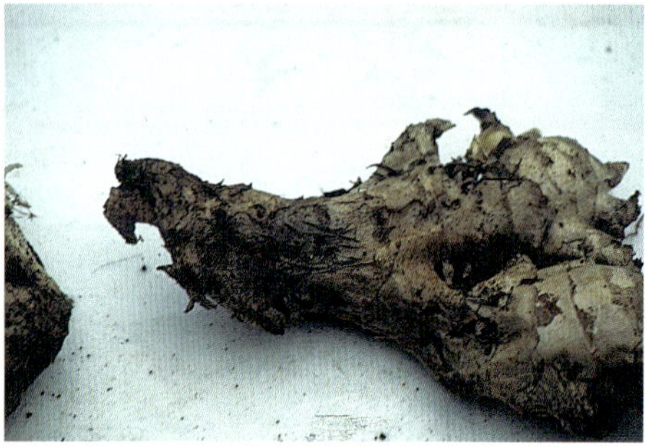

22C

22F

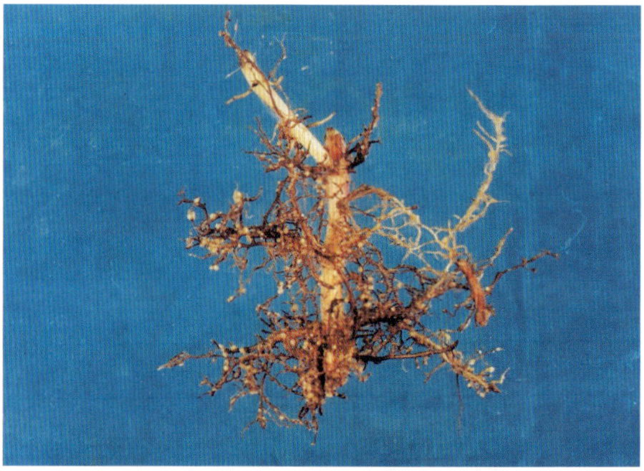

22E

Plate 22. (A) Symptoms of yellow or slow decline disease in black pepper caused by *Radopholus similis* (Photo: V.K. Sosamma). **(B)** Patch of dead and dying black pepper vines due to *Radopholus similis* in Bangka, Indonesia (Photo: J. Bridge). **(C)** Ginger rhizome infected with *Radopholus similis* in Fiji showing dark, shallow water-soaked lesions (Photo: J. Bridge). **(D)** Damage to turmeric rhizome due to *Pratylenchus* sp. infestation (Photo: V.K. Sosamma). **(E)** Root galls on *Withania somnifera* infested with *Meloidogyne* sp. (Photo: R. Pandey). **(F)** Galled roots of *Mentha arvensis* infested with *Meloidogyne* sp. showing large egg masses on surface of root (Photo: R. Pandey).

23A

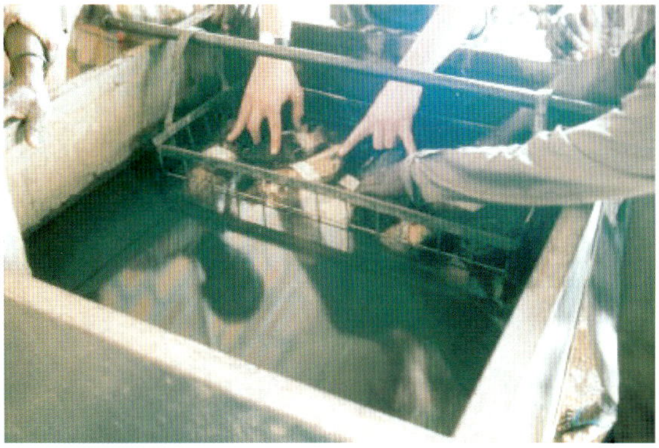

23B

23C

23D

23E

23F

Plate 23. (A) Coffee husks used to control root knot nematodes through stimulation of the antagonistic potential in the rhizosphere soil of short-cycle vegetables in the Philippines (Photo: R.A. Sikora). **(B)** Treatment of banana corms in a hot water bath to reduce *Radopholus similis* infestations, supplied by the plant protection unit in Tonga (Photo: P. Speijer). **(C)** Two species of *Crotalaria* used as an antagonistic crop and green manure to control *Meloidogyne incognita* on a medicinal crop in Brazil (Photo: R.A. Sikora). **(D)** Elevated beds planted to marigold (*Tagetes* sp.) as an antagonistic plant for incorporation as a green manure under plastic mulch to stimulate biofumigation for control of *Meloidogyne incognita* in Morocco (Photo: H. Kaak and R.A. Sikora). **(E)** Converted photograph taken initially with infrared photography over a sugarbeet field showing nests of damage caused by *Heterodera schachtii* on sugarbeet. **(F)** Solarization of vegetable beds with plastic mulch for root knot and wilt control in Jordan (Photo: H. Saleh).

24A

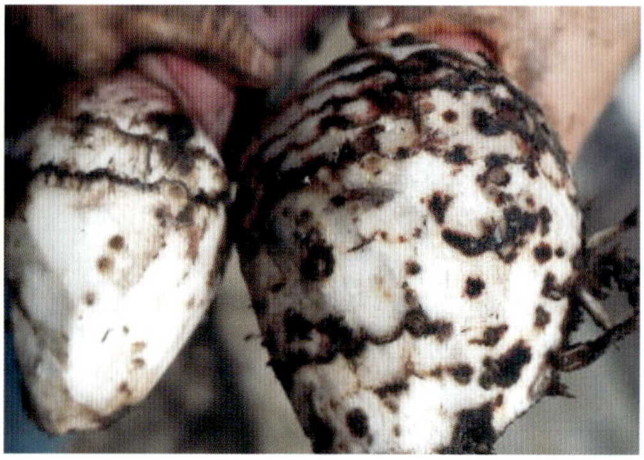

24B

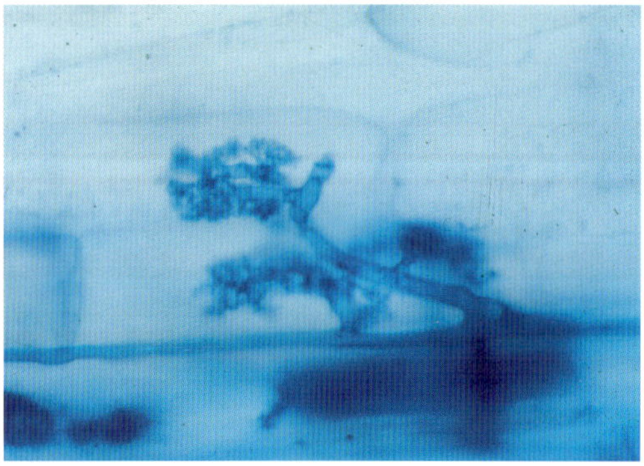

24C

24D

24E

24F

Plate 24. **(A)** Soil fumigation under plastic mulch for *Rotylenchulus reniformis* control in pineapple in Hawaii (Photo: R.A. Sikora). **(B)** Physical removal of *Radopholus similis* from banana corms in Tonga by paring, with dark spots being an indication of nematode infestation (Photo: P. Speijer and R.A. Sikora). **(C)** Arbuscule of an endotrophic mycorrhizal fungus in the roots of tomato, with known plant growth and health-promoting activity used for bio-enhancement of transplants for root knot control (Photo: R.A. Sikora). **(D)** Nematode-free tissue culture banana plants targeted for bio-enhancement with mutualistic fungal endophytes for *Radopholus similis* control in Costa Rica (Photo: L. Pocasangre, INIBAP & CATIE). **(E)** Highly susceptible and resistant groundnut cultivars growing in a field heavily infested with *Meloidogyne arenaria* in Texas, USA (Photo: J.L. Starr, Texas A&M University). **(F)** Tomato seedling showing grafted union between rootstock and shoot (Photo: R.A. Sikora, Taiwan).

2 Identification, Morphology and Biology of Plant Parasitic Nematodes*

David J. Hunt,[1] Michel Luc[2] and Rosa H. Manzanilla-López[3]
[1]*CABI Bioscience, Bakeham Lane, Egham, Surrey TW20 9TY, UK;* [2]*6 rue Boutard, 92200 Neuilly-sur-Seine, France;* [3]*Plant Nematode Interactions Unit, Rothamsted Research, Harpenden, Herts AL5 2JQ, UK*

Nematodes successfully colonize a greater variety of habitats than any other group of multicellular animals. Many species are free-living, feeding on bacteria or fungal spores, whereas others are predatory or parasitic in habit. The latter forms parasitize most groups of animals, including other nematodes, and a wide variety of algae, fungi and higher plants. However, despite such ecological diversity, nematodes are surprisingly similar in their structure.

This chapter starts with a brief, simplified account of the basic morphology, anatomy and bionomics of plant parasitic nematodes, followed by illustrated descriptions that concentrate on the diagnostic features of the most commonly occurring and/or most important plant parasitic genera referred to in the following chapters.

Morphology of Plant Parasitic Nematodes[1]

Plant parasitic nematodes almost invariably bear a mouth spear for penetrating plant cells, a feature that distinguishes them from the majority of other soil nematodes. It should be borne in mind, however, that non-phytoparasitic dorylaims also have a spear, as do many mycophagous, predatory or insect parasitic nematodes. The spear has evolved independently in each of the three major groups of plant parasitic nematodes. In the Tylenchida (including Tylenchina and Aphelenchina), the spear is also known as the **stylet**; in the Longidoridae (Dorylaimida), it is called the **odontostyle**; and in the Trichodoridae (Triplonchida), it is the **onchiostyle**. Tylenchs, the most speciose and important group of plant parasitic nematodes on a world scale, will be dealt with in most detail.

Tylenchs (Fig. 2.1A–J)

Tylenchs are basically bilaterally symmetrical, typically vermiform, animals that usually range from 0.2–1 mm in length. In some genera, the female loses the vermiform habit, becoming obese, even globose, in form.

*A revision of the chapter by M. Luc, D.J. Hunt and J.E. Machon.

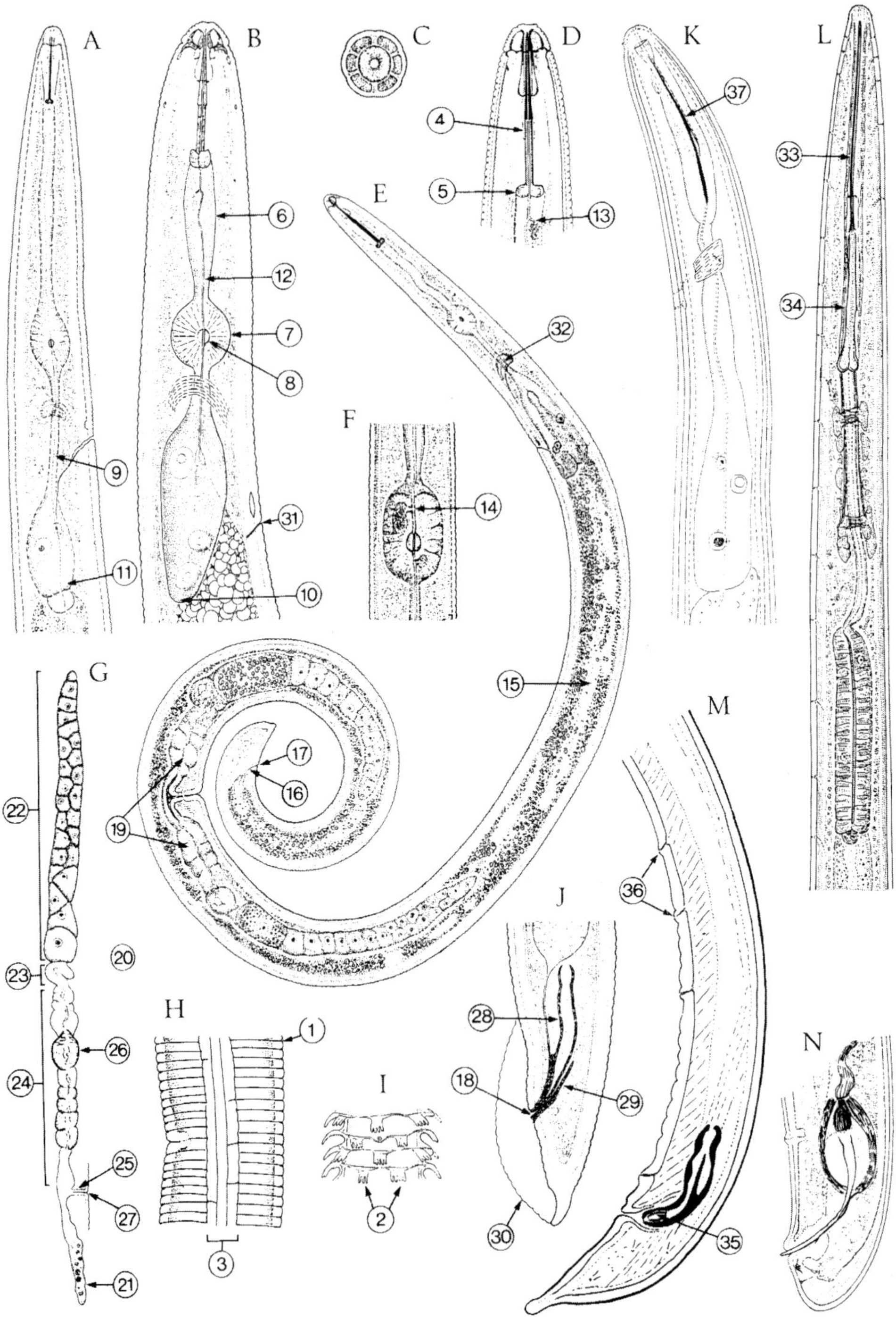

Fig. 2.1. Major diagnostic features of plant parasitic nematodes. Line drawings are for illustrative purposes only and are not to scale.

The **labial region**, when seen *en face* (Fig. 2.1C), is typically hexaradiate and has a central orifice, the mouth, through which the hollow stylet is protruded. Various sensory structures, including the **amphidial apertures**, occur on the labial region, which is often transversely annulated and usually separated from the body by a constriction. Internally, the labial region contains a sclerotized **framework** (or skeleton) to support the structure and for attachment of the stylet protractor muscles.

The body is enclosed in a cuticle, which is usually transversely annulated (H1) and may be ornamented with a variety of processes in the criconematid forms (I2). Longitudinal ridges occur in some species. Beneath the cuticle are the **hypodermis** and the longitudinal muscles, which are attached to four **chords** – longitudinal thickenings of the cuticle and hypodermis. The lateral chords are better developed than the ventral and dorsal ones and correspond externally to the **lateral field** which is marked by a number of **longitudinal lines** (H3) or **incisures**, the region between two incisures being known as a **band** or **ridge**. The central cavity of the nematode, the **pseudocoelom**, contains a viscous fluid, which acts as a **hydrostatic skeleton**. Suspended within the fluid are the three major organ systems – digestive, reproductive and excretory.

The digestive system comprises the **stylet**, **oesophagus**, **intestine** and **rectum**. The stylet (D4) is a protrusible cuticular tube, pointed anteriorly and with a subterminal aperture. It consists of an anterior conus attached posteriorly to a more or less cylindrical shaft, the latter generally swelling posteriorly to form three **basal knobs** (D5). Protractor muscles are attached to the knobs and extend anteriorly to the labial (or cephalic) skeleton.

The oesophagus (which is also referred to as the pharynx) comprises a narrow cylinder or procorpus (B6) which expands to form the **median bulb** (B7), a muscular swelling containing refringent valve plates (B8), before narrowing to the **isthmus** (A9) and then expanding into a glandular portion (B10, A11). There are three, one dorsal and two subventral, oesophageal glands which may form a bulb-like structure (A11) abutting the intestine or may be extended into a lobe overlapping the intestine (B10). Between the stylet base and the oesophago–intestinal junction runs a central tube, the **oesophageal lumen** (B12), through which glandular secretions and food pass. In Tylenchina, the dorsal oesophageal gland opens into the oesophageal lumen near the stylet base (D13), the two subventral glands opening within the median bulb, whereas in Aphelenchina all three glands open within the median bulb (F14). The **intestine** (E15) is a largely undifferentiated tube, which opens via the **rectum** (E16) at the **anus** (E17) or, in adult males, the **cloaca** (J18). In the males of certain genera, the digestive system is degenerate and non-functional.

The reproductive system in both sexes is tubular. The female genital system may be composed of two (E19), usually opposed, branches (**didelphic**) or reduced to a single branch (**monodelphic**). In monodelphy (G20), the posterior branch may be reduced to a **post-uterine sac** (G21) or be entirely absent, the other branch running anteriorly (**monoprodelphic**). Each branch has four major parts: **ovary** (G22), **oviduct** (G23), **uterus** (G24) and **vagina** (G25). There may also be a **spermatheca** (G26), a specialized uterine structure for storing sperm. The vagina opens to the exterior via the **vulva** (G27), a ventrally situated transverse slit in the middle or posterior section of the body. The male reproductive system is less variable. The single genital tube consists of a **testis**, **seminal vesicle** and **vas deferens** opening to the exterior with the intestine via a common aperture, the cloaca (J18). The copulatory organ consists of the paired **spicules** (J28) with a guiding piece, the **gubernaculum** (J29). The protrusible spicules are heavily cuticularized and serve to open the female vulva and channel sperm. The male tail often has cuticular expansions, the caudal alae (J30) or **bursa**, to assist in copulation.

The excretory system consists of a uninucleate **gland cell** connected via an **excretory canal** to the ventrally situated

excretory pore (B31). This pore is usually in the oesophageal region, but may be posteriorly located (e.g. *Tylenchulus*).

The nervous system consists of the **nerve ring** (E32), a circumoesophageal (sometimes circumintestinal) commisure, plus a network of nerves connected to body organs and various sensory structures. These sense organs are mostly in the labial region (**sensillae** and **amphids**), the oesophageal region (**cephalids**, **deirids**, **hemizonid** and **hemizonion**) and on the tail (**phasmids** and **caudalids**).

Longidoridae (Fig. 2.1L and M)

Compared with tylenchs, longidorids are much longer nematodes and range from 0.9 to over 12 mm in size. The cuticle is smooth and lateral fields are absent. The protrusible spear has a different origin from that of the tylenchs and is more properly called an odontostylet. It may be up to 300 μm long and consists of a needle-like **odontostyle** (L33) attached posteriorly to a cuticular extension, the **odontophore** (L34). A cuticularized **guiding ring** is located around the odontostyle. The oesophagus consists of a narrow anterior section and a posterior cylindroid expansion, which is both muscular and glandular. The female reproductive system is either didelphic or monodelphic; in the latter case, the anterior branch regresses and only the posterior branch remains (**opisthodelphic**). The male spicules are well developed and have lateral guiding pieces (M35). There is no gubernaculum or bursa, but a ventral series of sensory supplements (M36) run anteriorly from the cloaca. Some morphological features of tylenchs, such as excretory pore, phasmids, deirids and cephalids, are missing, whereas numerous somatic cuticular pores are present along the body.

Trichodoridae (Fig. 2.1K and N)

Trichodorids are rather plump, cigar-shaped nematodes, about 0.5–1.1 mm long and with a bluntly rounded labial region and tail. The cuticle is smooth and may swell enormously under the influence of acidic fixation. The curved spear is actually a mural tooth, and is properly referred to as an **onchiostyle** (K37). The oesophagus comprises a narrow cylindrical anterior section that swells gradually into a posterior bulboid expansion. The female genital system is usually didelphic, very exceptionally monodelphic. The male spicules are slightly curved and a weak bursa may be present. Ventral supplements occur.

Novel Approaches to Identification

Molecular methodologies in nematode identification and systematics have advanced tremendously in the last decade or so (see De Ley and Blaxter, 2002). Although widely used in systematics and phylogenetic studies, molecular techniques are also increasingly applicable to species identification, particularly so in morphologically conserved and/or speciose groups, such as the cyst nematodes, root knot nematodes, *Bursaphelenchus* and *Xiphinema*. Increasing attention is also being paid to other intractable groups, including the anguinids. In groups such as the heteroderids and meloidogynids, isozyme methodologies are also an important diagnostic tool.

Bionomics of Plant Parasitic Nematodes

Reproduction and development

Reproduction is usually either amphimictic (separate males and females) or parthenogenetic (males absent, very rare or nonfunctional), although hermaphrodism is also known. Eggs may be laid singly or stuck together in masses in a gelatinous matrix secreted by the female. Such egg masses are associated with species where the females swell and become sedentary, although some obese genera retain all the eggs within the body, the cuticle tanning on the death of the female to form a tough cyst. Egg sacs and cysts serve to protect the eggs.

Nematodes have four, exceptionally three (as in some longidorids), juvenile stages between the egg and adult, the intervening moults facilitating an increase in size. In tylenchs, the first stage juvenile, or J1, moults to the J2 within the egg, but in longidorids and trichodorids, it is the J1 that eclodes.

Environmental conditions

Although occupying many different ecological niches, nematodes are essentially aquatic animals. Plant parasitic nematodes require at least a film of water to enable locomotion and, as all species spend a greater or lesser proportion of their life within soil, its water content is a primary ecological factor. Although many species die in dry soils, others may survive in an anhydrobiotic state. Conversely, too much soil water may result in a lethal oxygen deficit, although certain genera (e.g. *Hirschmanniella*) thrive under such conditions.

Soil temperature is rarely a particularly important factor as it tends to remain reasonably stable in a given environment. Some tropical nematodes survive soil temperatures of 50°C, provided that sufficient time is available for them to enter anhydrobiosis.

Soil structure is influential as pore size affects the ease with which nematodes can move through the soil interstices. In general, sandy soils provide the best environment, soils with a high clay content or those with an excessively open texture inhibiting movement. However, saturated clay soils can be colonized successfully by certain specialized nematodes, including *Hirschmanniella* and some *Paralongidorus*. Soil pH may affect nematodes, but few data are available for tropical and subtropical species.

The maxim that 'where a plant is able to live, a nematode is able to attack it' is a good one. Nematodes are even able to attack the aerial parts of plants provided that the humidity is high enough to facilitate movement. Such conditions are provided in flooded rice fields where foliar species, such as *Aphelenchoides besseyi* and *Ditylenchus angustus*, can be devastating. Some *Bursaphelenchus* species, vectored by wood-boring insects, directly attack the trunk of coconut palm or pines. Other nematodes, such as some *Hirschmanniella* and *Halenchus* spp., attack algae and can live in seawater.

Hatching, host location and penetration

The eggs of many plant parasitic nematodes are deposited singly, either in the soil or within the plant tissues. Provided that other factors are favourable, they usually hatch irrespective of the presence of a host plant.

In the more advanced parasites, however, the eggs may be embedded in a gelatinous matrix to form an egg mass (e.g. *Meloidogyne*) or retained within the swollen female body, the cuticle of which tans to form a protective cyst (e.g. *Heterodera* and *Globodera*). Egg hatch in cyst nematodes is stimulated by root exudates from the host, a requirement that implies a restricted host range. Nematodes are attracted to plant roots by a variety of factors, which have yet to be fully elucidated. Such attractants can operate over considerable distances – up to 1 m, for example, in *Meloidogyne*.

There are three main types of parasitism (Fig. 2.2):

1. **Ectoparasitic** – the nematode remains in the soil and does not enter the plant tissues. It feeds by using the stylet to puncture plant cells; the longer the stylet, the deeper it can feed. The majority of ectoparasitic species remain motile, whereas some others, e.g. *Cacopaurus*, are permanently attached to the root by the stylet, which is deeply embedded in the plant tissue.
2. **Semi-endoparasitic** – only the anterior part of the nematode penetrates the root, the posterior section remaining in the soil phase.
3. **Endoparasitic** – the entire nematode penetrates the root. **Migratory endoparasites** retain their mobility and have no fixed feeding site within the plant tissue, whereas **sedentary endoparasites** have a fixed feeding site and induce a sophisticated trophic system of nurse cells or syncytia, thus allowing them to become obese and thereby lose their mobility.

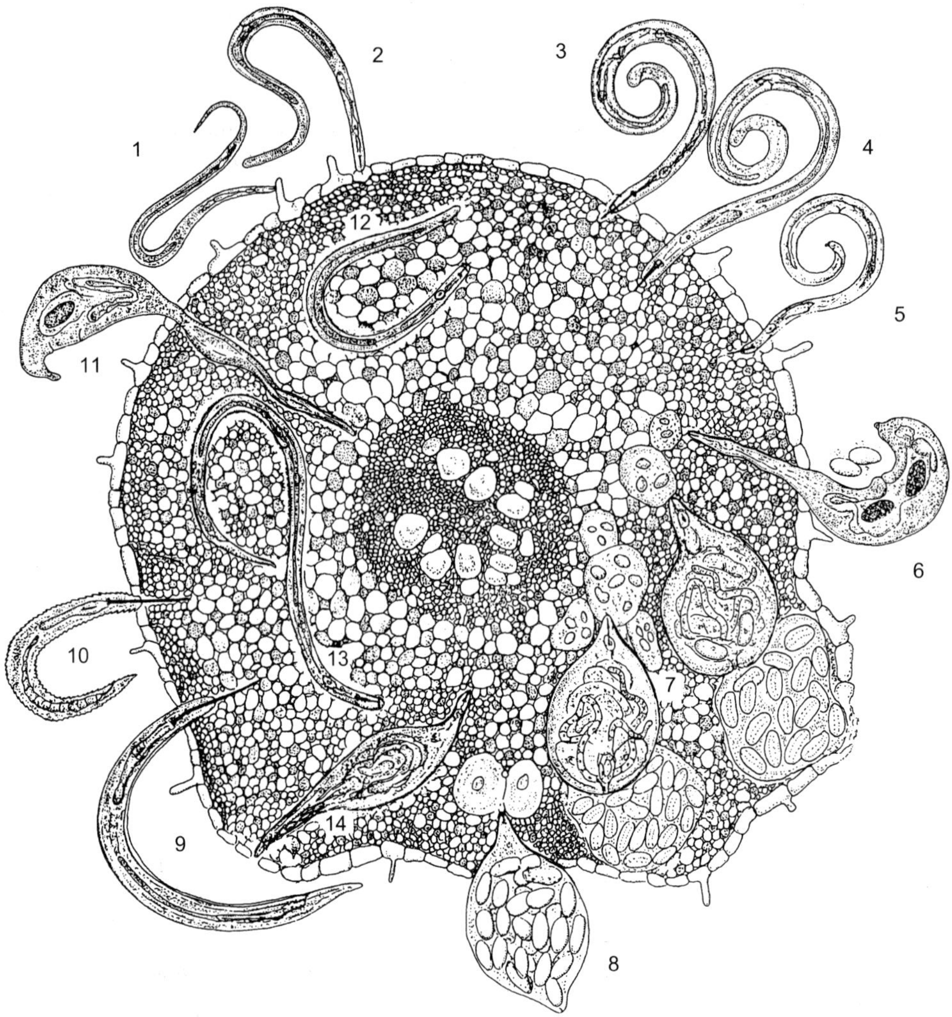

Fig. 2.2. Diagrammatic presentation of various types of tylenchid nematode feeding on root tissue. 1. *Ditylenchus*. 2. *Tylenchorhynchus*. 3. *Rotylenchus*. 4. *Hoplolaimus*. 5. *Helicotylenchus*. 6. *Rotylenchulus*. 7. *Meloidogyne*. 8. *Heterodera*. 9. *Hemicycliophora*. 10. *Criconemoides*. 11. *Tylenchulus*. 12. *Pratylenchus*. 13. *Hirschmanniella*. 14. *Nacobbus*. (Modified after Siddiqi, 1986.)

The above categories are not mutually exclusive as some genera may, depending on the host, be either semi-endoparasitic or migratory ectoparasitic, e.g. *Helicotylenchus*, whilst some sedentary parasites have only the anterior body embedded in the root (= **sedentary semi-endoparasites**), e.g. *Rotylenchulus* and *Tylenchulus*.

In *Meloidogyne* and *Heterodera/Globodera*, the J2 is the infective stage, but in ectoparasites and most migratory endoparasites any vermiform stage may feed on, or penetrate, the root (Fig. 2.3). Rarely, as in *Rotylenchulus*, the immature female is the infective stage, the non-feeding juveniles and males remaining in the soil.

Host reactions

As ectoparasites, e.g. *Tylenchorhynchus*, do not enter the plant, the damage they cause is usually limited to necrosis of those cells penetrated by the stylet. However, those species with longer stylets, such as *Xiphinema* or *Hemicycliophora*, can penetrate the tissues more deeply, thus killing more cells. Such nematodes tend to feed on meristematic tissue near the root tips, the concomitant damage resulting in galling or hooked roots and, if the growing point is destroyed, secondary root proliferation.

Endoparasites not only kill the cells they feed upon but, by burrowing through the root tissues, cause extensive destruction leading to cavitation and secondary infection. Successive generations of nematodes compound the damage, and it is not surprising that some of the most pathogenic nematodes belong to this group (*Pratylenchus*, *Radopholus* and *Hirschmanniella*).

Sedentary endoparasites have a sophisticated relationship with the host, involving transformation of root cells into a trophic system of nurse or transfer cells. The function of the trophic system is to operate as a nutrient sink so that the sedentary nematode is provided with a copious supply of nutrients, thus enabling it to increase enormously in size and thereby produce more eggs. In *Meloidogyne*, proliferation of the root cells is also incited, thus causing the characteristic galls.

Plants with the root system damaged by nematodes often show above-ground symptoms such as stunting, chlorosis, wilting, early senescence and reduced yield. These symptoms are a direct result of the impaired ability of the root system to deliver water and nutrients and thus may be confused with similar symptoms resulting from poor soil conditions and/or nutrient deficiencies.

The exact ways in which nematodes affect plants have yet to be fully elucidated and, besides impairing root function by physical damage, toxins may also be involved. An interesting case is 'Ontario peach decline' where a very low population of *Pratylenchus* can kill young trees. The nematodes metabolize the sugar part of cyanosides in the plant tissue and thus liberate the CNH radical which is highly toxic to the tree.

In nematology, a number of terms are used to describe the inter-relationships of host and parasite. Plants can be divided into **hosts** or **non-hosts** depending on whether nematode reproduction occurs. Non-hosts may be **immune**, i.e. no nematode penetration or reproduction, or **resistant**, i.e. allowing nematode penetration and a varying degree of parasitism, but not reproduction. Host plants are non-resistant or susceptible and can be good or poor hosts, depending on whether reproduction is high or low. Susceptible plants, which support the lowest levels of reproduction within a data set, have been referred to as partially resistant or even, in some cases (in an agronomical concept), as 'resistant'. Some resistant plants are used as 'trap crops' to attract the nematodes in the soil before sowing a crop susceptible to the nematode in question.

Variations in the ability of nematodes to reproduce on given plant species or cultivars are of great agricultural significance and are of two principal types. Nematode populations distinguished by their ability or inability to reproduce on designated plant species are known as host races. **Pathotypes** are variants of a host race or species, which are distinguished by their ability to reproduce on a designated host plant genotype (e.g. cultivar, line, etc.).

Tolerance refers to the amount of damage caused by the nematode to the plant and should not be confused with resistance (q.v.). A **tolerant host** suffers little damage even when heavily infected, whilst an **intolerant host** may be severely damaged, even if only lightly infested.

Survival

In the absence of a live host, nematodes may survive in the soil or in plant residues. Provided that the environment dries slowly, many nematodes are able to enter a reversible anhydrobiotic state when they are less susceptible to desiccation, temperature extremes and chemicals. In a number of genera, the eggs are the survival stage, being protected either in a gelatinous matrix (*Meloidogyne*, *Tylenchulus* and *Rotylenchulus*) or within the hardened cyst-like body of the dead female (*Heterodera* and *Globodera*). In the latter case, infective J2 nematodes may not hatch for several years after being laid. Anhydrobiosis is probably more common in tropical and subtropical areas than is currently realized and enables the organism to survive the dry season and also to nullify some non-chemical control methods, such as dry fallow. The record for longevity in the anhydrobiotic state is held by seed nematodes, such as *Anguina*, which have been recorded surviving for 39 years. A practical consequence of anyhydrobiosis is that extraction from dry soil requires a sufficient period of soaking for the nematodes to absorb water and thereby attain the active state.

Identification of the Major Genera

This section is intended to serve as a basic guide to the identification of the major parasitic genera found in tropical and subtropical agriculture. Each generic diagnosis has the major differential characters printed in bold. Genera are arranged according to systematic position (Table 2.1) rather than trophism. A full list of scientific authorities is given in Appendix B.

Table 2.1. Outline classification.

Order/suborder/superfamily	Family	Genus	Page
TYLENCHIDA			
Aphelenchina			
Aphelenchoidoidea	Aphelenchoididae	*Aphelenchoides*	19
		Bursaphelenchus	19
Tylenchina			
Tylenchoidea			
	Anguinidae	*Ditylenchus*	22
		Anguina[a]	22
	Belonolaimidae	*Tylenchorhynchus*	25
	Pratylenchidae	*Pratylenchus*	25
		Hirschmanniella	28
		Radopholus	28
		Nacobbus[a]	31
	Hoplolaimidae	*Helicotylenchus*	33
		Hoplolaimus	33
		Scutellonema	36
		Aorolaimus	36
		Aphasmatylenchus	36
		Rotylenchulus[a]	38
	Heteroderidae	*Heterodera*[a]	40
		Globodera[a]	40
	Meloidogynidae	*Meloidogyne*[a]	40
Criconematoidea	Criconematidae	*Criconemoides*	43
		Hemicycliophora	45
		Hemicriconemoides	45
	Tylenchulidae	*Tylenchulus*[a]	45
DORYLAIMIDA			
Dorylaimina			
Longidoroidea	Longidoridae	*Xiphinema*	48
		Longidorus	48
		Paralongidorus	48
TRIPLONCHIDA			
Diphtherophorina			
Trichodoroidea	Trichodoridae	*Trichodorus*	50
		Paratrichodorus	50

[a]Genera with obese sedentary females.

Aphelenchoides Fischer, 1894 (Aphelenchina, Aphelenchoididae)

Morphology: small to medium sized (0.4–1.2 mm), slender nematodes. **Females die straight or ventrally arcuate on heat relaxation, while the male tail curls ventrally to produce a 'walking-stick' shape. Labial region weakly sclerotized; stylet weak,** with or without basal swellings. **Oesophageal bulb well developed, spherical to rounded–rectangular in shape and more or less filling the body diameter. Dorsal oesophageal gland duct opening within bulb, just anterior to the valve plates. Oesophageal gland lobe overlapping intestine dorsally. Female: vulva posterior (60–75%); genital tract single, anteriorly directed. Tail medium conoid,** with or without terminal mucron(s). Male: tail medium conoid, **spicules** well developed, **thorn shaped. No bursa.**

Biology: ectoparasitic on leaves, stems and other parts of higher plants. Most species can also be readily cultured on various fungal hyphae. *Aphelenchoides besseyi* can withstand desiccation for several years. The life cycle is rapid and may be completed in as little as a week.

Major species: *Aphelenchoides* is a very speciose genus, the majority being fungal feeders. Several species, however, are also important phytoparasites, i.e. *A. arachidis*, *A. besseyi*, *A. fragariae* and *A. ritzemabosi.*

Distribution: *A. arachidis* is only currently recorded from groundnut (peanut) in northern Nigeria, but the other species are well distributed, with *A. besseyi* being found in most rice-growing areas.

Useful literature

CIH Descriptions of Plant-parasitic Nematodes, Sets 1–8. CAB International, Wallingford, UK (Set 1, No. 4; Set 3, No. 32; Set 8, No. 116).

Hunt, D.J. (1993) *Aphelenchida, Longidoridae and Trichodoridae: Their Systematics and Bionomics.* CAB International, Wallingford, UK.

Nickle, W.R. and Hooper, D.J. (1991) The Aphelenchina: bud, leaf, and insect nematodes. In: Nickle, W.R. (ed.) *Manual of Agricultural Helminthology.* Marcel Dekker, New York, pp. 465–507.

Bursaphelenchus Fuchs, 1937 (Aphelenchina, Aphelenchoididae)

= *Rhadinaphelenchus* J.B. Goodey, 1960

Morphology: the genus is similar in general respects to *Aphelenchoides*, although the male has **differently shaped spicules and cuticular alae (the 'bursa') on the tail tip.** In *B. cocophilus*, **both sexes are very slender (body length/body diameter = about 100). In addition, the female has an extremely long postvulval sac, very long, slightly tapering tail with a rounded tip, and a vulval flap.** The male tail tip bears a small cuticular flap (the 'bursa'), which is most easily visible in ventral view. **Dorsal limb of spicule elongate.**

Biology: mostly ectophoretic associates of various insects, including Coleoptera and Hymenoptera. There are two major phytoparasitic species, both being vectored by woodboring insects: *B. xylophilus*, which attacks pine trees; and *B. cocophilus* (formerly known as *Rhadinaphelenchus cocophilus*) which is parasitic in the stem of coconut palms, 10 g of tissue of which may contain 50,000 nematodes. *B. cocophilus* may also be found in cortical tissues of coconut roots. Infection often causes the development of a red or orange-red ring of tissue within the stem (hence the common name of 'red ring nematode'). The nematode is vectored by the *Rhynchophorus* palm weevil during oviposition, an infected palm dying in 2–4 months.

Major species: this is a large genus with many described species, although, of these, only two, *B. cocophilus* and *B. xylophilus*, are currently considered to be of major economic importance.

Taxonomic note: the red ring nematode has traditionally been placed in its own genus, i.e. *Rhadinaphelenchus.* Although this generic name may still be found in recent literature, the combination *B. cocophilus* is currently in more widespread use and is adopted herein.

Distribution: the genus is widespread, although *B. cocophilus* is restricted to the Caribbean, Central and South American regions. *B. xylophilus* is recorded from some tropical/subtropical regions, including Hong Kong and southern China, but mainly occurs in more temperate climates, e.g. Japan.

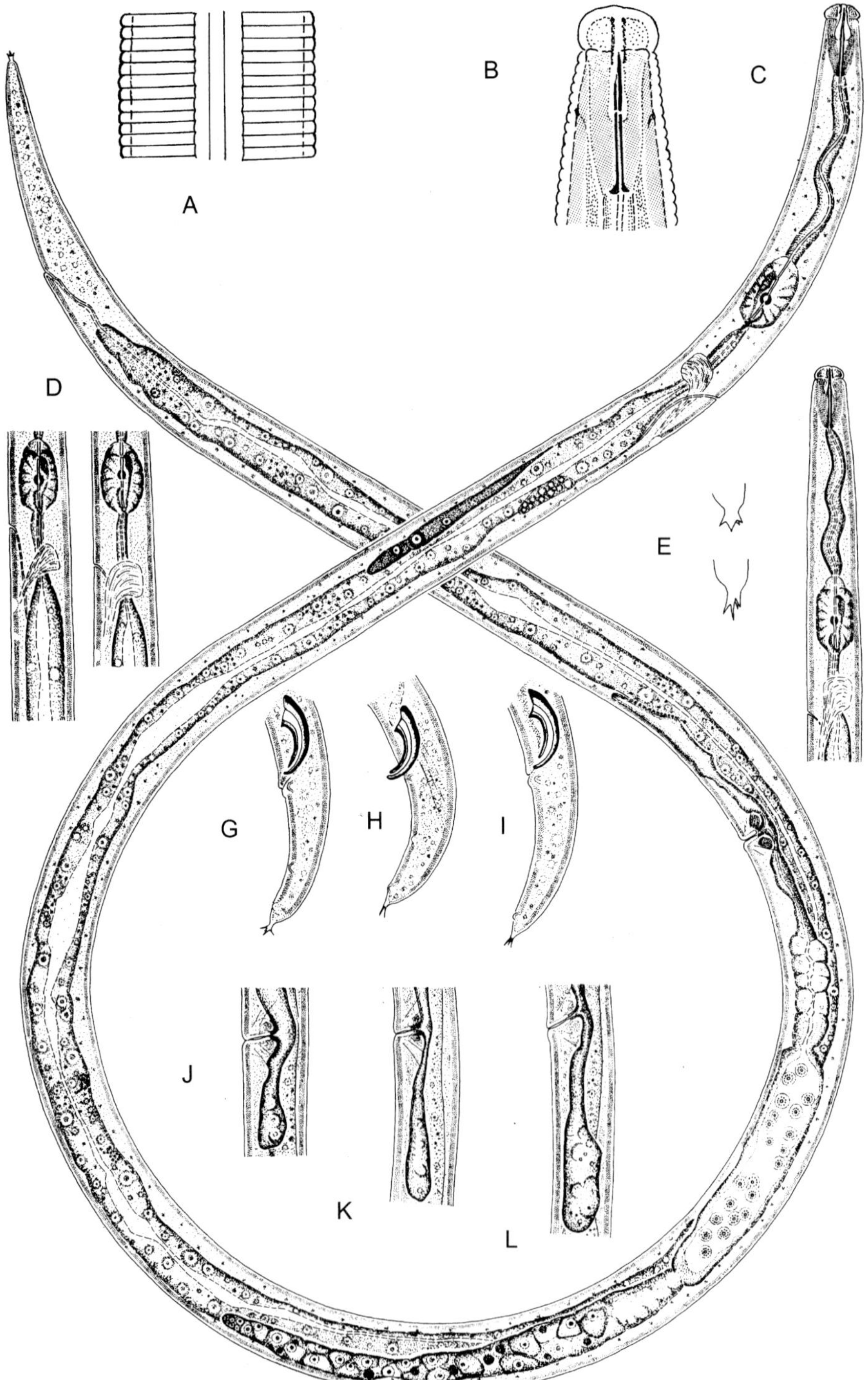

Fig. 2.3. *Aphelenchoides besseyi* (A) lateral field; (B) labial region; (C) entire female; (D) median bulb and excretory pore position; (E) female tail tips; (F) oesophageal region; (G–I) male tail region; (J–L) post-vulval sac. Line drawings are for illustrative purposes only and are not to scale.

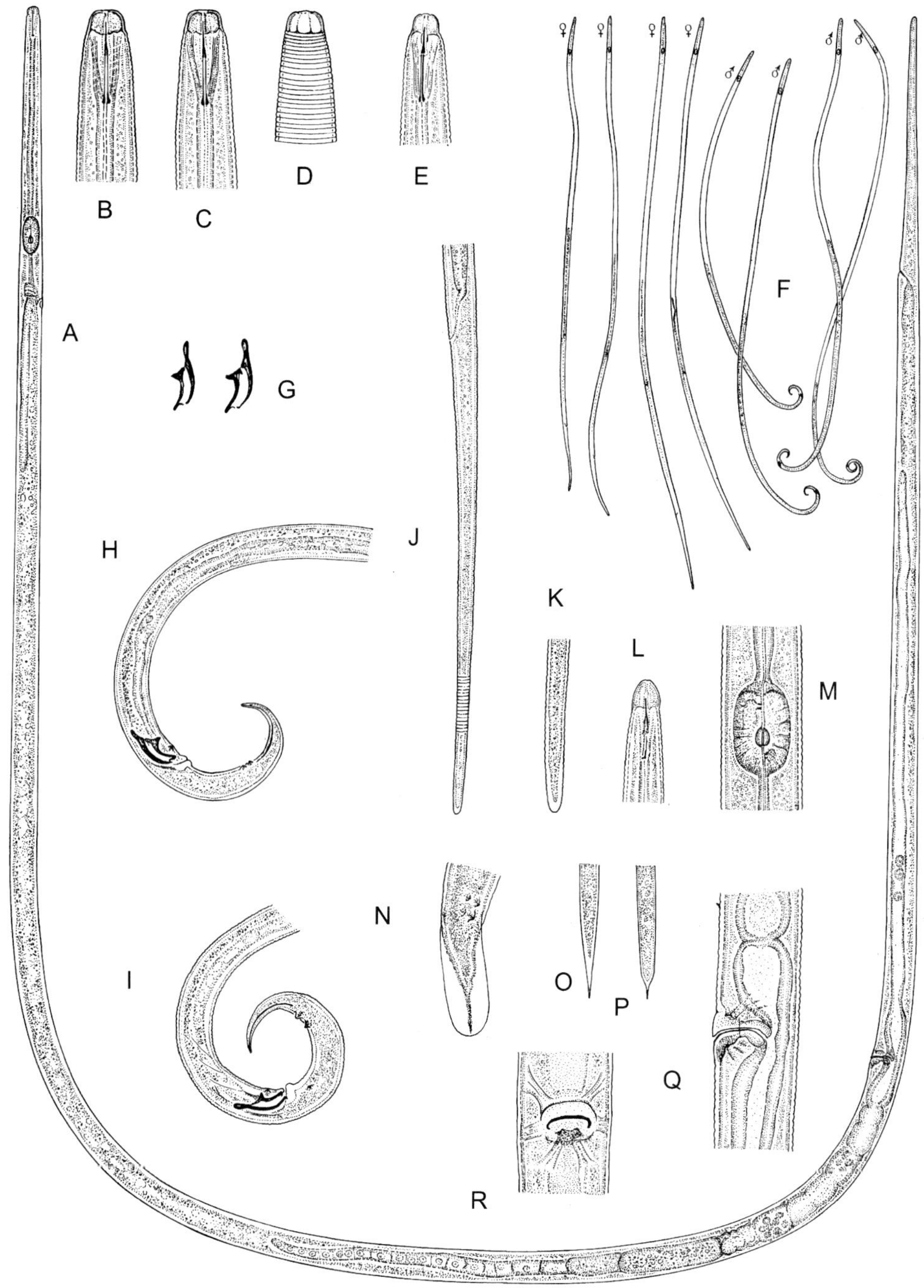

Fig. 2.4. *Bursaphelenchus cocophilus* (A) entire female; (B–D) female labial region; (E) male labial region; (F) entire females and males; (G) spicules; (H and I) male tail end; (J) female tail; (K) female tail tip; (L) juvenile labial region; (M) median bulb; (N) male 'bursal' flap; (O and P) juvenile tail tips; (Q) vulval region; (R) vulval slit in ventral view. Line drawings are for illustrative purposes only and are not to scale.

Useful literature

CIH Descriptions of Plant-parasitic Nematodes, Sets 1–8. CAB International, Wallingford, UK (Set 5, No. 72).

Dean, C.G. (1979) Red ring disease of *Cocos nucifera* L. caused by *Rhadinaphelenchus cocophilus* (Cobb, 1919) Goodey, 1960. An annotated bibliography and review. *Technical Communication No. 47*. CAB International, Wallingford, UK.

Hunt, D.J. (1993) *Aphelenchida, Longidoridae and Trichodoridae: Their Systematics and Bionomics*. CAB International, Wallingford, UK.

Nickle, W.R. and Hooper, D.J. (1991) The Aphelenchina: bud, leaf, and insect nematodes. In: Nickle, W.R. (ed.) *Manual of Agricultural Helminthology*. Marcel Dekker, New York, pp. 465–507.

Ditylenchus Filipjev, 1936 (Tylenchina, Anguinidae)

Morphology: slender nematodes dying straight or slightly curved ventrally on heat relaxation. Skeleton of labial region weakly sclerotized. Stylet of moderate strength and with small basal knobs. Oesophagus with a muscular median bulb; **isthmus gradually expanding to form the basal bulb**, which may extend as a lobe over the intestine. Female: **vulva well posterior. Genital tract single, anteriorly outstretched. Post-uterine sac present. Tail elongate, conoid.** Male: **bursa adanal,** not reaching tail tip. **Tail elongate, conoid.**

Biology: ectoparasites of plant stems and leaves but also found within the tissues. Infected stems and leaves are often stunted and deformed.

Major species: a large genus, most species of which are fungal feeders. Major phytoparasitic species include *D. angustus*, *D. dipsaci* and *D. africanus*.

Distribution: *D. angustus* is found in rice-growing areas of Bangladesh, Vietnam and other areas of Asia; *D. dipsaci* is restricted to the cooler regions of the tropics and subtropics, and *D. africanus* is so far known only from South Africa and Mozambique.

Confusable genus: *Aphelenchoides*, *Anguina* (juvenile stages).

Useful literature

CIH Descriptions of Plant-parasitic Nematodes, Sets 1–8. CAB International, Wallingford, UK (Set 1, No. 14; Set 5, No. 64).

Fortuner, R. (1982) On the genus *Ditylenchus* Filipjev, 1936 (Nematoda: Tylenchida). *Revue de Nématologie* 5, 17–38.

Sturhan, D. and Brzeski, M.W. (1991) Stem and bulb nematodes, *Ditylenchus* spp. In: Nickle, W.R. (ed.) *Manual of Agricultural Helminthology*. Marcel Dekker, New York, pp. 423–464.

Anguina Scopoli, 1777 (Tylenchina, Anguinidae)

Morphology: sexually dimorphic. Adult stages only found in plant galls, juveniles occurring in galls, plant tissue or soil, depending on stage of life cycle. General morphology similar to that of *Ditylenchus*. Female: obese, medium to large nematodes (1.5–5 mm) **dying spirally coiled** on heat relaxation. **Vulva very posterior with a single, anteriorly directed genital tract reflexed twice or more.** Numerous oocytes. Male: small to medium sized (1–2.5 mm) dying ventrally or dorsally (e.g. as in *A. tritici*) arcuate. **Testis well developed with one or more flexures. Bursa adanal.**

Biology: forming galls on stems, leaves or flowers of various plants. The J2 stage is found in the soil and feeds ectoparasitically on the plant tissues. The final moult takes place after gall formation, each female laying up to 2000 eggs. As the gall matures and dries, the J2 infectives slowly desiccate to an anhydrobiotic state and may survive many years.

Major species: *A. agrostis* complex, *A. tritici*.

Confusable genus: *Ditylenchus*, as the soil-dwelling juveniles look similar.

Useful literature

CIH Descriptions of Plant-parasitic Nematodes, Sets 1–8. CAB International, Wallingford, UK (Set 1, No. 13; Set 2, No. 20).

Brzeski, M.W. (1981) The genera of Anguinidae (Nematoda, Tylenchida). *Revue de Nématologie* 4, 23–34.

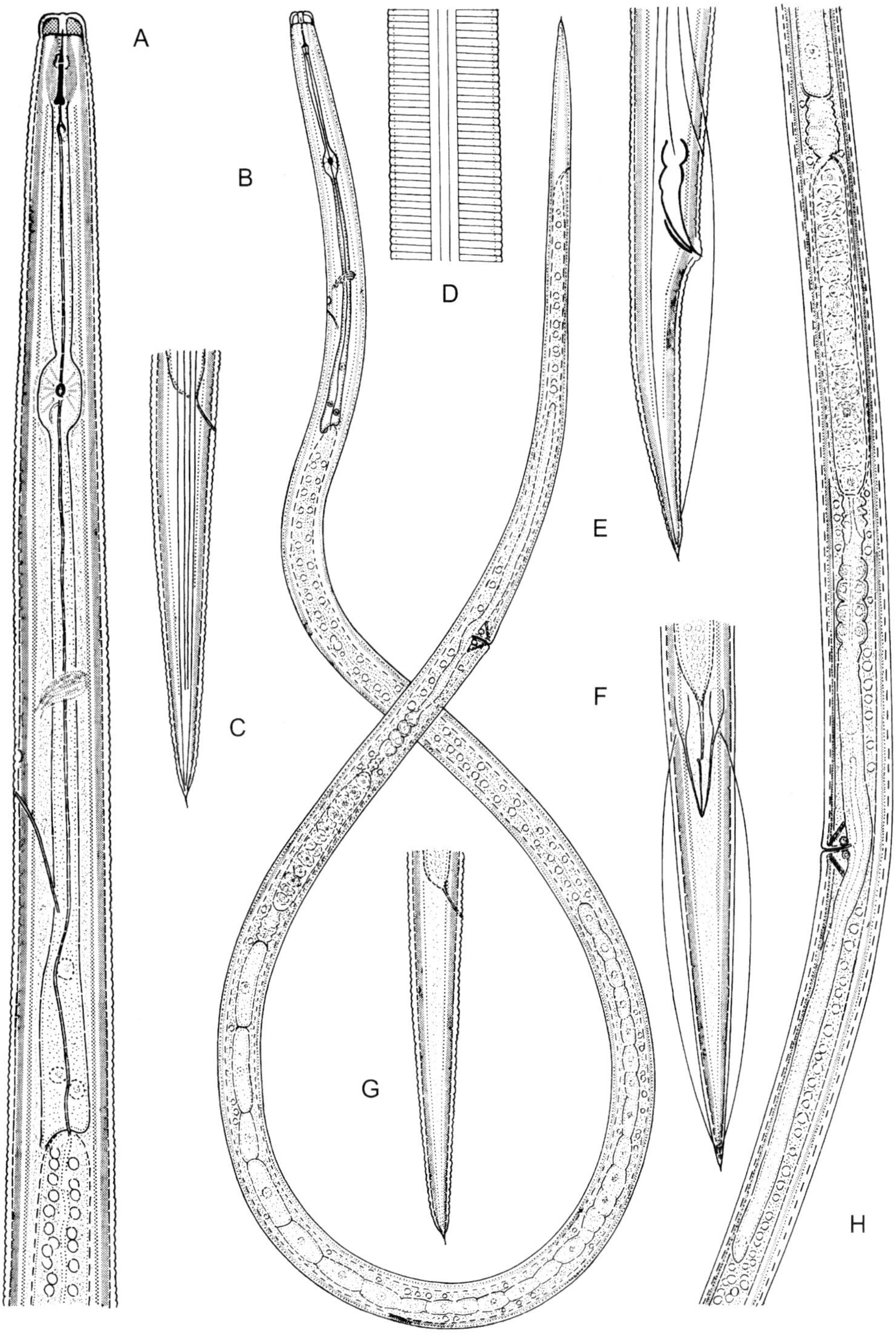

Fig. 2.5. *Ditylenchus angustus* (A) female oesophageal region; (B) entire female; (C and G) female tails; (D) lateral field; (E and F) male tails; (H) female genital tract. Line drawings are for illustrative purposes only and are not to scale.

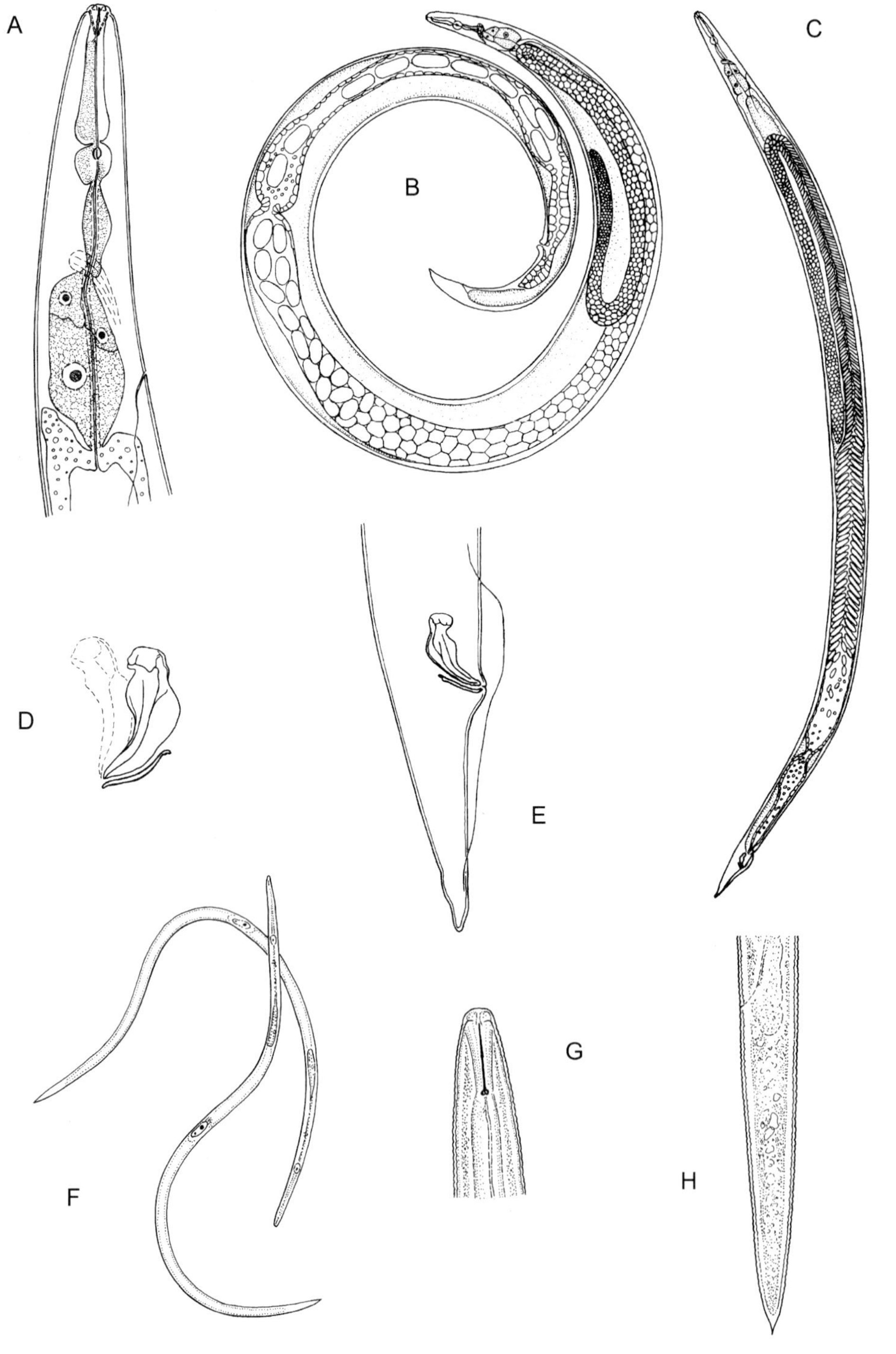

Fig. 2.6. *Anguina tritici* (A) female oesophagus; (B) entire female; (C) entire male; (D) male spicules; (E) male tail; (F) second stage juveniles; (G) J2 labial region; (H) J2 tail. (After Goodey, 1932; Thorne, 1949; Siddiqi, 1972.) Line drawings are for illustrative purposes only and are not to scale.

Chizov, V.N. and Subbotin, S.A. (1992) [Phytoparasitic nematodes of the subfamily Anguininae (Nematoda, Tylenchida). Morphology, trophic specialization, systematics.] *Zoologichesky Zhurnal* 69, 15–26 (In Russian).

Krall, E.L. (1991) Wheat and grass nematodes: *Anguina, Subanguina,* and related genera. In: Nickle, W.R. (ed.) *Manual of Agricultural Helminthology.* Marcel Dekker, New York, pp. 721–760.

Tylenchorhynchus Cobb, 1913 (Tylenchina, Belonolaimidae)

= *Telotylenchus, Dolichorhynchus, Trilineellus, Divittus, Morasinema, Tessellus, Neodolichorhynchus, Mulkorhynchus*

Morphology: small nematodes (rarely over 1 mm long), dying more or less straight or slightly curved ventrally on application of gentle heat. **No marked sexual dimorphism in form of anterior region.** Labial region rounded, continuous with body contour or slightly offset, with narrow annules and weak sclerotization. **Stylet slender, 15–30 μm long, moderately sclerotized with rounded, backwardly sloping, knobs.** Lateral field with two, three or four lines; cuticle sometimes divided into blocks. Oesophagus equally developed in both sexes; median bulb fusiform, moderately developed; oesophageal glands abutting the intestine or, very rarely, overlapping. Female: **vulva median with two equally developed genital tracts;** one directed anteriorly, one posteriorly. Spermatheca rounded. **Tail, about three anal body diameters long, conoid to subcylindrical, with rounded tip.** Male: tail elongate, conical–pointed, bursa extending to tail tip, trilobed in some species. Spicules slightly curved.

Biology: migratory ecto-, semi-ecto- or endoparasites. Most species are bisexual. Polyphagous. Not considered as being very important parasites. Well distributed in all climatic areas.

Major species: *T. annulatus, T. brassicae, T. mashhoodi.*

Confusable genera: *Trichotylenchus, Quinisulcius, Merlinius, Amplimerlinius.*

Useful literature

Anderson, R.V. and Potter, J.W. (1991) Stunt nematodes: *Tylenchorhynchus, Merlinius,* and related genera. In: Nickle, W.R. (ed.) *Manual of Agricultural Helminthology.* Marcel Dekker, New York, pp. 529–586.

CIH Descriptions of Plant-parasitic Nematodes, Sets 1–8. CAB International, Wallingford, UK (Set 6, No. 85).

Fortuner, R. and Luc, M. (1987) A reappraisal of Tylenchina (Nemata). 6. The family Belonolaimidae Whitehead, 1960. *Revue de Nématologie* 10, 183–202.

Pratylenchus Filipjev, 1936 (Tylenchina, Pratylenchidae)

Morphology: small nematodes (<1 mm long) dying slightly curved ventrally on application of gentle heat. No marked sexual dimorphism in form of anterior region. **Labial region strongly sclerotized, low, flattened,** usually appearing as a dark, flat cap under the stereomicroscope, **divided into two, three or four annules and continuous with the body contour. Stylet is approximately 20 μm or less in length** (i.e. less than three times as long as the labial region diameter), moderately sclerotized and with rounded or anteriorly concave knobs. Oesophagus equally developed in both sexes, median bulb well developed; **oesophageal gland lobes overlapping intestine ventrally. Female: vulva well posterior at 70–80% of body length; genital system with a single, anteriorly directed, tract** (monoprodelphic) and a variable post-vulval section which may show some differentiation, but which is never functional; spermatheca oval or round and usually filled with sperm in bisexual species. **Tail subcylindroid or more or less conoid with a broad to narrowly rounded or truncate terminus,** which may be smooth or annulated. Male: tail short, dorsally convex–conoid; **bursa extending to tail tip;** spicules slender, arcuate.

Biology: migratory endoparasites with all stages found in the root cortex. Low soil populations can be associated with high root populations. The nematodes feed mainly on cortex cells and form cavities

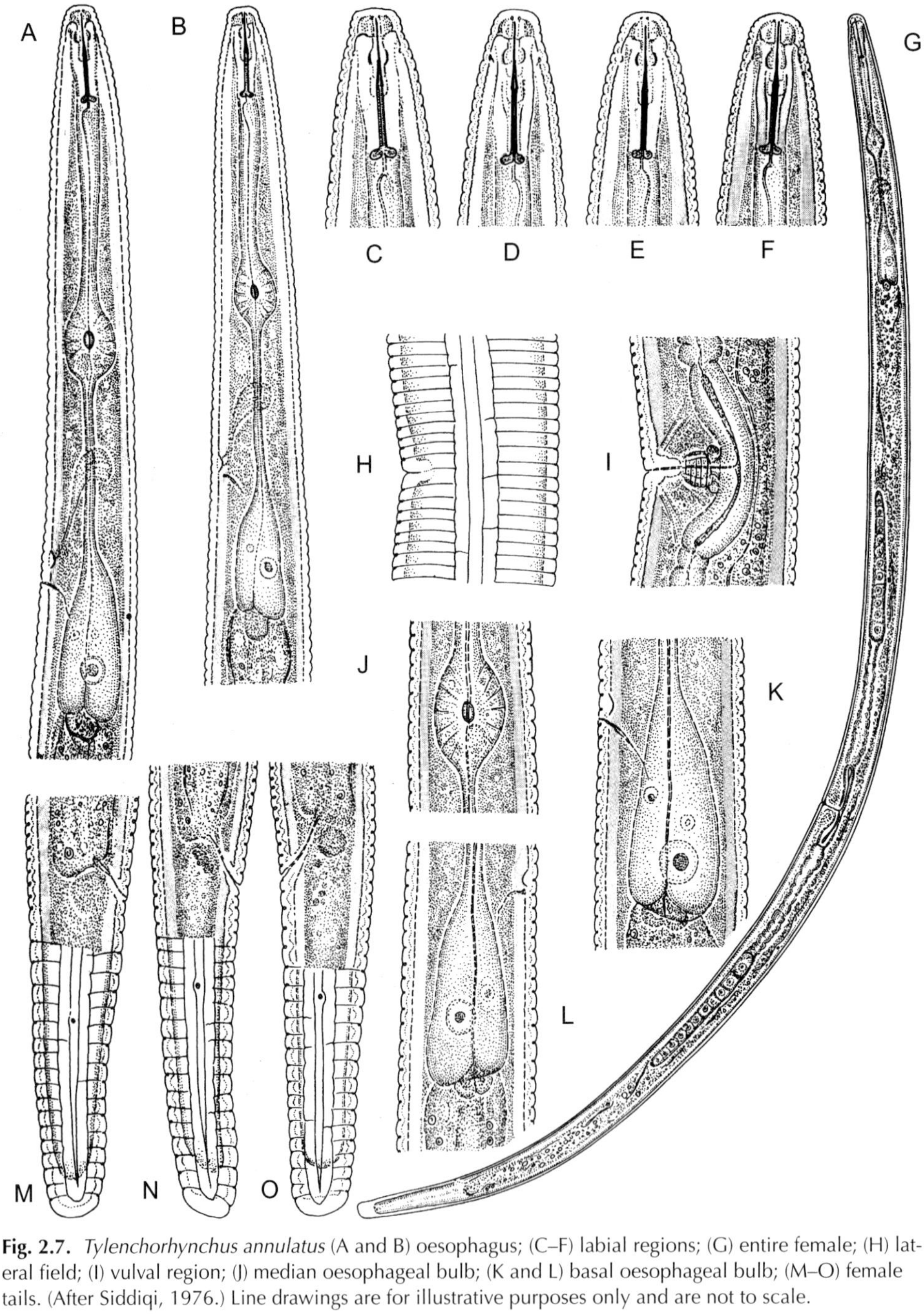

Fig. 2.7. *Tylenchorhynchus annulatus* (A and B) oesophagus; (C–F) labial regions; (G) entire female; (H) lateral field; (I) vulval region; (J) median oesophageal bulb; (K and L) basal oesophageal bulb; (M–O) female tails. (After Siddiqi, 1976.) Line drawings are for illustrative purposes only and are not to scale.

containing 'nests' or colonies of nematodes of all stages. Discoloration of affected tissues is usually pronounced. Above-ground symptoms of attack include chlorosis and stunting. Some species reproduce sexually while others are parthenogenetic. The life cycle may be completed in 3–4 weeks and the nematodes can survive in the absence of

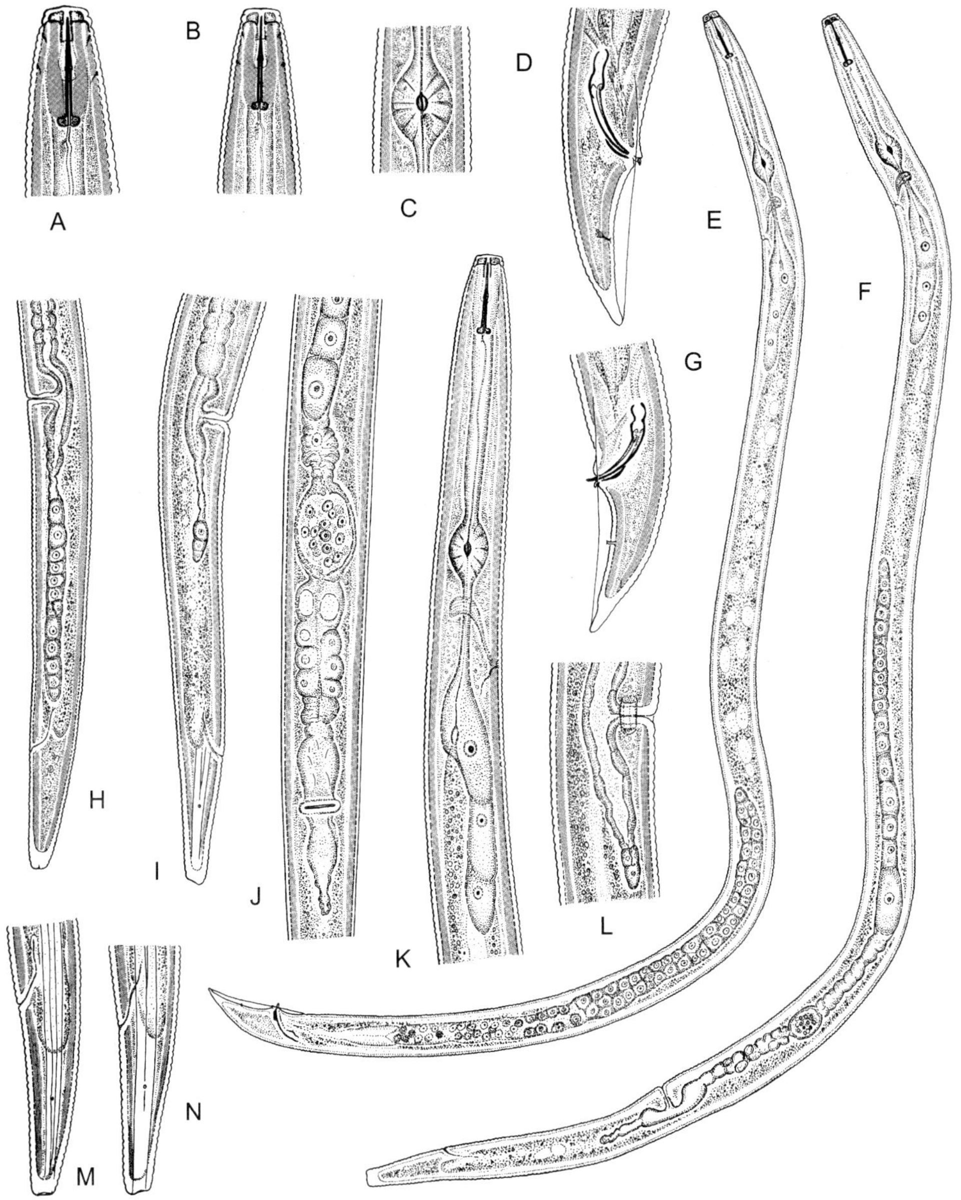

Fig. 2.8. *Pratylenchus coffeae* (A) female labial region; (B) male labial region; (C) median bulb; (D and G) male tail; (E) entire male; (F) entire female; (H and I) female posterior region; (J) female vulval region, ventral view; (K) oesophageal region; (L) vulval region; (M and N) female tails. (After Siddiqi, 1976.) Line drawings are for illustrative purposes only and are not to scale.

host plants for several months. Most important species are polyphagous, although *P. goodeyi* may be restricted to banana.
Major species: *P. brachyurus*, *P. coffeae*, *P. goodeyi*, *P. penetrans*, *P. zeae*.

Distribution: *P. brachyurus*, *P. coffeae* and *P. zeae* are widely distributed in tropical and subtropical areas; *P. penetrans* mainly in cooler regions of the tropics; *P. goodeyi* on banana in Crete and the Canary Islands

and in the cooler areas of Ethiopia, Kenya, Tanzania, Uganda and Burundi.
Confusable genus: *Radopholus*. Novices may confuse with *Aphelenchus avenae*, particularly in populations of the latter where there are abundant males.

Useful literature

Café Filho, A.C. and Huang, C.S. (1989) Description of *Pratylenchus pseudofallax* n.sp. with a key to species of the genus *Pratylenchus* Filipjev, 1936 (Nematoda: Pratylenchidae). *Revue de Nématologie* 12, 7–15.

CIH Descriptions of Plant-parasitic Nematodes, Sets 1–8. CAB International, Wallingford, UK (Set 1, No. 6; Set 2, No. 25; Set 6, Nos 77, 89; Set 8, No. 120).

Handoo, Z.A. and Golden, A.M. (1989) A key and diagnostic compendium to the species of the genus *Pratylenchus* Filipjev, 1936 (Lesion nematodes). *Journal of Nematology* 21, 202–218.

Loof, P.A.A. (1978) The genus *Pratylenchus* Filipjev, 1936 (Nematoda: Pratylenchidae): a review of its anatomy, morphology, distribution, systematics and identification. *Vaxskyddsrapporter, Jordbruk 5* Uppsala, Sweden.

Loof, P.A.A. (1991) The family Pratylenchidae Thorne, 1949. In: Nickle, W.R. (ed.) *Manual of Agricultural Helminthology*. Marcel Dekker, New York, pp. 363–421.

Hirschmanniella Luc & Goodey, 1963 (Tylenchina, Pratylenchidae)

Morphology: medium sized to long, slender nematodes (1.1–4 mm) **dying more or less straight or ventrally arcuate** on application of gentle heat. No marked sexual dimorphism in form of anterior region. **Labial region** continuous with body contour, **hemispherical or anteriorly flattened, annulated. Stylet strongly developed** (15–46 μm) with rounded basal knobs. **Oesophageal glands elongate and overlapping intestine in a long ventral lobe. Female: vulva median; genital system with two functional and equally developed genital tracts,** one anteriorly and one posteriorly directed; **tail elongate, conoid, terminal mucron often present. Male tail similar to female; bursa not reaching to tail tip,** spicules slender, arcuate.

Biology: migratory endoparasites, mainly of roots, but also corms and rhizomes, where they move freely through the tissues. Eggs are laid within the root, and development to the adult takes about 5–6 weeks. The genus is associated with aquatic environments – marsh, freshwater and marine. Most species are bisexual.
Major species: *H. mexicana* (= *caudacrena*), *H. imamuri*, *H. miticausa*, *H. mucronata*, *H. oryzae*, *H. spinicaudata*.
Distribution: the genus is distributed worldwide in suitable habitats, with *H. oryzae*, the major species, being widely distributed in the rice-growing areas of India, Bangladesh, Malaysia, Indonesia, the Philippines and Japan. It is also found in parts of Africa and South America.
Confusable genus: *Radopholus*.

Useful literature

CIH Descriptions of Plant-parasitic Nematodes, Sets 1–8. CAB International, Wallingford, UK (Set 2, No. 26; Set 5, No. 68).

Ebsary, B.A. and Anderson, R.V. (1982) Two new species of *Hirschmanniella* Luc & Goodey, 1963 (Nematoda: Pratylenchidae) with a key to nominal species. *Canadian Journal of Zoology* 60, 530–535.

Loof, P.A.A. (1991) The family Pratylenchidae Thorne, 1949. In: Nickle, W.R. (ed.) *Manual of Agricultural Helminthology*. Marcel Dekker, New York, pp. 363–421.

Radopholus Thorne, 1949 (Tylenchina, Pratylenchidae)

= *Neoradopholus*

Morphology: small nematodes (<1 mm long) dying more or less straight or slightly curved ventrally when heat relaxed. **Marked sexual dimorphism in form of anterior region: female labial region low, rounded, continuous or slightly offset from body contour; male labial region higher, often knob-like and more offset. Male labial sclerotization, stylet and oesophagus reduced;** female cephalic sclerotization strong, stylet and oesophagus well developed. Median bulb in female oesophagus well developed and **oesophageal glands**

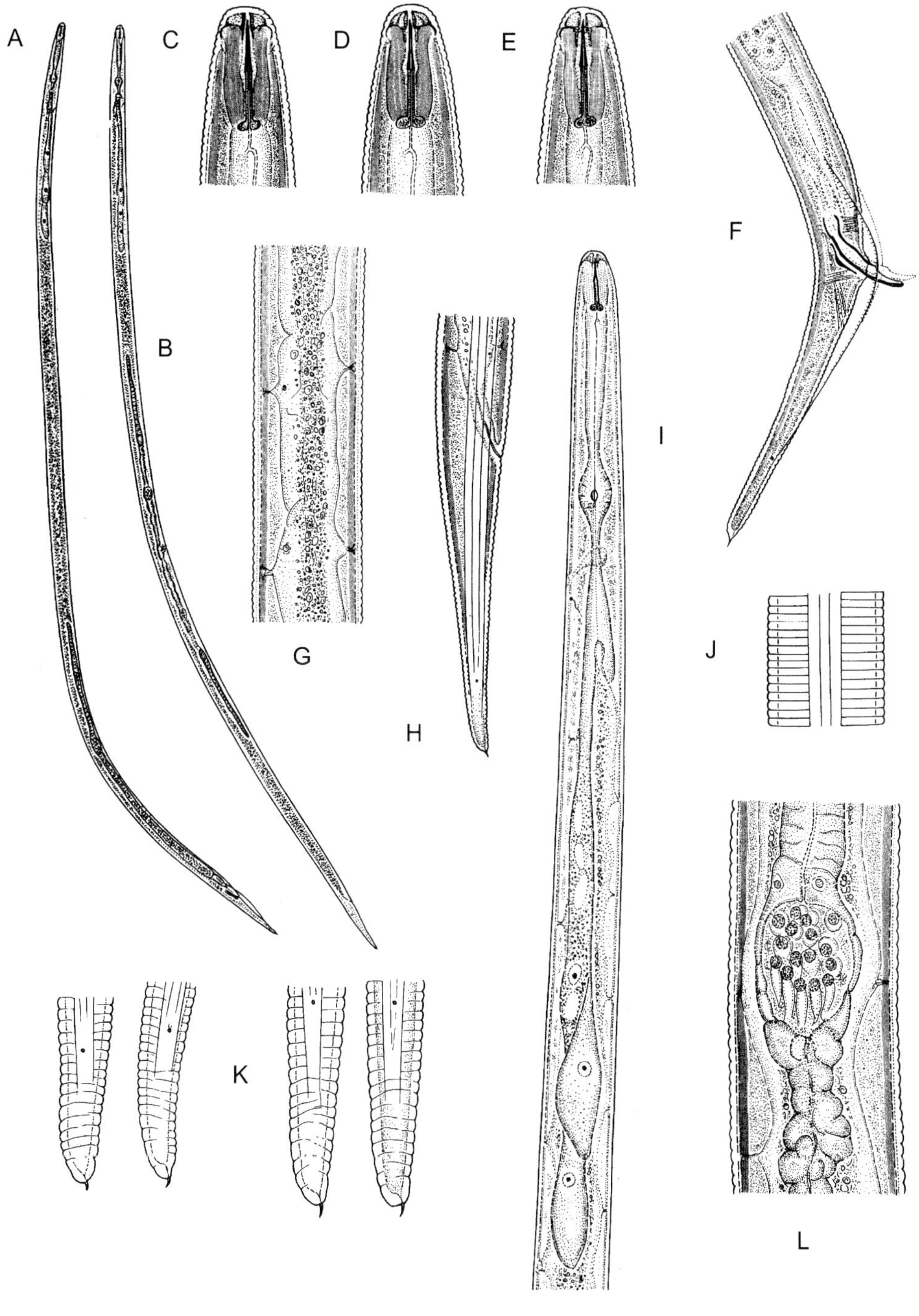

Fig. 2.9. *Hirschmanniella oryzae* (A) entire male; (B) entire female; (C–E) labial region; (F) male tail; (G) mid-body showing 'Thorneian cells'; (H) female tail; (I) oesophageal region; (J) lateral field; (K) female tail tips; (L) spermatheca with sperm. (After Siddiqi, 1973.) Line drawings are for illustrative purposes only and are not to scale.

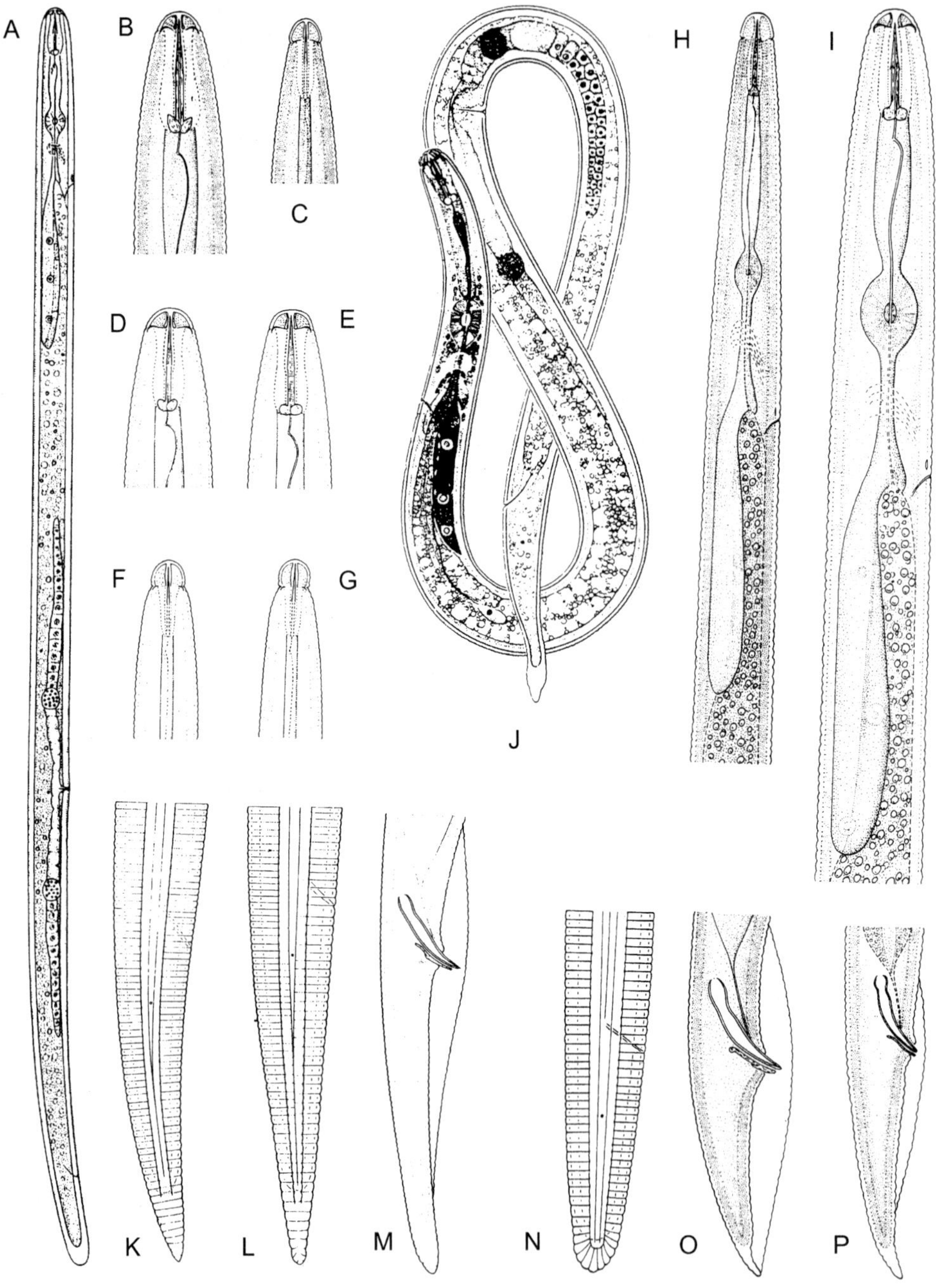

Fig. 2.10. (A) *Radopholus rotundiseminus*, (H, I and P) *R. vangundyi*, (B, C, N and O) *R. inaequalis* and (D–G and J–M) *R. similis*. (A) Entire female; (B, D and E) female labial region; (C, F and G) male labial region; (H) male oesophagus; (I) female oesophagus; (J) entire female; (K and L) female tails; (M) male tail; (N) female tail; (O) male tail; (P) male tail. Line drawings are for illustrative purposes only and are not to scale.

mostly overlapping intestine dorsally. Female: **vulva median, with two functional and equally developed genital tracts,** spermathecae rounded and with sperm in bisexual species; tail elongate, conoid (~60 μm long in *R. similis*). Male: tail elongate, conoid, ventrally arcuate; **bursa not reaching to tail tip in most species,** including *R. similis*; spicules slender, arcuate.

Biology: migratory endoparasites of root and corm/tuber tissues. In roots, the feeding activities are restricted to the cortex causing cavitation, discoloration and severe damage, allowing secondary invasion by other microorganisms. The adult male is non-feeding. The major species is *R. similis* which has two recognized host races or biotypes, one attacking banana and many other plants, but not citrus, the other (previously recognized as a separate species, *R. citrophilus*, by some authorities) attacking both citrus and banana as well as a variety of other plants. It is possible that *R. similis* includes a range of host races, current evidence also indicating a highly variable pathogenicity.

Major species: *R. similis, R. citri, R. bridgei, R. duriophilus, R. musicola.*

Distribution: the majority of species have been described from Australasia. However, *R. similis* has been introduced worldwide in tropical regions and occurs virtually everywhere that banana is grown. The citrus race of *R. similis* is only recorded from Florida.

Confusable genera: *Achlysiella, Pratylenchus, Hirschmanniella.*

Useful literature

CIH Descriptions of Plant-parasitic Nematodes, Sets 1–8. CAB International, Wallingford, UK (Set 2, No. 27).

Colbran, R.C. (1970) Studies of plant and soil nematodes. 15. Eleven new species of *Radopholus* Thorne and a new species of *Radopholoides* de Guiran (Nematoda: Tylenchoidea) from Australia. *Queensland Journal of Agricultural and Animal Sciences* 27, 437–460.

El-Badri, G.A.A., Geraert, E. and Moens, M. (1999) Morphological differences among *Radopholus* populations (Nematoda: Tylenchida) from banana in Africa. *Journal of Nematode Morphology and Systematics* 2, 1–16.

Loof, P.A.A. (1991) The family Pratylenchidae Thorne, 1949. In: Nickle, W.R. (ed.) *Manual of Agricultural Helminthology.* Marcel Dekker, New York, pp. 363–421.

Ryss, A. (1997) Computerized identification of species of the genus *Radopholus* (Tylenchida: Pratylenchidae). *Russian Journal of Nematology* 2, 137–142.

Sher, S.A. (1968) Revision of the genus *Radopholus* Thorne, 1949 (Nematoda: Tylenchoidea). *Proceedings of the Helminthological Society of Washington* 35, 219–237.

Nacobbus Thorne & Allen, 1944 (Tylenchina, Pratylenchidae)

Morphology: sexually dimorphic. Immature female (in soil or in roots). Vermiform, slender, 0.6–1 mm long. Labial area rounded, continuous with body contour. Labial sclerotization strong; stylet robust, with rounded basal knobs. Oesophagus with strong median bulb and strong valves; **oesophageal glands long, overlapping intestine dorsally. Vulva located posteriorly** (V = 90–95%); vulval lips not protruding. **Single anterior genital tract present. Tail short, rounded.** Mature females (in roots): **body saccate; anterior and posterior portions tapering.** Genital tract convoluted. Tail short. Male: similar to immature female, except for sexual characters. Spicules curved. Tail short; bursa reaching tail tip. Juveniles: uncoiled J4 resembles immature female.

Biology: in some species, the eggs are laid within a gelatinous matrix formed by the female. On hatching, the J2 invades a root, but does not form a fixed feeding site. Instead the juveniles migrate through the tissue and may even leave the root and enter another. The J3 and J4 stages are less mobile. After the final moult, the immature female may leave the root and enter another before taking up position near the vascular tissue and initiating a syncytial trophic system and gall formation. As the female develops, the posterior region extends towards the epidermis and an opening in the gall is formed through which the gelatinous matrix and eggs are extruded. In another species, *N. dorsalis*, the eggs are retained within the female body.

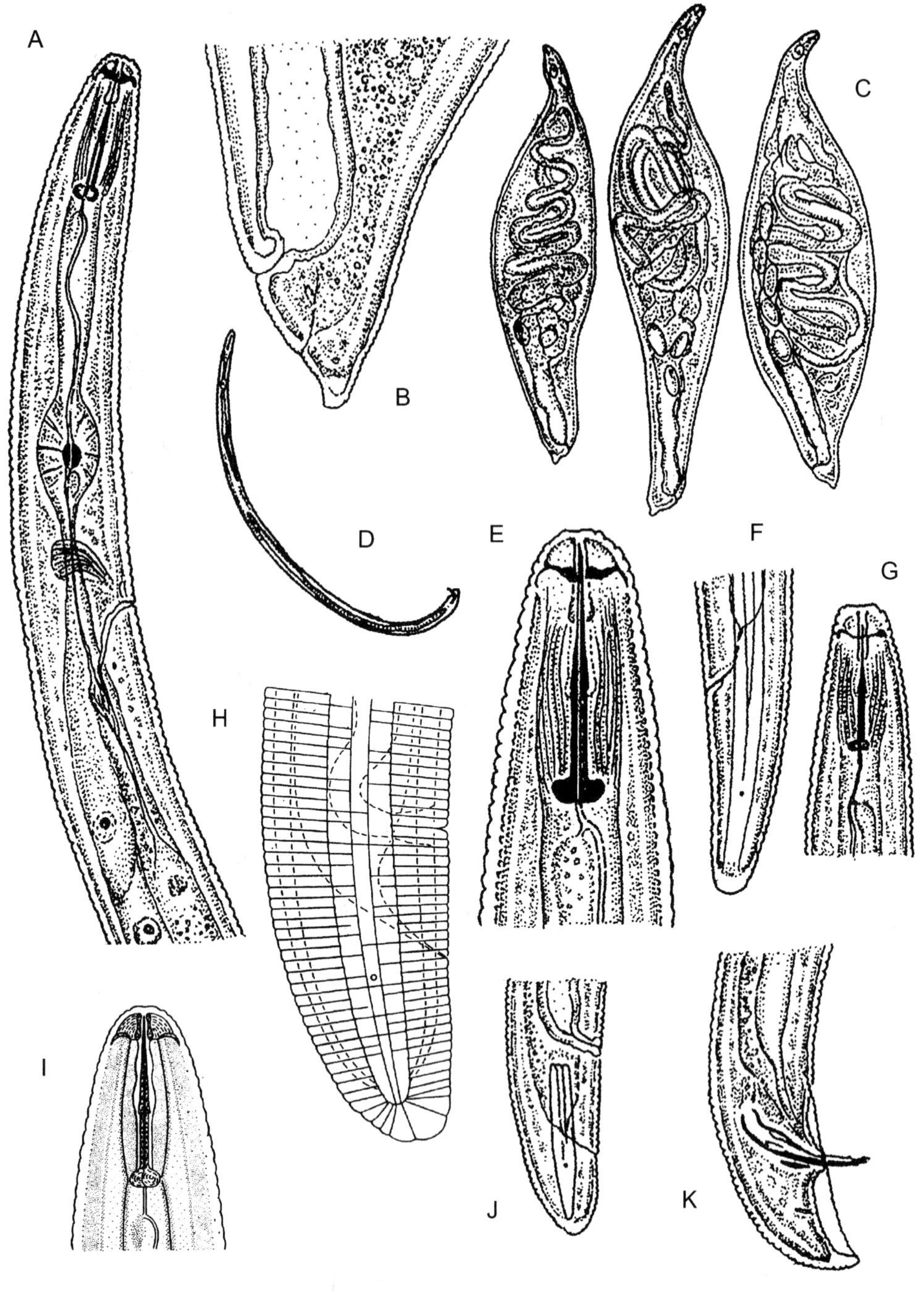

Fig. 2.11. *Nacobbus aberrans* (A) male oesophageal region; (B) tail region of mature female; (C) mature females; (D) entire male; (E and I) male labial region; (F) tail region of second stage juvenile; (G) labial region of second stage juvenile; (H and J) immature female posterior region; (K) male tail. Line drawings are for illustrative purposes only and are not to scale.

Major species: *N. aberrans*, *N. bolivianus*, *N. dorsalis*.
Distribution: indigenous to the Americas and only known to be established there.
Confusable genus: mature females may be confused with *Meloidogyne*. Under the stereomicroscope, immature vermiform females may be confused with *Meloidogyne* males, and the coiled J3 or J4 juveniles may be confused with *Helicotylenchus*.

Useful literature

CIH Descriptions of Plant-parasitic Nematodes, Sets 1–8. CAB International, Wallingford, UK (Set 8, No. 119).

Jatala, P. (1991) Reniform and false root-knot nematodes, *Rotylenchulus* and *Nacobbus* spp. In: Nickle, W.R. (ed.) *Manual of Agricultural Helminthology*. Marcel Dekker, New York, pp. 509–528.

Manzanilla-López, R.H., Costilla, M.A., Doucet, M., Franco, J., Inserra, R.N., Lehman, P.S., Cid del Prado-Vera, I., Souza, R.M. and Evans, K. (2003) The genus *Nacobbus* Thorne & Allen, 1944 (Nematoda: Pratylenchidae): systematics, distribution, biology and management. *Nematropica* 32, 149–227.

Reid, A., Manzanilla-López, R.H. and Hunt, D.J. (2003) *Nacobbus aberrans* (Thorne, 1935) Thorne & Allen, 1944 (Nematoda: Pratylenchidae); a nascent species complex revealed by RFLP analysis and sequencing of the ITS-rDNA region. *Nematology* 5, 441–451.

Helicotylenchus Steiner, 1945 (Tylenchina, Hoplolaimidae)

= *Rotylenchoides*, *Zimmermannia*

Morphology: small to medium sized nematodes (0.4–1.2 mm) **usually dying in a spiral** (rarely C-shaped) on heat relaxation. **Labial region conoid–rounded, rarely truncate, sclerotization moderate. Stylet well developed, usually 3–4 times the lip region diameter in length and with rounded or cup-shaped knobs. Opening of dorsal oesophageal gland duct 25–50% of stylet length posterior to knobs. Oesophageal gland lobe overlapping intestine mainly ventrally.** Female: vulva posterior (60–70%), **both genital tracts usually fully developed,** posterior branch rarely reduced and non-functional (= *Rotylenchoides*). **Tail short, usually dorsally convex–conoid or hemispherical.** A terminal projection or mucron may be present. **Phasmids small, dot-like.** Male: tail short, spicules well developed, arcuate. Bursa reaching tail tip.
Biology: ectoparasitic, semi-endoparasitic or endoparasitic nematodes of roots. All stages can be found in the root cortex, but migration through the tissues has not been reported. Small lesions are formed which become necrotic as secondary invasion proceeds. Polyphagous. Most species are parthenogenetic but one of the most common and most damaging species, *H. multicinctus*, is amphimictic.
Major species: *H. dihystera*, *H. erythrinae*, *H. mucronatus*, *H. multicinctus*, *H. pseudorobustus*.
Distribution: throughout the tropical and subtropical areas.
Confusable genus: *Rotylenchus* (has the dorsal oesophageal gland duct opening more anterior and dorsally overlapping gland lobe). J2 stage may be confused with *Rotylenchulus* juveniles.

Useful literature

Boag, B. and Jairajpuri, M.S. (1985) *Helicotylenchus scoticus* n.sp. and a conspectus of the genus *Helicotylenchus* Steiner, 1945 (Tylenchida: Nematoda). *Systematic Parasitology* 7, 47–58.

CIH Descriptions of Plant-parasitic Nematodes, Sets 1–8. CAB International, Wallingford, UK (Set 1, No. 9; Set 2, No. 23; Set 8, No. 109).

Fortuner, R. (1991) The Hoplolaiminae. In: Nickle, W.R. (ed.) *Manual of Agricultural Helminthology*. Marcel Dekker, New York, pp. 669–719.

Hoplolaimus von Daday, 1905 (Tylenchina, Hoplolaimidae)

= *Basirolaimus*, *Hoplolaimoides*

Morphology: nematodes of medium length (1–2 mm) dying slightly curved ventrally on application of gentle heat. **Labial region high, offset,** rounded and **with massive sclerotization.** Basal lip annule may be

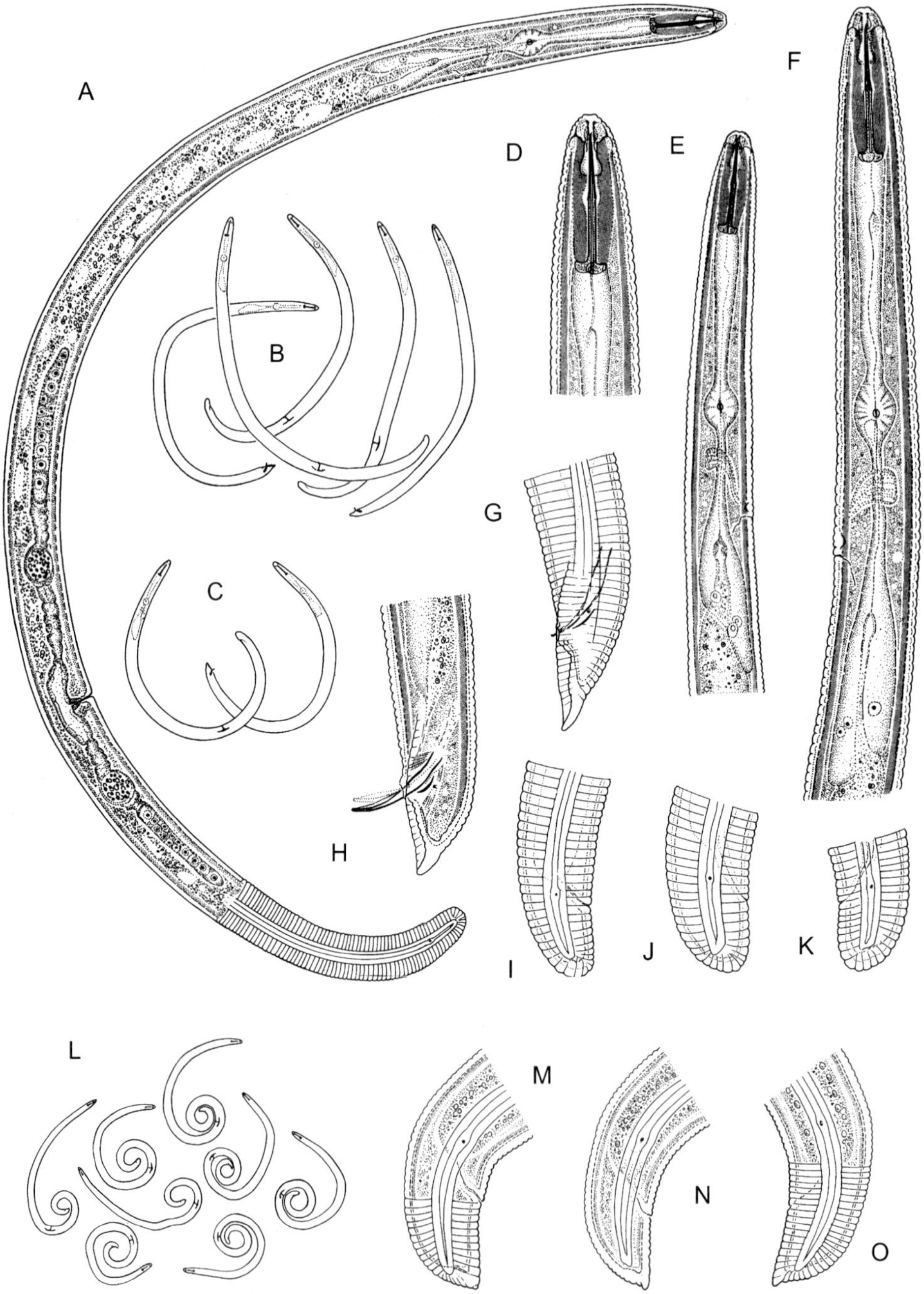

Fig. 2.12. *Helicotylenchus multicinctus* (A) entire female; (B) females; (C) males; (D) female labial region; (E) female oesophagus; (F) male oesophagus; (G and H) male tails; (I–K) female tails. *H. dihystera* (L) females; (M–O) female tails. (After Siddiqi, 1972, 1973.) Line drawings are for illustrative purposes only and are not to scale.

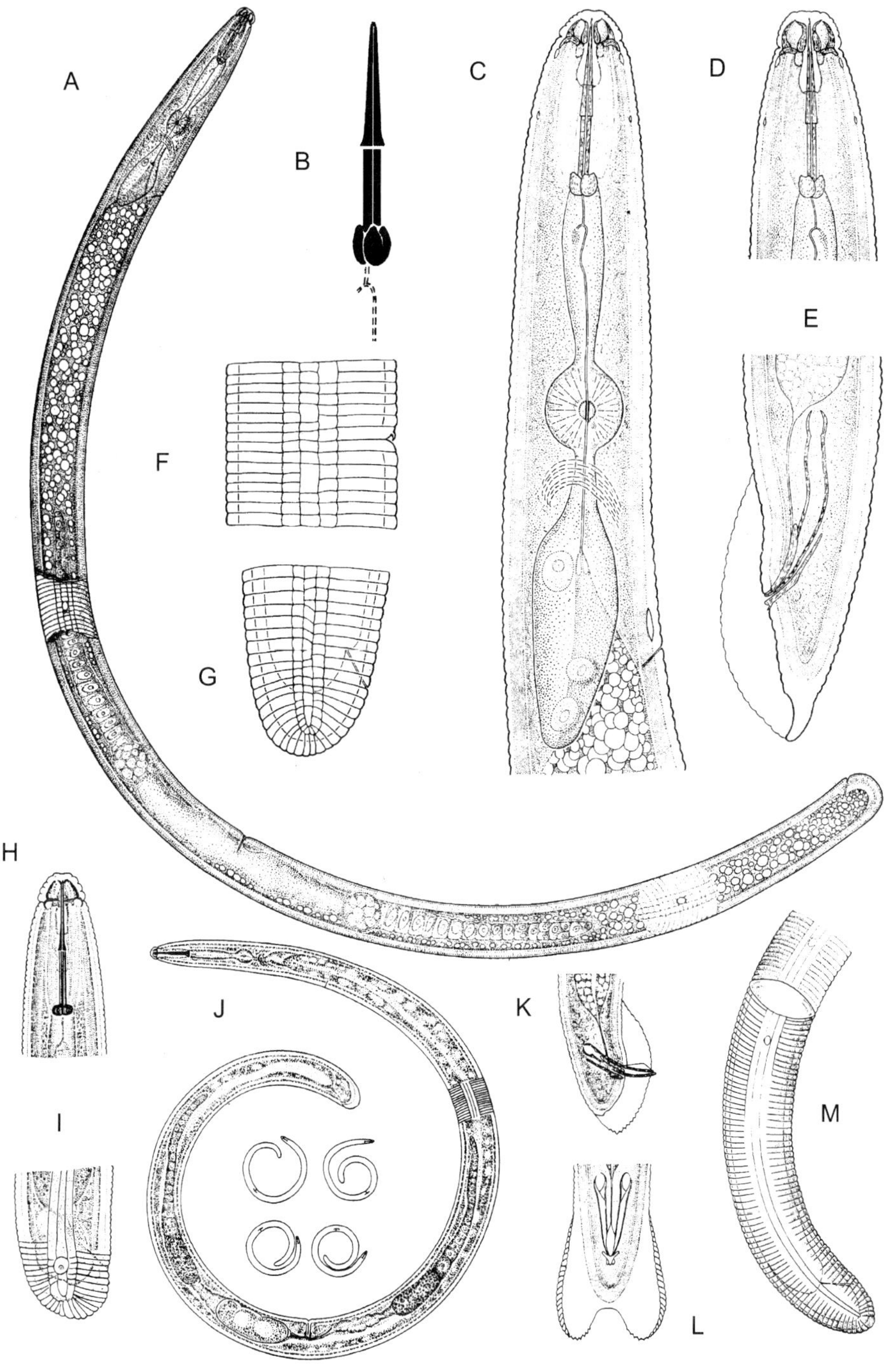

Fig. 2.13. *Hoplolaimus galeatus* (A) entire female; (C) female oesophagus; (D) male labial region; (E) male tail; (F) vulval region and lateral field; (G) female tail. *H. seinhorsti* (B) stylet and tulip-shaped knobs. *Scutellonema brachyurus* (H) labial region; (I) female tail; (J) adult females. *Aorolaimus luci* (K) male tail, lateral view; (L) male tail, ventral view; (M) female posterior region showing scutella. Line drawings are for illustrative purposes only and are not to scale.

divided into small squares. **Stylet massive, 40–50 μm long, with well developed basal knobs bearing anterior tooth-like projections.** Oesophagus well developed with a **dorsally overlapping gland lobe** containing either three or six (= *Basirolaimus*) nuclei. Female: vulva median, genital system consisting of two opposed tracts. Tail short, bluntly rounded. **Phasmids enlarged to form scutella, one being between anus and vulva and the other anterior to vulva.** Male: tail short, spicules well developed, arcuate. Bursa extending to tail tip. **Scutella situated at similar relative positions to female.**
Major species: *H. columbus*, *H. indicus*, *H. pararobustus*, *H. seinhorsti*.

Scutellonema Andrássy, 1958 (Tylenchina, Hoplolaimidae)

Morphology: small to medium-sized nematodes (0.3–1.5 mm) usually dying in a C-shape or open spiral. **Labial region with moderate sclerotization. Stylet of medium development with rounded knobs. Oesophagus with dorsal overlap.** Female: vulva median with two opposed genital tracts. Tail short, bluntly rounded. **Phasmids enlarged to form scutella which are opposite one another and either on, or very near, the tail.** Male: tail short, spicules well developed, arcuate. Bursa extending to tail tip.
Major species: *S. brachyurus*, *S. bradys*, *S. cavenessi*.

Aorolaimus Sher, 1964 (Tylenchina, Hoplolaimidae)

= *Peltamigratus*, *Nectopelta*

Morphology: similar to *Scutellonema* in general characters but female differs in having **scutella located well anterior to anus (yet posterior to vulva) and not opposite one another**. Males have scutella similarly arranged to the female and a large bursa which in many species is extended beyond the tail tip as two lobes.
Major species: *A. luci*.

Biology: all three genera are migratory endoparasites of roots and/or tubers. Most species are polyphagous. Reproduction can be amphimictic or parthenogenetic. *Scutellonema bradys* causes a serious dry rot of yam tubers.
Distribution: widespread in tropical and subtropical areas although *Aorolaimus* is more restricted to South America and parts of Africa.

Useful literature

Bittencourt, C. and Huang, C.S. (1986) Brazilian *Peltamigratus* Sher, 1964 (Nematoda: Hoplolaimidae), with descriptions of six new species. *Revue de Nématologie* 9, 3–24.

CIH Descriptions of Plant-parasitic Nematodes, Sets 1–8. CAB International, Wallingford, UK (Set 1, No. 10; Set 3, No. 33; Set 4, No. 54; Set 5, No. 66; Set 6, Nos 76, 81).

Fortuner, R. (1991) The Hoplolaiminae. In: Nickle, W.R. (ed.) *Manual of Agricultural Helminthology*. Marcel Dekker, New York, pp. 669–719.

Germani, G., Baldwin, J.G., Bell, A.H. and Wu, X.Y. (1985) Revision of the genus *Scutellonema* Andrássy, 1958 (Nematoda: Tylenchida). *Revue de Nématologie* 8, 289–320.

Aphasmatylenchus Sher, 1965 (Tylenchina, Hoplolaimidae)

Morphology: medium-sized nematodes (0.9–1.8 mm) assuming an open C-shape on heat relaxation. Weak sexual dimorphism in form of anterior region. **Annules** prominent, either smooth or, as **in *A. straturatus*, with numerous longitudinal striae dividing each annule into small blocks (corncob-like configuration).** Labial region offset from body contour, annulated, conoid with distinct labial disc. **Stylet strongly developed, less than three labial region diameters long** and with rounded basal knobs. **Oesophageal glands overlapping intestine in a mostly ventral lobe.** Intestinal fasiculi present, extending beyond rectum into tail. **Female: vulva median; genital system with two functional and equally developed genital tracts,** one anteriorly and one posteriorly directed; **tail cylindroid to conoid–rounded, phasmids absent. Male stylet and oesophagus less well developed**

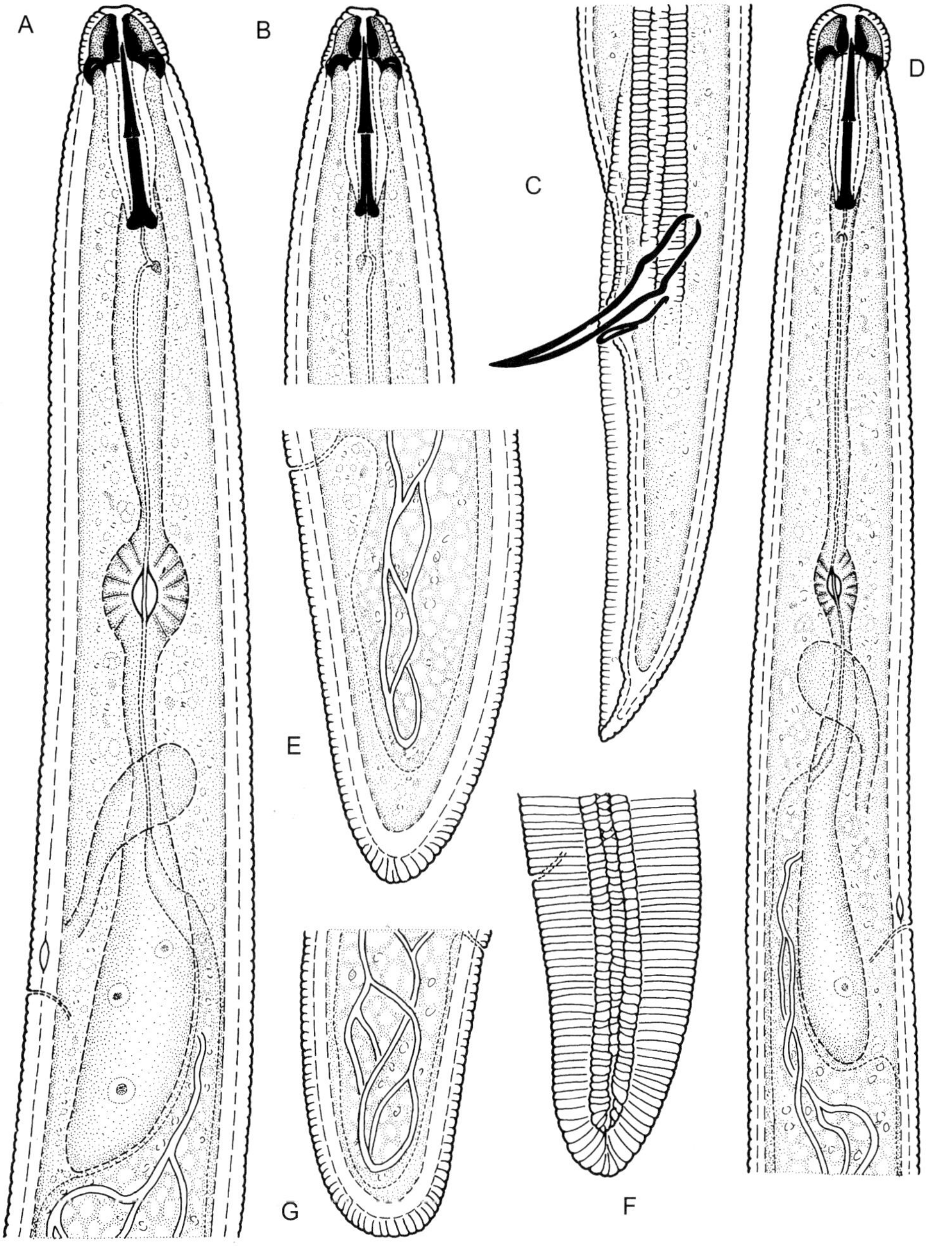

Fig. 2.14. *Aphasmatylenchus straturatus* (A) female anterior region; (B) female labial region; (C) male tail; (D) male anterior region; (E–G) female tail region showing intestinal fasiculi; (F) female tail region, surface view. (After Germani, 1977.) Line drawings are for illustrative purposes only and are not to scale.

than in female, tail elongate conoid, tapering to a pointed terminus; bursa reaching to tail tip. **Phasmids absent.** Spicules robust, arcuate.

Biology: usually migratory ectoparasites, although they may also be found inside roots. *A. straturatus* parasitizes legumes, including groundnut (where it causes

'voltaic chlorosis'), soybean, pigeonpea and cowpea. This species has also been associated with the Shea butter tree (*Butyrospermum parkii*) throughout Burkina Faso. The nematodes do not appear to be capable of entering an anhydrobiotic state, but migrate deeper into the soil horizon during the dry season. The type species, *A. nigeriensis*, was found in the rhizosphere of *Theobroma cacao* and *Hevea brasiliensis*.
Major species: *A. straturatus*, *A. nigeriensis*, *A. liberiensis*.
Distribution: the genus is predominantly found in the sahelian zone of West Africa where it has been recorded from Nigeria, Burkina Faso (formerly Upper Volta), Côte d'Ivoire, Mali, Liberia and Senegal. *A. nigeriensis* has also been recorded from tropical rainforest in French Guyana, South America.
Confusable genus: *Scutellonema*, *Aorolaimus*.

Useful literature

Baujard, P. and Martiny, M. (1995) Ecology and pathogenicity of the Hoplolaimidae (Nemata) from the sahelian zone of West Africa. 4. The genus *Aphasmatylenchus* Sher, 1965. *Fundamental and Applied Nematology* 18, 355–360.

CIH Descriptions of Plant-parasitic Nematodes, Sets 1–8. CAB International, Wallingford, UK (Set 7, No. 104).

Fortuner, R. (1991) The Hoplolaiminae. In: Nickle, W.R. (ed.) *Manual of Agricultural Helminthology*. Marcel Dekker, New York, pp. 669–719.

***Rotylenchulus* Linford & Oliveira, 1940 (Tylenchina, Hoplolaimidae)**

Morphology: sexually dimorphic. Immature female (free in soil): body vermiform, small (0.23–0.64 mm), dying ventrally arcuate on application of gentle heat. **Labial region continuous with body contour,** rounded to conoid, striated. Labial sclerotization of medium development. **Stylet of medium strength, with rounded basal knobs.** Oesophagus with well developed median bulb and valves; **dorsal oesophageal gland opening located well posterior to stylet base** (0.6–1.9 times stylet length); **oesophageal glands well developed with a long lateral overlap.** Vulva posteriorly situated (V = 58–72); vulval lips not protuberant. **Two genital tracts,** each with a double flexure. Tail conoid, with rounded terminus. **Mature female (on roots): swollen to kidney-shaped body.** Anterior part irregular. Vulval lips protruding. Genital tracts convoluted. Male: vermiform. Labial sclerotization, stylet and oesophagus reduced (median oesophageal bulb weak, without valves) but conspicuous. Spicules curved. Tail pointed. **Bursa not reaching tail tip.** Juvenile: resembling immature female, but shorter and lacking vulva and genital tracts.
Biology: the eggs are laid in a gelatinous matrix. On hatching, the juveniles moult to the immature female or male without feeding. The immature female is the invasive stage, but only the anterior section penetrates the root tissue, the posterior part remaining in the soil and becoming obese (i.e. a sedentary semi-endoparasite). About 50 eggs are deposited in a gelatinous matrix secreted by specialized vaginal cells.
Major species: *R. borealis*, *R. parvus*, *R. reniformis*.
Distribution: *R. reniformis* is almost ubiquitous in tropical and subtropical soils, although the other species appear to be more restricted in distribution.
Confusable genus: *Senegalonema*.

Useful literature

CIH Descriptions of Plant-parasitic Nematodes, Sets 1–8. CAB International, Wallingford, UK (Set 1, No. 5; Set 6, No. 83).

Dasgupta, D.R., Raski, D.J. and Sher, S.A. (1968) A revision of the genus *Rotylenchulus* Linford & Oliveira, 1940 (Nematoda: Tylenchidae). *Proceedings of the Helminthological Society of Washington* 35, 169–192.

Jatala, P. (1991) Reniform and false root-knot nematodes, *Rotylenchulus* and *Nacobbus* spp. (1991). In: Nickle, W.R. (ed.) *Manual of Agricultural Helminthology*. Marcel Dekker, New York, pp. 509–528.

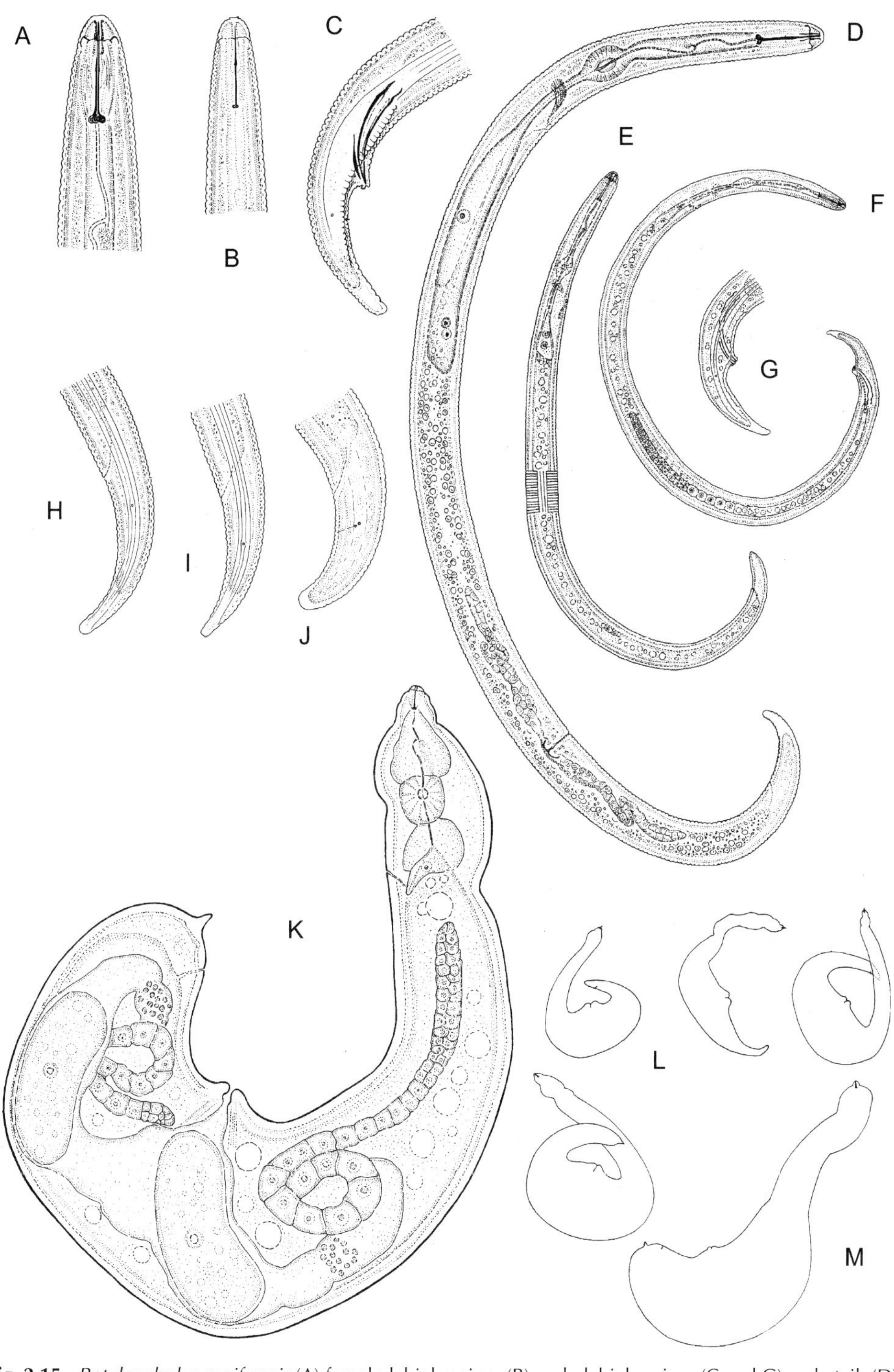

Fig. 2.15. *Rotylenchulus reniformis* (A) female labial region; (B) male labial region; (C and G) male tail; (D) entire immature female; (E) entire juvenile; (F) entire male; (H and I) immature female tails; (J) juvenile tail; (K and M) entire mature females. *R. parvus* (L) entire mature females. Line drawings are for illustrative purposes only and are not to scale.

Robinson, A.F., Inserra, R.N., Caswell-Chen, E.P., Vovlas, N. and Troccoli, A. (1997) *Rotylenchulus* species: identification, distribution, host range, and crop resistance. *Nematropica* 27, 127–180.

Heterodera Schmidt, 1871 (Tylenchina, Heteroderidae)

= *Bidera*, *Ephippiodera*

Morphology: sexually dimorphic. Female: obese, **lemon-shaped,** approximately 300 μm in diameter with a distinct neck and partially enclosed either in root tissue or in the soil. **Oral disc squarish, strongly offset.** Vulva subterminal, near anus. Cuticle thick, whitish at first, but tanning to a brownish-black colour as cyst matures. **Eggs retained within protective cyst. Vulva and anus located on a terminal cone with two translucent areas, the fenestrae, on either side of vulval slit.** Two convoluted genital tracts. **In young females, excretory pore visible at level of, or posterior to, median bulb valve plates.** Male: **vermiform with body often twisted through 180° on heat relaxation;** found free in soil. **Stylet and skeleton of labial region robust. Tail short, hemispherical. Spicules opening subterminally. No bursa.** Juvenile (J2): vermiform, 450–600 μm long with **stylet and labial region skeleton robust. Tail conical with hyaline area starting well before tail terminus.**

Globodera Skarbilovich, 1959 (Tylenchina, Heteroderidae)

Morphology: similar to *Heterodera* except for the **globose cyst. Vulva and anus not elevated on a terminal cone and vulval slit surrounded by a single, circular, fenestra.**
Biology: in most species, all the eggs are retained within the mature cyst, although in some species a voluminous external egg mass is present (e.g. *H. oryzae*). Eggs often hatch in response to root exudates from a host plant, although other hatching factors can be involved. The J2 emerges from the egg, invades a root and induces a feeding site composed of syncytial nurse cells. Root galling is not induced. The J2 swells and moults three times to form the adult female which enlarges rapidly, the posterior region bursting through the root epidermis. Males are more commonly produced when food is in short supply. They assume a vermiform state within the J4 cuticle before burrowing out of the root into the soil. Females produce several hundred eggs and, after death, the cuticle of the female tans to form a protective cyst.
Major species: *H. avenae*, *H. cajani*, *H. ciceri*, *H. glycines*, *H. latipons*, *H. sacchari*, *G. pallida*, *G. rostochiensis.*
Distribution: although the majority of *Heterodera* species are temperate in distribution, some species are present in tropical or subtropical crops, whereas *Globodera* species tend to be confined to cooler regions.
Confusable genera: *Afenestrata*, *Cactodera*, *Punctodera.* The J2 infectives can be confused with those of other genera of the same family and share some similarities with those of *Meloidogyne.*

Useful literature

Baldwin, J.G. and Mundo-Ocampo, M. (1991) Heteroderinae, cyst- and non cyst-forming nematodes. In: Nickle, W.R. (ed.) *Manual of Agricultural Helminthology.* Marcel Dekker, New York, pp. 275–362.

CIH Descriptions of Plant-parasitic Nematodes. Sets 1–8. CAB International, Wallingford, UK (Set 1, No. 2; Set 2, Nos 16, 17; Set 4, No. 48; Set 8, No. 118).

Golden, A.M. (1986) Morphology and identification of cyst nematodes. In: Lamberti, F. and Taylor, C.E. (eds) *Cyst Nematodes.* NATO ASI Series, Plenum Press, London, pp. 23–46.

Meloidogyne Goeldi, 1887 (Tylenchina, Meloidogynidae)

= *Hypsoperine*

Morphology: sexually dimorphic. Female: embedded in root tissue, globose, 0.3–0.7 mm in diameter **with a slender neck. Vulva subterminal near anus.** Cuticle whitish, thin, annulated. Stylet

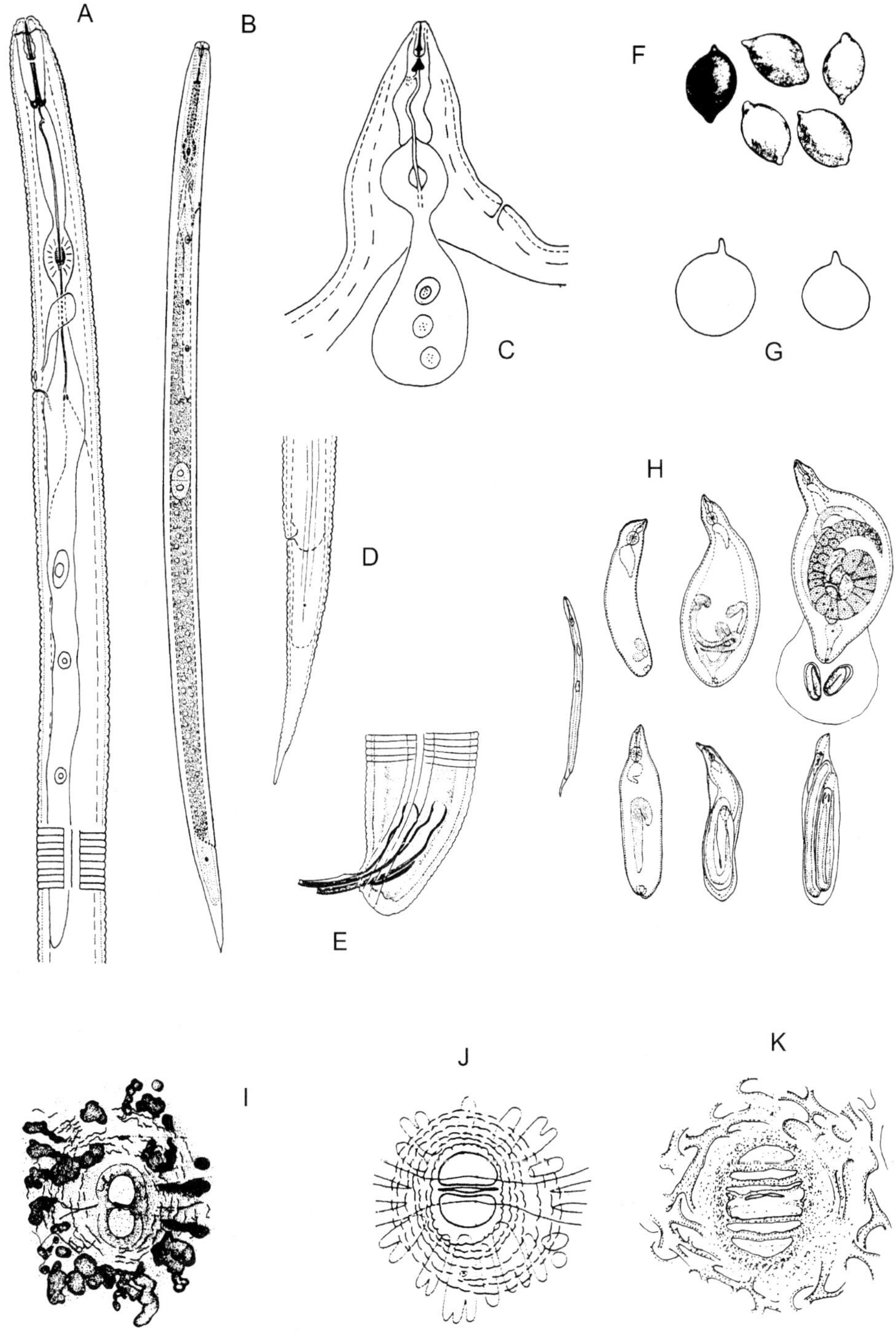

Fig. 2.16. *Globodera rostochiensis* (C) female anterior region; (G) entire cysts; (K) perivulval area. *Heterodera avenae* (E) male tail; (F) cysts; (I) perivulval area. *H. glycines* (J) perivulval area. *H. oryzae* (D) juvenile tail. *H. sacchari* (A) J2 oesophagus; (B) J2 infective juvenile. *H. schachtii* (H) developmental stages. Line drawings are for illustrative purposes only and are not to scale.

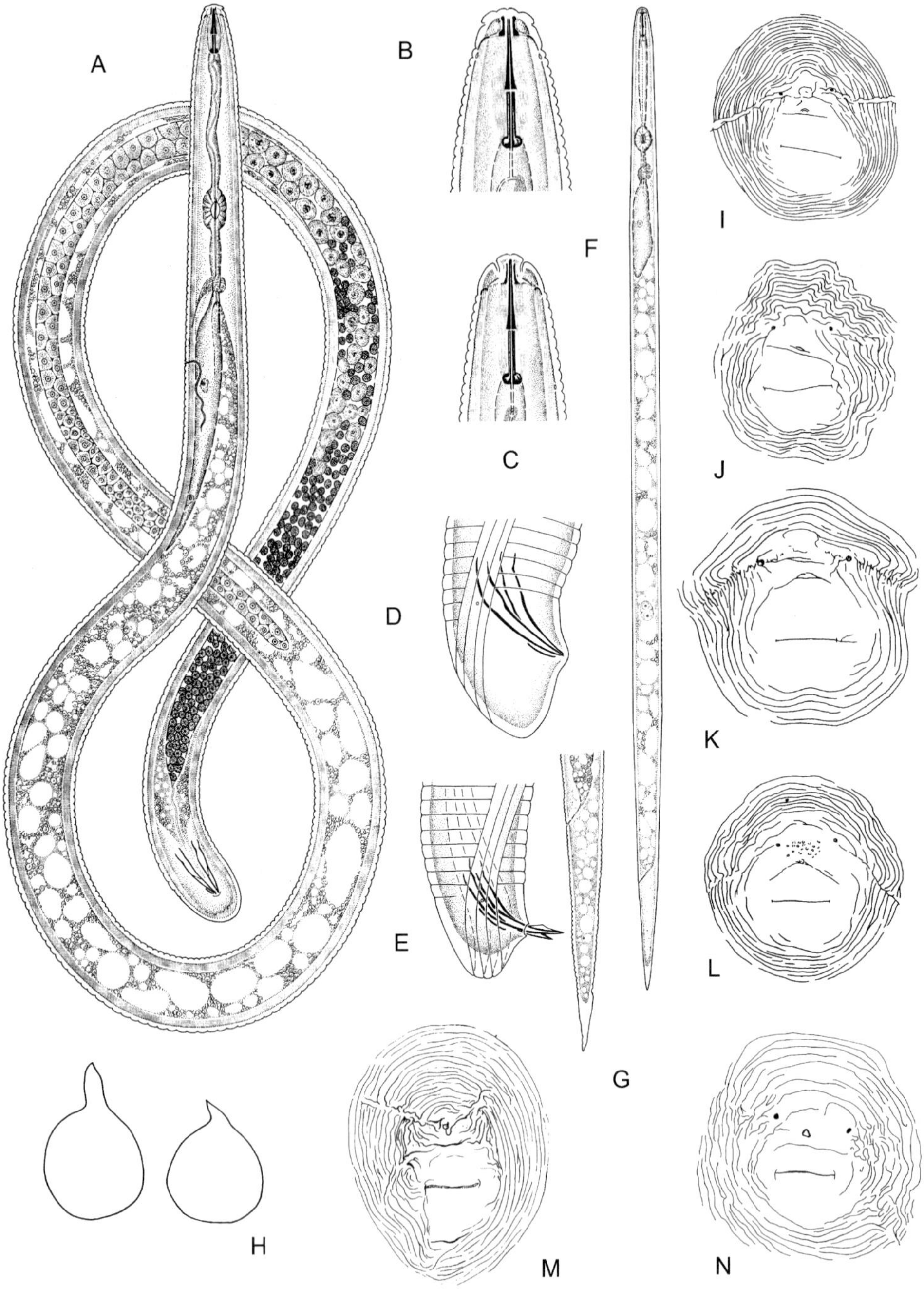

Fig. 2.17. *Meloidogyne incognita* (A) entire male; (B and C) male labial region; (D and E) male tail; (F) entire infective juvenile (J2); (G) J2 tail; (H) mature females. Perineal patterns (I) *M. javanica*; (J) *M. incognita*; (K) *M. arenaria*; (L) *M. hapla*; (M) *M. graminicola*; (N) *M. exigua*. Line drawings are for illustrative purposes only and are not to scale.

short, moderately sclerotized. Labial region skeleton weak. **Excretory pore located anterior to median bulb valve plates** and often near to stylet base. Two convoluted genital tracts. Eggs **deposited outside body in a gelatinous matrix.** Male: vermiform, free-living in soil, 1–2 mm long. **Body usually twisted through 180° along its length on heat relaxation. Stylet and labial region skeleton robust. Tail short,** hemispherical. Spicules robust. Bursa absent. Juveniles (J2): slender, vermiform, about 450 μm long. Stylet and labial region skeleton weakly sclerotized. **Tail conical with hyaline portion starting near tail tip.**
Biology: in most species, the eggs are retained within a gelatinous matrix outside the swollen female body. On hatching, the J2 invades a host root and induces a trophic system of giant cells. Cortical cells are induced to multiply to form the characteristic gall. The remainder of the life cycle is similar to that of *Heterodera*/*Globodera* except that in most species the females do not normally burst out of the root because of the surrounding gall tissue.
Major species: *M. arenaria*, *M. exigua*, *M. graminicola*, *M. hapla*, *M. incognita*, *M. javanica*, *M. mayaguensis.*
Distribution: widely distributed throughout the tropical and subtropical regions.
Confusable genera: *Nacobbus*. The J2 infective stage might be confused with those of *Heterodera/Globodera*, but has weaker cephalic sclerotization, a less robust stylet and a shorter hyaline region in the tail.

Useful literature

CIH Descriptions of Plant-parasitic Nematodes. Sets 1–8. CAB International, Wallingford, UK (Set 1, No. 3; Set 2, No. 18; Set 4, No. 49; Set 5, No. 62; Set 6, No. 87).

Eisenback, J.D. (1997) *Root-knot Nematode Taxonomic Database.* CAB International, Wallingford, UK (CD-ROM).

Eisenback, J.D. and Triantaphyllou, H.H. (1991) Root-knot nematodes: *Meloidogyne* species and races. In: Nickle, W.R. (ed.) *Manual of Agricultural Helminthology.* Marcel Dekker, New York, pp. 191–274.

Jepson, S.B. (1987) *Identification of Root-knot Nematodes* (Meloidogyne *Species).* CAB International, Wallingford, UK.

Sasser, J.N. and Carter, C.C. (eds) (1985) *An Advanced Treatise on* Meloidogyne. Vols 1 and 2. North Carolina State University, Raleigh, North Carolina.

Criconemoides Taylor, 1936 (Tylenchina, Criconematidae)

= *Criconemella, Macroposthonia, Mesocriconema, Xenocriconemella, Madinema, Seshadriella, Neobakernema, Crossonemoides*

Morphology: strong sexual dimorphism. Female: body 0.20–1 mm long, stout, dying straight or slightly curved, with rounded anterior end, and rounded to conical posterior part. **Cuticle provided with 42–200 prominent, retrorse annules,** with a **smooth or finely crenate posterior margin.** Labial area not well separated from rest of body, marked by one or two thinner annules. **Stylet strong, basal knobs with a forwardly directed process** (= anchor shaped). **Oesophagus with a strong median bulb which is fused with the procorpus; glands forming a small posterior bulb. Vulva posterior. One genital tract, extending anteriorly.** Spermatheca laterally situated. Male: body slender and short. Anterior end rounded. No stylet; oesophagus degenerate. Spicule short, slightly curved. Bursa weakly developed, exceptionally absent. Tail pointed. Juveniles: resembling female. Annules smooth to finely crenate (exceptionally with a row of scales) on posterior margin.
Biology: migratory ectoparasites on perennial crops, trees and vines. Males non-feeding. Most species are parthenogenetic. Only a few species have been proved to be harmful. Found in all geographic areas.
Major species: *C. axestis*, *C. onoensis*, *C. sphaerocephalus*, *C. xenoplax.*
Confusable genera: *Criconema*, *Discocriconemella*, *Hemicriconemoides.*
Taxonomic note: species of *Criconemoides* have also been commonly placed in one or more of the following genera: *Macroposthonia*, *Criconemella* or *Mesocriconema.* This situation can be confusing and must be borne in mind when consulting the

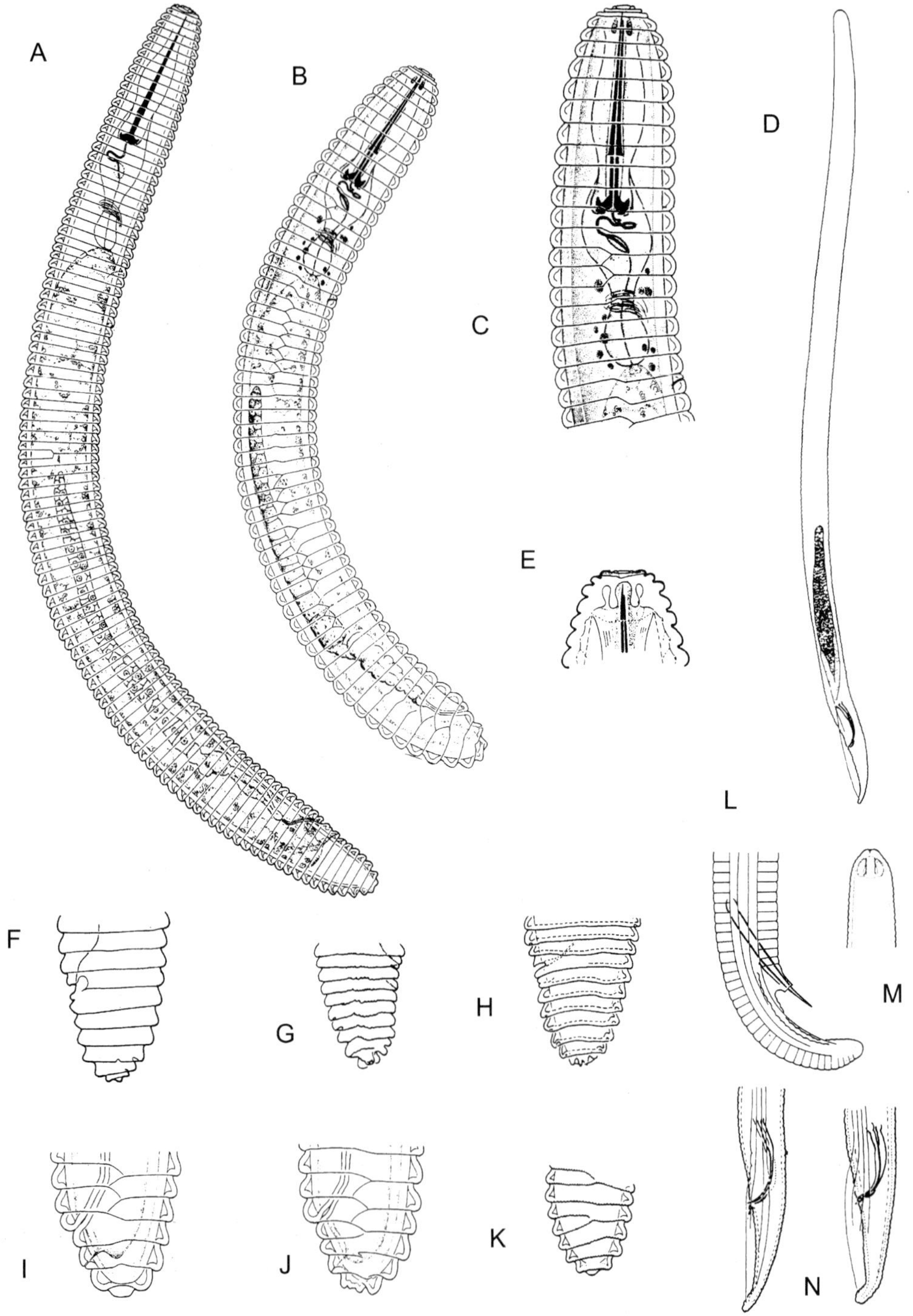

Fig. 2.18. *Criconemoides pseudohercyniensis* (D) entire male; (E) female labial region; (G) female tail; (N) male tails. *C. onoensis* (H) female tail. *C. sphaerocephalus* (B) entire female; (C) female oesophageal region; (I and J) female tails. *C. xenoplax* (A) entire female; (F) female tail; (K) juvenile tail; (L) male tail; (M) male labial region. Line drawings are for illustrative purposes only and are not to scale.

literature, both old and new. The ICZN has recently decreed that *Criconemoides* is the valid generic name for this assemblage of species.

Useful literature

CIH Descriptions of Plant-parasitic Nematodes, Sets 1–8. CAB International, Wallingford, UK (Set 1, No. 127; Set 2, No. 28).

Raski, D.J. and Luc, M. (1987) A reappraisal of Tylenchina (Nemata). 10. The superfamily Criconematoidea Taylor, 1956. *Revue de Nématologie* 10, 409–444.

Hemicycliophora de Man, 1921 (Tylenchina, Criconematidae)

= *Aulosphora*, *Colbranium*, *Loofia*

Morphology: strong sexual dimorphism. Female: body straight, or slightly ventrally curved, 0.6–1.9 mm long, stout. Anterior end rounded. Posterior end pointed, more rarely rounded. **Cuticle with detached sheath** (= 'double' cuticle); external layer marked by numerous (up to 400) prominent annules; **annules not retrorse.** No true lateral field, but cuticle may be variously ornamented (longitudinal lines, squares, dots, scratches, etc.). Labial area not separated from body, marked by 2–3 annules. **Stylet strong, long, with rounded basal knobs. Oesophagus with strong median bulb fused with procorpus; glands forming a small terminal bulb abutting intestine. Vulva posteriorly situated. One anteriorly directed genital tract;** spermatheca lateral. Anus and rectum vestigial. Post-vulval part generally conical with pointed terminus, more rarely cylindrical with rounded extremity. Male: slender, with simple cuticle (outer layer not detached). **No stylet. Oesophagus degenerate. Spicules strong, semi-circular to hook-shaped. Bursa adanal, well developed.** Tail long, conical, often angled ventral to body axis. Juveniles: resembling female.

Biology: as for *Criconemoides*.

Major species: *H. arenaria*, *H. parvana*, *H. typica*.

Confusable genera: *Caloosia*, *Hemicriconemoides*.

Useful literature

Brzeski, M.W. (1974) Taxonomy of Hemicycliophorinae (Nematoda, Tylenchida). *Zeszyty probl. Postep. Naukowe robn.* 154, 237–330.

Hemicriconemoides Chitwood & Birchfield, 1957 (Tylenchina, Criconematidae)

Morphology: strong sexual dimorphism. Female: similar in many ways to *Hemicycliophora*, but shorter (usually ~0.5 mm long) with fewer annules and with **very closely adpressed 'double' cuticle.** Stylet knobs with **anteriorly directed processes.** Tail short, conoid. **Juveniles resembling female but posterior margin of body annules ornamented with scales or short denticles.**

Biology: similar to *Criconemoides*.

Major species: *H. cocophillus*, *H. mangiferae*.

Confusable genera: *Caloosia*, *Hemicycliophora*.

Useful literature

CIH Descriptions of Plant-parasitic Nematodes, Sets 1–8. CAB International, Wallingford, UK (Set 7, No. 99).

Tylenchulus Cobb, 1913 (Tylenchina, Tylenchulidae)

Morphology: sexually dimorphic. Immature female (free in soil): body vermiform, ventrally curved posteriorly, small (<0.5 mm). Labial region rounded, continuous with body contour. **Labial sclerotization weak.** Stylet of medium development with rounded basal knobs. **Oesophagus with strong median bulb not well separated from procorpus; glands forming a basal bulb. Vulva very posteriorly situated; genital tract single, anteriorly outstretched. Excretory pore located very posteriorly and only slightly anterior to vulva.** Tail conical. **No anus or rectum.** Mature female: **anterior part embedded in root tissue,** irregular, slender, with thin cuticle. Posterior part,

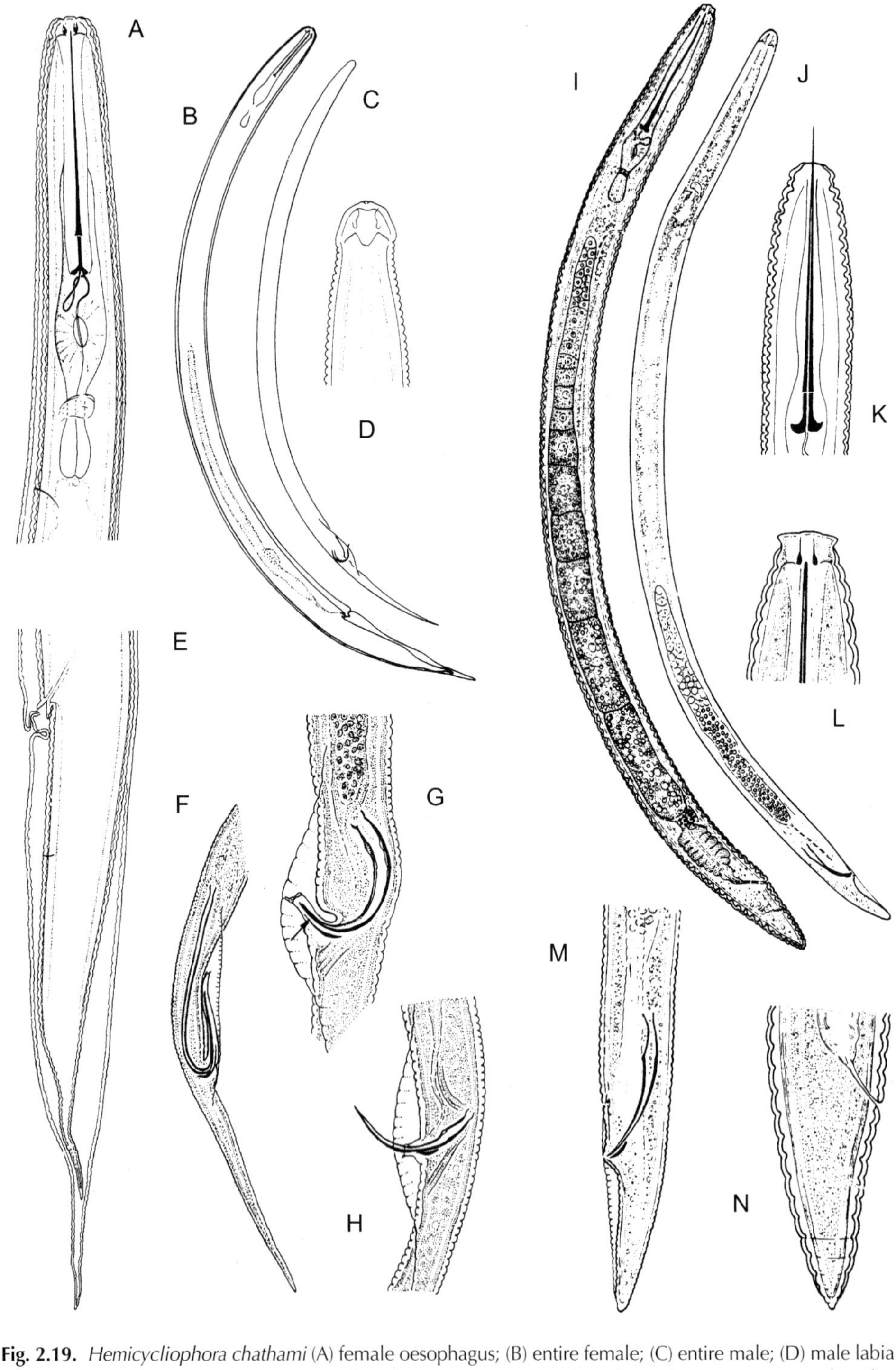

Fig. 2.19. *Hemicycliophora chathami* (A) female oesophagus; (B) entire female; (C) entire male; (D) male labial region; (E) female posterior region; (G) male tail. *H. penetrans* (F) male tail. *H. thienemanni* (H) male tail. *Hemicriconemoides mangiferae* (I) entire female; (J) entire male; (L) female labial region; (M) male tail; (N) female tail. *H. chitwoodi* (K) female stylet. Line drawings are for illustrative purposes only and are not to scale.

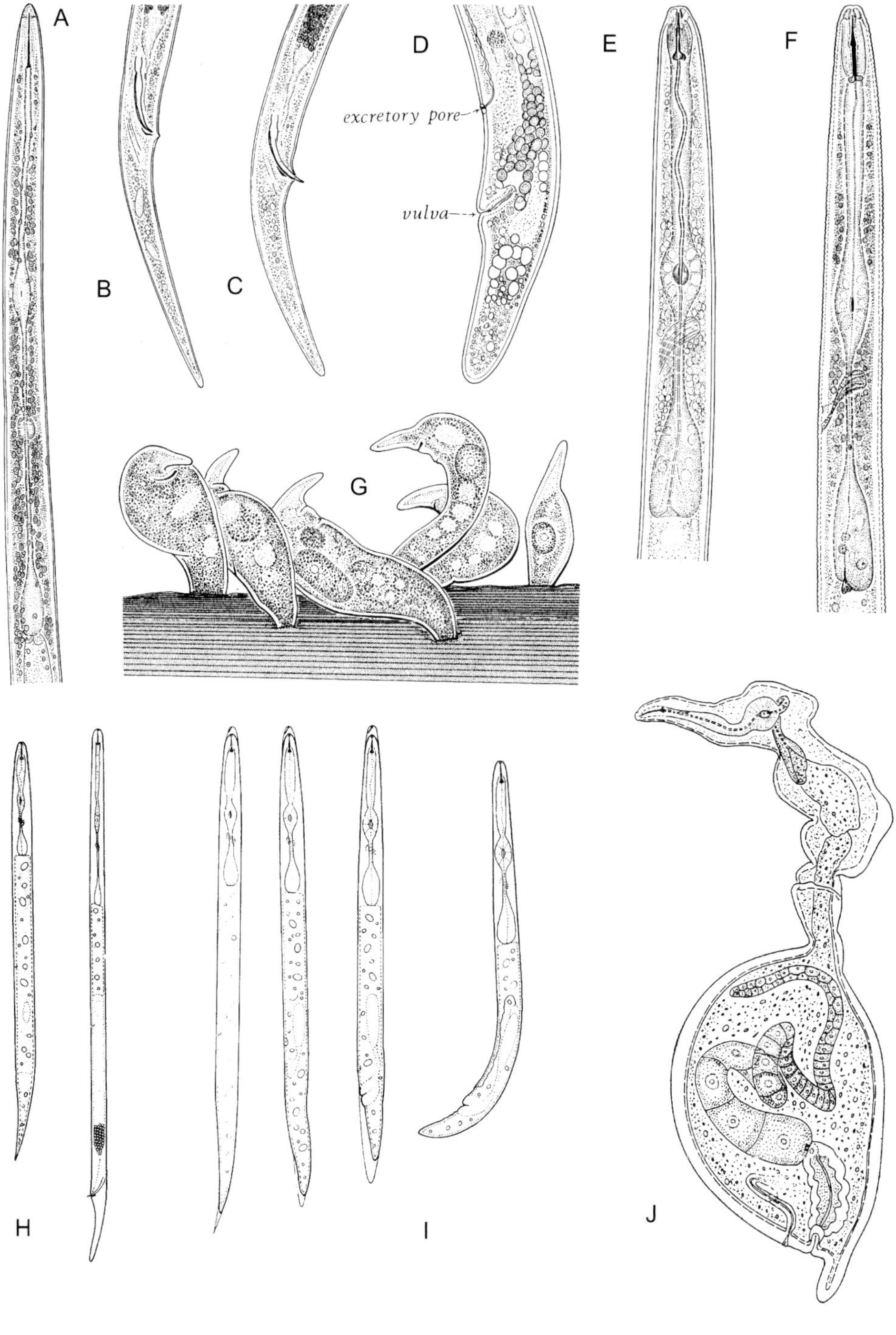

Fig. 2.20. *Tylenchulus semipenetrans* (A) male oesophagus; (B and C) male tails; (D) female posterior region; (E) immature female oesophagus; (F) juvenile oesophagus; (G) mature females attached to root; (H) development of male; (I) development of female; (J) entire female. Line drawings are for illustrative purposes only and are not to scale.

bursting out of root, swollen with very thick cuticle and a pointed postvulvar section; **excretory pore and vulva very posterior.** Excretory cell well developed, producing a gelatinous matrix. Genital tract convoluted, with several eggs. **No anus or rectum.** Male: body vermiform, short and slender. **Cephalic sclerotization, stylet and oesophagus reduced.** Spicules slightly curved. **No bursa.** Tail conical, pointed. Juvenile: body vermiform. Labial sclerotization, stylet and oesophagus similar to those of immature females. Tail long, pointed. Genital primordium differently shaped in male and female juveniles from J2 onwards.

Biology: the eggs are contained in a gelatinous matrix produced by the secretory/excretory cell. After hatching, male juveniles moult to the adult without feeding, whilst female juveniles feed on cortical cells. The immature female penetrates deeper into the root, the anterior end penetrating deep into the cortex whilst the posterior section, which becomes obese, remains outside the root. A highly sophisticated system of trophic nurse cells is initiated around the female labial region. (Note: a heavily infested citrus root, when carefully rinsed in water, retains a collar of earth adhering to the gelatinous egg sacs underneath.)

Major species: *T. semipenetrans*.

Distribution: found almost everywhere that citrus is grown on any scale and often causing a severe disease 'slow decline'.

Confusable genus: *Trophotylenchulus*.

Useful literature

CIH Descriptions of Plant-parasitic Nematodes, Sets 1–8. CAB International, Wallingford, UK (Set 3, No. 34).

Raski, D.J. (1991) Tylenchulidae in agricultural soils. In: Nickle, W.R. (ed.) *Manual of Agricultural Helminthology*. Marcel Dekker, New York, pp. 761–794.

Xiphinema Cobb, 1913 (Dorylaimina, Longidoridae)

Morphology: slender nematodes, 1.3–5 mm long. Labial region continuous or offset. **Amphidial aperture a broad slit leading back to a funnel-shaped pouch.** Stylet very long (60–250 μm) consisting of a needle-like **odontostyle with a forked base** attached to an **odontophore with three prominent basal flanges. Stylet guide appearing tubular with the 'guide ring' located in posterior half of odontostyle.** Oesophagus consisting of a long, narrow procorpus and a short, glandular bulb. Female: vulva usually at 40–50%, but may be more anterior. Usually two genital tracts present, but in some species the anterior tract is non-functional (mono-opisthodelphic or pseudo-mono-opisthodelphic) and reduced to varying degrees, or even entirely absent, in which case the vulva is more anteriorly located (V = 25%). Tail very variable from short and rounded to long filiform. Male: spicules very powerful, arcuate. Ventral supplements form a precloacal row.

Longidorus Micoletzky, 1922 (Dorylaimina, Longidoridae)

Morphology: similar to *Xiphinema* but body thinner and may be up to 11 mm long. **Amphids pouch-like and opening via a minute, inconspicuous pore. Odontostyle/odontophore junction not forked, odontophore lacking flanges** and odontostylet less strongly sclerotized. **Guide ring located in anterior half of odontostyle.**

Paralongidorus Siddiqi, Hooper & Khan, 1963 (Dorylaimina, Longidoridae)

= *Siddiqia*, *Inagreius*, *Longidoroides*

Morphology: similar to *Longidorus*, but **amphidial pouch stirrup-shaped** and **amphidial aperture broad and slit-like** as in *Xiphinema*.

Biology: long-lived, migratory ectoparasites attacking a wide variety of hosts. The favoured point of attack is at or near the root tip, resulting in hooked root tips and/or terminal galls. Attacked root systems are stunted, lack developed laterals and show necrosis at the feeding sites. *Xiphinema* tends to be more abundant under woody hosts, whereas *Longidorus* and *Para-*

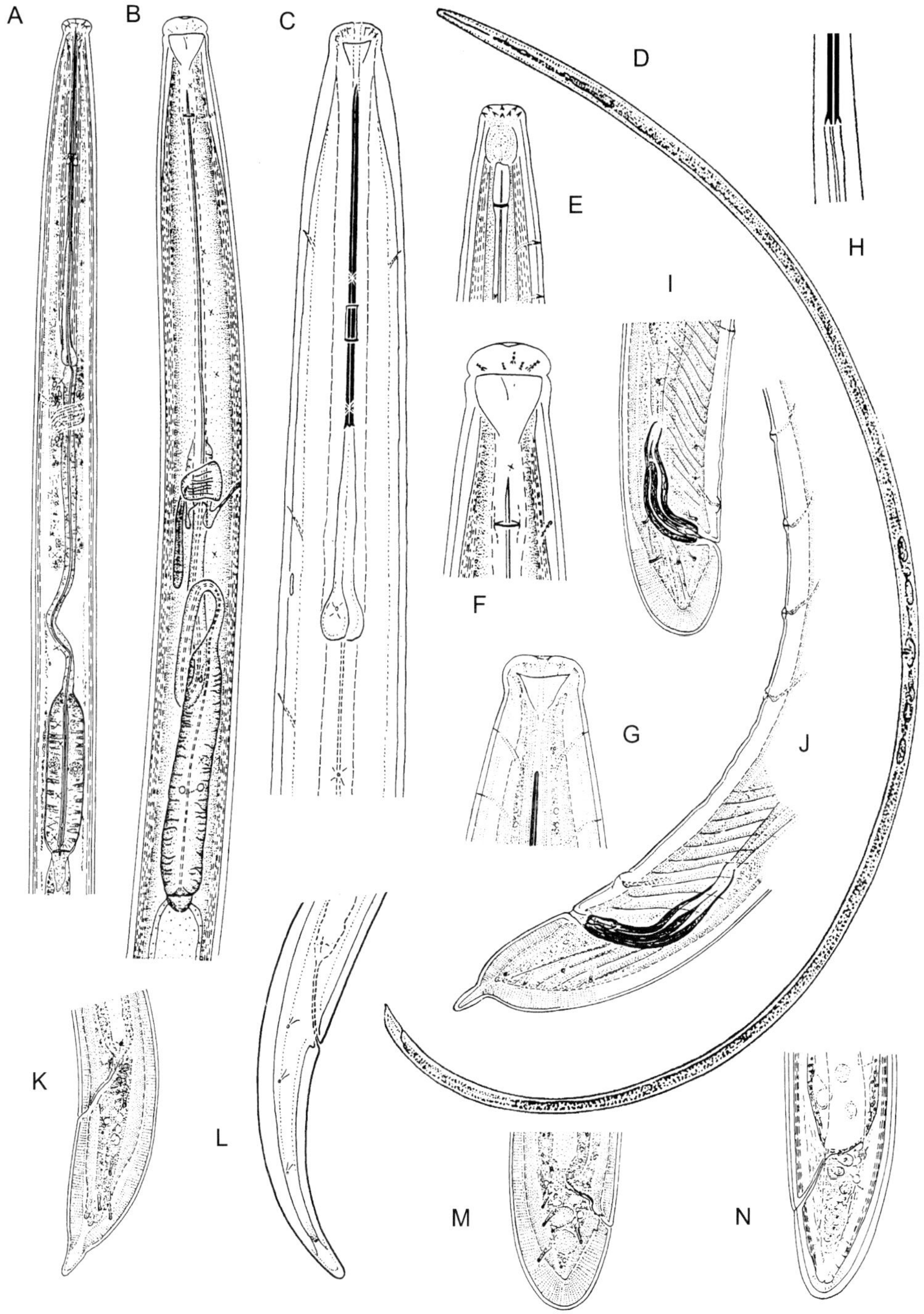

Fig. 2.21. *Longidorus fursti* (A) oesophagus; (N) female tail. *L. elongatus* (E) labial region. *Paralongidorus natalensis* (B) oesophagus; (F) labial region. *Xiphinema heynsi* (I) male tail; (M) female tail. *X. mammatum* (J) male tail. *X. neobasiri* (D) entire female; (G) labial region; (K) female tail. *X. savanicola* (C) oesophagus; (H) odontostyle/odontophore junction; (L) female tail. Line drawings are for illustrative purposes only and are not to scale.

longidorus are more common under non-woody plants, particularly grasses and cereals. Greatest populations are found below 30 cm. With few exceptions, sandy soils support higher populations than heavier soils. Some species have been shown to be virus vectors. Reproduction is amphimictic or parthenogenetic.
Major species: *X. americanum* group, *X. index*, *X. elongatum*, *L. africanus*, *L. laevicapitatus*, *P. australis*.
Distribution: *Longidorus* is mainly found in cooler areas whilst *Xiphinema* and *Paralongidorus* are more tropical.
Confusable genera: each other, *Paraxiphidorus*, *Xiphidorus*.

Useful literature

CIH Descriptions of Plant-parasitic Nematodes, Sets 1–8. CAB International, Wallingford, UK (Set 2, No. 29; Set 3, No. 45; Set 8, No. 117).

Hunt, D.J. (1993) *Aphelenchida, Longidoridae and Trichodoridae: Their Systematics and Bionomics*. CAB International, Wallingford, UK.

Loof, P.A.A. and Luc, M. (1993) A revised polytomous key for the identification of species of the genus *Xiphinema*, Cobb, 1913 (Nematoda: Longidoridae) with exclusion of the *X. americanum*-group: supplement 1. *Systematic Parasitology* 24, 185–189.

Loof, P.A.A., Luc, M. and Baujard, P. (1996) A revised polytomous key for the identification of species of the genus *Xiphinema*, Cobb, 1913 (Nematoda: Longidoridae) with exclusion of the *X. americanum*-group: supplement 2. *Systematic Parasitology* 33, 23–29.

Trichodorus Cobb, 1913 (Diphtherophorina, Trichodoridae)

Morphology: body stout, cigar shaped, 0.8–1.2 mm long. Cuticle smooth. Labial region continuous with body contour; papillae prominent. **Onchiostyle** (= stylet) **tripartite, curved.** Oesophagus **slender anteriorly with a posterior bulboid expansion.** Female: vulva median with **strong vaginal sclerotization, vagina** well developed, **extending into body for about half its diameter, one pair of lateral body pores present within one body diameter of vulva.** Two genital tracts. **Tail rounded, very short;** anus almost terminal. Male: spicules arcuate, gubernaculum present. **Protractor muscles conspicuous, of unusual form** and encapsulating spicule shafts. Ventral supplements present, **bursa** usually **absent or very small.**

Paratrichodorus Siddiqi, 1974 (Diphtherophorina, Trichodoridae)

= *Atlantadorus*, *Nanidorus*

Morphology: very similar to *Trichodorus* but **cuticle markedly swelling in response to acidic fixatives.** Female: **vulva with weak vaginal sclerotization, vagina weakly developed, extending into body for about a third of its diameter. No lateral body pores within one body diameter of vulva.** Male: **spicule protractor muscles inconspicuous. Bursa present.**
Biology: ectoparasitic on the roots of perennial and woody plants. The main area of attack is just behind the root tip, thereby restricting root elongation. The root tip is then attacked, as are the developing lateral root initials, resulting in the characteristic 'stubby root' system. Both genera are more common in light or sandy soils, and highest densities tend to occur at depths of 30–40 cm. Some species are known to be virus vectors and it is likely that the other species are potential vectors.
Major species: *T. primitivus*, *T. similis*, *T. viruliferus*, *P. minor*, *P. pachydermus*.
Distribution: worldwide. *Trichodorus* tends to occur more in temperate regions whilst *Paratrichodorus* is more tropical.
Confusable genera: *Monotrichodorus* (only one female genital tract) and each other.

Useful literature

CIH Descriptions of Plant-parasitic Nematodes, Sets 1–8. CAB International, Wallingford, UK (Set 1, No. 15; Set 4, No. 59; Set 6, No. 86; Set 7, No. 103; Set 8, No. 112).

Decraemer, W. (1991) Stubby root and virus vector nematodes: *Trichodorus*, *Paratrichodorus*,

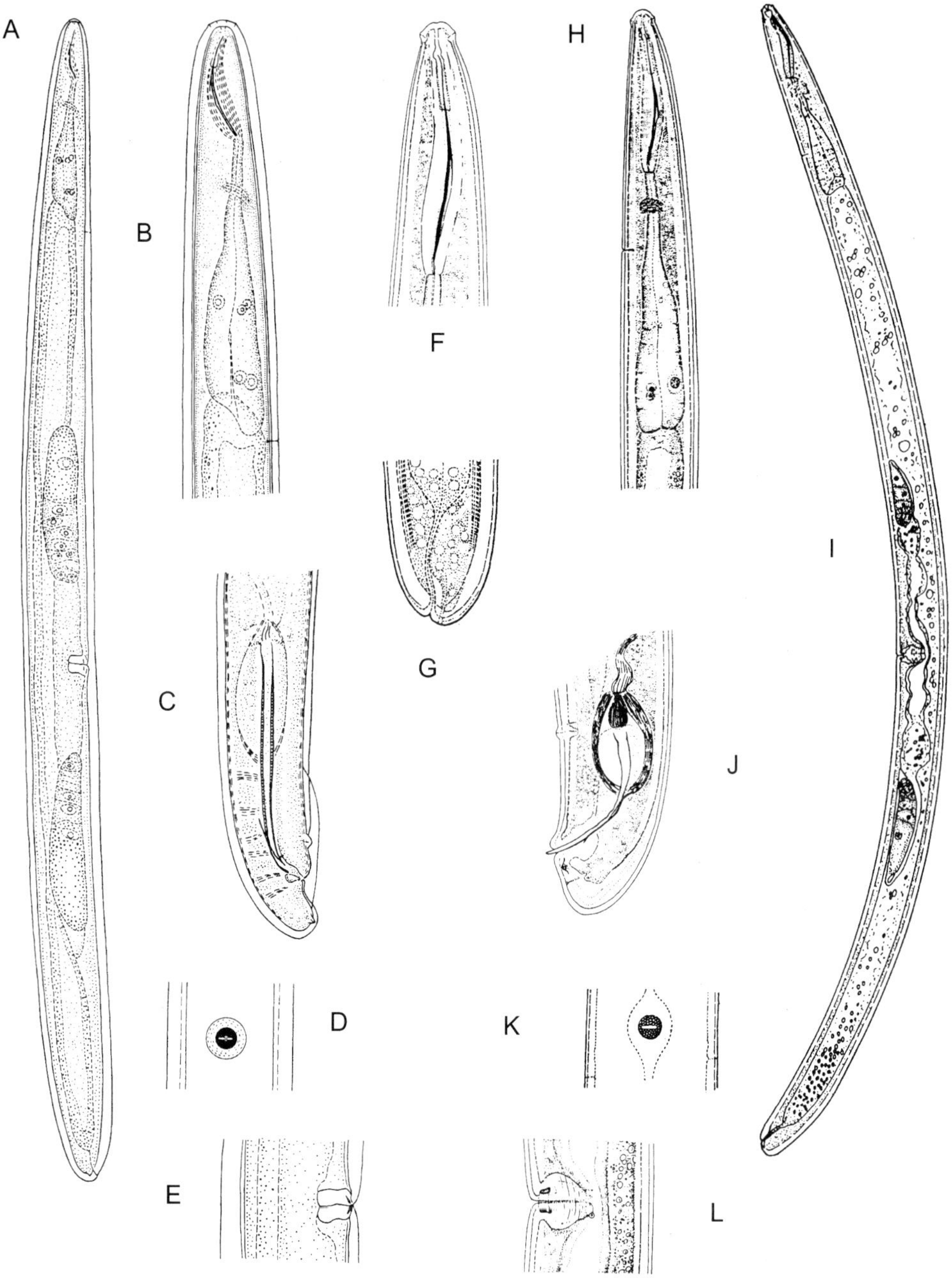

Fig. 2.22. *Paratrichodorus minor* (A) entire female; (B) oesophagus; (C) male tail; (D) vulva, ventral view; (E) vulva, lateral view. *Trichodorus primitivus* (F) labial region; (H) oesophagus; (J) male tail; (L) vulva lateral view. *T. similis* (G) female tail; (K) vulva, ventral view. *T. viruliferus* (I) entire female. Line drawings are for illustrative purposes only and are not to scale.

Allotrichodorus and *Monotrichodorus.* In: Nickle, W.R. (ed.) *Manual of Agricultural Helminthology.* Marcel Dekker, New York, pp. 587–625.

Decraemer, W. (1995) *The Family Trichodoridae: Stubby Root and Virus Vector Nematodes.* Kluwer Academic Publishers, Dordrecht, The Netherlands.

De Waele, D. and Brzeski, M.W. (1995) *Trichodorus altaicus* sp. n. (Nematoda: Trichodoridae) and a key to the species of the genus *Trichodorus. Fundamental and Applied Nematology* 18, 181–187.

Hunt, D.J. (1993) *Aphelenchida, Longidoridae and Trichodoridae: Their Systematics and Bionomics.* CAB International, Wallingford, UK.

Acnowledgements

Reproduction of illustrations from the following sources is gratefully acknowledged: *Cahier ORSTOM, Série Biologie*: Luc, M. 11, 5–131 (Fig. 2.18H). *Journal of Helminthology*: Fortuner, R. 44, 141–152 (Fig. 2.3C–L). *Journal of Nematology*: Sher, S.A. 2, 228–235 (Fig. 2.11H and I). *Nematologica*: Sher, S.A. 6, 155–169 (Fig. 2.13A and C–G); 9, 267–295 (Fig. 2.13K–M). *Phytopathology*: Raski, D.J. 40, 135–152 (Fig. 2.16H). *Phytophylactica*: Jacob, P.J.F. and Heyns, J. 14, 169–178 (Fig. 2.21A, B, F and N). (Reproduced under South Africa Government Printer's copyright authority 9017 of 5 July 1989.) *Proceedings of the Helminthological Society of Washington*: Dasgupta, D.R., Raski, D.J. and Sher, S.A. 35, 169–192 (Fig. 2.15A–N); Sher, S.A. 35, 219–237 (Fig. 2.10A–G and I–P). *Revue de Nématologie*: Luc, M. and Southey, J.F. 3, 243–269 (Fig. 1 R, N; Fig. 21 C, M); Siddiqi, M.R. 2, 51–64 (Fig. 2.1R and N; Fig. 2.21G, I–K and M); 3, 179–199 (Fig. 2.19F–H). *Soil and Freshwater Nematodes*: Goodey, J.B. Methuen. 544 pp. (Fig. 2.20J). *Systematic Parasitology*: Orton Williams, K.J. 8, 207–214 (Fig. 2.19A–E and G). (Reprinted by permission of Kluwer Academic Publishers.) *United States Department of Agriculture* (Fig. 2.10H; After N.A. Cobb, 1915). All other illustrations by courtesy of CABI Publishing.

Note

[1] General information on nematode morpho-anatomy and biology can be found in Dropkin (1980) and Maggenti (1981). In addition, excellent illustrated descriptions of various plant and insect parasitic nematodes, together with data on biology and classification, can be found in Siddiqi (2000).

General References

Anonymous (2003) *Crop Protection Compendium. Global Module.* CAB International, Wallingford, UK (CD-ROM).

CIH Descriptions of Plant-parasitic Nematodes, Sets 1–8. (1972–1984) CAB International, Wallingford, UK.

De Ley, P. and Blaxter, M. (2002) Systematic position and phylogeny. In: Lee, D.L. (ed.) *The Biology of Nematodes.* Taylor & Francis, London, pp. 1–30.

Dropkin, V.H. (1980) *Introduction to Plant Nematology.* John Wiley & Sons, New York.

Fortuner, R., Geraert, E., Luc, M., Maggenti, A.R. and Raski, D.J. (1987–1988) A reappraisal of Tylenchina (Nemata). *Revue de Nématologie* 10, 127–232, 409–444; 11, 159–188.

Hunt, D.J. (1993) *Aphelenchida, Longidoridae and Trichodoridae: Their Systematics and Bionomics.* CAB International, Wallingford, UK.

Kleynhans, K.P.N., Van den Berg, E., Swart, A., Marais, M. and Buckley, N.H. (1996) *Plant Nematodes in South Africa.* Plant Protection Research Institute Handbook No. 8, PPRI, Pretoria, South Africa.

Maggenti, A.M. (1981) *General Nematology.* Springer Verlag, New York.

Nickle, W.R. (1991) *Manual of Agricultural Helminthology.* Marcel Dekker, New York.

Siddiqi, M.R. (2000) *Tylenchida Parasites of Plants and Insects,* 2nd edn. CAB International, Wallingford, UK.

Starr, J.L., Cook, R. and Bridge, J. (2002) *Plant Resistance to Parasitic Nematodes.* CAB International, Wallingford, UK.

Taylor, C.E. and Brown, D.J.F. (1997) *Nematode Vectors of Plant Viruses.* CAB International, Wallingford, UK.

3 Methods for Extraction, Processing and Detection of Plant and Soil Nematodes*

David J. Hooper,[1] Johannes Hallmann[2] and Sergei A. Subbotin[3]

[1] *Formerly Entomology and Nematology Department, AFRC Institute of Arable Crops Research, Rothamsted Experimental Station, Harpenden, Herts AL5 2JQ, UK;* [2] *Biologische Bundesanstalt für Land- und Forstwirtschaft, Institut für Nematologie und Wirbeltierkunde, Toppheideweg 88, D-48161 Münster, Germany;* [3] *Institute of Parasitology of the Russian Academy of Sciences, Leninskii Prospect 33, Moscow 117071, Russia*

Introduction

Diagnosis of nematode damage requires methods for their extraction, handling and detection. The methods take advantage of size, density and motility of the nematodes to separate them from plant tissue and soil particles by means of sieving, centrifugation and filtration. Different methods allow different applications such as for diagnosis, determination of infestation levels, monitoring nematode populations and statutory testing for the presence of quarantine nematodes. Besides morphology and morphometrics, molecular techniques are increasingly used for rapid and accurate identification of nematodes. This chapter gives details of the most common methods. There are many modifications to the basic methods often determined by local supplies of equipment and operating conditions. Further references with excellent sections on methodology are: Ayoub (1980), Dropkin (1989), Hunt and De Ley (1996), Shurtleff and Averre (2000), Southey (1986), Thorne (1961) and Zuckerman *et al.* (1985).

Sampling

Most migratory plant parasitic nematodes are found around plant roots, and so rhizosphere samples are preferable. Badly stunted plants may have too small a root system to support many nematodes, and samples from nearby, less affected, plants may yield more specimens. Usually few nematodes occur in the top 5 cm of soil which can be discarded from samples. Soil samples are generally taken to a depth of 15–20 cm, but 60 cm might be appropriate for nematodes affecting tree crops and other deep-rooted perennials. Nematodes are not uniformly distributed in soil. Areas of nematode damage may be circular to oval or rectangular in outline; patches of poor growth may follow the rows. Sampling for stem and foliar nematodes should be from symptomatic plants. Soil samples and plant material to be examined for nematodes should be kept moist. Polythene bags are excellent containers for samples; soil and/or roots keep well in them but whole plants are best kept sep-

*A revision of the chapter by D.J. Hooper.

 Plant Parasitic Nematodes in Subtropical and Tropical Agriculture, 2nd Edition (eds M. Luc, R.A. Sikora, J. Bridge)

arate from soil. Plant tops usually decompose faster than roots and should be in separate bags if they are to be stored for more than a day or two. Warm storage above 20°C adversely affects nematodes from plants and soil, so samples should be kept cool, at around 5°C in temperate regions, 10–16°C in warmer regions of the middle latitudes, and 16–18°C in the tropics and subtropics. Although it is common practice to store samples in refrigerators, low temperature (~5°C) can adversely affect the recovery of some nematodes from tropical soils (Whyte and Gowen, 1974). For more information on sampling procedures, especially sample size and sampling intensity for different crops, see Shurtleff and Averre (2000).

Fixation of Plant Tissue and Soil

In most cases, plant tissue and soil samples will be processed for nematodes within a few days after sampling. However, fixation of plant tissue and soil can be useful in preventing population changes during extended storage and avoiding quarantine restrictions applicable to live material. Roots and shoot tissue can be fixed for storage, subsequent examination or staining by adding to them preferably hot (60–70°C) formal acetic (FA, 4:1) or 5% formalin (2% formaldehyde solution). Alternatively, fresh material can be put directly into hot lactophenol/lactoglycerol; this softens tissues and is particularly helpful in the recovery of *Meloidogyne* females from roots. For soil samples, Elmiligy and De Grisse (1970) mixed hot fixative (100 ml of 40% formaldehyde + 10 ml of glycerol + 890 ml of distilled water at ~80°C) with soil samples. Nematodes in soils treated by fixation are extracted using centrifugal flotation.

Materials for Nematode Extraction

Extraction and handling of plant parasitic nematodes mainly require basic materials which can be bought at the local market (e.g. sieves, dishes, flasks, filters, funnels and tubing) or made individually (e.g. nematode transfer pick, counting dishes, sieves and racks). Plastic or stainless steel is preferable for nematode extraction rather than brass/bronze gauze, rings or pans because metallic ions, especially copper, released into small volumes of static water can be toxic to nematodes, especially dorylaims (Pitcher and Flegg, 1968). However, brief contact with metal sieves as in the sieving technique does not appear to be harmful. Stainless steel sieves of various sizes are offered by several laboratory suppliers. Cheaper ones can be made by buying the wire gauze separately and fixing it between two vinyl rings cut from a drainpipe of 15–20 cm in diameter.

Several methods make use of the ability of mobile nematodes to pass through a filter, thus separating them from plant debris and soil particles. Cotton wool milk filters, wet-strength paper handkerchiefs and towels are suitable, as are various types of cotton cloth or muslin. Tissues containing odour or toxic substances should be avoided. It is necessary to select a filter that retains as much debris as possible but with sufficiently large pores for the nematodes to migrate through. For larger nematodes such as *Longidorus* spp., a nylon gauze of about 90 μm aperture, secured to a supporting ring, will often give a clean enough extract. Various grades of lingerie material, nylon or terylene, are also suitable. Supports to hold the sample above water level can be easily made by fixing wet-strength viscose or wire mesh between two vinyl rings cut from a drainpipe.

Direct Examination of Plant Material

Nematodes can usually be seen by examining small amounts of gently washed plant tissue such as roots, leaves, stems or seeds with a stereoscopic microscope at magnifications from 15 to 50× using transmitted and/or incident light. Examine the plant tissue in water in an open Petri dish or large watch glass, and tease it apart with strong mounted needles. Nematodes released from the tissues will float out and can be collected with a handling needle or fine pipette. Migratory endoparasites (e.g. *Aphelenchoides*, *Ditylenchus*, *Hirschmanniella*, *Pratylenchus*, *Radopholus* and *Bursaphelenchus* (*Rhadinaphelenchus*)) emerge in a few minutes and can be found moving about on the bottom of the dish. Sedentary endoparasitic nematodes (e.g. *Globodera*, *Heterodera*, *Meloidogyne* and *Nacobbus*) may be seen attached to the surface of roots or in dissected tissue. Semi-endoparasites (e.g. *Rotylenchulus* and *Tylenchulus*) and firmly attached ectoparasites can be seen attached to the surface of the roots. Since nematodes tend to migrate from damaged tissue, it is often worthwhile to re-examine the sample after a few hours.

To recover females of root knot nematodes (*Meloidogyne* spp.) from roots, carefully tease away the tissue with forceps and a fine needle to release the head and neck; avoid puncturing the body. Dissection and storage in 0.9% NaCl helps to avoid the osmotic effect of water, which tends to cause females to burst.

Staining of nematodes in plant tissue

Since nematodes are translucent and difficult to see in plant tissues, staining helps to visualize them. Plant material is gently washed free from soil or debris and any thick material should be sliced thinly before staining. Detection of *Meloidogyne* females can be facilitated by staining the roots in Phloxine B (0.15 g/l water) for 15–20 min, rinsing and examining them in water; the gelatinous matrix of the egg sac is stained red (Holbrook *et al.*, 1983) although a few species, e.g. *M. artiellia*, do not stain well.

When staining specimens within leaves, stems and roots, the plant tissue needs first to be cleared in diluted sodium hypochlorite bleach (5.25% NaOCl or Clorox) for about 4 min. Trial and error is needed to determine the right strength and incubation time of the bleach, e.g. tomato roots clear quickly, but coffee roots are extremely difficult to clear. Thoroughly rinse the roots on a 100 μm aperture sieve to remove all traces of the bleach, which inhibits staining by acid fuchsin. Transfer the plant material into a glass vial and cover it with the acid fuchsin solution (875 ml of lactic acid, 63 ml of glycerol, 62 ml of water, 0.1 g of acid fuchsin). Boil the solution for about 30 s in a microwave oven or on a hot plate in a ventilated area to avoid the vapour of lactic acid. Several small samples can be stained in one operation by wrapping each in a piece of muslin cloth. The plant tissue is allowed to cool in the stain before being transferred to a sieve (100 μm aperture) to gently wash off excess stain in running tap water. If destaining with tap water is not sufficient, transfer the tissue in equal volumes of glycerol and distilled water acidified with a few drops of lactic acid. Depending upon the type of material, differentiation may take from several hours to 2–3 days, but the stained nematodes should be seen eventually in largely unstained tissue. Alternatively, plant tissue can be stained in acidified lactoglycerol plus 0.05% acid fuchsin or 0.05% methyl blue stain for a few minutes (Bridge *et al.*, 1982).

Extraction from Plant Material

Most commonly used methods for the separation of nematodes from plant material rely on the activity of nematodes (e.g. modified Baermann funnel technique); they are therefore not suitable for extracting sluggish (e.g. *Criconemoides*, *Hemicycliophora* and *Xiphinema*) or sedentary nematodes (e.g. *Globodera*, *Heterodera*, *Meloidogyne*, *Rotylenchulus* and *Tylenchulus*), although the juveniles and males of such forms will usually be recovered. For the latter, maceration–filtration or the mistifier technique are more suitable. Comparing the efficiency of these three techniques to extract *Pratylenchus zeae* and *Hirschmaniella oryzae* from rice roots, Prot *et al.* (1993) found the maceration–filtration or mistifier techniques most efficient. Other, less often used methods include the incubation technique (Young, 1954; West, 1957). Nematode extraction from bulky plant substrates such as bulbs, corms or enlarged storage roots can present difficulties. In such cases, the plant tissue can be peeled and used for nematode extraction to provide reliable data (McSorley *et al.*, 1999). Always wash plant material free of debris and adhering soil particles before extraction.

Baermann funnel technique

The Baermann funnel technique uses a funnel of 10–15 cm in diameter with rubber tubing attached to the funnel stem and closed with a spring or screw clip. The funnel is placed in a suitable support and almost filled with tap water. Plant material containing nematodes is chopped into small pieces of about 1 cm length, placed in a square of muslin cloth, nylon gauze, etc., which is folded to enclose the material, and then gently submerged in the water in the funnel (Fig. 3.1A). Nematodes

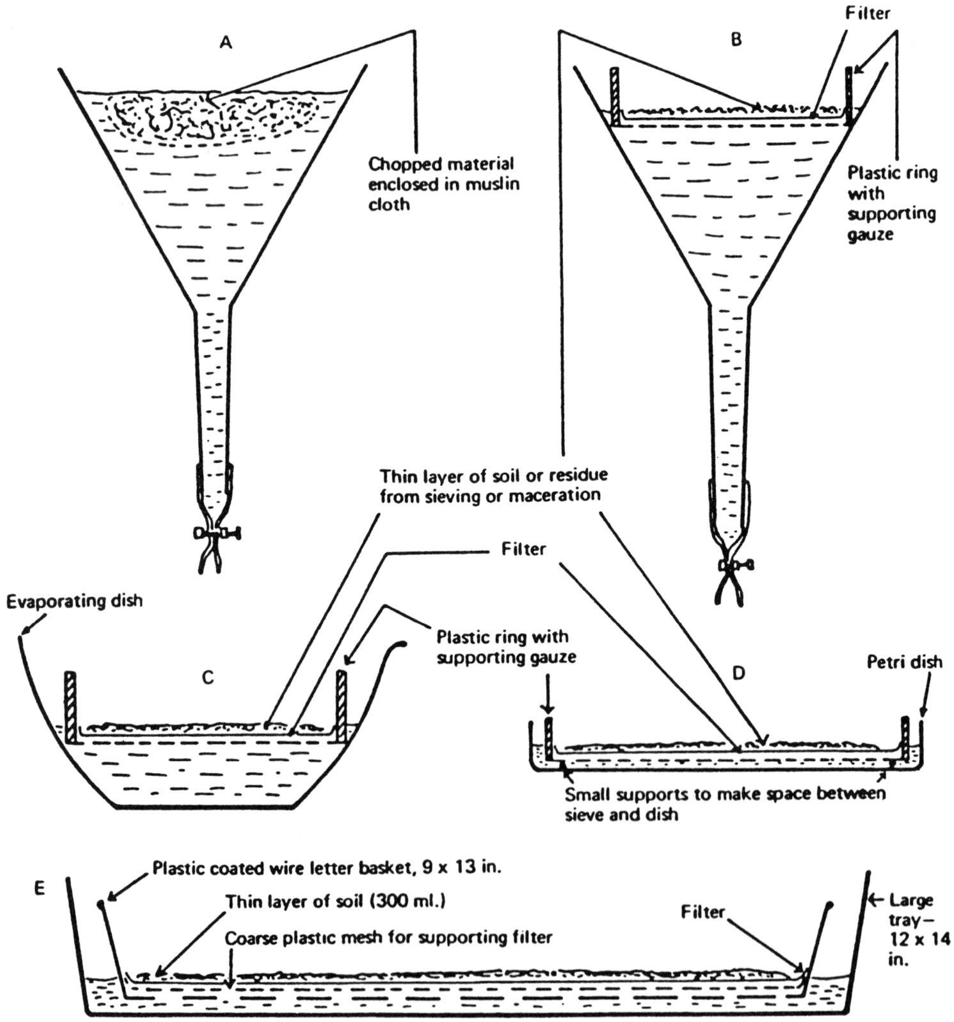

Fig. 3.1. Baermann funnel and modifications for extraction of active nematodes from chopped plant material, from thin layers of soil, or from residues obtained by sieving or maceration. The filter is a cotton wool milk filter, wet-strength facial tissue, coarse cotton cloth, or fine woven nylon or terylene cloth. Plastic rings are cut from perspex, polythene or vinyl tubes. The supporting gauze is muslin or nylon cloth held with an elastic band, or a coarse plastic mesh stuck or fused to the edge of the ring.

emerge from the tissues and sink to the bottom of the funnel stem. After 24–48 h, fully open the clamp to rapidly withdraw 5–10 ml of water containing the nematodes and transfer it to a shallow viewing dish for examination. This technique should not be used in its original form, as nematode recovery is less than 20% of that of other methods (Oostenbrink, 1970), mainly because of anaerobic conditions due to bacterial decay of the submerged organic matter and lack of oxygen at the base of the funnel stem. However, this technique has been modified in several ways to become a standard method for extraction of nematodes from plant tissue and soil.

Modifications of the Baermann funnel technique

Modifications of the Baermann funnel are widely used to extract active adult and juvenile nematodes (e.g. *Anguina*, *Aphelenchoides*, *Ditylenchus*, *Hirschmaniella*, *Pratylenchus* and *Radopholus*). Examples of modified Baermann techniques are illustrated in Fig. 3.1B–E. The funnel technique uses a supporting mesh (see Materials for nematode extraction) to hold the plant tissue partly submerged in water to avoid anaerobic decomposition (Fig. 3.1B). A milk filter or paper tissue is placed on the support and the chopped plant material is placed on it. Fill the funnel with tap water and set the sieve in the funnel to partly submerge the filter in the water. Samples must not be completely submerged in water. Drain off sufficient water if necessary. After 24–48 h, collect the nematode suspension as described above.

Using a shallow tray, dish or bowl (Whitehead and Hemming, 1965; Rodríguez-Kábana and Pope, 1981) instead of a funnel further improves oxygenation and reduces the number of nematodes remaining on the funnel wall (Fig. 3.1C and D). Similar to the above, a milk filter or paper tissue is placed on a support and the chopped plant material placed on it. A circle of muslin or paper tissue placed on top of the material will keep it moist and prevent it from floating. The support, with the material to be treated, is placed in a tray filled with tap water. Glass rods or small feet attached to the sieve ring are used to give a space of about 2 mm between the base of the sieve and the collecting tray. The material should be almost awash and, when it is not, more tap water should be added carefully between the outside of the support and the edge of the collecting dish. Avoid too large sample sizes; split the sample or use larger trays of 20–30 cm in diameter instead (Fig. 3.1E). Do not pour water over the sample to avoid washing debris through the filter. After 24–48 h, the support with the sample is gently removed and the contents of the dish transferred with a spray bottle to a beaker. The sample can be re-immersed in fresh tap water for further extraction of nematodes. Oxygenation, hence nematode extraction, can be improved by wetting the roots with tap water containing 1–3% H_2O_2 (Tarjan, 1967). H_2O_2 is often used for extracting migratory endoparasites from fleshy roots (e.g. banana), especially where high temperatures reduce oxygenation (P.R. Speijer, personal communication).

Mistifier technique (Seinhorst, 1950)

Nematodes recovered by this method are more active than by the previous methods because oxygenation is better, and sap and decomposition products from the material, which inactivate nematodes, are washed away. A fine mist of water is sprayed over the plant material. A spray nozzle, passing about 4.5 l/h is used. Most systems use an intermittent spray of say 1 min in every 10 min. Oil burner nozzles or gas jets can sometimes be adapted, and a water pressure of about 2.8 kg/cm^2 is usually required to give a suitable mist. The plant material to be treated is finely chopped into pieces 3–4 mm long and placed on a milk filter or tissue supported on a mesh set in a funnel as described for the modified Baermann technique (Fig. 3.2). Optimum sample size depends on sieve diameter and water flow

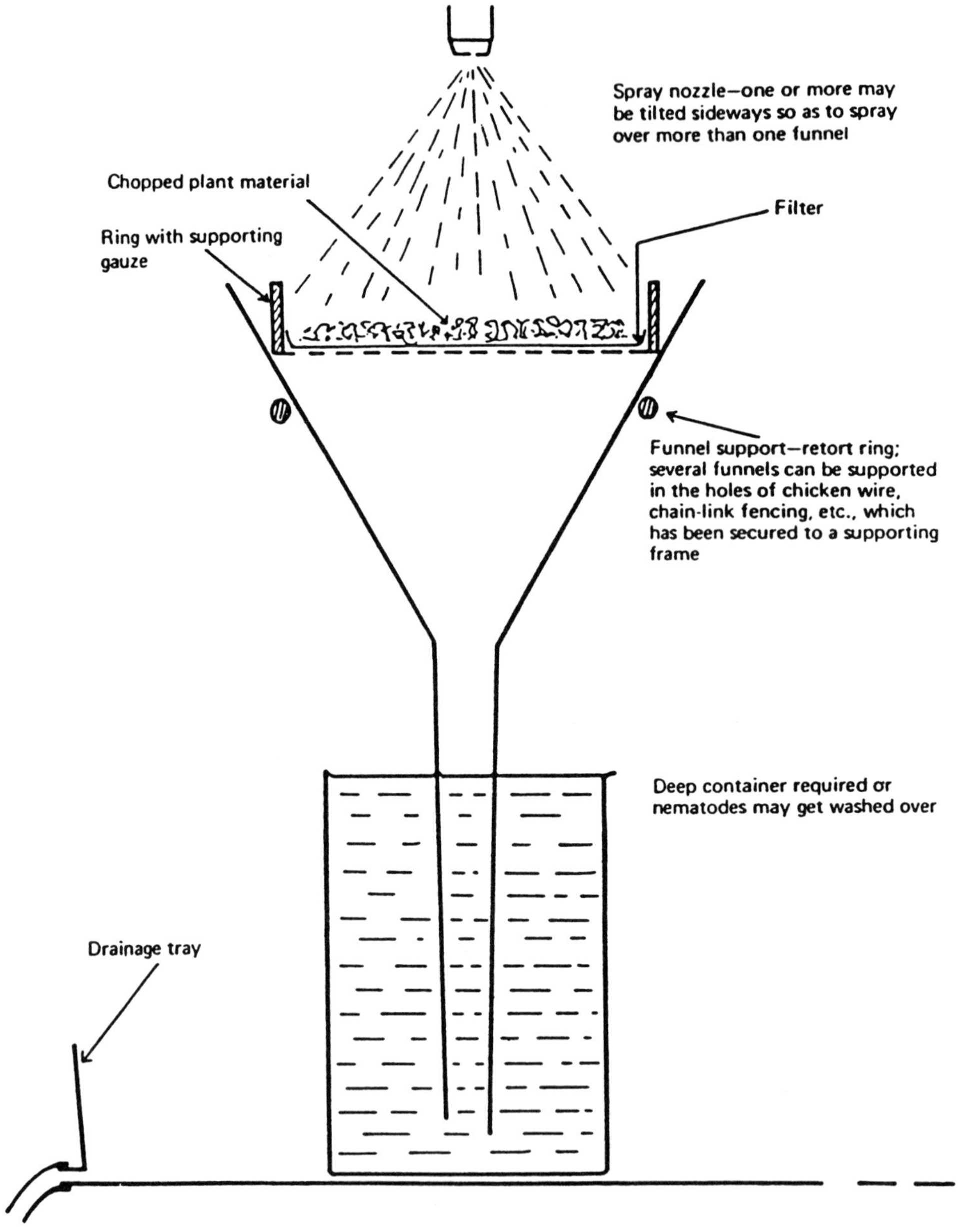

Fig. 3.2. Mist extraction of active nematodes from chopped plant material. The apparatus may be covered with plastic sheeting to prevent spread of the spray.

rate; increasing sample size can decrease the efficacy of extraction (De Waele *et al.*, 1987). Nematodes collected in the tube attached to the funnel stem can be released in a beaker for further examination. Compared with the modified Baermann techniques, plant tissue will decompose much more slowly, thus allowing prolonged extraction times of up to 2 weeks (e.g. freshly hatched *Meloidogyne* juveniles). Several funnels can be set up on a rack and one or two nozzles can supply all of them. The whole apparatus can be set up on the bench if enclosed with a polythene cover and left to stand on a drainage tray. For a more elaborate apparatus using collection trays instead of beakers, see Southey (1986). This method is suitable for recovering most active nematodes, but not for *Bursaphelenchus* (*Rhadinaphelenchus*), which swims and is lost in the overflowing water.

Maceration techniques

Maceration is used for extracting active nematodes as well as immobile stages of sedentary nematodes from bulbs, cloves, corms, storage roots, crowns, leaves and small plants. The plant material is chopped into lengths of 1 cm or less and then placed in about 100 ml of tap water and macerated in an electric mixer with revolving knife blades (e.g. common household blender, Waring blender, M.S.E. Atomix, Sunbeam domestic or Dormeyer blender). The maceration time required depends on the type of mixer used and, to some extent, on the type of plant material. Maceration needs to be continued long enough to give nematodes easy egress from the tissues but not to damage or render them immobile. For the extraction of eggs (e.g. *Meloidogyne* spp.) from root tissue, the sodium hypochlorite (NaOCl) technique described by Hussey and Barker (1973) is recommended. Comparing various maceration treatments, Stetina *et al.* (1997) achieved the most effective extraction of nematodes and eggs when maceration was preceded by 10 s in 0.5% NaOCl. Maceration methods in general are often quicker and more efficient than those described previously. However, the maceration action may release toxic substances from the plant material that can kill or immobilize nematodes. Toxic substances can be partially removed and extraction efficacy improved by pouring the macerated debris and water through the filter on the Baermann dish, removing the water in the dish and refilling the dish with tap water. Plant debris hindering nematode observation can be cleaned by one of the following processes.

Filtration

The nematode suspension is cleaned by active migration through a filter using the modified Baermann technique. Nematodes are collected after 24–48 h and examined under the microscope.

Flocculation–flotation (Escobar and Rodriguez-Kabana, 1980)

To extract *Radopholus similis* from banana, 25 g of roots are macerated in 100 ml of water. Then 250 ml of 1 M sucrose solution containing 12.5 μg/ml of the flocculating agent Separan NP10 are added and mixed. After standing for 2 min, the clear supernatant is poured through a 400 μm aperture sieve over one with 80 μm apertures; the sieves are sprayed with water and nematodes are washed from the 80 μm sieve into a counting dish for examination.

Centrifugal flotation (Coolen and D'Herde, 1972; Coolen, 1979)

The macerated plant sample is poured on to a 1200 μm aperture sieve resting in a funnel standing in a 500 ml centrifuge tube. The residue on the sieve is carefully washed with a spray before it is discarded. A 5 ml aliquot of kaolin powder is added to the extract in the centrifuge tube and the contents thoroughly mixed with a Vibromixer. Tubes are balanced and centrifuged for 4 min at 1500 *g*; the supernatant is poured off and the residue resuspended in sucrose, $ZnSO_4$ or $MgSO_4$ solution of specific gravity 1.18 and mixed with a Vibromixer or manually for at least 30 s. Tubes are balanced with the appropriate solution and centrifuged for 4 min at 1500 *g*. The supernatant is then poured on to a 5 μm aperture sieve, and the nematodes and eggs are collected in a beaker. De Waele *et al.* (1987) found that the efficiency of extraction of *Pratylenchus* from maize roots decreased with an increase in sample size, and so the root mass treated should be standardized for comparative studies.

Extraction of *Bursaphelenchus* (*Rhadinaphelenchus*) from coconut stem tissue

The material is chopped, well macerated (see above) and the suspension transferred to a 2 l conical flask which is then filled with water and allowed to stand for 30 min. The flask is then shaken and inverted with its neck in a vessel of water and the suspension allowed to settle for 30 min. The contents of the lower vessel are discarded and the flask contents are sieved four times through a 63 μm aperture sieve; the residue is washed off each time and collected in a beaker (after Fenwick, 1963).

Extraction from Soil

Nematode extraction from soil requires techniques different from plant tissue, except for the modified Baermann technique, which is widely used also to separate mobile nematode stages from soil. However, this technique is inefficient in recovering some large nematodes (e.g. *Longidorus* and *Xiphinema*) or nematodes with cuticular appendages (e.g. Criconematids). These are best extracted using sieving or elutriation techniques. Sieving or sieving plus filtering are quick methods for assessing all types of active, inactive and dead nematodes in soil, but they are not very quantitative as they are subject to much operator error. Elutriation techniques are very versatile methods capable of extracting wet cysts and vermiform nematodes from soil or root knot females from root debris providing the appropriate sized sieves and the correct flow rate of water are used. Flotation techniques give the most efficient and quickest extraction of active and sedentary nematodes from soil. Ideally, large centrifuge tubes (300–1000 ml) are preferable, but smaller tubes can be used especially when used in conjunction with a sieving technique. Other less frequently used techniques include the Seinhorst two-flask technique, which is a simple method giving a more efficient and cleaner extract than direct sieving (Seinhorst, 1955). A combination of techniques can improve accuracy of the assessment, as noted by Demeure and Netscher (1973) for *Meloidogyne* in a sandy clay soil.

Comparing the different techniques, Yen *et al.* (1998) found higher recovery rates of *Meloidogyne incognita*, *Pratylenchus coffeae*, *Aphelenchoides besseyi* and free-living nematodes when using the centrifugal flotation method and flotation–sieving technique than the modified Baermann funnel method. Griffiths *et al.* (1990) received significantly more nematodes by using flotation with colloidal silica (Ludox HS30) than by centrifugal flotation in sucrose, modified Baermann funnel extraction or repeated sieving. Comparing the modified Baermann technique with flotation–sieving, Rodríguez-Kábana and Pope (1981) extracted higher numbers of *Pratylenchus*, *Meloidogyne* and *Heterodera* with the modified Baermann method but *Helicotylenchus* and *Hoplolaimus* were higher for the flotation–sieving method. Nematode recovery, especially of specimens from roots (e.g. *Meloidogyne* and *Pratylenchus*), can be improved by incubating the soil sample at room temperature for 3–4 weeks prior to extraction.

Modified Baermann technique (Whitehead and Hemming, 1965)

The modified Baermann technique requires little labour and uses simple equipment. For soil samples up to 100 ml, flower pot dishes or plastic bowls of 10 cm in diameter can be used. For handling larger samples, the Baermann tray or dish technique is generally preferred over the Baermann funnel technique. A support to hold the soil above water level is made from a plastic-covered letter basket (22 × 32 cm) or a frame inside which is placed a coarse plastic mesh and on top of this a double layer of muslin cloth, paper tissue or milk filter. The basket is stood in a collecting tray (e.g. a large photographic dish, baking dish or glasshouse tray). Up to 300 ml of finely crumbled soil, passed through an 8 mm aperture sieve if necessary, is evenly spread in a thin layer over the filter in the basket. Water is carefully added down the inside edge of the collecting tray until the soil layer looks wet (Fig. 3.1E). To obtain a clean extract, it is important not to move the tray once the water has been added. Space can be saved by making a simple rack to hold the trays, and evaporation can be lessened by covering with polythene sheeting. Most nematodes will have collected on the floor of the tray after 24 h, but root knot juveniles from egg masses or some endoparasites from root fragments may take several days to emerge. The basket is then slowly and carefully removed and the nematode suspension from the tray beneath can be concentrated by pouring into a narrow beaker and leaving to settle for 4 h or more when the supernatant water can be syphoned off; or the extract can be concentrated in large (8 cm × 40 cm) glass cylinders having a funnel-like base fitted with a tap or pinch-cock (Whitehead and Hemming, 1965). Alternatively, the suspension can be concentrated quickly by passing it through a very fine sieve (< 25 μm aperture), washing the nematodes off the sieve and collecting them in a vessel.

Sieving technique (Cobb, 1918)

The sieving technique is also known as the 'bucket-sieving' method. Although crude, it is widely used as it enables the extraction of large numbers of both active and inactive nematodes in a relatively short time. Equipment required includes two plastic buckets (5 l), sieves of 15–20 cm diameter made with wire mesh (preferably stainless steel) of an aperture size of 2 mm, 710, 250, 125, 90, 63, 45 and 25 μm, respectively and tall 100 ml beakers for the residue from the sieves.

Usually only three or four of the set of sieves will be used for a particular sample, with the sieves selected to match the size of nematode it is hoped to extract, and to suit the type of soil involved. In general, sieve openings should be no greater than one-tenth of the nematode length. Most adults of large nematodes (e.g. *Anguina*, *Belonolaimus*, *Hirschmanniella*, *Longidorus* and *Xiphinema*) are caught on a 250 μm aperture sieve, adults of average-sized nematodes (e.g. *Aphelenchoides*, *Ditylenchus* and *Hemicycliophora*) on a 90 μm aperture sieve, and many juveniles and small adults (e.g. *Criconemoides*, *Paratrichodorus*, *Paratylenchus*, *Pratylenchus* and *Radopholus*) on a 63 μm aperture. A 45 μm or even 25 μm aperture sieve is used to recover small juveniles (e.g. *Meloidogyne*, *Heterodera* and most others). Ready-made sieves are expensive. Use sieves singly, never stack them and never attempt to work a sample through them all simultaneously, as this may reduce the efficiency of recovery. Fine sieves are easily clogged, but this can partially be avoided by pouring the suspension on a sieve inclined at an angle of about 30° to the horizontal; however, the number of nematodes caught on the sieve will also be reduced (Araya *et al.*, 1998). Gently patting the underside of the sieve into the water in the bucket below and lifting it in and out a few times will help to clear it. Sonicate sieves for cleaning. The method is as follows.

1. Mix the soil sample thoroughly and place a known volume of soil (100–500 ml) in bucket I and fill with about 1–4 l of water. Dry soils should be soaked for a few hours. The mixture is stirred to free nematodes from the soil and suspend them in the water. Flocculating agents, such as Separan NP10 (12.5 μg/ml), might be used to help to break up soil aggregates in heavy clay soils.

2. Let the mixture settle for 30–60 s and decant over a 2 mm aperture sieve into bucket II. Avoid pouring the sediment. Add less water to the sediment in bucket I and repeat this step 2–3 times to increase nematode recovery. Any sediment left in bucket I is then discarded and bucket I washed out. The sieve is rinsed over bucket II. The residue on this sieve may contain very large nematodes, but usually it can safely be discarded.

3. The contents of bucket II are stirred, allowed to settle for about 10 s and then poured through a 710 μm aperture sieve into the clean bucket I, leaving behind heavy soil particles to which more water is added and the process repeated, if desired. The sieve over bucket I is rinsed. The residue on this sieve may contain only a few large nematodes, but this often depends on how much debris is present. To collect the residue, hold the sieve over bucket I at a steep angle (35–45°) and direct a gentle stream of water on to its upper side to wash the nematodes to the bottom edge of the sieve. Small nematodes and eggs will be washed through the sieve into bucket I and recovered later. Transfer the nematodes on the sieve into a 250 ml beaker using a gentle stream of water, leaving behind any heavy particles.
4. Bucket II is cleaned and the process repeated using 250, 125 and 90 μm aperture sieves and collecting the residues, as described above. The residues of each sieve can be pooled in one beaker or kept separate in different beakers. If the contents of the beakers appear cloudy, it is because the residue on the sieve was rinsed inadequately. If necessary, the contents should be poured back on to the sieve and rinsed again over the bucket containing the remaining suspension before proceeding to the next sieve in the series. The contents of the collecting beakers are allowed to settle for 1–2 h and the supernatant liquid is carefully decanted or syphoned off leaving about 20 ml in the bottom. The material can be transferred to a viewing dish and examined.

Some workers shorten the whole procedure by transferring the soil suspension directly through a 1–2 mm aperture sieve to remove very coarse material followed by a 45 μm aperture sieve which collects the nematode specimens. This procedure is less suitable for larger sample sizes (> 250 ml) and heavy soil due to clogging of the fine sieve. Although this technique is less laborious, nematode losses may be higher. If the suspension still contains a significant amount of debris, further processing by centrifugal flotation or modified Baermann techniques will result in an almost clean nematode suspension. However, sluggish and inactive nematodes can be lost (e.g. *Longidorus*/*Xiphinema*).

Elutriation techniques

Elutriation techniques extract nematodes of defined size by using an upcurrent of water to separate them from soil particles and hold them in suspension. They give a cleaner extraction than that obtained by direct sieving, although they are not any more efficient. Flow rates can readily be adjusted to suit soil type and the size of nematode to be extracted. Of the models that have been developed (Seinhorst, 1956; Tarjan *et al.*, 1956; Oostenbrink, 1960), the No. III model of Oostenbrink is often used because it is robust and easily operated and cleaned. Oostenbrink (1960) or Southey (1986) should be consulted for details. Winfield *et al.* (1987) described a column elutriator for extracting nematodes and other small invertebrates, referred to as a Wye Washer. This equipment was shown to achieve extraction rates equal to or better than existing techniques and have operational advantages as soil samples up to 1 kg can be processed; however, water use and price are high.

The fluidizing column (Trudgill *et al.*, 1973) is a simple, robust and versatile elutriator which has been modified by several workers. The version (Figs 3.3 and 3.4) used at Rothamsted has an internal diameter of 7.5 cm and a column height of 42 cm above the disc. It is constructed from a plastic (perspex) cylinder which fits tightly into a short cylindrical base sealed by an O ring. The base contains a plastic sintered plate, and water is introduced beneath the plate through a side arm with a perforated end piece. By varying the water flow rate, preferably with a flow meter, all types and sizes of nematodes can be recovered. Up to 200 cm^3 of soil can be treated. The soil is mixed in water and passed through a coarse sieve of 8 mm aperture. The prepared sample is then added with the column about one-third full of water. The

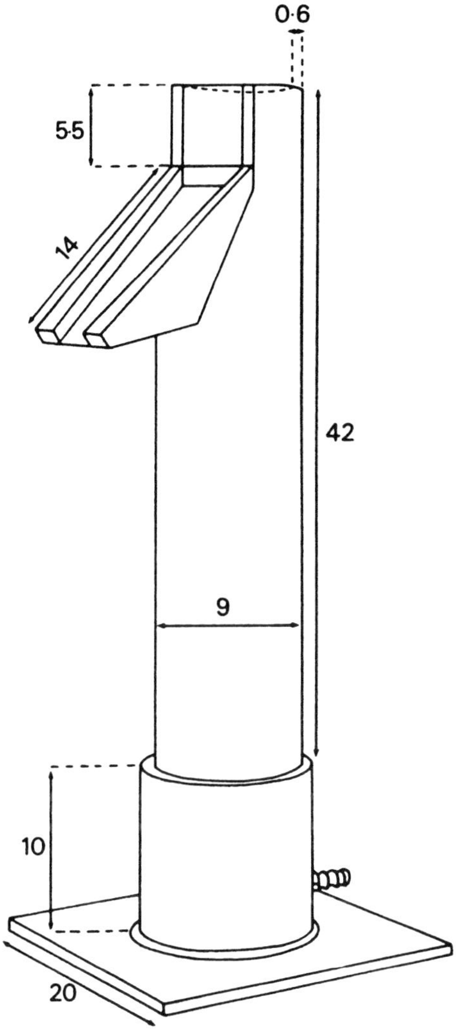

Fig. 3.3. Fluidizing column, with dimensions in cm (from Trudgill *et al.*, 1973, reproduced with permission from *Nematologica*).

Fig. 3.4. Fluidizing column in operation (photo: Rothamsted Experimental Station).

upward water flow, through the sintered plate, is adjusted to a rate of about half that required to wash over the nematodes and is allowed to run for 3 min to mix and fluidize the suspension, then for a further 3 min at the full rate to extract the desired nematodes. The overflow from the column is caught on a sieve or bank of sieves of appropriate size. In order to obtain reasonably clean extracts, the flow of water through the column needs careful control. Trudgill *et al.* (1973) give a terminal velocity (settling rate) of 0.11 cm/s for *Longidorus leptocephalus* adults and 0.01 cm/s for cyst nematode (heteroderid) juveniles. Thus, for a column with a 3.75 cm radius, the least flow to extract longidorids would be $\pi \times (3.75)^2$ (area of the disc) × 0.11 (settling rate) × 60 (seconds to minutes) = 291 ml/min; for heteroderid juveniles, the flow rate would be 29 ml/min. In practice, about twice these flow rates should be used to ensure a good recovery of nematodes. The apparatus should be run at approximately 300 or 30 ml/min for 3 min and then at 600 or 60 ml/min for longidorids or heteroderid juveniles, respectively. Longidorid adults would be caught on a 150 μm aperture sieve and heteroderid juveniles on one with 45 μm apertures. Extracts from the sieves can be concentrated and cleaned as described for the sieving technique. Much faster flow rates (3.5 l/min for 3 min then 7 l/min for 3 min) are required to extract heteroderid females and cysts from moist soils. The extract is caught on a 250 μm aperture sieve after passing through a 840 μm sieve to remove coarse debris.

Flotation techniques

Nematodes can be extracted from soil and organic debris by floating them out in a solution of specific gravity greater than their own. As the method does not rely on the mobility of nematodes, it is extremely useful for extracting sluggish forms such as criconematids as well as dead, moulting or fixed nematodes and eggs. Centrifugal flotation is generally a more efficient nematode extraction method than Baermann, sieving or elutriation techniques. Flotation is often used to clean extracts obtained by sieving or elutriation but can also be applied directly to soil samples. Solutions of sucrose, $MgSO_4$ or $ZnSO_4$ can be used. Sugar is the most used solute because it is cheap; however, Rodríguez-Kábana and King (1975) found that blackstrap molasses was even cheaper and, because of higher viscosity, more effective than sucrose for extracting nematodes. $MgSO_4$ does not have the stickiness of sugar, and $ZnSO_4$ has fewer osmotic effects but is more acid and toxic. Other manufactured solutes (Ludox, Ficoll and Percol) have advantages over $MgSO_4$ and $ZnSO_4$ but are more expensive (Viglierchio and Yamashita, 1983; Bloemers and Hodda, 1995). To reduce the osmotic stress by the solutes, nematodes should be rinsed with water as soon as possible to aid their recovery. A solution with a specific gravity of about 1.18 (673 g of sugar dissolved in water and made up to 1 l) is suitable for most nematodes; however, a more dense solution of specific gravity 1.25 (1210 g of sugar dissolved in water and made up to 1 l) is required for very long nematodes such as *Longidorus* and *Xiphinema* but also for nematode cysts. The specific gravity of a solution should be checked just prior to its use as changes in temperature and microbial activity can cause a considerable decrease in concentration. The suspensions recovered are usually so clean that they can be caught on very fine sieves of 5–20 μm aperture for direct counting.

Centrifugal flotation (Caveness and Jensen, 1955; Dunn, 1971)

The soil sample is mixed and passed through a 1 cm aperture sieve to remove stones or coarse debris. A total of 100–250 ml of soil is placed in a 800–1000 ml centrifuge tube and water added up to 2 cm from the tube brim. Kermarrec and Bergé (1971) recommend the addition of kaolin powder, 1 ml to 100 ml of suspension, before centrifuging to aid sedimentation and to give a more compact surface to the sediment pellet. The contents are thoroughly mixed using a Vibromixer or mechanical device. The tubes are balanced by adding water and centrifuged at about 1800 *g* for 4 min, after which the centrifuge must be carefully braked to avoid vibrations that will disturb the sediment pellet. The supernatant containing organic debris is discarded and the tube almost filled with the suspending solution (specific gravity 1.18) and stirred mechanically or Vibromixed to resuspend the pellet containing the nematodes. Tubes are balanced by adding more solution and re-centrifuged at 1800 *g* for 4 min. The supernatant is poured through a sieve of 53 μm aperture or less (e.g. 25 μm to avoid loss of smaller nematodes), quickly rinsed with tap water and collected in a beaker or counting dish. Alternatively, the supernatant can be poured into excess water (~1:5) to reduce osmotic stress on the nematodes. The relative centrifugal force represents the force on particles due to gravity: $g = 0.00001118 \times$ radius of centrifuge arm to tip of tube in cm $\times$ (speed in r.p.m.)^2.

Sieving/centrifugal flotation

Jenkins (1964) modified the technique of Caveness and Jensen (1955) to handle larger soil samples. A total of 100–500 ml of soil are washed through an 840 μm aperture sieve into a bucket, and made up to about 6 l with water. After stirring, the suspension is allowed to settle for 30 s before the supernatant is decanted through a 52 μm aperture sieve. The first bucket is refilled and the process repeated. The sievings are collected in two 50 ml centrifuge tubes which are balanced before spinning at 1800 *g* for 4–5 min. The supernatant is poured off and replaced by sucrose solution (specific gravity 1.18). The tubes are balanced, shaken, and spun for 0.5–1 min. The supernatant is poured through sieves of 53 μm aperture or less and the sievings are washed before collection in a beaker for examination. Extracts obtained by elutriation can also be cleaned using this Jenkins modification. Gooris and D'Herde (1972) and Demeure and Netscher (1973) described more elaborate methods for extracting *Meloidogyne* stages, including egg masses.

Flotation, flocculation/sieving (Byrd et al.*, 1966)*

In this method, flocculating chemicals are used instead of centrifugation to separate soil particles from suspension in 1.0 M (342 g/l solution) sucrose solution. Separan is an effective flocculating agent irrespective of soil type or pH. Ferric chloride ($FeCl_4$) can be used, but the concentration is critical and must be varied according to soil type and pH. This method takes only 1–3 min per sample and gives good yields of *Xiphinema*, trichodorids and spiral nematodes, but small forms such as *Criconemoides* may be trapped in the flocculated material and lost.

Fifty cm^3 of soil are placed in a 600 ml beaker and made up to 350 ml with 1.0 M sugar solution containing 12.5 μg/ml of Separan. This is stirred gently with a mechanical stirrer (1600 r.p.m.) for 20 s and then allowed to settle for 2–5 min. The nematode suspension is then decanted through a 355 μm aperture sieve set over one of 45 μm aperture or smaller. The residue on the sieves is rinsed and washed into a beaker; the contents are swirled, allowed to settle for a few seconds, and then poured back on to the 45 μm or smaller aperture sieve leaving behind heavier particles. The nematodes are then washed from the sieve into a beaker with about 25 ml of water.

Mishra *et al.* (1977) pooled soil extracts obtained by sieving into a beaker, mixed in 0.2% Separan CP-7 and, after allowing particles to settle for 1 min, decanted the supernatant through a 50 μm aperture sieve to recover the nematodes. The process is repeated three or more times on the residue left in the beaker. This modification avoids the use of a sucrose solution and, because sieved extracts only are treated, larger volumes of soil can be handled initially. Rush (1970) extracted *Xiphinema americanum* from soil using Separan without sucrose. Sieve aperture must be modified to nematode size.

Extraction of heteroderid cysts from dry soils

The saccate dead females, 'cysts', containing eggs of heteroderid nematodes float in water when they are dried. To extract cysts, the soil sample is air dried and passed through a 4 mm aperture sieve. A 100–1000 cm^3 sample of the dried soil is placed in a plastic bucket and made up to about 2–5 l with water, thoroughly stirred with a strong stream of water or manually. Allow the coarse material to sediment for 1–3 min. Any cysts present will float to the surface with other organic debris. Decant through a 2 mm aperture sieve over a 250 μm aperture sieve (a 100 μm aperture sieve may be needed to catch small cyst nematodes such as *Heterodera trifolii*). Repeat the process 2–3 times if necessary. Wash the residue on the sieves and collect the cysts on the 250 μm aperture sieve for further examination. Alternatively, the float can be poured on to a filter paper in a funnel, the water drained off, and the paper examined for cysts, most of which will occur along the 'tide mark' left at the upper water level (Shepherd, 1986).

Other methods commonly used to extract cysts from soil include the Fenwick can (Fenwick, 1940) and its modified version described by Oostenbrink (1950), Oostenbrinks's Model III elutriator (Oostenbrink, 1960), fluidizing column (Trudgill *et al.*, 1973), 'Schuiling' centrifuge (Hietbrink and Ritter, 1982) and

Wye Washer (Winfield *et al.*, 1987). See Shepherd (1985) for further details on these methods. Based on Riggs *et al.* (1997), sieving was more efficient than elutriation for extracting cysts. If cysts are to be used further as inoculum in biotests, note that the contents of *Globodera* but not *Heterodera* cysts will survive desiccation.

Storage

Many nematodes remain in good condition for several days when stored in shallow, fresh tap water at about 5–10°C. Contaminating bacteria can be suppressed by adding three drops of 5% streptomycin sulphate solution per 5 ml of nematode suspension. Tropical nematodes needed for live cultures or for experimental use should be stored at room temperature and aerated with an aquarium pump. For long-term storage (e.g. germplasm collection, maintenance of genetic lines, reference material or inoculum), nematodes can be stored in liquid nitrogen. Cryopreservation has been shown to work for several nematodes. For *Pratylenchus thornei*, the survival rate was 76% when nematodes were pre-treated in 14–17% glycerol for 5 days before storage in liquid nitrogen (Galway and Curran, 1995). Thawed nematodes were able to reproduce and infect carrot disc cultures. Similar survival rates were achieved by Beek *et al.* (1996) for *Meloidogyne hapla* and *M. chitwoodi* in liquid nitrogen after pre-treatment in 10% ethanediol for 2 h at room temperature and 40% ethanediol for 45 min on ice. Cysts of *Heterodera avenae* have been stored successfully at –18°C (Ireholm, 1996).

Examination of Nematode Suspensions

Direct examination

Extracted nematodes can be examined directly under a microscope to the genus level using viewing dishes or counting slides (Fig. 3.5). A good stereoscopic microscope with a range of magnifications 10× to 100×, a fairly flat field and good resolution are essential. Illumination by transmitted light should be as even as possible; small frosted strip-light tubes are suitable.

All or part of the extracted suspension, according to its density, is placed in an open counting dish and examined under the microscope. When samples are taken with a pipette, it should have a wide outlet to prevent debris clogging it. Petri dishes or flat-bottomed Syracuse watch glasses (Shurtleff and Averre, 2000) are often used, and a grid is etched, or scratched with a marking diamond, on the inside of the base to act as a guide when searching. Small disposable tissue culture plastic Petri dishes (5 cm in diameter) that have sloping sides can be used on which a grid is easily scratched with a needle (Fig. 3.5L). Merny and Luc (1969) describe an open plastic dish 5 ml capacity, with sloping sides to minimize the effect of the meniscus; the base is marked in 2 mm squares. Similar counting slides with 2 ml capacity were developed by Sikora for nematode counting on a compound microscope with moveable stage (Fig. 3.5H). A 10 ml capacity winding-track carved into a block of solid plastic was developed by Sikora and Nordmeyer and used to collect and count cysts in samples with debris (Fig. 3.5J). Some dishes have channels/ridges on the base which restrict the movement of nematodes: the Doncaster (1962) dish with concentric channels holds up to 40 ml. De Grisse (1963) moulded a rectangular dish with ridges, and Bridge (in Hooper, 1990; see Fig. 3.5F) designed a 5 ml plastic dish with a ridged base which is readily made by injection moulding. A counting slide primarily used for cysts is shown in Fig. 3.5G. Multichamber counting slides allow examination of several samples on one slide (Fig. 3.5B). The slide in Fig. 3.5K was made by removing squares from hardened parafilm in a Petri dish and then etching in a suitable grid. Touching the surface of the liquid with a needle dipped in detergent reduces surface tension and helps in nematode picking.

Fixed capacity, usually 1 ml, covered counting slide chambers are useful for routine counts when immediate access to nematodes within the suspension is not required. Examples are the Peters 1 ml or 2 × 0.5 ml counting slides made in glass by Hawksley (Fig. 3.5A and B) and the Fenwick multichamber slides which can be made in plastic (Doncaster *et al.*, 1967; Southey, 1986). To be sure of searching over the whole area of the dish, the space between the grid lines should be a little less than the field width of the microscope at the magnification being used. Thus, a dish with an extract containing large nematodes (*Xiphinema*, etc.), which would be examined at about 15× magnification, would have guide lines about 1 cm apart, whereas extracts containing average size nematodes would be examined at about

Fig. 3.5. Examples of counting slides/dishes: (A) Peter's 1 ml counting slide in glass as made by Hawksley; (B) multichamber counting slide in glass as made by Hawksley; (C) 1 ml counting slide made by MEKU; (D) 2 ml counting slide in plastic (made at BBA Münster); (E) microscope slide with ridges to hold a large cover slide, 1 ml volume (made by Sikora, Bonn University); (F) moulded plastic dish, 5 ml, with sloping sides and ridged grid (made at Rothamsted Experimental Station from a design by Bridge (in Hooper, 1990)); (G) glass ring, 38 mm, glued on a glass plate for counting cysts (made at BBA Münster); (H) 2 ml counting slide with sloping sides consisting of a 2 mm high plastic ring glued on a plastic plate of 75 × 37 mm (made at Bonn University from a design by Sikora); (I) 2 ml counting slide in plastic with a coverglass of 78 × 48 mm as the bottom to allow examination with an inverse microscope (made at BBA Münster); (J) 10 ml winding-track counting tray in plastic as made by Nordmeyer and Sikora (made at Bonn University); (K) multichamber counting slide with sloping sides made in paraffin within a 90 mm diameter plastic Petri dish (made at BBA Münster); (L) 50 mm diameter plastic tissue culture Petri dish marked for examination at 20–40×, base lines are cut with a plastic or glass writing knife into the lid (photo: BBA Münster).

50× and have lines about 3 mm apart. Some workers prefer to examine extracts in a dish with a thin base (e.g. a disposable plastic Petri dish) using the low/medium power objectives of an inverted, compound microscope when nematodes can be seen in more detail (Fig. 3.5C–E and I). Sikora etched markings lengthwise on to a glass microscope and then placed a large cover-slip on small supports glued to the slide to allow observation of nematodes in 0.2–0.5 ml samples at up to 400×. A hand tally counter or a bank of counters is a useful aid for counting different genera. For nematode identification to the species level, temporary or permanent slides have to be prepared, which includes handling of the nematodes.

Handling nematodes

There are various methods for handling nematodes. Small batches of nematodes can be selected and transferred from a suspension by using a fine pipette. The modified Hesling's device (Alam, 1990) or the suction device described by Sehgal and Gaur (1988) even allow the selection of individual specimens. However, in most cases, a handling needle is preferred, which is a dissecting needle handle to the end of which is attached with glue a nylon toothbrush bristle, sharpened bamboo splinter, eyebrow hair, fine wire or small wire loop. Old curved nylon toothbrush bristles are recommended as they can be tapered to the desired thickness with a sharp scalpel and they are not so easily damaged as other types. The quill and shaft of a moderate sized feather also make a convenient handling tool, the feather vane is removed and the thin end of the shaft shaped/sharpened; the thicker quill end can also be used, but the hollow core should be blocked off to prevent loss of nematodes up the quill by surface tension. Many beginners have difficulty in picking up nematodes with a bristle. To do this, the nematodes should be in shallow water, near the centre of the dish, and the lowest convenient microscope magnification should be used to give the greatest possible depth of focus and working distance. While viewed with the stereoscopic microscope, the handling needle is used to lift the nematode to the surface of the water, the bristle is then held immediately underneath the nematode and quickly flicked up so that the nematode is pulled out through the meniscus. Avoid using too fine and smooth a bristle as it will not have enough drag to bring the nematode up with it through the meniscus. The surface tension can be removed by adding a small drop of soap or detergent on a needle. Picking up fixed nematodes from glycerine is generally easier due to its higher viscosity.

Killing and fixing nematodes

For identification to the species level and permanent storage, nematodes must first be killed, fixed and properly mounted. A few specimens can be killed by transferring them to a drop of water on a 26 × 76 mm glass slide, which is then heated over a small flame for a few seconds until the nematodes suddenly straighten out. However, killing by placing the slide on a controlled hot plate at 65–70°C is most effective and prevents damage to specimens due to overheating. The specimen can be examined directly under the microscope, but in most cases will be transferred to fixative or fixed on the slide by adding an equal sized drop of double strength fixative (e.g. FA or TAF (1.5 ml of triethanolamine, 83.6%, Aqua dest, 14.9%, formalin, 35%)).

The following method is recommended for killing and fixing nematodes in one step: specimens are concentrated in about 3 ml of water in a 10 ml glass vial, either by centrifuging or by letting them settle and siphoning off the supernatant. The vial is shaken to disperse the nematodes. Threefold strength TAF or FA (or formal propionic (FP)) fixative 4:1 (preferably plus 2% glycerol) is heated to about 70–75°C and 6 ml are quickly added to the nematodes. This kills and fixes them in the one process (Seinhorst, 1966). The fixative can be heated in a small tube stood in water of the required temperature for a few minutes; preferably the temperature is monitored with a thermometer in the suspension. This method gives a very good fixation of glands and gonads. Nuclei tend to expand and are more easily seen. Although specimens appear rather dark as soon as they are fixed, processing to glycerol will eventually clear them. However, fixatives usually cause some shrinkage and/or distortion of the specimen (Grewal *et al.*, 1990).

Comparing the different methods, Grewal *et al.* (1990) found that killing and fixing with addition of hot (95°C) TAF produced the least affected specimens compared with FA 4:1 or FP 4:1. Chakrabarti and Saha (2001) came to similar conclusions using TAF at 50°C. The most lifelike specimens were produced when fixed in TAF and processed to glycerol by the slow method (outlined below) (Grewal *et al.*, 1990; Siddiqi, 2000).

Fixatives

Solutions of 5–10% formalin (2–4% formaldehyde), preferably plus 2% glycerol, are often used as fixative. Notice that due to toxic fumes, all work with formaldehyde must be done under the exhaust hood. The addition of a small amount of powdered $CaCO_4$ to the stock solution is recommended as this neutralizes the free formic acid that can cause darkening and granulation of tissues. Alternatively, the formic acid can be neutralized using triethanolamine as in TAF fixative (Courtney *et al.*, 1955). FA 4:1 and FP 4:1 are probably the most widely used fixatives that also allow long-term preservation. TAF is a commonly used fixative, as nematodes retain their lifelike appearance in it for several hours, but it is not a good long-term preservative, as some degeneration of the nematode cuticle can occur. However, specimens fixed in TAF and mounted in glycerol remain in good condition. Commonly used fixatives are:

Formalin: formalin (40% formaldehyde), 8 ml; distilled water up to 100 ml.

Formal acetic (FA) or formal propionic (FP) 4:1: formalin (40% formaldehyde), 10 ml; glacial acetic acid (or propionic acid), 1 ml; (glycerol, 2 ml); distilled water up to 100 ml.

As noted by Golden in Hooper (1970), the addition of 2% glycerol to the above means that nematodes can be brought directly from fixative to glycerol by slow evaporation (see below). Also as noted by Hooper (1987), nematodes stored in vials will eventually end up in glycerol should the fixative evaporate.

TAF: formalin (40% formaldehyde), 7 ml; triethanolamine, 2 ml; distilled water, 91 ml.

Nematodes will be spoiled if put alive into cold fixative. Alcoholic fixatives should be avoided as they usually shrink nematodes. Well-fixed specimens have a smooth outline, whereas distorted specimens are rarely worth keeping. Nematodes can be stored in formalin indefinitely. Vials containing them should be labelled with the identity of the nematode if known, source, locality, fixative used and date of fixation.

Processing and Mounting Nematodes

In fixed nematodes, much of the internal body contents, especially gonad structure, may be obscured by the granular appearance of the intestine. Specimens can be cleared by processing with lactophenol, lactoglycerol or glycerol, which are also suitable mountants. Although lactophenol has been widely used in the past, it is now recognized that phenol fumes are a danger to health. To avoid using phenol, Bridge *et al.* (1982) recommended the use of lactoglycerol. This is a solution of equal amounts of lactic acid, glycerol and distilled water, to which can be added 0.05% acid fuchsin or 0.05% methyl blue to stain the specimen if required. However, glycerine mounts are preferred. Several techniques exist that allow processing of the specimens through alcohol to glycerine with minimum time and effort (Hooper, 1987). Mounted specimens can deteriorate and the storage of some representatives in glycerol in vials is recommended.

Glycerol method

Most nematodes are best preserved in anhydrous glycerol. Transfer from the fixative to glycerol can follow a slow or rapid method. The former usually gives better preservation and is therefore recommended if time is not a limiting factor.

Slow method

Remove most of the fixative from preserved specimens in a small dish or deep glass block with a fine pipette, but take care not to inadvertently draw nematodes. Add 3–4 ml of the following solution: anhydrous glycerol, 2 ml; 96% ethanol, 1 ml; distilled water, 90 ml.

Cover the dish loosely and let the sample stand at room temperature for 2–3 weeks or until water and ethanol have all evaporated. The process can be speeded up in an oven at 30–40°C but the container needs to be well covered to ensure that the evaporation takes several days. If evaporation is too rapid, the nematodes shrink and become distorted. Golden (in Hooper, 1970) recommends the addition of a few drops of picric acid which helps to prevent clearing and fading of nematode stylets and the growth of moulds.

Rapid method (Seinhorst, 1962)

Fixed specimens are transferred to a small concave glass dish of 2–4 ml capacity containing about 0.5 ml of the following solution: 96% ethanol, 20 ml; glycerol, 1 ml; distilled water, 79 ml.

The dish with nematodes is placed into a closed glass vessel containing an excess (e.g. 1/10 volume of the vessel) of 96% ethanol. The dish is supported above the ethanol on a platform or grid. After a minimum of 12 h in an oven at 40°C, the specimens will be in a mixture of mainly ethanol, with some glycerol. The dish is removed from the vessel, excess ethanol can be withdrawn using a pipette, and a solution of five parts glycerol and 95 parts of 96% ethanol is added. The dish is then placed in a partly closed Petri dish in an oven at 40°C until the ethanol has evaporated. This should take at least 3 h; the nematodes are then in pure glycerol and should be mounted immediately in anhydrous glycerol. Note that nematodes processed to glycerol are very soft and should be handled carefully, preferably using a mounted eyebrow hair or similar soft bristle.

Mounting nematodes

The nematodes are best mounted on thin microscope glass slides (25 × 76 mm) using 19 mm diameter round coverslips. Cobb-type aluminium double coverglass slides (see Southey, 1986) allow examination from either side, but preparation is more laborious. Supports (e.g. stainless-steel wire, tungsten filaments of calibrated diameter, glass fibre or beads) about as thick as the nematode are used to prevent deformation of the specimens due to the weight of the coverglass.

Some important features of nematodes are most readily seen in freshly killed/fixed specimens mounted in TAF. Place the specimens plus similar sized supports in a small drop of fixative, add the coverglass on to it, blot off excess fixative from around the coverglass with a tissue and seal the coverglass. In spite of a good seal, nematode specimens in fixative usually start to dry out after a few days or undergo undesirable changes during prolonged storage. Process fixed nematodes to glycerine for long-term preservation.

For permanent mounts, a very small drop of anhydrous glycerol (heated for 4 h at 40°C in an oven) is placed in the centre of a clean microscope slide and nematodes of about equal diameter are transferred to it, using a handling needle, and arranged in the centre of the drop so that they are touching the slide surface, not floating. Three coverglass supports are arranged around the nematodes. Paraffin wax of melting point 60–65°C is used as seal but also provides additional support. A wax ring is prepared using a copper tube (15 mm in diameter, heatproof handle) heated in a flame, dipped in paraffin wax and applied to the centre of the slide surrounding the mountant. A clean coverglass (19 mm diameter circle No. 1) held with fine forceps is lowered on to the drop. A mounted needle held in the other hand can be used to help prevent the coverglass from sliding sideways when it is applied. It helps to prevent air bubbles from being trapped if the drop is kept as hemispherical as possible before applying the coverglass. The slide is placed on a hotplate at 65°C for a few seconds. As soon as the wax melts, press lightly with a mounted needle on the coverglass to make sure it has settled far enough; thick mounts prevent oil immersion objectives being used. The wax will set rapidly when the slide is placed on a cool surface. A secondary seal is desirable to prevent drying out and to prevent immersion oil dissolving the wax. Permount (Fisher Scientific), Corseal (Sabir *et al.*, 1997) or Glyceel (Bates, 1997) are excellent; nail varnish is a good substitute. The coverglass is ringed, using a small soft brush, with a thick but fairly narrow band of the sealant, making sure there is sufficient on the coverglass as well as on the slide. Repeat the process when the first ring has dried to give a good seal. The brush can either be kept in the sealant or in a tube of solvent (*n*-butyl acetate).

Instead of a wax ring, Siddiqi (2000) recommends the use of three small lumps of wax, each about the size of the mounting drop, arranged around the drop, and the coverglass is placed on the lumps and the slide then heated. The wax melts, allowing the coverglass to settle down, and confines the glycerol to the centre of the mount. It is important to retain a hemispherical drop of mountant before applying the coverglass or the wax may swamp the specimens.

Posterior cuticular patterns of *Meloidogyne* spp.

The cuticular markings surrounding the vulva and anus (posterior cuticular pattern or 'perineal' pattern) of females of *Meloidogyne* spp. are used in their identification (Taylor *et al.*, 1955; Franklin, 1962). Fresh or fixed galled roots are stained in cotton-blue lactophenol or lactoglycerol and allowed to differentiate. Females stained in fresh root material are preferable because their body contents are more easily removed (Franklin, 1962). About 20 females are dissected out and transferred, using fine-pointed forceps, to 45% lactic acid on a transparent plastic (e.g. perspex) slide or plastic Petri dish cover. Working at a magnification of at least 32×, preferably more, the swollen female is speared at the neck end with a very sharp, fine needle and held so that the posterior end can be cut off with an oculist's scalpel or sharp Borradaile needle. A hypodermic needle mounted on a handle also serves as a very good cutting tool. The inner tissue is carefully removed by lightly brushing with a flexible bristle. The cuticle is transferred to a drop of glycerol where it is trimmed to a size slightly greater than the pattern, which is then transferred to a drop of glycerol on a clean glass slide. The posterior patterns, outside uppermost, are arranged in one or two neat rows, and a coverglass is applied and sealed. Supports are optional. At least ten specimens from a population should be examined. The patterns can usually be seen satisfactorily at a magnification of about 500×, but, for species having small or indistinct patterns, an oil immersion objective and higher magnification may be needed.

As noted by Taylor (1987), the lip region shape and the position of the excretory pore in mature females are an aid to the identification of *Meloidogyne* spp. Gerber and Taylor (1988) give details of preparation and mounting so as to show the anterior end and perineal pattern on one specimen. The preparation is similar to that described above for perineal patterns only, but the mature female is pierced once or twice in the mid-body region and the body contents carefully squeezed out. The female is then orientated with the perineal pattern to one side and, using a fine scalpel or hypodermic needle, the posterior quarter of the body, without the pattern, is cut away, taking care not to damage the pattern. The prepared specimens are then mounted in glycerol with the cut opening underneath and the perineal pattern uppermost. For additional information on preparation methods for culturing and identification of *Meloidogyne* spp., see Barker *et al.* (1985) or Jepson (1987).

Vulval cones of cyst nematodes

The structure of the vulva, fenestra and associated internal structures as well as the general shape of cysts are used for identifying cyst nematodes (e.g. *Globodera* and *Heterodera*) (Hesling, 1978). Dry cysts should be soaked in water for up to 24 h before dissection. A moist cyst is placed on a perspex slide on the stage of a stereomicroscope and the posterior end cut off so that the fenestral area is in the centre of the cut piece. If necessary, the cut end is trimmed so that it is no more than 5–10 times the fenestral area. Using very fine forceps and a flexible probe (eyebrow or fine toothbrush bristle mounted on the end of a dissecting needle handle), any adhering body contents, e.g. eggs, are cleaned out taking particular care not to damage the structures associated with the vulva. Thick-walled and heavily pigmented species, bleached for a few minutes in H_2O_2, often have more visible structures. Avoid overbleaching. The cleaned vulval cones are washed in distilled water and then passed through 70, 95 and 100% ethanol to clove oil. After being cleared in clove oil, they are mounted in Canada balsam. The coverglass is supported with pieces of glass rod or broken coverglass thick enough to prevent crushing the specimen. Vulval cones may also be mounted in 'Euparal', after passage through 70% ethanol and isobutanol, or directly in glycerine and sealed.

A simpler method for the examination of the vulval cone of mature *Heterodera* cysts is described by Esser (1988). A block of 1.7% water agar (15 mm × 15 mm × 2 mm high) is put on a slide. A small 1 mm deep cavity slightly less than the diameter of the cyst is made on the agar block with a fine needle. The cyst is gently pushed into the cavity with the anterior end down until the vulva region of the cyst is at the same level as the agar surface. A small drop of water is added to a 15 mm coverslip which is inverted and dropped over the embedded cyst, which can then be viewed under the microscope. Correia and Abrantes (1997) describe an improved technique for mounting *Heterodera* cysts in glycerine agar.

Computerized systems

Image analysis systems can assist with the examination of nematode samples by counting nematodes in a suspension (Been *et al.*, 1996) or with automatic recognition of nematodes (Fernandez-Valdivia *et al.*, 1989). Furthermore, computerized keys can help with the identification of species (Viscardi and Brzeski, 1993, 1995). A wider application of image analysis is seen in special software for morphometrics on nematodes (e.g. Leica IM 500).

Molecular Diagnostics

Most methods of nematode diagnostics have some limitations. Species identification based on differences in morphological and morphometrical characters requires a lot of skill and is often inconclusive for individual nematodes. Isozyme or total protein analyses are relatively fast ways to identify root knot or cyst-forming nematode species. Differences in isozyme or protein patterns show significant consistency and are useful for species identification. However, reliable results can only be obtained with nematodes of specific developmental stage. DNA-based diagnostics do not rely on the express products of the genome and are independent of environmental influence or developmental stage. Recent progress in nematode diagnostics has been achieved due to introducing the polymerase chain reaction (PCR), a powerful method with widespread application in many biological fields (Fig. 3.6). A single nematode, egg or even a part of the nematode body could be identified using this technology. The majority of PCR-based

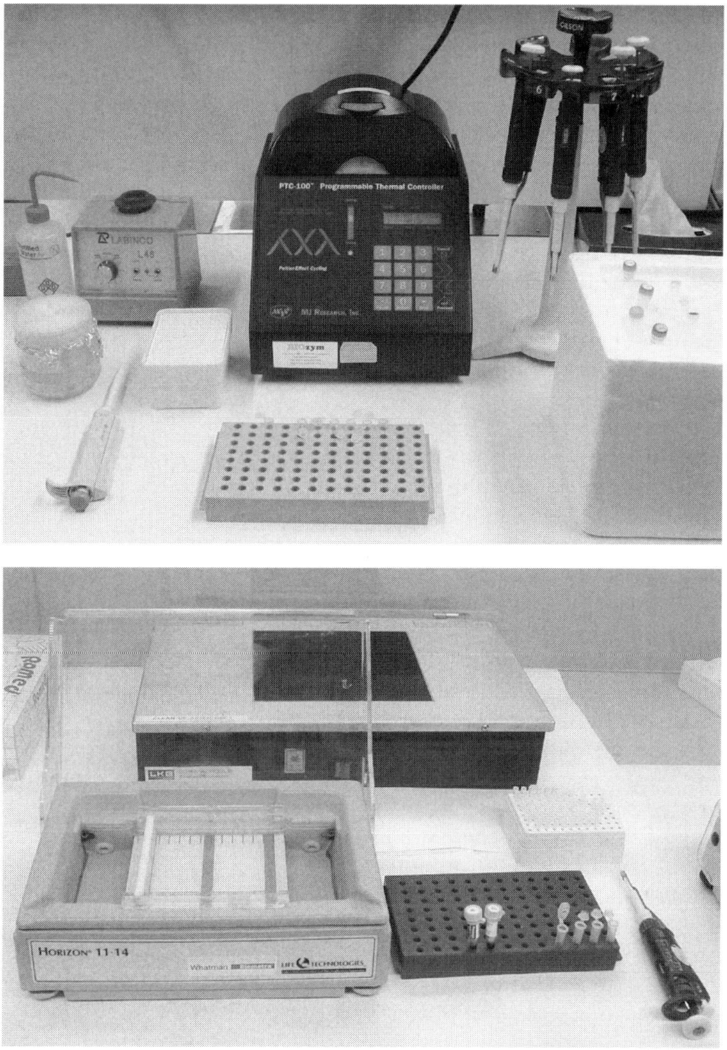

Fig. 3.6. Equipment required for PCR (top), electrophoresis and visualization of the PCR product on agarose gel (bottom).

techniques developed for nematode diagnostics indicate differences of the rRNA or mitochondrial DNA (mtDNA) gene sequences.

rRNA and mtDNA genes

The rRNA genes are arranged as tandem repeats with several hundred copies per genome. Each repeat includes the small subunit (SSU) gene, or 18S gene, the 5.8S gene and the large subunit (LSU) gene, or 28S gene, the spacer region between the subunit and 5.8S gene, called the internal transcribed spacers (ITS1 and ITS2), and between the gene cluster, called the intergenic spacer (IGS). In the root knot nematodes, the 5S gene is found in the IGS. The 18S gene evolves relatively slowly and is useful for comparison of distantly related groups, whereas ITS and IGS are considerably more variable and can be used to distinguish species or subspecies. Some regions of the 28S gene are also useful for species differentiation.

MtDNA is a circular double-stranded closed small structure, which is present in large copy numbers in the cell. Rapid evolution rates of specific genes in the mtDNA, which evolve ten times faster and more than nuclear genes, resulted in accumulated sequence polymorphism. This allows this molecule to be used as a useful marker for differentiation of nematode populations and of closely related species.

DNA extraction

The first step in molecular diagnostic procedures is the preparation of the template DNA (examples 1 and 2). Several protocols for the extraction of nucleic acids from nematodes are available (Curran *et al.*, 1985; Caswell-Chen *et al.*, 1992; Blok *et al.*, 1997). Some of these allow the isolation of microgram quantities of pure genomic DNA. However, because only small quantities of starting DNA are required for PCR amplification, simplified and rapid procedures generally can be used (Harris *et al.*, 1990; Subbotin *et al.*, 2000; Waeyenberge *et al.*, 2000; Floyd *et al.*, 2002). Using different extraction methods and commercial kits, nematode DNA can be obtained directly from soil samples (Nazar *et al.*, 1995; Waite *et al.*, 2003). Furthermore, extraction of DNA from formalin-fixed materials or nematodes embedded in glycerine on slides provides a new opportunity for molecular examination of reference materials (Thomas *et al.*, 2000).

Example 1: protocol for DNA extraction using proteinase K with Worm Lysis Buffer (WLB) (Waeyenberge *et al.*, 2000).

1. Pick a single or several nematodes and place in a 10 μl drop of double-distilled water on a glass slide under the dissecting microscope.
2. Cut nematodes into three or four pieces with a needle or scalpel.
3. Transfer worm bits with water to a sterile 0.2 ml Eppendorf tube containing 8 μl of WLB (500 mM KCl, 100 mM Tris–HCl pH 8.3, 15 mM $MgCl_2$, 10 mM dithiothreitol (DTT); 4.5% Tween-20) and 2 μl of proteinase K (600 μmg/ml).
4. Freeze at –80°C for 10 min.
5. Incubate at 65°C for 1 h and then heat at 95°C for 15 min.
6. Centrifuge for 1 min at maximum speed to remove debris. Use 1–4 μl of the supernatant in the PCR.

Example 2: protocol for DNA extraction using NaOH (Floyd *et al.*, 2002).

1. Pick individual nematodes directly into 20 μml of 0.25 M NaOH in a 0.2 ml Eppendorf tube and keep at room temperature from several minutes to several hours.
2. Heat the lysate for 3 min at 95°C.
3. Add 4 μl of HCl and 10 μl of 0.5 M Tris–HCl buffered at pH 8.0 to neutralize the base.
4. Add 5 μl of 2% Triton X-100.
5. Heat the lysate for 3 min at 95°C.
6. Use 0.5–2.0 μml of lysate for the PCR.

PCR

This enzymatic reaction allows *in vitro* amplification of target DNA fragments by up to a billionfold from complex DNA samples within a test tube. Any nucleic acid sequence can be detected by PCR amplification. The method requires a DNA template containing the region to be amplified, two oligonucleotide primers flanking this target region (Table 3.1), DNA polymerase and deoxyribonucleotide triphosphates (dNTPs) mixed in buffer containing magnesium ions ($MgCl_2$) (example 3). The PCR is performed in tubes with final volumes of 20–100 μl. The PCR procedure consists of a succession of three steps which are determined by temperature condition: template denaturation at 95°C for 3–4 min, primer annealing at 55–60°C for 1–2 min and extension at 72°C for 1–2 min. The PCR is carried out for 30–40 cycles in a thermocycler with programmed heating and cooling. Finally, PCR products are separated electrophoretically according to their size on agarose or polyacrylamide gels and visual-

Table 3.1. Universal primers frequently used for nematode diagnostics.

Code	Primer (5′–3′)	Amplified region	Reference
C2F3	GGT CAA TGT TCA GAA ATT TGT GG	3′ of COII to 16S	Powers and Harris (1993)
1108	TAC CTT TGA CCA ATC ACG CT	mitochondrial genes	
18S	TTG ATT ACG TCC CTG CCC TTT	ITS1 region of rDNA	Szalanski *et al.* (1997)
rDNA1.58S	GCC ACC TAG TGA GCC GCG CA		
18S	TTG ATT ACG TCC CTG CCC TTT	ITS1–5.8S–ITS2	Vrain *et al.* (1992)
26S	TTT CAC TCG CCG TTA CTA AGG	region of rDNA	
F194	CGT AAC AAG GTA GCT GTA G	ITS1–5.8S–ITS2	Ferris *et al.* (1993)
F195	TCC TCC GCT AAA TGA TAT G	region of rDNA	
SSU18A	AAA GAT TAA GCC ATG CAT G	18S gene of rDNA	Blaxter *et al.* (1998)
SSU26R	CAT TCT TGG CAA ATG CTT TCG		
D2A	ACA AGT ACC GTG AGG GAA AGT TG	D2–D3 expansion	De Ley *et al.* (1999)
D3B	TCG GAA GGA ACC AGC TAC TA	segments of	
TW81	GTT TCC GTA GGT GAA CCT GC	28S gene of rDNA	Joyce *et al.* (1994)
AB28	ATA TGC TTA AGT TCA GCG GGT	ITS1–5.8S–ITS2 region of rDNA	

ized by ethidium bromide under ultraviolet (UV) light or after silver staining. Once identified, nematode target DNA generated by PCR amplification can be characterized further by various analyses: restriction fragment length polymorphism (RFLP), single-strand conformation polymorphism (SSCP) or sequencing.

Example 3: PCR protocol.

1. Add a DNA suspension to the Eppendorf tube containing a PCR mixture with 5 μl of 10× PCR buffer, 10 μl of Q-solution, 1 μml of dNTP mixture (10 mM each) (*Taq* PCR Core Kit, Qiagen), 0.5 μl of each primer, 1 U of *Taq* polymerase, and double-distilled water to a final volume of 50 μl.
2. Put the tube in the PCR machine with the following thermal profile: an initial denaturation at 94°C for 4 min, 35 cycles of 94°C for 1 min, 55°C for 1.5 min, 72°C for 2 min and a final elongation step at 72°C for 10 min.
3. Run 2–5 μl of PCR product on a 0.8–1% agarose gel for 30–60 min at 90–100 V.

PCR-RFLP

Variation in sequences in PCR products can be revealed by restriction endonuclease digestion. The PCR product obtained from different species or populations can be digested by a restriction enzyme and the resulting fragment is separated by electrophoresis (example 4). If there is some difference in sequences situated within the restriction site of the enzyme, the digestion of the PCR products will lead to different electrophoretic profiles. It has been shown that comparison of restriction patterns derived from amplified ITS regions is a very useful approach to distinguish species and populations of *Aphelenchoides* (Ibrahim *et al.*, 1994), *Bursaphelenchus* (Hoyer *et al.*, 1998), cyst-forming nematodes (Thiéry and Mugniéry, 1996; Bekal *et al.*, 1997; Orui, 1997; Szalanski *et al.*, 1997; Subbotin *et al.*, 2000) (Fig. 3.7), *Ditylenchus* (Wendt *et al.*, 1993; Ibrahim *et al.*, 1994), *Nacobbus* (Reid *et al.*, 2003), *Pratylenchus* (Orui, 1996; Waeyenberge *et al.*, 2000), *Radopholus* (Fallas *et al.*, 1996), root knot nematodes (Zijlstra *et al.*, 1995; Schmitz *et al.*, 1998) and *Xiphinema* (Vrain *et al.*, 1992). Comparison of RFLP profiles from newly obtained samples with those from known species provide a quick tool for nematode identification. PCR-RFLPs are especially suited to identify nematodes of monospecific probes; this strategy does not allow mixed species populations to be identified.

Example 4: RFLP protocol.

1. Add 2–8 μl of PCR product to an

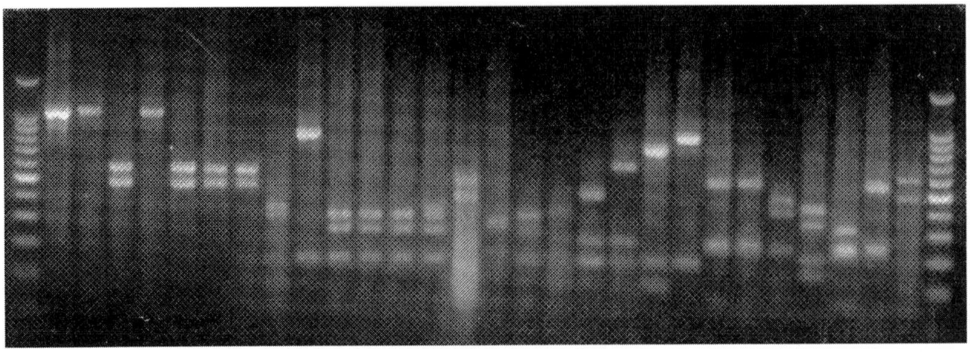

Fig. 3.7. RFLP patterns obtained after *Alu*I digestion of the amplified PCR product of the ITS-rDNA for cyst-forming nematodes. L, 100 bp DNA ladder; U, unrestricted PCR product; 1, 2, *H. avenae*; 3, *H. arenaria*; 4, *H. filipjevi*; 5, *H. aucklandica*; 6, *H. ustinovi*; 7, *H. latipons*; 8, *H. hordecalis*; 9, *H. schachtii*; 10, *H. trifolii*; 11, *H. medicaginis*; 12, *H. ciceri*; 13, *H. salixophila*; 14, *H. oryzicola*; 15, *H. glycines*; 16, *H. cajani*; 17, *H. humuli*; 18, *H. ripae*; 19, *H. fici*; 20, *H. litoralis*; 21, *H. carotae*; 22, *H. cruciferae*; 23, *Heterodera* sp.; 24, *H. cyperi*; 25, *H. goettingiana*; 26, *H. urticae*; 27, *Meloidodera alni* (Subbotin *et al.*, 2000).

Eppendorf tube containing 1.0 μl of 10× restriction enzyme buffer, 1 μl of restriction enzyme and double-distilled water to a final volume of 10 μl.

2. Put the tube in a water bath at 37°C (or other temperature required for digestion) for 6–12 h.

3. Centrifuge the tube for 30 s at maximum speed.

4. Run the reaction mixture on a 1.5% agarose gel in 1× TBE for 60–90 min at 90–100 V.

The restriction enzymes recommended for species identification are *Alu*I, *Ava*I, *Bsh*1236I, *Bsu*RI, *Cfo*I, *Hin*fI, *Mva*I, *Rsa*I and *Pst*I for cyst-forming nematodes, and *Alu*I, *Dra*I, *Hin*fI, *Msp*I, *Pvu*II and *Rsa*I for root knot nematodes.

PCR-SSCP

This technique has been applied successfully for rapid identification of cyst-forming nematodes and root knot nematodes from cultures and field samples (Clapp *et al.*, 2000). The distinguishing patterns obtained with PCR-SSCP are sequence dependent and utilize minor nucleotide differences across several hundred bases of sequences. It is a simple procedure where denatured, single-stranded PCR amplicons are separated electrophoretically in a non-denaturing polyacrylamide gel. The length, position and extent of self-complementary base pairs affect the conformation taken up by the molecules and thus their electrophoretic mobility. This effect is enhanced by minor length polymorphisms and increasing amounts of sequence variation. SSCP patterns are highly reproducible between gels and generate two markers from each DNA sequence present. The band patterns are compared with those obtained from controls or from pattern databases.

Sequencing

Direct sequencing of PCR products or sequencing of cloned PCR fragments provides full characterization of amplified target DNA. One of the first applications of PCR in plant nematology was presented by Ferris *et al.* (1993), who used the ITS rDNA sequences to establish the taxonomic and phylogenetic relationships of cyst-forming nematodes. The sequences of the ITS regions, fragments of 18S and 28S of rRNA genes, have been examined for a wide range of plant parasitic nematodes (Subbotin *et al.*, 2001b; Floyd *et al.*, 2002; Reid *et al.*, 2003). The comparison of newly obtained sequences from samples with those published or deposited in the GenBank is a most reliable approach for molecular identification. Increasing numbers of deposited nematode rDNA sequences as well as decreasing costs for sequence analyses will allow wider application of this still rather expensive procedure for routine nematode diagnostics in the future.

PCR with species-specific primers

PCR with specific primer combinations or multiplex PCR constitute a major development in DNA diagnostics and allow the detection of one or several species in a nematode mixture by a single PCR test, thus decreasing diagnostic time and costs. Species-specific primers are designed based on the broad knowledge of sequence divergence of the target DNA region in many populations of the same species and in closely related species. This knowledge allows the detection of populations with small differences in sequences, and avoids the amplification of an identical specific fragment in other species. The principle of this method is the alignment of the sequences from target and non-target organisms and the selection of primer mismatches to non-target organisms, but it shows sufficient homology for efficient priming and amplification of the target organism. In nematology, this diagnostic tool has been developed for

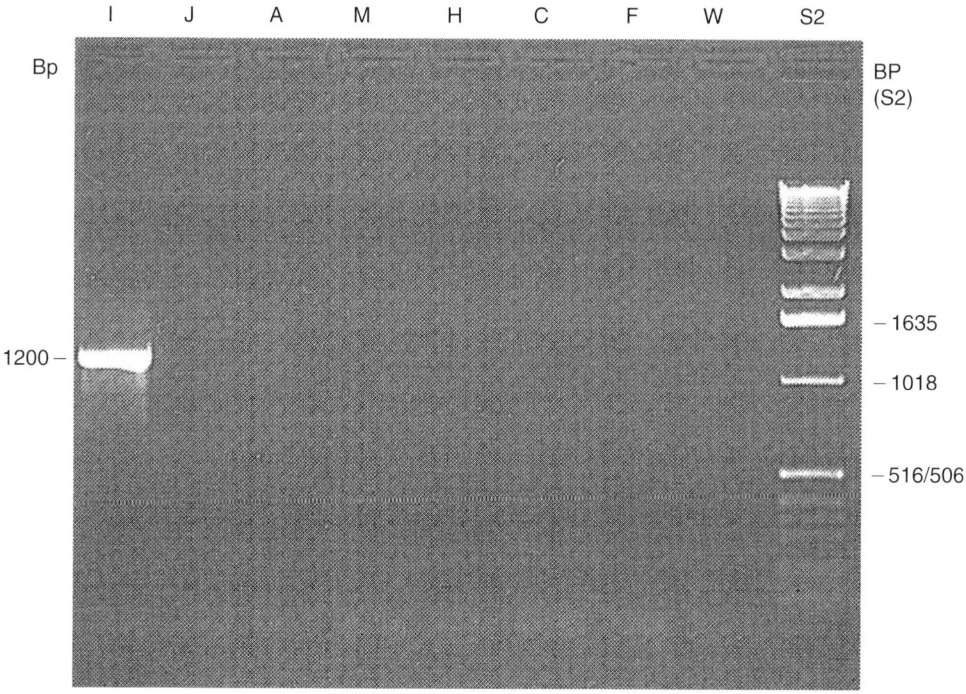

Fig. 3.8. Amplification product of PCR with species-specific primer Finc/Rinc for *Meloidogyne incognita*. I, *Meloidogyne incognita*; J, *M. javanica*; A, *M. arenaria*; M, *M. mayaguensis*; H, *M. hapla*; C, *M. chitwoodi*; F, *M. fallax*; W, no template DNA control; S, size marker (Zijlstra *et al.*, 2000).

Table 3.2. Species-specific primers developed for identification of cyst-forming and root knot nematodes.

Species	Primer set (5′–3′)	Amplicon length	Reference
Globodera pallida	PITSp4 ACA ACA GCA ATC GTC GAG ITS5 GGA AGT AAA AGT CGT AAC AAG G	265 bp	Bulman and Marshall (1997)
Globodera pallida	TGT CCA TTC CTC TCC ACC AG CCG CTT CCC CAT TGC TTT CG	768 bp	Fullaondo *et al.* (1999)
Globodera pallida	GGT GAC TCG ACG ATT GCT GT GCA GTT GGC TAG CGA TCT TC	238 bp	Mulholland *et al.* (1996)
Globodera rostochiensis	PITSr3 AGC GCA GAC ATG CCG CAA ITS5 GGA AGT AAA AGT CGT AAC AAG G	434 bp	Bulman and Marshall (1997)
Globodera rostochiensis	GCA AGC CCA GCG TCA GCA AC GAA CAT CAA CCT CCT ATC GG	315 bp	Fullaondo *et al.* (1999)
Globodera rostochiensis	GGT GAC TCG ACG ATT GCT GT GCA GTT GGC TAG CGA TCT TC	391 bp	Mulholland *et al.* (1996)
Heterodera glycines	GlyF1 TTA CGG ACC GTA ACT CAA rDNA2 TTT CAC TCG CCG TTA CTA AGG	181 bp	Subbotin *et al.* (2001a)
Heterodera schachtii	SHF6 GTT CTT ACG TTA CTT CCA TW81 GTT TCC GTA GGT GAA CCT GC	200 bp	Amiri *et al.* (2002)
Meloidogyne arenaria	TCG AGG GCA TCT AAT AAA GG GGG CTG AAT AAT CAA AGG AA	950 bp	Dong *et al.* (2001)
Meloidogyne arenaria	Far TCG GCG ATA GAG GTA AAT GAC Rar TCG GCG ATA GAC ACT ACA ACT	420 bp	Zijlstra *et al.* (2000)
Meloidogyne chitwoodi	MC3F CCA ATG ATA GAG ATA GGA AC MC1R CTG GCT TCC TCT TGT CCA AA	400 bp	Williamson *et al.* (1997)
Meloidogyne chitwoodi	C64 GAT CTA TGG CAG ATG GTA TGG A 1839 AGC CAA AAC AGC GAC CGT CTA C	900 bp	Petersen *et al.* (1997)
Meloidogyne chitwoodi	Fc TGG AGA GCA GCA GGA GAA AGA Rc GGT CTG AGT GAG GAC AAG AGT A	800 bp	Zijlstra (2000)
Meloidogyne exigua	Ex-D15-F CAT CCG TGC TGT AGC TGC GAG Ex-D15-R CTC CGT GGG AAG AAA GAC TG	562 bp	Randing *et al.* (2002)
Meloidogyne fallax	F64 TGG GTA GTG GTC CCA CTC TG 1839 AGC CAA AAC AGC GAC CGT CTA C	1100 bp	Petersen *et al.* (1997)
Meloidogyne fallax	Ff CCA AAC TAT CGT AAT GCA TTA TT Rf GGA CAC AGT AAT TCA TGA GCT AG	515 bp	Zijlstra (2000)
Meloidogyne hapla	GGC TGA GCA TAG TAG ATG ATG TT ACC CAT TAA AGA GGA GTT TTG C	1500 bp	Dong *et al.* (2001)
Meloidogyne hapla	MH0F CAG GCC CTT CCA GCT AAA GA MH1R CTT CGT TGG GGA ACT GAA GA	960 bp	Williamson *et al.* (1997)
Meloidogyne hapla	Fh TGA CGG CGG TGA GTG CGA Rh TGA CGG CGG TAC CTC ATA G	610 bp	Zijlstra (2000)
Meloidogyne incognita	TAG GCA GTA GGT TGT CGG G CAG ATA TCT CTG CAT TGG TGC	1350 bp	Dong *et al.* (2001)
Meloidogyne incognita	Inc-K14-F GGG ATG TGT AAA TGC TCC TG Inc-K14-R CCC GCT ACA CCC TCA ACT TC	399 bp	Randing *et al.* (2002)
Meloidogyne incognita	Finc CTC TGC CCA ATG AGC TGT CC Rinc CTC TGC CCT CAC ATT AGG	1200 bp	Zijlstra *et al.* (2000)
Meloidogyne javanica	CCT TAA TGT CAA CAC TAG AGC C GGC CTT AAC CGA CAA TTA GA	1650 bp	Dong *et al.* (2001)
Meloidogyne javanica	Fjav GGT GCG CGA TTG AAC TGA GC Rjav CAG GCC CTT CAG TGG AAC TAT AC	670 bp	Zijlstra *et al.* (2000)
Meloidogyne paranaensis	Par-C09-F GCC CGA CTC CAT TTG ACG GA Par-C09-R CCG TCC AGA TCC ATC GAA GTC	208 bp	Randing *et al.* (2002)

identification of cyst-forming and root knot nematodes (Table 3.2 (on p. 78) and Fig. 3.8 (on p. 77)), *Pratylenchus* (Uehara *et al.*, 1998), *Xiphinema* (Wang *et al.*, 2003) and *Ditylenchus* (Esquibet *et al.*, 2003). The multiplex PCR with specific primers for identification of several nematode targets in one assay is limited by the number of primer pairs that can be used in a single reaction and the number of bands that can be clearly identified without giving false-positive results. This technique requires precise optimization of the reaction conditions for the primer sets used simultaneously in the test.

Reverse dot-blot hybridization

This technique involves the use of PCR for simultaneous amplification and labelling of target DNA to generate digoxigenin-dUTP-labelled amplicons which are hybridized to specific immobilized oligonucleotide probes on a membrane. This approach can be used for simultaneous identification of many different nematodes from a single sample. Uehara *et al.* (1999) have demonstrated that this technology can be used for the identification of *Pratylenchus* species (Fig. 3.9).

RAPD-PCR

In contrast to the above-mentioned classical PCR method, the random amplified polymorphic DNA PCR (RAPD-PCR) or PCR with arbitrary primer (AP-PCR) does not require any information on the primer

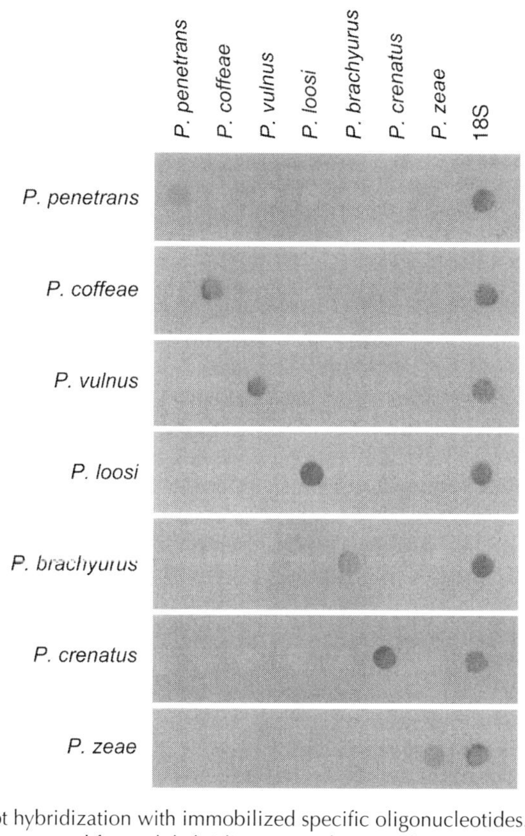

Fig. 3.9. Reverse dot-blot hybridization with immobilized specific oligonucleotides. The *Pratylenchus* species listed on the left were used for each hybridization (Uehara *et al.*, 1999).

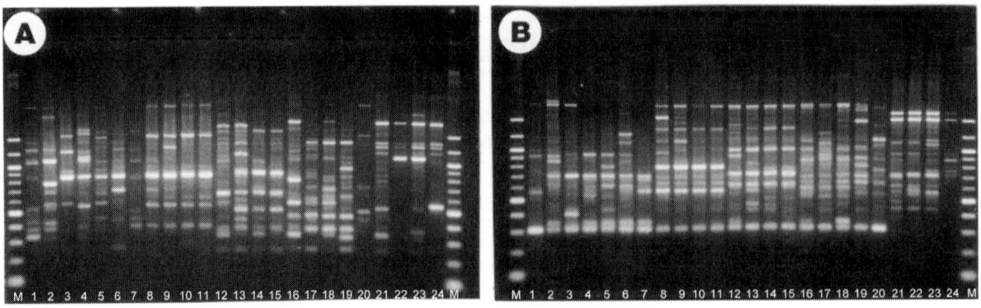

Fig. 3.10. RAPD patterns of 26 populations of the *Heterodera avenae* complex. Primers: A, A-16; B, A-18. Populations: 1, *H. avenae* (Taaken, Germany); 2, *H. avenae* (Santa Olalla, Spain); 3, *H. avenae* (Çukurova Ebene, Turkey); 4, *H. avenae* (Saudi Arabia); 5, *H. avenae* (Ha-hoola, Israel); 6, *H. avenae* (Israel); 7, *H. avenae* (near Delhi); 8, *H. australis* (South Australia, sample 3); 9, *H. australis* (Beulah, Australia); 10, *H. australis* (Victoria, Australia); 11, *H. australis* (Yorke Peninsular, Australia); 12, *H. mani* (Bayern, Germany); 13, *H. mani* (Heinsberg, Germany); 14, *H. mani* (Andernach, Germany); 15, *H. mani* (Germany); 16, *H. pratensis* (Missunde, Germany); 17, *H. pratensis* (Östergaard, Germany); 18, *H. pratensis* (Lindhöft, Germany); 19, *H. pratensis* (Lenggries, Germany); 20, *H. aucklandica* (One Tree Hill, New Zealand); 21, *H. filipjevi* (Saratov, Russia); 22, *H. filipjevi* (Akenham, England); 23, *H. filipjevi* (Torralba de Calatrava, Spain); 24, *H. filipjevi* (Selçuklu, Turkey). M, 100 bp DNA ladder (Biolab). (Source: Subbotin *et al.*, 2003.)

design. This PCR technology uses a single random primer about ten nucleotides long, approximately 50% GC rich and lacking any internal inverted repeats. By lowering the annealing temperature during the amplification cycle, the primer anneals at random in the genome, allowing the synthesis of highly polymorphic amplification products. RAPD-PCR distinguishes nematode species, subspecies and races and is used for root knot nematodes (Cenis, 1993; Blok *et al.*, 1997) and cyst-forming nematodes (Caswell-Chen *et al.*, 1992; Thiéry *et al.*, 1997) (Fig. 3.10). However, the reproducibility of the results is the most critical point for application of this technique for diagnostic purposes. Specific sequences for certain species or races, called SCARs (sequence characterized amplified regions), can be derived from RAPD fragments and further used to design species-specific primers.

AFLP

The amplified fragment length polymorphism (AFLP) technique has been developed by Vos *et al.* (1995) and is based on the selective amplification of genomic restriction fragments. AFLP involves three steps: (i) digestion of DNA with two restriction enzymes and ligation of specific adapters to the restriction fragments; (ii) PCR amplification of a subset of the restriction/adapter fragments under stringent conditions; and (iii) gel electrophoresis analysis of the amplified restriction fragments. The AFLP technique has several advantages over RAPD in that it produces results that are very reproducible and it has higher resolutions generating many more amplified fragments. AFLP fingerprinting has been applied successfully for the evaluation of inter- and intraspecific genetic variation of cyst-forming nematodes (Folkertsma *et al.*, 1996; Marché *et al.*, 2001) and root knot nematodes (Semblat *et al.*, 1998).

References

Alam, M.M. (1990) A modification of Hesling's device for rapid selection of nematodes from aqueous suspension. *Nematologica* 36, 254–256.

Amiri, S., Subbotin, S.A. and Moens, M. (2002) Identification of the beet cyst nematode *Heterodera schachtii* by PCR. *European Journal of Plant Pathology* 108, 497–506.

Araya, M., Jaén, R., Cheves, A. and Centeno, M. (1998) Effect of sieve washing time, mesh size, and mesh position, on nematode recovery from banana roots. *Agronomía Costarricense* 22, 205–210.

Ayoub, J.R. (1980) *Plant Nematology: an Agricultural Training Aid,* Revised edn. NemaAid Publications, Sacramento, California.

Barker, K.R., Carter, C.C. and Sasser, J.N. (1985) *An Advance Treatise on* Meloidogyne *Volume II: Methodology.* North Carolina State University and United States Agency for International Development, Raleigh, North Carolina.

Bates, J.W. (1997) The slide-sealing compound 'Glyceel'. *Journal of Nematology* 29, 565–566.

Been, T.H., Meijer, E.M.J., Beniers, A.E. and Knol, J.W. (1996) Using image analysis for counting larvae of potato cyst nematodes (*Globodera* spp.). *Fundamental and Applied Nematology* 19, 297–304.

Bekal, S., Gauthier, J.P. and Rivoal, R. (1997) Genetic diversity among a complex of cereal cyst nematodes inferred from RFLP analysis of the ribosomal internal transcribed spacer region. *Genome* 40, 479–486.

Blaxter, M., De Ley, P., Garey, J.R., Liu, L.X., Scheldeman, P., Vierstraete, A., Vanfleteren, J.R., Mackey, L.Y., Dorris, M., Frisse, L.M., Vida, J.T. and Thomas, W.K. (1998) A molecular evolutionary framework for the phylum Nematoda. *Nature* 392, 71–75.

Bloemers, G.F. and Hodda, M. (1995) A method for extracting nematodes from a tropical forest soil. *Pedobiologia* 39, 331–343.

Blok, V.C., Phillips, M.S., McNicol, J.W. and Fargette, M. (1997) Genetic variation in tropical *Meloidogyne* spp. as shown by RAPDs. *Fundamental and Applied Nematology* 20, 127–133.

Bridge, J., Page, S. and Jordan, S. (1982) An improved method for staining nematodes in roots. *Report of Rothamsted Experimental Station for 1981,* Part 1, 171.

Bulman, S.R. and Marshall, J.W. (1997) Differentiation of Australasian potato cyst nematode (PCN) populations using the polymerase chain reaction (PCR). *New Zealand Journal of Crop and Horticultural Sciences* 25, 123–129.

Byrd, D.W., Nusbaum, C.J. and Barker, K.R. (1966) A rapid flotation–sieving technique for extracting nematodes from soil. *Plant Disease Reporter* 50, 945–957.

Caswell-Chen, E.P., Williamson, V.M. and Wu, F.F. (1992) Random amplified polymorphic DNA analysis of *Heterodera cruciferae* and *H. schachtii* populations. *Journal of Nematology* 24, 343–351.

Caveness, F.E. and Jenson, H.J. (1955) Modification of the centrifugal-flotation technique for the isolation and concentration of nematodes and their eggs from soil and plant tissue. *Proceedings of the Helminthological Society of Washington* 22, 87–89.

Cenis, J.L. (1993) Identification of four major *Meloidogyne* spp. by random amplified polymorphic DNA (RAPD-PCR). *Phytopathology* 83, 76–78.

Chakrabarti, U. and Saha, M. (2001) Effect of killing, fixing and mounting methods on morphological clarity of *Tylenchorhynchus mashhoodi,* Siddiqi and Basir, 1959. *Indian Journal of Nematology* 31, 16–22.

Clapp, J.P., Van der Stoel, C.D. and Van der Putten, W.H. (2000) Rapid identification of cyst (*Heterodera* spp., *Globodera* spp.) and root-knot (*Meloidogyne* spp.) nematodes on the basis of ITS2 sequence variation detected by PCR-single-strand conformational polymorphism (PCR-SSCP) in cultures and field samples. *Molecular Ecology* 9, 1223–1232.

Cobb, N.A. (1918) *Estimating the Nema Population of the Soil.* Agricultural Technology Circular 1, Bureau of Plant Industry, United States Department of Agriculture, Washington, DC.

Coolen, W.A. (1979) Methods for the extraction of *Meloidogyne* spp. and other nematodes from roots and soil. In: Lamberti, F. and Taylor, C.E. (eds) *Root-knot Nematodes (*Meloidogyne *Species) Systematics, Biology and Control.* Academic Press, London, pp. 317–329.

Coolen, W.A. and d'Herde, C.J. (1972) *A Method for the Quantitative Extraction of Nematodes from Plant Tissue.* Ghent State Agriculture Research Centre.

Correia, F.J.S. and de Abrantes, I.M.O. (1997) An improved technique for mounting *Heterodera* cysts for light microscopy. *Nematologica* 43, 507–509.

Courtney, W.D., Polley, D. and Miller, V.L. (1955) TAF, an improved fixative in nematode technique. *Plant Disease Reporter* 39, 570–571.

Curran, J., Baillie, D.L. and Webster, J.M. (1985) Use of restriction fragment length differences in genomic DNA to identify nematode species. *Parasitology* 90, 137–144.

De Grisse, A. (1963) A counting dish for nematodes excluding border effect. *Nematologica* 9, 162.

De Ley, P., Félix, M.-A., Frisse, L.M., Nadler, S.A., Sternberg, P.W. and Thomas, W.K. (1999) Molecular and morphological characterisation of two reproductively isolated species with mirror-image anatomy (Nematoda: Cephalobidae). *Nematology* 1, 591–612.

Demeure, Y. and Netscher, C. (1973) Méthode d'estimation des populations de *Meloidogyne* dans le sol. *Cahiers ORSTOM Series Biol.* No. 21, 8–90.

De Waele, E., De Waele, D. and Wilken, R. (1987) Effect of root mass on the efficacy of methods for extracting root-lesion nematodes from maize roots. *Phytophylactica* 19, 473–474.

Doncaster, C.C. (1962) A counting dish for nematodes. *Nematologica* 7, 334–336.

Doncaster, C.C., Edwards, B.S. and Shepherd, A.M. (1967) A labour-saving method of making Fenwick multi chamber counting slides for nematodes. *Nematologica* 12, 644.

Dong, K., Dean, R.A., Fortnum, B.A. and Lewis, S.A. (2001) Development of PCR primers to identify species of root-knot nematodes: *Meloidogyne arenaria, M. hapla, M. incognita* and *M. javanica. Nematropica* 31, 271–280.

Dropkin, V.H. (1989) *Introduction to Plant Nematology*, 2nd edn. John Wiley & Sons, New York.

Dunn, R.A. (1971) Comparison of two centrifugal flotation techniques with Seinhorst elutriation for efficiency of nematode extraction from soil. *Journal of Nematology* 3, 308–309.

Elmiligy, L.A. and De Grisse, A. (1970) Effect of extraction technique and adding fixative to soil before storing on recovery of plant parasitic nematodes. *Nematologica* 16, 353–358.

Escobar, J. and Rodríguez-Kábana, R. (1980) Comparacion de un metodo de flotacion con uno de tamizado para la determinacion de *Radopholus similis* en raices de banana. *Nematropica* 10, 86–88.

Esquibet, M., Grenier, E., Plantard, O., Andaloussi, F.A. and Caubel, G. (2003) DNA polymorphism in the stem nematode *Ditylenchus dipsaci*: development of diagnostic markers for normal and giant races. *Genome* 46, 1077–1083.

Esser, R.P. (1988) A simple method for examination of the vulva area of mature cysts of *Heterodera* spp. *Journal of Nematology* 20, 497–498.

Fallas, G.A., Hahn, M.L., Fargette, M., Burrows, P.R. and Sarah, J.L. (1996) Molecular and biochemical diversity among isolates of *Radopholus* spp. from different areas of the world. *Journal of Nematology* 28, 422–430.

Fenwick, D.W. (1940) Methods for the recovery and counting of cysts of *Heterodera schachtii* from soil. *Journal of Helminthology* 18, 155–172.

Fenwick, D.W. (1963) Recovery of *Rhadinaphelenchus cocophilus* (Cobb 1919) Goodey, 1960 from coconut tissues. *Journal of Helminthology* 37, 11–14.

Fernandez-Valdivia, J., Castillo, P. and Gomez-Barcina, A. (1989) Use of image analysis for automatic recognition of nematodes. In: Fortuner, R. (ed.) *Nematode Identification and Expert System Technology*. Plenum Press, New York, pp. 293–299.

Ferris, V., Ferris, J.M. and Faghihi, J. (1993) Variation in spacer ribosomal DNA in some cyst-forming species of plant parasitic nematodes. *Fundamental and Applied Nematology* 16, 177–184.

Floyd, R., Abede, E., Papert, A. and Blaxter, M. (2002) Molecular barcodes for soil nematode identification. *Molecular Ecology* 11, 839–850.

Folkertsma, R.T., Rouppe Van der Voort, J.N.A.M., De Groot, K.E., Zandvoort, P., Schots, A., Goomers, F., Helder, J. and Bakker, J. (1996) Gene pool similarities of potato cyst nematode population assessed by AFLP analysis. *Molecular Plant–Microbe Interaction* 9, 47–54.

Franklin, M.T. (1962) Preparation of posterior cuticular patterns of *Meloidogyne* spp. for identification. *Nematologica* 7, 336–337.

Fullaondo, A., Barrena, E., Viribay, M., Barrena, I., Salazar, A. and Ritter, E. (1999) Identification of potato cyst nematode species *Globodera rostochiensis* and *G. pallida* by PCR using specific primer combination. *Nematology* 1, 157–163.

Galway, N.J. and Curran, J. (1995) Cryopreservation of *Pratylenchus* spp. *Journal of Nematology* 27, 483–485.

Gerber, K. and Taylor, A.L. (1988) A simple technique for mounting whole root-knot nematode females. *Journal of Nematology* 20, 502–503.

Gooris, J. and d'Herde, C.J. (1972) *A Method for the Quantitative Extraction of Eggs and Second Stage Juveniles of* Meloidogyne *spp. from Soil.* Ghent State Agricultural Research Centre.

Grewal, P.S., Richardson, P.N. and Wright, D.J. (1990) Effects of killing, fixing and mounting methods on taxonomic characters of parthenogenic adult female *Caenorhabditis elegans* (Nematoda: Rhabditidae). *Revue de Nématologie* 13, 437–444.

Griffiths, B.S., Boag, B., Neilson, R. and Palmer, L. (1990) The use of colloidal silica to extract nematodes from small samples of soil or sediment. *Nematologica* 36, 465–473.

Harris, T.S., Sandall, L.J. and Powers T.O. (1990) Identification of single *Meloidogyne* juveniles by polymerase chain reaction amplification of mitochondrial DNA. *Journal of Nematology* 22, 518–524.

Hesling, J.J. (1978) Cyst nematodes: morphology and identification of *Heterodera, Globodera* and *Punctodera.* In: Southey, J.F. (ed.) *Plant Nematology Reference Book.* Ministry of Agriculture Fisheries and Food, No. 407 (GDI), London, pp. 125–155.

Hietbrink, H. and Ritter, C.E. (1982) Separating cysts from dried soil samples by a new centrifugation and a flotation method. In: *Proceedings of XVIth International Symposium, European Society of Nematologists,* St Andrews, UK, pp. 28–29.

Holbrook, C.C., Knauft, D.A. and Dickson, D.W. (1983) A technique for screening peanut for resistance to *Meloidogyne arenaria. Plant Disease* 67, 957–958.

Hooper, D.J. (1970) Handling, fixing, staining and mounting nematodes. In: Southey, J.F. (ed.) *Laboratory Methods for Work with Plant and Soil Nematodes.* Ministry of Agriculture, Fisheries and Food Technical Bulletin 2, London, pp. 39–54.

Hooper, D.J. (1987) Observations on the curation of plant and soil nematodes. *Nematologica* 32, 312–321.

Hooper, D.J. (1990) Extraction and processing of plant and soil nematodes. In: Luc, M., Sikora, R.A. and Bridge, J. (eds) *Plant Parasitic Nematodes in Subtropical and Tropical Agriculture.* CAB International, Wallingford, UK, pp. 45–68.

Hoyer, U., Burgermeister, W. and Braasch, H. (1998) Identification of *Bursaphelenchus* species (Nematoda, Aphelenchoididae) on the basis of amplified ribosomal DNA (ITS-RFLP). *Nachrichtenblatt des Deutschen Pflanzenschutzdienstes* 50, 273–277.

Hunt, D.J. and De Ley, P. (1996) Nematodes in soil. In: Hall, G.S. (ed.) *Methods for the Examination of Organismal Diversity in Soils and Sediments.* CAB International, Wallingford, UK, pp. 227–240.

Hussey, R.S. and Barker, K.R. (1973) A comparison of methods for collecting inocula for *Meloidogyne* spp., including a new technique. *Plant Disease Reporter* 57, 1025–1028.

Ibrahim, S.K., Perry, R.N., Burrows, P.R. and Hooper, D.J. (1994) Differentiation of species and populations of *Aphelenchoides* and of *Ditylenchus angustus* using a fragment of ribosomal DNA. *Journal of Nematology* 18, 109–114.

Ireholm, A. (1996) Long-term storage of *Heterodera avenae* cysts. *Fundamental and Applied Nematology* 19, 357–361.

Jenkins, W.R. (1964) A rapid centrifugal flotation technique for separating nematodes from soil. *Plant Disease Reporter* 48, 692.

Jepson, S.B. (1987) *Identification of Root-knot Nematodes (*Meloidogyne *Species).* CAB International, Wallingford, UK.

Joyce, S.A., Reid A., Driver, F. and Curran, J. (1994) Application of polymerase chain reaction (PCR) methods to identification of entomopathogenic nematodes. In: Burnell, A.M., Ehlers, R.-U. and Masson, J.P. (eds) *COST 812 Biotechnology: Genetics of Entomopathogenic Nematode–Bacterium Complexes.* Proceedings of Symposium and Workshop, St Patrick's College, Maynooth, County Kildare, Ireland. Luxembourg, European Commission, DG XII, pp. 178–187.

Kermarrec, C. and Bergé, J.-B. (1971) Méthodes d'estimation des populations de nématodes des sols et des recines. *Bulletin of the Society of Ecology* 3, 151–169.

Marché, L., Valette, S., Grenier, E. and Mugniéry, D. (2001) Intra-species DNA polymorphism in the tobacco cyst-forming complex (*Globodera tabacum*) using AFLP. *Genome* 44, 941–946.

McSorley, R., Frederick, J.J. and McGovern, R.J. (1999) Isolation of *Meloidogyne incognita* from caladium corms. *Nematropica* 29, 245–248.

Merny, G. and Luc, M. (1969) Les techniques d'echantillonnage des peuplements de nématodes dans le sol. In: Lamotte, M. and Bourliere, F. (eds) *Problémes d'Ecologie: l'Echantillonnage des Peuplements Animaux des Milieux Terrestres.* Masson et Cie, Paris, pp. 237–272.

Mishra, S.D., Vijayalakshmi, K. and Seshadri, A.R. (1977) A rapid sieving–sedimentation method for extracting nematodes from soil. *Indian Journal of Nematology* 5, 259–260.

Mulholland, V., Carde, L., O'Donnell, K.J., Fleming, C.C. and Powers, T.O. (1996) Use of the polymerase chain reaction to discriminate potato cyst nematode at the species level. In: Marshall, G. (ed.) *Diagnostics in Crop Protection* (BCPC Monograph No 65), pp. 247–252.

Nazar, R.N., Robb, E.J. and Volossiouk, T. (1995) Direct extraction of fungal DNA from soil. In: Akkermans, A.D.L., van Elsas, J.D. and de Bruijn, F.J. (eds) *Molecular Microbial Ecology Manual.* Kluwer Academic Publishers Group, Dordrecht, The Netherlands, p. 488.

Oostenbrink, M. (1950) Het aardappelaaltje (*Heterodera rostochiensis* Wollenweber), een gevaarlijke parasite voor de eenzijdige aardappel-cultuur. *Verslagen en Mededelingen van de Plantenziektenkundige Dienst to Wageningen,* No. 115.

Oostenbrink, M. (1960) Estimating nematode populations by some selected methods. In: Sasser, J.N. and Jenkins, W.R. (eds) *Nematology. Fundamentals and Recent Advances With Emphasis* on *Plant-parasitic and Soil Forms.* University of North Carolina Press, Chapel Hill, North Carolina, pp. 85–102.

Oostenbrink, M. (1970) Comparison of techniques for population estimation of soil and plant nematodes. In: Phillipson, J. (ed.) *Methods of Study in Soil Ecology.* Proceedings of the Paris Symposium, UNESCO Paris, pp. 249–255.

Orui, Y. (1996) Discrimination of the main *Pratylenchus* species (Nematode: Pratylenchidae) in Japan by PCR-RFLP analysis. *Applied Entomology and Zoology* 31, 505–514.

Orui, Y. (1997) [Discrimination of *Globodera rostochiensis* and four *Heterodera* species (Nematoda: Heteroderidae) by PCR-RFLP analysis]. *Japanese Journal of Nematology* 27, 67–75.

Petersen, D.J., Zijlstra, C., Wishart, J., Blok, V. and Vrain, T.C. (1997) Specific probes efficiently distinguish root-knot nematode species using signature sequence in the ribosomal intergenetic spacer. *Fundamental and Applied Nematology* 20, 619–626.

Pitcher, R.S. and Flegg, J.J.M. (1968) An improved final separation sieve for the extraction of plant-parasitic nematodes from soil debris. *Nematologica* 14, 123–127.

Powers, T.O. and Harris, T.S. (1993) A polymerase chain reaction method for identification of five major *Meloidogyne* species. *Journal of Nematology* 25, 1–6.

Prot, J.C., Gergon, E.B. and Matias, D.M. (1993) Influence of extraction procedures from root samples on the recovery and infectivity of *Pratylenchus zeae* and *Hirschmanniella oryzae. Nematologia Mediterranea* 21, 133–137.

Randing, O., Bongiovanni, M., Carneiro, R.M.D.G. and Castagnone-Sereno, P. (2002) Genetic diversity of root-knot nematodes from Brazil and development of SCAR markers specific for the coffee-damaging species. *Genome* 45, 862–870.

Reid, A., Manzanille-López, R.A. and Hunt, D.J. (2003) *Naccobus aberrans* (Thorne, 1935) Thorne and Allen, 1944 (Nematoda: Pratylenchidae); a nascent species complex revealed by RFLP analysis and sequencing of the ITS-rDNA. *Nematology* 5, 441–452.

Riggs, R.D., Schmitt, D.P. and Mauromoustakos, A. (1997) Comparison of extraction and shipping methods for cysts and juveniles of *Heterodera glycines. Journal of Nematology* 29, 127–132.

Rodríguez-Kábana, R. and King, P.S. (1975) Efficiency of extraction of nematodes by flotation–sieving using molasses and sugar and by elutriation. *Journal of Nematology* 7, 54–59.

Rodríguez-Kábana, R. and Pope, M.H. (1981) A simple incubation method for the extraction of nematodes from soil. *Nematropica* 11, 175–185.

Rush, M.C. (1970) Transmission of tobacco ringspot virus from native host to *Cucumis sativus* by *Xiphinema americanum. Phytopathology* 60, 917–918.

Sabir, N. (1997) 'Corseal' a new coverslip ringing material for permanent slides. *Indian Journal of Nematology* 26, 123–124.

Schmitz, V.B., Burgermeister, W. and Braasch, H. (1998) Molecular genetic classification of Central European *Meloidogyne chitwoodi* and *M. fallax* populations. *Nachrichtenblatt des Deutschen Pflanzenschutzdienstes* 50, 310–317.

Sehgal, M. and Gaur, H.S. (1988) A new simple device for picking inactive and active nematodes. *Indian Journal of Nematology* 18, 145.

Semblat, J.-P., Wajnberg, E., Dalmasso, A., Abad, P. and Castagnone-Sereno, P. (1998) High resolution DNA fingerprinting of parthenogenetic root-knot nematodes using AFLP analysis. *Molecular Ecology* 7, 119–125.

Seinhorst, J.W. (1950) De betekenis van de toestand van de grond voor het optreden van aantasting door het stengelaaltje (*Ditylenchus dipsaci* (Kühn) Filipjev). *Tijdschrift over Plantenziekten* 56, 289–348.

Seinhorst, J.W. (1955) Een eenvoudige methode voor het afscheiden van aaltjes uit grand. *Tijdschrift over Plantenziekten* 61, 188–190.

Seinhorst, J.W. (1956) The quantitative extraction of nematodes from soil. *Nematologica* 1, 249–267.

Seinhorst, J.W. (1962) On the killing, fixation and transferring to glycerin of nematodes. *Nematologica* 8, 29–32.

Seinhorst, J.W. (1966) Killing nematodes for taxonomic study with hot F.A. 4:1. *Nematologica* 12, 178.

Shepherd, A.M. (1986) Extraction and estimation of cyst nematodes. In: Southey, J.F. (ed.) *Laboratory Methods for Work with Plant and Soil Nematodes.* HMSO Books, Norwich, UK, pp. 31–49.

Shurtleff, M.C. and Averre, C.W. III (2000) *Diagnosing Plant Diseases Caused by Nematodes.* APS Press, St Paul, Minnesota.

Siddiqi, M.R. (2000) *Tylenchida Parasites of Plants and Insects,* 2nd edn. CAB International, Wallingford, UK.

Southey, J.F. (ed.) (1986) *Laboratory Methods for Work with Plant and Soil Nematodes.* HMSO Books, Norwich, UK.

Stetina, S.R., McGawley, E.C. and Russin, J.S. (1997) Extraction of root-associated *Meloidogyne incognita* and *Rotylenchulus reniformis. Journal of Nematology* 29, 209–215.

Subbotin, S.A., Waeyenberge, L. and Moens, M. (2000) Identification of cyst forming nematodes of the genus *Heterodera* (Nematoda: Heteroderidae) based on the ribosomal DNA-RFLPs. *Nematology* 2, 153–164.

Subbotin, S.A., Peng, D. and Moens, M. (2001a) A rapid method for the identification of the soybean cyst nematode *Heterodera glycines* using duplex PCR. *Nematology* 3, 365–371.

Subbotin, S.A., Vierstraete, A., De Ley, P., Rower, J., Waeyenberge, L., Moens, M. and Vanfleteren, J.R. (2001b) Phylogenetic relationships within the cyst-forming nematodes (Nematoda, Heteroderidae) based on analysis of sequences from the ITS regions of ribosomal DNA. *Molecular Phylogenetic and Evolution* 21, 1–16.

Subbotin, S.A., Sturhan, D., Rumpenhorst, H.J. and Moens, M. (2003) Molecular and morphological characterisation of the *Heterodera avenae* species complex (Tylenchida: Heteroderidae). *Nematology* 5, 515–538.

Szalanski, A.L., Sui, D.D., Harris, T.S. and Powers, T.O. (1997) Identification of cyst nematodes of agronomic and regulatory concern with PCR-RFLP of ITS1. *Journal of Nematology* 29, 255–267.

Tarjan, A.C., Simanton, W.A. and Russell, E.E. (1956) A labor-saving device for the collection of nematodes. *Phytopathology* 46, 641–644.

Tarjan, A.C. (1967) Influence of temperature and hydrogen peroxide on the extraction of burrowing nematodes from citrus roots. *Plant Disease Reporter* 51, 1024–1028.

Taylor, A.L. (1987) Identification and estimation of root-knot nematode species in mixed populations. *Florida Department of Agriculture and Consumer Services Bulletin* 12, Gainesville, Florida.

Taylor, A.L., Dropkin, V.H. and Martin, G.C. (1955) Perineal patterns of root-knot nematodes. *Phytopathology* 45, 26–34.

Thiéry, M. and Mugniéry, D. (1996) Interspecific rDNA restriction fragment length polymorphism in *Globodera* species, parasites of Solanaceous plants. *Fundamental and Applied Nematology* 19, 471–479.

Thiéry, M., Fouville, D. and Mugniéry, D. (1997) Intra- and interspecific variability in *Globodera,* parasites of Solanaceous plants, revealed by random amplified polymorphic DNA (RAPD) and correlation with biological features. *Fundamental and Applied Nematology* 20, 495–504.

Thomas, W.K., Vida, J.T., Friese, L.M., Mundo, M. and Baldwin, J.G. (1997) DNA sequencing from formalin-fixed nematodes: integrating molecular and morphological approaches to taxonomy. *Journal of Nematology* 29, 250–254.

Thorne, G. (1961) *Principles of Nematology.* McGraw Hill, New York.

Trudgill, D.L., Evans, K. and Faulkner, G. (1973) A fluidising column for extracting nematodes from soil. *Nematologica* 18, 469–475.

Uehara, T., Mizukubo, T., Kushida, A. and Momota, Y. (1998) Identification of *Pratylenchus coffeae* and *P. loosi* using specific primers for PCR amplification of ribosomal DNA. *Nematologica* 44, 357–368.

Uehara, T., Kushida, A. and Momota, Y. (1999) Rapid and sensitive identification of *Pratylenchus* spp. using reverse dot blot hybridisation. *Nematology* 5, 549–555.

van der Beek, H.J.G., Veldhuis, W.B.J., Zijlstra, C. and van Silfhout, C.H. (1996) Preservation of *Meloidogyne hapla* and *M. chitwoodi* in liquid nitrogen: differences in response between populations. *Fundamental and Applied Nematology* 19, 227–234.

Viglierchio, D.R. and Yamashita, T.T. (1983) On the methodology of nematode extraction from field samples: density flotation techniques. *Journal of Nematology* 15, 444–449.

Viscardi, T. and Brzeski, M.W. (1993) DITYL: computerized key for species identification of *Ditylenchus* (Nematoda: Anguinidae). *Fundamental and Applied Nematology* 16, 389–392.

Viscardi, T. and Brzeski, M.W. (1995) Computerized key for eelworms identification. *Materialy Sesji Instytutu Ochrony Roślin* 35, 374–375.

Vos, P., Hogers, R., Bleeker, M., Reijans, M., van de Lee, T., Hornes, M., Frijters, A., Por, J., Peleman, J., Kuiper, M. and Zabeau, M. (1995) AFLP: a new technique for DNA fingerprinting. *Nucleic Acids Research* 23, 4407–4414.

Vrain, T.C., Wakarchuk, D.A., Lévesque, A.C. and Hamilton, R.I. (1992) Intraspecific rDNA restriction fragment length polymorphism in the *Xiphinema americanum* group. *Fundamental and Applied Nematology* 15, 563–573.

Waeyenberge, L., Ryss, A., Moens, M., Pinochet, J. and Vrain, T.C. (2000) Molecular characterization of 18 *Pratylenchus* species using rDNA restriction fragment length polymorphism. *Nematology* 2, 135–142.

Waite, I.S., O'Donnell, A.G., Harrison, A., Davies, J.T., Colvan, S.R., Ekschmitt, K., Dogan, H., Wolters, J.T., Bongers, T., Bongers, M., Bakonyi, G., Nagy, P., Papatheodorou, E.M., Stamou, G.P. and Boström, S. (2003) Design and evaluation of nematode 18S rDNA primers for PCR and denaturing gradient gel electrophoresis (DGGE) of soil community DNA. *Soil Biology and Biochemistry* 35, 1165–1173.

Wang, X., Bosselut, N., Castagnone, C., Voisin, R., Abad, P. and Esmenjaud, D. (2003) Multiplex polymerase chain reaction indentification of single individuals of the longidorid nematodes *Xiphinema index, X. diversicaudatum, X. vuittenezi,* and *X. italiae* using species primers from ribosomal genes. *Phytopathology* 93, 160–166.

Wendt, K.R., Vrain, T.C. and Webster, J.M. (1993) Separation of three species of *Ditylenchus* and some host races of *D. dipsaci* by restriction fragment length polymorphism. *Journal of Nematology* 25, 361–364.

West, J.A. (1957) Recommended changes in recovery techniques for burrowing nematode. *Plant Disease Reporter* 41, 600–602.

Whitehead, A.G. and Hemming, J.R. (1965) A comparison of some quantitative methods of extracting small vermiform nematodes from soil. *Annals of Applied Biology* 55, 25–38.

Whyte, E.B. and Gowen, S.R. (1974) Recovery of nematodes from banana roots and soil samples. *Nematropica* 4, 27–41.

Williamson, V.M., Caswell-Chen, E.P., Westerdahl, B.B., Wu, F.F. and Caryl, G. (1997) A PCR assay to identify and distinguish single juveniles of *Meloidogyne hapla* and *M. chitwoodi. Journal of Nematology* 29, 9–15.

Winfield, A.L., Enfield, M.A. and Forman, J.H. (1987) A column elutriator for extracting cyst nematodes and other small invertebrates from soil samples. *Annals of Applied Biology* 111, 223–231.

Yen, J.H., Lee, M.D., Chen, D.Y., Lin, C.Y. and Tsay, T.T. (1998) The comparison of three nematode-extraction methods on four selected nematodes. *Plant Protection Bulletin (Taipei)* 40, 153–162.

Young, T.W. (1954) An incubation method for collecting migratory endoparasitic nematodes. *Plant Disease Reporter* 38, 794–795.

Zijlstra, C. (2000) Identification *of Meloidogyne chitwoodi, M. fallax* and *M. hapla* based on SCAR PCR: a powerful way of enabling reliable identification of populations or individuals that share common traits. *European Journal of Plant Pathology* 106, 283–290.

Zijlstra, C., Lever, A.E.M., Uenk, B.J. and Vansilhout, C.H. (1995) Differences between ITS regions of isolates of root-knot nematodes *Meloidogyne hapla* and *M. chitwoodi. Phytopathology* 85, 1231–1237.

Zijlstra, C., Donkers-Venne, D.T.H.M. and Fargette, M. (2000) Identification of *Meloidogyne incognita, M. javanica* and *M. arenaria* using sequence characterised amplified region (SCAR) based PCR assays. *Nematology* 2, 847–853.

Zuckerman, B.M., Mai, W.F. and Krusberg, L.R. (1985) *Plant Nematology Laboratory Manual.* University of Massachusetts Agricultural Experiments Station, Amherst, Massachusetts.

4 Nematode Parasites of Rice*

John Bridge,[1] Richard A. Plowright[2] and Deliang Peng[3]
[1]*CABI Bioscience, UK Centre, Bakeham Lane, Egham, Surrey TW20 9TY, UK;*
[2]*29 Huntstile, Goathurst, Bridgwater, Somerset TA5 2DQ, UK;* [3]*Institute of Plant Protection, Chinese Academy of Agricultural Sciences, Beijing 100094, China*

Rice (*Oryza* spp.) is the most important food crop in the world, being the staple food for more than half of the world's population, predominantly in Asia where more than 90% of the world's rice is grown and consumed. It is a very versatile crop and there are many types of rice adapted to various environments and cultivation practices.

Essentially there are five major rice-growing environments (Khush, 1984), which have a profound impact on the plant parasitic nematode fauna and their concomitant damage.

1. Irrigated: About 53% of the world rice area is irrigated and provides up to 75% of the total world rice production. Irrigated (inundated) areas have good water control and rice is flooded throughout the growing season.
2. Rainfed lowland: Approximately 31% of the world rice area is planted in rainfed lowland areas. Rainfed lowlands have a wide variety of growing conditions related to depth and duration of standing water on the crop. The fields are bunded but are entirely dependent on rainfall.
3. Deepwater: Areas classified as deepwater occur in the river deltas of South and South-east Asia occupying about 3% of the world rice area. There is no water control, and flooding occurs only during part of the growing season when water depths vary to over 3 m.
4. Tidal wetlands: Tidal wetlands occur near sea coasts and inland estuaries and are directly or indirectly influenced by tides.
5. Upland: Upland rice is grown in soils without surface water accumulation. It is rainfed without any water control. Upland rice occupies approximately 13% of the world rice area and yields are generally low. Most rice in Africa and Latin America is upland.

Nematodes of Rice

Many genera and species of parasitic nematodes are associated with rice, but only some of these are known or suspected to cause yield loss (Table 4.1). They have diverse parasitic habits, but all cause mechanical damage and/or malfunctions of the physiological processes involved in

*A revision of the chapter by J. Bridge, M. Luc and R.A. Plowright.

Table 4.1. Plant nematode genera and species known or suspected to cause yield loss in rice and their means of spread.

Nematodes	Rice affected	Means of spread
Foliar parasites		
Ditylenchus angustus	Lowland and deepwater	Stem and panicles, soil
Aphelenchoides besseyi	Upland, irrigated, lowland and deepwater	Seed, stem and panicles, soil
Root parasites		
Criconemoides onoensis	Upland, irrigated and lowland	Soil
Heterodera elachista	Upland and irrigated	Soil and roots
H. oryzae	Upland and irrigated	Soil and roots
H. oryzicola	Upland and irrigated	Soil and roots
H. sacchari	Upland and irrigated	Soil and roots
Hirschmanniella belli	Irrigated, lowland and deepwater	Soil and roots
H. gracilis	Irrigated, lowland and deepwater	Soil and roots
H. imamuri	Irrigated, lowland and deepwater	Soil and roots
H. mexicana	Irrigated, lowland and deepwater	Soil and roots
H. mucronata	Irrigated, lowland and deepwater	Soil and roots
H. oryzae	Irrigated, lowland and deepwater	Soil and roots
H. spinicaudata	Irrigated, lowland and deepwater	Soil and roots
Hoplolaimus indicus	Upland and irrigated	Soil and roots
Meloidogyne graminicola	Upland, irrigated, lowland and deepwater	Soil and roots
M. hainanensis	Upland and irrigated	Soil and roots
M. incognita	Upland and irrigated	Soil and roots
M. javanica	Upland and irrigated	Soil and roots
M. arenaria	Upland and irrigated	Soil and roots
M. oryzae	Irrigated	Soil and roots
M. salasi	Upland and irrigated	Soil and roots
M. triticoryzae	Upland and irrigated	Soil and roots
Paralongidorus australis	Upland and irrigated	Soil
Pratylenchus brachyurus	Upland	Soil and roots
P. indicus	Upland	Soil and roots
P. pseudopratensis	Upland	Soil and roots
P. zeae	Upland	Soil and roots
Xiphinema ifacolum	Upland	Soil

plant development, resulting in poor growth and yield loss. Some species cause damage in all rice environments whilst others are more restricted (Table 4.1). Nevertheless, rice nematodes can be divided conveniently into two groups depending on their parasitic habits: the foliar parasites, feeding on stems, leaves and panicles; and the root parasites.

Foliar Parasites

Only two species, *Ditylenchus angustus* and *Aphelenchoides besseyi*, are known foliar parasites of rice, although others are suspected.

Ditylenchus angustus

D. angustus, the cause of 'ufra' (India) or 'Tiem Dot San' (Vietnam), occurs in Bangladesh, Burma, India, Madagascar, Malaysia, Thailand and Vietnam, mainly in major river deltas on both deepwater and lowland rice.

Symptoms of damage

During vegetative growth, symptoms of nematode damage are prominent white patches, or white speckles in a splash pattern at the bases of young leaves (Fig. 4.1 and Plate 1A). Brown stains may develop on leaves and sheaths and later intensify

Fig. 4.1. White patches on rice leaf base caused by *Ditylenchus angustus.* (Photo: J. Bridge.)

Fig. 4.2. Twisting and distortion of leaf bases caused by *Ditylenchus angustus.* (Photo: J. Bridge.)

to a dark brown colour; leaves inside such sheaths may be wrinkled. Young leaf bases are twisted, leaf sheaths distorted, and the lower nodes can become swollen with irregular branching (Fig. 4.2). After heading, infected panicles are usually crinkled with empty, shrivelled glumes, especially at their bases; the panicle head and flag leaf are twisted and distorted (Fig. 4.3 and Plate 1B). Panicles often remain completely enclosed within a swollen sheath or only partially emerge (Fig. 4.4) (Butler, 1913; Hashioka, 1963; Vuong and Rabarijoela, 1968; Cox and Rahman, 1980; Chakrabarti *et al.*, 1985). Dark brown patches of ufra-infected plants can be observed in the field normally after panicle initiation (Plate 1C). *D. angustus* can significantly reduce plant heights and photosynthetic rates in leaves (Ali *et al.*, 1997).

Biology and life cycle

D. angustus is an ectoparasite, feeding on young, foliar tissues. Nematodes in water invade rice within 1 h, but invasion varies

Fig. 4.3. Twisting and distortion of rice panicles and flag leaf caused by *Ditylenchus angustus*. (Photo: J. Bridge).

Fig. 4.4. Partial emergence of a rice panicle due to *Ditylenchus angustus*. (Photo: J. Bridge.)

with plant age, older plants being less easily invaded (Rahman and Evans, 1988). In deepwater rice seedlings, nematodes are found around the growing point, but in all parts of the plant in lowland rice. Nematodes are carried or migrate upwards to feed on newly forming tissues enclosed in the rolled leaf sheaths. They accumulate and feed on the primordia of the developing panicles; at harvest, they are coiled in a quiescent state mainly within the dried glumes of the lower spikelets on each panicle, but not within the grains. Activity and infectivity are resumed when water returns for the next rice crop. On deepwater rice in Bangladesh, Butler (1913) assumed that multiplication of *D. angustus* takes place between May, June and November with at least three generations. The greatest infection of rice occurs in the temperature range 27–30°C (Butler, 1913, 1919; Hashioka, 1963; Vuong and Rabarijoela, 1968; Vuong, 1969).

Survival and means of dissemination

Between crops, *D. angustus* remains active in ratoons, volunteer or wild rice

(Rathaiah, 1988) and other hosts. It also survives in a desiccated state in crop residues, mainly panicles enclosed or partially enclosed in leaf sheaths (Cox and Rahman, 1979b; Kinh, 1981). Nematodes can be reactivated in water after 7–15 months (Butler, 1913) but may not remain infective. There is an 'overwinter decay' of *D. angustus* in crop residues between rice crops (Cox and Rahman, 1979b), and populations rapidly decline after harvest. However, the different stages of *D. angustus* show no intrinsic ability to control water loss and survive severe desiccation. They are dependent on high humidities and/or protection by plant tissues for long-term survival (Ibrahim and Perry, 1993).

Nematodes in flooded soil are inactive in less than 4 months (Butler, 1913) and probably lose their infectivity in a much shorter period. However, infested soil dried for 6 weeks can produce ufra disease symptoms 2 months after planting rice (Cuc, 1982b). Soil from around diseased plants does not normally appear to produce the disease (Hashioka, 1963) and is a minor component in disease transmission and nematode survival.

Most *D. angustus* die after a few days in water, but survival for longer periods has been observed (Butler, 1919). Nematode death appears to occur in water, but even a relatively brief survival in water would allow *D. angustus* to spread by water flow to infect new plants (Hashioka, 1963; Sein and Zan, 1977). Long-distance transmission in runoff water, canals and rivers is possible. Nematodes can migrate from diseased to healthy plants in water, and by stem and leaf contact under high humidity (> 75% relative humidity) (Rahman and Evans, 1988).

D. angustus does not have an actual survival stage and cannot survive severe desiccation (Ibrahim and Perry, 1993). The nematodes can be found inside filled and unfilled spikelets of freshly harvested rice, but generally not in dried seed from infected plants (Butler, 1919; Hashioka, 1963; Sein, 1977b; Cuc and Giang, 1982), apart from one report from India (Prasad and Varaprasad, 2002), and dissemination in seed is therefore rare or unlikely.

Environmental conditions affecting parasitism

D. angustus is a parasite of deepwater, irrigated and lowland rice and requires at least 75% humidity to migrate on the foliage. Ufra disease is most severe in the wettest years and in the wettest areas of Bangladesh where the median rainfall exceeds 1.6 m (Cox and Rahman, 1980). In Vietnam, the disease is most severe in months of high rainfall or in fields with high water levels (Cuc and Kinh, 1981).

Hosts of *D. angustus*

Hosts are mainly confined to wild and cultivated species of deepwater and lowland rice (*Oryza sativa* var. *fatua*, *O. glaberrima*, *O. cubensis*, *O. officinalis*, *O. meyriana*, *O. latifiolia*, *O. perennis*, *O. eichingeri*, *O. alta*, *O. minuta*), but *Leersia hexandra* has also been found to support populations of the nematode (Hashioka, 1963; Vuong and Rabarijoela, 1968; Sein and Zan, 1977). Two other weeds, *Echinochloa colona* and *Sacciolepsis interrupta*, have also been found to be infected (Cuc, 1982a).

Disease complexes

The ufra nematode can increase the nitrogen content of rice plants and thus the plants become more susceptible to the plant pathogen *Pyricularia oryzae* (Mondal *et al.*, 1986). Foliar brown spots associated with the nematode could be secondary invasion sites for *Fusarium* and *Cladosporium* fungi (Vuong, 1969).

Economic importance

Ufra has a restricted distribution because of the unique environmental requirements of the nematode. It is often localized in a rice-growing region and does not always occur in the same fields every year. The worldwide and national yield losses caused by *D. angustus* are therefore seemingly low. In Bangladesh, for example, an annual yield loss of 4% (20% yield loss over 20% of the area) has been estimated on deepwater rice (Catling *et al.*, 1979). However, when it does occur, it is one of the most devastating of all diseases affecting rice (Cox and Rahman, 1980).

D. angustus has been a serious problem in Vietnam in the Mekong Delta. It can cause 50–100% loss of deepwater, irrigated and lowland rice, and, during 1974, hundreds of hectares of deepwater rice in one Province were totally lost (Cuc and Kinh, 1981). During 1982, 60,000–100,000 ha of rice in the Mekong Delta were affected by *D. angustus* (Catling and Puckridge, 1984) and, in Dong Thap Province, 10,000 ha were affected (Puckeridge, 1988). However, since that time, the rice areas damaged by *D. angustus* have been greatly reduced mainly because of the marked reduction in the cultivation of deepwater rice, which has been replaced by irrigated lowland rice cultivars (Prot, 1994a). Hashioka (1963) estimated that 500 ha of lowland rice in southern Thailand had yield losses of 20–90% caused by ufra. Rice in Assam and West Bengal, India has been found infected with *D. angustus*, with losses estimated at 10–30% in some areas (Pal, 1970; Rao *et al.*, 1986b). In Bangladesh, 60–70% of low lying areas covering about 200,000 ha were found to be infested with *D. angustus* (Mondal and Miah, 1987).

Serious yield losses can occur if transplanted rice seedlings are infected with *D. angustus*, even at low initial percentage infection. Yield losses varying from 1.26 to 3.94 t/ha have been recorded with 4–10% infected seedlings (Mondal *et al.*, 1988).

Management measures

Many different measures to control *D. angustus* have been suggested, some practical, others less feasible. Those likely to achieve the best results are destruction or removal of infested stubble and straw, crop rotation, control of weeds and volunteer rice, control of water flow, varietal resistance and escape cropping.

DESTRUCTION OR REMOVAL OF INFESTED STUBBLE AND STRAW. Burning of infested crop residues gives very effective control and has long been advocated (Butler, 1919). Thorough burning is essential, although it is not always possible where soil remains waterlogged after harvest or when a large proportion of the straw is removed for other purposes, e.g. for cattle fodder, leaving insufficient for effective burning (McGeachie and Rahman, 1983). Ploughing-in crop residues can reduce ufra as nematodes decline more rapidly in moist soil than in foliar remains (Butler, 1919). New growth in rattooning hills, following rice harvest, should be destroyed to prevent further multiplication of *D. angustus*. This is not always possible and depends on local resources and soil conditions.

CROP ROTATION. Growing a non-host crop such as jute in rotation with deepwater rice can reduce the incidence of ufra in fields where the rise of floodwater is not excessively fast (McGeachie and Rahman, 1983). Lowland transplanted rice rotated with mustard, another non-host, and jute is less affected by ufra than continuously cultivated rice (Miah and Rahman, 1985; Chakraborti, 2000b).

ELIMINATING OTHER HOSTS. Removal of volunteer and ratoon rice plants, wild rice and other host weeds will help prevent the carry-over of nematodes from one rice crop to the next (Hashioka, 1963; Sein and Zan, 1977).

CONTROLLING WATER FLOW. As nematodes can easily be spread in surface water, preventing river overflow into fields by improved bunding or banks could be beneficial (Sein and Zan, 1977).

RESISTANCE. A large number of deepwater and lowland rice cultivars have been tested against *D. angustus*. In Vietnam, four high-yielding local improved breeding lines (IR9129-393-3-1-2, IR9129-169-3-2-2, IR9224-117-2-3-1 and IR2307-247-2-2-3) and three cultivars (BKN6986-8, CNI-53 and Jalaj) are described as slightly infected (Kinh and Phuong, 1981; Kinh and Nghiem, 1982). A Burmese cultivar (B-69-1) from the Irawaddy Delta was tolerant of ufra disease (Sein, 1977a) and a Thailand cultivar (Khao Tah Ooh) was relatively less susceptible (Hashioka, 1963). Two cultivars in West Bengal, India (IR36 and IFT4094) were also less susceptible (Chakrabarti *et al.*, 1985). Complete resistance to *D. angustus* has been found in a wild rice, *Oryza subulata*, and a deepwater cv. RD-16-06 (Miah and Bakr, 1977b). The Rayada group of deepwater rice lines show the most promise because of their strong resistance. Rayada lines are highly resistant to *D. angustus* in Bangladesh, and others have shown moderate resistance (Rahman, 1987, 1994; Das and Sarmah, 1995). The cv. Rayada B3 has been shown to be both ufra resistant and high yielding (Das *et al.*, 2000). Resistance is partly mediated by a rapid necrotic response to nematode feeding (Plowright and Gill, 1994) and involves the increased levels of chlorogenic acid and sythesis of the rice phtytoalexin sakuranetin (Plowright *et al.*, 1996).

The cvs Padmapani and Digha are not attacked by *D. angustus* in areas of India and Bangladesh. It is suggested that they escape the disease because of their short growth duration (Mondal and Miah, 1987; Rathaiah and Das, 1987).

ESCAPE CROPPING. *D. angustus* survives for a limited period, and lengthening the overwinter period can reduce primary infection (Cox and Rahman, 1980; McGeachie and Rahman, 1983; Das and Bhagawati, 1992). This can be achieved with deepwater rice by using short duration cultivars or late sowing and transplanting. Manipulation of rice cropping patterns and cultivation techniques could be a useful means of control (McGeachie and Rahman, 1983). Since *D. angustus* enters the leaf sheath primarily at the water surface (Plowright and Gill, 1994), short periods of submergence of young seedlings can reduce infection by nematodes.

CHEMICAL. Chemicals such as carbofuran, mocap, hexadris monocrotophos, phenazine and benomyl have been used with some success, but their high cost and difficulties of correct application make them uneconomical and they have not been recommended for large-scale field use.

The greatest reduction in nematode populations and disease incidence has been achieved with carbofuran and benomyl, alone and in combination (Miah and Bakr, 1977a; Sein, 1977c; Cox and Rahman, 1979a; Rahman *et al.*, 1981; Miah and Rahman, 1985; Nguyen *et al.*, 1993; Mian *et al.*, 1994). Combined spraying of carbosulfan and triazophos has also proved effective (Das, 1996). The rates used are generally uneconomical.

The chemical Azadirachtin, produced from neem (*Azadirachta indica*), has had some success when used as an integrated application combining seed treatment, dipping seedling roots and foliar spray, or with cultural practices (Chakraborti, 1999, 2000a,b). Neem seed dust itself also gives good control of the nematode, as effective as carbofuran (Rahman, 1996).

Summary of management measures against D. angustus

The recommended management measures against *D. angustus* are broadly those put forward by the Deepwater Rice Management Project (Anonymous, 1987): (i) thorough burning of crop residues to eliminate all infested stem terminals; (ii) extending the overwintering period by delayed planting; and (iii) the use of shorter duration cultivars. The use of resistant cultivars, when they become available, should prove to be the most effective measure.

Methods of diagnosis

D. angustus is found in the foliage of growing plants (and crop residues) mainly near the growing points of leaves and inflorescences, and it is these portions of the plants that need to be sampled. Pieces of plant about 5 mm long are cut longitudinally to expose the innermost young leaves.

Nematodes can be extracted from plant pieces placed in a small container on a Baermann funnel or small tray with water and left for 24 h or overnight before examining the suspension (Chapter 3).

For immediate examination of material, the rolled leaves or young inflorescence can be teased apart in a Petri dish of water and observed directly. Nematodes are active in fresh material but will require some time to resume activity from dried panicles.

Aphelenchoides besseyi

Aphelenchoides besseyi is seed borne and causes the disease 'white tip'. It has been recorded in most rice-growing areas of the world (Ou, 1985) including Iran (Pedramfar *et al.*, 2001), Italy (Moretti, 1997; Cotoneo and Moretti, 2001) and Turkey (Ozturk and Enneli, 1997).

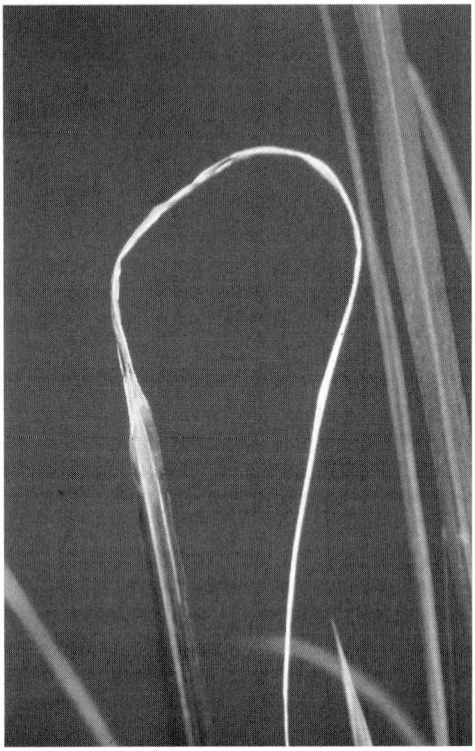

Fig. 4.5. White tip symptoms on rice leaf caused by *Aphelenchoides besseyi*. (Photo: J. Bridge.)

Symptoms

Susceptible plants can be symptomless, but in general yield loss only occurs in plants showing some symptoms. During early growth, the most conspicuous symptom is the emergence of the chlorotic tips of new leaves from the leaf sheath (Fig. 4.5). These tips later dry and curl, whilst the rest of the leaf may appear normal. The young leaves of infected tillers can be speckled with a white splash pattern, or have distinct chlorotic areas. Leaf margins may be distorted and wrinkled, but leaf sheaths are symptomless (Plate 1D).

Viability of infected seed is lowered, germination is delayed (Tamura and Kegasawa, 1959b) and diseased plants have reduced vigour and height (Todd and Atkins, 1958). Infected panicles are shorter, with fewer spikelets and a smaller proportion of filled grain (Dastur, 1936; Yoshii, 1951; Todd and Atkins, 1958).

In severe infections, the shortened flag-leaf is twisted and can prevent the complete extrusion of the panicle from the boot (Yoshii and Yamamoto, 1950a; Todd and Atkins, 1958). *A. besseyi* infestation reduces seed swelling (Togashi and Hoshino, 2001), the grain is small and distorted (Todd and Atkins, 1958) and the kernel may be discoloured and cracked (Uebayashi *et al.*, 1976) (Fig. 4.6). Infected plants mature late and have sterile panicles borne on tillers produced from high nodes.

Biology

When seed infected with *A. besseyi* is sown, the anabiotic nematodes rapidly become active and, during early growth, *A.*

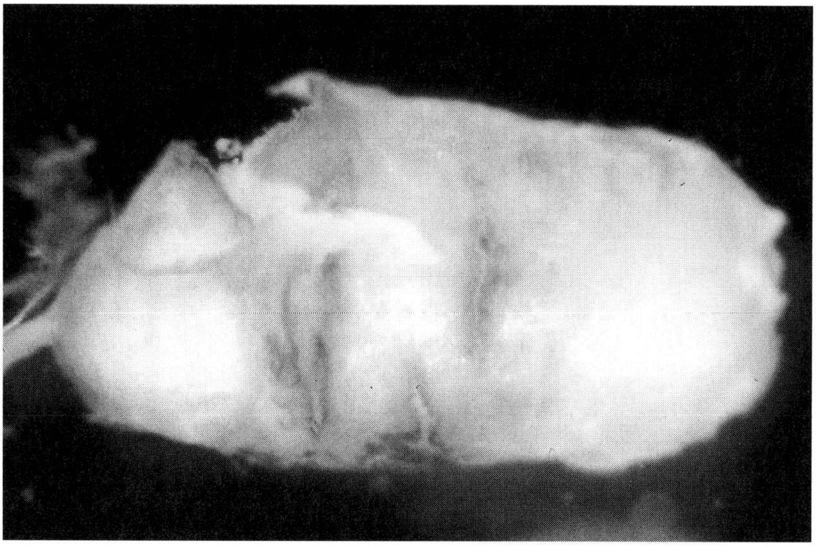

Fig. 4.6. Necrotic lesions on rice seed endosperm caused by *Aphelenchoides besseyi*. (Photo: R.A. Plowright.)

besseyi is found within the innermost leaf sheath, feeding ectoparasitically around the apical meristem (Yoshii and Yamamoto, 1950b; Goto and Fukatsu, 1952; Todd and Atkins, 1958). A rapid increase in nematode numbers takes place at late tillering (Goto and Fukatsu, 1952) and is associated with the reproductive phase of plant growth (Huang and Huang, 1972). Nematodes are able to enter spikelets before anthesis, within the boot, and feed ectoparasitically on the ovary, stamens, lodicules and embryo (Dastur, 1936; Huang and Huang, 1972). However, *A. besseyi* is more abundant on the outer surface of the glumes and enters when these separate at anthesis (Yoshii and Yamamoto, 1950b). As grain filling and maturation proceed, reproduction of the nematode ceases, although the development of J3 to adult continues until the hard dough stage (Huang and Huang, 1972). The population of anabiotic nematodes is predominantly adult female (Huang *et al.*, 1979). These nematodes coil and aggregate in the glume axis. More nematodes occur in filled grain than in sterile spikelets (Yoshii and Yamamoto, 1950b), and infected grain tends to occur more towards the middle of the panicle (Goto and Fukatsu, 1952).

A. besseyi is amphimictic (Huang *et al.*, 1979), and males are usually abundant; however, reproduction can be parthenogenetic (Sudakova and Stoyakov, 1967). The optimum temperature for oviposition and hatch is 30°C. At 30°C the life cycle is 10 ± 2 days and lengthens significantly at temperatures below 20°C (Huang *et al.*, 1972). No development occurs below 13°C (Sudakova, 1968).

Survival and dissemination

A. besseyi aggregate in the glume axis of maturing grain and slowly desiccate as kernel moisture is lost. They become anabiotic and are able to survive for 8 months to 3 years after harvest (Cralley, 1949; Yoshii and Yamamoto, 1950b; Todd, 1952; Todd and Atkins, 1958). Survival is enhanced by aggregation and a slow rate of drying (Huang and Huang, 1974), but the number (Yoshii and Yamamoto, 1950b; Sivakumar, 1987a) and infectivity (Cralley and French, 1952) of nematodes are reduced as seed age increases. It is ironic that good seed storage conditions probably prolong nematode survival. More nematodes survive in seeds stored with low moisture than in seeds at high moisture levels at most temperatures (Chaudhury and Chaudhury, 1996).

A. besseyi is not thought to survive long periods in soil between crops (Cralley and French, 1952; Yamada *et al.*, 1953), although anabiotic nematodes may survive on rice husks and plant debris. Sivakumar (1987b) found *A. besseyi* reproducing on *Curvularia* and *Fusarium* in straw after harvest.

The principal dispersal method for *A. besseyi* is seed. The inadvertent dissemination of infected seed must account for its worldwide distribution. On a local scale, *A. besseyi* can be transmitted in flood water in lowland rice (Tamura and Kegasawa, 1958; Uebayashi and Imamura, 1972), but the survival of nematodes in water decreases as temperature increases from 20 to 30°C (Tamura and Kegasawa, 1958). High seeding rates in infected seedbeds also facilitate local dispersal (Kobayashi and Sugiyama, 1977).

Environmental factors affecting parasitism

A. besseyi is able to infect rice in most environments, but infection and damage are generally greater in irrigated lowland and deepwater than in upland. In Brazil, da Silveira *et al.* (1977) found significantly more infestations in irrigated rice than in upland, and in Japan infection was greater in flooded conditions (Tamura and Kegasawa, 1959a).

A. besseyi is active and feeds at a relative humidity greater than 70% (Tikhanova, 1966) and, consequently, a high relative humidity during the reproductive phase of the crop is required for migration into the panicle (Sivakumar, 1987b) and favours symptom development (Dastur, 1936).

Other hosts of A. besseyi

The host range encompasses more than 35 genera of higher plants (Fortuner and Williams, 1975) although host races are thought to exist. The wild annual rice *O. breviligulata* A. Chev. and Roehr. and *O. glaberrima* Steud. are good hosts. Other important hosts include some common weeds of rice fields, e.g. *Cyperus iria* L., *Setaria viridis* Beauv. and *Panicum sanguinale* L. (Yoshii and Yamamoto, 1950b), and food crops such as *Dioscorea trifida* L. (yam), *Ipomoea batatas* (sweet potato), *Allium cepa* L. (onion), *Zea mays* L. (maize) and *Colocasia esculenta* L. (taro). It has also been found on chilli pepper (*Capsicum annuum* var. *longum*) in Sarawak (Hockland and Eng, 1997). In addition, many saprophytic and pathogenic fungi are good hosts, e.g. *Alternaria* spp., *Curvularia* spp., *Fusarium* spp., *Helminthosporium* spp., *Nigrospora* sp., *Sclerospora* sp. and *Botrytis cinerea*. Rao (1985) found that *A. besseyi* survived but did not multiply on the rice blast fungus, *Pyricularia oryzae*, and Iyatorni and Nishizawa (1954) reported that *A. besseyi* can feed and reproduce on the stem rot fungus *Sclerotium oryzae*.

Disease complexes

The involvement of *A. besseyi* in disease complexes is not widely researched. In Bangladesh, *A. besseyi* occurs with *D. angustus* (Timm, 1955) and *Meloidogyne graminicola*, but little is known of their associations. In pot tests, the effects of *A. besseyi* and *M. graminicola* on yield of flooded rice were additive, but *M. graminicola*-infected plants had more *A. besseyi*/seed at harvest than those with *A. besseyi* alone (Plowright, 1986).

A. besseyi appears to influence the symptom development of some fungal pathogens of rice such as *Sclerotium oryzae* (stem rot) and *Pyricularia oryzae* (blast) (Nishizawa, 1953a; Tikhanova and Ivanchenko, 1968; McGrawley *et al.*, 1984). *Curvularia lunata* in rice seed can cause a build-up of *A. besseyi* numbers and increases grain deformation (Rao *et al.*, 1994), and rice kernels infected by *A. besseyi* are predisposed to secondary infection by saprophytes such as *Enterobacter agglomerans* which causes black, wedge-shaped spots on grain (Nishizawa, 1976; Uebayashi *et al.*, 1976).

Economic importance and population-damage threshold levels

A. besseyi is widely distributed because of its dissemination in seed, but its importance varies between regions, countries and localities. Within a locality, the incidence and severity of the disease can change from year to year and are strongly influenced by cultural practices and local rice types.

Damage in a susceptible cultivar largely depends on the percentage of infested seed sown and the number of *A. besseyi*-infested seeds. Generally, population densities per seed number or weight are counted. Fukano (1962) determined an economic damage threshold density (300 live nematodes/100 seeds), which provides a useful basis for damage prediction since, in many countries, very little information on the pest status of *A. besseyi* exists.

Yield loss data for *A. besseyi* have been widely reported. In the 1950s, typical figures for susceptible cultivars in the USA were 17.5, 4.9 and 6.6% in different years (Atkins and Todd, 1959), and 10–30% in Japan (Yamada and Shiomi, 1950; Yoshii and Yamamoto, 1950a; Yoshii, 1951). *A. besseyi* has been controlled in the USA by seed treatment and resistant cultivars and is no longer a pest (Hollis and Keoboonrueng, 1984). *A. besseyi* also disappeared from Japan, but has re-occurred, the economic value of infected discoloured grain being reduced if infection exceeds 0.7% (Inagaki, 1985). In China, yield losses can be as high as 45% when plant infestation levels exceed 50% (Tsay *et al.*, 1998).

A. besseyi damage has been reported from deepwater rice in Bangladesh. More than 50% of fields are infected and the panicle weight of heavily infected plants (650 nematodes/100 seeds) was one-third that of less infected plants (112 nematodes/100 seeds) (Rahman and McGeachie, 1981; Rahman and Taylor, 1983). In contrast, local cultivars in Thailand appear to be tolerant of *A. besseyi* and no symptoms have been observed despite widespread infection (Buangsuwon *et al.*, 1971). Rao (1976) reported severe symptoms in the field in India, but accurate yield loss assessment is lacking. Muthukrishnan *et al.* (1974) observed that plants sometimes recover after early severe damage, and computed losses of 0.2–10%.

In Africa, *A. besseyi* is widespread, particularly in west and central Africa, Madagascar and the Comoro Islands (Barat *et al.*, 1969). White tip is very likely to be causing significant yield loss in the mangrove swamp rice of Sierra Leone, where the widely grown cultivars are very susceptible to *A. besseyi* (3000–10,000 *A. besseyi*/100 seeds), and the incidence and severity of the disease were said to be increasing (Fomba, 1984). Yield loss is also likely in Tanzania, where levels of infested seed are very high (2–82%) and average 68 *A. bessey*/infested seed (Taylor *et al.*, 1972), and in Madagascar where Vuong (1969) considered that all seed was infested above the Fukano (1962) threshold. In Nigeria, infestation levels can be 2–400 per 100 seeds, but were commonly less than 100 per 100 seeds (Babatola, 1984). In the former USSR, the yield loss of a susceptible cultivar was 54%. *A. besseyi*-infested seed (80%) gave rise to only 31% damaged plants in the field (Popova, 1984). Yield loss in central-west Brazil would seem unlikely with the infestation levels (10–140 per 100 seeds) given by Huang *et al.* (1977), unless grain has a high percentage infestation.

Management measures

Preventing dispersal of *A. besseyi* requires the elimination of nematodes from seed, e.g. by hot water or chemical seed treatments. Resistant cultivars and cultural methods have been used to reduce infection below damage thresholds, and tolerant cultivars avoid yield loss without nematode control. Stubble burning prevents transmission of *A. besseyi* in straw and chaff, but would have to be used in conjunction with other control measures.

HOT WATER TREATMENT. There are numerous references on the hot water treatment of rice seed (Cralley, 1949; Yoshii and Yamamoto, 1950c, 1951; Todd and Atkins,

1958; Borovkova, 1967). The most effective control requires seed to be pre-soaked in cold water for 18–24 h, then immersed in water at 51–53°C for 15 min. Higher temperatures (55–61°C for 10–15 min) are required if seed is not pre-soaked. The temperature and duration of treatment must he closely monitored, and after treatment the seed must be dried at 30–35°C or sun dried if stored, but otherwise can be sown directly in the field. For quarantine purposes, at the International Rice Research Institute, seed was soaked in cold water for 3 h followed by hot water at 52–57°C for 15 min. Simply water-soaking seeds followed by relatively rapid air-drying can in itself cause marked nematode mortality in seeds (Hoshino and Togashi, 2000).

CHEMICAL. Various chemical seed treatments have been used, sometimes to good effect (Ribeiro, 1977), but it is also reported that nematicide treatment of seeds has very little effect on nematode mortality within the seeds (Hoshino and Togashi, 2000). However, benomyl seed treatment and spraying with benomyl 1 or 15 days after transplanting can be used to protect rice plants from infestation by *A. besseyi* (Gergon and Prot, 1993). Chemical soil application is said to be effective (Rao, 1986a), although pre-harvest chemical treatments alone are reported to be only partially effective (Aleksandrova, 1981) and there is no evidence that chemical soil treatment is an economical proposition.

RESISTANCE AND TOLERANCE. Resistance to *A. besseyi* appears to be widespread. Cralley (1949) and Cralley and Adair (1949) first reported variations in susceptibility of rice to *A. besseyi* and listed the cvs Arkansas Fortuna, Nira 43 and Bluebonnet as resistant. In the USA, *A. besseyi* has been controlled principally through the use of resistant cultivars. Resistance to *A. besseyi* has been reported from Japan (Nishizawa, 1953b; Yamada *et al.*, 1953; Goto and Fukatsu, 1956), Korea (Park and Lee, 1976), India (Rao *et al.*, 1986a), Brazil (Oliveira, 1989), Russia (Popova *et al.*, 1994) and Italy (Orsenigo, 1954). Resistance to *A. besseyi* is said to be genetically controlled and carried by the Japanese cv. Asa-Hi (Nishizawa, 1953b).

Screening for resistance, based primarily on symptom expression, has commonly revealed symptomless but susceptible (i.e. tolerant) cultivars (Nishizawa, 1953b; Goto and Fukatsu, 1956), and there is a strong influence of environment on *A. besseyi* development and damage.

CULTURAL. Irrigating seedbeds (Yamada *et al.*, 1953) or direct seeding into water (Cralley, 1956) reduces infection. In these conditions, nematodes emerge and lose vigour before seed germination. High seedling rates in the seedbed (Kobayashi and Sugiyama, 1977) and high numbers of seedlings per hill (Yamada *et al.*, 1953) tend to increase infection by increasing the number of infection loci in the field. Such problems are thought to be responsible for the re-occurrence of *A. besseyi* in Japan (Inagaki, 1985). In the USA (Cralley, 1949) and Japan (Yoshi and Yamamoto, 1951; Yamada *et al.*, 1953), early planting presumably in cooler conditions reduced or eliminated *A. besseyi* infection. In Korea, rotating beans with rice decreases field populations of *A. besseyi* (Kim *et al.*, 1996).

Summary of management measures against A. besseyi

- Hot water treatment of seed. Probably the most effective and cheapest control measure.
- Resistant or tolerant cultivars.
- Early planting if rice season is preceded by a cooler period.
- Low seedbed planting densities.

Methods of diagnosis

Different sampling methods are used depending on the stage of crop growth. During early growth and tillering, *A. besseyi* is found in the base of the culm and between leaf sheaths. For immediate inspection, plant tissue is carefully teased in water to release nematodes. Plant tissue can be stained before examination, which is particularly useful for detecting low numbers. Alternatively, *A. besseyi* can be extracted from chopped tillers placed on a sieve, or directly in water.

During the reproductive phase, *A. besseyi* is progressively found on or in developing spikelets, and peak numbers are found at flowering. *A. besseyi* is recovered from spikelets and grain by soaking a known number in water for 24–48 h at 25–30°C. Quantitative extraction requires that the glumes are separated from the kernel yet remain in the extract. Better recovery is achieved from hulled grain, but extraction from unhulled grain is less tedious and is a practical method for detection of *A. besseyi* (e.g. for quarantine) especially if extraction time is extended to more than 2 days (Gergon and Mew, 1991).

The percentage of infested seed is a useful parameter, but extracting from individual seeds is time consuming. However, detailed analysis can be done on individual seeds. A method that achieves very good nematode recovery is splitting individual rice seeds and then transferring into single pipette tips. Tips containing a split seed are then singly placed upright in glass vials with water (Hoshino and Togashi, 1999). However, these same authors found that mass extraction of split seeds to determine low levels of nematodes was as efficient and far less laborious than the single seed method (Hoshino and Togashi, 2002).

Root Parasites

Meloidogyne species

Root knot nematodes, *Meloidogyne* spp., have been found on rice in many countries. Probably the most damaging species, *M. graminicola*, is distributed mainly in the countries of South and South-east Asia (Burma, Bangladesh, India (including Sikkim), Nepal, Pakistan, Sri Lanka, Laos, Thailand, Vietnam, Taiwan, Indonesia and the Philippines) and is likely to occur in other countries of the region. *M. graminicola* has also been reported on rice in the USA, Brazil and Colombia. *M. graminicola* is a damaging parasite on upland, lowland and deepwater rice. *M. oryzae* has only been found in Surinam, South America (Maas *et al.*, 1978) on irrigated rice. *M. hainanensis* is reported parasitizing rice in Hainan Island, China (Guo *et al.*, 1984; Liao and Feng, 1995), and a species described as *M. lini* is also reported from rice roots in China (Yang *et al.*, 1988). *M. triticoryzae* is identified as a parasite of both rice and wheat in the rice–wheat cropping systems of northern India (Gaur *et al.*, 1993; Gaur, 2003). Four species of *Meloidogyne* occur only on upland and hydromorphic rice: *M. incognita* (Costa Rica, Cuba, Egypt, Côte d'Ivoire, Nigeria, South Africa and Japan), *M. javanica* (Brazil, Egypt, Comoro Islands, Nigeria, Côte d'Ivoire and Ghana) (Coyne *et al.*, 1999), *M. arenaria* (Nigeria, Egypt and South Africa) and *M. salasi* (Costa Rica and Panama) (Lopez, 1984).

Symptoms

All *Meloidogyne* spp. can cause swellings and galls throughout the root system. Infected root tips become swollen and hooked, a symptom which is especially characteristic of *M. graminicola* and *M. oryzae* (Figs 4.7 and 4.9, Plate 1E). Galls caused by *M. salasi* also occur mostly on the root tips of rice (Sancho *et al.*, 1987).

Above-ground symptoms vary according to the type of rice and the species of *Meloidogyne*. In upland conditions and shallow intermittently flooded land, all species can cause severe growth reduction, unfilled spikelets, reduced tillering, chlorosis, wilting and poor yield (Babatola, 1984). Symptoms often appear as patches in a field.

M. graminicola is known to cause serious damage to deepwater rice. Prior to flooding, symptoms are the typical stunting and chlorosis of young plants. When flooding occurs, submerged plants with serious root galling are unable to elongate rapidly, and do not emerge above the water level (Bridge and Page, 1982). This causes death or drowning out of the plants, leaving patches of open water in the flooded fields.

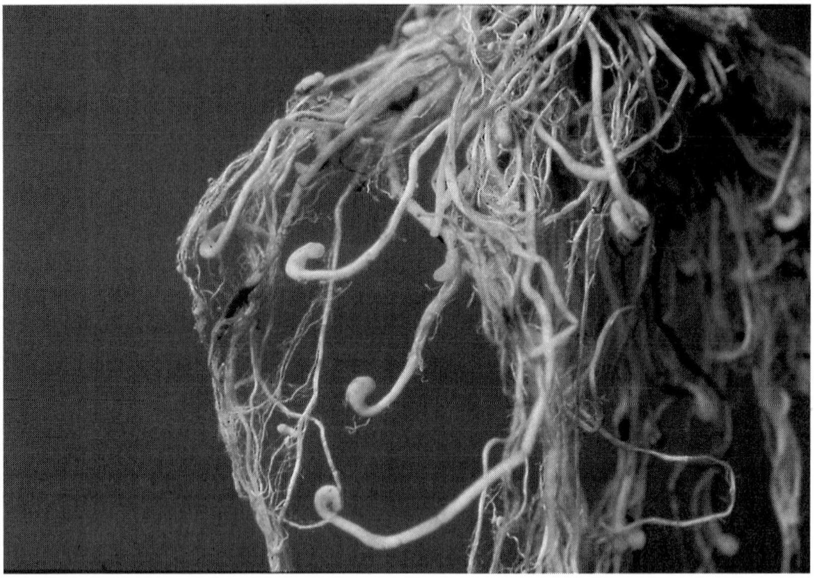

Fig. 4.7. Characteristic hooked root tip galls on rice caused by *Meloidogyne graminicola.* (Photo: J. Bridge.)

Biology and life cycle

The biology and life cycle of *M. incognita* and *M. javanica* on rice are similar to those described for other crops. The life cycle of *M. oryzae* is 4 weeks at a mean temperature of 27°C (Segeren-V.d. Oever and Sanchit-Bekker, 1984). *M. graminicola* from Bangladesh has a very short life cycle on rice of less than 19 days at temperatures of 22–29°C (Bridge and Page, 1982), and an isolate from the USA completed its cycle in 23–27 days at 26°C (Yik and Birchfield, 1979). In India, the life cycle of *M. graminicola* is reported to be 26–51 days depending on the time of year (Rao and Israel, 1973). Females and egg masses of *M. oryzae* are completely embedded in root tissues, and up to 50 females can be present in a single gall (Segeren-V.d. Oever and Sanchit-Bekker, 1984).

Infective, second-stage juveniles of *M. graminicola* invade rice roots in upland conditions just behind the root tip (Buangsuwon *et al.*, 1971; Rao and Israel, 1973). Females develop within the root, and eggs are laid mainly in the cortex (Roy, 1976a) (Plate 1F). Juveniles can remain in the maternal gall or migrate intercellularly through the aerenchymatous tissues of the cortex to new feeding sites within the same root (Bridge and Page, 1982). This behaviour appears to be an adaptation by *M. graminicola* to flooded conditions, enabling it to continue multiplying within the host tissues even when roots are deeply covered by water. Juveniles that migrate from rice roots in flooded soil cannot reinvade (Bridge and Page, 1982). *M. triticoryzae* behaves in a similar way in different water regimes and there is less invasion of rice roots when the soil is puddled (Chandel *et al.*, 2002b), and the females and egg sacs usually remain inside the root tissues (Chandel *et al.*, 2001).

M. triticoryzae produces three kinds of unhatched second stage juveniles: those that hatch freely in water; those that require rice root diffusates to stimulate hatch; and those that do not hatch even in the presence of root diffusates that are said to equate with a diapause (Gaur *et al.*, 2000).

Biological races

Rice cultivars are susceptible to race 1 of *M. arenaria* and races 2 and 4 of *M. incognita* (Ibrahim *et al.*, 1983).

Survival and means of dissemination

M. incognita, *M. javanica*, *M. arenaria* and *M. salasi* are parasites mainly of upland rice and survive in soil as eggs or juveniles, or on alternative hosts. They do not survive long periods in flooded soil. *M. oryzae* can survive in shallow flooded (< 10 cm) rice fields for relatively short periods (Segeren V.d. Oever and Sanchit-Bekker, 1984), but *M. graminicola* is well adapted to flooded conditions and can survive in waterlogged soil as eggs in egg masses or as juveniles for long periods. Numbers of *M. graminicola* decline rapidly after 4 months, but some egg masses can remain viable for at least 14 months in waterlogged soil (Roy, 1982). *M. graminicola* can survive in soil flooded to a depth of 1 m for at least 5 months (Bridge and Page, 1982); it cannot invade rice in flooded conditions but quickly invades when infested soils are drained (Manser, 1968). All *Meloidogyne* spp. can be spread in soil and on seedlings of other crop hosts planted to a field. Because *M. oryzae* and, especially, *M. graminicola* are found in flooded rice, there is the additional danger of dissemination in irrigation and runoff water.

Alternative hosts of Meloidogyne

M. incognita, *M. javanica* and *M. arenaria*, the most widespread root knot species, have numerous hosts other than rice. *M. graminicola* also has a wide host range which includes many of the common weeds of rice fields (Table 4.2). It is parasitic on both the *indica* and *japonica* races of *Oryza sativa* (Manser, 1971) and can also be a damaging parasite of vegetables, such as onion (Gergon *et al.*, 2001). A number of weeds and crops are also alternative hosts of *M. oryzae* (Maas *et al.*, 1978; Segeren-V.d. Oever and Sanchit-Bekker, 1984) and *M. salasi* (Lopez, 1984; Salazar and Quesada, 1999).

M. triticoryzae in India is known to reproduce on the crops wheat, barley, sorghum, soybean, okra, green gram, berseem (*Trifolium alexandrinum*) and some cultivars of potato, and also on the weed species *Cyperus rotundus*, *Echinochloa colonum*, *E. crus-galli*, *Leptochloa coloniculus* and *Phalaris minor* (Gaur and Sharma, 1998; Chandel *et al.*, 2002a).

Economic importance

M. incognita can cause poor seedling establishment and reduced yields in upland rice. Yields can decrease to 60% when 8000 eggs and juveniles/dm^3 of soil are present at sowing (Babatola, 1984). Significant yield reductions can occur in both upland and irrigated rice with *M. incognita* (Ibrahim *et al.*, 1972), but damage is generally more severe under upland conditions (Fademi, 1984). Damage to irrigated rice will occur where seedlings are raised in well-drained nursery soils. High initial soil populations of both *M. incognita* and *M. javanica* are necessary to cause yield loss in rice, and populations above 1000 eggs/plant are needed to reduce grain yield with *M. javanica* (Sharma, 1980) or as high as 35,000 eggs/plant to reduce growth by around 40% (Ferraz, 1995). Populations of 128 eggs and juveniles/cm^3 of Venezuelan isolates of *M. incognita* have been shown to kill rice plants (Greco *et al.*, 2000).

M. graminicola can cause economic yield loss in upland, lowland and deepwater rice. In upland rice, there is an estimated reduction of 2.6% in grain yield for every 1000 nematodes present around young seedlings (Rao and Biswas, 1973). The population levels which cause 10% loss in yield of upland rice are 120, 250 and 600 eggs/plant at 10, 30 and 60 days age of plants, respectively, in direct seeded crops (Rao *et al.*, 1986). In flooded rice, damage by *M. graminicola* is caused in nurseries before transplanting (Fig. 4.8, Plate 2A) – the tolerance limit of seedlings is less than one second stage juvenile/cm^3 of soil (Plowright and Bridge, 1990). Damage also occurs prior to flooding where rice is sown directly in well-drained soils. Experiments have shown that 4000 juveniles/plant of *M. graminicola* can cause destruction of up to 72% of deepwater rice plants by drowning out. Losses as high as this in the field are unlikely as natural root populations vary considerably (Bridge and Page, 1982).

Table 4.2. Hosts of *Meloidogyne graminicola.*

Abelmoschus esculentus (L.) Moench.	*Impatiens balsamina* L.
Ageratum conyzoides L.	*Imperata cylindrica* (L.) Beauv.
Allium cepa L.	*Lactuca sativa* L.
Alopecurus carolinianus Walt.	*Leersia hexandra* Sw.
Amaranthus viridis L.	*Leucas lavendulaefolia* J.E. Smith
Ammania petandra Roxb.	*Ludwigia repens* J.R. Forst.
Andropogon sp.	*Lycopersicon esculentum* Mill.
Avena sativa L.	*Murdannia nudiflora* (L.) Brenan
Beta vulgaris L.	*Musa* sp.
Blumea sp.	*Oplismenus compositus* (L.) Beauv.
Borreria hispida (L.) K. Schum.	*Oryza sativa* L.
B. ramosa (L.) Stapf.	*Oxalis corniculata* L.
Brassica juncea (L.) Czern. & Coss	*Panicum miliare* Lam.
B. oleracea L.	*P. miliaceum* L.
Catharanthus roseus (L.) Don	*P. repens* L.
Centella asiatica (L.) Urban	*Paspalum scrobiculatum* L.
Commelina benghalensis L.	*Pennisetum typhoides* (Burm. f.) Stapf & Hubbard
Colocasia esculenta (D) Schott	*P. pedicillatum* L.
Corchorus capsularis L.	*Petunia* sp.
Courtosia cyperoides Nees	*Phaseolus vulgaris* L.
Cucumis sativus L.	*Phlox drummondii* Hook.
Cymbopogon citratus (DC.) Stapf.	*Phyllanthus urinaria* L.
Cynodon dactylon (L.) Pers.	*Pisum sativum* L.
Cyperus brevifolis (Rottb.) Hassk.	*Poa annua* L.
C. compressus L.	*Portulaca oleracea* L.
C. deformis L.	*Ranunculus pusillus* Poir.
C. iria L.	*Rungia parviflora* Nees
C. pilosus Vahl.	*Saccharum officinarum* L.
C. procerus Rottb.	*Sacciolepsis indica* (L.) Chase
C. pulcherrimus Willd. ex Kunth	*Scirpus articulatus* L.
C. rotundus L.	*Scoparia dulcis* L.
Desmodium triflorum (L.) DC.	*Setaria italica* (L.) Beauv.
Digitaria longiflora (Retz.) Pers.	*Solanum melongena*
D. sanguinalis (L.) Scop.	*S. nigrum* L.
Echinochloa colonum (L.) Link	*S. sisymbriifolium* L.
E. crusgalli (L.) Beauv.	*Sorghum bicolor* (L.) Moench
Eclipta prostrata l.	*Sphaeranthus africanus* L.
Eleusine coracana (L.) Gaertn.	*Sphenoclea zeylanica* Gaertn.
E. indica (L.) Gaertn.	*Spinacea oleracea* L.
Eragrostis gangetica (Roxb.) Steud.	*Stellaria media* (L.) Cyrillo
E. plumosa Link	*Trifolium repens* L.
Euphorbia hirta L.	*Triticum aestivum* L.
Fimbristylis miliacea (L.) Vahl	*Urena lobata* L.
F. podocarpa Nees	*Vandellia* sp.
Fuirena glomerata Lam.	*Vernonia cinerea* (L.) Less
Glycine max (L.) Merr.	*Vicia faba* L.
Gnaphalium purpureum L.	*Vigna mungo* (L.) Heppner
Grangea maderaspatana Poir.	*V. radiata* (L.) Wilcz.
Hedyotis diffusa Willd.	*V. unguiculata* (L.) Walp.
Herminium sp.	*Zea mays* L.

Sources: Birchfield (1965); Buangsuwon *et al.* (1971); Manser (1971); Roy (1977a,b); Yik and Birchfield (1979); MacGowan and Langdon (1989).

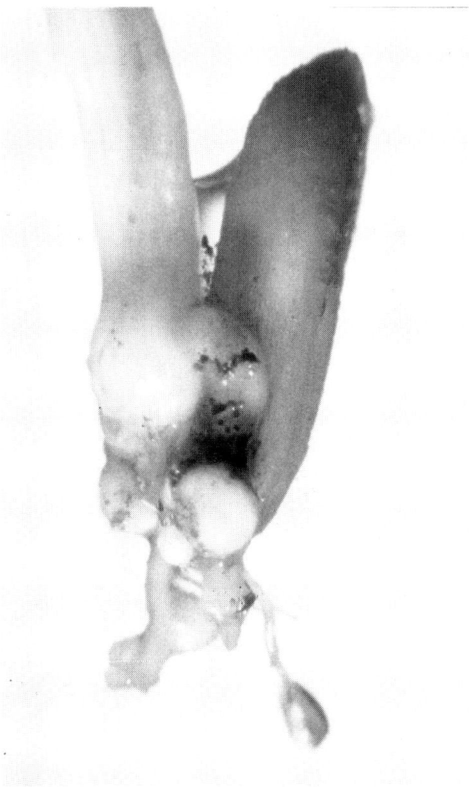

Fig. 4.8. *Meloidogyne graminicola* root galls on rice seedlings. (Photo: R.A. Plowright.)

Management measures

The recommended control of *Meloidogyne* on rice depends on the species. Flooding of soil even for relatively short periods will control or alleviate damage caused by *M. incognita*, *M. javanica* and *M. arenaria* and probably *M. salasi*, but continuous flooding would be necessary for *M. oryzae* and *M. graminicola*. Increasing soil fertility can compensate for some damage by the nematodes (Diomandé, 1984). Resistant cultivars hold out the most promise for effective and economic control, and some resistance to the different species has been found. Chemical control on the field scale is generally uneconomical particularly with low-yielding upland rice, but could be an economical proposition for nursery soils.

FLOODING. *M. incognita*, *M. javanica* and *M. arenaria* are not important parasites of lowland rice except in nursery seedlings, and can be controlled by flooding where this is possible. Although *M. oryzae* can survive some flooding, it can be controlled at depths greater than 10 cm (Segeren-V.d. Oever and Sanchit-Bekker, 1984). It is mainly a problem in the elevated areas of flooded rice fields where levelling is poor. *M. graminicola* will survive normal flooding, but damage to the crop can be avoided by raising rice seedlings in flooded soils thus preventing root invasion by the nematodes (Bridge and Page, 1982). Continuous flooding is highly effective in controlling *M. graminicola* in Vietnam (Kinh *et al.*, 1982). Similarly, in the Philippines, yield losses due to *M. graminicola* may be prevented or minimized when the rice crop is flooded early and kept flooded until a late stage of development (Soriano *et al.*, 2000).

There has been an increase in the incidence of *M. graminicola* damage in rice fields in the Philippines related to a decrease in the availability of water for agricultural use, and the nematode is mainly found in non-permanently flooded fields (Prot, 1994a; Prot *et al.*, 1994). Flooding the soil for 3 weeks prior to transplanting to control weeds has often been replaced with the more economical use of direct wet seeding in saturated but not flooded soils. Also, farmers are more likely to use intermittent rather than continuous flooding to save water. Both of these water management activities allow juveniles of *M. graminicola* to parasitize roots from the soil, which they are unable to do in continuously flooded conditions (Prot, 1994a). Puddling of soil prior to transplanting and prolonged early flooding reduces populations of both *M. graminicola* and *M. triticoryzae* in rice fields in India (Garg *et al.*, 1995).

RESISTANCE. A number of rice cultivars and breeding lines have been recorded as resistant to *Meloidogyne* species, although only a small number of these are truly resistant. Diomandé (1984) found that cultivars of *O. glaberrima* were resistant to *M. incognita*. Generally, cultivars of *O. sativa* were susceptible although some improved cvs, IRAT 109, IRAT 112, IRAT 133, IRAT 106, and a traditional cv., CG-18, also showed tolerance. Rice cvs IR 28, IR 459 and P24 are 'resistant' to *M. arenaria*, *M. javanica* and *M. incognita* (races 2, 3 and 4), and A95, Giza 171 and Giza 172 are 'resistant' to *M. incognita* (race 3) and *M. javanica* (Ibrahim *et al.*, 1983). The cvs IR 20, Ikong Pao Faro 21 and 27 support low populations of *M. incognita* in Nigeria (Babatola, 1980; Fademi, 1987). *O. glaberrima* is resistant to *M. graminicola*, and some progeny from interspecific crosses with *O. sativa* appear to be less susceptible (Plowright *et al.*, 1999). The majority of *O. sativa* cultivars are susceptible to *M. graminicola*. For example, all 80 cultivars tested in Laos were found to be susceptible (Manser, 1971). However, there are a number of cultivars from India, Thailand and the USA that are reported to be resistant to *M. graminicola* (Roy, 1973; Jena and Rao, 1974, 1976; Prasad *et al.*, 1979, 1986b; Yik and Birchfield, 1979; Chunram, 1981; Rao *et al.*, 1986b). It has been shown that tolerance levels of rice cultivars to *M. graminicola* are affected by whether the crop is grown in upland or flooded conditions (Tandingan *et al.*, 1996).

CROP ROTATION. Certain crops are resistant or poor hosts of *M. graminicola* and could be used in rotation to reduce nematode populations, e.g. castor, cowpea, sweet potato, soybean, sunflower, sesame, onion, turnip, *Phaseolus vulgaris*, jute and okra (Rao *et al.*, 1986a). Soil populations of *M. graminicola* are reduced when rice is preceded by the planting of mustard (*Brassica campestris* subsp. *oleifera*) and guzitil (*Guizotia abbysinica*) in Bangladesh (Rahman, 1990). Long rotations, greater than 12 months, will be needed to reduce *M. graminicola* soil populations to low levels. Introducing a fallow into the rotation will also give control of the nematodes but, to be effective, it needs to be a bare fallow free of weed hosts (Roy, 1978) and is therefore impractical in most circumstances. However, one weed, *Eclipta alba*, is toxic to *M. graminicola* and could be grown and incorporated into the field soil to kill the nematodes (Prasad and Rao, 1979b).

SOIL AMENDMENTS. The use of decaffeinated tea waste and water hyacinth compost has been suggested to control *M. graminicola* (Roy, 1976b), and some reduction in populations is reported following the incorporation of other chopped 'botanicals', *Polygonum*, *Ageratum*, *Mikania* and also water hyacinth (Das *et al.*, 1999).

SOIL SOLARIZATION. The method can be effective on a small scale, such as on nursery beds. It uses clear polyethylene sheets which are laid on the surface of the beds for a period of 3–4 weeks in sunny weather. It can give a reduction of populations of *Meloidogyne* spp. in rice beds of over 80% and improve seedling growth (Ganguly *et al.*, 1996).

CHEMICALS. Seed treatments, root dips, soil drenches and soil incorporation have been tested in experimental trials with varying success in India (Rao *et al.*, 1986a; Rahman and Das, 1994), but their practical and economic applicability have not been determined. Carbofuran and diazinon gave effective control of *M. graminicola* in Vietnam when applied to irrigation water (Kinh *et al.*, 1982), but this means of application has many dangers. Seed treatment with neem-based pesticides can reduce populations and damage caused by *M. graminicola* (Das and Deka, 2002).

Diagnosis

The presence and populations of *Meloidogyne* in rice roots can be determined by standard root staining techniques (Chapter 3). Root extractions will only isolate hatched juveniles and males, and a combination of root maceration and staining of a known weight of roots can be a more efficient and practical way of determining populations of sedentary females within roots. Assessing the severity of root damage by the amount of galling (root knot index) is a practical and speedy method, but can be difficult with rice. One useful rating system is to rate only the percentage of affected large roots with the root tip galls characteristic of *Meloidogyne* on rice (Diomandé, 1984) and, using this system, a rice root knot rating chart has been devised making use of the actual percentage of roots galled to determine and rate the extent of damage caused by *M. graminicola* on rice (Fig. 4.9), but could also be used for other root knot species.

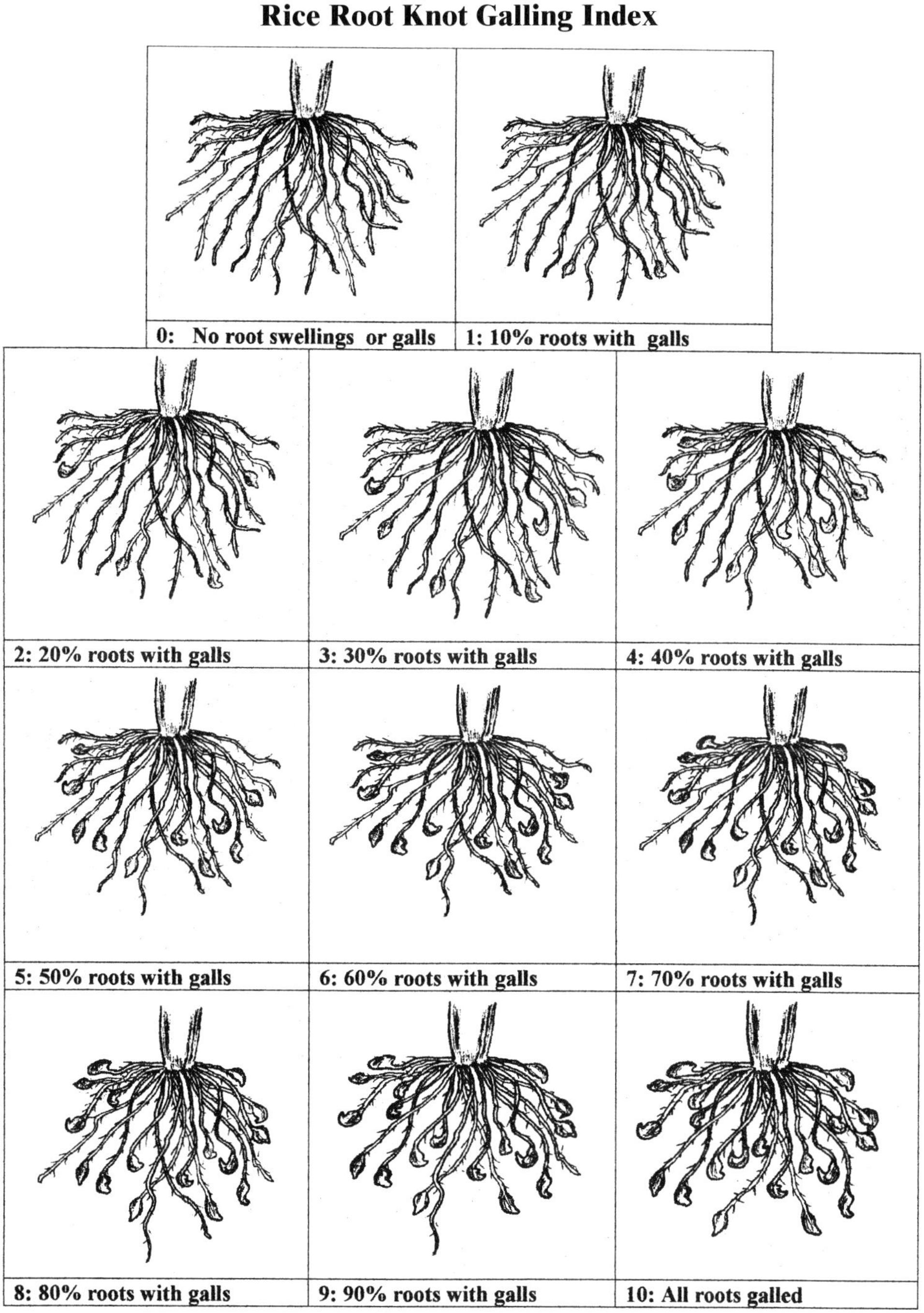

Fig. 4.9. Root knot rating chart to determine damage caused to rice roots by *Meloidogyne graminicola*. (J. Bridge.)

Hirschmanniella

A number of *Hirschmanniella* species, known collectively as rice root nematodes, are parasites of irrigated, lowland and deepwater rice (Table 4.1). They are found in flooded fields and occur in the majority of rice-growing regions. They are reported on rice from China, India (including Sikkim), Nepal, Pakistan, Bangladesh, Sri Lanka, Korea, Japan, the Philippines, Vietnam, Egypt, West Africa, Brazil, Portugal and, most recently, Iran (Minassian and Barooti, 1997). The most commonly recorded species is *H. oryzae*, but there was a tendency in the early literature for all *Hirschmanniella* spp. found in rice roots to be grouped under the name *H. oryzae* (Taylor, 1969). Seven species are reported to damage rice (*H. belli*, *H. gracilis*, *H. imamuri*, *H. mexicana* (= *H. caudacrena*), *H. mucronata*, *H. oryzae* and *H. spinicaudata*) (Table 4.1), whilst a further 12 species have been found in rice roots (*H. diversa*, *H. dubia*, *H. indica*, *H. kaverii*, *H. magna*, *H. mangaloriensis*, *H. marina*, *H. microtyla*, *H. nghetinhiensis*, *H. ornata*, *H. shamimi* and *H. thornei*). Seventeen species are known from rice in China alone (Li, 1987; Zhang, 1987; Wu *et al.*, 1995; Gao *et al.*, 1999; Wang and Pan, 1999; Liao *et al.*, 2000). Four species have been recorded from weeds in rice fields (*H. asteromucronata*, *H. furcata*, *H. obesa* and *H. truncata*).

Symptoms of damage

There are no easily identifiable above-ground symptoms of nematode damage in the field. Retardation of growth rate occurs especially in early growth, with a decrease in tillering. Yellowing of rice plants is observed occasionally (Plate 2B), and flowering can be delayed by up to 14 days. Roots invaded by *Hirschmanniella* spp. turn yellowish brown and rot (Van der Vecht and Bergman, 1952; Mathur and Prasad, 1972b; Muthukrishnan *et al.*, 1977; Babatola and Bridge, 1979; Fortuner and Merny, 1979; Hollis and Keoboonrueng, 1984; Khuong, 1987; Ichinohe, 1988).

Biology

Hirschmanniella spp. are migratory endoparasites of roots (Fig. 4.10, Plate 2C). The nematodes produce cavities and channels through the cortex which become

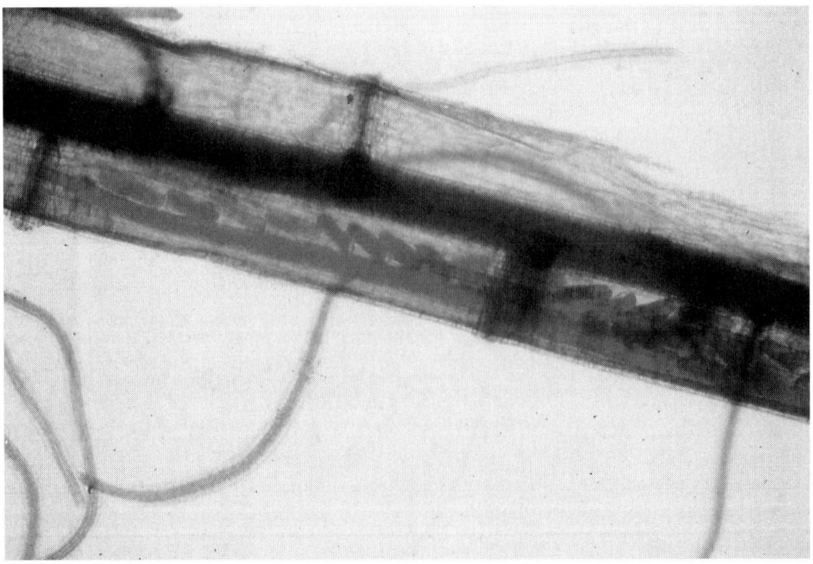

Fig. 4.10. *Hirschmanniella oryzae* female and eggs in roots of rice. (Photo: J. Bridge.)

necrotic for some distance into the root (Van der Vecht and Bergman, 1952; Mathur and Prasad, 1972b; Lee and Park, 1975; Babatola and Bridge, 1980; Hollis and Keoboonrueng, 1984).

Eggs of *H. oryzae* are deposited in the roots a few days after invasion, and hatching occurs 4–6 days after deposition (Van der Vecht and Bergman, 1952; Mathur and Prasad, 1972a). The life cycle is of variable length. In north India, it is suggested that there is only one generation of *H. oryzae* a year (Mathur and Prasad, 1972a); in Japan, two generations (Kuwahara and Iyatomi, 1970; Ou, 1985); and in Senegal, three generations (Fortuner and Merny, 1979). Maximum root populations occur between tillering and heading of the rice crop (Kuwahara and Ivatomi, 1970; Fortuner and Merny, 1979).

Survival and means of dissemination

H. oryzae survives between crops in weeds and other hosts (Table 4.3), in ratooning rice roots and in undecayed roots of rice stubble (Mathur and Prasad, 1973b; Feng, 1986; Ichinohe, 1988). *Hirschmanniella* spp. can also survive in soil. They survive longer in roots than in soil, but survival of root populations is shorter in flooded soil due to the more rapid decay of roots. Populations of *H. oryzae* decrease slowly in wet rice fields in the absence of a host, surviving for at least 7 months (Park *et al.*, 1970), and are eradicated after 12 months

Table 4.3. Hosts of *Hirschmanniella* spp. parasitic on rice.

Weeds	*F. miliacea* (L.) Vah
Ageratum congzoides	*Hydrolea zeylanica* (L.) Vahl[a]
Alternanthera sessilis R. Br	*Ischaeum rugosum* Salisb.
A. philoxenoides	*Ixeris denticulata*
Astragalus sinicus L.	*Leonurus artemisia*
Bidens bipinata	*Lindernia antipoda* (L.) Alston
Boerhavia diffusa[a]	*Ludwigia perennis* L.
Brachiaria ramosa (L.) Stapf[a]	*Mnesithia laevis* (Retz.) Kunth
Cleochars yokiscens	*Monochoria hastata* (L.) Solms[a]
Coryza canadesis	*M. vaginalis* (Burm. f.) Presl
Crogophora sp.[a]	*Nelumbo nucifera* Gaerm.
Cyperus difformis L.[a]	*Polygonum plebejum*[a]
C. elatus L.	*P. hydropiper*
C. nutans Vahl	*Scirpus articulatus* L.
C. iria L.	*Sesbania aculeata* L.[a]
C. procerus Rottb	*Sporobolus indicus*
C. pulcherrimus Willd. ex Kunth	*Vallisneria spiralis* L.
C. rotundus L.[a]	
Digitaria sanguinalis	**Crops**
Echinochloa colona (L.) Link[a]	*Oryza sativa* L.[a]
E. crus-galli (L.) Beauv.[a]	*Abelmoschus esculentus* (L.) Moench.
Eclipta alba (L.) Hassk.[a]	*Gossypium hirsutum* L.
E. prostrata	*Hordeum vulgare* L.
Eichhornia crassipes (Mart.) Solms	*Lycopersicon esculentum* Mill.
Eleocharis spiralis (Rottb.) Roem & Schult.[a]	*Pennisetum typhoides* (Burm. f.) Stapf & Hubbard
Eleusine indica (L.) Gaertn.[a]	*Saccharum officinarum* L.
Eragrostis pilosa (L.) Beauv.	*Triticum aestivum* L.
Fimbristylis ferruginea (L.) Vahl[a]	*Zea mays* L.
F. globulosa (Retz.) O. Kuntze	

[a]Plants supporting high nematode populations.
Sources: Van der Vecht and Bergman (1952); Kawashima (1963); Yamsonrat (1967); Mathur and Prasad (1973b); Babatola (1979); Mohandas *et al.* (1979); Venkitesan *et al.* (1979); Razjivin *et al.* (1981); Edward *et al.* (1985); Khuong (1987); Kumar (1990); Gao *et al.* (1998a, 1999).

(Fortuner and Merny, 1979). In dry conditions, survival is enhanced by quiescence (Fortuner and Merny, 1979), e.g. *H. oryzae* can survive for longer than 12 months in soils that are not continually wet (Muthukrishnan *et al.*, 1977). *H. oryzae*, *H. imamuri* and *H. spinicaudata* have also been shown to survive in anaerobic conditions over a wide range of pH (Babatola, 1981). In fallow field soil, populations of *H. oryzae* can survive high temperatures of 35–45°C and low temperatures of 8–12°C (Mathur and Prasad, 1973b).

Hirschmanniella is spread in irrigation and flood water, and in soil adhering to implements and field workers. Where there is a long history of rice cultivation, the nematodes are likely to be widespread. In Japan, for example, virtually every rice paddy is infested with either *H. imamuri* or *H. oryzae* (Ichinohe, 1988). The nematodes are also disseminated to the field in roots of rice seedlings from nurseries. *Hirschmanniella* spp. are unusual nematodes, being perfectly adapted to constant flooding (Fortuner and Merny, 1979).

Alternative hosts

Hirschmanniella spp. are parasites of a considerable number of rice field weeds (Van der Vecht and Bergman, 1952) mainly of the families Cyperaceae and Gramineae (Table 4.3). Few cultivated crops are hosts of *H. oryzae* in India (Mathur and Prasad, 1973b); however, some crop plants are hosts of *Hirschmanniella* spp. (Babatola, 1979).

Disease complexes

Necrotic areas develop around nematodes as they migrate and feed on cortical tissues, but diminish as nematodes penetrate deeper into the roots. This suggests a phoretic relationship between the rice root nematodes and soil microorganisms, as necrosis does not occur at all in the absence of these organisms (Babatola and Bridge, 1980). Similarly, 'root browning' of rice, caused mainly by soil microorganisms, is increased in the presence of *H. oryzae* (Lee and Park, 1975).

Economic importance

It has been estimated that *Hirschmanniella* spp. infest 58% of the world's rice fields, causing 25% yield losses (Hollis and Keoboonrueng, 1984). However, there are discrepancies in yield loss estimates around the world and suggestions that yield reductions occurring in the presence of *Hirschmanniella* are not always solely attributable to the nematodes. In Japan, for example, it has not always been possible to demonstrate high correlations between nematode population levels and yield reductions (Ichinohe, 1988). Similarly in the Côte d'Ivoire, where nematicide treatments against *H. spinicaudata* increased rice yields by 20–53%, there was no significant correlation between yields and nematode populations. The suggested explanation is that there is a bacteriological factor present which suppresses both nematodes and rice yields (Cadet and Quénéhervé, 1982). Contrasting evidence in Senegal in microplots has established that *H. oryzae* can cause a yield loss of 42% when fertilizers are not applied, with nematode populations at harvest of 3200–6000 nematodes/dm^3 of soil, and 5–30 nematodes/g of root. Even when rice is grown in the best conditions with adequate fertilizers, yield losses are 23%, with nematode populations at harvest of 1500–2500/dm^3 of soil and 90–410 nematodes/g of root (Fortuner, 1974, 1977, 1985).

Experiments with *Hirschmanniella* spp. have established varying degrees of yield loss. Inoculations of one and ten *H. oryzae*/g of soil caused 27 and 39.4% yield loss, respectively (Jonathan and Velayutham, 1987), and the numbers of panicles and grain weight were reduced by 16 and 32%, respectively, with a population level of 1200 *Hirschmanniella*/plant (Yamsonrat, 1967). *H. imamuri*, *H. oryzae* and *H. spinicaudata* reduced yields by 31–34.3% at population levels of 1000 nematodes/plant or 500 nematodes/dm^3 of soil (Babatola and Bridge, 1979). The yield of plants inoculated with 5000 *H. mucronata*/plant at 1 and 40 days was reduced by 50.6 and 45.6%, respectively (Panda and Rao, 1971). *H. oryzae* populations of 100/plant reduced grain yield by 35% (Mathur and Prasad, 1972b). In microplots, natural populations of 29–68 *H. oryzae*/500 cm^3 of soil at transplanting reduced grain weight by 13.8–19.2% (Venkitesan *et al.*, 1979).

In Vietnam, economic damage by *Hirschmanniella* spp. occurs when 40 or more nematodes are present in a rice hill 1 week after transplanting; equivalent after multiplication to 800 nematodes per hill at heading (Khuong, 1987). Yield losses caused by *Hirschmanniella* spp. are influenced by soil fertility (Fortuner and Merny, 1979), age of plant when infected (Panda and Rao, 1971), number of crops and flooding (Khuong, 1987), and seasonal climatic conditions (Mathur and Prasad, 1972b).

Management measures

Management of *Hirschmanniella* spp. has been achieved or recommended by various practices, in particular fallow, weed control, use of 'resistant' cultivars, rotation with non-host plants, chemical soil treatment of nurseries and fields, and chemical root dipping and seed coating.

CULTURAL PRACTICES. Yield losses due to *Hirschmanniella* spp. are greater in poor soils. It is, therefore, possible to reduce yield losses by improving the nutritional status of the soil (Mathur and Prasad, 1972b). Nematode populations decline in the absence of host plants, but a considerable percentage can survive depending on environmental conditions (Van der Vecht and Bergman, 1952; Mathur and Prasad, 1973a; Muthukrishnan *et al.*, 1977). Prolonged fallows might control *Hirschmanniella*, but the evidence suggests that fallows would need to be at least 12 months in wet conditions and longer in dry. They would also need to be free of other crop and weed hosts. The management of weeds, which are generally good hosts, will reduce nematode populations both in the absence of rice and during growth of the crop. Time of transplanting can be important and, in the Punjab, there is less build-up of *H. oryzae* when basmati rice is transplanted later in mid-July compared with mid-June (Randhawa *et al.*, 1991).

Rotation of crops is not possible in continuous rice cropping, but is often normal practice where a single wet season rice crop is followed by dry season crops. In fields with a single rice crop, populations of *Hirschmanniella* are always low in some localities (Khuong, 1987). This is due to a combination of dry soil and non-host dry season crops such as cowpea, pigeonpea, soybean, groundnut (peanut), sweet potato, sorghum, finger millet, tobacco, cabbage and onion against *H. oryzae*, *H. imamuri* and *H. spinicaudata* (Mathur and Prasad, 1973b; Babatola, 1979; Gao *et al.*, 1998a,b) and millet, cotton and wheat against *H. oryzae* in India (Mathur and Prasad, 1973b). Any of these or other non-host crops in rotation with rice should reduce the risk of *Hirschmanniella* damage, but their host status may vary with different nematode species.

Three green manure legume crops, *Sesbania rostrata*, *Sphenoclea zeylanica* and *Aeschynomene afraspera*, can give good, practical control with the additional benefit of increased soil nitrogen (Mohandas *et al.*, 1981; Germani *et al.*, 1983; Hendro *et al.*, 1992; Prot, 1992). The yield of rice following *Sesbania* was increased by 214% in micro plots compared with repeated rice cropping. *Sphenoclea* can give 99% control of *Hirschmanniella* spp. *Sesbania* appears to act as a trap crop (Germani *et al.*, 1983), while *Sphenoclea* produces toxic plant exudates (Mohandas *et al.*, 1981). Unfortunately, *S. rostrata* is a very good host of the rice root knot nematode, *M. graminicola*, and it should be used with caution for the management of rice nematodes (Prot, 1994b).

Oil cakes used as organic amendments, particularly those of castor (*Ricinus communis*) and neem, can significantly reduce populations of *H. oryzae* (Jonathan and Pandiarajan, 1991; Khan and Shaukat, 1998).

Other cultural measures to alleviate damage by *Hirschmanniella* spp. in Japan are: (i) early planting and (ii) direct sowing, which both reduce initial infection (Sato *et al.*, 1970; Nakazato *et al.*, 1964 quoted in Fortuner and Merny, 1979).

RESISTANCE. The majority of rice cultivars tested are good hosts of *Hirschmanniella* spp. These include cultivars from India, Korea, Japan, Nigeria, El Salvador, Iraq, Ecuador, Thailand and Vietnam. In Korea, all 270 cultivars tested were susceptible to *H. oryzae*, although six supported only low numbers (Park *et al.*, 1970). Cultivars supporting relatively low nematode numbers have been rated as 'resistant' (Arayaungsarit *et al.*, 1986; Rao *et al.*, 1986a). Some of these could be truly resistant, such as cv. TKM9 to *H. oryzae* from India (Ramakrishnan *et al.*, 1984). Because of the widespread occurrence of *Hirschmanniella* in rice fields, for example from all locations in Thailand (Yamsonrat, 1967) and virtually every rice paddy in Japan, it is possible that the rice cultivars which now grow best in paddies are those which are relatively resistant to, or tolerant of, *Hirschmanniella* spp. (Ichinohe, 1988).

CHEMICAL. High yield increases have been achieved using chemicals against *Hirschmanniella*, but there is little indication that chemical control is economical or practical except in special circumstances (Ichinohe, 1972).

Most of the available chemicals with nematicidal action have been applied with varying success against *Hirschmanniella*, especially in India (Edward *et al.*, 1985; Rao *et al.*, 1986a), and also in Japan (Ichinohe, 1988), Thailand (Taylor, 1969) and Côte d'Ivoire (Cadet and Quénéhervé, 1982). Chemical control has been attempted by application to field and nursery soil, as root dips and for soaking seeds. In field soil, various methods of application have been tried, including soil incorporation, application in standing water and 'mud ball' application (Prasad *et al.*, 1986a). Bare root dips for transplanted seedlings in a range of nematicides can reduce *H. oryzae* populations and increase yields (Lahan *et al.*, 1999).

Heterodera species

Four main cyst nematodes infecting rice are *Heterodera oryzicola*, *H. elachista*, *H. oryzae* and *H. sacchari*. *H. oryzicola* is found only on upland rice in Kerala State, India (Rao and Jayaprakash, 1978), and *H. elachista* specifically on upland rice in Japan (Okada, 1955). *H. oryzae* occurs on lowland rice in parts of the Côte d'Ivoire, Senegal (Fortuner and Merny, 1979), in Bangladesh (Page and Bridge, 1978), in Nepal (Sharma *et al.*, 2001) and Iran (Pedramfar *et al.*, 2001). *H. sacchari* occurs on upland and flooded rice throughout West Africa (Côte d'Ivoire, Ghana, Guinea, Benin, Togo, Nigeria and Liberia) (Babatola, 1984; Lamberti *et al.*, 1991; Coyne *et al.*, 1996, 1999; Coyne and Plowright, 2000). The Japanese *Heterodera* sp., first referred to by Okada (1955), was attributed to *H. oryzae* until being described as *H. elachista* by Ohshima (1974). *H. mothi* and *H. graminis* have also been found in fields in Nepal cropped with rice and wheat (Sharma *et al.*, 2001) and, in the same cropping system in India, another species described as *H. skohensis* has been found with the rice crop (Kaushal *et al.*, 2000). Rice is reported to be a good host of *H. sorghi* (Srivastava and Sethi, 1987).

Symptoms

The symptoms of infection by each species are similar. Root growth is suppressed, and infected roots turn brown or black. Lemon-shaped white females and brown cysts can be observed protruding from infected roots (Fig. 4.11, Plate 2D). Rice responds to *H. sacchari* by the proliferation of secondary roots, which have a compensatory function (Babatola, 1983a), but generally the reduced size and function of cyst nematode-infected roots leads to leaf chlorosis and slowed plant growth and development, i.e. stunting and reduced tillering. Seedlings are usually more vulnerable, and Jayaprakash and Rao (1984) have observed seedling death in patches heavily infested by *H. oryzicola*.

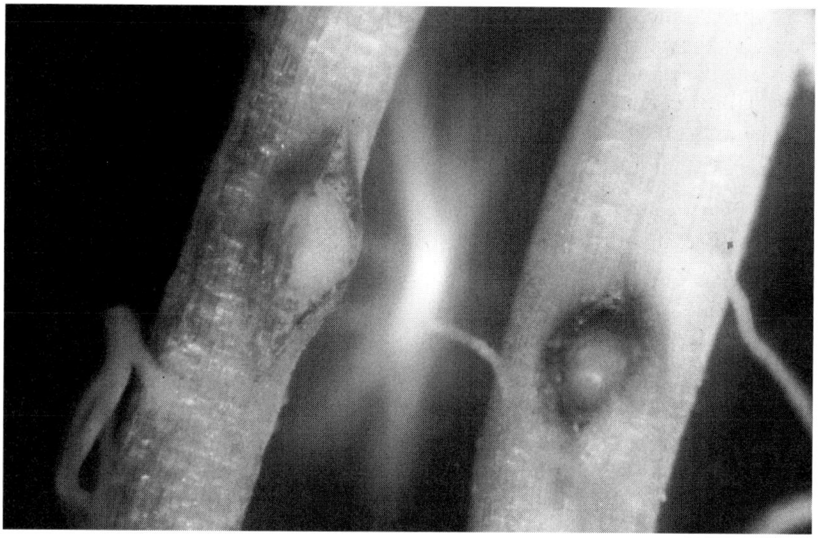

Fig. 4.11. *Heterodera oryzicola* cyst and white female emerging from roots of rice. (Photo: R.A. Plowright.)

Biology

H. oryzicola and *H. elachista* are parasites of upland rice, and *H. sacchari* is damaging only in upland rice (Babatola, 1983a) although it is also found in flooded conditions. *H. oryzae* differs by its adaptation to flooding, and second stage juveniles of *H. oryzae* can survive better in anaerobic than in aerobic water (Reversat, 1975).

The biology is as described in Chapter 2. Females of *H. oryzicola*, *H. elachista* and *H. oryzae* deposit many eggs into a large egg sac attached to the vulval cone. Juveniles in egg sacs hatch freely in water, but there is evidence that exudates from actively growing roots are required to stimulate hatch from cysts of *H. oryzicola* (Jayaprakash and Rao, 1982b) and *H. oryzae* (Merny, 1966). These differences in hatching behaviour indicate that J2s from later generation egg sacs invade rice during crop growth and that cysts are principally a means of survival (Plate 2E). In contrast, *H. sacchari* rarely has an egg sac and eggs hatch freely in water. *H. sacchari* also differs from the other rice cyst nematodes as it is a parthenogenetic triploid, the others being amphimictic. The life cycle of each species is complete in 24–30 days, which allows multiple generations depending on the duration of the crop; *H. oryzicola* is said to have 12 generations/year in continuous rice, while *H. oryzae*, *H. elachista* and *H. sacchari* have 2–3 generations per crop (Berdon and Merny, 1964; Merny, 1966, 1972; Netscher, 1969; Netscher *et al.*, 1969; Nishizawa *et al.*, 1972; Shimizu, 1977; Jayaprakash and Rao, 1982a; Sharma and Swarup, 1984). *H. oryzicola* is dependent on rice root diffusates to induce substantial egg hatch; this is not the case with *H. sacchari*, which will hatch in water (Ibrahim *et al.*, 1993).

Alternative hosts

H. oryzicola and *H. oryzae* have a narrow host range, with many wild and cultivated Gramineae being non-hosts (Merny and Cadet, 1978; Sharma and Swarup, 1984). *H. oryzicola* has some weed hosts, e.g. *Cynodon dactylon* and *Brachiara* sp. (Charles and Venkitesan, 1985); and some Cyperaceae, e.g. *Mariscus umbellatus* and *Kyllinga monocephala*, are hosts of *H. oryzae* and *H. oryzicola* (Merny and Cadet, 1978; Charles and Venkitesan, 1990). Surprisingly, banana is a good host of both nematodes (Taylor, 1978; Charles and Venkitesan, 1985, 1990). In this respect, *H.*

sacchari is again quite distinct as it has a wide host range, including many wild Cyperaceae and Gramineae indigenous of West African savannah and humid lowlands (Odihirin, 1975).

Economic importance

Because of their restricted distribution, cyst nematodes on rice are largely of local importance. Shimizu (1977) noted that damage by *H. elachista* varied between years, and this is likely to be true for the other species, as local climatic and edaphic factors, and cultural practices vary (see Coyne *et al.*, 1998). Shimizu (1971) considered that *H. elachista* was important in later growth (presumably grain filling and maturation) and could decrease yield by 7–19%. In India, higher yield losses (17–42%) are attributed to *H. oryzicola* (Kumari and Kuriyan, 1981). *H. oryzae* is a minor problem in Senegal and Côte d'Ivoire, and is replaced by *H. sacchari* in mixed populations; its importance on rice crops in Bangladesh requires assessment. *H. sacchari* populations in Côte d'Ivoire increased rapidly with intensive wet season rice cropping, leading to yield losses of 50% (Coyne and Plowright, 1998). Coyne and Plowright (2000) demonstrated that such losses in *O. sativa* were correlated with barely detectable pre-sowing nematode population densities.

Management measures

CULTURAL PRACTICES. Exploiting the narrow host range of *H. oryzicola*, *H. elachista* and *H. oryzae* through rotation with non-host crops is likely to be beneficial, e.g. rotation with soybean or sweet potato to control *H. elachista* has given yield improvements of 2.8- to 3.7-fold (Nishizawa *et al.*, 1972). However, the traditional, long fallow periods in forest and forest savannah of the Côte d'Ivoire did not clearly influence the prevalence of *Heterodera* (Coyne *et al.*, 1998). Experimentally, Coyne and Plowright (1998) controlled *H. sacchari* using solarization.

RESISTANCE. Rice cultivars vary in their susceptibility to *H. oryzae* (Merny and Cadet, 1978), *H. sacchari* (Babatola, 1983b) and *H. oryzicola* (Jayaprakash and Rao, 1983), but few have complete resistance. The African rice *O. glaberrima* is resistant to *H. sacchari* (Reversat and Destombes, 1998; Coyne *et al.*, 1999). The resistance is qualitative and inherited in progeny from interspecific crosses with *O. sativa* (Plowright *et al.*, 1999). Ashurst *et al.* (2001) and Amoussou (2002) showed that resistance was controlled by a single recessive gene identified as Hsa-10g by Lorieux *et al.* (2003). Microsatellite markers have been found linked to resistance and the resistance gene (Amoussou, 2002; Lorieux *et al.*, 2003).

Cultivars of *O. sativa* rarely have multiple nematode resistance; cvs Lalnakanda, CR143-2-2 and TKM6, although resistant to *H. oryzicola*, are susceptible to *M. graminicola* (Prasad *et al.*, 1986c). *O. glaberrima*, on the other hand, is resistant to both *H. sacchari* and *Meloidogyne* spp., but resistance is under different genetic control and is not inherited equally by the progeny of interspecific crosses with *O. sativa* (Plowright *et al.*, 1999).

Pratylenchus species

Ten species of root lesion nematodes have been reported on rice throughout the world. The most common are *Pratylenchus zeae*, found in Africa, North, Central and South America, Australia, South and South-east Asia and Egypt, and *P. brachyurus*, reported from Africa, South America, Pakistan and the Philippines. They occur predominantly on upland rice, and only *P. zeae* and *P. indicus*, a species found in India and Pakistan, have been reported to cause damage.

Symptoms

There are generally no specific aboveground symptoms of infection by *P. zeae* (Plowright *et al.*, 1990). However, the leaves of 22-day-old rice seedlings infected with *P. indicus* are said to yellow from the tip, wilt and dry up (Rao and Prasad, 1977). *Pratylenchus* spp. cause discrete lesions in the root cortex which become necrotic and coalesce as infection spreads. Root size and function are diminished, growth rate (either tillering or shoot extension) is reduced and plants become stunted. *Pratylenchus* sp. is said to be associated with a disease known as entorchamiento in Colombia; the symptoms are stunted growth, twisting and yellowing of leaves and proliferation of deformed secondary roots (Pardo and Munoz, 1994).

Biology

Population levels of *P. indicus* decline rapidly during the fallow periods and persist in low numbers (Prasad and Rao, 1978a). *P. zeae* can survive in a cultivated clean fallow for up to 6 months (Plowright *et al.*, 1989). Weed hosts of *P. zeae* are *Cynodon dactylon*, *Amaranthus spinosus*, *Dactylodenium aegyptium*, *Digitaria sanguinalis* and *Echinochloa* sp. (Fortuner, 1976).

Invasion by *P. zeae* takes place within 1 week of emergence, the life cycle being completed in about 30 days. *P. indicus* completes a life cycle in 33–34 days, and several overlapping generations occur on a single crop (Prasad and Rao, 1982a). The optimum temperature for *P. indicus* reproduction is 23–30°C, and peaks of population are always preceded immediately by rainfall (Prasad and Rao, 1979a). During crop growth, *P. zeae* is found mainly in rice roots, and soil population levels are generally low. Plowright *et al.* (1990) found that the rate of *P. zeae* reproduction was greatest after flowering, and numbers increased towards grain maturity. *P. zeae* migrates into soil from heavily infected necrotic roots (Plate 2F). *Pratylenchus* spp. are readily disseminated in soil and infected root material.

Economic importance

Despite the prevalence of *P. zeae* in upland rice, there is very little information on its pest status. However, in South-east Asian upland rice ecosystems, *Pratylenchus* spp. together with *Meloidogyne* spp. are potentially the most economically important nematode pests (Prot *et al.*, 1996). Plowright *et al.* (1990) have shown that rice yield can be increased 13–29% by control of *P. zeae*, but some cultivars may be tolerant of infection. The maximum yield reduction in the field was 30% with an infection of 1000 *P. zeae*/g of root at harvest, and higher nematode densities at harvest will not necessarily cause further yield loss. Martin (1972) reported that the growth of rice infected with more than 500 *Pratylenchus* sp. (probably *P. zeae*)/g of root was poor, and severely stunted plants had more than 3500 nematodes/g of root. Prasad and Rao (1978b) found that the yield of rice cv. Bala was reduced by 33% at final population densities of *P. indicus* up to 1625/g of root. The data suggest that *P. zeae* and *P. indicus* can cause yield loss in upland rice, but further studies are required.

Management measures

P. zeae can be managed effectively using chemicals, e.g. carbofuran (Plowright *et al.*, 1990; Sahoo and Sahu, 1993a). However, chemical control is undesirable in upland rice and requires economic appraisal. Control through crop rotation has been reported using poor or non-host crops such as *Vigna radiata* (L.) Wilczek (mung bean), *Vigna mungo* (L.) Hepper (black gram), *Vigna unguiculata* (L.) Walp (cowpea) and *Sesamum indicum* L. (sesame) (Prasad and Rao, 1978a). The yield of rice after rice in fields heavily infested with *P. zeae* was 37% lower than the yield of rice after cowpea, but two successive croppings with resistant legume crops are necessary to reduce nematode populations to a low level, and this rotation will protect only one rice crop from the nematode (Aung and Prot, 1990). However, *P. zeae* has a wide host range, and many of the food crops (mainly cereals) grown in upland rice cropping systems are good hosts (Table 4.4), as also are the many weeds and wild

Table 4.4. Some important hosts of *Pratylenchus zeae*.

Oryza sativa L.	*Vigna unguiculata* L. (Walp)
O. glaberrima Steud	*Lycopersicon esculentum* Mill
O. breviligulata A. Chev & Rocchr	*Ipomoea batatas* (L.)
Eleusine coracana (L.) Gaertn	*Glycine max* (L.) Merr
Sorghum bicolor (L.) Moench	*Arachis hypogaea* L.
Zea mays L.	*Saccharum* spp.
Triticum aestivum	*Solanum tuberosum* L.
Avena sativa L.	*Allium cepa* L.
Hordeum vulgare L.	*Lactuca sativa* L.
Secale cereale L.	*Nicotiana tabacum* L.
Amaranthus sp.	*Gossypium* spp.

rices found in upland rice fields (Sahoo and Sahu, 1993b). Fallow periods of a practical length will reduce but not eliminate damage by *P. zeae* to susceptible, intolerant cultivars.

Differences in susceptibility of rice cultivars and accessions to *P. zeae* (R.A. Plowright and D. Matias, unpublished data) and *P. indicus* (Prasad and Rao, 1982b) have been found, but no useful field resistance has yet been identified. Upland rice cultivars appear to differ in their tolerance of *P. zeae* (Plowright *et al.*, 1990); if this is a reliable and hereditable trait, then it will be useful for alleviating yield loss.

Criconemoides (*Criconemella*) and *Criconema*

Criconemoides (= *Criconemella*) spp. (*C. annulatus*, *C. curvatus*, *C. incisus*, *C. informis*, *C. obtusicaudatus*, *C. onoensis*, *C. ornatus*, *C. oryzae*, *C. palustris*, *C. paragoodeyi*, *C. rusticus*, *C. sphaerocephala* and *C. tescorum*) and *Criconema crassianulatum*, *C. corbetti*, *C. jaejuense* and *C. cardamomi* occur on upland and flooded rice in various areas of the world (Fortuner and Merny, 1979; Fortuner, 1981; De Waele and Van den Berg, 1988; Berg *et al.*, 1989; Lamberti *et al.*, 1991; Choi and Geraert, 1994; Sperandio and Amaral, 1994; Pedramfar *et al.*, 2001; Sharma *et al.*, 2001; Coyne and Plowright, 2002), but only *C. onoensis* has been shown to be harmful (Hollis and Keoboonrueng, 1984). *C. onoensis* is known to occur on rice in the USA, Guinea, Côte d'Ivoire, Mauritius, Surinam, Belize and India (Luc, 1970; Maas, 1970; Baclri, 1978; Hollis and Keoboonrueng, 1984; Chinappen *et al.*, 1988).

In flooded rice fields, *C. onoensis* causes no obvious symptoms but, in pot tests, the presence of 210 nematodes/dm^3 of soil can cause severe stunting and yellowing of plants (Hollis, 1977). Parasitized main and secondary roots are stunted, with lesions near club-shaped root tips. *C. onoensis* is ectoparasitic, feeding on or near root tips of both flooded and upland rice. In West Africa, *C. palustris* is more common than *C. onoensis* in flooded rice (Luc, 1970; Merny, 1970).

Dissemination of *C. onoensis* could result from transportation of infested soil and certainly by irrigation water in flooded rice. Survival is ensured by the presence of several permanent weed hosts belonging to the Cyperaceae and Gramineae such as *Cynodon dactylon*, *Paspalum hydrophilum*, *Cyperus iria*, *C. esculentus*, *C. haspan*, *C. articulatus*, *Fimbristylis milacea*, *Fuirena* sp. and *Eleocharis* spp. (Hollis, 1972a,b; Hollis and Joshi, 1976). Rice supports only low population densities because of root decay caused by early nematode attack (Hollis, 1977).

Aggressive Cyperaceae weeds are very susceptible to *C. onoensis* and may proliferate in the absence of the nematode. Thus chemical control of the nematode is effective only if rice fields are weeded. Hand removal is uneconomical and the combined use of nematicides and herbicides may be harmful to rice. However, the nematicide Furadan can be satisfactorily combined with herbicides containing the active ingredient 3,4 dichloro-propionanilide (Hollis and Keoboonrueng, 1984). The increase of rice yield after weeding and treatment with phenamiphos is about 17% (Hollis, 1977).

In Louisiana, *C. onoensis* decreased rice production in 1967 by 15% (Hollis *et al.*, 1968), and *C. onoensis* populations as high as 4200/l of soil may reduce yields of upland rice in Mauritius (Chinappen *et al.*, 1988).

Hoplolaimus

A number of lance nematodes (*Hoplolaimus* spp.) are found on upland rice; *H. indicus*, a migratory endoparasite, is reported to be a damaging parasite of rice in India and Nepal (Das and Rao, 1970; Sharma *et al.*, 2001), and another species, *H. clarissimus*, is associated with damage to rice in Togo where rice is cropped continuously on the same fields (Coyne *et al.*, 1996).

Damage by *H. indicus* is not always obvious in the field and, in the early seedling stage, is very similar to nitrogen deficiency. Leaves of seedlings infected by *H. indicus* are yellowish before turning brown and brittle with ash-coloured tips. Plants are stunted, with shortened upper internodes; new leaves can be curled. The symptoms can be less apparent in the latter stage of the crop (Banerji and Banerji, 1966; Das and Rao, 1970). Rice roots have brown lesions at invasion points. Cavities can be found in the cortex, cells lose their rigidity, vascular elements become distorted and roots become flaccid (Das and Rao, 1970; Ramana and Rao, 1975; Alam *et al.*, 1978).

There are few studies of the yield losses caused by *H. indicus* in the field, but, in pot experiments, initial population levels of 100–10,000 nematodes/plant can reduce numbers of tillers by 21.5–36.0% and reduce grain yields by 10.7–19.8% (Ramana and Rao, 1978).

Paralongidorus, Longidorus

Four species of *Paralongidorus* have been recorded on flooded rice: *P. oryzae* occurs in Nepal and India (Verma, 1973); and *P. lutensis* and *P. zenobiae* are found on deep-water rice in Bangladesh (Hunt and Rahman, 1991). *P. australis* is a recognized important parasite of rice locally in North Queensland, Australia (Stirling and Vawdrey, 1984). *Longidorus pisi* has been found on upland and irrigated rice in South Africa (Berg *et al.*, 1989).

In the field, *P. australis* causes poor growth, mainly in rice planted during the summer. The first symptoms appear 7–10 days after flooding and develop into patches of stunted yellow plants, of which many may die. Primary roots show brown necrotic tips, sometimes hooked or curled; secondary roots are shorter than normal, often with a forked appearance. The root system is severely reduced (Fig. 4.12), attacked roots being 1–5 cm long versus 15–20 cm in healthy plants (Stirling and Channon, 1986). Experimentally inoculating rice seedlings with 250–900 nematodes/plant produces symptoms of damage (Stirling, 1984). *P. australis* is an unusually long species, the smallest juveniles being 2–5 mm long and the adults often reaching 10 mm (Stirling and McCulloch, 1985). This inhibits movement in relatively dry or even fine-grained

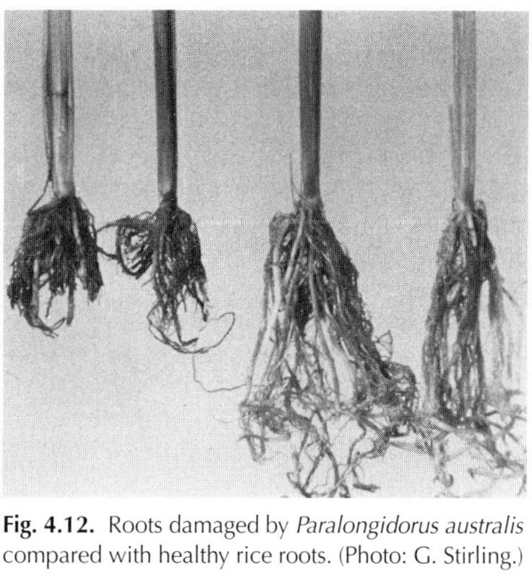

Fig. 4.12. Roots damaged by *Paralongidorus australis* compared with healthy rice roots. (Photo: G. Stirling.)

wet soils and restricts full activity to flooded conditions (Stirling, 1985). The nematode is able to survive in micro-aerobic and anaerobic soils. The life cycle is long, lasting 3–4 rice crops, i.e. about 2 years (Stirling and Shannon, 1986), with most of the active population in the top 25 cm of the soil. Optimal temperature for nematode development is 22–30°C. After harvest, the nematodes move deeper as the soil dries and become anabiotic. They can survive at least 14 months, resuming activity when the soil is flooded (Stirling, 1985). Being limited to flooded rice fields in a relatively narrow area, and with no other known host, the risk of dissemination of this nematode is low.

Control can be achieved by increasing the rate of nitrogenous fertilizer in combination with deep ploughing (> 40 cm) or by changing to moist cultivation rather than flooded in order to inhibit nematode movement (Stirling and Shannon, 1986). Delaying flooding after sowing decreases the degree of nematode damage (Stirling *et al.*, 1989). Control by dry fallow is effective but not normally appropriate because *P. australis* can remain anabiotic for several years. Crop rotation with maize, sorghum or soybeans may be a preferable substitute for fallow. No resistance has been found, but some rice cultivars are more tolerant than others (Stirling *et al.*, 1989).

Xiphinema

Xiphinema bergeri is very common in flooded rice fields of Senegal, Côte d'Ivoire, Guinea, Ghana and Gambia (Fortuner and Merny, 1973; Coyne *et al.*, 1999, 2000), and appears to be widespread in Western Africa; *X. rotundatum* has been found occasionally in Côte d'Ivoire (Merny, 1970).

Several species of *Xiphinema* have been recorded from the rhizosphere of upland rice: *X. insigne* and *X. orbum* in India, *X. nigeriense* and *X. oryzae* in Nigeria, *X. seredouense* in Guinea and *X. cavenessi* in Côte d'Ivoire. A total of 23 species of *Xiphinema* have been found associated with rice in West Africa (Coyne *et al.*, 2000). None of these species is known to be harmful. However, Lamberti *et al.* (1988, 1991) claim that *X. ifacolum* is pathogenic on upland rice in Liberia.

Other Nematodes

Many nematodes, in addition to those already discussed, are found with rice (Fortuner and Merny, 1979), but few of these are reported to be associated with damage and are probably of minor or local importance.

Tylenchorhynchus

Tylenchorhynchus spp. are very common in upland, lowland and deepwater rice throughout the world. They have been found infecting rice in Central and South America, Africa, the Middle East, South and South-eastern Asia, Malaysia and Australia. *T. annulatus* (syn. *martini*) has the widest distribution and is the main species found in irrigated rice fields. Other less commonly reported species on rice are *T. claytoni*, *T. mashoodi*, *T. elegans*, *T. crassicaudatus*, *T. clarus*, *T. nudus*, *T. karnalensis*, *T. iarius*, *T. oryzae*, *T. clavicaudatus* and *T. brassicae*. *T. annulatus*, *T. nudus* and *T. brassicae* can be pathogenic to rice in pot culture (Khan *et al.*, 1990; Haidar *et al.*, 1996; Khan and Shaukat, 2000), and damage is accentuated by an aggregation phenomenon known as 'swarming' (Joshi and Hollis, 1976). However, none of the above species has been shown consistently to cause damage to rice in the field.

Aorolaimus and *Scutellonema*

Aorolaimus nigeriensis has been found in large populations associated with chlorosis and stunting of rice in Togo (Coyne *et al.*, 1996). *Scutellonema brachyurum* and *S. clathricaudatum* are considered to be possible damaging parasites (Baqri and Ahmad, 2000; Coyne *et al.*, 2001).

Helicotylenchus and *Caloosia*

Helicotylenchus spp. are commonly found with rice throughout all rice-growing regions. Species recorded on rice include *H. dihystera*, *H. crenacauda*, *H. erythrinae*, *H. indicus*, *H. digitiformis*, *H. abunaamai*, *H. astriatus* and *H. egyptiensis*. Often, *Helicotylenchus* spp. are the most prevalent parasitic nematodes on rice as reported in West Africa (Coyne *et al.*, 2000) and India (Baqri and Ahmad, 2000), but there are few reports of associated damage to the crop. *H. abunaamai* has been observed feeding ectoparasitically on rice roots (Padhi and Das, 1984). Similarly, *Caloosia paxi* feeds ectoparasitically on upland rice roots and can arrest their apical growth (Rao and Mohanadas, 1976). It has been speculated that high populations of *H. dihystera* may affect rice yields in Liberia (Lamberti *et al.*, 1991) and Mauritius (Chinappen *et al.*, 1988).

Conclusions and Future Prospects

Most rice nematodes are potentially damaging, but their economic importance is strongly influenced by the environment. With some widespread nematodes, such as *A. besseyi*, the damage they cause is not proportional to their distribution; for others, such as *Hirschmanniella* spp., yield losses are probably underestimated. The damage caused by *D. angustus* can be devastating, but it has a limited distribution and its occurrence is unpredictable. Furthermore, as new rice cultivars are bred and regional cropping practices change, nematodes may emerge to be even more important. An ominous example of this is the spread of *D. angustus* from its traditional host, deepwater rice, to the more widely grown and globally important irrigated and lowland rice. Other new nematode problems are surfacing, e.g. *Paralongidorus* at present only known to be damaging in Australia and *Aorolaimus* in Africa. *Paralongidorus* in particular could be more widespread on rice and may have avoided detection as it is difficult to isolate.

Control of rice nematodes poses a number of problems, primarily because measures to control one nematode may increase the damage caused by another. This complicates the recommendation of cultural methods for nematode control on rice and other crops in a rice cropping system, e.g. flooding reduces or eliminates populations of *Pratylenchus*, *Hoplolaimus*, *Heterodera* and most *Meloidogyne* spp., but encourages *Hirschmanniella* spp. Significant reductions in populations of *Hirschmanniella* attacking rice and in soil populations of *Meloidogyne* spp. damaging vegetables can be achieved where irrigated or lowland rice is rotated with upland vegetable crops. However, this same system would increase damage to and yield loss of rice by *M. graminicola*. An accurate knowledge of the species present in a field is thus an important prerequisite for investigating such control methods. Chemical control of rice nematodes will rarely be economic or efficient, and the dangers and difficulties of applying nematicides in flooded rice are self-evident. In flooded soils, sulphur dioxide, produced by anaerobic bacteria, could be used as a form of nematode control, and some trials have proven the efficacy of such phenomena (Jacq and Fortuner, 1979). The difficulty is that rice seedlings may also be killed. More research on this and other similar techniques could be beneficial, but requires the cooperation of nematologists, agronomists and soil microbiologists. Cultivars with resistance or tolerance to nematodes could offer acceptable and economic control of rice nematodes, but there are few, if any ongoing rice nematode resistance breeding programmes. There is some information on the variations in the susceptibility of rice cultivars to most rice nematodes, but essentially very little is known about the mechanisms and inheritance of resistance. Progress is being made with some of the important rice nematodes, but a coordinated international effort is required by nematologists, agronomists and plant breeders to identify and transfer resistance to commercially acceptable rice cultivars.

Acknowledgements

We wish to acknowledge all those who have helped us in our work and in the preparation of this chapter, particularly the rice farmers, our colleagues and fellow nematologists in the rice-growing regions of the world.

References

Alam, M.M., Saxena, S.K. and Khan, A.M. (1978) Suitability of crops to certain ectoparasitic nematodes. *Acta Botanica Indica* 6 (Supplement), 205–208.

Aleksandrova, I.V. (1981) [Decontamination of seed rice from *Aphelenchoides besseyi* Christie by combined treatment with chemical and irradiation.] *Byulleten Vsesoytiznogo Instituta Gel'mintologii im. K. I. Skryabina* 31, 105.

Ali, M.R., Fukutoku, Y. and Ishibashi, N. (1997) Effect of *Ditylenchus angustus* on growth of rice plants. *Japanese Journal of Nematology* 27, 52–66.

Amoussou, P.-L., Ashurst, J., Bridge, J., Jones, M., Snape, J. and Koyama, M. (2002) Mapping of natural nematode resistance in rice for use by West African farmers (abstract). *Proceedings of the Plant, Animal and Microbes Genome Conference X.* 12–16 January 2002, San Diego, California.

Anonymous (1987) *Deep Water Rice Pest Management Project. Recognition of Ufra Disease.* International Rice Research Institute, Government of India and West Bengal.

Arayaungsarit, L., Panpitpat, R. and Puttirut, N. (1986) Reaction of rice varieties to root nematode *Hirschmanniella* spp. in the field. *International Rice Research Newsletter* 11, 14–15.

Ashurst, J., Bridge, J., Amoussou, P.-L. and Plowright, R. (2001) Testing for resistance to rice nematodes in interspecific crosses between the indigenous African rice, *Oryza glaberrima* and the Asian rice *O. sativa* (abstract). *Association of Applied Biologists Nematology Meeting.* December 2001.

Atkins, J.G. and Todd, E.H. (1959) White tip disease of rice. III. Yield tests and varietal resistance. *Phytopathology* 49, 189–191.

Aung, T. and Prot, J.-C. (1990) Effects of crop rotations on *Pratylenchus zeae* and on yield of rice cultivar UPL Ri-5. *Revue de Nématologie* 13, 445–447.

Babatola, J.O. (1979) Varietal reaction of rice and other food crops to the rice-root nematodes, *Hirschmanniella oryzae, H. spinicaudata* and *H. imamuri. Nematropica* 9, 123–128.

Babatola, J.O. (1980) Reactions of some rice cultivars to the root-knot nematode, *Meloidogyne incognita. Nematropica* 10, 5–9.

Babatola, J.O. (1981) Effect of pH, oxygen and temperature on the activity and survival of *Hirschmanniella* spp. *Nematologica* 27, 285–291.

Babatola, J.O. (1983a) Pathogenicity of *Heterodera sacchari* on rice. *Nematologia Mediterranea* 11, 21–25.

Babatola, J.O. (1983b) Rice cultivars and *Heterodera sacchari. Nematologia Mediterranea* 11, 103–105.

Babatola, J.O. (1984) Rice nematode problems in Nigeria; their occurrence, distribution and pathogenesis. *Tropical Pest Management* 30, 256–265.

Babatola, J.O. and Bridge, J. (1979) Pathogenicity of *Hirschmanniella oryzae, H. spinicaudata* and *H. imamuri* on rice. *Journal of Nematology* 11, 128–132.

Babatola, J.O. and Bridge, J. (1980) Feeding behaviour and histopathology of *Hirschmanniella oryzae, H. imamuri* and *H. spinicaudata* on rice. *Journal of Nematology* 12, 48–53.

Banerji, S.N. and Banerji, D.K. (1966) Occurrence of the nematode *Hoplolaimus indicus* in West Bengal. *Current Science* 35, 597–598.

Baqri, Q.H. (1978) Nematodes from West Bengal (India). VI. Species of Criconematoidea (Tylenchida). *Indian Journal of Nematology* 8, 116–121.

Baqri, Q.H. and Ahmad, N. (2000) Qualitative and quantitative studies of plant and soil inhabiting nematodes associated with rice crop in Sikkim, India. *Records of the Zoological Survey of India* 98, 137–148.

Barat, H., Delassus, M. and Huu-Hai-Vuong (1969) The geographical distribution of white tip disease of rice in tropical Africa and Madagascar. In: Peachey, J.E. (ed.) *Nematodes of Tropical Crops.* Technical Communication No. 40. Commonwealth Bureau of Helminthology, pp. 269–273.

Berdon, B.R. and Merny, G. (1964) Biologie d'*Heterodera oryzae* Luc and Berdon, 1961. 1. Cycle du parasite et réactions histologiques de l'hôte. *Revue de Pathologie Végétale et d'Entomologie Agricole de France* 43, 43–53.

Birchfield, W. (1965) Host parasite relations and host range studies of a new *Meloidogyne* species in Southern USA. *Phytopathology* 55, 1359–1361.

Borovkova, A.M. (1967) [*Aphelenchoides besseyi* in the USSR and its control.] In: Sveshnikova, N.M. (ed.) [*Nematode Diseases of Crops.*] Izdatelstvo 'Kolos', Moscow, pp. 153–157.

Bridge, J. and Page, S.L.J. (1982) The rice root-knot nematode, *Meloidogyne graminicola,* on deep water rice (*Oryza sativa* subsp. *indica*). *Revue de Nématologie* 5, 225–232.

Buangsuwon, D., Tonboon-eke, P., Rujirachoon, G., Braun, A.J. and Taylor, A.L. (1971) Nematodes. In: *Rice Diseases and Pests of Thailand,* English edition. Rice Protection Research Centre, Rice Department, Ministry of Agriculture, Thailand, pp. 61–67.

Butler, E.J. (1913) Diseases of rice. *Agricultural Research Institute, Pusa, Bulletin,* 34.

Butler, E.J. (1919) The rice worm (*Tylenchus angustus*) and its control. *Memoirs of the Department of Agriculture in India.* Botanical Series 10, 1–37.

Cadet, P. and Quénéhervé, P. (1982) Action des nématicides en rizière inondée contre *Hirschmanniella spinicaudata. Revue de Nématologie* 5, 93–102.

Catling, H.D. and Puckridge, D.W. (1984) Report on a visit to Vietnam, October, 1983. *Deepwater Rice* 3–4.

Catling, H.D., Cox, P.G., Islam, Z. and Rahman, L. (1979) Two destructive pests of deep water rice – yellow stem borer and ufra. *ADAB News* 6, 16–21.

Chakrabarti, H.S., Nayak, D.K. and Pal, A. (1985) Ufra incidence in summer rice in West Bengal. *International Rice Research Newsletter* 10 (1), 15–16.

Chakraborti, S. (1999) Non-synthetic chemical based management approaches for rice stem nematode. *Pest Management and Economic Zoology* 7, 161–165.

Chakraborti, S. (2000a) Evaluation of neem products on management of ufra disease of rice. *Indian Journal of Nematology* 30, 234–236.

Chakraborti, S. (2000b) Assessment of some management practices on rice stem nematode. *Indian Journal of Nematology* 30, 245–247.

Chandel, S.T., Gaur, H.S. and Alam, M.M. (2001) Histopathology of wheat roots infected with root-knot nematode, *Meloidogyne triticoryzae. Pakistan Journal of Nematology* 19, 67–70.

Chandel, S.T., Gaur, H.S. and Alam, M.M. (2002a) Population dynamics of the root-knot nematode *Meloidogyne triticoryzae* under five rice based cropping systems. *Archives of Phytopathology and Plant Protection* 35, 43–51.

Chandel, S.T., Gaur, H.S. and Alam, M.M. (2002b) Effect of tillage and water management on the population behaviour of root-knot nematodes, *Meloidogyne triticoryzae,* in rice crop. *Archives of Phytopathology and Plant Protection* 35, 195–200.

Charles, J.S. and Venkitesan, T.S. (1985) New hosts of *Heterodera oryzicola* Rao and Jayaprakash, 1978, in Kerala, India. *Indian Journal of Nematology* 14, 181–182.

Charles, J.S.K. and Venkitesan, T.S. (1990) Host records of the rice cyst nematode, *Heterodera oryzicola. Indian Journal of Nematology* 20, 222–224.

Chaudhury, R. and Chaudhury, R. (1996) Effect of storage temperature and moisture on the survival of the white tip nematode, *Aphelenchoides besseyi,* and infested paddy, *Oryza sativa,* seeds. *Annals of Plant Protection Sciences* 4, 16–21.

Chinappen, M., Lamberti, F., Ciancio, A. and Jokhun, P. (1988) Losses caused by concomitant infestation of *Criconemella onoensis* and *Helicotylenchus dihystera* on upland rice in Mauritius. *Nematologia Mediterranea* 16, 175–177.

Choi, Y.E. and Geraert, E. (1994) Systematic study of Criconematoidea from Korea I. Description of *Ogma koesanense* sp. n. and four new species for Korea. *Korean Journal of Applied Entomology* 33, 184–193.

Chunram, C. (1981) Progress report on the root-knot nematode studies in Thailand. *Proceedings of the 3rd Research Planning Conference on Root-knot Nematodes,* Meloidogyne *spp.* Region VI, 20–24 July, 1981, Jakarta, pp. 97–103.

Cotoneo, A. and Moretti, F. (2001) *Aphelenchoides besseyi* nelle sementi di riso: situazione nella campagna 2000. *Sementi Elette* 47, 26–28.

Coyne, D.L. and Plowright, R.A. (1998) Use of solarisation to control *Heterodera sacchari* and other plant parastic nematodes in the field: a modified technique for experimental purposes. *International Journal of Nematology* 8, 81–84.

Coyne, D.L. and Plowright, R.A. (2000) *Heterodera sacchari*: field population dynamics and damage to upland rice in Côte d'Ivoire. *Nematology* 2, 541–550.

Coyne, D.L., Plowright, R.A. and Fofana, I. (1996) Preliminary investigations of nematodes associated with rice in Guinea, Benin and Togo. *Afro-Asian Journal of Nematology* 6, 70–73.

Coyne, D.L., Plowright, R.A., Thio, B. and Hunt, D.J.H. (1998) Plant parasitic nematode diversity and prevalence in traditional upland rice in Ivory Coast: preliminary observations on the effects of cropping intensification. *Fundamental and Applied Nematology* 21, 723–732.

Coyne, D.L., Plowright, R.A., Twumasi, J. and Hunt, D.J.H. (1999) Prevalence of plant parasitic nematodes associated with rice in Ghana with a discussion of their importance. *Nematology* 1, 399–405.

Coyne, D.L., Hunt, D.J., Plowright, R.A. and Darboe, M.L.K. (2000) Further observations on nematodes associated with rice in Côte d'Ivoire. *International Journal of Nematology* 10, 123–130.

Coyne, D.L., Smith, M. and Plowright, R.A. (2001) Plant parasitic nematode populations on upland and hydromorphic rice in Côte d'Ivoire: relationship with moisture availability and crop development on a valley slope. *Agriculture, Ecosystems and Environment* 84, 31–43.

Cox, P.G. and Rahman, L. (1979a) Synergy between benomyl and carbofuran in the control of ufra. *International Rice Research Newsletter* 4 (4), 11.

Cox, P.G. and Rahman, L. (1979b) The overwinter decay of *Ditylenchus angustus. International Rice Research Newsletter* 4 (5), 14.

Cox, P.G. and Rahman, L. (1980) Effects of ufra disease on yield loss of deep water rice in Bangladesh. *Tropical Pest Management* 26, 410–415.

Cralley, E.M. (1949) White tip of rice. *Phytopathology* 39, 5.

Cralley, E.M. (1956) A new control measure for white tip. *Arkansas Farm Research* 5, 5.

Cralley, E.M. and Adair, C.R. (1949) Rice diseases in Arkansas in 1948. *Plant Disease Reporter* 33, 257–259.

Cralley, E.M. and French, R.G. (1952) Studies on the control of white tip of rice. *Phytopathology* 42, 6.

Cuc, N.-t.T. (1982a) New weed host of rice stem nematode identified in Vietnam. *International Rice Research Newsletter* 7 (3), 15.

Cuc, N.-t.T. (1982b) Field soil as a source of rice stem nematodes. *International Rice Research Newsletter* 7 (4), 15.

Cuc, N.-t.T. and Giang, L.T. (1982) Relative humidity and nematode number and survival in rice seeds. *International Rice Research Newsletter* 7 (4), 14.

Cuc, N.-t.T. and Kinh, D.N. (1981) Rice stem nematode disease in Vietnam. *International Rice Research Newsletter* 6 (6), 14–15.

Das, D., Saikia, M.K. and Sarmah, D.K. (1999) Comparative efficacy of botanicals and antagonistic fungi for the management of rice root-knot nematode, *Meloidogyne graminicola. International Journal of Tropical Agriculture* 17, 287–290.

Das, D., Sarma, N.K. and Das, P. (2000) Rayada B3 – a high yielding, ufra-resistant deepwater rice for Assam, India. *International Rice Research Notes* 25, 17.

Das, P. (1996) An integrated approach for management of rice stem nematode, *Ditylenchus angustus* in deep water rice in Assam. *Indian Journal of Nematology* 26, 222–225.

Das, P. and Bhagawati, B. (1992) Incidence of rice stem nematode, *Ditylenchus angustus*, in relation to sowing time of deep water rice in Assam. *Indian Journal of Nematology* 22, 86–88.

Das, P. and Deka, B.C. (2002) Efficacy of neem-based pesticides against *Meloidogyne graminicola* on rice as seed treatment. *Indian Journal of Nematology* 32, 204–205.

Das, P.K. and Rao, Y.S. (1970) Life history and pathogenesis of *Hoplolaimus indicus* in rice. *Indian Phytopathology* 23, 459–464.

Das, P. and Sarmah, N.K. (1995) Sources of resistance to rice stem nematode, *Ditylenchus angustus*, in deep-water rice genotypes. *Annals of Plant Protection Sciences* 3, 135–136.

da Silveira, S.G.P., Curi, S.M., Fernandes, C.D. and de Bona, A. (1977) Occurrencia do nematoide *Aphelenchoides besseyi* Christie, 1942, em areas produtoras de sementes de arroz, no estado de Sao Paulo. *Reunias de Nematologia* 2, 81–91.

Dastur, J.F. (1936) A nematode disease of rice in the Central Provinces. *Proceedings of the Indian Academy of Sciences*, Section B 4, 108–121.

De Waele, D. and van den Berg, E. (1988) Nematodes associated with upland rice in South Africa, with a description of *Hemicycliophora oryzae* sp.n. (Nemata: Criconematoidea). *Revue de Nématologie* 11, 45–51.

Diomandé, M. (1984) Response of upland rice cultivars to *Meloidogyne* species. *Revue de Nématologie* 7, 57–63.

Edward, J.C., Sharma, N.N. and Agnihothrudu, V. (1985) Rice root nematode (*Hirschmanniella* spp.) – a review of the work done in India. *Current Science* 54, 179–182.

Fademi, O.A. (1984) Influence of rate and time of carbofuran application to control root-knot nematodes in upland rice. *International Rice Research Newsletter* 9 (6), 22–23.

Fademi, O.A. (1987) Resistance of some rice varieties to the root-knot nematode (RKN) *Meloidogyne incognita. International Rice Research Newsletter* 12 (1), 11.
Feng, R.Z. (1986) [A preliminary study on the distribution, occurrence and fluctuation of rice root nematodes.] *Journal of Guangxi Agricultural College* 2, 57–62.
Ferraz, L.C.C.B. (1995) Danos causados a duas variedades de arroz pelonematoide das galhas *Meloidogyne javanica. Summa Phytopathologica* 21, 36–37.
Fomba, S.N. (1984) Rice disease situation in mangrove and associated swamps in Sierra Leone. *Tropical Pest Management* 30, 73–81.
Fortuner, R. (1974) Evaluation de dégats causés par *Hirschmanniella oryzae* (Van Breda de Haan, 1902) Luc & Goodey, 1963, nématode endoparasite des racines du riz irrigué. *L'Agronomie Tropicale, Nogent* 29, 708–714.
Fortuner, R. (1976) *Pratylenchus zeae. C. I. H. Descriptions of Plant-parasitic Nematodes.* Set 6, No. 77.
Fortuner, R. (1977) Fertilization du riz et dégats causés par le nématode *Hirschmanniella oryzae* (Van Breda de Haan) Luc & Goodey. *Procès-verbal de la Séance du ler Juin, 1977, Académie d'Agriculture de France* 58, 624–630.
Fortuner, R. (1981) Les nématodes associés au riz pluvial en Côte d'Ivoire. *L'Agronomie Tropicale, Nogent* 36, 70–77.
Fortuner, R. (1985) Nematode damage assessment problems and solutions for a rice nematode: *Hirschmanniella oryzae. California Plant Pest and Disease Report* 4, 2–7.
Fortuner, R. and Merny, G. (1973) Les nématodes parasites des racines associés au riz en Basse-Casamance et en Gambie. *Cahiers de l'ORSTOM, Serie Biologie* 21, 3–20.
Fortuner, R. and Merny, G. (1979) Root-parasitic nematodes of rice. *Revue de Nématologie* 2, 79–102.
Fortuner, R. and Orton Williams, K.J. (1975) Review of the literature on *Aphelenchoides besseyi* Christie, 1942, the nematode causing 'white tip' disease in rice. *Helminthological Abstracts, Series B, Plant Nematology* 44, 1–40.
Fukano, H. (1962) [Control method against rice white tip disease.] *Nogyo oyobi Engei* 37, 689–692.
Ganguly, A.K., Pankaj and Sirohi, A. (1996) Effect of soil solarization of rice nursery beds to suppress plant parasitic nematodes. *International Rice Research Notes* 21, 80–81.
Gao, X., Li, J. and Zhou, H. (1998a) Studies on the host range of *Hirschmanniella oryzae* and its associated weeds in Guangzhou, China. *International Journal of Nematology* 8, 92–93.
Gao, X., Zhou, H. and Feng, Z. (1998b) Effects of cultivation methods on *Hirschmanniella oryzae. Chinese Rice Research Newsletter* 6, 8–9.
Gao, X.B., Li, J.Q. and Zhou, H.J. (1999) Studies on host range of *Hirshchmanniella oryzae. Journal of Yunnan Agriculture University* 14 (Supplement), 117.
Garg, R.N., Gaur, H.S. and Singh, G. (1995) Effect of puddling and water regimes on soil physical properties, plant parasitic nematodes and performance of rice crop. *Annals of Plant Protection Sciences* 3, 121–126.
Gaur, H.S. (2003) *Root-knot Disease of Rice and Wheat: Problem and Management.* Division of Nematology, Indian Agricultural Research Institute, New Delhi, India.
Gaur, H.S. and Sharma, S.N. (1998) Studies on the host range of the root-knot nematode, *Meloidogyne triticoryzae,* among cultivated crops and weeds. *Annals of Plant Protection Sciences* 6, 41–47.
Gaur, H.S., Saha, M. and Khan, E. (1993) *Meloidogyne triticoryzae* sp. n. (Nematoda: Meloidogynidae) a root-knot nematode damaging wheat and rice in India. *Annals of Plant Protection Sciences* 1, 18–26.
Gaur, H.S., Beane, J. and Perry, R.N. (2000) The influence of root diffusate, host age and water regime on hatching of the root-knot nematode, *Meloidogyne triticoryzae. Nematology* 2, 191–199.
Gergon, E.B. and Mew, T.W. (1991) Evaluation of methods for detecting the nematode *Aphelenchoides besseyi* Christie in routine seed testing of rice. *Seed Science and Technology* 19, 647–654.
Gergon, E.B. and Prot, J.C. (1993) Effect of benomyl and carbofuran on *Aphelenchoides besseyi* on rice. *Fundamental and Applied Nematology* 16, 563–566.
Gergon, E.B., Miller, S.A. and Davide, R.G. (2001) Occurrence and pathogenicity of rice root-knot nematode (*Meloidogyne graminicola*) and varietal reaction of onion (*Allium cepa*). *Philippine Agricultural Scientist* 84, 43–50.
Germani, G., Reversat, G. and Luc, M. (1983) Effect of *Sesbania rostrata* on *Hirschmanniella oryzae* in flooded rice. *Journal of Nematology* 15, 269–271.
Goto, K. and Fukatsu, R. (1952) [Studies on white tip of rice plant caused by *Aphelenchoides oryzae* Yokoo. 11. Number and distribution of the nematode on the affected plants.] *Annals of the Phytopathological Society of Japan* 16, 57–60.

Goto, K. and Fukatsu, R. (1956) [Studies on the white tip of rice plant. III. Analysis of varietal resistance and its nature.] *Bulletin of the National Institute of Agricultural Sciences, Tokyo, Series C, Plant Pathology and Entomology* No. 6, 123–149.

Greco, N., Crozzoli, R., Lamberti, F. and Brandonisio, A. (2000) Pathogenic potential of the root-knot nematode, *Meloidogyne incognita,* on rice in Venezuela. *Nematologia Mediterranea* 28, 47–55.

Guo, Y.Y., Zhou, N.S., Wu, T.T. and Qiu, H.L. (1984) [An investigation and study on rice root-knot nematode on Hainan Island.] *Guangdong Agricultural Science* (*Guangdong Nongye Kexue*) No. 4, 32–34.

Haidar, M.G., Nath, R.P. and Thakur, S.K. (1996) Growth response of rice plant to stunt nematode, *Tylenchorhynchus nudus. Journal of Research, Birsa Agricultural University* 8, 163–164.

Hashioka, Y. (1963) The rice stem nematode *Ditylenchus angustus* in Thailand. *FAO Plant Protection Bulletin* 11, 97–102.

Hendro, M.E., Prot, J.-C. and Madamba, C.P. (1992) Population dynamics of *Hirschmanniella mucronata* and *H. oryzae* on *Sesbania rostrata, Aeschynomene afraspera* and rice cv. IR 58. *Fundamental and Applied Nematology* 15, 167–172.

Hockland, S. and Eng, L. (1997) *Capsicum annuum* v. *longum* a new host record for the rice white tip nematode, *Aphelenchoides besseyi. International Journal of Nematology* 7, 229.

Hollis, J.P. (1972a) Nematicide–weeds interaction in rice fields. *Plant Disease Reporter* 56, 420–424.

Hollis, J.P. (1972b) Competition between rice and weeds in nematode control tests. *Phytopathology* 62, 764.

Hollis, J.P. (1977) Loss in yield of rice caused by the ring nematode *Criconemoides onoensis* revealed by the elimination of the yellow nutsedge, *Cyperus esculentus. Nematologica* 23, 71–78.

Hollis, J.P. and Joshi, M.M. (1976) Weed effect on chemical control of ring nematodes in rice. *Annual Proceedings of the American Phytopathological Society* 1975, 44–45.

Hollis, J.P. and Keoboonrueng, S. (1984) Nematode parasites of rice. In: Nickle, W.R. (ed.) *Plant and Insect Nematodes.* Marcel Dekker, New York, pp. 95–146.

Hollis, J.P., Embabi, M.S. and Alhassan, S.A. (1968) Ring nematode disease of rice in Louisiana. *Phytopathology* 58, 728–729.

Hoshino, S. and Togashi, K. (1999) A simple method for determining *Aphelenchoides besseyi* infestation level of *Oryza sativa* seeds. *Journal of Nematology* 31, 641–643.

Hoshino, S. and Togashi, K. (2000) Effect of water-soaking and air-drying on survival of *Aphelenchoides besseyi* in *Oryza sativa* seeds. *Journal of Nematology* 32, 303–308.

Hoshino, S. and Togashi, K. (2002) Mass extraction method for determining *Aphelenchoides besseyi* density in *Oryza sativa* seeds. *Japanese Journal of Nematology* 32, 25–29.

Huang, C.S. and Huang, S.P. (1972) Bionomics of white-tip nematode, *Aphelenchoides besseyi* in rice florets and developing grains. *Botanical Bulletin of Academia Sinica* 13, 1–10.

Huang, C.S. and Huang, S.P. (1974) Dehydration and the survival of rice white tip nematode *Aphelenchoides besseyi. Nematologica* 20, 9–18.

Huang, C.S., Huang, S.P. and Lin, L.H. (1972) The effect of temperature on development and generation period of *Aphelenchoides besseyi. Nematologica* 18, 432–438.

Huang, C.S., Cupertino, F.P. and Marinelli, N.M. (1977) Incidence of white-tip nematode, *Aphelenchoides besseyi,* in stored rice seed from Central-West Brazil. *PANS* 23, 65–67.

Huang, C.S., Huang, S.P. and Chiang, Y.C. (1979) Mode of reproduction and sex ratio of rice white-tip nematode, *Aphelenchoides besseyi. Nematologica* 25, 244–260.

Hunt, D.J. and Rahman, L. (1991) Two new species of *Paralongidorus* (Nematoda: Dorylaimida) from deepwater in Bangladesh. *Afro-Asian Journal of Nematology* 1, 83–85.

Ibrahim, I.K.A., Ibrahim, I.A. and Rezk, M.A. (1972) Pathogenicity of certain parasitic nematodes on rice. *Alexandria Journal of Agricultural Research* 20, 175–181.

Ibrahim, I.K.A., Rezk, M.A. and Khalil, H.A.A. (1983) Resistance of some plant cultivars to root-knot nematodes, *Meloidogyne* spp. *Nematologia Mediterranea* 11, 189–192.

Ibrahim, S.K. and Perry, R.N. (1993) Desiccation survival of the rice stem nematode *Ditylenchus angustus. Fundamental and Applied Nematology* 16, 31–38.

Ibrahim, S.K., Perry, R.N., Plowright, R.A. and Rowe, J. (1993) Hatching behaviour of the rice cyst nematodes *Heterodera sacchari* and *H. oryzicola* in relation to age of host plant. *Fundamental and Applied Nematology* 16, 23–29.

Ichinohe, M. (1972) Nematode diseases of rice. In: Webster, J.M. (ed.) *Economic Nematology.* Academic Press, London, pp. 127–143.

Ichinohe, M. (1988) Current research on the major nematode problems in Japan. *Journal of Nematology* 20, 184–190.

Inagaki, H. (1985) The plant parasitic nematodes important in Japan and the related researches. *JARQ* 18, 194–201.

Iyatomi, K. and Nishizawa, T. (1954) [Artificial culture of the strawberry nematode, *Aphelenchoides fragariae*, and the rice white-tip nematode *Aphelenchoides besseyi*.] *Japanese Journal of Applied Zoology* 19, 8–15.

Jacq, V. and Fortuner, R. (1979) Biological control of rice nematodes using sulphate reducing bacteria. *Revue de Nématologie* 2, 41–50.

Jayaprakash, A. and Rao, Y.S. (1982a) Life history and behaviour of the cyst nematode, *Heterodera oryzicola* Rao and Jayaprakash, 1978 in rice (*Oryza sativa* L.). *Proceedings of the Indian Academy of Sciences, Animal Sciences* 91, 283–295.

Jayaprakash, A. and Rao, Y.S. (1982b) Hatching behaviour of the cyst-nematode *Heterodera oryzicola*. *Indian Journal of Nematology* 12, 400–402.

Jayaprakash, A. and Rao, Y.S. (1983) Reaction of rice cultivars against the cyst nematode, *Heterodera oryzicola*. *Indian Journal of Nematology* 13, 117–118.

Jayaprakash, A. and Rao, Y.S. (1984) Cyst nematode, *Heterodera oryzicola* and seedling blight fungus *Sclerotium rolfsii* disease complex in rice. *Indian Journal of Nematology* 14, 58–59.

Jena, R.N. and Rao, Y.S. (1974) Root-knot nematode resistance in rice. *Indian Journal of Genetics and Plant Breeding* 34, 443–449.

Jonathan, E.I. and Pandiarajan, P. (1991) Effect of organic amendments in controlling rice root nematodes. *International Rice Research Newsletter* 16, 25.

Jonathan, E.I. and Velayutham, B. (1987) Evaluation of yield loss due to rice root nematode *Hirschmanniella oryzae*. *International Nematology Network Newsletter* 4, 8–9.

Joshi, M.M. and Hollis, J.P. (1976) Pathogenicity of *Tylenchorhynchus martini* swarmers to rice. *Nematologica* 22, 123–124.

Kaushal, K.K., Sharma, S.B. and Singh, M. (2000) *Heterodera skohensis*, a new cyst nematode species (Nematoda: Heteroderidae) from Kangra valley of Himachal Pradesh, India. *International Journal of Nematology* 10, 67–70.

Kawashima, K. (1963) [Investigations on *Hirschmannia oryzae*. I. Varietal susceptibility to the nematode. II. Susceptibilities of weeds to the nematode. III. Durable effect of fungicides.] *Annual Report of Society Plant Protection, North Japan* 14, 111, 112–113, 158–159.

Khan, A. and Shaukat, S.S. (1998) Effect of three organic amendments and carbofuran on populations of *Hirschmanniella oryzae* and *Tylenchorhynchus*. *International Journal of Nematology* 8, 94–96.

Khan, A. and Shaukat, S.S. (2000) Effect of single and simultaneous inoculations of *Hirschmanniella oryzae* and *Tylenchorhynchus annulatus* to rice seedlings under laboratory conditions. *Sarhad Journal of Agriculture* 16, 197–200.

Khan, A., Jaffry, A.H. and Khan, M.A. (1990) Pathogenicity of *Tylenchorhynchus brassicae* to three cultivars of rice. *International Nematology Network Newsletter* 7, 42–43.

Khuong, N.B. (1987) *Hirschmanniella* spp. in rice fields of Vietnam. *Journal of Nematology* 19, 82–84.

Khush, G.S. (1984) Terminology for rice growing environments. In: *Terminology for Rice Growing Environments.* International Rice Research Institute, Los Baños, Philippines, pp. 5–10.

Kim, J.I., Han, S.C. and Choi, D.R. (1996) Investigation of plant parasitic nematodes in paddy–upland rotation. *RDA Journal of Agricultural Science, Crop Protection* 38, 387–391.

Kinh, D.-N. (1981) Survival of *Ditylenchus angustus* in diseased stubble. *International Rice Research Newsletter* 6 (6), 13.

Kinh, D.-N. and Nghiem, N.-T. (1982) Reaction of rice varieties to stem nematodes in Vietnam. *International Rice Research Newsletter* 7 (3), 6–7.

Kinh, D.-N. and Phuong, C.-D. (1981) Reaction of some deepwater rice varieties to *Ditylenchus angustus*. *International Rice Research Newsletter* 6 (6), 6–7.

Kinh, D.-N., Huong, N.M. and Ut, N.U. (1982) Root-knot disease of rice in the Mekong Delta, Vietnam. *International Rice Research Newsletter* 7 (4), 15.

Kobayashi, Y. and Sugiyama, T. (1977) [Dissemination and reproduction of white tip nematodes *Aphelenchoides besseyi* in mechanical transplantation rice culture.] *Japanese Journal of Nematology* 7, 74–77.

Kumar, S. (1990) Alternate hosts for rice root nematode *Hirschmanniella oryzae*. *International Nematology Network Newsletter* 7, 4.

Kumari, U. and Kuriyan, K.J. (1981) Cyst nematode, *Heterodera oryzicola*, on rice in Kerala. I. Estimation of loss in rice due to *H. oryzicola* infestation, in field conditions. *Indian Journal of Nematology* 11, 106.

Kuwahara, M. and Iyatomi, K. (1970) Studies on the bionomics of the rice root nematode *Hirschmanniella* imamuri Sher and *Hirschmanniella oryzae* (Soltwedel) Luc and Goodey, with special reference to its mode of life and population dynamics. *Japanese Journal of Applied Entomology and Zoology* 14, 117–121.

Lahan, K.K., Sinha, A.K. and Das, P. (1999) Evaluation of chemicals as bare root dip against rice root nematode, *Hirschmanniella oryzae. Indian Journal of Nematology* 29, 8–12.

Lamberti, F., Bleve-Zacheo, T., Tuopay, D.K., Ciancio, A. and Boiboi, J.B. (1988) Relationships between *Xiphinema ifacolum* and rice in Liberia. *Nematologia Mediterranea* 15 (1987), 303–314.

Lamberti, F., Ciancio, A., Tuopay, D.K., Boiboi, J.B., Vovlas, N., Bleve-Zacheo, T. and Elia, F. (1991) Nematode threats to rice in Liberia. *Nematologia Mediterranea* 19, 291–303.

Lee, Y.B. and Park, J.S. (1975) [Influence of rice root nematode (*Hirschmanniella oryzae*) on the root browning of rice.] *Korean Journal of Plant Protection* 14, 35–36.

Li, D.Z. (1987) Description of species of parasiting nematodes of genus *Hirshchmanniella* on rice in Sichuan province. *Journal of South-west Agricultural University* 9 (1), 33–35.

Liao, J.L. and Feng, Z.X. (1995) A root knot nematode, *Meloidogyne hainanensis* sp. nov. (Nematoda: Meloidogynidae) parasitizing rice in Hainan, China. *Journal of South China Agriculture University* 16 (3), 34–39.

Liao, J.L., Feng, Z.X., Li, S.M. and Hu, Y.M. (2000) Species of *Hirschmanniella* on rice and their distribution in China. *Nematologia Mediterranea* 28, 107–110.

Lopez, R. (1984) *Meloidogyne salasi* sp.n. (Nematoda: Meloidogynidae), a new parasite of rice (*Oryza sativa* L.) from Costa Rica and Panama. *Turrialba* 34, 275–286.

Lorieux, M., Reversat, G., Garcia Diaz, S.X., Denance, C., Jouvenet, N., Orieux, Y., Bourger, N., Pando Bahuon, A. and Ghesquire, A. (2003) Linkage mapping of Hsa-10g, a resistance gene of African rice to the cyst nematode, *Heterodera sacchari. Theoretical and Applied Genetics* 107, 691–696.

Luc, M. (1970) Contribution a l'étude du genre *Criconemoides. Cahiers de l'ORSTOM, Série Biologie* 11, 69–131.

Maas, P.W. (1970) Tentative list of plant parasitic nematodes in Surinam, with description of two new species of Hemicycliophorinae. *Bulletin van het Landbouwproefstation Suriname* No. 87.

Maas, P.W.T., Sanders, H. and Dede, J. (1978) *Meloidogyne oryzae* n.sp. (Nematoda: Meloidogynidae) infesting irrigated rice in Surinam (South America). *Nematologica* 24, 305–311.

MacGowan, J.B. and Langdon, K.R. (1989) Hosts of the rice root-knot nematode, *Meloidogyne graminicola. Nematology Circular, Gainesville* No. 172.

McGrawley, E.C., Rush, M.C. and Hollis, J.P. (1984) Occurrence of *Aphelenchoides besseyi* in Louisiana rice seed and its interaction with *Sclerotium oryzae* in selected cultivars. *Journal of Nematology* 16, 65–68.

Manser, P.D. (1968) *Meloidogyne graminicola* a cause of root knot of rice. *FAO Plant Protection Bulletin* 16, 11.

Manser, P.D. (1971) Notes on the rice root knot nematode in Laos. *FAO Plant Protection Bulletin* 19, 138–139.

Martin, G.C. (1972) Rhodesia. A nematode affecting rice. *FAO Plant Protection Bulletin* 20, 40–41.

Mathur, V.K. and Prasad, S.K. (1972a) Embryonic development and morphology of larval stages of the rice root nematode, *Hirschmanniella oryzae. Indian Journal of Nematology* 2, 146–157.

Mathur, V.K. and Prasad, S.K. (1972b) Role of the rice root nematode, *Hirschmanniella oryzae* in rice culture. *Indian Journal of Nematology* 2, 158–168.

Mathur, V.K. and Prasad, S.K. (1973a) Control of *Hirschmanniella oryzae* associated with paddy. *Indian Journal of Nematology* 3, 54–60.

Mathur, V.K. and Prasad, S.K. (1973b) Survival and host range of the rice root nematode, *Hirschmanniella oryzae. Indian Journal of Nematology* 3, 88–93.

McGeachie, I. and Rahman, L. (1983) Ufra disease: a review and a new approach to control. *Tropical Pest Management* 29, 325–332.

Merny, G. (1966) Biologie d'*Heterodera oryzae* Luc and Berdon, 1961. II. Rôle des masses d'oeufs dans la dynamique des populations et la conservation de l'espèce. *Annales des Epiphyties* 17, 445–449.

Merny, G. (1970) Les nématodes phytoparasites des rizieres inondées en Côte d'Ivoire. I. Les espèces observées. *Cahiers de l'ORSTOM, Serie Biologie* 11, 3–43.

Merny, G. (1972) Les nématodes phytoparasites des rizières inondées de Côte d'Ivoire. III. Etude sur a dynamique des populations de deux endoparasites: *Hirschmanniella spinicaudata* et *Heterodera oryzae. Cahiers de l'ORSTOM, Serie Biologie* 16, 31–87.

Merny, G. and Cadet, P. (1978) Penetration of juveniles and development of adults of *Heterodera oryzae* on different plants. *Revue de Nématologie* 1, 251–255.

Miah, S.A. and Bakr, M.A. (1977a) Chemical control of ufra disease of rice. *PANS* 23, 412–413.

Miah, S.A. and Bakr, M.A. (1977b) Sources of resistance to ufra disease of rice in Bangladesh. *International Rice Research Newsletter* 2 (5), 8.

Miah, S.A. and Rahman, M.L. (1985) Severe ufra outbreak in transplanted rice in Bangladesh. *International Rice Research Newsletter* 10 (3), 24.

Mian, I.H., Latif, M.A., Rahman, M.L. and Bahar, M.R. (1994) Nematicides and mustard oil cake to control ufra disease of rice. *Bangladesh Journal of Plant Pathology* 10, 5–10.

Minassian, V. and Barooti, S. (1997) First report on occurrence of *Hirschmanniella oryzae* in Iran. *Iranian Journal of Plant Pathology* 33, 31–32.

Mohandes, C., Rao, Y.S. and Sahu, S.C. (1981) Cultural control of rice root nematodes (*Hirschmanniella spp.*) with *Sphenoclea zeylandica*. *Proceedings of the Indian Academy of Sciences (Animal Sciences)* 90, 373–376.

Mondal, A.H. and Miah, S.A. (1987) Ufra problem in low lying areas of Bangladesh. *International Rice Research Newsletter* 12 (4), 29–30.

Mondal, A.H., Rahman, L., Ahmed, H.J. and Miah, S.A. (1986) The causes of increasing blast susceptibility of ufra infected rice plants. *Bangladesh Journal of Agriculture* 11, 77–79.

Mondal, A.H., Rahman, M.L. and Miah, S.A. (1988) Distribution and severity of rice seedling diseases in boro seedbeds in Bangladesh. *International Rice Research Newsletter* 13, 20.

Moretti, M. (1997) White tip: nuova malattia del riso in Italia causata dal nematode *Aphelenchoides besseyi*. *Informatore Agrario* 53, 47–51.

Muthukrishnan, T.S., Rajendran, G. and Chandrasekara, J. (1974) Studies on the white tip nematode of rice *Aphelenchoides besseyi* in Tamil Nadu. *Indian Journal of Nematology* 4, 188–193.

Muthukrishnan, T.S., Rajendran, G., Ramamurthy, V.V. and Chandrasekaran, J. (1977) Pathogenicity and control of *Hirschmanniella oryzae*. *Indian Journal of Nematology* 7, 8–16.

Netscher, C. (1969) L'ovogenèse et la reproduction chez *Heterodera oryzae* et *H. sacchari* (Nematoda: Heteroderidae). *Nematologica* 15, 10–14.

Netscher, C., Luc, M. and Merny, G. (1969) Description du mâle d'*Heterodera sacchari* Luc & Merny, 1963. *Nematologica* 15, 156–157.

Nguyen, T.T.C., Tran, V.P. and Prot, J.C. (1993) Efficacy of benomyl in controlling the ufra nematode in Vietnam. *International Rice Research Notes* 18, 37–38.

Nishizawa, T. (1953a) [On the relation between the rice nematode disease 'white tip' and the stem-rot of rice plants.] *Annals of the Phytopathological Society of Japan* 17, 137–140.

Nishizawa, T. (1953b) [Studies on the varietal resistance of rice plants to the rice nematode disease 'Senchu shingare byo' (V1).] *Bulletin of the Kyushu Agricultural Experimental Station* 1, 339–349.

Nishizawa, T. (1976) [Occurrence of abnormal rice kernels associated with the infestation of white tip nematode, *Aphelenchoides besseyi*.] *Japanese Journal of Nematology* 6, 73–79.

Nishizawa, T., Shimizu, K. and Nagashima, T. (1972) Chemical and cultural control of the rice nematode, *Heterodera oryzae* Luc et Berdon Brizuela, and hatching responses of the larvae to some root extracts. *Japanese Journal of Nematology* 2, 27–32.

Odihirin, R.A. (1975) Occurrence of *Heterodera* cyst nematode (Nematoda: Heteroderidae) on wild grasses in southern Nigeria. *Occasional Publication, Nigerian Society for Plant Protection* 1, 24–25.

Ohshima, Y. (1974) *Heterodera elachista* n.sp., an upland rice cyst nematode from Japan. *Japanese Journal of Nematology* 4, 51–56.

Okada, T. (1955) On morphological characters of the upland rice nematode, *Heterodera* sp. *Annual Meeting of the Japanese Society of Applied Entomology and Zoology* 11.

Oliveira, J.V. (1989) Avaliacao da resistencia de quatro genotipos de arroz irrigado ao nematoide *Aphelenchoides besseyi,* Christie, 1942. *Agronomia Sulriograndense* 25, 11–18.

Orsenigo, M. (1954) Suscettibilita di varieta Italiane di riso alla malattia della 'white-tip'. *Annali della Facolta di Agraria Universita Cattolica del S. Cuore, Milano* 51, 1–7.

Ou, S.H. (1985) *Rice Diseases*, 2nd edn. Commonwealth Mycological Institute, Kew, UK.

Ozturk, G. and Enneli, S. (1997) (Determination of *Aphelenchoides besseyi* (Aphelenchida: Aphelenchoididae), the white tip nematode, harmful on rice for the first time in Turkey) (In Turkish). *Turkiye Entomoloji Dergisi* 21, 129–132.

Padhi, N.N. and Das, S.N. (1984) Some aspects of the ecological behaviour of *Helicotylenchus abunaamai* Siddiqi, 1972. *Indian Journal of Nematology* 14, 25–29.

Page, S.L.J. and Bridge, J. (1978) *Plant Nematodes on Deep Water Rice in Bangladesh. Report on Visit to Bangladesh.* Ministry of Overseas Development, UK.

Pal, A.K. (1970) Survey on nematode infection of paddy and successful treatment with hexadrin. *Proceedings of the Indian Science Congress Association* 57, 496.

Panda, M. and Rao, Y.S. (1971) Evaluation of losses caused by the root-nematodes (*Hirschmanniella mucronata* Das) in rice (*Oryza sativa* L.). *Indian Journal of Agricultural Science* 41, 611–614.

Pardo, B.F. and Munoz, B.D. (1994) Agente causal del entorchamiento en el cultivo del arroz en los Llanos Orientales. *Arroz* 43, 16–22.

Park, J.S. and Lee, J.O. (1976) Varietal resistance of rice to the white-tipped nematode *Aphelenchoides besseyi. International Rice Research Newsletter* 1, 15.

Park, J.S., Han, S.C. and Lee, Y.B. (1970) [Studies on the ecology and feeding preference of *Hirschmanniella oryzae*.] *Research Reports of the Office of Rural Development, Korea, Plant Environment* 13, 93–98.

Pedramfar, H., Pourjam, E. and Kheiri, A. (2001) Plant parasitic nematodes associated with rice in Guilan Province, Iran. *Iranian Journal of Plant Pathology* 37, 285–301.

Plowright, R.A. (1986) The interaction of *Aphelenchoides besseyi* and *Meloidogyne graminicola* on rice cv. IR36. *18th International Symposium of European Society of Nematologists,* Antibes, p. 24.

Plowright, R.A. and Bridge, J. (1990) Effect of *Meloidogyne graminicola* (Nematoda) on the establishment, growth and yield of rice cv. IR36. *Nematologica* 36, 81–89.

Plowright, R.A. and Gill, J.R. (1994) Aspects of resistance in deepwater rice to the stem nematode *Ditylenchus angustus. Fundamental and Applied Nematology* 17, 357–367.

Plowright, R.A., Matias, D., Aung, T. and Mew, T.W. (1990) The effect of *Pratylenchus zeae* on the growth and yield of upland rice. *Revue de Nématologie* 13, 283–292.

Plowright, R.A., Grayer, R.J., Gill, J.R., Rahman, M.L. and Harborne, J.B. (1996) The induction of phenolic compounds in rice after infection by the stem nematode *Ditylenchus angustus. Nematologica* 42, 564–578.

Plowright, R.A., Coyne, D.L., Nash, P. and Jones, M.P. (1999) Resistance to the rice nematodes *Heterodera sacchari, Meloidogyne graminicola* and *M. incognita* in *Oryza glaberrima* and *O. glaberrima* × *O. sativa* interspecific hybrids. *Nematology* 1, 745–751.

Popova, M.B. (1984) [Principal factors determining the intensity and pathogenicity of *Aphelenchoides besseyi* infestation on rice.] *Byulleten' Vsesoyuznogo Instituta Gel'mintologfl im. K. I. Skryabina* No. 36, 39–41.

Popova, M.B., Zelenskii, G.L. and Subbotin, S.A. (1994) An assessment of resistance in cultivars of *Oryza sativa* L. to *Aphelenchoides besseyi* Christie, 1942. *Russian Journal of Nematology* 2, 41–44.

Prasad, J.S. and Rao, S.Y. (1978a) Influence of crop rotations on the population densities of the root lesion nematode *Pratylenchus indicus* in rice and rice soils. *Annales de Zoologie et Ecologie Animale* 10, 627–634.

Prasad, J.S. and Rao, S.Y. (1978b) Potentiality of *Pratylenchus indicus* the root lesion nematode as a new pest of upland rice. *Annales de Zoologie et Ecologie Animale* 10, 635–640.

Prasad, J.S. and Rao, Y.S. (1979a) Factors influencing the prevalence of the root-lesion nematode, *Pratylenchus indicus,* Das 1960 in rice. *Revue d'Ecologie et de Biologie du Sol* 16, 517–523.

Prasad, J.S. and Rao, Y.S. (1979b) Nematicidal properties of the weed *Eclipta alba* Hask (Compositae). *Revista di Parasitologia* 50, 87–90.

Prasad, J.S. and Rao, Y.S. (1982a) Studies on the life cycle of *Pratylenchus indicus,* the root lesion nematode in rice. *Oryza* 19, 209–211.

Prasad, J.S. and Rao, Y.S. (1982b) Screening of some rice cultivars against the root lesion nematode, *Pratylenchus indicus. Nematologia Mediterranea* 10, 215–216.

Prasad, J.S. and Varaprasad, K.S. (2002) Ufra nematode, *Ditylenchus angustus* is seed borne! *Crop Protection* 21, 75–76.

Prasad, J.S., Panwar, M.S. and Rao, Y.S. (1986a) Chemical control of the root nematode (*Hirschmanniella mucronata*) in rice. *Beitrage zur Tropischen Landwirtschaft und Veterinarmedizin* 24, 65–69.

Prasad, J.S., Panwar, M.S. and Rao, Y.S. (1986b) Screening of some rice cultivars against the root knot nematode *Meloidogyne graminicola. Indian Journal of Nematology* 16, 112–113.

Prasad, J.S., Panwar, M.S. and Rao, Y.S. (1986c) Reaction of some rice cultures to root-knot and cyst nematodes. *Oryza* 23, 59–61.

Prasad, K.S.K., Krishnappa, K. and Rao, Y.S. (1979) Response of some improved rice varieties to the development and reproduction of *Meloidogyne graminicola* Golden and Birchfield, 1965. *Mysore Journal of Agricultural Sciences* 13, 60–62.

Prot, J.-C. (1992) Use of green manure crops in control of *Hirschmanniella mucronata* and *H. oryzae* in irrigated rice. *Journal of Nematology* 24, 127–132.

Prot, J.-C. (1994a) Effects of economic and policy changes on status of nematode pests in Vietnam and the Philippines. *Fundamental and Applied Nematology* 17, 195–198.

Prot, J.-C. (1994b) The combination of nematodes, *Sesbania rostrata,* and rice: the two sides of the coin. *International Rice Research Notes* 19, 30–31.

Prot, J.-C., Soriano, I.R.S. and Matias, D.M. (1994) Major root parasitic nematodes associated with irrigated rice in the Phillipines. *Fundamental and Applied Nematology* 17, 75–78.

Prot, J.-C., Piggin, C., Courtois, B. and Schmit, V. (1996) Nematode pests in upland rice production systems. *IRRI Discussion Paper Series* No. 16, 239–245.

Puckeridge, D.W. (1988) DWR in S. Vietnam, July, 1988. *Deepwater Rice* 14, 6–7.

Rahman, M.F. and Das, P. (1994) Seed soaking with chemicals for reducing infestation of *Meloidogyne graminicola* in rice. *Journal of the Agricultural Science Society of North East India* 7, 107–108.

Rahman, M.L. (1987) Source of ufra-resistant deep water rice. *International Rice Research Newsletter* 12 (1), 8.

Rahman, M.L. (1990) Effect of different cropping sequences on root-knot nematode, *Meloidogyne graminicola,* and yield of deepwater rice. *Nematologia Mediterranea* 18, 213–217.

Rahman, M.L. (1994) New ufra-resistant rice lines. *International Rice Research Notes* 19, 16.

Rahman, M.L. (1996) Ufra disease management in rainfed lowland and irrigated rice. *Bangladesh Journal of Botany* 25, 13–18.

Rahman, M.L. and Evans, A.A.F. (1988) Studies on host–parasite relationships of rice stem nematode, *Ditylenchus angustus* (Nematoda: Tylenchida) on rice, *Oryza sativa,* L. *Nematologica* 33, 451–459.

Rahman, M.L. and McGeachie, I. (1982) Occurrence of white tip disease in deepwater rice in Bangladesh. *International Rice Research Newsletter* 7 (5), 15.

Rahman, M.L. and Taylor, B. (1983) Nematode pests associated with deepwater rice in Bangladesh. *International Rice Research Newsletter* 8 (5), 20–21.

Rahman, M., Sharma, B.R. and Miah, S.A. (1981) Incidence and chemical control of ufra in boro fields. *International Rice Research Newsletter* 6 (2), 12.

Ramakrishnan, S., Varadharajan, G. and Sutharsan, P.D. (1984) TKM9 is resistant to rice-root nematode. *International Rice Research Newsletter* 9 (5), 20.

Ramana, K.V. and Rao, Y.S. (1975) Histopathogenesis in rice roots infested by *Hoplolaimus indicus. Indian Journal of Nematology* 5, 232–234.

Ramana, K.V. and Rao, U.S. (1978) Evaluation of yield losses due to lance nematode, *Hoplolaimus indicus* Sher in rice. *Andhra Agriculture Journal* 24, 124–125.

Randhawa, N., Singh, I., Sakhuja, P.K. and Malhi, S.S. (1991) Effect of date of transplanting of basmati rice cultivars and spacing on the population build-up of rice root nematode, *Hirschmanniella oryzae. Indian Journal of Nematology* 21, 131–135.

Rao, J. (1985) Population build-up of white tip nematode (*Aphelenchoides besseyi* Christie, 1942) in different monoxenic cultures. *Oryza* 22, 45–49.

Rao, J., Kauraw, L.P. and Anand, P. (1994) Deterioration of seed quality due to interaction between *Aphelenchoides besseyi* and *Curvularia lunata* in rice. *Indian Phytopathology* 47, 92–93.

Rao, Y.S. (1976) Abnormal development of earheads in rice (*Oryza sativa* L.) due to the white tip nematode (*Aphelenchoides besseyi* Christie, 1942). *Current Science* 45, 560–561.

Rao, Y.S. and Biswas, H. (1973) Evaluation of yield losses in rice due to the root-knot nematode *Meloidogyne incognita. Indian Journal of Nematology* 3, 74.

Rao, Y.S. and Israel, P. (1973) Life history and bionomics of *Meloidogyne graminicola,* the rice root-knot nematode. *Indian Phytopathology* 26, 333–340.

Rao, Y.S. and Jayaprakash, A. (1977) Leaf chlorosis in rice due to a new cyst nematode. *International Rice Research Newsletter* 2 (4), 5.

Rao, Y.S. and Jayaprakash, A. (1978) *Heterodera oryzicola* n.sp. (Nematoda: Heteroderidae) a cyst nematode on rice (*Oryza sativa* L.) from Kerala State, India. *Nematologica* 24, 241–346.

Rao, Y.S. and Mohanadas, C. (1976) Occurrence of *Caloosia heterocephala* n.sp. (Nematoda: Hemicycliophoridae) on roots of rice (*Oryza sativa*). *Nematologica* 22, 227–234.

Rao, Y.S. and Prasad, J.S. (1977) Root-lesion nematode damage in upland rice. *International Rice Research Newsletter* 2 (1), 6–7.

Rao, Y.S., Prasad, J.S. and Panwar, M.S. (1986a) Nematode problems in rice: crop losses, symptomatology and management. In: Swarup, G. and Dasgupta, D.R. (eds) *Plant Parasitic Nematodes of India. Problems and Progress.* Indian Agricultural Research Institute, New Delhi, pp. 279–299.

Rao, Y.S., Prasad, J.S. and Panwar, M.S. (1986b) Stem nematode (*Ditylenchus angustus*) a potential pest of rice in Assam and West Bengal, India. *International Nematology Network Newsletter* 3 (4), 24–26.

Rathaiah, Y. (1988) Volunteer rice, host for diseases in DWR. *Deepwater Rice* 14, 1–2.

Rathaiah, Y. and Das, G.R. (1987) Ufra threatens deepwater rice in Majuli, Assam. *International Rice Research Newsletter* 12 (4), 29.

Razjivin, A., Fernandez, M., Ortega, J. and Quincosa, A. (1981) Nuevas especies de *Hirschmanniella* (Nematoda: Pratylenchidae), parasitas de plantas indeseables en plantaciones de arroz. *Poeyana* 216, 1–11.

Reversat, G. (1975) Etude préliminaire de la survie en anadrobiose des juveniles du nématode *Heterodera oryzae* (Tylenchida: Heteroderidae). *Comptes Rendus de l'Académie des Sciences, Paris Série D* 280, 2865–2868.

Ribeiro, A.S. (1977) Tratamento de sementes de arroz irrigado para controle da disseminacao do nematoide *Aphelenchoides besseyi.* In: *Anais, VIII Reuniano da Cultura do Arroz,* Rio Grande do Sul, Embressa Brasileria de Pesquisa Agropecuaria, pp. 138–142.

Roy, A.K. (1973) Reaction of some rice cultivars to the attack of *Meloidogyne graminicola. Indian Journal of Nematology* 3, 72–73.

Roy, A.K. (1976a) Pathological effects of *Meloidogyne graminicola* on rice and histopathological studies on rice and maize. *Indian Phytopathology* 29, 359–362.

Roy, A.K. (1976b) Effect of decaffeinated tea waste and water hyacinth compost on the control of *Meloidogyne graminicola* on rice. *Indian Journal of Nematology* 6, 73–77.

Roy, A.K. (1977a) Weed hosts of *Meloidogyne graminicola. Indian Journal of Nematology* 7, 160–163.

Roy, A.K. (1977b) Host suitability of some crops to *Meloidogyne graminicola. Indian Phytopathology* 30, 483–485.

Roy, A.K. (1978) Effectiveness of rotation with non-host or fallow on reducing infestation of *Meloidogyne graminicola. Indian Journal of Nematology* 8, 156–158.

Roy, A.K. (1982) Survival of *Meloidogyne graminicola* eggs under different moisture conditions *in vitro. Nematologia Mediterranea* 10, 221–222.

Sahoo, C.R. and Sahu, S.C. (1993a) Effect of different nematicides in control of the root lesion nematode, *Pratylenchus zeae* in rice. *Annals of Plant Protection Sciences* 1, 139–140.

Sahoo, C.R. and Sahu, S.C. (1993b) Weeds and wild rices as alternate hosts of *Pratylenchus zeae* in upland paddies. *Journal of Research APAU* 21, 51–52.

Salazar, L. and Quesada, M. (1999) Nuevo hallazgo relacionado con la distribucion geografica de *Meloidogyne salasi* en arroz y nuevos hospederos. *Agronomia Mesoamericana* 10, 103–105.

Sancho, C.L., Salazar, L. and Lopez, R. (1987) Efecto de la densidad inicial del inoculo sobre la patogenicidad de *Meloidogyne salasi* en tres cultivares de arroz. *Agronomia Costarricense* 11, 233–238.

Sato, T., Koyama, T. and Koshihara, T. (1970) [Relation between the occurrence of rice root nematodes (*Hirschmanniella oryzae* Luc & Goodey and *Hirschmanniella imamuri* Sher) and the cultivating practices of paddy rice plant.] *Bulletin of the Tohoku National Agricultural Experiment Station* No. 39, 207–219.

Segeren-V.d. Oever, H. and Sanchit-Bekker, M.L. (1984) Observations on *Meloidogyne oryzae* Maas. Sanders and Dede 1978 in irrigated rice in Suriname. *Surinaamse Landbouw* 32, 51–59.

Sein, T. (1977a) Varietal resistance to ufra disease in Burma. *International Rice Research Newsletter* 2 (2), 3.

Sein, T. (1977b) Seed-borne infection and ufra disease. *International Rice Research Newsletter* 2 (2), 6.

Sein, T. (1977c) Testing some pesticides against ufra disease. *International Rice Research Newsletter* 2 (2), 6.

Sein, T. and Zan, K. (1977) Ufra disease spread by water flow. *International Rice Research Newsletter* 2 (2), 5.

Sharma, R.D. (1980) Patogenicidade de um nematoide *Meloidogyne javanica* (Treub, 1885) Chitwood, 1949 ao arroz (*Oryza sativia* L.). *Sociedade Brasileira de Nematologia* No. 5, 91–98.

Sharma, S.B. and Swarup, G. (1985) *Cyst Forming Nematodes of India.* Cosmo Publications, New Delhi.

Sharma, S.B., Pande, S., Saha, M., Kaushal, K.K., Lal, M., Singh, M., Pokharel, R., Upreti, R.P. and Singh, K. (2001) Plant parasitic nematodes associated with rice and wheat based cropping systems in Nepal. *International Journal of Nematology* 11, 35–38.

Shimizu, K. (1971) [Influence of the rice cyst nematode on the yield of paddy rice.] *Annual Meeting of the Japanese Society of Applied Entomology and Zoology.* Fuchu, Tokyo, 7–9 April 1971, p. 32.

Shimizu, K. (1977) [Seasonal fluctuation and rhizospherical distribution of population of the upland rice cyst nematode, *Heterodera elachista*.] *Japanese Journal of Nematology* 7, 49–57.

Sivakumar, C.V. (1987a) Post-embryonic development *Aphelenchoides besseyi in vitro* and its longevity in stored rice seeds. *Indian Journal of Nematology* 17, 147–148.

Sivakumar, C.V. (1987b) The rice white tip nematode in Kanyakumari district Tamil Nadu, India. *Indian Journal of Nematology* 17, 72–75.

Soriano, I.R.S., Prot, J.-C. and Matias, D.M. (2000) Expression of tolerance for *Meloidogyne graminicola* in rice cultivars as affected by soil type and flooding. *Journal of Nematology* 32, 309–317.

Sperandio, C.A. and Amaral, A.S. (1994) Nematoides fitoparasitas associados ao arroz irrigado no Rio Grande do Sul. *Lavoura Arrozeira* 47, 3–5.

Srivastava, A.N. and Sethi, C.L. (1987) Additional host records of *Heterodera sorghi*. *Indian Journal of Nematology* 17, 342–343.

Stirling, G.R. (1984) A needle nematode (*Paralongidorus* sp.) causing poor growth of rice in North Queensland. *Australasian Plant Pathology* 13, 22–23.

Stirling, G.R. (1985) Observations on the biology and ecology of *Paralongidorus australis* in rice in North Queensland. *Nematologica* 31 (1986), 218–228.

Stirling, G.R. and McCulloch, J.S. (1985) *Paralongidorus australis* n.sp. (Nematoda, Longidoridae), causing poor growth of rice in Australia. *Nematologica* 30 (1984), 387–394.

Stirling, G.R. and Shannon, E. (1986) Needle nematodes. A potential problem for the Burdekin rice industry. *Queensland Agricultural Journal* 112, 258–261.

Stirling, G.R. and Vawdrey, L.L. (1985) Distribution of a needle nematode, *Paralongidorus australis*, in rice fields and areas of natural vegetation in North Queensland. *Australasian Plant Pathology* 14, 71–72.

Stirling, G.R., Vawdrey, L.L. and Shannon, E.L. (1989) Options for controlling needle nematode (*Paralongidorus australis*) and preventing damage to rice in northern Queensland. *Australian Journal of Experimental Agriculture* 29, 223–232.

Sudakova, M.I. (1968) [Effect of temperature on the life cycle of *Aphelenchoides besseyi*.] *Parazitologiya* 2, 71–74.

Sudakova, M.I. and Stoyakov, A.V. (1967) [Reproduction and duration of life of *Aphelenchoides besseyi* Christie, 1942.] *Zoologicheski Zhurnal* 46, 1079–1099.

Tamura, I. and Kegasawa, K. (1958) [Studies on the ecology of the rice nematode, *Aphelenchoides besseyi* Christie. II. On the parasitic ability of rice nematodes and their movement into hills.] *Japanese Journal of Ecology* 8, 37–42.

Tamura, I. and Kegasawa, K. (1959a) [Studies on the ecology of the rice nematode, *Aphelenchoides besseyi* Christie. IV. The injurious feature and population dynamics of nematodes in unhulled rice grain with special reference to the cultural environment of rice plant.] *Japanese Journal of Ecology* 9, 65–68.

Tamura, I. and Kegasawa, K. (1959b) [Studies on the ecology of the rice nematode, *Aphelenchoides besseyi* Christie, V. On the abnormal growth of rice plant and decrease in yield caused by rice nematode.] *Japanese Journal of Ecology* 9, 120–124.

Tandingan, I.C., Prot, J.-C. and Davide, R.G. (1996) Influence of water management on tolerance of rice cultivars for *Meloidogyne graminicola*. *Fundamental and Applied Nematology* 19, 189–192.

Taylor, A.L. (1969) Nematode parasites of rice. In: Peachey, J.E. (ed.) *Nematodes of Tropical Crops*. Technical Communication, No. 40, Commonwealth Bureau of Helminthology, St Albans, UK, pp. 264–268.

Taylor, D.P. (1978) Parasitism of banana by *Heterodera oryzae*. *Revue de Nématologie* 1, 165–169.

Taylor, D.P., Ngundo, B.N. and Othieno, S.M. (1972) Tanzania. Rice white tip nematode. *FAO Plant Protection Bulletin* 20, 41–42.

Tikhonova, L.V. (1966) [*Aphelenchoides besseyi* Christie, 1942 (Nematoda: Aphelenchoididae) of rice and its control.] *Zoologicheski Zhurnal* 45, 1759–1766.

Tikhanova, L.V. and Ivanchenko, Y.N. (1968) [The combined infection of rice with a fungus, *Piricularia* and a nematode, *Aphelenchus*.] *Trudy V.Yesoy,uznogo Instituta Gel'mintologii im. K. 1. Skryabina* 14, 287–290.

Timm, R.W. (1955) The occurrence of *Aphelenchoides besseyi* Christie, 1942 in deep water paddy of East Pakistan. *Pakistan Journal of Science* 7, 47–49.

Todd, E.H. (1952) Further studies on the white tip disease of rice. *Proceedings, Association of Southern Agricultural Workers, 49th Annual Convention*, p. 141.

Todd, E.H. and Atkins, J.G. (1958) White tip disease of rice. 1. Symptoms, laboratory culture of nematodes and pathogenicity tests. *Phytopathology* 48, 632–637.

Togashi, K. and Hoshino, S. (2001) Distribution pattern and mortality of the white tip nematode, *Aphelenchoides besseyi* (Nematoda: Aphelenchoidae), among rice seeds. *Nematology* 3, 17–24.

Tsay, T.T., Cheng, Y.H., Teng, Y.C., Lee, M.D., Wu, W.S. and Lin, Y.Y. (1998) Bionomics and control of rice white tip disease nematode, *Aphelenchoides besseyi*. *Plant Protection Bulletin Taipei* 40, 277–286.

Uebayashi, Y., Amano, T. and Nakanishi, I. (1976) [Studies on the abnormal rice kernel 'Kokutemnai'. Mechanisms of symptom development.] *Japanese Journal of Nematology* 6, 67–72.

van den Berg, E. and De Waele, D. (1989) Further observations on nematodes associated with rice in South Africa. *Phytophylactica* 21, 125–130.

Van der Vecht, J. and Bergman, B.H.H. (1952) Studies on the nematode *Radopholus oryzae* (Van Breda de Haan) Thorne and its influence on the growth of the rice plant. *Pemberitaarl Balai Besar Penielidikan Pertanian Bogor* No. 131.

Venkitesan, T.S., Nalinakumari, T. and Satykumar, J. (1979) Distribution of rice root nematode in Kerala. *Indian Journal of Nematology* 9, 55.

Verma, R.S. (1973) Two new species in the subfamily Longidorinae (Nematoda) from Uttar Pradesh, India, with a key to species of *Paralongidorus* Siddiqi *et al.*, 1963. *Zoologischer Anzeiger* 190, 170–174.

Vuong, H.H. (1969) The occurrence in Madagascar of the rice nematodes. *Aphelenchoides besseyi* and *Ditylenchus angustus*. In: Peachey, J.E. (ed.) *Nematodes of Tropical Crops*. Technical Communication No. 40. Commonwealth Bureau of Helminthology, St Albans, UK, pp. 274–288.

Vuong, H.H. and Rabarijoela, P. (1968) Note préliminaire sur la présence des nématodes parasites du riz à Madagascar: *Aphelenchoides besseyi* Christie 1942, *Ditylenchus angustus* (Butler, 1913) Filipjev, 1936. *L'Agronomie Tropicale, Nogent* 23, 1025–1048.

Wang, J.X. and Pan, C.S. (1999) Studies of rice root nematodes (*Hirshchmanniella* species). *Journal of Xiamen University (Nature Science)* 16 (2), 155–159.

Wu, H.P., Yang, R.Z. and Xie, Y.L. (1995) The identification and distribution of rice root nematode (*Hirschmanniella* spp.) in Anhui Province. *Journal of Anhui Agricultural University* 22 (3), 239–245.

Yamada, W. and Shiomi, T. (1950) [Studies on the rice white tip disease. II. Disease control with special reference to rice seed disinfection.] *Special Bulletin, Okayama Prefecture Agricultural Experiment Station* No. 47, 1–8.

Yamada, W., Shiomi, T. and Yamamoto, H. (1953) [Studies on the white tip disease of rice plants. III. Hibernation of the causal nematode, varietal resistance of rice plants to the disease, and effect of rice culture upon the disease occurrence.] *Special Bulletin, Okayama Prefecture Agricultural Experiment Station* No. 48, 27–36.

Yamsonrat, S. (1967) Studies on rice-root nematodes (*Hirschmanniella* spp.) in Thailand. *Plant Disease Reporter* 51, 960–963.

Yang, B.J., Hu, K.J. and Xu, B.W. (1988) A new species of root-knot nematode *Meloidogyne lini* n.sp. parasitizing rice. *Journal of Yunnan Agricultural University* 3, 11–17.

Yik, C.P. and Birchfield, W. (1979) Host studies and reactions of cultivars to *Meloidogyne graminicola*. *Phytopathology* 69, 497–499.

Yoshii, H. (1951) [On the growth habit and yield of rice plants affected with *Aphelenchoides oryzae*.] *Science Bulletin of the Faculty of Agriculture, Kyushu University* 12, 133–141.

Yoshii, H. and Yamamoto, S. (1950a) A rice nematode disease 'Senchu Shingare Byo'. I. Symptom and pathogenic nematode. *Journal of the Faculty of Agriculture, Kyushu University* 9, 209–222.

Yoshii, H. and Yamamoto, S. (1950b) A rice nematode disease 'Senchu Shingare Byo'. II. Hibernation of *Aphelenchoides oryzae*. *Journal of the Faculty of Agriculture, Kyushu University* 9, 223–233.

Yoshii, H. and Yamamoto, S. (1950c) A rice nematode disease 'Senchu Shingare Byo'. IV. Prevention of the present disease. *Journal of the Faculty of Agriculture, Kyushu University* 9, 293–310.

Yoshii, H. and Yamamoto, S. (1951) [On some methods for the control of the rice nematode disease.] *Science Bulletin of the Faculty of Agriculture, Kyushu University* 12, 123–131.

Zhang, S.S. (1987) Identification of rice root nematodes (*Hirshchmanniella* spp.) in Fujian, China. *Journal of Fujian Agricultural College* 16 (2), 155–159.

5 Nematode Parasites of Cereals*

Alexander H. McDonald[1] and Julie M. Nicol[2]

[1]*ARC-Grain Crops Institute, Private Bag X1251, Potchefstroom 2520, Republic of South Africa;* [2]*CIMMYT International, PK 39, Emek, Ankara, Turkey*

Cereals constitute the world's most important source of food. Amongst cereals, wheat, maize and rice occupy the most eminent positions in terms of production, acreage and source of nutrition, particularly in developing countries. Barley, sorghum, millet, oat, rye and the other edible grains, conversely, are restricted to specific growing regions and are limited in area under cultivation. It has been estimated that about 70% of the land cultivated for food crops is devoted to cereal crops. The contribution of individual crops to total world cereal production can be seen in Table 5.1. Cereals as a source of human nutrition and animal feed provide both calories and proteins. It has been estimated that wheat will produce more calories, proteins and essential amino acids from a hectare of arable land than the livestock that can be supported by that land (Johnson, 1984).

Although the introduction of new cultivars of wheat, maize, rice and other cereals has boosted agricultural output, the yield potential of the new cultivars has not been fully expressed and is often far below theoretical maximum yields. This disparity between actual and theoretical yield expression can be attributed to 'production constraints'. Attention has, therefore, been focused on minimizing these constraints to increase production. Although insect pests and diseases have long been recognized as important constraints affecting crop production, extensive research on the 'weak linkages' such as plant parasitic nematodes in the plant–pest system is lacking. As most nematodes live in the soil, they represent one of the most difficult pest problems to identify, demonstrate and control (Stirling *et al.*, 1988). Farmers, agronomists and pest management consultants commonly underestimate their effects, but it has been estimated that some 10% of the world crop production is lost as a result of plant nematode damage (Whitehead, 1998). It is pertinent also to consider in many of the cereal systems discussed in this chapter the interaction of nematodes with other plant pathogens, particularly soil-borne fungi, and in many cases the synergism which results in more damage than either pathogen alone.

Management of nematodes may be approached by using a complement of methods in an integrated pest management system or may involve only one of these methods. Some of the most commonly

*A revision of the chapter by G. Swarup and Carlos Sosa-Moss.

Table 5.1. Contribution of cereals in world food production.

	Total production (1000 t)								
Continent	Maize	Rice	Wheat	Barley	Sorghum	Millet	Oat	Rye	Total cereals
World	606,436	593,508	586,088	1135,227	58,321	27,273	25,825	21,033	2,073,993
Developed	332,354	25,618	314,747	112,658	15,425	1,328	23,447	19,898	845,475
Developing	274,082	567,891	271,340	22,568	42,896	25,945	2,377	1,136	1,208,235
Africa	42,185	17,020	16,604	3,476	19,859	13,423	143	32	114,847
Asia	163,953	540,917	253,661	19,941	12,045	12,544	1,181	1,057	1,007,994
Australia	376	1,348	20,642	5,893	1,829	51	1,226	21	32,014
Europe	71,174	3,186	191,338	85,900	713	935	16,655	19,308	403,220
North and Central America	271,780	11,373	84,012	18,305	18,964	282	5,399	515	411,317
South America	56,777	19,646	19,498	1,382	4,907	39	1,184	101	103,682

Sourced from FAO website. Figures represent 5 year average 1998–2002.

practised methods will be discussed, including crop rotation, use of resistant and tolerant cultivars or varieties, cultural practices and chemicals. It is important to stress that the most appropriate control method will be determined by the nematode involved and the economic feasibility of implementing a possible management practice. The purpose of this chapter is to provide an insight into the economically important nematodes on cereal grains other than rice, which is covered separately in Chapter 4. Information is presented here on their currently known distribution, biology and life cycle, damage potential, economic importance and management options that exist for their control.

The relationship between initial nematode density and crop yield is important in determining the economic impact of plant parasitic nematodes on a cereal crop. Cultivar selection and inherent environmental conditions affect crop loss at a specific population density. Economic threshold levels whereby these nematodes cause damage are not reported to a great extent in this review, as many of these numbers can be misleading. Very few published papers have used the same sampling and extraction protocols, hence nematode numbers presented are reflective of the sampling and extraction methodologies used by different authors and are therefore difficult to compare.

For further references and illustration of many of these nematodes, please refer to the reviews of Kort (1972), Griffin (1984), Sikora (1988), Rivoal and Cook (1993), De Waele and McDonald (2000), Kollo (2002) and Nicol (2002).

Wheat and Barley

Today, wheat (*Triticum aestivum*) is grown on more than 270 Mha, which is more land than any other commercial crop and continues to be the most important food grain source for humans. It is grown in most countries in the world in a range of climates and has a number of end uses, mostly human, but also feed. With the predicted 35% increase in population to 7.9 billion by the year 2025, it is clear that wheat demand will increase by about 48%; 584 Mt in 1995–1999 annual production to 860 Mt (Marathee and Gomez-MacPherson, 2001). These increases in production will occur in two ways: (i) by expanding the wheat area; and (ii) by improving the yield per unit area sown. It is expected that the rate of increase in production will slow down as there will be little new area available for cultivation or irrigation, and the gap between yield potential and actual yield will be smaller, particularly in developing countries (Curtis, 2002). It is expected by 2030 that over half the wheat production will be from developing countries (Marathee and Gomez-MacPherson, 2001).

Another closely related cereal, barley (*Hordeum vulgare*), is grown in nearly all cultivated areas of the temperate zones and in many subtropical areas in addition to the high altitude sections of the torrid zones of both hemispheres. The uses of barley are for human consumption, animal feed and the production of malt for beer. Barley is one of the most dependable crops where drought occurs.

Three other important cereals to mention are oat (*Avena sativa*), rye (*Secale cereale*) and triticale (wheat × rye), which are used predominantly for animal feed. Oat grows best in cool, moist regions of the world, has a high water requirement and is less sensitive than wheat or barley to soil conditions (Leonard and Martin, 1965). Rye is considered as one of the most resilient cereals where poor soils, severe winters and drought occur, and is used for grain, pasture or even a green manure crop (Leonard and Martin, 1965). Triticale is less well known; however, it is increasing in importance due to its wide adaptability in poor soils and climates and for use as animal feed. A general observation is that rye offers in many cases a high level of resistance against the cereal root nematodes. Consequently, this applies to some triticale cultivars, depending on the regions of rye chromosomes incorporated. These two crops offer great potential in rotational control of cereal nematodes, particularly where small grain cereals are the predominant feature of the cropping system.

The availability of fertilizer and water are the two most important abiotic factors regulating yields of most plants including small cereal cultivars. Crop yields are also profoundly influenced by other factors such as cultivar selection, pesticide use and management practices. These, in turn, affect nematode population fluctuations and the degree of economic loss. It is appreciated that plants undergoing stress, especially with regard to water and nutrients, are more likely to suffer damage from cereal nematodes that attack roots.

Nematodes of Wheat and Barley

Although quite a sizeable number of plant parasitic nematodes have been recorded associated with wheat and barley, only a few species can be regarded as economically important. The most important nematodes of these crops are: (i) cereal cyst nematodes, *Heterodera* species; (ii) the root lesion nematodes, *Pratylenchus* species; (iii) the ear cockle nematode, *Anguina tritici*; (iv) the root knot nematodes, *Meloidogyne* species; and (v) the stem nematode, *Ditylenchus dipsaci*.

Heterodera avenae

Distribution

The cereal cyst nematodes are a group of several closely related species and are considered to be one of the most important groups of plant parasitic nematodes on a worldwide basis. Recent taxonomic advances have reclassified some of the previously known strains. *H. avenae* Gotland strain has been renamed *H. filipjevi*. They have a global distribution, and the clear delineation of specific species to certain climatic zones is not well defined. Several of the species have been found in tropical and subtropical environments.

The most commonly recorded species of economic importance on cereals is *Heterodera avenae*, which has been detected in many countries, including Australia, Canada, Israel, South Africa, Japan and most European countries (Kort, 1972), as well as India (Sharma and Swarup, 1984; Sikora, 1988), China (Peng, China, personal communication) and countries within North Africa and western Asia, including Morocco, Tunisia, Libya and Pakistan (Sikora, 1988), Turkey (Rumpenhorst *et al.*, 1996) and recently Algeria (Mokabli *et al.*, 2001) and Saudi Arabia (Ibrahim *et al.*, 1999).

H. avenae is considered the principal species on temperate cereals (Rivoal and Cook, 1993), while another important cereal species, *H. latipons*, is essentially only Mediterranean in distribution, being found in Syria (Sikora and Oostendorp, 1986; Scholz, 2001), Israel (Kort, 1972; Mor *et al.*, 1992), Cyprus (Sikora, 1988), Turkey (Rumpenhorst *et al.*, 1996), Italy and Libya (Kort, 1972). However, it is also known to occur in northern Europe (Sabova *et al.*, 1988). Another species with an increasingly wide distribution is *H. filipjevi*, formerly know as Gotland strain of *H. avenae*. It has been found in Russia (Balakhnina, 1989; Subbotin *et al.*, 1996), Tadzhikistan (Subbotin *et al.*, 1996), Iran (Sturhan, 1996; Tanha Maafi *et al.*, 2003), India (Bishnoi and Bajaj, 2000, 2002), Sweden (Cook and Noel, 2002) and Turkey (Rumpenhorst *et al.*, 1996; Nicol *et al.*, 2002).

Other *Heterodera* species known to be of importance to cereals include *H. hordecalis* in Sweden, Germany and Britain (Andersson, 1974; Cook and York, 1982a; Sturhan, 1982), and in Iran (Maafi, 2003), *H. zeae*, which is found in India, Pakistan (Sharma and Swarup, 1984; Maqbool, 1988) and Iraq (Stephan, 1988), and various others including *H. mani*, *H. bifenestrata* and *H. pakistanensis*, as well as an unrelated species of cyst nematode, *Punctodera punctata* (Sikora, 1988).

Biology and life cycle

The life cycle of *H. avenae* involves only one generation during a cropping season, irrespective of geographic region, and the host range of this nematode is restricted to graminaceous plants. There is sexual dimorphism, with males remaining worm-like, whereas females become lemon-shaped and spend their life inside or attached to a root. Nematode infection often causes a 'bushy-knotted' appearance to the overall root system (Plate 3A). Eggs are retained within the female's body and, after the female has died, the body wall hardens to a resistant brown cyst, which protects the eggs and juveniles. The moment such cysts turn brown, juvenile emergence stops completely. The eggs within a cyst remain viable for several years (Kort, 1972).

The induction of dormancy appears to be correlated with the change in cyst colour as well as with increases in temperature. Juvenile emergence from eggs in brown cysts requires a period of dormancy of 2 or more months and is strongly regulated by an increase in temperature (Banyer and Fisher, 1971, 1976). Rajan (1984) notes that when suitable temperature conditions are available, emergence of juveniles may take place spontaneously. Often the periods of mass emergence from cysts coincide with a cropping season. *H. avenae* juveniles penetrate roots and pass through cells towards the stele where they induce the development of a feeding syncytium. Adult females are white and lemon-shaped, turning brown after death. Males are essential for fertilization.

Temperature, availability of moisture and root diffusate are important determinants of juvenile emergence. Emergence of *H. avenae* can take place at temperatures between 10 and 25°C, with the optimum between 20 and 22°C (Winslow, 1955; Swarup and Gill, 1972). The optimum for the Australian *H. avenae* population is 10°C (Brown, 1987). Fluctuating temperatures or alternate exposure of cysts to low and high temperatures stimulates *H. avenae* maximum emergence; release of juveniles at low temperatures of 10–15°C can be obtained with wheat and barley root diffusates. Root diffusate from 1-week-old barley seedlings stimulates emergence of juveniles from the cysts (Gill, 1967; Williams and Beane, 1972). To date, the other species are anticipated to have similar biology; however, few comparable studies have been conducted. Furthermore, ecotypes of *H. avenae* have been reported, which result in differences in hatching cycles from induction or suppression of dormancy (diapause) by different temperature conditions (Rivoal and Cook, 1993).

Survival

Encysted eggs of *H. avenae* survive for several years at 5°C when stored at low relative humidity (Kyrou, 1976; Meagher, 1982). Furthermore, eggs in cysts are quite susceptible to drying, with prolonged exposures markedly reducing juvenile emergence. However, populations present in the tropics that are exposed to prolonged dry summer conditions do not completely lose their viability. Even in the hot dry summers existing in Israel and India, juveniles in the cyst remain viable until suitable temperatures for emergence are reached (Minz, 1956). Under fallow, non-host or resistant cultivars populations decline by 70–80% annually by hatching and death of juveniles (Andersson, 1982).

Environmental factors

Many abiotic factors, e.g. fertility, pH, soil type and organic matter content, influence nematode population development and damage severity (Duggan, 1961). Moderate nematode population levels under favourable environmental conditions for plant growth may not cause as much damage as when plant growth is restricted by moisture stress or low fertility levels (Kornobis *et al.*, 1980). Increased nitrogen application is known to reduce the intensity of nematode damage to the crop, but at high nematode population levels this may no longer hold true (Germershauzen *et al.*, 1976).

H. avenae has been associated with economic levels of damage almost exclusively in light soils. Part of the reason for this association is that sand particles are optimal for nematode development and sand has a lower water-holding capacity. However, the nematode also thrives well in the slightly heavier soils of the western area of Rajastan, India. Irrespective of soil type, when the intensity of cropping exceeds a certain limit, damage is imminent (Kort, 1972).

Symptoms of damage

The symptoms associated with *H. avenae* damage are characterized by uneven patches of poor growing plants, randomly distributed throughout a field, and may vary in size from 1 to 100 m^2 or more (Fig. 5.1; Plate 3B). Damage to plants and the size and number of patches are directly related to nematode population levels as well as nematode distribution in the field. Under monoculture, the patches coalesce and damage can uniformly cover the entire field within 3–4 years. Severely infected plants remain stunted, 30–60 cm high. The leaves of cereal plants become pale, yellowish-green in colour, with thin and narrow leaf blades and generally fewer tillers. Ears, if formed, have very few grains.

Symptoms produced on roots are different, depending on the host. Wheat attacked by *H. avenae* shows increased root production such that the roots have a 'bushy-knotted' appearance (Plate 3A), with several females visible at each knot (Rivoal and Cook, 1993). Oat roots are shortened and thickened, while barley roots appear less affected. Other species of *Heterodera* also appear to produce host-specific symptoms on the roots of cereals. For example, in Israel, *H. latipons* did not produce knotted roots like *H. avenae* (Mor *et al.*, 1992).

Such root symptoms are recognizable within 1–2 months after sowing in tropical environments. Under European conditions, root division takes place at the points of juvenile invasion, giving an appearance of a knotted root system. In Australia, a much-branched root system is characteristic of infested wheat and barley and to a lesser extent oat. Tufting of roots may not be noticeable during field examination due to adhering soil (Holdeman and Watson, 1977). Wheat and barley in India are sown in the last 2 weeks of November or early December. The above-ground symptoms of damage can be seen within a month after sowing, becoming quite marked by the end of January. During this period, second stage juveniles are abundant in the soil. By mid-February, white females can be seen attached to roots.

Fig. 5.1. Uneven patchy growth of a wheat crop in field infested with *Heterodera avenae*. (Photo: R.A. Sikora.)

Pathotypes

Unfortunately, populations of the cereal cyst nematode are very heterogeneous for virulence but also differ in cereal species host range (Cook and Noel, 2002). The existence of pathotypes in *H. avenae* populations was noticed on barley cultivars as early as 1920 in Sweden. Results obtained with the International Test Assortment in 1972–1973 demonstrated the existence of more pathotypes than originally recognized, especially in subtropical regions. For instance, Barley 191, which is reported to be resistant to the known populations of *H. avenae* in Europe, is susceptible to *H. avenae* populations in Australia, Norway and India (Stoen, 1971; Brown, 1972; Mathur *et al.*, 1974). In these tests, it is quite difficult to make clear-cut distinctions between resistance and susceptibility based on the number of cysts alone. Pathotypes may also occur in mixtures, further complicating delineation of the pathotype in a particular sample. This and the inclusion of additional hosts other than those recommended in the International Test Assortment may be responsible for conflicting observations on pathotype numbers from India (Mathur *et al.*, 1974; Swarup *et al.*, 1979). Furthermore, in northern Europe, most populations reproduce well on oat, but in southern Europe, North Africa and part of Asia, oat are nonhosts to most populations (Cook and Noel, 2002). Although the pathotype scheme (Table 5.2) by Andersen and Andersen (1982) has the simplicity of being based on known R-genes or at least resistance sources, it suffers from underestimating the polymorphism of resistance and avirulence genes (Cook and Noel, 2002).

Previously Mathur *et al.* (1974) reported five pathotypes from soils of North India on the basis of host differentials, while Swarup *et al.* (1979) reported two pathotypes from India. Siddiqui and Hussain (1989) studied the Uttar Pradesh population, and reported two different populations on the basis of host differentials and designated them as pathotype 1 and 2. Andersen and Andersen (1982) concluded on the basis of literature that three pathotypes Ha21, Ha31 and Ha41 occur in India. Bekal *et al.* (1998) indicated that the Nazafgargh population of India might belong to pathotype Ha71. Recent studies by Bishnoi and Bajaj (2000, 2002) concluded that there is a *H. avenae* complex in India. They reported that the Delhi, Jaipur, Narnaul, Sirsa and Udaipur populations belong to pathotype Ha21, whereas Himachal Pradesh, Ambala and Punjab populations had their affinities to Ha31 and Ha41 pathotypes, respectively. On the basis of detailed morphological studies, eight populations could be distinguished in two different morphological groups. One group that comprised Delhi, Jaipur, Narnaul, Sirsa and Udaipur populations represented *H. avenae*, whereas, H.P., Ambala, and Punjab populations represented *H. filipjevi* (Madzhidov, 1981). Stelter (1984) designated them Hf31 and Hf41. It is clear from recent studies that pathotype-1 (H.P., Ambala and Punjab populations) of *H. avenae* reported earlier by Mathur *et al.* (1974) and Swarup *et al.* (1979) now represent *H. filipjevi*, whereas other populations are *H. avenae*.

Unfortunately, the presence of pathotypes with other *Heterodera* species are less well understood but are essential to understand the biology of the nematode and possible control options involving host resistance. The different species of *Heterodera* are difficult to differentiate easily and require a strong taxonomic understanding of morphological traits of cysts or juveniles. Recent molecular techniques such as restriction fragment length polymorphism (RFLP) of ribosomal DNA have enabled solid taxonomic differentiation among several entities of the cereal cyst nematode complex (Bekal *et al.*, 1997; Subbotin *et al.*, 2000). See Chapter 2 for more details on these techniques.

Damage potential and economic importance

The damage threshold (i.e. the given population of a pathogen to cause a given yield loss) of *Heterodera* will be determined by many environmental and genotypic factors,

Table 5.2. Pathotypes of cereal cyst nematodes defined by an International Test Assortment of cereal cultivars.

	H. avenae group		Ha1 group		Ha2 group		Ha3 group		*H. hordecalis*		*H. bifenestra*		
Pathotype	Ha11	Ha21	Ha31	Ha41	Ha51	Ha61	Ha71	Ha12	Ha13	Ha23	Ha33	Hh1	Hb1
Differential cereal species cultivar [R-gene][a]													
Barley													
Emir [+ ex Emir]	S	S	—	S	—	R	S	S	S	S	S	S	S
Ortolan [*Rha1*]	R	R	R	R	R	R	R	S	S	S	S	S	S
Siri [*Rha2* + ex Herta]	R	R	R	S	S	S	R	R	S	S	S	S	S
Moroco [*Rha3* +]	R	R	R	R	R	R	R	R	R	R	R	R	S
Varde	S	—	—	S	—	S	S	S	S	S	S	S	S
KVL191[*Rha2* +]	R	R	R	—	S	S	S	R	–	—	—	—	—
Bajo Aragon	R	—	—	R	—	R	R	R	S	S	S	S	(R)
Herta	S	S	R	—	R	—	R	S	S	—	—	—	—
Martin 403-2 [2 dom]	R	—	—	R	—	R	R	R	R	S	S	S	S
Dalmastische	(R)	—	—	S	—	R	(S)	S	S	(R)	S	(R)	S
La Estanzuela	—	—	—	—	—	—	S	—	—	(R)	—	(R)	S
Harlan 43	R	—	—	—	—	—	R	R	—	R	S	—	—
Oat													
SunII [minor genes]	S	R	R	R	R	S	R	S	S	S	S	R	S
Nidar	S	—	—	S	—	S	R	S	S	S	S	R	S
Pusa hybrid BS1 [1 dom]	R	R	—	R	R	R	R	R	S	R	S	R	S
Silva [> 1 gene]	(R)	—	—	R	—	(R)	R	(R)	(R)	(R)	S	R	S
A. sterilis 1376 [1–3 dom]	R	R	—	R	R	R	R	R	R	R	R	R	S
IGV.H 72-646	R	—	—	R	—	R	R	R	S	S	S	—	S
Wheat													
Capa	S	S	—	S	—	S	S	S	S	S	S	R	S
Loros [*Cre1* +]	R	R	—	R	—	(R)	R	R	(R)	S	S	R	(R)
Iskamish K-2-light	S	—	—	R	—	(R)	—	S	S	S	S	R	R
AUS 10894 [*Cre1* +]	R	—	—	R	—	R	S	R	(R)	S	S	R	(R)
Psathias	—	—	—	S	—	—	—	S	S	S	R	R	S

Modified from Cook and Rivoal (1998).

[a]Resistance genes are those in italics (*Rha1*, *Rha2*, *Rha3* define the pathotype groups); dom, dominant gene; +, additional gene(s) inferred; S, susceptible; R, resistant (< 5% new females on susceptible control); () intermediate; —, no documentation. Sourced from Rivoal and Cook (1993, 1998) and previously modified from Andersen and Andersen (1982) and their revision.

such as water and nutrient availability and tolerance and/or resistance reaction of a given cultivar or variety. Furthermore, interpretation of the damage threshold between specific nematological studies should be carried out with extreme caution, as very few studies are truly comparable, with inherent differences in sampling protocol, extraction procedure and nematode renumerification. It is for this reason that the studies conducted to date are only listed here. The reader should interpret these accordingly (Duggan, 1961; Stone, 1968; Dixon, 1969; Gill and Swarup, 1971; Meagher and Brown, 1974; Simon and Rovira, 1982; Handa *et al.*, 1985; Dhawan and Nagesh, 1987; Fisher and Hancock, 1991; Zancada and Althöfer, 1994; Al-Hazmi *et al.*, 1999; Ibrahim *et al.*, 1999).

Water stress is one of the key environmental conditions that can exacerbate damage caused by *H. avenae*. The number of juveniles penetrating host roots also has a direct bearing on the expression of damage. With increasing inoculum density, more juveniles penetrate the roots, but the percentage penetration decreases (O'Brien and Fisher, 1978). Gokte and Swarup (1984a) reported that an inoculum increase of 100 to 1000 eggs and juveniles/g of soil resulted in a fourfold increase in penetration, whereas the next tenfold increase caused only a twofold increase, affecting cyst production. The number of juveniles penetrating wheat roots increases linearly with increasing inoculum densities until a maximum is reached (O'Brien and Fisher, 1978).

H. avenae in the north-western part of India and in southern Australia is considered a major limiting factor of wheat and barley. Figures have been computed that suggest that for every 10 eggs/g of soil, there is a loss of 188 kg/ha in wheat and 75 kg/ha in barley (Duggan, 1961; Dixon, 1969). In the north-western part of India, significant yield increases of wheat and barley have been obtained after nematicidal treatments (Swarup *et al.*, 1976). Yield losses due to this nematode are 15–20% on wheat in Pakistan (Maqbool, 1988), 40–92% on wheat and 17–77% on barley in Saudi Arabia (Ibrahim *et al.*, 1999), and 20% on barley and 23–50% on wheat in Australia (Meagher, 1972). Staggering annual yield losses of £3 million in Europe and AUS$72 million in Australia have been calculated as being caused by *H. avenae* (Wallace, 1965; Brown, 1981). The losses in Australia are now greatly reduced due to their control with resistant and tolerant cultivars.

H. avenae and *H. zeae* are major pests of wheat and barley in Pakistan (Maqbool, 1988). In India, *H. zeae* is considered to be one of the most economically important nematodes attacking cereals (Sharma and Swarup, 1984). *H. avenae* has been associated with a severe disease present in India known as 'molya', which tends to occur only on the more temperate cereals such as barley and wheat, while tropical cereals such as sorghum and maize are non-hosts (Gill and Swarup, 1971; Sharma and Swarup, 1984).

Little is known about the economic importance of the species *H. latipons* even though it was first described in 1969 (Sikora, 1988). Recent studies by Scholz (2001) implicate yield loss with both barley and durum wheat with *H. latipons*. Field studies in Cyprus indicated a 50% yield loss on barley (Philis, 1988). Because the cysts are similar in size and shape, it is possible that previous findings of this recently described nematode species have been attributed erroneously to the economically important *H. avenae* (Kort, 1972). In West Asia and North Africa, *H. latipons* has been found on wheat and barley in four countries (Sikora, 1988). It has also been confirmed recently in Turkey (Rumpenhorst, 1996; Nicol *et al.*, 2002) and from several Mediterranean countries, associated with poor growth of wheat (Kort, 1972). Unfortunately this nematode has not been studied in detail and information on its host range, biology and pathogenicity is scarce, but it is suspected to be an important constraint on barley and durum wheat production in temperate, semi-arid regions (Sikora, 1988; Ismail, 2000, 2001; Scholz, 2001; Scholz and Sikora, 2004).

Similarly, *H. filipjevi* is most probably an economically important nematode on cereals due to its widespread distribution and previous misidentifications as *H. avenae* in the former USSR and also Sweden. Preliminary data from Turkey implicate yield losses up to 35% on common winter wheats (J.M. Nicol, unpublished). Given the increased recognition and incidence, these species are now being identified as a constraint to cereal production (Philis, 1988; Oztürk *et al.*, 1999; Scholz, 2001).

Management measures

In areas where *H. avenae* is responsible for economic losses, the best approaches have been adoption of crop rotation and use of resistant cultivars.

CHEMICAL. Although in the past it has been shown that low rates of non-fumigant nematicides provided effective and economical control under severe infestation conditions in nematode control programmes in Australia, India and Pakistan (Gurner *et al.*, 1980; Swarup, 1984; Maqbool, 1988), the present-day cost and environmental concerns associated with these chemicals do not make them a viable economic alternative for almost all farmers. However, their use in scientific experiments to understand the importance of these nematodes will remain vital.

CULTURAL PRACTICES. One of the most efficient methods of controlling *H. avenae* is with the use of grass-free rotations using non-host crops. In long-term experiments, non-host or resistant cereal frequencies of 50% (80% in lighter soils) keep populations below damaging thresholds (Rivoal and Besse, 1982; Fisher and Hancock, 1991). In Europe, a 4 year rotation can be practised for nematode control, but economic factors do not permit such long rotations in most subtropical and tropical countries.

In some countries, cereals are the prominent rotation; therefore, in order to understand and utilize cereal rotations, a clear understanding of species and pathotypes in a given region is necessary, but it is indeed possible to use rotational options. Oat is resistant in Australia but is susceptible in the UK (Cook and York, 1988), while most cultivars of rye are generally resistant. Within triticales, there are cultivars that have resistance that can be utilized (Cook and York, 1987).

Clean fallow can reduce population densities of the nematode, and 1–5 deep ploughings during hot summer months can cause reductions in nematode populations between 9.3 and 42.4%, with a corresponding yield increase of 4.4–97.5% (Mathur *et al.*, 1987), but are not always economically and environmentally sound. The decrease in population is attributable to killing of cyst contents (eggs and juveniles) due to desiccation by intense solar heat and hot winds.

RESISTANCE. Cultivar resistance is considered one of the best methods for nematode control and has been found to be successful in several countries such as Australia, Sweden and France on a farm scale (R. Rivoal, France, personal communication). However, it has also been observed that use of resistance, especially derived from single dominant genes, may cause disequilibrium in biological communities and possibly ecological replacement with other nematodes such as *Pratylenchus* (Lasserre *et al.*, 1994). Another potential concern is breakdown of resistance sources with repeated use. This has occurred in France with the resistant oat cultivar Panema with the appearance of a new *H. avenae* pathotype (Lasserre *et al.*, 1994). In Denmark, a new virulent pathotype of *H. avenae* was selected by growing the same barley cultivars over a long period with the *Rha2*, and the same with a related species, *H. filipjevi* in Sweden (Cook and Noel, 2002).

In order for cultivar resistance to be effective, a sufficient understanding of the number of species and pathotypes within species is essential. The International Test Assortment of barley, oat and wheat (Andersen and Andersen, 1982) offers classification of pathotype variation (Table 5.2). Although useful, a pathotype scheme for a species complex, based on interaction with three cereal genera, will not easily describe extensive variation in virulence (Rivoal and Cook, 1993). Furthermore, to date, there are few molecular or other diagnostic methods that can provide consistent and reliable pathotype and pathogenicity differentiation.

The extensive review by Rivoal and Cook (1993), revised by Nicol (2002) and presented here in Table 5.3, gives some indication of worldwide accessions of germplasm within oat, barley, triticale, rye, wheat and wild grass relatives that offer control of some of the nematode species and pathotypes and, where known, the genetic control and chromosome location. Some resistant cultivars simultaneously reduce populations of several European pathotypes (Williams and Siddiqi, 1972). Additional *Triticum* accessions have been identified that appear to possess high degrees of resistance to a broad array of *Heterodera* species and pathotypes. Fortunately, many of the sources implicate major gene inheritance, which makes selection for these relatively efficient.

In India, seven resistant cultivars (BH331, BH338, C-164, Rajkiran, RD 2035, RD 2052 and RD 2508) of barley were tested against Ambala, Sirsa, Jaipur and Udaipur populations of the *H. avenae* complex. None of the tested cultivars was found to be resistant against Ambala populations (*H. filipjevi*), whereas Rajkiran, C-164 and RD 2035 were resistant to Jaipur, Udaipur and Sirsa populations of *H. avenae* (Yadav, India, personal communicaton). The barley cultivars BH331 and BH338 were susceptible against both *H. avenae* and *H. filipjevi*. With wheat, a new variety, Raj MR-1, was developed and released recently by ARS Durgapura Jaipur in Rajasthan, which holds resistance to *H. avenae* from that region (Yadav, India, personal communication).

Molecular technology has also been applied to identify markers for various cereal cyst nematode resistance genes, using techniques such as RFLP and PCR (polymerase chain reaction), in both barley (Kretschmer *et al.*, 1997; Barr *et al.*, 1998) and wheat (Eastwood *et al.*, 1994a; Williams *et al.*, 1994; Ogbonnaya *et al.*, 1996; Lagudah *et al.*, 1998; Paull *et al.*, 1998). Furthermore, many of the wild grass relatives have been introgressed into a hexaploid wheat background for breeding purposes. Many of these have had molecular work applied to identify the location and the possibility of producing markers to the known gene(s). More details about introgressions, substitutions and molecular characterization of these materials can be found in McIntosh *et al.* (2001). Some of these markers are actively being implemented in marker-assisted selection (MAS) and pyramiding of gene resistance in Australian cereal breeding programmes against *H. avenae*, pathotype Ha13 (Jefferies *et al.*, 1997; Ogbonnaya *et al.*, 1998). This is an example where there is sufficient understanding of the biology of the pathogen and genetic control of the resistance so that both conventional breeding and the tools of molecular biology can be combined for the advancement of controlling this disease. Such potential exists for other nematodes, but will require a similar understanding and combining of related skills.

The utilization of these identified sources and possibly of other as yet unidentified sources of resistance is country specific and dependent on the number and types of *Heterodera* species and pathotypes that need to be controlled. For example, in Israel, all locally grown wheat and barley cultivars tested against *H. avenae* and *H. latipons* are excellent hosts. However, the oat cultivars tested were extremely poor hosts to *H. avenae* but good hosts to *H. latipons* (Mor *et al.*, 1992). In Mediterranean countries such as Algeria, Spain, Israel and southern France, oat appears generally to be a poor host for *H. avenae*, in comparison with northern Europe where it is considered to be a good host, suggesting a possibility that the nematode has developed host race types (R. Rivoal, France, personal communication). Many countries unfortunately have limited resources and/or expertise to establish this information, and current control methods are based on understanding the response of

Table 5.3. Principal sources of genes[a] used for wheat breeding resistance to cereal cyst nematode (*Heterodera avenae*) and root lesion nematode (*Pratylenchus thornei* and *P. neglectus*).

Cereal species	Cultivar or line	Origin	Genetic information R-gene(s)[b]	Response to pathotypes[c]	Use in cultivars	References[d]
Cereal cyst nematode						
Oat						
Avena sterilis	I376	?	1–3 major genes	R to all *Ha1*, *Ha2* and *Ha3*	UK	
Avena spp.	US1624 (CI3444)		Major gene	R to *Ha1* and *Ha2*, S to *Ha3*	Sweden, Denmark, UK	
Avena spp.	Avon and several Australian cvs	?		R in Australia (*Ha13*), S to *Ha1* and *Ha2*	Australia	
A. sativa	Panema	UK	1 dom, from I376	S. Australia	UK	
	Nelson	Sweden	1 dom, from C.I. 3444, allelic to Panema	—	N.W. Europe	
			2 dom		France	
A. byzantina	NZ Cape	New Zealand	?	S, UK	Australia	
	Mortgage Lifter	Australia	2 rec	—	—	
	TAMO 301, 302	Texas, USA	?	—	Australia	
	No 11527	?	?	R, Siberia		
Barley						
Hordeum spp.	Many cvs, e.g.	Northern Europe landraces	*Rha1*	R, to *Ha1* in many cvs	N. Europe cvs 1900–1950s	
	Emir		*Rha?*	R to *Ha61* (Norway, NL, India, Siberia)	Susceptible in most of Europe	
	North African accessions?	N. Africa	*Rha2*	R to *Ha1*, *Ha2* and S to *Ha3*	Cvs in Denmark, Sweden, UK	
	Morocco from N. Africa	Morocco	*Rha3*	R to *Ha1*, *Ha2* and *Ha3*	Not in cultivars	
	Galleon		Major gene	R to *Ha13*	Australia	
	Drost	Sweden	1 dom (*Rha1*)	many bred cvs		
	Ortolan	Germany	1 or 2 dom, allelic to *Rha1*	pR, Australia	N. Europe	
	ex. L.P.191	?N. Africa	1 dom, (*Rha2*)		N. Europe	
	ex. Morocco	N. Africa	not linked to *Rha1*		Australia	
	L.P. 191, Morocco		1, dom, (*Rha3*) allelic to *Rha2*		—	

	Athenais	Greece	1–2 dom		Australia	
	Nile, C.I.3576	Egypt			Australia	
	C.I. 8147	Turkey	1 dom, not *Rha1*	R, pathotype-1	Australia	
	Martin	Algeria	1 dom, similar to *Rha2*	(Delhi	Australia	
	C164, RD2052	India	1 dom, not *Rha1* 2 dom, ?similar to *Rha3* 1 dom	population)	India	
Wheat						
T. aestivum	Loros, AUS10894	?	***Cre1***[a] (formerly *Ccn1*), on chromosome 2BL.	pR to several pathotypes	N.W. Europe, Australia	Slootmaker *et al.* (1974); Bekal *et al.* (1998)
	Katyil	Australia	***Ccn1***	S, India	Australia	
	Festiguay	Australia	***Cre8*** (formerly *CreF*) on chromosome 7L? Recent analysis suggests 6B	pR in cvs Australia (*Ha13*)	Australia	Paull *et al.* (1998); Williams *et al.* (unpublished)
	AUS4930 = 'Iraq 48'	Iraq	Possibly identical genetic location to *Cre1*. Also resistance to *Pratylenchus thornei*	R to several pathotypes and CCN species and *Pratylenchus thornei*	Australia, France, CIMMYT Int. – under scientific evaluation	Bekal *et al.* (1998); Nicol *et al.* (1998, 1999, 2001); Green (personal communication); Lagudah (personal communication)
T. durum	Psathias 7654, 7655, Sansome, Khapli	?	?	S, to some pathotypes, pR to others	—	Rivoal *et al.* (1986)
Triticale						
Triticosecale	T701-4-6	Australia	*CreR* on chromosome 6RL		Australia	Asiedu *et al.* (1990); Dundas *et al.* (2001)
Secale cereale	R173 family		*CreR* on chromosome 6RL			Taylor *et al.* (1998)
	Driva	Australia	?	=Ningadhu in cv. Tabara	Australia	
	Salvo	Poland	?		UK	
Wild grass relatives						
Aegilops. tauschii	CPI 110813	Central Asia	*Cre4* on chromosome 2DL	R Australia	Australia synthetic hexaploid lines	Eastwood *et al.* (1994a); Rivoal *et al.* (2001)
A. tauschii	AUS18913	?	***Cre3*** on chromosome 2DL	R Australia	Australia advanced breeding lines	Eastwood *et al.* (1994a); Rivoal *et al.* (2001)
A. peregrina (*A. variabilis*)	1		*Cre(3S)* with (*Rkn2*) on chromosome 3S; *CreX*, not yet located			Barloy *et al.* (1996); Jahier *et al.* (1998); Rivoal *et al.* (2001); Barloy (unpublished); Lagudah (personal communication)

Continued

Table 5.3. *Contiinued.*

Cereal species	Cultivar or line	Origin	Genetic information R-gene(s)[b]	Response to pathotypes[c]	Use in cultivars	References[d]
A. longissima	18	?	?	R to four French pathotypes and *Meloidogyne naasi*	France	Bekal *et al.* (1998)
A. geniculata	79 MZ1, MZ61, MZ77, MZ124		?	R and pR to several pathotypes	France – under scientific evaluation	Bekal *et al.* (1998); Zaharieva *et al.* (2001)
A. triuncialis	TR-353	?	*Cre7* (formerly *CreAet*)	R and pR to several pathotypes	France – under scientific evaluation	Romero *et al.* (1998)
A. ventricosa	VPM 1		*Cre5* (formerly *CreX*), on chromosome 2AS	R to several pathotypes	Spain – under scientific evaluation	Jahier *et al.* (2001); Ogbonnaya *et al.* (2001b)
	11, AP-1, H-93-8		*Cre2* (formerly *CreX*) on genome N^v			Delibes *et al.* (1993); Andres *et al.* (2001); Rivoal *et al.* (2001)
	11, AP-1, H-93-8, H-93-35		*Cre6*, on chromosome $5N^v$			Ogbonnaya *et al.* (2001b); Rivoal *et al.* (2001)
Root lesion nematode						
T. aestivum	GS50a	Australia – reselection from Australian cultivar 'Gatcher'	Resistance to Pt			Thompson and Clewett (1986)
	AUS4930 = Iraq 48	Iraq		Resistance to Pt but also portrays resistance to CCN	Australia, CIMMYT – under scientific investigation	Nicol *et al.* (1998, 1999, 2001)
	Excalibur	Australia – reselection of commercial cultivar 'Excalibur'		Resistance to Pn (*Rlnn1*), on chromosome 7AL		Williams *et al.* (2002)
	Croc_1/*Ae. tausch.* (224)// Opata	Synthetic derivative		Resistance to Pt. Unknown from where resistance is derived		Nicol *et al.* (2001)
A. tauschii	CPI 110872			Resistance to Pt and Pn		Thompson (personal communication)
A. geniculata	MZ10, MZ61, MZ96, MZ144	Middle East and West Asia		Moderate resistance to Pt. Several also portray resistance to CCN		Zaharieva *et al.* (2002)

[a]See also differentials listed in Table 5.2.
[b]dom or rec, dominant or recessive genes. Characterized single gene; bold indicates a marker implemented in a commercial breeding programme; see Ogbonnaya *et al.* (2001b).
[c]R, resistant; pR, partially resistant; S, susceptible.
[d]Sourced from Rivoal and Cook (1993, 1998).
Pt, *Pratylenchus thornei*; Pn, *Pratylenchus neglectus*; ?, no published scientific studies conducted.

local cultivars to the pathogen(s). In order to make best use of existing research findings, greater collaboration between research institutions such as the Consultative Group of International Agricultural Research (CGIAR) and countries and research groups where the nematode is considered important is essential. The most recent reports by such collaborations are referred to by Rivoal *et al.* (2001), which offer a great start to unravelling the complex puzzle of *Heterodera* populations and existing knowledge of resistant sources and their possible use in controlling the cyst nematode in different regions of the world.

BIOLOGICAL CONTROL. As reviewed by Trudgill *et al.* (1992), most research on biological control has been done on three main types of agents, namely obligate parasites, facultative parasites and rhizosphere bacteria. Only agents that parasitize adult females and/or their eggs have been found so far to provide effective natural control. Natural enemies of *H. avenae*, mainly fungi, have been recognized for quite some time, but not as yet exploited as biological control agents for field application. Particularly in extensive field crops, the ability to manipulate such fungi to generate nematode-suppressive soils to control populations below economic threshold levels is very difficult. However, studies conducted in the 1980s in the UK demonstrated that despite cereal monoculture, populations of *H. avenae* were maintained below an economically damaging threshold by parasitic fungi, indicating natural suppression of the population (Kerry and Crump, 1977; Kerry, 1981; Kerry *et al.*, 1982a,b,c; Crump *et al.*, 1983). Dackman and Nordbring-Hertz (1985) found that cysts of *H. avenae* in Sweden where all eggs were parasitized commonly gave rise to pure colonies of egg parasites, while cysts in which only a portion of the eggs were infected gave rise to multiple opportunistic species. Sharma and Swarup (1988) detected *Pasteuria penetrans*, a bacterial parasite of juveniles, which may prove to be a promising agent for *H. avenae* control. The potential use of mutualistic fungal and bacterial endophytes applied to the seed to reduce nematodes may prove promising (Sikora, 2000; Pagdham and Sikora, 2004).

Currently, several commercial biological control products are available for the control of sedentary nematodes, including cyst and root knot. These include *P. penetrans* produced by Nematech Ltd, Tokyo, *Paecilomyces lilacinus* produced by Prophyta GmbH, Malchow, Germany, and a similar product developed by Biological Control Products, South Africa. Their use for controlling cereal cyst nematode on cereals is not reported in the literature, but they have been effective against other cyst nematodes in greenhouse trials (Kiewnick and Sikora, 2003). They are more commonly used on higher value, more intensive agricultural crops such as tomato. Trudgill *et al.* (1992) reinforce that the greatest value of biocontrol agents will be in combination with other control options.

Pratylenchus

Distribution

The genus *Pratylenchus* is a large group with many species affecting both monocots and dicots. Many of the species are morphologically similar, which makes them difficult to identify. At least eight species of lesion nematodes have been recorded for small grains (Rivoal and Cook, 1993). Four species, *P. thornei*, *P. crenatus*, *P. neglectus* and *P. penetrans*, have a worldwide distribution, especially in the temperate zones (Kort, 1972).

P. thornei is the most studied species and is a known parasite of cereals worldwide, being found in Syria (Greco *et al.*, 1984; Saxena *et al.*, 1988), the former Yugoslavia, Mexico and Australia (Fortuner, 1977), Canada (Yu, 1997), Israel (Orion *et al.*, 1982), Morocco (Ammati, 1987), Pakistan and India (Maqbool, 1988), Turkey (Nicol *et al.*, 2002), Algeria (Troccoli *et al.*, 1992) and Italy (Lamberti, 1981). Unfortunately, very little is known about the economic importance and distribution of the other species on cereals.

Biology and life cycle

Pratylenchus species are polycyclic, polyphagous, migratory root endoparasites, which are not confined to fixed places for their development and reproduction. Eggs are laid in the soil or inside plant roots. The nematode invades the tissues of the plant root, migrating and feeding inside a root. Secondary attack by fungi frequently occurs at these lesions. The life cycle is variable between species and environment, and ranges from 45 to 65 days (Agrios, 1988).

Environmental factors

P. thornei is active during the growing season and subsequently survives the period before the next crop (in a desiccated state if drought) until reactived by rainfall in the following rainy season (Grandison and Wallace, 1974). Work on a closely related species, *P. mediteraneus*, also found on cereals indicated that low soil moisture was a major ecological factor affecting nematode multiplication in Israel (Orion *et al.*, 1984). Studies by Glazer and Orion (1983) indicate that *P. thornei*, a species closely related to *P. mediteraneus*, was able to withstand desiccation for up to 7–8 months, remaining infective. Survival of *P. thornei* in 200 g soil samples was reduced by 80% by drying from 19.5 to 5% moisture and/or high temperatures (Baxter and Blake, 1968).

P. crenatus is more common in light soils, *P. neglectus* in loamy soils and *P. thornei* in heavier soil types (Kort, 1972). However, the work of Nicol (1996) and Nicol *et al.* (2002) suggests that both *P. thornei* and *P. neglectus* can occur in a range of soil types, and mixtures of the two species are not uncommon in southern Australia and the Central Anatolian Plateau of Turkey.

Nitrogen, commonly applied to cereals, has important effects on plant growth and populations of *P. thornei*. Van Gundy *et al.* (1974) report nitrogen to provide some level of control, but only when the population was near the economic threshold for damage. However, work by Doyle *et al.* (1987) with *P. thornei* and by Orion *et al.* (1984) with *P. mediteraneus* did not find any differences. Kimpinski (1972) found that the concentration of ammonium nitrate was correlated with fewer numbers and lower densities of *P. neglectus* in wheat roots. Potassium and phosphorus fertilizers did not significantly increase wheat yields in the *P. thornei*-infested fields (Doyle *et al.*, 1987), and no change was found in numbers of *P. neglectus* with the application of either of these (Kimpinski, 1972).

Symptoms of damage

Pratylenchus feed on and destroy roots, resulting in characteristic dark brown or black lesions on the root surface, hence their name 'lesion' nematodes (Fig. 5.2, Plate 3C). Above-ground symptoms of *Pratylenchus* on cereals, like other cereal root nematodes, are non-specific, with infected plants appearing stunted and unthrifty, sometimes with reduced numbers of tillers and yellowed lower leaves (Fig. 5.3).

Pathotypes

As reviewed by De Waele and Elsen (2002), biological diversity among populations of the same species has been reported in *P. brachyurus*, *P. goodeyi*, *P. loosi*, *P. neglectus*, *P. penetrans* and *P. vulnus*. To date, there is no record of differences within *P. thornei*. Furthermore, screening of identified resistant accessions in Australia, Mexico and Turkey with local populations reveals the resistance to pertain under greenhouse and field conditions. However, care should be taken to examine the reproductive fitness between root lesion nematode populations from the field and also in greenhouse studies to be sure of the usability of plant resistance reactions, as nematodes in culture collections for an extended period of time can lose their pathogenicity (De Waele and Elsen, 2002).

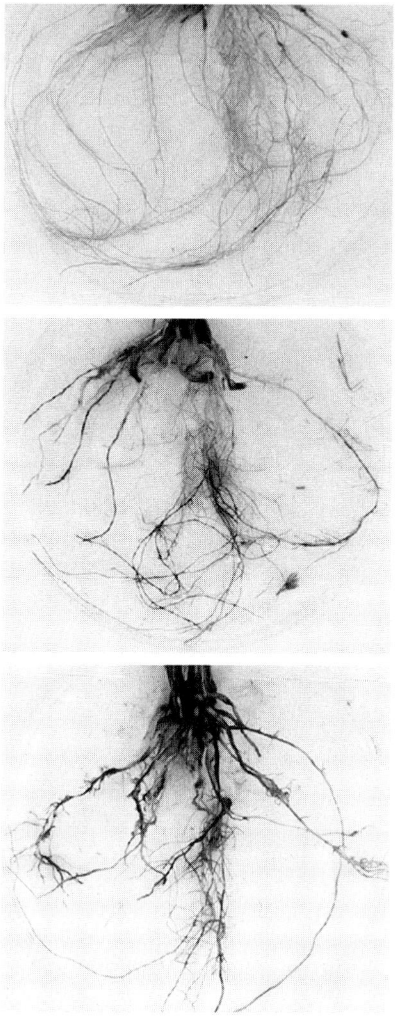

Fig. 5.2. Symptoms of root lesion nematode, *Pratylenchus thornei*, on susceptible wheat, showing extensive lesions, cortical degradation and reduction in both seminal and lateral root systems with increasing nematode density from top to bottom under natural field infestation. (Photo: J.M. Nicol, CIMMYT.)

Damage potential and economic importance

Although *Pratylenchus* is capable of multiplying for several generations during a single season, they spread only from plant to plant due to their relative immobility. The impact of plant parasitic nematodes on plant health and crop yield varies with biogeographic location, cropping sequence and intensity, cultivar selection, soil characteristics and nematode community structure (McKenry and Ferris, 1983). As mentioned previously, the economic threshold for plant damage will depend on many such factors, and interpretation of the damage threshold between specific nematological studies should be done with extreme caution, as very few studies are truly comparable. As stated earlier, there are inherent differences in sampling protocol, extraction procedure and nematode renumerification. It is for this reason that the studies conducted below are only listed. The reader should interpret these accordingly (Van Gundy *et al.*, 1974; Orion *et al.*, 1984; Doyle *et al.*, 1987; Lasserre *et al.*, 1994; Nicol *et al.*, 1999; Taylor *et al.*, 1999; Nicol and Ortiz-Monasterio, 2004).

Fig. 5.3. Winter wheat attacked by root lesion nematode, *Pratylenchus neglectus*, showing patchy distribution, reduced tillering and emergence of infected plants. (Photo: R. Rivoal and R. Cook.)

As previously mentioned, the most studied of these species on wheat is *P. thornei* and, somewhat less so, *P. neglectus* and *P. penetrans*. *P. thornei* is considered the economically most important species in at least three countries. Yield loss of 38–85% on wheat have been reported in Australia (Thompson and Clewett, 1986; Doyle *et al.*, 1987; Taylor and McKay, 1993; Eastwood *et al.*, 1994b; Nicol *et al.*, 1999; Taylor *et al.*, 1999), 10–40% in Mexico (Van Gundy *et al.*, 1974; Nicol and Ortiz-Monasterio, 2004) and 70% in Israel (Orion *et al.*, 1984). *P. thornei* appears to be associated with regions experiencing a Mediterranean climate. It is highly probable, given the distribution of this nematode, that similar losses may also be occurring in many other countries, but this has not been studied.

The other species of lesion nematodes where yield loss studies have been conducted (*P. neglectus* and *P. penetrans*) are not recognized as having a global distribution on cereals, and the current yield loss studies would suggest that the damage potential of these nematodes is not as great as that of *P. thornei*. In Australia, losses on wheat with *P. neglectus* ranged from 16 to 23% (Vanstone *et al.*, 1995; Taylor *et al.*, 1999), while in Canada *P. penetrans* losses were 10–19% (Kimpinski *et al.*, 1989). Yield loss work by Vanstone *et al.* (1998) in the field where both *P. thornei* and *P. neglectus* were present indicates losses between 56 and 74% on wheat. Studies by Sikora (1988) have identified *P. neglectus* and *P. penetrans* in addition to *P. thornei* on wheat and barley in northern Africa and all of these plus *P. zeae* in western Asia. Further work is necessary to determine the significance of these species in these regions.

Management measures

In several countries such as Australia, Israel and Mexico, research has explored different options to control *Pratylenchus* and has concluded that an integrated pest management approach is required. This implies both the use of varietal selection with an emphasis on resistance and tolerance, and avoiding rotations, which encourage multiplication, particularly susceptible wheat after wheat.

CHEMICAL. As with cereal cyst nematode, chemical control, although in most cases effective against root lesion nematodes, is not economically viable or environmentally acceptable with cereal crops.

CULTURAL PRACTICES. The use of crop rotation is a limited option for root lesion nematodes, due to the polyphagous nature of the nematode. Little is understood about the potential role of crop rotation in controlling these nematodes, although some field and laboratory work has been undertaken to better understand the ability of both *P. thornei* (Van Gundy *et al.*, 1974; O'Brien, 1983; Clewett *et al.*, 1993; Hollaway *et al.*, 2000) and *P. neglectus* (Vanstone *et al.*, 1993; Lasserre *et al.*, 1994; Taylor *et al.*, 1999, 2000) to utilize cereals and leguminous crops as hosts. Results from these studies indicate that hosting ability is both species- and cultivar-specific, with both legumes and cereals. As a result, it is essential that hosting ability studies are conducted with local/regional cultivars. It is possible, depending on crop rotation patterns and the population dynamics of nematodes, that resistant cultivars of cereals alone may not be sufficient to maintain nematode populations below economic levels of damage.

As with cereal cyst nematode, some triticale varieties such as Abacus and Muir in Australia are known to host fewer nematodes than with bread or durum wheats, and hence may offer some useful rotational options (Farsi *et al.*, 1995).

Cultural methods offer some control options, but are of limited effectiveness. To be of major significance, these need to be integrated with other control measures. Di Vito *et al.* (1991) found that mulching fields with polyethylene film for 6–8 weeks suppressed *P. thornei* populations by 50%. Van Gundy *et al.* (1974) found that delaying sowing of irrigated wheat by 1 month in Mexico gave maximum yields. In Australia, cultivation reduced populations of *P. thornei* (Thompson *et al.*, 1983; Klein *et al.*, 1987) and, in Israel, Orion *et al.* (1984) found that biannual fallowing reduced *P. mediteraneus* populations by 90% and increased grain yields by 40–90%. An 11 year management trial conducted in Queensland revealed that the topsoil of zero tillage fallow systems had higher *P. thornei* populations than mechanically cultivated treatments (Thompson *et al.*, 1983).

RESISTANCE. Unlike cereal cyst nematode, no commercially available sources of cereal resistance are available to *P. thornei*, although sources of tolerance have been used by cereal farmers in northern Australia for several years (Thompson *et al.*, 1997). As illustrated in Table 5.3, Thompson and Clewett (1986), Nicol *et al.* (1996, 1999) and Nicol (2002) identified wheat lines that have proven field resistance, and work is continuing to breed this resistance into suitable backgrounds. Recent work by Thompson and Haak (1997) identified 29 accessions from the D-genome donor to wheat, *Aegilops tauschii*, suggesting that there is future potential for gene introgression. Some of this material also contained the *Cre3* and other different, unidentified sources of cereal cyst nematode resistance genes conferring resistance to some cereal cyst nematode pathotypes. As with the cereal cyst nematode, molecular biology is being used to determine the genetic control, location and the subsequent identification of markers for resistance to both *P. thornei* and *P. neglectus*. Recent work with Australian germplasm referred to by McIntosh *et al.* (2001) reports the gene *Rlnn1* on chromosome 7AL, effective against *P. neglectus*, and two quantitative trait loci (QTLs) on chromosomes 2BS and 6DS*i*. No commercial sources of resistance are currently available for species of *Pratylenchus* that attack cereals.

BIOLOGICAL CONTROL. Successful biological control of *Pratylenchus* species is likely to be difficult due to their migratory behaviour. *Pratylenchus* species spend much of their lives in roots and tend to be found only in soil when their host plants are stressed, senescing or diseased, or when their hosts have been ploughed out after harvest (Stirling, 1991). Currently, several commercial biological control products are available for the control of nematodes, but their use for controlling lesion nematode on cereals is not reported in the literature. However, as mentioned previously, their application and use is more common on higher value, more intensive agricultural crops such as tomato. As stated for cyst nematode control, the greatest value of biocontrol agents will be in combination with other control options.

Anguina tritici

Distribution

Seed gall nematode (*Anguina tritici*), commonly known as 'ear cockle', is frequently found on small grain cereals where farm-saved seed is sown without use of modern cleaning systems. It is of historical importance since it is the first plant parasitic nematode recorded in the literature. Cereals are infected throughout western Asia and North Africa (Sikora, 1988), the Indian subcontinent, China, parts of Eastern Europe (Tesic, 1969; Swarup, 1986), Iraq (Stephan, 1988), Turkey (Yuksel *et al.*, 1980) and Pakistan (Maqbool, 1988). It has also been reported from most European countries, Russia, Australia, New Zealand, Egypt, Brazil and several areas in the USA. It has only been detected much later on barley in northern Iraq, where infestations reached 90% (Al-Talib *et al.*, 1986; Stephan, 1988).

Biology and life cycle

Nematode-infected seed galls, which may be present already in the soil or sown into the soil at planting with contaminated seed, become moist and soft, with soil moisture facilitating the release of juveniles. A single gall may contain over

10,000 dormant juveniles. Approximately 1 week after seed galls infected with nematodes are placed in the soil, juveniles can be traced in the growing point of a germinating plant. These juveniles move upward passively on the growing point as the plant grows. They do not exhibit any morphological change until approximately 2 months. Nematode morphological changes take place only when the juveniles penetrate a flower primordial after 2–3 months and then turn into adults. As a result, ovules and other flowering parts of a plant are transmuted into galls or 'cockles' (Fig. 5.5). Nematodes mature inside galls and females lay thousands of eggs from which juveniles hatch and remain dormant in seed. The total life cycle is completed in around 4 months (Swarup and Sosa-Moss, 1990). Temperature, humidity, planting depth and the source of galls are the major determinants in symptom expression. The nematode favours wet and cool weather (Kort, 1972).

Temperature, humidity and the source of galls are particularly important for development of yellow ear rot (Plate 3D). This nematode-vectored bacterial disease, vernacularly known as 'tundu' or 'tannan' in India, is also commonly found associated with the ear cockle nematode problem. The disease was first recorded from India by Hutchinson (1917), where the nematode is associated with a bacterium *Corynebacterium michiganense* pv. *tritici*. The bacterium is frequently present along with juveniles in galls and is responsible for expression of the disease. On its own, the bacterium is only capable of producing yellow streaks on leaves that run parallel to the veins. The nematode carries the bacterium to the growing point as an external body contaminant (Gupta and Swarup, 1972). Atmospheric temperatures between 5 and 10°C and a relative humidity of 95–100% favour multiplication of the bacterium in plants.

The bacterium multiplies very quickly under favourable environmental conditions, increasing its concentration in a plant and forming a thick, viscous fluid in which nematode juveniles are not able to survive. Under such conditions, emerging ears are totally sterile and are covered with yellow slime. Yellow ear rot requires a combination of 0.4 optical density of the bacterium and 10,000 nematode juveniles for maximum expression of the bacterial phase of the disease. However, under less favourable conditions for the bacterium, nematode juveniles survive to produce partial ear cockle and partial yellow ear rot symptoms. Economic losses associated with this combination are increased because of the lower price for infected grain (Rivoal and Cook, 1993).

Symptoms of damage

Symptoms of *A. tritici* attack may be indicated by small and dying plants with leaves generally twisted due to nematode infection (Swarup and Sosa-Moss, 1990). Infected ears are easily recognized by their smaller size and darkened colour compared with normal seeds, but infected seeds may be easily confused with bunt (*Tilletia tritici*). Under dry conditions, juveniles may survive for decades (Kort, 1972).

In both ear cockle and yellow ear rot, the first observable symptom is an enlargement of the basal stem portion near the soil base, visible in 3-week-old wheat seedlings. The emerging leaves are twisted (Fig. 5.4) and crinkled. Frequently, some leaves remain folded with their tips held near the growing point. These leaves, after about 30–45 days, straighten out and many appear normal, with faint ridges on the surface. In comparison with healthy seedlings, the affected plants are dwarfed, with a spreading habit. These symptoms are more clearly discernible on young seedlings and decrease with plant age. Under very low infestation levels, plants may not exhibit any visible symptoms, even though a few seed galls are produced in the ears, whereas severely infested plants may die without heading. Infested seedlings produce more tillers and grow faster than normal plants, but not necessarily with an increase in the number of ears (Swarup and Sosa-Moss, 1990).

Fig. 5.4. Twisted leaves of wheat caused by *Anguina tritici*. (Photo: R.A. Sikora.)

Furthermore, ears emerge roughly a month earlier in diseased plants. Such ears are short and broad, with very small or no awns on the glumes (Fig. 5.5). Nematode galls replace either all or some of the grains. In the yellow ear rot disease (Plate 3D), the characteristic feature is the production of a bright yellow slime- or gum-like substance on the abortive ears as well as leaves, which remains in contact with such ears while still in the boot leaf stage. Under humid conditions, the bacterial slime trickles down tissues (Swarup and Sosa-Moss, 1990) and upon drying it appears brown in colour. An infected spike is narrow and short, with wheat grains partially or completely replaced by slime. In the latter event, an emerging spike remains sterile. The stalk of an infected spike is always distorted.

Damage potential and economic importance

Worldwide, wheat, barley and rye are commonly attacked, but barley is less attacked in India (Paruthi and Gupta, 1987). In Iraq, ear cockle is an important pest on wheat, with infection ranging from 0.03 to 22.9% and causing yield losses up to 30% (Stephan, 1988). Barley is also attacked in Iraq and Turkey (Yuksel *et al.*, 1980; Al-Tabib *et al.*, 1986).

Fig. 5.5. *Anguina tritici*-infested ear head with ear cockle (right). (Photo: J. Bridge and D.J. Hunt.)

In Pakistan, ear cockle is a known pest on wheat and barley and is found in nearly all parts of the country, causing losses of 2–3%. However, in association with the yellow ear rot bacterium, it produces serious yield losses on wheat (Maqbool, 1988). In China, Chu (1945) found yield losses between 10 and 30% on wheat. A minimum population of 10,000 juveniles/kg of soil is essential for development of ear cockle. Disease intensity is greater when nematode galls are placed in soil at a depth of 2–6 cm than when placed deeper.

Management measures

SEED HYGIENE. Ear cockle can easily be controlled by seed hygiene. Clean, uninfected seed can be obtained either through use of certified seed or by cleaning infected seed using modern seed cleaning techniques, or by sieving and flotation (Singh and Agrawal, 1987). Although it has been eradicated from the Western hemisphere through adoption of this approach, it remains a problem on the Indian subcontinent, in western Asia and to some extent China (Swarup and Sosa-Moss, 1990).

Since ear cockles are the only source for perpetuation of both diseases, their removal from contaminated seed lots can completely eliminate both diseases. Galls are lighter in weight than wheat seed and can be easily discarded through a winnowing process or by flotation of contaminated seeds in 20% brine solution. It is important, however, to wash wheat seed after brine treatment two or three times in water to remove adhering salt particles, otherwise seed germination is impaired.

To dispense with salt treatment, Byars (1920) suggested pre-soaking contaminated seeds in water, then soaking them at either 50°C for 30 min, 52°C for 20 min, 54°C for 10 min or 56°C for 5 min. The principle is to reactivate quiescent juveniles before killing them with hot water. Leukel (1957) suggested pre-soaking galls for 4–6 h in water and then exposing them to hot water at 54°C for 10 min.

RESISTANCE AND ROTATIONS. For countries where hygiene practices are difficult to implement, host resistance and rotation offer some hope. The earliest record of a resistance source is the cultivar Kanred (Leukel, 1924) used in a breeding programme initiated by Shen *et al.* (1934). Crosses between Kanred and a highly susceptible wheat cultivar resulted in a few lines in the F_2 and F_3 generations free from nematode attack. Unfortunately, this work was not continued. However, since then, resistance to *A. tritici* has been identified in Iraq in both wheat and barley (Saleh and Fattah, 1990), and in Pakistan (Shahina *et al.*, 1989), and was sought in India (Swarup and Sosa-Moss, 1990). In Iraq, laboratory screening has identified sources of resistance in both wheat and barley (Stephan, 1988). Oat, maize and sorghum are considered to be non-hosts (Limber, 1976; Paruthi and Gupta, 1987) and, while they may offer some option for reducing populations by rotation, the disease is not completely controlled.

Meloidogyne

Distribution

Root knot nematodes are the most economically important group of plant parasitic nematodes worldwide, attacking nearly every crop grown (Sasser and Freckman, 1987). Several *Meloidogyne* spp. are known to attack cereals and tend to favour light soils and warm temperatures. Several species attack Poaceae in cool climates, including *M. artiellia*, *M. chitwoodi*, *M. naasi*, *M. microtyla* and *M. ottersoni* (Sikora, 1988). In warm climates, *M. graminicola*, *M. graminis*, *M. kikuyensis* and *M. spartinae* are important (Taylor and Sasser, 1978). In tropical and subtropical areas, *M. incognita*, *M. javanica* and *M. arenaria* are all known to attack cereal crops (Swarup and Sosa-Moss, 1990).

To date, only *M. naasi* and *M. artiellia* have been shown to cause significant damage to wheat and barley in the winter growing season in the subtropics (Sikora,

1988). The most important and most studied species of the root knot nematodes on cereals worldwide are described below. There is little information on most other species, many of which are of unknown importance.

M. naasi is reported from most northern European countries, the USA and the former USSR, occurring mostly in temperate climates (Kort, 1972). However, it has also been found in Iran on wheat (Kort, 1972), in the Mediterranean area on barley, in the Maltese islands (Inserra *et al*., 1975), and in New Zealand and Chile on small grains (Jepson, 1987). It is probably the most important root knot nematode affecting grain in most European countries (Kort, 1972). *M. naasi* does not appear to be widespread in temperate, or tropical and subtropical semi-arid regions such as western Asia and northern Africa (Sikora, 1988). *M. naasi* is a polyphagous nematode, reproducing on at least 100 species of plants (Gooris and D'Herde, 1977) including barley, wheat, rye, sugarbeet, onion and several broadleaf and monocot weeds (Kort, 1972). Generally Poaceae are considered to be better hosts (Gooris, 1968). In Europe, oat is a poor host compared with other cereals, whereas in the USA oat is an excellent host of *M. naasi* (Kort, 1972). Host races of *M. naasi* have been identified in the USA by using differential hosts (Michel *et al*., 1973), which makes control of this nematode more difficult.

Other species of root knot nematodes attacking cereals include *M. artiellia*, which has a wide host range including crucifers, cereals and legumes, especially chickpea (Ritter, 1972; Di Vito *et al*., 1985). It is known to reproduce well on cereals and severely damage legumes (Kyrou, 1969; Sikora, 1988). This nematode is chiefly known from Mediterranean Europe in Italy, France, Greece and Spain (Di Vito and Zacheo, 1987), but also West Asia (Sikora, 1988), Syria (Mamluk *et al*., 1983), Israel (Mor and Cohn, 1989) and western Siberia (Shiabova, 1981).

Populations of *M. graminicola* in rice–wheat rotation areas of South Asia have been observed to cause severe damage to wheat roots, under conditions of artificial inoculation (Soomro and Hague, 1992) and in soil bioassay tests of rice–wheat production fields in Bangladesh (Padgham *et al*., 2004) and India (Gaur and Sharma, 1999). Wheat varieties from India have been reported to support poor to excellent reproduction of *M. graminicola* (Roy, 1977), and all economically important wheat varieties from Bangladesh supported excellent reproduction of *M. graminicola* (Padgham, Germany, 2004, personal communication).

M. chitwoodi is a pest on cereals in the Pacific North West of the USA and is also found in Mexico, South Africa and Australia (Eisenback and Triantaphyllou, 1991). Many cereals, including wheat, oat, barley and maize, and a number of dicots, are known to be hosts (Santo and O'Bannon, 1981). The three species *M. incognita*, *M. javanica* and *M. arenaria* were found to be good hosts on a range of cereal cultivars including wheat, oat, rye and barley under greenhouse conditions (Johnson and Motsinger, 1989). *M. graminis* is not known to be widely distributed, being limited to the southern USA, where it is associated with cereals and, more often, turfgrasses (Eriksson, 1972).

Biology and life cycle

Root knot nematodes cause typical small-sized root galls on roots. Egg masses attached to the posterior end of protruding females are normally transparent, but darken on exposure to air, and can resemble cysts of *Heterodera avenae*. Young juveniles of *M. naasi* invade roots of cereals within 30–45 days of germination, after which small galls on root tips can be observed. *M. naasi* generally has one generation per season (Rivoal and Cook, 1993). Egg masses in galls survive in the soil. Eggs have a diapause, broken by increasing temperature after a cool period (Antoniou, 1989). In warmer regions on perennial or volunteer grass hosts, more than one generation per season is possible (Kort, 1972). Juveniles develop and females become almost spherical in shape. Females deposit eggs in an egg sac and usually appear 8–10 weeks after sowing and are found embedded in the gall tissue (Kort, 1972). Large galls may contain 100 or more egg-laying females (Rivoal and Cook, 1993).

Symptoms of damage

Towards the end of a growing season, galling of the roots, especially the root tips, is common. Galls are typically curved, horseshoe- or spiral-shaped (Kort, 1972).

Symptoms of *M. naasi* attack closely resemble those caused by *H. avenae*, with patches of poorly growing, yellowing plants that may vary in size from a few square metres to larger areas. Other root knot nematodes attacking cereals are suspected to produce similar symptoms, but most are much less studied than *M. naasi.*

Damage potential and economic importance

Information on the economic importance of root knot nematodes on cereals is limited to a few studied species. *M. naasi* can seriously affect wheat yield in Chile (Kilpatrick *et al.*, 1976) and Europe (Person-Dedryver, 1986). On barley, it has been known to cause up to 75% yield loss in California, USA (Allen *et al.*, 1970). It is also associated with yield loss in barley in France (Caubel *et al.*, 1972), Belgium (Gooris and D'Herde, 1977) and Great Britain (York, 1980). Severe losses can occur, with entire crops of spring barley lost in The Netherlands and France (Schneider, 1967). *M. naasi* damage is not known to be widespread in temperate semi-arid regions (Sikora, 1988).

Damage to wheat by *M. artiellia* is known from Greece, southern Israel and Italy (Kyrou, 1969; Mor and Cohn, 1989). In Italy, 90% yield losses on wheat have been recorded (Di Vito and Greco, 1988). *M. chitwoodi*, an important pathogen of potato, also damages cereals in Utah, USA (Inserra *et al.*, 1985) and Mexico (Cuevas and Sosa-Moss, 1990). In controlled laboratory studies, *M. incognita* and *M. javanica* have been shown to reduce plant growth of wheat (Abdel Hamid *et al.*, 1981; Roberts *et al.*, 1981; Sharma, 1981) and, similarly, *M. chitwoodi* (Nyczepir *et al.*, 1984). *M. incognita* is a known field problem on wheat in north-western India (Swarup and Sosa-Moss, 1990).

Management measures

Control methods for root knot nematodes have been investigated in more detail for the known economically important species *M. naasi.* Partial resistance was found in barley and also in *Triticum squarrosa* and *T. monococcum*, while full resistance was identified with *Hordeum chilense*, *H. jabatum*, *T. umbellulatum* and *T. variabile* (bread wheat) (Cook and York, 1982b; Roberts *et al.*, 1982; Person-Dedryver and Jahier, 1985). Resistance has also been expressed in *H. chilense* (Person-Dedryver *et al.*, 1990; Yu *et al.*, 1990).

Cultural management options for *M. naasi* include rotations, using poor or non-host crops (Cook *et al.*, 1986), and use of fallow during the hatching period (Allen *et al.*, 1970; Gooris and D'Herde, 1972).

Rotations offer some options for *M. artiellia*. Di Vito *et al.* (1985) were able to demonstrate that, although most legumes and Graminaceae are hosts, cowpea, lupin, sainfoin and maize could be considered non-hosts.

***Ditylenchus* spp.**

Distribution

The genus *Ditylenchus* comprises many species that are prevalent in a wide range of climatic conditions from temperate, subtropical to tropical, where moisture regimes enable nematode infection, multiplication and dispersal (Plowright *et al.*, 2002). *Ditylenchus dipsaci* is by far the most common and important species of stem nematode on cereals, particularly on oat and rye, and is widespread throughout western and central Europe, the USA, Canada, Australia, Brazil, Argentina, and North and South Africa (Plowright *et al.*, 2002).

Biology and life cycle

D. dispaci is a migratory endoparasite and invades foliage and the base of stems of cereal plants, where it migrates through tissues and feeds on adjacent cells. Reproduction continues inside a plant almost all year round, but is minimal at low temperatures. When an infected plant dies, nematodes return to the soil from where they infect neighbouring plants. Typical symptoms of stem nematode attack include basal swellings, dwarfing and twisting of stalks and leaves, shortening of internodes and many axillary buds, producing an abnormal number of tillers to give a plant a bushy appearance (Fig. 5.6, Plate 3E). Heavily infected plants may die in the seedling stage, resulting in bare patches in a field, while other attacked plants fail to produce flower spikes (Kort, 1972).

The nematodes are highly motile in soil and can cover a distance of 10 cm within 2 h (Kort, 1972), hence their ability to spread from one plant to another is rapid. There are a number of biological races or strains of *D. dipsaci*, which are morphologically indistinguishable but differ in host range. Kort (1972) stated that the rye strain is more common in Europe and the oat strain is more common in Britain. Rye strains attack rye and oats as well as several other crops, including bean, maize, onion, tobacco, clover and also a number of weed species commonly associated with the growth of cereals in many countries (Kort, 1972). The oat strain attacks oat, onion, pea, bean and several weed species, but not rye (Kort, 1972). Wheat is also attacked by *D. dipsaci* in central and Eastern Europe (Rivoal and Cook, 1993). The giant race of *D. dipsaci* is widely distributed throughout North Africa and the Near East on many crops and needs to be monitored for effects on cereals (see Chapter 8).

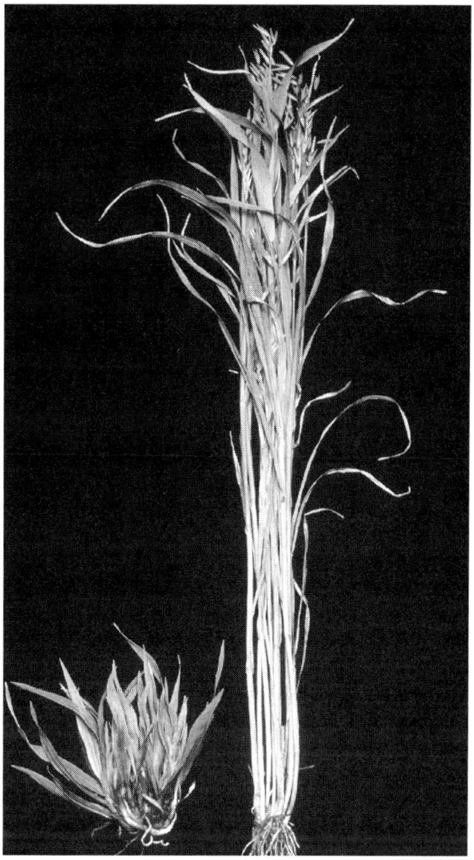

Fig. 5.6. Close up of stem nematode, *Ditylenchus dipsaci*, damage on susceptible oat indicating severe dwarfing, twisting of leaves, and an abnormal number of tillers giving the plant a bushy, stunted appearance. (Photo: S. Taylor.)

Damage potential and economic importance

Economic damage by *D. dipsaci* depends on a combination of factors such as host plant susceptibility, infection level of soil, soil type and weather conditions. This is complicated further by the extensive intraspecific variation, where more than 30 biological races are known to exist within *D. dipsaci* (Janssen, 1994). Furthermore, environmental conditions such as extended soil moisture content in the surface layer of soil provide optimum nematode activity, hence increasing the chance of a heavy attack. Economic damage is rarely associated with sandy soils, but soils with a clay base are more likely to be associated with damage (Kort, 1972). It is a problem with cereal crops growing on heavy soils in high rainfall areas (Griffin, 1984). The nematode is economically important on rye and oat but not on wheat and barley (Sikora, 1988). Although few studies have looked at the economic importance of this nematode, work on oat in England attributed a 37% yield loss to *D. dipsaci* (Whitehead *et al.*, 1983) and in Italy was considered an important factor in poor wheat yields, where damage caused by *D. dipsaci* was associated with the presence of *Fusarium* spp. (Belloni, 1954).

Management measures

The occurrence of different biological races or strains of *D. dipsaci* makes it a difficult nematode to control. The only economical and highly effective method is use of host resistance (Table 5.4).

As a result of the polyphagous nature of *D. dipsaci* being a pest on lucerne (alfalfa), red and white clover, pea, bean and bulbous species of the Liliaceae, including garlic, onion, tulip and narcissus, the use of crop rotation in some cropping systems is limited. However, within lucerne, red and white clover, oat, garlic, strawberry and sweet potato, resistant cultivars have been developed, as reviewed by Plowright *et al.* (2002).

In Britain, the most successful oat crop has resistance derived by the landrace cv. Grey Winter, which is controlled by a single dominant gene that is now bred into several commercial cultivars (Plowright *et al.*, 2002). In other oat, resistance may be derived from Uruguayan landraces. The wild oat, *Avena ludoviciana*, has more than one gene for resistance (Plowright *et al.*, 2002), whilst a number of other oat cultivars have been reported resistant (Whitehead, 1997), but many of these offer only partial resistance or tolerance.

Rotational combinations of non-hosts including barley and wheat offer some control for the rye and oat races of *D. dipsaci*. However, once susceptible oat crops have been damaged, rotations are largely ineffective (Rivoal and Cook, 1993).

Table 5.4. Crop cultivars and accessions resistant to stem nematode, *Ditylenchus dipsaci*.

Crop	Species	Cultivar/accession	Country	Reference
Lucerne	*Medicago sativa*	Vertus	Sweden	Cook and Yeates (1993)
		Nova	Australia	
		Washoe Lahontan	USA	
		Resistador II		
White clover	*Trifolium repens*	Line G49	New Zealand	Mercer and Grant (1995)
	Tolerant	Sebeda	New Zealand	West and Steele (1986)
		Katrina		
		Alice	UK	Cook and Evans (1988)
		Donna		
		Aran		
		Pronitro		
Rye	*Secale cereale*	Ottersum (landrace)	The Netherlands	Ritzema-bos (1922)
		Heertvelder		
Faba bean	*Vicia faba*	INRA 29H	France	Caubel and Le Guen (1983)
		Several		Gastel (1990); Hanounik *et al.* (1986)
		Souk el Arba Rharb (landrace)	Morocco	Schreiber (1977)
Red clover	*Trifolium pretense*	Sabtoron	UK	Plowright *et al.* (2002)
		Norseman		
Oat	*Avena sativa*	Grey Winter		
		Penirth		
		Anita	Belgium	Clamont (1985)
		Bettong	Australia	MacDaniel and Barr (1994)
	A. ludoviciana	Cc 4346	UK	Griffiths *et al.* (1957)

From Plowright *et al.* (2002).

Other nematodes

There are other plant parasitic nematodes, such as *Longidorus elongatus*, *Merlinius brevidens* and species of *Tylenchorhynchus* and *Paratrichodorus*, which have been found or are implicated potentially to cause yield loss on cereals, although their global distribution and economic importance to date have not been clearly defined. *Tylenchorhynchus nudus*, *T. vulgaris* and *M. brevidens* are responsible for poor growth in limited areas of the USA and India (Smolik, 1972; Upadhyaya and Swarup, 1981). *Paratrichodorus anemones* and *P. minor* are two species reported to cause damage to cereal crops in the USA, with wheat seeded early in autumn in sandy soils being highly susceptible to *P. minor*. Elekcioglu and Gozel (1997) clearly demonstrated field population dynamics in relation to wheat growth for the nematode complex *Pratylenchus thornei*, *Paratrophurus acristylus* and *Paratylenchus* species in Turkey, concluding that the importance of the two latter genera requires further investigation. Other cyst nematodes, such as *Punctodera puncata* and *Heterodera hordecalis*, have been described from roots of cereals in several countries, but their distribution and economic importance are unknown. These nematodes or nematode combinations can be found in the chapters reviewed by Kort (1972), Griffin (1984), Swarup and Sosa-Moss (1990) and Rivoal and Cook (1993).

Maize

Zea mays L. is one of the most important cereal crops used in the human diet in large parts of the world and an important feed component for livestock. In terms of total world production, maize on average over the last 5 years outranked paddy rice and wheat. Global production exceeds 600 Mt, with about 60% produced in the developed countries, particularly by the USA. China produces 27% of the world's maize and the rest is grown in countries of Latin America, Africa and southern Asia (Table 5.1), with a large proportion being produced in the tropics and subtropics.

Many plant pathogens and pests, including plant parasitic nematodes, cause considerable loss during crop development and aggravate plant damage under moisture and other stress conditions. Information on the importance of plant parasitic nematodes used to be very limited, but a significant number of publications on maize–nematode associations have appeared over the past decade. This implies increasing awareness of the importance of nematode damage to this very important food and fodder crop. Previously, the crop was commonly regarded as a non-host to several nematode species (Idowu and Fawole, 1990; Toida *et al.*, 1991; Rodríguez-Kábana, 1992), probably because yield losses may go unnoticed as a result of extensive root systems, inadequate control measures (Riekert, 1996; Koenning *et al.*, 1999) or lack of typical symptoms (Asmus *et al.*, 2000). Awareness that specific extraction (Riekert, 1995) and resistance assessment methods (Ibrahim *et al.*, 1993) may affect quantifications is a major factor when considering the importance of nematodes to a crop such as maize. The prominence of maize in the global and many local economies (Table 5.1) and as a staple food to millions emphasizes the fact that the impact of nematode parasitism on this crop should not be underestimated. The extensive use of maize in rotation systems further necessitates a profound knowledge of the crop's host status to economically important nematode species.

Nematodes of Maize

Over 60 nematode species have been found associated with maize in different parts of the world. Most of them have been recorded from roots and soil around maize roots, with information on the biology or pathogenicity of many of these species not readily available. The most important groups of plant parasitic nematodes demonstrated to be important limiting factors in maize production from all over the world are: (i) the root knot nematodes,

Meloidogyne species; (ii) the root lesion nematodes, *Pratylenchus* species; and (iii) the cyst nematodes, *Heterodera* species.

Meloidogyne

Distribution

Root knot nematodes, comprised of more than 50 species, are considered economically important on most crops in the world (Sasser, 1977; Hirschmann, 1985; Jepson, 1987). Some species have a worldwide distribution and have wide host ranges, while others are limited in distribution and are more host specific. Several races with differential host ranges occur within species (Sasser and Triantaphyllou, 1977; Kleynhans, 1991). It is important, therefore, to know the status and distribution of root knot nematodes on an important crop such as maize. *M. incognita* and *M. javanica* have been detected damaging maize in almost all maize-growing regions of the world. *M. africana* and *M. arenaria* have been recorded on maize in India (Krishnamurthy and Elias, 1967) and Pakistan (Maqbool, 1980, 1981). *M. arenaria* has also been reported by several authors from the USA and elsewhere as being associated with maize or that maize germplasm exhibits variable response to this root knot nematode species (Keetch and Buckley, 1984; Windham and Williams, 1987; Ibrahim *et al.*, 1993; Kinloch and Dunavin, 1993; Davis and Timper, 2000; Timper *et al.*, 2002). *M. chitwoodi* interaction with mycorrhizal fungi on maize was studied in a greenhouse (Estanol-Botello *et al.*, 1999), but maize is considered by some as a poor host to race 2 of this nematode species (Al-Rehiayani and Hafez, 1998), while reference to good host maize cultivars exists (Cardwell and Ingham, 1997). The root knot nematode species and races found in association with maize have very wide host ranges as would be evident from other chapters in this book and many other references. Weeds could also play an important role in root knot nematode-susceptible crop sequences (Meyer and Van Wyk, 1989).

Biology and life cycle

Completion of the life cycle of this group of nematodes varies with conditions and host, but could be approximately 20 days when conditions are optimal (Taylor and Sasser, 1978). These authors report that a single female could produce over 1000 eggs under optimal conditions, but this figure also varies greatly (Barker *et al.*, 1985). Under poor growing conditions, *M. javanica* juveniles may enter young roots, but fail to mature (Shepherd, 1981).

Symptoms

Above-ground symptoms include stunting, leaf chlorosis and patchy growth (Fig. 5.8). Root galls may be small or large, terminal or subterminal (Fig. 5.7) or further back along the root (Fig. 5.9). Typical gall symptoms may be totally absent (Becerra and Sosa-Moss, 1977; Idowu, 1981; Riekert,

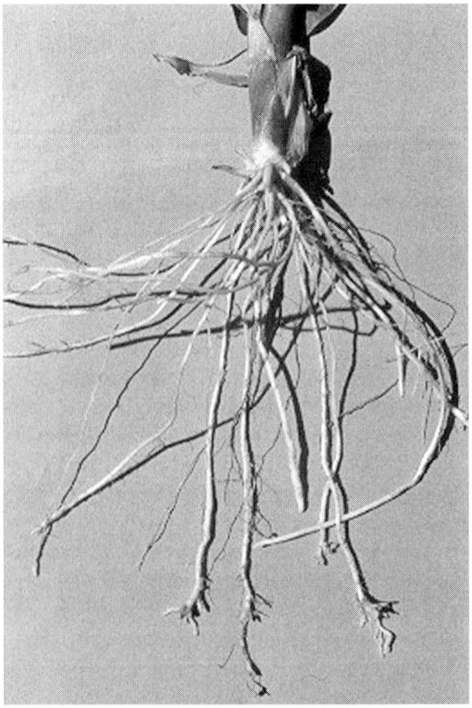

Fig. 5.7. *Meloidogyne* infection of young maize plants with galls and typical root tip branching. (Photo: A.H. McDonald.)

Fig. 5.8. Field symptoms on maize infested with *Meloidogyne*. (Photo: A.H. McDonald.)

Fig. 5.9. Severely galled maize roots. (Photo: A.H. McDonald.)

1995; Asmus *et al.*, 2000), and therefore it should be emphasized that maize often has mistakenly been considered a poor host or even immune to root knot nematodes. Histologically, *M. javanica* infection of maize roots shows typical multinucleated giant cell development in vascular tissue as well as embedded egg masses in inconspicuous galls, mostly close to root apexes (Asmus *et al.*, 2000).

Since root galls are often small or even lacking, a root system should be stained and examined for nematode penetration if root knot nematodes are suspected of being important or if juveniles are detected in the soil. Root tip galls can also be confused with galls produced by ectoparasites such as *Xiphinema* (Fig. 5.10; Plate 3F). Riekert (1995) modified an NaOCl extraction technique (Hussey and Barker, 1973; Hussey and Boerma, 1981) specifically for root knot nematode assessment on maize. Others used the same or similar methodology, particularly in maize resistance assessment of root knot nematodes (Williams and Windham, 1990; Davis and Timper, 2000). Gall indices (Johnson *et al.*, 1999) and staining methods (Windham and Williams, 1994a) are, however, also used.

Pathotypes

The four races of *M. incognita* and race 2 of *M. arenaria* sometimes reproduce well on maize, but some cultivars exhibit specificity to a specific race (Lopez, 1981; Oteifa and Elgindi, 1982; Williams and Windham, 1990; Ibrahim *et al.*, 1993; Windham and Williams, 1994b; Davis and Timper, 2000).

Damage potential and economic importance

Although root knot nematodes occur frequently in maize fields, information on economic losses is lacking. However, indirect observations when nematicides are applied in root knot-infected soils suggest that these nematodes are economically important in maize (Riekert, 1996; Riekert and Henshaw, 1998). In Jamaica (Hutton, 1976, 1981), greater root knot damage occurred when maize was sown after sugarcane. Failure to demonstrate yield reduction due to nematode parasitism in maize was explained by Dickson and McSorley (1990) as being a result of extensive root growth in this crop after the seedling stage. This is due to high fertilization and watering levels applied to this crop and it obscures measurable injury levels. Koenning *et al.* (1999) add a lack of adequate control measures on maize as a reason for ignorance of nematode damage on the crop. Goswami and Raychaudhuri (1978) studied the interaction between mosaic virus and *M. incognita* in pot trials. They found that the mosaic symptoms appeared earlier and nematode reproduction was greater when both pathogens were together than when alone. It remains an important aspect to be alert to root knot nematode infestation of maize, particularly in low input production conditions.

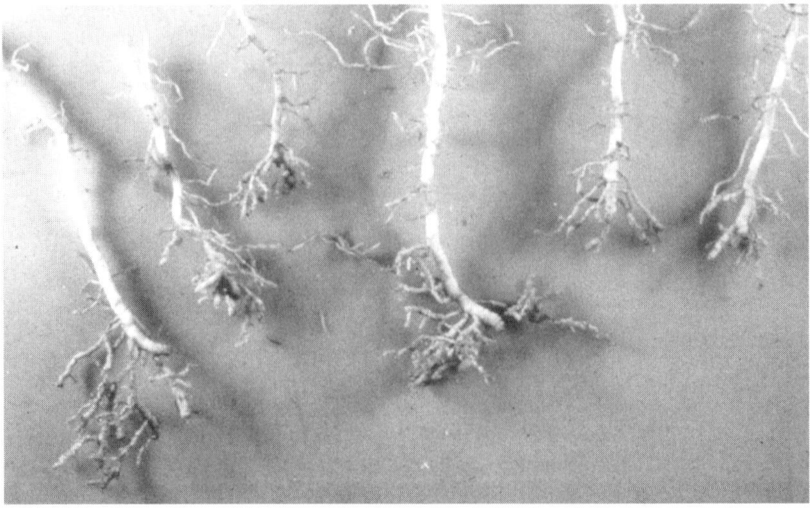

Fig. 5.10. *Xiphinema* root tip galling of maize. (Photo: B. Jacobsen and R.A. Sikora.)

Pratylenchus

Distribution

Lesion nematodes are cosmopolitan in maize fields (De Waele and Jordaan, 1988a; Tacconi *et al.*, 1988; Mizukubo *et al.*, 1990; Gao and Cheng, 1992; Prasad *et al.*, 1995; Koenning *et al.*, 1999) and are often associated with poor growth and yield reduction (Dickson and McSorley, 1990; Aflomi and Fawole, 1991; McDonald and Van den Berg, 1993; Meintjes, 1993). *Pratylenchus brachyurus*, *P. zeae* and *P. penetrans* are the most commonly encountered species in subtropical and tropical regions (De Waele and Jordaan, 1988a; Jordaan *et al.*, 1989; Lordello *et al.*, 1992), followed by *P. coffeae*, *P. delattrei*, *P. goodeyi* (Prasad *et al.*, 1995), *P. hexincisus*, *P. neglectus*, *P. pratensis*, *P. sefaensis* and *P. thornei* (Loof, 1978).

Lesion nematodes have wide host ranges (Loof, 1978), which can affect the selection of crop used to control the nematode in rotations. In addition, the presence of weed hosts in a field can strongly influence lesion nematode densities in maize fields (Egunjobi, 1974; Stradioto *et al.*, 1983; Jordaan and De Waele, 1988).

Biology and life cycle

Host plant, temperature and soil type are very important ecological factors for plant parasitic nematodes, but as *Pratylenchus* species are very polyphagous, environmental factors seem to dominate the host plant in this genus (Loof, 1978). The general biology and life cycle of this group of nematodes are described in Chapter 1. Temperature greatly affects the development and reproduction of *Pratylenchus*, e.g. *P. zeae*, *P. brachyurus* and *P. hexincisus* reproduce well at 30°C, whereas *P. penetrans* prefers lower temperatures of 20–24°C (Olowe and Corbett, 1976; Zirakparvar *et al.*, 1980). Frequently the optimum temperature for nematode development is correlated with the optimum temperature required for good plant growth (Olowe and Corbett, 1976). A similar effect was recorded by Dickerson *et al.* (1964), who found differences in the top weight of plants inoculated with *P. penetrans* over the uninoculated controls at 20°C, but not at 24°C.

Soil type (Loof, 1978) and tillage practices (Minton, 1986) have also been recorded to affect lesion nematode population dynamics. Most *Pratylenchus* species thrive well in a wide range of soil types, but for others a particular soil may be more suitable (Loof, 1978). Naganathan and Sivakumar (1975, 1976) reported higher population densities of *P. delattrei* in sandy clay loam soil than in any other soil type. Conversely, *P. hexincisus* is found in a wide range of soil types, but reproduces best in sandy soils (Swarup and Sosa-Moss, 1990).

Moisture is an important factor affecting the development and damage potential of *Pratylenchus* species (Corbett, 1976; McDonald *et al.*, 1987; McDonald and Van den Berg, 1993). In Nigeria, Egunjobi (1974) demonstrated pathogenicity of *P. brachyurus* on maize and found increased nematode development during the rainy season.

Symptoms of damage

Generally the nematode species, population density and environmental conditions affect symptom expression. Therefore, above-ground symptoms are not highly specific (Jepson, 1987). Nematode damage to fibrous root systems can result in destruction of cortical parenchyma and epidermis (Gao and Cheng, 1992), which may cause sloughing-off of the tissue and severe necrosis (Plate 3C). In addition, severe root pruning as well as proliferation of lateral roots may occur (Ogiga and Estey, 1975; Zirakparvar, 1980). *P. zeae* causes a mechanical breakdown of cells and necrosis of stellar and cortical tissues, resulting in formation of cavities (Olowe and Corbett, 1976; Olowe, 1977). Patel *et al.* (2002a) recorded considerable reduction in root and shoot weight, plant height and chlorophyll content, and an almost tenfold increase in *P. zeae* numbers in maize grown in pots. In contrast, *P. brachyurus* causes more necrosis than mechanical damage (Corbett, 1976). Damage by lesion nematodes can often be diagnosed by the presence of small lesions (Corbett, 1976; Fortuner, 1976) on the root surface.

Damage potential and economic importance

Nematode populations may increase considerably under continuous maize cropping, ultimately resulting in significant yield losses (Reversat and Germani, 1985; Maqbool and Hashmi, 1986). Yield loss estimates in maize due to *Pratylenchus* species are scarce, mostly as a result of confounding effects of other factors (Dickson and McSorley, 1990; Todd and Oakley, 1996; Koenning *et al.*, 1999). Smolik and Evenson (1987) found direct relationships between *P. hexincisus* and *P. scribneri* and maize yield loss, indicating that *P. hexincisus* was more damaging to dry land maize than was *P. scribneri* to irrigated maize. A questionnaire survey to agricultural research institutions in South Africa put *Pratylenchus* species second overall after root knot nematodes in terms of economic importance (Keetch, 1989). *Pratylenchus*, along with *Meloidogyne* and *Hoplolaimus* were the most frequently reported genera on maize in the USA (Koenning *et al.*, 1999).

In Nigeria, *P. brachyurus* has been reported to be responsible for 28.5% yield reduction. The reduction in yield was correlated with a 50% increase in nematode density (Egunjobi, 1974). Zirakparvar (1980) reported that *P. hexincisus* causes reductions in root and top weights of plants, while *P. pseudopratensis* reduced maize yield in Nigeria but had no effect on top weight and plant height (Afolami and Fawole, 1991).

Indirect evidence has been obtained with nematicides where the detected yield increases suggested that lesion nematodes are important limiting factors in maize cultivation (El-Meleigi, 1989; Riekert, 1996). Yield increases of 33–128% have been observed by Walters (1979) in South Africa following the application of nematicides, a 10–54% increase in the USA (Bergeson, 1978; Norton *et al.*, 1978) and a twofold increase in Brazil (Lordello *et al.*, 1983).

Precise evaluations of losses in maize caused by lesion nematodes are hampered by secondary infections of nematode lesions by fungi and bacteria (Egunjobi, 1974). Jordaan *et al.* (1987) found that the effect on maize plant growth is more severely affected under combined inoculation of *Pratylenchus zeae*, *P. brachyurus* and *Fusarium moniliforme* than when inoculated with nematodes alone. The effect is greater during the seedling stage, and Patel *et al.* (2002b) confirm that these plants are predisposed to *P. zeae* infection by this fungus. Although synergistic effects between *P. pratensis* and *F. moniliforme* on maize were also found by Revelo Moran *et al.* (1993), there was a reduction in nematode numbers in the presence of the fungus under both greenhouse and field conditions. Maize could serve as an inoculum reservoir for tobacco rattle virus, transmitted by *Paratrichodorus allius* to cause corky ringspot of potato (Motjahedi *et al.*, 2002). The complex nature of these interactions of nematodes and fungi on a crop such as maize discourages research on this topic.

Heterodera

Although more than nine species of cyst nematodes have been recorded associated with maize in subtropical and tropical countries, only three, i.e. *Heterodera zeae*, *H. avenae* and *Punctodera chalcoensis*, are considered economically important (Luc, 1986).

H. cajani, *H. delvii*, *H. gambiensis*, *H. graminis*, *H. oryzae* and *H. sorghi* have been recorded sporadically, but their role as parasites of maize remains uncertain (Koshy and Swarup, 1972; Merny and Cadet, 1978; Prasad *et al.*, 1980; Sharma and Swarup, 1984; Reversat and Germani, 1985). Swarup *et al.* (1964) from India first recorded *H. avenae* on maize in the subtropics. The nematode has also been reported in maize fields in Egypt (Ibrahim *et al.*, 1986). The worldwide distribution on cereals as well as information on the biology of this nematode species have been discussed in the section under wheat. It has been suggested that there are virulent and less virulent pathotypes in *H. avenae* populations with regard to their ability to parasitize maize (Saefkow and Lucke, 1979; Saefkow, 1983).

Heterodera zeae

Distribution

This nematode was first described from India by Koshy *et al.* (1970), where it is widely distributed (Sharma and Swarup, 1984). The nematode has been also reported from Egypt (Ibrahim *et al.*, 1976), Pakistan (Maqbool, 1980), the USA (Golden and Mulvey, 1983), Thailand (Chinnarsi *et al.*, 1995) and Rajastan (Aruna and Siddiqui, 1997a).

Koshy *et al.* (1970) originally reported barley (*Hordeum vulgare*) as a host for *H. zeae*. Srivastava and Swarup (1975) recorded *Setaria indica* and *Zea mexicana* as additional hosts (Sharma and Swarup, 1984). Oat (Ringer *et al.*, 1987), wheat (Shahina and Maqbool, 1990; Abadir *et al.*, 1994) and vetiver grass (*Vetiveria zizanioides*) (Bajaj and Gupta, 1994) have been added to the host list. Shahina and Maqbool (1990) regarded the non-graminaceous families Malvaceae, Compositae, Cruciferae and Cucurbitaceae as non-hosts. Ringer *et al.* (1987) performed a comprehensive host range test, which included several weed species.

Biology and life cycle

Temperature plays an important role in the biology of *H. zeae* (Koenning *et al.*, 1999). Srivastava (1980) found that the most favourable temperature for emergence of juveniles from cysts is 25°C, with 91% emergence. At temperatures of 10 or 15°C, only 10–20% of the juveniles emerge. However, other reports of optimum temperature ranges for reproduction from between 28 and 36°C have been published (Krusberg, 1988; Parihar and Yadav, 1992; Hashmi *et al.*, 1993b), but egg hatching is significantly slower as temperatures drop below 25–20°C (Hutzell and Krusberg, 1990). Temperature and level of maize hybrid susceptibility affect the population dynamics of *H. zeae* (Ismail *et al.*, 1994).

The life cycle is short, taking only 15–18 days (Srivastava and Sethi, 1985b; Hutzell and Krusberg, 1990). It has been speculated that the nematode may complete 6–7 generations during one crop season (Srivastava and Sethi, 1985a, 1986).

Generally, the nematode reproduces well in moderately light soils. The addition of clay to soil mixtures resulted in a proportional decline in nematode reproduction levels (Srivastava and Sethi, 1984a).

Symptoms of damage

H. zeae-infested plants exhibit poor and unthrifty growth and are stunted and pale green in colour (Koshy and Swarup, 1971).

Pathotypes

Three host races have been distinguished based on reproduction and host preference (Bajaj and Gupta, 1994).

Damage potential and economic importance

Though the pathogenicity of the nematode has been demonstrated on maize, data on economic damage to the crop are lacking (Koenning *et al.*, 1999). However, Srivastava and Sethi (1984b) showed that plant growth reductions are directly correlated with initial nematode population density. Hashmi *et al.* (1993b) found consistent reductions in maize plant weights in the presence of *H. zeae* at optimum temperature regimes for the nematode. Maize growth and yield are suppressed by 13–73% in the presence of *H. zeae*, and damage is more profound under hot and dry conditions (Krusberg *et al.*, 1997). Maize plant growth progressively declines with increasing initial population densities of this nematode (Ismael *et al.*, 1994).

Maize plants infected by a combined inoculum of *H. zeae* and *Cephalosporium maydis* show more late wilt symptoms than when inoculated with either alone (Singh and Siradhana, 1988).

Punctodera chalcoensis

Vázques (1976) surveyed maize fields in Mexico State during 1960 and recorded a cyst nematode, identified then as *Heterodera punctata*, on maize roots. Sosa-Moss (1965) demonstrated distinct morphological differences between the Mexican population and the original description of *H. punctata* (Thorne, 1928). He also reported that the Mexican population attacks maize instead of wheat and grasses, which are common hosts of *H. punctata*. The species was later redescribed as *Punctodera chalcoensis* (Stone *et al.*, 1976).

Distribution

P. chalcoensis is limited in distribution to Mexico where it is considered of extreme importance. The nematode has been given the local name of Mexican corn cyst nematode. Out of 300 graminaceous plants tested, only *Z. mays* and *Z. mexicana* (Teosinte) were considered hosts (Stone *et al.*, 1976).

Biology and life cycle

The nematode has one generation per year and survives winter in diapause (Sosa-Moss, 1987). The nematode survives and reproduces well in all soil types and causes severe damage on volcanic sandy soils.

Symptoms

Maize fields infested with this cyst nematode exhibit patches of stunted and chlorotic plants. Damage can be severe and is dependent on cultivar susceptibility, nematode density and adequate soil moisture levels in the latter part of the growing season.

In heavily infested sandy soils, plants are markedly stunted, with chlorotic leaves exhibiting pale colour stripes. It is important to distinguish these symptoms from those caused by the virus disease 'Rayado Fino' where the pale striped lines are in green leaves rather than in yellowish leaves as in the case of nematode infestation.

A maize root system is generally poorly developed when infected by this nematode. Two months after planting, corresponding to the initiation of the rainy season, large numbers of white females can be observed on the root surface.

Damage potential and economic importance

Under glasshouse conditions, Sosa-Moss and Gonzales (1973) obtained a reduction of about 60% in yield in heavily infested soils. Although yield loss in the field is considered to be high, experimental data are lacking.

Other nematodes associated with maize

Many other plant parasitic nematodes have been found associated with maize (De Waele and Jordaan, 1988a; Jordaan *et al.*, 1989; Koenning *et al.*, 1999). In most of these cases, their importance to maize production has not been determined (Koenning *et al.*, 1999). Of limited or local importance are species of *Belonolaimus*, *Criconemella* (McSorley and Dickson, 1990; Huang *et al.*, 1997), *Hoplolaimus*, *Tylenchorhynchus*, *Helicotylenchus* (Haidar and Nath, 1992), *Rotylenchulus*, *Longidorus*, *Paratrichodorus* (McSorley and Gallaher, 1994), *Ditylenchus* (Basson *et al.*, 1990; MacGuidwin and Slack, 1991), *Quinisulcius* (Stoyanov *et al.*, 1990) and *Radopholus* (Price, 1994; Fogain and Gowen, 1995). *Longidorus* and *Xiphinema* can cause severe root tip damage on sandy soils and yield loss, especially under moisture stress situations. *Belonolaimus longicaudatus* can cause severe losses to sweetcorn on sandy soils in Florida (Rhoades, 1977), and linear relationships between decrease in maize yield and increase in a *Belonolaimus* species population were found (Todd, 1989). Seed-borne specimens of *Ditylenchus dipsaci* and *Ditylenchus* juveniles were detected on maize (Tenente *et al.*, 2000). Knuth (2000) reports differential susceptibility of maize varieties to *D. dipsaci*, the nematode causing significant yield losses and affecting the rate of seedling development.

Management measures for maize nematodes

CHEMICAL. Utilization of nematicides is limited in most instances for economic or political reasons, as well as the fact that their application has led to inconsistent results (McDonald *et al.*, 1987; McDonald and De Waele, 1987a,b; Barnard *et al.*, 1989; Johnson *et al.*, 1990; Shahina and Maqbool, 1990; Johnson and Leonard, 1995; Riekert, 1996). Inoculation of soil planted to maize with effective nitrogen fixation agents after treatment with nematicides is recommended because *Azospirillum* species stimulate growth after treatment with nematicides (Fayez, 1990). Responsible use of chemical control of nematodes in maize could always be a useful production management tool, particularly when used in integrated nematode management systems (Barnard *et al.*, 1989; Johnson *et al.*, 1990; Johnson and Leonard, 1995). Most importantly, the effect of environment on treatment success (McDonald and De Waele, 1987b; Barnard *et al.*, 1989; Badra and Adesiyan, 1990) and possible carry-over benefits (Johnson and Leonard, 1995) must never be disregarded with a crop such as maize.

RESISTANCE. In a review of resistance of maize to plant parasitic nematodes, Jordaan and De Waele (1987) highlight a very important consideration when wanting to introduce resistance to nematodes in maize, which is that it may be in exchange for other commercially desirable or preferred traits (see also Williams *et al.*, 1990). On the other hand, pedigree breeding without selecting for nematode resistance may result in highly susceptible and intolerant crops, which could be very costly in any kind of production system. These authors distinguish between resistance to nematode reproduction and tolerance to the damage caused by nematodes. They describe tolerance as the ability of a crop to suffer no damage or yield reduction despite being moderately to heavily infected by nematodes. When trying to introduce resistance to nematodes in commercial hybrids, nematologists should always endeavour to work with plant breeders to ascertain that the end-product will be acceptable to producer, processor and consumer alike.

Many maize cultivars have been reported to be resistant to *Meloidogyne*, *Helicotylenchus* and *Paratrichodorus* (Johnson, 1975), *M. incognita* and *M. javanica* (Nishizawa, 1981; Oteifa and Elgindi, 1982; De Brito and Antonio, 1989; Ribeiro *et al.*, 2002) and *M. arenaria* (Sasser and Kirby, 1979). Windham and Williams (1994a) reported retarded *M. incognita* development or failure of juveniles to reach maturity in maize hybrids exhibiting some level of resistance. Windham and Williams (1987, 1988) screened 64 commercial maize hybrids for resistance to *M. arenaria*, *M. incognita* and *M. javanica*, and found resistance to *M. arenaria* among the hybrids as well as three inbred lines with resistance to *M. javanica*. Twenty-five commercial tropical hybrids were all more susceptible to *M. arenaria* than *M. incognita* in a greenhouse screening (Windham and William, 1994b). Davis and Timper (2000), however, conclude that maize is generally more resistant to *M. arenaria* than *M. incognita*, which could benefit maize–groundnut (peanut) rotation systems but not cotton–maize sequences. Lordello *et al.* (1989) found resistance to *M. javanica* in cultivars and claim immunity as a dominant trait in one, which they traced back to one parental line, IAC Ip365-4-1 (Lordello and Lordello, 1992).

Combining abilities of maize lines are important factors to consider, both for yield and for nematode resistance. Poerba *et al.* (1990) found general (additive resistance) and specific (single gene dominance) combining resistance to *M. javanica* in diallel crosses between maize inbred lines, with Mp307 the best source of resistance. A later study by Williams and Windham (1992) with inbred line diallel crosses on *M. incognita*, however, showed general combining ability to be a better source of variation. Mp307 remained the best source of resistance. Four more inbred lines subsequently have been registered as sources of resistance to *M. incognita* and *M. arenaria* (Williams and Windham, 1998). These lines have not been tested for combining ability for yield but are white kernel types. Open-pollinated maize varieties were also screened as possible sources of resistance to *M. incognita* in the USA (Aung *et al.*, 1990, 1991), and two, Old Raccoon and Tebeau, showed higher levels of resistance than the resistant check. Most of the screenings and selections mentioned above were done with the host races of the root knot nematode populations identified, which is highly recommended.

Jordaan and De Waele (1987) state that it could be more difficult to identify resistance to migratory than sedentary endoparasites, but cite several reports of resistance in maize to other nematode species. Lordello *et al.* (1985) have also identified several maize genotypes resistant to *Pratylenchus zeae* and *P. brachyurus*. Two wild maize species, *Zea diploperennis* and *Z. mexicana*, have been reported to be resistant against *Pratylenchus scribneri* and *Helicotylenchus pseudorobustus* (Norton *et al.*, 1985). Wicks *et al.* (1990) developed and registered a yellow maize line with resistance to *P. hexincisus* and *P. scribneri*, as well as to important fungal diseases of maize, ear rot and Tursicum. Good specific combining ability has been identified for this line. Cultivar reaction to *Helicotylenhus pseudodigonicus* varied in greenhouse screenings, and a positive relationship was found between nematode inoculum levels and percentage reduction of root and shoot growth of maize.

Windham and Lawrence (1992) report high levels of resistance in commercial maize hybrids to *Rotylenchulus reniformis* based on reproduction of the nematode. Hashmi *et al.* (1993a) demonstrated in greenhouse screening of inbred maize lines that resistance to *Heterodera zeae* exists in this crop. Variation in susceptibility to *H. zeae* in maize lines was also recorded by Aruna and Siddiqui (1997b). Singh and Patel (1999) report a maize variety resistant to *Tylenchorhynchus vulgaris*, and a variable genotype reaction to *T. zambiensis* was observed by Venditti and Noel (1995).

Although almost all of the above-mentioned authors recommend use of nematode resistance alone or in combination with other nematode management tools, the focus of introduction of resistance should be to produce a genotype with acceptable agronomic traits, durable resistance and affordable seed. Use of marker-assisted selection (Young and Mudge, 2002) should be considered, especially where single gene dominance is available. As this technology develops, it could assist breeders in introducing preferred genes from the male as well as female side to ensure sustainable heritability of these traits.

CULTURAL. Practices such as crop rotation, tillage, planting time, application of organic amendments and sanitation have been tested, and in many cases were demonstrated to be effective in reducing nematode populations. In most cases, maize was tested for its application as a non-host crop against root knot nematodes affecting other crops in the rotation. Therefore, little is actually known concerning their effects on root knot population densities in a maize crop. It should be stressed again that in some countries maize is damaged by root knot nematode and reproduction occurs even though typical root galls are not visible (Becerra and Sosa-Moss, 1977; Idowu, 1981).

Some recent studies on crop rotations or sequences where maize was involved point out dangers of ineffective crop choices due to susceptibility of maize (Florini and Loria, 1990; Gallaher *et al.*, 1991; Todd, 1991; Riekert and Henshaw, 1998; Hague *et al.*, 2002) to different nematode species. There are also cautions about targeting only one nematode species in a rotation system when other damaging species are also present (McSorley and Gallaher, 1992; McSorley and Dickson, 1995) and longer sequences of resistant crops before planting a susceptible crop are more effective (Johnson *et al.*, 1999; Chen *et al.*, 2001). While it is sensible to test the host suitability of all crops used in a system to all potentially important nematodes (Wang *et al.*, 2002), it must be understood that rotation alone may not be sufficient to prevent subsequent susceptible crops from suffering nematode damage. Additional control strategies such as nematode resistance should be integrated for effective management of plant parasitic nematodes (Kinloch and Dunavin, 1993). Radish as well as French and African marigold (*Tagetus patula* and *T. erecta*) reduce *Pratylenchus* species populations in maize-based rotations (Knuth, 2002). The wide host ranges of plant parasitic nematodes include several weed species (Salawu and Oyewo, 1999), which must be taken into consideration when control strategies are designed. Weeding of maize plots reduced populations of *Ditylenchus* species, *Heterodera* species and *Tylenchorhynchus clarus* (Youssef, 1998).

There are reports of highly effective nematode management and yield increase in crops where maize is used as a resistant rotational crop (Acosta *et al.*, 1991; Rodríguez-Kábana *et al.*, 1991; Davis *et al.*, 2003), particularly where maize is a non-host to cyst nematodes (Noel and Edwards, 1996). As well as direct crop effects on nematode populations, availability of plant nutrients such as nitrogen and phosphates also plays an important role, particularly in rotations with legumes (Bürkert *et al.*, 2001).

The increasing popularity of conservation tillage and no-till requires a good understanding of the effect of tillage on plant parasitic nematode populations, particularly where other nematode management practices are also used and where these tillage regimes are important for more reasons than nematode management. Minton (1986) gives an overview of tillage and accompanying factors affecting nematode populations. Sometimes tillage effects such as organic matter restitution and soil compaction apparently have little effect on nematode numbers (Esmenjaud *et al.*, 1990; McSorlley and Gallaher, 1993, 1994), while sometimes nematode populations are greatly influenced by tillage (Yeates and Hughes, 1990; Ivezic *et al.*, 2000; Sumner *et al.*, 2002). The important influence of environmental factors in these systems and inter-relationships cannot be overemphasized (Yeates and Hughes, 1990; Yeates *et al.*, 1993). Combining tillage and rotation systems in nematode management strategies is recommended (Cabanillas *et al.*, 1999).

In Mexico, it has been observed that early sowing dates, as well as adequate fertilization reduces damage caused to maize by *Punctodera chalcoensis* (Sosa-Moss and Gonzalez, 1973; Sosa-Moss, 1987). Krusberg *et al.* (1997) found no alleviation of damage by *H. zeae* to maize by fertilizer amendments, but Ivezic *et al.* (1996) obtained up to 60% reduction in nematode populations dominated by *P. thornei* in maize fields after application of high levels of potassium. Animal litter affected numbers of nematodes associated with maize (Sumner *et al.*, 2002), while compost affected densities of several nematode species associated with this crop, although large amounts were needed to induce responses (McSorley and Gallaher, 1996). McSorley and Gallaher (1997) ascribe the inconsistent performance of compost against plant parasitic nematodes on maize to the positive effect of the amendment on crop performance. More consistent effects are observed after prolonged application of compost, which improves soil organic matter content and water-holding capacity. Although there is no interaction between *Meloidogyne* species and organic amendment rates, population densities of *M. incognita* and *M. javanica* decrease with increasing residue rates, while maize plant growth increase (Albuquerque *et al.*, 2002).

BIOLOGICAL. The growth promotion effects of *Trichoderma* species on maize in the presence or absence of *M. arenaria* are not indicative of parasitism by the fungus on the nematode but could be as a result of compounds produced by the fungus that have a direct or indirect effect on nematodes (Windham *et al.*, 1989). Riekert and Tiedt (1994) provide evidence of *Arthrobotrys dactyloides* trapping of *M. incognita* juveniles on the surface of maize roots, but they regard the commercialization of nematode-trapping fungi as of limited use. Several species of nematode-trapping fungi were present in a maize–tomato rotation, although detection frequencies and population densities did not differ significantly between organically and conventionally treated plots (Timm *et al.*, 2001). Bourne (2001) obtained 50% reduction in numbers of *M. incognita* after application of *Pochonia chlamydosporia* in rotations with maize and susceptible crops, and Bourne and Kerry (1999) obtained significant control of *M. incognita*, *M. javanica* and *M. arenaria* in maize with application of this fungus. More than 50% control of *Pratylenchus* species was achieved with *Paecilomyces lilacinus* (Gapasin, 1995), and strains of *Pseudomonas* species inhibit invasion of *Meloidogyne* species and *Radopholus similis* in maize, tomato and banana roots (Aalten *et al.*, 1998). Mycorrhizal fungi of the genus *Glomus* reduce *M. chitwoodi* juvenile numbers on maize (Estanol-Botello *et al.*, 1999). None of these biocontrol agents can be used economically at the present time in cereal crops.

Sorghum

In terms of worldwide production, sorghum (*Sorghum bicolor* (L.) Moench) is the fifth most important cereal in the world. Sorghum is thought to originate from Africa (Maunder, 2002), and over the last 5 years Africa has had the highest production of all continents (Table 5.1). The crop is also very important in regions such as Asia, and Central and North America where production is fairly stable over most regions (FAO, 2002).

Sorghum is an important food and fodder crop of dry land agriculture and it is adapted to a wide range of environmental conditions, from semi-arid through temperate to high rainfall areas (Kollo, 2002). Sorghum is used in various forms of unleavened bread in India and Central America, as fermented bread in Sudan, Ethiopia and India, or as porridge in Africa and India. It is also boiled like rice and is used to produce alcoholic as well as non-alcoholic beverages in some African countries. In some parts of Africa, sorghum is also eaten as a vegetable. Green and dried fodder is an important roughage for cattle. Sorghum is also used for ethanol production in countries such as Brazil (Dahlberg and Frederiksen, 2000).

Nematodes of Sorghum

Although a number of nematode species have been recorded associated with this crop, little information is available on specific nematode problems. Increased yields after chemical treatment of soil where high population densities of specific nematode species were recorded provide indirect evidence of significant economic damage. Nematode damage to sorghum is most likely when the crop is cultivated in monoculture. From a global perspective, species in three genera could be considered more important: the root lesion nematode, *Pratylenchus*, the stunt nematode, *Tylenchorhynchus*, and the root knot nematode, *Meloidogyne* (De Waele and McDonald, 2000).

Pratylenchus

Root lesion nematode species are omnipresent and frequently reported to be associated with sorghum (De Waele and Jordaan, 1988b; Sharma and McDonald, 1990; De Waele and McDonald, 2000). Many reports are conflicting in terms of the damage potential of lesion nematodes on sorghum, which could be attributed to differences in factors such as cultivar, environment and infestation levels (Kollo, 2002). Some frequently reported lesion nematode species associated with sorghum are *P. zeae*, *P. brachyurus*, *P. crenatus*, *P. penetrans*, *P. coffeae*, *P. scribneri*, *P. goodeyi* and *P. hexincisus* (Motalaote *et al.*, 1987; De Waele and Jordaan, 1988b; Gallaher *et al.*, 1991; Todd, 1991; Prasad *et al.*, 1995; De Waele and McDonald, 2000). As a result of infestation by lesion nematodes, roots exhibit necrotic lesions. In heavily infested fields, plants appear stunted and chlorotic. *P. zeae* is reported to reduce uptake of nutrients and water from soil. The species also suppresses top and root growth of sorghum (Chevres-Roman *et al.*, 1971; Bee-Rodriguez and Ayala, 1977; Claflin, 1984; Cuarezma-Teran and Trevathan, 1985; Motalaote *et al.*, 1987). Several reports of interactions of lesion nematodes with plant pathogens were published (Bee-Rodriguez and Ayala, 1977; Kollo, 2002). Motalaote *et al.* (1987) reported differential susceptibility of sorghum genotypes to *P. zeae*. Sorghum is generally reported as a good rotation crop in potato and cereal (Florini and Loria, 1990), maize, soybean, sorghum (Gallaher *et al.*, 1991) and cereal, lucerne and soybean (Todd, 1991) cropping systems.

Tylenchorhynchus

The stunt nematodes *Tylenchorhynchus martini*, *T. nudus* and *Quinisulcius acutus* (Claflin, 1984; Cuarezma-Teran and Trevathan, 1985) have been recorded as associated with unthrifty growth of sorghum plants. Both *T. martini* and *T. nudus* increase in numbers under sorghum monoculture and cause damage at levels of 2000–5000 nematodes/250 cm^3 of soil. Yield increases by 55% after nematicide treatment where *T. martini* is the dominant nematode population (Hafez and Claflin, 1982). Similarly, *T. nudus* was reported to reduce plant growth by 10 and 56% in fertilized and unfertilized plots, respectively (Smolik, 1977). At least eight more species of *Tylenchorhynchus* have been reported as parasites of sorghum (Kollo, 2002).

Nematode feeding results in poorly developed root systems. Root tips may be short and become swollen. Stunted growth and decline of seedling vigour may be observed in severely infested fields (Claflin, 1984). Although 30% reduction in root fresh weight can be caused by stunt nematodes on sorghum, top growth is less affected (Kollo, 2002). Interactions with other plant pathogens and stunt nematodes are reported on sorghum (Sharma and McDonald, 1990).

Meloidogyne

Sorghum is a good host for a number of root knot nematode species. *M. incognita*, *M. arenaria*, *M. javanica*, *M. naasi* and *M. graminicola* are reported associated with sorghum (Sharma and McDonald, 1990; De Waele and McDonald, 2000; Kollo, 2002). *M. acronea* has been detected on sorghum in South Africa (Coetzee, 1956) and Malawi (Bridge *et al.*, 1976). In Malawi, three cultivars were shown to support high to moderate root populations of the local isolate. The nematode was responsible for delayed flowering and yield losses of 56% in sorghum cv. Lindse 555 in pot experiments (S.L.J. Page and J. Bridge, unpublished), with delayed flowering also observed in the field (Page, 1985). Specific races of *M. incognita* and *M. arenaria* are also reported to be better adapted to sorghum (Ibrahim *et al.*, 1993; Kollo, 2002), whilst only race 5 of *M. naasi* parasitizes on sorghum (Ediz and Dickerson, 1976). The nematode causes stunting and chlorosis of infested plants. The optimum soil temperature for development is 26°C, and the life cycle is completed in 34 days.

M. incognita infestation results in production of elongated swellings or discrete knots and proliferation of roots (Orr and Morey, 1978). Galls produced by *M. naasi* are similar but smaller than *M. incognita* galls (De Waele and McDonald, 2000), while *M. acronea* induces extensive root proliferation but inconspicuous root galls (Page, 1985). Screenings of sorghum genotypes against *Meloidogyne* species led to the conclusion that this crop generally is a poor host to root knot nematodes and therefore a suitable rotation option with more susceptible crops (Fortnum and Currin, 1988). Similarly, De Waele and McDonald (2000) have reported variable levels of resistance to be present in sorghum, with some reports of high susceptibility (McSorley and Gallaher, 1992) to resistant germplasm in Brazil (Sharma and McDonald, 1990; Ribeiro *et al.*, 2002). There are not many nematode control options for sorghum due to its low value and the poor conditions it is cultivated under in most parts of the world. Improvement of growing conditions and low-input management practices are therefore recommended (Kollo, 2002).

Other nematodes associated with sorghum

Many other plant parasitic nematode species have been associated with sorghum (De Waele and McDonald, 2000; Kollo, 2002). *Longidorus africanus* and *Heterodera zeae* (Lamberti, 1969; Singh *et al.*, 1979) are associated with sorghum and have been shown to be pathogenic in pot experiments. *Heterodera gambiensis* has been found associated with the crop only in Gambia (Merny and Netscher, 1976); however, damage was not observed in the field in subsequent survey work (Bridge *et al.*, 1978). *Criconemoides ornatus* and *C. sphaerocephala* reproduce well on sorghum (Gallaher *et al.*, 1991; McSorley and Gallaher, 1993), but they are not considered to be economically important (McSorley and Gallaher, 1992). Several weed species in addition to sorghum are good hosts to *Belonolaimus* species, which requires stringent weeding where sorghum is a rotation crop (Todd, 1991).

Millets

Millets are warm weather cereals with small grains. They comprise the following species: *Panicum miliare*, *P. miliaceum*, *P. scrobiculatum*, *Setaria italica*, *Echinochloa colosna*, *Digitaria exilis*, *Eragrostis teffi* and *Eleusine coracana* (finger millet), which are collectively known as small millets (Esele, 2002), and *Pennisetum glaucum*, which is pearl millet (Hash and Witcombe, 2002). These crops form an important staple food in India and several countries of Africa, the Near East and South Asia. Despite their importance (Table 5.1), there is little information on nematode associations with millets. Reasons could be that they are largely produced in a subsistence context, on marginal soils and under adverse climatic conditions (Hash and Whitcombe, 2002). They are grown almost exclusively for animal feed in developed countries (Kollo, 2002).

Pearl millet

Pearl or bulrush millet (*Pennisetum glaucum* (L.) R. Br.) is cultivated for grain and fodder in the arid regions of Africa, India and Asia and as a pasture in the USA. This crop is highly tolerant to stresses such as drought, low soil fertility and heat (Kollo, 2002). A number of plant parasitic nematode species have been recorded in the rhizosphere of the crop. Pearl millet is a host for both *M. incognita* and *M. javanica* (Handa *et al.*, 1971), whereas genotypes in Brazil are all resistant to *M. javanica* and *M. incognita* (Ribeiro *et al.*, 2002). In the north-western sector of India, *M. incognita* has been reported to be a field problem where it occurs in combination with *Sclerospora graminicola*. Appearance of symptoms of green ear disease caused by the fungus was advanced by about 2 weeks when root knot nematodes were present (Vaishnav and Sethi, 1978). Depending on cultivar, the crop is a poor/non-host for *Meloidogyne acronea* (Bridge *et al.*, 1976; Page, 1983). *M. arenaria* race 2 populations are suppressed by pearl millet in rotations with soybean, resulting in low gall indices on soybean (Kinloch and Dunavin, 1993). Millet in the former USSR is affected by *Longidorus elongatus*. The infested plants are stunted and chlorotic with shortened, thick and deformed roots, with yield reductions of 41% (Semkin, 1975).

In glasshouse tests, pearl millet proved to be a favourable host for *Tylenchorhynchus vulgaris* multiplication (Upadhyaya and Swarup, 1972). A report from the southern part of India also suggests that the reniform nematode, *Rotylenchulus reniformis* may be a problem on pearl millet (Seshadri, 1970). Several plant parasitic nematode species are considered of variable importance on millet in different countries. An association between a species of *Fusarium* and *Xiphinema* is reported from Zimbabwe (Sharma and McDonald, 1990). De Waele *et al.* (1998) found 16 plant parasitic nematode species associated with pearl millet in a survey of maize and millet in Namibia, and Hasan *et al.* (1998) provide a list of nematodes hosted by pearl millet, sorghum and maize. Van Biljon and Meyer (2000) found pearl millet to be a good host to *Pratylenchus delattrei* but not *P. zeae*, whereas pearl millet has good resistance to *P. penetrans* (Belair *et al.*, 2002). Kollo (2002) provides extensive lists of species and their reproductive potentials on pearl millet among some other crops. Variable levels of resistance in pearl millet breeding material against *M. incognita* and *M. arenaria* exist (Timper *et al.*, 2002).

Finger millet

The only nematodes of importance on finger or African millet, *Eleusine coracana*, are *Heterodera gambiensis* and *H. delvii*, both recorded on this crop in the southern part of India and Gambia (Bridge *et al.*, 1978). From the same area, *R. reniformis* is also reported to be a problem in the field (Seshadri, 1970; Krishna Prasad and Krishnappa, 1982). *P. penetrans* has a reproduction rate of 5.8 over the initial population on foxtail millet, highlighting the dangers of using susceptible crops in rotation systems (Belair *et al.*, 2002).

Conclusions

There are several genera and species of nematodes that are of economic importance to small grain cereals. Despite sustaining research activities during the past half-century, wheat and rice are the main cereal crops generally perceived to have major nematode problems. Our understanding of some nematodes such as the cereal cyst nematode, *H. avenae*, is much more extensive than others with respect to both biology and control measures, mainly in the form of host resistance. Others such as ear cockle nematode, *A. tritici*, are relatively easily controlled with the adoption of seed hygiene. Although maize nematode research increased significantly over the past decade, barley, sorghum and millets have not received the same attention, though in some areas nematodes may be responsible for economic damage to the crops. Previously, cereals were considered poor hosts of root knot nematodes, but it has become quite apparent that *Meloidogyne* species are very important, particularly on maize. Unfortunately, our knowledge is limited with respect to basic biology and control options for most of the other important nematodes described.

Management of nematodes in cereals has so far been dependent largely on the use of rotation and a limited number of resistant cultivars. The cost of chemicals is prohibitive, and in many cases environmentally unacceptable to the average cereal producer. In the future, our ability to reduce yield losses caused by nematodes will require a greater understanding of many basic questions about nematode biology and the application of appropriate control measures. As a consequence, it is inevitable that breeding for resistance and perhaps tolerance is the major strategy for long-term and environmentally sound control of these parasites, in association with the most appropriate integrated management practices. If in the future biological control proves effective under field conditions and acceptable on an economic basis, then it could be incorporated into integrated pest management systems. At the present time, its use is limited in scope. Although there was a significant increase in resistance studies and many useful sources were added, their use is dependent on commercialization. There is, however, still a great demand for improvement and adaptability of these cereal genotypes to tropical and subtropical conditions, where they are most needed and should be concentrated upon. To capitalize on this information, it is necessary to combine research efforts, particularly for some of the more complex nematodes with race and pathotype differences. Hence the need for global collaborative research programmes is great. Furthermore, the adoption of molecular tools to assist in both pathogen identification and plant breeding will become an integral part of future research developments and ultimate control of these important pests.

New challenges to nematologists in these fields come with trends such as the introduction of genetically modified crops, organic crop production and renewed focus on reduced tillage or no-till. The pesticide industry is under all kinds of pressure, and withdrawal of certain highly effective nematicides will have continued and increased impact on crop production. Simple nematode management technology will be replaced by complicated system management strategies, with increased demand for knowledge of the pest and its interaction with host and environment.

It should be mentioned here that countries with more developed research programmes should assist less fortunate countries with research facilities and manpower. It is in these countries where little is understood about the distribution, importance and control of nematodes where the net benefit of adopting appropriate control measures could be enormous and is considered of extreme humanitarian importance.

References

Aalten, P.M., Vitour, D., Blanvillain, D., Gowen, S.R. and Sutra, L. (1998) Effect of rhizosphere fluorescent *Pseudomonas* strains on plant-parasitic nematodes *Radopholus similis* and *Meloidogyne* spp. *Letters in Applied Microbiology* 27, 357–361.

Abadir, S.K., Ismail, A.E. and Kheir, A.M. (1994) Influence of preceding crops on population densities of *Heterodera zeae* on corn. *Annals of Agricultural Science, Moshotor* 32, 1035–1041.

Abdel Hamid, M.E., Ramadan, H.H., Salem, F.M. and Osman, G.Y. (1981) The susceptibility of different field crops to infestation by *Meloidogyne javanica* and *M. incognita acrita. Anzeiger für Schädlingskunde, Pflanzenschutz, Umweltschutz* 54, 81–82.

Acosta, N., Román, O., Vicente, N.E. and Sáchez, L.A. (1991) [Crop rotation systems and population levels of nematodes.] *Journal of Agriculture of the University of Puerto Rico* 75, 399–405.

Afolami, S.O. and Fawole, B. (1991) Effects of *Pratylenchus sefaensis* Fortuner 1973 on growth and yield of *Zea mays* L. cv. FARZ-7 under continuous cropping. *Plant and Soil* 138, 133–138.

Agrios, G.N. (1988) *Plant Pathology*, 3rd edn. Academic Press, London.

Albuquerque, P.H.S., Pedrosa, E.M.R. and Moura, R.M. (2002) Host nematode relationships in *Meloidogyne* spp. infested soil amended with stillage and filter cake. *Nematologia Brasileira* 26, 27–34 (abstract).

Al-Hazmi, A.S., Al-Yahya, F.A. and Abdul-Razig, A.T. (1999) Damage and reproduction potentials of *Heterodera avenae* under outdoor conditions. *Supplement to the Journal of Nematology* 31, 662–666.

Allen, M.W., Hart, W.H. and Baghott, K.V. (1970) Crop rotation controls the barley root-knot nematode at Tulelake. *California Agriculture* 24, 4–5.

Al-Rehiayani, S. and Hafez, S. (1998) Host status and green manure effect of selected crops on *Meloidogyne chitwoodi* race 2 and *Pratylenchus neglectus. Nematropica* 28, 213–230.

Al-Talib, N.Y., Al-Taae, A.K.M., Neiner, S.M., Stephan, Z.A. and Al-Baldawi, A.S. (1986) New record of *Anguina tritici* on barley from Iraq. *International Nematology Network Newsletter* 3, 25–27.

Ammati, M. (1987) Nematode status on food legumes in Morocco. In: Saxena, M.C., Sikora, R.A. and Srivastava, J.P. (eds) *Nematodes Parasitic to Cereals and Legumes in Temperate Semi-arid Regions.* International Centre for Agricultural Research in the Dry Areas (ICARDA), Syria, pp. 169–172.

Andersen, S. and Andersen, K. (1982) *Annual Report, 1981–1982.* Queensland Department of Primary Industries, Brisbane, Australia.

Andersson, S. (1974) *Heterodera hordecalis* n.sp. (Nematoda: Heteroderidae) a cyst nematode of cereals and grasses in southern Sweden. *Nematologica* 20, 445–454.

Andersson, S. (1982) Population dynamics and control of *Heterodera avenae* – a review with some original results. *EPPO Bulletin* 12, 463–475.

Andrés, M.F., Romero, M.D., Montes, M.J. and Delibes, A. (2001) Genetic relationships and isozyme variability in the *Heterodera avenae* complex determined by isoelectric focusing. *Plant Pathology* 50, 270–279.

Antoniou, M. (1989) Arrested development in plant parasitic nematodes. *Helminthological Abstracts Series B* 58, 1–9.

Aruna, P. and Siddiqui, A.U. (1997a) Population fluctuation of *Heterodera zeae* on maize and barley in Rajastan. *Journal of Mycology and Plant Pathology* 27, 321–322.

Aruna, P. and Siddiqui, A.U. (1997b) Evaluation of maize lines against corn cyst nematode *Heterodera zeae. Journal of Mycology and Plant Pathology* 27, 322–324.

Asiedu, R., Fisher, J.M. and Driscoll, C.J. (1990) Resistance to *Heterodera avenae* in the rye genome of triticale. *Theoretical and Applied Genetics* 79, 331–336.

Asmus, G.L., Ferraz, L.C.C.B. and Appezzato da Gloria, B. (2000) [Anatomical changes in corn (*Zea mays* L.) roots caused by *Meloidogyne javanica*.] *Nematropica* 30, 33–39 (abstract).

Aung, T., Windham, G.L. and Williams, W.P. (1990) Reproduction of *Meloidogyne incognita* on open-pollinated maize varieties. *Supplement to the Journal of Nematology* 22, 651–653.

Aung, T., Williams, W.P., Windham, G.L. and Davis, F.M. (1991) Characterization of open-pollinated varieties. *Research Report – Mississippi Agricultural and Forestery Experiment Station 16.*

Badra, T. and Adesiyan, S.O. (1990) Field corn response to granular pesticides: residues in grains and soils and nematode biological interactions. *Indian Journal of Nematology* 20, 1–9.

Bajaj, H.K. and Gupta, D.C. (1994) Existence of host races in *Heterodera zeae* Koshy *et al. Fundamental and Applied Nematology* 17, 389–390.

Balakhnina, V.P. (1989) Resistance of varieties of *Triticum durum* desf. and *Triticum aestivum* L. to the oat cyst nematode. In: *Gel'mintologiya segodnya: Problemy i perspektivy. Tezisy dokladov nauchnoi konferentsii*, Tom 2, 4–6 Aprelya 1989, Moscow, pp. 36–37.

Banyer, R.J. and Fisher, J.M. (1971) Effect of temperature on hatching of eggs of *Heterodera avenae. Nematologica* 17, 519–534.

Banyer, R.J. and Fisher, J.M. (1976) Effect of host and environment on some aspects of biology of *Heterodera avenae. Nematologica* 22, 298–305.

Barker, K.R., Schmitt, D.P. and Imbriani, J.L. (1985) Nematode population dynamics with emphasis on determining damage potential to crops. In: Barker, K.R., Carter, C.C. and Sasser, J.N. (eds) *An Advanced Treatise on* Meloidogyne. *Volume II: Methodology*. North Carolina State University Department of Plant Pathology and USAID, Raleigh, North Carolina, pp. 135–148.

Barloy, D., Martin, J., Rivoal, R. and Jahier, J. (1996) Genetic and molecular characterization of lines of wheat resistant to cereal cyst nematode *Heterodera avenae*. In: *Proceedings of the Third International Congress of Nematology*, 7–12 July 1996, Pointe à Pitre, Guadeloupe, French West Indies, p. 164 (abstract).

Barnard, A.J., McEwen, J., Hornby, D. and Beane, J. (1989) Effects of aldicarb, benomyl, dazomet, permethrin, pirimicarb, phorate and nitrogen fertilizer on maize (*Zea mays*) grown for 14 consecutive years on the same site. *Journal of Agricultural Science* 112, 339–349.

Barr, A.R., Chalmers, K.J., Karakousis, A., Manning, S., Lance, R.C.M., Lewis, J., Jefferies, S.P. and Langridge, P. (1998) RFLP mapping of a new cereal cyst nematode resistance locus in barley. *Plant Breeding* 117, 185–187.

Basson, S., De Waele, D. and Meyer, A.J. (1990) An evaluation of crop plants as hosts for *Ditylenchus destructor* isolated from peanut. *Nematropica* 20, 23–29.

Baxter, R.I. and Blake, C.D. (1968) *Pratylenchus thornei* – a cause of root necroses in wheat. *Nematologica* 14, 351–361.

Becerra, L.E.N. and Sosa-Moss, C. (1977) *Meloidogyne incognita* (Kofoid and White 1919) Chitwood 1949, atacando *Zea mays* (L), en México sin causar agallas radiculares. *Memorias IX Reunión de Nematólogos de los Trópicos Americanos*. OTAN-ONTA, Lima, Peru.

Bee-Rodriguez, D. and Ayala, A. (1977) Interaction of *Pratylenchus zeae* with four soil fungi on sorghum. *Journal of Agriculture, University of Puerto Rico* 61, 501–506.

Bekal, S., Jahier, J. and Rivoal, R. (1998) Host response of different triticeae to species of the cereal cyst nematode complex in relation to breeding resistant durum wheat. *Fundamental and Applied Nematology* 21, 359–370.

Belair, G., Fournier, Y., Dauphinais, N. and Dangi, O.P. (2002) Reproduction of *Pratylenchus penetrans* on various rotation crops in Quebec. *Phytoprotection* 83, 111–114.
Belloni, V. (1954) Comparsa in thalia di un anguillulosi del frumento e prove de lotta. *Notiziario sulle Malattie delle Piante* 27, 3–6.
Bergeson, G.N. (1978) Control of the lesion nematode (*Pratylenchus* spp.) in corn with carbofuran. *Plant Disease Reporter* 62, 295–297.
Bishnoi, S.P. and Bajaj, H.K. (2000) On the species and pathotypes of *Heterodera avenae* complex of India. Paper presented at *National Symposium of Nematological Society of India*, held at Orrisa University of Agriculture & Technology, Bhuneshwar, India, 11–14 Oct. 2000, p. 61.
Bishnoi, S.P. and Bajaj, H.K. (2002) Response of resistant barley cultivars to the Indian populations of *Heterodera avenae* complex. *Indian Journal of Nematology* 32 (2), 125–128.
Bourne, J.M. and Kerry, B.R. (1999) Effect of the host plant on the efficacy of *Verticillium chlamydosporium* as a biological control agent for root-knot nematodes at different nematode densities and fungal application rates. *Soil Biology and Biochemistry* 31, 75–84.
Bourne, J.M. (2001) Making a soil suppressive to root-knot nematodes by applications of *Verticillium chlamydosporium. Proceedings of the IOBC-WPRS Study Group 'Integrated Control of Soil Pests'.* Bulletin OILB-SROP, Bad Honnef, Germany, 24, 25–30.
Bridge, J., Jones, E. and Page, L.J. (1976) *Meloidogyne acronea* associated with reduced growth of cotton in Malawi. *Plant Disease Reporter* 60, 5–7.
Bridge, J., Terry, P.J., Tunstall, J.P. and Waller, J.M. (1978) A survey of crop pests, diseases and weeds of the Gambia. *Report UK Ministry of Overseas Development.*
Brown, R.A. (1981) Nematode diseases. In: *Economic Importance and Biology of Cereal Root Diseases in Australia.* Report to Plant Pathology Subcommittee of Standing Committee on Agriculture, Australia.
Brown, R.H. (1972) Chemical control on cereal cyst nematode in Victoria – a comparison of systemic and contact nematicides. *Australian Journal of Experimental Agriculture and Animal Husbandry* 12, 662–667.
Brown, R.H. (1987) Control strategies of low value crops. In: Brown, R.H. and Kerry, B.R. (eds) *Principles and Practice of Nematode Control in Crops.* Harcourt Brace Jovanovic, Sydney, Australia, pp. 351–387.
Bürkert, A., Bagayoko, M., Alvey, S. and Bationo, A. (2001) Causes of legume-rotation effects in increasing cereal yields across the Sudanian, Sahelian and Guinean zone in West Africa. In: Horst, W.J., Schenk, M.K., Bürkert, A., Claassen, N., Flessa, H., Frommer, W.B., Goldbach, H., Olfs, H.W. and Romheld, V. (eds) *Plant Nutrition – Food Security and Sustainability of Agro-ecosystems.* Kluwer Academic Publishers, Dordrecht, The Netherlands, pp. 972–973.
Byars, L.P. (1920) The nematode disease of wheat caused by *Tylenchus tritici. United States Department of Agriculture Bulletin 842.*
Cabanillas, H.E., Bradford, J.M. and Smart, J.R. (1999) Effect of tillage systems, soil type, crop stand and crop sequence on reniform nematodes after harvest. *Nematropica* 29, 137–146.
Cardwell, D.M. and Ingham, R.E. (1997) Reproduction of *Meloidogyne chitwoodi* on popcorn cultivars. *Supplement to the Journal of Nematology* 29, 657–661.
Caubel, G. and Le Guen, J. (1983) Variability of relationships between *Vicia* bean and stem nematode (*Ditylenchus dipsaci*) characterization of varietal resistance. *Proceedings of the 1st European Conference on Grain Legumes*, Angers, France, pp. 337–338.
Caubel, G., Ritter, M. and Rivoal, R. (1972) Observations relatives à des attaques du nématode *Meloidogyne naasi* Franklin sur céréales et graminées fourragères dans l'Ouest de la France en 1970. *Compte Rendus des Séances de l'Académie d'Agriculture de France* 58, 351–356.
Chen, S., Porter, P.M., Reese, C.D. and Stienstra, W.C. (2001) Crop sequence effects on soybean cyst nematode and soybean and corn yields. *Crop Science* 41, 1843–1849.
Chevres-Roman, R., Gross, H.D. and Sasser, J.N. (1971) The influence of selected nematode species and number of consecutive plantings of corn and sorghum on forage production, chemical composition of plants and water use efficiency. *Nematropica* 1, 40–41.
Chinnarsi, B., Tangchitsomkid, N. and Toida, Y. (1995) *Heterodera zeae* on maize in Thailand. *Japanese Journal of Nematology* 24, 35–38.
Chu, V.M. (1945) The prevalence of the wheat nematode in China and its control. *Phytopathology* 35, 288–295.

Claflin, L.E. (1984) Plant-parasitic nematodes affecting sorghum. In: Mughogho, L.K. and Rosenberg, G. (eds) *Sorghum Root and Stalk Rots: a Critical Review.* ICRISAT, Patancheru, India, pp. 53–58.

Clamont, G. (1985) Breeding for resistance to the cereal cyst nematode (*Heterodera avenae* Woll.) and to stem nematode (*Ditylenchus dipsaci* (Kühn) Fil.) in Belgium. *Comptes Rendus des Séances de l'Académie d'Agriculture de France* 71, 751–760.

Clewett, T.G., Thompson, J.P. and Fiske, M.L. (1993) Crop rotation to control *Pratylenchus thornei.* In: Vanstone, V.A., Taylor, S.P. and Nicol, J.M. (eds) *Proceedings of the* Pratylenchus *Workshop, Ninth Biennial Australasian Plant Pathology Conference*, Hobart, Tasmania, Australia.

Coetzee, V. (1956) *Meloidogyne acronea*, a new species of root-knot nematode. *Nature* 177, 899–900.

Cook, R. and Evans, D.R. (1988) Observations on resistance in white clover (*Trifolium repens* L.) to the stem nematode (*Ditylenchus dipasci* (Kühn) Filipjevi). *Journal of Agricultural Science* 110, 145–154.

Cook, R. and Noel, G.R. (2002) Cyst nematodes: *Globodera* and *Heterodera* species. In: Starr, J.L., Cook, R. and Bridge, J. (eds) *Plant Resistance to Parasitic Nematodes.* CAB International, Wallingford, UK, pp. 71–106.

Cook, R. and Yeates, G.W. (1993) Nematode pests of grassland and forage crops. In: Evans, K., Trudgill, D.L. and Webster, J.M. (eds) *Plant Parasitic Nematodes in Temperate Agriculture.* CAB International, Wallingford, UK, pp. 305–350.

Cook, R. and York, P.A. (1982a) Resistance of cereals to *Heterodera avenae*: methods of investigation, sources and inheritance of resistance. *EPPO Bulletin* 12, 423–434.

Cook, R. and York, P.A. (1982b) Genetics of resistance to *Heterodera avenae* and *Meloidogyne naasi.* In: *Proceedings of the 4th International Barley Genetics Symposium*, Edinburgh, UK.

Cook, R. and York, P.A. (1987) Variation in Triticales in reaction to cereal cyst nematode. *Annals of Applied Biology 110, Tests of Agrochemicals and Cultivars* 8, 158–159.

Cook, R. and York, P.A. (1988) Reaction of some European and Australian oat cultivars to cereal cyst nematode. *Annals of Applied Biology 112, Tests of Agrochemicals and Cultivars* 9, 84–85.

Cook, R., York, P.A. and Guile, C.T. (1986) Effects and control of cereal root-knot nematode in barley/grass rotations. In: *Proceedings of the 1986 British Crop Protection Conference – Pests and Diseases*, pp. 433–440.

Corbett, D.C.M. (1976) *Pratylenchus brachyurus. C.I.H. Descriptions of Plant-parasitic Nematodes.* Set 6, No. 89.

Crump, D.H., Sayre, R.M. and Young, L.D. (1983) Occurrence of nematophagous fungi in cysts. *Plant Disease* 67, 63–64.

Cuarezma-Teran, J.A. and Trevathan, L.F. (1985) Effects of *Pratylenchus zeae* and *Quinisulcius acutus* alone and in combination on sorghum. *Journal of Nematology* 17, 169–174.

Cuevas, O.Y.J. and Sosa-Moss, C. (1990) Host plants of *Meloidogyne chitwoodi* in the states of Tlaxcala and Puebla, Mexico. *Current Nematology* 1, 69–70.

Curtis, B.C. (2002) Wheat in the world. In: Curtis, B.C., Rajaram, R. and Gómez Macpherson, H. (eds) *Bread Wheat Improvement and Production.* FAO, Rome, pp. 1–18.

Dackman, D. and Nordbring-Hertz, B. (1985) Fungal parasites of the cereal cyst nematode *Heterodera avenae* in Southern Sweden. *Journal of Nematology,* 17, 50–55.

Dahlberg, J.A. and Frederiksen, R.A. (2000) Introduction. In: Frederiksen, R.A. and Odvody, G.N. (eds) *Compendium of Sorghum Diseases,* 2nd edn. The American Phytopathological Society, St Paul, Minnesota, pp. 1–4.

Davis, R.F. and Timper, P. (2000) Resistance in selected corn hybrids to *Meloidogyne arenaria* and *M. incognita. Supplement to the Journal of Nematology* 32, 633–640.

Davis, R.F., Koenning, S.R., Kemerait, R.C., Cummings, T.D. and Shurley, W.D. (2003) *Rotylenchulus reniformis* management in cotton with crop rotation. *Journal of Nematology* 35, 58–64.

De Brito, J.A. and Antonio, H. (1989) [Screening maize genotypes for resistance to *Meloidogyne javanica.*] *Nematologia Brasileira* 13, 129–137.

Delibes, A., Romero, D., Aguaded, S., Duce, A., Mena, M., Lopez-Brana, I., Andrés, M.F., Martin-Sanchez, J.-A. and Garcia-Olmedo, F. (1993) Resistance to the cereal cyst nematode (*Heterodera avenae* Woll.) transferred from the wild grass *Aegilops ventricosa* to hexaploid wheat by a stepping-stone procedure. *Theoretical and Applied Genetics* 87, 402–408.

De Waele, D. and Elsen, A. (2002) Migratory endoparasites: *Pratlylenchus* and *Radopholus* species. In: Starr, J.L., Cook, R. and Bridge, J. (eds) *Plant Resistance to Parasitic Nematodes.* CAB International, Wallingford, UK, pp. 175–206.

De Waele, D. and Jordaan, E.M. (1988a) Plant-parasitic nematodes on field crops in South Africa. 2. Maize. *Revue de Nématologie* 11, 65–74.

De Waele, D. and Jordaan, E.M. (1988b) Plant-parasitic nematodes on field crops in South Africa. 1. Sorghum. *Revue de Nématologie* 11, 203–212.
De Waele, D. and McDonald, A.H. (2000) Diseases caused by nematodes. In: Frederiksen, R.A. and Odvody, G.N. (eds) *Compendium of Sorghum Diseases*, 2nd edn. American Phytopathological Society, St Paul, Minnesota, pp. 50–53.
De Waele, D., McDonald, A.H., Jordaan, E.M., Orion, D., Van den Berg, E. and Loots, G.C. (1998) Plant-parasitic nematodes associated with maize and pearl millet in Namibia. *African Plant Protection* 4, 113–117.
Dhawan, S.C. and Nagesh, M. (1987) On the relationship between population densities of *Heterodera avenae*, growth of wheat and nematode multiplication. *Indian Journal of Nematology* 17, 231–236.
Dickerson, O.J., Darling, H.M. and Griffin, C.D. (1964) Pathogenicity and population trends of *Pratylenchus penetrans* on potato and corn. *Phytopathology* 54, 317–322.
Dickson, D.W. and McSorley, R. (1990) Interaction of three plant-parasitic nematodes on corn and soybean. *Supplement to the Journal of Nematology* 22, 783–791.
Di Vito, M. and Greco, N. (1988) Investigation on the biology of *Meloidogyne artiellia*. *Revue de Nématologie* 11, 221–225.
Di Vito, M. and Zacheo, G. (1987) Responses of cultivars of wheat to *Meloidogyne artiellia*. *Nematologia Mediterranea* 15, 405–408.
Di Vito, M., Greco, N. and Saxena, M.C. (1991) Effectiveness of soil solarization for control of *Heterodera ciceri* and *Pratylenchus thornei* on chickpeas in Syria. *Nematologica Mediterranea* 19, 109–111.
Di Vito, M., Greco, N. and Zacheo, G. (1985) On the host range of *Meloidogyne artiellia*. *Nematologia Mediterranea* 13, 207–212.
Dixon, G.M. (1969) The effect of cereal cyst eelworm on spring sown cereals. *Plant Pathology* 18, 109–112.
Doyle, A.D., McLeod, R.W., Wong, P.T.W., Hetherington, S.E. and Southwell, R.J. (1987) Evidence for the involvement of the root lesion nematode *Pratylenchus thornei* in wheat yield decline in northern New South Wales. *Australian Journal of Experimental Agriculture* 27, 563–570.
Duggan, J.J. (1961) The effect of cereal root eelworm on its hosts. *Irish Journal of Agriculture Research* 1, 7–16.
Dundas, I.S., Frappell, D.E., Crack, D.M. and Fisher, J.M. (2001) Deletion mapping of a nematode resistance gene on rye chromosome 6R in wheat. *Crop Science* 41, 1771–1778.
Eastwood, R.F., Lagudah, E.S. and Appels, R. (1994a) A directed search for DNA sequences tightly linked to cereal cyst nematode resistance genes in *Triticum tauschii*. *Genome* 37, 311–319.
Eastwood, R.F., Smith, A. and Wilson, J. (1994b) *Pratylenchus thornei* is causing yield losses in Victorian wheat crops. *Australasian Nematology Newsletter* 5, 2.
Ediz, S.A. and Dickerson, O.J. (1976) Life cycle, pathogenicity, histopathology and host range of race 5 of the barley root-knot nematode. *Journal of Nematology* 8, 228–232.
Egunjobi, O.A. (1974) Nematodes and maize growth in Nigeria. II. Effects of some amendments on populations of *Pratylenchus brachyurus* and on the growth and production of maize (*Zea mays*) in Nigeria. *Nematologia Mediterranea* 3, 5–73.
Eisenback, J.D. and Triantaphyllou, H.H. (1991) Root-knot nematode: *Meloidogyne* species and races. In: Nickle, W.R. (ed.) *Manual of Agricultural Nematology*. Marcel Dekker Inc., New York, pp. 191–274.
Elekcioglu, I.H. and Gözel, U. (1997) Effect of mixed population of *Paratrophurus acristylus*, *Pratylenchus thornei* and *Paratylenchus* sp. (Nematoda: Tylenchida) on yield parameters of wheat in Turkey. *International Journal of Nematology* 7, 217–220.
El-Meleigi, M.A. (1989) Effect of soil moisture, carbofuran and MDMV on *Fusarium moniliforme* and *Pratylenchus* spp., stalk rot and grain yield of maize. *Applied Agricultural Research* 4, 248–252.
Eriksson, K.B. (1972) Nematode diseases of pasture legumes and turf grasses. In: Webster, J.M. (ed.) *Economic Nematology*. Academic Press, New York, pp. 66–96.
Esele, J.P. (2002) Diseases of finger millet – a global overview. In: Leslie, J.F. (ed.) *Sorghum and Millets Diseases*. Iowa State University Press, Ames, Iowa, pp. 21–26.
Esmenjaud, D., Rivoal, R. and Marzin, H. (1990) Numbers of *Pratylenchus* spp. (Nematoda) in the field on winter wheat in different cereal rotations. *Nematologica* 36, 217–226.
Estanol-Botello, E., Ferrera-Cerrato, R., Sosa-Moss, C., Santizo-Rincon, J.A. and Quintero, L.R. (1999) [Interaction of the nematode *Meloidogyne chitwoodi* with three strains of the fungus *Glomus* sp. and the distribution of dry matter of young maize plants.] *Terra* 17, 17–25.
FAO (2002) FAO website. Available at: http://apps.fao.org/page/collections-subset=agriculture

Farsi, M., Vanstone, V.A., Fisher, J.M. and Rathjen, A.J. (1995) Genetic variation in resistance to *Pratylenchus neglectus* in wheat and triticales. *Australian Journal of Experimental Agriculture* 35, 597–602.

Fayez, M. (1990) Interaction of some nematicides with *Azospirillum lipoferum* and the growth of maize. *Zeitschrift für Pflanzenernährung und Bodenkunde* 153, 219–223.

Fisher, J.M. and Hancock, T.W. (1991) Population dynamics of *Heterodera avenae* Woll. in South Australia. *Australian Journal of Agricultural Research* 42, 53–68.

Florini, D.A. and Loria, R. (1990) Reproduction of *Pratylenchus penetrans* on potato and crops grown in rotation with potato. *Journal of Nematology* 22, 106–112.

Fogain, R. and Gowen, S. (1995) Pathogenicity on maize and banana among isolates of *Radopholus similis* from four producing coutries of Africa and Asia. *Fruits (Paris)* 50, 5–9.

Fortnum, B.A. and Currin, R.E. (1988) Host suitability of grain sorghum cultivars to *Meloidogyne* spp. *Annals of Applied Nematology* 2, 61–64.

Fortuner, R. (1976) *Pratylenchus zeae. C.I.H. Descriptions of Plant-parasitic Nematodes.* Set 6, no. 77.

Fortuner, R. (1977) *Pratylenchus thornei. C.I.H. Descriptions of Plant Parasitic Nematodes.* Set 7, no. 93.

Gallaher, R.N., McSorley, R. and Dickson, D.W. (1991) Nematode densities associated with corn and sorghum cropping systems in Florida. *Supplement to the Journal of Nematology* 23, 668–672.

Gao, X.B. and Cheng, H.R. (1992) Observations on infections of *Pratylenchus scribneri* in maize roots. *Nematologia Mediterranea* 20, 141–142.

Gapasin, R.M. (1995) Evaluation of *Paecilomyces lilacinus* (Thom.) Samson for the control of *Pratylenchus* sp. in corn. *Biocontrol* 1, 35–39.

Gastel, G. (1990) Resistenzprüfung von Ackerbohnen (*Vicia faba*) gegen das Stengelälchen (*Ditylenchus dipsaci*). Dissertation, Universität Hohenheim, Germany.

Gaur, H.S. and Sharma, S.N. (1999) Relative efficacy of bioassay and extraction of juveniles from soil for detection and estimation of population levels of the root-knot nematodes, *Meloidogyne graminicola* and *M. triticoryzae*. *Annals of Plant Protection Science* 7, 75–79.

Germershauzen, K., Kastner, A. and Schmidt, W. (1976) Observations on the threshold of damage by *Heterodera avenae* on cereals under different rates of nitrogen fertilization. *Vortragstagung zu Aktuellen Problemen der Phytonematologie* 11–30.

Gill, J.S. (1967) Studies on *Heterodera avenae,* the casual organism of 'molya' disease of wheat and barley. PhD Thesis, Punjab University, India.

Gill, J.S. and Swarup, G. (1971) On the host range of cereal cyst nematode, *Heterodera avenae,* the casual organism of 'molya' disease of wheat and barley in Rajasthan, India. *Indian Journal of Nematology* 1, 63–67.

Glazer, I. and Orion, D. (1983) Studies on anhydrobiosis of *Pratylenchus thornei. Journal of Nematology* 28, 333–338.

Gokte, N. and Swarup, G. (1984a) Relationship between initial inoculum levels of *Heterodera avenae* and penetration of juveniles in the roots of susceptible and resistant barley cultivars. *Indian Journal of Nematology* 14, 95–96.

Golden, A.M. and Mulvey, R.H. (1983) Redescription of *Heterodera avenae,* the corn cyst nematode, with SEM observations. *Journal of Nematology* 15, 60–70.

Gooris, J. (1968) Host plants and non-host plants of the Gramineae root-knot nematode *Meloidogyne naasi* Franklin. *Mededelingen van de Rijksfaculteit Landbouwwetenschappen te Gent* 33, 85–100.

Gooris, J. and D'Herde, C.J. (1972) Mode d'hivernage de *Meloidogyne naasi* Franklin dans le sol et lutte par rotation culturale. *Revue de l'Agriculture, Bruxelles* 25, 659–664.

Gooris, J. and D'Herde, C.J. (1977) *Study on the Biology of* Meloidogyne naasi *Franklin, 1965.* Ministry of Agriculture, Ghent, Belgium.

Goswami, B.K. and Raychaudhuri, S.P. (1978) Interaction of rootknot nematodes and viruses in maize and tobacco in India. *Third International Congress of Plant Pathology,* 16–23 August, Germany.

Grandison, G.S. and Wallace, H.R. (1974) The distribution and abundance of *Pratylenchus thornei* in fields of strawberry clover (*Trifolium fragiferum*). *Nematologica* 20, 283–290.

Greco, N., DiVito, M., Saxena, M.C. and Reddy, M.V. (1984) Investigation on the root lesion nematode *Pratylenchus thornei* in Syria. *Nematologia Mediterranea* 16, 101–105.

Griffin, G.D. (1984) Nematode parasites of alfalfa, cereals, and grasses. In: Nickle, W.E. (ed.) *Plant and Insect Nematodes.* Marcel Dekker Inc., New York, pp. 243–321.

Griffiths, D.J., Holden, J.H.W. and Jones, J.M. (1957) Investigations on resistance of oats to stem eelworm, *Ditylenchus dipsaci* (Kühn). *Annals of Applied Biology* 45, 709–720.

Gupta, P. and Swarup, G. (1972) Ear-cockle and yellow rot disease of wheat. II. Nematode–bacterial association. *Nematologica* 18, 320–324.

Gurner, P.S., Dube, A.J. and Fisher, J.M. (1980) Chemical control of cereal cyst-nematodes (*Heterodera avenae*) on wheat by a new low-volume applicator. *Nematologica* 26, 448–454.

Hafez, S.L. and Claflin, L.E. (1982) Control of plant parasitic nematodes on grain sorghum and yield response. *Fungicide-Nematicide Tests* 37, 198.

Hague, S.S., Overstreet, C. and Van Santen, E. (2002) Crop rotation effects on nematode populations. *Proceedings of the 25th Annual Southern Conservation Tillage Conference for Sustainable Agriculture.* Auburn University, Auburn, Alabama, pp. 156–160.

Haidar, M.G. and Nath, R.P. (1992) Distribution of plant parasitic nematodes in maize fields and seasonal variations of *Hoplolaimus indicus. Current Nematology* 3, 173–177.

Handa, D.K., Mathur, B.N. and Bhargava, L.P. (1971) Occurrence of root-knot on pearl millet. *Indian Journal of Nematology* 1, 244.

Handa, D.K., Mathur, R.L., Mathur, B.N. and Yadav, B.D. (1985) Estimation of losses in barley due to cereal cyst nematode in sandy and sandyloam soils. *Indian Journal of Nematology* 15, 163–166.

Hanounik, S.B., Halila, H. and Harrabi, M. (1986) Resistance in *Vicia faba* to stem nematode (*Ditylenchus dipsaci*). *FABIS Newsletter* 16, 37–39.

Hasan, N., Pandey, K.C., Bhaskar, R.B. and Hazra, C.R. (1998) Plant parasitic nematodes associated with major fodder crops in India. *Indian Journal of Nematology* 28, 100–103.

Hash, C.T. and Whitcombe, J.R. (2002) Gene management and breeding for downy mildew resistance. In: Leslie, J.F. (ed.) *Sorghum and Millets Diseases.* Iowa State University Press, Ames, Iowa, pp. 27–36.

Hashmi, G., Hashmi, S., Krusberg, L.R. and Heuttel, R.N. (1993a) Resistance in *Zea mays* to *Heterodera zeae. Supplement to the Journal of Nematology* 25, 820–823.

Hashmi, S., Krusberg, L.R. and Sardanelli, S. (1993b) Reproduction of *Heterodera zeae* and its suppression of corn plant growth affected by temperature. *Journal of Nematology* 25, 55–58.

Hirschmann, H. (1985) The genus *Meloidogyne* and morphological characters differentiating its species. In: Sasser, J.N. and Carter, C.C. (eds) *An Advanced Treatise on* Meloidogyne. Volume I: *Biology and Control.* North Carolina State University Department of Plant Pathology and USAID, Raleigh, North Carolina, pp. 79–93.

Holdeman, Q.L. and Watson, T.R. (1977) *The Oat Cyst Nematode,* Heterodera avenae, *a Root Parasite of Cereal Crops and Other Grasses.* Department of Food and Agriculture, California.

Hollaway, C.J., Taylor, S.P., Eastwood, R.F. and Hunt, C.H. (2000) Effect of field crops on density of *Pratylenchus* in southeastern Australia, Part 2: *P. thornei. Journal of Nematology* 32, 600–608.

Huang, X., Becker, J.O. and Huang, X. (1997) *In vitro* culture and feeding behaviour of *Belonolaimus longicaudatus* on excised *Zea mays* roots. *Journal of Nematology* 29, 411–415.

Hussey, R.S. and Barker, K.R. (1973) A comparison of methods of collecting inocula of *Meloidogyne* spp., including a new technique. *Plant Disease Reporter* 57, 1025–1028.

Hussey, R.S. and Boerma, H.R. (1981) A greenhouse screening procedure for root-knot nematode resistance in soybeans. *Crop Science* 21, 794–796.

Hutchinson, C.M. (1917) A bacterial disease of wheat in Punjab. *Memoirs of Department of Agriculture, India,* Bacteriological Series 1.

Hutton, D.G. (1976) Informe sobre el status de los problemas e investigaciones con *Meloidogyne* spp en Jamaica. *Memorias de la Conferencia Regional de Planeamiento del Proyecto Internacional* Meloidogyne, Panamá, pp. 11–22.

Hutton, D.G. (1981) Informe sobre el trabajo desarrollado en Jamaica para el Proyecto internacional desde Mayo de 1976 a Julio de 1978. *Memorias de la Segunda Conferencia de Planamiento del Proyecto Internacional* Meloidogyne *Region I,* Panamá, pp. 62–78.

Hutzell, P.A. and Krusberg, L.R. (1990) Temperature and the life cycle of *Heterodera zeae. Journal of Nematology* 22, 414–417.

Ibrahim, A.A.M., Al-Hazmi, A.S., Al-Yahya, F.A. and Alderfasi, A.A. (1999) Damage potential and reproduction of *Heterodera avenae* on wheat and barley under Saudi field conditions. *Nematology* 1, 625–630.

Ibrahim, I.K.A., Rezk, M.A. and Ibrahim, A.A.M. (1986) Occurrence of the cyst nematodes *Heterodera avenae, H. daverti* and *H. rosii* in northern Egypt. *Journal of Nematology* 18, 614 (abstract).

Ibrahim, I.K.A., Lewis, S.A. and Harsman, D.C. (1993) Host suitability of graminaceous crop cultivars for isolates of *Meloidogyne arenaria* and *M. incognita. Supplement to the Journal of Nematology* 25, 858–862.

Idowu, A.A. (1981) A review of root-knot nematode work on maize at National Cereals Research Institute Abadan, and prospects for future studies. *Proceedings of the 3rd Research Planning Conference on Root-Knot Nematodes,* Meloidogyne *spp.,* Ibadan, Nigeria, pp. 122–127.

Idowu, A.A. and Fawole, B. (1990) Effects of cowpea and maize root leachates on *Meloidogyne incognita* egg hatch. *Journal of Nematology* 22, 136–138.

Inserra, R.N., Lamberti, F., Volvas, N. and Dandria, D. (1975) *Meloidogyne naasi* nell'Italia Meridionale e Malta. *Nematologia Mediterranea* 3, 163–166.

Inserra, R.N., Vovlas, N., O'Bannon, J.H. and Griffin, G.D. (1985) Development of *Meloidogyne chitwoodi* on wheat. *Journal of Nematology* 17, 322–326.

Ismael, A.E., Abadir, S.K. and Kheir, A.M. (1994) Damage potential and reproduction of *Heterodera zeae* on corn under field conditions. *Bulletin of the Faculty of Agriculture, University of Cairo* 45, 231–242 .

Ismail, S., Schuster, R.-P. and Sikora, R.A. (2000) Factors influencing dormancy of the Mediterranean cereal cyst nematode *Heterodera latipons* on barley. *Communications in Agriculture and Applied Biological Sciences* 65, 529–535.

Ivezic, M., Samota, D., Raspudic, E. and Horvat, D. (1996) Population density of plant parasitic nematodes of maize at higher doses of potassium fertilizer. *Poljoprireda* 2, 29–32.

Ivezic, M., Raspudic, E. and Brmez, M. (2000) Structure of nematode communities in different agroecosystems in Croatia. *Helminthologia* 37, 165–169.

Jahier, J., Abelard, P., Tanguy, A.M., Dedryver, F., Rivoal, R., Khatkar, S. and Bariana, H.S. (2001) The *Aegilops ventricosa* segment on chromosome 2AS of the wheat cultivar VPM1 carries the cereal cyst nematode resistance gene *Cre5. Plant Breeding* 120, 125–128.

Jahier, J., Rivoal, R., Yu, M.Q., Abelard, P., Tanguy, A.M. and Barloy, D. (1998) Transfer of genes for resistance to cereal cyst nematode from *Aegilops variabilis* Eig to wheat. *Journal of Genetics and Breeding* 52, 253–257.

Janssen, G.J.W. (1994) The relevance of races in *Ditylenchus dipsaci* (Kühn) Filipjev, the stem nematode. *Fundamental and Applied Nematology* 17, 469–473.

Jefferies, S.P., Barr, A.R., Langridge, P., Chalmers, K.J., Kretschmer, P., Gianquitto and Karakousis, A. (1997) Practical application of molecular markers in barley breeding. In: *Proceeding of the 8th Australian Barley Technical Symposium,* September 1997, Gold Coast, Queensland, Australia.

Jepson, S.B. (1987) *Identification of Root-knot Nematodes (*Meloidogyne *Species).* CAB International, Wallingford, UK.

Johnson, A.W. (1975) Resistance of sweetcorn cultivars to plant-parasitic nematodes. *Plant Disease Reporter* 59, 373–376.

Johnson, A.W. and Leonard, R.A. (1995) Effects and carry-over benefits of nematicides in soil planted to a sweetcorn–squash–vetch cropping system. *Supplement to the Journal of Nematology* 27, 563–570.

Johnson, A.W. and Motsinger, R.E. (1989) Suitability of small grains as hosts of *Meloidogyne* species. *Supplement to the Journal of Nematology* 21, 650–653.

Johnson, A.W., Dowler, C.C., Glaze, N.C., Sumner, D., Chalfant, R.B., Golden, A.M. and Epperson, J.E. (1990) Effects of nematicides on nematode population densities and crop yield in a turnip–corn–pea cropping system. *Phytoprotection* 71, 55–63.

Johnson, A.W., Sumner, D.R., Windham, G.L. and Williams, W.P. (1999) Effects of resistant corn hybrid and fenamiphos on *Meloidogyne incognita* in a corn–squash rotation. *Journal of Nematology* 184–190.

Johnson, V.A. (1984) New horizons in cereal production. In: Gallagher, E.J. (ed.) *Cereal Production.* Butterworths, London, pp. 1–13.

Jordaan, E.M. and De Waele, D. (1987) Resistance of maize to plant-parasitic nematodes: a literature review. *Proceedings of the Seventh South African Maize Breeding Symposium, 1986.* Department of Agriculture and Water Supply, Pretoria, pp. 60–66.

Jordaan, E.M. and De Waele, D. (1988) Host status of five weed species and their effects on *Pratylenchus zeae* infestation of maize. *Journal of Nematology* 20, 620–624.

Jordaan, E.M., Loots, G.C., Jooste, W.J. and De Waele, D. (1987) Effects of root-lesion nematodes (*Pratylenchus brachyurus* Godfrey and *P. zeae* Graham) and *Fusarium moniliforme* Sheldon alone or in combination, on maize. *Nematologica* 33, 213–219.

Jordaan, E.M., De Waele, D. and Van Rooyen, P.J. (1989) Endoparasitic nematodes in maize roots in the Western Transvaal as related to soil texture and rainfall. *Journal of Nematology* 21, 356–360.

Keetch, D.P. (1989) A perspective of plant nematology in South Africa. *South African Journal of Science* 85, 506–508.

Kerry, B.R. (1981) Fungal parasites, a weapon against cyst nematodes. *Plant Disease* 65, 390–393.

Kerry, B.R. and Crump, D.H. (1977) Observations on fungal parasites of females and eggs of the cereal cyst nematode, *Heterodera avenae* and other cyst nematodes. *Nematologica* 23, 193–201.

Kerry, B.R., Crump, D.H. and Mullen, L.A. (1982a) Studies on the cereal cyst nematode, *Heterodera avenae* under continuous cereals, 1974–1978. I. Plant growth and nematode multiplication. *Annals of Applied Biology* 100, 477–487.

Kerry, B.R., Crump, D.H. and Mullen, L.A. (1982b) Studies on the cereal cyst nematode, *Heterodera avenae*, under continuous cereals, 1975–1978. II. Fungal parasitism of nematode females and eggs. *Annals of Applied Biology* 100, 489–499.

Kerry, B.R., Crump, D.H. and Mullen, L.A. (1982c) Natural control of the cereal cyst nematode, *Heterodera avenae*, by soil fungi at three sites. *Crop Protection* 1, 99–109.

Kilpatrick, R.A., Gilchrist, L. and Golden, A.M. (1976) Root knot on wheat in Chile. *Plant Disease Reporter* 60, 135.

Kimpinski, J. (1972) The ecology of *Pratylenchus minyus.* PhD Thesis, University of Adelaide, Adelaide, Australia.

Kimpinski, J., Anderson, R.V., Johnston, H.W. and Martin, R.A. (1989) Nematodes and fungal diseases in barley and wheat on Prince Edward Island. *Crop Protection* 8, 412–416.

Kinloch, R.A. and Dunavin, L.S. (1993) Summer cropping effects on the abundance of *Meloidogyne arenaria* race 2 and subsequent soybean yield. *Supplement to the Journal of Nematology* 25, 806–808.

Klein, T.A., McLeod, R.W. and Marshall, D.R. (1987) Northern wheat yield decline in relation to *Pratylenchus thornei. Australasian Plant Pathology Society Sixth Conference.* Adelaide, Australia.

Kleynhans, K.P.N. (1991) *The Root-knot Nematodes of South Africa.* Technical Communication No. 231. South African Department of Agricultural Development, Pretoria.

Knuth, P. (2000) [Infestation with nematodes – are there any tolerant varieties?] *Mais* 1, 28–31.

Knuth, P. (2002) [Nematodes in multiyear maize cultivation: pests often profit from poor growth conditions.] *Mais* 30, 107–109.

Koenning, S.R., Overstreet, C., Noling, J.W., Donald, P.A., Becker, J.O. and Fortnum, B.A. (1999) Survey of crop losses in response to phytoparasitic nematodes in the United States for 1994. *Supplement to the Journal of Nematology* 31, 587–618.

Kollo, I.A. (2002) Plant-parasitic nematodes of sorghum and pearl millet: emphasis on Africa. In: Leslie, J.F. (ed.) *Sorghum and Millets Diseases.* Iowa State University Press, Ames, Iowa, pp. 259–266.

Kornobis, S., Wolny, S. and Wilski, A. (1980) The effect of *Heterodera avenae* population density in the soil on the yield and plant weight of oats. *Zeszyty. Problemowe Postetow Nauk Rolniczych* 232, 9–17.

Kort, J. (1972) Nematode diseases of cereals of temperate climates. In: Webster, J.M. (ed.) *Economic Nematology.* Academic Press, London, pp. 97–126.

Koshy, P.K. and Swarup, G. (1971) Distribution of *Heterodera avenae, H. zeae, H. cajani,* and *Anguina tritici* in India. *Indian Journal of Nematology* 1, 106–111.

Koshy, P.K. and Swarup, G. (1972) Susceptibility of plants to pigeon-pea cyst nematode. *Indian Journal of Nematology* 2, 1.

Koshy, P.K., Swarup, G. and Sethi, C.L. (1970) *Heterodera zeae* n.sp. (Nematoda: Heteroderidae). A cyst-forming nematode on *Zea mays. Nematologica* 16, 511–516.

Kretschemer, J.M., Chalmers, K.J., Manning, S., Karakousis, A., Barr, A.R., Islam, A.K.M.R., Logue, S.J., Choe, W., Barker, S.J., Lance, R.M.C. and Langridge, P. (1997) RFLP mapping of the Ha2 cereal cyst nematode resistance gene in barley. *Theoretical and Applied Genetics* 94, 1060–1064.

Krishnamurthy, G.V.G. and Elias, N.A. (1967) Host range of *Meloidogyne incognita* causing root-knot on tobacco in Junsur, Mysore State. *Indian Phytopathology* 20, 274–277.

Krishna Prasad, K.S. and Krishnappa, K. (1982) Occurrence of *Heterodera gambiensis* from India. *Indian Journal of Nematology* 12, 405.

Krusberg, L.R. (1988) The corn cyst nematode, *Heterodera zeae,* in the United States. In: Maqbool, M.A., Ghaffar, A.M. and Krusberg, L.R. (eds) *Advances in Plant Nematology.* University of Karachi, Karachi, Pakistan, pp. 171–175.

Krusberg, L.R., Sardanelli, S. and Grybauskas, A.P. (1997) Damage potential of *Heterodera zeae* to *Zea mays* as affected by edaphic factors. *Fundamental and Applied Nematology* 20, 593–599.

Kyrou, N.C. (1969) First record of occurrence of *Meloidogyne artiellia* on wheat in Greece. *Nematologica* 15, 432–433.

Kyrou, N.C. (1976) Biological notes on *Heterodera avenae* Woll., 1924 studied on wheat in central Macedonia. *Annales de l'Institut Phytopathologique Benak New Series* 11, 187–192.

Lagudah, E.S., Moullet, O., Ogbonnaya, F., Eastwood, R., Appels, R., Jahier, J., Lopez-Brana, I. and Delibes de Castro, A. (1998) Cyst nematode resistance genes in wheat. In: *Proceedings from the 7th International Congress of Plant Pathology*, Edinburgh, UK.

Lamberti, F. (1969) Pathogenicity of *Longidorus africanus* on selected field crops. *Plant Disease Reporter* 53, 421–424.

Lamberti, F. (1981) Plant nematode problems in the Mediterranean region. *Helminthological Abstracts, Series B* 50, 145–166.

Lasserre, F., Rivoal, R. and Cook, R. (1994) Interactions between *Heterodera* and *Pratylenchus neglectus* on wheat. *Journal of Nematology* 26, 336–344.

Leonard, W.H. and Martin, J.H. (1965) *Cereal Crops*. Macmillan, London.

Leukel, R.W. (1924) Investigations on the nematode disease of cereals caused by *Tylenchus tritici*. *Journal of Agricultural Research* 27, 928–956.

Leukel, R.W. (1957) Nematode disease of wheat and rye. *Farmers' Bulletin*, USDA No. 1607.

Limber, D. (1976) Artificial infection of sweetcorn seedlings with *Anguina tritici* Steinbuch (1799) Chitwood, 1935. *Proceedings of the Helminthological Society of Washington* 43, 201–203.

Loof, P.A.A. (1978) The genus *Pratylenchus* Filipjev, 1936 (Nematoda: Pratylenchidae): a review of its anatomy, morphology, distribution, systematics and identification. *Vaxtskydds Rapporter Jordbruk 5*. Swedish University of Agricultural Sciences.

López, R. (1981) Informe sobre las investigaciones realizadas en la Universidad de Costa Rica como parte del 'Proyecto Internacional Meloidogyne'. *Memorias de la Segunda Conferencia Regional de Planeamiento del Proyecto Internacional Meloidogyne Región I, Panamá*, pp. 16–22.

Lordello, A.I.L. and Lordello, R.R.A. (1992) [Maize genotypes indicated for planting in areas infested with *Meloidogyne javanica* Pt.] *Agronômico* 44, 21–22.

Lordello, A.I.L., Lordello, R.R.A. and Sawazaki, E. (1989) [Resistance of maize genotypes to *Meloidogyne javanica* Pt.] *Nematologia Brasileira* 13, 71–79.

Lordello, R.R.A., Sawazaki, E., Lordello, A.I.L. and Sobrinho, J.A. (1983) Controle de *Pratylenchus* spp. em milho, com nematicidas sistemicos e con torta de mamona. *Soc. Brasil. Nemat. Public.*

Lordello, R.R.A., Lordello, A.I.L., Sawazaki, E. and Aloisi Sobinho, J. (1985) Reaco de genotipus de miltho a *Pratylenchus* spp. *Nematologia Brasileira* 9, 163–173.

Lordello, R.R.A., Lordello, A.I.L. and Sawazaki, E. (1992) [Population fluctuation and control of *Pratylenchus* spp. on corn.] *Summa Phytopathologica* 18, 146–152.

Luc, M. (1986) Cyst nematodes in equatorial and hot tropical regions. In: Lamberti, F. and Taylor C.E. (eds) *Cyst Nematodes*. Plenum Publishing, London, pp. 355–372.

MacDaniel, M.E. and Barr, A.R. (1994) Registration of Australian winter cereal cultivars. *Avena sativa* (oat) cv. Bettong. *Australian Journal of Experimental Agriculture* 34, 701.

MacGuidwin, A.E. and Slack, S.A. (1991) Suitability of alfalfa, corn, oat, red clover and snapbean as hosts for the potato rot nematode, *Ditylenchus destructor*. *Plant Disease* 75, 37–39.

Mamluk, O.F., Augustin, B. and Bellar, M. (1983) New record of cyst and root-knot nematodes on legume crops in the dry areas of Syria. *Phytopathologia Mediterranea* 22, 80.

Maqbool, M.A. (1980) Occurrence of eight cyst nematodes on some agricultural crops in Pakistan. Karachi University. *Journal of Science* 8, 103–108.

Maqbool, M.A. (1981) Occurrence of root-knot and cyst nematodes in Pakistan. *Nematologia Mediterranea* 9, 211–212.

Maqbool, M.A. (1988) Present status of research on plant parasitic nematodes in cereals and food and forage legumes in Pakistan. In: Saxena, M.C., Sikora, R.A. and Srivastava, J.P. (eds) *Nematodes Parasitic to Cereals and Legumes in Temperate Semi-arid Regions*. ICARDA, Aleppo, Syria, pp. 173–180.

Maqbool, M.A. and Hasmi, S. (1986) Population trends of parasitic nematodes in different cropping sequences and the effect on yield of corn cv. AZAM. *International Nematology Network Newsletter* 3, 34–38.

Marathee, J.-P. and Gomez-MacPherson, H. (2001) Future world supply and demand. In: Bonjean, A.P. and Angus, W.J. (eds) *The World Wheat Book: a History of Wheat Breeding*. Lavoisier Publishing, Paris, pp. 1107–1116.

Mathur, B.N., Arya, H.C., Mathur, R.L. and Handa, D.K. (1974) The occurrence of biotypes of the cereal cyst nematode (*Heterodera avenae*) in the light soils of Rajasthan and Haryana, India. *Nematologica* 20, 19–26.

Mathur, B.N., Handa, D.K. and Swarup, G. (1987) Effect of deep summer ploughings on the cereal cyst nematode, *Heterodera avenae* and yield of wheat in Rajasthan, India. *Indian Journal of Nematology* 17, 292–295.

Maunder, A.B. (2002) Sorghum worldwide. In: Leslie, J.F. (ed.) *Sorghum and Millets Diseases.* Iowa State University Press, Ames, Iowa, pp. 11–17.
McDonald, A.H. and De Waele, D. (1987a) Control strategies against nematode pests of maize: a literature review. *Proceedings of the Seventh South African Maize Breeding Symposium, 1986.* Department of Agriculture and Water Supply, Pretoria, pp. 55–60.
McDonald, A.H. and De Waele, D. (1987b) Effect of two nematicides on nematode populations associated with maize. *Phytophylactica* 19, 475–478.
McDonald, A.H. and Van den Berg, E. (1993) Effect of watering regimen on injury to corn and grain sorghum by *Pratylenchus* species. *Journal of Nematology* 25, 654–658.
McDonald, A.H., Luis, J.H., Loots, G.C. and De Waele, D. (1987) Chemical control of root-lesion nematodes (*Pratylenchus* spp.) on maize in South Africa. *Phytophylactica* 19, 478–483.
McIntosh, R.A., Devos, K.M., Dubcovsky, D. and Rogers, W.J. (2001) Catalogue of gene symbols for wheat. In: Slinkard, A.E. (ed.) *Proceedings of the 9th Wheat Genetics Symposium,* Volume 5. University Extension Press, University of Saskatchewan, Saskatoon, Canada.
McKenry, M.V. and Ferris, H. (1983) Nematodes, on challenging problems in plant health. In: Kommendahl, T. (ed.) *Challenging Problems in Plant Health.* American Phytopathological Society Press, St Paul, Minnesota, pp. 1267–1279.
McSorley, R. and Dickson, D.W. (1990) Vertical distribution of plant-parasitic nematodes in sandy soil under maize. *Plant and Soil* 123, 95–100.
McSorley, R. and Dickson, D.W. (1995) Effect of tropical rotation crops on *Meloidogyne incognita* and other plant-parasitic nematodes. *Supplement to the Journal of Nematology* 27, 535–544.
McSorley, R. and Gallaher, R.N. (1992) Comparison of nematode population densities on six summer crops at seven sites in North Florida. *Supplement to the Journal of Nematology* 24, 699–706.
McSorley, R. and Gallaher, R.N. (1993) Effect of crop rotation and tillage on nematode densities in tropical corn. *Supplement to the Journal of Nematology* 25, 814–819.
McSorley, R. and Gallaher, R.N. (1994) Effect of tillage and crop residue management on nematode densities on corn. *Supplement to the Journal of Nematology* 26, 669–674.
McSorley, R. and Gallaher, R.N. (1996) Effect of yard waste compost on nematode densities and maize yield. *Supplement to Journal of Nematology* 28, 655–660.
McSorley, R. and Gallaher, R.N. (1997) Effect of compost and maize cultivars on plant-parasitic nematodes. *Supplement to Journal of Nematology* 29, 731–736.
Meagher, J.W. (1972) *Cereal Cyst Nematode (*Heterodera avenae *Woll).* Studies on Ecology and Content in Victoria. Technical Bulletin 24, Department of Agriculture, Victoria.
Meagher, J.W. (1982) The effect of environment on survival and hatching of *Heterodera avenae. EPPO Bulletin* 12, 361–369.
Meagher, J.W. and Brown, R.H. (1974) Microplot experiments on the effect of plant hosts on populations of cereal cyst nematode (*Heterodera avenae*) and its potential as a pathogen of wheat. *Nematologica* 20, 337–346.
Meintjes, S. (1993) The interaction between carbofuran, maize and nematodes. *Phytophylactica* 25, 253–255.
Mercer, C.F. and Grant, J.L. (1995) Resistance of the white clover variety G49 and its parent lines to stem nematode (*Ditylenchus dipsaci*). *New Zealand Journal of Agricultural Research* 38, 495–499.
Merny, G. and Cadet, P. (1978) Penetration of juveniles and development of adults of *Heterodera oryzae* on different plants. *Revue de Nématologie* 1, 251–255.
Merny, G. and Netscher, C. (1976) *Heterodera gambiensis* n.sp. (Nematode: Tylenchida) parasite du mil et du sorgho en Gambie. *Cahiers ORSTOM, Serie Biologie* 11, 209–218.
Meyer, A.J. and Van Wyk, R.J. (1989) Susceptibility of rotation crops and weeds in tobacco fields to *Meloidogyne* species and host races. *Phytophylactica* 21, 205–207.
Michel, R.E., Malek, R.B., Taylor, D.P. and Edwards, D.I. (1973) Races of the barley root-knot nematode, *Meloidogyne naasi* I. Characterisation by host preference. *Journal of Nematology* 5, 41–44.
Minton, N.A. (1986) Impact of conservation tillage on nematode populations. *Journal of Nematology* 18, 135–140.
Minz, G. (1956) Cyst forming nematodes in Israel. *Plant Disease Reporter* 40, 971–973.
Mizukubo, T., Toida, Y., Keereewan, S. and Yoshida, M. (1990) *Pratylenchus subranjani* n.sp. (Nematoda: Pratylenhidae) from maize in Thailand. *Applied Entomology and Zoology* 25, 311–318.
Mokabli, A., Valette, S. and Rivoal, R. (2001) Différenciation de quelques espèces de nématodes à kystes des céréales et des graminées par électrophorèse sur gel d'acétate de cellulose. *Nematologia Mediterranea* 29, 103–108.

Mor, M. and Cohn, E. (1989) New nematode pathogens in Israel: *Meloidogyne* on wheat and *Hoplolaimus* on cotton. *Phytoparasitica* 17, 221.

Mor, M., Cohn, E. and Spiegel, Y. (1992) Phenology, pathogenicity and pathotypes of cereal cyst nematodes, *Heterodera avenae* Woll. and *H. latipons* (Nematoda: Heteroderidae) in Israel. *Nematologica* 38, 444–501.

Motalaote, B., Starr, J.L., Frederikson, R.A. and Miller, F.R. (1987) Host status and susceptibility of sorghum to *Pratylenchus* species. *Revue de Nématologie* 10, 81–86.

Motjahedi, H., Crosslin, J.M., Thomas, P.E., Santo, G.S., Brown, C.R. and Wilson, J.H. (2002) Impact of wheat and maize as rotational crops on corky ringspot disease of Russet Norkotah potato. *American Journal of Potato Research* 79, 339–344.

Naganathan, T.G. and Sivakumar, C.V. (1975) Host parasite relationships and influence of soil types on the lesion nematode *Pratylenchus delattrei* Luc, 1958 on maize. *Indian Journal of Nematology* 5, 162–169.

Naganathan, T.G. and Sivakumar, C.V. (1976) Control of the lesion nematode *Pratylenchus delattrei* Luc, 1958 on maize. *Indian Journal of Nematology* 6, 32–38.

Nicol, J.M. (1996) The distribution, pathogenicity and population dynamics of *Pratylenchus thornei* (Sher and Allen, 1954) on wheat in South Australia. PhD Thesis, University of Adelaide, Adelaide, Australia.

Nicol, J.M. (2002) Important nematode pests of cereals. In: *Bread Wheat Production and Improvement.* FAO Plant Production and Protection Series, FAO, Rome, Italy, pp. 345–366.

Nicol, J. and Ortiz-Monasterio, I. (2004) Effects of the root lesion nematode, *Pratylenchus thornei*, on wheat yields in Mexico. *Journal of Nematology* (in press).

Nicol, J.M., Davies, K.A. and Eastwood, R. (1998) AUS4930: a new source of resistance to *Pratylenchus thornei* in wheat. *Abstracts of 24th International Symposium of the European Society of Nematologists*, St Andrews, UK.

Nicol, J.M., Davies, K.A., Hancock, T.W. and Fisher, J.M. (1999) Yield loss caused by *Pratylenchus thornei* on wheat in South Australia. *Journal of Nematology* 31, 367–376.

Nicol, J.M., Rivoal, R., Trethowan, R.M., Van Ginkel, M., Mergoum, M. and Singh, R.P. (2001) CIMMYT's approach to identity and use resistance to nematodes and soil-borne fungi, in developing superior wheat germplasm. In: Bedö, Z. and Langö, L. (eds) *Wheat in a Global Environment.* Kluwer Academic Publishers, Dordrecht, The Netherlands, pp. 381–389.

Nicol, J., Rivoal, R., Vallette, S., Bolat, N., Aktas, H., Braun, H.J., Mergoum, M., Yildrim, A.F., Bagci, A., Eleckcioglu, H. and Yahyaoui, A. (2002) The frequency and diversity of the cyst and lesion nematode on wheat in the Turkish Central Anatolian Plateau. *Journal of Nematology* 4, 272 (abstract).

Nishizawa, J. (1981) Preliminary experiment on varietal reaction of corn to three *Meloidogyne* species. *Proceedings of the 3rd Research Planning Conference on Root-knot Nematodes,* Meloidogyne *spp. Region VI, 20–24.* North Carolina State University, Raleigh, North Carolina, pp. 55–58.

Noel, G.R. and Edwards, D.L. (1996) Population development of *Heterodera glycines* and soybean yield in soybean–maize rotations following introduction into a noninfested field. *Journal of Nematology* 28, 335–342.

Norton, D.C., Tollefson, J., Hinz, P. and Thomas, S.H. (1978) Corn yield increases relative to non-fumigant chemical control of nematodes. *Journal of Nematology* 10, 160–166.

Norton, D.C., Edwards, J. and Hinz, P.N. (1985) Nematode populations in maize and related species. *Maydica* 30, 67–74.

Nyczepir, A.P., Inserra, R.N., O'Bannon, J.H. and Santo, G.S. (1984) Influence of *Meloidogyne chitwoodi* and *M. hapla* on wheat growth. *Journal of Nematology* 16, 162–165.

O'Brien, P.C. (1983) A further study on host range of *Pratylenchus thornei. Australian Plant Pathology* 12, 1–3.

O'Brien, P.C. and Fisher, J.M. (1978) Factors influencing the number of larvae of *Heterodera avenae* within susceptible wheat and barley seedlings. *Nematologica* 24, 295–304.

Ogbonnaya, F., Moullet, O., Eastwood, R. and Lagudah, E.S. (1996) The development and application of a PCR based assay linked to the Cre3 gene locus, conferring resistance to cereal cyst nematode, *Heterodera avenae* in wheat. In: Richards, R., Wrigley, C., Rawson, H., Rebetzke, G., Davidson, J. and Brettell, R. (eds) *Proceedings of the 8th Assembly, Wheat Breeding Society of Australia*, pp. 148–152.

Ogbonnaya, F.C., Subrahmanyam, N.C., Moullet, O., de Majnik, J., Eagles, H.A., Brown, J.S., Eastwood, R.F., Kollmorgen, J., Appels, R. and Lagudah, E.S. (2001) Diagnostic DNA markers for cereal cyst nematode resistance in bread wheat. *Australian Journal of Agricultural Research* 52, 1367–1374.

Ogiga, I.R. and Estey, R.H. (1975) Penetration and colonization of *Brassica rapa* and *Zea mays* root lesions by *Pratylenchus penetrans. Phytoprotection* 56, 23–30.

Olowe, T. (1977) Histological changes in maize root induced by *Pratylenchus brachyurus* and *P. zeae* in the absence of other micro-organisms. *Nigeria Journal of Plant Protection* 3, 41–51.

Olowe, T. and Corbett, D.C.M. (1976) Aspects of the biology of *Pratylenchus brachyurus* and *P. zeae. Nematologica* 22, 202–211.

Orion, D., Krikun, J. and Amir, J. (1982) Population dynamics of *Pratylenchus thornei* and its effect on wheat in a semi arid region. *Abstracts of 24th International Symposium of the European Society of Nematologists,* St Andrews, UK, p. 48.

Orion, D., Amir, J. and Krikun, J. (1984) Field observations on *Pratylenchus thornei* and its effects on wheat under arid conditions. *Revue de Nématologie* 7, 341–345.

Orr, C.C. and Morey, E.D. (1978) Anatomical response of grain sorghum roots to *Meloidogyne incognita acrita. Journal of Nematology* 10, 48–53.

Oteifa, B.A. and Elgindi, D.M. (1982) Relative susceptibility of certain commercially important cultivars to existing biotypes of *Meloidogyne incognita* and *M. javanica* in Nile-Delta, Egypt. *Proceedings of the 3rd Research Planning Conference on Root-Knot Nematodes* Meloidogyne *spp. Region VII,* Coimbra, Portugal, pp. 157–169.

Oztürk, G., Yildirim, A.F. and Kepenekci, I. (1999) [Investigation on the effect of *Heterodera filipjevi* Madzhidov on cereal yields which is one of the important cyst nematodes in cereal planted areas in Konya province.] In: Ekiz, H. (ed.) *Orta Anadolu'da hububat tardotless imdotless indotless in sorunlardotless i ve cözüm yollardotless i Sempozyumu.* Konya, Turkey, 8–11 Haziran, pp. 477–482.

Padgham, J.L., Duxbury, J.M., Mazid, A.M., Abawi, G.S. and Hossain, M. (2004) Yield loss caused by *Meloidogyne graminicola* on lowland rainfed rice in Bangladesh. *Journal of Nematology* (in press).

Page, S.L.J. (1983) Biological Studies of the African Cotton Root Nematode, *Meloidogyne acronea.* PhD Thesis, University of London, UK.

Page, S.L.J. (1985) *Meloidogyne acronea. C.I.H. Description of Plant-parasitic Nematodes.* Set 8, no. 114.

Parihar, A. and Yadav, B.S. (1992) Studies on the life cycle of corn cyst nematode *Heterodera zeae* Koshy *et al.*, 1970 on maize (*Zea mays* L.). *Pakistan Journal of Nematology* 10, 99–101.

Paruthi, I.J. and Gupta, D.C. (1987) Incidence of tundu in barley and kanki in wheat field infested with *Anguina tritici. Harayana Agricultural University Journal of Research* 17, 78–79.

Patel, N.B., Patel, D.J. and Patel, A.D. (2002a) Effect of *Pratylenchus zeae* on maize. *Indian Phytopathology* 55, 333–334.

Patel, N.B., Patel, D.J. and Patel, A.D. (2002b) *Pratylenchus zeae* and *Fusarium moniliforme* interaction in maize. *Indian Journal of Nematology* 32, 80–81 (abstract).

Paull, J.G., Chalmers, K.J. and Karakousis, A. (1998) Genetic diversity in Australian wheat varieties and breeding material based on RFLP data. *Theoretical Applied Genetics* 96, 435–446.

Person-Dedryver, F. (1986) Incidence du nématode à galle *Meloidogyne naasi* en cultures céréalières intensives. *Dix Années d'Etudes Concertées INRA-ONIC-ITCF, 1973–1983.* INRA, Paris, pp. 175–187.

Person-Dedryver, F. and Jahier, J. (1985) Les céréales à paille hôtes de *Meloidogyne naasi* Franklin II. Variabilité du comportement multiplicateur ou résistant de variétés cultivées en France. *Agronomie* 5, 573–578.

Person-Dedryver, F., Jahier, J. and Miller, T.E. (1990) Assessing the resistance to cereal root-knot nematode, *Meloidogyne naasi,* in a wheat line with the added chromosome arm 1HchS of *Hordeum chilense. Journal of Genetics and Breeding* 44, 291–295.

Philis, I. (1988) Occurrence of *Heterodera latipons* on barley in Cyprus. *Nematologia Mediterranea* 16, 223.

Plowright, R.A., Caubel, G. and Mizen, K.A. (2002) *Ditylenchus* species. In: Starr, J.L., Cook, R. and Bridge, J. (eds) *Plant Resistance to Parasitic Nematodes.* CAB International, Wallingford, UK, pp. 107–139.

Poerba, Y.S., Windham, G.L. and Williams, W.P. (1990) Resistance of maize hybrids to *Meloidogyne javanica. Nematropica* 20, 169–172.

Prasad, J.S., Seshu Reddy, K.V. and Sikora, R.A. (1995) Hosts of the banana root-lesion nematode, *Pratylenchus goodeyi* in East Africa. *Nematologia Mediterranea* 23, 253–254.

Prasad, K.S.K., Krishnappa, K., Setty, K.G.H., Reddy, B.M.R. and Reddy, H.R. (1980) Susceptibility of some cereals to *ragi* cyst nematode, *Heterodera delvii. Current Research* 9, 114.

Price, N.S. (1994) Alternate cropping in the management of *Radopholus similis* and *Cosmopolites sordidus,* two important pests of banana and plantain. *International Journal of Pest Management* 40, 237–244.

Rajan (1984) Evaluation of cyst extraction techniques and their effect on biology of *Heterodera cajani, H. zeae* and *H. avenae.* PhD Thesis, IARI, New Delhi, India.

Revelo Moran, J.A., Cid Del Pado Vera, I., Zavaleta-Mejía, E., Renfro, B. and Zarate De Lara, G. (1993) [Influence of *Pratylenchus pratensis* on corn rot caused by *Fusarium moniliforme* var. *subglutinans* Es.] *Revista Mexicana de Fitopatología* 11, 13–18.

Reversat, G. and Germani, G. (1985) Recherches sur la pathogenie des nematodes associes auf mais fourrager au Sénegal. *Revue de Nématologie* 8, 27–30.

Ribeiro, N.R., Silva, J.F.V., Meirelles, W.F., Craveiro, A.G., Parentoni, S.N. and dos Santos, F.G. (2002) [Evaluation of maize, sorghum and millet genotypes for resistance against *Meloidogyne javanica* and *M. incognita* race 3. Pt.] *Revista Brasileira de Milho e Sorgo* 1, 102–106.

Riekert, H.F. (1995) A modified sodium hypochlorite technique for the extraction of root-knot nematode eggs and larvae from maize root samples. *African Plant Protection* 1, 41–43.

Riekert, H.F. (1996) Greenhouse assessment of maize growth and yield response to nematode control with aldicarb. *African Crop Science Journal* 4, 471–475.

Riekert, H.F. and Henshaw, G.E. (1998) Effect of soybean, cowpea and groundnut rotations on root-knot nematode build-up and infestation of dryland maize. *African Crop Science Journal* 6, 377–383.

Riekert, H.F. and Tiedt, L.R. (1994) Scanning electron microscopy of *Meloidogyne incognita* juveniles entrapped in maize roots by a nematode-trapping fungus *Arthrobotrys dactyloides*. *South African Journal of Zoology* 29, 189–191.

Ringer, C.E., Sardanelli, S. and Krusberg, L.R. (1987) Investigations of the host range of the corn cyst nematode, *Heterodera zeae*, from Maryland. *Annals of Applied Nematology* 1, 97–106.

Ritter, M. (1972) Role economique et importance des *Meloidogyne* en Europe et dans le bassin Mediterraneen. *OEPP/EPPO Bulletin* 2, 17–22.

Ritzema-bos, J. (1922) Het stengelaatje. *Tijdschrifft Plantenziekten* 28, 159–180.

Rivoal, R. and Besse, T. (1982) Le nématode à kyste des céréales. *Perspectives Agricoles* 63, 38–43.

Rivoal, R. and Cook, R. (1993) Nematode pests of cereals. In: Evans, K., Trudgill, D.L. and Webster, J.M. (eds) *Plant Parasitic Nematodes in Temperate Agriculture*. CAB International, Wallingford, UK, pp. 259–303.

Rivoal, R., Doussinault, G. and Jahier, J. (1986) Résistance au développement d'*Heterodera avenae* Woll. chez différentes espèces de *Triticum*. *Agronomie* 6, 759–765.

Rivoal, R., Bekal, S., Valette, S., Gauthier, J.P., Bel Hadj Fradj, M., Mokabli, A., Jahier, J., Nicol, J. and Yahyaoui, A. (2001) Variation in reproductive capacity and virulence on different genotypes and resistance genes of Triticeae, in the cereal cyst nematode species complex. *Nematology* 3, 581–592.

Rhoades, H.L. (1977) Effect of nonfumigant nematicides on the sting nematode, *Belonolaimus longicaudatus*, and yield of sweetcorn. *Soil and Crop Science Society of Florida Proceedings* 37, 218–220.

Roberts, P.A., Van Gundy, S.D. and McKinney, H.E. (1981) Effects of soil temperature and planting date of wheat on *Meloidogyne incognita*, reproduction, soil populations and grain yield. *Journal of Nematology* 13, 338–345.

Roberts, P.A., Van Gundy, S.D. and Waines, J.G. (1982) Reaction of wild and domesticated *Triticum* and *Aegilops* species to root-knot nematodes (*Meloidogyne*). *Nematologica* 28, 182–191.

Rodrígues-Kábana, R. (1992) Cropping systems for the management of phytonematodes. In: Gommers, F.J. and Maas, P.W.T. (eds) *Nematology from Molecule to Ecosystem: Proceedings of the Second International Nematology Congress*, Veldhoven, The Netherlands. European Society of Nematologists, Dundee, UK, pp. 219–233.

Rodrígues-Kábana, R., Weaver, D.B., Robertson, D.G., Weaver, C.F. and Carden, E.L. (1991) Rotations of soybean with tropical corn and sorghum for the management of nematodes. *Supplement to the Journal of Nematology* 23, 662–667.

Romero, M.D., Montes, M.J., Sin, E., Lopez-Brana, I., Duce, A., Martin-Sanchez, A.J., Andres, M.F. and Delibes, A. (1998) A cereal cyst nematode (*Heterodera avenae* Woll.) resistance gene transferred from *Aegilops triuncalis* to hexaploid wheat. *Theoretical and Applied Genetics* 96, 1135–1140.

Roy, A.K. (1977) Host suitability of some crops to *Meloidogyne graminicola*. *Indian Phytopathology* 30, 483–485.

Rumpenhorst, H.J., Elekçioglu, I.H., Sturhan, D., Öztürk, G. and Enneli, S. (1996) The cereal cyst nematode *Heterodera filipjevi* (Madzhidov) in Turkey. *Nematologia Mediterranea* 24, 135–138.

Sabova, M., Valocka, B. and Liskova, M. (1981) Effect of *Heterodera avenae* on some cereal cultivars under experimental conditions. *Ochrana Rostlin* 17, 191–197.

Sabova, M., Valocka, B., Liskova, M. and Vargova, V. (1988) The first finding of *Heterodera latipons* Franklin, 1969 on grass stands in Czecholovakia. *Helminthologia* 25, 201–206.

Saefkow, M. (1983) Pathotypenuntersuchungen mit Populationen von *Heterodera avenae* an Mais. *Zeitschrift für Pflanzenkrankheiten und Pflanzenschutz* 90, 337–344.

Saefkow, M. and Lucke, E. (1979) Die Entwicklung von *Heterodera avenae* in Maiswurzeln. *Nematologica* 25, 309–313.

Salawu, E.O. and Oyewo, A.C. (1999) New and known weed hosts of *Meloidogyne incognita* (Kofoid and White, 1919) Chitwood, 1949, a root-knot nematode in selected maize farms at Tanke, Ilorin, Nigeria. *Pakistan Journal of Nematology* 17, 97–101.

Saleh, H.M. and Fattah, F.A. (1990) Studies on the wheat seed gall nematode. *Nematologia Mediterranea* 18, 59–62.

Santo, G.S. and O'Bannon, J.H. (1981) Pathogenicity of the Columbia root-knot nematode (*Meloidogyne chitwoodi*) on wheat, corn, oat and barley in the Pacific North West. *Journal of Nematology* 13, 548–550.

Sasser, J.N. (1977) Worldwide dissemination and importance of root-knot nematodes, *Meloidogyne* spp. *Journal of Nematology* 9, 26–29.

Sasser, J.N. and Freckman, D.W. (1987) A world perspective on nematology: the role of the society. In: Veech, J.A. and Dickson, D.W. (eds) *Vitas on Nematology*. Society of Nematologists, Hyattsville, Maryland, pp. 7–14.

Sasser, J.N. and Kirby, M.F. (1979) *Crop Cultivars Resistant to Root-knot Nematodes,* Meloidogyne *species*. A Cooperative Publication of the Department of Plant Pathology, North Carolina State University and the US Agency for International Development.

Sasser, J.N. and Triantaphyllou, A.C. (1977) Identification of *Meloidogyne* species and races. *Journal of Nematology* 9, 283 (abstract).

Saxena, M.C., Sikora, R.A. and Srivastava, J.P. (1988) *Nematodes Parasitic to Cereals and Legumes in Temperate Semi-arid Regions*. ICARDA, Aleppo, Syria.

Schneider, J. (1967) Un nouveau nématode du genre *Meloidogyne* parasite des céréales en France. *Phytoma* 185, 21–25.

Scholz, U. (2001) Biology, Pathogenicity and Control of the Cereal Cyst Nematode *Heterodera latipons* Franklin on Wheat and Barley under Semiarid Conditions, and Interactions with Common Root Rot *Bipolaris sorokinana* (Sacc.) Shoemaker [Teleomorph: *Cochliobolus sativus* (Ito et Kurib.) Drechs. ex Dastur]. PhD Thesis, University of Bonn, Bonn, Germany.

Scholtz, U. and Sikora, R.A. (2004) Hatching behaviour and life-cycle of *Heterodera latipons* Franklin as affected by agro-ecological conditions existing in semi-arid growing regions. *Nematologica* 6, 245–256.

Schreiber, M.L. (1977) Lebenweise, Bedeutung und Bekämpfungsmöglischkeiten von *Ditylenchus dipsaci* (Kühn) Filipjevi an ackerbohnen *Vicia faba* L. in Morokko. Dissertation, Technical University of Berlin, Berlin, Germany.

Semkin, A.I. (1975) The effect of *Longidorus elongates* on the crop of millet. *Chelyabinsku Gosudarstvennyi Pedagogichesku Institut Voprosy Zooloqu USSR* 4, 3–8.

Seshadri, A.R. (1970) Nematology. In: *Agricultural Year Book – Vistas in Crop Yields*. Indian Council of Agricultural Research, New Delhi, pp. 370–411.

Shahina, F. and Maqbool, M.A. (1990) Host range studies of *Heterodera zeae* and *H. avenae*. *Pakistan Journal of Nematology* 8, 79–85.

Shahina, F., Abid, M. and Maqbool, M.A. (1989) Screening for resistance in corn cultivars against *Heterodera zeae*. *Pakistan Journal of Nematology* 7, 75–79.

Sharma, R.D. (1981) Pathogenicity of *Meloidogyne javanica* to wheat. In: *Trabalhes Apresentados V Reunas, Brasileira de Nematologia,* 9 B. Publicao No. 5.

Sharma, R.D. and Swarup, G. (1988) *Pathology of Cyst Nematodes*. Malhotra Publishing House, New Delhi.

Sharma, S.B. and McDonald, D. (1990) Global status of nematode problems of groundnut, pigeonpea, chickpea, sorghum and pearl millet, and suggestions for future work. *Crop Protection* 9, 453–458.

Sharma, S.B. and Swarup, G. (1984) *Cyst Forming Nematodes of India*. India Cosmo Publication 1, New Delhi.

Shen, T.H., Tai, S.E. and Chia, W.L. (1934) A preliminary report on the inheritance of nematode resistance and length of beak in a certain wheat cross. *Bulletin College of Agriculture and Forestry,* No. 19, University of Nanking.

Shepherd, J.A. (1981) Report to the third regional conference on root-knot nematode research held at the International Institute of tropical Agriculture, Ibadan, Nigeria. *Proceedings of the 3rd Research Planning Conference on Root-knot Nematodes,* Meloidogyne *spp. Region IV and Region V,* Ibadan, Nigeria, pp. 21–27.

Sikora, R.A. (1988) Plant parasitic nematodes of wheat and barley in temperate and temperate semi-arid regions – a comparative analysis. In: Saxena, M.C., Sikora, R.A. and Srivastava, J.P. (eds) *Nematodes Parasitic to Cereals and Legumes in Temperate Semi-arid Regions*. ICARDA, Aleppo, Syria, pp. 46–48.

Sikora, R.A. and Oostendorp, M. (1986) *Report: Occurrence of Plant Parasitic Nematodes in ICARDA Experimental Fields.* ICARDA, Aleppo, Syria.
Simon, A. and Rovira, A.D. (1982) The relation between wheat yield and early damage of root by cereal cyst nematode. *Australian Journal of Agriculture and Animal Husbandry* 22, 201–208.
Singh, D. and Agrawal, K. (1987) Ear cockle disease (*Anguina tritici* (Steinbuch) Filipjev) of wheat in Rajasthan, India. *Seed Science and Technology* 15, 777–784.
Singh, K.P., Bino, B.N. and Edward, J.C. (1979) Effect of *Heterodera zeae* on growth of *Sorghum vulgare* and on the rhizosphere microflora. *Indian Phytopathology* 32, 167.
Singh, S.D. and Siradhana, B.S. (1988) Effect of nematodes on late wilt of maize. *Indian Phytopathology* 41, 173–176.
Singh, U.S. and Patel, D.J. (1999) Evaluation of maize varieties/hybrids against stunt nematode, *Tylenchorhynchus vulgaris. Indian Journal of Nematology* 29, 89–90.
s'Jacob, J.J. (1962) Beobachtungen an *Ditylenchus radicicola* (Greeff). *Nematologica* 7, 231–234.
Slootmaker, L.A.J., Lange, R., Jochemsen, G. and Schepers, J. (1974) Monosomic analysis in breadwheat of resistance to cereal root eelworm. *Euphytica* 23, 497–503.
Smolik, J.D. (1972) Reproduction of *Tylenchorhynchus nudus* and *Helicotylenchus leiocephalus* on spring wheat and effect of *T. nudus* on growth of spring wheat. *Proceedings South Dakota Academy of Science* 51, 153–159.
Smolik, J.D. (1977) Effects of *Trichodorus allius* and *Tylenchorhynchus nudus* on growth of sorghum. *Plant Disease Reporter* 61, 855–858.
Smolik, J.D. and Evenson, P.D. (1987) Relationship of yields and *Pratylenchus* spp. population densities in dryland and irrigated corn. *Annals of Applied Nematology* 1, 71–73.
Soomro M.H. and Hague, N.G.M. (1992) Effect of *Meloidogyne graminicola* on root growth of graminaceous plants. *Nematologia Mediterranea* 20, 143–147.
Sosa-Moss, C. (1965) Recherches sur *Heterodera punctata* Thorn, parasite du mais au mexique. *Comptes Rendus du Huitieme Symposium International de Nematologie*, Antibes, France. E.J. Brill, Leiden, The Netherlands.
Sosa-Moss, C. (1987) Cyst nematodes in Mexico, Central and South America. *Nematologia Mediterranea* 15, 1–2.
Sosa-Moss, C. and González, C. (1973) Respuesta de maíz chalpueno fertilizado y no fertilizado a cuatro diferentes niveles de *Heterodera punctata* raza mexicana (Nematode: Heteroderidae). *Nematropica* 3, 13–14.
Srivastava, A.N. (1980) Studies on the maize cyst nematode *Heterodera zeae*. PhD Thesis, IARI, New Delhi, India.
Srivastava, A.N. and Sethi, C.L. (1984a) Effect of soil type on reproduction of *Heterodera zeae. Indian Journal of Nematology* 14, 72.
Srivastava, A.N. and Sethi, C.L. (1984b) Relationships of initial populations of *Heterodera zeae* with plant growth of maize and nematode reproduction. *Indian Journal of Nematology* 14, 110–114.
Srivastava, A.N. and Sethi, C.L. (1985a) On the larval penetration and biology of *Heterodera zeae* on maize. *Indian Journal of Nematology* 15, 18–20.
Srivastava, A.N. and Sethi, C.L. (1985b) Effect of temperature on larval emergence of *Heterodera zeae. Indian Journal of Nematology* 15, 242–243.
Srivastava, A.N. and Sethi, C.L. (1986) Seasonal population fluctuations of *Heterodera zeae, Rotylenchulus reniformis* and *Tylenchorhynchus vulgaris* in maize–cowpea–wheat rotation. *Indian Journal of Nematology* 16, 16–18.
Srivastava, A.N. and Swarup, G. (1975) Preliminary studies on some graminaceous plants for their susceptibility to the maize cyst nematode *Heterodera zeae* Koshy *et al.*, 1970. *Indian Journal of Nematology* 5, 257–259.
Stephan, Z.A. (1988) Plant parasitic nematodes on cereals and legumes in Iraq. In: Saxena, M.C., Sikora, R.A. and Srivastava, J.P. (eds) *Nematodes Parasitic to Cereals and Legumes in Temperate Semi-arid Regions.* ICARDA, Aleppo, Syria, pp. 155–159.
Stirling, G.R. (1991) *Biological Control of Plant Parasitic Nematodes. Progress, Problems and Prospects.* CAB International, Wallingford, UK.
Stoen, M. (1971) *Heterodera avenae* rase – og resistenundes-desokelser. *N.J.F. Kongressen Uppsala, Seksion IV* 101–102.
Stone, A.R., Sosa-Moss, C. and Mulvey, R.H. (1976) *Punctodera chalcoensis* n.sp. (Nematoda: Heteroderidae) a cyst nematode from Mexico parasitizing *Zea mays. Nematologica* 22, 381–389.

Stone, L.E.W. (1968) Cereal cyst nematode in spring barley damage assessments 1963–64. *Plant Pathology* 17, 145–150.

Stoyanov, D., Baicheva, O., Daskalova, Y. and Shtereva, L. (1990) Parasitic root nematodes in wheat and maize (Family Tylenchorhynchidae) [Bg, Ru]. *Rasteniev' dni Nauki* 27, 112–116.

Stradioto, M.F., Barbosa Ferraz, L.C.G. and Pitelli, R.A. (1983) Dinamica Populacional de *Pratylenchus brachyurus* em cultura de Milho (*Zea mays* L.) infestada por plantas daninhas. *Societa Brasil. Nemat. Public* 7, 99–115.

Sturhan, D. (1982) Distribution of cereal and grass cyst nematodes in the Federal Republic of Germany. *EPPO Bulletin* 12, 321–324.

Sturhan, D. (1996) Occurrence of *Heterodera filipjevi* (Madzhidov, 1981) Stelter, 1984 in Iran. *Pakistan Journal of Nematology* 14, 89–93.

Subbotin, S.A., Rumpenhorst, H.J. and Sturhan, D. (1996) Morphological and electrophoretic studies on populations of the *Heterodera avenae* complex from the former USSR. *Russian Journal of Nematology* 49, 29–38.

Subbotin, S.A., Waeyenberge, L. and Moens, M. (2000) Identification of cyst forming nematodes of the genus *Heterodera* (Nematoda: Heteroderidae) based on the ribosomal DNA-RFLP. *Nematology* 2, 153–164.

Sumner, D.R., Hall, M.R., Gay, J.D., MacDonald, G., Savage, S.I. and Bramwell, R.K. (2002) Root diseases, weeds, and nematodes with poultry litter and conservation tillage in a sweetcorn–snap bean double crop. *Crop Protection* 21, 963–972.

Swarup, G. (1986) Investigations on wheat nematodes. In: Tandon, J.P. and Sethi, A.P. (eds) *Twenty-five Years of Coordinated Wheat Research 1961–86*. Wheat Project Directorate, Indian Council of Agricultural Research, New Delhi, pp. 189–206.

Swarup, G. and Gill, J.S. (1972) Factors influencing emergence of larvae from cysts of *Heterodera avenae*. *Indian Journal of Experimental Biology* 10, 219–225.

Swarup, G. and Singh, K. (1961) Molya disease of wheat and barley in Rajasthan. *Indian Phytopathology* 14, 127–133.

Swarup, G. and Sosa-Moss, C. (1990) Nematode parasites of cereals. In: Luc, M., Sikora, R.A. and Bridge, J. (eds) *Plant Parasitic Nematodes in Subtropical and Tropical Agriculture*. CAB International, Wallingford, UK, pp. 109–136.

Swarup, G., Mathur, R.L., Seshadri, A.R., Raski, D.J. and Mathur, B.N. (1976) Response of wheat and barley to soil fumigation by DD and DBCP against 'molya' disease caused by *Heterodera avenae*. *Indian Journal of Nematology* 6, 150–155.

Swarup, G., Sethi, C.L., Seshadri, A.R. and Kaushal, K.K. (1979) On the biotypes of *Heterodera avenae*, the causal organism of 'molya' disease of wheat and barley in India. *Indian Journal of Nematology* 9, 164–168.

Swarup, G., Prasad, S.K. and Raski, D.J. (1996) Some *Heterodera* species from India. *Plant Disease Reporter* 48, 235.

Tacconi, R., Santi, R. and Pola, R. (1988) [*Pratylenchus neglectus* on corn (*Zea mays*) in Emilia Romagna.] *Informatore Fitopatologico* 38, 59–61.

Tanha Maafi, Z., Subbotin, S.A. and Moens, M. (2003) Molecular identification of cyst-forming nematodes (Heteroderidae) from Iran and a phylogeny based on ITS-rDNA sequences. *Nematology* 5, 99–111.

Taylor, A.L. and Sasser, J.N. (1978) *Biology, Identification and Control of Root-knot Nematodes (*Meloidogyne *species)*. North Carolina State University Department of Plant Pathology and USAID, Raleigh, North Carolina.

Taylor, C., Shepherd, K.W. and Langridge, P. (1998) A molecular genetic map on the long arm of chromosome 6R of rye incorporating the cereal cyst nematode resistance gene, *CreR*. *Theoretical and Applied Genetics* 97, 100–102.

Taylor, S.P. and McKay, A. (1993) Sampling and extraction methods for *Pratylenchus thornei* and *P. neglectus*. In: Vanstone, V.A., Taylor, S.P. and Nicol, J.M. (eds) *Proceedings of the* Pratylenchus *Workshop, Ninth Biennial Australasian Plant Pathology Conference*, Hobart, Tasmania, Australia.

Taylor, S.P., Vanstone, V.A., Ware, A.H., McKay, A.C., Szot, D. and Russ, M.H. (1999) Measuring yield loss in cereals caused by root lesion nematodes (*Pratylenchus neglectus* and *P. thornei*) with and without nematicide. *Australian Journal of Agricultural Research* 50, 617–622.

Taylor, S.P., Hollaway, G.J. and Hunt, C.H. (2000) Effect of field crops on population densities of *Pratylenchus neglectus* and *P. thornei* in southeastern Australia; Part 1: *P. neglectus*. *Journal of Nematology* 32, 591–599.

Tenente, R.C.V., Manso, E.C. and Gonzaga, V. (2000) [Nematodes detected in imported germplasm and their eradication during 1995 to 1998, Pt.] *Nematologia Brasileira* 24, 79–81.
Tesic, T. (1969) [A study on the resistance of wheat varieties to wheat eelworm (*Anguina tritici* Stein.).] *Savrenema Poljoprivreda* 17, 541–543.
Thompson, J.P. and Clewett, T.G. (1986) Research on root-lesion nematode. In: *Queensland Wheat Research Institute Biennial Report 1982–1984*. Queensland Department of Primary Industries, Queensland. Government, Wheat Research Institute, Toowoomba, Queensland, pp. 32–35.
Thompson, J.P. and Haak, M.I. (1997) Resistance to root-lesion nematode (*Pratylenchus thornei*) in *Aegilops tauschii* Coss., the D-genome donor to wheat. *Australian Journal of Agricultural Research* 48, 553–559.
Thompson, J.P., Mackenzie, J. and McCulloch, J. (1983) Root lesion nematode (*Pratylenchus thornei*) on Queensland wheat farms. *Proceedings of the 46th International Congress of Plant Pathology*, Melbourne, Australia (abstract).
Thompson, J.P., Brennan, P.S., Clewett, T.G. and Sheedy, J.G. (1997) *Disease Reactions. Root-lesion Nematode.* Northern Region Wheat Variety Trials 1996. Queensland Department of Primary Industries Publication, Australia.
Thorne, G. (1928) *Heterodera punctata* n.sp., a nematode parasitic on wheat roots from Saskatchewan. *Scientific Agriculture* 8, 707–710.
Timm, L., Pearson, D. and Jaffee, B. (2001) Nematode-trapping fungi in conventionally and organically managed corn–tomato rotations. *Mycologia* 93, 25–29.
Timper, P., Wilson, J.P., Johnson, A.W. and Hanna, W.W. (2002) Evaluation of pearl millet grain hybrids for resistance to *Meloidogyne* spp. and leaf blight caused by *Pyricularia grisea*. *Plant Disease* 86, 909–914.
Todd, T.C. (1989) Population dynamics and damage potential of *Belonolaimus* sp. on corn. *Supplement to the Journal of Nematology* 21, 697–702.
Todd, T.C. (1991) Effect of cropping regime on populations of *Belonolaimus* sp. and *Pratylenchus scribneri* in sandy soil. *Supplement to the Journal of Nematology* 23, 646–651.
Todd, T.C. and Oakley, T.R. (1996) Seasonal dynamics and yield relationships of *Pratylenchus* spp. in corn roots. *Supplement to the Journal of Nematology* 28, 676–681.
Toida, Y., Keereewan, S. and Puttirut, N. (1991) Assessment and prevention of mungbean damage caused by the root-knot nematode, *Meloidogyne javanica* in Thailand. *Japanese Journal of Nematology* 21, 6–10.
Troccoli, A., Lamberti, F. and Greco, N. (1992) *Pratylenchus* species occurring in Algeria (Nematoda Pratylenchidae). *Nematologia Mediterranea* 20, 97–103.
Trudgill, D.L., Kerry, B.R. and Phillips, M.S. (1992) Seminar: integrated control of nematodes (with particular reference to cyst and root knot nematodes). *Nematologica* 38, 482–487.
Upadhyaya, K.D. and Swarup, G. (1972) Culturing, host range and factors affecting multiplication of *Tylenchorhynchus vulgaris* on maize. *Indian Journal of Nematology* 2, 139–145.
Upadhyaya, K.D. and Swarup, G. (1981) Growth of wheat in the presence of *Merlinius brevidens* singly and in combination with *Tylenchorhynchus vulgaris*. *Indian Journal of Nematology* 11, 42–46.
Vaishnav, M.U. and Sethi, C.L. (1978) Interaction of root-knot and stunt nematodes with *Sclerospora graminicola* on bajra. *Indian Phytopathology* 31, 497–500.
Van Biljon, E.R. and Meyer, A.J. (2000) Reproduction of *Pratylenchus zeae* and *P. delattrei* on crops resistant to *Meloidogyne javanica* and *M. incognita* races 2 and 4. *African Plant Protection* 6, 43–45.
Van Gundy, S.D., Perez, B.J.G., Stolzy, L.H.A. and Thomason, I.J. (1974) A pest management approach to the control of *Pratylenchus thornei* on wheat in Mexico. *Journal of Nematology* 6, 107–116.
Vanstone, V.A., Nicol, J.M. and Taylor, S.P. (1993) Multiplication of *Pratylenchus neglectus* and *P. thornei* on cereals and rotational crops. In: Vanstone, V.A., Taylor, S.P. and Nicol, J.M. (eds) *Proceedings of the* Pratylenchus *Workshop, 9th Biennial Australasian Plant Pathology Conference*, Hobart, Tasmania, Australia.
Vanstone, V.A., Taylor, S.P., Evans, M.L., McKay, A.C. and Rathjen, A.J. (1995) Resistance and tolerance of cereals to root lesion nematode (*Pratylenchus neglectus*) in South Australia. In: *Proceedings of the 10th Biennial Conference, Australian Plant Pathology Conference*, Lincoln, New Zealand, August 1995, p. 40.
Vanstone, V.A., Rathjen, A.J., Ware, A.H. and Wheller, R.D. (1998) Relationship between root lesion nematodes (*Pratylenchus neglectus* and *P. thornei*) and performance of wheat varieties. *Australian Journal of Experimental Agriculture* 38, 181–188.
Vazquez, J.T. (1976) *Infestaciones de NematodosFitoparásitos Como Factor Limitante en la Producción de Maiz en el Antiplano Mexicano.* Folleto. DAGEN.

Venditti, M.E. and Noel, G.R. (1995) Comparative host suitability of selected crop species to *Tylenchorhynchus zambiensis*. *Nematropica* 25, 15–25.

Wallace, H.R. (1965) The ecology and control of the cereal root nematode. *Journal of Australian Institute of Agricultural Science* 31, 178–186.

Walters, M.C. (1979) The possible status of parasitic nematodes as limiting factors in maize production in South Africa. In: *Proceedings of the Second South African Maize Breeding Symposium.* Pietermaritzburg, 1976. Department of Agricultural Technical Services, Republic of South Africa 142, 112–118.

Wang, K.H., McSorley, R., Gallaher, R.N., Tubbs, R.S. and Van Santen, E. (2002) Effects of winter and fall cover crops on plant-parasitic nematode population development. *Proceedings of the 25th Annual Southern Conservation Tillage Conference for Sustainable Agriculture.* Auburn University, Auburn, Alabama, pp. 152–155.

West, C.P. and Steele, K.W. (1986) Tolerance of clover cultivars to stem nematode (*Ditylenchus dipsaci*). *New Zealand Journal of Agricultural Research* 14, 227–229

Whitehead, A.G. (1998) *Plant Nematode Control.* CAB International, Wallingford, UK.

Whitehead, A.G., Tite, D.J. and Fraser, J.E. (1983) Control of stem nematode *Ditylenchus dipsaci* (oat race) by aldicarb and resistant crop plants. *Annals of Applied Biology* 103, 291–299.

Wicks, Z.W. III, Smolik, J.D., Carson, M.L. and Scholten, G.G. (1990) Registration of SD101 parental line of maize. *Crop Science* 30, 242.

Williams, K.J., Fisher, J.M. and Langridge, P. (1994) Identification of RFLP markers linked to the cereal cyst nematode resistance gene (*Cre*) in wheat. *Theoretical and Applied Genetics* 83, 919–924.

Williams, K.J., Taylor, S.P., Bogacki, P., Pallotta, M., Bariana, H.S. and Wallwork, H. (2002) Mapping of the root lesion nematode (*Pratylenchus neglectus*) resistance gene *Rlnn1* in wheat. *Theoretical and Applied Genetics* 104, 874–879.

Williams, T.D. and Beane, J. (1972) Effect of temperature and root exudates on *H. avenae* hatch. *Report of Rothamsted Experiment Station for 1973*, pp. 171–172.

Williams, T.D. and Siddiqi, M.R. (1972) *Heterodera avenae. C.I.H. Descriptions of Plant-parasitic Nematodes.* Set 1, No. 2.

Williams, W.P. and Windham, G.L. (1990) Resistance of maize to *Meloidogyne arenaria* and *Meloidogyne javanica. Theoretical and Applied Genetics* 80, 810–812.

Williams, W.P. and Windham, G.L. (1992) Reaction of diallel cross of maize to *Meloidogyne incognita* under field conditions. *Field Crops Research* 30, 167–171.

Williams, W.P. and Windham, G.L. (1998) Registration of root-knot resistant maize germplasm lines Mp709, Mp710, Mp711, and Mp712. *Crop Science* 38, 563.

Williams, W.P., Davis, F.M. and Windham, G.L. (1990) Registration of Mp708 germplasm line of maize. *Crop Science* 30, 757.

Windham, G.L. and Lawrence, G.W. (1992) Host status of commercial maize hybrids to *Rotylenchulus reniformis. Supplement to the Journal of Nematology* 24, 745–748.

Windham, G.L. and Williams, W.P. (1987) Host suitability of commercial corn hybrids to *Meloidogyne arenaria* and *M. incognita. Annals of Applied Nematology* 1, 13–16.

Windham, G.L. and Williams, W.P. (1988) Reproduction of *Meloidogyne javanica* on corn hybrids and inbreds. *Annals of Applied Nematology* 2, 25–28.

Windham, G.L. and Williams, W.P. (1994a) Penetration and development of *Meloidogyne incognita* in roots of resistant and susceptible corn genotypes. *Journal of Nematology* 26, 80–85.

Windham, G.L. and Williams, W.P. (1994b) Reproduction of *Meloidogyne incognita* and *M. arenaria* on tropical corn hybrids. *Supplement to the Journal of Nematology* 26, 753–755.

Windham, G.L., Windham, M.T. and Williams, W.P. (1989) Effects of *Trichoderma* spp. on maize growth and *Meloidogyne arenaria* reproduction. *Plant Disease* 73, 493–495.

Winslow, R.D. (1955) The hatching response of some root eelworms of the genus *Heterodera. Annals of Applied Biology* 45, 19–36.

Yeates, G.W. and Hughes, K.A. (1990) Effect of three tillage regimes on plant and soil nematodes in oat/maize rotations. *Pedobiologia* 34, 379–387.

Yeates, G.W., Wardle, D.A. and Watson, R.N. (1993) Relationships between nematodes, soil microbial biomass and weed-management strategies in maize and asparagus cropping systems. *Soil Biology and Biochemistry* 25, 869–876.

York, P.A. (1980) Relationship between cereal root-knot nematode *Meloidogyne naasi* and growth and yield of spring barley. *Nematologica* 26, 220–229.

Young, N.D. and Mudge, J. (2002) Marker-assisted selection for soybean cyst nematode resistance. In: Starr, J.L., Cook, R. and Bridge, J. (eds) *Plant Resistance to Parasitic Nematodes*. CAB International, Wallingford, UK, pp. 241–252.
Youssef, M.M.A. (1998) Efficacy of some weed control treatments on nematode populations in a maize field. *Anzeiger für Schädelingskunde, Pflantzenschutz, Umweltschutz* 71, 159–160.
Yu, M.Q., Person-Dedryver, F. and Jahier, J. (1990) Resistance to root knot nematode, *Meloidogyne naasi* (Franklin) transferred from *Aegilops variabilis* Eig to bread wheat. *Agronomie* 6, 451–456.
Yu, Q. (1997) First report of *Pratylenchus thornei* from spring wheat in southern Ontario. *Canadian Journal of Plant Pathology* 19, 289–292.
Yüksel, H. and Güncan, Döken, M.T. (1980) The distribution and damage of bunts (*Tilletia* spp.) and wheat gall nematode [*Anguina tritici* (Steinbuch) Chitwood] on wheat in the eastern part of Anatolia. *Journal of Turkish Phytopathology* 9, 77–88.
Zaharieva, M., Monneveux, P., Henry, M., Rivoal, R., Valkoun, J. and Nachit, M.M. (2001) Evaluation of a collection of wild wheat relative *Aegilops geniculata* Roth. and identification of potential sources for useful traits. *Euphytica* 119, 33–38.
Zaharieva, M., Nicol, J., Romero, H. and Rivoal, R. (2002) Resistance of *Aegilops geniculata* Roth to cereal cyst nematode (*Heterodera avenae*) and root lesion nematode (*Pratylenchus thornei*). *Nematology* 4, 240 (abstract).
Zancada, M.C. and Althöfer, M.V. (1994) Effect of *Heterodera avenae* on the yield of winter wheat. *Nematologica* 40, 244–248.
Zirakparvar, M.E. (1980) Host range of *Pratylenchus hexincisus* and its pathogenicity on corn, soybean and tomato. *Phytopathology* 70, 549–573.
Zirakparvar, M.E., Norton, D.C. and Cox, C.P. (1980) Population increase of *Pratylenchus hexincisus* on corn as related to soil temperature and type. *Journal of Nematology* 12, 313–318.

6 Nematode Parasites of Solanum and Sweet Potatoes*

Maria I. Scurrah,[1] Björn Niere[2] and John Bridge[3]
[1]*Department of Nematology and Entomology, International Potato Centre, PO Box 5969, Lima, Peru;* [2]*Biologische Bundesanstalt für Land- und Forstwirtschaft, Toppheideweg 88, D-48161 Münster, Germany;*
[3]*CABI Bioscience UK Centre, Bakeham Lane, Egham, Surrey TW20 9TY, UK*

Root and tuber crops are the most important food commodities produced in many subtropical and tropical countries. World production figures for 2002 (FAO, 2003) show that root and tuber cultivation is increasing and are the key sources of carbohydrates in the tropical world, where clonal reproduction and poor soils gives them advantages for subsistence agricultures and are second only to cereals in total world supply. This chapter covers two of these important crops, the solanum potato (*Solanum tuberosum*) and the sweet potato (*Ipomoea batata*).

Potato

The solanum or Irish potato, *Solanum tuberosum* L., originating from the Andean highlands of South America, is a major food crop in 57 countries, which is more countries than any other single crop, with the exception of maize, and it is the only tuber crop produced in any significant amount in the developed countries.

While potato occupies fourth place in importance amongst the major food crops, in terms of dry matter production per hectare, it is the third highest on the list. It ranks first and third in the list of edible energy and protein production per hectare per day, respectively (Horton *et al.*, 1984).

In recent years, in subtropical and tropical countries, potato production has spread gradually out of its traditionally cool environment at higher altitudes into hotter and, generally, drier areas. It is increasingly grown as a winter crop in many irrigated, arid areas of large, commercial farms as better varieties have become available in developing countries (Fig. 6.1).

The production of this crop has been expanding to relatively warm and humid zones that are optimum for the development of many pathogens and pests, including nematodes.

Of the factors which adversely influence the production of potatoes from seed tubers or true potato seed (TPS), nematodes are amongst the most important pest constraints. Currently, the distribution of nematodes in most temperate potato-grow-

*A revision of part of the chapter by P. Jatala and J. Bridge.

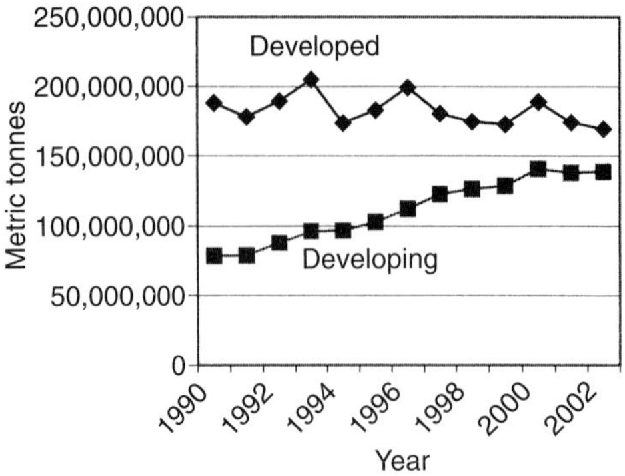

Fig. 6.1. World potato production: developed versus developing countries.

ing areas of the world is well known. While infestation in some countries may be regarded as minor, in other areas high infestations cause severe yield losses and/or affect quality.

Nematodes of Potato

Nematodes recognized as major parasites of potato are *Globodera* spp., *Meloidogyne* spp., *Nacobbus aberrans*, *Ditylenchus* spp. and *Pratylenchus* spp.

However, many other species are found associated with potato, such as *Belonolaimus longicaudatus*, *Atalodera* (=*Thecavermiculatus*) *andinas*, *Xiphinema* spp., *Rotylenchulus* spp., *Radopholus similis*, *Longidorus* spp., *Paratrichodorus* spp., *Trichodorus* spp. and *Paratylenchus* spp.; most of these have not been properly assessed.

Globodera

Potato cyst nematodes, *Globodera pallida* and *G. rostochiensis*, are the most important nematodes of potatoes and have received the greatest attention (Jensen *et al.*, 1979). They are mainly distributed in cooler areas of subtropical and tropical regions, as well as temperate regions of the world (see Chapter 22, map). They are believed to have evolved along with their principal hosts, potatoes, in the highlands of Peru and Bolivia. Brücher (1960) suggested that the Andes of northern Argentina may be their centre of origin, because wild *Solanum* species with resistant genes have been identified there, which still is a strong argument although resistant genes have also been identified in Bolivian and Peruvian species (Hawkes, 1994). The fact remains that they were introduced to Europe, probably in the mid or late 19th century, on South American potatoes imported for breeding purposes (Winslow, 1978a). They have since spread from there to most potato-growing areas in the tropical and subtropical zones, probably in soil adhering to seed potatoes exported from Europe initially to their colonies, e.g. Pakistan, India, South Africa, Oceania, North America, and later spreading with trade to the Mediterranean countries, northern and central South America and the Philippines. An introduction that may go back to the centre of origin is the Japanese population of potato cyst nematode: Yameda *et al.* (1972) reports viable cysts in Peruvian guano. However, only *G. rostochiensis* was identified, which is rarely found in most of Peru (Evans *et al.*, 1975). The most recent record comes from Indonesia, where in 2003 potato cyst nematode was detected (Siwi Indarti *et al.*, 2004).

Symptoms of damage

There are no specific above-ground symptoms of diagnostic value associated with potato cyst nematode infections. However, root injury causes stress and reduces the uptake of water and nutrients which in turn causes stunting, yellowing and other discoloration (Plate 4A), and wilting of the foliage under drought conditions. Early senescence and proliferation of lateral roots are often associated with nematode infection. Small immature females of white and yellow stages can be observed on the roots at flowering (Brown, 1969) (Fig. 6.2). Females of *G. rostochiensis* will go through a yellow stage, while G. *pallida* females remain white until dead (Guile, 1970) (Plate 4B). Females can sometimes be

Fig. 6.2. Females and cysts of *Globodera rostochiensis* on roots of potato. (Photo: J. Bridge.)

observed on the tuber surface (Franco, 1981). When females die, they become cysts, and their cuticles become brown or leathery, and contain as many as 500 nematode eggs.

Biology

Eggs inside cysts remain viable in soil for a long period of time; they contain second stage juveniles, which are the infective stage, and are stimulated to hatch by potato root exudates. Juveniles become active at 10°C and maximum root invasion takes place at 16°C (Franco, 1979). Invasion damages roots and stunts plants, in proportion to the field population of the nematodes. It has been shown that juveniles secrete an expansin that relaxes and breaks the bonds of cell walls, which explains the rapid penetration of nematodes into roots (Qin *et al.*, 2004). Juveniles then induce cells near their head to grow and accumulate nutrients; this is called a feeding site or syncytium. Male nematodes leave the roots after the final moult, whereas female nematodes become sedentary and moult to the adult female. The tail end of the mature, enlarged females ruptures the root tissue, but they remain attached to the root by their heads and protruding necks, which stay inserted in the root tissue. The fertilized females become large and subspherical and go through a sequence of colour change prior to dying on roots of potato and becoming cysts. It is these females that have grown so large on the surface of the roots that one can see them attached to the roots without magnification; Peruvian farmers call them the Spanish equivalent of nits. Potato cyst nematodes complete one generation during a growing season (Morris, 1971).

Species, pathotypes and virulence groups

Morphological, developmental and sterile offspring between races with white and yellow females gave enough reasons for Stone (1973) to divide potato cyst nematodes into two species, *Heterodera rostochiensis* and *H. pallida*. Later, potato cyst nematodes have been assigned to the genus *Globodera* (Mulvey and Stone, 1976). This divison has been substantiated with biochemical methods (for an overview, see Fleming and Powers, 1998). Two-dimensional gel electrophoresis (Bossis and Mugniéry, 1993) and molecular differences detected by internal transcribed spacer-restriction fragment length polymorphism (ITS-RFLP) (Thiéry and Mugniéry, 1996), specific polymerase chain reaction (PCR) products (Shields *et al.*, 1996) and random amplified polymorphic DNA (RAPD) analysis (Folkertsma *et al.*, 1994; Thiéry *et al.*, 1997) support the distinction of *G. pallida* and *G. rostochiensis*. Species-specific primers to detect the two species in mixtures have been designed (Bulman and Marshall, 1997; Fullaondo *et al.*, 1999).

Differential plants from breeding programmes are used to separate nematode populations that possess different virulence genes; these populations are called pathotypes for each species. Pathotypes of *G. pallida* and *G. rostochiensis* were designated initially based on their ability to reproduce on differential potato clones. Canto-Saenz and Scurrah (1977) and Kort *et al.* (1977) proposed international schemes for the identification of potato cyst nematode pathotypes (Table 6.1). Franco and Gonzalez (1990) later added an additional differential clone to distinguish pathotype P_6A (Table 6.1). While for some pathotypes a gene-for-gene interaction has been identified, some pathotypes have been established against differential clones of unknown genetic constitution. Resistance against pathotypes R_1A/Ro1 and R_1B/Ro4 (South American/European scheme) and pathotype P_1A/Pa1 is conferred by the major genes H_1 (from *S. tuberosum* ssp. *andigena*) and H_2 (from *S. multidissectum*), respectively. A gene-for-gene relationship was shown for the H_1 gene and inbred lines of *G. rostochiensis* (Janssen *et al.*, 1996). In this respect, only pathotypes R_1A/Ro1, R_1B/Ro4 and P_1A/Pa1 may be regarded as true pathotypes (Trudgill, 1985). Especially *S. vernei*-derived hybrids contain polygenic resistance to *G. pallida*, and expression of this type of resistance is of quantitative nature. Stone (1985) therefore proposed to abandon the term pathotype for potato cyst nematode populations defined against differential clones with unknown (polygenic or oligogenic) resistance genes. The other pathotypes in the Kort *et al.* (1977) scheme may be composed of different proportions of individuals carrying the same type of virulence genes. For populations belonging to those pathotypes, it was suggested to use the term virulence group (Anonymous, 1985; Mugniéry *et al.*, 1989). Scurrah and Franco (1985) also suspected different virulence groups in populations of pathotype P_5A from South America.

It appears that only part of the virulence genes present in the South American *Globodera* spp. populations were introduced to Europe. More variation is evident in *Globodera* spp. populations from the Andean region, which is the area where this parasite co-evolved with its host. Populations from South America can be distinguished from European populations by molecular methods (Grenier *et al.*, 2001) and virulence characteristics (Phillips and Trudgill, 1998).

Survival and dissemination

Second stage dormant juveniles inside eggs will remain viable in cysts for over 20 years in soils under severe environmental stress (Oostenbrink, 1966). They withstand temperatures of extreme cold (–15°C) and soil desiccation for long periods. A large portion of eggs will hatch only if they are stimulated by potato root exudates, but some eggs will hatch without the presence

Table 6.1. Differential hosts used for separating pathotypes of *Globodera rostochiensis* and *G. pallida* as proposed in South American and European schemes for the identification of potato cyst nematodes.

Differential host	*Globodera rostochiensis*					*Globodera pallida*						
South American scheme	R_1A	R_1B	R_2A	R_3A		P_1A	P_1B	P_2A	P_3A	P_4A	P_5A	P_6A
European scheme	Ro1	Ro4	Ro2	Ro3	Ro5	Pa1				Pa2	Pa3	
Solanum tuberosum ssp. *tuberosum*	+	+	+	+	+	+	+	+	+	+	+	
S. tuberosum ssp. *andigena* CPC 1673	–	–	+	+	+	+	+	+	+	+	+	
S. tuberosum ssp. *andigena* 280090.10										–	–	+
Solanum kurtzianum 60.21.19	–	+	–	+	+	+	+	–	+	+	+	
Solanum vernei 58.1642/4	–	+	–	–	+	+	+	+	–	+	+	
Solanum vernei 62.33.3	–	–	–	–	+	–	+	–	–	–	+	
Solanum vernei 65.346/19	–	–	–	–	–	+				+	+	
Solanum multidissectum P55/7	+	+	+	+	+	–	–	+	+	+	+	

+, indicates a multiplication rate (final population P_f /initial population P_i) > 1; –, indicates a multiplication rate (P_f/P_i) < 1. From: Canto-Saenz and Scurrah (1977); Kort *et al.* (1977); Franco and Gonzalez (1990).

of a host plant, which contributes to the natural decline of potato cyst nematode populations in the field. The rate of decline depends on many factors which are not well defined; however, soil moisture, temperature and non-host crops exert an influence on the cyst, which acts as a protective shell around the eggs. In Europe, the rate of natural decline over 1 year is estimated at between 20 and 30% (Turner, 1996; Trudgill *et al.*, 2003). In New Zealand, *G. pallida* in volcanic soil declined by 70%, whereas in Alluvial silt-loam the decline was only 31%. In the same soil, *G. rostochiensis* declined 60%, while in organic peat *G. pallida* declined 30% and *G. rostochiensis* 57% (Marshall, 1998). In dry areas of the Andes, the decline is less, and hence the traditional rotation of 7 years from one potato crop to the next, which is now found to be the ideal time required to bring populations down to below damage thresholds. Some crops are trap crops and elicit hatching, which accelerates the decline (Main *et al.*, 1999; Scholte, 2000).

Cyst nematodes are disseminated passively as cysts over long distances by movement of infested soil mostly clinging on to seed tubers, and for short distances by farm implements. Irrigation water can also disseminate the nematodes (Jones, 1970). The status of *G. pallida* and *G. rostochiensis* as quarantine pests and strong legal restrictions for the production of seed potato in many European countries have limited further spread from Europe to new places. However, within developing countries, internal long-distance distribution of the nematodes is a major problem, especially with the increase in cultivation of potato in Asia.

Environmental factors

The conditions which favour successful potato production are also favourable for nematode multiplication and survival. They flourish in cool soil temperatures, and high soil temperatures for prolonged periods will limit development and reproduction (Jones, 1970). Soil moisture of field capacity will enhance juvenile movement, while soil nutritional status has no effect on nematodes, other than that caused by crop performance. The nematodes tolerate the same soil pH that is tolerated by the potato plants (Jones, 1970).

Other hosts

Potato cyst nematodes are host specific and have a limited host range. Aubergine, tomatoes and a few solanaceous weeds are known to harbour the nematodes, but are not considered as efficient hosts (Evans and Stone, 1977).

Disease complexes

Potato cyst nematodes not only cause wounds in roots, but also provide entry sites for other organisms. This is of particular importance to fungi and bacteria. A greater yield loss was found when the fungus *Verticillium dahliae* was in the soil (Storey and Evans, 1987). Interactions have been reported between *G. pallida* and *Pseudomonas solanacearum* (Jatala *et al.*, 1976) and between *G. pallida* and *Verticillium dahliae* (Harrison, 1971; Franco and Bendezu, 1985).

Economic importance

High losses occur in areas of intensive potato cultivation. Yield losses of as high as 80% are not uncommon in some potato-growing areas of the tropics where infestation levels are high and continuous potato cultivation is practised, as in Bolivia (Franco *et al.*, 1998), although estimated losses in Bolivia are complicated by the potatoes also being affected by *Nacobbus aberrans*. Franco estimated high infestations in 35% of the fields in the departments of Chuquisaca and La Paz and with the market values at the time giving a yearly loss to farmers of US$16 million. Such estimations are lacking for other tropical areas, which vary from year to year and from area to area; however, the dryer the environment is, the higher the losses as damaged roots will be less effective in transporting moisture and nutrients from the soil. Generally, late cultivars exhibit less yield loss (tolerance) as new roots form after invasion, which does not happen with early cultivars. The long rotations imposed by this nematode make for difficulties in areas which are reliant on this tuber for their food and/or income. These areas, however, have been the powerhouses to develop resistant varieties such as in the starch-producing areas of The Netherlands. In developing countries, farmers have overused chemicals, and this practice needs to change in the future (Canto, 1996; Van Riel and Mulder, 1998).

Management measures

Clean planting material is the best way to control potato cyst nematodes and to restrict infestations of new land. Once potato cyst nematodes are introduced to a field, it is virtually impossible to eradicate the nematodes, and use of resistant varieties is the best means of managing them. Resistant varieties can reduce potato cyst nematode populations in the field by 60–90% (Mugniéry *et al.*, 1989). Nematodes hatch and invade the roots of resistant potato plants but cannot complete their cycle. A resistant variety should therefore reduce potato cyst nematode populations more than the natural annual decline occurring in the field. Resistance in these varieties will be most likely to be controlled by a major gene and is not influenced by the environment. In the absence of resistant varieties to certain pathotypes or virulence groups of potato cyst nematodes, especially of *G. pallida*, the term partially resistant varieties was introduced (Mugniéry *et al.*, 1989). Resistance in partially resistant varieties is controlled by several genes, and expression may be influenced by environmental factors, initial population density or the population of a particular pathotype or virulence group. Ideally, a partially resistant variety will not allow multiplication of the nematode.

The above-described pathotype schemes are in use for the classification of resistant or partially resistant potato varieties to potato cyst nematode to date (Anonymous, 2003; Baarveld *et al.*, 2003) and are useful not only to breeders but also to farmers who need to know what varieties can be grown with the potato cyst nematode populations found in their farms. From a practical point of view, varieties should be used as differentials rather than unadapted clones, which are difficult to maintain and grow. However, the problem that varieties may not be available indefinitely and the fact that varieties may be locally restricted remain concerns for the usage of such clones in pathotype schemes. Resistant clones developed in Europe tend to be susceptible in the Andes, and vice versa (Mugniery *et al.*, 1989; Franco, 1994). Resistant varieties have been released in Peru, to P_4A and P_5A *G. pallida* pathotypes which, as we stated, are quite different from the European pathotypes (Llontop and Franco, 1988; Bendezu, 1997). Therefore, varieties need to be tested against nematode populations that occur at the respective locations taking into account environmental conditions. India has also released a nematode-resistant variety for *G. rostochiensis* populations that thrive in the Nilgiris and Kodai hills; the variety, SON110, is also listed as resistant to late blight (Shekharmat, 1985).

The use of resistance has an immediate selective effect on the population, and thus growing susceptible varieties or varieties which carry different resistance genes should be part of a management system to slow down the selection of virulent pathotypes present in potato cyst nematode populations. Gene H_1 found in the late 1940s in clones of *S. andigena* collections is effective against population R_1A/Ro1 of *G. rostochiensis* and has remained effective against this population. The intensive use of varieties carrying this gene has resulted in a shift in the field populations from *G. rostochiensis* to *G. pallida* in Wales and England (Minnis *et al.*, 2002). Selection pressure exerted by resistant potato varieties within *G. pallida* was reported by Beniers *et al.* (1995). After 8 years of cultivation of a resistant variety, an increased virulence of the *G. pallida* population present was observed. Fitness was not affected by an increase in virulence (Turner, 1990; Beniers *et al.*, 1995). The detection of a new pathotype P_6A after cultivation of the potato variety Maria Huanca, resistant to pathotypes P_4A and P_5A, was reported from South America (Franco *et al.*, 1998). Artificial selection for virulence within populations of *G. pallida* on resistant potato clones was demonstrated by several authors (Turner, 1990; Pastrik *et al.*, 1995; Schouten and Beniers, 1997) and it was also shown that even susceptible potato varieties are not selectively neutral (Phillips *et al.*, 1998). This further demonstrates the variability between and within *G. pallida* populations and the difficulties in managing this species of potato cyst nematode. However, it should be noted that it takes several, repeated crops of potato for such a selection to occur.

Long-term rotations of up to 8 years may be needed to reduce nematode populations to below damaging densities (Evans and Haydock, 2000). All non-host crops reduce potato cyst nematode populations, depending on agro-ecological factors. Much work has been done on correct rotations which are location specific. Canto (1996) found in Cajamarca, Peru, higher yields after maize and barley. Also, in the higher areas of both Peru and Bolivia, lupins and faba beans have been reported as excellent cleaning crops grown in rotation and increasing yields of potatoes in the following seasons. Sikora (1984) developed a number of rotations for control of potato cyst nematode in the upland tropical growing areas of the Philippines. A combination of resistant, susceptible and early maturing potato cultivars integrated with non-host crops was used to suppress population densities. In addition, rotations were designed to take advantage of nematode diapause to escape damage and to trap late penetrating segments of the population.

Trap crops, which induce hatching but prevent the reproduction of the nematodes, are new and important tools to shorten rotations. In Europe, *S. sisymbriifolium* has been proposed as such a trap crop as nematode populations can be reduced by 70–80% (Scholte and Vos, 2000). However, this is not a commercial crop. In Bolivia, certain varieties of barley and oca (*Oxalis tuberosum*) actually produce greater hatch. Although these options are not yet being used by farmers, they could prove very useful (Franco *et al.*, 1999).

Nematicides are known to reduce nematode populations at low densities and give early crop protection (Whitehead, 1975; Evans and Haydock, 2000; Trudgill *et al.*, 2003), and they are still on the recommendation list of many countries although it is evident that at higher populations, they may not prevent multiplication of the nematodes (Trudgill *et al.*, 2003). As the nematicide breaks down, a rebound effect of the nematode population can be observed and populations may even increase. Granular nematicides are also less effective against *G. pallida* (Evans and Haydock, 2000). Consumer and environmental concerns are making farmers look at non-chemical alternatives; however, in many areas, chemical control is still a key control measure. Research in Ecuador has shown that it is imperative that correct application procedures are followed in terms of protective equipment as severe health problems have been documented of farmers or farm workers who do not follow label instructions (Yangen *et al.*, 2000).

Utilization of these various measures in an integrated management programme will help in keeping the populations below the damage threshold and reduce dissemination, as well as the emergence of new pathotypes.

Diagnosis

Early detection and identification is the first step to take proper action against potato cyst nematodes. The way to find out if symptoms of poor growth are caused by *Globodera* is to carefully uproot plants with as much root as possible and to examine these. White or yellow round females clinging to roots around flowering time is the best method to diagnose the presence of potato cyst nematode in an area (Fig. 6.2, Plate 4B). This is a very good method to map out areas of infestation during the growing season and has been found to be more efficient than soil sampling (Wood *et al.*, 1983).

Soil analysis for extraction of cysts will also provide an excellent means of diagnosis (Haydock and Perry, 1998). It is important, however, to note that it can take several years from the time of introduction until the nematodes become established and reach the detection level (Trudgill *et al.*, 2003). Soil sampling methods, especially for statutory soil sampling prior to planting of seed potato, should be sensitive enough to detect potato cyst nematodes even before visible symptoms could be observed. The detection level depends mainly on the number of cores, the sampling grid and the amount of soil taken per unit area (Been and Schomaker, 2000). Surveys for the detection of potato cyst nematodes are an important instrument to establish the presence of species and pathotypes/virulence groups in certain areas.

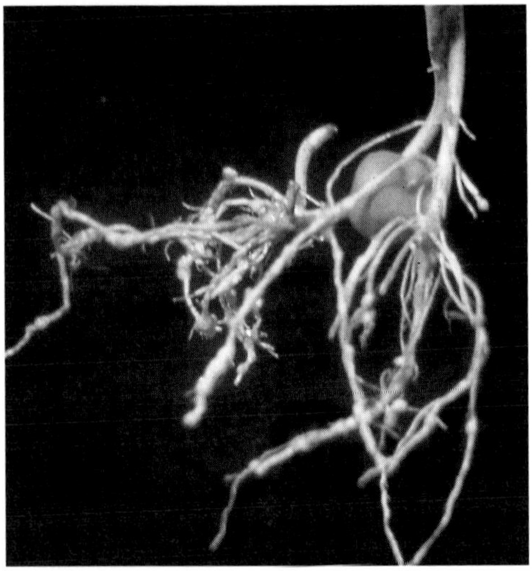

Fig. 6.3. Galls on the roots of potato caused by *Meloidogyne incognita* in Bolivia. (Photo: J. Bridge.)

Meloidogyne

Root knot nematodes are cosmopolitan in distribution, attacking almost all major crops and many weed species. Of the 80 species described, only seven have been associated with potato. Five species of *Meloidogyne* attacking potato are considered of global importance; *M. incognita* is the most widely distributed species in the tropics followed by *M. javanica* and *M. arenaria*, while *M. hapla*, *M. chitwoodi*, *M. fallax*, Karsen 1996 and *M. thamesi* are found principally in the cooler temperate regions (Taylor and Sasser, 1978; Brown and Mojtajedi, 2004).

Symptoms of damage

There are no specific above-ground symptoms. Infected plants exhibit stunting, yellowing, and tend to wilt under moisture stress. Infected roots will have galls or knots of various sizes and shapes (Fig. 6.3). Galling incidence and size are dependent upon nematode density, and the nematode species. *M. hapla* and *M. chitwoodi* galls are usually smaller than those caused by other species and have extensive lateral root formation. *M. incognita* have large and distinctive root galls. Infected tubers exhibit characteristic symptoms. Under favourable environmental conditions, tubers of all sizes can become infected (Jatala, 1975). Tubers infected with *M. incognita* have galls which give a warty appearance or can become deformed on the surface (Fig. 6.4, Plate 4C), *M. chitwoodi* causes pimple-like galls on tubers, and *M. hapla* does not cause distinct galled tubers, but can cause general swelling at high levels of infection. The depth of penetration of tubers varies but, depending on the tuber size, nematode females are usually found 1–2 mm below the skin feeding on vascular tissue (Jatala, 1975). All species produce necrotic spots in the region between tuber surface and the vascular ring (Plate 4D). This is tuber tissue reaction to the deposition of eggs and the gelatinous matrix.

Biology

The biology and life cycle of *Meloidogyne* species on potatoes follow the general patterns described for this genus (Chapter 2).

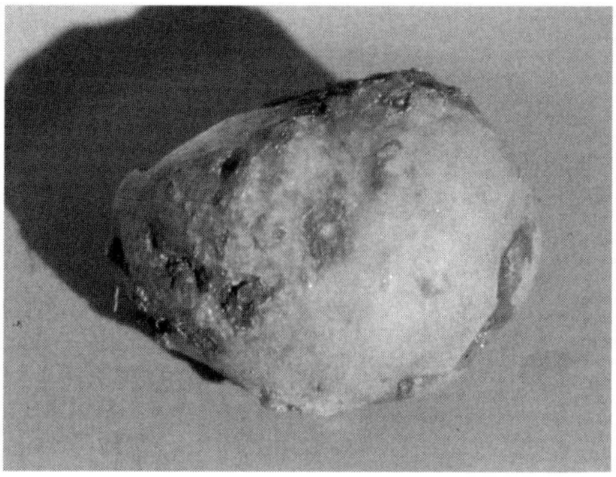

Fig. 6.4. Swellings on the surface of a potato tuber caused by *Meloidogyne incognita*. (Photo: J. Bridge.)

Both roots and tubers are infected; however, the first generation occurs mainly on the root systems, while the succeeding generations attack tubers. There are up to five generations on the susceptible host under favourable environmental conditions, but *M. incognita* can endure for up to 12 generations (Santos, 2001).

Races

There are several races of *Meloidogyne* species (see Chapter 9). All races of these nematodes attack potatoes in varying degrees.

Survival and dissemination

Since *Meloidogyne* species attack a large number of plant species, their population can be maintained on weeds and volunteer crops. However, in the absence of a suitable host, their populations are drastically reduced. Cold temperature reduces root knot populations and they overwinter usually in the form of eggs, although the ability of juveniles to go through anhydrobiosis may contribute to the survival of some *Meloidogyne* species. Infected tubers, plant parts and planting material, as well as movement of infested soil by farm machinery, and irrigation water are the main avenues of disseminating *Meloidogyne* species. Infected weeds and volunteer crops can also be sources of inoculum.

Environmental factors

M. incognita, *M. javanica* and *M. arenaria* develop better in higher temperatures and cannot withstand cool temperatures. Hence, they are of great economic importance in the tropics and warm temperate regions of the world. *M. hapla*, *M. chitwoodi* and *M. fallax*, on the other hand, are cool temperature nematodes and have an optimum temperature of 20°C (Taylor and Sasser, 1978). They are basically distributed in the northern part of North America and in Europe, but *M. chitwoodi* is also found in The Netherlands and Mexico (C. Sosa-Moss, personal communication).

Other hosts

Meloidogyne species have a wide host range and attack many agriculturally important crops and weeds. Most of the tuber-bearing *Solanum* species are susceptible to *Meloidogyne* species.

Disease complexes

Meloidogyne species often interact with other pathogenic organisms in development of disease complexes. Perhaps the most important interaction of these nematodes on potatoes is their association with *Ralstonia solanacearum* (Jatala *et al.*, 1975). Resistance of potatoes to bacterial wilt is broken in the presence of *M. incognita* (Jatala *et al.*, 1975; Jatala and Martin, 1979). Other interactions include their association with the *Verticillium* wilt organism and *Rhizoctonia solani*.

Economic importance

Although losses vary depending upon the cultivar and environmental conditions, they can reach 25% or more (Mai *et al.*, 1981). Loss consists of direct damage to the plant, as well as reduction in tuber quality. Infected tubers are economically undesirable and can serve as an inoculum source (Jatala, 1975). The finding of large commercial potato-growing areas in the Pacific Northwest USA and in The Netherlands, infected with *M. hapla*, *M. fallax* and *M. chitwoodi*, as well as the expansion of potato cultivation in warmer areas has increased the interest in controlling damage caused by root knot nematodes (Brown and Mojtajedi, 2004). A survey in Indonesia found that 97% of the fields destined for potato cultivation were heavily infected with root knot nematode (Suri and Jayasinghe, 2002).

Management measures

Since *Meloidogyne* species deposit their eggs in a gelatinous matrix (usually outside of the root surface) that is relatively unprotected, chemical control has been most successful in reducing their populations (Taylor and Sasser, 1978). The use of resistant cultivars and rotation with non-host crops are probably the most economical means for controlling *Meloidogyne* species. Resistant potato material with adaptation to warm temperatures of the tropics has not yet been developed. However, advanced clones arising from careful breeding of resistant *Solanum sparsipilum* in Tunisia show no galling, while var. Desirée has heavy galling (Berthou *et al.*, 1996). Similarly Brown *et al.* (1995, 1999) reported resistance from wild potatoes (*S. bulbocastanum* and *S. hougasii*) and their incorporation into cultivated germplasm. Utilization of these sources constitutes the most practical means of controlling root knot nematodes on potatoes.

Crop rotation with resistant crops will keep populations down, and there are several resistant tomato varieties as well as sweet potato varieties resistant to *M. incognita*.

Diagnosis

Sampling and extraction procedures are presented in Chapter 3. Additional methods of diagnosis include direct observation of roots and tubers and the use of a galling index (see Chapter 9). Staining the tuber and root tissues may aid in detection of nematodes. Bioassays with susceptible tomatoes and checking for root galling have proven reliable methods to determine if an area is infected with *Meloidogyne* before planting, although this can be a lengthy procedure as galling of the roots may not appear for 20–40 days. The simplest method is to take soil from the field and plant to seedling plants in pots.

Nacobbus aberrans

The false root knot nematode, *Nacobbus aberrans*, has been found associated with numerous crops and native plants in temperate and subtropical regions of North and South America. In North America, it is reported in the USA and Mexico. In the USA, it attacks sugarbeet and other field vegetable and weed hosts, but not potato (Thorne and Schuster, 1956; Inserra, 1983).

In Mexico, it causes economic loss in tomato, bean and chilli peppers. Nematodes from Hidalgo Morelos and Mexico are able to attack both sugarbeet and potato, thus differing from the race in the USA. In South America, *N. aberrans* is a problem in Argentina, northern Chile, Peru, Bolivia and Ecuador (Mai *et al.*, 1981; Manzanilla-Lopez *et al.*, 2002). With the exception of Ecuador, where it has not been reported on potatoes, *Nacobbus* is a major pest of potatoes and other vegetable field crops in these South American countries (Franco, 1994). Glasshouse populations have been reported from England (Franklin, 1959) and The Netherlands, probably from infected material introduced from the American continent. Although there is a report on the occurrence of this nematode in India, its presence cannot be confirmed (Sher, 1970), and there is also an unconfirmed report from China (Yin and Feng, 1981). It is considered to be the most important constraint to potato production in southern Peru and Bolivia (Mai *et al.*, 1981). The wide distribution of *N. aberrans* in the Americas has probably resulted from its host range, which includes many weeds, and from the passive transport of the nematode with propagative plant material such as seed potatoes and other tuber-forming hosts (Jatala and Scurrah, 1975).

Symptoms

Patches of poor growth are a common feature of affected crops. Above-ground symptoms include stunting, chlorotic leaves with rolled margins, and wilting. Root galls are produced by the false root knot nematodes, and normally the infected plants have few or no small feeder roots. Galls caused by *N. aberrans* can be mistaken for those caused by *Meloidogyne* species, but they differ in usually being more rounded and forming on the roots in a rosary-bead-like fashion (Fig. 6.5, Plate 4E) and, hence, the common name of rosary-bead nematode or 'Rosario' is given to *N. aberrans*. Potato root galls generally contain only one female. The number of galls increases when tubers begin to form. Although it does not cause easily recognizable symptoms on potato tubers, the tubers have spongy tissue in the lenticels. The large cells have an inflated appearance which later become flatter and suberized. *N. aberrans* usually penetrates the tubers to a depth of 1–2 mm below the skin (Mai *et al.*, 1981; Manzanilla-López *et al.*, 2002).

Biology and life cycle

N. aberrans has a combination of pratylenchid and heteroderid strategies for invasion, feeding and establishment in the host, reflected in the migratory behaviour of the juveniles, vermiform immature females and males, and the sedentary endoparasitic habit of the mature females. The first moult is within the egg; second stage juveniles emerge and invade small feeder roots. This stage feeds actively and supports the next two moults which hardly feed and are sluggish. They will then undergo an additional two moults before leaving the root system as pre-adults (Mai *et al.*, 1981; Baldwin and Cap, 1998; Manzanilla-López *et al.*, 2002). Under certain conditions, they remain in the root system in a quiescent stage for some time. The quiescent or dormancy stage can be reduced by drying or cooling factors. Once the pre-adults become active, they invade the root system and produce small necrotic lesions prior to gall formation. Production of necrotic lesions by juvenile invasion is not as frequent as those caused by pre-adults. A portion of those that leave the root system

Fig. 6.5. Small, round, bead-like galls on roots of potato caused by *Nacobbus aberrans*. (Photo: J. Bridge.)

become males. After the establishment of pre-adult females and gall formation, the nematodes develop to maturity, depositing a portion of their eggs in a gelatinous matrix on the root surface. Females often retain a portion of their eggs in their bodies in addition to depositing them in a gelatinous matrix. Pre-adults and juveniles also attack tubers, penetrating approximately 1–2 mm below the skin surface. There is no tuber galling or deformation associated with nematode infection. Depending upon the host, temperature and race of the nematode, generation time is usually between 25 and 30 days (Mai *et al.*, 1981). The presence of overlapping generations was first observed by Clark (1967). Under field conditions, there are three or four peaks of motile stages (Manzanilla-López *et al.*, 1998).

Races

N. aberrans can be separated into bean, potato and sugarbeet groups. The populations of each group have distinct host preferences and do not reproduce on graminaceous species or on leguminous species of the genera *Medicago* and *Lupinus*. Temperature requirements of *Nacobbus* are grounds for considering geographical races. For example, in the Andes, damage occurs in the high, cool Andes (15–18°C) and also in warmer temperatures in the subtropical lowlands of Argentina and Ecuador (20–26°C). Resistant varieties of potato have been found at high altitudes in La Paz, Bolivia which are susceptible at lower altitudes in Cochabamba (Franco *et al.*, 1998).

Survival and dissemination

False root knot nematodes are resistant to low temperatures, withstanding temperatures of –15°C. They can also survive in desiccated soil, a characteristic which makes this nematode quite unique in its biology (Jatala and Kaltenbach, 1979). Exposure of infested soil to cool temperatures for 2 weeks prior to planting with potatoes enhances infection and the severity of the nematode damage.

N. aberrans has a wide host range which includes at least 84 plant species, and many common weeds are good hosts (Manzanilla-López *et al.*, 1999). The weed *Aspergula arvensis* has been linked with the rapid spread of the nematode throughout Bolivia (Doucet *et al.*, 1994). Planting infected tubers, as well as movement of infested soil that adheres to potatoes and farm implements, is the major means of dissemination of this nematode. An examination of seed tubers bought in seed markets in Bolivia showed that 86% of the seed tubers harboured significant levels of *Nacobbus*, an important means of dissemination (Rojas *et al.*, 1997).

Environmental factors

False root knot nematodes have a wide temperature adaptability, surviving and reproducing most rapidly at a temperature range of 20–26°C. However, in the Andes, they are associated with potatoes at temperatures of 15–18°C and are not limited by soil types (Mai *et al.*, 1981). Periods of soil cooling and desiccation aid in revival of nematode activity during spring, causing subsequent root infection (Jatala and Kaltenbach, 1979). Thus *Nacobbus* is well adapted to survive extended periods of dry and cold, and this, added to its ability to colonize many weeds, makes it one of the most difficult nematodes to manage effectively.

Disease complexes

N. aberrans is often associated with *Meloidogyne* spp. and *Globodera* spp., when it seems it plays the role of a competitor, sometimes causing synergistic symptoms in plants. A relationship has been noted with *Synchitrium endobioticum* (Montalvo, 1993) and it often occurs together with *Spongospora subterranea* (Mai *et al.*, 1981).

Economic importance

N. aberrans plays an important role in reducing the yield of potatoes in Bolivia,

Argentina and Peru. A case study in Bolivia by Ramos *et al.* (1998) gives the total infected area of 131,330 ha with an estimated loss of US$51,775,119. The most severe losses occurred in the Department of Cochabamba due to higher nematode populations, thus showing the seriousness of the yield loss caused by this nematode as it affects the economies of potato-farming families in Bolivia.

Management measures

Immersion of seed potato in hot water and Clorox bleach destroys inoculum and is a recommended first step if the source of seed is questionable (Franco *et al.*, 1993). Nematicide tests focused on reducing galling and increasing yield found that, at normal rates, aldicarb, phenamiphos, carbofuran, oxamyl and CGA-12223 did not reduce galling. However, higher dosages (5 kg a.i./ha) reduced the number of galls from 552/plant down to 114/plant (Otazu *et al.*, 1985). Standard applications of commercial nematicide formulations do not reduce populations of the nematode (Manzanilla-López *et al.*, 2002).

Because of its extensive host range, control by crop rotations is difficult, although members of the Gramineae and most of the Leguminoseae are non-hosts (Mai *et al.*, 1981). A range of crops and varieties have been tested as potential trap crops for *N. aberrans* and, in general, the oats and ocas behaved as non-hosts, but some 46% of the oats, 100% of the quinoas and 8% of the isaños (mashuas) allowed invasion but not reproduction, and therefore are considered suitable as trap-crop plants (Main *et al.*, 1999).

Screening and breeding for resistance started in the 1970s. This has led to the cleaning of a native potato variety Gendarme identified as resistant which shows no galling to *Nacobbus*; however, some biological races of *Nacobbus* have been found to produce galling on this variety (Oros *et al.*, 1996).

Using various organic amendments increases yields but often also galling. However, chicken manure can also reduce galling. Other amendments such as *Brassica oleracea* reduced galling, but extremely large quantities are required (35 and 52 t/ha) (Manzanilla-López *et al.*, 2002).

Diagnosis

Sampling and extraction of *N. aberrans* from soils and roots are similar to those described for *Meloidogyne* spp. Diagnosis of symptoms on roots can be problematic and they often are mistaken for those caused by *Meloidogyne* spp. However, *N. aberrans* galls are characteristically formed on the lateral part of the roots, and the galls often occur in a bead-like fashion (Fig. 6.5, Plate 4E) with or without the presence of small root extensions from galls, as with *M. hapla*.

According to Montalvo *et al.* (1992), the best method to detect *Nacobbus* in fields is a bioassay consisting of growing a potato plant in moist soil and maintained in a closed transparent container (e.g. plastic bag) kept at 25°C in darkness, which can be assessed after 30 or 35 days. The development of galls on the roots is an indicator of the level of infection; this method also works for potato cyst nematode. Other methods tend to underestimate potential populations. Several reports have noted the potential danger that even initial low levels of infestation can lead to crop loss.

Ditylenchus

Potato rot or tuber nematode, *Ditylenchus destructor*, and potato stem nematode, *D. dipsaci*, have been reported from temperate climates, particularly eastern and western Europe. They also occur in North America and certain parts of South America (Mai *et al.*, 1981). However, the lack of economic damage or recognition of this pest from the potato fields in the tropics and subtropics is evident by the lack of extensive literature citations. Potato rot nematode occurs in many potato-producing countries, but the damage is only apparent in temperate zones.

Symptoms of damage

D. dipsaci is mainly a parasite of the foliage where it attacks leaves and petioles, causing shortened, thickened and malformed foliage. This nematode also injures tubers, producing conical pits often accompanied by skin splitting (Mai *et al.*, 1981).

D. destructor mainly damages tubers. The earliest below-ground symptoms are small, white, chalky or light-coloured spots just below the surface of the tuber. The symptoms become evident in the advanced stages of development when the tuber surface is marked by sunken, dark-coloured pits or skin cracks. Subsurface tissue will develop a brown, matted, wool-like appearance. As the affected areas coalesce, tissues darken and are invaded by bacteria and fungi.

The tuber skin becomes paper thin and cracks as the underlying tissue dries and shrinks. Under certain environmental conditions, bacterial wet rot may cause complete destruction (Mai *et al.*, 1981).

Biology

D. destructor enters small potato tubers through lenticels on the skin near eyes. Nematodes at first exist singly or in small numbers in the tissue just beneath the skin of the tubers, and small white lesions are present during early and mid-season tuber formation. More tuber tissue becomes involved as populations increase. The nematode continues to live and develop in harvested tubers (Winslow, 1978b; Mai *et al.*, 1981).

Survival, dissemination and host range

D. destructor has a wide host range, can survive on weeds, and on a wide range of soil-inhabiting fungi (Winslow, 1978b; Jensen *et al.*, 1979). It can also survive on infected tubers left in the field. Dissemination occurs by introduction of infected tubers and in soil adhering to seed pieces (Mai *et al.*, 1981). Irrigation water and cultivation by infested farm tools and machinery are other sources of inoculum dissemination. The nematode will survive in soils at temperatures as low as –28°C. However, major infestation will occur at 15–20°C and a rather high relative humidity of 90–100%. Apparently, high relative humidity is a very important factor in the establishment of the nematode. The nematode cannot survive under drought or low (< 40%) relative humidity (Winslow and Willis, 1972; Winslow, 1978b; Jensen *et al.*, 1979).

Economic importance and control

High yield losses occur in the areas where climatological conditions favour establishment of the potato rot nematodes. The effect of nematodes will manifest itself at harvest or storage when infected tubers will rot. The use of healthy tubers and soil fumigation are the most effective measures in controlling the nematodes. Rotation of potatoes with sugarbeet and other non-host crops can reduce nematode populations (Winslow, 1978b). Various cultural control programmes have contributed successfully to the management of these nematodes (Winslow and Willis, 1972; Winslow, 1978b; Jensen *et al.*, 1979).

Pratylenchus

Root lesion nematodes, *Pratylenchus* spp., are known to damage potatoes in the temperate, tropical and subtropical regions. *Pratylenchus crenatus*, *P. neglectus*, *P. thornei*, *P. scribneri*, *P. brachyurus*, *P. andinus*, *P. penetrans*, *P. coffeae*, *P. vulnus* and *P. flakkensis* are the most important species associated with potatoes (Jensen *et al.*, 1979; Mai *et al.*, 1981). High populations of lesion nematodes cause areas of poor growth; plants are less vigorous, turn yellow and cease to grow. Damage is often caused by direct feeding, and, usually, only cortical tissues are affected. Large nematode populations cause extensive lesion formation and cortex destruction of unsuberized feeder roots (Mai *et al.*, 1981).

Tubers are often attacked and small lesions are formed on the surface. Infected tubers are sources of nematode inoculum and aid in the survival of the nematodes. *Pratylenchus* spp. have a wide host range and are distributed extensively in the tropics, subtropics and temperate regions. Because of their extensive host range, crop rotations are not normally practical and should be developed with caution. These nematodes interact with a series of pathogenic organisms in development of disease complexes (Jensen *et al.*, 1979; Mai *et al.*, 1981). Soil fumigation and utilization of resistant potato clones have been identified (Dunn, 1973). Hot water treatment of infected tubers at 50°C for 45–60 min may also be an aid to reducing nematode spread (Koen, 1969; Yokoo and Matsunobu, 1975).

Other nematodes of potatoes

Although many other nematodes are reported to cause serious damage to potatoes, few are of global concern. Other important nematodes of potatoes in the tropics and subtropics are *Atalodera* (=*Thecavermiculatus*) *andina*, *Trichodorus* and *Paratrichodorus* spp. *T. andina* is an important nematode of potatoes in some Andean regions of Peru (Jatala, 1989). However, the extent of distribution and economic damage of this nematode to potatoes is not well documented. *Trichodorus* and *Paratrichodorus* spp. are of importance because of their involvement in the dissemination of potato viruses (Jensen *et al.*, 1979). In addition to their role in the transmission of viruses, they can also cause severe damage to the root system, leading to stunting and early senescense of the potato plant (Jensen *et al.*, 1979).

Other nematodes, such as *Belonolaimus longicaudatus*, *Radopholus similis* and *Rotylenchulus reniformis*, are also known to be of importance to potato production (Winslow, 1978b; Jensen *et al.*, 1979). However, they are generally not of any major global consequence to potato production.

Sweet Potato

Sweet potato, *Ipomoea batatas* (L) Lam., a native of tropical America, is more widely grown in developing countries than any other root crop. It is grown in tropical, subtropical and warmer temperate zones. Of all the world's root and tuber crops, sweet potato is second only to solanum potato in importance. Asia, especially China and Japan, accounts for the largest portion of sweet potato cultivation in the world (Chandra, 1994).

Taxonomically the *I. batatas* complex includes *I. trifida*, *I. littoralis* and *I. leucantha* within a single group on anatomical grounds. Although there are several other *Ipomoea* species consisting of an anatomically differentiated group of genomes comprised of diploids and tetraploids, their values are primarily for breeding research (Yen, 1982).

Sweet potato is a perennial herb with vine-like habits and variations in leaf form. The storage roots become swollen as the plant matures. It is vegetatively propagated and can be grown in relatively infertile soils with few inputs and can withstand periods of irregular drought and rainfall (Horton *et al.*, 1984). Storage roots can be left in the ground after maturity but, once harvested, they generally have a short storage life. Sweet potato ranks fourth and sixth on the list of dry matter production per hectare and edible energy production per hectare per day, respectively.

Nematodes of Sweet Potato

Although a large number of nematode species are associated with sweet potatoes, only a few are of economic concern. The most important nematode genera attacking sweet potatoes are *Meloidogyne* spp., *Rotylenchulus reniformis*, *Pratylenchus* spp. and *Ditylenchus* spp.

Meloidogyne

Root knot nematodes, *Meloidogyne* spp., are widely distributed in the tropics, subtropics and warmer temperate regions of the world. *M. incognita* is the most important species of the genus attacking sweet potatoes and has a wide global distribution. *M. arenaria*, *M. hapla* and *M. javanica* are also found infecting sweet potato, although it is a non-host to certain isolates of *M. javanica*. The distribution of *M. hapla* is limited to the cooler, temperate growing regions. In Japan, populations of *M. incognita* are the most pathogenic to sweet potato, but *M. arenaria* and *M. hapla* can also infect and reproduce on different cultivars; however, *M. javanica* does not complete its life cycle on the crop (Sano and Iwahori, 2002).

Symptoms

Meloidogyne species attack both roots (Fig. 6.6) and storage roots (Plate 4F), causing swellings or galls of different shapes, but they fail to induce the prominent galls on sweet potato as they do on many other crops. If the initial nematode population is high, they cause a pruning effect which can be overcome by vigorous growth and excessive lateral root production (Jatala, 1989). They also cause root tip necrosis in hypersensitive and resistant plants, while causing a somewhat general root necrosis in roots of susceptible cultivars. Physiological stresses associated with nematode parasitism can induce longitudinal cracking during development and swelling of the storage roots (Clark and Moyer, 1988). This root cracking can allow the establishment of secondary organisms and subsequent rotting (Lawrence *et al.*, 1986). Females can be observed on sliced storage roots and are usually associated with brown, necrotic cells around them (Plate 4F). Infected plants exhibit general symptoms of damage associated with poor root growth, such as yellowing, stunting and the tendency to wilt during the warmer periods of the day.

Biology

The life cycle of *M. incognita* and other root knot species on sweet potato follows the general pattern specific to this genus (Chapter 2). Feeder and storage roots are attacked at the same rate. Depth of penetration is dependent upon the time of penetration of storage roots. With a life cycle of 30–40 days, *M. incognita* can complete several generations during the growing season of the crop dependent upon the prevailing temperature (Jatala and Russell, 1972). *Meloidogyne* species do well in light, friable, sandy loam soil which constitute the major portion of the world's sweet potato-growing areas.

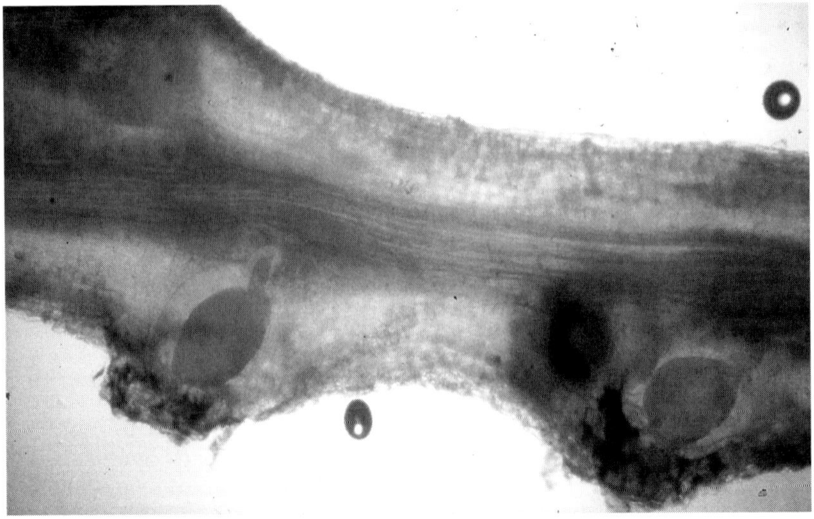

Fig. 6.6. Females of *Meloidogyne incognita* within sweet potato roots. (Photo: J. Bridge.)

Survival and dissemination

Meloidogyne juveniles and/or eggs survive in storage roots and can be disseminated in root, but not stem, propagative material. Irrigation water and unclean farm tools and machinery can aid dissemination of the nematodes. Nematodes can survive on many alternative weed hosts.

Economic importance

Meloidogyne species can reduce plant growth and yield. In South Africa, sweet potato cv. Blesbok is susceptible to both *M. incognita* and *M. javanica*, and the nematodes can cause over 11% decrease in the marketable yield due to a reduction in the storage roots produced (Kistner *et al.*, 1993). The observable damage to roots in the form of deep cracks greatly reduces the marketable value of sweet potato tubers. Tuber damage is of importance in assessing economic losses, and the nematode effects on quality and grade are of particular importance in developed countries (CIP, 1992; Johnson *et al.*, 1992; Sharma *et al.*, 1997). The yields of two *Meloidogyne*-resistant cultivars in Brazil ranged from 24 to 40 t/ha compared with 9 t/ha for other cultivars (da Silveira *et al.*, 1997). In addition, infected storage roots crack easily and the cracks provide the avenue for penetration and establishment of many secondary and/or pathogenic organisms affecting the quality of storage roots. An important economic factor in *Meloidogyne* infestation is its interaction with other pathogens in the establishment of disease complexes.

Management measures

Crop rotation and intercropping for reducing nematode populations are difficult with *Meloidogyne* species because of their extensive host range. A crop highly susceptible to root knot nematodes should be avoided in the cropping system. The antagonistic plants *Crotalaria juncea* and *C. spectabilis* have been tested against *M. incognita* (and *Pratylenchus coffeae*) in sweet potato fields in Japan, with some success (Torigoe, 1996).

Since sweet potato cultivation is generally conducted on a low cash input, the application of chemical control measures is usually cost prohibitive. Nevertheless, many organophosphates and carbamates, such as nemacur and aldicarb, are effective in controlling *Meloidogyne* species (Clark *et al.*, 1980; Gapasin, 1981). In the USA, pre-plant nematicide treatments of soil infested with *M. incognita* both doubled the yield of marketable sweet potato tubers and reduced the percentage of cracked tubers by over 40% (Hall *et al.*, 1988). Application of 1,3-dichloropropene at 56.12 l/ha, Vorlex (dichloropropene-dichloropropane) at 5.0 gal/ha or metam at 20.0 gal/ha to sweet potato fields in North Carolina significantly reduced root damage due to *M. incognita* (Averre *et al.*, 1993).

Resistance

Resistance at different levels has been found in numerous sweet potato cultivars in Japan, South America and the USA, although resistance can vary with different populations of the nematode. Cultivars Hernadez, Excel and Jewel are resistant to North Carolina populations of *M. incognita* race 3 and to *M. javanica*. These three cultivars plus two others, Beauregard and Porto Rico, are also resistant to *M. arenaria* race 2 (Cervantes-Flores *et al.*, 2002a). The virulence of nematode populations of the same host race varied among and within sweet potato genotypes although several clones showed resistance to all North Carolina *Meloidogyne* populations tested, suggesting that different genes could be involved in the resistance of sweet potato to root knot nematodes (Cervantes-Flores *et al.*, 2002b).

In Brazil, Rio Doce, Brazlandia Roxa and Paulistinha clones (de Freitas *et al.*, 2001) and cultivars Canuana and Palmas (da Silveira *et al.*, 1997) have shown high degrees of resistance to *M. incognita* and *M. javanica*. The cultivars Supresa, Arroba, Pira 1 and Coquino plus 21 clones have also shown degrees of resistance to *M. incognita* (races 1, 2 and 3) and *M. javanica* in Brazil (Peixoto *et al.*, 1998). In contrast, all of 27 selections of sweet potato from the Root and Tuber Germplasm Collection, University of Venezuela were susceptible to *M. incognita*, and only one, UVC-8, showed resistance (Montes *et al.*, 1998); three selections, UCV-2, UCV-7 and Catemaco, in Venezuela showed tolerance to *M. incognita* (Crozzoli *et al.*, 1994).

In India, three high yielding cultivars, Sree Vardhini, Sree Nandini and H268, and two short duration cultivars, Sree Rethna and Sree Bhadra, are highly resistant to the local populations of *M. incognita* (Mohandas and Palniswami, 1990; Mohandas *et al.*, 1996; Vimala and Rajendran, 1998). In Japan, many cultivars have been identified as resistant or slightly resistant to different populations of *M. incognita* (Sano *et al.*, 2002; Katayama *et al.*, 2003; Tamiya *et al.*, 2003), and control of root knot by crop rotation with resistant cultivars is recommended (Fukunaga and Iwahori, 2002). Improved sweet potato cultivars bred for the arid and saline soils of northern Chile that have resistance to local *Meloidogyne* spp. are Comensal, Costanero, Tacna, Yarada and ST87.030 (Gallo *et al.*, 2001). In China, the cultivars Lanshu 88, Xushu 18, Shenglibaiho, Lanruishao and 52-45 are resistant to *M. incognita* (Chen, 1993). Other cultivars carrying various degrees of resistance to *Meloidogyne* spp., particularly to *M. incognita*, are Heartogold, Norin no. 2, Norin no. 5, Nemagold, Ruby, Taihaku and Tirivan (Sasser and Kirby, 1979).

Biological control

Pasteuria penetrans, an obligate bacterial parasite of nematodes, has been used in Japan to control *M. incognita* on sweet potato. Soils treated with 2×10^{10} endospores/m^2 have lower populations of the nematode, and marketable yield is significantly higher (Tateishi, 1998, 1999). Over a long period of consecutive cropping, soils with *P. penetrans* had significantly fewer *M. incognita* juveniles present in the seventh and eighth cropping cycles and increased marketable yields of tubers (Tateishi and Sano, 2001).

Treatment of tubers

Hot water treatment of 65 min at 47°C (Burk and Tennyson, 1941) and hot air treatment of 4–8 h at 50°C (Martin, 1962) is effective in eliminating *Meloidogyne* from root propagative material. Similarly, chemical dip treatment of the propagation material in a solution of oxamyl or side dressing with nematicides at the time of planting will allow the establishment of the crop by providing early protection against nematodes (Rodriguez-Kabana *et al.*, 1978).

Diagnosis

Damage to roots can be assessed by rating the number of galls on roots, taking into account the root necrosis as they relate to the total root mass. The degree of storage root infection can be determined by slicing the roots at 0.5 cm thickness and observing the tissue for the presence of females. Staining the tissue will aid in detection of females with egg masses.

Rotylenchulus reniformis

Rotylenchulus reniformis, the reniform nematode, has been reported in most of the southeastern USA and many other tropical and subtropical areas of the world where sweet potatoes are grown (Martin, 1960; Birchfield and Martin, 1965; Fassuliot and Rau, 1967; Bird *et al.*, 1973; Brathwaite, 1977a,b; Gapasin and Valdez, 1979). It is commonly found on sweet potato in Japan and has been isolated from 60 to over 80% of sweet potato fields around Kyushu; it is considered to be a damaging pest of the crop in the area (Iwahori *et al.*, 2000, 2001). Infestations of fields by *R. reniformis* and *M. incognita* in Papua New Guinea are considered to be part of the reason for sweet potato yield decline (Hartemink *et al.*, 2000). *R. reniformis* is the most predominant nematode on sweet potato in Kerala, India (Ramakrishnan and Mohandas, 1996) and it commonly occurs in mixed populations with other species on sweet potato in Egypt (Kassab and Taha, 1990).

Infestation by *R. reniformis* may cause cracking of storage roots (Clark and Wright, 1983). The induced cracks are deep and the exposed surfaces are healed over by formation of callus and periderm. No juveniles and adults are found within the cracked sweet potatoes. The population level necessary for cracking may be very low and is probably less than that for yield reduction. Selection P-104 is reported to be resistant to cracking (Clark and Wright, 1983). *R. reniformis* populations in the USA restricted storage root growth of a susceptible cultivar but not shoot growth. Root necrosis occurs and becomes more pronounced as the numbers of the nematode increase (Walters and Barker, 1994).

R. reniformis may also interact with other pathogenic organisms, such as *Fusarium* spp., in development of disease complexes. Thomas and Clark (1983a) showed that *R. reniformis* and *M. incognita* were capable of inhibiting each other and becoming the predominant species in a sweet potato field. Glasshouse studies, however, showed that *R. reniformis* was inhibited and *M. incognita* became predominant in concomitant infection of sweet potato (Thomas and Clark, 1983b). Data on control of these nematodes on sweet potatoes are rather limited. Birchfield and Martin (1968) demonstrated that, under field conditions, reniform nematodes can be controlled by in-row treatment with some nematicides in the halogenated hydrocarbon group. Some nematicides in the organophosphate and carbamate group also showed good control of nematodes, resulting in improved quality and yields of sweet potatoes.

Another species, *R. variabilis*, was commonly found endoparasitic in sweet potato roots in Kenya (Njuguna and Bridge, 1998).

Pratylenchus spp.

The root lesion nematodes, *Pratylenchus* spp., most commonly found with sweet potatoes are *P. brachyurus* and *P. coffeae*, causing necrotic lesions of both feeder and storage roots. There does appear to be a certain degree of resistance to lesion nematodes in some of the existing sweet potato cultivars. Some local Peruvian cultivars, such as Nemanete and Bakongo, with resistance to *M. incognita* are known also to exhibit resistance to another species, *P. flakkensis* (Anguiz and Canto, 1991). In Japan around Kyushu, *Pratylenchus* spp. were found in 12–22% of sweet potato fields, with *P. coffeae* being the most predominant species (Iwahori *et al.*, 2000, 2001). *P. coffeae* is thought to cause serious losses of sweet potato in Japan, and there have been breeding programmes to identify a source of resistance to the nematode (Marumine and Sakamoto, 1979; Suzuki, 1989). Sweet potato populations of *P. coffeae* from different regions of Japan exhibited different reproduction rates and amount of root damage, some being very virulent. Using the polymerase chain reaction-restriction fragment length polymorphism (PCR-RFLP) technique revealed a distict polymorphism and suggested the presence of more than two species of *Pratylenchus* (Mizukubo and Sano, 1997). Because of their relatively large host range, control measures against *Pratylenchus* spp., such as rotation, may not be very effective.

Ditylenchus spp.

Ditylenchus dipsaci, the stem nematode, and *D. destructor*, the potato rot nematode are reported as serious pests of sweet potato in China (Jiang, 1990; Zhang, 1992; Wang and Zhao, 1994). They cause a brown to black necrotic layer within the storage root, often leading to complete decay, especially following secondary invasion by pathogenic fungi. Some cultivars of sweet potato have been found to be resistant, including Yushu 13, Lushu 78066 and Shengli 100 (Sun and Chen, 1994; Lin *et al.*, 1999; Yang *et al.*, 1999). In resistant cultivars, the xylem parenchyma cell walls are thicker and more lignified than in susceptible cultivars (Lin *et al.*, 1996).

Other nematodes

Other nematodes of possible importance to sweet potato production when present in large populations are *Paratrichodorus* spp., *Belonolaimus longicaudatus*, *Radopholus similis*, *Helicotylenchus* spp. and *Scutellonema* spp. In pot experiments in India, *R. similis* caused 72–84% reduction in the weights of sweet potato roots at an initial inoculum level of 10,000 nematodes/plant; the economic threshold level is said to 100 nematodes/plant (Koshy and Jasy, 1991).

References

Anguiz, R. and Canto, S.M. (1991) Reaccion de cultivares de comote (*Ipomoea batatas* (L.) Lam.) a *Pratylenchus flakkensis* Seinhorst. *Nematropica* 21, 197–201.

Anonymous (1985) Conclusions of the EPPO Workshop on Cyst Nematodes Münster, Germany, 26–28 June, 1984. *EPPO Bulletin* 15, 121–122.

Anonymous (2003) *Bundessortenamt – Beschreibende Sortenliste Kartoffeln.* Deutscher Landwirtschaftsverlag, Hannover, Germany.

Averre, C.W., Barker, K.R., Sorensen, K.A. and Westerbeek, P. (1993) Sweet potato root-knot nematode management with soil fumigants. *Fungicide and Nematicide Tests* 48, 213.

Baarveld, H., Peeten, H.M.G., Schipper, E., Schipper, J.K. and Delleman, J. (eds) (2003) *Netherlands Catalogue of Potato Varieties.* NIVAP, The Netherlands. Available at: www.nivap.nl

Baldwin, J.G. and Cap, G.B. (1992) Systematics of *Nacobbus*, the false root-knot nematode. In: Gommers, F. and Maas, P.W.-T.H. (eds) *Nematology from Molecule to Ecosystems.* European Society of Nematologists, Invergowrie, Dundee, UK, pp. 101–112.

Been, T.H. and Schomaker, C.H. (2000) Development and evaluation of sampling methods for fields with infestation foci of potato cyst nematodes (*Globodera rostochiensis* and *G. pallida*). *Phytopathology* 90, 647–656.

Bendezu, I. (1997) Comparison of genetic variablity in European and South American populations of potato cyst nematodes by variation in DNA and virulence towards plant resistance genes. PhD Thesis, University of Nottingham, UK.

Beniers, A., Mulder, A. and Schouten, H.J. (1995) Selection for virulence of *Globodera pallida* by potato cultivars. *Fundamental and Applied Nematology* 18, 497–500.

Berthou, F., Rouselle, P. and Mugniéry, D. (1996) Ex-*Solanum sparsipilum-chaconese* heat stable resistance to *Meloidogyne arenaria* in *Solanum tuberosum. Proceedings of III International Congress of Nematology*, La Guadeloupe, pp. 101–102.

Birchfield, W. and Martin, W.J. (1965) Effects of reniform nematode populations on sweet potato yields. *Phytopathology* 55, 497.

Birchfield, W. and Martin, W.J. (1968) Evaluation of nematicides for controlling reniform nematodes in sweet potatoes. *Plant Disease Reporter* 52, 127–131.

Bird, G.W., Crawford, J.L. and McGlohon, N.E. (1973) Distribution, frequency of occurrence and population dynamics of *Rotylenchulus reniformis* in Georgia. *Plant Disease Reporter* 57, 399–401.

Bossis, M. and Mugniéry, D. (1993) Specific status of six *Globodera* parasites of solanaceous plants studied by means of two-dimensional gel electrophoresis with a comparison of gel patterns by a computed system. *Fundamental and Applied Nematology* 16, 47–56.

Brathwaite, C.W.D. (1977a) Outbreaks and new records in Barbados. *FAO Plant Protection Bulletin* 25, 210.

Brathwaite, C.W.D. (1977b) The response of sweet potato to nematode infection. *Newsletter International Society Tropical Root Crops* 10, 32–38.

Brown, C.R. and Mojtahedi, H. (2004) Breeding potato for resistance to *Meloidogyne* species and trichodorid-vectored viruses. In: *Breeding the Solanacea* (in press).

Brown, C.R., Mojtahedi, H. and Santos, G.S. (1995) Introgression of resistance to Columbia and Northern root-knot nematodes from *Solanum bulbocastanum* into cultivated potato. *Euphytica* 83, 71–78.

Brown, C.R., Mojtahedi, H. and Santos, G.S. (1999) Genetic analysis of resistance to *Meloidogyne chitwoodi* introgressed from *Solanum hougasii* in cultivated potato. *Journal of Nematology* 31, 264–271.

Brown, E.B. (1969) Assessment of the damage caused to potatoes by potato cyst eelworm, *Heterodera rostochiensis* Woll. *Annals of Applied Biology* 63, 493–502.

Brucher, H. (1960) Über das Wild-Vorkommen des Nematoden *Heterodera rostochiensis* in Nord-Argentinien. *Die Naturwissenschaften* 47, 21–22.

Bulman, S.R. and Marshall, J.W. (1997) Differentiation of Australasian potato cyst nematode (PCN) populations using the polymerase chain reaction (PCR). *New Zealand Journal of Crop and Horticultural Science* 25, 123–129.

Burk, E.F. and Tennyson, G. (1941) Hot water treatment for control of nematodes in sweet potato seed roots. *Proceedings of the American Society for Horticultural Science* 39, 299–302.

Canto-Saenz, M. (1996) Integrated control of potato cyst nematode. In: *Progress Report Centro International de Papa (CIP).*

Canto-Saenz, M. and Scurrah, M. (1977) Races of potato cyst nematode in the Andean region and a new system of classification. *Nematologica* 23, 340–349.

Cervantes-Flores, J.C., Yencho, G.C. and Davis, E.L. (2002a) Efficient evaluation of resistance to three root-knot nematode species in selected sweet potato cultivars. *HortScience* 37, 390–392.

Cervantes-Flores, J.C., Yencho, G.C. and Davis, E.L. (2002b) Host reactions of sweet potato genotypes to root-knot nematodes and variations in virulence of *Meloidogyne incognita* populations. *HortScience* 37, 1112–1116.

Chandra, S. (1994) Contributions of tropical root crops to socioeconomic improvement of the developing world. In: Ofori, F. and Hahn, S.K. (eds) *Tropical Root Crops in a Developing Economy. Proceedings of the 9th Symposium of the International Society for Tropical Root Crops,* 20–26 October 1991, Accra, Ghana, pp. 149–163.

Chen, P. (1993) Study on the nematodes around rhizosphere of sweet potato. *Southwest China Journal of Agricultural Sciences* 6 (Supplement), 33–37.

CIP (1992) *Annual Report 1991.* CIP, Lima, Peru.

Clark, C.A. and Moyer, J.W. (1988) *Compendium of Sweet Potato Diseases.* APS, St Paul, Minnesota.

Clark, C.A. and Wright, V.L. (1983) Effect and reproduction of *Rotylenchulus reniformis* on sweet potato selections. *Journal of Nematology* 15, 197–203.

Clark, C.A., Birchfield, W., Yik, C.P. and Boniol, D.P. (1980) Application rates and methods for Nemacur 3SC for nematode control on sweet potato, 1979. *Fungicide and Nematicide Tests* 35, 222.

Clark, S.A. (1967) The development and life history of the false root knot nematode *Nacobbus serendipiticus. Nematologica* 13, 91–101.

Coursey, D.G. and Booth, R.H. (1977) Root and tuber crops. In: Leakey, C.L.A. and Willis, J.B. (eds) *Food Crops of the Lowland Tropics.* Oxford University Press, Oxford, pp. 75–96.

Crozzoli, R., Cati, F. and Vovlas, N. (1994) Respuesta de diez selecciones de batata al nematodo agallador *Meloidogyne incognita. Fitopatalogia Venezolana* 7, 50–54.

Da Ponte, J.J. and Franco, A. (1981) Manipueira, un nematicida nao convencional de comprovada potencialidade. In: *Trabalhos Apresentados Reuniae de Sociedade Brasileira de Nematologia, Piracicaba,* Brazil 5, pp. 25–33.

da Silveira, M.A., de Azevedo, S.M., Maluf, W.R., Campos, V.P. and Momente, V.G. (1997) Canuana e Palmas: novas cultivares de batata doce resistentes aos nematoides das galhas. *Horticultura Brasileira* 15, 122–123.

de Freitas, J.A., dos Santos, G.C., Souza, V.S. and de Azevedo, S.M. (2001) Resistencia de clones de batata doce, *Ipomoea batatas* L., aos nematoides causadores de galhas. *Acta Scientiarum* 23, 1257–1261.

Doucet, M., Ponce De Leon, E.L. and Franco, J. (1994) *Spergula arvensis* y su asociación con *Nacobbus aberrans* en el cultivo de papa en Bolivia. *Nematropica* 24, 69–72.

Dunn, R.A. (1973) Resistance in potato (*Solanum tuberosum*) to *Pratylenchus penetrans. Abstracts of the Papers of the International Congress of Plant Pathology (2nd).* Minneapolis, Minnesota, 5–12 September. American Phytopathological Society Inc. No. 0860, St Paul, Minnesota.

Evans, K. and Haydock, P.P.J. (2000) Potato cyst nematode management – present and future aspects. *Annals of Applied Biology* 59, 91–97.

Evans, K. and Stone, A.R. (1977) A review of the distribution and biology of the potato cyst nematodes *Globodera rostochiensis* and *G. pallida. PANS* 23, 178–189.

Evans, K., Franco, J. and Scurrah, M.M. (1975) Distribution of species of potato cyst nematodes in South America. *Nematologica* 21, 365–369.

Fassuliotus, G. and Rau, G.J. (1967) The reniform nematode in South Carolina. *Plant Disease Reporter* 51, 557.

Fleming, C.C. and Powers, T.O. (1998) Potato cyst nematode diagnostics: morphology, differential hosts and biochemical techniques. In: Marks, R.J. and Brodie, B.B. (eds) *Potato Cyst Nematodes: Biology, Distribution and Control.* CAB International, Wallingford, UK, pp. 91–114.

Folkertsma, R.T., Rouppe van der Voort, J.N.A.M., van Gent-Pelzer, M.P.E., de Groot, K.E., van den Bos, W.J., Schots, A., Bakker, J. and Gommers, F.J. (1994) Inter- and intraspecific variations between populations of *Globodera rostochiensis* and *G. pallida* revealed by random amplified polymorphic DNA. *Phytopathology* 84, 807–811.

Franco, J. (1979) Effect of temperature on hatching and multiplication of potato cyst nematodes. *Nematologica* 25, 237–244.

Franco, J. (1981) Potato cyst nematodes, *Globodera* spp. *Technical Information Bulletin 9.* International Potato Centre, Lima, Peru.

Franco, J. (1994) Problemas de nematodos en la produccion de papa en climas templados en la region andina. *Nematropica* 24, 179–195.

Franco, J. and Bendezú, E. (1985) Estudio del complejo *Verticillium dahliae* Kleb. y *Globodera pallida* Stone y su efecti en el comportamienti de algunos cultivares peruanos de papa. *Fitopatologia* 20, 21–27.

Franco, J. and Gonzalez, A. (1990) A new race of *Globodera pallida* attacking potatoes in Peru. *Revue de Nématologie* 13, 181–184.

Franco, J., Ramos, J., Oros, R., Main, G. and Ortuño, N. (1989) Economic losses caused by *Nacobbus aberrans* and *Globodera* spp. on the potato crop in Bolivia. *Revista Latino-americana de la Papa* 11, 41–66.

Franco, J., Montecinos, R. and Montalvo, R. (1993) Chemical treatment of tubers infected by *Nacobbus aberrans* to avoid nematode dispersal. *Nematropica* 23, 117.

Franco, J., Oros, R., Main, G. and Ortuño, N. (1998) Potato cyst nematodes (*Globodera* species) in South America. In: Marks, R.J. and Brodie, B.B. (eds) *Potato Cyst Nematodes: Biology, Distribution and Control.* CAB International, Wallingford, UK, pp. 239–269.

Franco, J., Main, G. and Oros, R. (1999) Trap crops as a component for the integrated management of *Globodera* spp. (potato cyst nematodes) in Bolivia. *Nematropica* 29, 53–62.

Franklin, M.T. (1959) *Nacobbus serendipiticus* n.sp., a root-galling nematode from tomatoes in England. *Nematologica* 4, 286–293.

Fukunaga, M. and Iwahori, H. (2002) Control of root-knot nematodes on sweet potato by crop rotation with resistant cultivar. *Kyushu Plant Protection Research* 48, 64–67. (In Japanese.)

Fullaondo, A., Barrena, E., Viribay, M., Barrena, I., Salazar, A. and Ritter, E. (1999) Identification of potato cyst nematode species *Globodera rostochiensis* and *G. pallida* by PCR using specific primer combinations. *Nematology* 1, 157–163.

Gallo, P., Escobar, H., Chavez, R., Jimenez, M., Torres, A. and Carrion, H. (2001) Introduccion y evaluacion de variedades mejoradas de camote (*Ipomoea batatas* L.) en las zonas arido salinas del norte de Chile. *IDESIA* 19, 57–70.

Gapasin, R.M. (1981) Control of *Meloidogyne incognita* and *Rotylenchulus reniformis* and its effect on the yield of sweet potato and cassava. *Annals of Tropical Research* 3, 92–100.

Gowen, S.R. (1985) *Report on Nematode Survey in the Solomon Islands and Vanuatu.* Report to the UK Overseas Development Administration, London.

Grenier, E., Bossis, M., Fouville, D., Renault, L. and Mugniéry, D. (2001) Molecular approaches to the taxonomic position of Peruvian potato cyst nematodes and gene pool similarities in indigenous and imported populations of *Globodera. Heredity* 86, 277–290.

Guile, C.T. (1970) Further observations on cyst colour changes in potato cyst eelworm pathotypes. *Plant Pathology* 19, 1–6.

Harrison, J.A.C. (1971) Association between the potato cyst nematode, *Heterodera rostochiensis* Woll. and *Verticillium dahliae* Kleb. in the early-dying disease of potatoes. *Annals of Applied Biology* 67, 185–193.

Hartemink, A.E., Poloma, S., Maino, M., Powell, K.S., Egenae, J. and O'Sullivan, J.N. (2000) Yield decline of sweet potato in the humid lowlands of Papua New Guinea. *Agriculture, Ecosystems and Environment* 79, 259–269.

Haydock, P.P. and Perry, J.N. (1998) The principles and practice of sampling for the detection of potato cyst nematodes. In: Marks, R.J. and Brodie, B.B. (eds) *Potato Cyst Nematodes: Biology, Distribution and Control.* CAB International, Wallingford, UK, pp. 61–74.

Hawkes, J.G. (1994) Origins of cultivated potatoes and species relationships. In: Bradshaw, J.E. and Mackay, G.R. (eds) *Potato Genetics.* CAB International, Wallingford, UK.

Horton, D., Lynam, J. and Knipscheer, H. (1984) Root crops in developing countries – an economic appraisal. *Proceedings of Sixth Symposium of the International Society for Tropical Root Crops.* International Potato Centre, Lima, Peru, 21–26 February 1983, pp. 9–39.

Inserra, R.N. (1983) Development and pathogenicity of the false root-knot nematode *Nacobbus aberrans* and its relationship with *Heterodera schachtii* and *Meloidogyne hapla* on sugarbeet. PhD thesis, Utah State University, Logan, Utah.

Iriiarte, L., Franco, J. and Ortuño, N. (1994) Efecto de la incorporación de cal, haba y tarwi en veerde sobre la población de *Globodera* spp y en el rendimiento del cultivo de papa. In: *III Reunion Nacinald e la Papa*, Cochabamba, Bolivia.

Iwahori, H., Sano, Z. and Torigoe, H. (2001) Distribution of the main plant-parasitic nematodes in sweet potato and taro fields in Kyushu and Okinawa, Japan. 2. Survey in several islands around Kyushu and the Nansei Islands. *Kyushu Plant Protection Research* 47, 112–117.

Janssen, G.J.W., Janssen, R., van Norel, A. and Hoogendorn, C.J. (1996) Expression of resistance to the root-knot nematodes, *Meloidogyne hapla* and *M. fallax* in wild *Solanum* spp. under field conditions. *European Journal of Plant Pathology* 102, 859–865.

Jatala, P. (1975) Root-knot nematodes (*Meloidogyne* species) and their effect on potato quality. In: *6th Triennial Conference of the EAPR,* 15–19 September 1975, Wageningen, The Netherlands.

Jatala, P. (1989) Nematodes in tuber and root crops and their management. In: *Proceedings of the 11th International Congress in Plant Protection,* Manila, Philippines.

Jatala, P. and Kaltenbach, R. (1979) Survival of *Nacobbus aberrans* in adverse conditions. *Journal of Nematology* 11, 303.

Jatala, P. and Martin, C. (1977) Interaction of *Meloidogyne incognita acrita* and *Pseudomonas scolanacearum* on *Solanum chacoense* and *S. sparsipilum*. *Proceedings of the American Phytopathological Society* 4, 178.

Jatala, P. and Russell, C.C. (1972) Nature of sweet potato resistance to *Meloidogyne incognita* and the effect of temperature in parasitism. *Journal of Nematology* 4, 1–7.

Jatala, P. and Scurrah, M.M. (1975) Mode of dissemination of *Nacobbus* spp. in certain potato growing areas of Peru and Bolivia. *Journal of Nematology* 7, 324–325.

Jatala, P., French, E.R. and Gutarra, L. (1975) Interrelationships of *Meloidogyne incognita acrita* and *Pseudomonas solanacearum* on potatoes. *Journal of Nematology* 7, 325.

Jatala, P., Gutarra, L., French, E.R. and Arango, J. (1976) Interaction of *Heterodera pallida* and *Pseudomonas solanacearum* on potatoes. *Journal of Nematology* 8, 289–290.

Jensen, H.J., Armstrong, L. and Jatala, P. (1979) *Annotated bibliography of nematode pests of potato.* Joint Publication of the International Potato Centre, Apartado 5969, Lima, Peru, and Oregon State University Agricultural Experiment Station, Corvallis, Oregon.

Jiang, Q.M. (1990) The damage and control of sweet potato (*Ipomoea batatas*) stem nematode (*Ditylenchus* spp.). *Bulletin of Agricultural Science and Technology* No. 1, 28.

Johnson, A.W., Dowler, C.C., Glaze, N.C., Chalfant, R.B. and Golden, A.M. (1992) Nematode numbers and crop yield in a fenamiphos-treated sweetcorn–sweet potato–vetch cropping system. *Journal of Nematology* 24, 533–539.

Jones, F.G.W. (1970) The control of the potato cyst nematode. *Journal of the Royal Society of Arts* 118, 179–199.

Kassab, A.S. and Taha, A.H.Y. (1990) Aspects of the host–parasite relationships of nematodes and sweet potato 1. Population dynamics and interaction of *Criconemella* spp., *Rotylenchulus reniformis, Tyenchorhynchus* spp. and endomycorrhiza. *Annals of Agricultural Science, Cairo* 35, 497–508.

Katayama, K., Tamiya, S., Kuranouchi, T., Komaki, K. and Nakatani, M. (2003) New sweet potato cultivar 'Quick Sweet'. *Bulletin of the National Institute of Crop Science* 3, 35–52.

Kistner, M.H., Daiber, K.C. and Bester, C. (1993) the effect of root-knot nematodes (*Meloidogyne* spp.) and dry land conditions on the production of sweet potato. *Journal of the Southern African Society for Horticultural Sciences* 3, 108–110.

Koen, H. (1969) Thermotherapeutic control of *Pratylenchus brachyurus* in potato tubers. *Phytophylactica* 1, 67–69.

Kort, J., Ross, H., Rumpenhorst, H.J. and Stone, A.R. (1977) An international scheme for identifying and classifying pathotypes of potato cyst nematodes *Globodera rostochiensis* and *G. pallida. Nematologica* 23, 333–339.

Koshy, P.K. and Jasy, T. (1991) Effect of the burrowing nematode, *Radopholus similis,* on sweet potato. *Journal of Root Crops* 17, 201–206.

Lawrence, G.W., Clark, C.A. and Wright, V.L. (1986) Influence of *Meloidogyne incognita* on resistance and susceptible sweet potato cultivars. *Journal of Nematology* 18, 59–65.

Lin, M.S., He, L.M., Wen, L. and Fang, Z.D. (1996) Mechanism of morphological structure of sweet potato resistance to potato rot nematode (*Ditylenchus destructor*). *Scientia Agricultura Sinica* 29, 8–12.

Lin, M.S., Wen, L. and Fang, Z.D. (1999) Potato rot nematodes and stem nematode disease of sweet potato. *Jiangsu Journal of Agricultural Sciences* 15, 186–190.

Llontop, J. and Franco, J. (1988) El nematode del quiste de la papa (Globoderaso.) en la region de la Libertad, Peru. Distribucion e identificiaon de escpecies y razas de *G. pallida*. *Fitopatologia* 23 (2), 49–64.

Mai, W.F., Brodie, B.B., Harrison, M.B. and Jatala, P. (1981) Nematodes. In: Hooker, W.J. (ed.) *Compendium of Potato Diseases*. American Phytopathological Society, St Paul, Minnesota, pp. 93–101.

Manzanilla-López, R.H., Cid del Prado, V.I., Evans, K., Alejo, C.J., Franco-Avila, E. and Carrillo-Fonseca, C. (1998) Population fluctuation of *Nacobbus aberrans* (Thorne 1935) Thorne & Allen, 1944 under field conditions in Chapingo, Mexico. *Nematropica* 28, 139.

Manzanilla-López, R.H., Hardin, S. and Evans, K. (1999) Morphometric study on twelve populations of *Nacobbus aberrans* (Thorne, 1935) Thorne & Allen 1944 (Nematoda: Pratylenchidae) from Mexico and South America. *Nematology* 1, 477–498.

Manzanilla-López, R.H., Costilla, M.A., Doucet, M., Franco, J., Inserra, R.N., Lehman, P.S., Cid del Prado-Vera, I., Souza, R.M. and Evans, K. (2002) The genus *Nacobbus* Thorne & Allen, 1944 (Nematoda: Pratylenchidae): systematics, distribution, biology and management. *Nematropica* 32, 149–227.

Main, G., Franco, J. and Ortuño, N. (1999) Los cultivos trampa como alternativa para reducir las pobaciones de *Nacobbus aberrans* y *Globodera* spp. en papa. *Fitopatologia* 34, 35–41.

Marshall, J.W. (1998) Potato cyst nematodes (*Globodera* species) in New Zealand and Australia. In: Marks, R.J. and Brodie, B.B. (eds) *Potato Cyst Nematodes: Biology, Distribution and Control*. CAB International, Wallingford, UK, pp. 347–394.

Martin, W.J. (1962) Elimination of root-knot nematodes from infested sweet potato roots and plants. *Plant Disease Reporter* 46, 21–23.

Marumine, S. and Sakamoto, S. (1979) Selection for resistance to the root lesion nematode (*Pratylenchus coffeae*) in sweet potato. *Kyushu Agricultural Research, Japan* No. 41, 48.

Mizukubo, T. and Sano, Z. (1997) *Pratylenchus coffeae* virulent races on sweetpotato. *Sweet Potato Research Front* No. 4, 2.

Minnis, S.T., Haydock, P.P.J., Ibrahim, S.K., Grove, I.G., Evans, K. and Russell, M.D. (2002) Potato cyst nematodes in England and Wales – occurrence and distribution. *Annals of Applied Biology* 140, 187–195.

Mohandas, C. and Palaniswami, M.S. (1990) Resistance in sweet potato (*Ipomoea batatas* L.) to *Meloidogyne incognita* (Kofoid & White) Chitwood in India. *Journal of Root Crops* 16, 148–149.

Mohandas, C., Palaniswami, M.S., Kurup, G.T., Potty, V.P., Padmaja, G., Kabeerathumma, S. and Pillai, S.V. (1996) Resistance in tuber crops to *Meloidogyne incognita* (Kofoid and White, 1919). In: Palaniswami, M.S., Kurup, G.T., Potty, V.P., Padmaja, G. and Kabeerathumma, S. (eds) *Tropical Tuber Crops: Problems, Prospects and Future Strategies*. Science Publishers, Inc., Enfield, New Hampshire, pp. 366–369.

Montalvo, R., Montecinos, R. and Franco, J. (1992) Métodos de laboratorio para el diagnóstico de *N. aberrans* en muestras de suelo. In: *Informe Anual IBTA-PROINPA 1991–92*. Cochabamba, Bolivia, pp. 139–140.

Montes, L., Crozzoli, R. and Vargas, G. (1998) Respuesta de selecciones de batata al nematodo agallador *Meloidogyne incognita* en Venezuela. *Nematropica* 28, 113–117.

Morris, R.F. (1971) Distribution and biology of the golden nematode, *Heterodera rostochiensis* in Newfoundland. *Nematologica* 17, 370–376.

Mugniéry, D., Phillips, M.S., Rumpenhorst, H.J., Stone, A.R., Treur, A. and Trudgill, D.L. (1989) Assesssment of partial resistance of potato to, and pathotype and virulence differences in potato cyst nematode. *Bullettin OEPP* 19, 7–25.

Mulvey, R.H. and Stone, A.R. (1976) Description of *Punctodera matadorensis* n.gen., n.sp. (Nematoda Heteroderidae) from Saskatchewan with list of species and generic diagnosis of *Globodera* (n. rank), *Heterodera*, and *Sarisodera*. *Canadian Journal of Zoology* 54, 772–785.

Njuguna, L.K. and Bridge, J. (1998) Plant parasitic nematodes of Irish potatoes (*Solanum tuberosum*) in Central Province and sweet potatoes (*Ipomoea batatus*) in Central, Nyanza and Coast Provinces of Kenya. *International Journal of Nematology* 8, 21–26.

Oostenbrink, M. (1966) Major characteristics of the relation between nematodes and plants. *Mededelingen Landbouwhogeschool Wageningen* 66, 1–46.

Oros, R., Ortuño, N. and Franco, J. (1996) Estudio y distribución de razas. Estudio de razas de *N. aberrans* en el cultivo de papa. In: *Informe Anual 1995–96 IBTA-PROINPA*. Cochabamba, Bolivia, pp. 18–21.

Otazu,V., Hoopes, R., Caero, G. and Huayta, U.I. (1985) El rosario de la papa causado por *Nacobbus aberrans* (Thorne 1935) Thorne and Allen 1994, su efecto en el rendimiento y algunos aspectos que inciden en su propagación y prevalencia en Bolivia. *Fitopatologia* 20, 65–70.

Pastrik, K.H., Rumpenhorst, H.J. and Burgermeister, W. (1995) Random amplified polymorphic DNA analysis of a *Globodera pallida* population selected for virulence. *Fundamental and Applied Nematology* 18, 109–114.

Peixoto, J.R., Ferraz, F.M., Santos, L.C., de Angelis, B. and Juliatti, F.C. (1998) Selecao de genotipos de batata doce resistentes ao nematoide das galhas (*Meloidogyne* spp.). *Fitopatologia Brasileira* 23, 51–57.

Phillips, M.S. and Trudgill, D.L. (1998) Variation of virulence, in terms of quantitative reproduction of *Globodera pallida* populations, from Europe and South America, in relation to resistance from *Solanum vernei* and *S. tuberosum* ssp. *andigena* CPC 2802. *Nematologica* 44, 409–423.

Phillips, M.S., Blok, V.C., Ploeg, A. and Harrower, B.E. (1998) Studies on an artificially fragmented population of potato cyst nematode *Globodera pallida*. *Nematologica* 44, 655–666.

Qin, L., Kudla, U., Roze, E.H.A., Goverse, A., Popeijus, H., Nieuwland, J., Overmars, H., Jones, J.T., Schots, A., Smant, G., Bakker, J. and Helder, J. (2004) Plant degradation: a nematode expansin acting on plants. *Nature* 427, 30.

Ramakrishnan, S. and Mohandas, C. (1996) Density and frequency of various plant parasitic nematodes in sweet potato growing areas of Kerala. *Journal of Root Crops* 22, 112–114.

Ramos, J., Franco, J., Ortuño, N., Oros, R. and Main, G. (1998) Incidencia y severidad de los nematodos parsitos del cultivo de la papa, *Nacobbus aberrans* y *Globodera* spp. y determinaciones de especies de *Globodera* spp. en Bolivia. In: *IBTA-PROINPA*. Cochabamba, Bolivia.

Rodriguez-Kabana, R., Turner, J.L. and Ingram, E.G. (1978) Tratamiento de raíces de batata con el nematocida sistémico oxamy para el control de nematodos fitoparásitos. *Nematropica* 8, 26–31.

Rojas, F., Franco, J. and Ortuño, N. (1997) Las ferias agrícolas: fuente de diseminación de *Nacobbus aberrans*. *Revista Latinoamericana de la Papa* 9/10, 35–48.

Salaman, R. (1949) *The History of Social Influence of the Potato.* Cambridge University Press, London.

Sano, Z. and Iwahori, H. (2002) Host suitability of the sweet potato, *Ipomoea batatas*, to major root-knot nematodes, *Meloidogyne* species. *Sweet Potato Research Front* No.13.

Sano, Z., Iwahori, H., Tateishi, Y. and Kai, Y. (2002) Differences in the resistance of sweet potato cultivars and lines to *Meloidogyne incognita* populations. *Japanese Journal of Nematology* 32, 77–86.

Santos, G.S. (2001) Root knot nematodes. In: Stevenson W.R. *et al.* (eds) *Compendium of Potato Diseases*, 2nd edn. APS Press, St Paul, Minnesota, pp. 51–53.

Sasser, J.N. and Kirby, M.F. (1979) *Crop Cultivars Resistant to Root-knot Nematodes,* Meloidogyne *Species.* International Meloidogyne Project Contract No. AID/ta-c-1234. North Carolina State University, Raleigh, North Carolina.

Scholte, K. (2000) Screening of non-tuber bearing Solanaceae for resistance to and induction of juvenile hatch of potato cyst nematodes and their potential for trap cropping. *Annals of Applied Biology* 136, 239–246.

Scholte, K. and Vos, J. (2000) Effect of potential trap crops and planting date in soil infected with potato cyst nematode and root knot nematodes. *Annals of Applied Biology* 137, 153–164.

Scurrah, M. and Franco, J. (1985) Breeding for resistance of *Globodera pallida* at CIP. *EPPO Bulletin* 15, 167–173.

Schouten, H.J. and Beniers, J.E. (1997) Durability of resistance to *Globodera pallida* I. Changes in pathogenicity, virulence and aggressiveness during reproduction on partially resistant potato cultivars. *Phytopathology* 87, 862–867.

Sharma, S.B., Price, N.S. and Bridge, J. (1997) The past, present and future of plant nematology in International Agricultural Research Centres. *Nematological Abstracts* 66 No. 3, 119–142.

Shekharmat, G.S. (1985) *20 Years of Potato Research All India Coordination Potato Improvement Project.* Central Potato Research Institute, Shimla, Delhi.

Sher, S.A. (1970) Revision of the genus *Nacobbus* Thorne and Allen, 1944 (Nematoda: Tylenchoidea). *Journal of Nematology* 2, 228–235.

Shields, R., Fleming, C.C. and Stratford, R. (1996) Identification of potato cyst nematodes using the polymerase chain reaction. *Fundamental and Applied Nematology* 19, 167–173.

Sikora, R.A. (1984) Importance of diapause, resistant and early maturing potato cultivars, chicken manure and non-host crops for *Globodera rostochiensis* integrated control in the upland tropics. *Mededelingen Faculteit Landbouwwetenschappen Gent* 49, 613–620.

Siwi Indarti, Bambang, R.T.P., Mulyadi and Triman, B. (2004) First record of potato cyst nematode *Globodera rostochiensis* in Indonesia. *Australasian Plant Pathology* 33, 325–326.

Stone, A.R. (1973) *Heterodera pallida* n.sp. (Nematoda: Heteroderidae), a second species of potato cyst nematode. *Nematologica* 18, 591–606.

Stone, A.R. (1985) Co-evolution of potato cyst nematodes and their hosts: implications for pathotypes and resistance. *EPPO Bulletin* 15, 130–137.

Storey, G.W. and Evans, K. (1987) Interactions between *Globodera pallida* juveniles, *Verticillium dahliae* and three potato cultivars, with descriptions of associated histopathologies. *Plant Pathology* 36, 192–200.

Sun, F.R. and Chen, H.C. (1994) New sweet potato variety with high resistance to stem nematode – Lushu 78066. *Crop Genetic Resources* No. 4, 58.

Suri, F.V. and Jayasinghe, U. (2002) Survey of potato fields for root knot nematode in Ngablak, Central Java. In: Fuglie, K. (ed.) *Progress in Potato and Sweetpotato Research in Indonesia.* Indonesian Agency for Agricultural Research, Jakarta and Development (IAARD), Ministry of Agriculture and International Potato Center, Bogor, Indonesia, pp. 89–91.

Suzuki, M. (1989) New summer crop cultivars registered by the Ministry of Agriculture, Forestry and Fisheries in 1989. *Japanese Journal of Breeding* 39, 499–506.

Tamiya, S., Nakatani, M., Komaki, K., Katayama, K. and Kuranouchi, T. (2003) New sweet potato cultivar, 'Purple Sweet Lord'. *Bulletin of the National Institute of Crop Science* No. 4, 29–43.

Taeishi, Y. (1998) Suppression of *Meloidogyne incognita* and yield increase of sweet potato by field application of *Pasteuria penetrans. Japanese Journal of Nematology* 28, 22–24.

Tateishi, Y. (1999) Biological control of *Meloidogyne incognita* on sweetpotatoes for table use by *Pasteuria penetrans. Sweet Potato Research Front* 9, 4.

Tateishi, Y. and Sano, Z. (2001) *Pasteuria penetrans* for controlling *Meloidogyne incognita* on sweet potato cultivars with different susceptibility to the root-knot nematode. II. Nematode suppressive effects during the 7th to the 9th cropping seasons. *Kyushu Plant Protection Research* 47, 118–122.

Taylor, A.L. and Sasser, J.N. (1978) *Biology, identification and control of root-knot nematodes* (Meloidogyne *species*). International Meloidogyne Project. North Carolina State University Graphics.

Thiéry, M. and Mugniéry, D. (1996) Interspecific rDNA restriction fragment length polymorphism in *Globodera* species, parasites of solanaceous plants. *Fundamental and Applied Nematology* 19, 471–479.

Thiéry, M., Fouville, D. and Mugniéry, D. (1997) Intra- and interspecific variability in *Globodera,* parasites of solanaceous plants, revealed by random amplified polymorphic DNA (RAPD) and correlation with biological features. *Fundamental and Applied Nematology* 20, 495–504.

Thomas, R.J. and Clark, C.A. (1983a) Effects of concomitant development on reproduction of *Meloidogyne incognita* and *Rotylenchulus reniformis* on sweet potato. *Journal of Nematology* 15, 215–221.

Thomas, R.J. and Clark, C.A. (1983b) Population dynamics of *Meloidogyne incognita* and *Rotylenchulus reniformis* alone and in combination and their effects on sweet potato. *Journal of Nematology* 15, 204–211.

Thorne, G. and Schuster, M.L. (1956) *Nacobbus batatiformis* n.sp. (Nematoda: Tylenchida), producing galls on the roots of sugar beets and other plants. *Proceedings of Helminthological Society of Washington* 23, 128–134.

Torigoe, H. (1996) Correlation between the suppressive effects of several enemy plants and the recovery of *Pratylenchus coffeae* and *Meloidogyne incognita* populations in taro and sweet potato fields in southern Kyushu. *Proceedings of the Association for Plant Protection of Kyushu* 42, 83–88.

Trudgill, D.L. (1986) Potato cyst nematodes: a critical review of the current pathotyping scheme. *EPPO Bulletin.*

Trudgill, D.L., Elliott, M.J., Evans, K. and Phillips, M.S. (2003) The white potato cyst nematode (*Globodera pallida*) – a critical analysis of the threat in Britain. *Annals of Applied Biology* 143, 73–80.

Turner, S. (1990) The identification and fitness of virulent potato cyst-nematode populations (*Globodera pallida*) selected on resistant *Solanum vernei* hybrids for up to eleven generations. *Annals of Applied Biology* 117, 385–397.

Turner, S.J. (1996) Population decline in potato cyst nematodes (*Globodera rostochiensis, G. pallida*) in field soils in Northern Ireland. *Annals of Applied Biology* 129, 315–322.

Van Riel, H. and Mulder, A. (1998) Potato cyst nematodes (*Globodera species*) in Western Europe. In: Marks, T.J. and Brodie, B.B. (eds) *Potato Cyst Nematodes: Biology, Distribution and Control.* CAB International, Wallingford, UK, pp. 274–298.

Vilsoni, F. and Heinlein, M. (1982) Host range and reproductive capacity of the reniform nematode *Rotylenchulus reniformis* on some crop cultivars in Fiji. *Fiji Agricultural Journal* 44, 61–66.

Vilmala, B. and Rajendran, P.G. (1998) 'Sree Rethna' and 'Sree Bhadra' – two promising short duration varieties of sweet potato. *Journal of Root Crops* 24, 25–30.

Walters, S.A. and Barker, K.R. (1994) Reproductive and damage potentials of two populations of *Rotylenchulus reniformis* on sweet potato and related comparisons with *Meloidogyne javanica* on tomato. *Supplement to the Journal of Nematology* 25, 830–835.

Wang, Z.M. and Zhao, Y.E. (1994) The seriousness of stem nematode (*Ditylenchus* sp.) on sweet potato in Henan Province. *Technology and Extension of Plant Protection* No. 2, 40.

Whitehead, A.G. (1969) The distribution of root-knot nematodes (*Meloidogyne* spp.) in tropical Africa. *Nematologica*, 315–333.

Whitehead, A.G. (1975) Chemical control of potato cyst nematode. *Research Review* 8, 17–23.

Winslow, R.D. (1978a) An overview of the important nematode pests of potato. *Developments in Control of Nematode Pests of Potato. Report of the 2nd Nematode Planning Conference.* International Potato Centre, Lima, Peru, pp. 16–30.

Winslow, R.D. (1978b) Other nematode pests: importance and control. *Developments in the Control of Nematode Pests of Potato. Report of the 2nd Nematode Planning Conference.* International Potato Centre, Lima, Peru, pp. 138–153.

Winslow, R.D. and Willis, R.J. (1972) Nematode diseases of potatoes. In: Webster, J.M. (ed.) *Economic Nematology*. Academic Press, London, pp. 17–48.

Wood, F.H., Foot, M.A., Dale, P.S. and Barber, C.J. (1983) Relative efficiency of plant sampling and soil sampling in detecting the presence of low potato cyst nematode infestations. *New Zealand Journal of Experimental Procedure* 11, 271–273.

Yameda, E., Takakura, S. and Tezuka, H. (1972) On the occurrence of the potato cyst nematodes, *Heterodera rostochiensis*, Japan. *Japanese Journal of Nematology* 2, 12–15.

Yang, G.H., Kang, Z.H., Xiao, L.H., Hou, L.X., Yang, X.P. and Gu, J.H. (1999) Breeding of a new starch processing type sweet potato variety Yushu 13 with high yield and multi-resistance. *Journal of Henan Agricultural Sciences* 3, 3–5.

Yangen, D., Cole, D., Crissman, C. and Sherwood, S. (2003) *Human Health, Environmental, and Economic Effects of Pesticide Use in Potato Production in Ecuador.* Centro International de Papa (CIP).

Yen, D.E. (1982) Sweet potato in historical perspective. In: Villareal and Griggs, T.D. (eds) *Sweet Potato, Proceedings of the First International Symposium.* AVRDC, Tainan, Taiwan, pp. 17–30.

Yin, K.C. and Feng, Z.X. (1981) The investigation of plant nematodes. *Acta Phytohylactica-Sinica* 8, 122–123.

Yokoo, T. and Matsunobu, K. (1975) On thermal death point of the potato lesion nematode *Pratylenchus pratensis. Agricultural Bulletin, Saga University* 5, 97–102.

Zhang, J.Y. (1992) The occurrence and control of stem nematode (*Ditylenchus*) on sweet potato. *Hebei Agricultural Science* 1, 25.

7 Nematode Parasites of Tropical Root and Tuber Crops (Excluding Potatoes)*

John Bridge,[1] Danny L. Coyne[2] and Charles K. Kwoseh[3]
[1]*CABI Bioscience UK Centre, Bakeham Lane, Egham, Surrey TW20 9TY, UK;*
[2]*International Institute of Tropical Agriculture, Ibadan, Nigeria;* [3]*Department of Crop Science, University of Science and Technology, Kumasi, Ghana*

Root and tuber crops are the most important food commodities produced in many subtropical and tropical countries. World production figures for 2002 (FAO, 2002) show that root and tuber crops are the most important source of carbohydrates in the tropical world and are second only to cereals in total world supply. They all produce starchy storage organs that are modified stems or roots, generally referred to as rhizomes, corms or tubers.

The origin and history of root and tuber crops are well documented (Coursey and Haynes, 1970; Coursey and Booth, 1977; Leon, 1977). However, the actual contribution and potential of these crops in the world's food supply are poorly understood. In addition to Solanum potato (*Solanum tuberosum*) and sweet potato (*Ipomoea batatas*) (Chapter 6), the other most widely grown root and tuber crops are cassava (*Manihot esculenta*), yams (*Dioscorea* spp.), taro (*Colocasia esculenta*) and tannia (*Xanthosoma* spp.). A further 27 root and tuber crops have been described, many of which are not grown on a wide scale, but are of local importance (Kay, 1987).

Cassava

Cassava, *Manihot esculenta* Crantz, is a perennial woody shrub of the Euphorbiaceae family. It originated in tropical America although its exact area of original domestication is not known with certainty. There is a major centre of diversity of *Manihot* spp. in Brazil and a secondary centre in Mesoamerica (Cock, 1984). Its occurrence in the wild state is not known and its evolution as a species is directly linked to selection under cultivation by man (Horton *et al.*, 1984). However, there are a large number of wild *Manihot* spp. with which cassava can be readily crossed (Cock, 1984). From the Americas, cassava spread first to Africa and then to Asia. In Africa, separate introductions were made into the West, first through the Congo Basin, and the East (Jones, 1959).

There are two main groups – sweet and bitter cassavas. The enlarged storage roots have hydrocyanic glycosides in varying quantities depending on age, variety, and environmental conditions such as soil moisture and temperature (Nartey, 1977). Some varieties have customarily been des-

*A revision of part of the chapter by P. Jatala and J. Bridge.

 Plant Parasitic Nematodes in Subtropical and Tropical Agriculture, 2nd Edition (eds M. Luc, R.A. Sikora, J. Bridge)

ignated as sweet or bitter, purportedly in relation to their cyanogenic glucoside content. However, analysis of various parts of the plants of bitter and sweet varieties at different stages of growth can show comparable levels (Nartey, 1977). Cassava was selected originally for its enlarged roots, and ability to propagate from stem cuttings and erect plant type (Jennings, 1976). Although it is grown principally for its swollen roots (storage roots), its leaves are also eaten, particularly in parts of Africa, such as in the countries of the Congo Basin.

Because of its long growth period, its cultivation is limited primarily to the tropics and subtropics where it can be planted at any time of the year providing there is sufficient moisture for stem cuttings to take root. Once mature, storage roots can remain in the ground and be harvested from between 6 and 48 months after planting (Nweke *et al.*, 2002). It is the most widely grown root crop across varying agroclimatic conditions (Flach, 1982). It has the ability to produce economic yields under relatively marginal soil and rainfall conditions, and has the highest carbohydrate yield per unit of land and labour. It is compatible with a variety of associated crops and has essentially been recognized as a small farm and subsistence crop, requiring minimal cash input for production. However, to supply increasing urbanization and consumer demand, and with access to mechanized cassava processing equipment, the role of cassava is changing and it is being cultivated on an increasingly larger scale (Nweke *et al.*, 2002). In Asia, cassava is an important source of starch for industrial use. Nigeria produces more cassava than any other country, while Thailand is the biggest exporter of the crop.

Nematodes of Cassava

As with many tropical crops, a wide range of nematode species have been reported associated with cassava, from many different geographical areas. The nematode species associated with cassava are presented in various reports, the most comprehensive of which include those by Hogger (1971), Caveness (1980), McSorley *et al.* (1983b), Bridge *et al.* (1991), Ray *et al.* (1992) and Coyne *et al.* (2003). Although the list of nematodes is extensive, the majority of the nematode species appear, however, of limited importance, with little evidence of significant effect on the crop. Indeed, some reports are contradictory. The plant parasitic nematodes most frequently found associated with cassava are *Meloidogyne incognita*, *M. javanica*, *Pratylenchus brachyurus*, *Rotylenchulus reniformis*, *Helicotylenchus erythrinae* and *H. dihystera*. *M. incognita* and *M. javanica* are probably the most important nematodes, followed by *P. brachyurus*, *Helicotylenchus* spp. and *R. reniformis*, as they are found in abundance around the roots of cassava. Most of these nematodes may interact with other pathogenic organisms in the development of disease complexes. Most data relating to nematodes of cassava relate to diagnostic and distribution studies, with some information from screening studies and a limited amount from pathogenicity work, largely in pots.

Meloidogyne *species*

Root knot nematodes are by far the nematodes most commonly associated with cassava. They have been reported on cassava across Africa (Benin, Burundi, Cameroon, Côte d'Ivoire, Ghana, Kenya, Malawi, Mozambique, Niger, Nigeria, Tanzania, Togo, Uganda, Zambia and Zimbabwe); Asia (India, Malaysia, Taiwan and Thailand); the Pacific (Fiji and the Philippines); and the Americas (Antigua, Belize, Brazil, Colombia, Democratic Republic of Congo, Dominican Republic, Hawaii, Honduras, Peru, Puerto Rico, Trinidad and Tobago, Venezuela and the USA). *M. incognita* and *M. javanica* are the most important. *M. arenaria* and *M. hapla* are also reported (Tanaka *et al.*, 1979; Coyne *et al.*, 2003), although they are not of major concern. Numerous studies additionally make reference to *Meloidogyne* sp.

This may either be that identification to species was not attempted, or because the nematodes do not conform to the specifications of identified species. Furthermore, as with many crops, different *Meloidogyne* species can often occur in combination in the same situation.

Symptoms of damage

The typical knotting of the feeder and fine filamentous roots occurs and is the most obvious feature of *Meloidogyne* spp. infection (e.g. Bridge *et al.*, 1991). Such galling damage is common across cassava-growing areas but can vary considerably in the level of galling observed (Fig. 7.1, Plates 5A and C). However, the naturally 'knobbly' and rough texture of the feeder roots can disguise nematode damage (Coyne, 1995) (Fig. 7.2). The long duration over which cassava can remain in the ground and the common 'piece-meal' method of harvesting also mean that nematode-affected root systems may decompose in the ground or are not exposed at harvest for observation. In comparison with the damage reported on roots, less common and rarely documented is nematode damage to the storage roots themselves. In Kenya, severe damage to a small number of cassava germplasm lines (~1%) was observed in a breeder's selec-

Fig. 7.1. Galling of individual cassava roots infested with *Meloidogyne incognita*. (Photo: J. Bridge.)

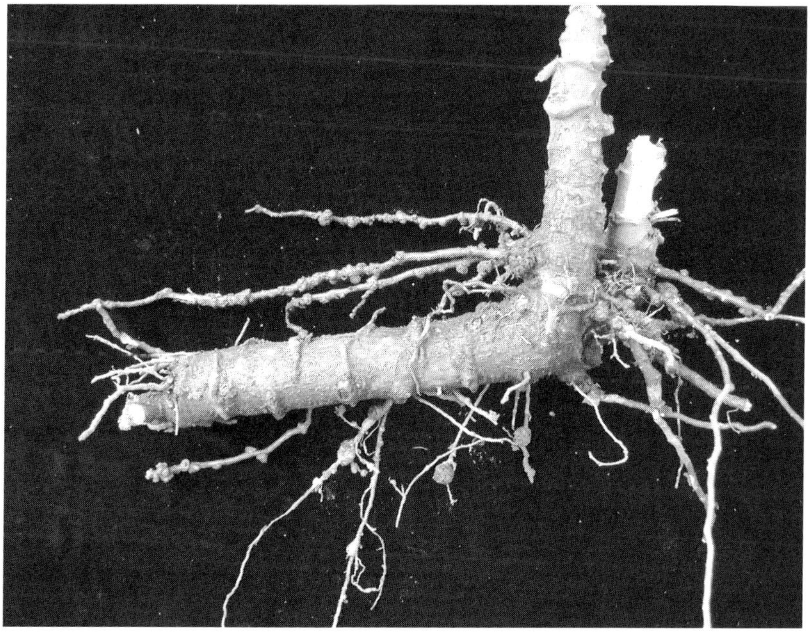

Fig. 7.2. Root system of cassava infested with *Meloidogyne incognita*. (Photo: D. Coyne.)

tion trial (Coyne *et al.*, 2004) (Fig. 7.3, Plate 5B). *M. incognita*, *M. javanica* and an unidentified *Meloidogyne* sp. were recovered from the cassava tissue. 'Bubbling' of the storage root surface occurred. In some cases, the surface was flaky in appearance with high levels of necrosis apparent under the surface, when thin sections were cut away, where the nematodes had infected the tissue. Other reports of extensive storage root deformation have been received from Mozambique (H.A.M. Van den Oever, 2000, Mozambique, personal communication) and limited observations of slight galling damage on tubers from Uganda (D.L. Coyne, unpublished data). Otherwise, *Meloidogyne* spp. associated with cassava concern solely the feeder (and fine) roots, with reports stating that damage does not occur to storage roots (Caveness, 1981; Coyne and Talwana, 2000; Makumbi-Kidza, 2001). Root knot nematodes do not appear to be directly related to rotting of storage tubers, although they are expected to lead to postharvest deterioration where storage roots are infected and galled. Above-ground symptoms of *Meloidogyne* spp. damage are not normally obvious. Under light infestation, increased aerial growth has been recorded (Caveness, 1982) and plant height observed to be unaffected following inoculation with *M. incognita* (Makumbi-Kidza *et al.*, 2000). Stem height and weight reduction (Gapasin, 1980, 1981; Caveness, 1981, 1982; Talwana *et al.*, 1997a) and reduced sprouting and establishment of cuttings, however, have been associated with high *Meloidogyne* populations (Talwana *et al.*, 1997a; Makumbi-Kidza *et al.*, 2000). In areas of Uganda, dead and dying cassava plants were associated with severe *M. incognita* infestation (Bridge *et al.*, 1991).

Although cassava contains cyanogenic glucosides, which probably form an element of the overall plant defence mechanism, there is little evidence to suggest that they are related to nematode damage or defence. In one study, assessment of root knot infection on 11 cassava cultivars was found to be unrelated to cyanide content (de Freitas and de Moura, 1986). However, Makumbi-Kidza (2001) found that two of ten clones assessed had higher cyanogenic

Fig. 7.3. Deformed and knobbly cassava root (left) due to feeding of *Meloidogyne incognita*. (Photo: D. Coyne.)

potential in storage roots in *M. incognita*-inoculated soil compared with non-inoculated soil. The remaining eight clones showed no difference between treatments. Makumbi-Kidza (2001) also showed that *M. incognita* egg mass formation on the feeder roots of selected clones was negatively correlated to the cyanogenic potential of the mother clone storage roots.

Disease complexes

There is little documented evidence that root knot nematodes form associations with other pests or pathogens on cassava. Galling and mechanical damage of roots by nematodes facilitate the entry and development of secondary pathogens, which will probably lead to increased levels of root necrosis and consequently reduced root weights compared with uninfected plants, as observed in some studies (Gapasin, 1980; Crozzoli and Hidalgo, 1992; Talwana *et al.*, 1997a; Coyne and Talwana, 2000). Preliminary data from studies in Nigeria have shown that the presence of *M. incognita* substantially increases the incidence and severity of damage to storage roots by *Botryodiplodia theobromae*, one of the main causal agents of root rot in Nigeria (Dixon *et al.*, 2003). However, the presence of *B. theobromae* resulted in reduced levels of galling on cassava roots (D.L. Coyne, unpublished data). Bridge *et al.* (1991), in Uganda, associated a possible secondary fungal root rot with severe nematode infestation in farmers' fields. The extent to which disease complexes occur, however, has been little investigated, and information is scarce, although nematode-infected roots are reportedly more susceptible to rot organisms (Théberge, 1985).

Economic importance

Under certain circumstances, root knot nematodes can be serious pests of cassava. However, while numerous pot and microplot studies have clearly demonstrated the highly pathogenic nature of *Meloidogyne* spp. on cassava (e.g. Caveness, 1981; Crozzoli and Parra, 1999; Coyne and Talwana, 2000; Makumbi-Kidza *et al.*, 2000), data on their economic impact on cassava are scarce and can be contradictary. Caveness (1982) showed that *Meloidogyne* spp. could cause 87% yield loss under heavy attack, with losses as high as 98% recorded in experimental plots (Théberge, 1985). However, evidence supporting similar or consistent levels of damage under farm conditions is limited, and quantification of the severe damage reported by Bridge *et al.* (1991) in Uganda is largely lacking. Gapasin (1980) concluded that initial populations of *Meloidogyne* spp. sufficiently large to cause injury to cassava are unlikely to occur naturally. Coyne and Namaganda (1994), however, observed root knot nematode galling damage to cassava roots on 94% of 88 fields examined in Uganda. Of those damaged, 17% were severely galled. Later, Coyne and Talwana (2000) related galling damage of cassava roots (cv. Ebwanateraka) from plants in farmers' fields in Uganda negatively with yield ($P = 0.05$). By extrapolating data (albeit crudely) from the two studies, Coyne (2003) estimated that 17% of Uganda cassava producers were losing 66% of their cassava to *Meloidogyne* spp. nematodes. In pot studies, Crozzoli and Parra (1999) established that the tolerance limit for aerial dry weight and root fresh weight on a cultivar (Tempranita) slightly affected by *M. incognita* race 2 was 1.0 J2/ml of soil. *Meloidogyne* spp. damage to cassava appears to be most important, in terms of crop yield response, at or before tuber initiation (Makumbi-Kidza *et al.*, 2000), a period when the crop is also most vulnerable to water stress (Ekanayake *et al.*, 1998). Therefore, it is possible that *Meloidogyne* spp. infection occurring after tuber initiation may lead to visually detectable galling damage, but not to yield reduction. This is possibly why there is difficulty in relating galling damage to yield from the field. Makumbi-Kidza *et al.* (2000) also determined that production loss by *M. incognita* was through a reduction in storage root number as opposed to a reduced weight of individual storage root.

In addition to the direct losses of both quantity and quality of the cassava crop, there is the added effect of reduced stem height and weight associated with high *Meloidogyne* populations (Gapasin, 1980, 1981; Caveness, 1981, 1982). This decreases the quality of the planting material available for the following season. Furthermore, the presence of *Meloidogyne* spp. at planting has been shown to suppress sprouting of cuttings (Talwana *et al.*, 1997a; Makumbi-Kidza *et al.*, 1999), suggesting that yields can be severely reduced through prevention of emergence.

Management measures

Considering the limited demonstration of damage by *Meloidogyne* spp. to cassava in the field, there has been little focus on management of the nematodes in cassava cropping systems, with the exception of varietal screening studies. Utilization of resistant cultivars on an international and national basis appears the most realistic and economical means of nematode management. However, in localized situations, particular management practices such as rotation systems, intercropping, fallowing, mulching and the use of nematicidal or antagonistic cover crops and by-products may be appropriate. Cassava cultivars differ considerably in their response to root knot nematodes (Da Ponte *et al.*, 1980; Caveness, 1981, 1982; Saka, 1982; Nwauzor and Nwankwo, 1989; Crozzoli and Hidalgo, 1992; Talwana *et al.*, 1997b; Coyne and Talwana, 2000; Makumbi-Kidza, 2001; Coyne *et al.*, 2004). Some cultivars have been recorded as immune, while others are highly susceptible. The *Meloidogyne* species screened against, however, has not always been identified, or has involved concomitant species. The differences in reaction of cultivars is no doubt due not only to the different species, races or pathotypes of *Meloidogyne* including combinations of species/pathotypes in the same location, but also to different population densities (McSorley *et al.*, 1983b). Caveness (1980), when screening cassava lines, determined that *M. incognita* was more aggressive than *M. javanica*.

Although yield increases have been obtained in Latin America with nematode control following soil fumigation (Da Ponte and Franco, 1981), the economic value of this is questionable (Hillocks and Wydra, 2002). Gapasin (1981) also reported that pre-plant application of the nematicides aldicarb, carbofuran and bunema increased yield. However, in field experiments in Uganda, Kenya and Nigeria, pre-plant application of phenamiphos or carbofuran in infested soils appears to have made little difference to yield, compared with untreated plots (D.L. Coyne, unpublished data). Neither did Diomandé (1982) obtain any yield improvement following fumigation with dibromochloropropane (DBCP) to control *M. javanica* in Côte d'Ivoire. Cassareep, a by-product of the cassava industry, was apparently effective in controlling *M. incognita* and *M. javanica* on cassava (Da Ponte and Franco, 1981). It is important to note that, as cassava production moves into monoculture and new high yielding cultivars are released, nematodes have the potential of becoming limiting factors in production in areas where the crop is being introduced.

***Pratylenchus* species**

P. brachyurus is probably the second most important nematode parasite of cassava after root knot. It occurs on the crop across cassava-growing locations around the world (McSorley *et al.*, 1983b). Other lesion nematodes have been associated with cassava, but not nearly to the extent of *P. brachyurus*. *P. pseudopratensis* is found in Nigeria (IITA, 1978), *P. zeae* in the Philippines (Timm, 1965) and *P. coffeae* in Java (de Fluiter and Mulholland, 1941 in McSorley *et al.*, 1983b) and Nigeria (Coyne, unpublished). Cassava is an excellent host and, in Ganavé, Togo, *P. brachyurus* was the most common nematode occurring on cassava, with population densities up to 400/g of root (DeGuiran, 1965). The nematode was attributed to contributing to a gradual yield decline over several years of production. However, in a field experiment at the same location, soil fumigation with DBCP improved yield by 8.5% for aerial growth and 7.9% for storage roots. In a greenhouse experiment in Brazil, an eightfold population increase in *P. brachyurus* density was observed after 3 months on cv. IAC-105.66 (Charchar and Huang, 1981). Zem (1979), however, reported that *P. brachyurus* caused no obvious damage to the crop in Brazil. Considerable variability in the reaction of cassava cultivars to *P. brachyurus* (Luc, 1971; Corbett, 1976) suggests that management of *P. brachyurus* may be possible through the use of tolerant or resistant cultivars. De Guiran (1965) described cultivars as highly susceptible and resistant depending on the development of *P. brachyurus* after 3 months on 42 cultivars.

Other nematodes of cassava

Despite the frequent occurrence of many other nematode species on cassava, there is little evidence of economic damage being caused by them. Some commonly occurring nematodes, such as *Aphelenchoides* spp. and *Aphelenchus avenae*, are primarily viewed as fungivorous. Their presence in high densities around plant roots, such as 15,000 in 10 g of root + 100 ml of soil (Coyne *et al.*, 2003), may possibly be related to fungal contamination of the roots, as observed by Bridge *et al.* (1991), or as part of the decomposition process. *R. reniformis*, although regularly associated with cassava, was found to decline under cassava (McSorley *et al.*, 1983b). *Scutellonema bradys* is reported from Nigeria (Caveness, 1967b), Ghana (Addoh, 1971) and Togo (Luc and de Guiran, 1960). While *S. bradys* causes substantial damage to yam production, no damage is reported for cassava. Cassava is described as an excellent host for *Scutellonema clathricaudatum*, along with *Helicotylenchus microcephalus* (Caveness, 1967a). In Uganda, cassava was host to at least six species of *Scutellonema* (*S. brachyurus*, *S. clathricaudatum*, *S. magniphasmum*, *S. paralabiatum*, *S. unum* and *Scutellonema* sp.) (Coyne *et al.*, 2003), and to four in Nigeria (*S. aberrans*, *S. bradys*, *S. cavenessi* and *S. clathricaudatum*) (F.E. Caveness, unpublished) but without causing any obvious damage. *Heterodera* spp. have not hitherto been associated with cassava but, in recent exploratory work in Cameroon, over 10% of fields surveyed contained *Heterodera* spp. juveniles from soil around cassava roots (Tambe, 1999), while a small number of cassava fields also contained *Heterodera* spp. juveniles in a survey in the Democratic Republic of Congo (D.L. Coyne, unpublished data).

Yams

Yams, *Dioscorea* spp., are probably one of the oldest food crops known to man (Alexander and Coursey, 1969). Their large-scale cultivation as food crops is restricted largely to three main areas of the world: West Africa, the Pacific area (including Japan), and the Caribbean, but is also of importance in parts of eastern Africa and tropical America. The majority of yams are produced in West Africa, where they are steeped in cultural history and revered as a cultural symbol of fertility. They are an essential element of marriages for instance in many West African cultures. Yam is the second most important root and tuber crop in the world and contributes more than 200 dietary calories everyday for over 60 million people (Nweke, 1991).

The genus *Dioscorea* consists of over 600 species, but only ten of these are important food yams: *D. rotundata* Poir., *D. cayenensis* Lam., *D. dumetorum* (Kunth) Pax., *D. hispida* Dennst., *D. alata* L., *D. esculenta* (Lour.) Burk., *D. bulbifera* L., *D. opposita* Thunb., *D. japonica* Thunb. and *D. trifida* L. In addition to the edible yams, a number of *Dioscorea* species have been commercially grown to provide a source of diosgenin, which is used in the manufacture of oral contraceptives, sex hormones and cortisone (Coursey, 1967; Purseglove, 1972; Kay, 1987).

Some yams produce single, large tubers, while others produce many small tubers. Yams can also form bulbils in the leaf axils as in *D. bulbifera* and some cultivars of *D. rotundata* and *D. alata*. Most yams have good storage qualities and can survive for periods of 3–4 months or longer. Therefore, they are relied upon for local food security and income generation. Yams are normally vegetatively propagated from whole, small tubers (seed tubers/seed yams), portions of tubers (setts) or bulbils. The small seed tubers can be formed by cutting and removing the main tuber during the growing season. They can also be produced by the use of 'minisetts' or 'microsetts' cut from tubers (International Institute of Tropical Agriculture, 1984). Yams can be monocropped but are more often intercropped. The ideal growing conditions are a long rainy season with rainfall of at least 1500 mm, a temperature of 30°C, and deep, loose, fertile soils (Coursey, 1972).

Nematodes of Yams

Many different nematode species have been found associated with yams. The nematodes of particular importance are endoparasites of roots and tubers. Those known to cause serious damage by mainly reducing tuber yield and quality are *Scutellonema bradys*, *Pratylenchus coffeae*, *Pratylenchus sudanensis* and *Meloidogyne* spp.

Scutellonema bradys

The yam nematode, *S. bradys*, is the cause of a decay of yam tubers known as 'dry rot disease'. It is found in many yam-growing areas of the world, having been reported from West Africa (Benin, Burkina Faso, Cameroon, Côte d'Ivoire, The Gambia, Ghana, Guinea, Mali, Nigeria, Senegal and Togo), the Caribbean (Barbados, Cuba, Dominica, Dominican Republic, Guadeloupe, Guatemala, Haiti, Jamaica, Martinique and Puerto Rico), Brazil, Venezuela (Crozzoli and Parra, 1991), Korea (Park *et al.*, 1998) and India.

Symptoms of damage

Dry rot of yams, which is directly associated with *S. bradys*, occurs in the outer 1–2 cm of tubers (Fig. 7.4, Plate 5D). The initial stage of dry rot consists of cream and light yellow lesions below the outer skin of the tuber.

There are no external symptoms at this stage. As the disease progresses, it spreads into the tuber, normally to a maximum depth of 2 cm but sometimes deeper. In these later stages of dry rot, infected tissues first become light brown and then turn dark brown to black. External cracks appear in the skin of the tubers and parts can flake off exposing patches of dark brown, dry rot tissues (Fig. 7.5, Plate 5E). The most severe symptoms of dry rot are seen in mature tubers especially during storage, when it is often associated with general decay of tubers. Dry rot, however, can also develop to quite an advanced stage without being visually obvious, causing deterioration of the tissue underneath an intact periderm and appearing healthy. Only once the surface is removed with a knife or thumbnail is the underlying damage revealed.

Fig. 7.4. Dry rot disease caused by *Scutellonema bradys* in the outer part of yam (*Dioscorea rotundata*) tuber (left) compared with healthy tuber (right). (Photo: J. Bridge.)

Fig. 7.5. Dry rot disease and flaking off of epidermis exposing dark diseased tissue underneath in yam (*Dioscorea alata*) tubers caused by *Scutellonema bradys*. (Photo: J. Bridge.)

No foliar symptoms have been observed on yams growing in soil infested with *S. bradys*.

Biology and life cycle

S. bradys is a migratory endoparasite present in yam soils, roots and tubers (Plate 6A). *S. bradys* is a vermiform nematode when mature, measuring about 1 mm in length, and has a well developed, stout stylet for puncturing cells. All active stages are infective. It invades the young, developing tubers through the tissues of the tuber growing point, alongside emerging roots and shoots, through roots and also through cracks or damaged areas in the tuber skin (Bridge, 1972).

Nematodes feed intracellularly in tuber tissues, resulting in rupture of cell walls, loss of cell contents and the formation of cavities (Goodey, 1935; Bridge, 1973; Adesiyan *et al.*, 1975a). They are mainly confined to the subdermal, peridermal and underlying parenchymatous tissues in the outer 1–2 cm of tuber. *S. bradys* continues to feed and reproduce in yams stored after harvesting. Populations can increase nine- to 14-fold in *D. rotundata* tubers over a 5–6 month storage period, and five- to eightfold in *D. alata* and *D. cayenensis*, respectively, over the same period (Bridge, 1973; Adesiyan, 1977). In tubers with partial dry rot, more nematodes are found in the oldest, apical portions, adjacent to the stems (Adesiyan, 1977). In Martinique, the highest rate of multiplication of *S. bradys* occurs within the tuber after it has been harvested and coincides with the initiation of tuber dormancy (Cadet and Quénéhervé, 1994).

S. bradys is also morphologically similar to two other species, *S. cavenessi* and *S. clathricaudatum*, which it has been suggested may all be synonomous with each other (Baujard and Martiny, 1995). Molecular assessment of *S. bradys* from within Nigeria and Benin has shown that substantial polymorphic variation does exist between different populations and between individuals within a population (V.M. Williamson and D.L. Coyne, unpublished data) but how this relates to biology is unknown.

Survival and dissemination

No true survival stage is known with *S. bradys*, but populations are maintained in the absence of yams probably on other host plants. Sizeable populations of the nematode can be found in soil at the beginning of the yam-growing season (Obigbesan and Adesiyan, 1981; Adesiyan and Badra, 1982).

Yams are propagated from whole tubers or pieces of tuber, which are the principal means of dissemination of *S. bradys*. Comparatively low populations of the nematodes in tubers do not produce external symptoms of damage (Bridge, 1973) and thus the risk of dissemination by this means is greater. Infested seed tubers rather than soil are probably the main source of nematode inoculum in yam fields.

Environmental factors affecting parasitism

Nematodes in stored tubers are affected by storage conditions. Populations of *S. bradys* increase at twice the rate in tubers stored at 22–32°C and relative humidity 40–85% compared with those in tubers stored at 16–18°C and relative humidity 80–85% (Adesiyan, 1977).

Other hosts

The most commonly grown food yams are all hosts of *S. bradys* and susceptible to dry rot disease. In West Africa, the *Dioscorea* species known to be attacked are *D. alata*, *D. bulbifera*, *D. cayenensis*, *D. dumetorum*, *D. esculenta* and *D. rotundata* (Baudin, 1956; Caveness, 1967a; Smit, 1967; Bridge, 1982). Wild *Dioscorea* spp. from forest soils in Nigeria and Cameroon have been shown to support populations of *S. bradys* and to cause dry rot in tubers (Bridge, 1982; Bridge *et al.*, 1995). Also the wild yam, *D. praehensilis*, from the Republic of Guinea is reported to be susceptible to *S. bradys* (Kwoseh, 2000). *D. alata*, *D. bulbifera*, *D. cayenensis*, *D. rotundata*, *D. trifida* and *D. transversa* are hosts of *S. bradys* in the Caribbean (Decker *et al.*, 1967; Ayala and Acosta, 1971; Belliard and Kermarrec, 1978; Kermarrec *et al.*, 1987), *D. cayenensis* in Brazil (Moura *et al.*, 1978) and *D. alata* in India (Nadakal and Thomas, 1967). There are many other crop and weed hosts of *S. bradys* (Luc and de Guiran, 1960; Adesiyan, 1976b; Bridge, 1982), but most plants are relatively poor hosts in comparison with yams. Sesame and cowpea support high root populations, and melon can increase soil populations. *S. bradys* also occurs on other root and tuber crops, such as cassava (Missah and Peters, 2001), *Xanthosoma* sp., *Colocasia esculenta* and *I. batatas* (Kermarrec *et al.*, 1987), although none appears to be a particularly good host.

Disease complexes

Dry rot disease can be caused by *S. bradys* in the absence of other organisms (Bridge, 1973; Adesiyan *et al.*, 1975a), although it has been suggested that the disease is caused by a bacterium, *Corynebacterium* sp., in association with *S. bradys*, which acts as a wounding agent (Ekundayo and Naqvi, 1972). The more extensive, internal decay of tubers known as 'wet rot', 'soft rot' or 'watery rot' is associated with fungal and bacterial pathogens (Adeniji, 1970; Ogundana *et al.*, 1970; Ekundayo and Naqvi, 1972). This general decay of tubers, which is a serious problem in stored yams, is increased when tubers are wounded or damaged (Adeniji, 1970; Ogundana *et al.*, 1970). The damage caused by nematodes can predispose the tubers to invasion by decay organisms, resulting in complete rotting of the tubers (Goodey, 1935). The principal fungi causing internal tuber decay are *Botryodiplodia theobromae* and *Fusarium* sp., although other fungi and a bacterium, *Erwinia* sp., are frequently isolated from decaying tissues (Coursey, 1967; Adeniji, 1970; Ogundana *et al.*, 1970; Ekundayo and Naqvi, 1972; Moura *et al.*, 1976; Demeaux *et al.*, 1982). Nematodes and fungi are found together in the transitional stage between dry rot and wet rot, but nematodes do not occur in the 'late wet rot' stage deep in the tubers (Adesiyan *et al.*, 1975a).

In the West Indies, *S. bradys* infrequently occurs together in the same tubers with *P. coffeae*; however, the most usual situation is infestation by one species only. The establishment of one species in tuber tissues apparently prevents concomitant infection by the other species (Castagnone-Sereno and Kermarrec, 1988). When both species are present, *P. coffeae* dominates over *S. bradys* (Acosta and Ayala, 1976a). Consequently, *S. bradys* is viewed as less of a yam problem in Caribbean islands such as Martinique and Guadeloupe, where it is seen to be displaced by *P. coffeae* (P. Cadet and P. Quénéhervé, personal communication).

Economic importance

The primary importance of *S. bradys* is in the direct damage it causes to the tubers, but the relationships between this damage and loss in total yield is difficult to determine (Wood *et al.*, 1980). However, weight differences between healthy and diseased tubers harvested from the field have been estimated to be 20–30% in Côte d'Ivoire (Smit in Bridge, 1982) and 0–29% in Nigeria (Wood *et al.*, 1980). Weight reduction due to moisture loss is more likely to occur in late harvested tubers left in dry soil (Bridge, 1982). Water loss from tubers continues during storage and is significantly greater in tubers infected with *S. bradys* compared with healthy tubers (Adesiyan *et al.*, 1975b; Cadet and Quénéhervé, 1994).

Dry rot of yams alone causes a marked reduction in the quality, marketable value and edible portions of tubers, and these reductions are more severe in stored yams. When dry rot is followed by wet rot in stored yams, losses of whole tubers can be as high as 80–100% (Adesiyan and Odihirin, 1975), but losses certainly increase with duration of storage. The degree of preharvest damage to tubers by *S. bradys* varied from 0 to 40% in Nigeria (Wood *et al.*, 1980). About 46.6% of IITA yam germplasm screened were naturally infested with *S. bradys* (Kwoseh, 2000). Also, almost 47% of all tubers on sale in Nigerian markets were infested with *S. bradys* (Bridge, 1973), and both dry rot and wet rot diseases of tubers have been observed in all Nigerian yam barns and markets sampled (Adesiyan and Odihirin, 1977). Nematode infection contributes to long-term storage losses, which have been estimated as 50% (Coursey, 1967).

Populations in the outer peelings of rotted yam tubers can average 100,000 nematodes (Adesiyan *et al.*, 1975a) and can exceed 300,000 nematodes/50 g of tuber peelings (Bridge, 1973). Low populations of the nematode produce only discrete areas of yellow necrotic tissues or dry rot internally, and populations in excess of 1000 nematodes/50 g of tuber peelings are necessary to produce observable, external symptoms of damage (Bridge, 1973). Preharvest losses due to *S. bradys* have received relatively little attention in comparison with postharvest impact. Cadet and Daly (1996) found that nematicide treatment of seed infested with *S. bradys* gave 14–15% yield increase, but this was not significant. K. Green (2001, Nigeria, personal communication) established that inoculation of high levels of *S. bradys* resulted in no differences in yield compared with uninoculated plants but that weight reduction during storage was 30% greater after 2 months in the inoculated plants (which had 188 *S. bradys*/g of tuber peel), while 68% were discarded, compared with 30% of the uninoculated plants.

Management measures

The management measures that can be used are: (i) controlling nematodes in field soil by chemical and cultural means; (ii) use of nematode-free planting material or treatment of seed tubers and setts prior to planting to reduce or eliminate nematodes from propagative material; and (iii) treatment of tubers after harvesting to prevent storage losses.

CULTURAL. Keeping fallow land free of all host plants is a suggested control of *S. bradys* in Cuba (Decker *et al.*, 1967), but this is not always economical or practical.

Rotation of crops to control *S. bradys* is also not always an appropriate option as yams are often grown as the first crop in a rotation after fallow. However, as demand for land increases through demographic pressure, fallow periods reduce and cropping systems change, using non- or poor crop hosts in rotations or as intercrops will help reduce soil populations (Adesiyan, 1976) as will the use of non-host cover crops. The cover crops *Aeschynomene histrix*, *Pueraria phaseoloides* and *Mucuna pruriens* (*utilis*) significantly suppressed *S. bradys* populations in tubers and plots compared with plots without cover crops (Claudius-Cole *et al.*, 2003). Cover crops such as *Tagetes* species, *Stylosanthes gracilis*, *Centrosema* species, *Aspilia latofolia* and groundnut (peanut) have been recommended for use to lower nematode populations and restore fertility for yam production in Nigeria (Atu and Ogbuji, 1983). Mulching has also been reported generally to reduce nematode populations compared with pre-planting levels in the soil (IITA, 1976). Crops that are known to support high populations of *S. bradys* such as cowpea, sesame, green gram, pigeonpea, kenaf, okra, tomato and melon should be avoided.

Yams are frequently intercropped, sometimes with as many as five other crops (Coursey, 1967). If these crops are hosts of *S. bradys*, they will encourage build-up of nematode densities, increasing the chances of damage to the tubers. As an example, populations of *S. bradys* significantly increased in yams intercropped with cowpea in Nigeria (Atu, 1991). Non-hosts of *S. bradys* should be used, where possible, to reduce the chances of damage to the tubers. Similarly, weed control and the exclusion of weed hosts of *S. bradys*, such as *Eupatorium*, *Synedrella* and *Chromolaena*, from around yams will help to reduce nematode damage (Adesiyan, 1976). The use of nematode-free propagative material is by far the most appropriate means of preventing nematode damage. Seed tubers showing symptoms of dry rot (cracking and flaking) should not be used for planting. The presence of dry rot in tubers without external symptoms can be determined by scraping away sections of tuber skin, or by the use of tuber pieces rather than whole tubers, enabling the grower to examine for dry rot symptoms before planting. Pieces from different parts of the tubers often contain varying population levels of *S. bradys* (Adesiyan, 1977). Any foliar material used for propagative material will be completely free of *S. bradys*. Yams, such as *D. bulbifera* and some forms of *D. alata*, can be readily propagated from bulbils or aerial tubers. A number of yams, such as *D. alata*, *D. rotundata* and *D. dumentorum*, can be produced from vine cuttings (Coursey, 1967). Even true seed can be used for propagating *D. rotundata* (Sadik and Okereke, 1975). Although these methods of propagation are not a practical means of producing ware tubers, they can be used to produce nematode-free seed tubers.

The method used to produce large numbers of seed tubers from relatively few yams by growing 'microsetts' or 'minisetts' cut from mature tubers (International Institute of Tropical Agriculture, 1984) will effectively produce nematode-free propagative material as long as clean, healthy 'mother seed yams' are selected. Likewise, the use of tissue-cultured plantlets can provide pest- and disease-free planting material. However, tissue-cultured plantlets can be relatively expensive, and it will probably be some time before systems are in place to deliver these high-tech alternatives routinely. IITA, however, makes extensive use of tissue-cultured material when providing newly bred material to national research programmes, in order to comply effectively with phytosanitary regulations (IITA, 2000). The use of wood ash to coat yam setts before planting is a traditional practice amongst some yam growers and can enhance tuber formation, but does not markedly decrease numbers of nematodes in tubers. Mixing cow dung in yam mounds before planting at a rate of 1.5 kg per mound (1886.3 kg/ha) can increase yields of tubers and significantly decrease nematode numbers (Adesiyan and Adeniji, 1976). Other organic manures may have a similar effect on nematode populations in yam mounds. The use of neem has also been investigated; nematode management has been observed and yields increased following the application of neem powder at 2.5 t/ha to the soil (Onalo *et al.*, 2001). NPK fertilizer can reduce *S. bradys* populations in tubers of *D. alata* to a very low level. In contrast, nitrogen alone can increase both populations of *S. bradys* and the percentage of infested tubers of *D. rotundata*, whereas phosphorus alone can decrease the percentage of infested tubers. These results support observations by farmers in certain yam-growing areas of Nigeria that yams fertilized with nitrogen alone do not store well, but yams fertilized with mixtures that contain phosphorus store longer (Adesiyan and Adeniji, 1976). However, the conditions under which experiments are conducted and the species of yams and cultivars used are likely to have a substantial influence on the results obtained. For example, *S. bradys* populations increased on *D. rotundata* but not *D. alata* or *D. cayenensis* following application of high rates of nitrogen combined with phophorus (Obigbesan and Adesiyan, 1981). High rates of nitrogen alone led to an increase in nematode populations, especially on *D. rotundata*.

HOT WATER TREATMENT. Hot water treatment (HWT) can reduce or eliminate *S. bradys* from tubers. While farmers appreciate the benefits of the treatment, the cost and the access to resources (e.g. firewood), the labour requirements and the expense of heating equipment, and the difficulties of maintaining constant temperatures are the main prohibitive factors against its widespread use by farmers. However, it is feasible for small-scale operations and for establishing nematode-free planting material (Speijer, 1996; Meerman and Speijer, 2001).

Most studies have shown that a water temperature of 50–55°C for up to 40 min gives the best control of *S. bradys* without damaging tubers. The age of the tuber, the species of *Dioscorea* and cultivar being treated, and the severity of infestation of the tubers will affect nematode control by HWT (Ayala and Acosta, 1971; Bridge, 1975; Acosta and Ayala, 1976; Adesiyan and Adeniji, 1976; Castagnone-Sereno, 1988). There is also some indication that HWT of tubers can adversely affect sprouting ability of seed pieces cut into minisetts (D.L. Coyne, unpublished data). The time of treatment can be critical. *D. rotundata* tubers treated immediately after harvesting rot completely, but those treated after a storage of 2–6 months show little sign of deterioration, although those treated soon after dormancy has broken are slower to sprout (Bridge, 1975; Adesiyan and Adeniji, 1976).

RESISTANCE AND TOLERANCE. Resistance to *S. bradys* has not been found in any of the landraces or accessions examined in two of the main yam species, *D. alata* and *D. rotundata* (Kwoseh, 2000; Kwoseh *et al.*, 2002). There is no firm evidence of complete resistance to *S. bradys* in yams, and all the main food yams (*D. alata*, *D. bulbifera*, *D. cayenensis*, *D. esculenta* and *D. rotundata*) are susceptible to damage. All cultivars of *D. alata*, *D. cayenensis* and *D. rotundata* that have been examined in West Africa were susceptible to infection by *S. bradys* (Adesiyan, 1977; Bridge, 1982). However, variations in relative susceptibility have been reported, and *D. dumetorum* is generally considered to be less readily invaded than other species. Two *D. dumetorum* accessions and *D. cayenensis* var. Afun screened were confirmed to support low levels of *S. bradys* reproduction and without damage (Kwoseh, 2000; Kwoseh *et al.*, 2002). In Puerto Rico, a casual observation suggests that *D. alata* cv. Florido is not susceptible to nematode attack (Ayala and Acosta, 1971). *S. bradys* resistance in yams is not likely to be controlled by recessive genes as hybrids derived from susceptible parents are also susceptible (Kwoseh, 2000).

CHEMICAL. Chemical control of *S. bradys* on yams has had some success, but information on the economics of this means of control is lacking for large-scale use. DD and 1,3-dibromochloropropane (DBCP) applied as soil treatments have, at best, only produced moderate yield increases and control of *S. bradys* (Anonymous, 1964; Ayala and Acosta, 1971). Four granular nematicides (aldicarb, oxamyl, carbofuran and miral or isazophos) applied as post-plant treatments in yam mounds 2 weeks after planting at a rate of 2 kg a.i./ha reduced soil populations of *S. bradys* to very low levels, with remarkable yield increases recorded. There was some accumulation of toxic residues in harvested tubers (Adesiyan and Badra, 1982).

Chemotherapy of tubers as a practical means of nematode control for yam growers could be an economical proposition. In West Africa, farmers often mix pesticides together with wood ash to coat seed yam pieces. However, the product used and rates of application depend on the availability of pesticides and cash, while the pesticides themselves may have been 'diluted' by traders or their shelf life may have expired. Significant increases in yield have been obtained by soaking tuber pieces of *D. alata* infected with *S. bradys* for 30 min in 1000 ppm a.i. aqueous solutions of the nematicides DD, carbofuran and oxamyl; the disinfectants calcium hypochlorite and formalin; and nitrogenous fertilizers ammonium sulphate and calcium nitrate. Tuber pieces are drained and air dried before planting. All treatments reduced *S. bradys* populations in tuber tissues, but none of them eliminated nematodes from the yams (Badra and Caveness, 1979). Soaking tubers in oxamyl prior to planting is recommended to control both *S. bradys* and *P. coffeae* (Castagnone-Sereno, 1988). Coating yam seed tubers with liquid ethoprophos and cadusafos does control populations of *S. bradys* in yam tissues. Although this does not produce yield improvement in the field, it does greatly reduce the infestation and storage rot of tubers. Three times as much untreated tuber yield was lost compared with treated tubers, and 80% of seed tubers harvested from the treated plants were nematode free compared with only 30% from the untreated plants (Cadet and Daly, 1996). It is also possible to use readily available household disinfectants as tuber dips to control *P. coffeae* in yams (Hutton, 1998; see below), and this could also be effective in controlling *S. bradys*.

BIOLOGICAL CONTROL. Some investigators are optimistic about the potential of biological control of nematodes. Biological control of nematodes was reported to be about a decade away from practical reality (Sterling, 1992); however, to date, there is no report in the literature about the use of biological agents for the control of *S. bradys*.

INTEGRATED MANAGEMENT. There is the need to formulate an adaptable package that will suit the circumstances of the majority of yam farmers. However, a minimum use of chemical nematicides to lower high populations of nematodes in the soil, and management of these lowered populations with nematode-free planting material obtained by hot water dip is suggested. Sanitation methods, HWT and nematicides as preplant dips have been used to achieve some control of parasitic nematodes (Swennen, 1990). The use of carefully selected nematode-free tubers combined with a fallow period will effectively reduce nematode problems to yam cultivation, e.g. *S. bradys* in *D. alata* (Bridge, 1978; Caveness, 1992).

Diagnosis

Assessment of the incidence and extent of dry rot disease in yam tubers can be done by direct observation. In tubers without obvious external symptoms of damage, it will be necessary to scrape away the surface layers, or section tubers to determine the presence of dry rot.

Nematodes will be found in soil and roots, which can be sampled, particularly at the end of the growing season. However, most nematodes will be found in tuber tissues, and sampling of these is the most appropriate means of assessing populations and the importance of *S. bradys.* Peelings of a known thickness (1 or 2 cm) are cut from tubers. These are chopped finely, teased apart or preferably macerated before placing on a support tissue or sieve in water (see Chapter 3). Thirty to 50% of nematodes will emerge from tissues in the first 3 days, but they will continue migrating from the tissues for over 20 days.

Pratylenchus coffeae

P. coffeae is widely distributed on many different crops throughout the tropics. It is recorded as a parasite of yams in Barbados, Jamaica, Puerto Rico and Belize (Ayala and Acosta, 1971; Brathwaite, 1977; Coates-Beckford and Brathwaite, 1977; Bridge *et al.*, 1996), Brazil (Moura *et al.*, 2001), China (Huang *et al.*, 1994), Taiwan (Tsay *et al.*, 1994) and in the Pacific islands of Papua New Guinea, Fiji, Niue, Tonga, Vanuatu and Solomon Islands (Bridge, 1988). *P. coffeae* is the cause of tuber dry rot disease of yams, known locally in Jamaica as 'burn'.

Symptoms of damage

The dry rot symptoms caused by *P. coffeae* in yam tubers are indistinguishable from those caused by *S. bradys* (Plate 5F). Brown, irregular dry rot extends 1–2 cm into the outer tissues of *D. rotundata* tubers (Acosta, 1974), but can occur as deep as 5 cm in *D. alata* tubers (Bridge and Page, 1984). The dry rot can be more pronounced in the oldest apical portions of the tubers adjacent to the vines (Acosta, 1974), or even restricted to these portions in newly harvested tubers (Bridge and Page, 1984). External symptoms observed on tubers of *D. alata*, *D. cayenensis* and *D. rotundata* are deep cracks, a corky appearance, exposed dark brown rotted areas, and diseased tubers being spongy to the touch (Thompson *et al.*, 1973; Acosta and Ayala, 1975; Bridge and Page, 1984). Necrosis or rotting caused by *P. coffeae* has also been observed in tubers of *D. esculenta* (Bridge and Page, 1984) and *D. trifida* (Hickling, 1974).

Above-ground symptoms of damage are not as obvious. Vines from tubers severely infected with *P. coffeae* are shorter and unthrifty (Coates-Beckford *et al.*, 1978). Planting material with a high proportion of dry rot can result in non-sprouting of tubers and poor stands in yam fields (Coates-Beckford and Brathwaite, 1977).

Biology

P. coffeae is a migratory endoparasite of yam roots and tubers. It is assumed to have a life cycle of 3–4 weeks on *Dioscorea* spp. (Thompson *et al.*, 1973), and the general behaviour of *P. coffeae* in yam tubers is probably very similar to that of *S. bradys*.

No information is available on whether *P. coffeae* of yams is a separate biological race from those that are important parasites of other crops, although this possibility does exist. An isolate of *P. coffeae* from banana in Ghana multiplied in roots of yams, but tubers had none or very low nematode populations and did not have any associated dry rot symptoms (Kwoseh, 2000).

P. coffeae reproduces and multiplies in stored yams and is disseminated in seed tubers. Numbers have been seen to increase from 185 nematodes/g of tuber tissue at harvest to 1450/g at planting (Kermarrec *et al.*, 1988). Hot water treatment for 45 min at 45°C increased yields by 23% in Jamaica (Hutton *et al.*, 1982). It can also be introduced into yam fields in the roots and plant tissues of other crops. The nematodes can survive in field soil between yam crops on other hosts. Soil contaminating machinery, tools, reusable containers, shoes, boots and clothing, animal hooves and fur are easy means of dispersal within and between fields (Adesiyan *et al.*, 1990).

Temperature can have a marked effect on nematodes. During storage, at ambient temperatures of 24–31°C, *P. coffeae* populations can rise to very high levels (939/g), but in tubers stored at 12–13°C the numbers of nematodes remain very low (< 1/g) (Thompson *et al.*, 1973).

Other hosts

P. coffeae is a parasite of *D. alata*, *D. cayenensis*, *D. esculenta*, *D. rotundata* and *D. trifida*. It has also been found associated with *D. bulbifera* in the Pacific (Orton Williams, 1980). In addition to yams, *P. coffeae* has an enormous host range covering almost all plant families.

Disease complexes

Dry rot of yams caused by *P. coffeae* is associated with other soft and wet rots in stored tubers (Coates-Beckford and Brathwaite, 1977; Bridge and Page, 1984). It is likely that similar inter-relationships between nematodes and other organisms that have been described or suspected with *S. bradys* also occur with *P. coffeae*.

Economic importance

P. coffeae is important as a parasite of the tubers, reducing their edible portions, marketable value and, particularly, their storage qualities. Where the nematode occurs, it can be very widespread. In Jamaica,

67–100% of *D. rotundata* and *D. cayenensis* tubers were found to be infected with *P. coffeae* (Thompson *et al.*, 1973), and the nematode is reported to cause considerable losses to the same yam species in Guadeloupe (Kermarrec *et al.*, 1988). Over 50% of *D. alata* tubers examined in Papua New Guinea had obvious signs of dry rot and were infested with *P. coffeae* sometimes in numbers in excess of 60,000 nematodes/50 g of tissues (Bridge and Page, 1984). *P. coffeae* has been found to cause 30–100% disease incidence on Chinese yam (Haung *et al.*, 1994).

Yield reduction, as measured by weight of tubers, mainly results from planting seed tubers infested with *P. coffeae*. However, yield reduction in relation to numbers of high-quality tubers produced can occur when *P. coffeae* is present initially in the soil. Soil populations of 600 *P. coffeae* per plant of *D. rotundata* can produce significant tuber damage, and 1000 nematodes per plant can cause complete deterioration and severe reduction in tuber quality. However, neither of these populations causes reduction in total weight of harvested tubers (Acosta and Ayala, 1975, 1976a). If seed tubers are badly affected by dry rot, they can be so weakened that sprouting does not occur (Coates-Beckford and Brathwaite, 1977).

Management measures

The management options that have been described against *S. bradys* are, in most cases, applicable to control of *P. coffeae*. The main exception is in the use of crop rotations because of the different host range of *P. coffeae*.

CULTURAL. Using plant material that is free of nematodes is an effective means of controlling or reducing damage by *P. coffeae*, as detailed for *S. bradys*. As with *S. bradys*, central or distal tuber pieces, which generally contain the least *P. coffeae*, are recommended for propagative material (Acosta, 1974).

P. coffeae has an extremely wide and varied host range, and there are few reports of resistant crops against the yam isolates of *P. coffeae*, making it difficult to recommend any effective crop rotation practices. However, in Puerto Rico, rotating *D. alata* cvs Kinabayo, Florido and Gunung with the highly susceptible *D. rotundata* cv. Habanero significantly reduced dry rot and improved the quality and yield of the Habanero yams (Oramas Nival and Rodriguez, 2002). The weeds *Rottboellia exalta* and *Setaria barbata*, commonly found in yam plantations in Guadeloupe, are excellent hosts for *P. coffeae*, and it is recommended that they are removed (Kermarrec *et al.*, 1988).

PHYSICAL. The theoretical, but not always practical, control of *P. coffeae* in yam tubers can be achieved by HWT similar to that for *S. bradys*. Immersion of tubers in hot water can markedly reduce tuber populations of *P. coffeae* but rarely eliminates them without damaging the tuber. Hot water at 46–52°C for 15–30 min has been recommended for control of *P. coffeae* in *D. rotundata* tubers (Acosta and Ayala, 1976b). Use of seed tubers with extreme dry rot should be avoided as the treatment of these is less effective. Treatments in water at 51°C for 15–35 min have also effectively suppressed populations of *P. coffeae* and dry rot in *D. rotundata* and *D. cayenensis* tubers as well as increasing vine growth (Coates-Beckford *et al.*, 1978; Kermarrec *et al.*, 1988). However, HWT can cause severe physiological damage (Thompson *et al.*, 1973; Coates-Beckford *et al.*, 1977).

RESISTANCE AND TOLERANCE. It is suggested that *D. alata* cv. Florido is not susceptible to attack by *P. coffeae* (or *S. bradys*) in Puerto Rico (Ayala and Acosta, 1971). *D. esculenta* is possibly less susceptible to *P. coffeae* because of its different growth habit (Bridge and Page, 1984).

CHEMICAL. Chemical treatments of tubers prior to planting or storage have been tested for control of *P. coffeae*. No treatment with chemicals has been found to completely eliminate nematodes from tubers, but oxamyl dips can greatly reduce *P. coffeae* populations in tubers (Oramas Nival, 2002).

Field treatments to control *P. coffeae* are reported to be successful but, as with *S. bradys* on yams, the economics of their use in different situations have not been determined. Aldicarb as a single application at planting at a rate of 5.4 kg a.i./ha can give 72% control of *P. coffeae* (and *Rotylenchulus* sp.) and significantly increase high-quality tuber yields of *D. rotundata* in Puerto Rico. This nematicide is more effective than carbofuran and fensulfothion (Roman *et al.*, 1984a). Significant increases in yield of *D. rotundata* have also been obtained by a combination of foliar and seed tuber treatments with oxamyl (Roman *et al.*, 1984b).

It has been found in Jamaica that the simple and readily available household disinfectants, 'Dettol', 'Jeyes Fluid' and bleach can be as effective as the nematicide oxamyl in controlling *P. coffeae* and dry rot when used as tuber dips prior to planting (Hutton, 1998).

Meloidogyne species

The root knot nematodes, *Meloidogyne* spp., have been found on yams in Africa (Benin, Burkina Faso, Côte d'Ivoire, Ghana, Mali, Nigeria, Tanzania, Togo and Uganda), the Caribbean (Jamaica, Martinique, Puerto Rico and Trinidad), the Pacific (Fiji, Kiribati, Niue, Papua New Guinea and Western Samoa), Brazil, Costa Rica, Guatemala, China, Korea and Japan. The species of *Meloidogyne* identified as parasites of yams are *M. arenaria*, *M. hapla*, *M. incognita* and *M. javanica*, but worldwide *M. incognita* is the most important. *M. incognita* was the most prevalent and widespread nematode on yam in Ghana (Kwoseh, 2000).

Symptoms of damage

Meloidogyne spp. cause typical knotting or galling of yam roots. In addition, nematodes parasitizing the tubers produce galls in the outer tuber tissues, giving rise to abnormal, warty or knobbly tubers. In older tubers, dark necrotic spots can be observed in the outer tuber tissues surrounding individual females. Internal rotting of tubers has also been found associated with *Meloidogyne* spp. in certain yam species. Sprouting from galled tubers can be reduced or suppressed, and root proliferation from galls on tubers ('crazy root symptoms') can occur (Schieber, 1961; Jenkins and Bird, 1962; Bridge, 1973; Kermarrec, 1974; Adesiyan and Odihirin, 1978; Nwauzor and Fawole, 1981).

Foliar symptoms on food yams are observed occasionally. Early yellowing, leaf fall and termination of vine growth have been seen on *D. rotundata* infected with *M. incognita*, but infection only rarely reduces total tuber yield of these yams (Adesiyan and Odihirin, 1978; Nwauzor and Fawole, 1981; Atu *et al.*, 1983). *M. incognita* produces obvious galling on tubers of *D. trifida* (Kermarrec, 1974) and on *D. rotundata*, *D. alata* and *D. praehensilis*, as well as intraspecific yam hybrids (Kwoseh, 2000). Seedlings of 'medicinal' yams (*D. composita* Hemsl., *D. floribunda* Mart. et Gal. and *D. spiculiflora* Hemsl.) used for the production of cortisone precursors can be severely stunted or killed by *M. arenaria* and *M. incognita*, mainly the latter, with foliar chlorosis and leaf dieback (Schieber and Lassman, 1961; Jenkins and Bird, 1962; Bruhn and Koeh, 1963).

Biology and life cycle

The behaviour of *Meloidogyne* in yam roots is similar to that in other crops (Chapter 2), but in tubers there are some unusual features.

The life cycle of *M. incognita* in *D. rotundata* or *D. alata* tubers is 35 days (Nwauzor and Fawole, 1981). In *D. alata*, most nematodes are concentrated to a depth of 2 mm, with none beyond the 8 mm depth; in *D. rotundata*, they are concentrated at depths between 4 and 6 mm, with few at 14 mm (Nwauzor and Fawole, 1981). Females and egg masses produced in tuber tissues of *D. composita*, *D. floribunda* and *D. spiculiflora* become surrounded by lignified cells, preventing migration of hatched juveniles into surrounding tissues and causing their death (Bruhn and Koch, 1962; Jenkins and Bird, 1962; Koch, 1975). In *D. rotundata*, a similar host reaction occurs with *M. incognita*, which either kills or decreases juvenile and egg populations in stored tubers (Bridge, 1973; Nwauzor and Fawole, 1981). *M. hapla* develops in tubers of *D. batatas* (= *D. opposita*) until eggs are produced, and these only hatch when the tuber decays (Kawamura and Hirano, 1961).

Races

Host races are known in *Meloidogyne*, but it has not been determined which races, if any, are peculiar to yams. *M. incognita* race 2 is reported to infest *D. alata*, *D. bulbifera*, *D. cayenensis*, *D. esculenta* and *D. rotundata* in Nigeria (Atu *et al.*, 1984).

Survival and dissemination

Where *Meloidogyne* juveniles and/or eggs survive in stored tubers, they will be disseminated in propagative material. However, *Meloidogyne* species have extremely wide host ranges, and damaging populations will come from field soil having survived on other weed hosts, or be introduced into yam fields on infested seedlings of other crops.

Hosts

Susceptible yam hosts of *M. incognita* are *D. alata*, *D. bulbifera*, *D. cayenensis*, *D. composita*, *D. esculenta*, *D. floribunda*, *D. praehensilis*, *D. rotundata*, *D. spiculiflora* and *D. trifida*; hosts of *M. javanica* are *D. alata*, *D. opposita* and *D. rotundata*, and *D. opposita* (= *D. batatas*) is a host of *M. arenaria* in China (Gao *et al.*, 2000) and *M. hapla* in Japan (Kawamura and Hirano, 1961) and Korea (Park *et al.*, 1998). In addition to yams, *Meloidogyne* spp. have a very wide host range on weeds and crop plants.

Disease complexes

Yam tubers infested with *Meloidogyne* spp. are more prone to fungal and/or bacterial rot during storage than tubers free of the nematodes (Schieber, 1961; Schieber and Lassmann, 1961; Badra *et al.*, 1980; Nwauzor and Fawole, 1981).

Economic importance

Meloidogyne spp. adversely affect the marketable value of tubers because of the unappealing, warty appearance, and they are associated with rot of stored yams.

M. incognita completely destroyed a crop of *D. trifida* in Martinique at soil populations of 30,000 juveniles/100 g of soil (Kermarrec, 1974), and in Nigeria a combination of root knot nematodes and *S. bradys* caused the abandonment of large areas of yam farms (Adesiyan and Odihirin, 1977). Yields of yams severely infested with *M. arenaria* can be reduced by 24–80% in China (Gao, 1992). *M. javanica* populations of 30,000 nematodes per plant can reduce yields of *D. opposita* by over 50% (Nishizawa, 1973). Lower populations (5000 nematodes per plant) of both *M. incognita* and *M. javanica* significantly reduce yields of *D. alata* but not of *D. cayenensis* or *D. rotundata* (Adesiyan and Odihirin, 1978). Even populations of *M. incognita* as low as 100 juveniles per plant are said to reduce tuber yields of *D. rotundata* in India (Mohandas and Ramakrishnan, 1997). Other results suggest that reduction in yield is not the important part of nematode damage with *D. rotundata* as both relatively low and very high populations of *M. incognita* and *M. javanica* (100,000 and 156,000 eggs or juveniles per plant) do not appreciably decrease tuber weights (Acosta and Ayala, 1975; Nwauzor and Fawole, 1981; Atu *et al.*, 1983; Atu and Ogbuji, 1986).

The tuber quality as it relates to marketable value is often of primary importance in determining the economic damage caused by root knot nematodes. The proportion of yams with galled tubers collected from yam barns and markets in

Nigeria can be as high as 90% for *D. alata* and 70% for *D. rotundata* (Adesiyan and Odihirin, 1978). It is estimated that there is a reduction of 39–52% in the price of galled tubers compared with healthy ones (Nwauzor and Fawole, 1981). In Nigeria, the economic threshold at which control measures should be initiated is suggested to be the point at which 40% or more of tubers are galled. This is based on differences in market value between infected and healthy tubers. Experimentally, this has been shown to occur when soil populations of *M. incognita* at planting are 50–250 eggs per plant (Atu *et al.*, 1983).

Other losses caused by *M. incognita* and *M. javanica* in stored tubers are reduction in the edible portion (more peel has to be removed), a weight loss and an increase in the number of rotted tubers in both *D. alata* and *D. rotundata* (Nwauzor and Fawole, 1981).

Management measures

There are a few specific control measures that can be used against root knot nematodes, but in general many of those described above for other yam nematodes can be applied.

CULTURAL. The carry-over of high populations of nematodes in seed tubers is not as serious a problem with *Meloidogyne* as it is with the dry rot nematodes, but it does occur (Nwauzor and Fawole, 1981), and the use of obviously galled tubers for propagative material should be avoided. Local practices need to be changed, for example, in Nigeria, where most farmers deliberately keep galled tubers for use as planting material because of the low selling price (Nwauzor and Fawole, 1981). In Ghana, farmers remarked that knotted yam tubers were only observed on the fourth consecutive crop (Kwoseh, 2000). Therefore, continuous cropping of yams on the same piece of land season after season should be avoided.

Crop rotation will be difficult for *Meloidogyne* spp. management because of their very wide host range, but crops highly susceptible to root knot nematodes should be excluded from a cropping system. Severe damage to yam seedlings can occur when yams are grown alongside, or immediately after, a root knot-susceptible crop (Bridge, 1982). In Nigeria, intercropping highly susceptible crops such as okra, pumpkin and yam bean (*Sphenostylis stenocarpa*) with yam increases the damage by *M. incognita* to *D. rotundata* tubers (Atu and Ogbuji, 1986).

PHYSICAL. HWT can be used to control *Meloidogyne* spp. in tubers. As before, the economics and the success of the method will depend on many factors including species and age of yam tubers, nematode densities and depth of infestation. Dipping tubers of *D. alata*, *D. rotundata* and *D. floribunda* in water at 50–51°C for 30 min can effectively eliminate *Meloidogyne* (mainly *M. incognita*) from galled tubers (Hawley, 1956; Nwauzor and Fawole, 1981).

RESISTANCE AND TOLERANCE. The only yam species consistently found to be resistant to attack by *M. incognita* is the cluster yam, *D. dumentorum* (Caveness, 1979; Nwauzor and Fawole, 1981; Atu *et al.*, 1984; Kwoseh, 2000). *D. alata* cv. Obunenyi is reported to be resistant to *M. incognita* in Nigeria (Atu *et al.*, 1984), and *D. cayenensis* can be resistant to *M. incognita* and *M. javanica* (Adesiyan and Odihirin, 1978; Nwauzor and Fawole, 1981) although at least two cultivars of *D. cayenensis*, Oku and Apani, are known to be susceptible to *M. incognita* (Atu *et al.*, 1984). *D. esculenta* cv. Sree Latha is resistant to *M. incognita* in India (Mohandas *et al.*, 1996).

In pots in Ghana, one cultivar of *Dioscorea* sp., three *D. alata*, one *D. cayenensis* and nine *D. rotundata* cultivars obtained from farmers in the Ejura district were all highly susceptible to *M. incognita* (Hemeng, 1978). Two yam lines planted in a field naturally infested with *M. javanica* in Uganda were found to be uninfected at harvest, while most lines were heavily infected, suggesting the existence of resistance to root knot nematodes (Mudiope *et al.*, 1998). Lowe (1992) also reported that other lines of *D. rotundata* were not attacked by *M. incognita* race 2 and *M. javanica*.

CHEMICAL. In Nigeria, some farmers have used carbofuran granules applied to yam planting stations at a rate of 3 kg a.i./ha to control nematodes in fields infested with *Meloidogyne* (Atu and Ogbuji, 1986). This is reported to be an economical proposition when over 40% of yam tubers are found to be galled (Atu *et al.*, 1983). Granular oxamyl at rates of 3 or 6 kg a.i./ha applied at planting and at three 4-week intervals can control *M. javanica* on *D. rotundata*. In the presence of both *M. javanica* and *P. brachyurus*, tuber yields can be increased by over 40% when granular oxamyl at 3 kg a.i./ha applied at planting is combined with subsequent applications of calcium nitrate or ammonium sulphate incorporated at three 4-week intervals, each 60 kg of nitrogen. These treatments also reduce the incidence of rot in stored yams associated with the nematodes (Badra *et al.*, 1980).

INTEGRATED MANAGEMENT. Yam farmers intercrop yams with other crops for security in case of crop failure, thus it is important to know the host status of component crops/weeds in the yam cropping system and also of improved fallow crops. Choice of the right intercrops would reduce build-up of nematodes in farmers' fields, thus increasing their profits from quality produce. Also, detailed studies including the economics of HWT of seed yams or yam setts could help to develop a reasonably cheap and practical management package for yam farmers.

Diagnosis

Sampling and extraction of *Meloidogyne* spp. from yam roots and soil are as described in Chapter 3. Damage to tubers can be assessed by rating the number of galls or percentage area of tubers covered in galls. Population counts of juveniles hatched from eggs in the outer tuber layers can be done by the standard methods for extraction from plant tissues. Estimating populations of females in the outer tissues requires cutting the part of the tuber to be sampled into thin slices. Nematodes can be removed manually by teasing the tissue under a microscope, or the slices can be stained in the normal way and nematodes counted directly whilst embedded in the tissues.

Pratylenchus sudanensis

P. sudanensis was first observed on yam in 1993 during field studies in Uganda (Coyne *et al.*, 2003) where it was present at a mean density of 468 per 10 g of root + 100 ml of soil and is since reported as the dominant nematode species occurring on yam in Uganda, at up to 300 nematodes/g of tuber (Mudiope *et al.*, 2001). Although yam is not a key staple crop in Uganda, it is locally important within the country and region and contributes to crop diversity. The nematode has since been associated with cracked tubers (Mudiope *et al.*, 2003), which is associated with, as yet, an undefined condition, which results in rapid tuber deterioration (N. Wanyera, Uganda, 2002, personal communication). Inoculation with as few as 100 and as many as 10,000 *P. sudanensis* per plant resulted in 28 and 52% dead roots, compared with 3% in uninoculated plants in pot studies (Mudiope *et al.*, 2003). Host range studies of *P. sudanensis* in Sudan showed that 20 plant species acted as favourable hosts, especially cotton (cvs Barakat and Barac(67)B), sorghum millet (cv. Dwarf White Milo), pigeonpea (cv. Local) and Lubia bean (cv. Local) (Saadabi, 1985). Wheat (cv. Giza 155) and groundnut (cv. Ashford) were considered poor or non-hosts. *P. sudanensis* is morphologically similar to *Pratylenchus pseudopratensis* and, during the survey studies in 1993 (Coyne *et al.*, 2003), there was sometimes difficulty in differentiating some specimens between the two species (D.J. Hunt, UK, 2003, personal communication), and it was suggested that perhaps a morphological continuum exists between the two identified species, which may be better delineated using molecular analysis.

Other nematode parasites of yams

Other species of *Pratylenchus* are known to be parasites of yam. *P. brachyurus* has been found in tubers, roots and yam soil in Nigeria (Caveness, 1967b), Côte d'Ivoire (Miege, 1957), Guatemala (Jenkins and Bird, 1962), Fiji and Tonga (Bridge, 1988).

Radopholus cf. *similis* has been found causing dry rot of yam tubers in Papua New Guinea (Bridge and Page, 1984) and in New Caledonia. The dry rot disease is similar to that caused by *P. coffeae* and *S. bradys*, but diseased tissues tend to be lighter brown in colour (Bridge and Page, 1984). *R. similis* has also been found infesting tubers in Fiji (Butler and Vilsoni, 1975) and yam roots in the Solomon Islands (Bridge, 1988).

Aphelenchoides besseyi, a foliar nematode, is known to occur in large populations in the foliage and tubers of *D. trifida* in Guadeloupe associated with drying and blackening of the foliage, and wasting and cracking of tubers with internal decay (Kermarrec and Anais, 1973).

A 'black scurf-like syndrome' of Chinese yam, *D. opposita*, was shown to be caused by *Paratrichodorus porosus* in Japan (Nishizawa, 1973). Symptoms of the disease are blackening, cracking and corkiness of the tuber tips. The disease increases in severity with successive planting of yams. *P. porosus* also reduces the weight of the tubers and greatly inhibits their elongation, resulting in small rounded rather than long thin tubers.

Of the remaining nematodes associated with yams, the only other species identified as parasites of yam roots or tubers are *R. reniformis*, *S. clathricaudatum* and *H. dihystera*, although neither *S. cavenessi* nor *S. clathricaudatum*, both known to occur in West Africa, was found infecting stored yam tubers in Ghana (Kwoseh, 2000).

Taro

Taro (*Colocasia esculenta* (L.) Schott.), also known as cocoyam, dasheen and eddoe, is grown throughout the tropics, subtropics and warmer regions of the temperate zone. It belongs to the Araceae family, which comprises about 110 genera and over 2000 species, and is believed to have originated in South-east Asia. It is mostly a staple food or subsistence crop but is grown commercially in some countries. There are two botanical varieties of *Colocasia*, the 'eddoe type' *C. esculenta* var. *antiquorum*, which has a relatively small corm surrounded by large well-developed cormels, and the 'dasheen type' *C. esculenta* var. *esculenta*, which has a large central corm and numerous but small cormels. They can be grown in dry upland or flooded areas, depending on the type and cultivar (Plate 6B). They grow best with daily average temperatures of 20–27°C and rainfall of 2500 mm/year or more (Purseglove, 1972; Kay, 1987).

Taros are propagated vegetatively using whole corms or cormels, pieces of corms or the leaf-bearing tops of mature corms (the lower 30–50 cm of the petiole with the top 1–2 cm of the corms). They can be grown in flat wet areas, steep hillsides where rainfall is sufficient, or in 'patches' or pits in swampy areas (Purseglove, 1972; Kay, 1987).

Nematodes of Taro

The nematodes known to be damaging parasites of taro are *Meloidogyne* spp., *Hirschmanniella miticausa* and *P. coffeae*. Other nematodes found associated with tissue damage or present in high populations on the crop are *Radopholus* sp. and *Rotylenchulus reniformis*.

Meloidogyne species

The root knot nematodes, *Meloidogyne* spp. (*M. incognita*, *M. javanica* and *M. arenaria*), have been reported on *Colocasia* from Cuba (Lorenzo and Fernandez, 1982), Puerto Rico (Ayala, 1969), Trinidad (Brathwaite, 1972a), Florida (Byars, 1917; McSorley *et al.*, 1983a), Hawaii (Parris, 1940; Sipes and Arakaki, 1997), East Africa (Whitehead, 1969), Nigeria (Caveness, 1967), Ghana (Addoh, 1971), the

Philippines (Timm, 1965), Papua New Guinea (Bridge and Page, 1984), Niue, Western Samoa, Tonga, Fiji (Orton Williams, 1980; Fliege and Sikora, 1981), the Solomon Islands (Gowen, 1985), Taiwan (Huang *et al.*, 1972), Japan (Iwahori *et al.*, 2001), Egypt (Byars, 1917) and India (Nirula, 1959). *M. hapla* was found additionally on *Colocasia* in Uganda in low densities (Coyne *et al.*, 2003).

Symptoms of damage

Both *M. incognita* and *M. javanica* can cause galling of roots and corms. On young feeder roots, galls are small and irregular. Infested older roots become thickened with large swellings, although the symptoms are not always obvious. On corms, nematodes cause blister-like swellings, which later become large round or oblong galls, 2–15 mm in diameter, deforming the corms. Such infested corms are known to rot in storage. Nematodes can be present in yellow areas of variable size internally even though external symptoms are not present on the corms. The above-ground symptoms occur in patches in the field. Affected plants are stunted and unhealthy with yellowed leaves, which can turn brown and die (Nirula, 1959; Srivastava *et al.*, 1971; Brathwaite, 1972b; Lorenzo and Fernandez, 1982).

Survival and means of dissemination

Meloidogyne spp. can be carried over from one *Colocasia* crop to the next in a wide range of other host crops and weeds. As the nematodes feed and reproduce in corm tissues, they can be disseminated in corms and cormels if infested material is used for propagation.

Environmental factors affecting parasitism

Root knot nematodes are especially serious on the eddoe type or upland taro, *C. esculenta* var. *antiquorum*; *Meloidogyne* populations could be suppressed when taro is grown in very wet or flooded conditions (McSorley *et al.*, 1983a).

Economic importance

Losses caused by *Meloidogyne* have been described as severe in India where local farmers have in the past had to abandon cultivation of *Colocasia* because of the nematodes (Srivastava *et al.*, 1969). It is suggested that *Colocasia* (and *Xanthosoma*) are more tolerant of *M. incognita* than other crops, and high pre-plant populations of the nematode have to be present in field soil for damage to occur (McSorley *et al.*, 1983a). The malformation of corms due to galling reduces their marketable value (Srivastava *et al.*, 1971). The yield of the susceptible cultivar Sree Pallavi was significantly reduced by 21% following inoculation with 1000 second stage juvenile *M. incognita* in pots in India (Mohandas and Palaniswami, 1990).

Management measures

Use of nematode-free planting material will prevent dissemination into the field; seed corms or cormels should be free of any external symptoms of root knot damage. Selecting planting material from land with no previous records of nematode attack will reduce the risk of damage. Root knot can be controlled in corms by dipping in hot water at 50°C for 40 min (Byars, 1917), but this is unlikely to be an economic measure for large-scale farming.

Most root knot damage to taro is likely to occur if the crop is grown in field soils with high populations of *Meloidogyne* present. Planting taro intercropped with, or after, susceptible crops should be avoided. Rotating with a range of poor or non-host cover crops (barley, *Panicum maximum*, *Neonotonia wightii*, marigold, sesame or sunnhemp) incorporated into the ground before planting taro can reduce soil populations of *M. javanica* (Sipes and Arakaki, 1997).

The number of contradictory reports on damage by *Meloidogyne* may be due to the different host reactions of the many taro cultivars that are grown worldwide (McSorley *et al.*, 1983a). One cultivar, 'Dodare', in Japan was found to be completely resistant to both *M. incognita* and *M. javanica* (Inagaki, 1981), while cv. 'Samra' in Fiji is described as moderately susceptible to these two species (Kirby, 1977). In India, cv. C9 is classed as immune to *M. incognita* (Mohandas *et al.*, 1996). The cvs Mana Ulaulu and Piko Ulaulu are possible sources of partial resistance and/or tolerance to *M. javanica* in Hawaii (Sipes *et al.*, 1995).

Diagnosis

Standard methods for the extraction of nematodes from soil and roots can be used (Chapter 3). Assessing *Meloidogyne* populations in corms and the damage they cause can be done in a similar way to that used for yam tubers.

Hirschmanniella miticausa

H. miticausa is the causal organism of a taro corm rot disease known as 'miti-miti' in the Solomon Islands. The disease and nematode have been reported from four islands in the Solomon Islands group (Mortimer *et al.*, 1981) and the highlands of Papua New Guinea (Bridge and Page, 1984). A *Hirschmanniella* sp. has also been recorded associated with taro in Taiwan (Huang *et al.*, 1972).

Symptoms of damage

The initial foliar symptoms of miti-miti disease are wilting of the older leaves, which eventually become chlorotic, while the new central leaf, instead of bending, remains straight. Taro plants with the disease die prematurely as a result of corm damage.

Corms with the disease, cut longitudinally, at first show red streaks radiating from the base of the corm. These later become irregular, 1–10 mm wide, zones of dry brown rot, with the advancing diseased tissues remaining red (Plates 6C and D). The basal portions of severely diseased corms are often completely decayed due to a brown soft rot (Fig. 7.6). The numbers of cormels are reduced in plants with the disease (Mortimer *et al.*, 1981; Bridge *et al.*, 1983).

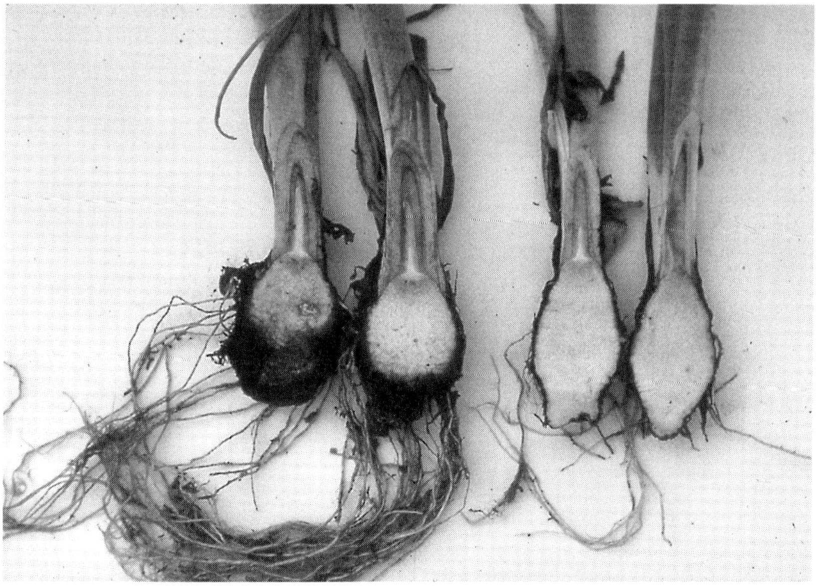

Fig. 7.6. Miti-miti disease of taro (*Colocasia esculenta*) corms caused by *Hirschmanniella miticausa* (left) plus secondary rot. (Photo: J. Bridge.)

Biology

H. miticausa is a migratory endoparasite. In growing taro plants, the highest populations occur in the corms with less in roots and relatively few in surrounding soil. Nematodes are found in, or immediately around, red necrotic tissues of the corm in the basal portion; relatively small numbers occur in the white centre tissues, and nematodes are rarely found in the crown (the top 1 cm). Numbers of nematodes commonly exceed 1000/10 g and can be over 3000/10 g of corm tissue.

The nematode is disseminated in diseased corm planting material. Other hosts are not known, but the nematode probably can survive for some period of time in field soil without hosts. It is found causing miti-miti disease of taro in dryland soils, rainfed mountain slopes and in flooded swamp pits.

Disease complexes

Nematode activity in corm tissues probably predisposes the corms to invasion of secondary pathogens, causing the extensive outer soft rot invariably associated with the disease. Fungi isolated from areas of soft rot in corms with miti-miti are *Corticum solani*, *Pythium vexans*, *Fusarium solani* and *F. oxysporum* (Bridge *et al.*, 1983).

Economic importance

Miti-miti disease renders taro corms inedible and, when severe, can destroy almost all consumable corm tissues of the crop. In parts of the Solomon Islands, the disease is so devastating that taro cultivation has been almost entirely abandoned, particularly where continuous cultivation has occurred in swamp pits (Patel *et al.*, 1984).

Management measures

The disease is at present restricted to those areas of the Pacific where taro is a subsistence crop. This limits the control measures that can be recommended, particularly the use of expensive nematicides.

Planting material infested with *H. miticausa* is the main source of inoculum in new land. Nematodes can be eliminated from normal planting material (corm top and 40 cm of leaf base) by immersing in hot water at 50°C for 15 min without damaging the tissues (Mortimer *et al.*, 1981). Because of the difficulties of treatment, it cannot be generally recommended to taro growers, but it could be used to establish a source of nematode-free planting material.

The most practical measure for small growers is to completely remove all nematodes from planting material manually. Nematodes rarely occur in the top few centimetres of the corm. Trimming the corm top back to white, healthy tissues will ensure that most, if not all, planting material is free of nematodes (Mortimer *et al.*, 1981). Planting corms or cormels, as compared with corm tops, will increase the risk of spreading nematodes.

Where taro is grown on hillsides, there is a risk of nematodes being carried downhill in runoff water. This can be avoided by making new plantings uphill from old taro gardens (Mortimer *et al.*, 1981).

These hygiene measures cannot be used in areas where there is intensive and continuous taro production such as in swamp pits in parts of the Solomon Islands. Where this occurs, the only practical solution is the use of resistant cultivars. One such resistant cultivar has been identified, a taro that occurs wild and is used only when other foods are scarce, and crosses between this taro and high yielding cultivars are possible (Patel *et al.*, 1984).

Diagnosis

H. miticausa is a large nematode and is most efficiently extracted from soil by a sieving and sedimentation method (Chapter 3). However, as most nematodes are found in plant tissues, their extraction from corms will give the most accurate assessment of their presence and population levels using a standard tissue extraction method.

Pratylenchus coffeae

The lesion nematode, *P. coffeae*, has been found parasitic on taro in Papua New Guinea (Bridge and Page, 1984), Fiji (Kirby *et al.*, 1980; Orton Williams, 1980), the Solomon Islands (Mortimer *et al.*, 1981) and in the warmer parts of Japan (Inagaki, 1985; Iwahori *et al.*, 2001). However, it is reported causing injury to taro only in Japan (Nishizawa and Ohshima, 1972; Oashi, 1984; Inagaki, 1985; Yamada, 2001).

Symptoms of damage

P. coffeae has consistently been found to be associated with a disease of taro in Japan causing poor plant growth, root decay and reduced number of cormels. Two months after planting, roots turn brown and then rot. This is followed by stunted top growth and, in serious cases, withering and death of the leaves 5 months after planting. The disease is most commonly seen in fields with continuous taro cultivation (Oashi, 1984). In Papua New Guinea, *P. coffeae* causes localized necrosis of root and corm tissues (Bridge and Page, 1984).

Biology

All stages of *P. coffeae* are found in roots, corms and in soil around taro. Highest populations occur in roots and soil, with less in the 'skin' of the corms (Oashi, 1984).

Economic importance

Field trials have shown that, by controlling *P. coffeae* in seed corms and field soil, yields of corms can be increased threefold. The most serious damage and highest nematode populations occur where taro is cultivated continuously, although there is a suggestion that nematodes may not be the only cause of problems with continuous taro cultivation (Oashi, 1984).

Management measures

The suggested management measures against *P. coffeae* on taro include disinfection of seed corms, reduction of soil populations and crop rotation (Oashi, 1984).

It is recommended that seed corms are selected from healthy parent plants and all roots are removed before planting. In Japan, nematodes can be eliminated from corms by soaking in a disinfectant ('cartap aqueous solution') for 30 min (Iwahashi, 1977), but chemical residues may be a problem.

Lowest populations of *P. coffeae* are found in soils that have been flooded previously, and planting taro in rice paddy field soil compared with dry, upland soil reduces the risk of damage. Combining disinfection of the seed corms with cultivation in paddy soil can almost eliminate nematodes from the crop, increasing corm germination and yields.

In Japan, taro is a comparatively low income crop, and the use of nematicides is thought to be uneconomic although nematicides will give some control of *P. coffeae* (Torigoe, 1993). Crop rotation is considered a more appropriate control measure. Soil populations of *P. coffeae* are decreased in land planted to groundnut, marigold, radish and *Stevia rabaudiana* Cav., but the nematodes increase to large numbers as soon as taro is cultivated, and a rotation of 2 or more years between taro crops is necessary (Oashi, 1984; Torigoe, 1994; Yamada, 2001). Use of antagonistic plants such as *Crotalaria juncea* and *C. spectabilis* in rotation can be effective against *Pratylenchus* (and *Meloidogyne*) (Torigoe, 1996).

Diagnosis

Determining the presence of nematodes in association with diseased plants will require sampling and extraction from soil and plant tissues. It will not always be possible to obtain a direct association between visible root damage symptoms and nematode numbers as *P. coffeae* can be found in superficially healthy, white roots (Oashi, 1984).

Other nematodes of taro

R. reniformis has been recorded associated with *Colocasia* in Puerto Rico

(Ayala, 1969), Taiwan (Huang *et al.*, 1972), Fiji, Western Samoa, the Solomon Islands, Tonga (Orton Williams, 1980; Fliege and Sikora, 1981) and Florida. Although high population levels of *R. reniformis* (1767 nematodes/100 cm^3 of soil) can be found with *Colocasia*, no effect on yield was noted in Florida (McSorley *et al.*, 1983a).

An undescribed *Radopholus* sp. is reported from necrotic tissues of taro corms and roots in Papua New Guinea (Bridge and Page, 1984). *Radopholus* spp. have been found associated with taro in Fiji, Tonga and Western Samoa (Kirby *et al.*, 1980; Orton Williams, 1980).

Aphelenchoides besseyi is recorded in large numbers from taro corms with rot (Bridge and Page, 1984).

Xanthosoma

There are about 40 species of *Xanthosoma* with the common names of tannia, tanier, yautia, malanga and new cocoyam. They can be confused with the genus *Colocasia* because of their similar botany but are distinguished by their different leaves.

Xanthosoma is native of tropical America but has spread widely throughout the tropical world. Some species are grown for their edible tubers or leaves; others can be grown for their ornamental foliage. The most widely grown edible species is *X. sagittifolium* (L.) Schott., others are *X. atrovirens* Koch and Bouchd, *X. violaceum* Schott, *X. caracu* Koch and Bouchd and *X. brasiliense* Engl. They can grow to a height of 2 m. A corm is produced which bears up to ten or more lateral cormels (Purseglove, 1972; Kay, 1987).

Tannias are propagated vegetatively from pieces of main corm, cormels or the tops of the main corm plus 20–30 cm of leaves. They can be grown in pure stands but are more often intercropped with tree crops and other plants. They require well-drained soils and cannot withstand waterlogging, and prefer an average annual rainfall of 140–200 cm (Purseglove, 1972; Kay, 1987).

Nematodes of *Xanthosoma*

Comparatively little information is available on the importance of nematodes associated with *Xanthosoma*. Only *Meloidogyne* spp., *R. reniformis* and *P. coffeae* are reported to cause damage to the crop.

Meloidogyne species

Four species of *Meloidogyne* have been found with *Xanthosoma*: *M. arenaria* is reported from Cuba (Decker and Casamayor, 1966) and Tanzania (Runkulatile *et al.*, 1990); *M. incognita* from Puerto Rico (Roman, 1978), Nigeria (Caveness *et al.*, 1981), Cuba (Decker and Casamayor, 1966) and Papua New Guinea (Bridge and Page, 1984); and *M. javanica* from Fiji and Tonga (Orton Williams, 1980), Venezuela (Crozzoli *et al.*, 1995), Colombia (Navarro and Barriga, 1975), Tanzania (Runkulatile *et al.*, 1990) and Florida, USA (McSorley *et al.*, 1983a). *Meloidogyne* spp. are also reported on tannia from Kiribate and Western Samoa in the Pacific (Orton Williams, 1980) and from Trinidad (Brathwaite, 1972b). *M. incognita*, *M. javanica*, *M. arenaria* and *M. hapla* were all found occurring on *Colocasia* in Uganda, with *M. hapla* in greater mean densities (Coyne *et al.*, 2003).

M. incognita has been found in high populations causing galling and roughening of the surface of *Xanthosoma* corms (Acosta, 1979). Similarly, *M. javanica* can cause obvious corm damage (Orton Williams, 1980). *M. arenaria* has been shown to cause severe galling and malformation of *X. sagittifolium* corms (Decker and Casamayor, 1966). *Meloidogyne* has also been reported in association with stunting and yellowing of *Xanthosoma* plants, with nematode galls localized at root tips (Roman, 1978). However, most findings suggest that *Xanthosoma* spp. are generally tolerant of *Meloidogyne* except when pre-plant populations are high (McSorley *et al.*, 1983a). Initial soil populations of 5000 *M. incognita* juveniles/l of soil can reduce corm weight of *X. sagittifolium*, but nematode populations decline to only 14/l of soil at harvest, suggesting that the crop is a very poor host (Caveness *et al.*, 1981). In Tanzania, however, *X. sagittifolium* was found to be a good host of both *M. javanica* and *M. arenaria* where they were recorded occurring in densities up to 6570 J2/g of dry root tissue (Runkulatile and Teri, 1990). Infected roots showed black to brown lesions and stubby root characteristics but did not cause gall formation.

It has been suggested that *M. incognita* is involved in a *Xanthosoma* root rot disease in Papua New Guinea (Bridge and Page, 1984).

Rotylenchulus reniformis

The reniform nematode *R. reniformis* is reported on *Xanthosoma* spp., sometimes in high populations, in the Pacific islands of Fiji, Kiribati, Western Samoa, Tonga (Orton Williams, 1980) and Papua New Guinea (Bridge and Page, 1984), also from Puerto Rico (Ayala and Ramirez, 1964), Trinidad (Brathwaite, 1972b) and Florida (McSorley *et al.*, 1983a).

Soil populations of 400 *R. reniformis*/100 cm^3 of soil can cause reduction in root weight and a 26% reduction in dry weight of marketable cormels of *X. caracu*. The same population levels did not affect the yield of *X. atrovirens* (McSorley *et al.*, 1983). Populations of 100–1000 nematodes/100 cm^3 of soil have been found associated with small root lesions on *X. sagittifolium* (Brathwaite, 1972b). In Fiji, *R. reniformis* occurred in 80% of *X. sagittifolium* plantings (Orton Williams, 1980), but tannia was a non-host for the nematode in a host range trial (Vilsoni and Heinlein, 1982).

The amount and type of damage caused by *R. reniformis* will depend on the species and cultivars of *Xanthosoma*, as well as populations of the nematode present in the soil. Nematode control has been recommended only in sites heavily infested by *R. reniformis* but not where populations are low (McSorley *et al.*, 1983a).

Pratylenchus

Pratylenchus spp. have been recorded on *X. violaceum* in Honduras (Pinochet and Ventura, 1980) and on *X. sagittifolium* in Fiji, Tonga and Western Samoa (Orton Williams, 1980). In Fiji, *P. coffeae* was found associated with 50% of *Xanthosoma* plants examined, occasionally present in the outer corm layers in areas around the margin of blackened, rotted tissue.

Helicotylenchus multicinctus

H. multicinctus was associated with cocoyam in Ghana (Addoh, 1971) and was found regularly in the roots of *Xanthosoma* plants in Uganda (Coyne *et al.*, 2003). However, no damage has been associated with *H. multicinctus*.

Other Root and Tuber Crops

There are over 27 species of minor root and tuber crops that are of local importance in several tropical and subtropical regions of the world (Kay, 1987). Nematological information is not available for most of these crops. Those crops on which some nematological investigations have been done are giant taro (*Alocasia* spp.), giant swamp taro (*Cyrtosperma chamissonis*), Chinese water chestnut (*Eleocharis dulcis*), and crops in certain tropical regions of Central and South America, oca *Oxalis tuberosa*, olluco, *Ullucus tuberosus*, arracacha, *Arracacia xanthorrhiza*, and mashua, *Tropaeolum tuberosum*, which constitute the basic diet of the population.

Giant taro

Giant taros (*Alocasia* spp.) are grown for their large edible corms. The most common species is *A. macrorrhiza* (L.) G.Don.

A number of plant parasitic nematodes have been isolated from around *Alocasia* plants, but there is no information on their importance. Most records come from the

Pacific (Orton Williams, 1980). Two species of *Meloidogyne*, *M. javanica* and *M. arenaria*, are reported causing root galls on *Alocasia* sp. in southern Africa (Martin, 1969).

Swamp taro

The swamp taro, *Cyrtosperma chamissonis* (Schott) Merr., is a crop of the Pacific grown in flooded swamp land for its large edible corms.

There are very few records of plant parasitic nematodes associated with *Cyrtosperma*. *Criconemoides denoudeni*, *C. onoensis*, *H. dihystera*, *Meloidogyne* sp. and *P. coffeae* have been found around plants in Fiji (Orton Williams, 1980). However, there is now strong evidence that a corm rot of swamp taro is caused by the burrowing nematode, *R. similis*, in the Pacific islands of Yap, Palau and Guam (Jackson, 1987). *R. similis* has been consistently isolated from roots and corms with the disease. Corms have small shallow holes, no more than 1–2 cm deep for the most part, except in severe instances when the entire basal part of the corm is decayed. Beneath these, the rot is brown and superficial but sometimes extending as narrow channels deep into the centre of the corm (G.V.H. Jackson, personal communication) (Plates 6E and F).

Chinese water chestnut

Chinese water chestnut (*Eleocharis dulcis* Burm.f.Trin.ex Hensch) is commercially cultivated in South-east Asia, the Pacific and southern USA for its edible corms. *Dolichodorus heterocephalus*, the awl nematode, is reported to reduce growth of the crop in the USA (Tarjan, 1952).

Oca and olluco

Oca, *Oxalis tuberosa* Molina, is an important crop of the cold areas of the Andes, grown at elevations of over 3000 m from Bolivia to Venezuela as a minor crop, considered of less importance than potato, but more important than olluco. There are several kinds of oca: the bitter, which has white tubers, and the sweet, with tubers of various colours. Because of the high content of calcium oxalate in the tubers, they can only be eaten after days of exposure to sun.

Olluco, *Ullucus tuberosus* Caldas, is endemic to the Andes and constitutes one of the staple food crops in the region from Bolivia to Colombia, and is an important crop after potato and oca. Tubers vary in shape and colour. As in potatoes and oca, they are often dehydrated and made into chuno (frozen, thawed and dehydrated). It replaces potatoes in certain zones of the cold altiplano where the excess humidity becomes a limiting factor to potato production.

Several nematode species are known to be associated with oca and olluco (Jatala, 1989). *Atalodera* (=*Thecavermiculatus*) *andina* and *Nacobbus aberrans* are quite widely distributed in the areas of oca and olluco cultivation (Aztocaza Perez, 1980; Jatala, 1989). Although roots of these crops are severely infected by *A. andina* and *N. aberrans*, the economic importance of these nematodes as production constraints is not well known. Reactions of these crops to *A. andina* and *N. aberrans* indicate the possibility of an available resistant gene base. *Meloidogyne* species are often found in association with *A. andina* and *N. aberrans* on the roots of these crops. This nematode, however, does not constitute a major concern in production. Because of the fact that these are primarily small farm crops with limited economic input for production, chemical control of nematodes is not practised.

Arracacha

Arracacha, *Arracacia xanthorrhiza* Bancroft, belongs to the family Apiaceae (Umbelliferae) and could be one of the earliest domesticated crops in the American continent as it is found in burial sites. It is a perennial herb, native of areas from

Mexico to Peru. Its fleshy tubers have an agreeable flavour and constitute an important food item amongst the people of Central America and the Andean regions of South America. It has various names throughout Latin America: Árracacham zanhoria blanca in Peru, Manidquina Salsa in Brazil and Peruvian carrot. It originates in the Andean foothills and grows best at altitudes between 1000 and 2500 m. From the Andean region of South America, its centre of origin, it was introduced successfully to mountainous regions of Brazil and Central America and, recently, to India and eastern Africa. Colombia is, however, probably the largest producer of this crop. Because of its rusticity and excellent quality starch, it is a rapidly expanding crop not only of small farmers but also into a commercial crop, with over 16,000 ha and average yields of 10 t/ha in several provinces in Brazil. It is not only appreciated in the fresh market, but the flour is especially used for baby food and cakes.

Nematodes

Of the nematode species attacking arracacha in South America, *Meloidogyne* spp. are of major importance (Jimenez *et al.*, 2001; Henz, 2002). Severe and early root infection inhibits the development of tubers. Infected plants exhibit general symptoms of stunting, yellowing and a tendency to wilt readily during the hot and dry period (P. Jatala, unpublished). In Brazil, both roots and tubers of arracacha can be severely infested by *M. hapla* and *M. incognita*. Galling of roots makes the tubers unmarketable, resulting in a 100% loss.

Arracacha favours a very high multiplication rate of *Meloidogyne* due to its very long (10 month) vegetative cycle, which favours more generations of the nematode. Thus farmers have to be very careful as to what crop follows arracacha, and two consecutive crops of arracacha are to be avoided. Prior to planting, poor host crops to the nematode should be planted, such as mucuna (*Stylosanthes guianensis*) and *Styzolobium* spp.

Corm disinfection can be effective. The corms should first be washed in running water, then dipped for 15 min in a 1% sodium hypochlorite solution of potable water. After this treatment, the corms need to be dried under ambient temperatures. Few sources of resistance have been detected, although cultivars with white roots are considered to show more resistance than those with yellow roots.

The lesion nematode, *Pratylenchus penetrans*, also causes small to large lesions of these organs, which can penetrate deep into the tuberous root (de Mendes *et al.*, 2001). Suggested control is by nursery soil treatment and the use of nematode-free planting material (Lordello, 1981).

Mashua

Mashua or aflu, *Tropaeolum tuberosum* Ruiz and Pav., probably originated in the altiplano zones of Peru and Bolivia. It is an annual crop that produces cone-shaped tubers similar to oca in form and colour. It is the least popular of the tubers and root crops of the region. The tubers are not palatable when eaten raw. They must be cured by the sun prior to cooking. They are also dehydrated to form chuno, as in potatoes, oca and olluco.

Of the nematode species attacking this crop, *N. aberrans* and *Meloidogyne* spp. are of major importance, and *N. aberrans* can become a limiting factor to production (Jatala, 1989). However, its economic damage to mashua production has not been documented throughout the range of its production.

Although chemicals are successful in controlling the nematodes, the fact that mashua is a small farm crop with minimal economic input means that no control measures are taken to reduce nematode attack.

Conclusions

In general, tuber and root crops constitute the major food source for a great part of the world's population. Assessment of nematode damage to minor tuber and root crops and their economic importance in production systems needs to receive greater attention. Although some tuber and root crops are the basic staple diet of the majority of the world population, nematological information regarding these crops is lacking. Similarly, a better understanding of the importance of several minor tuber and root crops and their utilization in the cropping systems may alleviate some of the food shortage experienced in many developing countries.

References

Acosta, N. (1974) Depth penetration of phytoparasitic nematodes in yam tubers. *Nematropica* 4, 7–11.

Acosta, N. (1979) *Meloidogyne incognita* in tanier corms in Puerto Rico. *Organization of Tropical American Nematologists Newsletter* 11 (1), 14.

Acosta, N. and Ayala, A. (1975) Pathogenicity of *Pratylenchus coffeae, Scutellonema bradys, Meloidogyne incognita* and *Rotylenchulus reniformis* on *Dioscorea rotundata. Journal of Nematology* 7, 1–6.

Acosta, N. and Ayala, A. (1976a) Effects of *Pratylenchus coffeae* and *Scutellonema bradys* alone and in combination on Guinea yam (*Dioscorea rotundata*). *Journal of Nematology* 8, 315–317.

Acosta, N. and Ayala, A. (1976b) Hot water and chemical dips for nematode control in tubers of *Dioscorea rotundata. Journal of Agriculture of the University of Puerto Rico* 60, 395–402.

Addoh, P.G. (1971) The distribution and economic importance of plant parasitic nematodes in Ghana. *Ghana Journal of Agricultural Science* 4, 21–32.

Adeniji, M.O. (1970) Fungi associated with storage decay of yam in Nigeria. *Phytopathology* 60, 590–592.

Adesiyan, S.O. (1976) Host range studies of the yam nematode, *Scutellonema bradys. Nematropica* 6, 60–63.

Adesiyan, S.O. (1977) Penetration and multiplication of *Scutellonema bradys* in yams (*Dioscorea* spp.). *Nematologia Mediterranea* 5, 313–317.

Adesiyan, S.O. and Adeniji, M.O. (1976) Studies on some aspects of yam nematode (*Scutellonema bradys*). *Ghana Journal of Agricultural Science* 9, 131–136.

Adesiyan, S.O. and Badra, T. (1982) Granular nematicides for control of the yam nematode, *Scutellonema bradys,* and relevant residues in raw tubers. *Journal of Nematology* 14, 213–216.

Adesiyan, S.O. and Odihirin, R.A. (1977) Plant parasitic nematodes associated with yam tubers in Mid-West State, Nigeria. *Nigerian Journal of Plant Protection* 3, 178–179.

Adesiyan, S.O. and Odihirin, R.A. (1978) Root knot nematodes as pests of yams (*Dioscorea* spp.) in southern Nigeria. *Nematologica* 24, 132–134.

Adesiyan, S.O., Odihirin, R.A. and Adeniji, M.O. (1975a) Histopathology studies of the yam tuber (*Dioscorea rotundata* Poir) infected with *Scutellonema bradys* (Steiner and Le Hew). *International Biodeterioration Bulletin* 11, 48–55.

Adesiyan, S.O., Odihirin, R.A. and Adeniji, M.O. (1975b) Economic losses caused by the yam nematode, *Scutellonema bradys,* in Nigeria. *Plant Disease Reporter* 59, 477–480.

Adesiyan, S.O., Caveness, F.E., Adeniji, M.O. and Fawole, B. (1990) *Nematode Pests of Tropical Crops.* Heinemann Educational Books (Nigeria) Limited.

Alexander, J. and Coursey, D.G. (1969) The domestication of the yams. In: Ucko, P.J. and Dimbleby, C.W. (eds) *The Domestication and Exploitation of Plants and Animals.* Gerald Duckworth, London, pp. 405–425.

Anonymous (1964) Nematode control trials with yam. *Annual Report for the Ministry of Agriculture Nigeria for 1962–63.*

Atu, U.G. (1991) Nematode problems in cropped reclaimed gully site at Owerri, Nigeria. *Afro-Asian Journal of Nematology* 1, 15–18.

Atu, U.G. and Ogbuji, R.O. (1986) Root-knot nematode problems with intercropped yam (*Dioscorea rotundata*). *Phytoprotection* 67, 35–38.

Atu, U.G., Odurukwe, S.O. and Ogbuji, R.O. (1983) Root-knot nematode damage to *Dioscorea rotundata. Plant Disease* 67, 814–815.

Atu, U.G., Okoli, O.O. and Ogbuji, R.O. (1984) Reactions of yams (*Dioscorea* spp.) to *Meloidogyne incognita* race 2. *International Nematology Network Newsletter* 1, 10–12.

Ayala, A. (1969) Nematode problems in Puerto Rican agriculture. In: Abad-Ramos, J. (ed.) *Proceedings of the Symposium on Tropical Nematology.* Agricultural Experimental Station, University of Puerto Rico, pp. 135–145.

Ayala, A. and Acosta, N. (1971) Observations on yam (*Dioscorea alata*) nematodes. *Nematropica* 1, 39–40.
Ayala, A. and Ramirez, C.T. (1964) Host-range, distribution and bibliography of the reniform nematode, *Rotylenchulus reniformis*, with special reference to Puerto Rico. *Journal of Agriculture of the University of Puerto Rico* 48, 140–161.
Aztocaza Perez, E.P. (1981) Nematodo Esférico Atacando Cultivos Andinos en el Perú. MS Thesis, Universidad Nacional Agraria, La Molina, Lima, Peru.
Badra, T. and Caveness, F.E. (1979) Chemotherapy of *Dioscorea alata* for disinfestation of *Scutellonema bradys*. *Nematropica* 9, 135–137.
Badra, T., Steele, W.M. and Caveness, F.E. (1980) The employment of a non-fumigant nematicide for control of root-knot and lesion nematodes on yams and crop preservation in storage. *Nematropica* 10, 81–85.
Baudin, P. (1956) Les maladies parasitaire des ignames en Côte d'Ivoire. *Revue de Mycologie, Paris 21* Supplement No. 2, 87–111.
Baujard, P. and Martiny, B. (1995) Ecology and pathogenicity of the *Hoplolaimidae* (Nemata) from the Sahelian zone of West Africa. 1. Field studies on *Scutellonema cavenessi* Sher, 1964. *Fundamental and Applied Nematology* 18, 261–269.
Belliard, A. and Kermarrec, A. (1978) Présence du nématode de l'igname (*Scutellonema bradys*) dans les tubercules de *Dioscorea trifida* en Republique Dominicaine. *Nouvelles Agronomiques des Antilles et de la Guyane* 4, 49–51.
Brathwaite, C.W.D. (1972a) *Colocasia esculenta* a new host of *Meloidogyne incognita* in Trinidad. *Plant Disease Reporter* 56, 618.
Brathwaite, C.W.D. (1972b) Preliminary studies of plant parasitic nematodes associated with selected crops at the University of the West Indies. *Plant Disease Reporter* 56, 1077–1079.
Brathwaite, C.W.D. (1977) Outbreaks and new records in Barbados. *FAO Plant Protection Bulletin* 25, 210.
Bridge, J. (1972) Nematode problems with yams (*Dioscorea* spp.) in Nigeria. *PANS* 1, 89–91.
Bridge, J. (1973) Nematodes as pests of yams in Nigeria. *Mededelingen Faculteit Landbouwwetenschappen Gent* 38, 841–852.
Bridge, J. (1975) Hot water treatment to control plant parasitic nematodes of tropical crops. *Mededelingen Faculteit Landbouwwetenschappen Gent* 40, 249–259.
Bridge, J. (1978) Pest control in tropical root crops. *PANS* 4, 234.
Bridge, J. (1982) Nematodes of yams. In: Miege, J. and Lyonga, S.N. (eds) *Yams. Ignames.* Clarendon Press, Oxford, pp. 253–264.
Bridge, J. (1988) Plant-parasitic nematode problems in the Pacific islands. *Journal of Nematology* 20, 173–183.
Bridge, J. and Page, S.L.J. (1984) Plant nematode pests of crops in Papua New Guinea. *Journal of Plant Protection in the Tropics* 1, 99–109.
Bridge, J., Mortimer, J.J. and Jackson, G.V.H. (1983) *Hirschmanniella miticausa* n.sp. (Nematoda: Pratylenchidae) and its pathogenicity on taro (*Colocasia esculenta*). *Revue de Nématologie* 6, 285–290.
Bridge, J., Otim-Nape, G.W. and Namaganda, J.M. (1991) The root-knot nematode *Meloidogyne incognita*, causing damage to cassava in Uganda. *Afro-Asian Journal of Nematology* 1, 116–117.
Bridge, J., Price, N.S. and Kofi, P. (1995) Plant parasitic nematodes of plantains and other crops in Cameroon, West Africa. *Fundamental and Applied Nematology* 18, 251–260.
Bridge, J., Hunt, D.J. and Hunt, P. (1996) Plant parasitic nematodes of crops in Belize. *Nematropica* 26, 111–119.
Brucher, H. (1960) Über das Wild-Vorkommen des Nematoden *Heterodera rostochiensis* in Nord-Argentinien. *Die Naturwissenschaften* 47, 21–22.
Bruhn, C. and Koch, W. (1962) Beobachtungen fiber den Befall mittelamerikanischer Dioscoreenarten durch *Meloidogyne incognita* (Kofoid and White). *Nematologica* 8, 307–312.
Bruhn, C. and Koch, W. (1963) Control of root-knot nematodes in *Dioscorea* tubers. *Phytopathology* 53, 24.
Butler, L. and Vilsoni, F. (1975) Potential hosts of the burrowing nematode in Fiji. *Fiji Agricultural Journal* 37, 38–39.
Byars, L.P. (1917) A nematode disease of the dasheen and its control by hot water treatment. *Phytopathology* 7, 66.
Cadet, P. and Daly, P. (1996) Use of nematicides to produce yam planting material free of *Scutellonema bradys* in Martinique (French West Indies). *Crop Protection* 15, 187–195.
Cadet, P. and Quénéhervé, P. (1994) Fluctuations naturelles de *Scutellonema bradys* (Nematoda: Hoplolaimidae) au cours de la croissance et du stockage de l'igname (*Dioscorea alata*) à la Martinique. *Nematologica* 40, 587–600.

Castagnone-Sereno, P. (1988) Désinfestation des semences d'igname par thermo ou chimiothérapie: efficacité nématicide et consequences agronomiques. *Turrialba* 38, 337–340.

Castagnone-Sereno, P. and Kermarrec, A. (1988) Association between *Pratylenchus coffeae* and *Scutellonema bradys* in yam tubers under agronomic conditions in the French West Indies. *Nematropica* 18, 155–157.

Caveness, F.E. (1967a) Shadehouse host ranges of some Nigerian nematodes. *Plant Disease Reporter* 51, 115–119.

Caveness, F.E. (1967b) *Nematology Studies 1960–1965.* End of tour progress report. Ministry of Agriculture and Natural Resources, Nigeria.

Caveness, F.E. (1980) Plant parasitic nematodes on cassava. In: Ezumah, H.C. (ed.) *Proceedings of the Workshop on Cassava Production and Extension in Central Africa.* International Institute of Tropical Agriculture, Mbanza-Ngungu, Zaire, 4, 83–106.

Caveness, F.E. (1981) Root-knot nematodes on cassava. *Annual Report, IITA,* 64–65.

Caveness, F.E. (1982) Root-knot nematodes as parasites of cassava. *IITA Research Briefs* 3, 2–3.

Caveness, F.E., Wilson, J.E. and Terry, R. (1981) Root-knot nematodes on tannia (*Xanthosoma sagittifolium*). *Nematologia Mediterranea* 9, 201–203.

Charchar, J.M. and Huang, C.S. (1981) Circulo de hospedeiras de *Pratylenchus brachyurus.* 3. *Plantas diversas Fitopatalogia Brasileira* 6, 469–473.

Claudius-Cole, A.O., Asiedu, R. and Fawole, B. (2004) Cultural control of *Scutellonema bradys* on *Dioscorea* spp. In: Akporoda, M.O. and Ekanayake, I.J. (eds) *Proceedings of the International Symposium for Root and Tuber Crops,* 12–16 November 2001, Ibadan (in press).

Coates-Beckford, P.L. and Brathwaite, C.W.D. (1977) Comparison of various treatments for the control of *Pratylenchus coffeae* in yam. *Nematropica* 7, 20–26.

Coates-Beckford, P.L., Hutton, D.G. and Jones, A.G. (1978) Effect of various treatments for the control of nematodes in the initial growth of yam. *Nematropica* 8, 6–7.

Cock, J.H. (1984) Cassava. In: Goldsworthy, P.R. and Fisher, N.M. (eds) *The Physiology of Tropical Field Crops.* John Wiley and Sons Ltd, Chichester, UK, pp. 529–550.

Corbett, D.C.M. (1976) *Pratylenchus brachyurus. C.I.H. Descriptions of Plant-parasitic Nematodes.* Set 6, No. 89.

Coursey, D.G. (1967) *Yams.* Longmans, London.

Coursey, D.G. and Booth, R.H. (1977) Root and tuber crops. In: Leakey, C.L.A. and Willis, J.B. (eds) *Food Crops of the Lowland Tropics.* Oxford University Press, Oxford, pp. 75–96.

Coursey, D.G. and Haynes, P.H. (1970) Root crops and their potential as food in the tropics. *World Crops* 22, 261–265.

Coyne, D.L. (1995) Nematode parasites of cassava. *African Crop Science Journal* 2, 355–359.

Coyne, D.L. (2004) Nematodes: pre- and post-harvest pests of root and tuber crops? A brief overview. In: Akporoda, M.O. and Ekanayake, I.J. (eds) *Proceedings of the International Symposium for Root and Tuber Crops,* 12–16 November 2001, Ibadan pp. 522–526.

Coyne, D.L. and Namaganda, J. (1994) Root-knot nematode, *Meloidogyne* spp., incidence on cassava in two areas of Uganda. *Roots* 1, 12–13.

Coyne, D.L. and Talwana, L.A.H. (2000) Reaction of cassava cultivars to root-knot nematode (*Meloidogyne* spp.) in pot experiments and farmer-managed field trials in Uganda. *International Journal of Nematology* 10, 153–158.

Coyne, D.L., Talwana, H.A.L. and Maslen, N.R. (2003) Plant-parasitic nematodes associated with root and tuber crops in Uganda. *African Plant Protection* 9 (2), 87–98.

Coyne, D.L., Khizzah, W. and Whyte, J. (2004) Root knot nematode damage to cassava in Kenya. *Roots* 9, 3–6.

Crozzoli, P.R. and Hidalgo, S.O. (1992) Response of ten cassava cultivars to the nematode *Meloidogyne incognita. Fitopatologia Venezolana* 5, 20–22.

Crozzoli, P.R. and Parra, M.D.M. (1991) Deteccion del nematodo *Scutellonema bradys,* causante de la pudricion seca del name en Venezuela. *Fitopatologia Venezolana* 4, 26.

Crozzoli, P.R. and Parra, N. (1999) Response of ten cassava cultivars to the nematode *Meloidogyne incognita. Nematologia Mediterranea* 27, 95–100.

Crozzoli, P.R., Moreno, N. and Torres, J.F. (1995) El nematodo *Meloidogyne javanica* asociado a danos en el cultivo del ocumo (*Xanthosoma sagttifolium*) en Venezuela. *Fitopatologia Venezolana* 8, 18.

Da Ponte, J.J. and Franco, A. (1981) Manipueira, un nematicida nao convencional de comprovada potencialidade. In: *Trabalhos Apresentados Reuniae de Sociedade Brasileira de Nematologia, Piracicaba,* Brazil 5, 25–33.

Da Ponte, J.J., Torres, J. and Simplicio, M.E. (1980) Behaviour of cassava cultivars in relation to root-knot nematodes. In: Lordello, L.G.E. (ed.) *IV Reuniao Brasiliera de Nematologia*, 16–20 July 1979, Sao Paulo, Publicacao No. Sociedade Brasiliera de Nematologia. Piracicaba, SP, Brazil, 4, pp. 107–133.

Decker, H. and Casamayor, G.R. (1966) Algunas observaciones sobre la presencia de nemátodos formadores de agallas en las raices (*Meloidogyne* spp.) en Cuba. *CENTRO, Boletin de Ciencias y Tecnologia de la Universidad Central de las Villas* 2, 19–29.

Decker, H., Casamayor, G.R. and Bosch, D. (1967) Observaciones sobre la presencia del nemátodo *Scutellonema bradys* en al tuberculo de ñame en la provincia de Oriente (Cuba). *CENTRO, Boletin de Ciencias Tecnología de la Universidad Central de Las Villas* 2, 67–70.

de Freitas, O.M.B.L. and de Moura, R.M. (1986) Response of cultivars of cassava (*Manihot esculenta* Crantz) to *Meloidogyne incognita* and *M. javanica* (Nematoda, Heteroderidae) and comparisons with hydrogen cyanide content. *Nematologia Brasileira* 10, 109–131.

De Guiran, G. (1965) Nematodes associes au manioc dans le sud du Togo. In: *Comptes Rendues des Ttravaux du Congres de la Protection Culture Tropicales*, Marseilles, pp. 677–680.

Demeaux, M., Babacauh, K.D. and Vivier, P. (1982) Problèmes posés par la conservation des ignames en Côte d'Ivoire et essais de techniques pour les résoudre. In: Miège, J. and Lyonga, S.N. (eds) *Yams. Ignames.* Clarendon Press, Oxford, pp. 320–328.

de Mendes, M.L., Pria, M.D., Geus-Bouwman, D.M., Vicente, F.R. and Tomaselli, F. (2001) Ocorrencia de *Pratylenchus penetrans* (Cobb, 1917) Chitwood and Oteifa, 1952, em mandioqhinha salsa (*Accacacia xanthorrhiza*) no municipio de Castro, PR. *Nematologia Brasileira* 25, 85–87.

de Tanaka, M.A.S., Chalfoun, S.M. and de Abreu, M.S. (1979) Doencas de mandioca en seu controle. *Informe Agropecuario* 5, 70–78.

Diomande, M. (1982) Root-knot nematodes on upland rice (*Oryza sativa* and *O. glaberrima*) and cassava (*Manihot esculenta*) in Ivory Coast. In: *Proceedings of the 3rd Research Planning Conference on Root-knot Nematodes,* Meloidogyne *spp., Regions IV and V (International* Meloidogyne *Project).* North Carolina State University Graphics, Raleigh, North Carolina, pp. 37–45.

Dixon, A.G.O., Bandyopadhyay, R., Coyne, D., Ferguson, M., Ferris, R.S.B., Hanna, R., Hughes, J., Ingelbrecht, I., Legg, J., Mahungu, N., Manyong, V., Mowbray, D., Neuenschwander, P., Whyte, J., Hartmann, P. and Ortiz, R. (2003) Cassava: from poor farmers' crop to pacesetter of African rural development. *Chronica Hortica* 43, 8–15.

Ekanayake, I.J., Osiru, D.S.O. and Porto, M.C.M. (1998) Physiology of cassava. *IITA Research Guide No. 55,* 3rd edn. International Institute of Tropical Agriculture, Ibadan, Nigeria.

Ekundayo, J.A. and Naqvi, S.H.Z. (1972) Preharvest microbial rotting of yams (*Dioscorea* spp.) in Nigeria. *Transactions of the British Mycological Society* 58, 15–18.

FAO (Food and Agriculture Organization) (2002) *FAOSTAT Agriculture Data.* FAO Statistical Databases. Available at: apps.fao.org

Flach, M. (1982) Ecological competition among the main moisture-rich starchy staples in the tropics and subtropics. *Proceedings of the 5th International Symposium on Tropical Root and Tuber Crops.* 17–21 September 1979, Los Banos, Laguna, the Philippines, pp. 345–375.

Fliege, F.H. and Sikora, R.A. (1981) Occurrence and distribution of plant-parasitic nematodes in W. Samoa. *Alafua Agricultural Bulletin* 6, 33–41.

Gao, Q.K. (1992) The observation of occurrence of root-knot nematodes on Chinese yam. *Chinese Vegetables* 5, 24–25.

Gao, Q.K., Zhao, H.H., Xu, M.E. and Liu, W.Z. (2000) Preliminary study on the root-knot nematode disease of yam. *Acta Phytopathologica Sinica* 30, 162–165.

Gapasin, R.M. (1980) Reaction of golden yellow cassava to *Meloidogyne* spp. inoculation. *Annals of Tropical Research* 2, 49–53.

Gapasin, R.M. (1981) Control of *Meloidogyne incognita* and *Rotylenchulus reniformis* and its effect on the yield of sweet potato and cassava. *Annals of Tropical Research* 3, 92–100.

Goodey, T. (1935) Observations on a nematode disease of yams. *Journal of Helminthology* 13, 173–190.

Gowen, S.R. (1985) *Report on Nematode Survey in the Solomon Islands and Vanuatu.* Report to UK Overseas Development Administration, London.

Hawley, W.O. (1956) Hot water treatment for the control of root-knot nematodes on *Dioscorea floribunda. Plant Disease Reporter* 40, 1045–1046.

Hemeng, O.B. (1978) Root-knot nematodes (*Meloidogyne incognita*) as serious pests of yam (*Dioscorea* spp.). *Ghana Journal of Agriculture* 11, 131–133.

Henz, G.P. (2002) Doencas da mandioqhinha salsa e sua situacao atual no Brasil. *Horticultura Brasileira* 20, 135–144.

Hickling, G.E. (1974) Nematode pests of yams in Jamaica. In: *Proceedings of the 12th Annual Meeting of the Caribbean Food Crops Society, Jamaica.* Ministry of Agriculture, Jamaica, pp. 63–67.
Hillocks, R.J. and Wydra, K. (2002) Bacterial, fungal and nematode diseases. In: Hillocks, R.J., Thresh, J.M. and Belloti, A.C. (eds) *Cassava: Biology, Production and Utilization.* CAB International, Wallingford, UK, pp. 261–280.
Hogger, C.H. (1971) Plant parasitic nematodes associated with cassavas. *Tropical Root and Tuber Crops Newsletter* 4, 4–9.
Horton, D., Lynam, J. and Knipscheer, H. (1984) Root crops in developing countries – an economic appraisal. *Proceedings of Sixth Symposium of the International Society for Tropical Root Crops.* 21–26 February 1983, International Potato Centre, Lima, Peru, pp. 9–39.
Huang, C.S., Tsai, Y.P., Tu, C.C., Lin, Y.Y. and Huang, S.P. (1972) *Plant parasitic nematodes in Taiwan.* Monograph Series No. 1. Institute of Botany Academia Sinica Taipei, Taiwan.
Huang, W.H., Gao, X.B. and Lu, J.H. (1994) Pathogen identification and pathogenicity of the Chinese yam root rot disease caused by lesion nematodes. *Journal of South China Agricultural University* 15, 35–38.
Hutton, D.G. (1998) Use of household disinfectants to suppress *Pratylenchus coffeae* and dry rot of yellow yam (*Dioscorea cayenensis*). *Tropical agriculture* 75, 49–52.
Hutton, D.G., Wahab, A.H. and Murray, H. (1982) Yield response of yellow yam (*Dioscorea cayenensis*) after disinfesting planting material of *Pratylenchus coffeae. Turrialba* 32, 493–496.
Inagaki, H. (1981) Introductory reports of studies on *Meloidogyne* spp. in Japan. In: *Proceedings of the Third Research Planning Conference on Root-knot Nematodes,* Meloidogyne *spp. Region VI.* Jakarta, Indonesia, pp. 40–49.
Inagaki, H. (1985) The plant parasitic nematodes important in Japan and the related researches. *JARQ* 18, 194–201.
IITA (International Institute of Tropical Agriculture) (1976) *Annual Report for 1975.* Ibadan, Nigeria.
IITA (International Institute of Tropical Agriculture) (1978) *Annual Report for 1977.* Ibadan, Nigeria.
IITA (International Institute of Tropical Agriculture) (1984) A system to increase seed yam production. *Research Highlights for 1983.* Ibadan, Nigeria, pp. 103–107.
IITA (International Institute of Tropical Agriculture) (2000) *Annual Report 2000.* IITA, Ibadan, Nigeria.
Iwahashi, T. (1977) [Control of root-lesion nematode of dasheen, I: soaking method of seed tuber.] *Kyushu Agricultural Research* 38, 81 (in Japanese).
Iwahori, H., Sano, Z. and Torigoe, H. (2001) Distribution of the main plant-parasitic nematodes in sweet potato and taro fields in Kyushu and Okinawa, Japan. *Kyushu Plant Protection Research* 47, 112–117.
Jackson, G.V.H. (1987) *Corm Rot of* Cyrtosperma *in Guam.* Report to the Federated States of Micronesia and Palau.
Jatala, P. (1989) Nematodes in tuber and root crops and their management. *Proceedings of the 11th International Congress in Plant Protection.* Manila, the Philippines.
Jatala, P. and Kaltenbach, R. (1979) Survival of *Nacobbus aberrans* in adverse conditions. *Journal of Nematology* 11, 303.
Jenkins, W.R. and Bird, G.W. (1962) Nematodes associated with wild yam, *Dioscorea* sp., with special reference to the pathogenicity of *Meloidogyne incognita incognita. Plant Disease Reporter* 46, 858–860.
Jennings, D.L. (1976) Cassava. In: Simmonds, N.W. (ed.) *Evolution of Crop Plants.* Longmans, Green & Co. Ltd, London.
Jimenez, N.P., Crozzoli, R.P., Montilla, J.C. and Carrasco, A. (2000) El nematodo *Meloidogyne hapla* asociado con danos en el cultivo del apio (*Arracacia xanthorrhza*) en el Estado Merida, Venezuela. *Fitopatologia Venezolana* 13, 42–43.
Jones, W.O. (1959) *Manioc in Africa.* Stanford University Press, Stanford, California.
Kawamura, T. and Hirano, K. (1961) [Host–parasite relationship of *Meloidogyne hapla* Chitwood on yam (*Dioscorea batatas* Decne).] *Annals of the Phytopathological Society of Japan* 26, 7–15 (in Japanese).
Kay, D.E. (revised by Gooding, E.G.B.) (1987) *Crop and Product Digest No. 2 – Root Crops,* 2nd edn. Tropical Development and Research Institute, London.
Kermarrec, A. (1974) Les nématodes de l'igname dans les Antilles. *Nouveau Maraich. et Viv de VINRA* 7/8, 95–105.
Kermarrec, A. and Anais, A. (1973) Presencia en Guadalupe de *Aphelenchoides besseyi* Christie 1942 (Nematoda, Aphelenchoididae) en foliage del mapuey: *Dioscorea trifida* L. *Turrialba* 23, 389–393.
Kermarrec, A., Castagnone-Sereno, P., Degras, L., Anais, A. and Denon, D. (1987) Nouvelle distribution *Scutellonema bradys* (Tylenchida: Hoplolaiminae) dans le Caraibe. Le cas des Antilles Francaises. *Medelingen van de Faculteit Landbouwwetenschappen Rijksuniversiteit Gent* 52, 617–624.

Kermarrec, A., Degras, L. and Anais, A. (1988) Une grave maladie parasitaire de d'igname Gross-Caille yams (*Dioscorea cayenensis, D. rotundata*) due au nématode des lesions racinaires (*Pratylenchus coffeae*). *Bulletin Agronomique Antilles Guyane* 7, 36–38.
Kirby, M.F. (1977) Control of root knot nematodes in Fiji. *Fiji Agricultural Journal* 39, 87–95.
Kirby, M.F., Kirby, M.E., Siddiqi, M.R. and Loof, P.A. (1980) Fiji nematode survey report: plant parasitic nematode distributions and host associations. *Bulletin No. 68, Ministry of Agriculture and Fisheries, Fiji.*
Koch, W. (1975) Control of diseases, pests and weeds in cultivating steroid plants. In: *International Plant Protection Congress (8th), Moscow, 21–16 August 1975. Reports and Informations. Section VI.* Moscow, Russia, pp. 239–244.
Kwoseh, C.K. (2000) Identification of resistance to major nematode pest of yams (*Dioscorea* spp.) in West Africa. PhD Thesis, University of Reading, UK.
Kwoseh, C.K., Plowright, R.A. and Bridge, J. (2002) The yam nematode: *Scutellonema bradys*. In: Starr, J.L., Cook, R. and Bridge, J. (eds) *Plant Resistance to Nematodes*. CAB International, Wallingford, UK, pp. 221–228.
Leon, J. (1977) Origin evolution, and early dispersal of root and tuber crops. In: Cock, L., MacIntyre, R. and Graham, M. (eds) *Proceedings of the Fourth Symposium of the International Society for Tropical Root Crops*. CIAT, Cali, Colombia, pp. 20–36.
Lordello, L.G.E. (1981) *Nematóides das Plantas Cultivadas*, 6th edn. Livraria Nobel, SA.
Lorenzo, E. and Fernandez, E. (1982) Reporte de nematodos noduladores en la malanga 'Isléna Japonesa' (*Colocasia esculenta*) en Cuba. *Ciencia y Técnica en la Agricultura. Proteccion de Plantas* 5, 31–38.
Lowe, J. (ed.) (1992) Nematological research at IITA 1969–1988: a summary of investigations conducted by Field E. Caveness. *Plant Health Research Monograph 2*. International Institute of Tropical Agriculture, Ibadan, Nigeria.
Luc, M. (1971) Nematode parasites of food crops in West Africa. Paper presented at a Seminar on Agricultural Research in West Africa, University of Ibadan, 24–28 May 1971.
Luc, M. and De Guiran, G. (1960) Les nématodes associés aux plantes de l'ouest Africain. Liste preliminaire. *Agronomie Tropicale* 15, 434–449.
Makumbi-Kidza, N.N. (2001) Studies on the distribution, pathogenicity and control of the root-knot nematode (*Meloidogyne incognita*) on cassava (*Manihot esculenta*). PhD Thesis, University of Bonn, Germany.
Makumbi-Kidza, N.N., Speijer, P.R., Ekanayake, I.J. and Sikora, R.A. (1999) Effect of *Meloidogyne incognita* on sprouting of two cassava (*Manihot esculenta* Crantz) cultivars. *Journal of Root Crops* 25, 175–177.
Makumbi-Kidza, N.N., Speijer, P.R. and Sikora, R.A. (2000) The influence of *Meloidogyne incognita* on growth and storage-root formation of young cassava, *Manihot esculenta* Crantz, plants. *Supplement to the Journal of Nematology* 32 (4S), 475–477.
Martin, G.C. (1969) Plants infected with root-knot nematodes (*Meloidogyne* spp.) in Rhodesia, Malawi and Zambia. *Technical Bulletin No. 8*, Rhodesia Ministry of Agriculture, Salisbury.
McSorley, R., O'Hair, S.K. and Parrado, J.L. (1983a) Nematodes associated with the edible aroid genera *Xanthosoma* and *Colocasia* and their effects on yield. *Nematropica* 13, 165–180.
McSorley, R., O'Hair, S.K. and Parrado, J.L. (1983b) Nematodes of cassava, *Manihot esculenta* Crantz. *Nematropica* 13, 261–287.
Meerman, J.C. and Speijer, P.R. (2001) *Proceedings of the 7th Triennial Symposium of the International Society for Tropical Root Crops – Africa Branch*, 11–17 October 1998. Cotonou, Benin, pp. 620–622.
Miège, J. (1957) Influences de quelques caractères des tubercules semences sur la levée et le rendement des ignames cultivées. *Journal d'Agriculture Tropicale et de Botanique Appliquée* 4, 315–342.
Missah, A. and Peters, J.C. (2001) Plant parasitic nematodes associated with soils, intercrops and tubers of yam in Ghana. *Proceedings of the 7th Triennial Symposium of the International Society for Tropical Root Crops – Africa Branch*, 11–17 October 1998. Cotonou, Benin, pp. 639–644.
Mohandas, C. and Palaniswami, M.S. (1990) Effect of *Meloidogyne incognita* on three popular cultivars of taro. *Journal of Root Crops: ISRC National Symposium 1990, Special*, pp. 207–209.
Mohandas, C. and Ramakrishnan, S. (eds) (1997) Pathogenic effect of root-knot nematode, *Meloidogyne incognita* on African white yam, *Dioscorea rotundata. Indian Journal of Nematology* 27, 233–236.
Mohandas, C., Palaniswami, M.S. and Pillai, S.V. (1996) Resistance in tuber crops to *Meloidogyne incognita* (Kofoid and White, 1919). In: Kurup, G.T., Palaniswami, M.S., Potty, V.P., Padmaja, G. and Kabeerathumma, S. (eds) *Tropical Tuber Crops: Problems, Prospects and Future Strategies*. Science Publishers, Enfield, New Hampshire, pp. 366–369.

Mortimer, J.L., Bridge, J. and Jackson, G.V.H. (1981) *Hirschmanniella* sp., an endoparasitic nematode associated with miti-miti disease of taro corms in the Solomon Islands. *FAO Plant Protection Bulletin* 29, 9–11.

Moura, R.M., Pio-Ribeiro, G., Coêlho, R.S.B. and Silva-Junior, J.N. (1976) *Penicillium sclerotigenum* Yamamoto, principal fungo causador de podridão em tuberas de inhame (*Dioscorea cayenensis* Lam.) no Estado de Pernambuco. *Fitopatologia Brasileira* 1, 67–76.

Moura, R.M., Coêlho, R.S.B. and Pio-Ribeiro, G. (1978) Estudo etio1ógico e efeito de 1,2-dibromo-3cloropropano no controle à caxa preta do inhame (*Dioscorea cayenensis* Lam.). *Fitopatologia Brasileira* 3, 47–53.

Moura, R.M., Pedrosa, E.M.R. and Guimaraes, L.M.P. (2001) Novos dados sobre a etiologia da easca preta do inhame no nordeste do Brasil. *Nematologia Brasileira* 25, 235–237.

Mudiope, J., Speijer, P.R., Maslen, N.R. and Adipala, E. (1998) Evaluation of yam host-plant response to root-knot nematodes in Uganda. *African Plant Protection* 4, 119–122.

Mudiope, J., Speijer, P.R., Maslen, N.R. and Adipala, E. (2001) *Pratylenchus*, the dominant genus affecting yam (*Dioscorea* spp.) in Uganda. *Proceedings of the 7th Triennial Symposium of the International Society for Tropical Root Crops – Africa Branch*, 11–17 October 1998, Cotonou, Benin, pp. 637–638.

Mudiope, J., Coyne, D.L., Adipala, E. and Sikora, R.A. (2004) Pathogenicity of the lesion nematode, *Pratylenchus sudanensis* (Loof & Yassin) in pots. *Proceedings of the VIth African Crop Science Conference*, 12–17 October 2003, Nairobi, Kenya.

Nadakal, A.M. and Thomas, N. (1967) Observations of nematodes associated with dry rot of *Dioscorea alata* L. *Science and Culture* 33, 142–143.

Nartey, F. (1977) *Manihot esculenta* (cassava): cyanogenesis, ultra-structure and seed germination. PhD Thesis, University of Copenhagen, Munksgaard, Copenhagen, Denmark.

Navarro, R. and Barriga, R. (1975) Identifications of *Meloidogyne* spp., root-knot nematodes, in Columbia. *Proceedings of the American Phytopathological Society* 2, 77–78.

Nirula, K.K. (1959) Root knot nematode on *Colocasia*. *Current Science* 28, 125–126.

Nishizawa, T. (1973) Pathogenicity of *Trichodorus porosus* to Chinese yam. *Japanese Journal of Nematology* 3, 33–37.

Nishizawa, T. and Ohshima, Y. (1972) [Annual changes in the nematode fauna in experimental fields with continuous or rotational cropping with 4 crops.] *16th Annual Meeting of the Japanese Society of Applied Entomology and Zoology*, Shizuoka, Japan, p. 120 (in Japanese).

Nwauzor, E.C. and Fawole, B. (1981) Root-knot nematodes on yams in eastern Nigeria. *Proceedings of the 3rd Research Planning Conference on Root-knot Nematodes,* Meloidogyne *spp., Regions IV and V,* Ibadan, Nigeria, pp. 161–167.

Nwauzor, E.C. and Nwankwo, O.C. (1989) Reaction of some cassava cultivars/lines to root-knot nematode, *Meloidogyne incognita* race 2. *International Nematology Network Newsletter* 6, 40–42.

Nweke, F.I., Ugwu, B.O., Asadu, C.L.A. and Ay, P. (1991) Production costs in the yam-based cropping systems of south-western Nigeria. *Resource and Crop Management Division. Research Monograph No. 6.* IITA, Ibadan, Nigeria, p. 29.

Nweke, F.I., Spencer, D.S.C. and Lynam, J.K. (2002) *The Cassava Transformation: Africa's Best Kept Secret.* Michigan State University Press, East Lansing, Michigan.

Oashi, K. (1984) [On the damage and ecological control of *Pratylenchus coffeae* infesting taro in south Kyushu.] *Japanese Journal of Nematology* 14, 61–62 (in Japanese).

Obigbesan, G.O. and Adesiyan, S.O. (1981) The influence of fertilizer nutrients on population levels of *Scutellonema bradys* in three yam species. *Nematologia Mediterranea* 9, 21–28.

Ogundana, S.K., Naqvi, S.H.Z. and Ekundayo, J.A. (1970) Fungi associated with soft rot of yams (*Dioscorea* spp.) in storage in Nigeria. *Transactions of the British Mycological Society* 54, 445–451.

Onalo, J.I., Asiedu, R. and Adisiyan, S.O. (2001) Control of the yam nematode (*Scutellonema bradys*) with neem fruit powder. *Proceedings of the 7th Triennial Symposium of the International Society for Tropical Root Crops – Africa Branch,* 11–17 October 1998, Cotonou, Benin, pp. 634–637.

Oramas Nival, D. (2002) Seed treatment of yam (*Dioscorea rotundata*) with oxamyl L during storage for the control of *Pratylenchus coffeae*. *Journal of Agriculture of the University of Puerto Rico* 86, 169–172.

Oramas Nival, D. and Rodriguez, J.A. (2002) Rotation of *Dioscorea rotundata* cv. Habanero with *D. alata* cultivars for the control of dry-rot disease. *Journal of Agriculture of the University of Puerto Rico* 86, 123–130.

Orton Williams, K.J. (1980) Plant-parasitic nematodes of the Pacific. *Technical Report, Volume 8, UNDP/FAO-SPEC Survey of Agricultural Pests and Diseases in the South Pacific.* Commonwealth Institute of Helminthology, St Albans, UK.

Park, S.D., Khan, Z., Kim, S.J., Kim, K.J. and Min, K.K. (1998) Occurrence and distribution of plant parasitic nematodes in yam (*Dioscorea batatas*) fields in Korea. *International Journal of Nematology* 8, 141–144.

Parris, G.K. (1940) A check-list of fungi, bacteria, nematodes and viruses occurring in Hawaii, and their hosts. *Plant Disease Reporter Supplement* 121, 1–91.

Patel, M.Z., Saelea, J. and Jackson, G.V.H. (1983) Breeding strategies for controlling diseases of taro in Solomon Islands. *Proceedings of the Sixth Symposium of the International Society for Tropical Root Crops.* Lima, Peru, pp. 143–149.

Pinochet, J. and Ventura, O. (1980) Nematodes associated with agricultural crops in Honduras. *Turrialba* 30, 43–47.

Purseglove, J.W. (1972) *Tropical Crops. Monocotyledons.* Longmans, London.

Ray, S., Sahoo, N.K. and Mohanty, K. (1992) Plant parasitic nematodes associated with tuber crops in Orissa. *Journal of Root Crops* 18, 70–72.

Roman, J. (1978) *Fitonematologia Tropical.* Universidad de Puerto Rico, Estacion Experimental Agricola, Rio Piedras, Puerto Rico.

Roman, J., Oramas, D. and Green, J. (1984a) Nematicide evaluation for the control of the nematodes of yam (*Dioscorea rotundata* Poir). *Journal of Agriculture of the University of Puerto Rico* 68, 157–162.

Roman, J., Oramas, D. and Green, J. (1984b) Use of oxamyl for the control of nematodes in yam (*Dioscorea rotundata* Poir). *Journal of Agriculture of the University of Puerto Rico* 68, 383–386.

Runkulatile, H.H.E. and Teri, J.M. (1990) Plant-parasitic nematodes associated with cocoyams (*Xanthosoma sagitifolium*) in Morogoro region of Tanzania. In: *Proceedings of the Workshop on the Global Status of and Prospects for Integrated Pest Management of Root and Tuber Crops in the Tropics,* 25–30 October 1987, Ibadan, Nigeria, pp. 193–195

Saadabi, A.M. (1985) Studies on the biology and control of some root-lesion nematodes with emphasis on *Pratylenchus sudanensis,* n.sp. Loof and Yassin, 1970 on cotton in the Gezira area. *International Nematology Network Newsletter* 2, 11–13.

Sadik, S. and Okereke, O.U. (1975) A new approach to improvement of yam, *Dioscorea rotundata. Nature* 254, 134–135.

Saka, V.W. (1982) International *Meloidogyne* project report in Malawi. In: *Proceedings of the 3rd Research Planning Conference on Root-knot Nematodes.* Meloidogyne *spp., Regions IV and V (International* Meloidogyne *Project).* North Carolina State University Graphics, Raleigh, North Carolina, pp. 31–36.

Schieber, E. (1961) Parasitic nematodes on *Dioscorea* in Guatemala. *Plant Disease Reporter* 45, 425.

Schieber, E. and Lassmann, D.K. (1961) Root-knot nematode on *Dioscorea* in Guatemala. *Plant Disease Reporter* 45, 981.

Sipes, B.S. and Arakaki, A.S. (1997) Root-knot nematode management in dryland taro with tropical cover crops. *Supplement to the Journal of Nematology* 29, 721–724.

Sipes, B.S., Nelson, S.C. and Arakaki, A.S. (1995) *Meloidogyne javanica* damage to dryland taro cultivars. *Afro-Asian Journal of Nematology* 5, 141–147.

Smit, J.J. (1967) Nematodes. In: Coursey, D.G. (ed.) *Yams.* Longmans, London, pp. 115–120.

Speijer, P.R. (1996) Hot water treatment: a possible strategy for farmer groups and NGOs. In: Green, K.R. and Florini, D.A. (eds) Pests and pathogens of yams in storage: a workshop report. *African Journal of Root and Tuber Crops* 1, 38–42.

Srivastava, A.S., Gupta, R.L., Singh, B. and Ram, S. (1969) Damage and control of *Meloidogyne javanica* (Treub) in *Colocasia* (*C. antiquorum*). *Labdev Journal of Science and Technology, Life Sciences* 7, 306–307.

Srivastava, A.S., Verma, R.S. and Pandey, R.C. (1971) Comparative efficacy of different nematicides tested against root-knot nematode infesting *Colocasia* (*C. anticorum*). *Labdev Journal of Science and Technology, Life Sciences* 9, 142–144.

Sterling, G.R., Gommers, F.J. and Maas, P.W.T. (1992) Outlook for biocontrol. In: *Nematology from Molecule to Ecosystem: Proceedings of the Second International Nematology Congress,* 11–17 August, Veldhoven, The Netherlands. European Society of Nematologists, pp. 257–265.

Swennen, R. (1990) *Plantain Cultivation Under West African Conditions: a Reference Manual.* IITA, Ibadan, Nigeria.

Talwana, L.A.H., Speijer, P.R., Adipala, E. and Maslen, N.R. (1997a) The effect of preplant population densities of root-knot nematodes on the establishment of cassava cuttings. *African Crop Science Conference Proceedings* 3, 1239–1242.

Talwana, L.A.H., Speijer, P.R., Adipala, E. and Maslen, N.R. (1997b) Early screening of cassava for resistance to root-knot nematodes. *Nematropica* 27, 19–25.

Tambe, T.J. (1999) A survey of the plant parasitic nematodes of cassava (*Manihot esculenta* Crantz) in the tropical rainforest of the South West Province, Cameroon. MSc Thesis, Faculty of Science, University of Buea, Cameroon.

Tarjan, A.C. (1952) Awl nematode injury on Chinese water chestnuts. *Phytopathology* 42, 114.

Théberge, R.L. (1985) *Common African Pests and Diseases of Cassava, Yam, Sweet Potato and Cocoyam.* International Institute of Tropical Agriculture, Ibadan, Nigeria, pp. 26–29.

Thompson, A.K., Been, B.O. and Perkins, C. (1973) Nematodes in stored yams. *Experimental Agriculture* 9, 281–286.

Timm, R.W. (1965) *A Preliminary Survey of the Plant Parasitic Nematodes of Thailand and the Philippines.* SEATO Thai Sambhand Printing Press, Bangkok.

Torigoe, H. (1993) Control of the root-lesion nematode (*Pratylenchus coffeae*) on taro (*Colocasia esculenta*) by combined application of 1,3-dichloropropene with dazomet or with oxamyl. *Proceedings of the Association for Plant Protection of Kyushu* 39, 94–97.

Torigoe, H. (1994) Control of the root-lesion nematode (*Pratylenchus coffeae*) on taro (*Colocasia esculenta*) by crop rotation with peanut and/or crops. *Proceedings of the Association for Plant Protection of Kyushu* 40, 113–118.

Torigoe, H. (1996) Correlation between the suppressive effects of several enemy plants and the recovery of *Pratylenchus coffeae* and *Meloidogyne incognita* populations in taro and sweet potato fields in southern Kyushu. *Proceedings of the Association for Plant Protection of Kyushu* 42, 83–88.

Tsay, T.T., Cheng, Y.H., Lin, Y.Y. and Cheng, C.F. (1994) Nematode diseases of root and tuber crops in Taiwan. *Plant Protection Bulletin Taipei* 36, 225–238.

Vilsoni, F. and Heinlein, M. (1982) Host range and reproductive capacity of the reniform nematode *Rotylenchulus reniformis* on some crop cultivars in Fiji. *Fiji Agricultural Journal* 44, 61–66.

Whitehead, A.G. (1969) The distribution of root-knot nematodes (*Meloidogyne* spp.) in tropical Africa. *Nematologica* 15, 315–333.

Wood, T.G., Smith, R.W., Johnson, R.A. and Kornolafe, P.O. (1980) Termite damage and crop loss studies in Nigeria – pre-harvest losses to yams due to termites and other soil pests. *Tropical Pest Management* 26, 355–370.

Yamada, M. (2001) Methods of control of injury associated with continuous vegetable cropping in Japan: crop rotation and several cultural practices. *JARQ* 35, 39–45.

Zem, A.C. (1979) New observations on nematodes and their hosts in the state of Bahia, Brazil. *Summa Phytopathologica* 5, 19–20 (abstract).

8 Nematode Parasites of Food Legumes*

Richard A. Sikora,[1] Nicola Greco[2] and João Flávo Velosa Silva[3]
[1]*Institut für Pflanzenkrankheiten der Universität Bonn, Nussallee 9, D-53115 Bonn, Germany;* [2]*Istituto per la Protezione delle Piante, Sezione di Bari, CNR, Via G. Amendola 165/A, 70126 Bari, Italy;* [3]*Embrapa Soybean, Phytopathology and Nematology, PO Box 231, 86001-970, Londrina, PR, Brazil*

The family Leguminosae, with approximately 650 genera and 18,000 species, is the third largest family of flowering plants. Although legumes are found throughout the world, the greatest diversity exists in the tropics and subtropics. The family is divided into three subfamilies: Caesalpinioideae with approximately 2800 species, mainly trees of tropical savannas and forests; Mimosoideae with about 2800 species, mostly small trees and shrubs of semi-arid tropical and subtropical regions; and Papilionoideae with about 12,000 species, containing the majority of food legumes and herbs, with a worldwide distribution (NAS, 1979; Purseglove, 1983).

Archaeological excavations have demonstrated that lentil, chickpea, lupin, string bean, broad bean, kidney bean, pea and soybean, among others, have played an important role as essential foods in the ancient civilizations of China, India, the Americas and the Near East as far back as 7000 BC (Brothwell and Brothwell, 1969). Of the more than 18,000 known legume species, fewer than 20 are of worldwide economic importance as food crops. However, over 200 have been considered important on a regional, local or future basis (NAS, 1979). For practical purposes, legume crops are often grouped under a variety of names including: legumes, pulses, grain legumes or beans. The use of any one term can be misleading, because these crops have a multitude of uses. These plants can be used as a grain, vegetable, green manure, pasture, cover crop to reduce erosion or as a source of fodder, cooking oil, or protein supplement, as well as for raw material in the food processing industry. Therefore, we have decided to use the broader term 'food legumes' for the crops discussed in this chapter. The main climatic zones, uses, distribution and relative economic importance of the major food legumes are presented in Table 8.1.

Legumes rank second to cereal crops in degree of nutritional importance for humankind. In many countries, they are the major source of protein, often containing 2–3 times more protein than cereals. It has been estimated that 80% of the protein in the diet of many tropical and subtropical countries is derived from vegetable products, among which food legumes predominate. In India, where in excess of 10 Mt are

*A revision of the chapter by R.A. Sikora and N. Greco.

Table 8.1. Common name, growing zone, uses, distribution and importance of food legumes in tropical and subtropical climatic areas.

	Climatic zones[a]				Uses				
Common names	T	ST	UT	D/SA	Grain	Veg.	Animal[b]	Distribution	Importance
Adzuki bean		x			x		x	Worldwide, China, Japan, SE Asia	++
Black gram, urd	x	x	x		x			India	++
Broad bean, faba bean	x	x			x	x	x	Worldwide	+++
Catjang bean	x	x				x	x	Worldwide, SE Asia	++
Chickpea	x	x		x	x	x	x	Worldwide	+++
Cowpea	x	x		x	x	x	x	Worldwide	+++
Grass pea, chickling pea		x	x		x		x	Worldwide	+
Haricot, kidney, bush, French, string bean	x	x			x	x		Worldwide	+++
Horse bean	x			x	x		x	Asia, Africa	+
Horse gram	x			x	x		x	Asia, Africa	+
Hyacinth bean, lablab	x	x		x	x		x	SE Asia	++
Lentil		x	x	x	x		x	Worldwide	+++
Lima bean, butter bean	x	x			x	x		Worldwide	++
Lupin, tarwi		x			x		x	N. and S. America, Mediterranean	+
Moth bean	x	x		x	x		x	Worldwide, India, S. America	+
Mung bean, green gram	x	x			x	x	x	Worldwide, India, China, SE Asia	+++
Pea		x	x		x	x	x	Worldwide	+++
Pigeonpea, red gram	x	x		x	x	x	x	Worldwide	++
Rice bean, red bean	x	x			x		x	Worldwide, SE Asia	++
Soybean, soya	x	x			x	x		USA, China, Brazil	+++
Winged bean	x				x	x	x	Worldwide, SE Asia	+

Brothwell and Brothwell (1969); NAS (1979); Ward *et al.* (1981).

[a]T, tropical; ST, subtropical; UT, upland tropics; D/SA, dry/semi-arid tropics.

[b]Animal = fodder, green manure, protein supplement or straw.

consumed every year, they supply the only high protein component of the diet (Kay, 1979). Legumes are the cheapest and most direct form of protein. They can be transported easily when dried and can be stored for long periods of time at room temperature without losing substantially on nutritional content.

The land area in food legume production, yield/ha and overall production are given in Table 8.2. The figures on production by crop and continent (Table 8.3) demonstrate the importance of the different food legumes in Asia where roughly 50% are consumed. The importance of plant parasitic nematodes, insects and diseases as well as abiotic constraints are reflected in the low per hectare yields in tropical agriculture when compared with yields in temperate agriculture (Table 8.2).

The symbiotic relationship between legumes and nitrogen-fixing *Rhizobium* bacteria gives these crops an economic advantage over crops requiring fertilizer. Part of the fixed nitrogen remains in the soil within crop residues after harvest, thus improving soil fertility. Food legumes are, therefore, an important component in tropical cropping systems, where they are rotated with such nutrient-demanding crops as rice and maize. In the subtropics, where soils are often deficient in organic matter, legumes can be used as green manure.

Cultivation techniques

Food legumes are adapted to a wide range of climatic conditions, particularly warm climates. Their deep root system favours survival during periods of drought, making them important crops in the semi-arid and dry regions of the tropics. In addition, a number of species grow well in moist climatic areas and are important crops in the humid tropics. Many food legumes are adapted to a wide range of soil types, high temperatures, low nutrient levels, alkalinity, acidity and high salt concentrations, making them important crops in marginal areas and in subsistence agriculture. Their capacity to grow under poor soil conditions may also be related to their ability to form symbiotic relationships with endomycorrhizal fungi, which are known to increase plant tolerance to a wide range of abiotic and biotic stress factors (Dehne, 1987) and in some cases to nematode infection (Sikora, 1981; Hussey and Roncadori, 1982).

Methods of cultivation vary greatly between climatic regions and within individual countries. The majority are sown by drilling or broadcasting either as a single crop or interplanted with other crops. When intercropped, the main crop is planted in rows and the legumes are broadcast after the main crop has been established (Kay, 1979; Ward *et al.*, 1981).

Table 8.2. Worldwide production of food legumes according to region in 2003.[a]

Region	Total area (1000 ha)	Yield (kg/ha)	Production (1000 t)
World	65,618	798	52,385
Africa	17,745	473	8,392
N. and C. America	5,981	841	6,777
S. America	4,422	1,133	3,436
Asia	31,526	744	23,445
Europe	3,741	2,053	7,680
Oceania	2,203	1,205	2,655
Former USSR	3,741[b]	1,372[b]	7,594

[a]FAO Statistics (2003).
[b]Average 1989–1991.

Table 8.3. World food legume production in t × 1000 in 2001.[a]

Continent	Bean	Broad bean	Dry pea	Chickpea	Lentil	Pigeonpea	Cowpea	Total legumes[b]	Soybean
Africa	2,311	1,175	266	329	77	261	3,533	8,392	999
N. and C. America	2,766	43	3,061	650	726	23	43	6,777	80,899
S. America	3,129	101	94	26	18	2	36	3,428	69,058
Asia	8,012	1,795	1,844	4,768	2,028	2,763	163	23,445	23,491
Europe	513	306	4,983	66	39	–	34	7,680	2,087
Oceania	40	200	407	240	181	–	3	2,655	105
Australia	40	20	356	240	180	–	3	2,597	105
World total	16,772	3,680	10,650	6,063	3,070	3,049	3,777	52,385	176,639

[a]FAO Statistics (2002).

[b]Includes other grain legumes; soybean is not included.

Nematodes of Food Legumes

Many plant parasitic nematodes have been found associated with legume crops (Goodey *et al.*, 1965; Sitaramaiah *et al.*, 1971; Bridge, 1981; Mani *et al.*, 1982; Ali and Askary, 2001). Those affecting forage, pasture and legumes grown mainly for cooking oil have been the subject of other review articles (Eriksson, 1972; Griffin, 1984; Schmitt and Noel, 1984; Sikora, 1987). The identification of races, biology and complex disease inter-relationships of the cyst nematodes affecting legumes in the *Heterodera trifolii* complex group were discussed by Sikora and Maas (1986).

Only those nematodes that are known to cause yield loss will be covered in this chapter; those that are only known to parasitize the crops and complete their life cycle on the plant will not be discussed in detail.

When food legumes are cultivated in semi-arid areas under rainfed conditions or in the dry season after the monsoon rains, infected plants are often severely damaged. Nematodes induce vascular disorders and reduce root penetration of the soil profile, increasing the negative impact moisture stress exerts on plant health.

Plant parasitic nematodes also affect plant vigour in some food legumes by suppressing *Rhizobium* root nodulation and nitrogen-fixing activity. Complex inter-relationships between nematodes and soil-borne fungal pathogens also play a significant role in reducing yield. The importance of these complex inter-relationships has received only minor attention.

Management measures, in the vast majority of cases where nematodes have been shown to be limiting factors, have not been developed adequately, leaving the farmer to his traditional cropping systems and ultimate poor yield. Furthermore, the effects of traditional multicropping rotation patterns on nematode population dynamics and crop growth are lacking for many parts of the tropical and subtropical zones. Many of these traditional cropping systems may be effective in checking nematode damage.

Although many breeding lines have shown different degrees of resistance to important nematodes, only a handful of resistant cultivars are available to the farmer. In many cases, the techniques used for testing have led to misinterpretation of results, with re-testing often failing to produce good sources of resistance for breeding programmes. Because breeding lines have little value to the grower, we have decided not to list all the lines tested. Lists of cultivars with resistance in food legumes have been compiled (Armstrong and Jensen, 1978; Sasser and Kirby, 1979; Bridge, 1981).

Black Gram, Urd, Mash

Black gram (*Vigna mungo* (L.) Hepper, syn. *Phaseolus mungo* L.), also known as urd or mash, probably originated in India and is a bushy annual common in Asia, Africa and America. The plant, which is very similar to mung bean, is resistant to high temperatures and is reasonably drought resistant. It is often grown intercropped with cotton, maize or sorghum.

Rotylenchulus reniformis has been detected on gram in Puerto Rico (Ayala and Ramirez, 1964). In India, *Heterodera cajani*, *R. reniformis* and *Tylenchorhynchus mashhoodi* have been found associated with the crop (Sitaramaiah, 1984) and are considered to be of economic importance. The root knot nematodes *Meloidogyne incognita* and *M. javanica* are known to infect black gram in Brazil (Freire *et al.*, 1972). Root knot has also been detected on the crop in India (Nadakal, 1964). In the Rajasthan area of India, *M. incognita* was found in 54% of the 176 fields sampled (Datta *et al.*, 1987).

Economic threshold level

M. incognita and *R. reniformis* were shown to cause significant growth reductions at 1 juvenile/cm^3 of soil in pot tests (Mishra and Gaur, 1981). Growth reduction increased with level of infestation, and both nematodes reduced the number of *Rhizobium* nodules. Zaki and Bhatti (1986) reported that *H. cajani* at 1 juvenile/g of soil did not affect shoot growth, but caused reductions in root weight. Gupta and Yadav (1979) in pot studies showed that plant growth was significantly reduced by *R. reniformis* at densities of more than 2 nematodes/g of soil. Damage threshold densities for *M. incognita* have been set at 1–2 juveniles/g of soil (Mahapatra *et al.*, 1999). *M. incognita* has been shown to have a negative effect on *Rhizobium* (Chahal and Chahal, 1987; Chahal *et al.*, 1988).

Root knot nematodes also interact with soil pathogens to cause more disease than when present alone. Wilting caused by *Fusarium pallidoroseum* increased greatly when in combination with *M. incognita* (Swain and Kar, 1994), and a reduction in plant growth and nodulation was dectected when *M. javanica* was associated with *Rhizoctonia bataticola* (Fazal *et al.*, 1998).

Management measures

In a study with five crop sequences designed specifically to control *M. incognita*, black gram after mustard and rice was the most effective in reducing root knot galling (Mahanta and Phukan, 1990).

The use of non-hosts and paddy rice in cropping systems will be the most effective and economical means of preventing damage. Resistant cultivars are not available, but moderate levels of resistance have been detected in some lines to *R. reniformis* (Routaray *et al.*, 1986; Midha and Trivedi, 1988). Resistance has been found in a number of lines to *M. javanica* and *M. incognita* (Handa, 1990; Kamalwanshi *et al.*, 2000), and to *H. cajani* (M.R. Siddiqui *et al.*, 1999).

Research has been conducted in the greenhouse that indicates that soil, and in some cases seed, treatment with arbuscular mycorrhizae can reduce root knot infection (Sankaranarayanan and Sundarababu, 1998; Bornali *et al.*, 2002). Plant health-promoting rhizobacteria combined with organic amendments gave good control of *M. javanica* and root pathogens (Siddiqui *et al.*, 2001a), whereas seed and soil treated with the egg pathogen *Paecilomyces lilacinus* gave good control of *M. incognita* and, when combined with other management tools, of *H. cajani* (Latha *et al.*, 2000).

Furthermore, studies have shown that organic amendments such as neem, castor and rice leaves as well as sawdust, chicken manure and oil cakes can reduce damage caused by *R. reniformis* and *M. incognita* (Anjum *et al.*, 1996; Bornali *et al.*, 1998). Seed treatment with neem-based formulations of seed kernel and seed coat has been shown to give significant control of both nematodes (Vijayalakshmi *et al.*, 1999). Nematicides as seed treatments have also been shown to be effective in reducing root knot (Kalita and Phukan, 1993).

The economic reality of using any of these approaches under practical field conditions on a crop with a low profit margin needs consideration when conducting research of this kind and in making recommendations to growers.

Broad Bean

Broad bean (*Vicia faba* L.), also known as faba, field, common, horse, tick and Windsor bean, is a subtropical or temperate crop that is probably native to the Mediterranean region or South-west Asia. It is grown in the winter season in the subtropics. The dried seeds are eaten as a porridge or consumed after baking as Foul in the Middle East, and the immature seeds also are often eaten as a vegetable after boiling. The seeds are also widely used as livestock and poultry feed. The crop is sometimes used as green manure and the dried residues as animal fodder.

A wide range of plant parasitic nematodes have been found associated with *V. faba*, but only a few are of widespread economic importance in the tropical and subtropical zones (Hooper, 1983b). In most cases, nematode damage occurs in the cooler winter growing seasons in the subtropics or in the upland tropical zones.

Ditylenchus

The stem nematode, *D. dipsaci*, is the most important nematode on broad bean in subtropical and temperate growing areas. The nematode has been detected attacking broad bean in many countries bordering the Mediterranean Sea, including Syria, Jordan, Turkey, France, Tunisia, Algeria, Morocco, Cyprus, Spain, Italy and Greece. Because of the nematode's worldwide distribution, it should be considered a potential pest in most areas where broad bean is grown (Hooper, 1972; Lamberti, 1981; Greco and Di Vito, 1987; Sellami, 1998; Abbad and Bachikh, 2001; Troccoli and Di Vito, 2002).

Biology

The stem and bulb nematode is a migratory endoparasite that feeds on stem, petiole, leaf, pod and seed tissue (see Chapter 2). The nematode does not cause damage to the root. Soil-borne *D. dipsaci* fourth stage juveniles penetrate the young seedling below the soil surface after germination. Damage is often more severe when seed-borne populations are already present in the tissue at planting. Cool, moist conditions for example, when present during the winter growing season in the Mediterranean region, favour nematode infection and disease development. As temperatures rise during the growing season, nematode development is often retarded, symptoms can disappear and the plant seems to recover.

Survival and means of dissemination

The fourth stage juvenile can withstand desiccation for many years. The nematodes often clump together to form 'nematode wool' when the plant tissue begins to dry. This wool can often be observed on the seeds in heavily infested pods. The presence of infective fourth stage juveniles in seed as well as in plant debris is important in the passive dissemination of the nematode over long distances. *D. dipsaci* is seed borne in broad bean, lucerne, onion, clovers and teasel.

The nematode in this desiccated stage can survive passage through pigs and cattle on infested seed (Palmisano *et al.*, 1971). Augustin (1985) was unable to detect passage of the nematode on infested straw in sheep.

Although nematode soil densities seem to decrease rapidly, Seinhorst (1956a) and Wilson and French (1975) showed that the nematode can survive for years without a host plant. However, many weeds and grasses are host for the nematode and may play an important role in its survival in the absence of cultivated plants. Nematode survival and damage are greater in heavy soils as compared with sandy soils (Seinhorst, 1956b). Hooper (1972) estimated that the nematode will die out within 8 years in the absence of a host, a rare occurrence in present day agriculture.

Races

Races of *D. dipsaci* normally cannot be identified morphologically nor with molecular techniques, with one exception, those attacking broad bean. Broad bean is attacked by the normal 'oat race' (1.2–1.4 mm adult or fourth stage juvenile body length) in temperate regions and by the 'giant race' (1.5–1.7 mm length) in the subtropical semi-arid regions of the Mediterranean. These two races can also be distinguished on the basis of chromosome number, $2n = 24$ in the 'oat race', versus double that number in the 'giant race'. Recently, these two races were also distinguished by molecular methods (Esquibet *et al.*, 1998). There are other races that can attack broad bean, but they are of marginal importance only in temperate regions.

The fact that the 'giant race' causes damage in England (Hooper, 1983a) and can survive under environmental conditions existing in Germany (D. Sturhan, Germany, 1989, personal communication) warrants closer examination of imported broad bean seeds originating from subtropical growing areas.

Symptoms

Although Hooper (1983a) suggested that the two races could be tentatively identified by the symptoms produced – the more severe symptoms being induced by the 'giant race' – he considered measurement of body length a more exact means of identification. The nematode can induce stem swelling and deformation of stem tissue (Fig. 8.1) or lesions which turn reddish-brown then black depending on cultivar and environmental factors. The lesions envelop the stem and increase in length, often advancing to the edge of an internode (Plate 7A). Leaf and petiole necrosis is also common under heavy infestations, but can be confused with symptoms produced by fungal leaf pathogens. Newly formed pods take on an even, dark brown appearance (Hooper, 1983a). Seeds infested with the nematode are darker, distorted, smaller in size and may have speckle-like spots (Fig. 8.2; Plate 7B) on the surface (Schreiber, 1977; Hooper, 1983b; Augustin, 1985). The percentage of seeds infested increases with infestation levels and is greatest when nematode-contaminated seed is used for sowing. Heavy infestations often kill the main shoot, which stimulates secondary tiller formation (Plate 7A). These newly formed shoots are often free from infection. The nematodes are found under the testa in depressions on either side of the radicle, causing necrotic patches, visible when the

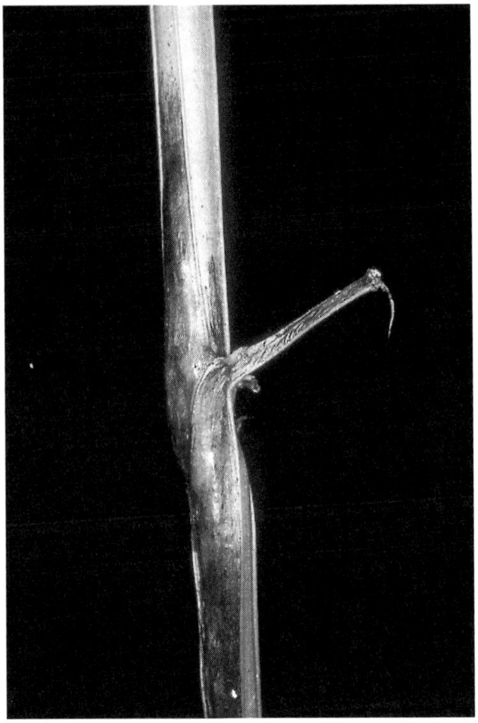

Fig. 8.1. Darkened and swollen stem, typical of *Ditylenchus dipsaci* 'giant race' infection of *Vicia faba*. (Photo: J. Bridge.)

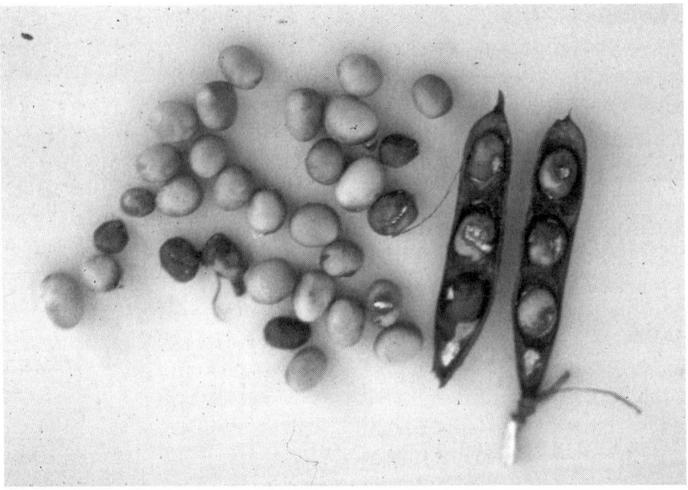

Fig. 8.2. Deformed and blackened seed and pods of *Vicia faba* infested with *Ditylenchus dipsaci* 'giant race'. (Photo: J. Bridge.)

testa is removed (Hooper, 1983b). Over 10,000 juveniles can be found in one infested seed.

It should be noted that Caubel and Leclercq (1989b) observed that two types of symptoms developed following infection, i.e. swelling and shortening of the inoculated axillary bud in resistant plants and necrotic lesions surrounding the inoculation site in susceptible lines.

Economic threshold level

Hooper (1983a) in field trials showed that the 'giant race' was more damaging to broad bean than the 'oat race', common to Europe, when *D. dipsaci*-infested straw was incorporated into the field. The 'giant race' caused 100 and 63% and the 'oat race' 82 and 1.3% stem and seed infection, respectively. The economic threshold level is not known for the 'giant race' on broad bean. The threshold levels for the 'oat race' on onion, celery and carrot is 2 nematodes/100 g of soil (Decker, 1969).

Other hosts

Although *D. dipsaci* has over 450 host plants (Hooper, 1972), the host range of the 'giant race' seems to be more limited. The 'giant race' is usually very damaging to wild oats, and some Moroccan populations also to pea (M. Di Vito, Italy, 2003, personal communication). Eleven out of 60 weed species found in fields of broad bean in Morocco were infested with the 'giant race'. In addition to weeds, *Avena sterilis*, *Vaccaria pyramidata* and *Verbena supia* were good hosts, as well as the economically important parasitic weed of broad bean *Orobanche crenata* (Abbad and Bachikh, 2001). The 'oat race' has also been reported to attack chickpea, pea and lentil in the Mediterranean basin, whereas the 'giant race' multliplied on faba bean, but only infested the stems of lentil and vetch (Caubel *et al.*, 1998c). Certain weeds serve as hosts for the 'oat race' (Green, 1981) and the 'giant race' of *D. dipsaci* (Augustin and Sikora, 1989a) and are important in maintaining high soil densities of the nematode.

In South Australia, the 'oat race' also affected emergence of canola, *Brassica napus*. Seedlings when inoculated showed typical symptoms of damage, whereas both tolerance and resistance were observed in mature plants (Taylor and Szot, 2000).

Management measures

Prevention of introduction by establishment of quarantine laws should be promoted. Seeds can be easily examined by the techniques outlined below and in Chapter 3. Rotation of 4 years with non-host crops and weed control of other hosts is required for successful control. Rotations of 3 years will reduce damage significantly compared with 2 years between crops. Caubel *et al.* (1998b) demonstrated that attacks by the 'oat race' are always associated with previous cropping of broad bean, pea or fodder beet, but not maize.

Fumigation has been used to eradicate the nematode from infested seeds, but will not give 100% control when high infestation exists (Powell, 1974; Augustin, 1985). Soil treatment with non-fumigant nematicides will prevent seed infestations and can be used to protect breeding material (Augustin and Sikora, 1984; Augustin, 1985).

Resistance to the 'giant race' is known from Egypt (B.A. Oteifa, Egypt, 1997, personal communication), where the nematode was not detected in a survey by Augustin (1985). The nematode was also reported on a local Moroccan cultivar by Schreiber (1977) and in Syria by Hanounik *et al.* (1986).

Good levels of resistance to the 'giant race' have been detected in breeding lines from ICARDA and INRA (Caubel *et al.*, 1998a). In a study of 250 accessions of broad bean from the INRA collection in Morocco, eight landraces from the Maghreb region and seven accessions from other origins were moderately resistant to the 'giant race'. The resistant lines from ICARDA showed a susceptible reaction and indicate that pathotypes may exist within the 'giant race' population (Abbad and Sellami, 1998).

Caubel and Leclercq (1989a) reported that INRA 29H was resistant and the ICARDA lines BPL 1696 and 1827 as well as FLIP 84-154 were intermediately resistant to the stem nematode. Abbad (2001) later reported that eight resistant lines from ICARDA and line INRA 29H were susceptible to a population from Dar Bouazza, Morocco, which indicates important variation within the 'giant race' that will need to be recognized in future resistant management and breeding programmes.

It should also be noted that the production of uninfested tillers after the main stem is killed by the nematode may be confused with resistance (Hooper, 1983a).

Heterodera

The pea cyst nematode, *Heterodera goettingiana*, is an important parasite of broad bean in many temperate regions. The nematode is a limiting factor in the cool growing season in some countries of subtropical North Africa, West Asia, Italy and Spain (Stone and Course, 1974). The nematode causes stunting in heavily infested fields (Fig. 8.3; Plate 7C).

Other hosts

Most host plants are in the tribe Vicieae of the family Leguminosae. *Pisum sativum*, *Lathyrus* species and species of *Vicia* as well as *Glycine max* are considered economically important hosts for *H. goettingiana* (Jones, 1950; Winslow, 1954). Soybean is a summer crop, and in the summer due to high temperatures *H. goettingiana* is for the most part quiescent. Therefore, it is a host but soybean would not be atttacked under field conditions. *Lens culinaris* Medic. is reported a host in the literature, but the authors believe this is by mistaken identity in that Tedford and Inglis (1999) found this crop to be a very poor to non-host.

Lentil was reported a host for *H. goettingiana* in the Irbid area of Jordan, but there is no evidence that the nematode was properly identified. In the same area in the early 1990s, lentil was infested by *H. ciceri* which was probably the nematode observed earlier on lentil (N. Greco, Italy, 2003, personal communication). In addition, many weeds are considered good hosts and are responsible for maintaining populations in the absence of susceptible crop plants.

Fig. 8.3. Broad bean crop showing a patch of stunted plants in a field infested with *Heterodera goettingiana*. (Photo: N. Greco.)

Biology

The biology and development of this cyst nematode are similar to those described for the other cyst nematodes in this chapter, and in Chapter 2. *H. goettingiana* only completes one generation per growing season if only cysts and not egg masses are produced, but multiple generations can be produced on both broad bean (Hooper, 1983a) and garden pea (Greco *et al.*, 1986a) sown in early autumn when egg masses are formed. Survival in the absence of a host has been reported to exceed 10 years (Brown, 1958).

Economic threshold level

The threshold level of broad bean to the nematode is 0.8 eggs/g of soil, with complete crop failure occurring at 64 eggs/g of soil (Greco *et al.*, 1991) The crop is, however, less susceptible to damage than pea. Growing the crop every 4 years in infested fields caused crop failure under temperate climatic conditions (Brown, 1958), but resulted in good crop stand under Mediterranean conditions (Di Vito and Greco, 1986).

Management measures

Effective control can be obtained by crop rotation with non-host crops. On uninfested land, Hooper (1983a) recommended reducing legume crops to once in 4 years. Where severe infestations are known, longer rotations are required (Brown, 1958).

Cicer arietinum L., *G. max* (L.), *Lupinus albus* L., *Medicago sativa* L., *Phaseolus vulgaris* L. and a number of clover species were found resistant to the nematode (Di Vito *et al.*, 1980; Tedford and Inglis, 1999).

Nematicides have been shown to be effective in controlling *H. goettingiana* on peas, but have not been examined on broad bean. Oxamyl at 6 g a.i./100 m row, applied in furrows, increased yield of pea and was considered to be economical, even though the nematode population increased tenfold after harvest (Green *et al.*, 1981). Nematicides, however, cannot be used economically for control of this nematode on broad bean.

Other nematodes of broad bean

There are a number of other nematodes that parasitize broad bean in the tropics and subtropics that are of local, limited or unknown importance.

The root knot nematodes *M. incognita*, *M. javanica*, *M. arenaria* and *M. artiellia* are known to attack broad bean in the tropics and subtropics (Goodey *et al.*, 1965). Damage caused by root knot nematodes has been observed in Italy (N. Greco and M. Di Vito, Italy, 1989, personal communication), Zimbabwe, Malawi, East Africa, Libya and Iraq (Hooper, 1983a). The symptoms of damage and methods of diagnosis are the same as those described for other legumes in this chapter. There also is an indication that the nematode can reduce nodulation (El Bahrawy and Salem, 1989). Control is usually accomplished by rotation with non-host crops, especially cereals. Care should be taken in selecting rotation crops, because of the nematodes' wide host range and known variability in the genus. Resistance is not known and nematicides are too expensive for practical use.

Some species of *Pratylenchus* cause extensive necrosis of the root tissue and yield loss in the subtropics and tropics (Trocccoli *et al.*, 2002). The impact of this group of nematodes to broad bean, however, has not been determined. However, broad bean is a good host for *P. neglectus*, *P. penetrans*, *P. pinguicaudatus* and *P. thornei* (Di Vito *et al.*, 2002a). Some broad bean lines were found resistant to them (Di Vito *et al.*, 2002b). In pot experiments, the tolerance limit of this legume to *P. neglectus* and *P. thornei* was 2 nematodes/cm^3 of soil (Di Vito *et al.*, 2000).

The burrowing nematode *Radopholus similis* has been shown to reproduce on broad bean only in India (Sosamma and Koshy, 1977). The reniform nematode, *R. reniformis*, has only been reported on broad bean in Pakistan and is of unknown importance (Timm, 1956). Hooper (1983a) discussed the distribution and importance of stunt nematodes in the family Tylenchorhynchidae; the nematodes are of limited economic importance.

Chickpea

Chickpea (*Cicer arietinum* L.), also known as gram and bengal gram, originated from Turkey and Syria around 5450 BC (Saxena, 1987). Production is concentrated in Asia where 84% of the world's crop is grown, with India accounting for about 57% of the area in cultivation. Other countries with extensive cultivation are: Pakistan, Myanmar (Burma), Iran, Ethiopia, Mexico, Canada and Australia. In the Mediterranean basin, chickpea is an important crop in Turkey, Syria, Morocco, Tunisia, Spain and Portugal. Although green pods and shoots of chickpea are also used for vegetables in India, this legume is used mainly as dried grains which are boiled, mashed or roasted, and used for flour in various foods. A minor portion is used as animal feed.

Two types of chickpea are commonly grown: (i) Desi – small-seeded with a brown seed coat common to India and used for flour, and 'dhal', an important split-pea vegetable, and to a lesser extent as animal feed; and (ii) Kabuli – large-seeded with a thin, light-coloured seed coat and usually consumed whole in West Asia.

Chickpea is moderately resistant to drought and sensitive to low temperature, therefore it is cultivated as a winter crop in India, Pakistan and Australian coastal areas and as a spring crop in Turkey, Syria, Ethiopia and Canada, with winter chickpea yielding nearly double the amount of spring chickpea. It can be cultivated successfully in areas with a minimum annual rainfall of 300 mm. Supplementary irrigation may double yields. Chickpea is irrigated in the Nile Valley of Egypt and Sudan, due to a lack of sufficient rainfall, and in India in areas whose soils have low water-holding capacity (Saxena, 1987).

Chickpea infested by nematodes are in general stunted, with chlorotic foliage. They flower poorly and give rise to few and small pods that are often empty. Senescence sets in earlier in heavily infested plants. The root system is reduced in size, *Rhizobium* nodulation is suppressed and the roots can show extensive necrosis. Since these symptoms are not specific, close examination of the root system is required for proper diagnosis. The nematodes associated with chickpea have been reviewed by Sharma (1985) and Ali (1995).

Meloidogyne

The species *M. javanica*, *M. incognita* and *M. arenaria* damage chickpea in India (Mathur *et al.*, 1969; Nath *et al.*, 1979) and *M. arenaria* in Ghana (Edwards, 1956). *M. javanica* was also found on chickpea in the Ethiopian highland. Infected chickpea have heavily galled roots (Fig. 8.4) which may rot. The concomitant presence of *M. incognita* and *M. javanica* may enhance the severity of the soil-borne fungus *Fusarium oxysporum* f.sp. *ciceri* (Siddiqui and Mahmood, 1994; Maheshwari *et al.*, 1997; Charu Jain and Trivedi, 1998). Moreover,

Fig. 8.4. Root of chickpea showing galls caused by infestation of *Meloidogyne incognita*. (Photo: N. Greco.)

chickpea lines may lose their resistance to the fungus when infested with these root knot nematodes (Maheshwari *et al.*, 1995; Rao and Krishnappa, 1996). The nematodes may also infest and develop on *Rhizobium* nodules which senesce earlier (Vovlas *et al.*, 1998).

In the subtropical semi-arid Mediterranean basin, damage is conspicuous when chickpeas are planted in sandy–loam soils in late summer or early autumn. Conversely, crop injury is minimized when chickpeas are sown from late autumn into the winter season. Soil temperatures suitable for nematode attack and development are not reached until late spring, allowing the plant to escape the damaging early root invasion process. For this reason, root knot nematodes, although important on other summer crops, do not constitute a problem in the Mediterranean basin.

The nematodes, however, are a serious problem in tropical zones. In India, Upadhyay and Dwivedi (1987) treated field plots infested with 4.6 *M. incognita* juveniles/cm^3 of soil with carbofuran and observed increases in yield of 40%. Yield losses of 31–37% were detected in nematicide trials when *M. incognita* was present at 2.5 juveniles/g of soil (Reddy, 1985), and Ali (1995) reported yield reduction up to 60%.

Economic threshold level

In pot experiments, the growth of chickpea was negatively affected when soil populations of *M. incognita* (Nath *et al.*, 1979) and *M. javanica* (Srivastava *et al.*, 1974) exceeded 0.2 juveniles/g of soil. Ahmad and Husain (1988) detected reductions in shoot length and total plant weight at densities of 1 juvenile/g of soil in pot studies. However, under field conditions, yield losses differ greatly between countries. This variation is caused by differences in soil type, environmental factors existing during the growing season in the different climatic zones and complex disease interrelationships. Therefore, field studies are required to estimate tolerance limits and make yield loss assessments.

Management measures

Crop rotation, including fallow, currently is used to control root knot on chickpea. Rotation is complicated by the wide host range of species of *Meloidogyne.* Nevertheless, groundnut (peanut) and winter cereals are non-hosts for *M. incognita* and *M. javanica*, and cotton is a non-host for *M. incognita* and *M. arenaria.* Saka and Carter (1987) listed hosts and non-hosts of *M. incognita.*

Sowing in late autumn, when soil temperature drops below 18°C, and harvesting in spring can limit or prevent nematode reproduction (Roberts *et al.*, 1981). Chickpea also should not be planted in early autumn in fields planted in the previous season to a summer host plant. In India, postponing sowing to late autumn has also been shown to suppress yield loss (Gaur *et al.*, 1979).

Weeds are often excellent hosts for root knot; therefore, good weed control can be important to a rotation programme under both non-host and fallow conditions.

Organic amendments have been incorporated into infested soils for control purposes. Attempts have also been made to control root knot nematodes in greenhouse trials with sawdust (Singh and Sitaramaiah, 1971) and plant leaves (Kaliram and Gupta, 1982). Although some nematode control and increased plant growth was obtained, the use of these materials in the field often is not practical on an expanded scale because of poor farmer access to the material, costs of transport, or the large amounts needed for adequate control.

Although nematicides and soil solarization are effective, they cannot be used to control nematodes economically on chickpea.

Soil treatment with several nematode antagonistic fungi and mycorrhizae has given promising results under controlled conditions, as did seed coating with nematicides, nematicidal active plant extracts, fungus filtrates and rhizobacteria, alone or in combination with other control means (Siddiqui and Mahmood, 1993, 1995; Charu Jain and Trivedi, 1997;

Siddiqui *et al.*, 2001a,b). However, the use of most of these new control options needs to be validated under field conditions.

Resistance

Chickpea lines and a few cultivars recently have been identified as resistant to root knot nematodes. The breeding material and cultivars available have poor agronomic characteristics and are presently of little importance to practical agriculture. However, Sharma *et al.* (1995) reported that the tolerant cvs N 31, N 59 and ICCC 42 performed better in a field infested with a mixed population of *M. incognita* and *M. javanica* in India.

Meloidogyne artiellia

This root knot nematode causes yellowing and stunting of plants and severe losses in yield (Fig. 8.5). *M. artiellia* was first reported from cabbage in England (Franklin, 1961) and later on chickpea in Spain, Italy (Greco, 1984), Syria (Greco *et al.*, 1992a), Turkey (Di Vito *et al.*, 1994b) and North Africa (Di Vito *et al.*, 1994a). The nematode differs significantly from the previously mentioned species of *Meloidogyne* in both morphometrics and ecology. Galls produced on chickpea by Syrian populations of the nematode are indistinct and almost totally absent in Italian populations. The most obvious symptom of nematode attack is the presence of large egg masses on the roots. Because of their size, they can be confused with cyst nematode females when observed with the naked eye (Plate 7D).

Other hosts

The nematode has a wide host range. Di Vito *et al.* (1985) found many cruciferous, cereal (except oat and maize) and leguminous crops (except lentil, haricot bean, cowpea, lupin, soybean and sainfoin) as good or very good hosts for the nematode. All species in the Solanaceae, Rosaceae, Linaceae, Compositae, Cucurbitaceae, Chenopodiaceae and Umbelliferae were poor or non-hosts.

Fig. 8.5. A chickpea crop showing yellowing and stunting of plants infested by *Meloidogyne artiellia* in Syria. (Photo: N. Greco.)

Biology

Investigations by Di Vito and Greco (1988a) demonstrated that second stage juveniles can invade chickpea roots at 10°C, but at this temperature adult stages were not formed after 66 days. Nematode development was also retarded at 30°C. In Italy and Syria, large egg masses can be observed in early April on the roots of chickpea sown the previous autumn, and in early May on spring-sown chickpea. Juveniles may hatch soon after the completion of embryogenesis.

The presence of a combination of insufficient rainfall and high temperature in spring in the Mediterranean basin often causes poor root growth which limits juvenile emergence from newly produced eggs. This interplay of biotic and abiotic factors is responsible for limiting the nematode to only one generation per growing season. However, if rainfall occurs late in the season, eggs hatch immediately and second stage juveniles survive during dry and hot summers in an anhydrobiotic condition (Di Vito and Greco, 1988a). The nematode seems to be adapted to a wide range of environmental conditions and develops well in a large variety of soil types including those containing 30–40% clay.

Consistent damage is caused to chickpea in Syria where this crop is rotated with durum hard wheat and barley, both good hosts for the nematode. A survey conducted in the 1980s (Greco *et al.*, 1992a) revealed that 13% of chickpea fields in the Aleppo province, northern Syria, were infested.

Economic threshold level

Microplot experiments have shown that chickpea is highly susceptible to nematode attack when population densities exceed 0.14 and 0.016 eggs/cm^3 of soil for winter- and spring-sown crops, respectively (Di Vito and Greco, 1988b).

Management measures

The parasite can be effectively controlled by rotating chickpea with non-host crops. In the Mediterranean area, cotton, sugarbeet, potato, oat, maize, lentil, tomato and melon are poor or non-host crops suitable for *M. artiellia* control programmes. The length of the rotation should be designed to reduce soil densities below threshold levels, which generally requires a 2–4 year period with non-host crops.

Although nematicides have been shown to be effective experimentally, they cannot be used economically on the crop. No attempts have been made to screen chickpea cultivars and lines for resistance to this nematode. Resistance was found in one accession each of *Cicer bijugum*, *C. chorassanicum* and *C. judaicum*, and two each of *C. pinnatifidum* and *C. reticulatum* (Di Vito *et al.*, 2001b). Only *C. reticulatum* is compatible with *C. arietinum* and can be used in breeding programmes.

Heterodera

A cyst nematode infesting chickpea was found in Syria by Mamluk *et al.* (1983) and was observed as the causal agent of severe chickpea decline in the Idleb province and other areas in the north of the country (Greco *et al.*, 1992a). The nematode was described as *Heterodera ciceri* by Vovlas *et al.* in 1985. The nematode belongs to the *H. trifolii* group and differs from *H. trifolii* in having abundant males, different host range and distinct morphological characteristics (Vovlas *et al.*, 1985; Sikora and Maas, 1986). The nematode has also been detected in several areas of Turkey (Di Vito *et al.*, 1994b) as well as in the Irbid Governorate of Jordan and the Beka'a Valley in Lebanon (N. Greco, 2004, unpublished).

Other hosts

The host range is confined to members of Leguminosae (Greco *et al.*, 1986b). The nematode reproduces well on chickpea, lentil, pea and grasspea (*Lathyrus sativus* L.) and poorly on *Vicia* spp., haricot bean, lupin and lucerne. However, a Syrian and a Turkish population also reproduced well on lucerne and *Medicago rigidula* (Di Vito *et al.*, 2001a). Broad bean and several clovers are very poor or non-hosts. In tests with plants in 13 botanical families, the nematode produced a few females only on carnation.

Biology

In comparative studies, nematode eggs hatched better at 15–25°C (Kaloshian *et al.*, 1986a) when stimulated by root leachates from pea (27–33%) than from the other host plants (maximum 16.5%) (Greco *et al.*, 1992b). Among the artificial hatching agents, the largest egg hatch occurred in a 3 mM zinc chlorine solution (maximum 58%). Although the nematode invades chickpea roots at 8°C, development only occurs at temperatures of 10°C and above (Kaloshian *et al.*, 1986a). Root invasion is suppressed at 30°C. Females may protrude a small gelatinous matrix, which is void of eggs (Kaloshian *et al.*, 1986b). In the field, large numbers of lemon-shaped white females (Plate 7E) can be seen at the beginning of April or 2 weeks later on the roots of winter- and spring-sown chickpeas, respectively. Cysts usually appear 14–16 days later (Greco *et al.*, 1988a) after an accumulation of 370 day degrees above the basal temperature of 10°C (Kaloshian *et al.*, 1986b).

Economic threshold level

The tolerance limit of chickpea to *H. ciceri* is 1 egg/cm^3 of soil. Yield losses of 20 and 50% can be expected in fields infested with eight or 16 eggs of the nematode/cm^3 of soil, respectively. Complete crop failure occurs in fields infested with ≥ 60 eggs/cm^3 of soil (Greco *et al.*, 1988a). Under field conditions, severe chickpea decline can be observed from the end of April onwards. At harvest, the protein content of chickpea grain produced in infested fields is significantly reduced, thus lowering the nutritional value of the grain.

Management measures

Since this nematode has a rather narrow host range, it can be controlled effectively by crop rotation (Saxena *et al.*, 1992). An annual decline of 50% of the nematode population using non-host crops has been reported (Saxena *et al.*, 1992). These results demonstrated that short 3–4 year rotations are effective in reducing the nematode densities to or below the tolerance limit.

Resistance

None of the nearly 10,000 chickpea lines screened showed resistance to *H. ciceri* (Di Vito *et al.*, 1996; Thompson *et al.*, 2000). However, resistance to the nematode was found in lines of *C. bijugum*, *C. pinnatifidum* and *C. reticulatum*. Because *C. reticulatum* can be crossed with *C. arietinum*, a research programme to introgress the resistance to *H. ciceri* in kabuli type cultigens is in progress at the International Center for Agricultural Research in the Dry Areas (ICARDA), Aleppo, Syria (Di Vito *et al.*, 1996; Malhotra *et al.*, 2002).

Another cyst nematode, *H. swarupi* (Sharma *et al.*, 1998), was described from roots of chickpea in Rajasthan, India. The nematode belongs to the *Heterodera schachtii* group, is close to *H. cajani* and also can infect pigeonpea. Nematode females turn yellow and produce an egg mass with eggs. Recently, *H. swarupi* has been detected in several other districts of Rajasthan, even in large numbers, but its impact on chickpea yield has not been assessed (Ali and Sharma, 2003).

Pratylenchus

Root lesion nematodes are migratory endoparasites that cause large cavities and necrosis in the cortex of chickpea roots (Fig. 8.6; Plate 7F). Eggs are deposited in the cavities within the root. Several generations may develop in a growing season, each taking about 1 month, and large numbers of specimens can be extracted from the roots at the early flowering stage

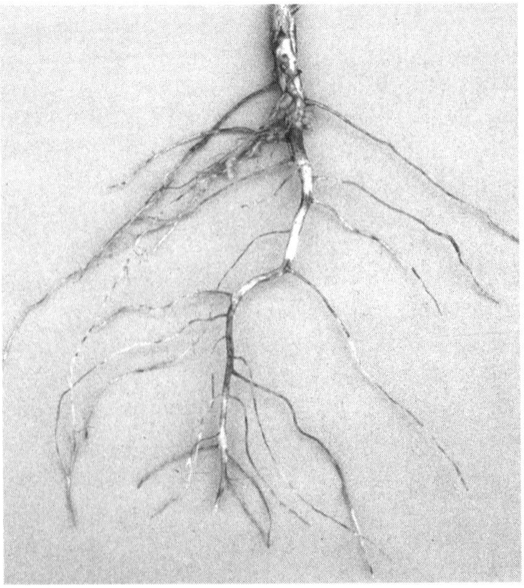

Fig. 8.6. Roots of chickpea exhibiting necrotic lesions caused by a lesion nematode *Pratylenchus* sp. (Photo: N. Greco.)

of the plants. Plant growth is reduced further through root damage caused by interrelationships with soil-borne root pathogens and adverse effects on *Rhizobium* nodulation. The reduced root system decreases plant resistance to drought conditions, which makes these nematodes important in the dry areas in both the semi-arid and dry regions of the world. In the absence of a host crop, *Pratylenchus* survive in the soil as eggs, juveniles or adults. In dry areas, they survive in an anhydrobiotic condition (Glazer and Orion, 1983).

The damage these nematodes cause in the field generally is not as severe as that caused by root knot and cyst nematodes. However, severe symptoms of infestation were observed in Turkey, Lebanon and in countries in North Africa. Because they are found in most fields on a worldwide basis, they are undoubtedly responsible for significant yield loss. Yield losses of 25 and 75% in winter- and spring-sown chickpea, respectively, were observed in Syria in a field infested with *P. thornei* (Greco *et al.*, 1988b).

The most important lesion nematode is *P. thornei*, which has a cosmopolitan distribution. In the Mediterranean region, the nematode was detected in 72% of chickpea fields in Syria (Greco *et al.*, 1992a), 61% in Turkey (Di Vito *et al.*, 1994b), 92% in southern Spain (Castillo *et al.*, 1996) and 28–61% in North Africa (Di Vito *et al.*, 1994a). Under field conditions in Syria, the tolerance limit of chickpea to the nematodes was 0.03 specimens/cm^3 of soil, with yield loss of 58% at 2 specimens/cm^3 of soil (Di Vito *et al.*, 1992). In India, population densities $\geq$ 0.1/g of soil were responsible for significant growth reduction, while densities of $\geq$ 4/g of soil also reduced germination (Walia and Seshadri, 1985a). The reaction of chickpea cultivars may differ, and some can be tolerant (Castillo *et al.*, 1998). The nematode appears to reproduce well on cool season crops and poorly on warm season crops (Di Vito *et al.*, 1992, 2002a).

Other species of root lesion nematodes found infesting chickpea in the Mediterranean region are *P. mediterraneus*, *P. neglectus*, *P. penetrans* and, seldomly, *P. crenatus*, *P. pratensis*, *P. pinguicaudatus* and *P. zeae* (Greco *et al.*, 1992a; Di Vito *et al.*, 1994a,b). However, the impact of these species on chickpea has not been assessed.

In Australia, both *P. thornei* and *P. neglectus* are widespread in wheat fields and they damage chickpea when this pulse is rotated with winter cereals. *P. thornei* appears to be present mostly in the clay soils of the northern grain regions of Australia, while *P. neglectus* prefers the rather light soils of the southern part of the country (Thompson *et al.*, 2000). In nematicide trials, yield increases of 25–60% were observed (Thompson *et al.*, 2000), which gives an indication of the level of loss that can be incurred when the nematodes are present.

Management measures

Specific management measures have not been developed for lesion nematodes on chickpea. Most species of *Pratylenchus* have wide host ranges; therefore, control by rotation is problematic. This is especially true in rotations with winter cereals which are often good hosts for the lesion nematode. However, rotation of cool season with warm season crops would be a satisfactory approach to control *P. thornei*.

Although chemical control is not an economically acceptable management measure, it has been demonstrated that split applications of aldicarb at 10 kg a.i./ha at sowing and after seed germination will control *P. thornei* and increase yield (Greco *et al.*, 1988b). Seed treatment with aldicarb, carbofuran and fensulfothion gave satisfactory control of the nematode in pot tests (Walia and Seshadri, 1985b), whereas under field conditions aldicarb failed to control the nematode (Greco *et al.*, 1988b). At the present time, there are no cultivars with resistance to root lesion nematodes available. However, resistance to *P. thornei* was reported in several lines and accessions of cultivated (Ali and Ahmad, 2000) and wild chickpeas (Di Vito *et al.*, 2001b).

Rotylenchulus

The reniform nematode, *R. reniformis*, has been found associated with chickpea mainly in India (Rashid *et al.*, 1973; Ali, 1995) and also in Ghana (Edwards, 1956). Another reniform nematode, *R. macrosoma*, occurs in chickpea fields in Syria, but it has never been found in the roots of this pulse. *R. reniformis* survives in the soil in the juvenile and adult male stages. Immature females penetrate the root and become established in the endodermis (Rebois *et al.*, 1975). The kidney-shaped females produce a gelatinous matrix that covers the female body in which about 50 eggs are laid. Soil adhering to this matrix often can hamper detection of the female on the root surface.

Economic threshold level

Mahapatra and Padhi (1986) demonstrated in greenhouse tests that population densities of ≥ 0.5 nematodes/g of soil reduce plant growth, and that growth reductions of 80% occur at 10 nematodes/g of soil.

Management measures

Rotations designed to reduce nematode densities are difficult to develop because of the nematode's wide host range. The only acceptable recommendation is to avoid growing chickpea in heavily infested fields and to test local crops for non-host status before suggesting alternative cropping systems. However, in India, paddy rice reduces populations of several nematodes, including *R. reniformis* (Haidar *et al.*, 2001). Although nematicides are effective in control of *R. reniformis*, they are not an economical alternative on chickpea.

Other nematodes of chickpea

Several other nematode species have been found associated with chickpea (Ali, 1995). In South Australia, the 'oat race' of *D. dipsaci* is considered a severe problem on chickpea and pea. Young plants are very susceptible to the nematode, while adult plants are resistant (Thompson *et al.*, 2000). Species of *Amplimerlinius, Aorolaimus, Helicotylenchus, Merlinius, Criconemoides, Paratrophurus, Pratylenchoides, Tylenchus, Tylenchorhynchus* and *Zygotylenchus* were commonly found associated with chickpea in Mediterranean countries (Greco *et al.*, 1992a; Di Vito *et al.*, 1994a,b; Castillo *et al.*, 1996). In India, *Helicotylenchus indicus, H. sharafati, Hoplolaimus dimorphicus* (Mulk and Jairajpuri, 1974, 1975), *Tylencholaimus asymmetricus* (Khan and Ahmad, 1994), *Tylenchorhynchus vulgaris* (Gill and Swarup, 1977), *T. cicerus* (Kakar *et al.*, 1995) and many others (Ali, 1995; Ali and Sharma, 2003) were detected. Species of *Tylenchus, Scutellonema* and *Aphelenchoides* were observed in Sudan (El Tigani *et al.*, 1970), and *Tylenchorhynchus annulatus, Helicotylenchus digonicus* and *Hoplolaimus indicus* in Pakistan (Maqbool, 1986). With the exception of *H. indicus* and *T. vulgaris*, the pathogenicity of these nematodes on chickpea has not been demonstrated. Sartaj *et al.* (1999) observed significant damage to chickpea caused by 500 *H. indicus* specimens per plant, and Gill and Swarup (1977) demonstrated that densities of *T. vulgaris* ranging from 10 to 20,000/500 g of soil caused increasing reductions in plant growth. Control measures have never been developed for these marginal pests.

Cowpea

Cowpea (*Vigna unguiculata* (L.) Walp. aggreg.) is known in the dry grain form as black-eyed pea, southern bean, China pea and marble pea, and in the green pod form as yard-long bean, asparagus bean, Bodi bean and snake bean. It is an annual plant with a great deal of varietal variation, including climbing, bushy prostrate and erect forms that probably originated in Africa or South-east Asia. Although the plant is used mainly for dried seeds, it is also used as a vegetable, pot herb and green manure. It is a hot weather crop well adapted to the semi-arid regions and hot humid growing regions.

It is usually grown under rainfed conditions on well drained soil (Kay, 1979). It is often intercropped with cereals, especially sorghum and millet, and can be planted without land preparation.

Meloidogyne

Root knot nematodes are serious pests of cowpea on a worldwide basis. *M. incognita* and *M. javanica* are the major species found on cowpea in most growing regions. Other important species are *M. arenaria* reported from Brazil, Cyprus and the USA; *M. hapla* from Brazil; *M. ethiopica* from Tanzania; *M. africana* from East Africa; *M. kikuyensis* from Kenya; and *M. mayaguensis* from Florida, USA, also attacking the cv. Iron Clay resistant to *M. incognita* (Brito *et al.*, 2003a,b).

Whereas *M. incognita* was widespread in Georgia cowpea fields, causing an estimated 5–10% yield loss, all other species detected were sporadic in occurrence, with losses estimated at below 1% (Toler *et al.*, 1963). In California, *M. javanica* and *M. incognita* are considered serious pests (Thomason and McKinney, 1960). Robinson (1961) reported the common occurrence of *M. javanica* in Australia. *M. arenaria* present in soil taken from groundnut (peanut) fields caused severe damage to cowpea in Alabama, USA.

Symptoms

Symptoms of damage induced by root knot include patches of stunted and yellowed plants (Fig. 8.7). Severe damage can lead to reduced numbers of leaves and buds.

Economic threshold level

In India, the threshold level, determined in glasshouse studies in sterilized soil, was 100 juveniles of *M. incognita*/500 g of soil (Sharma and Sethi, 1975), with significant yield reduction occurring at 2 juveniles/cm^3 of soil (Sarmah and Sinha, 1995). Visual symptoms of damage first occurred at 1000 and 10,000 juveniles/500 g of soil. *M. javanica* densities of 1000 or 10,000/500 g of soil caused growth reductions in pot tests (Gupta, 1979). At high densities, severe root galling occurs (Fig. 8.8; Plate 8A). In Venezuela, the tolerance limit of the susceptible cv. Manuare was 0.03 eggs and juveniles/cm^3 of soil for *M. incognita* race

Fig. 8.7. Poor growth of cowpea infested with *Meloidogyne javanica* in Nigeria. (Photo: J. Bridge.)

Fig. 8.8. *Meloidogyne incognita* galls on cowpea in Nigeria. (Photo: J. Bridge.)

2, while the resistant cv. Ojito Negro tolerated up to 0.74 eggs and juveniles/cm^3 of soil. Maximum yield reductions were 72% for the susceptible cultivar and only 20% for the resistant cultivar (Crozzoli *et al.*, 1997, 1999).

Disease complexes

The presence of heavy infestations of *M. javanica* on a cowpea cultivar tolerant to wilt caused by *F. oxysporum* f.sp. *tracheiphilum* caused increased wilting when compared with the susceptible cultivar (Thomason *et al.*, 1959). Moreover, Roberts *et al.* (1995) observed increased wilting of a susceptible cowpea cultivar with the concomitant presence of *M. incognita* and the wilt fungus, but nematode infection did not increase the wilting of wilt-resistant genotypes. Interactions were also observed between *M. incognita* and *Macrophomina phaseolina* (Devi and Goswami, 1992) and *Rhizoctonia solani* (Kassab and Ali, 1996).

High densities of *M. incognita* have also been shown to lead to poor nodulation and decreased nitrogen levels in the plant (Sharma and Sethi, 1976a; Abedinia, 1978; Ali *et al.*, 1981). In these studies, root knot galls were found on nodules, and nodules were also produced on the surface of nematode galls. The symbiotic inter-relationship was not affected at low population densities. Taha and Kassab (1980) reported that *M. javanica* when inoculated simultaneously with *Rhizobium* did not affect nodulation.

Management measures

Crop rotation can be an efficient means of controlling root knot nematodes in this crop. Proper selection and placement of non-host crops and resistant cultivars in rotation with susceptible cultivars can lead to control and yield increase. The wide host range of the three major species of root knot and the poorly understood host spectrum of most other species requires careful selection and testing prior to development of rotation schemes. Proper selection of non-hosts is also required because of the presence of races within the genus *Meloidogyne*.

da Ponte (1972) recommended rotations with graminaceous crops or *Crotalaria*. Populations of root knot decreased greatly in fallowed plots when *C. spectabilis* Roth. was grown as a weed-free cover crop as compared with the control (Rhoades, 1964). Mulching with cowpea foliage was also highly effective in suppressing populations (Rhoades and Forbes, 1986).

Egunjobi *et al.* (1986) showed that *M. javanica* populations were lower when cowpea and maize were grown under mixed rather than under sole cropping systems. The results suggested that this cropping system could be used for control of the nematode. Castillo *et al.* (1976) reported that one crop of paddy rice was sufficient to effectively reduce root knot nematode infestations in succeeding susceptible legume crops. The reduction was even greater than with rotations with non-host crops. Rotation of cowpea with winter crops, such as rye and narrow-leafed lupin, may decrease populations of *M. incognita* and *Pratylenchus* spp. (Wang *et al.*, 2002). Dukes *et al.* (1979) demonstrated that resistant cultivars were more effective than non-fumigant nematicides in reducing root knot damage.

Organic amendments have been used to suppress root knot nematode populations on a number of crops (Singh and Sitaramaiah, 1966). Neem cake incorporation in the previous crop caused a reduction in the density of all nematodes in the soil on the following cowpea crop (Jain and Hasan, 1986). Cocoa pod husks incorporated at 6000 kg/ha caused 28% reductions in galling and 6.7% increases in yield (Egunjobi, 1985; Egunjobi and Olaitan, 1986).

Although fumigant and non-fumigant nematicides reduce root knot densities and can cause significant increases in yield, their use on cowpea is not economical, unless cowpea is grown for the production of green pods used as vegetables. Several investigations demonstrated the efficacy of seed treatments with nematicides or nematode antagonists, plant extracts (especially *Azadirachta indica* and *Calotropis procera*), foliar sprays and soil applications of these extracts. However, more field tests are required to confirm the suitability of these treatments.

Resistance

Thomason and McKinney (1960) reported that all 44 cowpea cultivars and plant introductions tested showed some resistance to *M. incognita*, but were moderately to highly susceptible to *M. javanica*. Satisfactory levels of resistance to the three major root knot species were not found in 362 lines evaluated by Caveness (1965) in Nigeria. Amosu (1974) and Ogbuji (1978) reported a number of cultivars with some resistance to *M. incognita*. Of 241 lines tested in Nigeria, four were considered resistant and 28 moderately resistant to *M. incognita* (Caveness, 1979). He considered the lack of good sources of resistance critical for crop improvement breeding programmes. Bridge (1987) listed known cultivars and breeding lines with moderate to high levels of resistance to the various root knot species attacking the crop. Sharma and Sethi (1976b) reported that 15 lines and three cultivars were resistant to *M. incognita*. In field trials with 104 lines and cultivars, 11 showed high degrees of resistance to a population mixture of *M. javanica* and *M. incognita* (Patel *et al.*, 1977). Yield increases from three cultivars resistant to *M. incognita* ranged from 19 to 69% (Dukes *et al.*, 1979). Hadisoeganda and Sasser (1982) reported that variability in susceptibility exists to species of root knot and to *M. incognita* races 1, 2 and 3. All lines tested were, however, resistant to *M. incognita* race 4. Of 289 lines screened for resistance to *M. incognita* in India, 93 exhibited some degree of resistance (Singh and Reddy, 1982). Sasser and Hartman (1985) reported that of the 27 lines tested, most were resistant to *M. hapla* and *M. incognita*, moderately resistant to *M. javanica* and mostly susceptible to races of *M. arenaria*. A coordinated effort is needed to evaluate these lines again for sources of resistance. During the last decade, more cowpea lines (Subramaniyan *et al.*, 1997; Devi *et al.*, 1999), breeding material (Thies, 2000) and cultivars (Rodriguez *et al.*, 1996) resistant to *Meloidogyne* spp. have been identified. Recently, other cowpea cultivars having good agronomic traits and resistance to *M. incognita*, viruses, several fungal pathogens and insects have been released or identified. Among the most promising are the cvs Texas Pinkeye, Purple Hull (Miller and Scheuring, 1994), Carolina Crowder, Better Snap, Tender Cream (Fery and Duke, 1992, 1995b, 1996) and California Blackeye 27 (Ehlers *et al.*, 2000) all in the USA, Pampo and Otília in Brazil (da Ponte *et al.*, 1993) and Ojito Negro in Venezuela (Crozzoli *et al.*, 1995). Wang and McSorley (2002) reported the cvs Colossus, California Blackeye No 5, Iron Clay Magnolia Blackeye, Mississippi Purple, Mississippi Silver, Tennesse Brown and Zippercream as poor or non-hosts for *M. incognita*, with some being resistant to other *Meloidogyne* species. These resistant cultivars can be rotated with susceptible crops for the management of *M. incognita* (Ogallo *et al.*, 1999).

Mutiple genes seem to control resistance to *M. incognita* in cowpea. Most of the cultivars contain the resistant gene *RK*, as in the cultivar Mississippi Silver, but Fery and Dukes (1995a) found that three resistant lines (US 566, US 567 and US 568) possess a gene conditioning the resistance which is allelic to the *RK* gene. Moreover, Roberts *et al.* (1996) stated that the accession IT84S-2049, from Africa, contains a new dominant resistant gene (*RK*) conferring resistance to several populations of *M. incognita* and *M. javanica*, including populations virulent to cultivars containing the *RK* gene.

Heterodera

The cyst nematode *H. cajani* has been found associated with cowpea in a number of regions of India (Koshy and Swarup, 1971b) and has been detected on cowpea in Egypt (Aboul-Eid and Ghorab, 1974). The host range is limited to the Leguminosae or Pedaliaceae (Sharma and Swarup, 1984). Although the nematode seems to be widespread in India, crop loss assessment data are lacking (Luc, 1985). In a glasshouse study, an Egyptian population retarded emergence of leaves and retarded and reduced the number of flowering buds, flowers, growing pods and yield (Aboul-Eid and Ghorab, 1974).

Economic threshold level

Shoot length was reduced in glasshouse experiments when the population density ranged between 10 and 20 juveniles/100 g of soil (Sharma and Sethi, 1975; Zaki and Bhatti, 1986). Both root and shoot length were reduced at nematode densities of 100 juveniles/100 g of soil (Sharma and Sethi, 1975).

Disease complexes

The nematode can complete its life cycle on nodular tissue and can reduce the number of *Rhizobium* nodules (Sharma and Sethi, 1975). Cowpea growth was not affected when *Rhizoctonia bataticola* was inoculated prior to, simultaneously with or after *H. cajani* in glasshouse tests (Walia and Gupta, 1986).

Management measures

The most effective management measure for cyst nematodes is rotation with non-host crops. Cowpea rotated with paddy rice may be less affected by the nematode because of the negative effect of flooding on nematode densities. Although nematicides have been shown to suppress nematode attack, in general they cannot be used economically on this crop. The efficacy of seed treatments with neem products (Devi, 2000) and soil incorporation of the bacteria *Pasteuria penetrans* (Singh and Dhawam, 1994) and the fungi *Pochonia chlamidosporia* (syn. *Verticillium chlamydosporium*) and *P. lilacinus* (Preeti and Trivedi, 2000) has been demonstrated in the laboratory but needs confirmation under field conditions.

Resistance

In India, screening demonstrated that the cv. Rituraj was highly resistant, the cvs Bandel and Pusa Komal were resistant (Devi, 2001) and the cv. Barsati Mutant was tolerant to the nematode (Sharma and Sethi, 1976b).

H. glycines and *H. schachtii* have been reported on cowpea but at present are of unknown economic importance on the crop. *Heterodera vigni* reported from cowpea is now recognized as a junior synonym of *H. cajani*. Nine cultivars of cowpea tested for susceptibility to *H. glycines* were resistant to the nematode (Epps, 1969).

Rotylenchulus

The reniform nematode, *R. reniformis*, has been found associated with cowpea in India and the USA. Yield losses were detected when soil was treated with 1,3-dichloropropene (1,3-D) or ultra-high-frequency electromagnetic energy (Heald *et al.*, 1974). Crop loss assessment, however, is still needed to determine the true importance of the nematode on the crop, because of the broad-spectrum activity of the fumigant and electromagnetic energy.

Races

Dasgupta and Sehadri (1971) divided the nematode populations into two races on their ability to parasitize cowpea, castor or cotton, with race A reproducing on all three hosts and race B only on cowpea. More recently, Rao and Ganguly (1996) added millet and mustard to the above list of test plants and found that of six Indian populations of the nematode all reproduced on cowpea, castor and cotton, one did not reproduce on millet and mustard, one reproduced on millet and not on mustard, one reproduced on mustard and not on millet, and one reproduced on both millet and mustard.

Economic threshold level

The nematode reduced emergence by 7–9 days and seedling density by 6–11% at densities of 1 nematode/g of soil in glasshouse studies (Nanjappa *et al.*, 1978). A significant reduction in height, and fresh shoot and root weights was observed in pot tests with 1000 juveniles per plant (Gupta and Yadav, 1980).

Management measures

The narrow host range of the nematode, especially that of race B, should allow excellent control with crop rotation. For example, nematode densities were suppressed when cowpea was grown intercropped with maize (Egunjobi *et al.*, 1986). Although breeding lines have been found with resistance to the nematode, commercial cultivars are not yet available (Thakar and Patel, 1984).

Soil solarization was considered an effective method for reducing nematode densities to a depth of 15 cm (Heald and Robinson, 1987), but this method and soil treatments with nematicides probably cannot be used economically on this crop, unless it is destined for the fresh vegetable market in major cities where the produce could attract a high price.

Other nematodes of cowpea

Hoplolaimus seinhorsti, an endoparasitic nematode, was shown to cause severe damage to cowpea in Nigeria. The nematode induced marked necrosis in both the lateral and secondary lateral roots in field plot studies in Nigeria (Bridge, 1973). After 9 weeks, most of the lateral feeder roots were very badly rotted or missing. The number of nematodes increased to a maximum of 1110/root system after 5 weeks.

Haricot Bean

Haricot bean (*Phaseolus vulgaris* L.), also known as French, common, kidney, string, salad, runner or snap bean originated in Mexico between 2300 and 4000 BC. It is the most widely cultivated food legume (Table 8.3). In 2000, approximately 27 Mha were in production. Among the food legumes, *P. vulgaris* is the most uniformly distributed crop in the world and the main food legume in the Americas, where it is of great agricultural importance, especially in Brazil, Mexico and the USA. In Asia, haricot beans are cultivated extensively in India, with 36% of the world acreage. Extensive plantings also exist in China, Indonesia, Iran, Myanmar (Burma), North Korea, Pakistan, Thailand, Turkey and Vietnam. In Africa, the main producers are Burundi, Cameroon, Congo, Ethiopia, Rwanda, Tanzania and Uganda. In Europe, with only 2% of world acreage, this pulse is only of importance in Albania, Belarus, Greece, Italy, Moldova, Poland, Rumania, Spain, Ukraine and the former Yugoslavia. Nearly all countries in the tropics and subtropics produce *P. vulgaris* for dried grains which are eaten whole or mashed mainly in soup.

In addition to dried grain, 0.7 Mha are used for fresh green seeds, whole pods or are canned or frozen. In several countries, beans also are cultivated in glasshouses for the high value fresh vegetable market.

Phaseolus spp. are sensitive to low temperature; therefore, in the subtropics, they are cultivated during the warm seasons and sown early in the spring or in summer after the winter crop. The crop is therefore infected with many nematode species that have higher temperature optimums. The crop is grown as a sole crop, semi-climbing and as a climbing bean in relay systems with maize. Beans are often grown intercropped with maize.

Heterodera

The soybean cyst nematode *H. glycines*, besides infesting soybean, also attacks *Phaseolus* spp. This is important because *Phaseolus* beans are often rotated with soybean. Crop loss due to *H. glycines* infestations on haricot beans have been reported mainly in the USA (Noel, 1982).

Symptoms

Nematode attack is similar to that observed on soybean. In glasshouse tests, haricot bean was less susceptible than soybean to *H. glycines* (Abawi and Jacobsen, 1984). Data on yield loss incurred in the field are lacking. The level of invasion and reproduction of *H. glycines* on haricot bean is similar to or larger than that encountered on soybean (Abawi and Jacobsen, 1984; Melton *et al.*, 1985).

H. glycines must be considered a potential problem on haricot bean in areas where the nematode occurs, especially if it is rotated with soybean or other host crops. Abawi and Jacobsen (1984) postulated that because of the larger root size of haricot bean compared with that of soybean, reproduction rates of *H. glycines* on the former would be larger under field conditions and thus lead to larger soil population densities.

Management measures

The control measures devised for control of *H. glycines* on soybean should also be used when dealing with this nematode on haricot beans. There seems to be large variation in cultivar susceptibility to the nematode. The cvs Kentucky Wonder Pole and Kentucky Wonder Improved Rust Resistant are resistant to *H. glycines* (Melton *et al.*, 1985) and should be recommended to avoid yield losses and reduce nematode population densities.

Meloidogyne

M. incognita, *M. javanica* and *M. arenaria* appear to be the most common root knot species of haricot beans and have been reported causing damage in the Americas, Africa and Asia. There is probably no country in the tropics and subtropics in which beans are not affected by root knot nematodes. *M. mayaguensis* has been reported to damage haricot bean in Florida, USA (Brito *et al.*, 2003b). This nematode is probably more widespread than is thought in tropical America, South Africa and West African countries (Brito *et al.*, 2003a). Moreover, *M. chitwoodi* and *M. hapla* damage the crop in northern USA (Hafez and Sundararaj, 1999), and *M. brasiliensis* has potential to do so in Brazil (Charchar and Eisenback, 2002).

Symptoms

Although symptoms of nematode attack on aerial parts are similar to those caused by these nematodes on other crops, gall size on the roots of *Phaseolus* spp. is variable and may be nearly undetectable (Blazey *et al.*, 1964). In the latter case, the only visible symptom on the roots is the presence of large egg masses. However, severe galling was observed in Brazil (Lordello and De Oliveira Santos, 1960) and in Chile (Fig. 8.9). Due to the large number of types and cultivars of haricot bean and to the presence of root knot races (see Chapter 9), the intensity of damage caused by *Meloidogyne* spp. varies greatly.

Disease complexes

M. incognita will reduce the number and nitrogen-fixing efficiency of bacterial nodules on roots of haricot bean (Singh and Reddy, 1981; Mohanty *et al.*, 2001) and has been shown to increase the severity of *Macrophomina phaseolina* (Al Hazmi, 1985). Hutton *et al.* (1972) and France and Abawi (1995) detected increased wilting by *Fusarium solani* f.sp. *phaseoli* on beans attacked by *M. arenaria*, *M. javanica* and *M. incognita*, and found that resistance to the fungus can be lost in the presence of *M. incognita* infection. Extreme fungal root rotting is often associated with root knot damage (Fig. 8.10; Plate 8B).

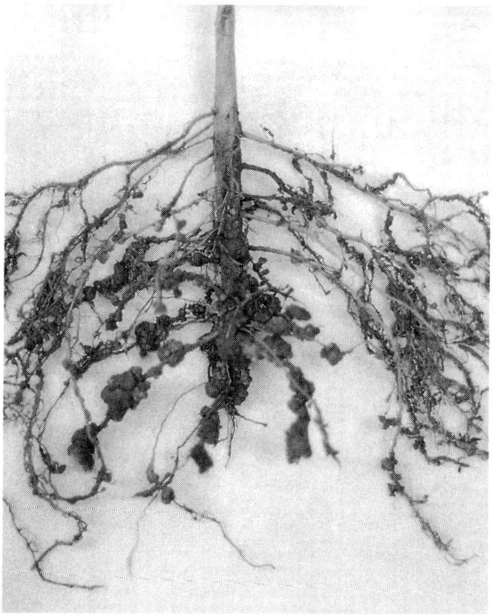

Fig. 8.9. Roots of haricot bean heavily galled by root knot nematodes, *Meloidogyne* spp., in Chile. (Photo: N. Greco.)

Economic threshold level

The extent of yield loss caused by *Meloidogyne* spp. to haricot bean has not been assessed under field conditions. The information available was derived from yields obtained in nematicide trials or pot experiments. Sharma (1981) observed significant growth reduction in soil infested with *M. javanica* at 1 egg/g of soil and a reduction of 82% at 10 eggs/g soil in glasshouse experiments. In pot experiments, Crozzoli *et al.* (1997) found tolerance limits of three haricot bean cultivars to *M. incognita* of 0.02–0.03 eggs/cm^3 of soil and that yield of green pods was reduced to 35–53% in soil infested with 4 eggs/cm^3 of soil.

Fig. 8.10. *Meloidogyne incognita*: galling and root rotting of haricot bean roots in the Philippines due to the interaction between nematodes and soil fungi. (Photo: R. Sikora.)

Management measures

Abiotic stress caused by adverse environmental factors and inter-relationships between root knot nematodes and other soil-borne pathogens are responsible for severe damage under field conditions (Fig. 8.10; Plate 8B). Planting time certainly plays an important role on the amount of yield losses. Most species of *Meloidogyne* found in the tropics and subtropics would be unable to invade bean roots initially if the crop is sown at the end of winter or early in the spring, when soil temperatures are below 15°C. Escape from early root penetration would give the plant a head start. Yield would increase because the larger root system could withstand the damage caused by delayed nematode invasion. Moreover, these beans would be harvested by the end of spring or early in summer, thus limiting the number of nematode generations produced (often to only one) and overall population densities. Sowing bean late in spring or in summer would cause early nematode invasion, the development of multiple generations, severe damage and high soil densities. Destruction of the infested roots of the preceding crop, as soon as possible, is also suggested to accelerate the decline of the nematode soil population density.

Root knot nematodes can be controlled satisfactorily with nematicides at the same rates suggested on other crops. Application of nematicides on 30–35 cm wide bands would reduce treatment costs. Seed treatment with oxamyl at 3–10% (w/v) or carbofuran 3% (w/w) prevented development of *M. incognita* in glasshouse tests (Rodriguez-Kabana *et al.*, 1976; Mohan and Mishra, 1993). Efficacy under field conditions was not determined.

Haricot beans have rather short growing seasons and, therefore, reduced rates of nematicides may be sufficient to give control and reduce possible environmental impact. The use of nematicides which move systemically into the plant on beans grown for the fresh vegetable market, because of the short growing season, must be closely monitored.

In countries with sufficient solar energy levels, root knot nematodes can be effectively controlled by a 4–8 weeks solarization, assuming that the land will remain uncropped during the summer. Control is even higher when this method is used in the glasshouse. Solarization is lethal to other soil-borne pathogens and weeds but is only effective in the upper soil layers (maximum 30 cm deep) and does not reach nematodes that may migrate up to the crop. The combined use of solarization and heated water increases soil penetration and efficacy (Saleh *et al.*, 1988). The costs involved, however, may limit the use of this technology for haricot bean production.

The incorporation into the soil of organic amendments (Acosta *et al.*, 1995; Ibrahim and Ibrahim, 2000; Sharma and Singh, 2001) has proved to give satisfactory nematode control, and this material should be cheap and easily available. The use of biological agents, including *Pasteuria penetrans*, is also showing promise but needs to be confirmed under field situations (Triviño and Gowen, 1996; dos Santos and Ferraz, 2000).

When beans are grown for green pod or green seed production, roots should be destroyed as soon as possible after harvest to prevent further nematode development on roots remaining in the soil.

Resistance

The breeding lines B-3864 (Fassuliotis *et al.*, 1967) and B-4175 (Wyatt *et al.*, 1980), both resistant to *M. incognita*, were derived from the Mexican line PI 165426. Further selection enabled Wyatt *et al.* (1983) to release the cv. Nemasnap, the first bush snap bean cultivar resistant to *M. incognita*. Moreover, the cvs Alabama N1, Carioca, Manoa Wonder and Riotibagi were found to be resistant to one or more species of warm season root knot nematodes. Alabama N1 and PI 165426 also possess resistance to *M. hapla* race A (Chen and Roberts, 2003). More cultivars resistant to *M. incognita* are reported by Blazey *et al.* (1964). In Brazil, Ribeiro and Ferraz (1983) tested 49 cultivars and lines and found that 37-R, Honduras-35, 51051 and Rajado Ag. 496 could be considered resistant to *M. javanica* although data were variable. In Kenya, the cvs Kahuti, Red Haricot, Rono, Saginaw and Kiburn were resistant to local populations of *M. incognita* and *M. javanica* (Ngundo, 1977). Germplasm material and cultivars have been screened for resistance to root knot nematodes (Omwega *et al.*, 1989; Mullin *et al.*, 1991b; Hafez and Sundararaj, 1999), resulting in more information on sources of resistance to these nematodes in haricot bean. The resistance to *M. incognita* in haricot bean was considered to be linked to two independent genes in the cvs Springwater Half Runner and Wingard Wonder according to Blazey *et al.* (1964), and to three pairs of recessive genes according to Hartman (1971) in the cv. Alabama No 1. Omwega *et al.* (1990a) conversely found that a single dominant gene (*Me1*) was responsible for resistance to *M. javanica*, *M. incognita* race 1 and *M. arenaria* race 1 in bean lines derived from the landraces G2618 and G1805. However, the reaction of the known resistant lines to different *Meloidogyne* species, populations or races may vary. Moreover, different sources of resistance are not equally heat stable, and heat stability also differs with nematode species and race (Omwega *et al.*, 1990b; Mullin *et al.*, 1991a; Sydenham *et al.*, 1997). Most of the resistant germplasm lines are available at Centro International de Agricultura Tropicale (CIAT), Cali, Colombia. However, before undertaking a breeding programme, it is suggested that the most suitable resistance source is selected on the basis of its reaction to local populations of root knot nematodes and environmental conditions.

Assuming that the mentioned resistant cultivars have good agronomic attributes and are suitable to local climates, they should be integrated into management systems in areas infested with root knot nematodes. Attention should also be paid to resistance management that requires rotating with different sources of resistance or periodically with susceptible cultivars to prevent resistance-breaking race formation.

Rotylenchulus

The reniform nematode, *R. reniformis*, also damages haricot bean, especially, but not only, in southern USA and tropical American countries (Tarte, 1971). This nematode also reduces *Rhizobium* root nodulation, and seed protein content, and increases the severity of the fungus *F. solani*. Investigations on yield loss and control have been reported (McSorley, 1980; McSorley *et al.*, 1981; McSorley and Pohronenzy, 1984). Nematode threshold levels, however, have not been determined.

Satisfactory nematode control was obtained with six foliar sprays of oxamyl at 0.56 kg a.i./ha combined with a soil drench of 2.24 kg a.i./ha of the same chemical, furrow application of 2.5 kg a.i./ha of carbofuran (Brancalion and Lordello, 1981), and pre-plant fumigation with 120–240 l of DD/ha (Thames and Heald, 1974).

Rotations with cotton should be avoided, because Thames and Heald (1974) demonstrated that pre-plant soil populations of *R. reniformis* following cotton were ten times higher than following grain sorghum. Information on resistant cultivars is scarce.

Pratylenchus

Several lesion nematodes have been reported on haricot bean causing extensive root necrosis and yield reduction. Among them, *P. scribneri* (Thomason *et al.*, 1976) and *P. penetrans* (Elliot and Bird, 1985) have been shown to reduce plant growth when soil populations exceed 0.5 nematodes/cm^3 of soil. The cvs Saginaw, Gratiot and Kentwood were tolerant to *P. penetrans*. It should be noted that *P. penetrans* reduced arbuscular mycorrhiza, *Glomus fasciculatum*, levels. The latter is important in phosphorus uptake by the root system. Although *P. penetrans* reproduction was not affected by mycorrhiza, the presence of the fungus symbiont reduced the severity of nematode damage. This indicates that mycorrhizal fungi are important in regulating nematode populations in haricot bean (Elliot *et al.*, 1984).

The cosmopolitan species *P. neglectus*, *P. alleni*, *P. brachyurus* and *P. thornei* infect haricot bean. Moreover, populations of *P. zeae* from Brazil, Malawi and Mozambique, *P. pinguicaudatus* from North Africa, and the banana lesion nematode, *P. goodeyi*, from Uganda were found to reproduce on haricot bean. Their importance in crop production is unknown.

Means of management suggested for root lesion nematodes on other crops should work satisfactorily on haricot bean.

Other nematodes of haricot bean

The false root knot nematode, *Nacobbus aberrans*, another sedentary endoparasitic nematode, is found in the Americas, and populations of the nematode from the states Puebla, Guanajato, Zacatecas and San Luis Potosi in Mexico damage haricot bean, with yield losses up to 36% having been reported (Lehman, 1985; Toledo *et al.*, 1993; Manzanilla-Lopez *et al.*, 2002; I. Cid del Prado-Vera, Mexico, 2003, personal communication). Infected roots show large galls similar to those of *Meloidogyne* spp. Therefore, close observation is required for correct diagnosis. *N. aberrans* seems to be less pathogenic than root knot. One generation requires 36 days at 25°C. The nematode has a wide host range, including sugarbeet, tomato, potato, pepper and many cruciferous plants and a variety of weeds. The wide host range complicates the development of effective rotation systems for control purposes. The nematode reproduces well on a number of different soil types, and damage is not restricted to sandy soils as is the case with most root knot species. Nematode populations from different areas may have different host ranges, indicating the possible existence of races or pathotypes. The populations damaging haricot bean in Mexico attack only this crop and chilli pepper and are classified as belonging to 'the bean group' (Manzanilla-Lopez *et al.*, 2002). Moreover, the cvs Amarillo Calpan, Bayo Mecentral, Negro San Luis and Rio Grande are resistant to Mexican populations of the nematode (I. Cid del Prado-Vera, Mexico, 2003, personal communication).

Foliar damage caused by *Aphelenchoides ritzemabosi* was sometimes observed on haricot bean following lucerne in Wyoming, USA. The nematode is common in lucerne fields along with *D. dipsaci*. It persisted up to 27 months in dried bean leaves, thus facilitating its persistence. However, this nematode is not considered to cause economic loss unless environmental conditions are very suitable (Franc *et al.*, 1996). *Belonolaimus longicaudatus*, *B. gracilis*, *Hoplolaimus galeatus*, *Zygotylenchus guevarai*, *Helicotylenchus dihystera*, *Tylenchorhynchus acutus* and *Dolichodorus heterocephalus* have also been reported from haricot bean. The potato rot nematode, *Ditylenchus destructor* (MacGuidin and Slack, 1991), and *Hemicycliophora poranga* (Chitambar, 1993) have potential to damage haricot bean in the USA and *Heterodera cajani* in India (Jain *et al.*, 1994) as this legume has been shown to be a good host for these nematodes.

Yield increases have been obtained following the application of nematicides in infested fields. Studies on their threshold levels and the exact extent of yield loss associated with these nematodes have not been conducted. These nematodes often occur concomitantly with economically important species, e.g. *H. glycines*, *R. reniformis* and species of *Meloidogyne* and *Pratylenchus*. Nematicides suggested for the control of the latter are usually effective against nematodes of lesser importance in the same field.

Lentil

Lens culinaris Medic. is a small-seeded legume that has been cultivated since ancient times in the Mediterranean region and more recently in Asia and in the Americas. India with 34%, Canada with 19% and Turkey with 12% of total world production are the largest growers of lentil. The crop is also important in Syria, Bangladesh, Iran, Pakistan, China, Ethiopia, Morocco, Spain, Chile, the USA and Australia. Lentil is a winter crop normally rotated with cereals and cultivated from sea level to more than 3000 m elevation. It is moderately resistant to low temperature and drought, but yields poorly in wet soils. Lentil is used mainly for human consumption in soup, roasted as a snack and for baking flour. The straw has a high nutritional value and is commonly used as animal fodder.

Heterodera

H. ciceri is a major limiting factor affecting lentil production in North Syria and is the only cyst nematode known to damage lentil in the field. The nematode causes severe stunting and yellowing which can be observed early in April.

Economic threshold level

Lentil is less susceptible than chickpea to this cyst nematode. The tolerance limit (Greco *et al.*, 1988a) on lentil was 2.5 eggs/cm^3 of soil compared with 1 egg/cm^3 for chickpea. Yield losses of 20% occurred in fields infested with 20 eggs/cm^3 of soil, but up to 50% when population densities exceeded 64 eggs/cm^3. Lentil produced on fields infested with *H. ciceri* also contained less protein. *H. ciceri* reproduction in the field was similar to that on chickpea at low population densities. Lower reproductive rates, however, were obtained at $\geq$ 2 eggs/cm^3 of soil, due to lower numbers of new cysts produced and reduced number of eggs per cyst (Greco *et al.*, 1988a).

Ditylenchus

D. dipsaci, the stem nematode, has been reported on lentil in Syria (Greco and Di Vito, 1987) and was isolated from the base of stems showing brownish necrotic lesions. Although the impact of the nematode on crop growth has not been measured, it can be assumed that *D. dipsaci* could damage lentil if late winters and early springs are cool and moisture levels are high.

Avoiding rotations with other host plants for the nematode, wider row spacing and proper weed control should be adequate to limit damage caused by the stem nematode. Augustin and Sikora (1989a) reported on the importance of weeds in Syria on population dynamics of the 'giant race' of *D. dipsaci*.

Other nematodes of lentil

Among other nematodes occasionally found in the rhizosphere of lentil are *Helicotylenchus mucronatus* (Mulk and Jairajpuri, 1974) and *M. javanica* (Prakash, 1981) in India and *M. incognita* in Pakistan (Maqbool, 1986). Interaction between *M. javanica* and *F. oxysporum* f.sp. *lentis* has been observed (Ali and Dwivedi, 2001). However, the root knot nematode species should not constitute a problem, because lentil is a winter crop and low temperatures are unfavourable for the development of these two species. Populations of *Pratylenchus mediterraneus*, *P. neglectus*, *P. penetrans*, *P. pinguicaudatus*, *P. thornei* and *Pratylenchoides leiocauda*, from the Mediterranean basin, can infest lentil, but their impact on the crop has not been assessed (Di Vito *et al.*, 1994a,b, 2002a). *R. reniformis* is reported on lentil in India (Fazal *et al.*, 1995).

Moth Bean

Moth bean (*Phaseolus aconitifolius* Jacq. syn. *P. trilobus* Ait.), also known as dew and mat bean, is a perennial or annual creeping legume native to India, Pakistan and Myanmar. It is of importance in the semi arid regions where it is eaten whole after frying, split as dhal or used for flour. It has also been planted in California and Texas in the USA.

The crop has been reported to be a host for *H. glycines* (Riggs and Hamblen, 1962) and attacked by root knot nematodes (Bessey, 1911) in the USA. *M. incognita* and *R. reniformis* have been shown to cause significant reductions in plant growth in glasshouse pot tests at levels of ≤ 1 juveniles/g of soil (Mishra and Gaur, 1981). In similar tests, Zaki and Bhatti (1986) detected reduction in growth caused by *H. cajani* when plants were inoculated with 10 juveniles/kg of soil.

Resistance was detected in two lines tested in microplots (Hasan and Jain, 1986).

Mung Bean

Mung bean (*Phaseolus aureus* Roxb, syn. *Vigna radiata* (L.) Wilczek var. *radiata*), also known as green or golden gram, probably originated in India. It is an annual, warm temperature crop that can be planted in both the main growing season and as a mid-season crop. It is an important grain crop and is probably best known when used as a vegetable in the form of bean sprouts. It is often rotated with rice where it is planted directly into the stubble by broadcasting, or it is intercropped with cereals. Mung bean is tolerant to alkaline and saline growing conditions.

Meloidogyne

All four major species of root knot nematodes have been shown to parasitize mung bean. Species of *Meloidogyne* are a serious problem in India, Thailand, the Philippines and the USA (Bridge, 1981). *M. javanica* has been shown to cause damage to the crop in the Philippines (Castillo, 1975).

Prasad *et al.* (1971) evaluated field damage and noted that the nematode had a greater impact on grain formation than on pod setting. Root knot nematodes caused severe galling of the root system, chlorosis and stunting. *M. incognita* caused significant reductions in plant growth, nodulation and nitrogen content of the shoot and root (Hussaini and Seshadri, 1975; Inderjit Singh *et al.*, 1977).

Although no apparent differences in shoot growth were noticed after 2 months, when 14-day-old plants were inoculated with 0, 10, 25, 50 or 100 *M. javanica* egg masses (Catibog and Castillo, 1975), the severe root galling produced indicated that inoculation at planting would have resulted in greater losses. Losses of 28% were measured in a field infested with a mixed population of *M. incognita* and *R. reniformis* (Castillo *et al.*, 1977).

Management measures

Standard rotations, especially those including paddy rice, probably limit the degree of damage caused by nematodes on this crop. The extent to which root knot nematodes affect the crop in multiple cropping situations is not known.

Yield increases of 68% were obtained in field trials when aldicarb was applied at 1.5 kg a.i./ha (Yein *et al.*, 1977; Sultan *et al.*, 1985). Seed treatment with neem cake and neem oil reduced *M. incognita* penetration by 75 and 64%, respectively (Vijayalakshmi and Goswami, 1986). Neither treatment was shown to be an economically feasible approach to control.

Although a number of breeding lines have been shown to be moderately resistant to *M. incognita* in India (Mathur *et al.*, 1973; Hussaini and Seshadri, 1976), cultivars with good agronomic characteristics are not available.

Rotylenchulus

R. reniformis is considered to be an important pest of mung bean in the Philippines (Castillo, 1975). Control measures have not been developed for the nematode. Patel and Thakar (1985) reported that two breeding lines were moderately resistant to the nematode. Castillo *et al.* (1978) showed that flooding for 30 days effectively reduced population levels in pot tests.

Other nematodes of mung bean

Mung bean has been reported to be a suitable host for the soybean cyst nematode *H. glycines* (Epps and Chambers, 1959). The nematode caused severe stunting on two cultivars.

Pea

Pisum sativum L., or garden pea, is a food legume used as both a dried grain and a fresh vegetable. Pea was originally cultivated for grain, and only in the 16th century did the use of fresh seeds become popular. In the last few decades, pea has probably become the most common frozen vegetable in the USA and in Europe. Europe, including the former USSR, accounts for 33% of the world pea acreage, China for 12% and India 11%. Small amounts are grown in Burundi, Ethiopia, the USA, Peru, Pakistan, Denmark, France, Hungary, the UK and Australia. Only 0.8 Mt are devoted to the production of green peas for the frozen food industry. Pea straw is also used for livestock feeding.

Heterodera

The cyst nematodes *H. goettingiana*, *H. trifolii* (Mulvey and Anderson, 1974) and *H. ciceri* (Greco *et al.*, 1986b) reproduce well on garden pea. No damage by the latter two species has been reported on pea in the subtropical regions of the Mediterranean where both species occur. In Mongolia, a population of *H. glycines* was found to reproduce on pea (Zhang, 1995). The most noxious cyst nematode affecting pea is *H. goettingiana*. This cyst nematode is widespread in Europe and the Mediterranean basin. In 1992, severe infestations of pea crops by *H. goettingiana* were also observed in western Washington state in the USA (Handoo *et al.*, 1994).

Infested fields show patches in which garden peas are stunted, chlorotic (Fig. 8.11) and have few flowers, which produce small and often empty pods. Symptoms of nematode infestations are very evident at flowering. Heavily infected plants have large numbers of swollen females on the surface of roots (Fig. 8.12). The root systems are reduced in size, and exhibit poor nodulation. Additional applications of fertilizer may not lessen damage. Damage is amplified by an inter-relationship of *H. goettingiana* with the soil-borne fungus *F. oxysporum* f.sp. *pisi* (Garofalo, 1964). In dry areas, pea suffers greatly from drought due to the reduced size and efficiency of the root system. Senescence also tends to occur earlier.

Fig. 8.11. Patch of stunted and yellow garden pea in a field heavily infested with *Heterodera goettingiana*. (Photo: N. Greco.)

Fig. 8.12. Roots of peas heavily infested with white females of *Heterodera goettingiana*. (Photo: N. Greco.)

Economic threshold level

The extent of damage caused by the nematode varies with cultivar and environmental conditions. However, Greco *et al.* (1991) reported a tolerance limit of garden pea to *H. goettingiana* of 0.5 eggs/cm^3 of soil, with 20–50% yield losses expected at between 3 and 8 eggs/cm^3 of soil. Complete crop failure occurs at densities of ≥ 32 eggs/cm^3 of soil.

Other hosts

H. goettingiana reproduces well on garden pea (*P. sativum*), field pea (*P. arvense* L.), broad bean (*Vicia faba* L.), vetch (*Vicia* spp.) and grass pea (*Lathyrus sativus* L.). Reproduction on other cultivated leguminous species is negligible. Several wild species of *Vicia* and *Lathyrus* (Jones, 1950; Winslow, 1954) are also hosts and are responsible for maintaining high soil densities even in the absence of host crops.

Biology

The time required by juveniles to reach the adult stage is strongly influenced by temperature and can take 7 weeks in winter and only 2 weeks in spring (Greco *et al.*, 1986a). *H. goettingiana*, having a minimum temperature for development of 4.4°C, can penetrate and develop on pea during the winter season (Beane and Perry, 1984). On garden pea sown in mid-autumn, females are formed by the end of autumn or in early winter. In this season, soil temperature is below 15°C, and the females protrude egg masses containing 100–150 eggs. When peas are sown from late autumn throughout early spring, females occur in the spring. By then, soil temperature may exceed 15°C and low moisture availability is common. Therefore, egg masses will not be protruded, or they will be small and empty. While eggs in egg masses hatch promptly when suitable environmental conditions exist (15–20°C and adequate soil moisture), no substantial hatch occurs in new cysts during the first 2 months. Egg hatch is suppressed at 25°C and therefore no root invasion would occur during the warm season. In England, one generation per year was reported on garden pea and two on broad bean (Jones, 1950). In the subtropical climate of the Mediterranean region, only one generation is completed on pea sown from late autumn onwards, but two to three generations if pea is sown in early autumn. In the latter case, egg masses could be produced and a high reproduction rate expected (Greco *et al.*, 1986a).

Management measures

Management of *H. goettingiana* varies with crop type. Cultivation of early pea for green pod production usually gives high return and therefore the use of nematicides is economical. Nematode control can be obtained by fumigating the soil 3–4 weeks before sowing, with DD or a mixture of DD and methyl isothiocyanate at 100–300 l/ha, depending on the degree of soil infestation (Di Vito *et al.*, 1973). Similar results would be expected using 1,3-D and other acceptable fumigants. Granular nematicides, such as aldicarb, fenamiphos, oxamyl and carbofuran, at 5–10 kg a.i/ha also give satisfactory nematode control and increased yield (Di Vito *et al.*, 1973; Whitehead *et al.*, 1979). Improved control is achieved by incorporating these non-fumigant nematicides into the top 10–15 cm of soil at sowing only, or at sowing and again after emergence. Granular nematicides must enter the soil solution to become effective and, therefore, irrigation may be required prior to and/or after treating the soil in semi-arid areas.

Soil solarization could be an alternative method for cyst nematode control on high value crops (Greco *et al.*, 1985). Mulching irrigated soil with thin (30–50 μm) polyethylene sheets for 4–8 weeks can reduce *H. goettingiana* in regions with sufficient solar energy assuming that the field can remain free of crops for the required time. However, solarization and non-fumigant nematicides usually are less effective than fumigants.

None of the above methods is economically acceptable when garden peas are grown for the production of dried grain. Rotating pea with non-host crops for a 3–6 year period will reduce nematode densities to non damaging levels, assuming an annual population decline of 50% (Di Vito and Greco, 1986).

Meloidogyne

Garden pea is a good host for root knot nematodes even though reports on infestations are limited. *M. incognita* was reported on pea in India (Reddy, 1985). There is little doubt that this and the other warm season root knot nematodes can be important parasites of peas in the tropics. *M. artiellia* has potential to damage pea in the Mediterranean area, and *M. brasiliensis* was found infesting pea in Brazil (Charchar and Eisenback, 2002). In the subtropics, pea is mostly grown as a winter crop and therefore damage caused by root knot nematodes would be negligible, unless, however, pea is sown early in autumn after a summer host crop, in which case pea growth would be reduced at an early stage. Above-ground symptoms of nematode attack are similar to those outlined for *H. goettingiana*. The roots exhibit large galls, are reduced in size and *Rhizobia* nodulation is reduced. Interaction of *M. incognita* with *F. oxysporum* f.sp. *pisi* has been demonstrated (Z.A. Siddiqui *et al.*, 1999). The tolerance limit of pea to *M. incognita* was about 0.5 eggs/g of soil (Siddiqui *et al.*, 1995) and probably less to *M. javanica*.

Peas escape nematode attack in the subtropics, when sowing is postponed to mid-autumn, or when temperatures drop. In other areas, seed treatment with 1% aldicarb, fenamiphos, carbofuran, carbosulfan and neem-based products has been shown to increase yield (Mani and Sethi, 1984; Mojumder *et al.*, 2002). Soil treatments with fumigant and non-fumigant nematicides, although effective, are uneconomical on this crop. Resistance to *M. incognita* was found in a few lines of pea, but it was not confirmed.

Other nematodes of garden pea

D. dipsaci damages garden pea in several countries (Hooper, 1972; Thompson *et al.*, 2000). Infected plants show extensive brownish and necrotic lesions on the stems (Fig. 8.13) and leaf chlorosis. These symptoms can be confused with those produced by other nematodes and diseases. *D. dipsaci* damages epidermal, cortical parenchyma and external phloem tissue, thereby adversely affecting translocation processes. In Australia, severe damage is

Fig. 8.13. Peas showing stem necrosis caused by infestation of *Ditylenchus dipsaci* in Italy. (Photo: N. Vovlas.)

caused by *D. dipsaci* on pea at the seedling stage, and a 30% reduction in seedling emergence has been observed (Thompson *et al.*, 2000). Infected pods are distorted and contain few seeds, which in turn may also be infected. It is not known whether the nematode can survive for a long time within grains as is typical on other crops. However, pea seed infestation is much less than in broad bean (Knuth, 1993).

In the subtropics, attacks of *D. dipsaci* are more severe on garden pea sown in autumn, and symptoms become more obvious throughout late winter and early spring. The same control measures suggested for this nematode on broad bean should also be adopted on pea. The root lesion nematodes, *Pratylenchus crenatus* and *P. penetrans*, have been found in association with pea decline. Mediterranean populations of *P. neglectus*, *P. penetrans*, *P. pinguicaudatus* and *P. thornei* have potential to damage pea (Di Vito *et al.*, 2002), whereas pea appears to be resistant or tolerant to Australian populations of *P. thornei* and *P. neglectus* (Thompson *et al.*, 2000). Symptoms caused by these nematodes are similar to those observed on other crops. *Pratylenchus* spp. are also known to break down plant resistance to *Fusarium* wilt (Oyekan and Mitchell, 1971). *R. reniformis* is found worldwide and damages pea, especially in India where a tolerance limit of 0.1 nematodes/g of soil was estimated (Vats and Dalal, 1998). The nematode reduces *Rhizobium* nodulation and may interact with *F. oxysporum* f.sp. *pisi* (Vats and Dalal, 1997). In Brazil, *Helicotylenchus dihystera* is considered a severe constraint of wheat and pea (Sharma *et al.*, 1993).

Pigeonpea

Pigeonpea (*Cajanus cajan* (L.) Mill.), also known as red gram, Congo pea and no-eyes pea, originated in Africa around 2000 BC. Pigeonpea is a woody, short-lived perennial shrub that reaches a height of up to 3.5 m. It is grown in both the tropics and subtropics and is very common in India where over 80% of the world crop is grown and consumed. The drought-resistant crop is often intercropped with cereals in India and Africa especially in semi-arid regions. The crop, which is usually planted as an annual and grown for dried grain, is used for dhal (decorticated split seed) in a variety of foods. In other countries, the green seeds are eaten as a substitute for, or in preference to, green peas. A large number of plant parasitic nematode species have been found associated with pigeonpea on a worldwide basis (Sharma, 1985). The vast majority are of limited economic importance. However, recent research work has shown that significant yield loss is exerted on the crop by some species of plant parasitic nematodes.

Heterodera

The cyst nematode *H. cajani* described by Koshy (1967) was first recorded on pigeonpea in India by Swarup *et al.* (1964). The nematode subsequently has been reported attacking the crop in a number of states in

India (Sharma and Swarup, 1984). The exact distribution and frequency of occurrence within the country, however, have not been determined. The nematode was detected in only seven out of 471 fields examined by Koshy and Swarup (1971a) and more recently has been detected in a large number of experimental fields in central India. The nematode is more prevalent on vertisol rather than alfisol soils.

Sharma *et al.* (1992) reviewed the nematodes of pigeonpeas, their biology and control.

Symptoms

In the field, yellowing and stunting have been observed; the former varies with plant genotype. In glasshouse tests, plants infected with 1000 or 5000 juveniles/500 cm^3 of sterilized soil were stunted with smaller internodes and leaves. Chlorosis, however, was not very apparent. Stunting was directly related to initial nematode density (S.B. Sharma, India, 1988, personal communication).

Other hosts

More than 105 plant species belonging to 58 genera in the families Leguminosae and Pedaliaceae are known hosts (Koshy and Swarup, 1972). Important hosts are chickpea, horse gram, hyacinth bean, soybean, tepary bean, moth bean and a number of species in the genera *Phaseolus* and *Vicia*.

Economic threshold level

Field densities have been shown to range from two to 130 cysts/500 cm^3 of soil. The highest numbers were detected on perennial plants or in fields cropped successively for 3–4 years (S.B. Sharma, 1988, India, personal communication). Plants associated with high cyst densities growing in vertisol soils were stunted and frequently chlorotic. Symptoms of damage seemed to be more prevalent in the Kharif crop planted in the autumn. Initial populations of 5 juveniles/100 cm^3 of soil were found to affect plant growth. Zaki and Bhatti (1986) reported that 100 juveniles/kg of soil caused significant reductions in growth in pot trials. Nematicide treatment led to grain yield increases of 20–25% over the controls (Sharma *et al.*, 1993).

Biology

At a soil temperature of 29°C, the nematode completes one generation in 16 days (Koshy and Swarup, 1971a). Optimum temperature for emergence is 28°C, with distinct reductions in emergence at 25°C (Sharma and Swarup, 1984). The largest number of juveniles emerged between August and October. An initial density of 1 juvenile/cm^3 of soil caused a 14–24% reduction in plant growth. The tolerance limit in the field was estimated at 2.6 eggs and juveniles/cm^3 of soil at sowing time (Sharma *et al.*, 1993).

Disease complexes

Wilt intensity caused by *Fusarium udum* increased significantly when combined with *H. cajani* in greenhouse tests. The pigeonpea lines used, however, reacted differently to the nematode–fungus combination. In one instance, the pathogenic effects of the nematode on plant growth were negated in the presence of the fungus (S.B. Sharma, India, 1988, personal communication).

Although *H. cajani* females have also been observed attached to *Rhizobium* nodules, nothing is known about the effects of the inter-relationship on plant health.

Management measures

Strategies for control of the nematode will have to stress rotation and resistance. Rotation with cereal crops, especially millet, probably limit nematode damage in most established rotation schemes. *Echinocloa colona* (barnyard millet), *Paspalum scorbiculatum* (Kodo millet), *Setaria italica* (Italian millet), *Chionachne* spp., *Trilobachne* spp. and *Zea mexicana* (teosinte) were shown to be non-hosts (Sharma and Swarup, 1984) and could be used effectively in crop rotation patterns.

Nematicides have been tested effectively against *H. cajani* (Patel *et al.*, 2000). In most cases, they are used to demonstrate impact on yield. The cost of these products limits their use in the field. Zaki and Bhatti (1986) attempted control using seed treatment with non-fumigant nematicides, which could reduce costs substantially. Although they were effective in reducing nematode populations, plant growth was also suppressed. Solarization has also been shown to reduce nematode densities. Neither control measure, however, can be used on this crop on an economic basis. Seed treatment with neem-based products has also been shown to reduce penetration of the nematode (Devi, 2000; Dibakar *et al.*, 2000; Vijayalakshmi *et al.*, 2001) and may be of regional importance depending on cost factors. Field testing, however, is lacking.

A new strain of *P. penetrans*, an obligate bacterial parasite, has been shown to reduce nematode penetration and development, and could be important in developing suppressive soils when commercial inoculum becomes available (Singh and Dhawan, 1993, 1994).

Resistance

Many of the pigeonpea types and genotypes grown are unimproved landraces. This germplasm should serve as a basis for the development of nematode-resistant cultivars with good agronomic characteristics. A number of lines have been reported to be resistant to the nematode; however, retesting has not always substantiated the results (Devi, 1998). Variation in testing techniques and in reporting the degree of resistance must be more closely monitored to avoid improper designation of the level of resistance.

Meloidogyne

M. javanica was found on pigeonpea in Puerto Rico (Ayala, 1962a), Brazil (Lordello and Arruda, 1956) and Malawi (Reddy *et al.*, 1993). Pigeonpea was shown to be highly susceptible to a population of *M. arenaria* taken from groundnut fields in Alabama, USA (Rodriguez-Kabana and Ingram, 1978). The nematode causes significant amounts of galling on the root system, leading to reduced growth and overall yield (Plate 8D). Plant growth was significantly reduced in pot tests at initial densities of 100 juveniles/500 g of soil (Pathak *et al.*, 1985). Salam and Khan (1986) reported that *M. javanica* caused increased wilt in plants affected by *F. oxysporum* f.sp. *udum*, and Dwivedi *et al.* (1992) demonstrated the same for *M. incognita*. Field trials with nematicides where galling was reduced 53–61% demonstrated a 14.2% avoidable yield loss due to a mixed population of *M. incognita* and *M. javanica* (Patel and Patel, 1993). Seed treatment with neem-based products reduced nematode infection (Dibakar *et al.*, 2000).

Many accessions and cultivars have been shown to have resistance to *M. incognita* (Wani and Alam, 1995; Suhail *et al.*, 2001). A number of breeding lines have been shown to be highly resistant to both *M. incognita* and *M. javanica*, but are susceptible to *Fusarium udum* (Patel *et al.*, 1987). Acosta *et al.* (1986) reported that all cultivars tested were susceptible to *M. javanica*. Siddiqui *et al.* (1991) detected resistance to *M. arenaria* race 2 as well as to *M. incognita* and *M. javanica*, but not to *M. arenaria* race 1.

Rotylenchulus

Linford and Oliveira (1940) in Hawaii were the first to report *R. reniformis* on pigeonpea. It has since been reported attacking the crop in Puerto Rico, Jamaica and India. The nematode causes yellowing of new leaves, progressive dieback of twigs and main stems, and premature death of many plants in Jamaica (Hutton and Hammerton, 1975). Galls are not produced on the root system as with root knot nematode. However, the females embedded in their egg masses on the surface of the root are diagnositic for infection (Plate 8C). Although the root system was reduced in size, extensive necrosis was not observed.

Root death seemed to be caused by excessive infection of the root tip (Ayala, 1962b). Jain and Sharma (1996) showed that the nematode increased the intensity of *F. udum* wilt on the wilt-resistant cv. ICPL 270 in the fields.

Thakar and Yadav (1985a) reported significant reductions in plant weight at 1000 or 10,000 nematodes/700 g of soil in pot tests on susceptible or resistant cultivars, respectively. Suppression of growth was also detected at densities of 100 nematodes/500 g of soil (Pathak *et al.*, 1985).

The nematode also reproduced on *Rhizobium* nodules (Ayala, 1962c). In a glasshouse experiment, race A caused marked reductions in total plant fresh weight after 30 days at a density of 142 nematodes/100 g of soil (Thakar and Yadav, 1985a).

Pigeonpea lines have been shown to be moderately resistant to the nematode (Thakar and Yadav, 1985b; Patel *et al.*, 1987; Suhail *et al.*, 2001). Resistance was reported in the accessions Pusa-33, 78, 84 and 85 in India (Ahmad, 1992). Field testing at ICRISAT showed that some germplasm has tolerance to the nematode (Sharma *et al.*, 2000).

Other nematodes of pigeonpea

Germani (1972) reported that *Aphasmatylenchus straturatus* was associated with stunted and chlorotic pigeonpea in Upper Volta. *Hoplolaimus seinhorsti* has been found associated with poor plant growth in India. Pigeonpea was a poor host for *Pratylenchus zeae* in Malawi (Jones and Hillocks, 1995).

In pot experiments, wilt caused by *F. udum* was not affected by simultaneous or sequential inoculation of *Tylenchorhynchus vulgaris*, *Helicotylenchus indicus* or *Hoplolaimus indicus* (Hasan, 1984).

Soybean

Glycine max (L.) Merr., originally confined to temperate zones, is becoming more important in many tropical and subtropical regions, especially in Brazil, South America, the Far East and, more recently, Africa. Whole soybeans have not always been accepted as a food legume in many countries, because of the development of an objectionable flavour during processing. Technology now exists that allows use of the whole bean in many foods (Hinson and Hartwig, 1977), and varieties more adapted to human consumption have been developed. Most soybean, however, is still processed for oil, high protein meal animal supplement, soy flour, soybean milk and curd. The average cultivar grown in north America contains 40% protein and 21% oil on a dry weight basis.

The crop can be grown successfully under a wide range of temperature conditions as long as adequate amounts of moisture are available during the seed development period (Hinson and Hartwig, 1977). A growing season with little or no moisture stress for about a 120 day period produces near maximum yields. Although soybean is usually drilled in rows, it probably can be intercropped successfully with cereals. In Asia, the seed is often inserted into the hills remaining after the rice harvest. Minimum tillage is effective, but requires adequate equipment and herbicide application. A major factor limiting adaptation to the humid and subhumid tropics is that seeds lose their viability and rapidly degrade in storage.

Meloidogyne

Root knot nematodes *M. incognita*, *M. javanica* and *M. arenaria* are important factors limiting soybean production. According to Schmitt and Noel (1984), the latter two species are becoming more important in warmer climatic regions. This is probably related to the introduction of the crop into new growing regions where cropping patterns have favoured these two species. They are likely to become important pests wherever soybean is grown.

Root knot nematodes cause varying degrees of stunting, chlorosis and in some cases early senescence, depending on the initial population density. Losses can often be related to intensity of galling, which is also dependent on initial population densities. Galls on the root system are typical of root knot infection, but can be confused with *Rhizobium* nodules by inexperienced observers.

Losses of 90% due to *M. incognita* have been reported from Florida (Kinloch, 1974) and Brazil (J. Silva, Brazil, 2004, personal communication). The level of damage is lower in North Carolina when compared with Florida, indicating that temperature affects crop loss intensity (Schmitt and Noel, 1984).

Economic threshold level

Kinloch (1982) showed that plant growth is inversely proportional to initial population density. Environmental factors, especially moisture, have a strong influence on the level of crop loss, with higher yield associated with increased moisture availability (Barker, 1982). Others factors, such as soil compaction, potassium deficiency and low amounts of organic matter, also contribute to overall loss. Losses incurred at a specific threshold level are therefore highly variable.

Disease complexes

Goswami and Agarwal (1978) in pot tests showed that yield reductions were greater when *M. incognita* was present with *F. oxysporum* or *F. solani* than when inoculated singly.

Management measures

The use of crop rotation is hampered by the wide host range of all three root knot species. With the exception of grasses, few alternative non-host crops exist. There are differences in nematode reproduction between varieties of the same cultivated species. The use of nematicides is not an economically acceptable means of controlling this nematode on soybean.

Resistance

The use of resistant varieties is the most economical means of controlling root knot nematodes in soybean. A number of cultivars are available that are resistant to *M. incognita*, *M. javanica* and *M. arenaria* (Armstrong and Jensen, 1978; Sasser and Kirby, 1979). The majority of resistant varieties come from the cv. Bragg. Some *H. glycines* sources of resistance also have resistant genes effective against root knot nematodes.

Heterodera

The soybean cyst nematode *H. glycines* is a major limiting factor in semi-arid regions of the USA and has been reported to occur in China, the former Soviet Union, Colombia, Korea, Indonesia, Egypt, Argentina, Brazil (Noel, 1985) and Italy (Manachini, 2000). The nematode causes severe stunting and yellowing of the foliage and, in extreme cases, plant death (Fig. 8.14; Plate 8E). Yield losses can range from 10 to 80% depending on rainfall, soil fertility, the presence of other diseases and nematode density (Jacobsen *et al.*, 1983).

Races

Several race classification schemes have been proposed using host differentials (Golden *et al.*, 1970; Inagaki, 1979; Riggs and Schmitt, 1988). The problems associated with race designation have been discussed elsewhere (Schmitt and Noel, 1984; Noel, 1985). The two classifications most commonly used are those proposed by Riggs and Schmitt, (1988) and Niblack *et al.* (2002) as shown in Table 8.4.

Biology

Optimum temperature for emergence and penetration is 24°C and for development 28–31°C. There is little or no development at 15°C or below, or at 33°C or above (Schmitt and Noel, 1984). The nematode is

Fig. 8.14. Severe damage to soybean in a field in Brazil due to *Heterodera glycines*. (Photo: J. Silva.)

Table 8.4. Bioassay for race characterization of *Heterodera glycines*; reaction/reproduction on cultivar or line.

Race	Pickett	Peking	PI88788	PI90763
1	–	–	+	–
2	+	+	+	–
3	–	–	–	–
4	+	+	+	+
5	+	–	+	–
6	+	–	–	–
7	–	–	+	+
8	–	–	–	+
9	+	+	–	–
10	+	–	–	+
11	–	+	+	–
12	–	+	–	+
13	–	+	–	–
14	+	+	–	+
15	+	–	+	+
16	–	+	+	+

Reproduction of *Heterodera glycines* on the four test cultivars/lines of less than 10% of the susceptible control cultivar Lee is considered negative and above 10% positive.

reported to have a diapause stage (Ross, 1963) which may reduce spontaneous emergence at a given time of year. The nematode is also susceptible to desiccation (Slack and Hamblen, 1961). The percentage survival of eggs and juveniles decreases with increasing temperature from northern to southern growing regions of the USA (Noel, 1985). The reduction is considered to be due to the influence of temperature on nematode activity and increased biological control through soil pathogens and parasites.

The nematode will complete 6–7 generations per season in temperate growing areas, with the greatest increase in density occurring in the first generation (Lawn and Noel, 1986).

Economic threshold level

Noel (1984) reported that, on silt loam soils with 2% organic matter, economic losses were incurred when densities were 699 or more eggs and juveniles or 12 cysts containing viable eggs in 250 cm^3 of soil.

Other hosts

Noel (1985) reported that other hosts of economic importance were: adzuki bean (*Phaseolus angularis* Wright), haricot bean and some species of *Lespedeza* and *Melilotus*. Monocotyledonous species have not been reported to be hosts.

Disease complexes

The nematode will severely reduce *Rhizobium* nodule weight and the level of nitrogen fixation (Lehman *et al.*, 1971).

Management measures

Rotation with non-host crops for 2 years (Fig. 8.15) will reduce populations sufficiently to allow planting of susceptible cultivars (Schmitt and Noel, 1984). Resistant cultivars are effective against some races of the nematode (Fig. 8.16; Plate 8F). The use of resistant cultivars (Wrather *et al.*, 1984) and possibly tolerant cultivars (Boerma and Hussey, 1984) in the rotation would increase the effectiveness of integrated control programmes. Problems associated with rotation management have been discussed by Noel (1985). Nematicides are not used for control of this nematode on a field scale. Sources of resistance have been given by Tisselli *et al.* (1980).

Fig. 8.15. Effect of rotation with maize–groundnut–soybean (left) versus monoculture of soybean (right) on soybean growth in a field infested with the soybean cyst nematode *Heterodera glycines*. (Photo: D. Schmitt.)

Fig. 8.16. Growth differences between soybean cultivars Clark-63 (susceptible, left) and Custer (resistant, right) infected with soybean cyst nematode, *Heterodera glycines*. (Photo: R.A. Sikora.)

Other nematodes of soybean

Rotylenchulus reniformis can cause stunting and chlorosis on soybean. The nematode has been found attacking soybean in a number of tropical and subtropical countries (Schmitt and Noel, 1984). Rotation with non-host crops for 2 or more years is an effective control measure. The wide host range of this nematode requires careful selection of rotation crops. Resistant cultivars are available (Birchfield *et al.*, 1971; Lim and Castillo, 1979). Some soybean cultivars resistant to *H. glycines* developed from Peking can also be resistant to *R. reniformis*.

Hoplolaimus columbus has been shown to cause damage in the south-eastern USA. High densities of a *Hoplolaimus* sp. were also detected in the rhizosphere of soybean in India (Sikora, 1972). *Belonolaimus longicaudatus*, which is also limited to the south-eastern USA, will cause stunting, chlorosis and wilting. The nematode is usually controlled with crop rotation.

P. brachyurus and other *Pratylenchus* species have been found attacking soybean in most growing regions. They can cause stunting, leaf yellowing and yield loss depending on soil densities at planting. Yield losses are linearly related to *P. brachyurus* densities in a sandy–clay loam soil (Schmitt and Barker, 1981). Control of these nematodes is hampered by wide host ranges and the presence of multiple species in a field. The lesion nematodes are also known to increase damage caused by root-rotting fungi, which may further reduce yield.

Winged Bean

Winged bean (*Psophocarpus tetragonolobus* (L.) D.C.), also known as Goa bean, asparagus pea, four-angled bean, Manila bean and princess pea, originated in Asia or Africa. It is a perennial crop grown as an annual for green immature pods, seeds, tubers and leaves in the humid tropics. The crop is resistant to high temperatures and is often intercropped with sweet potato, taro, banana, sugarcane and vegetables. It can be grown as a dry season crop with irrigation, but is not drought resistant.

Meloidogyne

Root knot nematodes have been shown to cause serious damage to winged bean in a number of tropical countries. *M. incognita* has been reported on the crop in Papua New Guinea (Price and Linge, 1979), India (Singh *et al.*, 1979), Okinawa (Teruya *et al.*, 1984) and Nigeria (Whitehead, 1969). *M. javanica* caused damage in Papua New Guinea (Price and Linge, 1979), Brazil (Lordello and de Almeida, 1979) and Okinawa (Teruya *et al.*, 1984). Root knot nematodes are considered the most widely distributed pests of winged bean in Papua New Guinea.

A '*Meloidogyne–javanica–incognita–arenaria* species complex' was responsible for severe galling to roots and tubers in the Côte d'Ivoire (Fortuner *et al.*, 1979). Species of *Meloidogyne* have also been reported from Mauritius (de Sorney, 1913) and the Philippines (Fajardo and Palo, 1933).

The distribution of the two major species attacking winged bean is influenced by temperature. *M. incognita* seems to be more predominant in the warmer coastal regions of Papua New Guinea and at lower altitudes in East Africa, whereas *M. javanica* is common in the highlands and higher altitudes (Whitehead, 1969; Price and Linge, 1979). These observations are supported by the fact that hatching of local populations occurs in a temperature range of 25–30°C for *M. incognita* and 20–30°C for *M. javanica* (Price and Linge, 1979). In the field, the juveniles penetrate the root within 1 week and females and galls develop after 4 weeks (Linge, 1976).

In the Côte d'Ivoire, the root knot nematode species complex caused heavy root galling and tuber galling so severe that they were unsuitable for consumption. An estimated 50–70% of the tubers failed to develop. Damage to the tubers was observed even at very low initial infestation levels (Fortuner *et al.*, 1979). Damage seems to be more severe on winged bean grown in the dry season (Khan, 1976; Price and Linge, 1979).

No attempts have been made to develop control measures for root knot nematodes on this crop. Resistance to *M. incognita* has not been detected in the lines screened to date (Duncan *et al.*, 1979; Singh *et al.*, 1979; Valdez, 1981; Phukan and Hazarika, 1985). Breeding lines with resistance to *M. javanica* have been found (Valdez, 1981).

Other nematodes of winged bean

A number of plant parasitic nematodes of unknown importance have been found associated with winged bean (Teruya *et al.*, 1984; Bridge, 1987).

Nematode Parasites of Other Food Legumes

A large number of food legumes have not been discussed in detail in this chapter. Most of these crops were considered to be of local importance. In some cases, only a few reports of nematodes associated with the crop were found. The plant parasitic nematodes that have been found associated with these food legumes have been compiled from major lists (Table 8.5) and are not considered complete. Species of root knot nematodes, cyst nematodes (*Heterodera*) and lesion nematodes parasitize many of these crops. The stem and bulb nematode, *D. dipsaci*, the reniform nematode and *Belonolaimus* cause severe damage on many food legumes and are most probably important on the crops listed. The species that have been reported to attack a number of these crops and that may be economically important are *H. glycines*, *M. arenaria*, *M. incognita*, *M. javanica* and *R. reniformis.*

Conclusions and Future Prospects

For many of the food legumes discussed, there is a definite lack of information on the presence and distribution of plant parasitic nematodes within the major growing regions. In some cases, survey work has only been conducted near research stations, with a complete lack of survey data on nematode distribution and frequency of occurrence in the major growing regions. Crop loss assessment has not been conducted in the majority of cases where important plant parasitic nematodes are known to occur. This situation has not changed over the past 15 years.

Food legumes are not high value cash crops; therefore, control is often limited to rotation with non-host crops. Resistance is important, but in many crops is not known or has not been transferred to cultivars suitable for farmer use.

The development of rotations for nematode management in temperate regions, where one crop per year is grown, is reasonably easy to formulate. In the tropics and subtropics, however, intercropping, and sequential and relay cropping, involving the production of 2–4 crops in 1 year, is common practice (Steiner, 1982; Ruthenberg, 1983). Designing rotations for nematode management under these conditions is a challenge to nematology. Bridge (1987) suggested a number of approaches to nematode control in cropping systems, and suggestions for integrated management are made in Chapter 22.

In some cases, nematicides have been suggested for nematode control on these crops, although their use is debatable due to high cost. In addition, a multitude of biological control agents have been tested and in some cases recommended for field use without reflection of the cost–benefit relationship to these low value crops.

Research on the influence of different cropping systems and the long-term effects of crop rotations on nematode population dynamics and yield loss has not been conducted. Whereas data on intercropping systems have demonstrated that crop yield can be increased in legume–cereal intercrop situations, the effects of intercropping on damage caused by plant parasitic nematodes have not been ascertained.

Rotation, especially with non-host crops and where possible in a paddy rice cropping system, could be an efficient method of controlling nematodes in the subsequent

Table 8.5. Plant parasitic nematodes associated with food legumes of local or limited importance in tropical and subtropical climatic areas (Goodey *et al.*, 1965; Mani *et al.*, 1982; Sitaramaiah, 1984; Saka and Carter, 1987).

	Adzuki bean	Catjang bean	Cluster bean	Grass pea	Horse gram	Hyacinth bean	Jack bean	Lima bean	Lupin, pearl	Lupin, white	Moth bean	Rice bean	Runner bean	Sword bean	Tepary bean	Velvet bean
Belonolaimus spp.								◆								
Ditylenchus dipsaci			◆										◆			
Helicotylenchus spp.				◆			◆									
Heterodera cajani			◆		◆	◆	◆					◆			◆	
H. glycines	◆	◆		◆		◆	◆	◆	◆	◆	◆	◆			◆	
H. goettingiana				◆					◆							
H. lespedezae	◆															
H. schachtii	◆	◆		◆												
H. trifolii				◆												
Hirschmanniella mucronata					◆											
Hoplolaimus spp.		◆	◆	◆		◆					◆					
Longidorus spp.							◆									
Meloidogyne spp.	◆		◆	◆	◆	◆	◆	◆	◆	◆	◆	◆	◆	◆	◆	
M. arenaria		◆				◆				◆			◆			
M. hapla		◆														
M. incognita		◆	◆	◆	◆	◆	◆	◆		◆		◆	◆	◆		
M. javanica			◆		◆	◆		◆		◆	◆	◆	◆			◆
Pratylenchus brachyurus						◆	◆	◆			◆					
P. coffeae				◆				◆								
P. pratensis											◆					
P. scribneri				◆				◆								
P. vulnus								◆			◆					
Radopholus similis			◆	◆						◆		◆	◆	◆	◆	
Rotylenchulus reniformis						◆		◆		◆						
Scutellonema spp.						◆										
Trichodorus spp.											◆					
Tylenchorhynchus spp.			◆	◆	◆	◆					◆				◆	
Xiphinema spp.						◆										

legume crop. Dry fallow in the semi-arid subtropics is also effective in reducing population densities. Research, however, is needed to determine if these observations are valid in all situations. The use of trap crops, that act as green manures and control components, should be looked at as an alternative control measure.

Nematicides are still too costly for the vast majority of food legumes. The development of a new generation of nematicides that are safe and effective and that could be used as seed dressing could allow their incorporation into nematode management systems.

There are a number of publications that list resistant cultivars and lines of food legumes (Armstrong and Jensen, 1978; Sasser and Kirby, 1979; Tisselli *et al.*, 1980; Bridge, 1981). In many instances, however, screening for resistance or tolerance has not been initiated. In other cases, known sources of resistance, because of inadequate methodology, have led to false interpretation of results. Coordination of the screening process is needed if good resistance or tolerance is to be developed in many of the important food legumes.

Diagnosis

Root knot nematodes

Species of root knot nematodes can usually be recognized by the presence of root galls, which with most species affecting food legumes in tropical and subtropical climates are large. To the untrained eye, root knot galls often resemble *Rhizobium* nodules. The latter, however, are distinct knots of root tissue attached to the surface of the root which can be easily detached from the root surface, whereas galls are swellings arising on all sides of the root.

Above-ground symptoms vary from stunting to chlorosis. Plants may wilt when exposed to moisture stress and in cases involving inter-relationships with fungal wilt diseases. In some plants, early senescence has been reported.

Cyst-forming nematodes

The presence of white lemon-shaped or round females, 0.4–0.8 mm in length, attached to the root surface is the most characteristic symptom of this group of nematodes. Knowledge of the day-degrees, the sum of temperature above the minimum temperature needed for activity, that coincides with appearance of adult females on the root surface can be used to simplify detection in field survey work. The presence of white females on the root surface is a simultaneous verification of parasitism.

The presence of cysts in soil samples is an indication that a cyst nematode problem is present in the cropping system; it does not indicate which crop or weed is being parasitized. Cyst colour varies greatly from white to dark brown. Colour can be species specific, but usually indicates cyst ages, with dark brown an indication of an old cyst.

The extraction of cysts from a predetermined quantity of soil and determination of the total number of eggs and juveniles found in the extracted cysts is the most exact measure used to determine nematode densities and to study population dynamics.

Stem and bulb nematode

Wallace (1962) demonstrated that the stem nematode migrates to the soil surface after rain. The date selected for soil sampling and the depth of sampling, therefore, are important in determining nematode densities when only looking at nematodes in the upper soil layers.

On broad bean, leaf spot symptoms caused by fungal diseases can be confused with necrosis induced by the stem nematode. The spots on infested seed cannot be used as a diagnostic characteristic because they can be caused by insect damage and water spotting (Plate 7B).

For routine studies and experimentation, Hooper (1983a) suggested soaking 150 g of seed in 500 ml of water overnight. To prevent introduction of the nematode into nematode-free areas, a high level of nematode extraction accuracy from seed is necessary. Augustin and Sikora (1989b) suggested first soaking and then maceration of the seed and extraction on a modified Baermann tray (see Chapter 3).

Lesion nematodes

Species of *Pratylenchus* cause distinct small brown to black lesions on the root surface of many food legumes (Plate 7F). They can often be seen with a simple magnifying lens in the field or with a field microscope. In extreme cases, the lesions coalesce to form large necrotic lesions. The nematodes can be extracted as outlined in Chapter 3.

References

Abawi, G.S. and Jacobsen, B.J. (1984) Effect of initial inoculum densities of *Heterodera glycines* on growth of soybean and kidney bean and their efficiency as host under greenhouse conditions. *Phytopathology* 74, 1470–1474.

Abbad, A.F. (2001) Screening of *Vicia faba* for resistance to the 'giant race' of *Ditylenchus dipsaci* in Morocco. *Nematologia Mediterranea* 29, 29–33.

Abbad, A.F. and Bachikh, J. (2001) Studies on the host range of *Ditylenchus dipsaci* in Morocco. *Nematologia Mediterranea* 29, 51–57.

Abbad, A.F. and Sellami, S. (1998) Screening of *Vicia faba* resistance to the giant race of *Ditylenchus dipsaci* in the Maghreb region. In: *3rd European Conference on Grain Legumes. Opportunities for High Quality, Healthy and Added-value Crops to Meet European Demands.* Valladolid, Spain, pp. 122–123.

Abedinia, M. (1978) Host–parasite relationships of the root knot nematode *Meloidogyne incognita* and cowpea. PhD Thesis, University of Reading, Reading, UK.

Aboul-Eid, H.Z. and Ghorab, A.I. (1974) Pathological effects of *Heterodera cajani* on cowpea. *Plant Disease Reporter* 58, 1130–1133.

Acosta, N., Vicente, N. and Toro, J. (1986) Susceptibility of pigeon pea (*Cajanus cajan*) cultivars and lines to *Meloidogyne javanica. Nematropica* 16, 1–10.

Acosta, N., Vargas, R., Román, O., Vicente, N. and Sánchez, L. (1995) *Mucuna deringiana* incorporada vs. no incorporada al suelo y rendimento en siembras subsiguientes de tomate, habichuelka y maiz. *Journal of Agriculture of the University of Puerto Rico* 79, 65–74.

Ahmad, S. (1992) Susceptibility of pigeon pea accessions to *Rotylenchulus reniformis. Tests of Agrochemicals and Cultivars* 13, 114–115.

Ahmad, S. and Husain, S.I. (1988) Effect of root knot nematodes on qualitative and quantitative characters of chickpea. *International Nematology Network Newsletter* 5, 12–13.

Al Hazmi, A.S. (1985) Interaction of *Meloidogyne incognita* and *Macrophomia phaseolina* in a root-rot disease complex of french bean. *Phytopathologische Zeitschrift* 113, 311–316.

Ali, M.A., Trabulsi, I.Y. and Abd-Elsamea, M.E. (1981) Antagonistic interaction between *Meloidogyne incognita* and *Rhizobium leguminosarum* on cowpea. *Plant Disease* 65, 432–435.

Ali, S.S. (1995) *Nematode Problems in Chickpea.* Indian Institute of Pulses Research, Kanpur, India.

Ali, S.S. and Ahmad, R. (2000) Screening of chickpea germplasm against nematodes. *International Chickpea and Pigeonpea Newsletter* 7, 8–9.

Ali, S.S. and Askary, T.H. (2001) Taxonomic status of phytonematodes associated with pulse crops. *Current Nematology* 12, 75–84.

Ali, S.S. and Dwivedi, R.P. (2001) Effect of interaction between *Fusarium oxysporum* f.sp. *lentis* and *Meloidogyne javanica* on lentil. *Indian Journal of Pulse Research* 14, 71–73.

Ali, S.S. and Sharma, S.B. (2003) Nematode survey of chickpea production areas in Rajasthan, India. *Nematologia Mediterranea* 31, 147–149.

Amosu, J.O. (1974) The reaction of cowpea (*Vigna unguiculata* L. Walp) to the root knot nematode (*Meloidogyne incognita*) in Western Nigeria. *Nigerian Agricultural Journal* 11, 165–169.

Anjum, A., Alam, M.M. and Ahmad, A. (1996) Efficacy of some organic amendments against reniform nematode, *Rotylenchulus reniformis* on blackgram. *Annals of Plant Protection Sciences* 4, 122–125.

Armstrong, J.M. and Jensen, H.J. (1978) *Indexed Bibliography of Nematode-resistance in Plants.* Station Bulletin 639, Agricultural Experiment Station Oregon State University, Corvallis, Oregon.

Augustin, B. (1985) Biologie, Verbreitung und Bekämpfung des Stengelälchens, *Ditylenchus dipsaci* (Kühn), Filipjev and *Vicia faba* L. in Syrien und anderen Ländern des Nahen Ostens und Vorderafrikas. PhD Thesis, University of Bonn, Germany.

Augustin, B. and Sikora, R.A. (1984) Influence of aldicarb and fenamiphos on the population dynamics of soil or seed-borne *Ditylenchus dipsaci* and yield of *Vicia faba*. *Proceedings of the International Congress of Nematology*, 5–10 August 1984, Guelph, Canada, 7 (abstract).

Augustin, B. and Sikora, R.A. (1989a) Studies on the host range of the normal and giant faba bean race of *Ditylenchus dipsaci*. *Nematologia Mediterranea* 17, 63–66.

Augustin, B. and Sikora, R.A. (1989b) Zur Methodik des Nachweises von *Ditylenchus dipsaci* in Körnerlegumenosen-saatgut. *Gesunde Pflanzen* 5, 189–192.

Ayala, A. (1962a) Occurrence of the nematode *Meloidogyne javanica* on pigeonpea roots in Puerto Rico. *Journal of Agriculture, University of Puerto Rico* 46, 154–156.

Ayala, A. (1962b) Pathogenicity of the reniform nematode on various hosts. *Journal of Agriculture, University of Puerto Rico* 46, 73–82.

Ayala, A. (1962c) Parasitism of bacterial nodules by the reniform nematode. *Journal of Agriculture, University of Puerto Rico* 46, 67–69.

Ayala, A. and Ramirez, C.T. (1964) Host range, distribution and bibliography of the reniform nematode, *Rotylenchulus reniformis* with special reference to Puerto Rico. *Journal of Agriculture, University of Puerto Rico* 48, 142–161.

Barker, K. (1982) Influence of soil moisture, cultivar, and population density of *Meloidogyne incognita* on soybean yield in microplots. *Journal of Nematology* 14, 429 (abstract).

Beane, J. and Perry, R.N. (1984) Observation on the invasion and development of the pea cyst-nematode *Heterodera goettingiana* in winter field beans (*Vicia faba*). *Annals of Applied Biology* 105, 503–508.

Bessey, E.A. (1911) Root knot and its control. *Bulletin Bureau Plant Industry, US Department of Agriculture*, No. 217.

Birchfield, W., Williams, C., Hartwig, E.E. and Brister, L.R. (1971) Reniform nematode resistance in soybeans. *Plant Disease Reporter* 55, 1043–1045.

Blazey, D.A., Smith, P.G., Gentile, A.G. and Miyagawa, S.T. (1964) Nematode resistance in the common bean. *Journal of Heredity* 55, 20–22.

Boerma, H.R. and Hussey, R.S. (1984) Tolerance to *Heterodera glycines* in soybean. *Journal of Nematology* 16, 289–296.

Bornali, M., Aparajita, B., Mahanta, B. and Borah, A. (1998) Effect of organic amendments for the management of *Meloidogyne incognita* on greengram and blackgram. *Journal of the Agricultural Science Society of North East India* 11, 73–76.

Bornali, M., Phukan, P.N. and Mahanta, B. (2002) Effect of *Glomus fasciculatum* and *Meloidogyne incognita* on blackgram. *Annals of Plant Protection Sciences* 10, 407–408.

Brancalion, A.M. and Lordello, L.G.E. (1981) Controle quimico de *Rotylenchulus reniformis* em algodeiro. *Sociedade Brasilieira de Nematologica* 5, 191–195.

Bridge, J. (1973) *Hoplolaimus seinhorsti*, an endoparasitic nematode of cowpea in Nigeria. *Plant Disease Reporter* 57, 798–799.

Bridge, J. (1981) Nematodes. In: Ward, A., Mercer, S.L. and Howe, V. (eds) *Pest Control in Tropical Grain Legumes*. Centre for Overseas Pest Research, London, pp. 111–125.

Bridge, J. (1987) Control strategies in subsistence agriculture. In: Brown, J. and Kerry, B. (eds) *Principles and Practice of Nematode Control in Crops*. Academic Press, Australia, pp. 389–420.

Brito, J., Inserra, R., Lehman, P. and Dixon, W. (2003a) The root knot nematode *Meloidogyne mayaguensis* Rammah and Hirshmann, 1988 (Nematoda: Tylenchida). In: Pest Alert, Website of Florida Department of Agriculture and Consumer Service, Division of Plant Industry, Gainesville, Florida. Available at: www.doacs.state.fl.us/~pi/enpp/nema/mayaguensis.html

Brito, J.A., Stanley, J.D., Mendes, M.L. and Dickson, D.W. (2003b) Host status of selected plant species to *Meloidogyne mayaguensis* from Florida. *Abstracts XXXV Annual Meeting of Organization of Nematologists of Tropical America*, 21–25 July, Guayaquil, Ecuador, p. 13.

Brothwell, D.R. and Brothwell, P. (1969) *Food in Antiquity*. Thames and Hudson, London.

Brown, E.B. (1958) Pea root eelworm in the eastern counties of England. *Nematologica* 3, 257–268.

Castillo, M.B. (1975) Plant parasitic nematodes associated with mung bean, soybean and peanut in the Philippines. *Philippine Agriculturist* 59, 91–99.

Castillo, M.B., Arceo, M.B. and Litsinger, J.A. (1976) Nematodes in cropping patterns. IV. Populations of plant parasitic nematodes associated with cropping patterns under different rice-growing environments in Manaoag, Pangasinan. *Philippine Phytopathology* 12, 24–29.

Castillo, M.B., Alejar, M.S. and Litsinger, J.A. (1977) Pathological reactions and yield loss of mung bean to known populations of *Rotylenchulus reniformis* and *Meloidogyne acrita*. *Philippine Agriculturist* 61, 12–24.

Castillo, M.B., Alegar, M.S. and Harwood, R.R. (1978) Effect of flooding on pot populations of *Rotylenchulus reniformis* and *Meloidogyne incognita. Philippine Phytopathology* 12, 11–16.
Castillo, P., Gomez-Barcina, A. and Jiménez-Diaz, R.M. (1996) Plant parasitic nematodes associated with chickpea in southern Spain and effect of soil temperature on reproduction by *Pratylenchus thornei. Nematologica* 42, 211–219.
Castillo, P., Vovlas, N. and Jiménez-Diaz, R.M. (1998) Pathogenicity and histopathology of *Pratylenchus thornei* populations on selected chickpea genotypes. *Plant Pathology* 47, 370–376.
Catibog, C.S. and Castillo, M.B. (1975) Pathogenicity of *Meloidogyne javanica* on mung bean *Phaseolus aureus* Roxb. *Philippine Agriculturist* 59, 189–195.
Caubel, G. and Leclercq, D. (1989a) *Vicia faba* lines resistant to the giant race of stem nematode (*Ditylenchus dipsaci* (Kuehn) Fil:). *FABIS Newsletter* 25, 45–48.
Caubel, G. and Leclercq, D. (1989b) Estimation de la resistance a la race geante de *Ditylenchus dipsaci* par les symptomes chez la feverole (*Vicia faba* L.). *Nematologica* 35, 216–224.
Caubel, G., Bossis, M., Esquibet, M., Esnault, R., Le-Guen, J., Maurin, G., Romane, G. and Le-Guen, J. (1998a) Resistance of *Vicia faba* to stem nematode, *Ditylenchus dipsaci.* In: *3rd European Conference on Grain Legumes. Opportunities for High Quality, Healthy and Added-value Crops to Meet European Demands,* Valladolid, Spain, p. 237.
Caubel, G., Esnault, R., Lefloch, D., Maurin, G. and Le-Guen, J. (1998b) Incidence of crop rotations on stem nematode, *Ditylenchus dipsaci,* attacking faba bean (*Vicia faba*). In: *3rd European Conference on Grain Legumes. Opportunities for High Quality, Healthy and Added-value Crops to Meet European Demands,* Valladolid, Spain, p. 266.
Caubel, G., Andaloussi, F.A., Bekal, S., Bossis, M., Esquibet, M., Sellami, S. and Tivoli, B. (1998c) Variabilite de populations du nematode des tiges, *Ditylenchus dipsaci* sur quelques legumineuses a grosses graines. In: Caubel, G. (ed.) *Les Legumineuses Alimentaires Mediterraneennes. Contraintes Biotiques et Potentialites de Developpement,* Rennes, France, *Colloques-de-l'INRA* 88,173–186.
Caveness, F.E. (1965) *End of Tour Progress Report on Nematology Project.* Ministry of Agriculture and Natural Resources, Western Region, Nigeria.
Caveness, F.E. (1979) Cowpea, lima bean, cassava, yams and *Meloidogyne* spp. in Nigeria. In: Lamberti, F. and Taylor, C.E. (eds) *Root Knot Nematodes.* Academic Press, New York, pp. 295–300.
Chahal, P.P.K. and Chahal, V.P.S. (1987) Adverse effect of *Meloidogyne incognita* on the functioning of nodules of mungbean (*Vigna radiata*). *Nematologia Mediterranea* 15, 13–19.
Chahal, V.P.S., Chahal, P.P.K. and Kapoor, S. (1988) Effect of *Meloidogyne incognita* on symbiosis between *Rhizobium* and mash (*Vigna mungo*). *Crop Improvement* 15, 95–97.
Charchar, J.M. and Eisenback, J.D. (2002) *Meloidogyne brasiliensis* n.sp. (Nematoda: Meloidogynidae), a root knot nematode parasitising tomato cv. Rossol in Brazil. *Nematology* 4, 629–643.
Charu Jain and Trivedi, P.C. (1997) Nematicidal activity of certain plants against root knot nematode, *Meloidogyne incognita* infecting chickpea, *Cicer arietinum. Annals of Plant Protection Sciences* 5, 171–174.
Charu Jain and Trivedi, P.C. (1998) Effect of the interaction between *Meloidogyne incognita, Fusarium oxysporum* and *Rhizoctonia bataticola* on chickpea. *Journal of Phytological Research* 11, 95–97.
Chen, P. and Roberts, P. (2003) Virulence in *Meloidogyne hapla* differentiated by resistance in common bean (*Phaseolus vulgaris*). *Nematology* 5, 39–47.
Chitambar, J.J. (1993) Host range of *Hemicycliophora porangia. Fundamental and Applied Nematology* 16, 557–561.
Crozzoli, R., Higuera, A. and Casassa, A.M. (1995) Respuesta de algunos materiales de frijol al nemátodo *Meloidogyne* spp. *Revista de la Facultad de Agronomía, Universidad del Zulia* 12, 485–490.
Crozzoli, R., Greco, N., Suárez, A.C. and Rivas, D. (1997) Patogenecidad del nematodo agallador, *Meloidogyne incognita,* en cultivares de *Phaseolus vulgaris* y *Vigna unguiculata. Nematropica* 27, 61–67.
Crozzoli, R., Greco, N., Suárez, A.C. and Rivas, D. (1999) Patogenecidad del nematodo agallador, *Meloidogyne incognita,* en *Vigna unguiculata. Nematropica* 29, 99–103.
da Ponte, J.J. (1972) Doencas do feijoeiro macássar, *Vigna sinensis* Endl. no nordest Brasileiro. *Boletim da Sociedad e Cearense de Agronomia* 13, 1–12.
da Ponte, J.J., Alves, M.E., Almeida, D.M. and Câmara Do Vale, C. (1993) Atributo de resistência à meloidoginose e de produtividade de um novo cultivar de caupi. *Nematologia Brasileira* 17, 161–167.
Dasgupta, D.R. and Seshadri, A.R. (1971) Races of the reniform nematode, *Rotylenchulus reniformis* Linford and Oliveira, 1940. *Indian Journal of Nematology* 1, 21–24.

Datta, S., Trivedi, P.C. and Tiagi, B. (1987) Nematodes of guar and mung in some areas of Rajasthan, India. *International Nematology Network Newsletter* 4, 12–16.

Decker, H. (1969) *Phytonematologie.* VEB Deutscher Landwirtschaftsverlag, Berlin, Germany.

Dehne, W.W. (1987) Zur Nutzung der VA Mykorrhiza als Antistreßfaktor. *Angewandte Botanik* 61, 135–143.

de Sorney, P. (1913) *Les Plantes Tropicales Alimentaires et Industrielles de la Famille des Legumineuses.* P. 3. Paris Challamel.

Devi, L.S. (1998) Evaluation of pigeonpea cultivars for resistance to pigeonpea cyst nematode, *Heterodera cajani. Indian Journal of Nematology* 28, 232–233.

Devi, L.S. (2000) Effect of neem products against *Heterodera cajani* on pigeon pea and cowpea. *Indian Journal of Nematology* 30, 250–252.

Devi, L.S. (2001) Resistance of *Heterodera cajani* in cowpea. *Annals of Plant Protection Sciences* 9, 361–362.

Devi, L.S., Dwivedi, S. and Devi, H.J. (1999) Screening of greengram, blackgram and cowpea varieties for their resistance against *Meloidogyne incognita. Annals of Plant Protection Sciences* 7, 226–227.

Devi, T.P. and Goswami, B.K. (1992) Interrelationships of root knot nematodes *Meloidogyne incognita* and *Macrophomina phaseolina* with cowpea. *Annals of Agricultural Research* 13, 267–268.

Dibakar, D., Mishra, S.D. and Das, D. (2000) Effect of neem seed powder and neem based formulations as seed coating against *Meloidogyne incognita, Heterodera cajani* and *Rotylenchulus reniformis* infecting pigeon pea. *Current Nematology* 11, 13–23.

Di Vito, M., Greco, N. and Zaccheo, G. (1985) On the host range of *Meloidogyne artiellia. Nematologia Mediterranea* 13, 207–212.

Di Vito, M. and Greco, N. (1986) The pea cyst nematode. In: Lamberti, F. and Taylor, C.E. (eds) *Cyst Nematodes.* Plenum Publishing Co., New York, pp. 321–332.

Di Vito, M. and Greco, N. (1988a) Investigation on the biology of *Meloidogyne artiellia. Revue de Nématologie* 11, 221–225.

Di Vito, M. and Greco, N. (1988b) The relationship between initial population densities of *Meloidogyne artiellia* and yield of winter and spring chickpea. *Nematologia Mediterranea* 16, 163–166.

Di Vito, M., Lamberti, F. and Inserra, R. (1973) Prove preliminari di lotta chimica contro *Heterodera goettingiana* Liebscher, grave parassita del pisello nell'Italia Meridionale. *Nematologia Mediterranea* 1, 125–138.

Di Vito, M., Greco, N. and Lamberti, F. (1980) Comportamento di popolazione di *Heterodera goettingiana* su specie diverse di leguminose. *Informatore Fitopatologico* 30 (5), 7–10.

Di Vito, M., Greco, N. and Saxena, M.C. (1992) Pathogenicity of *Pratylenchus tornei* on chickpea in Syria. *Nematologia Mediterranea* 20, 71–73.

Di Vito, M., Greco, N., Halila, H.M., Mabsoute, L., Beniwal, S.P.S., Saxena, M.C., Singh, K.B. and Solh, M.B. (1994a) Nematodes of cool-season food legumes in North Africa. *Nematologia Mediterranea* 22, 3–10.

Di Vito, M., Greco, N., Oreste, G., Saxena, M.C., Singh, K.B. and Kusmenoglu, I. (1994b) Plant parasitic nematodes of legumes in Turkey. *Nematologia Mediterranea* 22, 245–251.

Di Vito, M., Singh, K.B., Greco, N. and Saxena, M.C. (1996) Sources of resistance to cyst nematode in cultivated and wild *Cicer* species. *Genetic Resources and Crop Evolution* 43, 103–107.

Di Vito, M., Zaccheo, G. and Catalano, F. (2000) Effect of *Pratylenchus neglectus* and *P. thornei* on the growth of faba bean. *Nematologia Mediterranea* 28, 261–265.

Di Vito, M., Greco, N., Malhotra, R.S., Singh, K.B., Saxena, M.C. and Catalano, F. (2001a) Reproduction of eight populations of *Heterodera ciceri* on selected plant species. *Nematologia Mediterranea* 29, 79–90.

Di Vito, M., Troccoli, A., Zaccheo, G. and Catalano, F. (2001b) Sources of resistance to the nematodes *Meloidogyne artiellia* and *Pratylenchus thornei* in wild *Cicer* species. In: *Proceedings of the 4th European Conference on Grain Legumes*, 8–12 July 2001, Cracow, Poland, pp. 70–71.

Di Vito, M., Catalano, F. and Zaccheo, G. (2002a) Reproduction of six populations of *Pratylenchus* from the Mediterranean region on selected plant species. *Nematologia Mediterranea* 30, 103–105.

Di Vito, M., Catalano, F., Zaccheo, G., Sadici, M. and Kharrat, M. (2002b) Response of lines of faba bean to six populations of root lesion nematodes (*Pratylenchus* spp.). *Nematologia Mediterranea* 30, 107–109.

dos Santos, M.A. and Ferraz, S. (2000) Efficiência de cinco isolados de fungos nematófagos no controle de *Meloidogyne* spp. ao longo de três cultivos sucessivos. *Nematologia Brasileira* 24, 193–201.

Dukes, P.D., Fery, R.L. and Hamilton, M.G. (1979) Comparison of plant resistance and a nematicide for control of southern root knot nematodes in southernpeas, *Vigna unguiculata. Phytopathology* 69, 526–527.

Duncan, L.W., Caveness, F.E. and Perez, A.T. (1979) The susceptibility of winged bean (*Psophocarpus tetragonolobus*) to the root knot nematodes, *Meloidogyne incognita*, race 2, and *M. javanica*. *Tropical Grain Legume Bulletin, IITA* 15, 30–34.

Dwivedi, R.K., Upadhyay, K.D. and Dwivedi, K. (1992) Interaction of root-knot nematode, *Meloidogyne incognita* and fungus, *Fusarium oxysporum* f.sp. *udum* on pigeon pea. *Indian Journal of Nematology* 22, 96–100.

Edwards, E.E. (1956) Studies on resistance to root knot nematode of the genus *Meloidogyne* Goeldi, 1887. *Proceedings of the Helminthological Society of Washington* 23, 112–118.

Egunjobi, O.A. (1985) Effects of cocoa pod husks soil amendment on cowpea infestation by *Meloidogyne* spp. *Pakistan Journal of Nematology* 3, 99–103.

Egunjobi, O.A. and Olaitan, J.O. (1986) Response of *Meloidogyne incognita*-infected cowpea to some agro-waste soil amendments. *Nematropica* 16, 33–34.

Egunjobi, O.A., Akonde, P.T. and Caveness, F.E. (1986) Interaction between *Pratylenchus sefaensis, Meloidogyne javanica* and *Rotylenchulus reniformis* in sole and mixed crops of maize and cowpea. *Revue de Nématologie* 9, 61–70.

Ehlers, J.D., Hall, A.E., Patel, P.N., Roberts, P.A. and Matthews, W.C. (2000) Registration of 'California Blackeye 27' cowpea. *Crop Sciences* 40, 854–855.

El Bahrawy, S.A. and Salem, F.M. (1989) Interaction between *Rhizobium leguminosarum* and *Meloidogyne javanica* nematode in broad bean under nematicide application. *Zentralblatt für Mikrobiologie* 144, 279–281.

Elliot, A.P. and Bird, G.W. (1985) Pathogenicity of *Pratylenchus penetrans* to navy bean (*Phaseolus vulgaris* L). *Journal of Nematology* 17, 81–85.

Elliot, A.P., Bird, G.W. and Safir, G.R. (1984) Joint influence of *Pratylenchus penetrans* and *Glomus fasciculatum* (Phycomyceta) on the ontogeny of *Phaseolus vulgaris*. *Nematropica* 14, 111–119.

El Tigani, M., El Amin, M. and Siddiqi, M.R. (1970) Incidence of plant parasitic nematodes in the Northern Fung Area, The Sudan. *FAO Plant Protection Bulletin* 18, 102–106.

Epps, J.M. (1969) Nine varieties of southern peas resistant to the soybean cyst nematode. *Plant Disease Reporter* 53, 245.

Epps, J.M. and Chambers, A.Y. (1959) Mung bean (*Phaseolus aureus*), a host of the soybean cyst nematode (*Heterodera glycines*). *Plant Disease Reporter* 43, 981–982.

Eriksson, K.B. (1972) Nematode diseases of pasture legumes and turf grasses. In: Webster, J.M. (ed.) *Economic Nematology*. Academic Press, London, pp. 66–96.

Esquibet, M., Bekal, S., Castagnone-Sereno, P., Gauthier, J.P., Rivoal, R. and Caubel, G. (1998) Differentiation of normal and giant *Vicia faba* populations of the stem nematode *Ditylenchus dipsaci*: agreement between RAPD and phenotypic characteristics. *Heredity* 81, 291–298.

Fajardo, R.G. and Palo, M.A. (1933) The root knot nematode, *Heterodera radicicola* (Greef) Müller, of tomato and other plants in the Philippine Islands. *Philippine Journal of Science* 51, 457–484.

FAO (2003) *FAO Production Statistics*. Available at: apps.fao.org/faostat/

Fassuliotis, G., Hoffman, J.C. and Deakin, J.R. (1967) A new source of root knot nematode resistance in snap bean. *Nematologica* 13, 141.

Fazal, M., Siddiqui, M.R., Imran, M., Shah, N.H. and Siddiqui, Z.A. (1995) Susceptibility of some lentil cultivars to reniform nematodes, *Rotylenchulus reniformis*, and their effect on peroxidase activity and protein content. *Annals of Plant Protection Sciences* 3, 41–45.

Fazal, M., Bhat, M.Y., Imran, M. and Siddiqui, Z.A. (1998) A disease of blackgram involving *Meloidogyne javanica* and *Rhizoctonia bataticola*. *Nematologia Mediterranea* 26, 63–66.

Fery, R.L. and Dukes, P.D. (1992) 'Carolina Crowder' southernpea. *HortScience* 27, 1335–1337.

Fery, R.L. and Dukes, P.D. (1995a) Registration of US-566, US-567, and US-568 root knot nematode resistant cowpea germplasm lines. *Crop Science* 35, 1722.

Fery, R.L. and Dukes, P.D. (1995b) 'Bettersnap' southernpea. *HortScience* 30, 1318–1319.

Fery, R.L. and Dukes, P.D. (1996) 'Tender Cream' southernpea. *HortScience* 31, 1250–1251.

Fortuner, R., Fauquet, C. and Lourd, M. (1979) Diseases of the winged bean in Ivory Coast. *Plant Disease Reporter* 63, 194–199.

Franc, G.D., Beaupré, C.M.S., Gray, F.A. and Hall, R.D. (1996) Nematode angular leaf spot of dry bean in Wyoming. *Plant Disease* 80, 476–477.

France, R.A. and Abawi, G.S. (1995) Interaction between *Meloidogyne incognita* and *Fusarium oxysporum* f.sp. *phaseoli* on selected bean genotypes. *Journal of Nematology* 26, 464–474.

Franklin, M.T. (1961) A British root knot nematode, *Meloidogyne artiellia* n.sp. *Journal of Helminthology*, R.T. Leiper Supplement, 85–92.

Freire, F.C. das O., Diógenes, A.M. and da Ponte, J.J. (1972) Nematodes das galhas *Meloidogyne javanica* e *M. incognita,* parasitando leguminosas forrageiras. *Revista da Sociedade Brasileira de Fitopatologia* 5, 27–32.

Garofalo, F. (1964) Fenomeni di correlazione tra *Heterodera goettingiana* e *Fusarium oxysporum* (Schl.) Snyd. et Hans. Nell'avvizzimento delle piante di pisello e di lupino. *Bollettino dei Laboratorio Sperimentale dell'Osservatorio Fitopatologico, Torino* 27, 33–48.

Gaur, H.S., Mishra, S.D. and Sud, U.C. (1979) Effect of date of sowing on the relation between the population density of the root knot nematode, *Meloidogyne incognita* and the growth of three varieties of chickpea, *Cicer arietinum. Indian Journal of Nematology* 9, 152–159.

Germani, G. (1972) Une chlorose des légumineuses de Haute-Volta lliée à la présence d'un nématode. *Comptes Rendus des Séances Hebdomadaires de l'Académie d'Agriculture de France* 58, 202–204.

Gill, J.S. and Swarup, G. (1977) Pathogenic effect of *Tylenchorynchus vulgaris* on gram. *Indian Journal of Nematology* 7, 155–156.

Glazer, I. and Orion, D. (1983) Studies on anhydrobiosis of *Pratylenchus thornei. Journal of Nematology* 15, 333–338.

Golden, A.M., Epps, J.M., Riggs, R.D., Duclos, L.A., Fox, J.A. and Bernard, R.L. (1970) Terminology and identity of infraspecific forms of the soybean cyst nematode (*Heterodera glycines*). *Plant Disease Reporter* 54, 544–546.

Goodey, J.B., Franklin, M.T. and Hooper, D.J. (1965) *The Nematode Parasites of Plants Catalogued Under their Hosts.* CAB, Farnham Royal, UK.

Goswami, B.K. and Agarwal, D.K. (1978) Interrelationships between species of *Fusarium* and root knot nematodes, *Meloidogyne incognita* in soybean. *Nematologia Mediterranea* 6, 125–128.

Greco, N. (1984) Presenza di *Meloidogyne artiellia* su cece in Italia. *Nematologia Mediterranea* 12, 235–238.

Greco, N. and Di Vito, M. (1987) The importance of plant parasitic nematodes in food legume production in the Mediterranean Region. In: Saxena, M.C., Sikora, R.A. and Srivastava, J.P. (eds) *Proceedings of a Workshop on Nematodes Parasitic to Cereals and Legumes in Temperate Semi-arid Regions,* 1–5 March 1987, Larnaca, Cyprus, pp. 29–45.

Greco, N., Brandonisio, A. and Elia, F. (1985) Control of *Ditylenchus dipsaci, Heterodera carotae* and *Meloidogyne javanica* by solarization. *Nematologia Mediterranea* 13, 191–197.

Greco, N., Di Vito, M. and Lamberti, F. (1986a) Studies on the biology of *Heterodera goettingiana* in Southern Italy. *Nematologia Mediterranea* 14, 23–39.

Greco, N., Di Vito, M., Reddy, M.V. and Saxena, M.C. (1986b) Effect of Mediterranean cultivated plants on the reproduction of *Heterodera ciceri. Nematologia Mediterranea* 14, 193–200.

Greco, N., Di Vito, M., Saxena, M.C. and Reddy, M.V. (1988a) Effect of *Heterodera ciceri* on yield of chickpea and lentil and development of this nematode on chickpea in Syria. *Nematologica* 34, 98–114.

Greco, N., Di Vito, M., Saxena, M.C. and Reddy, M.V. (1988b) Investigation on the root lesion nematode, *Pratylenchus thornei,* in Syria. *Nematologia Mediterranea* 16, 101–105.

Greco, N., Ferris, H. and Brandonisio, A. (1991) Effect of *Heterodera goettingiana* population densities on the yield of pea, broad bean and vetch. *Revue de Nématologie* 14, 619–624.

Greco, N., Di Vito, M. and Saxena, M.C. (1992a) Plant parasitic nematodes of cool season food legumes in Syria. *Nematologia Mediterranea* 20, 37–46.

Greco, N., Di Vito, M. and Nombela, G. (1992b) The emergence of juveniles of *Heterodera ciceri. Nematologica* 38, 514–519.

Green, C.D. (1981) The effects of weeds and wild plants on the reinfestation of land by *Ditylenchus dipsaci* (stem and bulb nematode) and on the stability of its populations. In: Thresh, J.M. (ed.) *Pests, Pathogens and Vegetation.* Pitman, London, pp. 217–224.

Green, C.D., Williamson, K., Dennis, E.B. and McBurney, T. (1981) The effect of oxamyl on the growth of peas attacked by the pea cyst nematode. *Annals of Applied Biology* 97, 303–309.

Griffin, G.D. (1984) Nematode parasites of alfalfa, cereals, and grasses. In: Nickle, W.R. (ed.) *Plant and Insect Nematodes.* Marcel Dekker Inc., New York, pp. 243–323.

Gupta, D.C. (1979) Studies on the pathogenicity and relative susceptibility of some varieties of cowpea [*Vigna unguiculata* (L) Walp.] against *Meloidogyne javanica. Forage Research* 5, 141–145.

Gupta, D.C. and Yadav, B.S. (1979) Studies on the pathogenicity of reniform nematode, *Rotylenchulus reniformis* to urd, *Vigna mungo* (L) Wilczek. *Indian Journal of Nematology* 9, 48–50.

Gupta, D.C. and Yadav, B.S. (1980) Pathogenicity of *Rotylenchulus reniformis* on cowpea. *Nematologia Mediterranea* 8, 91–93.

Hadisoeganda, W.W. and Sasser, J.N. (1982) Resistance of tomato, bean, southern pea, and garden pea cultivars to root knot nematodes based on host suitability. *Plant Disease* 66, 145–150.
Hafez, S.L. and Sundararaj, P. (1999) Screening of commercial cultivars of *Phaseolus vulgaris* against *Meloidogyne chitwoodi* and *M. hapla*. *International Journal of Nematology* 9, 163–167.
Haidar, M.G., Askary, T.H. and Nath, R.P. (2001) Nematode population as influenced by paddy based cropping sequences. *Indian Journal of Nematology* 31, 68–71.
Handa, D.K. (1990) Reaction to urdbean cultivars and losses caused by *Meloidogyne javanica*. *International Nematology Network Newsletter* 7, 20–21.
Handoo, Z.A., Golden, A.M., Chitwood, D.J., Haglund, W.A., Inglis, D.A., Santo, G.S., Baldwin, J.G. and Williams, D.J. (1994) Detection of pea cyst nematode (*Heterodera goettingiana*) in pea in western Washington. *Plant Disease* 78, 831.
Hanounik, S.B., Halila, H. and Harrabi, M. (1986) Resistance in *Vicia faba* to stem nematodes (*Ditylenchus dipsaci*). *FABIS Newsletter (Faba Bean Information Service, ICARDA)* 16, 37–39.
Hartman, R.W. (1971) Inheritance of resistance to root knot nematodes (*Meloidogyne incognita*) in beans (*Phaseolus vulgaris* L). *Journal of the American Society of Horticultural Sciences* 96, 344–347.
Hasan, A. (1984) Synergism between *Heterodera cajani* and *Fusarium udum* attacking *Cajanus cajan*. *Nematologia Mediterranea* 12, 159–162.
Hasan, N. and Jain, R.K. (1986) Screening mothbean (*Vigna aconitifolius* Jacq) germplasm for resistance to root-knot nematode (*Meloidogyne incognita*). *Agricultural Science Digest, India* 4, 198–199.
Heald, C.M. and Robinson, A.F. (1987) Effects of soil solarization on *Rotylenchulus reniformis* in the lower Rio Grande Valley of Texas. *Journal of Nematology* 19, 93–103.
Heald, C.M., Menges, R.M. and Wayland, J.R. (1974) Efficacy of ultra-high frequency (UHF) electromagnetic energy and soil fumigation on the control of the reniform nematode and common purslane among southern peas. *Plant Disease Reporter* 58, 985–987.
Hinson, K. and Hartwig, E.E. (1977) Soybean production in the tropics. *FAO Plant Production and Protection Paper*, No. 4.
Hooper, D.J. (1972) *Ditylenchus dipsaci*. *CIH Description of Plant-parasitic Nematodes*, Set 1, No. 14. CIH, St Albans, UK.
Hooper, D.J. (1983a) Observations on stem nematode, *Ditylenchus dipsaci*, attacking field beans, *Vicia faba*. *Rothamsted Report for 1983* 2, 239–260.
Hooper, D.J. (1983b) Nematode pests of *Vicia faba* L. In: Hebblethwaite, P.D. (ed.) *The Faba Bean (*Vicia faba *L).* Butterworths, London, pp. 347–370.
Hussaini, S.S. and Seshadri, A.R. (1975) Interrelationships between *Meloidogyne incognita* and *Rhizobium* sp. on mung bean (*Phaseolus aureus*). *Indian Journal of Nematology* 5, 189–199.
Hussaini, S.S. and Seshadri, A.R. (1976) Resistance in some mung bean (*Vigna radiata*) varieties and breeding lines to the root knot nematode, *Meloidogyne incognita*. *Indian Journal of Nematology* 6, 131–137.
Hussey, R.S. and Roncadori, R.W. (1982) Vesicular-arbuscular mycorrhizae may limit nematode activity and improve plant growth. *Plant Disease* 66, 9–14.
Hutton, D.G. and Hammerton, J.L. (1975) Investigating the role of *Rotylenchulus reniformis* in a decline of pigeon pea. *Nematropica* 5, 24 (abstract).
Hutton, D.G., Wilkinson, R.E. and Mai, W.F. (1972) Effect of two plant-parasitic nematodes on *Fusarium* dry root rot of beans. *Phytopathology* 63, 749–751.
Ibrahim, A.A.M. and Ibrahim, I.K.A. (2000) Evaluation of non-chemical treatments in the control of *Meloidogyne incognita* on common bean. *Pakistan Journal of Nematology* 18, 51–57.
Inagaki, H. (1979) Race status of five Japanese populations of *Heterodera glycines*. *Japanese Journal of Nematology* 9, 1–4.
Inderjit Singh, V.P.S., Chahal, P.K., Sakhuja, P.K. and Chohan, J.S. (1977) Effect of different levels of *Meloidogyne incognita* in the presence or absence of *Rhizobium phaseoli* on *Phaseolus aureus*. *Indian Journal of Nematology* 7, 172–174.
Jacobsen, B.J., Edwards, D.I., Noel, G.R. and Shurtleff, M.C. (1983) The soybean cyst nematode problem. *Report on Plant Diseases, University of Illinois Cooperative Extension Service*, No. 501.
Jain, K.C. and Sharma, S.B. (1996) Loss of *Fusarium* wilt resistance in a pigeon pea line ICPL 270 in reniform nematode infested soil at ICRISAT Asia Center. *International Chickpea and Pigeonpea Newsletter* 3, 90.
Jain, R.K. and Hasan, N. (1986) Efficacy of 'Neem' cake on fodder production, photosynthetic pigments and nematodes associated with oats and its residual effect on cowpea. *Indian Journal of Nematology* 16, 98–100.

Jain, S.K., Dave, G.S. and Jain, A.K. (1994) Host range of cyst nematode *Heterodera cajani* in cold weather. *Indian Phytopathology* 47, 177–178.

Jones, F.G.W. (1950) Observations on the beet eelworm and other cyst forming species of *Heterodera*. *Annals of Applied Biology* 37, 304–440.

Jones, M.L. and Hillocks, R.J. (1995) Host status for *Pratylenchus zeae* of food crops and associated weed species in Malawi. *Afro Asian Journal of Nematology* 5, 120–126.

Kakar, G., Khan, E. and Siddiqi, M.R. (1995) Two new species belonging to genera *Tylenchorhynchus* Cobb, 1913 and *Telotylenchus* Siddiqi, 1960. *Annals of Plant Protection Sciences* 3, 149–154.

Kaliram, D.C. and Gupta, D.C. (1982) Efficacy of plant leaves, nematicides and fertilizer alone and in combination against *Meloidogyne javanica* infecting chickpea (*Cicer arietinum* L.). *Indian Journal of Nematology* 12, 221–225.

Kalita, D.N. and Phukan, P.N. (1993) Efficacy of four chemicals as seed treatment for the control of *Meloidogyne incognita* on blackgram. *Current Nematology* 4, 149–151.

Kaloshian, I., Greco, M. and Saad, A.T. (1986a) Hatching of cysts and infectivity of *Heterodera ciceri* on chickpea. *Nematologia Mediterranea* 14, 129–133.

Kaloshian, I., Greco, M., Saad, A.T. and Volvas, N. (1986b) Life cycle of *Heterodera ciceri* on chickpea. *Nematologia Mediterranea* 14, 135–145.

Kamalwanshi, R.S., Anis, K., Srivastava, A.S. and Khan, A. (2000) Varietal reaction of blackgram against *Meloidogyne incognita*. *Annals of Plant Protection Sciences* 8, 274–275.

Kassab, A.S. and Ali, M.K. (1996) Interactions among *Meloidogyne incognita, Rotylenchulus reniformis, Rhizoctonia solani* and *Rhizobium* on cowpea. *Annals of Agricultural Science* 41, 521–531.

Kay, D.E. (1979) *Food Legumes.* TPI Crop and Product Digest No. 3. Tropical Pest Institute, London.

Khan, T.N. (1976) Papua New Guinea: a centre of genetic diversity in winged bean (*Psophocarpus tetragonolobus* (L.) DC). *Euphytica* 25, 693–706.

Khan, T.N. and Ahmad, W. (1994) Description of *Ischiodorylaimus paraugandanus* sp.n. and *Tylencholaimus asymmetrichus* sp.n. (Nematoda: Dorylaimida) from India. *Indian Journal of Nematology* 24, 83–87.

Kinloch, R.A. (1974) Response of soybean cultivars to nematicidal treatments of soil infested with *Meloidogyne incognita*. *Journal of Nematology* 6, 7–11.

Kinloch, R.A. (1982) The relationship between soil populations of *Meloidogyne incognita* and yield reduction of soybean in the coastal plain. *Journal of Nematology* 14, 162–167.

Knuth, P. (1993) Stengelälchen (*Ditylenchus dipsaci*) in Saatgut von Ackerbohnen und Erbsen, Untersuchungen und Ergebnisse der Jahre 1987 bis 1991 in Baden-Württemberg. *Gesunde Pflanzen* 45, 250–254.

Koshy, P.K. (1967) A new species of *Heterodera* from India. *Indian Phytopathology* 20, 272–274.

Koshy, P.K. and Swarup, G. (1971a) Investigations on the life history of the pigeon-pea cyst nematode, *Heterodera cajani*. *Indian Journal of Nematology* 1, 44–51.

Koshy, P.K. and Swarup, G. (1971b) Distribution of *Heterodera avenae, H. zeae, H. cajani*, and *Anguina tritici* in India. *Indian Journal of Nematology* 1, 106–111.

Koshy, P.K. and Swarup, G. (1972) Susceptibility of plants to pigeon-pea cyst nematode, *Heterodera cajani*. *Indian Journal of Nematology* 2, 1–6.

Lamberti, F. (1981) Plant nematode problems in the Mediterranean region. *Helminthological Abstracts, Series B, Plant Nematology* 50, 145–166.

Latha, T.K.S., Rajeswari, E. and Narasimhan, V. (2000) Management of root-rot disease complex through antagonists and chemicals. *Indian Phytopathology* 53, 216–218.

Lawn, D.A. and Noel, G.R. (1986) Field interrelationships among *Heterodera glycines, Pratylenchus scribneri* and three other nematode species associated with soybean. *Journal of Nematology* 18, 98–106.

Lehman, P.S. (1985) *Nacobbus,* the false root knot nematode. Florida Department of Agriculture and Consumer Services, Division of Plant Industry. *Nematology Circular,* No. 119.

Lehman, P.S., Huisingh, D. and Barker, K.R. (1971) The influence of races of *Heterodera glycines* on nodulation and nitrogen-fixing capacity of soybean. *Phytopathology* 61, 1239–1244.

Lim, B.K. and Castillo, M.B. (1979) Screening soybeans for resistance to reniform nematode disease in the Philippines. *Journal of Nematology* 11, 275–282.

Linford, M.B. and Oliveira, J.M. (1940) *Rotylenchulus reniformis* nov. gen., n.sp., a nematode parasite of roots. *Proceedings Helminthological Society of Washington* 7, 35–42.

Linge, D.S. (1976) Studies on the root knot of winged-bean (*Psophocarpus tetragonolobus* (L) Dc.) in Papua New Guinea. Thesis, Department of Agriculture, University of Papua New Guinea.

Lordello, L.G.E. and Arruda, H.V. (1956) Nematoides parasitando Guandu. *Bragantia* 15, 5–6.
Lordello, L.G.E. and de Oliveira Santos, C.F. (1960) Incidencia de nematodes en cultura de feijao. *O Biologico* 26, 213–217.
Lordello, R.R.A. and de Almeida, L.D. (1979) Feijao alado, nova hospedeiro de *Meloidogyne javanica. Revista Agricultura, Piracicaba* 54, 136.
Luc, M. (1985) Cyst nematodes in equatorial and hot tropical regions. In: Lamberti, F. and Taylor, C.E. (eds) *Cyst Nematodes.* Plenum Press, New York, pp. 355–372.
MacGuidwin, A.E. and Slack, S.A. (1991) Suitability of alfalfa, corn, oat, red clover, and snapbean as hosts for the potato rot nematode, *Ditylenchus destructor. Plant Disease* 75, 37–39.
Mahanta, B. and Phukan, P.N. (1990) Effect of crop rotations on the development of *Meloidogyne incognita* on blackgram. *Indian Journal of Nematology* 20, 227–229.
Mahapatra, B.C. and Pahdi, N.N. (1986) Pathogenicity and control of *Rotylenchulus reniformis* on *Cicer arietinum. Nematologia Mediterranea,* 14, 287–290.
Mahapatra, S.N., Swain, P.K. and Narain, A. (1999) Pathogenicity and varietal reaction of blackgram against *Meloidogyne incognita* and *Fusarium oxysporum. Indian Journal of Nematology* 29, 13–18.
Maheswari, T.U., Sharma, S.B., Reddy, D.D.R. and Haware, M.P. (1995) Co-infection of *Fusarium oxysporum* f.sp. *ciceri* and *Meloidogyne javanica. Journal of Nematology* 27, 649–653.
Maheswari, T.U., Sharma, S.B., Reddy, D.D.R. and Haware, M.P. (1997) Interaction of *Fusarium oxysporum* f.sp. *ciceri* and *Meloidogyne javanica* on *Cicer arietinum. Journal of Nematology* 29, 117–126.
Malhotra, R.S., Singh, K.B., Di Vito, M., Greco, N. and Saxena, M.C. (2002) Registration of ILC 10765 and ILC 10766 chickpea germplasm lines resistant to cyst nematode. *Crop Science* 42, 1756.
Mamluk, O.F., Augustin, B. and Bellar, M. (1983) New records of cyst and root knot nematodes on legume crops in the dry areas of Syria. *Phytopathologia Mediterranea* 22, 80.
Manachini, D. (2000) First report of *Heterodera glycines* Ichinoe on soybean in Italy. *Bollettino di Zoologia Agraria e Bachicoltura, Serie II* 32, 261–267.
Mani, A. and Sethi, C.L. (1984) Plant growth of chickpea as influenced by initial inoculum levels of *Meloidogyne incognita. Indian Journal of Nematology* 14, 41–44.
Mani, A., Koshy, P.K. and Swarup, G. (1982) Nematodes associated with grain and forage legumes. *FAO Technical Document,* No. 129.
Manzanilla-Lopez, R.H., Costilla, M.A., Doucet, M., Franco, J., Inserra, R.N., Lehmann, P.S., Cid del Prado-Vera, I., Souza, R.M. and Evans, K. (2002) The genus *Nacobbus* Thorne & Allen, 1944 (Nematoda: Pratylenchidae): systematics, distribution, biology and management. *Nematropica* 32, 149–227.
Mathur, B.N., Handa, D.K. and Singh, H.G. (1969) Note on the occurrence of *Meloidogyne arenaria* as a serious pest of *Cicer arietinum. Madras Agricultural Journal* 56, 744.
Mathur, R.L., Handa, D.K. and Mathur, B.N. (1973) Resistance of varieties of cowpea, green gram and guar to root knot nematodes (*Meloidogyne javanica* and *M. incognita*). *Indian Journal of Mycology and Plant Pathology* 3, 182–183.
Maqbool, M.A. (1986) *Classification and Distribution of Plant Parasitic Nematodes in Pakistan.* National Nematological Research Centre, University of Karachi, Pakistan.
McSorley, R. (1980) Effect of *Rotylenchulus reniformis* on snap bean and methods for control by oxamyl. *Nematropica* 10, 89–95.
McSorley, R. and Pohronezny, K. (1984) Cost-effectiveness of nematode control by fumigation with SMDC on Rockdale soils. *Proceedings of Soil and Crop Science Society of Florida* 43, 188–192.
McSorley, R., Pohronezny, K. and Stall, W.M. (1981) Aspects of nematode control in snap bean with emphasis on the relationship between nematode density and plant damage. *Proceedings of the Florida State Horticultural Society* 94, 134 136.
Melton, T.A., Jacobsen, B.J., Hagedorn, S.J. and Noel, G. (1985) Comparative host suitabilities of snap bean to soybean cyst nematode (*Heterodera glycines*). *Plant Disease* 69, 119–122.
Midha, R.L. and Trivedi, P.C. (1988) Reaction of some black-gram varieties to the reniform nematode *Rotylenchulus reniformis. International Nematology Network Newsletter* 5, 18–19.
Miller, J.C. and Scheuring, D.C. (1994) 'Texas Pinkeye Purple Hull' cowpea. *HortScience* 29, 926–927.
Mishra, S.D. and Gaur, H.S. (1981) Pathogenicity of *Meloidogyne incognita* and *Rotylenchulus reniformis* to mothbean, *Vigna aconitifolius* Jacq. *Indian Journal of Nematology* 11, 228–230.
Mohan, S. and Mishra, S.D. (1993) Management of *Meloidogyne incognita* infecting frenchbean through seed coating with chemicals. *Indian Journal of Nematology* 23, 92–94.
Mohanti, K.C., Dash, L. and Swain, S.C. (2001) Effect of *Meloidogyne incognita* on nodulation and root biochemistry in French bean. *Annals of Plant Protection Sciences* 9, 81–86.

Mojumder, V., Pamkaj and Singh, J. (2002) Comparative efficacy of seed coating of pea with neem based pesticides and carbosulfan against root-knot, reniform and other plant parasitic nematodes. *Annual of Plant Protection Sciences* 10, 404–406.

Mulk, M.M. and Jairajpuri, M.S. (1974) Nematodes of leguminous crops in India. II. Five new species of *Helicotylenchus* Steiner, 1945 (Hoplolaimidae). *Indian Journal of Nematology* 4, 212–221.

Mulk, M.M. and Jairajpuri, M.S. (1975) Nematodes of leguminous crop in India. III. Three new species of *Hoplolaimus* Daday, 1905 (Hoplolaimidae). *Indian Journal of Nematology* 5, 1–8.

Mullin, B.A., Abawi, G.S. and Pastor-Corrales, M.A. (1991a) Modification of resistance expression of *Phaseolus vulgaris* to *Meloidogyne incognita* by elevated temperatures. *Journal of Nematology* 23, 182–187.

Mullin, B.A., Abawi, G.S., Pastor-Corrales, M.A. and Kornegay, J.L. (1991b) Reaction of selected bean pure lines and accessions to *Meloidogyne* species. *Plant Disease* 75, 1212–1216.

Mulvey, R.H. and Anderson, R.V. (1974) *Heterodera trifolii. CIH Description of Plant-parasitic Nematodes*, Set 4, No. 46. CIH, St Albans, UK.

Nadakal, A.M. (1964) Studies on plant parasitic nematodes of Kerala. III. An additional list of plants attacked by root knot nematode, *Meloidogyne* sp. (Tylenchoidea: Heteroderidae). *Journal of Bombay Natural History Society* 61, 467–469.

Nanjappa, C.K., Panda, M. and Seshadri, A.R. (1978) Effects of *Meloidogyne incognita* and *Rotylenchulus reniformis* on the emergence and survival of seedlings. In: *Abstracts of the Third International Congress of Plant Pathology*, München, Germany, p. 150.

NAS (National Academy of Science) *ad hoc* panel of the Advisory Committee on Technology innovation (1979) *Tropical Legumes: Resources for the Future.* National Academy of Sciences, Washington, DC.

Nath, R.P., Banerjee, A.K., Haider, M.G.I. and Singh, B.K. (1979) Studies on the nematodes of pulse crops in India. I. Pathogenicity of *Meloidogyne incognita* on gram. *Indian Phytopathology* 32, 28–31.

Ngundo, B.W. (1977) Screening of bean cultivars for resistance to *Meloidogyne* spp. in Kenia. *Plant Disease Reporter* 61, 991–993.

Niblack, T.L., Arelli, P.R., Noel, G.R., Opperman, C.H., Orf, J., Schmitt, D.P., Shannon, J.G. and Tylka, G.L. (2002) A revised classification scheme for genetically diverse populations of *Heterodera glycines. Journal of Nematology* 34, 279–288.

Noel, G.R. (1982) Soybean cyst nematode in commercial snap beans. *Plant Disease* 66, 520–522.

Noel, G.R. (1984) Relating numbers of soybean cyst nematodes to crop damage. In: *Proceedings of the 5th Cyst Nematode Workshop*, p. 17.

Noel, G.R. (1985) The soybean cyst nematode. In: Lamberti, F. and Taylor, C.E. (eds) *Cyst Nematodes.* Plenum Press, New York, pp. 257–268.

Ogallo, J.L., Goodell, P.B., Eckert, J.W. and Roberts, P.A. (1999) Management of root knot nematodes with resistant cotton cv. Nemx. *Crop Science* 39, 418–421.

Ogbuji, R.O. (1978) Reactions of cowpea to *Meloidogyne incognita* in eastern Nigeria. *PANS* 24, 32–34.

Omwega, C.O., Thomason, I.J., Roberts, P.A. and Waines, J.G. (1989) Identification of new sources of resistance to root-knot nematodes in *Phaseolus. Crop Science* 29, 1463–1468.

Omwega, C.O., Thomason, I.J. and Roberts, P.A. (1990a) Effect of temperature on expression of resistance to *Meloidogyne* spp. in common bean (*Phaseolus vulgaris*). *Journal of Nematology* 22, 446–451.

Omwega, C.O., Thomason, I.J. and Roberts, P.A. (1990b) A single dominant gene in common bean conferring resistance to three root knot nematode species. *Phytopathology* 80, 745–748.

Oyekan, P.O. and Mitchell, J.E. (1971) Effect of *Pratylenchus penetrans* on the resistance of pea variety to *Fusarium* wilt. *Plant Disease Reporter* 55, 1032–1035.

Palmisano, A.M., Tacconi, R. and Trotti, G.C. (1971) Sopravvivenza di *Ditylenchus dipsaci* (Kühn) Filipjev (Nematoda: Tylenchidae) al processo digestiva nei suini, equini e bovini. *Redia* 52, 725–737.

Patel, B.A., Chavda, J.C., Patel, S.T. and Patel, D.J. (1987) Reaction of some pigeonpea lines to reniform nematode (*Rotylenchulus reniformis*). *International Pigeonpea Newsletter* 6, 57–59.

Patel, D.S. and Thakar, N.A. (1985) Reaction of some mungbean varieties to the reniform nematode, *Rotylenchulus reniformis* Linford and Oliveira, 1940. *Indian Journal of Nematology* 15, 239.

Patel, G.A. and Patel, D.J. (1993) Avoidable yield losses in pigeonpea cv. Pusa Ageti due to *Meloidogyne javanica. International Pigeonpea Newsletter* 17, 26–27.

Patel, G.A., Shah, H.M. and Patel, D.J. (1977) Screening of cowpea cultivars against root knot nematodes. *Indian Journal of Nematology* 7, 169–170.

Patel, R.G., Patel, B.A., Vyas, R.V., Patel, H.R. and Patel, D.J. (2000) Effectiveness of phenamiphos for management of *Heterodera cajani* in pigeonpea. *Indian Journal of Nematology* 30, 252–253.

Pathak, K.N., Nath, R.P. and Gaider, M.G. (1985) Effect of initial inoculum levels of *Meloidogyne incognita* and *Rotylenchulus reniformis* on pigeonpea and their interrelationship. *Indian Journal of Nematology* 15, 177–179.

Phukan, P.N. and Hazarika, M.H. (1985) Reaction of some winged bean varieties to root knot nematode, *Meloidogyne incognita. Indian Journal of Nematology* 15, 246.

Powell, D.F. (1974) Fumigation of field beans against *Ditylenchus dipsaci. Plant Pathology* 23, 110–113.

Prakash, A. (1981) *Lens culinaris* – a new host for root knot nematode, *Meloidogyne javanica* in India. *National Academic Science Letters* 4, 459.

Prasad, S.K., Chawla, M.L., Kumar, S. and Saxena, H.P. (1971) Root knot nematode, *Meloidogyne javanica* (Treub, 1885) Chitwood, 1949 and stem borer, *Stomopteryx nertaria* Meyrick affecting green gram, *Phaseolus aureus. Indian Journal of Entomology* 33, 55–60.

Preeti, B. and Trivedi, P.C. (2000) Assessment of nematophagous fungi and neem cake compatibility against *Heterodera cajani* on cowpea. *Indian Phytopathology* 53, 478–479.

Price, T.V. and Linge, D.S. (1979) Studies on the root knot nematodes of winged bean (*Psophocarpus tetragonolobus*) in Papua New Guinea. *Tropical Agriculture* 56, 345–352.

Purseglove, J.W. (1983) *Tropical Crops – Dicotyledons.* Longman Group Limited, London.

Rao, G.M.V.P. and Ganguly, S. (1996) Host preference of 6 geographical isolates of reniform nematode, *Rotylenchulus reniformis. Indian Journal of Nematology* 26, 19–22.

Rao, V.K. and Krishnappa, K. (1996) Interaction effects of *Meloidogyne incognita* and *Fusarium oxysporum* f.sp. *ciceri* on root knot disease and wilt incidence in chickpea cultivars. *Indian Journal of Nematology* 26, 117–121.

Rashid, A., Khan, F.A and Khan, A.M. (1973) Plant parasitic nematodes associated with vegetables, fruits, cereals and crops in North India. *Indian Journal of Nematology* 3, 8–23.

Rebois, R.V., Madden, P.A. and Eldridge, B.J. (1975) Some ultrastructural changes induced in resistant and susceptible soybean roots following infection by *Rotylenchulus reniformis. Journal of Nematology* 7, 122–139.

Reddy, P.P. (1985) Estimation of crop losses in peas due to *Meloidogyne incognita. Indian Journal of Nematology* 15, 226.

Reddy, M.V., Subrahmanyam, P., Likoswe, A.A., Saka, V.W. and Msuku, W.A.B. (1993) Pigeon pea diseases in Malawi. *FAO Plant Protection Bulletin* 41, 101–106.

Rhoades, H.L. (1964) Effect of *Crotalaria spectabilis* and *Sesbania exaltata* on plant nematode populations and subsequent yield of snap beans and cabbage. *Proceedings of the Florida State Horticultural Society* 77, 233–237.

Rhoades, H.L. and Forbes, R.B. (1986) Effects of fallow, cover crops, organic mulches, and fenamiphos on nematode populations, soil nutrients, and subsequent crop growth. *Nematropica* 16, 141–151.

Ribeiro, C.A. and Ferraz, S. (1983) Resistencia varietal do feijoerio (*Phaseolus vulgaris* L.) a *Meloidogyne javanica. Sociedade Brasileira de Nematologia* 7, 261–270.

Riggs, R.D. and Hamblen, M.L. (1962) Soybean–cyst nematode host studies in the family Leguminosae. *Report of the Arkansas Agricultural Experiment Station, University of Arkansas,* Serial 110, 1–18.

Riggs, R.D. and Schmitt, D.P. (1988) Complete characterization of the race scheme for *Heterodera glycines. Journal of Nematology* 20, 392–395.

Roberts, P.A., Van Gundy, S.D. and McKinney, H.E. (1981) Effect of soil temperature and planting date of wheat on *Meloidogyne incognita* reproduction, soil population, and grain yield. *Journal of Nematology* 13, 378–345.

Roberts, P.A., Frate, C.A., Matthews, W.C. and Osterli, P.P. (1995) Interactions of virulent *Meloidogyne incognita* and *Fusarium* wilt on resistant cowpea genotype. *Phytopathology* 85, 1288–1295.

Roberts, P.A., Mattheus, W.C. and Ehlers, J.D. (1996) New resistance to virulent root knot nematodes limited to the *RK* locus of cowpea. *Crop Science* 36, 889–894.

Robinson, P.E. (1961) Root knot nematodes and legume nodules. *Nature* 189, 506–507.

Rodríguez, I., Rodríguez, M.G., Sánchez, L. and Iglesias, A. (1996) Expression of resistance to *Meloidogyne incognita* in cowpea cultivars (*Vigna unguiculata*). *Revista de Protección Vegetal* 11, 63–65.

Rodriguez-Kabana, R. and Ingram, E.G. (1978) Susceptibilidad del gandul a especies de fitonematodos en Alabama. *Nematropica* 8, 32–34.

Rodriguez-Kabana, R., King, P.S. and Truelove, B. (1976) A seed treatment method for control of plant parasitic nematodes. *Proceedings of the American Phytopathological Society* 3, 297.

Routaray, B.N., Sahoo, H. and Das, S.N. (1986) Evaluation of green gram and black gram varieties against reniform nematode, *Rotylenchulus reniformis. Indian Journal of Nematology* 16, 27–29.

Ross, J.P. (1963) Seasonal variation of larval emergence from cysts of the soybean cyst nematode. *Phytopathology* 53, 608–609.

Ruthenberg, H. (1983) *Farming Systems in the Tropics*, 3rd edn. Clarendon Press, Oxford.

Saka, V.E. and Carter, C.C. (1987) Hosts and nonhosts of the root knot nematode *Meloidogyne incognita.* Cooperative Publication of the Department of Plant Pathology, North Carolina State University and the United States Agency for International Development, Raleigh, North Carolina.

Salam, M.A. and Khan, M.W. (1986) Reaction of some cultivars of pigeon pea against *Fusarium udum* and *Meloidogyne javanica. International Nematology Network Newsletter* 3, 16–17.

Saleh, H., Abu-Gharbieh, W.I. and Al-Banan, L. (1988) Effect of solarization combined with solar-heated water on *Meloidogyne javanica. Nematologica* 34, 290–291.

Sankaranarayanan, C. and Sundarababu, R. (1998) Effect of *Rhizobium* on the interaction of vesicular-arbuscular mycorrhizae and root-knot nematode on blackgram (*Vigna mungo*). *Nematologia Mediterranea* 26, 195–198.

Sarmah, B. and Sinha, A.K. (1995) Pathogenicity of *Meloidogyne incognita* on cowpea. *Plant Health* 1, 12–14.

Sartaj, A., Verma, T.N. and Alam, M.M. (1999) Inoculum potential of spiral nematode, *Helicotylenchus indicus* in relation to growth of chickpea. *Annals of Plant Protection Sciences* 7, 231–233.

Sasser, J.N. and Kirby, M. (1979) *Crop Cultivars Resistant to Root Knot Nematodes,* Meloidogyne *Species.* Department of Plant Pathology, North Carolina State University, Raleigh, North Carolina.

Sasser, J.N. and Hartman, K.M. (1985) Evaluation of some pigeonpea lines for resistance to the root knot nematodes, *Meloidogyne* spp. *International Pigeonpea Newsletter* 4, 44–45.

Saxena, M.C. (1987) Agronomy of chickpea. In: Saxena, M.C. and Singh, K.B. (eds) *The Chickpea.* CAB International, Wallingford, UK, pp. 207–232.

Saxena, M.C., Greco, N. and Di Vito, M. (1992) Control of *Heterodera ciceri* by crop rotation. *Nematologia Mediterranea* 20, 75–78.

Schmitt, D.P. and Barker, K.R. (1981) Damage and reproduction potentials of *Pratylenchus brachyurus* and *P. penetrans* on soybean. *Journal of Nematology* 13, 327–332.

Schmitt, D.P. and Noel, G.R. (1984) Nematode parasites of soybeans. In: Nickel, W.R. (ed.) *Plant and Insect Nematodes.* Marcel Dekker, New York, pp. 13–59.

Schreiber, E.R. (1977) Lebensweise, Bedeutung und Bekämpfungsmöglichkeiten von *Ditylenchus dipsaci* (Kühn) Filipjev an Ackerbohne (*Vicia faba* L.) in Marokko. Dissertation, Technischen Universität, Berlin, D83 No. 79.

Seinhorst, J.W. (1956a) Population studies on stem eelworms (*Ditylenchus dipsaci*). *Nematologica* 1, 159–164.

Seinhorst, J.W. (1956b) Biologische Rassen van het stengelaaltje *Ditylenchus dipsaci* (Kühn) Filipjev en hun Waardplaten. *Tijdskrift voor Plantenziekten* 62, 179–188.

Sellami, S. (1998) Importence der nématode des tiges (*Ditylenchus dipsaci*) sur fève en Algerie. In: Tivoli, B. and Caubel, G. (eds) *Les Legumineuses Alimentaires Mediterraneennes. Contraintes Biotiques et Potentialites de Developpement.* Rennes, France, *Colloques de INRA* 88, 157–163.

Sharma, N.K. and Sethi, C.L. (1975) Effects of initial inoculum levels of *Meloidogyne incognita* and *Heterodera cajani* on cowpea and on their population development. *Indian Journal of Nematology* 5, 148–154.

Sharma, N.K. and Sethi, C.L. (1976a) Interrelationship between *Meloidogyne incognita, Heterodera cajani* and *Rhizobium* sp. on cowpea (*Vigna sinensis* (L.) Savi). *Indian Journal of Nematology* 6, 117–123.

Sharma, N.K. and Sethi, C.L. (1976b) Reaction of certain cowpea varieties to *Meloidogyne incognita* and *Heterodera cajani. Indian Journal of Nematology* 6, 99–102.

Sharma, R.D. (1981) Pathogenicity of *Meloidogyne javanica* to bean (*Phaseolus vulgaris* L.). *Sociedade Brasilerira de Nematologia* 5, 137–144.

Sharma, R.D., Da Silva, D.B. and Castro, L.H.R. (1993) Efeito de *Helicotylenchus dihystera* sobre trigo e ervilha cultivados em solo provenientes de trêssistema des de preparo. *Nematologia Brasileira* 17, 85–95.

Sharma, S.B. (1985) *A World List of Nematode Pathogens Associated with Chickpea, Groundnut, Pearl Millet, Pigeon Pea, and Sorghum.* ICRISAT, Patancheru, Andhra Pradesh, India.

Sharma, S.B. and Swarup, G. (1984) *Cyst Forming Nematodes of India.* Cosmo Publications, New Delhi, India.

Sharma, S.B., Smith, D.H. and McDonald, D. (1992) Nematode constraints of chickpea and pigeonpea production in the semiarid tropics. *Plant Disease* 76, 868–874.

Sharma, S.B., Nene, Y.L., Reddy, M.V. and McDonald, D. (1993) Effect of *Heterodera cajani* on biomass and grain yield of pigeon pea on vertisol in pot and field experiments. *Plant Pathology* 42, 163–167.

Sharma, S.B., Mohiuddin, M., Reddy, M.V., Singh, O., Rego, T.J. and Singh, U. (1995) Tolerance in chickpea to *Meloidogyne javanica. Fundamental and Applied Nematology* 18, 197–203.

Sharma, S.B., Siddiqui, M.R., Fazul Rahaman, P., Ali, S.S. and Ansari, M.A. (1998) Description of *Heterodera swarupi* sp.n. (Nematoda: Heteroderidae), a parasite of chickpea in India. *International Journal of Nematology* 8, 111–116.

Sharma, S.B., Jain, K.C. and Lingaraju, S. (2000) Tolerance to reniform nematode (*Rotylenchulus reniformis*) race A in pigeonpea (*Cajanus cajan*) genotypes. *Annals of Applied Biology* 136, 247–252.

Sharma, S.K. and Singh, I. (2001) Effect of soil amendment with cruciferous oil seed crops on *Meloidogyne incognita. Plant Disease Research* 16, 277–279.

Siddiqui, I.A., Shaukat, S.S. and Ehteshamul, H.S. (2001a) Use of plant growth-promoting rhizobacteria (PGPR) and soil organic amendments for the management of root diseases complex of urdbean. *Agrobotanica* 54, 65–70.

Siddiqui, I.A., Ehteshamul, H.S., Zaki, M.J., Ghaffar, A. and Shaukat, S.S. (2001b) Control of chickpea root knot complex with *Pseudomonas aeruginosa* and selected pesticides. *International Journal of Nematology* 1, 85–91.

Siddiqui, M.R., Hyder, A., Pramila, G.A. and Gupta, P. (1999) Evaluation of greengram and blackgram germplasm against pigeonpea cyst nematode. *Journal of Mycology and Plant Pathology* 29, 135–136.

Siddiqui, Z.A. and Mahmood, I. (1993) Biological control of *Meloidogyne incognita* race 3 and *Macrophomina phaseolina* by *Paecylomices lilacinus* and *Bacillus subtilis* alone and in combination on chickpea. *Fundamental and Applied Nematology* 16, 215–218.

Siddiqui, Z.A. and Mahmood, I. (1994) Interaction of *Meloidogyne javanica, Rotylenchulus reniformis, Fusarium oxysporum* f.sp. *ciceri* and *Bradyrhizobium japonicum* on the wilt disease complex of chickpea. *Nematologia Mediterranea* 22, 135–140.

Siddiqui, Z.A. and Mahmood, I. (1995) Management of *Meloidogyne incog*nita race 3 and *Macrophomina phaseolina* by fungus filtrates and *Bacillus subtilis* in chickpea. *Fundamental and Applied Nematology* 18, 71–76.

Siddiqui, Z.A., Husain, S.I. and Fazal, M. (1991) Response of twenty varieties of pigeonpea (*Cajanus cajan* L.) against *Meloidogyne javanica. Current Nematology* 2, 139–142.

Siddiqui, Z.A., Mahmood, I. and Ansari, M.A. (1995) Effect of different levels of *Meloidogyne incognita* on the growth of pea in the presence and absence of *Rhizobium. Nematologia Mediterranea* 23, 249–251.

Siddiqui, Z.A., Mir, R.A. and Mahmood, I. (1999) Effects of *Meloidogyne incognita, Fusarium oxysporum* f.sp. *pisi, Rhizobium* sp., and different soil types on growth, chlorophyll, and carotenoid pigments of pea. *Israel Journal of Plant Sciences* 47, 251–256.

Sikora, R.A. (1972) Plant parasitic nematodes associated with sugarbeet and soybean in India. *Indian Journal of Nematology* 2, 200–201.

Sikora, R.A. (1981) Interaction between plant parasitic nematodes, plant roots, and vesicular arbuscular mycorrhizae. In: *Proceedings of Symposium on Biological and Chemical Interactions in the Rhizosphere,* Stockholm, Sweden, pp. 115–135.

Sikora, R.A. (1987) Analysis of plant parasitic nematodes affecting pasture and forage legumes in temperate semiarid regions. In: Saxena, M.C., Sikora, R.A. and Srivastava, J.P. (eds) *Nematodes Parasitic to Cereals and Legumes in Temperate Semi-arid Regions.* Proceedings of a Workshop, Larnaca, Cyprus. International Center for Agricultural Research in Dry Areas (ICARDA), Aleppo, Syria, pp. 13–28.

Sikora, R.A. and Maas, P.W.T. (1986) An analysis of the *Heterodera trifolii* complex and other species in the *schachtii* group attacking legumes. In: Lamberti, F. and Taylor, C.E. (eds) *Cyst Nematodes.* Plenum Press, New York, pp. 293–313.

Singh, B. and Dhawan, S.C. (1993) A new bacterial strain of *Pasteuria penetrans,* its host range and effect of temperature on spore attachment to second-stage juveniles of pigeon-pea cyst nematode, *Heterodera cajani. Indian Journal of Nematology* 20, 161–166.

Singh, B. and Dhawan, S.C. (1994) Effect of *Pasteuria penetrans* on the penetration and multiplication of *Heterodera cajani* in *Vigna unguiculata* roots. *Nematologia Mediterranea* 22, 159–161.

Singh, D.B. and Reddy, P.P. (1981) Influence of *Meloidogyne incognita* infestation on *Rhizobium* nodule formation in french bean. *Nematologia Mediterranea* 9, 1–5.

Singh, D.B. and Reddy, P.P. (1982) Cultivars of cowpea resistant to root knot nematode in India. *Tropical Pest Management* 28, 185–186.

Singh, D.B., Parvatha, R. and Rajendran, R. (1979) Reaction of certain winged bean varieties to the root knot nematode, *Meloidogyne incognita. Indian Journal of Nematology* 9, 43–45.
Singh, R.S. and Sitaramaiah, K. (1966) Incidence of root knot of okra and tomato in oil-cake amended soil. *Plant Disease Reporter* 5, 668–672.
Singh, R.S. and Sitaramaiah, K. (1971) Control of root knot through organic and inorganic amendments of soil: effect of sawdust and inorganic nitrogen. *Indian Journal of Nematology* 1, 80–84.
Sitaramaiah, K. (1984) *Plant Parasitic Nematodes of India.* Today and Tomorrow's Printers and Publishers, New Delhi, India.
Sitaramaiah, K., Singh, R.A., Singh, K.P. and Sikora, R.A. (1971) *Plant Parasitic and Soil Nematodes of India.* U.P. Agriculture University Experiment Station Bulletin, No. 3.
Slack, D.A. and Hamblen, M.L. (1961) The effect of various factors on larval emergence from cysts of *Heterodera glycines. Phytopathology* 51, 350.
Sosamma, V.K. and Koshy, P.K. (1977) Additional hosts of the burrowing nematode, *Radopholus similis,* infesting coconut palm in South India. *Plant Disease Reporter* 61, 760–761.
Srivastava, A.S., Upadhyay, K.D. and Singh, G. (1974) Effect of root knot nematode, *Meloidogyne javanica,* on gram crop. *Indian Journal of Nematology* 4, 248–251.
Stone, A.R. and Course, J.A. (1974) *Heterodera goettingiana. CIH Descriptions of Plant-parasitic Nematodes,* Set 4, No. 47. Commonwealth Institute of Helminthology, St Albans, UK.
Steiner, K.G. (1982) *Intercropping in Tropical Smallholder Agriculture with Special Reference to West Africa.* Gesellschaft für Technische Zusammenarbeit (GTZ) Publication No. 137.
Subramaniyan, S., Faziullah, A.K., Rajendran, G. and Vadivelu, S. (1997) Reaction of some mung, cowpea, lablab and tomato genotypes to the reniform and root knot nematodes. *Indian Journal of Nematology* 27, 130–131.
Suhail, A., Alam, M.M. and Anver, S. (2001) Reaction of pigeonpea accessions to root-knot nematode *Meloidogyne incognita* and reniform nematode *Rotylenchulus reniformis. International Chickpea and Pigeonpea Newsletter* 8, 41–42.
Sultan, M.S., Singh, I. and Sharma, S.K. (1985) Efficacy of some nematicides for the control of root knot nematode infesting mungo. *Nematologia Mediterranea* 13, 257–259.
Swain, P.K. and Kar, A.K. (1994) Disease complex involving *Meloidogyne incognita* and *Fusarium pallidoroseum* on black gram. *Orissa Journal of Agricultural Research* 7, 22–25.
Swarup, G., Prasad, S.K. and Raski, D.J. (1964) Some *Heterodera* species from India. *Plant Disease Reporter* 48, 235.
Sydenham, G.M., McSorley, R. and Dunn, R.A. (1997) Effects of temperature on resistance in *Phaseolus vulgaris* genotypes and on development of *Meloidogyne* species. *Journal of Nematology* 29, 90–103.
Taha, A.H.Y. and Kassab, A.S. (1980) Interrelations between *Meloidogyne javanica, Rotylenchulus reniformis* and *Rhizobium* sp. on *Vigna sinensis. Journal of Nematology* 12, 57–62.
Tarte, R. (1971) Estudios sobre la distribution y poblationes de nematodes en fincas hortalizas intensamente cultivades. *Turrialba* 21, 34–37.
Taylor, S. and Szot, D. (2001) First record of damage to canola caused by the oat race of stem nematode (*Ditylenchus dipsaci*). *Australasian Plant Pathology* 29, 153.
Tedford, E.C. and Inglis, D.A. (1999) Evaluation of legumes common to the Pacific Northwest as host for the pea cyst nematode, *Heterodera goettingiana. Journal of Nematology* 31, 155–163.
Teruya, R., Abe, J., Nagamine, M. and Nakamura, H. (1984) Nematological survey on winged bean fields in Okinawa with special reference to varietal difference in susceptibility to root knot nematodes. *Japan Agricultural Research Quarterly* 18, 142–157.
Thakar, N.A. and Patel, C.C. (1984) Screening of few cowpea varieties against reniform nematode, *Rotylenchulus reniformis* Linford and Oliveira, 1940. *Indian Journal of Nematology* 14, 204.
Thakar, N.A. and Yadav, B.S. (1985a) Comparative pathogenicity of the reniform nematode *Rotylenchulus reniformis* on a susceptible and a resistant variety of pigeonpea. *Indian Journal of Nematology* 15, 167–169.
Thakar, N.A. and Yadav, B.S. (1985b) Screening pigeonpeas for their resistance to reniform nematodes. *International Pigeonpea Newsletter* 4, 42–43.
Thames, W.H. and Heald, C.M. (1974) Chemical and cultural control of *Rotylenchulus reniformis* on cotton. *Plant Disease Reporter* 58, 337–341.
Thies, J.A. (2000) Reactions of regional cooperative Southernpea (Cowpea) breeding lines, cultivars, and standard checks to southern root knot nematode race 3, 1999. *Biological and Cultural Tests for Control of Plant Diseases* 15, 195.

Thomason, I.J. and McKinney, H.E. (1960) Reaction of cowpeas, *Vigna sinensis,* to root knot nematodes, *Meloidogyne* spp. *Plant Disease Reporter* 44, 51–53.
Thomason, I.J., Erwin, D.C. and Garber, M.J. (1959) The relationship of the root knot nematode, *Meloidogyne javanica,* to Fusarium wilt of cowpea. *Phytopathology* 49, 602–606.
Thomason, I.J., Rich, J.R. and Omelia, F.C. (1976) Pathology and histopathology of *Pratylenchus scribneri* infecting snap bean and lima bean. *Journal of Nematology* 8, 347–352.
Thompson, J.P., Greco, N., Eastwood, R., Sharma, S.B. and Scurrah, M. (2000) Integrated control of nematodes of cool season food legumes. In: Knight, R. (ed.) *Linking Research and Marketing Opportunities for Pulses in the 21st Century (Proceedings of the Third International Food Legumes Research Conference).* Kluwer Academic Publishers, Dordrecht, The Netherlands, pp. 491–506.
Timm, R.W. (1956) The reniform nematode, a root parasite of *Vicia faba. Pakistan Review of Agriculture* 2, 135.
Tisselli, I., Sinclair, J.B. and Hymowitz, T. (1980) Sources of resistance to selected fungal, bacterial, viral and nematode disease of soybeans. *International Agricultural Publications INTSOY,* Series No. 18.
Toledo, J.C., Sosa-Moss, C. and Zavaleta-Mejía, E. (1993) Gama de hospedero de cinco poblaciones mexicanas de *Nacobbus aberrans. Nematropica* 23, 105–108.
Toler, R.W., Tompson, S.S. and Barber, J.M. (1963) Cowpea (Southern pea) diseases in Georgia, 1961–1962. *Plant Disease Reporter* 47, 746–747.
Triviño, C.G. and Gowen, S.R. (1996) Deployment of *Pateuria penetrans* for the control of root knot nematodes in Ecuador. *Brighton Crop Protection Conference: Pest and Diseases* 1, 389–392.
Troccoli, A. and Di Vito, M. (2002) Root lesion and stem nematodes associated with faba bean in North Africa. *Nematologia Mediterranea* 30, 79–81.
Upadhyay, K.D. and Dwivedi, K. (1987) Analysis of crop losses in pea and gram due to *Meloidogyne incognita. International Nematology Network Newsletter* 4, 6–7.
Valdez, R.B. (1981) Screening tomato, eggplant and winged bean for resistance to *Meloidogyne incognita* and *M. javanica.* In: *Proceedings of the 3rd Research Planning Conference on Root-knot Nematodes,* Meloidogyne *spp.* July 1981, Jakarta, Indonesia. North Carolina State University Graphics, Raleigh, North Carolina, pp. 82–88.
Vats, R. and Dalal, M.R. (1997) Interaction between *Rotylenchulus reniformis* and *Fusarium oxysporum* f.sp. *pisi* on pea (*Pisum sativum* L.). *Annals of Biology (Ludhiana)* 13, 239–242.
Vats, R. and Dalal, M.R. (1998) Pathogenicity of reniform nematode, *Rotylenchulus reniformis,* on pea (*Pisum sativum* L.). *Indian Journal of Nematology* 28, 93–95.
Vijayalakshmi, K. and Goswami, B.K. (1986) Effect of seed treatments with neem cake and neem oil on the germination of moong (green gram) and its vulnerability to root knot nematodes. *International Nematology Network Newsletter* 3, 8–9.
Vijayalakshmi, M., Rajagopal, R., Mojumder, V., Raman, R. and Singh, P.P. (1999) Comparative efficacy of neem based nematicidal products for management of nematode pests of pulse crops. In: Saxena, R.C. (ed.) *Azadirachta-indica-A.-Jus.* pp. 223–234.
Vijayalakshmi, M., Mishra, S.D. and Mojumder, V. (2001) Effect of neem products as seed coating of pigeonpea seed on *Heterodera cajani. Pesticide Research Journal* 13, 103–105.
Vovlas, N., Greco, N. and Di Vito, M. (1985) *Heterodera ciceri* sp.n. (Nematoda: Heteroderidae) on *Cicer arietinum* L. from northem Syria. *Nematologia Mediterranea* 13, 239–252.
Vovlas, N., Castillo, P. and Troccoli, A. (1998) Histology of nodular tissue of three leguminous hosts infected by three root knot nematode species. *International Journal of Nematology* 8, 105–110.
Walia, K.K. and Gupta, D.C. (1986) Effect of the fungus, *Rhizoctonia bataticola* on the population development of pigeon-pea cyst nematode, *Heterodera cajani* on cowpea (*Vigna unguiculata*). *Indian Journal of Nematology* 16, 131–132.
Walia, R.K. and Seshadri, A.R. (1985a) Pathogenicity of the root-lesion nematode *Pratylenchus thornei* on chickpea. *International Chickpea Newsletter* 12, 31.
Walia, R.K. and Seshadri, A.R. (1985b) Chemical control of *Pratylenchus thornei* on chickpea through seed treatment. *International Chickpea Newsletter* 13, 32–34.
Wallace, H.R. (1962) Observations on the behaviour of *Ditylenchus dipsaci* in soil. *Nematologica* 7, 91–101.
Wang, K.H. and McSorley, R. (2002) Management of nematodes with cowpea cover crops. Available at: agroecology.ifas.ufl.edu/cowpea.htm
Wang, K.H., McSorley, R., Gallaher, R.N. and Tubbs, R.S. (2002) Effects of winter and fall cover crops on plant-parasitic nematode population development. In: van Santen, E. (ed.) *Making Conservation Tillage*

Conventional: Building a Future on 25 Years of Research. Proceedings of the 25th Annual Southern Conservation Tillage Conference for Sustainable Agriculture. Alabama Agricultural Experiment Station, Auburn University, Auburn, Alabama, pp. 152–155.

Wani, A.H. and Alam, M.M. (1995) Susceptibility of some pigeonpea varieties to the root-knot nematode, *Meloidogyne incognita*. *Annals of Plant Protection Sciences* 3, 182–183.

Ward, A., Mercer, S.L. and Howe, V. (1981) *Pest Control in Tropical Grain Legumes.* Centre for Overseas Pest Research, London.

Whitehead, A.G. (1969) The distribution of root knot nematodes (*Meloidogyne* spp.) in tropical Africa. *Nematologica* 15, 315–333.

Whitehead, A.G., Bromilow, R.H., Tite, D.J., Finch, P.H., Fraser, J.E. and French, E.M. (1979) Incorporation of granular nematicides in soil to control pea cyst-nematode, *Heterodera goettingiana*. *Annals of Applied Biology* 92, 81–91.

Wilson, W.R. and French, N. (1975) Population studies of stem eelworm (*Ditylenchus dipsaci*) in microplots, field plots and fields in north of England under different crop rotations. *Plant Pathology* 24, 176–183.

Winslow, R.D. (1954) Provisional lists of host plants of some root eelworm (*Heterodera* spp). *Annals of Applied Biology* 41, 591–605.

Wrather, J.A., Anand, S.C. and Dropkin, V.H. (1984) Soybean cyst nematode control. *Plant Disease* 68, 829–833.

Wyatt, J.E., Fassuliotis, G., Johnson, A.W., Hoffman, J.C. and Deakin, J.R. (1980) B4175 root knot nematode resistant snap bean breeding line. *HortScience* 15, 530.

Wyatt, J.E., Fassuliotis, G., Hoffman, J.C. and Deakin, J.R. (1983) 'Nemasnap' snap bean. *HortScience* 18, 776.

Yein, B.R., Singh, H. and Chabra, H.K. (1977) Effect of pesticides and fertilizers singly and in combination on the root knot nematode infesting mungo. *Indian Journal of Nematology* 7, 117–122.

Zaki, F.A. and Bhatti, D.S. (1986) Pathogenicity of pigeon-pea cyst nematode, *Heterodera cajani* Koshy, 1967 on some pulse crops. *Indian Journal of Nematology* 16, 30–35.

Zhang, D.S. (1995) Paulownia and pea as two additional hosts of the soybean nematode (*Heterodera glycines*). *Acta Phytopathologica Sinica* 25, 275–278.

9 Nematode Parasites of Vegetables*

Richard A. Sikora[1] and Emilio Fernández[2]
[1]*Soil Ecosystem Phytopathology and Nematology, Institut für Pflanzenkrankheiten, Universität Bonn, Nussallee 9, D-53115 Bonn, Germany;*
[2]*Instituto de Investigaciones de Sanidad Vegetal, Calle 110 no. 516, Entre BY5F Playa Miramar, Havana, Cuba*

Vegetables are one of the most important components of our daily diet as well as a high value cash crop for small and large growers alike. Vegetables, especially the leaf vegetables, are rich in protein, vitamins, minerals and fibre, and vegetables are a major source of protein in the humid tropics.

Mass transportation and modern processing has made many of these often highly perishable foods – which were previously only available on a seasonal basis in local markets or in restricted growing regions – readily available both nationally and internationally. Many vegetables that were once only of regional importance are now standard produce on markets throughout the world.

The major producers of vegetables in the tropics in order of importance are: Asia, Africa, South America and Central America. A significant amount of vegetable production also takes place in the subtropics on all continents. The types of vegetables grown are numerous, and full coverage is beyond the scope of this chapter. Many of the important crops that can be used as vegetables, for example taro and the leaves of cassava, have been discussed under root and tuber crops (Chapter 7). Similarly, many of the crops covered under food legumes (Chapter 8), such as garden pea, mung bean, broad bean and haricot bean, which also are often considered to be vegetables, will not be dealt with here. In Table 9.1, the overall level of production of a number of vegetable crops in these four tropical regions is tabulated.

In most areas of the world, vegetable consumption and production have expanded rapidly in the past two decades, with production significantly outpacing population growth in the four regions listed since 1990 when the first version of this book was published (Table 9.2). There have also been significant shifts in the amounts of specific vegetables produced in a region.

Surprisingly, the total amount of 'fresh vegetables' produced for the market as a percentage of total production has actually decreased slightly since 1990. This decrease is probably due to the loss of agricultural land to expanding metropolitan areas and the cost of long-distance transport and storage in tropical and subtropical climates (Table 9.3).

*A revision of the chapter originally written by Caspar Netscher and Richard A. Sikora.

Table 9.1. Area in 1000 ha, yield in metric tonnes and total production in 1000 metric tonnes for select vegetables in regions with large tropical and subtropical climates.[a]

	Africa			C. America			S. America			Asia		
Vegetable	Area	Yield	Product.	Area	Yield	Product.	Area	Yield	Product.	Area	Yield	Product.
Cabbages	86	17	1,485	22	14	327	59	9	498	2,260	20	44,909
Lettuce	15	20	299	13	21	271	15	13	193	587	19	11,144
Tomato	609	20	12,452	82	28	2,336	149	44	6,628	2,323	25	57,330
Cauliflower	13	20	241	22	12	253	5	16	75	632	19	12,117
Squash, gourds	227	8	1,788	39	12	473	5	137	7	858	18	26,469
Cucumbers and gherkins	4	16	72	1,729	18	3,167	94	13	717	857	14	11,557
Aubergines	46	19	940	3	24	60	1	19	9	1,506	17	26,000
Spinach	4	17	58	2	11	20	1	17	13	702	14	9,869
Chillies and peppers	268	8	1,989	146	13	1,814	29	14	397	970	15	14,056
Green onions and shallots	38	13	466	45	25	1,131	22	5	113	106	20	2,126
Dry onions	281	14	4,012	19	14	260	160	21	3,416	1,971	17	32,575
Garlic	32	12	367	6	8	47	45	8	346	902	12	10,722
Carrots	74	13	952	18	25	432	46	21	935	507	19	9,749
Maize	375	4	1,413	19	10	186	86	8	704	133	6	790

[a]From Anonymous (2003) FAOSTAT database at: appjs.fao.org/faostat

Increased production is associated with major advances in production and processing technology. In addition, modern breeding methods supported by new molecular techniques are making major strides to shorten the development time for cultivars with plant resistance to nematodes, insects, diseases and abiotic stresses, as well as in improving the nutritive value of vegetable crops.

Vegetable production in all tropical and subtropical areas is highly dependent on good nematode control. In many cases, nematode control is a pre-requisite to successful production. For example, vegetable crops grown for the fresh market have relied heavily on good control, with soil fumigation a standard practice. This is reflected by the fact that 75% of the soil fumigant methyl bromide, the most effective fumigant on the market, is used for soil treatment, with over 40% for vegetable production (Anonymous, 1998a,b).

Cultivation Techniques

Depending on demographic structure and economic development of a region, vegetable production in the subtropics and tropics varies from gathering of fruits, leaves and tubers found amongst the natural vegetation and various forms of multiple cropping to large-scale highly technical commercial field production.

The increase in the importance of vegetables is especially evident in countries with rapidly expanding populations, e.g. Africa and Asia, where large amounts of land near urban centres are devoted to vegetable production and where production since 1990 has expanded by 32 and 50%, respectively (Table 9.2).

Protected cultivation of vegetables in these peri-urban areas using plastic mulches, tunnels or plastic greenhouses has expanded significantly in many countries in the tropics and subtropics both for domestic use and for export to large and often lucrative overseas markets. The area in protected cultivation has increased drastically in many countries in the past 15 years. The largest greenhouse producing area in the world is in the Mediterranean region of Europe, with 100,000 ha of vegetables grown in greenhouses and 299,879

Table 9.2. Comparison of regional population and changes in total area, yield and production for all vegetables and melons between 1990 and 2002 in regions with large tropical and subtropical climates and percentage increase.[a]

	Population (number)			Area (1000 ha)			Yield (t/ha)			Production (1000 t)		
	1990	2002	%	1990	2002	%	1990	2002	%	1990	2002	%
Asia[b]	3,100,917	3,775,948	18	19,293	34,239	44	15	17	22	299,003	595,319	50
S. America	296,170	357,329	17	1,143	1,331	14	13	15	13	14,291	20,267	29
Africa	622,440	832,089	25	3,745	4,906	24	9	10	10	33,130	48,465	32
C. America	111,449	139,941	20	571	691	17	13	16	19	7,676	11,264	32

[a]From Anonymous (2003) FAOSTAT database at: appjs.fao.org/faostat
[b]Data for Asia only from 1992 onwards available.

Table 9.3. Comparison of fresh vegetable production as a percentage of total vegetable production between 1990 and 2002 in 1000 metric tonnes.[a]

	1990			2002		
	Total	Fresh	Percentage	Total	Fresh	Percentage
Asia	299,003	116,787	39	595,319	206,120	35
S. America	14,291	2,934	21	20,267	3,485	17
Africa	33,130	9,570	29	48,465	12,408	26
C. America	7,676	421	6	11,264	509	5

[a]From Anonymous (2003) FAOSTAT database at: appjs.fao.org/faostat

ha produced under plastic tunnels and mulches, with Spain having 46,000 ha and Italy 61,775 ha under greenhouse production. Japan, China and Turkey as well as many countries in North Africa also have significant areas under protected cultivation (Hanan, 1998). It is important to note that 5000 ha of the greenhouse production is in soilless culture (Cantliffe and Vansickle, 2003).

Peri-urban agriculture, or production in and around large metropolitan areas, has expanded rapidly to meet the demands for fresh vegetables in large urban centres worldwide. In large-scale peri-urban and intensive commercial production operations, plant protection is often highly developed and uses cutting edge technology, while in developing countries small resource-limited growers often lack coordinated plant protection support, leading to insufficient crop pest management.

Nematodes of Vegetables

Plant parasitic nematodes are an extremely important limiting factor in vegetable production, and in many areas a major factor requiring extensive use of pesticides. The role plant parasitic nematodes play in limiting vegetable production, however, depends to a large extent on the farming system employed. In general, nematodes will be less important under more extensive and varied growing systems typical of shifting cultivation and multiple intercrop farming systems in subsistence agriculture, as well as in widely spaced rotations of some commercial farming systems. However, nematodes are very important in more intensive production systems, for example in protected cultivation where mono-cropping is practised, or in field production systems where soil fumigation is followed by sequential cropping of a series of susceptible hosts (Taylor, 1976).

This was observed in Senegal where crops grown under local cropping conditions were not parasitized by root knot while neighbouring irrigated vegetable fields were heavily infested (Netscher, 1978). However, pressure on land and available resources has shifted production in many countries from small multiple crop production units toward more intensive production systems even on a small farm scale. Peri-urban production of vegetables for local city markets has increased to enormous levels in the past 20 years. Similarly, export of high value vegetables from tropical and subtropical production zones to satisfy the highly lucrative spring, autumn and winter markets in temperate zones around the world has had a major impact on vegetable production, and has resulted in development of major nematode problems. This has led to dependence on the use of soil fumigant nematicides, especially the highly effective and broad-spectrum fumigant methyl bromide. The loss of methyl bromide in 2005 due to environmental problems associated with ozone degradation (Anonymous, 1998a,b) has stimulated vast amounts of research to find effective alternatives.

From an ecological standpoint, crops grown in shifting cultivation and in the other multiple intercropping systems common to rural subtropical and tropical areas still have much in common with the natural flora. The distribution of important plant parasitic nematodes associated with the natural vegetation is clustered. The distributions of the species which survive the drastic shift to multiple inter-cropping are also heterogeneous even if polyphagous species are present. Extensive damage by nematodes, therefore, is extremely rare in the crops produced directly after clearing. Exceptions to the rule occur in those instances where nematode-infested planting material in the forms of seedlings or tubers is used for planting (Bridge, 1987).

Nematode infestations are promoted by the lack of quality nematode- and disease-free planting material. In many cases, these seedlings are produced under suboptimal conditions and are often infested with nematodes, insects and diseases (Singh *et al.*, 2000). Since commercial nurseries producing high quality vegetable seedlings often do not exist in Africa and Asia, local farmers using traditional methods produce their own planting material. The rule and not the exception is poor quality seedlings, infested with pests and diseases.

Multiple intercropping systems, although initially reflecting the natural flora, will promote nematode population build-up with time. The extent of the increase will depend on the type of nematodes initially present and on the percentage of susceptible plants per unit area (Noe and Sikora, 1990). Surveys in Niger and Benin showed that seedlings in small subsistence grower's nursery beds are often infested with root knot (R.A. Sikora, Bonn, 2004, unpublished data), and therefore within a very short time span the entire farm is threatened.

Damage intensity usually increases slowly with time in the multiple intercropping system, as compared with the rapid increase in damage encountered in large-scale vegetable production where monoculture or near monoculture is practised. Large differences also exist between the plant parasitic nematode communities of tropical and temperate regions where vegetable crops have been recorded as a host for at least one of the most frequently occurring species of root knot nematodes, *Meloidogyne incognita*, *M. javanica* and *M. arenaria*. Important temperate parasites such as *Ditylenchus dipsaci* and species of *Heterodera* are only of local importance in the warm tropics, but can be a problem in the cooler seasons in the subtropics and on vegetables grown at higher altitudes. Conversely, root knot nematodes that predominate in tropical regions are uncommon in temperate regions (Taylor and Sasser, 1978). Greater crop damage is to be expected in warmer regions and in summer crops than in cooler growing regions or in the upland tropics (Noe and Sikora, 1990).

Root knot nematodes, which increase to damaging levels within a few seasons under susceptible crops, are so common in subtropical and tropical vegetable production that frequently they are taken to represent 'nematodes' in general. Other economically important nematode species, in particular cyst nematodes but also *Rotylenchulus reniformis* and *Paratrichodorus minor*, are simply overlooked, because of a lack of distinct symptoms, and are often neglected by plant protection agencies. Nematodes such as *Heterodera schachtii*, *Nacobbus aberrans*, *Belonolaimus longicaudatus*, *Xiphinema* spp. and *Tylenchorhynchus brassicae* have, however, been shown to be serious pests. Feldmesser *et al.* (1971) estimated that loss in yield caused by all plant parasitic nematodes on 24 vegetables crops in the USA was approximately 11%.

Meloidogyne

Although over 90 species of *Meloidogyne* have been described to date, four species are of particular economic importance to vegetable production, *M. incognita*, *M. javanica*, *M. arenaria* and *M. hapla*. Out of 1000 root knot populations collected in 75 countries, 53% were identified as *M. incognita*, 30% as *M. javanica*, 8% as *M. arenaria*, 8% as *M. hapla* and 2% as *M. exigua* or other species (Taylor and Sasser, 1978).

M. incognita, *M. javanica*, *M. arenaria* and *M. hapla* have the widest host ranges. *M. incognita* and *M. javanica* are commonly found in the tropics, whereas *M. arenaria*, which is also found sporadically in the tropics, is more common in the subtropics. *M. hapla*, a species common in the temperate regions, can occasionally be found in the cooler upland tropics. In this chapter, *M. incognita* var. *acrita* Chitwood, 1949, later promoted to specific rank (Esser *et al.*, 1976; Jepson, 1987), is synonymized with *M. incognita* (Triantaphyllou and Sasser, 1960). It is important to detect a number of new species, often in mixed populations with *M. incognita*, that will be economically important limiting factors in vegetable production in the future.

M. chitwoodi, which has a wide host range and attacks many vegetable crops, has been detected attacking vegetable crops in North and South America as well as in Africa and Europe (Fig. 9.1). The nematode causes severe galling often at the root tip, and will be important in cool season vegetables in the upland tropics if spread is not limited by quarantine. *M. chitwoodi* has been declared a quarantine pest in Europe where it causes severe damage to a broad spectrum of vegetable crops and in particular potato. The nematode can be disseminated on infected tubers or other forms of planting material and attacks a wide range of vegetable crops as well as cereal crops often used in rotations to control root knot. The nematode is discussed in detail in Chapter 6. Two races have been reported for this important species (Santo and Pinkerton, 1985) whereby most carrot cultivars are considered moderate to good hosts and lucerne a non-host of race 1, whereas most carrot cultivars are non-hosts and lucerne a good host for race 2.

M. mayaguensis, a new species first described from Puerto Rico (Rammah and Hirschmann, 1988), is now considered to be one of the most pathogenic root knot species known. It has been detected in Cuba, Senegal, South Africa and Brazil, and most recently in Florida (Fig. 9.2). Because *M. mayaguensis* reacts similarly to *M. incognita* race 4 in standard differential host tests (Brito *et al.*, 2004a), its presence and importance may have been underestimated. Re-examination of *M. incognita* race 4 populations using modern molecular techniques may be required to obtain proper identification of this new species. Important is the fact that the nematode has the ability to overcome a number of genes that code for nematode resistance: the *Mi* gene in tomato, the *N* gene in pepper (Brito *et al.*, 2004b) as well as genes for resistance in soybean and sweet potato (Brito *et al.*, 2004d). The species has a wide host range, attacking, amongst others, bell pepper, tomato, beet, cabbage, broccoli, aubergine, celery, parsley, watermelon and pumpkin as well as tobacco, guava and coffee (Rodriguez, 2000; Anonymous, 2001; Rodriguez *et al.*, 2003). Low reproduction was detected on garden bean, horse bean and potato. In Brazil, cowpea and *Crotalaria juncea*, the latter often considered a root knot antagonistic plant, were also shown to be susceptible, whereas groundnut (peanut), maize and *Crotalaria spectabilis* were non-hosts (Guimaraes *et al.*, 2003). In Florida, cultivars of carrot and collard also were shown to be non-hosts (Brito *et al.*, 2004c).

M. floridensis is a new species described from Florida that was originally considered to be *M. incognita* (Nyczepir *et al.*, 1998; Handoo *et al.*, 2004). The nematode reproduces on root knot-resistant peach rootstocks and has been shown to parasitize tomato, watermelon and cotton, but not tobacco, pepper, groundnut, verbena, aubergine, squash or basil (Kokalis-Burelle and Nyczepir, 2004). The importance of this new species on a worldwide basis is unknown, but its similarity to *M. incognita* indicates that it may prove to be widespread. The main species of *Meloidogyne* found parasitizing vegetables are listed by crop in Table 9.4.

Symptoms of damage

The presence of galls on the root system is the primary symptom associated with *Meloidogyne* infection. In galls formed by

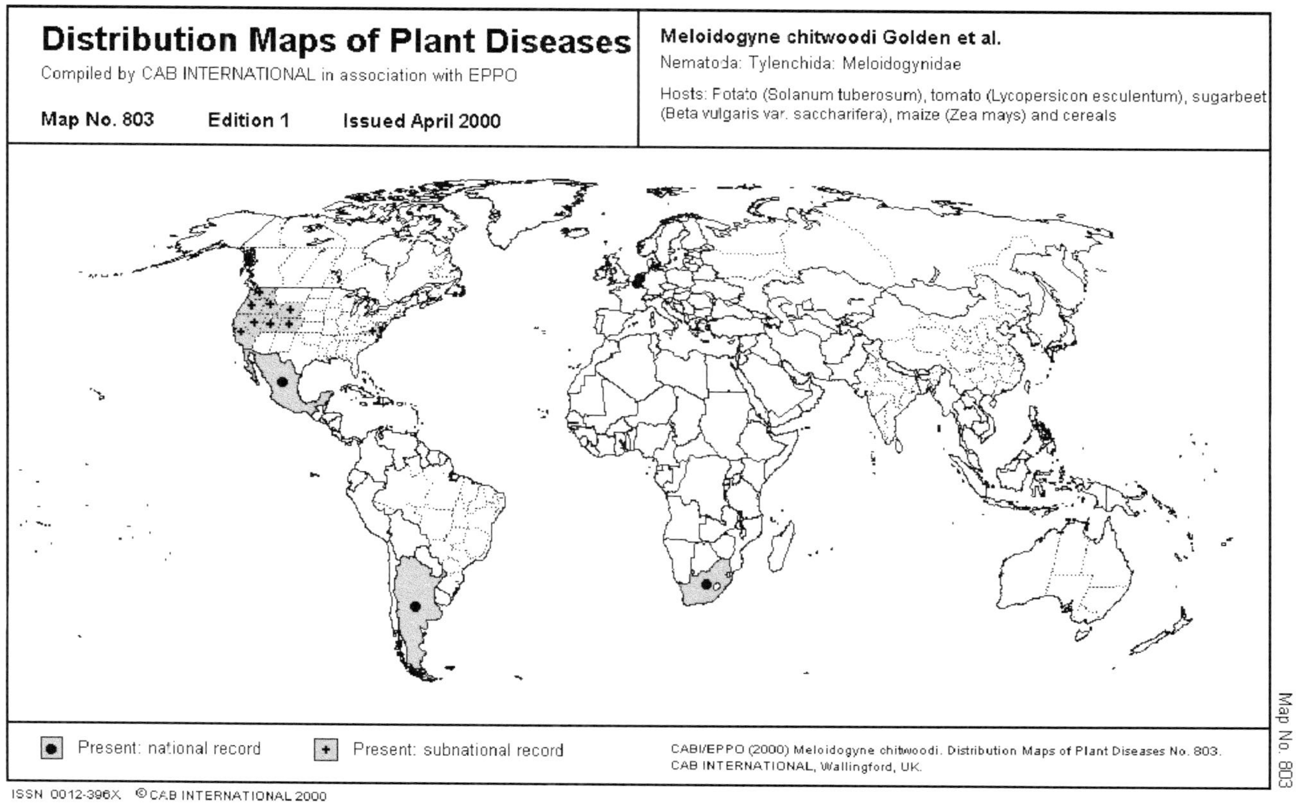

Fig. 9.1. Distribution map of *Meloidogyne chitwoodi* (Anonymous, 2001).

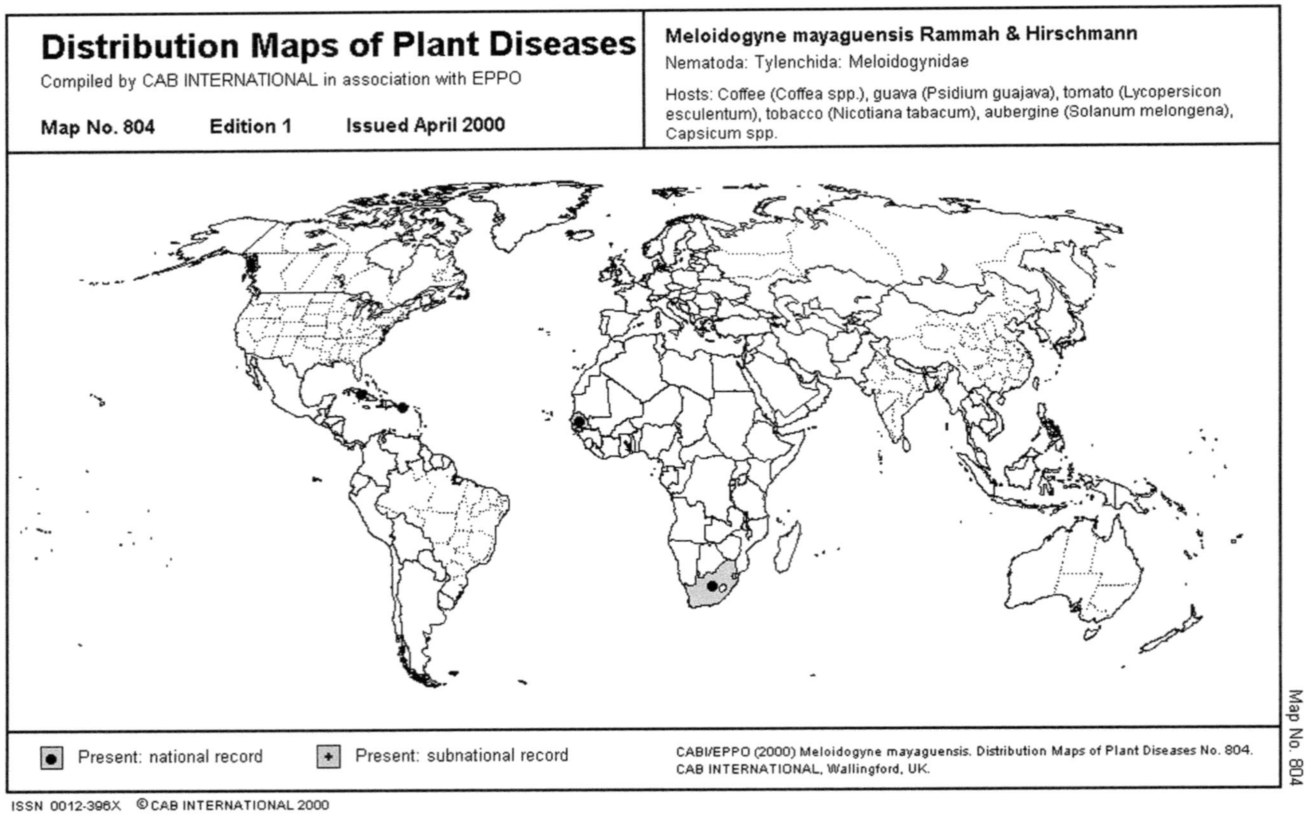

Fig. 9.2. Distribution map of *Meloidogyne mayaguensis* (Anonymous, 2001).

Table 9.4. Root knot nematodes, *Meloidogyne* species, associated with major vegetable crops in the subtropics and tropics.

		arenaria	*incognita*	*javanica*	*chitwoodi*	*hapla*	*mayanguensis*	*floridensis*
Allium asacolonium	Shallot		•	•				
A. cepa	Onion	•	•	•	•			
A. porrum	Leek		•	•				
A. sativum	Garlic		•					
A. schocnoprasum	Chives			•				
Amaranthus hybridus	Spinach (bajem)		•	•				
A. viridis	African spinach			•				
Apium graveolens	Celery		•	•			•	
Basella alba	Spinach		•	•			•	
Beta vulgaris	Beetroot	•	•		•	•	•	
Brassica chinensis	Chinese cabbage		•					
B. nigra	Black mustard	•						
B. oleracea var. *acephale*	Kaie	•		•				•
B. oleracea var. *botrytis*	Cauliflower	•		•			•	
B. oleracea var. *capitata*	Cabbage	•	•	•		•	•	
Capsicum annuum	Sweet pepper, chilli	•	•				•	•
C. frutescens	Cayenne pepper	•	•	•				
Celosia argentea	African spinach	•	•	•		•		
Citrullis vulgaris	Watermelon	•	•	•			•	
Cucumis meta	Melon	•	•	•				
C. sativus	Cucumber	•	•	•				•
Cucurbita maxima	Squash	•	•	•				•
C. pepo	Pumpkin	•	•	•			•	
Daucus carota	Carrot	•	•	•	•	•		
Ipomea reptans	'Spinach' (kangkung)					•		
Lactuca sativus	Lettuce	•	•	•		•	•	
Lagenaria siceraria	Bottle gourd	•		•				
L. vulgaris	Calabash	•	•	•				
Luffa cylindrica	Spange gourd		•	•				
Lycopersicon esculentum	Tomato	•	•	•	•		•	•
Momordica charantia	Balsam pear	•	•				•	
Petroselinum crispum	Parsley	•	•			•	•	•
Sechium edule	Chayotte			•				
Solanum melongena	Aubergine	•	•	•			•	•
S. tuberosum	Potato	•	•	•	•	•	•	
S. nigrum	Black nightshade	•	•	•				

one female, a swelling of the central cylinder, highly deformed vascular elements and the spherical part of the female surrounded by the cortical parenchyma can be easily observed at low magnification in stained roots (Plate 9A). The size and form of the gall depend on the species involved, the number of nematodes in the tissue, host and plant age. In cucurbits, the roots react to the presence of *Meloidogyne* by the formation of extremely massive, fleshy galls (Fig. 9.3), whereas in most other vegetables, galls are small to large and firm (Fig. 9.4; Plate 9B). In the upland tropics where temperate species are often detected or where temperatures reduce the growth and number of life cycles of warm temperature species small galls develop (Fig. 9.5). Root knot infections in young plants often lead to hooking of the tap root due to the presence of females on one side of the cortex (Fig. 9.6; Plate 9A).

Of importance in diagnosis of infection is the fact that in some cases galls are not visible at all, for example with *M. artiellia*, where galling is not induced and the exposed females and egg masses resemble cysts on the root surface. This nematode is also a serious parasite on legumes (see Chapter 8).

Similarly, symptoms of root knot on monocotyledonous crops such as onion and leek are very discrete, the main symptom being the presence of the protruding egg masses on the root surface. In some instances in the past, these crops were actually considered to be non-hosts. Galls on sweet and chilli pepper as well as on sweetcorn also are frequently small.

The symptoms caused by *M. hapla* on vegetables differ from those produced by most other species in that only small, more or less spherical galls are produced with profuse root branching originating from the gall

Fig. 9.3. Massive galls produced by *Meloidogyne javanica* on cucurbit roots in India (R.A. Sikora).

Fig. 9.4. Typical large firm galls, here on tomato, as produced by species of *Meloidogyne* on most vegetable crops in the tropics and subtropics (R.A. Sikora).

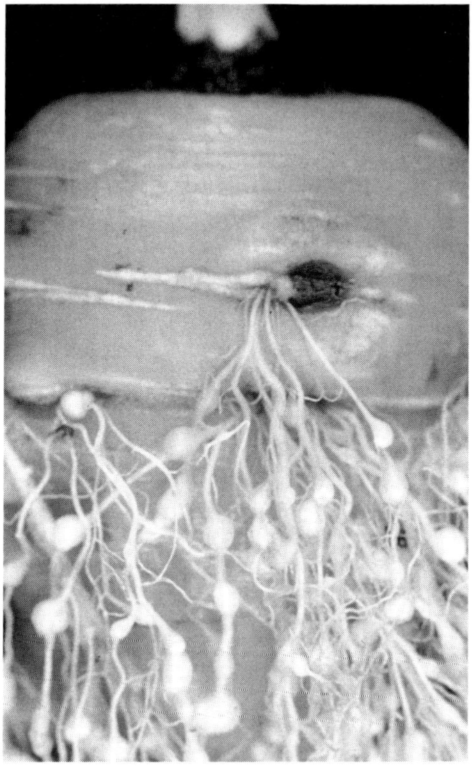

Fig. 9.5. Small galls on carrot produced by *Meloidogyne hapla*, that usually contain a single female (R.A. Sikora).

Fig. 9.6. Curved or hooked taproot on a cabbage seedling due to *Meloidogyne incognita* infection in Benin (R.A. Sikora).

tissue causing a 'bearded root' system (Plate 9C). When plants are severely infected by *Meloidogyne*, the normal root system is reduced to a limited number of severely galled roots with a completely disorganized vascular system. Rootlets are almost completely absent (Plate 9D). The roots are seriously hampered in their main functions of uptake and transport of water and nutrients. Plants wilt rapidly, especially under dry growing conditions, and are often stunted. Growth is retarded and leaves may be chlorotic (Plate 9E and F). In Thailand, wilting often occurs in non-chlorotic plants and has given rise to the term 'green wilt disease' (S. Sontirat, Bonn, 1990, personal communication). In cases where seedling infection has taken place, numerous plants die in the seedbed and seedlings do not survive transplanting. In those plants that do survive transplanting to the field, flowering and fruit production are strongly reduced. In addition to poor growth, severe losses in quality are caused by infection of the taproot which results in forking of the taproot and tuber deformation (Fig. 9.7; Plate 9C). Tuber crops such as carrot, beets, celery and radish can sustain immense losses due to poor marketability of deformed tubers. Tuber infection also makes long-term storage impossible in that these tap roots begin to rot due to fungal infection associated with nematode gall degradation. It should be noted that in some cases nematode reproduction and growth continue after harvest even in cool storage. Root tip galling and tuber galling are often associated with the presence of *M. chitwoodi* on vegetables which can cause severe root tip stunting (Fig. 9.8). Under high initial population densities or as the season advances, the galls are usually invaded by fungal pathogens and deleterious bacteria that induce severe root rotting. Such root rotting syndromes can cause even more yield

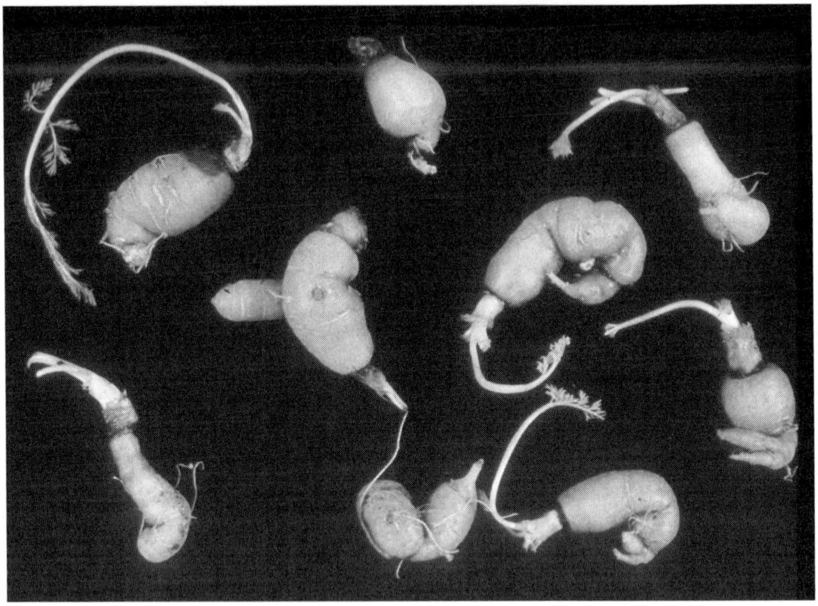

Fig. 9.7. Deformed taproots of carrot due to early root infection with *Meloidogyne incognita* (R.A.Sikora).

Fig. 9.8. Typical root tip galling due to *Meloidogyne chitwoodi* infection, here on leek, *Allium porrum* (courtesy of the Plant Protection Service Wageningen, The Netherlands).

loss than that caused when the nematode occurs alone (Plate 9D). In severe cases, the firm stele of the primary root is the only remnant of the original intact root system.

Biology and life cycle

There are optimum temperatures for different phases of the life cycle of *M. javanica* (Ferris and Van Gundy, 1979). The optimum temperature range for an Australian population was 25–30°C and that for a California population between 32 and 34°C. Dao (1970) demonstrated that populations adapt to local climatic conditions. Optimum temperatures for nematode development of most species important in the tropics correspond to those found in vegetable-growing regions, a factor ensuring serious root knot infestations. The distribution of major species found in the tropics and subtropics is determined by temperature. Temperature optima for *M. hapla* are at least 5°C lower than for the other major species in the tropics. *M. hapla* is therefore limited to the upland tropics and temperate growing regions. *M. incognita*, *M. javanica* and *M. arenaria* occur in areas with an average temperature of 36°C or lower in the warmest month. *M. hapla*, conversely, occurs in areas having a temperature as low as –15°C during the coldest month, but is limited to regions with an average high of less than 27°C during the warmest month (Taylor *et al.*, 1982). Cuadra (1983) reported that *M. incognita* could develop eight or more generations per year on tomato, with the length of the life cycle dependent on soil temperature. One cycle was completed in 19 days at 30.6°C versus 43 days at 21.8°C. Survival of eggs and juveniles of *M. javanica* decreased strongly when submitted to a temperature of 45°C for 3 h (Demeure, 1978).

Soil texture and structure are directly related to water-holding capacity and aeration and influence nematode survival, emergence and disease severity. Sikora (1989), studying paddy rice–vegetable cropping systems, detected severe root knot damage on vegetables grown in sandy soils after paddy, but a total absence in heavy clay soils after paddy. Soil type and soil pH have also been shown to influence nematode distribution (Taylor *et al.*, 1982). Soil type may also influence the types of crops grown, thereby affecting nematode distribution, population build-up and damage intensity. Nematode movement within a field or raised bed is also affected by soil type. For example, juveniles in sandy soils are able to move horizontally and vertically over distances of up to 75 cm in 9 days (Prot, 1977). Prot and Van Gundy (1981), however, found that migration decreased with increasing clay content of the soil, with no migration in soils with more than 30% clay. The effect of soil pH on root knot varies greatly. *Meloidogyne* species survive and reproduce at pH levels ranging from 4.0 to 8.0 (Ferris and Van Gundy, 1979). Emergence of *M. javanica* was greatest between 6.4 and 7.0 and inhibited below pH 5.2 (Wallace, 1966). Many tropical soils are very acid (pH of 4.5 is rather common), a fact that does not seem to prevent *Meloidogyne* build-up to extremely high densities.

Sasser (1954) proposed a method for the identification of the four major species, *M. incognita*, *M. javanica*, *M. arenaria* and *M. hapla*, based on the reaction of four hosts. The host differentials were expanded to include a tobacco cultivar with resistance to many *M. incognita* populations following the discovery of physiological races within *Meloidogyne* species (Taylor and Sasser, 1978).

It soon became evident that within species, great physiological variability existed. Riggs and Winstead (1959) demonstrated that when populations of *M. incognita* and *M. arenaria* were inoculated to resistant cultivars of tomato, enough selection pressure was exerted by the cultivar that within a short time resistant breaking populations called 'B races' were created. Sasser (1966) found that when different populations of the same species were inoculated to certain hosts, they often reacted differently. Thus certain populations of *M. incognita* parasitized cotton while others did not. In the same way, two categories of *M. arenaria* populations could be distinguished using groundnut as a differential host. When a resistant cultivar of tobacco, NC 95, was included in the host range, the situation became still more complicated; according to the reactions on the two differential hosts, cotton and tobacco, *M. incognita* populations could be split into four races. From these and other observations (Southards and Priest, 1973), it became evident that in contrast to other genera of parasitic nematodes, such as *Heterodera*, the identification of root knot did not automatically give exact indications of the host range of that population.

The use of host differentials (Hartman and Sasser, 1985) allows determination of the four main species and races of *Meloidogyne* (Table 9.5). Based on the results obtained with several hundred *Meloidogyne* populations, Sasser (1979a) concluded that there is considerable uniformity in host response and that resistance-breaking races are not common. In studies in Cuba with over 200 root knot populations from a wide spectrum of plants, *M. incognita* races 1, 2 and 3, *M. arenaria* race 2 as well as *M. javanica* and *M. hapla* were detected (Fernández *et al.*, 2001). However, Southards and Priest (1973) demonstrated that host differentials could react differently to populations of the same species. The development of resistance-breaking pathotypes on resistant tomato cultivars, as discussed later in this

Table 9.5. Differential host test identification of the most common *Meloidogyne* species and races (Hartman and Sasser, 1985).

	Tobacco	Cotton	Pepper	Watermelon	Groundnut	Tomato
M. incognita						
Race 1	–	–	+	+	–	+
Race 2	+	–	+	+	–	+
Race 3	–	+	+	+	–	+
Race 4	+	+	+	+	–	+
M. arenaria						
Race 1	+	–	+	+	+	+
Race 2	+	–	–	+	–	+
M. javanica	+	+	–	+	–	+
M. hapla	+	–	+	–	+	+

Cotton, cv. Deltapine; tobacco, cv. N.C.95; pepper, cv. Early California Wonder; watermelon, cv. Charleston Gray; groundnut, cv. Florunner; tomato, cv. Rutgers.
(–) Indicates a resistant host; (+) indicates a susceptible host.

chapter, further demonstrates the great genetic variability within this genus. Further complicating identification is the fact that many populations are composed of more than one species (Netscher, 1978; Fargette, 1987; Noling, 2003).

From one point of view, identification of *Meloidogyne* to species has little practical importance to vegetable growers, since most vegetables are susceptible to the major species encountered in the tropics. Furthermore, resistance is either non-existent in many crops or, if available, too expensive for most subsistence growers. Amaranthus, celosia, beetroot, Swiss chard, lettuce, most cabbages, cauliflower, most cucurbits, beans, peas, tomato, potato, aubergine, okra, carrot and many other vegetables have all been reported to be hosts of *M. arenaria*, *M. incognita* and *M. javanica* (see also Chapter 8 for other hosts). In some cases, these crops are also considered non-hosts or poor hosts depending on the root knot populations present. Accurate species identification of *Meloidogyne* can be important in the correct selection of non-host crops for rotation purposes or for use of a resistant cultivar when available.

Survival and means of dissemination

Root knot nematodes are obligate parasites, therefore, the absence of suitable host plants for prolonged periods ultimately leads to their disappearance. In the absence of susceptible crops, however, they often survive on weed hosts. In general, conditions favourable for plant growth will also be favourable for *Meloidogyne* reproduction. de Guiran and Demeure (1978) found that the optimum moisture level for emergence of *M. incognita* juveniles was slightly above field capacity. If under conditions optimum for emergence host plants are absent, juveniles will deplete their energy reserves in the soil and eventually die. Although nematode populations rapidly decline, a proportion of the eggs in the egg mass are in diapause and ensure perpetuation of the species (de Guiran, 1979; de Guiran and Villemin, 1980).

Under adverse environmental conditions, emergence and juvenile activity are reduced, thus increasing the chances of survival. Survival is influenced mainly by moisture content of the soil and to a lesser extent by temperature. High temperatures are often associated with low soil moisture content, whereas in the cases of waterlogged or inundated soils, high temperatures rarely occur. Juveniles and eggs survive periods of moisture stress in a state of anhydrobiosis. Egg masses collected from dry soils will contain empty eggs and anhydrobiotic eggs with second stage juveniles in diapause.

In field soil, the number of juveniles decreased from an initial infestation of approximately 10,000 nematodes/dm^3 of soil to zero after 12 weeks, when the soil was gradually dried (de Guiran, 1979). Similar effects were found in the dry season in Senegal (Demeure, 1977). Nematodes could not be detected in the top 20 cm of the soil at the end of the dry season. The number of nematodes in the 20–40 cm horizon, where available soil moisture was slightly higher, reached 0.9% of the initial population.

Dissemination takes place when juveniles or eggs are transported from infested to uninfested areas. Wind-borne dissemination of root knot nematodes has been reported (Orr and Newton, 1971) and occurs in regions where windstorms occur. This is probably a major factor during the monsoon season in Asia and on the Indian subcontinent. Spread with irrigation water has been demonstrated in the USA (Faulkner and Bolander, 1970) and in Spain (Tobar and Palacios, 1974), and definitely affects infield distribution. Root knot juveniles most probably spread into new fields by moving in runoff water into rivers and irrigation canals that then were tapped by farmers downstream. Dispersal in runoff water produced during rainstorms is another source of infield spread. Soil adhering to animals, footwear and agricultural implements also spreads infestations. Dispersal over great distances and over international borders occurs by movement of infested plants. The movement of the new species mentioned above therefore needs to be limited by good quarantine measures. Farms are often infested and damage maintained and intensified by growers using infested planting material.

Dissemination in peri-urban farms was observed to be due to the presence of root knot on old roots in household compost and in some commercially available organic amendments containing residues of infested weeds (Fernández *et al.*, 1994). Contamination, however, was not detected in animal-based organic matter such as earthworm humus or chicken manure.

Disease complexes

Many examples of disease complexes are known (Pitcher, 1963; Powell, 1971a,b; Taylor, 1979; Webster, 1985). Tomato plants wilt more quickly and can be killed when *Fusarium oxysporum* is present simultaneously (Plate 9F). Resistance in tomato cultivars to fungal wilt caused by *F. oxysporum* f.sp. *lycopersici* was reduced in the presence of *Meloidogyne* (Jenkins and Coursen, 1957; Sidhu and Webster, 1977). Conversely, Abawi and Barker (1984) did not detect any synergistic effect of *M. incognita* or Fusarium wilt on either resistant or susceptible tomato. Field studies on the importance of complex disease inter-relationships to crop production are scarce, and many of the experimental techniques used are considered inadequate (Wallace, 1983; Sikora and Carter, 1987). Many plants are susceptible to weak fungal pathogens only in the seedling stage. However, when simultaneously present with *Meloidogyne*, these fungi may increase damage to mature plants.

Valdez (1978) reported that damage to the root system caused by root knot nematode attack was responsible for increases in the intensity of bacterial wilt. Wilt is known to be more severe in root knot nematode-resistant tomato and aubergine cultivars in the presence of the nematode, with wilt developing 1–3 weeks earlier than with the bacterium alone.

Bacterial wilt of tomato caused by *Ralstonia solancearum* was positively correlated with *M. javanica* infection in the field on the Island of Youth in Cuba (Stefanova and Plumas, 1988). *M. incognita* race 1 was shown to increase wilt caused by both *R. solanacearum* and *F. oxysporum* f.sp. *lycopersici* on resistant tomato cultivars when inoculated simultaneously (Chindo *et al.*, 1991). Similar results were obtained by Deberdt *et al.* (1999) but only at high versus low temperatures. Their results indicated that at least one gene governing part of the bacterial wilt resistance is closely linked or allelic to the *Mi* gene in tomatoes for root knot resistance.

An interaction between root knot nematodes and bacterial canker caused by *Corynebacterium michiganense* has also been reported (de Moura *et al.*, 1975). The weight of the roots and shoots of tomato plants was more strongly reduced when secondary microbial invasion existed following inoculation with *M. incognita* than when aseptic juveniles were added (Mayol and Bergeson, 1970). Furthermore, Van Gundy *et al.* (1977) demonstrated that leachings of nematode-infected plants applied to tomato inoculated with *Rhizoctonia* resulted in the appearance of severe rot (Plate 9D), when compared with the controls. This again shows the intricacy of microbial interactions with nematodes in the rhizosphere and their importance for root health. Suppression of such disease complexes, which are very common in the tropics, by the control of *Meloidogyne* could increase yields significantly.

Economic importance

Estimations of vegetable crop losses in the tropics (Sasser, 1979b) ranged from 17 to 20% on aubergine, 18 to 33% on melon and 24 to 38% on tomato. In intensive commercial production, where sequential cropping of one susceptible crop after another is practised with up to four crops per year, the lack of effective root knot control would lead to total crop failure. The role *Meloidogyne* plays in crop loss is also difficult to ascertain in cases where crops are suffering from simultaneous attack by fungi, viruses, insects and other nematodes, a situation very common in tropical countries. Nematicide trials have been used to demonstrate losses associated with *M. incognita* infestations on a number of crops (Lamberti, 1979b). Crop loss due to this nematode ranged from 30 to 60% on aubergine and 50% on cantaloupe and watermelon. Nematode damage in protected cultivation, where susceptible crops are repeatedly planted in the same soil, often in the same elevated bed, is severe. Vegetable production in protected cultivation is often highly dependent on fumigant nematicides. In a survey of vegetable crops grown in plastic polytunnel greenhouses along coastal areas of north-east Spain, 50% of the 66 sites surveyed were infested with *M. incognita*, *M. arenaria* or *M. javanica* (Verdejo *et al.*, 1997).

In the USA, yield on plots infested with *M. incognita* and treated with DD-MENCS and planted with beans, summer squash, okra or cucumber increased 128, 180, 507 and 1175%, respectively (Johnson, 1985). These figures demonstrate clearly the economic impact these nematodes have on vegetable production in intensive agricultural production systems. Root knot nematodes, when present, also cause severe crop loss in multiple cropping systems even at the subsistence farming level. Crop loss assessment under these conditions is lacking and is needed to demonstrate the true impact of nematodes on vegetable production in small-scale subsistence farming systems (Noe and Sikora, 1990).

Economic threshold level

M. arenaria and *M. incognita* tolerance limits, or the population density at which damage is first observed, vary greatly with vegetable crop (Seinhorst, 1965; Barker and Olthof, 1976; Barker *et al.*, 1985; Di Vito *et al.*, 1986; Ferris *et al.*, 1986). The relationship between initial population densities of *M. incognita* race 1 and yield of susceptible and resistant tomato showed a tolerance limit of 0.55 eggs and juveniles/cm^3 soil for both types of tomato, whereas for artichoke seedlings the tolerance limit was 1.1, and for cabbage 0.5 (Di Vito *et al.*, 1991a,b). The wide variation in tolerance limits reflects the great difference in plant response to nematode infection as well as the influence of soil type and environmental conditions on disease development and severity (Ferris *et al.*, 1986).

In the San Joaquin Valley of California, USA, the number of juveniles in samples taken from sandy loam soils has been used for estimating potential yield loss in processing tomato production areas (Table 9.6). These figures are given here to be used as guidelines for estimating possible loss in other growing regions. Environmental fac-

Table 9.6. Effect of root knot nematode populations on processing tomato yield in San Joaquin Valley, California, USA, in a sandy loam soil (Anonymous, 1985).

Number of root knot juveniles/kg soil		
Autumn samples	Spring samples	Percentage of normal yield
≤ 160	≤ 25	100
310	50	98
620	100	95
940	150	91
1250	200	88
1560	250	85
1870	300	82
2190	350	79
2500	400	77
2810	450	74
3120	500	72
3440	550	69
3750	600	67
4060	650	65
4370	700	63
4690	750	61
5000	800	60
5310	850	58
5620	900	56
5920	950	55
6250	1000	53

tors, soil types and cropping sequences will affect damage threshold levels; therefore, one should be cautious when using these figures.

Techniques and strategies of root knot management

The variation in the manner in which vegetables are grown, that ranges from large-scale commercial production systems to shifting cultivation, prevents the development of one management strategy applicable to all situations. For example, the subsistence farmer frequently utilizes a mixture of local crops and cultivars of a crop to ensure that he or she has a minimum yield and usually will not or cannot follow modern management recommendations. These farmers often will not use a nematicide for economic reasons, and will not grow an unfamiliar nematode-resistant cultivar or do not have access to such planting material. On the other hand, a commercial plantation manager will not hesitate to utilize resistant cultivars or nematicides to protect a valuable crop (Radewald *et al.*, 1987; Noling and Becker, 1994; Noling, 2003). In the first case, crop improvement is more difficult or even impossible to implement; in the latter case, it is available to most growers having access to the technology.

A number of strategic reviews have been published that concentrate on specific regions or on nematode management in vegetable production (Johnson and Fassuliotis, 1984; Netscher and Sikora, 1990; Noling and Becker, 1994; Johnson, 1998; Sikora, 2002) and should be referred to for additional information. It should be noted that many of the techniques used for control of *Meloidogyne* on vegetables are used to control other plant parasitic nematodes affecting a wide array of crops (Nickel, 1984; Brown and Kerry, 1987; Barker *et al.*, 1998; Whitehead, 1998). This is especially important where multiple species of economically important nematodes affect the vegetable crop (Anonymous, 2004). In addition, many of these tools also limit infections by other soil-borne pests and disease including weeds, e.g. soil fumigation. To be effective, however, it is absolutely necessary to combine as many components as possible into a management system.

In commercial production of high value fresh vegetables, reliance on fumigant and non-fumigant nematicides is still the preferred method of management, especially where more than one vegetable crop is grown sequentially per year. In many cases, methyl bromide is used because of its broad spectrum of activity toward a wide range of pests and diseases, effective root knot nematode control, the short interval between treatment and planting and the increased number of crops per season it allows. However, with the inevitable loss of methyl bromide in 2005 or thereafter, alternative nematode management strategies are required (Anonymous, 1998a) and either have been developed or are in the process of being developed (Anonymous, 1998b; Sikora, 2001; Sikora *et al.*, 2004). Conversely, commercial production of vegetables for canning or processing is not as dependent on fumigation.

In the multiple cropping system used in small-scale farm situations and in extensive field production, where the use of nematicides is either not economically feasible or non-existent, effective alternatives need to be logically selected for management programmes based on economics and reliability.

Once large populations of *Meloidogyne* have developed in a field, it is virtually impossible to eradicate them completely from the soil. It is also difficult to maintain populations at sufficiently low levels without the use of effective management tools used in a logical ordered system. For example, although *M. javanica* densities were reduced to low levels – following either two non-hosts, or a resistant cultivar or a poor host – and aubergine yield increased significantly, nematode population density rose to high levels at the end of the first season (Netscher, 1981a). Fumigants also need to be used annually for effective control of root knot in vegetable production systems devoid of other tools in the management system. Root knot nematode control is a 'never ending battle' and, with the loss of methyl bromide, it will be dependent on a thorough knowledge of many aspects of practical nematology.

Physical methods of nematode management

QUARANTINE. Control strategies should be preventive rather than curative in nature and aimed from the onset at preventing the build-up of high population densities. Quarantine, if practised correctly, can add greatly by preventing introduction of a pest into a country or local region. The introduction of economically important nematodes such as *Belonolaimus*, *Nacobbus* and *Radopholus*, as well as important species of root knot have been excluded in the past and can be avoided further by good quarantine.

Important to vegetable production is the recent detection of highly damaging species of root knot: *M. chitwoodi*, *M. mayaguensis* and *M. floridensis* (Rammah and Hirschmann, 1988; Handoo *et al.*, 2004). In order to protect local production, effective quarantine laws and of course border inspections are needed for all three species. The distribution of *M. chitwoodi* and *M. mayaguensis* is presented in Figs 9.1 and 9.2; *M. floridensis* is presently limited in distribution to Florida in the USA.

At the national level, monitoring systems can be used to prevent local spread of nematodes by close scrutiny of commercial vegetable nurseries and nurseries on large production farms. For example, in Cuba, soil and all organic amendments targeted for use in vegetable seedbeds and/or nurseries are sampled either by bioassays with indicator plants or through soil extraction in plant protection laboratories (E. Fernández, unpublished).

FALLOW. Bare fallow is an effective means of managing root knot especially when it can be used in the hot, dry summer months between crops where alternative weed hosts are seldom a problem (Johnson and Fassuliotis, 1984; Brown and Kerry, 1987; Netscher and Sikora, 1990). In areas where climate is characterized by a prolonged and severe hot, dry season, fallow during the dry season, with soil tillage to dry the soil, followed by non-hosts during the wet season will result in significant reductions in *Meloidogyne* populations (Duc, 1980). Johnson and Fassuliotis (1984) reported that effective control could be obtained using summer fallows in hot, dry weather in arid areas or by withholding irrigation. Preventing alternative weed hosts from growing by repeated ploughing and disking at 2–4 week intervals or through the use of broad-spectrum herbicides is also a necessity. Of course bare fallow has to be economical and acceptable to the grower; therefore, it is most effective when other control techniques, i.e. root destruction or tillage, are used simultaneously. Under some conditions, fallowing has given equal or better control than rotation with non-host or fallow with cover crops (Kinloch and Dunavin, 1993). The negative effects on soil conservation also limit the use of bare fallow in many countries.

ROOT DESTRUCTION. Because root knot can survive and reproduce on the roots left in the soil after harvest, galled roots should be eliminated by uprooting and destruction. The spread of the nematode to the follow-up crop will be retarded and the overall population density reduced. It has been estimated that, when soil temperatures are high, each month that the root system survives causes a tenfold increase in root knot nematode densities (Anonymous, 2004). Root knot, for example, can even survive and reproduce in excavated roots and tubers over many weeks in such crops as tomato, pepper and even in small pieces of sweet potato tubers.

SOIL TILLAGE. Johnson *et al.* (1983) reported that standard tillage practices did not have significant affects on nematode densities in intensive vegetable cropping systems. However, where economical, repeated tilling of the soil at regular intervals for 30 days during hot and dry seasons between crops can significantly reduce root knot nematode densities in the upper horizons due to desiccation of eggs and juveniles. Tillage also eliminates alternative weed host and volunteer plants from the previous crop (Johnson and Fassuliotis, 1984; Perez, 1990). Mounding up of solanaceous crops such as tomato and pepper with upper horizon soil 30 days after transplanting led to the production of adventitious roots on the buried stems that improved plant vigour and offset some root knot damage to the lower root system (E. Fernández, unpublished).

FLOODING. Root knot densities drop significantly when soils are flooded for prolonged periods of time and, therefore, often are not considered to be severe problems in the dry season in tropical regions where paddy rice is a normal component of the rotation system. Thames and Stauer (1953) demonstrated that constant flooding of rice fields for 3 months gives acceptable control of root knot nematode for two succeeding vegetable crops. Root knot nematode densities were lower on susceptible dry season crops in paddy rice rotations than in upland areas in the Philippines (Castillo *et al.*, 1976b).

Sikora (1989) showed that the degree of root knot damage to processing tomato crops in the Philippines was less severe in rotations of paddy rice–tomato than in rotations without paddy rice. The level of galling decreased significantly with increasing clay content of the soil, indicating that soil type plays an active role in population reduction under flooded conditions. Similar effects of paddy rice cropping patterns were noted in northern Java, Indonesia (C. Netscher, France, 1989, personal communication).

In Florida, flooding alternated with drying during the summer has been recommended for vegetables grown on muck soils to reduce root knot nematode densities, with crops grown in unflooded fields more frequently damaged (Overmann, 1964). Noling (2003) stated that alternating

2–3 week cycles of flooding with drying seems to be more effective than long, continuous flooding cycles.

Root knot juveniles are killed after exposure to anaerobic conditions that begin in the soil a few days after flooding (Padgham, 2003). However, the susceptibility of root knot eggs to anaerobic conditions over time has not been studied. The impact of intermittent flooding with different lengths of aerobic and anaerobic conditions needs to be examined to optimize control. Combining flooding and solarization has also been examined as a means of control (Sotomayor *et al.*, 1999). It should be noted that concerns about water conservation would limit the use of this management tool in some countries (Noling and Becker, 1994). In addition, availability of water and the ability to control water levels are also a limiting factor in many areas where vegetables are grown.

ORGANIC AMENDMENTS. For simplicity, organic amendments is used here to mean all incorporated organic material added to the soil, in most cases in a dried state. Organic amendments added as fresh crop residue and grown in standard rotations, e.g. break, cover, trap, antagonistic or green manure crops, are discussed below.

It is a well known fact that incorporation of large amounts of organic material into the soil reduces root knot densities. Oil cakes, coffee husks, neem, marigold leaves, crustacean skeletons, sawdust, urea, chicken manure and bagasse amongst others have been used with some success (Singh and Sitaramaiah, 1966, 1967; Sikora *et al.*, 1973a; Muller and Gooch, 1982; Stirling, 1991; Sikora, 1992). Control may be due to: (i) toxic compounds present in the organic material as in neem; (ii) non-toxic compounds such as residual sugar in bagasse; (iii) toxic metabolites produced during microbial degradation; or (iv) enhancement of nematode antagonists.

Chitin amendments have received much interest in the past as an organic amendment in that they stimulate the antagonistic potential in soil toward nematodes (Main *et al.*, 1982; Rodriguez-Kabana *et al.*, 1987; Spiegel *et al.*, 1987; Galper *et al.*, 1990). Chitin originating from the crustacean industry in combination with waste products from the paper industry also has been used to reduce root knot nematodes (Culbreath *et al.*, 1985).

Organic amendments can increase the number and density of antagonists of root knot nematodes thereby increasing the overall antagonistic potential in the soil (Sikora, 1990; Stirling, 1991; Fernández *et al.*, 1998b). For example, the application of organic plant revitalizers based on renewable compounds extracted from palm oil and first used in the detergent industry were shown to stimulate rhizosphere microbial activity up to 19-fold and simultaneously reduce root knot galling (Mulawarman *et al.*, 2000, 2001).

Neem-based oil cakes and related products have been studied intensely in India for control of root knot nematodes. They have been used alone (Singh and Sitaramaiah, 1966, 1967) and combined with biocontrol agents (Naik *et al.*, 1998), for example with *Pochonia chlamydosporia* (syn. *Verticillium chlamydosporium*) (Rao *et al.*, 1998a,b) or with *Trichoderma harzianum* and *Glomus fasciculatum* (Reddy *et al.*, 1998), with reports of significant levels of control. However, little has been done to demonstrate the true economics and practical impact of this technology at the grower level. Although the use of organic amendments for effective nematode control is often limited by availability and in some cases by the large quantities needed, they will reduce nematode population densities to different degrees. In addition to their effects on nematode density, organic amendments also improve soil structure and water-holding capacity, reduce diseases and limit weed growth which ultimately leads to a stronger plant and improved tolerance to nematode attack.

SOLARIZATION AND SOIL HEATING. The lethal temperature for control of plant parasitic nematodes is considered to be around 45°C. Heating the soil either with dry or steam heat has been used for many years in protected cultivation to manage root knot nematodes, but the high cost of heating oil has limited its use drastically.

Soil solarization with plastic mulches, which leads to the development of lethal temperatures in the soil, is being used in some countries for control of root knot and soil-borne diseases (Katan, 1980; Whitehead, 1998). The technique is most effective in regions where high levels of solar energy are available for long periods of time. However, the limited depth to which lethal heat actually penetrates into the soil often restricts control to the upper 5–10 cm layer. Therefore, besides solar energy, root architecture of the crop to be grown and the depth of root knot infestations are important in selecting this approach. Manipulating root growth so that the root system remains shallow and in the upper horizon through controlled drip irrigation might increase the effectiveness of solarization in the early stages of plant growth.

The use of solarization could be effective in contained raised bed production units, as used in many peri-urban production systems. In addition, black plastic mulch (Abu-Gharbieh *et al.*, 1987) as opposed to clear plastic has been shown to give effective control, and combining a mulch with solar-heated water supplied through drip irrigation increased lethal temperature soil penetration and nematode control (Saleh *et al.*, 1988; Abu-Gharbieh *et al.*, 1991). In many climatic regions and in subsistence agriculture, the costs of using plastic mulches are limiting factors except for eliminating nematodes from soil in seedbeds (Bridge, 1987).

Solarization applied in the summer in Morocco before the next tomato in plastic greenhouses led to a 99% reduction in *M. javanica* densities when compared with the controls (Eddaoudi and Ammati, 1995). Similar results were obtained in India following solarization for 6 weeks in the summer months, with reductions in *M. incognita* and *Pythium aphanidermatum* (Reddy *et al.*, 2001). When a resistant cultivar followed solarization, production per plant was improved twofold over the susceptible variety in solarized soil. Solarization for 40–60 days during the Mediterranean summer gave yield increases equal to that of methyl bromide treatment (Noto, 1994).

Solarization reduced root knot, *Verticillium* wilt and weeds in autumn crops in Florida, even though climatic conditions are not considered ideal for soil solarization (Overman and Jones, 1986). Similar results were obtained in Cuba in peri-urban agriculture and in small farm production units using solarization under suboptimum conditions between July and September (Fernández and Labrada, 1995). Whether the use of solarization under suboptimal conditions is always effective and economical needs to be ascertained for each situation.

Soil solarization combined with dazomet or calcium cyanamide gave good control of root knot and increased tomato yield (Fiume and Parisi, 1995). Similarly, solarization together with carbofuran increased tomato yields 96% and solarization with neem cake 52%, coupled with a significant reduction in *M. javanica* (Sharma *et al.*, 1996). Solarization for 2–4 weeks, combined with cadusafos or fenamiphos, was considered a sustainable alternative to methyl bromide fumigation in greenhouse tomato in Cyprus (Ioannou *et al.*, 2002).

BIOFUMIGATION. This term normally refers to suppression of soil-borne pests and pathogens by the release in soil of biocidal compounds, principally isothiocyanates, when glucosinolates in cruciferous crop residues are hydrolysed (Kirkegaard *et al.*, 1998). The loss of traditional soil fumigants generated interest in breeding brassicas such as canola or fodder crops for simultaneous use in pest and disease control. Soil amended with fresh or dried cruciferous residues, at 38°C day and 27°C night temperatures, reduced *M. incognita* galling by 95–100% after 7 days incubation, with a simultaneous reduction in *Sclerotium rolfsii* and *Pythium ultimum* in controlled environment tests (Stapleton *et al.*, 1998). It should be noted here that many cruciferous plants are good hosts of some important species of *Meloidogyne*.

The term biofumigation is now used more freely whenever volatile substances are produced through microbial degradation of organic amendments that results in significant toxic activity toward a soil-borne pest or disease (Anonymous, 1998b; Bellow *et al.*, 1998). Biofumigation under these circumstances is greatest when there is an optimum combination of organic matter, high soil temperatures and adequate moisture to promote microbial activity leading to toxin production. In tropical and subtropical production systems, plastic mulch and drip irrigation improve the effectiveness of biofumigation.

Transporting organic amendments to the field or incorporating cover crops that produce large amounts of biomass into the soil together with plastic mulch and/or drip irrigation can increase the level of control attained significantly. The release of toxic compounds already present in antagonistic plants used as amendments, e.g. neem, marigold and castor, or the production of toxic compounds due to microbial fermentation of nutrient-rich organic amendments, e.g. velvet bean, sunnhemp or elephant grass, should lead to significant levels of nematode control.

Biofumigation using fresh marigold as an amendment is used effectively in root knot management in protected cultivation in Morocco. Mature plants are incorporated into beds during the summer, the beds outfitted with drip irrigation and then covered with plastic mulch for solarization and biofumigation (R.A. Sikora and H. Kaak, unpublished). Stirling and Stirling (2003) suggested incorporating brassicas into the soil combined with irrigation to control *M. javanica* in ginger. Biofumigation was considered not feasible in non-irrigated pineapple soils also affected by root knot nematode. When poultry droppings or pelleted slaughterhouse waste at 1 kg/m^2 was applied to soil under plastic tunnels in Italy and in Sicily, Fusarium wilt was controlled at a depth of 15–30 cm after 36 days of biofumigation (Primo and Cartia, 2001). Control was also obtained in Uruguay with *M. incognita* following biofumigation with chicken manure and rice hulls (Leon *et al.*, 2000). Control due to any form of biofumigation is probably the result of multifaceted mechanisms including: (i) non-host or trap cropping depending on the host status of the plant used; (ii) lethal temperature due to solarization; (iii) nematicidal action of toxic by-products produced during the degradation of organic matter; and (iv) stimulation of antagonists in the soil after biofumigation.

Cropping-based nematode management systems

Crop management is designed to attain high yield while simultaneously reducing nematode, insect, disease and weed problems, reduce erosion and improve soil fertility. In the tropics and subtropics, vegetable production systems are extremely diverse, with production over a 12 month growing season varying in structure from: (i) sequential cropping of 2–5 susceptible vegetable crops in one field without a break crop; (ii) rotation of one or more vegetable crops with a non-host; (iii) production of one vegetable crop and one cover crop or a weed fallow; and (iv) multiple cropping with vegetables intercropped with non-host crops.

Each production system has different requirements when it comes to combating root knot nematode infestations. In addition, the rotation crops used by a grower are planted for different reasons, with the type of rotation crop varying greatly between the tropics and subtropics. Selection is often dependent on the main cash crop in the cropping system.

Rotation crops are used to:

- suppress weed growth;
- prevent soil erosion;
- improve soil organic matter levels;
- improve water-holding capacity;
- raise the nitrogen concentration;
- control nematodes.

Nematode control achieved with crop management is attained by mechanisms including: starvation, trap cropping, antagonism, stimulation of soil antagonistic potential and/or different degrees of biofumigation. Conversely, in commercial production, where fumigation is the backbone of the cropping system and sequential cropping of susceptible vegetable crops is practised, rotation may not even be considered as a tool.

HEALTHY TRANSPLANTS. All crop nematode management-related strategies are useless if transplants are infested with root knot, since early root infection leads to severe crop loss. Only nematode-free seedlings should be selected for transplanting. Nurseries must be free of root knot nematodes in order to reduce dissemination into root knot-free production areas. Seedbeds should be selected on sites which previously were not planted to host plants. To reduce contamination, wherever possible, seedbeds should be planted for dry season crops on land normally flooded during the wet season, e.g. in previous paddy fields (Bridge, 1987; Sikora, 1988).

Chemical disinfestation is a common and effective practice in large production operations, whereas other methods must be considered for subsistence farming. Fumigant nematicides could be used in nurseries even in the case of traditional farming systems, because of the small amount needed and low impact on the environment. It should be noted that root knot juveniles move up to 1 m in 7 days; therefore, if a raised bed is exposed at the bottom to underlying infested soil, infection of the seedling will occur.

Soil can be heated in drums or on old sheets of metal over open fires before being added to trays, plastic bags or pots formed from banana leaves for seedling production in subsistence agriculture. Solarization of small quantities of soil under sandwiched pieces of plastic can also be effective. Heating soil in direct sunlight and drying reduces root knot densities drastically and can be effective for small farmers (Fernández *et al.*, 1994).

Pouring boiling water on to the surface of beds is a seemingly impractical method but it is very effective in eliminating root knot nematodes from tomato seedbeds and is used by farmers in Bolivia. The water is heated on wood fires immediately alongside the seedbeds. It is a method recommended by Centro de Investigación Agrícola Tropical (CIAT), Santa Cruz, Bolivia (P. Franco, CIAT, 2003, personal communication).

Small growers also should be trained to identify galling on transplants in order to eliminate diseased plants as well as to identify infested soil. Selecting seedbeds on sites never planted to susceptible host plants would be optimum. However, care must be taken since weed hosts could be present in such sites.

The production of seedlings using floating tray technology will eliminate root knot infection if the potting substrate is nematode free. However, treatment of the water with other pesticides can be prohibitive. Seedlings for soilless culture-based protected cultivation of vegetables also need to be nematode free. The introduction of root knot into the closed water circulation system is not expected and therefore is often overlooked, which leads to rapid spread in the greenhouses. Control may require thorough cleaning of all containers and pipes with the need for new nematode-free seedlings. Control in the irrigation water has been attained with heat and UV radiation (Runia, 1995; Hallmann *et al.*, 2004a).

NON-HOST CROPS. Non-host crops are defined here as crops harvested for marketing purposes as opposed to cover crops used for soil conservation, animal grazing or direct nematode control. Rotation with non-host crops of any type is the most important technique used for root knot management worldwide (Nusbaum and Ferris, 1973; Netscher and Sikora, 1990; Barker, 1991; Rodriguez-Kabana, 1992; Johnson, 1998). Many types of rotations have been proposed to reduce the impact of root knot nematodes in vegetable cropping systems (Page, 1979; Johnson and Fassuliotis, 1984; Sikora *et al.*, 1988).

A number of rotations are used effectively in the tropics, especially in Asia, that are predominantly composed of cruciferous crops moderately resistant or tolerant to root knot nematodes, together with a smaller number of highly susceptible crops (Fig. 9.9). Rotations designed in this manner can be used effectively to reduce root knot nematode densities even in high intensity sequential plantings. Rotations using moderately resistant or tolerant crops together with highly susceptible vegetable crops have been used in Mauritania, Malawi, Bangladesh and Niger for control of root knot. Vegetables considered moderately susceptible or tolerant to root knot were: cabbage, cauliflower and onion in Mauritania (Netscher and Luc, 1974), all cruciferous crops, onion and leek in Malawi (Bridge and Page, 1977) and broccoli, cauliflower, cabbage and onion in Bangladesh. Amaranthus and chilli were considered resistant in Bangladesh (Page, 1979), and onion and amaranthus were moderately resistant in Niger (Sikora *et al.*, 1988). Fernández *et al.* (1998a) separated crops into four groups: very susceptible, tomato, aubergine, lettuce, melon, cucumber, squash, okra; moderately susceptible, cabbage, cauliflower; slightly susceptible, onion, garlic; and resistant, mint, sesame, sorghum (Fernández *et al.*, 1998a). Classification of crops by this general reaction to infection seems to be independent of the *Meloidogyne* species concerned, but can vary from one population of a species to another (Netscher, 1970).

Taking advantage of these differences, Kanwar and Bhatti (1993, 1994) recommended a rotation cycle dominated by vegetables for control of *M. javanica*: tomato, onion, resistant tomato and okra. They also suggested a 1 year rotation cycle based on tomato, garlic, ridge gourd (*Luffa acutangola*) for *M. javanica*, and a rotation of cauliflower, garlic and brown sarsan (*Brassica campestris* ssp. *oleifera*), the latter effective in reducing nematode densities. It should be noted that these differences in susceptibility to root knot have been used to improve biocontrol efficacy of *Pochonia chlamydosporia* (Bourne and Kerry, 1999; Bourne, 2001; Kerry and Hidalgo-Diaz, 2004).

Root knot nematodes, however, are extremely polyphagous, therefore, relatively few non-host plants are available for control through crop rotation. Unfortunately, there are many reports of *Meloidogyne* populations parasitizing plants which have been reported non-hosts, an important factor in developing rotation-based control systems (Netscher and Taylor, 1979). Groundnut, for example, is often considered a non-host of *M. incognita* and *M. javanica* (Netscher, 1975). However, it is attacked by *M. javanica* in Zimbabwe (Martin, 1956) and the USA (Minton *et al.*, 1969) and is tolerant to *M.*

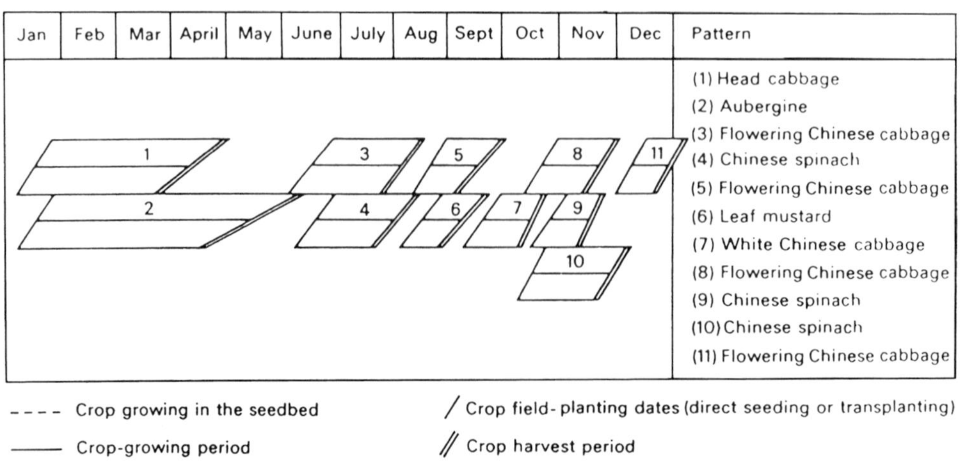

Fig. 9.9. Intensive vegetable rotation scheme with relay and sequential cropping of good and poor host plants often used in Asia (Ruthenberg, 1983, cited in Netscher and Sikora, 1990).

javanica in Bangladesh (Page, 1979). Therefore, recommended use in one country needs retesting in another.

Plants considered good host plants of a *Meloidogyne* species in one part of the world are not necessarily hosts to all populations of that species (Southards and Priest, 1973). Two races of *M. arenaria* were identified using groundut, previously considered a non-host, as a differential host (Sasser, 1966). Netscher (1970) showed that different populations within a species could be characterized by different virulence to a host. Lamberti (1979a) obtained similar results on tomato with 12 populations of *M. incognita* in southern Italy, with the level of galling differing up to fourfold on tomato depending on the crop origin of the initial population (Netscher and Sikora, 1990). Because of this large variation in host status within species of root knot, all crops being considered for rotation must be tested for host status to local populations before rotation schemes are recommended for the field. For example, there were differences in susceptibility in maize to *M. javanica*, with only eight of 34 tested cultivars actually having negative effects on population or a reproduction factor below 1 (Asmus *et al.*, 1995).

Care must be taken with regards to variation in nematode populations and to the composition of root knot species present in a field. Sometimes the *Meloidogyne* populations are composed of several species that may require different approaches for control. It should be noted that detection of species that make up less than 5% of the population is difficult. The detection of new species of root knot with host ranges very similar to that of old established species (*M. chitwoodi*, *M. mayaguensis* and *M. floridensis*) gives some indication of the diversity in populations being dealt with in the field. Therefore, low levels of genetic diversity may affect any crop in any field.

The fact that the minimum temperature required for *M. incognita* development in the root is significantly lower than the minimum 'activity threshold' of 18°C for *M. incognita* second stage juveniles (Roberts *et al.*, 1981) has been used to alter the date of planting for control of root knot. Changing the normal date of planting to coincide with low soil temperature was considered an important control tactic on carrots (Roberts, 1987) and could be used to limit nematode damage on vegetables in cool upland tropical regions.

Nut and shade trees used for food, windbreaks, building material or firewood vary in their status as root knot hosts. Neem (*Azadirachta indica*), cashew nut (*Anacardium occidentale*) and eucalyptus (*Eucalyptus camaldulensisare*) cultivars are usually considered resistant to *Meloidogyne* (Netscher, 1981b). Conversely, local trees as well as plants being selected for windbreaks, e.g. the baobab tree, *Adansonia digitata* (Taylor *et al.*, 1978), or *Prosopis julifiora* (Netscher and Luc, 1974), can be good hosts. Fruit trees such as papaya, which are often intercropped along or in the middle of small farmer vegetable fields, are often good hosts and serve as constant reservoirs for infections in these fields (Plate 9E). Furthermore, roots of some non-host crops can react to root knot penetration with local necrosis and, in the case of very high initial nematode densities, roots can be badly damaged and crop loss encountered.

A rotation of sesame, maize, groundnut, sorghum, cabbage, velvet bean and then resistant sweet potato was effective in controlling *M. incognita* in Cuba (Fernández *et al.*, 1992, 1998a). Root knot densities on tomato after sesame were reduced up to 75% as compared with rotation with sweet potato. Acosta *et al.* (1991) demonstrated that the yields of tomato from fields previously planted with maize were significantly higher than those with continuous tomato or tomato treated with a granular nematicide.

TRAP CROPS. In trap cropping, a good host crop is planted for a short duration of time to ensure good nematode penetration and then the developing sedentary juveniles in the root tissue are killed by root removal from the soil or by destruction of the root tissue by physical means or with herbicides. Trap cropping, which was developed originally to control cyst nematodes in sugarbeet, has been suggested for management of nematodes in vegetable crops by Potter and Olthof (1993).

Short cycle, susceptible crops are often used in traditional vegetable rotations, where they control root knot, often without the farmer's knowledge of the concept of trap cropping. In West Africa, black nightshade *Solanum nigrum*, which is used as a leafy vegetable, is continually grown sequentially in the same raised bed on a 3–4 week cycle. Because poisonous properties develop with age, only young plants are eaten. The crop must be harvested with the root system attached, giving the consumer information as to plant age. Root removal from the soil after 3 weeks ensures trapping and root knot death before egg laying is initiated (R.A. Sikora, unpublished data).

In Cuba, lettuce (*Lactuca sativa*) and radish (*Raphanus sativus*) are used as trap crops for root knot in management programmes in organic peri-urban production. The lettuce is harvested with the shoot and root system intact after 30–32 days growth. The roots are discarded before marketing, resulting in trapping and death of large numbers of root knot juveniles (Cuadra *et al.*, 2000).

COVER CROPS. Cover crops in this section are considered to be non-hosts of root knot that are used mainly to protect the soil from erosion, to suppress weed growth between major vegetable crops and to give some nematode control. They may also be used for animal fodder or grazing or as a green manure crop. Cover crops reduce root knot numbers inactively by being non-hosts. However, they also increase microbial activity after incorporation into the soil, which leads to increased numbers of antagonists and the microbial formation of nematicidal compounds in the soil. A number of non-host crops such as velvet bean (*Mucuna pruriens*), horse bean (*Canavalia ensiformis*) and jointvetch (*Aeschynomene americana*) have been tested for use as cover crops in the southern USA for nematode control (McSorley *et al.*, 1994 a,b).

A 2 year rotation with *Mucuna deeringiana*, as a cover crop ploughed into the soil 3 months after planting, gave effective control of *M. incognita* races 1 and 4 on tomato, beans or maize and strong yield increases in tomato (Acosta *et al.*, 1991, 1995). Quénéhervé *et al.* (1998) demonstrated that *Mucuna pruriens* can be used as a practical rotation crop for *M. incognita* control, when planted 3 months prior to a short-term vegetable crop such as lettuce.

Oil radish has been shown to be effective as a green manure crop towards *M. chitwoodi* race 2 (Rehiayani and Hafez, 1998). Rape as a green manure was shown to be more effective than velvet bean as a green manure in reducing *M. arenaria* on squash as well as in reducing the viability of *M. incognita* eggs (Crow *et al.*, 1996).

Some fodder and green manure crops considered to be non-hosts to species of *Meloidogyne*, which could be used in developing rotations, are listed in Table 9.7. Differences, however, in susceptibility between cultivars of the fodder grass *Panicum maximum*, considered a non-host of the more common tropical root knot nematodes, have been detected in South African populations of *M. incognita* (van der Linde, 1956).

The use of elephant grass, *Pennisetum purpureum*, as mulch or the cultivation of *Brachiaria plantaginea* led to significant reductions in galling over continuous tomato. Plant growth was increased the most in the *P. purpureum* treatment (Matsumoto *et al.*, 2002). This plant produces large amounts of biomass and as a mulch greatly stimulates microbial activity in the soil. In Alabama, the winter cover crops *Vica sativa*, *V. villosa* and *Trifolium incarnatum* incorporated into the soil before okra did not have a significant effect on *M. arenaria* or *R. reniformis*, nor on yield (Guertal *et al.*, 1998).

ANTAGONISTIC CROPS. Plants antagonistic to nematodes are those that are considered to produce antihelminthic compounds (Grainge and Ahmed, 1988; Jairajpuri *et al.*, 1990). These crops contain toxic substances with different modes of action

Table 9.7. Reaction of some fodder crops and green manures, often considered non-hosts, to *Meloidogyne* species.

Plant	*M. arenaria*	*M. javanica*	*M. incognita*
Aeschynome	–	–	+
Arachis hypogaea	+*	+	+
Crotalaria fulva	–	+	+
Crotalaria grahamiana	–	+	+
Crotalaria retusa	–	+	+
Crotalaria usaramoensis	–	–	+
Eragrostis curvula	–	+	+
Glycine javanica	–	–	+
Indigofera hirsuta	–	–	+
Panicum maximum	–	+	+
Stylosanthes gracilis	–	+	+

+, resistant; –, not tested; *, susceptible to many populations.

(Pandey *et al.*, 2003). The mechanisms responsible for control are often poorly understood and many tests have only been conducted *in vitro* with plant extracts. The production and active release of toxic substances while the crop is growing or after incorporation into the soil is usually responsible for control. These plants then are in a category similar to those listed above under biofumigation.

Marigold, sunnhemp, castorbean, partridge pea, asparagus and sesame have been studied extensively for nematode control activity. Six cover crops were used in rotations to control *M. incognita* on tomato, including the antagonistic plants castorbean, marigold, mustard, sesame and sunnhemp, with all crops leading to reduced root knot densities. Marigold had the greatest negative effect (Swamy *et al.*, 1995). Sesame and castor have been tested for use in the southern USA for nematode control with some success (McSorley *et al.*, 1994a,b). Sunnhemp is often used as a cover crop and green manure and sometimes considered to be an antagonistic crop for root knot nematode control. *Crotalaria longirostrata*, when grown as a cover crop and then incorporated into the soil, reduced *M. incognita* and *M. arenaria* galling of tomato (Villar and Zavaleta, 1990), with incorporation more effective than simultaneous interplanting of the two crops. The results suggested that toxic byproducts of microbial degradation were involved in control and not toxic exudates from the plant itself. *M. incognita* reproduction was equally reduced on *Crotalaria spectablilis* compared with tomato plants with the *Mi* gene (Esparrago *et al.*, 1999).

The best studied antagonistic plants are species in the genus *Tagetes* known to produce terthieny and derivatives of bithienyl that are toxic to root knot (Uhlenbrock and Bijloo, 1959; Varma *et al.*, 1978; Zavaleta-Mejia *et al.*, 1993). Sellami and Cheifa (1997) reported that *T. erecta*, grown 2.5 months prior to tomato, reduced root knot densities in greenhouses. El Hamawi and Mohamed (1990) showed that concomitantly planting *T. erecta* L. with tomato, green bean or cowpea in the greenhouse had only a slight effect on galling and no effect on *M. incognita* infection. Castro *et al.* (1990) demonstrated that crop rotation and soil incorporation of *T. erecta* resulted in significant reductions in *M. incognita* root galling and increased yield.

Ploeg (1999) demonstrated that *Tagetes patula*, *T. erecta*, *T. signata* and a *Tagetes* hybrid reduced galling in a subsequent susceptible tomato crop compared with the tomato–tomato rotation. In field tests, *T. patula* var. Single Gold and *Tagetes hybrid* var. Polynemao increased tomato yield 50% over a fallow treatment. Marigold Single Gold consistently reduced nematode infestation and galling and in part the results were comparable with methyl iodide fumigation (Ploeg, 2002).

RESISTANCE. The use of resistant cultivars is an elegant, economical and environmentally safe method for controlling root knot nematodes (Netscher and Mauboussin, 1973; Netscher and Sikora, 1990). Comprehensive reviews of most aspects of resistance to *Meloidogyne* have been published and should be consulted for more detailed information (Fassuliotis, 1979; Cook and Evans, 1987; Roberts, 1992; Johnson, 1998; Williamson, 1998; Hussey and Janssen, 2002). However, there are few sources of resistance amongst vegetable crops susceptible to *Meloidogyne*. Resistance has been found in pepper and bean cultivars and was incorporated into tomato via an embryo culture of a hybrid between a resistant line of *Lycopersicum peruvianum* and tomato. In most cases, the genetic basis for resistance is determined by one major gene (Gilbert and McGuire, 1956; Williamson, 1998). However, Hendy *et al.* (1985) reported the presence of five dominant genes which, when present in one genotype, protect against *M. incognita*, *M. javanica* and *M. arenaria*.

The *Mi* gene that confers resistance to *M. incognita*, *M. javanica* and *M. arenaria*, but not *M. hapla*, in tomato has been introduced into many cultivars following detection and the hybridization of *Lycopersicon esculentum* with the wild species *L. peruvianum* (Johnson and Fassuliotis, 1984). It is important that resistance genes that differ from *Mi* in properties and genetic position have been identified in *L. peruvianum* and will help broaden the base of root knot resistance (Williamson, 1998).

Mi-based resistance is used extensively on a worldwide basis for root knot control both at the commercial level and in home gardens. Radewald estimated that about 30% of processing tomatoes in California carried the *Mi* gene for resistance to root knot (cited in Koenning *et al.*, 1994) and in 2004 approximately 90% of fresh market tomato in California are estimated to have the *Mi* gene (I. Kaloshian, California, 2004, personal communication).

However, these cultivars are often not available to poor subsistence farmers. Resistance to root knot based on the *Mi* gene has been combined with resistance to the two fungal wilt pathogens *Fusarium* and *Verticillium* and to other major diseases. The root knot–*Fusarium oxysporum* wilt complex can be controlled by growing cultivars resistant to either the fungus or the nematode, or both. Conversely, the root knot–*Rhizoctonia solani* root rot complex, which is responsible for severe losses in the tropics and subtopics, can only be suppressed by controlling *Meloidogyne*, because of the lack of resistance to the fungus.

The small fruited hot peppers *Capsicum frutescens* L. var. *longum* are resistant to the major species of root knot, but not to *M. hapla* (Johnson, 1998). A number of root knot-resistant cultivars of bell pepper, *C. frutescens*, have been released in the USA. They are homozygous for the dominant *N* resistant gene toward *M. incognita*, *M. javanica* and races 1 and 2 of *M. arenaria* (Fery *et al.*, 1998; Thies and Fery, 2000a). The cultivar 'Charleston Belle' has been field tested with good results (Thies *et al.*, 2004) and resistance has been shown to hold up well under high soil temperatures that often negatively affect nematode resistance in other horticultural crops (Thies and Fery, 2000b).

Resistance has also been found in aubergine, where it was originally detected in *Solanum sisymbrifolium*, a closely related relative. Several wild species of *Cucumis* with resistance to root knot have also been detected (Fassuliotis, 1979). However, genetic barriers make it extremely difficult to introduce the resistance of the wild species into cultivated species. Modern molecular techniques such as protoplast culture and somatic hybridization may make it possible to create viable hybrids, and attempts are being made to develop interspecific hybrids (Starr *et al.*, 2003).

To date, little or no progress has been made in developing resistance to root knot in the Cucurbitaceae. However, most cucumber varieties seem to be more resistant to *M. hapla* and *M. arenaria* than to *M. incognita* or *M. javanica* (Johnson, 1998).

Resistant oil radish genotypes of *Raphanus sativus*, a green manure crop that has been used effectively to control sugarbeet cyst nematodes, were found to be very effective in reducing *M. hapla* and *M. incognita* numbers in the greenhouse and microplot tests (Bunte and Muller, 1996; Bunt *et al.*, 1997).

Resistant cultivars of any crop have an impact on nematode soil densities similar in intensity to that obtained with many soil fumigants in that they reduce soil population significantly. The major difference is that resistant cultivars do not reduce or eliminate the antagonistic potential in the soil which causes an enormous rebound of the root knot population during the season after fumigation.

The results of research in the southern USA showed that double cropping susceptible cucumbers using cucumber transplants versus direct sowing, when grown after a nematode-resistant tomato crop, was effective in improving cucumber yields in *M. incognita*-infested soils (Hanna *et al.*, 1994, 1996). This approach was more effective than a nematicide applied through drip irrigation for managing *M. incognita* (Colyer *et al.*, 1998). Similar results were obtained by double cropping cucurbit crops after the root knot-resistant bell pepper 'Charleston Belle' (Thies *et al.*, 2004). Cucumber yields were 87% heavier and the number of fruit 85% higher when planted after the resistant pepper cultivar. Squash yield increased 55% and the number of fruit 50% over the fields previously planted to susceptible pepper.

Lists of plants reported resistant to nematodes in general (Armstrong and Jensen, 1978) and crop cultivars resistant to species of *Meloidogyne* (Sasser and Kirby, 1979; Netscher and Sikora, 1990; Whitehead, 1998) have been compiled elsewhere. Lists of cultivars resistant to root knot nematodes, however, should be used with caution, because some of the cultivar reactions are often based on a limited number of field observations. Such tests are also not a guarantee that a cultivar is resistant to all populations of *Meloidogyne*.

GRAFTING. One of the most effective and innovative techniques developed for root knot control is the grafting of commercially valuable crop varieties on to nematode- and disease-resistant rootstocks. Although grafting has been practised since the 1920s in Japan and Korea, it has only recently become highly regarded in protected cultivation in the region for disease control. In Japan, 59% of the cucumber, tomato, aubergine, watermelon and melon grown in protected cultivation are tube grafted on to rootstocks of various types, because of increased vigour and tolerance or resistance to pests and diseases. Grafting robots have been developed to produce grafted plugs in nurseries (Oda, 1999).

The technique can be used effectively to control a number of diseases and root knot, and in many cases circumvents the long process needed to breed root knot resistance into all commercially acceptable cultivars (Black *et al.*, 2002). Depending on the price of production, it can be very effective in both field and protected cultivation of vegetables. It should be noted that nematode pathotypes can develop on these rootstocks, therefore, resistance management must be incorporated into these cropping systems.

Solanum torvum, which has been shown to have a high level of resistance to *M. incognita* and *M. arenaria*, but is a poor host for *M. javanica*, has been used successfully as a rootstock for aubergine (Dunay and Dalmasso, 1985). When the shoots of aubergine were grafted on to the rootstocks of *S. torvum*, *S. aethiopicum*, *S. sysimbriifolium*, *Cyphomandra betacea* (tamarillos), tomato line NR 62 or tomato cv. 'Giallo de Castellana' and compared with plants maintained on their own roots, the *Solanum* and tomato rootstocks all reduced plant susceptibility to *Meloidogyne*, with *S. torvum* the best combination for both control and yield (Porcelli *et al.*, 1990). Additional trials showed that aubergine grafted on to *S. torvum* rootstocks having resistance to root knot and soil-borne pathogens can compete with soil fumigation, regarding both control and yield increases (Morra *et al.*,

1992). Of seven wild species of *Solanum* tested, three were found to be resistant to *M. incognita*, i.e. *S. sisymbriifolium*, *S. torvum* and *S. toxicarium* (Mian *et al.*, 1995a). These species, when used as rootstocks, not only reduced galling on tomato, but also reduced bacterial wilt of aubergine caused by *Ralstonia solanacearum* (Mian *et al.*, 1995b).

Granges and Leger (1996) showed that when susceptible tomato were grafted on to rootstocks having resistance to species of *Meloidogyne* and various root pathogens, yield increased 50 and 30% at the beginning and end of harvest when compared with the non-grafted plants, respectively. Additional soil steaming did not increase productivity of the grafted plants. The highest profit margin was obtained when plants were grafted with two stems per nematode-resistant rootstock and planted at half the standard density. Susceptible tomato cultivars grafted on to the nematode-resistant rootstocks also produced significant yield increases and *M. incognita* control (Morra *et al.*, 1997). In tests conducted in Spain, grafted tomato held an intermediate place in both level of control and effect on yield increase between the resistant and susceptible cultivar (Sorribas *et al.*, 2004). The results showed that rootstocks used for grafting were only partially resistant to the nematode (S. Verdejo, Spain, 2004, personal communication). Augustin *et al.* (2002) reported that temperatures between 18 and 25°C did not affect the quality of rootstock resistance to *M. arenaria*, but that there are differences in the level of nematode resistance in rootstocks recommended for use.

The grafting of nematode-resistant rootstocks of sweet pepper, *Capsicum annuum*, on desirable but susceptible cultivars of sweet pepper, led to increased yields, the cost of which was only justified at high nematode infestations (Morra *et al.*, 2003).

Cucumbers grafted on to pumpkins initially had 71% fewer galls with a slight increase in galling over the season (Liu *et al.*, 1998). Of course these root knot-susceptible rootstocks will lead to high nematode populations over time and therefore their use needs to be integrated in an overall root knot management programme. Grafting in many countries could prove to be an alternative management approach, especially where temperature does not affect the genes controlling resistance.

RESISTANCE-BREAKING RACES. Resistant cultivars of crops susceptible to *Meloidogyne* do not necessarily protect the crop against all species of the genus. In addition, races may exist which are able to break resistance. The *Mi* gene does not confer immunity to *M. incognita* and *M. javanica* (Roberts and Thomason, 1986). Resistance-breaking races have been selected out of field populations of *M. incognita*, *M. javanica* and *M. arenaria* (Riggs and Winstead, 1959; Sauer and Giles, 1959). Root knot populations which were capable of attacking resistant cultivars have been detected even though they had previously never been exposed to the cultivars (Sikora *et al.*, 1973b; Netscher, 1977; Prot, 1984; Fargette, 1987; Berthou *et al.*, 1989). Resistance-breaking races were also selected from single egg mass populations of *M. incognita* and *M. javanica* in laboratory experiments (Triantaphyllou and Sasser, 1960; Netscher, 1977). Resistant cultivars, therefore, should be used judiciously and with caution or should be tested in advance for efficacy (Roberts *et al.*, 1986).

Kaloshian *et al.* (1996) detected an *M. incognita* population attacking an *Mi*-resistant tomato cultivar in California. Eddaoudi *et al.* (1997) reported that of 20 populations of *Meloidogyne* collected in two Moroccan vegetable-growing areas, nine populations were able to break resistance on the cultivar 'Darus' and six were virulent on 'VFN8'. In a survey of randomly selected populations of *M. incognita* and *M. javanica* from Crete, all *M. incognita* populations were avirulent toward *Mi* gene-resistant tomato, whereas three populations of *M. javanica* were identified as virulent (Tzortzakakis *et al.*, 1999). The results demonstrate the need for country-specific studies for the presence of *Mi* gene-resistance-breaking populations of root knot in order to optimize the use of resistance in nematode management programmes.

RESISTANCE MANAGEMENT. In many vegetable-growing regions, the *Mi* gene for resistance cannot be used due to its sensitivity to high temperatures. Dropkin (1969) showed that at 28°C the resistant cv. 'Nematex' was highly resistant to *M. incognita*, whereas at 32°C it was susceptible. In India, as well as in Senegal, a breakdown in resistance due to high soil temperatures has been observed (Sikora *et al.*, 1973b; Berthou *et al.*, 1989). In areas with high temperatures, cultural practices such as appropriate watering and mulching may reduce soil temperature to counteract and prevent loss of resistance. However, plastic mulches used for fumigation, solarization and plastic tunnels may elevate soil temperature above 28°C if planting is made directly through the plastic tarp.

Resistance management strategies need to be a major part of any vegetable production system wherever cultivars with the *Mi* gene for resistance to root knot are used, not only because of temperature inactivation, but also because of development of resistant breaking races over time.

The results of field research in the southern USA showed that a double cropping of cucumbers, using cucumber transplants versus direct seeding, after a nematode-resistant tomato crop, is effective in improving cucumber yields in root knot-infested soils (Hanna *et al.*, 1994, 1996). In addition, this cropping system was shown to be more effective than applying a granular nematicide through drip irrigation for managing *M. incognita* (Colyer *et al.*, 1998). Cropping systems of this nature allow the production of multiple cycles of high value crops and give simultaneous protection of resistant germplasm.

WEED CONTROL. The effect of any management programme can be seriously compromised if susceptible weeds are present in the field. Therefore, proper weed control contributes greatly to nematode management and effective crop improvement. Weeds, often good hosts of root knot nematodes, and multiple host species are commonly found in vegetable fields (Noling, 2003). In Table 9.8 a partial list of some weed hosts of important root knot nematodes of vegetables is given. If weed hosts are not controlled by proper management, they can sustain root knot populations even under non-hosts in a rotation. Such weeds are important hosts for root knot between the rows of fumigated beds in commercial production. Migration of significant levels of root knot into the beds over time leads to damage in follow-up crops (J. Noling, Florida, 2004, personal communication). Damage may even be higher due to loss of antagonistic potential in the bed due to fumigation. With the loss of methyl bromide, weed diversity and density may increase significantly. Since many weeds are good hosts of root knot, this could increase damage to a vegetable crop if alternative weed control is not practised properly.

A unique nematode management approach that takes advantage of weed growth has been developed in Costa Rica. All weeds, hosts and non-hosts, that grow in the rainy season before the next major vegetable crop are incorporated into the soil, drip irrigation added and the beds mulched with plastic. This combination leads to optimum moisture, high temperatures and a biofumigation that gives effective weed control and root knot management simultaneously. Planting is then done through the mulch into biofumigated soil at low nematode densities (R. Garron, Costa Rica, 2004 personal communication).

Chemical

Nematicides used in control of root knot nematodes are either fumigants, which are usually liquids and enter the soil water solution from a gas phase, or non-fumigants that are granular or liquid compounds, which are water soluble. In most cases, the fumigants are broad-spectrum contact nematicides effective against juveniles and eggs as well as other pests, diseases or weeds. Non-fumigant nematicides have either contact or nematistatic and systemic activity against nematodes and often against insects. In most cases, the

Table 9.8. Comparison of selected weed hosts of important root knot nematodes attacking vegetables.

	Meloidogyne species					
Weed	*incognita*	*javanica*	*arenaria*	*hapla*	*chitwoodi*	*mayaguensis*
Amaranthus albus	•	•				
Amaranthus retroflexus	•	•	•	•		
Ajuga reptans				•		•
Anthemis arvensis				•	•	
Atriplex papula	•					
Capsella bursa-pastoris	•	•	•	•		
Chenopodium album	•	•	•	•		
Cirsium arvense	•	•		•	•	
Clerodendrum ugandense	•					•
Convolvulus arvensis	•	•				
Cyperus rotundus	•	•				
Digitaria sanguinalis	•	•	•			
Erigeron spp.	•	•		•		
Erodium cicutarium	•			•		
Galinsoga ciliata				•		
Galinsoga parviflora	•	•	•	•	•	
Lamium amplexicaule	•			•	•	
Medicago arabica		•	•			
Poa annua	•		•			
Polygonum persicaria	•			•		
Portulaca oleracea	•	•	•	•		
Rumex crispus		•	•			
Senecio vulgaris	•		•	•		
Setaria verticillata	•	•	•			
Solanum nigrum	•	•	•	•	•	
Sonchus oleraceus	•	•	•	•		
Sonchus tenerrimus	•	•	•			
Stellaria media	•	•	•	•		
Taraxacum officinale	•			•	•	
Tibouchina elegans						•

From: Goodey *et al.* (1965); Fernández *et al.* (1993); Barcelo *et al.* (1997); Brito *et al.* (2004d); Zoon *et al.* (2004).

mechanism of action is associated with suppression of nematode mobility during the period when adequate concentrations are in the soil solution. The non-fumigant nematicides are not effective against the eggs of nematodes and in most cases do not kill the juveniles at the concentrations now being recommended for use. They give the plant a 'head start' by delaying nematode penetration during the highly sensitive seedling or post-transplant stage of plant development. There are a number of sources that give excellent reviews on the use of the most common fumigant and non-fumigant nematicides for a broad array of nematodes and crops and they should be consulted for more detail (Johnson, 1985; Hague and Gowen, 1987; Whitehead, 1998; Anonymous, 2004). In Appendix A, fumigant and non-fumigant nematicides are listed.

Fumigant nematicides are generally more effective in controlling root knot nematodes and in increasing crop yield than are non-fumigant nematicides. Because fumigant nematicides have a broader spectrum of activity, they control soil insects, fungal diseases and weeds in addition to other plant parasitic nematodes. This broad spectrum of activity also decreases the need for additional pesticide inputs, reducing overhead costs associated with crop production. Most of the fumigant nematicides listed in Appendix A have been shown to be highly effective in control programmes designed to reduce *Meloidogyne* losses in vegetables (Lamberti, 1979b; Johnson, 1985; Whitehead, 1998). They are used extensively for nematode control in large-scale production systems, and some vegetables grown on a large-scale basis in infested areas can only be produced economically together with fumigant application (Radewald *et al.*, 1987; Noling and Becker, 1994). In some growing areas, fumigants are applied under plastic mulch and the vegetables are planted through the mulch usually in raised beds. It should be noted that in many areas, soil temperatures may be too high for effective use of resistant cultivars.

Due to the multiple effects of nematodes, weeds and soil fungi on tomato production in many growing areas, a broad-spectrum fumigant is essential, especially where multiple susceptible crops are grown sequentially. Methyl bromide or combinations of this fumigant with chloropicrin applied under plastic mulch is the mainstay of growers having these multiple pest problems worldwide. The loss of methyl bromide due to side effects on ozone (Anonymous, 1998a) has stimulated a search for alternatives, with a number of fumigants studied in detail giving acceptable results (Noling and Becker, 1994; Jones *et al.*, 1996; Locascio *et al.*, 1997; Anonymous, 1998b; Csinos *et al.*, 2000). However, it should be noted that there are strong movements to keep methyl bromide on the market under critical use exemptions.

There are a number of fumigants that are in line to replace methyl bromide. Metham sodium, for example, is an effective fumigant that is released more slowly than methyl bromide and, therefore, could be applied by drip irrigation or even centre pivot overhead systems. This, however, will affect how vegetables are grown in some areas and may require retooling production systems. 1,3-Dichloropropene gives excellent control of nematodes, but requires chloropicrin to increase activity toward pathogens and weeds (Hague and Gowen, 1987; Noling and Becker, 1994; Whitehead, 1998). Limitations on use due to possible side effects on the environment are expected. Methyl iodide, another fumigant being looked at, and methyl bromide in two field trials produced at least 161 and 181% more marketable carrots without *M. incognita* damage than plants in control plots, indicating that the former may be an effective alternative to methyl bromide (Hutchinson *et al.*, 1999).

When used as directed, fumigants will give excellent nematode control and increase yield significantly. Because registration requirements and efficacy vary with country and crop, no attempt will be made here to list those still being used for the control of root knot nematodes in vegetables.

The majority of small farmers, especially those living at the subsistence level, cannot use fumigants because of a lack of capital for equipment, the nematicides or application.

The granular and/or liquid formulations of contact and/or systemic nematicides are suitable for commercial use as well as for use on small farms. The growers, however, must be made aware of proper handling and application techniques as well as time of application, since they are toxic to humans and the environment when improperly used. Non-fumigant nematicides are often not as effective as fumigants in increasing yields because they do not have broad-spectrum activity and in most cases only inactivate nematodes for short periods of time. Therefore, repeated applications are needed in multiple cropping vegetable systems. This is often uneconomical, environmentally questionable and can lead to biodegradation over time (Mojtahedi *et al.*, 1991; Stirling *et al.*, 1992).

A number of granular nematicides (phorate, aldicarb, carbofuran, oxamyl, thionazin, terbufos, isazophos, aldoxycarb, cloethocarb, ethoprophos, fenamiphos, cadusafos and avermectins) are effective against root knot nematodes on vegetable crops under field and plastic house conditions (Cadet, 1990; Basile *et al.*, 1993; Lamberti *et al.*, 1993; Philis, 1994; Verma *et al.*, 1994; Sasanelli *et al.*, 1996; Whitehead, 1998). Fosthiazate and DiTura, the latter a biological nematicide derived from the fermentation of a nematode parasitic isolate of the fungus *Myrothecium* sp. (Warrior *et al.*, 1999), are newer compounds that have been shown to control root knot (Toki and Imai, 1994).

Granular nematicides are either applied broadcasted over the soil surface and incorporated into the soil before planting or banded into or over the plant furrow. It is important that users realize human and environmental toxicity can occur and that the presence of residues in the harvested crop is possible if treatment recommendations are not followed.

Liquid formulations allow application by surface and drip irrigation (Overman, 1974; Johnson, 1985; Hague and Gowen, 1987; Whitehead, 1998; Anonymous, 2004), with the latter of extreme importance to vegetable production. Application through drip irrigation places the material directly in the rhizosphere and allows treatment at will or treatment when necessary during the growing season. It also allows splitting or extending application over specific time intervals to coincide with optimum control. For example, oxamyl applied to tomato by drip irrigation was more effective than granular nematicides applied at transplanting in controlling root knot (Philis, 1994; Russo *et al.*, 2003).

Dip treatment or treatment of transplants in nurseries (Ahuja, 1978; Mateille and Netscher, 1985; Cayrol *et al.*, 1993; Jansson and Rabatin, 1998) have also been effective. For example, Siddiqui *et al.* (1998) showed that dip treatment of seedlings of aubergine and tomato with fenamiphos significantly reduced *M. incognita* galling. In hydroponic systems, phenamiphos showed good results in reducing root knot (Pérez *et al.*, 1990).

Efforts are being made to develop formulations that allow seed treatment for nematode control that would greatly reduce the dose needed on a per hectare basis, reduce the environmental impact and reduce crop residues. In many short cycle vegetable crops and in crops with a taproot that may only need protection of 4–5 weeks, this could be an important treatment form. Pelleting seeds with carbofuran was effective toward stem nematode on broad bean (Schiffers *et al.*, 1985). Townshead (1990) showed that seed of carrot and tomato coated with oxamyl resulted in reduced galling by *M. hapla*. Treatment of bottle and bitter gourd with carbofuran, fenamiphos or phorate reduce *M. incognita* and increased yields (Siddiqui *et al.*, 1993).

Avermectins are macrocyclic lactones produced by the actinomycete *Strepomyces avermitilis* that have broad-spectrum antihelminthic activity (Cayrol *et al.*, 1993). They recently have been shown to be active against root knot as root dips and as seed treatments on a number of crops (Jansson and Rabatin, 1998; Jayakumar *et al.*, 2002; Rideout and Long, 2004; Smith-Becker and Becker, 2004). Effective control of *M. hapla* was obtained with low doses of abamectin as a seed treatment on pelleted and non-pelleted seeds of tomato (Abawi *et al.*, 2003), with root galling and number of eggs/g root significantly reduced after 6 weeks. Abamectin was also effective in controlling *M. incognita* as a seed treatment of cucumber (Becker *et al.*, 2003; Becker and Hofer, 2004). The reduction of nematode damage with seed treatment reduces costs and environmental impact and is a promising approach.

Fumigant and non-fumigant nematicides are effective components of management programmes for root knot. They will remain important in cropping situations where alternatives are not available or not effective. With the loss of methyl bromide, the use of other fumigants, cocktails of compounds and/or non-fumigant nematicides will increase. New nematicides that can be applied safely and if possible effectively to the seed or seedling or through drip irrigation would be a major step forward.

Biological

Four approaches are now important for management of root knot nematodes with antagonists in vegetable production: (i) inundative application of fungal pathogens, parasites or predatory fungi that infect eggs, juveniles or adults in the soil or on the root surface (Jatala, 1986; Kerry, 1987; Stirling, 1991; Kerry and Evans, 1996; Atkins *et al.*, 2004); (ii) field inoculation and management of the obligate bacterial parasite *Pasteuria penetrans* (Oostendorp and Dickson, 1991; Stirling, 1991; Trivino and Gowen, 1996; Gowen *et al.*, 1998); (iii) promotion of the naturally occurring antagonistic potential in soils with amendments or crop rotation (Sieverding, 1991; Sikora, 1992; Sikora *et al.*, 1994; Pyrowolakis *et al.*, 2002); and (iv) biological enhancement of transplants or planting material with plant health-promoting rhizosphere- or endorhiza-associated bacteria or fungi (Sikora, 1992, 1997; Sikora and Hoffmann-Hergarten, 1993; Hallmann and Sikora, 1994a,b; Sikora *et al.*, 2003).

Recent success in the development of fungal antagonists has coincided with significant advances in fermentation and formulation technology (Silman *et al.*, 1993; Lüth, 2000, 2004; Kiewnick, 2001). This has led to development of microbial products almost devoid of organic carrier material – a problem that has limited practical use in the past (Sikora, 1992). Modern solid-state fermentation allows economic production, storage, shipment and field application of fungal antagonists (Lüth and Eiben, 2003).

Paecilomyces lilacinus, which is predominantly a fungal egg pathogen, has been marketed for use in the Philippines and South Africa (Kiewnick, 2004). Effective biocontrol of root knot in the field has been reported on vegetables and other crops in a number of countries (Davide and Zorillla, 1983; Cabanillas and Barker, 1989; Lara *et al.*, 1996; Aceret *et al.*, 1999; Holland, 2001; Holland *et al.*, 2003; Kiewnick and Sikora, 2003). *P. lilacinus* strain-251 is presently licensed by biotech companies in Germany and South Africa for mass production and nematode control, and is being registered for the European and USA markets (Brückner, 2004; Kiewnick, 2004). Brückner (2004) recommended a split application programme for vegetable crops which involved: soil treatment with 4 kg/ha of the product 7 or 14 days before planting, then drenching seedling plugs 1 day prior to transplanting in a solution containing 10 g of *P. lilacinus* per 100 plants, followed by additional soil treatment in the field at 4–6 week intervals with 4 kg/ha as needed. Some form of root monitoring would be required. A total dose of 10–14 kg/ha of crop provided effective control and was considered economical.

The availability of a high quality commercial product that can be applied as a wettable powder to the soil, transplants or even through drip irrigation systems could have a significant impact on root knot control in specific types of vegetable cropping systems.

Pochonia chlamydosporia has been studied extensively both in the laboratory and under field conditions for root knot control (De Leij and Kerry, 1991; Hidalgo, 1999; Kerry, 2000, 2001; Atkins *et al.*, 2003, 2004). Bourne (2001) and Bourne and Kerry (1999) demonstrated that application of *P. chlamydosporia* in a rotation of less susceptible crops such as kale, beans and cabbage led to large reductions in root knot in the subsequent tomato crop, because egg masses are more exposed to the fungus on the rhizosphere of these crops. The application of *P. chlamydosporia* with arbuscular mycorrhizal fungi in the seedling stage was also shown to increase root knot control and plant growth (Rao *et al.*, 1997b), and applying the fungus with neem at transplanting improved efficacy (Rao *et al.*, 1998a).

An indigenous isolate of *P. chlamydosporia* from Cuba is mass produced in a biphasic process on rice in small-scale solid-state fermentation plants (Hidalgo, 1999; Kerry and Hidalgo-Diaz, 2004). Fermentation results in production of approximately 10^6 chlamydospores/g of medium. The spores are separated from the medium and, when applied in rotations with poor hosts (Atkins *et al.*, 2004), gave effective control of root knot on subsequent tomato in peri-urban organoponic vegetable production.

Trichoderma harzianum, which is known to be effective against fungal diseases, also has activity toward root knot nematodes. *Trichoderma harzianum* and *T. koningii* increased plant growth and reduced *M. arenaria* on maize under controlled conditions (Windham *et al.*, 1989). Control of root knot with an Indian strain of *T. harzianum* was enhanced by adding the antagonist to soil amended with neem cake (Rao *et al.*, 1997a). Both *T. harzianum* and *T. lignorum* increased plant growth and reduced *M. javanica* galling on tomato and aubergine in soil treated with the fungi 18 days prior to planting in greenhouse tests (Spiegel and Chet, 1998; Sharon *et al.*, 2001). Single treatments of *T. harzianum* and *T. virides* were effective at low initial root knot densities in one-cycle vegetable crops grown in organoponics (Perez, 2001). Two commercial strains of *T. harzianum* from the Middle East, applied to seedling plugs 1 week before transplanting and also broadcasted at 10 g/m furrow 1 week before planting, reduced *M. javanica* galling by 20–30% on aubergine, tomato and pepper and led to increased yield in two field trails near Jericho. The decrease in galling was not as great as with a non-fumigant nematicide (H. Saleh and A.A. Dababat, Palestine Authority, 2002, unpublished data).

Pasteuria penetrans is an obligate parasite of a number of important plant parasitic nematodes, in particular *Meloidogyne* (Birchfield and Antonpoulos, 1976; Stirling, 1991; Chen and Dickon, 1998). The spore form can resist both drought and exposure to non-fumigant nematicides (Mankau and Prasad, 1972). Stirling and Wachtel (1980) produced large numbers of spores by inoculating tomato with infected *Meloidogyne* juveniles. Dried tomato roots were then milled into a powder containing *Pasteuria* spores, a method that can be adapted to produce inoculum for small growers. The parasite can also be increased in root knot-infested fields by growing tolerant or moderately resistant crops (Oostendorp and Dickson, 1991; Gowen *et al.*, 1998; Gowen and Pembroke, 2001, 2004). Reincorporation of the roots of root knot host plants over three cycles led to spore populations in the soil that prevent root knot invasion (Melki *et al.*, 1998). The parasite seems to be more effective on warm soils and soils low in organic matter, which characterizes most tropical soils where root knot is a problem. Solarization and/or soil amendments may reduce the efficacy of *P. penetrans* (Freitas *et al.*, 2000a). Conversely, chloropicrin, found in many fumigant nematicides, was shown to have bactericidal effects on *P. penetrans* (Freitas *et al.*, 2000b). Metam sodium, on the other hand, did not adversely affect the parasite, which is important in intensive production systems where this fumigant may replace methyl bromide. Combining the parasite with plant resistance, oxamyl and solarization has also been investigated (Tzortzakakis and Gowen, 1994). Advances in fermentation of the parasite may make this antagonist available on a large scale to growers for management of root knot (Hewlett *et al.*, 2002).

The 'antagonistic potential' in a soil and its management has been considered a means of reducing the impact of root knot and other nematodes (Sikora, 1990, 1992; Stirling, 1991). Specific components of this potential can be measured and changes monitored using simple bioassays (Rodriguez-Kabana *et al.*, 1994; Sikora *et al.*, 1994; Terhardt *et al.*, 1997; Pyrowolakis *et al.*, 2002). Managing the 'antagonistic potential' to improve suppressiveness requires knowledge of the microbial communities (Vilich and Sikora, 1998) and more specifically antagonists conducive to management. Arbuscular mycorrhizal fungi are management-sensitive antagonists that are present in all soils and that react favourably to certain crop rotations (Smith, 1987; Sieverding, 1991; Sikora, 1995). A number of rotation crops such as vetch, clover, maize, bahiagrass and pearl millet, that are poor or non-hosts of root knot and decrease nematodes in rotations, also increase mycorrhizal densities in soils, which can then have positive effects on the following root knot-susceptible crop (Sumner *et al.*, 1999; Jothi *et al.*, 2000; Jothi and Rajeswari, 2001). It should be noted that crop rotation might also favour non-effective mycorrhiza species (Sieverding, 1991). The overall 'antagonistic potential' of a soil can be manipulated, for example using organic amendments and green manures of various forms (Singh and Sitaramaiah, 1966, 1967; Sikora *et al.*, 1973a). Amendments stimulate many antagonists and also lead to the production of toxic metabolites and/or biofumigants that together are responsible for the effectiveness of this management practice.

Biological enhancement of seeds, and vegetable transplants with antagonistic microorganisms, e.g. arbuscular mycorrhizal fungi, mutualistic fungal endophytes, plant health-promoting rhizosphere or mutualistic endophytic bacteria, has been shown to increase plant resistance and/or tolerance to root knot infection (Sikora, 1990, 1991, 1997; Sikora and Hoffmann-Hergarten, 1993; Hallmann and Sikora, 1994a,b; Schuster *et al.*, 1995; Hallmann, 2001; Hallmann *et al.*, 2004b).

Biological enhancement has been attained using antagonistic rhizobacteria as seed dressings (Sikora, 1988; Oostendorp and Sikora, 1989) and through application by drip irrigation systems (Zavaleta-Meija and Van Gundy, 1982). Tomato and pepper transplant production substrate treated with different formulations of plant growth-promoting rhizobacteria caused highly significant increases in tomato and pepper growth, vigour and survival in the field, with some formulations reducing the numbers of root knot galls on pepper (Kokalis-Burelle *et al.*, 2002). Rhizobacteria added to tomato 'seedling' trays under commercial nursery conditions caused significant reductions in root knot galling and increased tomato growth and yield (Reddy *et al.*, 2000). Some rhizobacteria, especially those that grow endophytically, have been shown to induce resistance (Hasky-Guenter *et al.*, 1998). Some of the latter isolates also have activity toward root knot and simultaneous activity toward Fusarium wilt of tomato (Hauschild *et al.*, 2000). Endophytic bacteria recently have been shown to significantly reduce root knot infection (Munif *et al.*, 2000) and induce systemic resistance in tomato (Munif *et al.*, 2001). Mahdy *et al.* (2000) also demonstrated significant levels of control with rhizosphere and endophytic bacteria against different species of *Meloidogyne* (Mahdy *et al.*, 2001).

Enhancement of plants with arbuscular mycorrhizal fungi, apart from providing plants with nutrients, reduces penetration and development of a number of root knot nematodes on a range of vegetable crops (Sikora, 1978, 1995; Hussey and Roncadori, 1982; Smith, 1987). An important strategy for effective control in vegetable crops is enhancement of seedlings to attain high mycorrhizal root colonization densities at transplanting. Mycorrhizal inoculum is commercially available for this purpose in many countries. Combining mycorrhizal fungi with plant health-promoting rhizobacteria during seedling development led to increased fungal colonization and root knot control in tomato at transplanting (Raimann and Sikora, 2003). The simultaneous use of bacteria that promote mycorrhizal root colonization and reduce root knot penetration increases efficacy (Sikora and Raimann, 2004).

Endophytic fungi are prime antagonists for use in biological enhancement of transplants for root knot control in one-cycle cropping systems. Hallmann and Sikora (1994a,b) showed that *M. incognita* galling was reduced by 50% in tomato inoculated with a mutualistic fungal endophyte. Root knot control with a non-pathogenic *F. oxysporum* isolate was significantly higher than with *Trichoderma* species in greenhouse tomato (A.A. Dababat and R.A. Sikora, Bonn, 2004, unpublished data). Mutualistic endophytic fungi have an advantage over arbuscular mycorrhizae in that they can be produced by liquid- or solid-state fermentation. They actively colonize the growth substrate as well as the endorhiza of the root occupied by the nematode (Sikora *et al.*, 2003).

Treatment of fumigated, biofumigated or solarized soil with biologically enhanced transplants would increase overall control, due to the lack of competitive microbial activity in this soil. To be effective, however, biological enhancement requires the existence of either commercial biocontrol products, as is the case with mycorrhizal fungi, that can be used by small or large commercial nursery production units that supply enhanced seedlings to growers. In some countries, for example in Cuba, antagonists are produced on a large scale centrally (Hidalgo, 1999; Fernández *et al.*, 2000; Perez, 2001) and enhanced seedlings are the result of inoculation in organoponic production. In many countries, large commercial growers produce their own high quality seedlings, and in some places large food store chains supply contract growers with clean and healthy seedlings. These could be inoculated before transport to the grower's fields with antagonists to increase yield and reduce pesticide use.

Summary of management measures

The principles and main components of effective control programmes and integrated pest control in vegetables as well as other crops have been discussed in this chapter as well as in good review articles (Taylor and Sasser, 1978; Johnson and Fassuliotis, 1984; Brown and Kerry, 1987; Johnson, 1998; Anonymous, 2004).

Control, as stated earlier, varies greatly between production systems (field, peri-urban organoponics, protected cultivation), with climatic conditions, production seasons and between countries with different eating habits and customs.

The main aspects we consider important for both field and protected cultivation of vegetables are summarized below.

- Prevent introduction through good quarantine.
- Build protected cultivation structures on root-knot-free land.
- Utilize nematode-free growth substrates for seedling production.
- Plant certified root-knot-free transplants.
- Introduce antagonists into the seedling production system.
- Always view previously infested land as infested.
- Develop farmer-based root gall monitoring protocols to estimate infestations.
- Design rotations that prevent the build-up of high nematode densities.
- Use nematicides judiciously based on monitoring previous crops.
- Take judicious advantage of resistant cultivars when available.
- Monitor soil temperature to prevent breakdown of *Mi* gene resistance.
- Rotate resistant and susceptible cultivars to prevent resistance-breaking pathotypes.
- Challenge all new crops with local populations for host status.
- Use root grafting of resistant root stocks if economical.
- Tolerant root stocks should be used if adaptable where resistance is lacking.
- Use paddy rice or controlled flooding for control.
- Destroy galled roots after harvest.
- Incorporate nematode desiccation through tillage-supported soil drying.
- Time planting for cooler periods to reduce infection.
- Biofumigate with incorporated organic amendments.
- Use organic matter to stimulate antagonistic potential.
- Trap crop if exact timing for crop destruction can be ensured.
- Introduce solarization in sun energy-rich regions.

Methods of diagnosis

The scattered or clustered distribution of most nematodes in the field makes reliable estimation of occurrence and/or population density extremely difficult. Due to the presence of egg masses, the spatial distribution of root knot is very heterogeneous. Techniques have been developed for extraction that are based on the fact that the egg masses remain intact in the soil either free or attached to host roots or root fragments (Dickson and Strubel, 1965; Byrd *et al.*, 1972; Gooris and d'Herde, 1972). After separating the organic matter from the soil using sieving or elutriation techniques, eggs are liberated from egg masses either chemically (Byrd *et al.*, 1972) or mechanically (Gooris and d'Herde, 1972). Demeure and Netscher (1973) observed egg masses present in the coarse sandy soil fraction and also suggested incubation of this fraction.

Even if the methods of extraction are sufficiently reliable, it is still virtually impossible to determine whether or not land is free from root knot, even when the results of soil analyses are negative. The majority of the methods used will not always detect egg masses in fields with low to moderate root knot infestation levels. Accuracy can be increased by increasing the volume of the soil sample taken from the field as well as the number of cores taken per unit area and by extracting greater quantities of soil than the usual 100–250 cm^3 recommended. The accuracy of the extraction method used in determining population densities is extremely important in estimating threshold levels. Barker (1985a,b) discusses sampling and extraction techniques and lists their relative efficiency.

Another problem, related to determination of population densities in sandy soils, is the migration of juveniles over substantial distances from the plant (Prot and Netscher, 1978).

Bioassay techniques, in which susceptible plants growing in the field are uprooted and examined for the presence of galls after a period of 3–6 weeks, constitute a means to evaluate the presence of infestations of soils with greater accuracy than soil analysis (McSorley and Parrado, 1983).

An accurate evaluation of root knot infestations in a field can be obtained at the end of the vegetative cycle of a susceptible crop. Plants are systematically uprooted and scored for severity of root galling, thereby giving an accurate estimation of the severity and the distribution of *Meloidogyne* in a field. This is the only method available for workers lacking basic nematological extraction equipment. A number of different root knot indices have been proposed (Zeck, 1971; Bridge and Page, 1980; Barker, 1985b). The root gall index proposed by Bridge and Page (1980) is typical of those often used in the field (Fig. 9.10).

In commercial vegetable production, growers could use random field root monitoring of gall intensity on a regular basis at harvest to determine future damage potential in the next crop (Noling, 2003). The number of plants pulled from the soil can vary from a few up to 25 plants depending on a grower's willingness to take low or high risk in the following crop. This gives the grower a fair impression of the root knot nematode situation in his field and aids in making decisions on control strategies for the next crop. Yield losses and root gall indices have a linear relationship, which varies in degree as to crop and environmental conditions (Barker *et al.*, 1981). A nomograph of root knot galling indices is shown in Fig. 9.11, to aid in comparison of the results of different indices used in the literature.

Rotylenchulus

After *Meloidogyne*, the reniform nematode, *Rotylenchulus reniformis*, is the most important nematode affecting vegetables. The nematode attacks over 100 plant species including many vegetable crops and is a limiting factor in vegetable production, but is often neglected or overlooked where it occurs concomitantly with *Meloidogyne*. The nematode has been detected in more than 36 countries (Heald and Thames, 1982). It has been recorded in Hawaii where it was first described (Linford and Oliviera, 1940), and in the southern USA, Mexico, the Caribbean, South America, the Middle East, most of Africa, India, South-east Asia and the Pacific.

Symptoms of damage

Above-ground symptoms include stunting and leaf curling (Singh and Khera, 1979). Root necrosis and cortical necrosis have

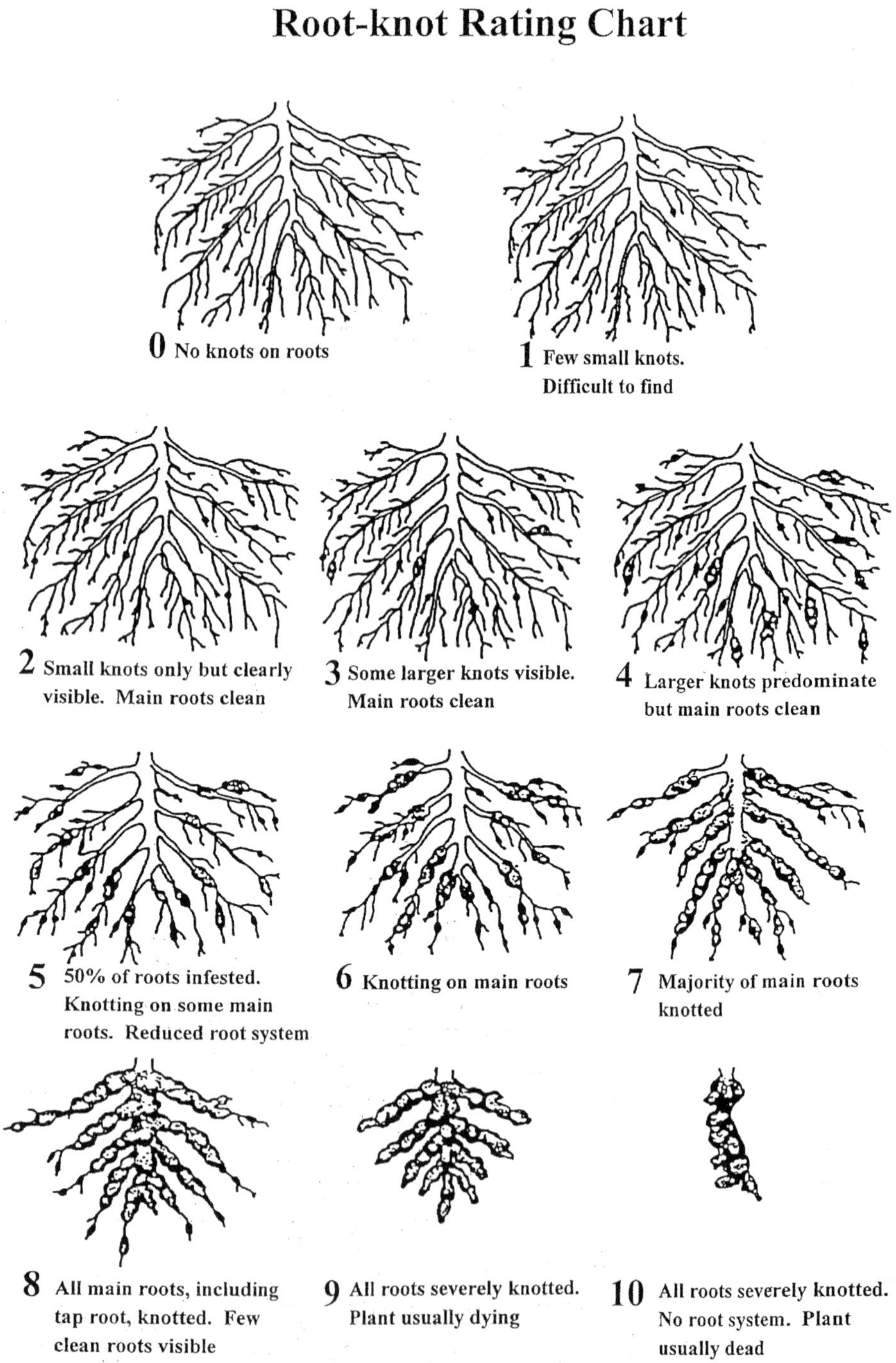

Fig. 9.10. Root galling rating scheme for evaluation of *Meloidogyne* infestation (Bridge and Page, 1980).

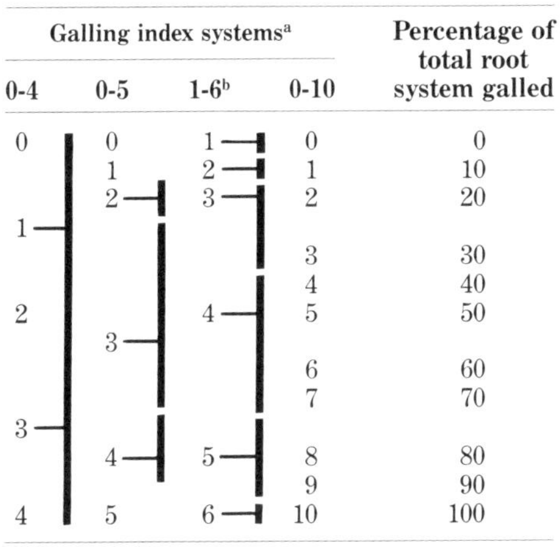

Fig. 9.11. Nomograph of root knot galling indices for *Meloidogyne* spp. (Barker, 1978).

been observed following infection. Cantaloupe growing in heavily infested soil was badly stunted and yields were greatly reduced (Heald, 1975). Leaf chlorosis can be produced (Bridge, 1983). Females and their adhering egg masses can be easily observed under the dissecting microscope (Fig. 9.12). Soil adhering to the gelatinous egg masses often gives them a dark appearance, aiding in detection.

Biology

Immature females penetrate the root and become sedentary. Galls are not produced. The life cycle is completed on okra in 24–29 days (Sivakumar and Seshadri, 1971). The existence of amphimictic and parthenogenetic races of *R. reniformis* has been demonstrated by Hirschmann and Triantaphyllou (1964).

The reniform nematode can survive in moist soil in the absence of hosts for 7 months, and for 6 months in dry soil. After 4 months, 84% of the nematodes were still alive (Sivakumar and Seshadri, 1979). Stoyanov (1971) reported that *R. reniformis* was able to survive 29 months in the absence of host plants.

The intensity of brinjal mosaic virus and okra yellow vein mosaic was promoted on plants parasitized by *R. reniformis* (Sivakumar and Merzainudeen, 1973; Naqvi and Alam, 1975). Charcoal rot caused by *Macrophomena phaseolina* on cantaloupe was significantly higher when the roots were infested with the reniform nematode (Carter, 1980).

Economic importance

Tomato yield was reduced following inoculation with 100 juveniles/plant (Singh and Khera, 1979). Snake gourd (*Trichosanthus dioica*) plants inoculated with 1000 nematodes were stunted and had smaller leaves than controls, and the roots were brown and showed cortical necrosis (Nath *et al.*, 1979). The nematode has been shown to damage a number of vegetable crops. Yield increases on okra, tomato, lettuce and squash of 19, 15, 57 and 69% were obtained with granular nematicides, respectively (Heald, 1978).

Management

CULTURAL. A 2 year rotation of cotton with sorghum was as effective as fumigation in reducing the nematode (Thames and Heald, 1974). Rotations which include soybeans resistant to the nematode also reduce densities (Gilman *et al.*, 1978). Nematode densities have also been reduced in rotations with maize, sugarcane and Pangola grass (Heald and Thames, 1982). A number of other crops are also known to be resistant to the nematode, including finger millet, groundnut, chillies, sugarcane and other grasses (Armstrong and Jensen, 1978; Bridge, 1983). Soil amendments such as animal manure and cotton seed cakes have been used with success to control the reniform nematode (Badra *et al.*, 1979). In glasshouse experiments, groundnut was a poor host of two populations of *R. reniformis* (Germani, 1978). Short periods of flooding of tomato in pot experiments reduced populations of the reniform nematode (Castillo *et al.*, 1976a). The nematode was also eradicated from infested soil following treatment with 50°C hot water for 5 min (Heald and Wayland, 1975).

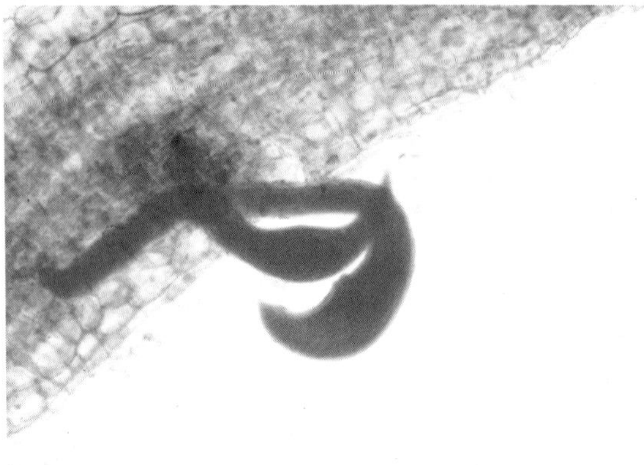

Fig. 9.12. Typical kidney-shaped females of *Rotylenchulus reniformis* on roots of tomato (R.A. Sikora).

Quénéhervé *et al.* (1998) suggested the use of *Mucuna pruriens* over *Tagetes erecta* and *Brachiaria decumbens* as a 3 month rotation crop prior to cultivation of short-term vegetable crops such as lettuce. A 2 year rotation with *Mucuna deeringiana* ploughed into the soil 3 months after planting versus non-incorporated was the most effective control measure for reniform nematodes on tomato, with yield increases of 176% (Acosta *et al.*, 1995). A number of other winter cover crops, incorporated into the soil before okra, did not have a major effect on *R. reniformis* or on yield (Guertal *et al.*, 1998).

Populations of *R. reniformis* dropped 86% following soil solarization and this was considered to provide season-long control on tomato even under conditions of abundant rainfall and extended cloud cover (Chellelmi *et al.*, 1994).

BIOLOGICAL. Very little has been done regarding biocontrol of this nematode, although it would be a good target for antagonists. *Paecilomyces lilacinus* reduced nematode densities and off-set damage to tomato at midseason and at harvest in field and greenhouse microplots (Walters and Barker, 1994). Sitaramaiah and Sikora (1982) were able to demonstrate that the penetration and reproduction of *R. reniformis* on tomato and cucumber were significantly reduced in the presence of the endomycorrhizal fungus *Glomus fasciculatum*. The results suggested that crop rotation to increase mycorrhizae might be important in regulating population densities.

RESISTANCE. There are only a few reports concerning resistance in vegetables to *R. reniformis*. In Egypt, the tomato cv. VFN 8 was shown to be moderately resistant to the reniform nematode (Oteifa and Osman, 1974). Balsubramanian and Ramakrishnan (1983) found that the tomato cvs Kalyanpur Sel 1 and Sel 2 were immune to the reniform nematode, while lines EC 118272 and EC 118276 were resistant.

CHEMICAL. A wide range of fumigant and non-fumigant nematicides is effective in controlling *R. reniformis* (Heald and Thames, 1982). The combination of nematicides and neem cake increased the yield of tomato and reduced reniform densities in field trials (Anitha *et al.*, 1998). Rich and Bird (1973) were able to reduce nematode penetration by a single foliar application of oxamyl. However, McSorley (1980) could not demonstrate effective nematode control following 6 weekly sprays with oxamyl on snap bean. All granular nematicides tested in Martinique reduced *R. reniformis* densities in tomato plots (Cadet, 1990).

Soil solarization combined with carbofuran increased tomato yields 96% and solarization with neem cake increased them 52% and reduced nematode densities

(Sharma *et al.*, 1996). In tests with cucumber, the growth and yield in soil mulched with clear plastic for 5 weeks were significantly greater than those in non-mulched soil and were related to lower reniform nematode densities and not changes in soil fertility (Coates *et al.*, 1998).

Nacobbus

There are at present three recognized species of the genus, *N. aberrans*, *N. bolivianus* and *N. dorsalis* (Manzanilla-Lopez *et al.*, 2002). *N. aberrans* is the species most commonly recorded, but it is known that it is a species complex (Reid *et al.*, 2003). One or more of these species are found in North, Central and South America and have also been detected in glasshouses in Europe. *N. aberrans* has been reported from cabbage, phaseolus beans, turnip, sweet pepper, chilli pepper, squash gourd, lettuce, tomato, *Cucumis sativus* and *Daucus carota*.

Symptoms of damage

The nematode produces galls similar in size to *M. hapla*. The galls are characteristically produced in strands or a bead-like fashion along the root (Plate 10A). The penetration of juveniles and immature females into the root can cause root necrosis (Bridge, 1983). Stunting, poor growth and chlorosis are typical above-ground symptoms associated with the endoparasitic nematode. Yield reduction can be significant (Schuster *et al.*, 1965). *N. aberrans* is an important pathogen in Mexico particularly on pepper, beans and tomato (Roman, 1978; Velasquez-Valle, 2001; Manzanilla-Lopez *et al.*, 2002).

The galls of *Nacobbus* spp. are often overlooked or mistaken for those produced by root knot nematodes, *Meloidogyne* species, because of the similarity in gall form. Galls only occur in the presence of the adult females.

Vargas *et al.* (1996) in greenhouse split-root tests demonstrated that *N. aberrans* can break down plant resistance to *Phytophthora capsici* in *Capsicum annuum*.

Biology

The females vary greatly in shape and will produce an egg sac that extends to the outside of the root (Clark, 1967; Johnson and Fassuliotis, 1984). According to Prasad and Webster (1967), the nematode completes a life cycle in 36 days at 25°C and in 43 days at 20 or 30°C. The genetic variability in *N. aberrans* and the existence of distinct geographical differences (Reid *et al.*, 2003) suggest that races may exist.

Control

Nacobbus can be controlled with both fumigant and non-fumigant nematicides. However, crop rotation with non-host crops is effective and more economical. Gomes (1973) reported that *Erodium cicutariuim* and *Brassica campestris* were not susceptible and hybrids of *Solanum andigenum* were resistant to the nematode. Bridge (1983) listed melon, squash, watermelon, groundnut, soybean, lucerne, oat, barley, rye, sorghum, wheat, maize, onion, okra, cotton, sunflower, *Phaseolus* spp., sesame, winged bean and rice as non-host crops that could be used in rotation. Because of the possible existence of races, retesting each crop with local populations is suggested as a necessary precaution.

Intercropping tomato with *Tagetes erecta*, independent of the planting date of *T. erecta* or spacing, showed a reduction in *N. aberrans* infection (Zavaleeta and Gomez, 1995).

Resistance to *N. aberrans* has been identified and confirmed in a range of *Lycopersicon* germplasm accessions, including those that possess genes for resistance to root knot nematodes (Veremis *et al.*, 1997).

Methods of diagnosis

The nematode can be easily detected by examining the root system during the growing season. Attention should be paid to the size of the galls and their orientation along the root system. If they are small and form bead-like strands along the root, they should be examined for *Nacobbus* females either by teasing out the females or by staining. However, when *Nacobbus* occurs in concomitant populations with *Meloidogyne*, it is often difficult to distinguish the different root galls.

Cyst Nematodes

Globodera

Globodera rostochiensis

The potato cyst nematode *G. rostochiensis* will infect and damage tomato and aubergine. The potato cyst nematode has been found infesting tomato in North, Central and South America (Bridge, 1983). The nematode is also present in Pakistan, India, the Mediterranean basin, South Africa and the Philippines. Symptoms include chlorosis, stunting and general poor growth. Detailed studies on yield losses and control, however, have not been reported for either crop.

Heterodera

Heterodera schachtii

This nematode has been found in Mexico (Sosa-Moss, 1986), the USA and Canada (Miller, 1986), Iraq (Stephan, 1986), Libya (Edongali, 1986), Senegal (Luc and Netscher, 1974) and Gambia (Bridge and Manser, 1980). The nematode causes significant losses on cruciferous crops. Yield reductions of 50% or more have been measured on Brussels sprouts, cabbage, broccoli and cauliflower when population densities are high (Miller, 1986). The nematode also attacks kale, Chinese cabbage, red beet, swede, spinach and turnip (Anonymous, 1987).

The sugarbeet cyst nematode is often found together with the cabbage cyst nematode, *H. cruciferae*, and, since cysts on the root system of both nematode species look similar (Plate 10B), proper identification is necessary in selecting control measures. Approximately 2–4 eggs/g of soil is used as a rough guideline for damage threshold levels in the Imperial Valley in California, USA (Anonymous, 1987). The nematode is controlled by long rotations or with fumigant nematicides (Lear *et al.*, 1966; Anonymous, 1987). Winter season crops and crops grown at higher altitudes are not damaged as severely.

Heterodera cruciferae

The cabbage cyst nematode, *H. cruciferae*, has been detected in California (Siddiqui *et al.*, 1973) and Libya (Edongali and Dabaj, 1982). The nematode causes significant damage to cruciferous crops in California, where it often occurs together in the same fields with *H. schachtii* (Anonymous, 1987). Although the nematode has many common hosts with the sugarbeet cyst nematode, its host range is somewhat smaller (Johnson and Fassuliotis, 1984). Seedlings infested with the nematode are stunted and exhibit interveinal chlorosis or leaf reddening (McCann, 1981). Cauliflower curd quality is reduced at 75 eggs/g of soil (Sykes and Winfield, 1966) and cabbage are severely stunted at 20 cysts/100 g of soil (McCann, 1981). Control is usually accomplished by crop rotation with non-host plants or by pre-planting fumigation (Anonymous, 1987).

Cactodera

Cactodera amaranthi

This cyst nematode has been found attacking spinach in central Mexico (Sosa-Moss, 1986), on *Amaranthus viridis* in Cuba (Stoyanov, 1972) and was detected in Florida (G. Rau, unpublished, cited in Luc, 1986). The host range of the nematode is limited to *A. viridis*, *A. spinosus* and *A. retroflexus* (Luc, 1986). Golden and Raski (1977) discussed the biology of the nematode.

METHODS OF DIAGNOSIS. All these endoparasitic, sedentary nematodes produce cysts on the surface of the root system at specific times in their life cycle (Plate 10B). The presence of cyst nematodes can be determined by carefully removing growing plants at different intervals during the growing season and examination of the roots with a hand lens. The detection of cysts imbedded in the root tissue is a clear sign of pathogenicity. Cysts can also be extracted from the soil using the techniques described in Chapter 3. The time of cyst appearance on the root surface is determined mainly by temperature. The cysts will also vary in colour from white through beige to dark brown. Cyst production and detection will also vary depending on the number of life cycles produced, e.g. the potato cyst nematode only has one generation per year, whereas the cabbage and sugarbeet cyst nematodes have many generations in a cropping season.

Ditylenchus

The onion race of the stem nematode, *D. dipsaci*, can cause severe damage to species of *Allium*, and especially garlic, in the winter season and in the cooler upland tropical and subtropical regions. The nematode is important on many crops in temperate regions of the world (Decker, 1969; Barker, 1991; Potter and Olthof, 1993). The nematode is a problem on lucerne in the subtropical regions of the USA, but does not seem to affect other crops in the region. The nematode has also been shown to cause severe injury to *Vicia faba* during the cool rainy winter growing season in the subtropical regions of North Africa (Saxena *et al.*, 1987). Vegetables growing in the warm tropics or during the summer season in the subtropics are not attacked. The nematode has been reported attacking species of *Allium* in a number of subtropical and tropical countries: Mexico, Venezuela, Ecuador, Peru, Colombia and the Dominican Republic, and various countries in the Mediterranean, Asia and the Pacific (Bridge and Hunt, 1986). In Morocco, the onion race of *D. dipsaci* was reported as causing severe injury to garlic, onion and peas, with infestation rates ranging from 55 to 100%.

Symptoms of damage

Penetration of onion leaves by this endoparasite causes leaf deformation and leaf swellings or blister-like areas on the surface (Plate 10C). The leaves grow in a disorderly fashion and often hang as if wilted. As the season progresses, they become chlorotic (Decker, 1969). Young plants can be killed when high infestations exist. Infected onions become swollen (bloat) and the bulbs may rot during storage (Bridge and Hunt, 1986). The inner scales of the bulb are usually more severely attacked than the outer scales. As the season advances, the bulbs become soft and, when cut open, show browning of the scales in concentric circles (Fig. 9.13). Conversely, *D. dipsaci* on garlic does not induce deformation or swellings, but causes leaf yellowing and death (Decker, 1969).

Biology

The fourth stage juveniles penetrate the stem and leaf tissue through the stomata. Egg laying begins at temperatures of 1–5°C with the optimum at 13–18°C. *D. dipsaci* completes one generation in 19–23 days at 15°C. Nematode activity stops at 36°C. The nematode prefers the cool moist climatic conditions existing in the upland tropics and wet winter seasons in the subtropics. *D. dipsaci* can parasitize plants on both heavy and light soils, although a higher incidence of infestation seems to occur on heavy soils.

Races

Although many races of *D. dipsaci* have been described (Sturhan, 1969), little to nothing is known about the race spectrum in those countries in the tropics where the nematode has been detected. It should be noted that most crops are usually attacked simultaneously by populations containing a mixture of races, which often makes determination of threshold levels difficult. The host range of many races has not been adequately determined. In Israel, two distinct races were identified, with one infecting onion and garlic but not *Phalaris*

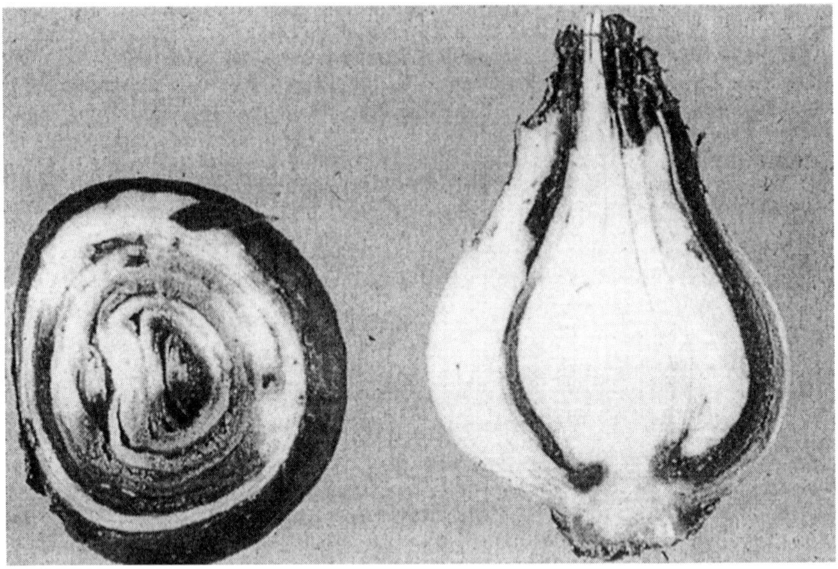

Fig. 9.13. *Ditylenchus dipsaci* induced browning of onion scales in cross- and longitudinal section of bulbs (R.A. Sikora from an AID Infodienst Bonn slide series).

canariensis, whereas the second infected *Phalaris* and oat but not onion or garlic (Aftalion and Cohn, 1990).

Survival and means of dissemination

The nematode can survive in the soil without a host plant for more than 1 year, and the fourth juvenile stage can survive in anabiosis for many years. The nematode can be disseminated by transportation in infested bulbs, plant residue and adhering soil. Seed-borne infections are also responsible for long-distance dissemination in onion, broad bean, beet and lucerne. Other hosts and weeds are responsible for maintaining infestations between onion and garlic. Bulbs harbouring light infestations will survive storage, and increase the level of infestation and losses in the following season when used as planting material. The nematode also attacks many weeds (Augustin and Sikora, 1989) present in field crops and these must be examined for host status, since high nematode densities can be maintained on these hosts. Abbad and Bachikh (2001) detected the nematode in 11 of 60 weed species growing in *Vicia faba* fields in Morocco, including the parasitic plant *Orobanche crenata*. A number of weed hosts lacking symptoms of infection were detected in fields after garlic cultivation in Brazil (Fonseca *et al.*, 1999).

Economic threshold level

According to Seinhorst (1956), the economic threshold level for onion is reached when ten or more nematodes are detected in 400 cm^3 of soil.

Management

Rotations with non-host crops for 3 years can be an effective means of control once the host range for a specific population or race is determined. Resistant cultivars of onion and garlic have not been developed for the commercial market (Bergquist and Riedei, 1972).

Fumigant nematicides are effective in reducing nematode infestation levels in the field but will not eradicate the nematode from the soil. The use of solarization combined with fumigation and granular nematicides increased the marketable yield of onion bulbs between 90 and 100% over the control (Lamberti *et al.*, 2001). However, solarization did not always improve results over individual chemical treatments nor did the use of non-fumigant nematicides after fumigation. Good control was attained in Italy with solarization for 4 weeks in July or August, with 1,3-dichloropropene or metham sodium in the autumn, or fenamiphos just before transplanting in January or February (Lamberti *et al.*, 2000). Greco *et al.* (1992) reported that more onion plants survived *D. dipsaci* infection and larger yields were obtained with a combination of solarization with 1,3-dicholoropropene. Non-fumigant nematicides were also effective in increasing marketable onion bulb yield and in reducing nematode densities in the soil (Greco *et al.*, 1992; Jaehn and Kimoto, 1995; Sasanelli *et al.*, 1995). The stem nematode can also be controlled in infested onion and garlic seed by treatment with methyl bromide (Hague, 1968; Infante and Sosa-Moss, 1971). The nematode can be controlled in onion bulbs by dipping in hot water at 44–45°C for 3 h (Bridge and Hunt, 1986). Temperature and time ratios are important for control and may vary with crop and cultivar. Formaldehyde was used until recently for control in onion bulbs and garlic seed cloves but has been removed from use for environmental and toxicological reasons. Jaehn (1995) reported that treatment of bulbs for 60 min in hot water at 49–50°C eradicated *D. dipsaci* from peeled seed bulbs of garlic.

Hot water dips without additives only partially controlled *D. dipsaci* when a warming pre-soak dip at 38°C for 30–60 min duration was followed by a hot water dip at 49°C for 15–30 min (Roberts and Matthew, 1995). Exposure to 49°C for 30 min caused slight retardations in emergence, but had no effect on the crop. Abamectin at 10–20 ppm as a 20 min hot dip at 49°C or as a 20 min cool dip at 18°C following a 20 min hot water dip was highly effective. Sodium hypochlorite in a 1.052–1.313% aqueous solution as a 20 min hot dip was highly effective in controlling *D. dipsaci*. Both treatments were non-toxic to the plant. Immersion of infected garlic cloves in solutions containing abamectin, however, led to yields equal to that of the uninfested controls and 56% higher than the untreated infested bulbs in trials conducted in California (Becker, 1999). In addition, 93% of the bulbs were nematode free compared with 46% in the nematode-infested control.

Gamma radiation at 0.1–0.5 kGy did not affect *D. dipsaci* in irradiated onion bulbs (Ignatowicz, 1998).

Methods of diagnosis

The presence of *D. dipsaci* can be easily determined by submerging small amounts of seed, stem, leaf or bulb tissue in water overnight to allow the active stages to escape. Detection in soil is more difficult because of the low population levels normally present. Consistent detection was obtained when sampling was done 1 week before harvesting by taking plants adjacent to plants showing symptoms (Jaehn and Kimoto, 1994).

Pratylenchus* and *Radopholus

Ten species of the lesion nematode, *Pratylenchus*, have been found in the rhizosphere or roots of vegetable crops: *P. brachyurus*, *P. barkati*, *P. dasi*, *P. coffeae*, *P. delattrei*, *P. loosi*, *P. singhi*, *P. thornei* and *P. zeae*. All species of *Pratylenchus* should be considered of potential importance when encountered in root tissue. Lesion nematodes are important parasites of many crops and are known to form disease complexes with many different soil-borne root-rotting fungi, thereby increasing root damage (Plate 10D). *P. brachyurus* and *P. zeae* have been detected in large numbers in the roots of vegetables. Little is known, however, about their impact on vegetable production. The over-riding importance of *Meloidogyne* in vegetable production, and the resulting lack of research on other plant parasitic nematode species, has limited our knowledge as to the exact importance of lesion nematodes in vegetable production.

The closely related burrowing nematode, *Radopholus*, has been detected in a number of vegetable crops, including kale, radish, tomato, aubergine, okra, carrot, onion, African spinach, watermelon, melon, calabash, pumpkin and squash. Crop loss studies have not been conducted.

Control

Lesion nematodes can be controlled with fumigant and non-fumigant nematicides, although this is probably not practical on an economic basis. Many species of *Pratylenchus* have wide host ranges, making the development of rotations difficult. A list of plants which have been reported to be resistant to the various species of *Pratylenchus* has been compiled by Armstrong and Jensen (1978). Ornat *et al.* (1999) recommended short-term clean fallow combined with root destruction between successive crops in intensive vegetable production for *P. neglectus* and root knot control. Yard waste compost did not affect final *Pratylenchus* densities in sweetcorn, yellow squash or okra in field tests, and in some cases reduced yield (McSorley and Gallaher, 1995).

Methods of diagnosis

Lesion nematodes produce small dark necrotic lesions on the root surface on many crops, which is the result of interrelationships with soil-borne fungal pathogens. The presence of lesions is a good indication that lesion nematodes are causing damage. The presence of the nematode should then be determined by extraction from the root tissue.

Belonolaimus

The sting nematodes, *B. gracilis*, *B. longicaudatus*, *B. euthychilus*, *B. maritimus* and *B. nortoni*, are common plant parasitic nematodes in the subtropical regions of the lower Coastal Plain of the south-eastern USA from Virginia to Florida and along the Gulf Coast into Texas. Note that the genus *Ibipora* found in Brazil is considered to be identical to *Belonolaimus*. Physiological races of *B. longicaudatus* have been detected (Abu-Gharbieh and Perry, 1970).

Symptoms

Damaged plants are stunted, chlorotic and wilt prematurely, with severe damage leading to plant death (Fig. 9.14). Nematode feeding induces stubby roots and necrotic lesions, which can expand to girdle the root. Perry and Rhoades (1982a) stated that 'infested areas consist of spots that vary in size and shape, but the boundary between diseased and healthy plants usually is fairly well defined'. Although disease complex associations have been detected on other hosts, they have not been observed on vegetable crops.

Biology

The nematodes are obligate parasites that cause damage to vegetables by feeding ectoparasitically on or near the root tip. The ectoparasite completes one generation within 28 days at an optimum temperature of 28–30°C.

Survival and means of dissemination

There is no definite survival stage in the life cycle of the nematode, with all stages of development present in the rhizosphere. The nematode may have been spread to many warmer regions of the world on golf course Bermuda grass sod (Perry and Rhoades, 1982a) but, because of its dependency on extreme sandy soil (Thames, 1959; Brodie and Quattlebaum, 1970), establishment has probably only occurred in a limited number of instances. The nematode seems to be most damaging on irrigated light soils, because of the nematode's requirement for uniform soil moisture, sandy soil and temperatures of 25–30°C for survival and multiplication.

Other hosts

The nematode causes severe damage to most agricultural crops including many wild plants and most vegetable crops

Fig. 9.14. Severely stunted celery due to *Belonolaimus longicaudatus* infestations in Florida (H. Rhoades).

(Graham and Holdeman, 1953; Good and Thornton, 1956; Robbins and Barker, 1973; Williams, 1974). Forage grasses and turf are also damaged by the nematode, whereas tobacco and watermelon are considered non-hosts. Because of the presence of races, variation in host range between populations should be expected.

Economic importance

Belonolaimus longicaudatus is the only species of the genus that has been shown to cause serious crop loss to vegetables. The species has been considered responsible for greater yield loss to vegetables in Florida than any other single plant pest of any type (Perry and Rhoades, 1982a). The nematode is highly pathogenic and even a single specimen in a soil sample can indicate that severe damage to a vegetable crop can occur. The sting nematode has been shown to damage a wide range of crops including okra, onion, celery, beetroot, cabbage, pepper, cucumber, pumpkin and carrot.

Management

CULTURAL. The addition of organic amendments that alter soil conditions has been shown to suppress the nematode because of its extreme sensitivity to changes in soil environmental conditions (Heald and Burton, 1968). Rotations designed to reduce population densities are difficult to select because of the wide host range, lack of resistant cultivars and possible presence of races in the species. A number of non-hosts are listed by Armstrong and Jensen (1978). Perry and Norden (1964) developed successful rotations using groundut, bahia grass and maize, although only the latter is a non-host throughout the nematode's range. The nematode did not reproduce on *Crotalaria spectabilis* in glasshouse tests (Rhoades, 1964) and, in the field, a summer cover crop of hairy indigo prevented a population increase (Rhoades, 1976a; Rhoades and Forbes, 1986). Fallowing and summer cover crops also reduced populations and increased yield (Rhoades, 1983). In field experiments, high populations developed on *Tagetes patula*, whereas low build-up was detected on joint vetch, *Aeschynomene americana* (Rhoades, 1980).

CHEMICAL. Nematicides are effective and have been widely used to control this nematode (Williams, 1974; Perry and Rhoades, 1982a). Good control has been obtained with pre-plant fumigant and non-fumigant nematicide treatment of cabbage and onion (Rhoades, 1969, 1971) and with both granular and transplant water application of non-fumigant nematicides on cabbage (Rhoades, 1976b). Johnson and Dickson (1973) obtained improved results when the nematicides were applied at planting as compared with pre-plant or post-plant treatments.

Methods of diagnosis

The nematode is an ectoparasite and can be easily extracted from sandy soils with modified Baermann dishes or sieving and elutriation techniques (see Chapter 3).

Trichodorus and *Paratrichodorus*

Species of stubby root nematodes, *Trichodorus* and *Paratrichodorus*, have been found throughout the world associated with vegetable crops. *Paratrichodorus minor* is considered an important limiting factor on vegetables grown in light soils in the subtropical regions (Perry and Rhoades, 1982b). *P. minor* attacks a wide range of vegetable crops and most other cultivated crop plants (Rohde and Jenkins, 1957; Perry and Rhoades, 1982b). *P. mirzai* and *T. viruliferus* are considered important on carrot and pepper, respectively. The stubby root nematodes prefer sandy or sandy-loam soils, but can occur in high numbers in organic soils (Perry and Rhoades, 1982b). This is probably true for all species in the two genera. The nematodes are ectoparasites feeding mainly on the root tip where damage suppresses elongation of the root and is responsible for the stubby root symptoms associated with these nematodes (Fig. 9.15; Plate 10E). The amount of damage to the root system varies with the vegetable crop attacked, but is characterized by reduced size and fewer, shorter rootlets (Johnson and Fassuliotis, 1984). The roots become discoloured and necrotic as the season advances. Netscher (1970) reported that *P. minor* caused a 50% reduction in root weight of tomato.

Plant growth is retarded and the foliage on stunted plants may become chlorotic (Christie and Perry, 1951). Some vegetables wilt when exposed to moisture stress. The nematodes cause severe crop losses to a variety of vegetable crops including onion, tomato, pepper, aubergine, beet, broccoli, Brussels sprout, cabbage, cauliflower, Chinese cabbage, radishes, swede, turnips, endive, lettuce and spinach (Anonymous, 2001, 2004).

Fig. 9.15. 'Stubby root' symptoms caused by the feeding of *Paratrichodorus minor* on maize (D.W. Dickson, in SON Slide Set-1).

Some species in these genera are also important vectors of virus diseases of horticultural crops. The plants, therefore, encounter both direct damage due to nematode feeding and indirect damage due to virus infection. *P. teres*, *P. pachydermus* and *T. similis*, for example, are the main vectors of tobacco rattle tobravirus that infects several ornamental bulb crops in Europe (Zoon *et al.*, 2002). With globalization and the movement of tubers and bulbs from producing countries in the upland tropics to consumers, these vectors could become of major importance.

Control by crop rotation is difficult in most cases because of the wide host range of this nematode. Ploughing reduced nematode numbers only moderately, whereas black radish decreased densities greatly. Some green manure crops increased nematode densities (Koot and Molendikk, 1997). *Crotalaria spectabilis* has been shown to be a non-host of the nematode and when used as a cover crop will reduce nematode densities (Rhoades, 1964). *Asparagus officinalis* var. *altilis* L. has also been shown to be resistant to attack, the resistance being induced by the production of a highly toxic glycoside (Rohde and Jenkins, 1958).

Fumigant and non-fumigant nematicides are effective in reducing initial damage and in giving the vegetable crop a head start on the nematode. However, it has been shown that nematode populations build up quickly (Perry, 1953). Some of the carbamate and phosphate non-fumigant nematicides exhibit longer durations of control than the fumigants (Rhoades, 1967).

Flooding for 2 weeks reduced populations significantly and the effect was improved by flooding followed by 2 weeks of drying (Overman, 1964). Yard waste organic amendments did not affect *P. minor* densities in the field (McSorley and Gallaher, 1995). Populations of *P. minor* on tomato following solarization were similar to that achieved with methyl bromide and chloropicrin fumigation in Florida (Chellelmi *et al.*, 1994).

Longidorus, Paralongidorus and *Xiphinema*

These nematodes have been shown to be potential problems in local areas. They can cause severe damage especially on sandy soils and are probably often overlooked wherever root knot nematodes predominate.

Xiphinema ifacolum increased fourfold on aubergine and tomato, and six- to tenfold on okra within 4 months. The nematode reduced okra and pepper growth and yield in the field, whereas *X. longicaudatum* severely depressed the growth of aubergine even though it seemed to be a poor host for the nematode (Lamberti *et al.*, 1992). *Longidorus africanus* caused damage to lettuce in the subtropical regions of southern California. Patchy growth and wilted seedlings were observed together with leaf margin chlorosis (Radewald *et al.*, 1969). Nematode feeding caused a reduction in elongation of the taproot and root tip swelling, typical of damage by a number of species of *Longidorus* and *Xiphinema* on other crops. Carrot and lettuce seedlings were shown to be highly sensitive to early attack by *L. africanus* (Huang and Ploeg, 2001). Delaying damage in the seedling stage was considered important in reducing damage in the field. *L. israelensis*, a parthenogenetic species, caused arrested root growth, root tip galling and deformed and forked taproots (Plate 10F). The nematode migrated to depths of 20–40 cm to survive hot, dry summer conditions (Peneva *et al.*, 1998). *L. vineacola* was reported to cause damage to celery in Israel (Cohn and Auseher, 1971).

Although viruliferous *X. americanum* have been found associated with watermelon, virus transmission does not seem to be a major problem in melon or vegetables (McGuire, 1982). The application of increasing amounts of sewage sludge to sandy soils led to a decrease in *X. basiri* and an increase in okra plant growth (Paulraj and Ramulu, 1994). Yard waste compost, however, did not affect *Xiphinema* densities in Florida field trials (McSorley and Gallaher, 1995).

Other Nematodes of Vegetables

Stunt nematodes are often found associated with vegetables. Twenty-two species of *Tylenchorhynchus* (three formerly named *Telotylenchus* and two *Quinisulcius*) and four species of *Merlinius* have been found in the rhizosphere of vegetable crops. With the exception of *Tylenchorhynchus brassicae*, none of the other species has been shown to be of significant economic importance on vegetable crops. *T. mashoodi* has been considered to be of potential importance on tomato.

T. brassicae has been detected in India, the Sultanate of Oman (Waller and Bridge, 1978) and Egypt (Oteifa and Eisharkawi, 1965). The nematode is a serious problem on most cruciferous crops and, when high populations of this nematode occur, growth is negatively affected (Khan, 1969). Of 22 vegetables, cabbage and cauliflower were the most suitable hosts. Large differences in the response of cultivars to the stunt nematode exist. One cauliflower cultivar and two accessions were considered resistant to *T. brassicae* in greenhouse tests (Pasha and Tiyagi, 1992). When the nematode was associated with *Rhizoctonia solani*, the emergence of vegetable seedlings was strongly reduced (Khan and Saxena, 1969).

The awl nematode, *Dolichodorus heterocephalus*, can cause damage to vegetables, especially on wet, sandy soils. In Florida, the nematode causes severe damage to tomato and celery, with losses on heavily infested soil often exceeding 50% (Tarjan *et al.*, 1952; Perry, 1953; Johnson and Fassuliotis, 1984). The nematode causes stubby root symptoms and severe root necrosis, indicating a long association with root-rotting fungi. The nematode can also attack the base of the hypocotyl where necrotic tissues can be observed (Johnson and Fassuliotis, 1984).

Spiral nematodes, *Helicotylenchus* spp. and *Scutellonema* spp., are commonly found in vegetable crops. Although more than 14 species of *Helicotylenchus* and three of *Scutellonema* have been detected in the rhizosphere of the various vegetable crops, none has been shown to be of economic importance in the field.

Species of *Hoplolaimus*, *Aorolaimus* (syn. *Peltamigratus*) and *Zygotylenchus* have been found in soil samples from vegetable crops. Their importance to vegetable production has not been determined.

Six species of ring nematode, *Criconemoides* (syn. *Criconemella* and *Macroposthonia*), have been detected in the rhizosphere of a wide range of vegetables. These nematodes are known to increase to high numbers in many subtropical soils and have been implicated as important limiting factors on a number of perennial crops and could be important on vegetables.

Future Prospects

Vegetable production is increasing in most subtropical and tropical countries in contrast to forced reductions in production being experienced in Western Europe and North America. This is seen in the large increase in land in vegetable production in the tropics and subtropics over the 1990s (Tables 9.2 and 9.3) – much of this related to cheaper production and upgraded transport. Similarly, this increase in the subtropics and tropics is associated with increased needs for inputs in the form of fertilizers and pesticides, components being reduced in Western Europe and North America, because of a reduction in farm subsidies and public awareness of the impact of agricultural inputs on the environment. Emphasis must be placed on preventing the spread of new and important species to uninfested areas. This will be important as globalization expands and the movement of fresh produce increases. The need for taxonomists with molecular skills to identify some of the very closely related species that occur simultaneously in the field has become evident.

Determination of threshold levels will be required to aid in selection of specific control measures for pest management programmes. Whether precision farming technology can help in this direction is still unknown. Vegetables are often attacked simultaneously by a multitude of different plant parasitic nematodes. This requires an expanded view of threshold levels, involving the effects of all the species involved. This is especially important where alternatives to fumigation are not available. Therefore, when determining damage intensity in the field, composite threshold levels, which include the inter-relationships between all economically important nematode species, will be needed.

The loss of methyl bromide for root knot management has negatively affected many commercial growers around the world. At the same time it has simultaneously generated a great deal of research to find effective alternatives – many of them discussed above in this chapter. It has also demonstrated the need to increase positions and research support in nematology. Alternatives cannot be found just by wishing and talking about farmer needs.

The development of resistant cultivars is playing an important role today and will increase in importance in the future, as is the use of grafting nematode-resistant root stocks. The development of transgenic cultivars is still not a reality, but could become an important tool for effective nematode management. The detection of resistant breaking races of *Meloidogyne* and new species able to break known genes for resistance underline the need to stress resistance management.

If environmentally safe nematicides are available in the future, an increase in use can be expected in many subtropical and tropical vegetable-growing regions, if only because of the loss of methyl bromide and the toxicity of many non-fumigant nematicides still on the market. The use of seed coating, root drenches and monitored application through drip irrigation is new to nematology and will reduce many negative side effects of the past. A new generation of nematicides that are both effective and safe is urgently needed.

There will always be an imbalance in the availability of pesticides between commercial growers and resource-limited growers, with the latter in most cases excluded for cost reasons. In these regions, resistance and cropping systems research must be strengthened. Success with antagonistic plants, grafting, biofumigation, trap cropping and management of antagonistic potential in the field must be looked at more closely for integrated approaches that are economically acceptable.

Biological control is an alternative that has become a reality with the development of cost-effective solid-state fermentation equipment and new formulations that reduce costs of transport and facilitate application. Commercial production of biologically enhanced seed and transplants needs to be promoted for early root protection. Suitable antagonists exist; the technology is available.

The 'all or nothing approach' to nematode control, or 'fumigate them', is a thing of the past. We need to improve 'nematode management' and maintain these pests at or below threshold levels. 'Living with them' is the concept of the future.

References

Abawi, G.S. and Barker, K.R. (1984) Effects of cultivars, soil temperature and population levels of *Meloidogyne incognita* on root necrosis and *Fusarium* wilt of tomatoes. *Phytopathology* 74, 333–338.

Abawi, G.S., Ludwig, J.W., Morton, H.V. and Hofer, D. (2003) Efficacy of abamectin as a seed treatment against *Meloidogyne hapla* and *Pratylenchus penetrans*. *Journal of Nematology* 35, 321–322.

Abbad, A.F. and Bachikh, J. (2001) Studies on the host range of *Ditylenchus dipsaci* in Morocco. *Nematologia Mediterranea* 29, 51–57.

Abu-Gharbieh, W.I. and Perry, V.G. (1970) Host differences among Florida populations of *Belonolaimus longicaudatus* Rau. *Journal of Nematology* 2, 209–216.

Abu-Gharbieh, W.I., Saleh, H.M. and Abu-Blan, H.A. (1987) Effect of 'solarization' using different thickness of black plastic tarping on soil-borne pathogens. Workshop, *Advances in Agriculture Through Plastic House Technology*, Amman, Jordan.

Abu-Gharbieh, W.I., Saleh, W.I.H. and Al Banna, L. (1991) Application of solar-heated water for soil solarization. *FAO Plant Production and Protection Paper* 109, 69–77.

Aceret, P.S., Trevorrow, P., Akiew, S., Tonello, P., DeFaveri, C. and Visini, J. (1999) Field evaluation of nematicidal products to control root-knot nematode (*Meloidogyne* spp.) of tobacco in Queensland. In: *12th Biennial Conference of the Australasian Plant Pathology Society*, Canberra. Conference Handbook, p. 217.

Acosta, N., Roman, O., Vicente, N.E. and Sachez, L.A. (1991) Sistemas de rotacion de cosechas y los niveles poblacionales de nematodos. *Journal of Agriculture of the University of Puerto Rico* 75, 399–405.

Acosta, N., Vargas, R., Roman, O., Vicente, N. and Sanchez, L. (1995) *Mucuna deeringiana* incorporada vs. no incorporada al suelo y rendimento en siembras subsiquientes de tomate, habichuelk y maiz. *Journal of Agriculture of the University of Puerto Rico* 79, 65–74.

Aftalion, B. and Cohn, E. (1990) Characterization of two races of the stem bulb nematode (*Ditylenchus dipsaci*) in Israel. *Phytoparasitica* 18, 229–232.

Ahuja, S. (1978) The effect of root-dip treatments on infestation of brinjal by *Meloidogyne incognita*. *Nematologia Mediterranea* 6, 133–134.

Anitha, B. and Subramanian, S. (1998) Management of the reniform nematode *Rotylenchulus reniformis* in tomato. In: Mehta, U. (ed.) *Proceedings of the Third International Symposium of Afro-Asian Society of Nematologists*. Sugarcane Breeding Institute, Coimbatore, India, pp. 249–250.

Anonymous (1985) Nematodes. In: *University of California Statewide Integrated Pest Management Project. Integrated Pest Management for Tomatoes.* Division of Agriculture and Natural Resources. Publication 3274, pp. 61–65.

Anonymous (1987) Nematodes. In: *University of California Statewide Integrated Pest Management Project. Integrated Pest Management for Cole Crops and Lettuce.* Division of Agriculture and Natural Resources. Publication 3307, pp. 81–88.

Anonymous (1998a) *Methyl Bromide – Getting Ready for the Phase Out.* United Nations Environment Programme, Industy and Environment. Ozone Action Programme, Paris.

Anonymous (1998b) *Report of the Methyl Bromide Technical Options Committee – 1998 Assessment of Alternatives to Methyl Bromide.* United Nations Environment Programme, Ozone Secretariat, Nairobi, Kenya.

Anonymous (2001) *Crop Protection Compendium, Global Module*, 3rd edn. CAB International, Wallingford, UK.

Anonymous (2003) *FAO Agricultural Production Statistics.* Available at: apps.fao.org

Anonymous (2004) Institute of Food and Agricultural Sciences IFAS *Florida Nematode Control Guide.* Entomology and Nematology Department and Cooperative Extension Service, University of Florida, Gainesville, Florida. Available at: edis.ifas.ufl.edu/CV112

Armstrong, J.M. and Jensen, H.J. (1978) *Indexed Bibliography of Nematode-resistance in Plants.* Corvallis, Oregon. *Agricultural Experiment Station Bulletin* 639.

Asmus, G.L., Andrade, P.J.M., Lafourcade, A.G. and Melo, A.P.J. (1995) Reacao de cultivares de milho (*Zea mays*) ao nematoide de galhas (*Meloidogyne javanica* (Treub) Chitwood). *Pesquisa em Andamento – EMBRAPA – Centro de Pesquisa Agropecuaria do Oeste* 1, 1–5.

Atkins, S.D., Hidalgo-Diaz, L., Kalisz, H., Mauchline, T.H., Hirsch, P.R. and Kerry, B.R. (2003) Development of a new management strategy for the control of root-knot nematodes (*Meloidogyne* spp.) in organic vegetable production. *Pest Management Science* 59, 183–189.

Augustin, B. and Sikora, R.A. (1989) Studies on the host range of the normal and giant faba bean races of *Ditylenchus dipsaci* in Syria and Germany. *Nematologia Mediterranea* 17, 63–66.

Augustin, B., Graf, V. and Laun, N. (2002) Temperature influencing efficiency of grafted tomato cultivars against root-knot nematode (*Meloidogyne arenaria*) and corky root (*Pyrenochaeta lycopersici*). *Zeitschrift für Pflanzenkrankheiten und Pflanzenschutz* 109, 371–383.

Badra, T., Salem, M.A. and Oteifa, B.A. (1979) Nematicidal activity of some organic fertilizers and soil amendments. *Revue de Nematologie* 2, 29–36.

Balasubramanian, P. and Ramakrishnan, C. (1983) Resistance to the reniform nematode *Rotylenchulus reniformis* in tomato. *Nématologia Mediterranea* 11, 203–204.

Barcelo, P., Sorribas, F.J., Ornat, C. and Verdejo-Lucas, S. (1997) Weed hosts to *Meloidogyne* spp. associated with vegetable crops in northeast Spain. In: Albajes, R. and Carnero, A. (eds) *Integrated Control in Protected Crops 'Mediterranean Climate'*. Avignon, France. *IOBC/WPRS Bulletin*, pp. 89–93.

Barker, K.R. (Chairman) (1978) Determining nematode population responses to control agents. In: Zehr, E. (ed.) *Methods for Evaluating Plant Fungicides, Nematicides, and Bactericides.* American Phytopathological Society, St Paul, Minnesota, pp. 114–125.

Barker, K.R. (1985a) Sampling nematode communities. In: Barker, K.R., Carter, C.C. and Sasser, J.N. (eds) *An Advanced Treatise on Meloidogyne, Volume I, Methodology.* North Carolina State University Graphics, Raleigh, North Carolina, pp. 3–17.

Barker, K.R. (1985b) Nematode extraction and bioassays. In: Barker, K.R., Carter, C.C. and Sasser, J.N. (eds) *An Advanced Treatise on Meloidogyne, Volume I, Methodology.* North Carolina University Graphics, Raleigh, North Carolina, pp. 19–35.

Barker, K.R. (1991) Rotation and cropping systems of nematode control: the North Carolina experience – introduction. *Journal of Nematology* 23, 342–343.

Barker, K.R. and Olthof, T.H.A. (1976) Relationships between nematode population density and crop responses. *Annual Review of Phytopathology* 14, 327–353.

Barker, K.R., Todd, F.A., Shane, W.W. and Nelson, L.A. (1981) Interrelationships of *Meloidogyne* species with flue-cured tobacco. *Journal of Nematology* 13, 67–79.

Barker, K.R., Schmitt, D.P. and Imbriani, J.L. (1985) Nematode population dynamics with emphasis on determining damage potential to crops. In: Sasser, J.N. and Carter, C.C. (eds) *An Advanced Treatise on Meloidogyne Volume I. Methodology.* North Carolina University Graphics, Raleigh, North Carolina, pp. 135–148.

Barker, K.R., Pederson, G.A. and Windham, G.L. (eds) (1998) *Plant and Nematode Interactions.* American Society of Agronomy, Madison, Wisconsin.

Basile, M., Carella, A. and Basile, A.C. (1993) Efficacia delle somministrazioni localizzate di fenamiphos al terreno nella lotta contro il nematode galligeno: *Meloidogyne incognita* su pomodoro. *Informatore Fitopatologico* 43, 57–59.

Becker, J.O. and Hofer, D. (2004) Efficacy comparison between seed-coated and soil-applied nematicides in root-knot nematode-infested cucumber fields. *Phytopathology Supplement* 94, 7.

Becker, J.O., Slaats, G. and Hofer, D. (2003) Cucumber seed coating with abamectin guards against early root damage by root-knot nematodes. *Phytopathology Supplement* 94, 149.

Becker, W.F. (1999) The effect of abamectin on garlic infected by *Ditylenchus dipsaci. Nematologia Brasileira* 2, 1–8.

Bello, A. (1998) Biofumigation and integrated crop management. In: Bello, A., González, J.A., Arias, M. and Rodríguez-Kábana, R. (eds) *Alternatives to Methyl Bromide for the Southern European Countries.* DG-XI, EU, CSIC, Madrid, pp. 99–126.

Bergquist, R.R. and Riedei, R.M. (1972) Screening onion (*Allium cepa*) in a controlled environment for resistance to *Ditylenchus dipsaci. Plant Disease Reporter* 56, 329–331.

Berthou, F., Ba-Diallo, A. and de Guiran, G. (1989) Caracterisation chez les nematodes *Meloidogyne* Goeldi (Tylenchida) de types virulents viv a vis du gene *Mi* de la tomate dans deux zolles maraicheres au Senegal. *Agronomie* 9, 877–885.

Birchfield, W. and Antonpoulos, A.A. (1976) Scanning electron microscope observations of *Duboscquia penetrans* parasitizing root-knot larvae. *Journal of Nematology* 8, 272–283.

Black, L.L., Wu, D.L., Wang, J.F., Kalb, T., Abbass, D. and Chen, J.H. (2002) Grafting tomatoes for production in the hot-wet season. *AVRDC International Cooperators' Guide,* Shanhua, Taiwan.

Bourne, J.M. (2001) Making a soil suppressive to root-knot nematodes by applications of *Verticillium chlamydosporium.* In: Sikora, R.A. (ed.) *Integrated Control of Soil Pests.* Proceeding of the IOBC-WPRS-Tri-trophic Interactions in the Rhizosphere and Root Health. Bad Honnef, Germany 1999. *Bulletin OILB-SROP* 25–30.

Bourne, J.M. and Kerry, B.R. (1999) Effect of the host plant on the efficacy of *Verticillium chlamydosporium* as a biological control agent of root-knot nematodes at different nematode densities and fungal application rates. *Soil Biology and Biochemistry* 31, 75–84.

Bridge, J. (1983) Nematodes. In: *Pest Control in Tropical Tomatoes.* Centre for Overseas Pest Research, Overseas Development Administration, London, pp. 69–84.

Bridge, J. (1987) Control strategies in subsistence agriculture. In: Brown, R.H. and Kerry, B.R. (eds) *Principles and Practice of Nematode Control in Crops.* Academic Press, New York, pp. 389–420.

Bridge, J. and Hunt, D. (1986) Nematodes. In: Tropical Development and Research Institute and Office of Overseas Development Administration. *Pest Control in Tropical Onions.* Tropical Development and Research Institute, London, pp. 65–77.

Bridge, J. and Manser, P.D. (1980) The beet cyst nematode in tropical Africa. *Plant Disease* 64, 1036.

Bridge, J. and Page, S.L.J. (1977) *An Assessment of the Importance and Control of Plant-parasitic Nematodes in Malawi.* ODM report on the visit to Malawi, 2 January to 21 April 1977. Ministry of Overseas Development, Ascot, UK.

Bridge, J. and Page, S.L.J. (1980) Estimation of root-knot nematode infestation levels on roots using a rating chart. *Tropical Pest Management* 26, 296–298.

Brito, J., Stanley, J., Cetintas, R., Powers, T.O., Inserra, R.N., McAvoy, E.J., Mendes, M.L., Crow, W.T. and Dickson, D.W. (2004a) Identification and host preference of *Meloidogyne mayaguensis* and other root-knot nematodes from Florida, and their susceptibility to *Pasteuria penetrans*. *Journal of Nematology* 36, 308–309 [abstract].

Brito, J.A., Stanley, J.D., Cetintas, R., DeVito, M., Thies, J.A. and Dickson, D.W. (2004b) Reproduction of *Meloidogyne mayaguensis* from Florida on root-knot nematode resistant tomato and pepper. *Journal of Nematology* 36, 308 [abstract].

Brito, J.A., Stanley, J.D., Mendes, M.L. and Dickson, D.W. (2004c) Host status of selected plant species to *Meloidogyne mayaguensis* from Florida. *Nematropica* 33, 99[abstract].

Brito, J.A., Inserra, R., Lehman, P. and Dixon, W. (2004d) Available at: www.doacs.state.fl.us/pi/enpp/nema/m-ayaguensis.html

Brodie, B.B. and Quattlebaum, B.H. (1970) Vertical distribution and population fluctuations of three nematode species as correlated with soil temperature, moisture and texture. *Phytopathology* 60, 1286.

Brown, R.H. and Kerry, B.R. (eds) (1987) *Principles and Practice of Nematode Control.* Academic Press, New York.

Brückner, S. (2004) Effectiveness and application strategies of the biological nematicide Bioact®WG. In: *Proceedings of an International Workshop on Development of Biocontrol Agents of Diseases for Commercial Applications in Food Production Systems.* Sevilla, Spain, p. 89.

Bunte, R. and Muller, J. (1996) Einfluss resistenter Ölrettich-Genotypen auf die Abundanzdynamick von *Meloidogyne hapla* und *M. incognita. Zeitschrift für Pflanzenkrankheiten und Pflanzenschutz* 103, 527–534.

Bunte, R., Muller, J. and Friedt, W. (1997) Genetic variation and response to selection for resistance to root-knot nematodes in oil radish (*Raphanus sativus* ssp. *oleiferus*). *Plant Breeding* 116, 263–266.

Byrd, D.W., Ferris, H. and Nusbaum, C.J. (1972) A method for estimating number of eggs of *Meloidogyne* spp. in soil. *Journal of Nematology* 4, 266–299.

Cabanillas, F. and Barker, K.R. (1989) Impact of *Paecilomyces lilacinus* inoculum level and application time on control of *Meloidogyne incognita* on tomato. *Journal of Nematology* 21, 115–120.

Cadet, P. (1990) Influence de quelques nematicides sur les rendements et la multiplication des nematodes parasites et la tomate a la Martinique. *Bulletin Agronomique des Antilles et de la Guyane* 10, 42–48.

Cantliffe, D.J. and Vansickle, J.J. (2003) Competitiveness of the Spanish and Dutch greenhouse industries with the Florida fresh vegetable industry. *University of Florida Extension Service.* Available at: edis.ifas.ufl.edu

Carter, W.W. (1980) Interaction of *Rotylenchulus reniformis* and *Macrophomina phaseolina* in charcoal rot of cantaloupe. *Journal of Nematology* 12, 217.

Castillo, M.B., Alegar, M.S. and Harwood, R.R. (1976a) Effect of flooding on pot populations of *Rotylenchulus reniformis* and *Meloidogyne incognita. Philippine Phytopathology* 12, 11–16.

Castillo, M.B., Arceo, M.B. and Litsinger, J.A. (1976b) Nematodes in cropping patterns. IV. Populations of plant parasitic nematodes associated with cropping patterns under different rice-growing environments in Manaoag, Pangasinan. *Philippine Phytopathology* 12, 24–29.

Castro, A.A.E., Zavaleta, M.E., Cid del Prado, V.I. and Zamudio, G.V. (1990) Rotacion e incorporacion de *Tagetes erecta* L. para el manejo de *Meloidogyne incognita* (Kofoid and White) Chitwood en el coltivo de tomate (*Lycopersicon esculentum* Miill.) en tecamachalco, Puebla. *Revista Mexicana de Fitopatologia* 8, 173–180.

Cayrol, J.C., Djian, C. and Frankowski, J.P. (1993) Efficacy of abamectin B1 for the control of *Meloidogyne arenaria. Fundamentals of Applied Nematology* 16, 239–246.

Chellemi, D.O., Olson, S.M., Scott, J.W., Mitchell, D.J. and McSorley, R. (1994) Reduction of phytoparasitic nematodes on tomato by soil solarization and genotype. *Journal of Nematology* 25, 800–805.

Chen, Z.X. and Dickson, D.W. (1998) Review of *Pasteuria penetrans*: biology, ecology and biological control potential. *Journal of Nematology* 30, 313–340.

Chindo, P.S., Khan, F.A. and Erinle, I.D. (1991) Reaction of three tomato cultivars to two vascular diseases in presence of the root-knot nematode, *Meloidogyne incognita* race 1. *Crop Protection* 10, 62–64.

Chitwood, M.G. (1949) 'Root knot nematodes' Part I. A revision of the genus *Meloidogyne* Goeldi, 1887. *Proceedings of the Helminthological Society of Washington* 16, 90–104.

Christie, J.R. and Perry, V.G. (1951) A root disease of plants caused by a nematode of the genus *Trichodorus. Science* 113, 491–493.

Clark, S.A. (1967) The development and life history of the false root-knot nematode, *Nacobbus serendipiticus. Nematologica* 13, 91–101.
Coates, B.P.L., Cohen, J.E., Ogle, L.R., Prendergasts, C.H. and Riley, D.M. (1998) Mulching soil to increase yield and manage plant parasitic nematodes in cucumber (*Cucumis sativus* L.) fields: influence of season and plastic thickness. *Nematropica* 28, 81–93.
Cohn, E. and Auscher, R. (1971) Seasonal occurrence of *Longidorus vineacola* on celery in Israel and its control. *Israel Journal of Agricultural Research* 21, 23–25.
Colyer, P.D., Kirkpatrick, T.L.L., Vernon, P.R., Barham, J.D. and Bateman, R.J. (1998) Reducing *Meloidogyne incognita* injury to cucumber in a tomato–cucumber double-cropping system. *Journal of Nematology* 30, 226–231.
Cook, R. and Evans, K. (1987) Resistance and tolerance. In: Brown, R.H. and Kerry, B.R. (eds) *Principles and Practice of Nematode Control in Crops.* Academic Press, New York, pp. 179–231.
Crow, W.T., Gueertal, E.A. and Rodriguez-Kabana, R. (1996) Responses of *Meloidogyne arenaria* and *M. incognita* to green manures and supplemental urea in glasshouse culture. *Journal of Nematology – Supplement* 28, 648–654.
Csinos, A.S., Sumner, D.R., Johnson, W.C., Johnson, A.W., McPherson, R.M. and Dowler, C.C. (2000) Methyl bromide alternatives in tobacco, tomato and pepper transplant production. *Crop Protection* 19, 39–49.
Cuadra, R.M. (1983) Número de generaciones y ciclo biológico de *Meloidogyne incognita* (Kofoid y White) Chitwood en Cuba. *Ciencia Agricultura* 20, 3–10.
Cuadra, R.M., Cruz, X. and Fajardo, J.F. (2000) Cultivos de ciclo corto como plantas trampas para el control del nematodo agallador. *Nematropica* 30, 241–246.
Culbreath, A.K., Rodrigues-Kabana, R. and Morgan-Jones, G. (1985) The use of hemicellulosic waste matter for reduction of the phytotoxic effect of chitin and control of root-knot nematodes. *Nematropica* 15, 49–75.
Dao, D.F. (1970) Climatic influence of temperature on distribution pattern of plant parasitic and soil inhabiting nematodes. *Mededelingen van de Landbouwhogeschool, Wageningen* 70, 187.
Davide, R.G. and Zorilla, R.A. (1983) Evaluation of a fungus, *Paecilomyces lilacinus* (Thom.) Samson, for the biological control of the potato cyst nematode *Globodera rostochiensis* Woll. as compared with some nematicides. *Philippine Agriculture* 66, 397–404.
Decker, H. (1969) *Phytonematologie.* VEB Deutscher Landwirtschaftsverlag, Berlin.
de Guiran, G. (1979) A necessary diapause in root-knot nematodes. Observations on its distribution and inheritance in *Meloidogyne incognita. Revue de Nématologie* 2, 223–231.
de Guiran, G. and Demeure, Y. (1978) Influence du potentiel hydrique des sols sur les masses d'oeufs de *Meloidogyne incognita. Revue de Nématologie* 1, 119–134.
de Guiran, G. and Villemin, M.A. (1980) Specificite de la diapause embryonnaire des oeufs de *Meloidogyne incognita* (Nematoda). *Revue de Nématologie* 3, 115–121.
Deberdt, P., Quénéhervé, P., Darrasse, A. and Prior, P. (1999) Increased susceptibility to bacterial wilt in tomatoes by nematode galling and the role of the *Mi* gene in resistance to nematodes and bacterial wilt. *Plant Pathology* 48, 408–414.
De Leij, F.A.A.M. and Kerry, B.R. (1991) The nematophagous fungus *Verticillium chlamydosporium* as a potential biological control agent for *Meloidogyne arenaria. Revue de Nématologie* 14, 157–164.
de Moura, R.M., Echandi, E. and Powell, N.T. (1975) Interaction *of Corynebacterium michiganense* and *Meloidogyne incognita* on tomato. *Phytopathology* 65, 1332–1335.
Demeure, Y. (1977) Persistance de l'infestation d'un sol par *Meloidogyne* sp. en saison seche au Senegal. *Cahiers ORSTOM, Serie Biologie* 11, 167–172.
Demeure, Y. (1978) Les Causes de la Survie de Certains Nematodes Phytoparasites (*Scutellonema cavenessi* et *Meloidogyne* spp.) Pendant la Saison Seche dans le Sahel Senegalais. These presentee devant l'Universite Claude Bernard (Lyon I).
Demeure, Y. and Netscher, C. (1973) Methode d'estimation des populations de *Meloidogyne* dans le sol. *Cahiers ORSTOM, Serie Biologie* 21, 85–90.
Dickson, D.W. and Strubel, F.B. (1965) A sieve-straining technique for extraction of egg masses of *Meloidogyne incognita* from soil. *Phytopathology* 50, 215–217.
Di Vito, M., Greco, N. and Carella, A. (1986) Effect of *Meloidogyne incognita* on the yield of eggplant. *Journal of Nematology* 18, 487–490.
Di Vito, M., Zaccheo, G. and Di Vito, M. (1991a) Population density of root-knot nematode, *Meloidogyne incognita* and growth of artichoke (*Cynara scolymus*). *Advances in Horticultural Science* 5, 81–82.

Di Vito, M., Cianciotta, V., Zaccheo, G. and Di Vito, M. (1991b) The effect of population densities of *Meloidogyne incognita* on yield of susceptible and resistant tomato. *Nematologia Mediterranea* 19, 265–268.

Dropkin, V. (1969) The necrotic reaction of tomatoes and other hosts resistant to *Meloidogyne*: reversal by temperature. *Phytopathology* 59, 1632–1637.

Duc, T.M. (1980) *Systeme de Cultures Irriguees a Partir des Eaux Sous-terraines.* Collection CIEH-CEFIGRE Bamako.

Dunay, M.C. and Dalmasso, A. (1985) Multiplication de *Meloidogyne javanica, M. incognita,* et *M. arenaria* sur divers *Solanum. Revue de Nématologie* 8, 312–334.

Eddaoudi, M. and Ammati, M. (1995) Study on the effects of single and combined soil solarization and the use of resistant tomato varieties on the populations of *Meloidogyne javanica* and the production of tomato in South Morocco. *Afro-Asian Journal of Nematology* 5, 28–33.

Eddaoudi, M., Ammati, M. and Rammah, A. (1997) Identification of the resistance breaking populations of *Meloidogyne* on tomatoes in Morocco and their effect on new sources of resistance. *Fundamental and Applied Nematology* 20, 285–289.

Edongali, E.A. (1986) Cyst forming nematodes in Libya. *NATO Advanced Study Institute, Series A: Life Sciences* 121, 389.

Edongali, E.A. and Dabaj, M. (1982) Cauliflower (*Brassica oleracea* cv. *capitata*) cyst nematode in Libya. *Libya Journal of Agriculture* 11, 205.

El Hamawi, M.H. and Mohamed B.E. (1990) The effect of marigold plants *Tagetes erecta* on infection of some vegetable crops with the root-knot nematode (*Meloidogyne incognita,* Chitwood 1949). *Bulletin of Faculty of Agriculture, University of Cairo* 41, Supplement, 1013–1021.

Esparrago, G., Barreiro, J.M., Ruales, C. and Bieche, B.J. (1999) Comparison of the reproduction of *Meloidogyne* populations on roots of *Crotalaria spectablilis* and processing tomato varieties with the *Mi* gene. *Acta Horticulturae* 487, 267–270.

Esser, R.P., Perry, V.G. and Taylor, A.L. (1976) A diagnostic compendium of the genus *Meloidogyne* (Nematoda: Heteroderidae). *Proceedings of the Helminthological Society of Washington* 43, 138–150.

Fargette, M. (1987) Use of the esterase phenotype in taxonomy of the genus *Meloidogyne.* 2. Esterase phenotypes in West African populations and their characterization. *Revue de Nématologie* 10, 45–55.

Fassuliotis, G. (1979) Plant breeding for root-knot nematode resistance. In: Lamberti, F. and Taylor, C.E. (eds) *Root-knot Nematodes* (Meloidogyne *species). Systematics, Biology and Control.* Academic Press, London, pp. 425–453.

Faulkner, L.R. and Bolander, W.J. (1970) Acquisition and distribution of nematodes in irrigation waterways of the Columbian Basin in Eastern Washington. *Journal of Nematology* 2, 362–367.

Feldmesser, J., Edwards, D.I., Epps, J.M., Heald, C.M., Jenkins, W.R., Jensen, H.J., Lear, B., McBeth, C.W., Nigh, E.L. and Perry, V.G. (1971) Estimated crop losses from plant-parasitic nematodes in the United States. Comm. Crop Losses. *Special Publication No. 1.* Society of Nematologists, Hyattsville, Maryland.

Fernández, E. and Labrada, R. (1995) Experiencia en el uso de la solarización en Cuba. In: *Memorias del taller 'Solarización del Suelo',* Escuela Agrícola Panamericana 'El Zamorano' Honduras, 18–21 Septiembre, pp. 5–6.

Fernández, E., Pérez, A., Lorenzo, E. and Vinent, E. (1992) Efectividad del uso del ajonjolí (*Sesanum indicum*) como cultivo intercosecha contra *Meloidogyne incognita. Revista de Protección Vegetal* 7, 39–42.

Fernández, E., Acosta, O., García, Y. and Cuadra, R. (1993) Manejo integrado de los nematodos del género *Meloidogyne* en el cafeto. *VIII Forum de Ciencia y Técnica.* INISAV, Cuba.

Fernández, E., Bernal, B., Vázquez, L., Gandarilla, H., Hernández, R., González, G., Echeverría, S., Amat, Z., Vázquez, T. and Fajardo, J.L. (1994) Manejo integrado de plagas en organopónicos. *IX Forum de Ciencia y Técnica.* INISAV, Cuba.

Fernández, E., Pérez, M., Gandarilla, H., Vázquez, R., Fernández, M., Paneque, M., Acosta, O. and Basterrechea, M. (1998a) Guía para disminuir infestaciones de *Meloidogyne* spp mediante el empleo de cultivos no susceptibles. *Boletín Técnico.* INISAV, Cuba 4, No. 3.

Fernández, E., Hernández, R., López, M. and Gandarilla, H. (1998b) Nematodos parásitos del banano y plátano. Manejo y Lucha biológica. *Boletín Técnico.* INISAV, Cuba 4, No. 5.

Fernández, E., Sánchez, L., Mena, J., Hidalgo, L. and Montes, M. (2000) Alternativas de control de nematodos en Cuba: situación actual y perspectivas. *Nematropica* 30, 127.

Fernández, E., Pérez, M., Gandarilla, H., Vázquez, R., Fernández, M., Paneque, P., Acosta, O. and Basterrechea, M. (2001) Rango de hospedantes de *Meloidogyne* spp dentro de los cultivos económicos de Cuba. *Nematropica* 31, 132.

Ferris, H. and Van Gundy, S.D. (1979) *Meloidogyne* ecology and host interrelationships. In: Lamberti, F. and Taylor, C.E. (eds) *Root-knot Nematodes (*Meloidogyne *species). Systematics, Biology and Control.* Academic Press, London, pp. 205–230.

Ferris, H., Ball, D.A., Beem, L.S. and Gudmundson, L.A. (1986) Using nematode count data in crop management decisions. *California Agriculture* 40, 12–14.

Fery, R.L., Dukes, P.D. Sr and Thies, J.A. (1998) 'Carolina Wonder' and 'Charleston Belle': southern root-knot nematode resistant bell peppers. *HortScience* 33, 900–902.

Fiume, F. and Parisi, B. (1995) La difesa del pomodoro in coltura protetta contro nematodi galligeni (*Meloidogyne* Goeldi) e *Pyrenochaeta lycopersici* (Schneider et Gerlach) in associazione. *Informatore Fitopatologico* 45, 44–50.

Fonseca, H.S., Jaehn, A. and Silva, M. de F.A. (1999) Association of *Ditylenchus dipsaci* with weeds harvested after the garlic crop. *Nematologia Brasileira* 23, 100–102.

Freitas, L.G., Mitchell, D.J., Dickson, D.W. and Chellemi, D.O. (2000a) Soil solarization and organic amendment effects on *Pasteuria penetrans. Nematologia Brasileira* 24, 133–146.

Freitas, L.G., Dickson, D.W., Mitchell, D.J. and Hewlett, T.E. (2000b) Infectivity and suppression of *Pasteuria penetrans* to *Meloidogyne arenaria* race 1 in tomato following soil fumigation. *Nematologia Brasileira* 24, 157–166.

Galper, S., Cohn, E., Spiegel, Y. and Chet, I. (1990) Nematicidal effect of collagen-amended soil and the influence of protease and collagenase. *Revue de Nématologie* 13, 67–71.

Germani, G. (1978) Tests preliminaires de sensibilite de deux cultures de tomate et d'un cultivar d'arachide a deux souches de *Rotylenchulus reniformis. Revue de Nématologie* 1, 11–12.

Gilbert, J.C. and McGuire, D.C. (1956) Inheritance of resistance to severe root-knot from *Meloidogyne incognita* in commercial type tomatoes. *American Society of Horticulture Science Proceeding* 68, 437–442.

Gilman, D.G., Jones, J.E., Williams, C. and Birchfield, W. (1978) Cotton–soybean rotation for control of reniform nematodes. *Louisiana Agriculture* 21, 10–11.

Golden, A.M. and Raski, D.J. (1977) *Heterodera thornei,* n.sp. (Nematoda: Heteroderidae) and a review of related species. *Journal of Nematology* 9, 93–112.

Gomes, T.J. (1973) Contribution to the study of infestation and dispersion of Cobb's false root-knot nematode *Nacobbus serendipiticus* Franklin in Peru. *Nematropica* 3, 4.

Good, J.M. and Thornton, B.D. (1956) Relative increases of populations of sting nematode, *Belonolaimus gracilis,* on six winter legumes. *Plant Disease Reporter* 40, 1050–1053.

Goodey, J.B., Franklin, M.T. and Hooper, D.J. (1965) *The Nematode Parasites of Plants Catalogued under Their Hosts.* Commonwealth Agricultural Bureaux International, UK.

Gooris, J. and de Herde, C.J. (1972) *A Method for the Quantitative Extraction of Eggs and Second Stage Larvae of* Meloidogyne *Populations from Soil.* Publication of the State Nematology and Entomology Research Station, Merelbeke, Belgium.

Gowen, S.R. and Pembroke, B. (2004) *Pasteuria penetrans* and the integrated control of root-knot nematodes. In: Sikora, R.A., Gowen, S., Hauschild, R. and Kiewnick S. (eds) *Multitrophic Interactions in Soils and Integrated Control.* Proceeding of the IOBC-WPRS-Bad Honnef, Germany 2003. *IOBC-WPRS Bulletin* 27, 75–77.

Gowen, S.R., Bala, G., Madulu, J., Mwageni, W. and Trivino, C.G. (1998) Field evaluation of *Pasteuria penetrans* for the management of root-knot nematodes. *Brighton Conference, Pests and Diseases* 3, 755–760.

Graham, R.W. and Holdeman, Q.L. (1953) The sting nematode, *Belonolaimus gracilis,* on cotton and other crops in South Carolina. *Phytopathology* 43, 434–439.

Grainge, M. and Ahmed, S. (1988) *Handbook of Plants with Pest-control Properties.* John Wiley and Sons, New York.

Granges, A. and Leger, A. (1996) Essai comparatif de plants de tomates greffes a une ou deux tetes et de desinfection du sol a la vapeur en serre. *Revue Suisse de Viticulture, d'Arboriculuture et d'Horticulutre* 28, 389–392.

Greco, N., D'Addabbo, T., Stea, V. and Brandonisio, A. (1992) The synergism of soil solarization with fumigant nematicides and straw for the control of *Heterodera carotae* and *Ditylenchus dipsaci. Nematologia Mediterranea* 20, 25–32.

Guertal, E.A., Sikora, E., Hagan, A.K. and Rodrigues-Kabana, R. (1998) Effect of winter cover crops on populations of southern root-knot and reniform nematodes. *Agriculture, Ecosystems and Environment* 70, 1–6.

Guimaraes, L.M.P., de Moura, R.M. and Pedrosa, E.M.R. (2003) *Meloidogyne mayaguensis* parasitism on different plant species. *Nematologia Brasileira* 27, 139–145.

Hague, N.G.M. and Gowen, S.R. (1987) Chemical control of nematodes. In: Brown, R.H. and Kerry, B.R. (eds) *Principles and Practice of Nematode Control.* Academic Press, New York, pp. 131–173.

Hallmann, J. (2001) Plant interactions with endophytic bacteria. In: Jeger, M.J. and Spence, N.J. (eds) *Biotic Interactions in Plant–Pathogen Associations.* CAB International, Wallingford, UK, pp. 87–119.

Hallmann, J. and Sikora R.A. (1994a) Influence of *Fusarium oxysporum,* a mutualistic fungal endophyte, on *Meloidogyne incognita* infection of tomato. *Zeitschrift für Pflanzenkrankheiten und Pflanzenschutz* 101, 475–481.

Hallmann, J. and Sikora, R.A. (1994b) Occurrence of plant parasitic nematodes and non-pathogenic species of *Fusarium* in tomato plants in Kenya and their role as mutualistic synergists for biological control of root-knot nematodes. *International Journal of Pest Management* 40, 321–325.

Hallmann, J., Hänisch, D., Braunsmann, J. and Klenner, M. (2004a) Plant parasitic nematodes in soilless culture systems. *Nematologica* (in press).

Hallmann, J., Faupel, A., Krechel, A., Sikora, R.A. and Berg, G. (2004b) Endophytic bacteria and biological control of nematodes. In: Sikora, R.A., Gowen, S., Hauschild, R. and Kiewnick, S. (eds) *Multitrophic Interactions in Soil.* Proceeding of the International Organization for Biological Control of Noxious Animals and Plants workshop. Bad Honnef, Germany 2003. *IOBC-WPRS Bulletin* 27, 83–94.

Hanan, J.J.H. (1998) *Greenhouses: Advanced Technology for Protected Horticulture.* CRC Press, Boca Raton, Florida.

Handoo, Z.A., Nyczepir, A.P., Esmenjaud, D., van der Beek, J.G., Castagnone-Sereno, P., Carta, L.K., Skantar, A.M. and Higgins, J.A. (2004) Morphological, molecular and differential-host characterization of *Meloidogyne floridensis* n.sp. (Nematoda: Meloidogynidae), a root-knot nematode parasitizing peach in Florida. *Journal of Nematology* 36, 20–35.

Hanna, H.Y., Colyer, P.D., Kirkpatrick, T.L., Romaine, D.J. and Vernon, P.R. (1994) Improving yield of cucumbers in nematode-infested soil by double-cropping with a resistant tomato cultivar, using transplants and nematicides. *Proceedings of the Florida State Horticultural Society* 106, 163–165.

Hanna, H.Y., Colyer, P.D., Kirkpatrick, T.L., Romaine, D.J. and Vernon, P.R. (1996) An economical, environmental-friendly approach to reducing nematode damage to cucumbers in infested soil. *Louisiana Agriculture* 39, 20–21.

Hartman, K.M. and Sasser, J.N. (1985) Identification of *Meloidogyne* species on the basis of differential host test and perineal-pattern morphology. In: Barker, K.R., Carter, C.C. and Sasser, J.N. (eds) *Advanced Treatise on* Meloidogyne. *Volume II Methodology.* North Carolina University Graphics, Raleigh, North Carolina, pp. 69–78.

Hasky-Günther, K., Hoffmann-Hergarten, S. and Sikora, R.A. (1998) Resistance against the potato cyst nematode *Globodera pallida* systemically induced by the rhizobacteria *Agrobacterium radiobacter* (G12) and *Bacillus sphaericus* (B43). *Fundamental and Applied Nematology* 21, 511–517.

Hauschild, R., Hallmann, J. and Sikora, R.A. (2000) *Fusarium oxysporum* and *Meloidogyne incognita* on tomato can be controlled by antagonistic rhizobacteria. *Communications in Agricultural and Applied Biological Sciences* 65, 527–531.

Heald, C.M. (1975) Pathogenicity and histopathology of *Rotylenchulus reniformis* infecting cantaloupe. *Journal of Nematology* 7, 149–152.

Heald, C.M. (1978) Effect of the reniform nematode on vegetable yields. *Plant Disease Reporter* 62, 902–904.

Heald, C.M. and Burton, G.W. (1968) Effect of organic and inorganic nitrogen on nematode populations on turf. *Plant Disease Reporter* 52, 46–48.

Heald, C.M. and Thames, W.H. (1982) The reniform nematode, *Rotylenchulus reniformis.* In: Riggs, R.D. and Editorial Committee, Southern Regional Research Committees S-76 and S-154 (eds) *Nematology in the Southern Region of the United States. Southern Cooperative Series Bulletin* 276, 139–143.

Heald, C.M. and Wayland, J.R. (1975) Factors affecting the control of *Rotylenchulus reniformis* with UHF electromagnetic energy. *Journal of Nematology* 7, 171–174.

Hendy, H., Pochard, E. and Dalmasso, A. (1985) Transmission heriditaire de deux nouvelles sources de resistance aux nematodes du genre *Meloidogyne* chez le piment *Capsicum annum* L. *Comptes Rendus de l'Academie d'Agriculture* 817–822.

Hewlett, T., Gerber, J.F., Smith, K.S. and White, J.H. (2002) *In vitro* culture of *Pasteuria penetrans*. *Nematology* 4, 152–153.

Hidalgo, L. (1999) Potencialidades de Cepas Autóctonas de *Verticillium chlamydosporium* (Goddard) Como Agente de Control Biológico de *Meloidogyne* spp. PhD Thesis, Universidad Agraria de La Habana 'Fructuoso Rodríguez Paez', Habana, Cuba.

Hirschmann, H. and Triantaphyllou, A.C. (1964) Reproduction in plant and soil nematodes. *Annual Review of Phytopathology* 2, 57–80.

Holland, R.J. (2001) *Paecilomyces lilacinus* as a biocontrol agent. PhD Thesis, Macquarie University, Australia.

Holland, R.J., Williams, K.L. and Nevalainen, K.M.H. (2003) *Paecilomyces lilacinus* strain Bioact 251 is not a plant endophyte. *Australasian Plant Pathology* 32, 473–478.

Huang, X. and Ploeg, A.T. (2001) Effect of plant age and *Longidorus africanus* on the growth of lettuce and carrot. *Journal of Nematology* 33, 2–3.

Hussey, R.S. and Janssen, G.J.W. (2002) Root-knot nematodes: *Meloidogyne* species. In: Starr, J.L., Cook, R. and Bridge J. (eds) *Plant Resistance to Parasitic Nematodes*. CAB International, Wallingford, UK, pp. 43–70.

Hussey, R.S. and Roncadori, R.W. (1982) Vesicular-arbuscular mycorrhizae may limit nematode activity and improve plant growth. *Plant Disease* 66, 9–14.

Hutchinson, C.M., McGiffen, M.E. Jr, Ohr, H.D., Sims, J.J. and Becker, J.O. (1999) Evaluation of methyl iodide as a soil fumigant for root-knot nematode control in carrot production. *Plant Disease* 83, 33–36.

Ignatowicz, S. (1998) Control of pests of stored onion with irradiation used for inhibition of onion sprouting. *Biuletyn Warzywniczy* 48, 65–76 (in Polish).

Infante, M. and Sosa-Moss, C. (1971) Chemical control of *Ditylenchus dipsaci* (Kühn) Filipjev in garlic seed. *Nematropica* 1, 13–14.

Ioannou, N., Ioannou, M. and Paroussi, G. (2002) Integrated management of soil-borne pathogens of greenhouse tomato in Cyprus. In: Voyiatzis, D. and Paroussis, E. (eds) *Proceedings of the Second Balkan Symposium on Vegetable and Potatoes*. Thessaloniki, Greece 2000. *Acta Horticulturae* 579, 433–438.

Jaehn, A. (1995) Termoterapia de adho para erradicac (cedilla) ao de *Ditylenchus dipsaci*. *Nematologia Brasileira* 19, 93–96.

Jaehn, A. and Kimoto, T. (1994) Amostragem de bulbos de alho em campos infestados por *Ditylenchus dipsaci*. *Nematologia Brasileira* 18, 35–41.

Jaehn, A. and Kimoto, T. (1995) Tratamento do alho-semente com abamectina para erradicacao de *Ditylenchus dipsaci*. *Nematologia Brasileira* 19, 97–99.

Jairajpuri, M.S., Alam, M.M. and Ahmad, I. (1990) *Nematode Biocontrol*. CBS Publishers, New Delhi.

Jansson, R.K. and Rabatin, S. (1998) Potential of foliar, dip and injection applications of avermectins for control of plant-parasitic nematodes. *Journal of Nematology* 30, 65–75.

Jatala, P. (1986) Biological control of plant parasitic nematodes. *Annual Review of Phytopathology* 24, 453–489.

Jayakumar, J., Rajendran, G. and Ramakrishnan, S. (2003) Avermectin, a novel nematicide for root-knot control in tomato. *Proceedings of a National Symposium on Biodiversity and Management of Nematodes in Cropping Systems for Sustainable Agriculture*. Jaipur, India 2002. Indian Agricultural Research Institute, pp. 154–157.

Jenkins, W.R. and Coursen, B.W. (1957) The effect of root-knot nematodes *Meloidogyne incognita acrita* and *Meloidogyne hapla* on *Fusarium* wilt of tomato. *Plant Disease Reporter* 41, 182–186.

Jepson, S.B. (1987) *Identification of Root-knot Nematodes (*Meloidogyne *species)*. CAB International, Wallingford, UK.

Johnson, A.W. (1985) The role of nematicides in nematode management. In: Sasser, J.N. and Carter, C.C. (eds) *An Advanced Treatise on* Meloidogyne. Volume I. *Biology and Control*. North Carolina University Graphics, Raleigh, North Carolina, pp. 249–267.

Johnson, A.W. (1998) Vegetable crops. In: Baker, K.R., Pederson, G.A. and Windham, G.L. (eds) *Plant and Nematode Interactions*. American Society of Agronomy, Madison, Wisconsin, pp. 595–636.

Johnson, A.W. and Fassuliotis, G. (1984) Nematode parasites of vegetable crops. In: Nickle, W.R. (ed.) *Plant and Insect Nematodes*. Marcel Dekker Inc., New York, pp. 323–372.

Johnson, A.W., Dowler, C.C., Glaze, N.C. and Sumner, D.R. (1983) Effect of intensive cropping systems and pesticides on nematodes and crop yields. *USAD-ARS Agricultural Research Results* ARR-S-14. New Orleans, Louisiana.

Johnson, J.T. and Dickson, D.W. (1973) Evaluation of methods and rates of application of three nematicide insecticides for control of the sting nematode on corn. *Proceedings Soil Crop Science of Florida* 32, 171–173.

Jones, J.P., Gilreath, J.P., Overman, A.J. and Noling, J.W. (1996) Control of soil-borne diseases of mulched tomato by fumigation. *Proceeding of the Florida State Horticultural Society* 108, 201–203.

Jothi, G. and Rajeswari, S. (2001) Management of root-knot nematode in brinjal by using VAM and crop rotation with green gram and pearl millet. *Journal of Biological Control* 15, 77–80.

Jothi, G., Mani, M.P. and Rajeswari, S. (2000) Management of *Meloidogyne incognita* on okra by integrating non-host and endomycorrhiza. *Current Nematology* 11, 25–28.

Kaloshian, I., Williamson, V.M., Miyao, G., Lawn, D.A. and Westerdahl, B.B. (1996) 'Resistant-breaking' nematodes identified in California tomatoes. *California Agriculture* 50, 18–19.

Kanwar, R.S. and Bhatti, D.S. (1993) Management of root-knot nematode (*Meloidogyne javanica*) by poor host crops in three crop rotations. *International Journal of Pest Management* 39, 304–308.

Kanwar, R.S. and Bhatti, D.S. (1994) Crucifers and garlic (*Allium sativum*) as rotational crops to manage root-knot nematode (*Meloidogyne javanica*) in vegetable crops. *Indian Journal of Agricultural Sciences* 64, 339–342.

Katan, J. (1981) Solar heating (solarization) of soil for control of soilborne pests. *Annual Review of Phytopathology* 19, 211–236.

Kerry, B.R. (1987) Biological control. In: Brown, R.H. and Kerry, B.R. (eds) *Principles and Practice of Nematode Control in Crops.* Academic Press, New York, pp. 233–263.

Kerry, B.R. (2000) Rhizosphere interactions and the exploitation of microbial agents for the biological control of plant parasitic nematodes. *Annual Review of Phytopathology* 38, 423–441.

Kerry, B.R. (2001) Exploitation of the nematophagous fungal *Verticillium chlamydosporium* Goddard for the biological control of root-knot nematodes (*Meloidogyne* spp.). In: Butt, T.M., Jackson, C. and Magan, N. (eds) *Fungi as Biocontrol Agents.* CAB International, Wallingford, UK, pp. 155–167.

Kerry, B.R. and Evans, K. (1996) New strategies for the management of plant parasitic nematodes. *American Phytopathological Society* 134–152.

Kerry, B. and Hidalgo-Diaz, L. (2004) Application of *Pochonia chlamydosporia* in the integrated control of root-knot nematodes on organically grown vegetable crops in Cuba. In: Sikora, R.A., Gowen, S., Hauschild, R. and Kiewnick, S. (eds) *Multitrophic Interactions in Soil and Integrated Control.* Proceedings of the International Organization for Biological Control of Noxious Animals and Plants, Bad Honnef, Germany 2003. *IOBC-WPRS Bulletin* 27, 123–126.

Khan, A.M. (1969) *Studies on Plant Parasitic Nematodes Associated with Vegetable Crops in Uttar Pradesh.* Final Technical Report of Grant No. FG In 225. Allighar Muslim University, India, Project No. A7-CR-65.

Khan, A.M. and Saxena, S.K. (1969) Effect of *Rhizoctonia solani* and *Tylenchorhynchus brassicae* on the emergence of cauliflower seedlings. *All Indian Nematology Symposium,* 1967, The Indian Agricultural Research Institute, New Delhi.

Kiewnick, S. (2001) Advanced fermentation and formulation technologies for fungal antagonists. In: Sikora, R.A. (ed.) *Integrated Control of Soil Pests.* Proceedings of the International Organization for Biological Control of Noxious Animals and Plants workshop. *Tri-trophic Interactions in the Rhizosphere and Root Health.* Bad Honnef, Germany, 1999. *IOBC-WPRS Bulletin* 24, 77–79.

Kiewnick, S. (2004) Biological control of plant parasitic nematodes with *Paecilomyces lilacinus,* strain 251. In: Sikora, R.A., Gowen, S., Hauschild, R. and Kiewnick, S. (eds) *Multitrophic Interactions in Soil and Integrated Control.* Proceedings of the International Organization for Biological Control of Noxious Animals and Plants, Bad Honnef, Germany 2003. *IOBC-WPRS Bulletin* 27, 133–136.

Kiewnick, S. and Sikora, R.A. (2003) Efficacy of *Paecilomyces lilacinus* (strain 251) for the control of root-knot nematodes. *Communications in Agriculture and Applied Biological Sciences* 68, 123–128.

Kinloch, R.A. and Dunavin, L.S. (1993) Summer cropping effects on the abundance of *Meloidogyne arenaria* race 2 and subsequent soybean yields. *Supplement to the Journal of Nematology* 25, 806–809.

Kirkegaard, J.A., Sarwar, M., Matthiessen, J.N., Thomas, G. and Monteiro, A.A. (1998) Assessing the biofumigation potential of crucifers. *Acta Horticulturae* 459, 105–111.

Koenning, S.R., Overstreet, C., Noling, J.W., Donald, P.A., Becker, J.O. and Fortnum, B.A. (1999) Survey of crop losses in response to phytoparasitic nematodes in the United States for 1994. *Journal of Nematology* 31, 587–618.

Kokalis-Burelle, N. and Nyczepir, A.P. (2004) Host range studies for *Meloidogyne floridensis. Journal of Nematology* 36, 328 [abstract].

Kokalis, B.N., Vavrina, C.S., Rosskopf, E.N. and Shelby, R.A. (2002) Field evaluation of plant growth-promoting rhizobacteria amended transplant mixes and soil solarization for tomato and pepper production in Florida. *Plant and Soil* 238, 257–266.

Koot, P.A.C. and Molendikk, L.P. (1997) Good control of *Trichodorus* nematode. *PAV Bulletin Akkrbouw* 29–31.

Lamberti, F. (1979a) Economic importance *of Meloidogyne* spp. in subtropical and Mediterranean climates. In: Lamberti, F. and Taylor, C.E. (eds) *Root-knot Nematodes (*Meloidogyne *species) Systematics, Biology and Control.* Academic Press, London, pp. 341–357.

Lamberti, F. (1979b) Chemical and cultural control. In: Lamberti, F. and Taylor, C.E. (eds) *Root-knot Nematodes (*Meloidogyne *Species) Systematics, Biology and Control.* Academic Press, London, pp. 403–423.

Lamberti, F., Ciancio, A., Boilboi, J.B., Tuopay, D.K., Bleve, Z.T. and Elia, F. (1992) Pathogenicity and reproduction of two species of *Xiphinema* on selected vegetable crops in Liberia. *Nematologia Mediterranea* 20, 113–123.

Lamberti, F., Ekanayake, H.M.R.K., Sasanelli, N. and Zacheo, F. (1993) Control of root-knot nematodes in Sri Lanka. *Nematologia Mediterranea* 21, 227–234.

Lamberti, F., Sasanelli, N., D'Addabbo, T., D'Aloisio, V., Carrella, A. and de Cosmis, P. (2000) Recenti acquisizioni nella lotta contro il nematode *Ditylenchus dipsaci* su cipolla. *Nematologia Mediterranea* 28, 17–25.

Lamberti, F., Sasanelli, N., D'Addabbo, T., D'Aloisio, V. and de Cosmis, P. (2001) Chemical treatments and soil solarization for the control of the stem nematode, *Ditylenchus dipsaci* on onion. *Nematologia Mediterranea* 29, 149–152.

Lara, J., Acosta, N., Betancourt, C., Vincente, N. and Rodrigues, R. (1996) Biological control of *Meloidogyne incognita* in tomato in Puerto Rico. *Nematropica* 26, 143–152.

Lear, B., Miyagawa, S.T., Johnson, D.E. and Atlee, C.B. Jr (1966) The sugarbeet cyst nematode associated with reduced yields of cauliflower and other vegetable crops. *Plant Disease Reporter* 50, 611–612.

Leon, L. de, Branchero, L., Lopez-Perez, J.A. and Bello, A. (2000) Control de *Meloidogyne incognita* en cultivo de tomate en Uruguay. *Boletin de Sanidad Vegetal* 26, 401–407.

Linford, M.B. and Oliveira, J.M. (1940) *Rotylenchulus reniformis* novum genus; new species, a nematode parasite of roots. *Proceedings of the Helminthological Society of Washington* 7, 35–42.

Liu, M.T., Xu, R.F., Wu, Q.S. and Chang, Q.X. (1998) Controlling nematodes by grafting cucumbers. *China Vegetables* 5, 36 pp. (in Chinese).

Locascio, S.J., Gilreath, J.P., Dickson, D.W., Kucharek, R.A., Jones, J.P. and Noling, J.W. (1997) Fumigant alternatives to methyl bromide for polyethylene-mulched tomato. *HortScience* 32, 1208–1211.

Luc, M. (1986) Cyst nematodes in equatorial and hot tropical regions. *NATO Advanced Study Institute, Series A: Life Sciences* 121, 355–372.

Luc, M. and Netscher, C. (1974) Presence of the sugar-beet nematode at Dakar. *FAO Plant Protection Bulletin* 22, 24.

Lüth, P. (2000) Fungicides from living organisms – problems and possibilities from the point of view of a small business. *Nachtrichtenblatt Deutsche Pflanzenschutzdienst* 52, 222–225.

Lüth, P. (2004) Cost effective production of fungal biocontrol agents. *Proceedings of an International Workshop on Development of Biocontrol Agents of Diseases for Commercial Applications in Food Production Systems.* Seville, Spain, 116 (abstract).

Lüth, P. and Eiben, U. (2003) Solid-state fermenter and method for solid-state fermentation. USA Patent US 6,620,614 B1.

Mahdy, M., Hallmann, J. and Sikora, R.A. (2000) Biological control of different species of the root-knot nematode *Meloidogyne* with the plant health promoting rhizobacteria *Bacillus cereus* S18. *Communications in Agricultural and Applied Biological Sciences* 65, 545–549.

Mahdy, M., Hallmann, J. and Sikora, R.A. (2001) Influence of plant species on the biological control activity of the antagonistic rhizobacteria *Rhizobium etli* strain G12 toward the root-knot nematode, *Meloidogyne incognita. Communications in Applied Biological Sciences* 66, 655–662.

Main, I.H., Godoy, G., Shelby, R.A., Rodriguez-Kabana, R. and Morgan-Jones, G. (1982) Chitin amendments for control of *Meloidogyne arenaria* in infested soil. *Nematropica* 12, 71–84.

Mankau, R. and Prasad, N. (1972) Possibilities and problems in the use of a sporozoan endoparasite for biological control of plant parasitic nematodes. *Nematropica* 2, 7–8.

Manzanilla-Lopez, R.H., Costilla, M.A., Doucet, J., Franco, J., Inserra, R.N., Lehman, P.S., Cid del Prado-Vera, I., Souza, R.M. and Evans, K. (2002) The genus *Nacobbus* Thorne & Allen, 1944 (Nematoda: Pratylenchidae): systematics, distribution, biology and management. *Nematropica* 32, 149–227.

Martin, G.C. (1956) The common root-knot nematode (*Meloidogyne javanica*) and crop rotation. 5. Cotton (*Gossypium hirsuitum* var. 9L34). *Rhodesian Farmer* 27, 33.

Mateille, T. and Netscher, C. (1985) Temporary protection of eggplant from *Meloidogyne incognita* by minute quantities of isazophos and aldicarb applied at seedling stage. *Revue de Nématologie* 8, 41–44.

Matsumoto, M.N., Goulart, A.M.C., Homechin, M., Sakamoto, M.M.Y. and Santiago, D.C. (2002) Mulching with *Pennisetum purpureum* and other cultural practices for management of *Meloidogyne javanica* on tomato under greenhouse conditions. *Nematologia Brasileira* 26, 101–104.

Mayol, P.S. and Bergeson, G.B. (1970) The role of secondary invaders in *Meloidogyne incognita* infection. *Journal of Nematology* 2, 80–83.

McCann, J. (1981) Threshold populations of *Heterodera cruciferae* and *H. schachtii* causing damage to cabbage seedlings. *Plant Disease Reporter* 65, 264–266.

McGuire, J M. (1982) Nematode transmission of viruses. In: Riggs, R.D. and Editorial Committees, Southern Regional Research Committees S-76 and S-154 (eds) *Nematology in the Southern Region of the United States. Southern Cooperative Series Bulletin* 276, 190–192.

McSorley, R. (1980) Effect *of Rotylenchulus reniformis* on snap bean and methods for control by oxamyl. *Nematropica* 10, 89–95.

McSorley, R. and Gallaher, R.N. (1995) Effect of yard waste compost on plant parasitic nematode densities in vegetable crops. *Journal of Nematology* 27, 545–549.

McSorley, R. and Parrado, J.L. (1983) A bioassay sampling plant for *Meloidogyne incognita. Plant Disease* 67, 182–184.

McSorley, R., Dickson, D.W. and de Brito, J.A. (1994a) Host status of selected tropical rotation crops to four populations of root-knot nematodes. *Nematropica* 24, 45–53.

McSorley, R., Dickson, D.W., de Brito, J.A. and Hochmuth, R.C. (1994b) Tropical rotation crops influence nematode densities and vegetable yield. *Journal of Nematology* 26, 308–314.

Melki, K.C., Giannakou, I.O., Pembroke, B. and Gowen, S.R. (1998) The cumulative build-up *of Pasteuria penetrans* spores in root-knot nematode infested soil and the effect of soil applied fungicides on its infectivity. *Fundamental and Applied Nematology* 21, 679–683.

Mian, I.H., Ali, M. and Akhter, R. (1995a) Grafting on *Solanum* rootstocks to control root-knot of tomato and bacterial wilt of eggplant. *Bulletin of the Institute of Tropical Agriculture, Kyushu University* 18, 41–47.

Mian, I.H., Akhter, R. and Islam, M.N. (1995b) Resistance of tomato and wild *Solanum* to *Meloidogyne incognita. Bulletin of the Institute of Tropical Agriculture, Kyushu University* 18, 33–40.

Miller, L.I. (1986) Economic importance of cyst nematodes in North America. *NATO Advanced Study Institute, Series A: Life Sciences* 121, 373–385.

Minton, N.A., McGill, J.F. and Golden, A.M. (1969) *Meloidogyne javanica* attacks peanuts in Georgia. *Plant Disease Reporter* 53, 668.

Mojtahedi, H., Santo, G.S. and Pinkerton, J.N. (1994) Efficacy of ethoprop on *Meloidogyne hapla* and *M. chitwoodi* and enhanced biodegradation in soil. *Journal of Nematology* 23, 272–279.

Morra, L., Mennella, G. and Amore, R. (1992) Innesto della melanzana (*Solanum melongena* L.) quale mezzo per la difesa da patogeni tellurici e l'aumento dell rese. *Colture Protette* 21, 85–93.

Morra, L., Correale, A., Bigotto, M. and Restaino, F. (1997) Valutazione di tre portinnesti per il pomodoro in serra. *Colture Protette* 26, 69–73.

Morra, L., Bigotto, M., Fesa, G., Santonicola, L., Apicella, A., Tonini, A. and Nervo, G. (2003) Valutazione di portinnesti per il paperone in Campania. *Colture Protette* 32, 83–91 (in Italian).

Mulawarman, Hallmann, J., Schuster, R.-P., Bell, D., Kopp-Holtwiesche, B. and Sikora R.A. (2000) Terrapy®, a natural product with plant growth enhancing effect and nematode control. *Communications in Agricultural and Applied Biological Sciences* 65, 537–544.

Mulawarman, Hallmann, J., Bell, D., Kopp-Holtwiesche, B. and Sikora R.A. (2001) Effects of natural products on soil organisms and plant health enhancement. *Communications in Agricultural and Applied Biological Sciences* 66, 609–617.

Muller, R. and Gooch, P.S. (1982) Organic amendments in nematode control. An examination of the literature. *Nematropica* 12, 319–326.

Munif, A., Hallmann, J. and Sikora R.A. (2000) Evaluation of the biocontrol activity of endophytic bacteria from tomato against *Meloidogyne incognita. Communications in Agricultural and Applied Biological Sciences* 65, 471–480.

Munif, A., Hallmann, J. and Sikora, R.A. (2001) Induced systemic resistance of selected endophytic bacteria against *Meloidogyne incognita* on tomato. *Communications in Applied Biological Sciences* 66, 663–669.

Naik, B.G., Nanjegowda, D., Ravi, K. and Reddy, P.P. (1998) In: Kumar, N.K.K. and Verghese, A. (eds) *Advances in IPM for Horticultural Crops*. Proceedings of the First National Symposium on Pest

Management in Horticultural Crops: Environmental Implications and Thrusts, Bangalore, India, 1997, pp. 321–422.

Naqvi, S.Q.A. and Alam, M.M. (1975) Influence of brinjal mosaic virus in the populations of *Tylenchorhynchus brassicae* and *Rotylenchulus reniformis* around eggplant roots. *Geobios* 2, 120–121.

Nath, R.P., Haidar, M.G., Akhter, S.W. and Prasad, H. (1979) Studies on the nematodes of vegetables in Bihar. Effect of the reniform nematode *Rotylenchulus reniformis* on *Trichosanthes dioica. Indian Journal of Nematology* 6, 175–177.

Netscher, C. (1970) Les nematodes parasites des cultures maraicheres au Senegal. *Cahiers ORSTOM Serie Biologie* 11, 209–229.

Netscher, C. (1975) Studies on the resistance of groundnut to *Meloidogyne* spp. in Senegal. *Cahiers ORSTOM Serie Biologie* 10, 227–232.

Netscher, C. (1977) Observations and preliminary studies on the occurrence of resistance breaking biotypes of *Meloidogyne* spp. on tomato. *Cahiers ORSTOM Serie Biologie* 11, 173–178.

Netscher, C. (1978) Morphological and physiological variability of species of *Meloidogyne* in West Africa and implications for their control. *Mededelingen Landbouwhogeschool Wageningen* 46.

Netscher, C. (1981a) Preliminary studies on the control of *Meloidogyne* species. *Proceedings of the 3rd Research Planning Conference on Root-knot nematodes* Meloidogyne *species. Regions IV and V*, Ibadan, Nigeria, pp. 247–253.

Netscher, C. (1981b) Arbres resistants au *Meloidogyne* spp.: Utilization comme brise vent au Senegal. *Agronomie Tropicale Nogent* 36, 175–177.

Netscher, C. and Luc, M. (1974) Nematodes associes aux cultures maraicheres en Mauritanie. *Agronomie Tropicale Nogent* 29, 697–701.

Netscher, C. and Mauboussin, J.C. (1973) Resultats d'un essai concernant l'efficacite comparee d'une variete resistance et de certains nematicides contre *Meloidogyne javanica. Cahiers ORSTOM, Serie Biologie* 21, 97–102.

Netscher, C. and Sikora, R.A. (1990) Nematode parasites of vegetables. In: Luc, M., Sikora, R.A. and Bridge, J. (eds) *Plant Parasitic Nematodes in Subtropical and Tropical Agriculture.* CAB International, Wallingford, UK, pp. 237–283.

Netscher, C. and Taylor, D.P. (1979) Physiologic variation within the genus *Meloidogyne* and its implications on integrated control. In: Lamberti, F. and Taylor, C.E. (eds) *Root-knot Nematodes (*Meloidogyne *species). Systematics, Biology and Control.* Academic Press, London, pp. 269–294.

Nickel, W.R. (ed.) (1984) *Plant and Insect Nematodes.* Marcel Dekker, New York.

Noe, J. and Sikora, R.A. (1990) Effects of tropical climates on the distribution and host-parasite relationship of plant parasitic nematodes. In: Luc, M., Sikora, R.A. and Bridge, J. (eds) *Plant Parasitic Nematodes in Subtropical and Tropical Agriculture.* CAB International, Wallingford, UK, pp. 583–597.

Noling, J.W. (2003) Nematodes and their management. Horticultural Sciences Department, UF/IFAS, Florida Extension Service, ENY450 7178. Available at: edu.ifas.ufl.edu/CV112p

Noling, J.W. and Becker, J.O. (1994) The challenge of research and extension to define and implement alternatives to methyl bromide. *Journal of Nematology, Supplement* 26, 573–586.

Noto, G. (1994) Soil solarization in greenhouse: effects on tomato crop. *Acta Horticulturae* 357, 237–242.

Nusbaum, C.J. and Ferris, H. (1973) The role of cropping systems in nematode population management. *Annual Review of Phytopathology* 11, 423–440.

Nyczepir, A.P., Esmenjaud, D. and Eisenback, J.D. (1998) Pathogenicity of *Meloidogyne* sp (FL-isolate) on *Prunus* in the southeastern United States and France. *Journal of Nematology* 30, 509 (abstract).

Oda, M. (1999) Grafting of vegetables to improve greenhouse production. *Extension Bulletin – Food and Fertilizer Technology Center, Taipei.*

Oostendorp, M. and Sikora, R.A. (1989) Seed treatment with antagonistic rhizobacteria for the suppression of *Heterodera schachtii* early root infection of sugar beet. *Revue de Nématologie* 12, 77–83.

Oostendorp, M., Dickson, D.W. and Mitchell, D.J. (1991) Population development of *Pasteuria penetrans* on *Meloidogyne arenaria. Journal of Nematology* 23, 58–64.

Ornat, C., Verdjo, L.S., Sorribas, F.J. and Tzortzakakis, E.A. (1999) Effect of fallow and root destruction on survival of root knot and root-lesion nematodes in intensive vegetable cropping systems. *Nematropica* 29, 5–16.

Orr, C.C. and Newton, O.H. (1971) Distribution of nematodes by wind. *Plant Disease Reporter* 55, 61–63.

Oteifa, B.A. and El-Shakawi, S.E. (1965) Species of some Egyptian parasitic nematodes on onion. *Bulletin of the Zoological Society of Egypt* 20, 1632–1637.

Oteifa, B.A. and Osman, A.A. (1974) Host–parasite relations of *Rotylenchulus reniformis* on *Lycopersicon esculentum. Simposia Internacional (XII) de Nematologia. Sociedad Europe de Nematologos,* Septiembre 1974, Granada, Spain, pp. 78–79.

Overman, A.J. (1964) The effect of temperature and flooding on nematode survival in fallow sandy soil. *Soil Crop Science Society of Florida Proceedings* 24, 141–149.

Overman, A.J. (1974) Nematicides in linear drip irrigation for full-bed mulch of tomato. *Proceedings of the Soil and Crop Science Society of Florida* 34, 197–200.

Overman, A.J. and Jones, J.P. (1986) Soil solarization, reaction, and fumigation effects on double-cropped tomato under full-bed mulch. *Proceedings Florida State Horticultural Society* 99, 315–318.

Padgham, J.L., Abawai, G.S. and Duxbury, J.M. (2003) Survival and infectivity of *Meloidogyne graminicola* in flooded and non-flooded soils. *Nematologia Mediterranea* 31, 225–230.

Page, S.L.J. (1979) *An Assessment of the Importance and Control of Plant-parasitic Nematodes of Vegetable Crops in Bangladesh.* ODM. Report of a visit to Bangladesh. Ministry of Overseas Development, Ascot, UK.

Pandey, R., Sikora, R.A., Kalra, A., Singh, H.B. and Pandey, S. (2003) Plants and their products act as major nematode inhibitory agents. In: Trivedi, P.C. (ed.) *Nematode Management in Plants.* Scientific Publishers, Jodhpur, India, pp. 103–131.

Pasha, M.J. and Tiyagi, S.A. (1992) Responses of some cauliflower cultivars/accessions to stunt nematode *Tylenchorhynchus brassicae. Test of Agrochemicals and Cultivars* 13, 92–93.

Paulraj, C. and Ramulu, U.S. (1994) Effect of soil application of low levels of urban sewage sludge on the uptake of nutrients and yield of certain vegetables. *Journal of the Indian Society of Soil Science* 42, 485–487.

Peneva, V., Orion, D., Shlevin, E., Bar Eyal, M. and Brown, D.J.F. (1998) *Longidorus israelensis* sp.n. (Nematoda: Dorylaimoidea), a parasite of carrot in Israel. *Fundamental and Applied Nematology* 21, 715–721.

Pérez, J. (1990) Determinación de los períodos de laboreo necesarios para disminuir la infestación de *Meloidogyne incognita* (Kofoid et White) Chitwood en suelos arcillosos. *Agrotecnia de Cuba* 15, 57–63.

Pérez, J., Lorenzo, E., Pérez, A. and Fernández, E. (1990) Comprobación en condiciones de producción del tratamiento a hidropónicos con nematicidas para el combate de *Meloidogyne* spp. *Ciencia y Técnica Agricultura Protección de Plantas* 13, 52–65.

Pérez, J.M. (2001) *Trichoderma,* Alternativas para el Control Biológico de Nematodos en el Marco de una Agricultura Sostenible. MSc Thesis. Universidad. Central 'Martha Abreu', Facultad de Ciencias Agropecuarias, Villa Clara, Cuba.

Perry, V.G. (1953) The awl nematode, *Dolichodorus heterocephalus,* a devastating plant parasite. *Proceedings of the Helminthological Society of Washington* 20, 21–27.

Perry, V.G. and Norden, A.J. (1964) Some effects of a cropping sequence on populations of certain plant nematodes. *Proceedings Soil and Crop Scientific Society of Florida* 23, 116–121.

Perry, V.G. and Rhoades, H.L. (1982a) The genus *Belonolaimus.* In: Riggs, R.D. and Editorial Committees, Southern Regional Research Committees S-76 and S-154 (eds) *Nematology in the Southern Region of the United States. Southern Cooperative Series Bulletin* 276, 144–149.

Perry, V.G. and Rhoades, H.L. (1982b) The trichodorid nematodes. In: Riggs, R.D. and Editorial Committees, Southern Regional Research Committees S-76 and S-154 (eds) *Nematology* in *the Southern Region of the United States. Southern Cooperative Series Bulletin* 276, 183–186.

Philis, J. (1994) Use of dazomet and oxamyl for controlling the root-knot nematode *Meloidogyne javanica* in glass houses. *Nematologia Mediterranea* 2, 241–243.

Pitcher, R.S. (1963) Interrelationships of nematodes and other pathogens of plants. *Helminthological Abstracts* 34, 1–17.

Ploeg, A.T. (1999) Greenhouse studies on the effect of marigolds (*Tagetes* spp.) on four *Meloidogyne* species. *Journal of Nematology* 31, 6–69.

Ploeg, A.T. (2002) Effects of selected marigold varieties on root-knot nematodes and tomato and melon yields. *Plant Disease* 86, 505–508.

Porcelli, S., Morra, L., del Piano, L. and D'Amore, R. (1990) Osservazioni sull'affinita ed comportamento vegeto produttivo di innesti tra melanzana ed altre Solanaceae. *Colture Protette* 19, 75–80.

Potter, J.W. and Olthof, T.H.A. (1993) Nematode pests of vegetable crops. In: Evans, K., Trudgill, D.L. and Webster, J.M. (eds) *Plant Parasitic Nematodes in Temperate Agriculture.* CAB International, Wallingford, UK, pp. 171–207.

Powell, N.T. (1971a) Interactions between nematodes and fungi in disease complexes. *Annual Review of Phytopathology* 9, 253–274.
Powell, N.T. (1971b) Interaction of plant parasitic nematodes with other disease causing agents. In: Zuckermann, B.M., Mai, W.F. and Rohde, R.A. (eds) *Plant Parasitic Nematodes* Volume II. Academic Press, New York, pp. 119–136.
Prasad, S.K. and Webster, J.M. (1967) Effect of temperature on the rate of development of *Nacobbus serendipiticus* in excised tomato roots. *Nematologica* 13, 85–90.
Primo, P.D. and Cartia, G. (2001) Solarizzazione e biofumigazione per il contenimento di *Fusarium oxysporum* f.sp. *melonis. Informatore Agrario* 57, 55–58.
Prot, J.C. (1977) Amplitude et cinetique des migrations de nematode *Meloidogyne javanica* sous l'influence d'un plant de tomate. *Cahiers ORSTOM Serie Biologie* 11, 157–166.
Prot, J.C. (1984) A naturally occurring resistance breaking biotype of *Meloidogyne incognita* on tomato: reproduction and pathogenicity on tomato cultivars Roma and Rossol. *Revue de Nématologie* 7, 23–28.
Prot, J.C. and Netscher, C. (1978) Improved detection of low population densities of *Meloidogyne. Nematologica* 24, 129–132.
Prot, J.C. and Van Gundy, S.D. (1981) Effect of soil texture and the clay component on migration of *Meloidogyne incognita* second stage juveniles. *Journal of Nematology* 13, 213–219.
Pyrowolakis, A., Westphal, A., Sikora, R.A. and Becker, J.O. (2002) Identification of root-knot suppressive soils. *Applied Soil Ecology* 19, 51–56.
Quénéhervé, P., Topart, P. and Martiny, B. (1998) *Mucuna pruriens* and other rotational crops for control of *Meloidogyne incognita* and *Rotylenchulus reniformis* in vegetables in polytunnels in Martinique. *Nematropica* 28, 19–30.
Radewald, J.D., Osgood, J.W., Mayberry, K.S., Paulus, A.O. and Shibuya, F. (1969) *Longidorus africanus* a pathogen of head lettuce in the Imperial Valley of southern California. *Plant Disease Reporter* 53, 381–384.
Radewald, J.D., McKenry, M.V., Roberts, P.A. and Westerdahl, B.B. (1987) The importance of soil fumigation for nematode control. *California Agriculture* 41, 16–17.
Rammah, A. and Hirschmann, H. (1988) *Meloidogyne mayaguensis* n.sp. (Meloidogynidae), a root-knot nematode from Puerto Rico. *Journal of Nematology* 20, 58–69.
Rao, M.S., Reddy, P.P. and Nagesh, M. (1997a) Management of root-knot nematode, *Meloidogyne incognita* on tomato by integration of *Trichoderma harzianum* with neem cake. *Zeitschrift für Pflanzenkrankheiten und Pflanzenschutz* 104, 423–425.
Rao, M.S., Kerry, B.R., Gowen, S.R., Bourne, J.M. and Reddy, P.P. (1997b) Management of *Meloidogyne incognita* in tomato nurseries by integration of *Glomus deserticola* with *Verticillium chlamydosporium. Zeitschrift für Pflanzenkrankheiten und Pflanzenschutz* 104, 419–422.
Rao, M.S., Reddy, P.I.P., Nagesh, M. and Reddy, P.P. (1998a) Integrated management of *Meloidogyne incognita* on tomato using oil cakes and a bio agent, *Verticillium chlamydosporium.* In: Kumar, N.K.K. and Verghese, A. (eds) *Advances in IPM for Horticultural Crops.* Proceedings of the First National Symposium on Peat Management in Horticultural Crops: Environmental Implications and Thrusts, Bangalore, India, 1997, pp. 343–348.
Rao, M.S., Reddy, P.P. and Nagesh, M. (1998b) Integrated management of *Meloidogyne incognita* on tomato using *Verticillium chlamydosporuim* and *Pasteuria penetrans. Pest Management in Horticultural Ecosystems* 4, 32–35.
Reddy, G.S., Rao, V.K., Sitaramaiah, K. and Chalam, R.V. (2001) Soil solarization for control of the soil-borne pathogen complex due to *Meloidogyne incognita* and *Pythium aphanidermatum. Indian Journal of Nematology* 31, 136–138.
Reddy, M.S., Dawkins, R.A. and Kloepper, J.W. (2000) Evaluation of biological preparations on root-knot severity and yield of tomato in Alabama – 1999. *Biological and Cultural Tests for Control of Plant Diseases* 15, 129.
Reddy, P.P., Nagesh, M., Devappa, V. and Kumar, M.V.V. (1998) Management of *Meloidogyne incognita* on tomato by integration of *Trichoderma harzianum, Glomus fasciculatum* and neem cake. In: Kumar, N.K.K. and Verghese, A. (eds) *Advances in IPM for Horticultural Crops.* Proceedings of the First National Symposium on Peat Management in Horticultural Crops: Environmental Implications and Thrusts, Bangalore, India, 1997, pp. 349–352.
Rehiayani, A.S. and Hafez, S. (1998) Host status and green manure effect of selected crops on *Meloidogyne chitwoodi* race 2 and *Pratylenchus neglectus. Nematropica* 28, 213–230.

Reid, A., Manzanilla-Lopez, R.H. and Hunt, D.J. (2003) *Nacobbus aberrans* (Thorne, 1935) Thorne & Allen, 1944 (Nematoda: Pratylenchidae); a nascent species complex revealed by RFLP analysis and sequencing of the ITS-rDNA region. *Nematology* 5, 441–451.

Reimann, S. and Sikora, R.A. (2003) Managing the mycorrhizosphere – an approach to sustainable agriculture after the phase out of methyl bromide. *Communications in Agriculture and Applied Biological Sciences* 68, 129–134.

Rhoades, H.L. (1964) Effect *of Crotalaria spectabilis* and *Sesbania exaltata* on plant nematode populations and subsequent yield of snap beans and cabbage. *Proceedings of Florida State Horticultural Society* 77, 233–237.

Rhoades, H.L. (1967) Efficacy of some experimental nematicides applied in-the-row on vegetables. *Proceedings of Florida State Horticultural Society* 80, 156–161.

Rhoades, H.L. (1969) Nematicide efficacy in controlling sting and stubby-root nematodes attacking onions in central Florida. *Plant Disease Reporter* 53, 728–730.

Rhoades, H.L. (1971) A comparison of pre-plant and post-plant nematicides for controlling sting nematodes. *Soil and Crop Science Society of Florida* 31, 260–262.

Rhoades, H.L. (1976a) Effect of *Indigofera hirsuta* on *Belonolaimus longicaudatus, Meloidogyne incognita* and *M. javanica* and subsequent crop yields. *Plant Disease Reporter* 60, 384–386.

Rhoades, H.L. (1976b) Comparison of granular and transplant water application of nonfumigant nematicides for controlling the sting nematode, *Belonolaimus longicaudatus,* on cabbage. *Soil and Crop Science Society of Florida* 36, 205–206.

Rhoades, H.L. (1980) Relative susceptibility of *Tagetes patula* and *Aeschynomene americana* to plant nematodes in Florida. *Nematropica* 10, 116–120.

Rhoades, H.L. (1983) Effects of cover crops and fallowing on populations of *Belonolaimus longicaudatus* and *Meloidogyne incognita* and subsequent crop yields. *Nematropica* 13, 9–16.

Rhoades, H.L. and Forbes, R.B. (1986) Effects of fallow, cover crops, organic mulch and phenimaphos on nematode populations, soil nutrients and subsequent growth. *Nematropica* 16, 141–151.

Rich, J.R. and Bird, G.W. (1973) Inhibition of *Rotylenchulus reniformis* penetration of tomato and cotton roots with foliar applications of oxamyl. *Journal of Nematology* 5, 221–224.

Rideout, S.L. and Long, D.H. (2004) STAN: a novel approach to nematode control in cotton. *Proceedings of the 43rd Annual Meeting Society of Nematologists.* Colorado, USA, Abstract 79.

Riggs, R.D. and Winstead, N.N. (1959) Studies on resistance in tomato to root-knot nematodes and on the occurrence of pathogenic biotypes. *Phytopathology* 49, 716–724.

Robbins, R.T. and Barker, K.R. (1973) Comparisons of host range and reproduction among populations of *Belonolaimus longicaudatus* from North Carolina and Georgia. *Plant Disease Reporter* 57, 750–754.

Roberts, P.A. (1987) The influence of planting date of carrot on *Meloidogyne incognita* reproduction and injury to roots. *Nematologica* 33, 335–342.

Roberts, P.A. (1992) Current status of the availability, development and use of host plant resistance to nematodes. *Journal of Nematology* 24, 213–227.

Roberts, P.A. and Matthew, W.C. (1995) Disinfection alternatives for control of *Ditylenchus dipsaci* in garlic seed cloves. *Journal of Nematology* 27, 448–456.

Roberts, P.A. and Thomason, J. (1986) Variability in reproduction of isolates of *Meloidogyne incognita* and *M. javanica* on resistant tomato genotypes. *Plant Disease* 70, 547–551.

Roberts, P.A., Van Gundy, S.D. and McKinney, H.E. (1981) The effects of soil temperature and planting date of wheat on *Meloidogyne incognita* reproduction, soil populations and grain yield. *Journal of Nematology* 13, 338–345.

Roberts, P.A., May, D. and Metthew, W.C. (1986) Root knot resistance in processing tomatoes. *California Agriculture* 40, 24–26.

Rodríguez, M.G. (2000) Identificación y Caracterización de *Meloidogyne mayaguensis* (Nemata: Meloidogynidae) en el Cafeto en Cuba. PhD Thesis. Universidad Agraria de La Habana 'Fructuoso Rodríguez', Havana, Cuba.

Rodríguez, M.G., Lourdes, S. and Rowe, J. (2003) Host status of agriculturally important plant families to the root-knot nematode *Meloidogyne mayaguensis* in Cuba. *Nematropica* 33, 125–130.

Rodriguez-Kabana, R. (1992) Cropping systems for the management of phytonematodes. In: Gommers, F.J. and Maas, P.W.T. (eds) *Nematology from Molecule to Ecosystem.* Proceedings of the European Society of Nematologists Veldhoven, Kekker and Huisman, The Netherlands, pp. 219–233.

Rodriguez-Kabana, R., Morgan-Jones, G. and Chet, I. (1987) Biological control of nematodes: soil amendments and microbial antagonists. *Plant and Soil* 100, 237–247.

Rodriguez-Kabana, R., Kokalis-Burelle, N., Kiewnich, S., Schuster, R.-P. and Sikora, R.A. (1994) Alginate screens for delivery of root-knot nematode inoculum and evaluation of microbial interactions. *Plant and Soil* 164, 147–154.

Rohde, R.A. and Jenkins, W.R. (1957) Host range of a species of *Trichodorus* and its host–parasite relationships on tomatoes. *Phytopathology* 47, 295–298.

Rohde, R.A. and Jenkins, W.R. (1958) Basis for resistance of *Asparagus officinalis* var. *altilis* L. to the stubby root nematode *Trichodorus christiei* Allen 1957. *Maryland Agriculture Experiment Station Bulletin A-97.*

Roman, J. (1978) *Fitonematologia Tropical.* Estacion Experimental Agricola, Universidad de Puerto Rico.

Ruthenberg, H. (1983) *Farming Systems in the Tropics,* 3rd edn. Oxford University Press, Oxford.

Runia, W.T. (1995) A review of possibilities for disinfection of recirculation water from soilless cultures. *Acta Horticulturae* 382, 221–229.

Russo, G., Loffredo, A., Fiume, G. and Fiume, F. (2003) Efficacy of oxamyl, alone or integrated against nematodes. *Informatore Agrario* 59, 56–58.

Saleh, H., Abu-Gharbieh, W.I. and Al-Banna, L. (1988) Effect of solarization combined with solar-heated water on *Meloidogyne javanica. Nematologica* 34, 290–291.

Santo, G.S. and Pinkerton, J.N. (1985) A second race of *Meloidogyne chitwoodi* discovered in Washington. *Plant Disease* 69, 361.

Sasanelli, N., D'Aloisio, V., Basile, M. and Rana, G.L. (1995) Control of *Ditylenchus dipsaci* on onion by chemical treatments with cadusafos, fenamiphos and aldicarb. *Afro-Asian Journal of Nematology* 5, 24–27.

Sasanelli, N., D'Addabbo, T., Basile, M. and Carella, A. (1996) Effect of cadusafos on the reproduction of *Meloidogyne incognita. Afro-Asian Journal of Nematology* 6, 36–39.

Sasser, J.N. (1954) Identification and host–parasite relationships of certain root-knot nematodes (*Meloidogyne* spp.). *Maryland Agricultural Experiment Station Bulletin A-77.*

Sasser, J.N. (1966) Behaviour of *Meloidogyne* species from various geographical locations on ten host differentials. *Nematologica* 12, 98–99.

Sasser, J.N. (1979a) Pathogenicity and host ranges in *Meloidogyne* species. In: Lamberti, F. and Taylor, C.E. (eds) *Root-knot Nematodes (*Meloidogyne *Species). Systematics, Biology and Control.* Academic Press, London, pp. 256–268.

Sasser, J.N. (1979b) Economic importance of *Meloidogyne* in tropical countries. In: Lamberti, F. and Taylor, C.E. (eds) *Root-knot Nematodes (*Meloidogyne *Species). Systematics, Biology and Control.* Academic Press, London, pp. 359–374.

Sasser, J.N. and Kirby, M.F. (1979) *Crop Cultivars Resistant to Root-knot Nematodes* Meloidogyne *Species.* A cooperative publication of the Department of Plant Pathology, North Carolina State University and the United States Agency for International Development, Raleigh, North Carolina.

Sauer, M.R. and Giles, J.E. (1959) A field trial with root-knot resistant tomato variety. *CSIRO Australian Irrigation Research Station Technical Paper* 3, 1–10.

Saxena, M.C., El-Moneim, A.A., Mamluk, O. and Hanounik, S.B. (1987) A review of nematology research in ICARDA. In: Saxena, M.C., Sikora, R.A. and Srivastava, J.P. (eds) *Nematodes Parasitic to Cereals and Legumes in Temperate Semi-arid Regions.* Proceedings of a Workshop, Larnaca, Cyprus 1987. ICARDA, Aleppo, Syria, pp. 69–84.

Schiffers, B.C., Fraselle, J. and Jaumin, L. (1985) Comparaison d'efficacite contre le *Ditylenchus dipsaci* (Kühn) Fil. de 4 traitements nematicides incorpores aux enrobages de semences de feverolle (*Vicia faba* L). *Mededelingen Landbouwhogeschool Wageningen* 50, 797–807.

Schuster, M.L., Sandstedt, R. and Estes, L.W. (1965) Host–parasite relations of *Nacobbus batatiformis* and the sugar beet and other hosts. *Journal of the American Society of Sugar Beet Technology* 13, 523–537.

Schuster, R.-P., Sikora, R.A. and Amin, N. (1995) Potential of endophytic fungi for the biological control of plant parasitic nematodes. *Communications in Agriculture and Applied Biological Sciences* 60, 1047–1052.

Seinhorst, J.W. (1956) Biologische Rassen von het stengelaaltje *Ditylenchus dipsaci* (Kühn) Filipjev en hun waardplanten. *Tijdschrift over Planteziekten* 62, 179–188.

Seinhorst, J.W. (1965) The relation between nematode density and damage to plants. *Nematologica* 11, 137–154.

Sellami, S. and Cheifa, H. (1997) Effet de *Tagetes erecta* contre les *Meloidogyne* sous abri-plastique. *Communications in Agriculture and Applied Biological Sciences* 62, 737–740.

Sharma, R., Rai, R.R. and Farhat, S. (1996) Integration of soil solarization, carbofuran and neem cake applications for nematode management on tomato. *Indian Journal of Plant Protection* 24, 110–114.

Sharon, E., Bar-Eyal, M., Chet, I., Herrere-Estrella, A., Kleifeld, O. and Spiegel, Y. (2001) Biological control of the root-knot nematode *Meloidogyne javanica* by *Trichoderma harzianum*. *Phytopathology* 91, 687–693.

Siddiqui, I.A., Sher, S.A. and French, A.M. (1973) *Distribution of Plant Parasitic Nematodes in California.* California Department of Food and Agriculture, Sacramento, California.

Siddiqui, M.A., Azam, M.F. and Saxena, S.K. (1993) Seed treatment with some chemicals for the control of the root-knot nematode on bottle gourd and bitter gourd. *Current Nematology* 4, 11–14.

Siddiqui, M.A., Atam, M.F. and Saxena, S.K. (1998) Evaluation of root dip treatments with nematicides on development of root galls and yield (*Meloidogyne incognita*) in tomato and aubergine. *Tests of Agrochemicals and Cultivars* 19, 8–9.

Sidhu, G. and Webster, J.M. (1977) Predisposition of tomato to the wilt fungus (*Fusarium oxysporum lycopersici*) by the root-knot nematode (*Meloidogyne incognita*). *Nematologica* 23, 436–442.

Sieverding, E. (1991) *Vesicular–Arbuscular Mycorrhiza Management in Tropical Agrosystems.* German Agency for Technical Cooperation (GTZ), Eschborn, Germany.

Sikora, R.A. (1978) Einfluss der endotrophen Mykorrhiza (*Glomus mosseae*) auf das Wirt-Parasit-Verhältnis von *Meloidogyne incognita* in Tomaten. *Zeitschrift für Pflanzenkrankheiten und Pflanzenschutz* 85, 197–202.

Sikora, R.A. (1988) Interrelationship between plant health promoting rhizobacteria, plant parasitic nematodes and soil microorganisms. *Communications in Agriculture and Applied Biological Sciences* 53, 867–878.

Sikora, R.A. (1989) *Importance of plant parasitic and entomopathogenic nematodes to vegetable production in the Philippines.* GTZ Report, Eschborn, Germany.

Sikora, R.A. (1990) Management of the soil antagonistic potential in agro-ecosystems for nematode control. *Proceeding II. International Nematology Congress,* Veldhoven, The Netherlands, pp. 249–256.

Sikora, R.A. (1991) The concept of plant-health promoting rhizobacteria for the biological control of plant parasitic nematodes. In: Keel, C., Koller, B. and Defago, G. (eds) Plant growth-promoting rhizobacteria – progress and prospects. Proceedings of the II International Workshop on Plant Growth-promoting Rhizobacteria. Interlaken, Switzerland, 1990. *WPRS Bulletin* XIV 8, 3–10.

Sikora, R.A. (1992) Management of the antagonistic potential in agricultural ecosystems for the biological control of plant parasitic nematodes. *Annual Review of Phytopathology* 30, 245–270.

Sikora, R.A. (1995) Vesicular–arbuscular mycorrhizae: their significance for biological control of plant parasitic nematodes. *Biocontrol* 1, 29–33.

Sikora, R.A. (1997) Biological system management in the rhizosphere: an inside-out/outside-in perspective. *Communications in Agriculture and Applied Biological Sciences* 62, 105–112.

Sikora, R.A. (ed.) (2001) *Integrated Control of Soil Pests.* Proceedings of the 1999 Workshop: Tri-trophic Interactions and Root Health. Bad Honnef, Germany. *IOBC/WPRS Bulletin* 24.

Sikora, R.A. (2002) Strategies for biological system management of nematodes in horticultural crops: fumigate, confuse or ignore them. *Communications in Agricultural and Applied Biological Sciences* 67, 5–18.

Sikora, R.A. and Carter, W.W. (1987) Nematode interactions with fungal and bacterial pathogens – fact or fantasy. In: Veech, J.A. and Dickson, D.W. (eds) *Vistas on Nematology.* Society of Nematologists, Hyattsville, Maryland, pp. 307–312.

Sikora, R.A. and Hoffmann-Hergarten, S. (1993) Biological control of plant parasitic nematodes with plant-health-promoting rhizobacteria. In: Lumsden, R.D. and Vaughn, F.L. (eds) *Pest Management: Biological Based Technologies.* Proceedings of the Beltsville Symposium XVIII, USDA, Beltsville, Maryland, 2–6 May 1993. American Chemical Society, Washington, DC, pp. 166–172.

Sikora, R.A., Singh, R.S. and Sitaramaiah, K. (1973a) Control of root-knot through organic and inorganic soil amendments. 3. Effect of rice husk and sugarcane bagasse. *Haryana Journal of Horticultural Science* 2, 123–127.

Sikora, R.A., Sitaramaiah, K. and Singh, R.S. (1973b) Reaction of root-knot nematode resistant tomato cultivars to *Meloidogyne javanica* in India. *Plant Disease Reporter* 57, 141–143.

Sikora, R.A., Reckhaus, P. and Adamou, I. (1988) Presence, distribution and importance of plant parasitic nematodes in irrigated agricultural crops in Niger. *Communications in Agricultural and Applied Biological Sciences* 53, 821–834.

Sikora, R.A., Schuster, R.-P. and Kiewnick, S. (1994) Indexing biodiversity and antagonistic potential in agricultural soils of Madagascar. *Communications in Agriculture and Applied Biological Sciences* 59, 781–790.

Sikora, R.A., Niere, B. and Kimenju, J. (2003) Endophytic microbial biodiversity and plant nematode management in African agriculture. In: Neuenschwander, P., Borgemeister, C. and Langewald, J. (eds) *Biological Control in IPM Systems in Africa.* CAB International, Wallingford, UK, pp. 179–192.

Sikora, R.A., Gowen, S., Hauschild, R. and Kiewnick S. (eds) (2004) *Multitrophic Interactions in Soil.* Proceedings of the 2003 workshop: multitropic interactions in soil and integrated control. Bad Honnef, Germany. *IOBC-WPRS Bulletin* 27.

Silman, R.W., Bothast, R.J. and Schisler, D.A. (1993) Production of *Colletotrichum truncatum* for use as a mycoherbicide: effect of culture drying and storage on recovery and efficacy. *Biotechnical Advances* 11, 561–575.

Singh, D.P., Rashid, M.M. and Razzaque, M.A. II (2000) *Development of Entrepreneurship in Vegetable Nursery and Seed Business in Bangladesh.* AVRDC-USAID Workshop Proceedings 1999 Dhaka, Bangladesh.

Singh, R.S. and Sitaramaiah, K. (1966) Incidence of root-knot of okra and tomatoes in oil-cake amended soil. *Plant Disease Reporter* 50, 668–672.

Singh, R.S. and Sitaramaiah, K. (1967) Effect of decomposing green leaves, sawdust and urea on the incidence of root-knot to okra and tomato. *Indian Phytopathology* 20, 349–355.

Singh, R.V. and Khera, S. (1979) Pathogenicity of *Rotylenchulus reniformis* on brinjal (*Solanum melongena* L). *Indian Journal of Nematology* 9, 117–124.

Sitaramaiah, K. and Sikora, R.A. (1982) Effect of the mycorrhizal fungus *Glomus fasciculatus* on the host parasite relationship of *Rotylenchulus reniformis. Nematologica* 28, 412–419.

Sivakumar, C.V. and Meerzainuddeen, M. (1973) The effect of yellow vein mosaic disease of okra on the reniform nematode, *Rotylenchulus reniformis. South Indian Horticulture* 21, 68–69.

Sivakumar, C.V. and Seshadri, A.R. (1971) Life history of the reniform nematode, *Rotylenchulus reniformis,* Linford and Oliveira, 1940. *Indian Journal of Nematology* 1, 7–20.

Sivakumar, C.V. and Seshadri, A.R. (1979) Longevity of the reniform nematode, *Rotylenchulus reniformis* in host free soil. *Indian Journal of Nematology* 6, 138–144.

Smith, G.S. (1987) Interactions of nematodes with mycorrhizal fungi. In: Veech, J.A. and Dickson, D.W. (eds) *Vistas on Nematology.* Society of Nematologists, Hyattsville, Maryland, pp. 292–300.

Smith-Becker, J. and Becker, J.O. (2004) Early protection of cotton seedlings by abamectin seed coating against root-knot nematodes. *Proceeding Abstracts of the 43rd Annual Meeting Society of Nematologists.* Colorado, USA. Abstract No. 174, 80.

Sorribas, J., Ornat, C. and Bunol, J. (2004) Management of *Meloidogyne javanica* by using grafted tomatoes and *Mi* resistant gene cultivars. *Phytoma-Espana* (in press).

Sosa-Moss, C. (1986) Cyst nematodes in Mexico, Central and South America. *NATO Advanced Study Institute, Series A: Life Sciences* 121, 397–398.

Sotomayor, D., Allen, L.H. Jr, Chen, Z., Dickson, D.W. and Hewlett, T. (1999) Anaerobic soil management practices and solarization for nematode control in Florida. *Nematropica* 29, 153–170.

Southards, C.J. and Priest, M.F. (1973) Variation in pathogenicity of seventeen isolates of *Meloidogyne incognita. Journal of Nematology* 5, 63–67.

Spiegel, Y. and Chet, I. (1998) Evaluation of *Trichoderma* spp. as a biocontrol agent against soilborne fungi and plant-parasitic nematodes in Israel. *Integrated Pest Management Reviews* 3, 169–175.

Spiegel, Y., Chet, I. and Cohn, E. (1987) Use of chitin for controlling plant parasitic nematodes. II. Mode of action. *Plant and Soil* 98, 337–345.

Stapleton, J., Duncan, R.A. and Johnson, R. (1998) Soil disinfestation with cruciferous amendments and sublethal heating: effects on *Meloidogyne incognita, Sclerotium rolfsii* and *Pythium ultimum.* Proceedings of the 7th International Congress of Plant Pathology, Edinburgh, UK. *Plant Pathology* 47, 737–742.

Starr, J.L., Cook, R. and Bridge, J. (eds) (2003) *Plant Resistance to Plant Parasitic Nematodes.* CAB International, Wallingford, UK.

Stefanova, M. and Plumas, G. (1988) Asociación de *Pseudomonas solanacearun* y *Meloidogyne javanica* en plantaciones de tomate de la Isla de la Juventud. *Ciencia y Técnica Agricultura. Protección de Plantas* 11, 87–94.

Stephan, Z.A. (1986) Host status of cyst nematodes in Iraq. *NATO Advanced Study Institute, Series A: Life Sciences* 121, 391.

Stirling, A.M., Stirling, G.R. and Macrae, I.C. (1992) Microbial degradation of fenamiphos after repeated application to a tomato-growing soil. *Nematologica* 38, 245–254.

Stirling, G.R. (1991) *Biological Control of Plant Parasitic Nematodes.* CAB International, Wallingford, UK.

Stirling, G.R. and Stirling, A.M. (2003) The potential of *Brassica* green manure crops for controlling root-knot nematode (*Meloidogyne javanica*) on horticultural crops in a subtropical environment. *Australian Journal of Experimental Agriculture* 43, 623–630.

Stirling, G.R. and Wachtel, M.F. (1980) Mass production of *Bacillus penetrans* for biological control of root knot nematodes. *Nematologica* 26, 308–312.

Stoyanov, D. (1971) Control de los nematodos parasitos de plantanos par medio de rotaciones y su duracion en tierra gin hospedos. *Revista de Agricultura Cuba* 4, 75–81.

Stoyanov, D. (1972) *Heterodera amaranthi* n. sp. (Tylenchida: Heteroderidae); un nematodo formador de quistes en Cuba. *Poeyana* 97, 1.

Sturhan, D. (1969) Das Rassenproblem bei *Ditylenchus dipsaci. Mitteilung der Biologischen Bundesanstalt für Land- und Forstwirtschaft* 136, 87–98.

Sumner, D.R., Minton, N.A., Brenneman, T.B., Burton, G.W. and Johnson, A.W. (1999) Root diseases and nematodes in Bahiagrass–vegetable rotations. *Plant Disease* 83, 55–59.

Swamy, S.D.R., Reddy, P.P., Jegowda, D.N. and Swamy, B.C.N. (1995) Management of *Meloidogyne incognita* in tomato nursery by growing trap/antagonistic crops in rotation. *Current Nematology* 6, 9–12.

Sykes, G.B. and Winfield, A.L. (1966) Studies on *Brassica* cyst nematode *Heterodera cruciferae. Nematologica* 12, 530–538.

Tarjan, A.C., Lownsbery, B.F. and Hawley, W.G. (1952) Pathogenicity of some plant-parasitic nematodes from Florida soils. I. The effect of *Dolichodorus heterocephalus* Cobb on celery. *Phytopathology* 42, 131–132.

Taylor, A.L. and Sasser, J.N. (1978) *Biology, Identification and Control of Root-knot Nematodes (*Meloidogyne *spp.).* North Carolina State University Graphics, Cooperative Publication of Department of Plant Pathology, North Carolina State University and US Agency for International Development, Washington, DC.

Taylor, A.L., Sasser, J.N. and Nelson, L.A. (1982) *Relationship of Climate and Soil Characteristics to Geographical Distribution of* Meloidogyne *Species in Agricultural Soils.* North Carolina State University Graphics, Cooperative Publication of the Department of Plant Pathology, North Carolina State University and US Agency for International Development, Washington, DC.

Taylor, C.E. (1979) *Meloidogyne* interrelationships with microorganisms. In: Lamberti, F. and Taylor, C.E. (eds) *Root-knot Nematodes (*Meloidogyne *Species). Systematics, Biology and Control.* Academic Press, London, pp. 375–398.

Taylor, D.P. (1976) Plant nematology problems in tropical Africa. *Helminthological Abstracts Series B, Plant Nematology* 45, 269–284.

Taylor, D.P., Netscher, C. and Germani, G. (1978) *Adansonia digitata* (Baobab), a newly discovered host for *Meloidogyne* sp. and *Rotylenchulus reniformis.* Agricultural implications. *Plant Disease Reporter* 62, 276–277.

Terhardt, J., Vilich, V. and Sikora, R.A. (1997) Description of microbial communities of rhizosphere soil based on the phospholipid fatty-acid spectrum analysis of microorganisms. In: Dehne, H.-W. *et al.* (eds) *Diagnosis and Identification of Plant Pathogens.* Kluwer Academic Publishers, Dordrecht, The Netherlands, pp. 345–348.

Thames, W.H. Jr (1959) Plant parasitic nematode populations of some Florida soils under cultivated and natural conditions. *Dissertation Abstracts* 20, 1109–1110.

Thames, W.H. and Heald, C.M. (1974) Chemical and cultural control of *Rotylenchulus reniformis* on cotton. *Plant Disease Reporter* 58, 337–341.

Thames, W.H. and Stauer, W.N. (1953) A preliminary trial of lowland culture rice in rotation with vegetable crops as a means of reducing root-knot nematode infestations in the Everglades. *Plant Disease Reporter* 37, 187–192.

Thies, J.A. and Fery, R.L. (2000a) Characterization of resistance conferred by the *N* gene to *Meloidogyne arenaria* races 1 and 2, *H. hapla* and *M. javanica* in two sets of isogenic lines of *Capsium annuum* L. *Journal of the American Society of Horticultural Science* 125, 71–75.

Thies, J.A. and Fery, R.L. (2000b) Heat stability of resistance to *Meloidogyne incognita* in Scotch Bonnet peppers (*Capsicum chinense* Jacq.). *Journal of Nematology* 32, 356–361.

Thies, J.A., Davis, R.F., Mueller, J.D., Fery, R.L., Langston, D.B. and Miller, G. (2004) Double-cropping cucumbers and squash after resistant bell pepper for root-knot nematode management. *Plant Disease* 88, 589–593.

Tobar, A. and Palacios, F. (1974) Nematode dispersal by irrigation water. *Symposia Internacional (XII) de Nematologia. Sociedad Europe de Nematologos,* Septiembre 1974, Granada, Spain, pp. 103–105.

Toki, R. and Imai, O. (1994) Nemathorin® (fosthiazate IKI-1145, ASC-66824), a new nematicide. *Agrochemicals Japan* 65, 13–15.

Townshead, J.L. (1990) Growth of carrot and tomato from oxamyl-coated seed and control of *Meloidogyne hapla*. *Journal of Nematology* 22, 170–175.

Triantaphyllou, A.C. and Sasser, J.N. (1960) Variation in perineal patterns and host specificity of *Meloidogyne incognita*. *Phytopathology* 50, 724–735.

Trivino, C.G. and Gowen, S.R. (1996) Deployment of *Pasteuria penetrans* for control of root-knot nematodes in Ecuador. *Pests and Diseases* 4, 389–392.

Tzortzakakis, E.A. and Gowen S.R. (1994) Evaluation of *Pasteuria penetrans* alone and in combination with oxamyl, plant resistance and solarization for control of *Meloidogyne* spp. on vegetables grown in greenhouses in Crete. *Crop Protection* 13, 455–462.

Tzortzakakis, E.A., Blok, V.C., Phillips, M.S. and Trudgill, D.L. (1999) Variation in root-knot nematode (*Meloidogyne* spp.) in Crete in relation to control with resistant tomato and pepper. *Nematology* 1, 499–506.

Uhlenbrock, J.A. and Bijloo, J.D. (1959) Nematicidal substance in marigold. *Tagetes erecta* Linn. *Recueil des Travaux Chimique des Pays Bas Belgium* 78, 239.

Valdez, R.B. (1978) Nematodes attacking tomato and their control. *1st International Symposium on Tropical Tomato, AVRDC Publication 78/59*, Taiwan, pp. 136–152.

Valesquez-Valle, R. (2001) Root-knot nematodes affecting vegetables and other crops in north central Mexico. *Revista Mexicana de Fitopatologia* 19, 107–109.

Van der Linde, W.J. (1956) The *Meloidogyne* problem in South Africa. *Nematologica* 1, 177–183.

Van Gundy, S.D., Kirkpatrick, J.D. and Golden J. (1977) The nature and role of metabolite leakage from root knot nematode galls and infection by *Rhizoctonia solani*. *Journal of Nematology* 9, 113–121.

Vargas, E.M.T., Zavaleta, M.E. and Hernandez, A.A.M. (1996) Breaking of resistance to *Phytophthora capsici* Leo. in chilli (*Capsicum annuum* L.) Serrano CM-34 by *Nacobbus aberrans* Thorne and Allen. *Nematropica* 26, 159–166.

Varma, M.K., Sharma, H.C. and Pathak, V.N. (1978) Efficacy of *Tagetes patula* and *Sesamum orientale* against root knot of eggplant. *Plant Disease Reporter* 62, 274–275.

Verdejo, L.S., Ornat, C., Sorribas, F.J. and Albajes, R. (1997) Management of root-knot nematodes in protected crops of northeast Spain. In: Albajes, R. and Carnero, A. (eds) *Integrated Control in Protected Ccrops 'Mediterranean Climate'*. Meeting Proceedings, Tenerife, Canary Islands. *Bulletin OILB-WPRS* 20, 94–98.

Veremis, J.C., Cap, G.B. and Roberts, P.A. (1997) A search for resistance in *Lycopersicum* spp. to *Nacobbus aberrans*. *Plant Disease* 81, 217–221.

Verma, A.C., Anwar, A. and Kumar, A. (1994) Efficacy of basamid Gr. against *Meloidogyne incognita* in a tomato nursery. *Afro-Asian Journal of Nematology* 4, 73–75.

Vilich, V. and Sikora, R.A. (1998) Diversity in soil-borne microbial communities: a tool for biological system management of root health. In: Boland, G.J. and Kuykendall, L.D. (eds) *Plant–Microbe Interactions and Biological Control*. Marcel Dekker, Inc., New York, pp. 1–15.

Villar, E.M.J. and Zavaleta, M.E. (1909) Efecto de *Crotolaria longirostrata* Hook y Arnott sobre nematodeos agalladores (*Meloidogyne* spp.). *Revista Mexicana de Fitopatologia* 8, 166–172.

Wallace, H.R. (1966) Factors influencing the infectivity of plant parasitic nematodes. *Proceedings of the Royal Society of London Series B* 164, 592–614.

Wallace, H.R. (1983) Interactions between nematodes and other factors on plants. *Journal of Nematology* 15, 221–227.

Waller, J.M. and Bridge, J. (1978) Plant disease and nematodes in the sultanate of Oman. *PANS* 24, 136–140.

Walters, S.A. and Barker, K.R. (1994) Efficacy of *Paecilomyces lilacinus* in suppressing *Rotylenchulus reniformis* on tomato. *Journal of Nematology Supplement* 26, 600–605.

Warrior, P., Rehberger, L.A., Beach, M., Grau, P.A., Kirfman, G.W. and Conley, J.M. (1999) Commercial development and introduction of DiTera®, a new nematicide. *Pesticide Science* 55, 376–379.

Webster, J.M. (1985) Interaction of *Meloidogyne* with fungi on crop plants. In: Sasser, J.N. and Carter, C.C. (eds) *An Advanced Treatise on* Meloidogyne. Volume I. *Biology and Control*. North Carolina University Graphics, Raleigh, North Carolina, pp. 183–192.

Whitehead, A.G. (1998) *Plant Nematode Control*. CAB International, Wallingford, UK.

Williams, K.J.O. (1974) *Belonolaimus longicaudatus*. *CIH Description of Plant-parasitic Nematodes*. Set 3, No. 40.

Williamson, V.M. (1998) Root-knot nematode resistance genes in tomato and their potential for future use. *Annual Review of Phytopathology* 36, 277–293.

Windham, G.L., Windham, M.T. and Williams, W.P. (1989) Effects of *Trichoderma* spp. on maize growth and *Meloidogyne arenaria* reproduction. *Plant Disease* 73, 493–495.

Zavaleta, M.E. and Gomez, R.O. (1995) Effect of *Tagetes erecta* L.–tomato (*Lycopersicon esculentum* Mill.) intercropping on some tomato pests. *Fitopatologia* 30, 35–46.

Zavaleta-Mejia, E. and Van Gundy, S.D. (1982) Effects of rhizobacteria on *Meloidogyne* infection. *Journal of Nematology* 15, 475–476.

Zavaleta-Mejia, E., Castro, A.E. and Zamudio, V. (1993) Effecto del cultivo e incorporaticion de *Tagetes erecta* L. sobre la problacion e infection de *Meloidogyne incognita* (Koford and White) Chitwood in chilli (*Capsicum annum* L.). *Nematologica* 23, 49–56.

Zeck, W.M. (1971) A rating scheme for field evaluation of root-knot nematode infestations. *Pflanzenschutz Nachrichten. Bayer AG* 24, 141–144.

Zoon, F.C., van Bruggen, A.S., de Heij, A., Asjes, C.J., van den Ende, J.E., van Bruggen, A.S., de Heij, A., van den Ende, J.E. and Littlejohn, G. (2002) Effect of green manure crops and organic amendments on incidence of nematode-borne tobacco rattle virus. In: Venter, R. and Lombard, C. (eds) *Proceedings of the Eighth International Symposium on Flower Bulbs*, Cape Town, South Africa, pp. 287–292.

Zoon, F., de Heij, A. and Kok, H.C.J. (2004) Weed hosts of *Meloidogyne chitwoodi* and *M. fallax*. Poster European Society of Nematology. Congress Abstracts. Available at: esn-symposium.ba.cnr.it

10 Nematode Parasites of Peanut*

Don W. Dickson[1] and Dirk De Waele[2]

[1]*Entomology and Nematology Department, Building 970, Natural Area Drive, PO Box 110620, University of Florida, Gainesville, FL 32611-0620, USA;* [2]*Katholieke Universiteit Leuven, Laboratory of Tropical Crop Improvement, Kasteelpark Arenberg 13, 3001 Leuven, Belgium*

The cultivated groundnut, or peanut (*Arachis hypogaea* L.), is an annual, self-pollinating, herbaceous legume native to South America (Hammons, 1982). It is a geotropic plant that produces its pods (fruits) underground. Flowering begins 4–6 weeks after planting and extends over a period of several weeks. Within about 1 week after the flowers are fertilized, pointed needle-like structures, carpophores, commonly called 'pegs', develop, elongate and grow into the soil to a depth of 2–7 cm. Upon entering the soil, the fertilized ovaries located behind the tip of the peg enlarge rapidly and pod growth begins. Two to four seeds are formed within a pod, but the number of seeds formed per pod depends on the groundnut variety. The length of time necessary for pod development to maturity may vary with cultivar and environmental conditions, e.g. cv. Florunner requires 63–70 days from the time the ovary begins enlarging to maturity (Williams and Drexler, 1981).

Groundnut was listed as one of the 20 crop plants that stand between man and starvation (Wittwer, 1981). Seeds from groundnut are rich in calories and contain 25% protein. They may be boiled, broiled, roasted, fried, ground into peanut butter, or crushed for oil. Groundnut-containing foods such as peanut butter, salted groundnuts, candies and snack-type crackers and cookies are popular because of their unique roasted groundnut flavour (McWatters and Cherry, 1982). However, this crop is grown primarily for cooking and salad oil. Oil extraction also produces a protein-rich by-product that may be used for human consumption if processed from an edible-grade groundnut; otherwise, it is used for animal feed.

Currently, groundnut is cultivated on all six continents, with major production in over 32 countries. Eight countries, the Peoples Republic of China, India, the USA, Indonesia, Argentina, Senegal, Zaire and Myanmar, produce 72% of the world supply. In 2003, approximately 31.6 Mt were produced on 22.4 Mha. The highest average yields per hectare are produced in the USA (2.81 t/ha) and the People's Republic of China (2.59 t/ha). Other countries have much lower yields (Anonymous, 1999). Production is dis-

*A revision of the chapter by N.A. Minton and P. Baujard.

tributed generally in tropical and subtropical regions of the world. Regions with loose, friable, sandy soils and warm temperatures are ideal for groundnut production.

Nematodes of Groundnut

Plant nematodes are primary parasites of groundnut in all production regions of the world. Based on a worldwide survey of nematologists, annual losses caused by all nematodes to groundnut were estimated at 12% and monetary losses were estimated at US$1.03 billion (Sasser and Freckman, 1987). The primary nematode parasites of groundnut include *Meloidogyne* spp., *Pratylenchus brachyurus*, *Belonolaimus longicaudatus*, *Criconemoides ornatus*, *Aphelenchoides arachidis*, *Aphasmatylenchus straturatus*, *Scutellonema cavenessi*, *Tylenchorhynchus brevilineatus* and *Ditylenchus africanus*. Each will be discussed in this chapter. Several other nematode species have been found associated with groundnut (Sharma, 1985), but are not included herein.

Meloidogyne

Three major *Meloidogyne* spp. parasitize groundnut and each is capable of causing severe suppression of groundnut yields and fruit quality. Two of the species, *M. arenaria* (common name is groundnut root knot nematode) and *M. javanica* (common name is Javanese root knot nematode), are highly virulent pathogens of groundnut, whereas *M. hapla* (common name is northern root knot nematode) causes less damage but nonetheless is an important disease-inducing agent of groundnut. All three species occur on groundnut worldwide (Sasser, 1977). *M. arenaria* and *M. javanica* are common in warm and hot regions of the world, whereas *M. hapla* occurs only in cooler regions. A new species of root knot nematode causing disease on groundnut in Texas, USA was recently described as *M. haplanaria* (Eisenback *et al.*, 2003).

In the USA, *M. arenaria* is the dominant *Meloidogyne* species parasitizing groundnut in Alabama, Florida, Georgia and Texas, while patchy occurrences have been reported in North Carolina, South Carolina and Virginia. In other regions of the world, *M. arenaria* is reported damaging groundnut in Zimbabwe (Martin, 1958), Israel (Orion and Cohn, 1975), Egypt (Ibrahim and El-Saedy, 1976a), India (Sharma *et al.*, 1978; Dhurj and Vaishnav, 1981; Sakhuja and Sethi, 1985c), Taiwan (Cheng and Tu, 1980; Cheng *et al.*, 1981) and China (Zhang, 1985).

M. javanica was first reported parasitizing groundnut in Zimbabwe (Martin, 1958). Although the species is highly virulent on groundnut, it is less frequently encountered on groundnut in the USA than *M. arenaria*. It is only reported parasitizing groundnut in Florida (Cetintas, 2003), Georgia (Minton *et al.*, 1969b) and Texas (Tomaszewski *et al.*, 1994). The species is also known to occur on groundnut in Egypt (Ibrahim and El-Saedy, 1976b), Brazil (Lordello and Gerin, 1981; Carneiro *et al.*, 2003) and India (Sakhuja and Sethi, 1985b).

M. hapla is encountered more frequently on groundnut in the more northern latitudes. However, the nematode may be encountered at higher elevations in tropical regions (Eisenback and Triantaphyllou, 1991). *M. hapla* has been reported infecting groundnut in all groundnut-producing states in the USA with the exception of Florida (Dickson, 1998). It is frequently encountered infecting groundnut in North Carolina, Virginia and Oklahoma. The species parasitizes groundnut in Israel (Minz, 1956), South Africa (Van der Linde, 1956), Australia (Colbran, 1958; Saint-Smith *et al.*, 1972), Zimbabwe (Martin, 1961), Japan (Mitsui *et al.*, 1976), Korea (Choi, 1981) and China (Yin and Feng, 1981; Yang, 1984; Zhang, 1985).

Infection and histopathology

Second stage juveniles (J2) of *Meloidogyne* spp. enter and damage groundnut roots, pegs and pods. Upon entering root tips, they cause only slight mechanical injury, except when large numbers enter in a limited area. *M. arenaria* J2 are capable of infecting groundnut roots the second day following plant inoculation (Minton, 1963). As a result of the juvenile feeding on vascular cells, large, multinucleate giant cells develop by the eighth day. Hyperplasia (increase in cell numbers) is observed in tissue adjacent to juveniles. Hyperplasia and hypertrophy (increase in cell size) result in the disorganization of vascular tissue and the formation of galled tissue. The parenchymatous cells associated with developing juveniles at the periphery of the stele multiply and grow out into the cortex. The enlarging juveniles crush adjacent cortical cells as they grow during development. Elongation of severely galled roots is slowed, resulting in a stunted root system. A major consequence of root knot nematode development and giant cell formation is the malformation of the xylem elements and the inhibition of secondary growth of the xylem and phloem tissues. As a consequence, infected roots do poorly in taking up nutrients and water.

Symptoms of damage

Noticeable above- and below-ground symptoms of root knot nematodes on groundnut can be observed as early as 45–75 days after planting, but most severe symptoms are observed after 90–120 days. Above-ground symptoms of root knot disease may be subtle or very conspicuous, especially as the crop nears maturity. The degree of symptoms depends on the growing environment as well as the population density of root knot nematode juveniles at the time of planting. In some cases, stunting of young plants may be severe (Fig. 10.1). The general characteristics of diseased groundnut plants are typical of other plants infected by these nematodes. As the crop nears maturity, heavily infected plants may be severely stunted, showing symptoms of chlorosis, incipient wilting, nutrient deficiencies, or even death when conditions are hot and dry. Symptoms are distributed in patches of varying sizes (Fig. 10.2). Infected plants exhibit a rusty, yellowish and mot-

Fig. 10.1. Severely stunted groundnut plant caused by early infection by *Meloidogyne arenaria* (right) compared with a healthy plant (left). (Photo: D.W. Dickson.)

Fig. 10.2. Field symptoms showing patchy distribution of damage caused by groundnut plants infected by *Meloidogyne arenaria*. (Photo: D.W. Dickson.)

tled appearance. If drought occurs near the end of the season, the severity of root knot disease is accentuated and weakened plants die (Fig. 10.3A and B). Early season symptoms include stunted plants that fail to cover the soil between rows (Fig. 10.4A and B). The slowly dying and browning plants present a mottled effect among the greener plants, but even when such plants show conspicuous galling their neighbouring plants in apparent vigour are usually also infected (Machmer, 1951). In China, *M. arenaria*-infected plants may become yellow and stunted as early as 40 days after planting (Zhang, 1985).

Second stage juveniles infect groundnut plants soon after they germinate, but noticeable galling and egg masses are not apparent on the roots until 55–90 days after planting. The characteristic symptom on roots is the abnormal swelling (galls or knots) (Fig. 10.5); however, these are often difficult for the novice to see. Galled tissue on roots may attain a diameter larger than that of normal adjacent roots, but, because of the abundance of nodules containing

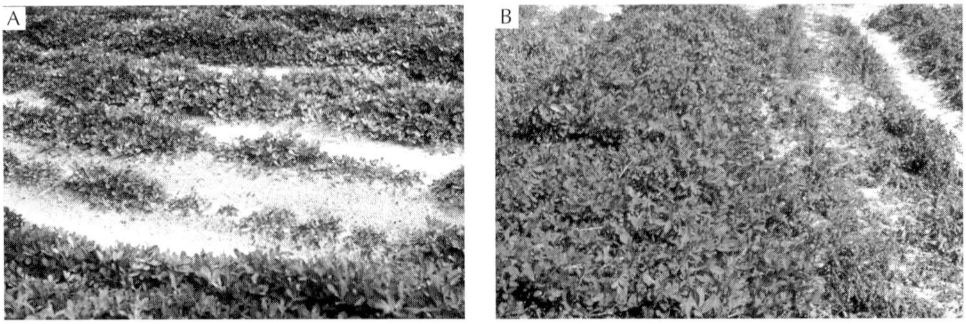

Fig. 10.3. Near-harvest field symptoms showing (A) stunted, yellow and (B) dying groundnut plants that are infected by *Meloidogyne arenaria*. (Photo: D.W. Dickson.)

Fig. 10.4. (A) Early season field symptoms showing *Meloidogyne arenaria* damage. (B) Stunted plants fail to cover the soil between rows. (Photo: D.W. Dickson.)

nitrogen-fixing bacteria (*Bradyrhizobium* (*Arachis*) sp.), the amount of galling is difficult to determine. This is distinct from what happens on tomato or cucumber where galling is evident on roots 2–3 weeks after planting. Groundnut root galls are small and generally discrete, whereas galls on other host crops may be large and become coalesced. Nematode galls can be distinguished from nodules containing nitrogen-fixing bacteria (Plate 11A). Nodules are distinctive, round swellings attached to the root and are easily detached, whereas nematode galls are swellings that constitute a part of the fibrous root system and cannot be removed without destroying the integrity of the root. Also, because second stage nematodes can infect nodules in some instances, galls may appear on the nodules, and vice versa. Root knot juveniles may also infect pegs and pods after blooming and initiation of pod set (generally ~45 days after planting). Galling on pegs and pods is distinctive and more easily seen than that on roots but, interest-

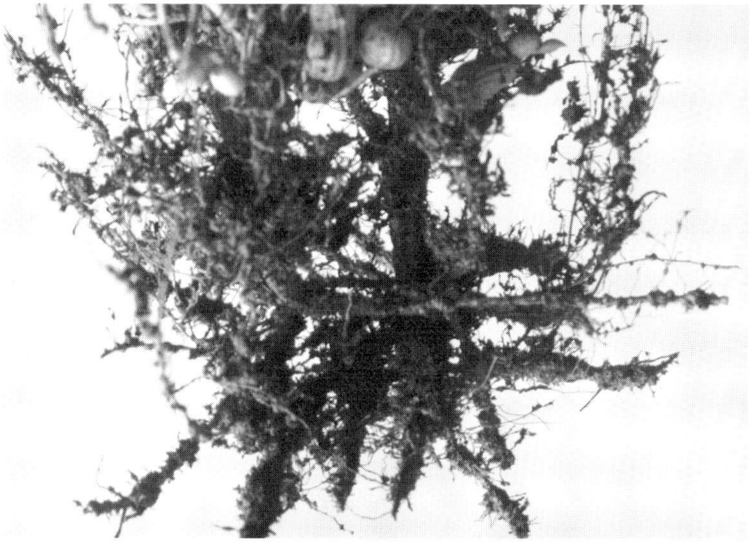

Fig. 10.5. Groundnut roots galled by *Meloidogyne arenaria*. (Photo: D.W. Dickson.)

Fig. 10.6 (A–C) Groundnut pegs and pods showing extensive galling caused by *Meloidogyne arenaria*. (Photo: D.W. Dickson.)

ingly, it does not always appear even though roots are galled (Fig. 10.6; Plate 11B). Situations where galling on pegs and pods is extensive generally result in a large reduction in potential yield. Yield potential is lost because pegs are weakened and easily fall off during harvest, pod formation is aborted, or damaged pods fail to produce seeds. Galling induced by *M. hapla* is distinctively different from that caused by *M. arenaria* or *M. javanica*. The former results in smaller galls with some root proliferation above galls that results in a denser root system (Sasser, 1954).

The most obvious sign of root knot nematodes is the observation of female nematodes in galled roots, pegs or pods. Females are globose, approximately the size of a typewritten period on a page (800 μm length × 500 μm wide), pearly white in colour, and have sharp pointed necks and heads off to one side that are generally visible. An egg mass is generally extruded from the vulva end of each female at or near the root surface. This positions the egg mass on the outside surface of galled tissue, which facilitates both egg hatch and secondary infection of roots by the freshly hatched juveniles. Egg masses are about 1 mm in width, appear as brownish masses adhering to galled tissue and contain up to 300–500 eggs each. They are generally plentiful along galled tissue. Staining with food colouring (Thies *et al.*, 2002) or Phloxine B (Dickson and Struble, 1965) enables one to see them more readily.

M. hapla symptoms are generally less severe than those caused by either *M. arenaria* or *M. javanica*. Above-ground symptoms of *M. hapla* may be difficult to detect because this pathogen causes less stunting or chlorosis. The most severe symptoms generally indicate large population densities of infective juveniles in the soil. Severity over a field varies, depending on the variations in soil population densities and soil type. Sandy areas within a field often show the most severe symptoms. Plants with light infections usually do not show stunting or chlorosis. Typically, the only indication of root knot disease on such plants is galls on roots, pegs and pods.

Biological races

Among the root knot nematode species that infect groundnut, there are two races reported for *M. arenaria* (Taylor and Sasser, 1978) and four proposed for *M. javanica* (Carneiro *et al.*, 2003). Of the two races of *M. arenaria*, race 1 infects groundnut and race 2 does not; whereas, of the four proposed races of *M. javanica*, races 3 and 4 infect groundnut, races 2 and 4 infect pepper, and race 1 infects neither. Host races of *M. arenaria* and *M. javanica* are morphologically indistinguishable (Sasser, 1979a; Osman *et al.*, 1985; Carneiro *et al.*, 1998, 2003), thus their separation depends on their reaction on differential host plants. Most *M. javanica* populations do not reproduce on groundnut (Taylor and Sasser, 1978). The worldwide distribution of populations of *M. javanica* that infects groundnut is listed above.

Survival and means of dissemination

Plant nematodes are moved by humans, animals, water, wind and any other means that move soil or infected parts of plants. Humans and animals that track across fields infested with plant nematodes may potentially spread nematodes to uninfested fields via soil adhering to their feet. Important among dispersal methods is the movement of soil from infested fields transported on all types of farming implements. Any type of cultivation equipment with soil adhering will move nematodes, but groundnut producers must look to their diggers and combines as a principal means of moving root knot nematodes. Plant nematodes may be dispersed from movement of freshly dug pegs, pods or roots; however, developmental stages of nematodes generally do not survive in these plant parts when they are well dried. Interestingly, wind and water play a major role in the dispersal of nematodes. Second stage juveniles of *Meloidogyne* spp. were among 28 genera recovered from dust traps placed 2 m above the ground in western Texas (Orr and Newton, 1971). Dispersal by surface runoff water and by irrigation also occurs (Faulkner and Bolander, 1966; Meagher, 1967; Sauer, 1968). Refuse from packing and processing plants that has not been thoroughly dried may harbour viable eggs and infective juveniles.

Environmental factors affecting parasitism

Temperature is considered the most important environmental factor affecting *Meloidogyne* spp. survival and parasitism, and the lower and upper temperatures for survival are approximately 0–5°C and 35–40°C, respectively (Taylor and Sasser, 1978). In general, the optimum temperature for survival of eggs and juveniles is 10–15°C (Bergeson, 1959; Thomason *et al.*, 1964). The optimum temperature for hatching of *M. hapla* and *M. javanica* is 25 and 30°C, respectively (Bird and Wallace, 1965). *M. javanica* had a significantly higher hatch rate at 30°C than *M. hapla*. At 14.3°C, the life cycle of *M. javanica* requires 56 days, whereas at 26.1°C only 21 days are required (Milne and Du Plessis, 1964).

There is general agreement that *Meloidogyne* spp. damage is greater in sandy soils than in soils with a large percentage of clay. In China, the incidence and severity of *M. arenaria* on groundnut were related to soil texture (Zhang, 1985).

Soil moisture is necessary to sustain activities of *Meloidogyne* spp. In moist soils of 40–60% field capacity, J2 are active and move through the soil in a film of water surrounding soil particles. In dry soils, they become inactive and die through desiccation (Van Gundy, 1985). In wet soils, hatching may be inhibited and juvenile movement slowed by lack of oxygen. All activity of *M. javanica* increased as the oxygen concentrations increased from 0.2 to 21%, and it was concluded that a favourable environment would be provided when moist soils drain rapidly and allow oxygen concentrations to increase above 10% (Baxter and Blake, 1969). *M. arenaria* was less damaging in fields that had a high water table than in well-drained fields (Zhang, 1985). Also, *M. arenaria* is less damaging to a groundnut crop that follows a flooded crop than in fields that are not flooded. The infectivity of J2 of root knot nematodes may be reduced when exposed to low temperatures (–8 to 20°C) in saturated soil as compared with exposure in soils with 51 cm moisture tension (Vrain, 1978).

Disease complexes

There is a great need for more detailed studies on disease complexes involving root knot nematodes on groundnut. This is especially true regarding interactions involving *Sclerotium rolfsii* Sacc. (southern blight) and *Cylindrocladium parasiticum* Crous, Wingefield & Alfenas (Cylindrocladium black rot (CBR)), which are two important soil-borne diseases that appear fairly frequently with root knot nematodes in many agricultural fields used for groundnut production (Melouk and Backman, 1995). The incidence of southern blight, which causes groundnut stem rot (Brenneman *et al.*, 1995), is reduced when crop rotation or nematicides are used to manage *M. arenaria* infection on groundnut (Rodríguez-Kábana *et al.*, 1982a, 1994). There have been few attempts to document a disease complex involving *M. arenaria* and southern blight even though they occur together frequently, and field observations and other studies suggest infection of groundnut by *M. arenaria* increases the incidence of southern blight. Recently, an attempt to show an interaction between *M. arenaria* and *S. rolfsii* in microplots was negative (Starr *et al.*, 1996).

Significant positive correlations between final populations of microsclerotia of *C. parasiticum* and *M. hapla* in a field test indicated that this nematode could affect CBR development (Diomandé and Beute, 1981b). It was also found that two populations of *M. arenaria* enhanced development of CBR on CBR-resistant groundnut (Diomandé *et al.*, 1981). Others show an increased severity of black rot on susceptible groundnut cultivars infected by either *M. arenaria* or *M. hapla* (Culbreath *et al.*, 1992).

A synergistic interaction in groundnut

pod rot and damping off occurred when *Pythium myriotylum* Drechs. was combined with *Fusarium solani* (Mart.) Sacc. and *M. arenaria* (Garcia and Mitchell, 1975a,b). Groundnut plants inoculated with *F. solani* mycelium and *M. arenaria* wilted sooner after inoculating than when *F. solani* was inoculated alone (Patel *et al.*, 1985). Results of a 2 year pot study showed that the presence of *M. arenaria* had no effect on the incidence of *Aspergillus flavus* Link in groundnut seeds (Minton and Jackson, 1967); however, after 1 year, the incidence of *A. flavus* was greater in shells of plants inoculated with both organisms than with *A. flavus* alone. In a microplot study, the incidence of *A. flavus* was greater in seeds of plants inoculated with *A. flavus* and *M. hapla* than in seeds of plants inoculated with only *A. flavus* (Minton *et al.*, 1969a). Aflatoxin was not detected in seeds of any treatment and was present in only one shell sample each of *A. flavus*- or *A. flavus* plus *M. hapla*-inoculated plants.

The interaction of concomitant populations of *M. arenaria* races 1 and 2, and *M. hapla* on groundnut was investigated (Hirunsalle *et al.*, 1995a). Race 2 of *M. arenaria* tended to depress *M. arenaria* race 1 development on groundnut, whereas *M. hapla* had little effect on the latter. In mixed (1:1) populations of *M. arenaria* races 1 and 2, race 1 was dominant on groundnut (Hirunsalle *et al.*, 1995b). A 2 year crop rotation with a poor host for race 1 lowered the number of nematodes; however, they increased rapidly when groundnut was reintroduced. *M. arenaria* race 1 had a greater rate of reproduction than *M. arenaria* race 2 in a 2 year rotation of groundnut following resistant tobacco, and in a 3 year groundnut rotation following 2 years of resistant tobacco. Although the reproductive potentials of *M. arenaria* and *M. hapla* are similar, interaction studies between them on five groundnut genotypes showed that *M. arenaria* had greater infection capacity and caused more crop damage than *M. hapla* (Hirunsalle *et al.*, 1995c). Thus, *M. arenaria* is more competitive than *M. hapla* on groundnut. Infection and reproductive potentials and crop damage induced by mixed populations were similar to those induced by *M. arenaria* alone.

Economic importance and population damage threshold levels

Yield suppression by plant nematodes is difficult to estimate because damage is seldom confined to a single nematode species (Sasser *et al.*, 1970, 1975a). Also, damage caused by low to moderate densities of plant nematodes often goes unnoticed. Where damaging levels of *M. arenaria* or *M. javanica* occur, more than 50% of yield potential can be lost. Even 100% losses have been observed in sections of severely infested fields; however, because of the uneven distribution of plant nematodes, losses over large fields may average less than 50%. In three nematicide efficacy trials conducted in groundnut fields heavily infested with *M. arenaria*, the most effective nematicide treatments increased yields an average of 83% (Dickson and Hewlett, 1988a). Estimated potential yield losses due to *M. arenaria* generally are less, ranging from 0.5% in Oklahoma to 5% in Alabama, whereas that for *M. hapla* ranged from 0.3% in Georgia to 5% in North Carolina (Anonymous, 1987). Suppression of yields by *Meloidogyne* spp. in West Africa and south-eastern Asia were estimated at 15% (Sasser, 1979b).

Some estimates of the percentage of groundnut fields infested by *Meloidogyne* spp. have been reported. In Alabama, Georgia and Texas, approximately 41, 10 and 26%, respectively, of the fields surveyed were found to be infested (Motsinger *et al.*, 1976; Ingram and Rodríguez-Kábana, 1980; Wheeler and Starr, 1987). Even larger estimates, up to 40%, have been reported in these southern states including Florida (Sturgeon, 1986). In the Punjab, India, *Meloidogyne* spp. juveniles were present in an average of 47% of soil samples collected from three groundnut-growing districts (Sakhuja and Sethi, 1985c). Galling on groundnut due to *Meloidogyne* spp. was noted in 31% of locations sampled. In

Egypt, 65% of soil and root samples collected from fields with poor-growing groundnut contained *Meloidogyne* spp. (Ibrahim and El-Saedy, 1976a). *M. javanica* was the dominant species, with *M. arenaria* present in a few of the root samples. Seventy-five percent of soil samples collected around groundnut plants in Guyana contained *Meloidogyne* spp. (Singh, 1972). A large percentage, up to 61%, of groundnut grown on 6200 ha in Leizhou Peninsula, Republic of China, was infected with *M. arenaria* (Zhang, 1985).

Several scientists have reported on the economic damage level of plant nematodes on groundnut. Advisories for damaging levels of plant nematodes are usually based on numbers of juveniles in soil because most extraction procedures do not recover nematode eggs from soil (Garcia, 1976; Rodríguez-Kábana *et al.*, 1986). However, timing of sampling is critical because population densities of *M. arenaria* juveniles in soil at planting time are usually relatively low or near undetectable levels. Hence, for grower advisory purposes, it is usually best to determine population densities as soon after harvest as practical. Once root knot disease is observed on groundnut, the problem will continue to increase unless the nematode is suppressed by natural biological antagonists or other causes (Dickson *et al.*, 1994).

In India, groundnut plants were stunted when inoculated with one *M. javanica* egg/cm^3 of soil (Sakhuja and Sethi, 1985b). A reduction of 27% in shoot length and 55% in dry shoot weight was obtained when plants were inoculated with 8 eggs/cm^3 of soil. In nematicide experiments, yields are usually negatively correlated with numbers of *Meloidogyne* juveniles in the soil. Regression analysis on data from 16 groundnut experiments in Alabama indicated that yields were negatively related to numbers of *M. arenaria* juveniles in the soil determined near harvest (Rodríguez-Kábana *et al.*, 1982b). On the basis of a linear regression model, it was determined that groundnut yield loss in microplots was 8.6% for each tenfold increase in initial population density of *M. hapla* juveniles in soil (Rickard *et al.*, 1977). A significant negative relationship between initial population densities of *M. arenaria* in microplot tests and groundnut yields was reported (Wheeler and Starr, 1987). A linear model estimated a 10% yield loss with initial populations of 44–83 eggs and juveniles/500 cm^3 of soil. In Florida, the damage threshold was estimated to be as low as a single juvenile per 100 cm^3 of soil (McSorley *et al.*, 1992). An inoculum density of 1000 *M. arenaria* juveniles/kg of soil caused a reduction of groundnut plant shoot growth, shoot weight and root length of 23.9, 33.1 and 31.9%, respectively (Dhurj and Vaishnav, 1981).

Management

Where root knot disease of groundnut exists, generally some means of management of the disease is required for profitable groundnut production. Each field should be evaluated based on the history of nematode damage before determining what management tactics to employ. The first line of defence should be preventing further development of the disease by reducing spread. Management of root knot disease is very difficult and costly once it becomes established, in terms of both time devoted to developing management tactics and resources that must be allocated.

CROP ROTATION. One of the most effective means of management is crop rotation, which includes plants that are non-hosts or resistant to *M. arenaria*, *M. hapla* or *M. javanica*. When the cash value for groundnut is low, this may be the only management tactic that can be used profitably. The objective is for groundnut to follow poor or non-hosts for root knot nematodes, such as cotton, maize, small grains and pasture grasses (Bailey, 1988; Hagan, 1988; Dickson and Melouk, 1995; Dunn and Dickson, 1995); however, it is important to realize that use of rotation varies with the nematode species present, cultivar of rotational crop and economics of growing a rotation crop.

Growing tropical forages for several years, e.g. bahiagrass (*Paspalum notatum* Flegge), has long been recognized as one of the best rotations to precede groundnut (Norton *et al.*, 1977; Rodríguez-Kábana *et al.*, 1994). Rotations of 3 or more years out of groundnut and other favourable hosts are better than 1 or 2 year rotations (Dickson and Hewlett, 1989); however, for such a rotation to work successfully, one must manage weeds in the forage grasses. A few common weeds that occur frequently in planting of forage grasses include hairy indigo (*Indigofera hirsuta* L.), alyceclover (*Alysicarpus vaginalis* (L.) DC.) and morning-glory (*Ipomoea* spp.), each of which is a good host for *M. arenaria*. Two weeds common in groundnut fields, coffee senna (*Cassia occidentalis* L.) and sicklepod (*C. obtusifolia* L.), are reported as hosts for the groundnut root knot nematode (Machmer, 1951). However, this is variable and needs further study for clarification. Another tropical forage, coastal bermudagrass (*Cynodon dactylon* (L.) Pers.), has been grown for decades as a forage and rotation crop for the management of root knot nematodes in flue-cured tobacco, sweet potato and other vegetables (Burton and Johnson, 1987), and has been proposed as a general solution to problems caused by root knot nematodes (Anonymous, 1989). Other data, however, show that 3 year rotations with coastal bermudagrass failed to increase groundnut yields or decrease densities of *M. arenaria* on groundnut (Rodríguez-Kábana *et al.*, 1994). Also, bermudagrass failed to reduce the incidence of southern blight (*S. rolfsii*), whereas both bahiagrass and cotton did.

Cotton is a good rotation crop with groundnut in situations where a single producer grows these two crops. *M. arenaria*, *M. hapla* and *M. javanica* do not live on cotton, whereas *M. incognita* races 3 and 4 that infect cotton do not live on groundnut (Sasser and Carter, 1982; Rodríguez-Kábana *et al.*, 1994). Groundnut yields were increased and densities of *M. arenaria* were reduced following 1 year of cotton (Rodríguez-Kábana *et al.*, 1987).

The susceptibility ratings of a number of plant species to *M. arenaria*, *M. hapla* and *M. javanica* are available (Sasser, 1954). Some of these may be used effectively in rotation with groundnut when *M. arenaria* or *M. hapla* is present. A checklist of more than 450 cultivars in 13 botanical families of crop plants reported to carry resistance to at least one *Meloidogyne* species is available (Sasser and Kirby, 1979). This list may serve as a useful guide for selecting cultivars to grow in rotation with groundnut. Seven of 30 crop plants tested in Taiwan were resistant (non-host) to *M. arenaria* (Cheng *et al.*, 1981). Although plants are listed as resistant or non-hosts, care must be exercised in their selection because all cultivars of a crop do not respond in the same way. A good example is maize (*Zea mays* L.), which is reported as an excellent rotational crop with groundnut. In recent years, some cultivars have been shown to support relatively high population densities of *M. arenaria* and *M. javanica* (Baldwin and Barker, 1970; Norse, 1972; Windham and Williams, 1994). Conversely, maize is resistant to *M. hapla* (Sasser, 1954; Baldwin and Barker, 1970), thus it is a suggested rotational crop for managing *M. hapla* on groundnut in Queensland, Australia (Broadley, 1981; Vance, 1981). Yet growing maize in a groundnut rotation is better than continuous groundnut or other good hosts of root knot nematodes, such as soybean, tobacco or vegetables, all of which are excellent hosts of root knot nematodes, except when root knot nematode-resistant soybean or tobacco is planted. Grain sorghum (*Sorghum vulgare* Pers.) appears to suppress root knot nematode population densities several-fold better than maize (Rodríguez-Kábana and Touchton, 1984; McSorley and Gallaher, 1991).

Some unusual crops such as partridge pea (*Cassia fasciculata* Michx.) and American jointvetch (*Aeschynomene americana* L.) suppressed population densities of *M. arenaria* in soil when either crop was grown for 2 years in rotation with groundnut (Rodríguez-Kábana *et al.*, 1991). Partridge pea reduced *M. arenaria* population densities more than jointvetch, but groundnut yields were increased after 2 years of planting jointvetch. Sesame (*Sesamum indicum* L.), castor (*Ricinus communis* L.), hairy indigo (*I. hirsuta* L.) and bahiagrass (*P. notatum* Flugge) were also listed as promising crops for managing *M. arenaria* in groundnut (Rodríguez-Kábana and Morgan-Jones, 1987). No *M. arenaria* galling was observed on American jointvetch, castor, cotton (*Gossypium hirsutum* L.), crotalaria (*Crotalaria spectabilis* Roth), sorghum–sudangrass (*Sorghum bicolor* × *S. sudanense* (Piper) Stapf) or resistant soybean (*Glycine max* L.) when grown in microplots, whereas hairy indigo supported only a low level of galling (McSorley *et al.*, 1994). It is important to note that hairy indigo, a common weed in production fields, is often observed heavily galled by *M. arenaria* in numerous groundnut fields in Florida (D.W. Dickson, personal observation). The reason for the variation in results with hairy indigo is not known.

The use of flood fallowing in conjunction with crop rotations may effectively reduce root knot nematode damage (Thames and Stoner, 1953; Zhang, 1985); however, in most instances, this is not practical. Rotating a winter small grain crop with groundnut can help prevent growth of weeds that are hosts of groundnut nematodes; however, since some small grain cultivars may also support low population densities if grown during warm weather, planting should be delayed until cool weather when nematode development and reproduction are reduced (Dunn and Dickson, 1995).

Crop rotation should not be expected to reduce a root knot nematode population abruptly because: (i) some of the nematode population will survive the winter without a host; (ii) most crop plants can support at least some nematode reproduction; and (iii) most fields have some weeds that support nematode reproduction. Rotation is a far better tool to help keep relatively low population densities from becoming too high, or for gradually reducing high population densities over several years (Dunn and Dickson, 1995).

OTHER CULTURAL METHODS. Destruction of roots of host crops that precede groundnut in a rotation to interrupt reproduction will reduce the potential for damage. Ploughing soils several weeks before applying nematicides and planting groundnut encourages the decay of live plant roots that protect nematodes from their enemies or from nematicides that are applied to the soil (Dunn and Dickson, 1995). Drying soils after they have been turned may reduce plant nematodes (Zhang, 1985). Clean fallowing for prolonged periods of time may also be effective.

In China, growers who fertilize well, especially with organic fertilizers, have fewer problems with root knot nematodes than growers who use less fertilizer (Zhang, 1985). *M. arenaria* was less damaging in soils with high water tables than in well-drained soils. This nematode was also less serious in irrigated than in non-irrigated fields (Zhang, 1985).

RESISTANCE TO ROOT KNOT NEMATODES IN GROUNDNUT. The most exciting prospect on the horizon for nematode management is the development of groundnut cultivars with root knot nematode resistance. Just a few years ago, this was considered improbable (Miller, 1972b; Minton and Hammons, 1975; Holbrook *et al.*, 1983). Thousands of groundnut introductions were screened without identifying resistance to *M. arenaria*. However, resistance was reported to *M. hapla* (Castillo *et al.*, 1973; Subrahmanyam *et al.*, 1983) and to *M. javanica* (Sakhuja and Sethi, 1985a). Then, in 1986, resistance to *M. arenaria* was reported in *A. glabrata* Benth., a rhizomatous groundnut that is not cross-compatible with *A. hypogaea* (Baltensperger *et al.*, 1986). However, a major breakthrough came in 1989 when resistance to *M. arenaria* was found in 21 *Arachis* spp. and two interspecific hybrids (Nelson *et al.*, 1989). They also reported resistance to *M. hapla* in two *Arachis* spp. and one of the interspecific hybrids. A systematic search was made of the *A. hypogaea* germplasm collection for useful sources of resistance (Holbrook and Noe, 1992). Several lines with moderate levels of resistance were identified, although none of them had the high levels of resistance observed in the wild groundnut species.

There are groups of scientists in the USA currently engaged in efforts to develop further resistance to *Meloidogyne* spp. in groundnut. These scientists are focusing on a continued systematic search of available germplasm collections of *A. hypogaea* for useful sources of resistance (Stephenson *et al.*, 1995). Resistance identified to date suppresses population densities of *M. arenaria* by 40–60% (Noe *et al.*, 1992). Yield potential of three lines with moderate resistance appears to be equal to that of the susceptible standards in the absence of nematode pressure and superior to the susceptible standard in the presence of damaging levels of *M. arenaria* (Holbrook *et al.*, 1995). Other groundnut scientists are focusing on the introgression of resistance from wild *Arachis* spp. into cultivated groundnut. Because *A. hypogaea* is an allotetraploid, whereas most wild species are diploids, introgression of nematode resistance genes from the wild species into cultivated groundnut is a difficult process. The effort to date has focused on the development of resistance to *M. arenaria* and has followed a diploid route to the introgression of resistant genes into *A. hypogaea* (Simpson, 1991). Three wild species were used to develop a nematode-resistant complex hybrid (TxAG-6 (Simpson *et al.*, 1993)) that is cross-compatible with *A. hypogaea*. All three wild species, *A. batizocoi* Krapov. & WC Greg., *A. cardenasii* Krapov. & WC Greg. and *A. diogoi* (= *A. chancoensis*) Hoehne, are resistant to *M. arenaria*, and *A. cardenasii* is also resistant to *M. hapla* (Nelson *et al.*, 1989). A backcrossing programme that uses cv. Florunner as the recurrent parent has been used for the introgression of resistant genes to *M. arenaria* from TxAG-6 into two released cultivars. The first cv. COAN was selected from the fifth backcross generation (Simpson and Starr, 2001) and the second cv. NemaTAM (Simpson *et al.*, 2003) was selected from the seventh backcross generation. Both resistant cultivars suppress nematode reproduction by more than 90% and have significantly greater yield potential than susceptible cultivars in grower fields where nematode population densities exceed the damage threshold density. Unfortunately, in fields not infested with root knot nematodes, yields of COAN are 10–20% less than those of the highest yielding susceptible cultivars, and yields of NemaTAM are statistically similar but numerically less than those of susceptible cultivars (Church *et al.*, 2000; Starr *et al.*, 2002).

Both random amplified polymorphic DNA (RAPD; Burow *et al.*, 1996) and restriction fragment length polymorphism (RFLP; Choi *et al.*, 1999) markers linked to the resistance locus in breeding lines derived from TxAG-6 (including COAN and NemaTAM) have been identified. These reports indicated that resistance in these genotypes is inherited as a single dominant gene that was derived from *A. cardenasii*. The RFLP markers were used to select individuals homozygous for resistance from a segregating population in the development of NemaTAM in a process that was more efficient than the more traditional progeny testing (Church *et al.*, 2000).

Other groups of scientists have focused on development of resistance to *M. arenaria* in Virginia market-type groundnut, using *A. cardenasii* as the source of resistance (Stalker *et al.*, 1994). High levels of resistance were introgressed into several breeding lines using a hexaploid route. This effort has reached the point where resistant lines are being screened for yield potential in field tests. Two RAPD markers linked to the resistance genes were identified (Stalker *et al.*, 1995). In lines with high levels of resistance, there is segregation in a 3:1 ratio indicative of a single dominant gene. Segregation patterns are more complex in lines with moderate levels of resistance (Stalker *et al.*, 1995). These programmes are complementary in that although each has used a different approach for introgression of genes, they may have resulted in the identification of different resistance genes. There is evidence that resistance to *M. arenaria* in *A. cardenasii* is conditioned by multiple, dominant, major genes (Starr and Simpson, 1991). Additionally, resistance in the F_2 generation from TxAG-6 × *A. hypogaea* and a derived BC3 population segregates as one dominant gene and one recessive gene (Church *et al.*, 2000). Further, resistance was recovered in the F_3 from susceptible F_2 individuals in a 1:3 (resistant:susceptible) ratio consistent with the F_2 being heterozygous for the recessive resistance gene. It has not yet been determined which of the three wild *Arachis* species used in the development of TxAG-6 contributed this recessive gene.

With the confirmation of parasitism of groundnut by populations of *M. javanica* from Egypt (Tomaszewski *et al.*, 1994), India (Sharma *et al.*, 1995) and the USA (Minton *et al.*, 1969; Abdel-Momen and Starr, 1997; Cetintas *et al.*, 2003), there is increased interest in this nematode among the resistance breeding programmes. Resistance to *M. javanica* is present in early generation breeding lines (TxAG-7 (Thomaszewski *et al.*, 1994) and TP-233 (Abdel-Momen *et al.*, 1998)). Data from several lines derived from the fourth backcross generation, however, suggest that genes conditioning resistance to *M. javanica* differ from those that condition resistance to *M. arenaria* (Abdel-Momen *et al.*, 1998). COAN (Simpson and Starr, 2001) and NemaTAM (Simpson *et al.*, 2003) are resistant to *M. arenaria* and *M. javanica*, which suggests that if different genes condition resistance to each nematode species, these genes may be tightly linked and segregate together in segregating populations. Other investigators have also reported resistance in groundnut to *M. javanica* (Sakhuja and Sethi, 1985a).

The widespread deployment of a single gene for resistance to *M. arenaria* is likely to lead to the eventual selection of nematode populations with increased virulence on that source of resistance and/or to a shift in nematode species. Increased virulence of some *M. incognita* populations on the *Mi-1* gene in tomato following repeated planting of resistant cultivars has been observed (Kaloshian *et al.*, 1996). Similarly, changes in virulence have been documented for cyst nematodes in response to use of specific resistance genes (Turner, 1990; Young, 1992). Shifts in root knot nematode species predominance due to introduction of specific resistance genes have been documented for tobacco (Fortnum *et al.*, 1984). The availability of multiple genes for resistance to the major species of root knot nematodes infecting groundnut is likely to be an important asset in that it will allow development of gene deployment systems to enhance the durability of resistance currently being developed.

CHEMICAL. Nematicides are one of the most reliable and efficient methods of managing important nematode diseases of groundnut (Figs 10.7 and 10.8). In cases of severely infested fields, it may be the only effective choice, especially where crop rotation or other cultural means of management that yield economic returns cannot be employed. There are currently two general types of nematicides available, fumigants and non-fumigants. The former is formulated as a liquid that, when applied into the soil profile, volatilizes to form a gas that is distributed uniformly through soil pore spaces, contacting and killing nematodes. Currently the only fumigant nematicide remaining on the market is 1,3-dichloropropene (1,3-D). This product must be applied about 7 days pre-plant to avoid possible phytotoxicity, and may be applied as a plough-down treatment (Fig. 10.9) as soil is prepared for planting or by chisel injection at least 7 days pre-plant (Fig. 10.10). Other compounds, including 1,3-D, that were used effectively for plant nematode management were 1,2-dibromo-3-chloropropane (DBCP), 1,2-dibromethane (ethylene dibromide (EDB)) and 1,2-dichloropropane 1,3-dichloropropene (DD). These compounds were recommended for nematode management of groundnut during the 1950s through to the early 1980s (Miller, 1951; Good *et al.*, 1958; Miller and Duke, 1961). DBCP was the principal nematicide used in the southern USA in groundnut production regions during the 1960s to mid 1970s. The compound was low cost and highly efficacious as a row treatment at relatively low dosages of 5–10 l/ha. However, because of human toxicology problems and environmental issues, it

Fig. 10.7. Field plots in Florida, USA treated with 1,3-D soil fumigant as a plough-down application. (Photo: D.W. Dickson.)

Fig. 10.8. Field plots in Florida, USA treated with aldicarb non-fumigant nematicide as an at-plant application. (Photo: D.W. Dickson.)

Fig. 10.9. Plough-down application of the soil fumigant 1,3-D. (Photo: D.W. Dickson.)

Fig. 10.10. Application equipment for soil injection of 1,3-D soil fumigant by chisel injection. Six chisels are arranged on the tool bar, each spaced 30 cm apart. The cylinder of 1,3-D is pressurized by nitrogen gas. (Photo: D.W. Dickson.)

was suspended by the US Environmental Protection Agency in 1978. Thereafter, EDB became the chemical of choice until 1983, when it too was suspended because of issues similar to those that impacted DBCP. In the mid 1970s DD was voluntarily withdrawn from the market by its manufacturer, thus leaving groundnut producers with only one choice for a fumigant nematicide, 1,3-D. Consequently, in recent years, much effort has been placed on determining more efficacious methods of applying 1,3-D. The compound is highly volatile, thus it is essential that it be properly applied and sealed (Minton and Csinos, 1986; Rodríguez-Kábana and Robertson, 1987; Riegel *et al.*, 2000a,b). New fumigation equipment has become available in the USA that ensures superior application of 1,3-D (Anonymous, 2001a,b). Relatively high rates (84–112 l/ha) of 1,3-D applied broadcast are required in fields heavily infested with the groundnut root knot nematode (Plate 11C; Dickson and Hewlett, 1988a). Also, in such fields, a combination treatment of 1,3-D and a non-fumigant applied post-plant at peg initiation, e.g. aldicarb, surpassed the performance of either applied alone (Fig. 10.11) (Dickson and Waites, 1978; Rodríguez-Kábana *et al.*, 1985).

Several non-fumigant compounds having both nematicidal and insecticidal properties were introduced in the late 1950s (Dickson and Smart, 1971; Minton and Morgan, 1974; Sasser *et al.*, 1975b). Non-fumigant nematicides are essentially all organophosphate or carbamate pesticides that are formulated as liquids or granules, either of which may be applied in the planting furrow or directly onto the soil surface and incorporated by tillage equipment. The active ingredient depends entirely on water for redistribution; thus, excessive rainfall or irrigation may cause a premature loss of the active ingredient from a root zone. Non-fumigant compounds that are currently labelled or have been labelled for use on groundnut include aldicarb, carbofuradan, ethoprop, fenamiphos and oxamyl. All of these have been evaluated for management of most major groundnut nematodes under various cultural conditions. Research on these compounds has been done in the USA (Minton and Morgan, 1974; Dickson and Waites, 1978, 1982; Rodríguez-Kábana *et al.*, 1981, 1982a; Minton *et al.*, 1984; Rodríguez-Kábana and

Fig. 10.11. Application equipment for a post-plant application (peg initiation) of a non-fumigant nematicide. The granular nematicide is spread in rows over vines by a 14 inch wide row bander. The burlap bags knock the granules off vines to the soil surface. (Photo: D.W. Dickson.)

King, 1985), India (Singh and Sakhuja, 1984), Australia (Colbran, 1968; Broadley, 1981) and China (Zhang, 1985).

BIOLOGICAL MANAGEMENT. Progress has been made over the past 20 years in the identification of possible biological control organisms that offer exciting possibilities for the future management of plant nematode diseases of groundnut. One of these organisms, *Pasteuria penetrans*, a bacterial endospore-forming obligate parasite of root knot nematodes (Fig. 10.12), has been reported as a suppressive agent for *M. arenaria* in groundnut fields in Florida and Georgia, USA (Minton and Sayre, 1989; Dickson *et al.*, 1994). Plant nematode population density suppression in soils is a concept that has been studied only recently. Suppressive soils are defined as those in which disease development is suppressed even though the pathogen is introduced in the presence of a susceptible host (Huber and Schneider, 1982). There are only a few documented reports of plant nematode-suppressive soils, with most

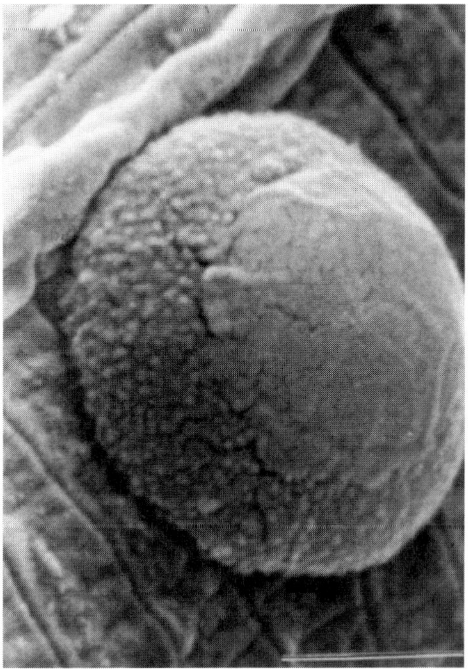

Fig. 10.12. Endospore of *Pasteuria penetrans* attached to the cuticle of a second stage juvenile of *Meloidogyne arenaria*. (Photo: D.W. Dickson.)

regarding fungal antagonists (Gair *et al.*, 1969; Stirling and Mankau, 1979; Jaffee and Zehr, 1982; Kerry, 1982); however, others have reported on suppressive soil sites infested with *P. penetrans* and root knot nematodes (Mankau, 1980; Stirling and White, 1982; Bird and Brisbane, 1988; Minton and Sayre, 1989; Dickson *et al.*, 1991, 1994).

It has been demonstrated that when *P. penetrans* is introduced into a soil containing high densities of *M. arenaria*, the bacterium will amplify to suppressive levels within 3 years (Oostendorp *et al.*, 1990, 1991) or sooner if high densities of endospores (100,000/g of soil) are added (Chen *et al.*, 1996). Others have also reported on the suppressiveness of this bacterium to plant nematodes (Mankau, 1975; Stirling, 1984; Brown *et al.*, 1985; Dube and Smart, 1987). Groundnut may be an ideal crop for amplifying *P. penetrans* to suppressive densities because it is grown in a hot climate and is a long season crop. Both conditions favour development of *P. penetrans* (Hatz and Dickson, 1992; Serracin *et al.*, 1997). Also, methods for harvesting the groundnut crop that include digging plants, drying on the soil surface, and then combining pods, leave behind root residues which most probably aid in the spread of endospores.

Over a period of 3–5 years, groundnut pods and pegs may be totally free of visible galling where *P. penetrans* occurs. Once J2 densities are reduced, *P. penetrans* density may also diminish due to a decrease in an available nematode host; however, because of the apparent persistence of endospores in soils, their disappearance or extinction may be delayed. The long-term persistence and suppressiveness of *P. penetrans* against *M. arenaria* race 1 were investigated in a suppressive site following 9 years of continuous cultivation of bahiagrass (*Paspalum notatum* cv. Pensacola var. Tifton 9), rhizomal groundnut (*Arachis glabrata* cv. Florigraze), weed fallow and 4 years of continuous groundnut. The percentages of J2 with endospores attached and endospore-filled females increased over the 4 year period, but the level of soil suppressiveness previously reported was not obtained (Cetintas and Dickson, 2005). Yet, testing the soil for its nematode suppressiveness clearly showed a higher level of suppressiveness than was observed in the field. After 4 years of continuous groundnut, roots, pods and pegs were heavily galled, and yields were low. This was probably due to the discovery of *M. javanica* infecting groundnut in this field. *M. javanica* was a non-host of the *P. penetrans* isolate in this field.

Methods of diagnosis

SAMPLING. Diagnosing root knot nematode damage on groundnut can best be done by periodic field observations and examinations of roots, pegs and pods in conjunction with plant nematode extraction or soil bioassays. A soil bioassay entails growing a root knot nematode-susceptible crop, e.g. groundnut, in soil collected from a field site suspected of having harmful nematodes. Characteristic foliage symptoms and galling of underground plant parts are clues that plants are diseased by root knot nematodes. The type of gall on the roots and pods may be a useful indicator of the *Meloidogyne* spp. present (Sasser, 1954). For an estimation of densities of nematodes in the soil, cores of soil must be taken (Barker *et al.*, 1986). Soil samples should be collected at or near harvest to determine the maximum population density. Soil bioassays are reported to be useful for establishing the level of infestations during winter or early spring months when population densities are low (Ingram and Rodríguez-Kábana, 1980).

EXTRACTION. *Meloidogyne* juveniles and eggs may be extracted from soil and roots using standard nematological laboratory procedures (Chapter 3). Females may be excised from root or pod tissues to allow for detailed morphological examination to assist with species identification. Also, individual excised females may be identified based on isozyme phenotypes, e.g. esterase and malate dehydrogenase band-ing profiles resolved by polyacrylamide gel electrophoresis (Dickson *et al.*, 1971; Esbenshade and Triantaphyllou, 1985, 1990).

Determining the relationship of nematode populations to crop loss

A measure of nematode involvement in potential groundnut yield losses may be determined by correlating numbers of *Meloidogyne* juveniles per unit of soil or root knot nematode indices with yields in nematicide-treated and untreated soil. Negative relationships were found between yields and the initial soil population density of *M. hapla* (Rickard *et al.*, 1977) and *M. arenaria* (Dhurj and Vaishnav, 1981; Wheeler and Starr, 1987; Koenning and Barker, 1992), as well as the final population density of *M. arenaria* in soil (Rodríguez-Kábana *et al.*, 1982b). Root knot nematode indices at harvest were correlated with yield for *M. arenaria* and *M. hapla* (Minton and Morgan, 1974).

In microplot studies in Florida, the damage threshold of *M. arenaria* on groundnut was estimated to be as low as a single juvenile per 100 cm^3 of soil (McSorley *et al.*, 1992). Based on studies in Texas, more than 10% of the groundnut fields from five major producing counties have *M. arenaria* population densities that exceed the level that causes 10% yield suppression (Wheeler and Starr, 1987). Models for aiding in the prediction of potential yield losses for a wide range of environmental conditions are not available.

Pratylenchus brachyurus

The lesion nematode, *Pratylenchus brachyurus*, is a major nematode parasite of groundnut, with a distribution mainly in the warmer groundnut production regions of the world (Loof, 1964). The species was first reported on groundnut in Alabama, USA in 1942 (Steiner, 1949), and is now known to parasitize groundnut in most of the groundnut-producing states in the USA, and several other countries of the world including Egypt (Oteifa, 1962), Australia (Colbran, 1968) and Zimbabwe (Anonymous, 1973). One other lesion nematode species, *P. coffeae*, was reported parasitizing groundnut in India (Chabra and Mahajan, 1976). Taxonomic separation of species of *Pratylenchus* is difficult because they exhibit little morphological diversity (Roman and Hirschmann, 1969).

Symptoms of damage

Distinct field symptoms of *P. brachyurus* damage on groundnut are difficult to discern. Severely infected groundnut plants may be stunted and chlorotic, but this is rare. Heavy infection by the lesion nematode is reported to cause extensive discoloration of below-ground plant parts and reduced root systems and pod weights. Above-ground symptoms may include slight stunting with unthrifty, yellow-green foliage (Miller and Duke, 1961; Boswell, 1968). The most obvious symptom of lesion nematode damage on groundnut is small, purplish-brown to black lesions that form on the groundnut shell (Fig. 10.13; Plate 11D) (Good *et al.*, 1958; Boswell, 1968). The plant nematode-induced lesions are described as giving pods a speckled appearance, and are conspicuous to the trained eye (Miller and Duke, 1961). These lesions have distinct boundaries and first appear as small brown tunnels in the shell and have the potential later to coalesce to form larger lesions. When coalesced, they are difficult to separate from those induced by other soil microbes. Secondary soil-borne pathogens may enter these lesions causing them to increase in size, or the infected pegs and pods may rot. Infection of pegs by *P. brachyurus* has been correlated with a peg rot condition resembling the peg rot disease caused by *Sclerotium rolfsii* (Good *et al.*, 1958). Combinations of fungal- and nematode-induced lesions may occur, but this has received little study (Good *et al.*, 1958). *P. brachyurus* can be found in roots and pegs, as well as shells of mature pods, but the nematode is more numerous in shell tissue. The lesions on mature pods are 'purplish-brown' and have

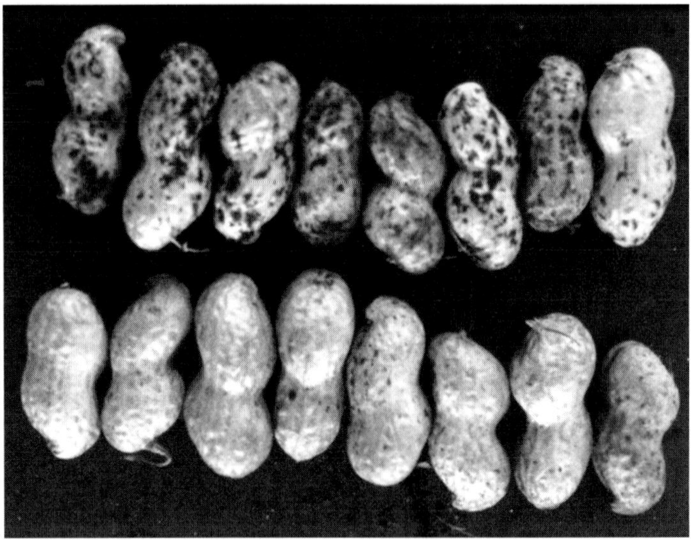

Fig. 10.13. Lesions on groundnut pods caused by *Pratylenchus brachyurus* infection (top), compared with non-infected pods (bottom). (Photo: D.W. Dickson.)

a somewhat darker colour with distinct boundaries as compared with those induced by a soil–microbial complex (Good *et al.*, 1958).

Symptoms of pod lesions may vary depending on type of groundnut or cultivar, e.g. they may be less conspicuous on Virginia-type groundnut than on Spanish and Runner types (Good *et al.*, 1958; Minton *et al.*, 1970). *P. brachyurus* feeding within the pegs weakens them, resulting in pod loss at harvest (Good *et al.*, 1958; Boswell, 1968; Jackson and Sturgeon, 1973). Also, microorganisms that colonize damaged pods may penetrate the shell and damage the seed, thus the yield, as well as the quality and value of the crop may be reduced (Good *et al.*, 1958).

Biology and life history

P. brachyurus is a migratory endoparasite that infects groundnut roots, pegs and pods, and feeds within the parenchymatous tissues. All life stages of the nematode except the egg and first stage juvenile (found inside the egg shell) are infective. These life stages including eggs are found within parasitized plant tissue, and the second, third and fourth stage juveniles, and adults are fusiform shaped (Chapter 2). They remain mobile, hence they are capable of migrating within plant tissue. The nematode is capable of moving to new infection sites on groundnut roots, pegs and pods. The nematode reproduces rapidly, which results in roots, pegs and pods containing thousands of nematodes; however, few lesion nematodes will be detected in soil surrounding roots (Steiner, 1945; Boyle, 1950). *P. brachyurus* are most numerous in mature shells, where they form dark-coloured necrotic lesions (Good *et al.*, 1958). The nematodes remain viable in these infected shells and serve as a source of inoculum even after the shells have been cured and stored over winter. The nematode may penetrate anywhere along roots, pegs and pods and move from old infection sites to induce new infection sites. Females lay eggs singly inside lesions or outside the plant tissue in soil. Apparently, the pod shell tissue is more favourable for reproduction, with 6–8 times greater numbers occurring in it as compared with equal portions of root tissue (Good *et al.*, 1958). Irrigation events markedly increase lesion nematode in groundnut pegs (Good and Stansell, 1965).

The presence or absence of host races within *P. brachyurus* has not been documented; however, their existence has been suggested by variation in numbers of *P. brachyurus* extracted from roots of citrus seedlings when inoculated with different nematode isolates (O'Bannon and Tomberlin, 1970). Field observations have also suggested behavioural differences within *P. brachyurus* populations, particularly on groundnut (Payan and Dickson, 1988). However, attempts to separate races of *P. brachyurus* failed to discern behavioural differences among four populations tested on seven species of crop plants (Payan and Dickson, 1988). Population densities on these seven crop plants of two nematode populations from groundnut were not different from two other nematode populations that originated from soybean or maize.

Survival and means of dissemination

P. brachyurus may overwinter in groundnut plant debris left in the soil (Graham, 1951; Good *et al.*, 1958; Feldmesser and Rebois, 1965). Since the nematode is polyphagous, it may survive and overwinter in live roots of many winter crops and weeds as well as in dead tissues. In South Africa, 66% of *P. brachyurus* from potato and maize were found in the soil organic matter at the end of winter although the organic matter constituted only 0.3% of the soil (Koen, 1967). *P. brachyurus* was recovered from groundnut shells that were stored at 24°C for 3, 6 and 28 months (Boswell, 1968).

P. brachyurus may be disseminated in many of the same ways as *Meloidogyne* spp. Since this is a migratory parasite and it infects most underground plant structures, it can be transported in infected roots and other underground plant parts in the soil. Generally, the major method of spread is by human activity, involving movement of plant material and soil and tillage equipment. Groundnut shells that are used to mulch soil or that are processed by grinding for use as diluents in certain preparations may carry live nematodes (Good *et al.*, 1958; Colbran, 1968). Also, water movement across infested fields as a result of either rainfall or irrigation may transport the nematode.

Environmental factors affecting parasitism

The distribution and parasitism of *P. brachyurus* are temperature related, consequently the nematode is restricted to warmer regions of the world (Loof, 1964). Reproduction in root and shell tissue of groundnut was greatest at 26°C (Boswell, 1968). Soil types and moisture may also affect parasitism of groundnut by *P. brachyurus* (Endo, 1959; Good and Stansell, 1965; Boswell, 1968).

Disease complexes

Disease complexes involving *P. brachyurus* and other soil microorganisms that would produce a peg rot have been suggested (Good *et al.*, 1958). *P. brachyurus* and *S. rolfsii* were frequently found occurring together as pathogens. Lesions of groundnut pods were found to contain both *P. brachyurus* and mycelium of fungi, most notably *Rhizoctonia solani*, *Fusarium* spp. and *Penicillium* spp. (Boswell, 1968). It is reported that lesions on roots, pods and pegs allow fungi and bacteria to enter damaged cells (Jackson and Sturgeon, 1973). The result is a peg–pod rot that weakens the stem to such an extent that the attached pod is lost during harvest. There is some indication that the presence of *P. brachyurus* is related to an increase of *Aspergillus flavus* in groundnut shells but not in seeds (Jackson and Minton, 1968).

Economic importance and population damage threshold levels

P. brachyurus is only occasionally associated with severe groundnut yield loss, thus damage by this nematode is often overlooked. Consequently, damage estimates for this nematode may be low since it has been reported in a large percentage of the groundnut production areas in the USA and in other countries. The following are reports of the percentages of fields sampled that were infested by *P. brachyurus* in the USA: Alabama 84% (Ingram and Rodríguez-Kábana, 1980); Georgia 17% (Motsinger *et al.*, 1976); Texas 16% (Wheeler and Starr, 1987); and South Carolina 14% (Alexander, 1963). Damage has also been reported in Florida (Dickson and Smart, 1971) and Arkansas (Jackson and Sturgeon, 1973). In Egypt, it was found that *P. brachyurus* infested 81% of groundnut fields (Oteifa, 1962), but in a later survey only 10% of samples were found to contain the nematode (Ibrahim and El-Saedy, 1976a). *P. brachyurus* occurs in groundnut fields in a variety of soils in South Burnett, Australia (Colbran, 1968); it is also widespread throughout Atherton Tablelands in North Queensland, Australia and was absent only in soils that had recently been brought into cultivation (Broadley, 1981). *Pratylenchus* sp. was found in 5000 soil samples collected from groundnut fields in Guyana (Singh, 1972).

Population damage thresholds for *P. brachyurus* have not been well defined. Numbers of *P. brachyurus* per gram of shell have been correlated with yields (Good *et al.*, 1958; Boswell, 1968; Minton and Morgan, 1974). A significant yield increase was reported in fumigant nematicide-treated plots in which there were 242 *P. brachyurus*/g of shell as compared with the untreated plots that had 2771/g of shell (Boswell, 1968). In Georgia, a significant yield increase was obtained in fumigated plots in which there were 127 *P. brachyurus*/g of shell compared with 2280/g of shell in untreated plots (Minton and Morgan, 1974).

In greenhouse and microplot studies, high initial population densities damage groundnut substantially; however, the nematode does not increase greatly on the groundnut cv. Florunner. Damage in the field is probably caused by high initial population densities that carried over from a previous crop, e.g. maize. In Texas, USA, the nematode did not increase on cotton or soybean cultivars (Thames, 1982). In lesion nematode-infested sites, up to 19% of the pods have been left in the ground at digging compared with sites where the nematode was controlled (Good *et al.*, 1958).

Management

Potential yield losses on groundnut caused by *P. brachyurus* are considered to be relatively small in relation to the amount of groundnut acreage infested. Hence, few management tactics specifically targeting this nematode are employed, except in certain areas where severe infestations and crop losses are known to occur. Growers who produce groundnut for green boiling or roasting in the shell generally must seek production fields known to be free of *P. brachyurus* infestations. In such crop uses, unsightly blemishes on the shells caused by lesion nematode are not acceptable.

Cultural practices

Generally, crop rotations for management of *P. brachyurus* in groundnut are not effective because of its wide host range that includes many agricultural crops and weeds, and because there are few alternative cash crops available for use in rotations with groundnut (Endo, 1959; Koen, 1967; Porter *et al.*, 1984). Population densities of *P. brachyurus* were greater in maize than in groundnut in a maize–groundnut rotation (Good *et al.*, 1954). *P. brachyurus* was also present in soil in rotations that included lupin (*Lupinus hirsutus* L.), oat (*Avena sativa* L.) and native grasses. However, their numbers were greater in lupin than in oat or native grasses (Good *et al.*, 1954). Fallowing for 6 weeks (May to June) or 9 months (May to March) in the southern USA reduced population densities of *P. brachyurus* in soil to undetectable levels (Brodie and Murphy, 1975).

Timely harvesting increases yield and value of groundnut in fields heavily infested with *P. brachyurus* (Good *et al.*, 1958; Boswell, 1968). However, larger groundnut yields were reported from *P.*

brachyurus-infested soil that had been fumigated, irrigated and harvested earlier than normal than for non-irrigated groundnut (Good and Stansell, 1965). Although irrigated groundnut yielded more, *P. brachyurus* was tenfold more numerous in shell tissue in irrigated than in non-irrigated plots.

RESISTANCE. No commercial groundnut cultivar possesses useful levels of resistance to *P. brachyurus*. Six groundnut cultivars were reported to be equally infected with *P. brachyurus*, but lesion symptoms were not as conspicuous on two of them (Minton *et al.*, 1970). Two groundnut plant introductions, PI290606 and PI295233, were reported to be resistant (Smith *et al.*, 1978). This work was confirmed and an additional resistant plant introduction, PI365553, was identified (Starr, 1984).

CHEMICAL. Where severe infestations of lesion nematodes occur, nematicide applications may be beneficial. Nematicides that control *Meloidogyne* spp. also control *P. brachyurus* (Good and Stansell, 1965; Boswell, 1968; Jackson and Sturgeon, 1973; Minton and Morgan, 1974).

Methods of diagnosis

SAMPLING. Both soil and below-ground plant parts can be assayed to determine population densities of *P. brachyurus*. Soil samples should be collected in such a manner as to obtain roots, pegs and pods. Shells usually yield more *P. brachyurus* per unit weight of tissue than roots. Soil samples should be collected shortly before or after harvest when soil population densities are likely to be at their greatest. The use of bioassays to establish the level of infestation in soils (Boswell, 1968) may be helpful if samples are collected during the winter or early spring when population densities are low.

EXTRACTION. *P. brachyurus* adults and juveniles may be extracted from roots by incubating roots in a mist chamber and from soil using standard nematological extraction procedures (Chapter 3).

Belonolaimus longicaudatus

The sting nematode, *Belonolaimus longicaudatus*, is primarily distributed in sandy soils along the Atlantic Coastal Plain from Connecticut and New Jersey to Florida, USA. There are also a few field sites infested west of the Mississippi River, namely in Arkansas, Kansas, Nebraska, Oklahoma, Texas and most recently California. However, with the exception of Oklahoma, there are no reports of sting nematode being a problem on groundnut in these states. *B. longicaudatus* is reported to occur in association with groundnut in most of the groundnut-producing states (Owens, 1951; Holderman, 1955; Rau, 1958; Wheeler and Starr, 1987), but loss estimates are only reported for Virginia, North Carolina and Oklahoma (Anonymous, 1987). In North Carolina, 16 counties, of which eight are major producers of groundnut, are known to be infested with *B. longicaudatus* (Cooper *et al.*, 1959). In Virginia, only a small percentage of the groundnut acreage has a problem with *B. longicaudatus*. There are no reports of *B. longicaudatus* causing damage to groundnut outside the USA.

Symptoms

B. longicaudatus is an ectoparasitic plant nematode that feeds at root tips and along succulent roots as well as on young pegs and pods. The most obvious symptom is the greatly abbreviated root system, which results in dwarfed, chlorotic plants. Plant growth may be uneven in heavily infested fields, and erratic stands may occur. Yield and quality of groundnut may be severely reduced.

Races

B. longicaudatus was first described from Florida (Rau, 1958), where the species is considered as one of the state's most important plant pathogens. The host range of *B. longicaudatus* is extensive and includes agronomic, horticultural and ornamental crops (Perry and Rhoades,

1982; Smart and Nguyen, 1991; Bekal and Becker, 2000). Several studies suggest the existence of physiological races of *B. longicaudatus* with different host ranges (Abu-Gharbieh and Perry, 1970; Robbins and Barker, 1973). Additionally, populations of *B. longicaudatus* from North Carolina and Georgia were found to differ morphologically from each other, and from Rau's description of *B. longicaudatus* (Robbins and Hirschmann, 1974). Matings between these two populations resulted in a few, non-fertile offspring (Robbins and Hirschmann, 1974). This led them to suggest that these populations may be different species. There are other scientists who question the nematode's taxonomic status because of reported variations in morphology and host specificity among different isolates (Owens, 1951; Perry and Norden, 1963; Good, 1968; Abu-Gharbieh and Perry, 1970; Rau and Fassuliotis, 1970; Robbins and Barker, 1973; Duncan *et al.*, 1996). Populations of this nematode are definitely pathogenic on groundnut in North Carolina and Virginia (Owens, 1951; Cooper *et al.*, 1959), but not in Georgia (Good, 1968) or Florida (Dickson, 1998). A population of *B. longicaudatus* collected from a groundnut-growing region in north central Florida did not cause damage or reproduce well on groundnut, whereas a population from a non-groundnut-growing region in central Florida did (Perry and Norden, 1963). There are instances in Florida where groundnut shows classic symptoms of *B. longicaudatus* damage, but the number of plants involved is always very small and disease development diminishes as the growing seasons progresses (D.W. Dickson, unpublished).

Means of dissemination

B. longicaudatus dispersal appears to be affected by certain biological and environmental restraints. It is suggested that soil texture, soil temperature and moisture are critical for the nematode's development (Perry, 1965; Robbins and Barker, 1974). The nematode may be disseminated by any means that will transport infested soil, such as farm equipment, animals, water, and transplants that have soil attached.

Environmental factors affecting parasitism

The limited distribution of *B. longicaudatus* suggests that its ecological requirements may be very specific. Fine-textured soils are believed to inhibit its movement and reproduction (Thames, 1959). There is minimal reproduction in soils with less than 80% sand content or more than 10% clay content (Robbins and Barker, 1974). In Virginia, *B. longicaudatus* is found only in the A-horizon of soils with a sand content of 84–94% (Miller, 1972a). Greater numbers of this nematode were also reported in the upper 20 cm of soil compared with a 20–40 cm depth, even though both layers contained 94–95% sand (Perez *et al.*, 2000).

Soil temperature and moisture have a large influence on the life cycle and reproductive rate in Florida, Georgia and North Carolina populations (Perry, 1964; Boyd and Perry, 1971; Robbins and Barker, 1974; Smart and Nguyen, 1991). In Florida, *B. longicaudatus* reproduced better at 29°C than at 27°C, but was greatly reduced at 35°C (Perry, 1964; Boyd and Perry, 1971). The reproduction of the Georgia population was greatest at 30°C, whereas reproduction of the North Carolina population was reduced at 30°C (Robbins and Barker, 1974). In Florida, populations either die or migrate downward when soil temperatures at 2.5 cm below the bare soil surface reach 39.5°C or higher (Boyd and Perry, 1971). The optimum soil moisture for reproduction was reported to be 7% (Robbins and Barker, 1974). In a greenhouse study, the life cycle *of B. longicaudatus* was completed in about 28 days (Smart and Nguyen, 1991). *In vitro* cultivation on excised maize roots has made it possible to complete more detailed information on the life cycle and behavioural characteristics of a California isolate of *B. longicaudatus* (Huang and Becker, 1997, 1999). The life cycle was completed in 1 month at 26–27°C and in 24 days at 28°C (Huang and Becker, 1997). All juvenile stages as well as the adults fed on root meristems, with mating occurring after males and females completed their last moult.

Economic importance and population damage threshold levels

Economic losses for groundnut in the USA due to *B. longicaudatus* are not great despite the extreme damage this nematode is capable of inflicting. Losses have only been reported for North Carolina (0.30%), Oklahoma (0.25%) and Virginia (0.50%) (Anonymous, 1987). Increases in yields of as much as 109–400% compared with untreated controls were obtained in North Carolina nematicide trials in which the average population density of *B. longicaudatus* ranged from about 10 to 43 nematodes/100 cm^3 of soil (Cooper *et al.*, 1959; Sasser *et al.*, 1960). The economic threshold level varied from two to five *B. longicaudatus*/130 cm^3 of soil, depending on the nematicide used.

Management

No commercial groundnut cultivar is resistant to *B. longicaudatus*. The nematode has a wide host range and only a few crop plants such as small grain, tobacco (*Nicotiana tabacum* L.) and watermelon (*Citrullus vulgaris* Schrad.) have reduced population densities when grown in rotation with groundnut (Holderman and Graham, 1953; Bailey, 1988). The use of nematicides has been the major means of management. Both fumigant and non-fumigant nematicides have in the past given excellent control and increased groundnut yields (Cooper *et al.*, 1959; Sasser *et al.*, 1960, 1975b).

Methods of diagnosis

SAMPLING. Early season seedling damage from *B. longicaudatus* is apparent especially if population densities are high. Above-ground symptoms will include severely stunted plants that appear in scattered portions of the field, and examination of roots of seedlings for damage as well as assessments of population densities in the soil are suggested. Soil samples should be collected using procedures recommended for recovery of ectoparasitic plant nematodes (Chapter 3).

EXTRACTION. The extraction of *B. longicaudatus* from soil may be done by using any one of a number of standard extraction procedures (Chapter 3).

Determining the relationship of populations to crop loss

The effects of *B. longicaudatus* on groundnut are reflected in plant growth, yield and quality (Cooper *et al.*, 1959; Sasser *et al.*, 1975a). Significant negative correlations of number of nematodes in the soil with yield and growth may be obtained during most of the growing season.

Criconemoides ornatus

Criconemoides ornatus (commonly called the groundnut ring nematode) was first reported associated with groundnut in Georgia (Boyle, 1950; Machmer, 1953). It is now known to occur in a large percentage of the groundnut production regions of the USA (Alexander, 1963; Minton *et al.*, 1963; Motsinger *et al.*, 1976; Ingram and Rodríguez-Kábana, 1980; Wheeler and Starr, 1987). *Criconemoides* spp. have been reported in Burkina Faso (Germani and Dhéry, 1973), Egypt (Ibrahim and El-Saedy, 1976a) and Gambia (Merny *et al.*, 1974).

Symptoms

A chlorotic condition of groundnut, called ‘groundnut yellows’, was reported in Georgia in a soil heavily infested with a species of *Criconemoides* (Machmer, 1953). Although the species involved was never identified, it was probably *C. ornatus*. The ‘yellows disease’ symptom was duplicated in microplots by using freshly extracted, greenhouse-grown inoculum of

C. ornatus (Barker *et al.*, 1982). Although yellowing of groundnut has been observed and duplicated in microplots, this is not a common symptom. In fact, it is probably rarely seen.

As few as 178 freshly introduced *C. ornatus*/500 cm^3 of soil will stunt groundnut. Roots, pegs and pods of plants growing in microplots in soil heavily infested with *C. ornatus* are severely discoloured with brown necrotic lesions (Minton and Bell, 1969). Small necrotic lesions caused by this nematode are often superficial, but necrosis in large lesions usually extends deep into the tissues. The primordia of many lateral roots and young roots are killed, which results in reduced numbers of lateral roots. Pod yields from nematode-infected plants are reduced by about one-half.

Survival and means of dissemination

Information relative to factors affecting survival of *C. ornatus* is limited. Little research has been done to determine soil type preferences, but survey results suggest the nematode favours lighter sandy soils (Barker, 1974). Population densities of the nematode decline rapidly in the presence of poor hosts. Because *C. ornatus* is an ectoparasite, dissemination occurs primarily via soil transported on farm equipment, on the feet of humans or animals, and in water.

Environmental factors affecting parasitism

The environmental factors affecting the parasitism of groundnut by *C. ornatus* have received little attention. The previous crop as well as geographic areas in North Carolina affect the occurrence and activity of *Criconemoides* spp. (Barker, 1974). The Coastal Plain region, with warm, sandy soils, has a greater abundance of *Criconemoides* spp. than the Piedmont and Mountain regions, each with soils that are cooler and contain more loam and clay. The frequency of occurrence of *Criconemoides* spp. on groundnut (54%) was greater than that for any other crop.

Disease complexes

Greenhouse studies in North Carolina revealed an interaction (enhancement of CBR) between *Cylindrocladium crotalariae* and *C. ornatus* on CBR-susceptible Florunner, but not on CBR-resistant NC 3033 groundnut cultivars (Diomandé and Beute, 1981a). The severity of CBR on Florunner was increased when the density of *C. ornatus* was 10/15 cm diameter clay pot and *C. crotalariae* was 0.25 and 2.5 microsclerotia/cm^3 of soil. Significant positive correlations between *C. ornatus* and *C. crotalariae* indicated that this nematode can affect CBR development in the field (Diomandé and Beute, 1981b).

Economic importance and population damage threshold limits

Damage to groundnut due to *C. ornatus* in the field is subtle, and low levels of damage may go undetected. Also, *C. ornatus* is seldom present alone, but usually occurs in polyspecific nematode communities. Therefore, losses due to only *C. ornatus* have not been well defined. Pod yield in a microplot experiment (Minton and Bell, 1969) was reduced by about a half in heavily inoculated soil. In a field experiment in which the soil was infested with five genera of nematodes in addition to *C. ornatus*, population densities of *C. ornatus* were negatively correlated with groundnut growth index and pod yield (Sasser *et al.*, 1975a).

Based on a linear regression model, it was determined that groundnut yield loss in microplots was 19% for each tenfold increase in initial populations of *C. ornatus* in the soil (Rickard *et al.*, 1977). As few as 178 *C. ornatus*/500 cm^3 of soil in a microplot experiment caused a significant yield loss (Barker *et al.*, 1982). In a second microplot experiment, the *C. ornatus* that reproduced the previous year on tobacco (a poor host) did not affect groundnut yields (Barker *et al.*, 1982). It was concluded that many of the nematodes present in the soil in the spring following tobacco may have been dead, since tobacco is a poor host. Therefore, the previous host may affect the infectivity of the nematodes present in the soil and present an important problem for nematode advisory programmes.

Management

Since losses due to *C. ornatus* have not been well defined, recommendations for control of this nematode when present as the primary pathogen are seldom made. Also, there is no known resistant commercial groundnut cultivar. Certain crops such as cotton, soybean, maize and sorghum grown in rotation with groundnut may reduce population levels (Good, 1968; Johnson *et al.*, 1974; Kinloch and Lutrick, 1975). Nematicides, both fumigant and non-fumigant, are considered effective against ring nematode (Minton and Morgan, 1974).

Methods of diagnosis

SAMPLING AND EXTRACTION. Evaluating soil population densities is the major means of diagnosing possible *C. ornatus* damage to groundnut. *C. ornatus* may be extracted from the soil using one of several methods, but the modified centrifugal flotation method is most useful because of the sluggish nature of this plant nematode (Chapter 3). Methods that depend on nematodes being active are poor choices for extraction of ring nematodes.

Determining the relationship of populations to crop losses

Even though *C. ornatus* is a weakly pathogenic nematode, negative correlations of population densities with yield and plant growth often suggest plant damage (Minton and Morgan, 1974; Sasser *et al.*, 1975a). Soil assays made early in the season (55–73 days after planting) may be more meaningful than assays made near harvest (Sasser *et al.*, 1975a).

Aphelenchoides arachidis

Aphelenchoides arachidis, the testa nematode, was described from northern Nigeria on groundnut (Bos, 1977a,b). This is the only country in the world to date where this nematode pest of groundnut occurs. A significant level of infestation occurs in only a limited area around Samaru, a low level of infestation at Kadawa and in one groundnut sample from Gwoza.

Symptoms

A. arachidis is a parasite of pods, testae, roots and hypocotyls, but not the cotyledons, embryos or other parts of the plant (Bos, 1977a; Bridge *et al.*, 1977). Seed coats were discoloured when more than 2000 *A. arachidis*/testa were present (Bridge *et al.*, 1977) (Plate 11E). Heavily infested seeds, examined immediately after removal from fresh, mature pods, are a light brown, have translucent testae and dark vascular strands within the testae. After infested seeds are dried, testae are often wrinkled and are darker brown than in non-infested seeds (Plate 11E). Nematodes are found mainly in the subepidermal parenchymatous layer, and around the tracheids of the testa. Testae infested with *A. arachidis* are thicker and more uneven than normal testae. Nematodes are found in subepidermal parenchyma cells where walls are broken and cells enlarge. The epidermal layer of the seed coat is reduced in infested testae, and the basal tissue, including the aleurone layer, is disorganized. Infested seeds of cv. Spanish 205 weighed less than healthy seeds, but nematode damage has little effect on seed germination.

Biology and life cycle

The nematode is a facultative endoparasite of groundnut (Bridge *et al.*, 1977). It also feeds ectoparasitically on groundnut roots and on two fungi, *Macrophomina phaseolina* (Tassi) Goid. and *Botrytis cinerea* Pers., that have been associated with seeds on agar plates. *A. arachidis* were found in the parenchymatous tissues of the testa, root cortex and hypocotyl, but not in the central stele or vascular bundles (Bridge *et al.*, 1977). Pods are invaded 10 days after the fruiting pegs penetrate into the soil, but the number of nematodes in pods does not increase rapidly until after 30 days, with largest numbers present at about day 60. All stages of the nematode, including eggs, were found throughout the testae, but at the end of the growing season heavily infested testae of mature seeds contained mainly juvenile stages with few adults. Testae showing no external symptoms contained mostly adults and eggs, often arranged along the vascular elements of the seed coats.

Biotypes

It is suggested that there are two biotypes of *A. arachidis*, one occurring on cereals and one on both cereal and groundnut (Bos, 1977b).

Survival and means of dissemination

A. arachidis survives desiccation in stored groundnut pods for 12 months (Bridge *et al.*, 1977). The second, third and fourth stage juveniles were extracted from dried testae and shells with no particular stage predominating, but adults were found alive only occasionally in either testae or shells of stored pods. No active nematodes were extracted from infested pods sun-dried in the field before storage. Volunteer plants in an infested field contained many adult nematodes, which suggests that they continue to develop to maturity under natural conditions in pods left in the ground during the dry season in Nigeria. Unless appropriate precautions are taken, *A. arachidis* has the potential to become a serious pest worldwide because it can be disseminated in infested seeds (Bridge *et al.*, 1977).

Disease complexes

Infestation of groundnut seeds by *A. arachidis* in field experiments predisposed seeds to invasion by fungi (McDonald *et al.*, 1979). Nematode-infested seeds had higher levels of fungal infection (*R. solani*, *S. rolfsii*, *Macrophomina phaseolina* and *Fusarium* spp.) than those that appeared to the eye to be nematode-free. Both rates of seedling emergence and total emergence are slightly lower for nematode-infested seeds than for clean seeds.

Economic importance and population damage threshold levels

Groundnut yields are not decreased by *A. arachidis*; however, the confectionery groundnut is devalued by the nematode because the nematode causes shrivelled and discoloured seeds (Bridge *et al.*, 1977). Severe infestation of groundnut with *A. arachidis* not only has an adverse effect on the appearance and size of seed, but it also may predispose seeds to an invasion by fungi that may lead to reduced seed emergence (McDonald *et al.*, 1979). Because its distribution is limited to Nigeria, *A. arachidis* has not caused major economic losses but, if it should become established in other groundnut-producing regions of the world, it could possibly become a major economic pest.

Management

Only limited information is available on management of *A. arachidis* on groundnut. No field-applied treatments have been reported, but a number of preventive measures are effective against further spread of the nematode. Immersing infested seed in four times their volume of water heated to 60°C and allowing to cool for 5 min gives complete control of the nematode without affecting germination (Bridge, 1975; McDonald and Misari, 1976; Bridge *et al.*, 1977). In northern Nigeria, very dry conditions make it possible to sun-dry pods after harvest in order to reduce the number of nematodes in pods (Bridge *et al.*, 1977). In more humid areas, sun-drying of pods may not be effective. Shelling groundnut before planting will also eliminate the tissues in which most of the nematodes occur and survive best (Bridge *et al.*, 1977).

Aphasmatylenchus straturatus

Aphasmatylenchus straturatus was described in 1970 from around roots of groundnut in south-west Burkina Faso, West Africa near Niangoloko village (Germani, 1970). It has not been reported to occur outside of Burkina Faso.

Symptoms

A. straturatus causes interveinal chlorosis, stunting, a poorly developed root system, reduction of *Rhizobium* nodules on roots and losses in potential groundnut yields (Germani and Dhéry, 1973; Germani and Luc, 1982a,b).

Biology and life cycle

The nematode is a migratory endo-ectoparasite on groundnut. Field observations indicate that it spends the dry season at a depth of 40–60 cm in the soil adjacent to roots of the karite (*Butyrospermum parkii* L.) tree or in the roots of this tree.

Groundnut is interplanted with the karite tree in many fields in Burkina Faso and, at the beginning of the rainy season, the nematode moves from tree roots and enters groundnut roots. The nematode is most abundant in early-maturing cultivars about 40 days after seeding and in late-maturing cultivars about 70 days after seeding. Approximately 100–110 days after seeding, the nematode leaves the groundnut roots and returns to roots of the karite tree. *A. straturatus* does not become anhydrobiotic.

Economic importance and population damage threshold levels

Disease symptoms may occur in the field when as few as 600 nematodes/dm^3 of soil are present, but approximately 2000 nematodes/dm^3 of soil are required to induce symptoms in the greenhouse. Potential yield reductions attributed to *A. straturatus* are estimated to range from 30 to 70%. In 1971, *A. straturatus* was estimated to infest approximately 4% of the groundnut production area of Burkina Faso, but had increased to 25% within 3 years. Since this nematode also parasitizes other economically important leguminous plants grown in Burkina Faso (Germani and Dhéry, 1973), its rapid spread poses a threat to groundnut and other legumes.

Management

There is little information available on the management of *A. straturatus* on groundnut, but nematicides (DBCP) applied at planting in the past gave satisfactory control (Dhéry *et al.*, 1975).

Methods of diagnosis

Soil samples for extraction of *A. straturatus* must be collected in the root zone of groundnut or karite trees during the dry season. If samples are collected in the root zone of groundnut, they should be taken at a depth of 0–20 cm, but if collected in the root zone of karite trees during the dry season, they should be taken at a depth of 40–60 cm.

Scutellonema cavenessi

Scutellonema cavenessi was described from northern Nigeria (Sher, 1964) but has since been found associated with most cultivated plants in Senegal and Mali. In Senegal, *S. cavenessi* was associated with poor growth of groundnut (Germani, 1979b, 1981b).

Symptoms

Foliage of groundnut plants grown in soil infested with *S. cavenessi* was chlorotic (Germani, 1979b). *S. cavenessi* is associated with reducing the number of lateral roots and *Rhizobium* nodules. Chlorosis was reduced in plots treated with DBCP, which also reduced population densities of *S. cavenessi*. Chlorosis was associated with a reduced level of nitrogen fixation and less total nitrogen in pods and foliage (Germani, 1979b). Application of the fumigants DBCP and EDB reduced the nematode population densities and increased vine and pod yield, the number and weight of *Rhizobium* nodules, the nitrogen and phosphorus content of foliage and seeds, and the level of endomycorrhizae infestation (Germani, 1979b, 1981b; Germani *et al.*, 1981, 1982, 1985; Germani and Reversat, 1982, 1983).

Biology and survival

In Senegal, *S. cavenessi* showed seasonality in activity (Demeure, 1978a; Demeure *et al.*, 1980). This nematode is active during the rainy season, but as the dry season progresses and the humidity of the soil drops to about 0.2%, nematodes 0–25 cm deep in the soil enter into a state of anhydrobiosis, in which they remain until the next rainy season.

Economic importance and population damage thresholds

S. cavenessi is distributed throughout the groundnut production area of Senegal, but the extent of potential crop loss has not been fully evaluated. Nevertheless, in experimental plots, nematicides have increased yields of pods from 20 to 220% and of vines from 40 to 270% (Germani *et al.*, 1985).

Management

There are no known plant cultivars resistant to *S. cavenessi.* Furthermore, all crops grown in rotation with groundnut in the Sahelian zone of Senegal are susceptible to this nematode. Bare fallow between crops of groundnut provided excellent management (Duncan, 1986) but, because of the high cost, this practice is not practical in the Sahelian zone. In the past, EDB and DBCP were the only nematicides tested that gave practical control. These nematicides, when used at 20 kg a.i./ha, gave excellent control and yield increases (Germani and Gautreau, 1976; Germani, 1979a,b, 1981a; Duncan and Baujard, 1986; Baujard *et al.*, 1987). There was also a residual effect of the nematicide on other crops grown in treated fields the following year. The fumigant nematicides were applied in or near the row with an animal-drawn injector metered with a ground-driven peristalic pump that applies a uniform rate as the apparatus is drawn across the field.

Methods of diagnosis

Soil samples for nematode assays should be collected in the groundnut root zone 25 cm deep using standard sampling and extraction techniques (Chapter 3). However, if samples are taken during the dry season when the nematode is in the anhydrobiotic state, the soil should be moistened if they are to be extracted by elutriation or Baermann techniques, otherwise they should be extracted by the centrifugal flotation method (Demeure, 1978b; Duncan, 1986; Duncan and Baujard, 1986).

Tylenchorhynchus brevilineatus

Tylenchorhynchus brevilineatus was first observed causing damage to groundnut in 1976 in the Kalahasti area of Andhra Pradesh State, India (Reddy *et al.*, 1984). The disease caused by this nematode is known as ‘Kalahasti malady’. Since 1976, the disease has been widespread in the Kalahasti area, but has also been observed in Nellore District in Andhra Pradesh (Reddy *et al.*, 1984). This nematode has not been reported as a pathogen of groundnut in other parts of the world.

Symptoms of damage

Disease symptoms are characterized by small pods and a brownish-black discoloration of the pod surfaces (Reddy *et al.*, 1984). Small, brownish-yellow lesions appear on the pegs and pod stalks and on young, developing pods. Lesion margins are slightly elevated because of host cell proliferation around them. The length of pod stalks is greatly reduced, and in advanced stages of the disease the pod surface becomes completely discoloured, but seeds from diseased pods are healthy. Discoloration is also observed on roots but is less severe than on pods.

Pathogenicity tests in the greenhouse corroborated field observations (Reddy *et al.*, 1984). Groundnut plants inoculated with 500 *T. brevilineatus*/12 cm diameter pot were severely stunted and had reduced root systems. Lesions were present on the roots but were not extensive. Pods were severely discoloured and small, but seeds from the discoloured pods were healthy. Brownish-yellow lesions were observed on individually inoculated pods after 15 days. The number of lesions increased and extensive discoloration was observed by 30 days after inoculation.

Management

Aldicarb and carbofuran non-fumigant nematicides provided control of *T. brevilineatus* when applied to groundnut 20 days post-plant. These nematicides reduced soil population densities of *T. brevilineatus* and the percentage of diseased pods (Reddy *et al.*, 1984). There was also an increase in plant height, pod yields, and pod and kernel weights.

Ditylenchus africanus

Ditylenchus africanus, the groundnut pod nematode, was originally described as *D. destructor*, the potato rot nematode. It was first reported damaging groundnut in the Transvaal Province of South Africa in 1987 (Jones and De Waele, 1988). A subsequent survey revealed the presence of this nematode in seven major groundnut-producing regions (De Waele *et al.*, 1988), where 75% of 877 seed samples that graded 'damaged' were infected. An average of 160 nematodes/seed was recovered. This nematode has not been reported on groundnut in other parts of the world.

Symptoms of damage

D. africanus was isolated from roots, pegs, shells and groundnut seeds (De Waele *et al.*, 1988). Visible symptoms are not apparent on roots, but seed show blemishes and premature germination before harvest (Fig. 10.14; Plate 11F). Infected pods of cv. Sellie are black, resembling black hull

Fig. 10.14. *Ditylenchus africanus*-infected seed (right), compared with non-infected seed (left). (Photo: D.W. Dickson.)

caused by *Chalara elegans* Nag Raj & Kendrick. Approximately 40–60% of the pods and seeds are destroyed in heavily infested fields. *D. africanus* is present in both hulls and seeds, which results in a lower quality grade and reduced groundnut yield.

In greenhouse pathogenicity tests (De Waele *et al.*, 1988), nematodes were present in the peg, exocarp and endocarp, testa and embryo, and on the cotyledons. The first symptom to develop was brown necrotic tissue at the pod base at the juncture of the peg and pod. The surface of infected tissue was dark brown and had a corky appearance. The most distinct symptom of advanced disease was dark brown to black discoloration of veins that extended longitudinally in the exocarp just beneath the pod surface. Infected pods lacked the lustre of healthy pods and appeared dead. Infected seeds were usually shrunken and the micropyles were dark brown to black. The testae were flaccid, had dark vascular strands and were easily removed. The inner layer of the testa had a distinct yellow discoloration. Infected embryos were usually olive green to brown instead of having the normal colourless to yellow appearance. The extent of potential yield losses caused by this nematode and research relative to its management has not been reported.

Biology and life history

D. africanus develops from egg to adult in 8 days at 25°C. At 28–30°C, egg hatching starts at around 3 days. By the 6th day, 90% of eggs have hatched. The nematode is able to enter a state of anhydrobiosis with about one-third of the anhydrobiotic nematodes becoming active after rehydration to invade hulls and seeds of a newly planted crop. Although *D. africanus* is present in the roots of groundnut and in soil, 90% of the total population at harvest is found in pods. Infestations up to 97,000 nematodes/pod are not uncommon for groundnut grown in the field.

D. africanus enters the immature pegs and pods of groundnut at the peg connection; however, the infective stage is unknown. It subsequently invades the parenchymatous region of the hull exocarp and the endocarp, and eventually the seed testa. The nematode causes malformation of the cells of infected tissues, cell wall breakage and cell collapse. Damage appears to be caused by enzymatic activity. The entire parenchyma region in some testae is destroyed. In immature pods, *D. africanus* may move across the fibrous region of the mesocarp into the hull of the endocarp. In mature pods, the lignification of the fibrous mesocarp at around 105 days is a barrier to penetration of the inner pod tissues. Nematodes artificially inoculated after 105 days are no longer able to cause damage to the seed. Increased numbers of eggs and anhydrobiotic forms of the nematode are found in the hull tissues and eggs are found in the seed testae of late harvested pods (~189 days after planting). Eggs, some containing first stage juveniles, have been observed in the parenchymatous mesocarp of the hull. Also, egg numbers increase in the seed testa of late harvested pods. Both occurrences may indicate the onset of survival mechanisms of the nematode. Apparently, all life stages can be found in the hull. It appears that eggs and anhydrobiotic forms are involved in winter survival in decaying hulls and stubble, whereas eggs are the important survival stage in stored seed.

Economic importance

About 200,000 ha of groundnut are grown annually in South Africa, with *D. africanus* being present in all major groundnut production areas of the country. Greenhouse damage potential studies showed that at 250 nematodes/3-l pot, 10–25% of seeds germinated into second generation seedlings before harvest, and fresh weight of harvested seed was suppressed 20–50%.

Other nematodes

A worldwide list of nematode pathogens associated with groundnut has been compiled (Sharma, 1985). The list is extensive and includes many genera and species that have not been proven to cause economic damage to groundnut. Additional research may demonstrate that some of these species are, in fact, pathogenic and pose a serious threat to groundnut production, while others may feed on groundnut but cause little or no economic damage.

The possibility that two nematodes that are not considered serious pathogens of groundnut interact with a virus to cause disease has been suggested. The clump disease of groundnut, caused by a virus, was eliminated in Senegal by treating the soil with DD (Merny and Mauboussin, 1973). It was suggested that one or more nematodes were acting as a vector and pointed out that *Longidorus pisi* was present in soil samples. In India, the disease was reduced in field experiments by 97 and 84% with DBCP and aldicarb, respectively (Singh and Sakhuja, 1984). Soil samples collected from the rhizosphere of diseased plants always contained *Paralongidorus citri*. Both nematodes are capable of transmitting plant viruses.

Conclusions and Future Prospects

Potential groundnut yield losses due to plant nematodes occur in every major groundnut production region of the world. With estimated loss projected at 12% (Sasser and Freckman, 1987), it is apparent that improved strategies are badly needed to reduce these losses.

Meloidogyne spp. are the most important plant nematodes damaging groundnut in most regions of the world, but, in some regions, such as in West Africa, other species may be more serious. In Senegal, for instance, *Meloidogyne* spp. do not damage groundnut and the crop is often rotated with vegetables to suppress *M. arenaria* population densities. A number of nematodes such as *A. arachidis*, *A. straturatus*, *S. cavenessi*, *T. brevilineatus* and *D. africanus* are reported to cause serious damage to groundnut in isolated regions of Africa and Asia, but not in other regions of the world. *B. longicaudatus* is a pathogen of groundnut in only certain regions of the USA. Questions have been raised as to why these nematodes have been reported damaging groundnuts only in these areas and what is the probability of their becoming pathogens in other regions of the world.

Nematode management in the past, particularly in industrialized countries, was based to a great extent on chemical control. In these countries, the loss of the fumigants DBCP, EDB and DD because of concerns for improved environmental protection and human safety has led to their suspension or withdrawal from the market. Potential for nematicides to cause water contamination, human and health considerations, as well as the increased cost of applying chemicals has increased the urgency to seek safer and more economical chemicals and to develop other means of management.

Resistant cultivars can be the best and most economical means of managing nematodes. Although until relatively recently resistance had not been identified or been incorporated into commercial cultivars, there has been a breakthrough in the search for resistance to the root knot species *M. arenaria* and *M. hapla*, and researchers in the USA are engaged in promising efforts to develop further resistance to *Meloidogyne* spp. Expanded utilization of cultural practices such as crop rotations, cover crops, trap crops, fallowing and flooding, organic amendments and other tactics that aid in reducing nematode damage is necessary for the maintenance of economical groundnut production. More research is needed on understanding and predicting naturally suppressive soils, so as to aid producers in capitalizing on this important and greatly underutilized tactic of nematode management. Efforts to prevent the spread of nematodes through sanitation and quarantine in extreme situations may contribute to future containment of nematode problems.

Nematologists and advisors to growers in the future will be challenged to devise more effective management schemes that will yield quality groundnuts and an economical return to the producers, while protecting the safety of the consumer and environment.

References

Abdel-Momen, S.M. and Starr, J.L. (1997) Damage functions for three *Meloidogyne* species on *Arachis hypogaea* in Texas. *Journal of Nematology* 29, 478–483.

Abdel-Momen, S.M., Simpson, C.E. and Starr, J.L. (1998) Resistance of interspecific *Arachis* breeding lines to *Meloidogyne javanica* and an undescribed *Meloidogyne* species. *Journal of Nematology* 30, 341–346.

Abu-Gharbieh, W.I. and Perry, V.G. (1970) Host differences among Florida populations of *Belonolaimus longicaudatus* Rau. *Journal of Nematology* 2, 209–216.

Alexander, P.M. (1963) Stylet-bearing nematodes associated with various plants in South Carolina, 1962–1963. *Plant Disease Reporter* 47, 978–982.

Anonymous (1973) Plant nematology. In: *Secretary for Agriculture Report for the Period 1 October 1971 to 30 September 1972.* Salisbury, Rhodesia.

Anonymous (1987) Bibliography of estimated crop losses in the United States due to plant-parasitic nematodes. *Annals of Applied Nematology* (*Journal of Nematology* vol. 19, Supplement) 1, 6–12.

Anonymous (1989) *Board of Agriculture, National Research Council. Alternative Agriculture.* National Academy Press, Washington, DC.

Anonymous (1999) USDA–NASS agricultural statistics 1999. Statistics of oilseeds, fats, and oils. Available at: www.usda.gov/nass/pubs/agr99/acro99.htm

Anonymous (2001a) Application technology for injecting Telone. *Methyl Bromide Alternatives* 7, 5–6.

Anonymous (2001b) Yetter's new rig makes Telone application easy. *Florida Grower* 94, 20.

Bailey, J.E. (1988) Peanut disease control. In: *North Carolina Agricultural Extension Service, North Carolina State University, Peanuts, 1988, AG-331*, 51–64.

Baldwin, J.G. and Barker, K.R. (1970) Host suitability of selected hybrids, varieties and inbreds of corn to populations of *Meloidogyne* spp. *Journal of Nematology* 2, 345–350.

Baltensperger, D.D., Prine, G.M. and Dunn, R.A. (1986) Root knot nematode resistance in *Arachis glabrata. Peanut Science* 13, 78–80.

Barker, K.R. (1974) Influence of geographic area and previous crop on occurrence and density of plant-parasitic nematodes in North Carolina. *Plant Disease Reporter* 11, 991–995.

Barker, K.R., Schmitt, D.P. and Campos, V.P. (1982) Response of peanut, corn, tobacco, and soybean to *Criconemella ornata. Journal of Nematology* 14, 576–581.

Barker, K.R., Townshend, J.L., Bird, G.W., Thompson, I.J. and Dickson, D.W. (1986) Determining nematode population responses to control agents. In: Hickey, K.D. (ed.) *Methods for Evaluating Pesticides for Control of Plant Pathogens.* APS Press, St Paul, Minnesota, pp. 283–296.

Baujard, P., Duncan, L.W., Pariselle, A. and Sarr, E. (1987) Étude des effets de quatre nématicides fumigants sur les nématodes et l'arachide au Sénégal. *Revue de Nématologie* 10, 355–360.

Baxter, R.I. and Blake, C.D. (1969) Oxygen and the hatch of eggs and migration of larvae of *Meloidogyne javanica. Annals of Applied Biology* 63, 191–203.

Bekal, S. and Becker, J.O. (2000) Population dynamics of the sting nematode in California turfgrass. *Plant Disease* 84, 1081–1084.

Bergeson, G.B. (1959) The influence of temperature on the survival of some species of the genus *Meloidogyne* in the absence of a host. *Nematologica* 4, 344–354.

Bird, A.F. and Brisbane, P.G. (1988) The influence of *Pasteuria penetrans* in field soils on the reproduction of root-knot nematodes. *Revue de Nématologie* 11, 75–81.

Bird, A.F. and Wallace, H.R. (1965) The influence of temperature on *Meloidogyne hapla* and *M. javanica. Nematologica* 11, 581–589.

Bos, W.S. (1977a) *Aphelenchoides arachidis* n.sp. (Nematoda: Aphelenchoidea) an endoparasite of the testa of groundnuts in Nigeria. *Zeitschrift für Pflanzenkrankheiten und Pflanzenschutz* 84, 95–99.

Bos, W.S. (1977b) A preliminary report on the distribution and host-range of the nematode *Aphelenchoides arachidis* Bos in the North of Nigeria. *Samaru Agricultural Newsletter* 19, 21–23.

Boswell, T.E. (1968) Pathogenicity of *Pratylenchus brachyurus* to Spanish peanut. PhD Dissertation, Texas A&M University, College Station, Texas.

Boyd, F.T. and Perry, V.G. (1971) Effects of seasonal temperatures and certain cultural treatments on sting nematodes in forage grass. *Proceedings of the Soil and Crop Science Society of Florida* 30, 360–365.

Boyle, L.W. (1950) Several species of parasitic nematodes on peanuts in Georgia. *Plant Disease Reporter* 34, 61–62.

Brenneman, T.B., Sumner, D.R., Baird, R.E., Burton, G.W. and Minton, N.A. (1995) Suppression of foliar and soilborne peanut diseases in bahiagrass rotation. *Phytopathology* 85, 948–952.

Bridge, J. (1975) Hotwater treatment to control plant parasitic nematodes of tropical crops. *Mededelingen van de Faculteit Landbouwwetendchappen Rijksuniversiteit Gent* 40, 249–259.

Bridge, J., Bos, W.S., Page, L.J. and McDonald, D. (1977) The biology and possible importance of *Aphelenchoides arachidis*, a seed-borne endoparasitic nematode of groundnuts from northern Nigeria. *Nematologica* 23, 253–259.

Broadley, R.A. (1981) Distribution and control of root-knot and lesion nematodes on peanuts in North Queensland. *Australian Journal of Experimental Agriculture and Animal Husbandry* 21, 223–226.

Brodie, B.B. and Murphy, W.S. (1975) Population dynamics of plant nematodes as affected by combination of fallow and cropping sequences. *Journal of Nematology* 7, 91–92.

Brown, S.M., Kepner, J.L. and Smart, G.C. Jr (1985) Increased crop yields following application of *Bacillus penetrans* to field plots infested with *Meloidogyne incognita*. *Soil Biology and Biochemistry* 17, 483–486.

Burow, M.D., Simpson, C.E., Paterson, A.H. and Starr, J.L. (1996) Identification of RAPD markers in peanut (*Arachis hypogaea*) associated with root-knot nematode (*Meloidogyne arenaria*) resistance, derived from *A. cardenasii*. *Molecular Breeding* 2, 369–379.

Burton, G.W. and Johnson, A.W. (1987) Coastal bermudagrass rotations for control of root-knot nematodes. *Journal of Nematology* 19, 138–140.

Carneiro, R.M.D.G., Castagnone-Sereno, P. and Dickson, D.W. (1998) Variability among four populations of *Meloidogyne javanica* from Brazil. *Fundamental and Applied Nematology* 21, 319–326.

Carneiro, R.M.D.G., Carneiro, R.G., Das Neves, D.I. and Almeida, M.R.A. (2003) Nova raça de *Meloidogyne javanica* detectada em *Arachis pintoi* no Estado do Paraná. *Nematologia Brasileira* 27, 219–221.

Castillo, M.B., Morrison, L.S., Russell, C.C. and Banks, D.J. (1973) Resistance to *Meloidogyne hapla* in peanut. *Journal of Nematology* 5, 281–285.

Cetintas, R. and Dickson, D.W. (2005) Persistence and suppressiveness of *Pasteuria penetrans* to *Meloidogyne arenaria* race 1. *Journal of Nematology* (in press).

Cetintas, R., Lima, R.D., Mendes, M.L., Brito, J.A. and Dickson, D.W. (2003) *Meloidogyne javanica* on peanut in Florida. *Journal of Nematology* 35, 433–436.

Chabra, H.K. and Mahajan, R. (1976) *Pratylenchus coffeae*, the root-lesion nematode in groundnut and its control by granular nematicides. *Nematologia Mediterranea* 4, 241–242.

Chen, Z.X., Dickson, D.W., McSorley, R., Mitchell, D.J. and Hewlett, T.E. (1996) Suppression of *Meloidogyne arenaria* race 1 by soil application of endospores of *Pasteuria penetrans*. *Journal of Nematology* 28, 159–168.

Cheng, Y.H. and Tu, C.C. (1980) Identification of *Meloidogyne arenaria* of peanut in Taiwan. *Journal of Agricultural Research of China* 29, 47–53.

Cheng, Y.H., Lo, W.C. and Tu, C.C. (1981) Studies on host range of *Meloidogyne arenaria* and its chemical control. *Research Bulletin of Tainan Dais* 15, 55–64.

Choi, K., Burow, M.D., Church, G., Burow, G., Paterson, A.H., Simpson, C.E. and Starr, J.L. (1999) Genetics and mechanism of resistance to *Meloidogyne arenaria* in peanut germplasm. *Journal of Nematology* 31, 283–290.

Choi, Y.E. (1981) The root-knot nematodes, *Meloidogyne* spp., in Korea. In: *Proceedings of the 3rd Research Planning Conference on Root-knot Nematodes,* Meloidogyne *spp., Region VI,* Jakarta, Indonesia.

Church, G.T., Simpson, C.E., Burow, M.D., Paterson, A.H. and Starr, J.L. (2000) Use of RFLP markers for identification of individuals homozygous for resistance to *Meloidogyne arenaria* in peanut. *Nematology* 2, 575–580.

Colbran, R.C. (1958) Studies of plant and soil nematodes. 2. Queensland host records of root-knot nematodes (*Meloidogyne* spp.). *Queensland Journal of Agricultural Science* 15, 101–136.

Colbran, R.C. (1968) Nematodes – important pests of peanuts. *Division of Plant Industry, Department of Primary Industries. Queensland, Australia, Advisory Leaflet* No. 955.

Cooper, W.E., Wells, J.C., Sasser, J.N. and Bowery, T.G. (1959) The efficacy of preplant and postplant applications of 1,2-dibromo-3-chloropropane on control of the sting nematode, *Belonolaimus longicaudatus*. *Plant Disease Reporter* 43, 903–908.

Culbreath, A.K., Beute, M.K., Shew, B.B. and Barker, K.R. (1992) Effects of *Meloidogyne hapla* and *M. arenaria* on black rot severity in new *Cylindrocladium*-resistant peanut genotypes. *Plant Disease* 76, 352–357.

Demeure, Y. (1978a) Influence des températures élevées sur les états actifs et anhydrobiotiques du nématode *Scutellonema cavenessi*. *Revue de Nématologie* 1, 13–19.

Demeure, Y. (1978b) Résistance à la sécheresse en zone sahélienne du nématode phytoparasite *Scutellonema cavenessi* Sher, 1963. *Cahiers de l'ORSTOM, Série Biologie* 10, 283–292.

Demeure, Y., Netscher, C. and Quénehervé, P. (1980) Biology of the plant parasitic nematode *Scutellonema cavenessi* Sher, 1964: reproduction, development and life cycle. *Revue de Nématologie* 3, 213–225.

De Waele, D., Jones, B.L., Bolton, C. and Van den Berg, E. (1988) *Ditylenchus destructor* in hulls and seeds of peanut. *Journal of Nematology* 21, 10–15.

Dhéry, M., Germani, G. and Giard, A. (1975) Résultats de traitements nématicides contre la chlorose et le rabougrissement de l'arachide en Haute Volta. *Cahiers de l'ORSTOM, Série Biologie* 10, 161–161.

Dhurj, I. and Vaishnav, M.U. (1981) Pathogenicity of root-knot nematode on groundnut. *Indian Journal of Nematology* 11, 217–218.

Dickson, D.W. (1998) Peanut. In: Barker, K.R., Pederson, G.A. and Windham, G.L. (eds) *Plant and Nematode Interactions.* American Society of Agronomy, Crop Science Society of America, and Soil Science Society of America, Madison, Wisconsin, pp. 523–566

Dickson, D.W. and Hewlett, T.E. (1988) Efficacy of fumigant and nonfumigant nematicides for control of *Meloidogyne arenaria* on peanut. *Annals of Applied Nematology* (*Journal of Nematology* Vol. 20, Supplement) 2, 95–101.

Dickson, D.W. and Hewlett, T.E. (1989) Effect of bahiagrass and nematicides on *Meloidogyne arenaria* on peanut. *Journal of Nematology* 21, 671–676.

Dickson, D.W. and Melouk, H.A. (1995) Management of nematode pests. In: Melouk, H.A. and Shokes, F.M. (eds) *Peanut Health Management.* APS Press, St Paul, Minnesota, pp. 83–86.

Dickson, D.W. and Smart, G.C. Jr (1971) Control of *Meloidogyne arenaria* and *Pratylenchus penetrans* on peanut with foliar applications of a systemic nematicide. *Journal of Nematology* 3, 307–308.

Dickson, D.W. and Struble, F.B. (1965) A sieving–staining technique for extraction of egg masses of *Meloidogyne incognita* from soil. *Phytopathology* 55, 497.

Dickson, D.W. and Waites, R.E. (1978) Efficacy of at-plant and additional at-pegging applications of nematicides for control of *Meloidogyne arenaria* on peanut. *Proceedings of the American Peanut Research and Education Association* 10, 51 (abstract).

Dickson, D.W. and Waites, R.F. (1982) Evaluation of nematicides for managing the peanut root-knot nematode on peanut, 1980. *Fungicide and Nematicide Tests, American Phytopathological Society* 37, 195–196.

Dickson, D.W., Huisingh, D. and Sasser, J.N. (1971) Dehydrogenases, acid and alkaline phosphatases, and esterases for chemotaxonomy of selected *Meloidogyne, Ditylenchus, Heterodera* and *Aphelenchus* spp. *Journal of Nematology* 3, 1–16.

Dickson, D.W., Mitchell, D.J., Hewlett, T.E., Oostendorp, M. and Kannwischer-Mitchell, M.E. (1991) Nematode-suppressive soil from a peanut field. *Journal of Nematology* 23, 526.

Dickson, D.W., Oostendorp, M., Giblin-Davis, R.M. and Mitchell, D.J. (1994) Control of plant-parasitic nematodes by biological antagonists. In: Rosen, D., Bennett, F.D. and Capinera, J.L. (eds) *Pest Management in the Subtropics. Biological Control – A Florida Perspective.* Intercept, Andover, UK, pp. 575–601.

Diomandé, M. and Beute, M.K. (1981a) Effects of *Meloidogyne hapla* and *Macroposthonia ornata* on Cylindrocladium black rot on peanut. *Phytopathlogy* 71, 491–496.

Diomandé, M. and Beute, M.K. (1981b) Relations of *Meloidogyne hapla* and *Macroposthonia ornata* populations to Cylindrocladium black rot in peanuts. *Plant Disease* 65, 339–342.

Diomandé, M., Black, M.C., Beute, D.K. and Barker, K.R. (1981) Enhancement of *Cylindrocladium crotalaria* root rot by *Meloidogyne arenaria* (race 2) on a peanut cultivar resistant to both pathogens. *Journal of Nematology* 13, 321–327.

Dube, B. and Smart, G.C. Jr (1987) Biological control of *Meloidogyne incognita* by *Paecilomyces lilacinus* and *Pasteuria penetrans*. *Journal of Nematology* 19, 222–227.

Duncan, L.W. (1986) Effect of bare fallow on plant parasitic nematodes in the Sahelian zone of Senegal. *Revue de Nématologie* 9, 75–81.

Duncan, L.W. and Baujard, P. (1986) Influence of nematicide placement depth and time of application on treatment efficacy on the Sahelian zone of Senegal. *Revue de Nématologie* 9, 135–139.

Duncan, L.W., Noling, J.W., Inserra, R.N. and Dunn, D. (1996) Spatial patterns of *Belonolaimus* spp. among and within citrus orchards on Florida's central ridge. *Journal of Nematology* 28, 352–359.

Dunn, R.A. and Dickson, D.W. (1995) Peanut nematode management. *Document RF-NG016.* Florida Cooperative Extension Service, University of Florida, Gainesville, Florida.

Eisenback, J.D., Bernard, E.C., Starr, J.L., Lee, T.A. and Tomaszewski, E.K. (2003) *Meloidogyne haplanaria* n.sp. (Nematoda: Meloidogynidae), a root-knot nematode parasitizing peanut in Texas. *Journal of Nematology* 35, 395–403.
Eisenback, J.D. and Triantaphyllou, H.H. (1991) Root-knot nematodes: *Meloidogyne* species and races. In: Nickle, W.R. (ed.) *Heteroderinae, Cyst and Non-cyst-forming Nematodes. Manual of Agricultural Nematology.* Marcel Dekker, New York, pp. 191–274.
Endo, B.Y. (1959) Responses of root-lesion nematodes, *Pratylenchus brachyurus* and *P. zeae* to various plants and soil types. *Phytopathology* 49, 417–421.
Esbenshade, P.R. and Triantaphyllou, A.C. (1985) Use of enzyme phenotypes for identification of *Meloidogyne* species. *Journal of Nematology* 22, 10–15.
Esbenshade, P.R. and Triantaphyllou, A.C. (1990) Isozyme phenotypes for the identification of *Meloidogyne* species. *Journal of Nematology* 17, 6–20.
Faulkner, L.R. and Bolander, W.J. (1966) Occurrence of large nematode populations in irrigation canals of south central Washington. *Nematologica* 12, 591–600.
Feldmesser, J. and Rebois, R.V. (1965) Temperature and moisture effects on *Pratylenchus brachyurus. Nematologica* 11, 37–38.
Fortnum, B.A., Krausz, J.P. and Conrad, N.G. (1984) Increasing incidence of *Meloidogyne arenaria* on flue-cured tobacco in South Carolina. *Plant Disease* 68, 244–245.
Gair, R., Mathias, P.L. and Harvey, P.N. (1969) Studies of cereal nematode populations and cereal yields under continuous or intensive culture. *Annals of Applied Biology* 63, 503–512.
Garcia, R.M. (1976) Vertical distribution and survival stages of *Meloidogyne arenaria* in Florida with peanut as a host crop. MS Thesis, University of Florida, Gainesville, Florida.
Garcia, R. and Mitchell, D.J. (1975a) Synergistic interactions of *Pythium myriotylum* with *Fusarium solani* and *Meloidogyne arenaria* in pod rot of peanut. *Phytopathlogy* 65, 832–833.
Garcia, R. and Mitchell, D.J. (1975b) Interactions of *Pythium myriotylum* with *Fusarium solani, Rhizoctonia solani,* and *Meloidogyne arenaria* in pre-emergence damping-off of peanut. *Plant Disease Reporter* 59, 665–669.
Germani, G. (1970) *Aphasmatylenchus straturatus* n.sp. (Nematode: Hoplolaimidae) from West Africa. *Proceedings of the Helminthological Society of Washington* 37, 48–51.
Germani, G. (1979a) Nematicide as a tool to study the impact of nematodes on plant productivity. In: Mongi, H.O. and Huxley, P.A. (eds) *Soil Research in Agroforestry: Proceedings of an Expert Consultation.* Nairobi, Kenya, pp. 297–313.
Germani, G. (1979b) Action directe et rémanente d'un traitement nématicide du sol sur 2 cultivars d'arachide au Sénégal. *Oléagineux* 34, 399–404.
Germani, G. (1981a) Étude au champ de l'évolution des populations du nématode *Scutellonema cavenessi* et de la cinétique de la fixation de N2 sur 3 cultivars d'arachide. *Oléagineux* 36, 247–249.
Germani, G. (1981b) Pathogenicity of the nematode *Scutellonema cavenessi* on peanut and soybean. *Revue de Nématologie* 4, 203–208.
Germani, G. and Dhéry, M. (1973) Observations et expérimentations concernant le role des nématodes dans deux affections de l'arachide en Haute Volta: la 'chlorose' et le 'clump'. *Oléagineux* 28, 235–242.
Germani, G. and Gautreau, J. (1976) Résultats agronomiques obtenus par des traitements nématicides sur arachide au Sénégal. *Cahiers de l'ORSTOM, Série Biologie* 11, 193–202.
Germani, G. and Luc, M. (1982a) Études sur la 'chlorose voltaique' des légumineuses due au nématode *Aphasmatylenchus straturatus* Germani. I. *Revue de Nématologie* 5, 139–146.
Germani, G. and Luc, M. (1982b) Études sur la 'chlorose voltaique' des légumineuses due au nématode *Aphasmatylenchus straturatus* Germani. *Revue de Nématologie* 5, 195–199.
Germani, G. and Reversat, G. (1982) Effet sur les rendements de l'arachide au Sénégal de deux produits nématicides, DBCP et EDB, et d'un amendement organique. *Oléagineux* 37, 521–524.
Germani, G. and Reversat, G. (1983) Effect du dibromochloropropane sur quelques espèces de nématodes reviviscents, parasites de l'arachide au Sénégal. *Revue de Nématologie* 6, 73–78.
Germani, G., Ollivier, B. and Diem, H.G. (1981) Interaction of *Scutellonema cavenessi* and *Glomus mosseae* on growth and N_2 fixation. *Revue de Nématologie* 4, 277–280.
Germani, G., Cuany, A. and Merny, G. (1982) L'analyse factorielle des correspondances appliquée a l'influence de deux nématodes sur la croissance de l'arachide et la fixation symbiotique de l'azote. *Revue de Nématologie* 5, 161–168.
Germani, G., Baujard, P. and Luc, M. (1985) *La Lutte Contre les Nématodes dans le Bassin Arachidier Sénégalais.* ORSTOM, Dakar.

Good, J.M. (1968) Relation of plant parasitic nematodes to soil management practices. In: Smart, G.C. and Perry, V.G. (eds) *Tropical Nematology*. University of Florida Press, Gainesville, Florida, pp. 113–118.

Good, J.M. and Stansell, J.R. (1965) Effect of irrigation, soil fumigation, and date of harvest on *Pratylenchus brachyurus* (Godfrey). *Nematologica* 11, 38–39.

Good, J.M., Robertson, W.K. and Thompson, L.C. Jr (1954) Effect of crop rotation on populations of meadow nematode, *Pratylenchus leiocephalus,* in Norfolk loamy, fine sand. *Plant Disease Reporter* 38, 178–180.

Good, J.M., Boyle, L.W. and Hammons, R.D. (1958) Studies of *Pratylenchus brachyurus* on peanuts. *Phytopathology* 48, 530–535.

Graham, T.W. (1951) Nematode root rot of tobacco and other plants. *South Carolina Agricultural Experiment Station Technical Bulletin* 390.

Hagan, A. (1988) Disease and nematode control. In: *Alabama Cooperative Extension Service, Auburn University. 1988 Peanut Pest Management, Weed, Insect, Disease and Nematode Control Recommendations. Circular ANR-360*, pp. 5–6.

Hammons, R.O. (1982) Origin and early history of the peanut. In: Patte, H.E. and Young, C.T. (eds) *Peanut Science and Technology*. American Peanut Research and Education Society, Yoakum, Texas, pp. 1–20.

Hatz, B. and Dickson, D.W. (1992) Effect of temperature and attachment, development, and interactions of *Pasteuria penetrans* on *Meloidogyne arenaria. Journal of Nematology* 24, 512–521.

Hirunsalle, A., Barker, K.R. and Beute, M.K. (1995a) Infection, reproduction potential, and root galling by root-knot nematode species and concomitant populations on peanut and tobacco. *Journal of Nematology* 27, 172–177.

Hirunsalle, A., Barker, K.R. and Beute, M.K. (1995b) Effects of peanut–tobacco rotations on population dynamics of *Meloidogyne arenaria* in mixed race populations. *Journal of Nematology* 27, 178–188.

Hirunsalle, A., Barker, K.R. and Beute, M.K. (1995c) Effects of peanut genotypes on *Meloidogyne* species interactions. *Journal of Nematology* 27, 189–199.

Holbrook, C.C. and Noe, J.P. (1992) Resistance to peanut root-knot nematode (*Meloidogyne arenaria*) in *Arachis hypogaea. Peanut Science* 19, 35–37.

Holbrook, C.C., Knauft, D.A. and Dickson, D.W. (1983) A technique for screening peanut for resistance to *Meloidogyne arenaria. Plant Disease* 67, 957–958.

Holbrook, C.C., Noe, J.P., Gorbet, D.W. and Stephenson, M.G. (1995) Evaluation of peanut breeding lines with resistance to the peanut root-knot nematode. *Proceedings of the American Peanut Research and Education Society*, Charlotte, North Carolina, 11–14 July 1995, 27, p. 24

Holderman, Q.L. (1955) The present known distribution of the sting nematode, *Belonolaimus gracilis,* in the Coastal Plain of the Southeastern United States. *Plant Disease Reporter* 39, 5–8.

Holderman, Q.L. and Graham, T.W. (1953) The effect of different plant species on the population trends of the sting nematode. *Plant Disease Reporter* 37, 497–500.

Huang, X. and Becker, J.O. (1997) *In vitro* culture and feeding behavior of *Belonolaimus longicaudatus* on excised *Zea mays* roots. *Journal of Nematology* 29, 411–415.

Huang, X. and Becker, J.O. (1999) Life cycle and mating behavior of *Belonolaimus longicaudatus* in gnotobiotic culture. *Journal of Nematology* 31, 70–74.

Huber, D.M. and Schneider, R.W. (1982) The description and occurrence of suppressive soils. In: Schneider, R.W. (ed.) *Suppressive Soils and Plant Disease.* APS Press, St Paul, Minnesota, pp. 1–7.

Ibrahim, I.K.A. and El-Saedy, M.A. (1976a) Plant parasitic nematodes associated with peanuts in Egypt. *Egyptian Journal of Phytopathology* 8, 31–35.

Ibrahim, I.K.A. and El-Saedy, M.A. (1976b) Development and pathogenesis of *Meloidogyne javanica* in peanut roots. *Nematologica Mediterranea* 4, 231–234.

Ingram, E.G. and Rodríguez-Kábana, R. (1980) Nematodes parasitic on peanuts in Alabama and evaluation of methods for detection and study of population dynamics. *Nematropica* 10, 21–30.

Jackson, C.R. and Minton, N.A. (1968) Pod invasion by fungi in the presence of lesion nematodes in Georgia. *Oléagineux* 23, 531–534.

Jackson, E.E. and Sturgeon, R.V. Jr (1973) Effects of nematicides upon root lesion nematode populations. *Journal of the American Peanut Research and Education Association* 5, 178–181.

Jaffee, B.A. and Zehr, E.I. (1982) Parasitism of the nematode *Criconemella xenoplax* by the fungus *Hirsutella rhossiliensis. Phytopathology* 72, 1378–1381.

Johnson, A.W., Dowler, C.C. and Hauser, E.W. (1974) Seasonal population dynamics of selected plant parasitic nematodes on four monocultured crops. *Journal of Nematology* 6, 187–190.

Jones, B.L. and De Waele, D. (1988) First report of *Ditylenchus destructor* in pods and seeds of peanut. *Plant Disease* 72, 453.

Kaloshian, I., Lawn, D.A., Williamson, V.M., Westerdahl, B.B. and Miyao, G. (1996) 'Resistance-breaking' nematodes identified in California tomatoes. *California Agriculture* 50, 18–19.

Kerry, B.R. (1982) The decline of *Heterodera avenae* populations. *European Plant Protection Organization Bulletin* 12, 491–496.

Kinloch, R.A. and Lutrick, M.C. (1975) The relative abundance of nematodes in an established field crop rotation. *Proceedings of the Soil and Crop Science Society of Florida* 34, 192–194.

Koen, H. (1967) Notes on the host range, ecology and population dynamics of *Pratylenchus brachyurus*. *Nematologica* 13, 118–124.

Koenning, S.R. and Barker, K.R. (1992) Relative damage functions and reproduction potentials of *Meloidogyne arenaria* and *M. hapla* on peanut. *Journal of Nematology* 24, 187–192.

Loof, P.A.A. (1964) Free-living and plant-parasitic nematodes from Venezuela. *Nematologica* 10, 201–300.

Lordello, A.I.L. and Gerin, M. (1981) Nematóide javanes parasitando raízes e nódules de amendoim. *Revista de Agricultura Piracicaba* 56, 278.

Machmer, J.H. (1951) Root-knot of peanut. I. Distribution. *Plant Disease Reporter* 35, 364–366.

Machmer, J.H. (1953) *Criconemoides* sp., a ring nematode associated with peanut 'yellows'. *Plant Disease Reporter* 37, 156.

Mankau, R. (1975) *Bacillus penetrans* n.comb. causing a virulent disease of plant-parasitic nematodes. *Journal of Invertebrate Pathology* 26, 333–339.

Mankau, R. (1980) Biological control of *Meloidogyne* populations by *Bacillus penetrans* in West Africa. *Journal of Nematology* 12, 230 (abstract).

Martin, G.C. (1958) Root-knot nematodes (*Meloidogyne* spp.) in the Federation of Rhodesia and Nyasaland. *Nematologica* 3, 332–349.

Martin, G.C. (1961) Plant species attacked by root-knot nematodes (*Meloidogyne* spp.) in the Federation of Rhodesia and Nyasaland. *Nematologica* 6, 130–134.

McDonald, D. and Misari, S.M. (1976) Disease and pests of groundnuts and their importance in crop exchange. Interafrican Symposium. The role of plant protection in crop improvement in Africa. Ibadan, Nigeria, 7–12 October, 1974. *African Journal of Plant Protection* 1, 75–82.

McDonald, D., Bos, W.S. and Gumel, M.H. (1979) Effects of infections of peanut (groundnut) seed by the testa nematode *Aphelenchoides arachidis*, on seed infection by fungi and on seeding emergence. *Plant Disease Reporter* 63, 464–467.

McSorley, R. and Gallaher, R.N. (1991) Nematode population changes and forage yields of six corn and sorghum cultivars. *Supplement to the Journal of Nematology* 23, 673–677.

McSorley, R., Dickson, D.W., Candanedo-Lay, E.M., Hewlett, T.E. and Frederick, J.J. (1992) Damage functions for *Meloidogyne arenaria* on peanut. *Journal of Nematology* 24, 193–198.

McSorley, R., Dickson, D.W., Brito, J.A. and Hochmuth, R.C. (1994) Tropical rotation crops influence nematode densities and vegetable yield. *Journal of Nematology* 26, 308–314.

McWatters, K.H. and Cherry, I.P. (1982) Potential food uses of peanut seed proteins. In: Pattee, H.E. and Young, C.T. (eds) *Peanut Science and Technology*. American Peanut Research and Education Society, Yoakum, Texas, pp. 689–736.

Meagher, J.W. (1967) Observations on the transport of nematodes in subsoil drainage and irrigation water. *Australian Journal of Experimental Agriculture and Animal Husbandry* 7, 577.

Melouk, H.A. and Backman, P.A. (1995) Management of soilborne fungal pathogens. In: Melouk, H.A. and Shokes, F.M. (eds) *Peanut Health Management*. APS Press, St Paul, Minnesota, pp. 75–82.

Merny, G. and Mauboussin, J.C. (1973) Action possible des nématodes dans le rabougrissement ou 'clump' de l'arachide. *Nematologica* 19, 406–408.

Merny, G., Fortuner, R. and Luc, M. (1974) Plant parasitic nematodes in Gambia. *L'Agronomie Tropicale, Nogent* 29, 702–707.

Miller, L.I. (1951) A report on the effect of ethylene dibromide soil treatment on root-knot control, nodulation, and yield of peanuts. *Virginia Journal of Science* 2, 109–112.

Miller, L.I. (1972a) The influence of soil texture on the survival of *Belonolaimus longicaudatus*. *Phytopathology* 62, 670–671 (abstract).

Miller, L.I. (1972b) Resistance of plant introductions of *Arachis hypogaea* to *Meloidogyne hapla*, *Meloidogyne arenaria*, and *Belonolaimus longicaudatus*. *Virginia Journal of Science* 23, 101 (abstract).

Miller, L.I. and Duke, G.B. (1961) Peanut nematode disease control. *Virginia Agricultural Experiment Station Bulletin* 520.

Milne, D.L. and Du Plessis, D.P. (1964) Development of *Meloidogyne javanica* (Treub) Chitwood on tobacco under fluctuating soil temperatures. *South African Journal of Agricultural Science* 7, 673–680.

Minton, N.A. (1963) Effects of two populations of *Meloidogyne arenaria* on peanut roots. *Phytopathology* 53, 79–81.

Minton, N.A. and Bell, D.K. (1969) *Criconemoides ornatus* parasitic on peanuts. *Journal of Nematology* 1, 349–351.

Minton, N.A. and Csinos, A.S. (1986) Peanut response to 1,3-D in *Meloidogyne arenaria* and *Sclerotium rolfsii* infested soil. *Proceedings of the American Peanut Research and Education Society* 18, 62 (abstract).

Minton, N.A. and Hammons, R.O. (1975) Evaluation of peanut for resistance to the peanut root-knot nematode, *Meloidogyne arenaria. Plant Disease Reporter* 59, 944–945.

Minton, N.A. and Jackson, C.R. (1967) Invasion of peanut pods by *Aspergillus flavus* and other fungi in the presence of root-knot nematodes. *Oléagineux* 22, 543–546.

Minton, N.A. and Morgan, L.W. (1974) Evaluation of systemic and nonsystemic pesticides for insect and nematode control in peanuts. *Plant Science* 1, 91–98.

Minton, N.A. and Sayre, R.M. (1989) Suppressive influence of *Pasteuria penetrans* in Georgia soils on reproduction of *Meloidogyne arenaria. Journal of Nematology* 21, 574–575.

Minton, N.A., Cairns, E.J., Minton, E.B. and Hopper, B.E. (1963) Occurrence of plant-parasitic nematodes in Alabama. *Plant Disease Reporter* 47, 743–745.

Minton, N.A., Bell, D.K. and Doupnik, B. Jr (1969a) Peanut pod invasion by *Aspergillus flavus* in the presence of *Meloidogyne hapla. Journal of Nematology* 1, 318–320.

Minton, N.A., McGill, J.F. and Golden, A.M. (1969b) *Meloidogyne javanica* attacks peanuts in Georgia. *Plant Disease Reporter* 53, 668.

Minton, N.A., Hammons, R.O. and Parham, S.A. (1970) Infection of shell and peg tissues of six peanut cultivars by *Pratylenchus brachyurus. Phytopathology* 60, 472–474.

Minton, N.A., Csinos, A.S. and Bell, D.K. (1984) Effects of six nematicides and two fungicides, applied in various combinations on peanuts. *United States Department of Agriculture, Agricultural Research Service, ARS-3.*

Minz, G. (1956) The root-knot nematode, *Meloidogyne* spp., in Israel. *Plant Disease Reporter* 40, 798–801.

Mitsui, Y., Yoshida, T., Okamoto, K. and Ishii, R. (1976) Relationship between nematode-trapping fungi and *Meloidogyne hapla* in peanut field. *Japanese Journal of Nematology* 6, 47–55.

Motsinger, R.E., Crawford, J.L. and Thompson, S.S. (1976) Nematode survey of peanuts and cotton in southwest Georgia. *Peanut Science* 3, 72–74.

Nelson, S.C., Simpson, C.E. and Starr, J.L. (1989) Resistance to *Meloidogyne arenaria* in *Arachis* sp. germplasm. *Supplement to the Journal of Nematology* 21, 654–660.

Noe, J.P., Holbrook, C.C. and Minton, N.A. (1992) Field evaluation of susceptibility to *Meloidogyne arenaria* in *Arachis hypogaea* plant introductions. *Supplement to the Journal of Nematology* 24, 712–716.

Norse, D. (1972) Nematode populations in a maize–groundnut–tobacco rotation and the resistance of maize varieties to *Meloidogyne javanica. Tropical Agriculture, Trinidad* 49, 355–360.

Norton, A.J., Perry, V.G., Martin, F.G. and NeSmith, J. (1977) Effect of age of bahiagrass sod on succeeding peanut crops. *Peanut Science* 4, 71–74.

O'Bannon, J.H. and Tomberlin, A.T. (1970) *Pratylenchus* sp. as citrus pathogens. *Phytopathology* 60, 1540.

Oostendorp, M., Dickson, D.W. and Mitchell, D.J. (1990) Host range and ecology of isolates of *Pasteuria* sp. from the southeastern United States. *Journal of Nematology* 22, 525–531.

Oostendorp, M., Dickson, D.W. and Mitchell, D.J. (1991) Population development of *Pasteuria penetrans* on *Meloidogyne arenaria. Journal of Nematology* 23, 58–64.

Orion, D. and Cohn, E. (1975) Crop rotation for control of certain nematodes. *Phytoparasitica* 3, 69 (abstract).

Orr, C.C. and Newton, O.H. (1971) Distribution of nematodes by wind. *Plant Disease Reporter* 55, 61–63.

Osman, H.A., Dickson, D.W. and Smart, G.C. Jr (1985) Morphological comparisons of host races 1 and 2 of *Meloidogyne arenaria* from Florida. *Journal of Nematology* 17, 279–285.

Oteifa, B.A. (1962) Species of root-lesion nematodes commonly associated with economic crops in the Delta of the U.A.R. *Plant Disease Reporter* 46, 572–575.

Owens, J.V. (1951) The pathological effects of *Belonolaimus gracilis* on peanuts in Virginia. *Phytopathology* 41, 29 (abstract).

Patel, H.R., Vaishnav, M.U. and Dhurj, I.U. (1985) Interaction of *Meloidogyne arenaria* and *Fusarium solani* on groundnut. *Indian Journal of Nematology* 15, 98–99.

Payan, L.A. and Dickson, D.W. (1988) Host specificity of four populations of *Pratylenchus brachyurus. Annals of Applied Nematology* (*Journal of Nematology* Vol. 20, Supplement) 2, 140–143.

Perez, E.E., Weingartner, D.P. and McSorley, R. (2000) Niche distribution of *Paratrichodorus minor* and *Belonolaimus longicaudatus* following fumigation on potato and cabbage. *Journal of Nematology* 32, 343–348.

Perry, V.G. (1964) Nematode host-parasitic relationships. *Florida Agricultural Experiment Station Annual Report 1964*, pp. 115–116.

Perry, V.G. (1965) Host–parasite relationships of certain nematodes and crop plants in Florida. *University of Florida Agricultural Experiment Stations Annual Report*, Gainesville, Florida, pp. 115–116.

Perry, V.G. and Norden, A.J. (1963) Some effect of a cropping sequence on populations of certain plant nematodes. *Proceedings of the Soil and Crop Science Society of Florida* 23, 116–121.

Perry, V.G. and Rhoades, H.L. (1982) The genus *Belonolaimus*. In: Riggs, R.D. (ed.) *Nematology in the Southern Region of the United States. Bulletin 276*. Agricultural Experiment Station, University of Arkansas, Fayettville, Arkansas, pp. 144–149.

Porter, D.M., Smith, D.H. and Rodríguez-Kábana, R. (1984) Introduction. In: Porter, D.M., Smith, D.H. and Rodríguez-Kábana, R. (eds) *Compendium of Peanut Diseases*. APS Press, St Paul, Minnesota, pp. 1–4.

Rau, G.J. (1958) A new species of sting nematode. *Proceedings of the Helminthological Society of Washington* 25, 95–98.

Rau, G.J. and Fassuliotis, G. (1970) Equal-frequency tolerance ellipses for population studies of *Belonolaimus longicaudatus*. *Journal of Nematology* 2, 84–92.

Reddy, D.D.R., Subrahmanyam, P., Sankara Reddy, G.H., Raja Reddy, C. and Siva Ras, D.V. (1984) A nematode disease of peanut caused by *Tylenchorhynchus brevilineatus*. *Plant Disease* 68, 526–529.

Rickard, D.A., Barker, K.R. and Wells, J.C. (1977) Effects of *Macroposthonia ornata* and *Meloidogyne hapla* on yield and value of peanut. *Journal of Nematology* 9, 281 (abstract).

Riegel, C., Dickson, D.W., Peterson, L.G. and Nance, J.L. (2000a) Rate response of 1,3-dichloropropene for nematode control in spring squash in deep sand soils. *Supplement to the Journal of Nematology* 32, 524–530.

Riegel, C., Dickson, D.W., Shaw, L.N., Peterson, L.G. and Whitty, E.B. (2000b) Comparison of different chisel types for 1,3-dichloropropene fumigation in deep sandy soils. *Nematropica* 31, 289–293.

Robbins, R.T. and Barker, K.R. (1973) Comparisons of host range and reproduction among populations of *Belonolaimus longicaudatus* from North Carolina and Georgia. *Plant Disease Reporter* 57, 750–754.

Robbins, R.T. and Barker, K.R. (1974) The effects of soil type, particle size, temperature, and moisture on reproduction of *Belonolaimus longicaudatus*. *Journal of Nematology* 6, 1–6.

Robbins, R.T. and Hirschmann, H. (1974) Variation among populations of *Belonolaimus longicaudatus*. *Journal of Nematology* 6, 87–94.

Rodríguez-Kábana, R. and King, P.S. (1985) Evaluation of selected nematicides for control of *Meloidogyne arenaria* in peanut: a multi-year study. *Nematropica* 15, 155–164.

Rodríguez-Kábana, R. and Morgan-Jones, G. (1987) Novel rotations and organic materials show promise for management of nematodes. *Alabama Agricultural Experiment Station, Highlights of Agricultural Research* 34, 13.

Rodríguez-Kábana, R. and Robertson, D.G. (1987) Control of *Meloidogyne arenaria* in peanut with 1,3-D: relative efficacy and application depth. *Nematropica* 17, 17–29.

Rodríguez-Kábana, R. and Touchton, J.T. (1984) Corn and sorghum as rotational crops for management of *Meloidogyne arenaria* in peanut. *Nematropica* 14, 26–36.

Rodríguez-Kábana, R., King, P.S. and Pope, M.H. (1981) Comparison of in-furrow applications and banded treatments for control of *Meloidogyne arenaria* in peanuts. *Nematropica* 11, 53–67.

Rodríguez-Kábana, R., Shelby, R.A., King, P.S. and Pope, M.H. (1982a) Application time and effectiveness of four systemic nematicides against *Meloidogyne arenaria* on Florunner peanut. *Nematropica* 12, 85–96.

Rodríguez-Kábana, R., Williams, J.C. and Shelby, R.A. (1982b) Assessment of peanut yield losses caused by *Meloidogyne arenaria*. *Nematropica* 12, 279–288.

Rodríguez-Kábana, R., Weaver, C.F. and King, P.S. (1985) Combinations of 1,3-D and aldicarb for management of *Meloidogyne arenaria* in peanuts. *Nematropica* 15, 93–106.

Rodríguez-Kábana, R., Weaver, C.F., Robertson, D.G. and Snoddy, E.L. (1986) Population dynamics of *Meloidogyne arenaria* juveniles in a field with Florunner peanut. *Nematropica* 16, 185–196.

Rodríguez-Kábana, R., Ivey, H. and Backman, P.A. (1987) Peanut–cotton rotations for the management of *Meloidogyne arenaria*. *Journal of Nematology* 19, 484–486.

Rodríguez-Kábana, R., Robertson, D.G., King, P.S. and Wells, L. (1991) American jointvetch and partridge pea for the management of *Meloidogyne arenaria* in peanut. *Nematropica* 21, 97–103.

Rodríguez-Kábana, R., Kokalis-Burelle, N., Robertson, D.G., King, P.S. and Wells, L.W. (1994) Rotations with coastal bermudagrass, cotton, and bahiagrass for management of *Meloidogyne arenaria* and southern blight in peanut. *Supplement to the Journal of Nematology* 26, 665–668.

Roman, J. and Hirschmann, H. (1969) Morphology and morphometrics of six species of *Pratylenchus*. *Journal of Nematology* 1, 363–386.

Saint-Smith, J.H., McCarthy, G.J.P., Rawson, J.E., Langford, S. and Colbran, R.C. (1972) Peanut growing. *Queensland Agricultural Journal* 98, 639–644.

Sakhuja, P.K. and Sethi, C.L. (1985a) Screening of groundnut germplasm for resistance to root-knot nematode, *Meloidogyne javanica*. *Indian Journal of Nematology* 15, 129–130.

Sakhuja, P.K. and Sethi, C.L. (1985b) Growth of groundnut as influenced by different inocula of *Meloidogyne javanica*. *Indian Journal of Nematology* 15, 135–137.

Sakhuja, P.K. and Sethi, C.L. (1985c) Frequency of occurrence of various plant parasitic nematodes and root-rot fungi on groundnut in Punjab. *Indian Journal of Nematology* 15, 191–194.

Sasser, J.N. (1954) Identification and host–parasite relationships of certain root-knot nematodes (*Meloidogyne* spp.). *University of Maryland, Agricultural Experiment Station, Technical Bulletin A-77.*

Sasser, J.N. (1977) Worldwide dissemination and importance of the root-knot nematodes, *Meloidogyne* spp. *Journal of Nematology* 9, 26–29.

Sasser, J.N. (1979a) Pathogenicity, host ranges, and variability in *Meloidogyne* spp. In: Lamberti, F. and Taylor, C.E. (eds) *Root-knot Nematodes (*Meloidogyne *Species): Systematics, Biology, and Control.* Academic Press, New York, pp. 257–268.

Sasser, J.N. (1979b) Economic importance of *Meloidogyne* in tropical countries. In: Lamberti, F. and Taylor, C.E. (eds) *Root-knot Nematodes (*Meloidogyne *Species): Systematics, Biology and Control.* Academic Press, New York, pp. 359–374.

Sasser, J.N. and Carter, C.C. (1982) Root-knot nematodes (*Meloidogyne* sp.): identification, morphological and physical variation, host range, ecology, and control. In: Riggs, R.D. (ed.) *Nematology in the Southern Region of the United States. Bulletin 276.* Agricultural Experiment Station, University of Arkansas, Fayettville, Arkansas, pp. 21–32.

Sasser, J.N. and Freckman, D. (1987) A world perspective on nematology: the role of the society. In: Veech, J.A. and Dickson, D.W. (eds) *Vistas on Nematology.* Society of Nematologists, Hyattsville, Maryland, pp. 7–14.

Sasser, J.N. and Kirby, M.F. (1979) Crop cultivars resistant to root-knot nematodes, *Meloidogyne* spp., with information on seed sources. *Cooperative Publication, Department of Plant Pathology, North Carolina State University and US Agency of International Development.* Raleigh, North Carolina.

Sasser, J.N., Cooper, W.E. and Bowery, T.G. (1960) Recent developments in the control of sting nematode, *Belonolaimus longicaudatus*, on peanuts with 1,2-dibromo-3-chloropropane and EN 18133. *Plant Disease Reporter* 44, 733–737.

Sasser, J.N., Barker, K.R. and Nelson, L.A. (1970) Relative effects of eight nematode species on growth, yield and value of peanut following chemical soil treatment. Proceedings of the Second International Congress of Parasitology, Washington, DC. *Journal of Parasitology* 56, 300–301.

Sasser, J.N., Barker, K.R. and Nelson, L.A. (1975a) Correlations of field populations of nematodes with crop growth responses for determining relative involvement of species. *Journal of Nematology* 7, 193–198.

Sasser, J.N., Barker, K.R. and Nelson, L.A. (1975b) Chemical soil treatment for nematode control on peanut and soybean. *Plant Disease Reporter* 59, 154–155.

Sauer, M.R. (1968) Nematodes in an irrigated vineyard. *Nematologica* 14, 457–458.

Serracin, M., Schuerger, A.C., Dickson, D.W. and Weingartner, D.P. (1997) Temperature-dependent development of *Pasteuria penetrans* in *Meloidogyne arenaria*. *Journal of Nematology* 29, 228–238.

Sharma, N.K., Sharma, S.K. and Midha, S.K. (1978) Root-knot disease of groundnut in Punjah. *Indian Farming* 28, 19–21.

Sharma, S.B. (1985) A world list of nematode pathogens associated with chickpea, groundnut, pearl millet, pigeonpea and sorghum. *Pulse Pathology Progress Report 42. International Crops Research Institute for the Semi-Arid Tropics* 5–8.

Sharma, S.B., Smith, D.H. and McDonald, D. (1995) Host races of *Meloidogyne javanica*, with preliminary evidence that the 'groundnut race' is widely distributed in India. *International Arachis Newsletter* 15, 43–45.

Sher, S.A. (1964) Revision of the Hoplolaiminae (Nematoda). III. *Scutellonema*, Andrassy, 1958. *Nematologica* 9, 421–443.

Simpson, C.E. (1991) Pathways for introgression of pest resistance into *Arachis hypogaea* L. *Peanut Science* 18, 22–26.

Simpson, C.E. and Starr, J.L. (2001) Registration of 'COAN' peanut. *Crop Science* 41, 918.

Simpson, C.E., Nelson, S.C., Starr, J.L., Woodard, K.E. and Smith, O.D. (1993) Registration of TxAG-6 and TxAG-7 peanut germplasm lines. *Crop Science* 33, 14–18.

Simpson, C.E., Starr, J.L., Church, G.T., Burow, M.D. and Paterson, A.H. (2003) Registration of 'NemaTAM' peanut. *Crop Science* 43, 1561.

Singh, I. and Sakhuja, P.K. (1984) Clump disease of groundnut and its control with nematicides. *Indian Journal of Nematology* 14, 54–55.

Singh, N.D. (1972) Plant-parasitic nematodes associated with some economic crops in Guyana. *Plant Disease Reporter* 56, 1059–1062.

Smart, G.C. Jr and Nguyen, K.B. (1991) Sting and awl nematodes: *Belonolaimus* spp. and *Dolichodorus* spp. In: Nickle, W.R. (ed.) *Manual of Agricultural Nematology*. Marcel Dekker, New York, pp. 627–667.

Smith, O.D., Boswell, T.E. and Thames, W.H. (1978) Lesion nematode resistance in peanuts. *Crop Science* 18, 1008–1011.

Stalker, H.T., Garcia, G.M., Shew, B.B., Beute, M.K., Isleib, T.G. and Kochert, G. (1994) Introgression from *A. cardenasii* to *A. hypogaea*. *Proceedings of the American Peanut Research and Education Society* 26, 33.

Stalker, H.T., Shew, B.B., Garcia, G.M., Beute, M.K., Barker, K.R., Holbrook, C.C., Noe, J.P. and Kochert, G.A. (1995) *Meloidogyne arenaria* resistance in advanced-generation *Arachis hypogaea* × *A. cardenasii* hybrids. *Proceedings of the American Peanut Research and Education Society* 27, 24.

Starr, J.L. (1984) Expression of resistance in peanuts, *Arachis hypogaea*, to *Pratylenchus brachyurus*: impact on screening for resistance. *Journal of Nematology* 16, 404–406.

Starr, J.L. and Simpson, C.E. (1991) Segregation of resistance to *Meloidogyne arenaria* in progeny of interspecific hybrids. *Proceedings of the American Peanut Research and Education Society* 23, 23.

Starr, J.L., Shim, M.Y., Lee, T.A. Jr and Simpson, C.E. (1996) Additive effects of *Meloidogyne arenaria* and *Sclerotium rolfsii* on peanut. *Journal of Nematology* 28, 99–106.

Starr, J.L., Morgan, E.R. and Simpson, C.E. (2002) Management of peanut root-knot nematode, *Meloidogyne arenaria*, with host resistance. Online. *Plant Health Progress* doi:10.1094/PHP-2002–1221–01-HM.

Steiner, G. (1945) Meadow nematodes as the cause of root destruction. *Phytopathology* 35, 935–937.

Steiner, G. (1949) Plant nematodes the grower should know. *Proceedings of the Soil Science Society of Florida, 1942, 4-B*, 72–117.

Stephenson, M.G., Holbrook, C.C., Noe, J.P. and Anderson, W.F. (1995) Evaluation of additional sources of resistance to the peanut root-knot nematode in the cultivated species of peanut. *Proceedings of the American Peanut Research and Education Society* 27, 25.

Stirling, G.R. (1984) Biological control of *Meloidogyne javanica* with *Bacillus penetrans*. *Phytopathology* 74, 55–60.

Stirling, G.R. and Mankau, R. (1979) Mode of parasitism of *Meloidogyne* and other nematode eggs by *Dactylella oviparasitica*. *Journal of Nematology* 11, 282–288.

Stirling, G.R. and White, A.M. (1982) Distribution of a parasite of root-knot nematodes in South Australian vineyards. *Plant Disease* 66, 52–53.

Sturgeon, R.V. (1986) Peanut disease loss estimates for major peanut producing states in the United States in 1984 and 1985. *Proceedings of the American Peanut Research and Education Society* 18, 24–26.

Subrahmanyam, P., Ghanekar, A.M., Nolt, B.L., Reedy, D.V.R. and McDonald, D. (1983) Resistance to groundnut diseases in wild *Arachis* species. *Proceedings of an International Workshop on Cytogenetics of Arachis*, 31 October–2 November 1983, ICRISAT Centre, Patancheru, India.

Taylor, A.L. and Sasser, J.N. (1978) *Biology, Identification and Control of Root-knot Nematodes* (Meloidogyne *species)*. A cooperative publication of the Department of Plant Pathology, North Carolina Sate University and USAID. North Carolina State University Graphics, Raleigh, North Carolina.

Thames, W.H. (1982) The genus *Pratylenchus*. In: Riggs, R.D. (ed.) *Nematology in the Southern Region of the United States. Bulletin 276*. Agricultural Experiment Station, University of Arkansas, Fayetteville, Arkansas, pp. 108–126.

Thames, W.H. Jr (1959) Plant parasitic nematode populations of some Florida soils under cultivated and natural conditions. *Dissertation Abstract* 20, 1109–1110.

Thames, W.H. Jr and Stoner, W.N. (1953) A preliminary trial of lowland culture rice in rotation with vegetable crops as a means of reducing root-knot nematode infestations in the Everglades. *Plant Disease Reporter* 37, 187–192.

Thies, J., Merrill, S.B. and Corley, E.L. (2002) Red food coloring stain: new, safer procedures for staining nematodes in roots and egg masses on root surfaces. *Journal of Nematology* 34, 179–181.

Thomason, I.J., Van Gundy, S.D. and Kirkpatrick, J.D. (1964) Mobility and infectivity of *Meloidogyne javanica* as affected by storage time and temperature in water. *Phytopathology* 54, 192–195.

Tomaszewski, E.K., Khalil, M.A.M., El-Deeb, A.A., Powers, T.O. and Starr, J.L. (1994) *Meloidogyne javanica* parasitic on peanut. *Journal of Nematology* 26, 436–441.

Turner, S.J. (1990) The identification and fitness of virulent potato-cyst nematodes (*Globodera pallida*) selected on resistant *Solanum vernei* hybrids for up to eleven generations. *Annals of Applied Biology* 117, 385–397.

Van der Linde, J.W. (1956) The *Meloidogyne* problem in South Africa. *Nematologica* 1, 177–183.

Van Gundy, S.D. (1985) Ecology of *Meloidogyne* spp. – emphasis on environmental factors affecting survival and pathogenicity. In: Sasser, J.N. and Carter, C.C. (eds) *An Advanced Treatise on* Meloidogyne *Biology and Control Vol. I.* University of North Carolina Press, Raleigh, North Carolina, pp. 178–182.

Vance, P.N. (1981) Peanut growing in the South Burnett. *Queensland Agricultural Journal* 107, 201–213.

Vrain, T.C. (1978) Influence of chilling and freezing temperatures on infestivity of *Meloidogyne incognita* and *M. hapla*. *Journal of Nematology* 10, 177–180.

Wheeler, T.A. and Starr, J.L. (1987) Incidence and economic importance of plant-parasitic nematodes on peanut in Texas. *Peanut Science* 14, 94–96.

Williams, E.J. and Drexler, J.S. (1981) A non-destructive method for determining peanut pod maturity. *Peanut Science* 8, 134–141.

Windham, G.L. and Williams, W.P. (1994) Reproduction of *Meloidogyne incognita* and *M. arenaria* on tropical corn hybrids. *Supplement to the Journal of Nematology* 26, 753–755.

Wittwer, S.H. (1981) The 20 crops that stand between man and starvation. *Farm Chemicals* 144, 17–28.

Yang, B. (1984) The identification of 15 root-knot nematode populations. *Acta Phytopathologica Sinica* 11, 107–112.

Yin, K.C. and Feng, Z.X. (1981) A preliminary survey on the parasitic nematodes of agricultural crops. *Phytopathologica* 8, 111–126.

Young, L.D. (1992) Problems and strategies associated with long-term use of nematode resistant cultivars. *Journal of Nematology* 24, 228–233.

Zhang, Y. (1985) Occurrence and control of peanut root-knot disease in non-irrigated sloping fields of Zhangjiang District. *Agricultural Science of Guangdong Province* 6, 48–49.

11 Nematode Parasites of Citrus*

Larry W. Duncan
University of Florida, IFAS, Citrus Research and Education Center, 700 Experiment Station Road, Lake Alfred, FL 33850, USA

Citrus is grown in more than 125 countries in a belt within 35° latitude north or south of the equator. The major limiting factor to citrus production is a requirement that the occurrence of freezing temperatures be of very short duration. Within the family Rutaceae, the genera *Citrus* (oranges, mandarins, pomelos, grapefruit, lemons, limes and citrons), *Fortunella* (kumquats) and *Poncirus* (trifoliate oranges) contain the principal commercial species (Swingle and Reese, 1967). Citrus production worldwide exceeded 96 Mt in 2002. Approximately 68% of the world's citrus production is consumed as fresh fruits, and about 11% of total production is used in international trade (Anonymous, 2002).

Citrus spp. are naturally deep-rooted plants (Ford, 1954a,b), and optimum growth requires deep, well-drained soils because roots will not grow into or remain in saturated zones. Nevertheless, trees can be well managed in areas with high water tables if grown on beds. Citrus grows well under any rainfall regime provided that adequate soil moisture can be maintained. Irrigation of citrus is commonly practised by a variety of methods that range from orchard flooding to low-volume drip or microsprinkler systems. In areas with sporadic rainfall, the ability to manage soil moisture is critical for good production, particularly during the period when fruit are set after the first seasonal flower bloom (Sites *et al.*, 1951). There is a tendency at present in the USA and elsewhere to increase early returns by planting higher density orchards with shorter life expectancies due to such diseases as citrus blight, tristeza and greening (Hearn, 1986).

Citrus Nematodes

Numerous nematode species are associated with the citrus rhizosphere (Cohn, 1972; Duncan, 1999). Few, however, have been shown to be of economic importance. With the notable exception of *Tylenchulus semipenetrans*, most nematode species capable of damaging mature citrus tend to be regional or local problems, due either to edaphic conditions or to the natural distribution of a particular nematode. Because the aetiology of specific nematode diseases of citrus affects management recommendations, the recognized nematode pathogens are discussed completely in separate sections.

*A revision of the chapter by L.W. Duncan and E. Cohn.

Tylenchulus semipenetrans

The 'citrus nematode', *Tylenchulus semipenetrans*, is aptly named since it occurs in all citrus-producing regions of the world and limits production of citrus fruits under a wide range of environmental and edaphic conditions. In the major citrus-producing regions, various surveys estimated that the nematode infests from 24–60% (Florida and California) to as many as 70–90% (Brazil, Spain, Texas and Arizona) of commercial orchards. Similar statistics are reported worldwide (Van Gundy and Meagher, 1977; Heald and O'Bannon, 1987; Esser *et al.*, 1993; Sorribas *et al.*, 2000; de Campos *et al.*, 2002). Expansion of citrus into new citrus areas presents an important opportunity to reduce the incidence of *T. semipenetrans*. For example, the incidence of the nematode is declining in Florida citrus as orchards are relocated southward to avoid freeze damage to trees. New orchards are planted in non-infested soil with trees certified free of the nematode. The widespread use of nematode-resistant rootstocks in older orchards is also reducing the economic importance of this nematode in Florida (Duncan *et al.*, 1994b).

Tylenchulus semipenetrans was first detected on citrus roots in California in 1912 and named and described during the next 2 years (Cobb, 1913, 1914). The nematode causes the disease 'slow decline' of citrus. The primary effect of *T. semipenetrans* in newly infested sites is a gradual reduction in tree quality so that over a period of years infested trees are smaller, less vigorous and less productive than normal. The name 'slow decline' is less appropriate when young trees are replanted into heavily infested soil where effects on tree growth may be noted soon after planting.

Symptoms

Symptom development depends on overall orchard conditions. Infested trees growing under otherwise optimum conditions may yield somewhat less fruit while appearing quite healthy. As conditions become less suitable for tree growth, the effects of citrus nematode parasitism are more apparent (Van Gundy and Martin, 1961; Van Gundy *et al.*, 1964; Heald and O'Bannon, 1987). In new citrus plantings, symptom development progresses slowly as nematode populations develop to high levels (Cohn, 1965b). Symptoms are those associated with poor root development. Leaves are smaller and may become chlorotic. In saline conditions, excessive sodium may accumulate in leaves (Mashela and Nthangeni, 2002). Wilting occurs earlier during periods of water stress, and leaf drop is more pronounced, producing exposed branch terminals.

Heavily infected feeder roots are slightly thicker than healthy roots and have a dirty appearance due to soil particles that adhere to gelatinous egg masses on the root surface (Fig. 11.1). Symptoms may not be apparent on lightly infected root systems so that infected nursery stock may easily go undetected. Feeder roots decay faster due to loss of integrity at the epidermis and at feeding sites in the cortex, resulting in invasion by secondary organisms (Schneider and Baines, 1964; Cohn, 1965b; Hamid *et al.*, 1988). This may be expressed as lesions on lightly infected roots, while heavy infections result in cortical sloughing and root death.

Biology and ecology

The biology of *T. semipenetrans* is described in Chapter 2. The life cycle is regulated by host phenology interacting with geographic variation and temporal changes in the soil environment. Most studies report one (Prasad and Chawla, 1966; Bello *et al.*, 1986; Sorribas *et al.*, 2000), two (Vilardebo, 1964; O'Bannon *et al.*, 1972; Salem, 1980; Baghel and Bhatti, 1982; Duncan *et al.*, 1993; Al Hinai and Mani, 1998; Sorribas *et al.*, 2000; Galeano, 2002) or three (Hamid *et al.*, 1988) distinct periods of active population growth per year, although no seasonality was evident during a survey in Israel (Cohn, 1966). When conditions are otherwise favourable, populations will increase between temperatures of 20 and 31°C, with maximum

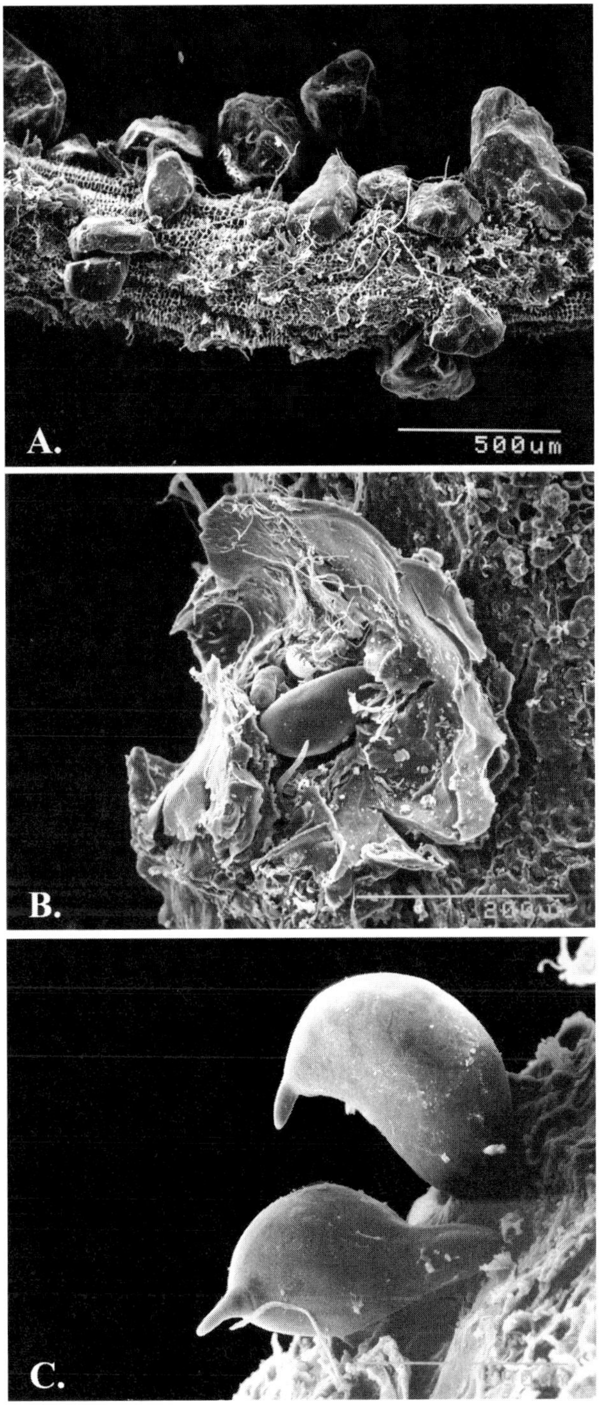

Fig. 11.1. Scanning electron micrographs of *Tylenchulus semipenetrans* on citrus roots. Sand grains adhere to the gelatinous matrix of the egg mass giving the roots a dirty appearance (A); when sand is gently removed, the gelatinous matrix (desiccated from the fixation process) is seen surrounding the female, eggs and hatched juveniles (B); removal of the egg mass reveals the posterior ends of two females (C).

development at 25°C and very slow development at the extremes (O'Bannon *et al.*, 1966). Low winter temperatures frequently regulate the population growth of the nematode (Duncan *et al.*, 1993). Summer soil temperatures in places such as Egypt, Texas, Oman and Spain approach the upper limit of this range and often correspond to population decline of the nematode (Salem, 1980; Davis, 1984; Al Hinai and Mani, 1998; Sorribas *et al.*, 2000). Similarly, in Arizona and Florida, population growth was slow on young trees until canopies developed sufficiently to shade the soil and provide optimum soil temperatures (Reynolds and O'Bannon, 1963a).

Soil moisture is often inversely related to population growth of *T. semipenetrans* (Duncan *et al.*, 1993; Sorribas *et al.*, 2000; Galeano, 2002) even though, compared with many plant parasitic nematodes, *T. semipenetrans* has little capacity for anhydrobiotic survival and nematode numbers decline quickly when trees become drought stressed (Van Gundy and Martin, 1961; Van Gundy *et al.*, 1964; Tsai and Van Gundy, 1988). Nevertheless, populations in extremely dry parts of the rhizosphere can either grow rapidly or decline precipitously, depending on whether part or all of the root system is affected by drought (Fig. 11.2). Hydraulic lift of water deep in soil to drier surface soil horizons via the root xylem (Caldwell *et al.*, 1991) creates an environment highly favourable for population growth of *T. semipenetrans* (Duncan and El-Morshedy, 1996). It is not known whether this is due to increased oxygen (Van Gundy *et al.*, 1962), passive movement of nematodes deeper in soil with precipitation, increased activity of natural enemies or other factors (Sorribas *et al.*, 2000). *T. semipenetrans* may have experienced less selection pressure for anhydrobiotic survival through co-evolution exclusively with deep-rooted woody perennials. The potential importance of hydraulic lift for *T. semipenetrans* is consistent with reports that peak population densities of this nematode in subtropical regions tend to be bimodal, occurring in the dry months that precede and follow the summer rainy season (Toung, 1963; Prasad and Chawla, 1966; O'Bannon *et al.*, 1972) and the observation in Mediterranean climates that higher population densities tended to occur under drip compared with flood irrigation (Sorribas *et al.*, 2000).

The phenology of citrus growth and development also affects population growth of *T. semipenetrans*. When soil temperature and moisture are not limiting, fibrous root growth alternates with growth of new leaves. Flushes of new fibrous roots permit increased population growth on young roots that are most suitable for penetration and development of *T. semipenetrans* (Cohn, 1964; O'Bannon *et al.*, 1972). In California, three annual flushes of root growth corresponded to three distinct peaks of numbers of *T. semipenetrans* females (Hamid *et al.*, 1988). The effect of the nematode on the normal pattern of root growth was demonstrated by reducing the nematode populations with oxamyl. Trees heavily infected by nematodes initiated 66% more new roots, but the root mass was reduced by 30%, due to the demand by the nematode for carbohydrates (Hamid *et al.*, 1988). The amount of carbohydrate available to nematodes is seasonal, decreasing markedly during the summer. Starch is an important nutrient for *T. semipenetrans* (Cohn, 1965a; Plate 12A), and the concentrations of starch and some sugars in fibrous roots were highly correlated with seasonal population density of the nematode (Duncan and Eissenstat, 1993; Duncan *et al.*, 1993). The concentrations of phenolic and lignin secondary compounds in citrus roots also vary seasonally and have been shown to be inversely related to *T. semipenetrans* population growth (Van Gundy and Kirkpatrick, 1964; Duncan *et al.*, 1993).

T. semipenetrans is broadly adapted to most edaphic conditions common to citriculture. The nematode will survive in any soil whose texture is suitable for citrus, although, unlike many nematode parasites, development in pot studies is often less rapid in sandy soils. Moderate amounts of clay and silt (Van Gundy *et al.*, 1964; Davide, 1971; Bello *et al.*, 1986) and

Fig. 11.2. The effect of soil water potential and hydraulic lift on population growth of *Tylenchulus semipenetrans*. Citrus seedlings were grown in double vertical pots (photograph) in which the top pot was infested with nematodes. Three treatments consisted of irrigating both pots (no drought), only the bottom pot (local drought) or neither pot (uniform drought). Hydraulic lift of water from the lower to the upper pot could occur only under the local drought treatment. Despite similar soil water potential under uniform and local drought in the upper pot (top graph panel), nematode population growth (lower two panels) was favoured by dry soil combined with hydraulic lift. (Redrawn from Duncan and El-Morshedy, 1996.)

organic matter (O'Bannon, 1968) favour infection and development. Populations develop best at pH 6.0–8.0; however, at less optimum pH, the nematode is also pathogenic to citrus (Van Gundy and Martin, 1962; Bello *et al.*, 1986; El-Borai *et al.*, 2003). Although *T. semipenetrans* population growth is not favoured by saline soil solutions (Kirkpatrick and Van Gundy, 1966), population density and damage to trees are often very high in orchards irrigated with saline water (Machmer, 1958; Cohn *et al.*, 1965). Mashela *et al.* (1992a,b) demonstrated that both resistant and susceptible citrus seedlings exposed temporarily to salinity and then grown under non-saline conditions were predisposed to higher nematode reproduction and suffered greater nematode damage than seedlings grown without prior exposure to salinity. Similar conditions occur during the rainy season in orchards irrigated with salinized water during the dry season. Increased arginine synthesis leading to a reduction in phenylalanine ammonia lyase was demonstrated in salt-stressed citrus plants and may result in fewer phenolic compounds for defence against nematodes (Dunn *et al.*, 1998). *T. semipenetrans*-infected trees accumulated higher concentrations of Na and Cl in leaves, reduced concentrations in roots, and experienced greater nutrient deficiencies (particularly K) in both leaves and roots than did non-infected trees under salinity (Van Gundy and Martin, 1961; Milne and Willers, 1979; Mashela and Nthangeni, 2002). A demonstrated increased rate of carbon flow to nematode-infected roots is consistent with an osmotic-based mechanism proposed to explain the variable affect of the nematode on concentrations of different elements in roots and leaves (Mashela and Nthangeni, 2002).

Reproductive rates of different biotypes of the nematode obviously vary with rootstock (O'Bannon and Hutchinson, 1974). Even on susceptible commercial rootstocks, reproduction rates may differ considerably (Davide, 1971; O'Bannon *et al.*, 1972). While the scion is reported not to influence resistance or susceptibility of a rootstock, it does influence the general quality of the root system in terms of nematode development (Kirkpatrick and Van Gundy, 1966; Bello *et al.*, 1986). Nematode morphology is also affected to some degree by the host species of citrus (Das and Mukhopadhyaya, 1985). Tree nutrition influences population levels (Martin and Van Gundy, 1963; Mangat and Sharma, 1981).

Biotypes and rootstock resistance

Physiological races or biotypes of *T. semipenetrans* exist based on host suitability (Baines *et al.*, 1969a,b). Three biotypes are commonly recognized (Inserra *et al.*, 1980; Gottlieb *et al.*, 1986; Verdejo-Lucas *et al.*, 1997). A 'Citrus' biotype was described from populations found throughout the US citrus-growing regions and Italy. It reproduces poorly on *P. trifoliata* but will reproduce on *Citrus* spp. and on the hybrids 'Carrizo' and 'Troyer' citrange as well as on olive (*Olea europeae*), grape (*Vitis vinifera*) and persimmon (*Diospyros* spp.). The 'Poncirus' biotype, found in California, reproduces on most citrus including *P. trifoliata*, and on grape, but not olive. A 'Mediterranean' biotype is similar to the 'Citrus' biotype, except that it does not reproduce on olive. It is found throughout the Mediterranean region, South Africa and perhaps India. Populations of a reported 'Grass' biotype that do not infect citrus have since been assigned to the species *Tylenchulus graminis* and *T. palustris* (Inserra *et al.*, 1988).

Since the biotypes vary by geographic region, so do suitably resistant cultivars. Within citrus, a number of cultivars of *P. trifoliata* are resistant to most populations of *T. semipenetrans* (Inserra *et al.*, 1988; Verdejo-Lucas and Kaplan, 2002). Resistant hybrids of *P. trifoliata* also provide acceptable rootstocks in some regions (Gottlieb *et al.*, 1986; Spiegel-Roy *et al.*, 1988; Verdejo-Lucas *et al.*, 2000). Swingle citrumelo (*C. paradisi* × *P. trifoliata*) is a commercially acceptable rootstock with a high degree of resistance to most populations of *T. semipenetrans*. It is also resistant to tristeza virus and tolerant of *Phytophthora nicotianae*, and is widely planted in Florida; however, it is intolerant of calcareous soils. Several hybrids of *P. trifoliata* × various mandarin (*C. reticulata*) rootstocks have inherited high resistance to *T. semipenetrans*, grow well in calcareous soils and are being evaluated for use in Spain (Verdejo-Lucas *et al.*, 2003). Selections of Poorman orange (*Citrus* × hybrid of undetermined origin) × *P. trifoliata* hybrids exhibiting combined resistance to *Phytophthora citrophthora* and tristeza were found to be highly resistant to more than one biotype of the nematode (Gottlieb *et al.*, 1986; Spiegel-Roy *et al.*, 1988). *Severinia buxifolia* is a citrus relative with a high degree of resistance to the citrus nematode which could become a source of germplasm in intergeneric breeding programmes.

Factors identified as responsible for resistance of citrus to *T. semipenetrans* population development include host cell hypersensitivity, wound periderm formation, compounds in root tissues which are toxic to the nematode, and unidentified factors which result in low rhizoplane nematode levels early during the infection process (Van Gundy and Kirkpatrick, 1964; Kaplan and O'Bannon, 1981; Galeano *et al.*, 2003), decreased female fecundity and a higher proportion of males (Verdejo-Lucas *et al.*, 2000). Resistance inherited from *P. trifoliata* is thought to be a dominant and oligogenic trait (Hutchinson, 1985). Eleven random amplified polymorphic DNA (RAPD) markers associated with resistance have been reported and are being evaluated to facilitate identification of resistance in breeding programmes (Ling *et al.*, 2000).

Other hosts

In general, the citrus nematode has a narrow range of host genera. Although 75 rutaceous species (mainly citrus and citrus hybrids) support the nematode, only a few non-rutaceous hosts have been identified, the most important of which are grape, olive and persimmon.

Disease interactions

Although early studies revealed no measurable disease synergism between *Fusarium solani* and *T. semipenetrans* when co-introduced in soil with citrus seedlings (Van Gundy and Tsao, 1963), subsequent work suggested that the nematode may increase the pathogenicity of this fungus (O'Bannon *et al.*, 1967; Labuschagne *et al.*, 1989; Walker and Morey, 1999). *P. nicotianae* is a more virulent pathogen of citrus roots than *F. solani* and frequently occurs in combination with *T. semipenetrans*. Levels of *P. nicotianae* in soil increased in a field trial when *T. semipenetrans* were controlled with nematicides (Graham and Duncan, 1997). Subsequently it was shown that pre-infection of citrus roots by *T. semipenetrans* can reduce the rate of infection by *P. nicotianae* (El-Borai *et al.*, 2002). Additional study of this unusual nematode–disease relationship is warranted to determine its significance for management recommendations.

Economic importance and population damage threshold levels

Although *T. semipenetrans* influences citrus yields differently under various circumstances, guidelines have been published to help interpret soil sample results. It was estimated in California that soil stages (juveniles/100 g of soil) below 800 represent a non-damaging population level (Garabedian *et al.*, 1984). Orchards with levels greater then 1600 may respond economically to nematicide treatment, and at levels above 3600 treatments may improve yield substantially. Populations were estimated during the peak growth period of May–July. Females per g of root are also used in California to define damage levels, with counts of less than 300, more than 700 and more than 1400 representing low, moderate and high ranges, respectively. The threshold was approximately 850 juveniles/100 cm^3 of soil when populations were measured during periods of low population development. Grapefruit yields in Texas orchards, some of which were treated with nematicides, were according to the equation:

$$\text{Yield} = 160.3\ e^{-.0000429x}$$

where yield is kg/tree and x = nematodes/100 cm^3 of soil (Timmer and Davis, 1982). Factors important in determining threshold levels are discussed in the sections on methods of diagnosis below.

When citrus is sold on the fresh fruit market, larger size fruit obtain premium prices. Because *T. semipenetrans* often reduces fruit size, the nematode can be of greater economic importance in orchards where the fruit is marketed fresh rather than for processing (Philis, 1989; McClure and Schmitt, 1996).

Methods of diagnosis

SAMPLING AND EXTRACTION. Because nematodes are aggregated in soil and along roots, sample size can be reduced by sampling during seasons of peak population size and in locations of highest feeder root and nematode concentration (Nigh, 1981a; Duncan, 1986). Stratification of orchards into areas of healthy and unhealthy trees may also improve sample precision (Scotto la Massèse, 1980). Seasonal variation in numbers of nematode life stages in the soil and roots are in the order of three- to tenfold (Salem, 1980; Baghel and Bhatti, 1982; Duncan *et al.*, 1993; Sorribas *et al.*, 2000). Thus, for comparative purposes, it is helpful to sample during the same season each year, preferably when peak populations are attained. Similarly, feeder roots and nematodes are more abundant beneath the tree canopy than at the dripline or in rows between trees (Nigh, 1981b; Davis, 1985; Duncan, 1986). Low volume irrigation systems concentrate root and nematode populations even further in the wetted zones.

Most published work on sample size indicates that accurate estimation of the population level of *T. semipenetrans* is costly. Five samples, each consisting of 12 cores (2.5 × 30 cm) of soil, were required to estimate population levels to within 20% of the true mean in a Texas grapefruit orchard (Davis, 1984). In Florida, between 30 and 75 cores were necessary to estimate population levels in 2 ha areas of various orchards within 40% of the true mean (McSorley and Parrado, 1982b; Duncan *et al.*, 1989, 1994a). Despite its low precision, sampling is valuable since the majority of population estimates are well above or below management threshold levels. Some laboratories suggest that samples be obtained to a depth of at least 60 cm (Van Gundy, 1984) although, in a study conducted in a shallow-rooted citrus orchard, the population levels in the first 30 cm of soil were used to predict the population level in the first 60 cm of the soil horizon (Duncan, 1986). Fibrous root mass density and density of root stages of the nematode can also be obtained from soil samples. For a given sample size, sample precision for root stages of the nematode is less than that for soil stages (Duncan *et al.*, 1993).

Laboratories frequently determine infestation levels as nematodes per unit soil weight or volume. Juveniles and males of *T. semipenetrans* can be separated from soil by most conventional methods. Unfortunately, extraction efficiencies are rarely reported and so it is often difficult to make direct comparisons between laboratories. For some soils, techniques based on Baermann funnel principles appear to be similar in efficiency to techniques employing density flotation if the layer of soil extracted is relatively thin (Nigh, 1981b; McSorley and Parrado, 1982a). However, other authors report major differences in efficiency of the two approaches (Galeano, 2002). A disadvantage to quantifying soil stages is that a given population level may represent a different parasitic burden depending on the root mass density of the tree (Scotto La Massèse, 1980; Duncan, 1986). Nematodes hatching from root samples are easily obtained (Young, 1954; Cohn *et al.*, 1965; McClure and Schmitt, 1996) and females per unit root can also be determined by extraction (Baines *et al.*, 1969b; Duncan *et al.*, 1993) or direct counts on stained roots (Davis and Wilhite, 1985).

ESTIMATING CROP LOSS. Economic loss assessment in mature, perennial crops is complicated by the fact that the difference in yields between nematode-infested and non-infested trees is due to long-term, cumulative stress. The nematodes on the root system affect the current crop; however, infested trees may also be smaller and less healthy due to previous effects of parasitism. As trees decline, they tend to support fewer nematodes. Other soil-borne factors frequently contribute to tree decline in addition to nematodes such as *Phytophthora* spp., salinity, poor soil drainage, drought and nutrient deficiency. Moreover, if stresses are removed, citrus trees allocate carbohydrate to vegetative growth before fruit growth (Eissenstat and Duncan, 1992). Thus, yields may (McClure and Schmitt, 1996) or may not increase in the first year following nematode management (Duncan, 1989; Le Roux *et al.*, 1991).

Two approaches have been widely employed for citrus nematode crop loss assessment: (i) nematode populations have been reduced with nematicides and subsequent yields monitored; and (ii) the relationship between nematode infestations and yields has been examined. Both approaches have limitations. It is evident from the bulk of reported evidence that the citrus nematode can reduce tree health and fruit yield and quality, but it is often not clear to what extent other factors may have influenced the results of these studies. When orchards are treated with nematicides, rhizosphere organisms in addition to nematodes are affected (Baines *et al.*, 1962, 1966; Mankau, 1968; Milne and du Toit, 1976; O'Bannon and Nemec, 1978). In the case of systemic chemicals, above-ground pests and other fauna associated with the tree may also be affected (Milne and De Villiers, 1977; Childers *et al.*, 1987). Chemical treatments may also directly affect plant development negatively (Cohn *et al.*, 1968; Timmer, 1977) or positively (Wheaton *et al.*, 1985). Similarly, relating crop yields to nematode infestation levels can be confounded by unmeasured edaphic variables that affect both nematode and tree. There are no reports of experiments comparing the growth and yield of citrus trees in the field that are randomly inoculated with *T. semipenetrans*. Such experiments would provide important information and are feasible because the nematode moves very slowly in an orchard if unaided by flowing water or other cultural practices (Duncan *et al.*, 1995).

Experiments in which nematicide treatments resulted in significant citrus yield increases have been widely reported (Baines, 1964; Yokoo, 1964; Cohn *et al.*, 1965; Oteifa *et al.*, 1965; Philis, 1969; O'Bannon and Tarjan, 1973; Vilardebo *et al.*, 1975; Davide and Dela Rose, 1976; Milne and Willers, 1979; Timmer and Davis, 1982; Childers *et al.*, 1987; Duncan, 1989; Le Roux *et al.*, 1991, 1998). Treatment responses in these and other experiments ranged from none to several hundred per cent increase in fruit from treated trees. Although yield response to nematicide treatment is often positive, results have been erratic. Good yield responses have been measured following treatments which did not reduce population levels (Davis *et al.*, 1982) and, in some cases, consistent, strong reduction of populations has not resulted in measurable yield response (Davis and Wilhite, 1985; Stirling and Wachtel, 1985). Such results indicate that we do not adequately understand the effects of some nematicide treatments, the damaging level of *T. semipenetrans* or the interaction of the nematode with other debilitating factors. On average, the reported yield increase in response to nematicide treatment has been of the order of 15–30%.

Studies relating tree quality and yield to nematode infestation level report similar findings. Under uniform soil conditions within orchards (Reynolds and O'Bannon, 1963b; Scotto la Massèse, 1980; Coelho *et al.*, 1983) or considering specific varieties between orchards (Davide, 1971), the highest levels of soil stages of *T. semipenetrans* frequently were measured beneath trees with only moderate symptoms. Healthy trees supported smaller populations that had not yet caused significant damage, while the reduced root systems of severe decline trees were incapable of supporting high nematode populations. In Israel, the average tree quality index declined with nematode infestation level beyond a specific threshold level (40,000 nematodes/g of root weight) (Cohn *et al.*, 1965). Citrus fruit yield has also been negatively correlated with infestation level (Willers, 1979; Timmer and Davis, 1982; Childers *et al.*, 1987). A Florida orchard was identified in which randomly distributed trees were infested or not infested by *T. semipenetrans*. Average soil texture, levels of salinity and nutrients, density of *P. nicotianae* and tree decline symptoms did not differ for infested or non-infested trees. However, leaf area, fibrous root mass density and fruit yield of infested trees were 32, 8 and 22% lower, respectively, than those from non-infested trees (Duncan *et al.*, 1995).

Management measures

Methods commonly employed to control *T. semipenetrans* depend on local conditions and focus on: (i) excluding the pest; (ii) minimizing losses through crop management; and (iii) reducing population levels of the pest.

EXCLUSION. Most citrus-growing regions have few serious nematode pests so that exclusion of *T. semipenetrans* from orchards is a realistic goal to preclude the perennial expense of nematode management. Occasional introductions of *T. semipenetrans* into non-infested orchards do not negate the value of a conscientious sanitation programme, since the nematode migrates very slowly by its own power (Meagher, 1967; Tarjan, 1971; Baines, 1974; Duncan *et al.*, 1995). In the absence of flooding and particularly with the use of low volume irrigation, trees may remain uninfected for long periods, despite the existence of nematodes on adjacent trees. Exclusion of *T. semipenetrans* is relatively simple in most newly planted orchards and in non-infested existing orchards. Since the host range of the nematode is limited to only a few non-rutaceous plant species, infestation usually results from movement of infected planting stock (Van Gundy and Meagher, 1977) or from contaminated equipment (Tarjan, 1956). Programmes to approve and monitor nursery sites and certify that nursery stock is nematode free have been highly effective in limiting the distribution of *T. semipenetrans* (Milne, 1982; Lehman, 1995). The Florida nursery certification programme was estimated to have saved growers US$33 million in 1994 by reducing yield losses from *T. semipenetrans* that would have otherwise occurred from the spread of this nematode (Lehman, 1995). Such programmes focus on: (i) continuous monitoring through soil sampling; (ii) isolating nursery locations to avoid runoff water from infested orchards; and (iii) security to prevent contaminated planting media or equipment from entering the nursery area. Separate equipment for use in infested and non-infested orchards may be feasible in some cases, otherwise equipment must be disinfested continually prior to movement into non-infested orchards (Esser, 1984). Irrigation with some forms of surface water such as canals and rivers has been found to represent a serious source of inter-orchard contamination by *T. semipenetrans* and *P. nicotianae* (Cohn *et al.*, 1976), particularly since pests can be widely spread in a short time. Irrigation water can be decontaminated through the use of settling ponds and filtration systems, but the procedures require careful maintenance (Cohn, 1976).

CROP MANAGEMENT. A key concept for successful management of *T. semipenetrans* is that of the limiting factor (Thomason and Caswell, 1987). Vigorous orchards in which nematode population densities exceed management thresholds are those in which nematode management is most likely to be profitable. Although citrus nematode may sometimes exacerbate damage caused by other stresses (Labuschagne and Kotze, 1988; Mashela and Nthangeni, 2002), citrus trees that are damaged by *Phytophthora* spp., poor drainage, salinity, frequent drought or other problems are unlikely to respond consistently to management of just *T. semipenetrans*. Therefore, it is important to ensure that orchards are managed properly in all respects, before investing in nematode management tactics.

DIRECT MANAGEMENT OF NEMATODE POPULATIONS. Direct suppression of citrus nematode populations relies on the use of resistant rootstocks, nematicidal chemicals, or physical methods such as solarization. The commercially acceptable resistant rootstock Swingle citrumelo is now widely planted in Florida and, combined with nursery certification, has appreciably reduced the occurrence of *T. semipenetrans* (Lehman, 1995). Resistance management appears to be an important consideration, because the Poncirus biotype occurs in regions with widespread use of *P. trifoliata* rootstocks (Baines *et al.*, 1969b). Resistance-breaking biotypes were detected on Swingle citrumelo in a Florida nursery (Duncan *et al.*, 1994b). When resistant rootstocks are used to replant an entire orchard, they are challenged only by nematodes that remain from the previous trees. However, if resistant rootstocks are used to replace unthrifty individual trees in orchards with susceptible rootstocks, the opportunity to break resistance increases because the resistant rootstock is challenged continuously by nematodes supported by the adjacent susceptible rootstocks (Duncan *et al.*, 1994b; Verdejo-Lucas *et al.*, 2003).

Nematicides are broadly classified by whether they are used prior to, or following, planting. The most effective pre-plant nematicides in citrus are fumigants such as methyl bromide, metam sodium and 1,3-dichloropropene. Previously, dibromochloropropane (DBCP) was widely used to control citrus nematodes until it was banned in most countries for health and environmental reasons, and methyl bromide is being phased out for the same reason. The fumigants act directly on nematodes as contact poisons. Pre-plant fumigation of old orchard sites with histories of citrus nematode infestation can be important to prevent the rapid infection of young trees (Baines *et al.*, 1956, 1966; O'Bannon and Tarjan, 1973; Le Roux *et al.*, 1998). Citrus nematodes are well adapted to survive in the absence of plants (Cohn, 1966; Van Gundy *et al.*, 1967) and have been detected in fields for as long as 9 years after the removal of citrus (Baines *et al.*, 1962; Hannon, 1964). Sorribas *et al.* (2003) demonstrated that trees on resistant rootstocks grow more quickly than on susceptible rootstocks in non-fumigated soil infested by *T. semipenetrans*, but not in soil fumigated with 1,3-dichloropropene. Net income from increased yield during years 4–8 after planting was 46–101% higher in plots fumigated with methyl bromide in South Africa (Le Roux *et al.*, 1998). Fumigants can also adversely affect young tree growth under some conditions (Cohn *et al.*, 1968; Milne, 1974). It is important to observe proper intervals between treatment and planting to avoid phytotoxicity. In nurseries which experience frequent or very thorough fumigation, mycorrhizal fungi may be nearly eradicated (O'Bannon and Nemec, 1978). To avoid phosphorus deficiency, replanted nursery stock should be mycorrhizal or seedbeds should be reinoculated with endomycorrhizal fungi. This problem is seldom encountered when replanting orchards since plants in fumigated sites are quickly invaded by fungi from adjacent soil if they are not mycorrhizal at the time of transplanting (Graham, 1988).

Pre-plant solarization of soil can also be highly beneficial to subsequent growth of citrus, but the reasons are unresolved. Increased tree growth and yield in response to solarization in South Africa are more likely to have resulted from early control of *P. nicotianae* than control of the citrus nematode (Cronje *et al.*, 2002). Indeed, the increased growth of trees due to solarization resulted in generally higher numbers of *T. semipenetrans* in solarized plots as long as 10 years following solarization.

Post-plant nematicides in citrus are generally carbamate or organophosphate, acetylcholinesterase inhibitors. Some post-plant citrus nematicides are translocated systemically within the tree and suppress insects and mites (both pest and beneficial species) in addition to nematodes. Thus, like many pesticides, some of these nematicides have the potential to disrupt biological control in the canopy of the tree. Oxamyl and fenamiphos have basipetal movement from the point of application, which provides a higher level of nematode control in the deeper soil profiles, compared with aldicarb which moves upward into the canopy (O'Bannon and Tarjan, 1979). All of the nematicides used in citrus are incorporated in the soil either mechanically or with irrigation for efficacy and safety. They are inappropriate for small farms that lack proper, safe application equipment.

Nematicide placement, application timing and application history are important considerations. Nematicides in large commercial citrus orchards are applied in bands down the tree rows or through low volume irrigation systems. Since the abundance of nematodes and feeder roots in the upper soil horizons declines quickly with distance from the trunk, nematicide bands, even for systemic products, are most effective when they are applied as much as possible beneath the tree canopy (Nigh, 1981a; Duncan, 1986, 1989). Applications through low volume irrigation systems deliver nematicides to areas of highest root and nematode abundance. Where population levels and root growth are seasonal, treatment should precede periods when nematodes actively invade new roots (Hamid *et al.*, 1988). Splitting the maximum allowable nematicide dose for multiple applications within a season can markedly increase efficacy. The life cycle of the citrus nematode was disrupted by three applications of cadusaphos, made at 60 day intervals, to the extent that nematodes were not detected on roots or in the soil for up to 4 years (Le Roux, 1995; McClure and Schmitt, 1996). Although less effective than cadusaphos, other nematicides also perform best on such a schedule. However, the profitability of nematicide use cannot be assessed from studies of 2 or 3 years duration, because continuous use can reduce the effectiveness of nematicides as a result of accelerated microbial degradation (Smelt *et al.*, 1996; Johnson, 1998).

Consideration of possible environmental effects should be part of a decision on whether to treat the soil with nematicides. As a class of pesticides, nematicides have been heavily restricted in recent years due to environmental contamination and possible health effects. Under certain conditions of soil type, precipitation rate and water table level, the potential for groundwater contamination exists for most chemicals that are applied to the soil. The treatment of nematode pests in citrus orchards resulted in contamination of large numbers of drinking water wells with several pesticides, some of which have subsequently been banned for use in the USA and elsewhere (Kaplan, 1988).

Additional nematode parasites of citrus

Nematodes other than *T. semipenetrans* currently known to be capable of damaging citrus tend to be very limited in distribution. Accordingly, with the exception of burrowing nematodes, considerably less is known about the relationship between other nematode species and citrus. Both migratory endoparasites (lesion and burrowing nematodes) and sedentary endoparasites (root knot nematodes), as well as a number of species of ectoparasitic nematodes can damage citrus. Additionally, there are nematode species commonly found in the citrus rhizosphere for which insufficient information exists to determine their pathogenic potential.

Radopholus similis and *R. citri*

Spreading decline is a severe disease of citrus caused by the citrus race of *Radopholus similis* that is only encountered on Florida's central ridge of deep sandy soils. *R. similis* is commonly called the burrowing nematode because of its extensive tunnelling through root tissue as a migratory endoparasite. The disease was first described in 1928 and the causal organism was identified in 1953 (Suit and DuCharme, 1953). The name of the disease is descriptive of the rapid progression of decline in infested groves which can reach 15 m/year. The citrus race of *R. similis* also parasitizes banana, but is distinct from the more widespread banana race for which citrus is not a host (DuCharme and Birchfield, 1956).

In 1984, the citrus race of *R. similis* was renamed *R. citrophilus* and designated as a sibling species to *R. similis*. The taxonomic change was based on putative differences in chromosome number, isozyme patterns, mating behaviour, host preference (Huettel *et al.*, 1984) and later morphology (Huettel and Yaegashi, 1988). Subsequent research by several laboratories failed to confirm the previous work and provided convincing evidence based on karyotype identity, morphological and genetic similarity and reproductive compatibility that *R. citrophilus* is a junior synonym of *R. similis* (Kaplan and Opperman, 1997, 2000; Kaplan *et al.*, 1997, 2000; Valette *et al.*, 1998; Elbadri *et al.*, 2002). An interesting aspect of the recent work on *Radopholus* systematics is that, compared with many nematode species, little intraspecific variation exists in the DNA sequences of genomic regions useful in taxonomy. The genetic similarity among *R. similis* populations worldwide may result from its wide host range combined with its recent dissemination worldwide on banana from its centre of origin somewhere in Australasia (Kaplan, 1994b; Fallas *et al.*, 1996; Machon and Bridge, 1996; Marin *et al.*, 1998).

R. citri was discovered in citrus roots in Indonesia (Bridge *et al.*, 1990; Hahn *et al.*, 1994; Machon and Bridge, 1996). The pathogenicity of *R. citri* was demonstrated and the nematode is associated with declining trees in Indonesia, but its economic importance in the region has not yet been characterized.

Symptoms

Spreading decline is generally distinguishable from other major decline diseases such as citrus blight in that large contiguous groups of trees are affected and expansion of the diseased area is rapid. Forced water uptake in the trunk of the tree (Graham *et al.*, 1983) is indistinguishable from normal trees and is another rapid preliminary method to determine whether a tree may be infected with *R. similis* rather than suffering from citrus blight. Decline trees have sparse foliage, particularly high in the canopy during the early stages of symptom development. Leaves and fruit are small, and fewer mature fruit remain on trees. Branch ends are bare and eventually entire branches die. Affected trees wilt rapidly during periods of low soil moisture particularly during the periods of drought that tend to occur in the winter and spring in Florida. It is during these periods that disease progression is most rapid.

Symptoms on roots are most apparent below 25–30 cm so that evidence of damage to the abundant shallow portion of the root system may be lacking (Ford, 1952, 1953). The most obvious symptom in the root system is the reduction in the quantity of feeder roots in the deeper soil profiles. At depths of 25–50 cm, 75% of the root system may remain, but below this level the root system is almost totally destroyed. Since mature citrus growing on the deep sands of the ridge may establish as much as half of the feeder roots between 1 and 6 m, destruction of the deep root system on a large tree accounts for the drought-related above-ground symptoms during periods of moisture stress. Infected feeder roots develop dark lesions at the points of nematode entry and activity which expand and coalesce as secondary pathogens destroy these tissues. Nematodes may burrow in a section of root for several weeks, completely destroying the phloem and much of the cortex, girdling the central cylinder (DuCharme, 1959; Plate 12B). On larger roots, the lesions can form callused margins (Feder and Feldmesser, 1956). The nematode penetrates the region of elongation and root tips can become swollen due to hyperplasia and stubby if terminals are penetrated (Feder and Feldmesser, 1956; DuCharme, 1959, 1968).

Biology

R. similis on citrus has a life cycle of 18–20 days under optimum conditions (DuCharme and Price, 1966), permitting population levels to increase rapidly when conditions are favourable (DuCharme and Suit, 1967). Following root penetration, mature females begin to lay eggs at an average rate of nearly two per day and eggs hatch in 2–3 days. In gnotobiotic culture, colonies initiated with single females attained average population levels of more than 11,000 individuals in less than 3 months, although rhizosphere competitors restrict population growth in orchards far below such a level (DuCharme and Price, 1966). The nematodes normally reproduce sexually; however, females that do not mate after a period of time reproduce as hermaphrodites (Brooks and Perry, 1962; Kaplan and Opperman, 2000). Mature males do not feed and comprise 0–40% of the population, averaging about 10% (DuCharme and Price, 1966). The nematode remains within the root until forced by overcrowding and decay to migrate.

Survival and means of dissemination

R. similis does not survive for long periods in the absence of host roots (DuCharme, 1955). In field trials in which root material was excluded, the nematode could not be detected in samples after 6 months (Tarjan, 1961). However, under more natural experimental conditions, the nematode has been detected up to 14 months under bare fallow conditions (Hannon, 1963), and unconfirmed reports suggest as long as 2 years (Suit *et al.*, 1967). Large root fragments that remain buried in soil after tree removal may help support populations during fallow.

The nematode is spread in contaminated rootstock (Poucher *et al.*, 1967), machinery (Tarjan, 1956), subsoil water (DuCharme, 1955) and it migrates rapidly along developing root systems. In orchards, the spreading decline disease is reported to move as much as 15 m/year (Poucher *et al.*, 1967), while in greenhouse tests, movement of about a quarter to a third of that rate has been measured (Feldmesser *et al.*, 1960; O'Bannon and Tomerlin, 1969a; Tarjan, 1971).

Host range

R. similis is remarkably polyphagous, attacking more than 250 plants in 15 families outside of the Rutaceae (Ford *et al.*, 1960). Within the citrus and closely related genera, more than 1200 species, varieties and hybrids have been screened for resistance or tolerance to *R. similis* (Ford and Feder, 1961; O'Bannon and Ford, 1976). Three varieties of citrus, Ridge Pineapple sweet orange, Estes rough lemon and Milam lemon, and a *P. trifoliata* × citrus hybrid, Carrizo citrange, have been released as rootstocks since 1958. Although data on tolerance under field conditions are very limited, all of the rootstocks subsequently have been shown to support biotypes of *R. similis* capable of breaking resistance (Kaplan and O'Bannon, 1985). In the case of Carrizo citrange, considerable variability exists within the progeny for susceptibility to burrowing nematodes (Kaplan, 1986); however, a breeding line known as Kuharski Carrizo has been identified in which resistance appears to be stable (Kaplan, 1994a).

Environmental factors affecting parasitism

The biology of *R. similis* related to citrus is strongly influenced by edaphic conditions. The nematode is found in citrus-growing regions of Florida other than the ridge, but populations do not develop to damaging levels. This is probably related to interactions between soil temperature, moisture and root growth periodicity. The cardinal temperature for *R. similis* is 24°C, and development occurs between 12 and 32°C. Optimum temperatures occur for the longest periods each year in the deeper soil horizons where the highest level of reproduction is known to occur. Highest absolute populations in soil samples are found in the late summer–early autumn period when optimum temperatures combine with an annual cycle of root growth to support population increase. As the root growth cycle declines later in the autumn, infected roots begin to die and soil populations begin to decline even though the nematodes recovered per unit of root tend to be highest in the late autumn (DuCharme, 1967, 1969). The temperature extremes in the surface soil horizon are nearer the limits for development of *R. similis* during the period of root growth which may partly explain low population development in surface roots. The nematode does not have a known resting stage so that moisture deficits which are more commonly encountered in the shallow horizons may also inhibit development in this zone (Tarjan, 1961).

Soil texture is also an important determinant in the spreading decline disease cycle. The nematode is more pathogenic to citrus in pot studies in sandy than loamy soils (O'Bannon and Tomerlin, 1971). Movement of *R. similis* is highest in light textured soil (Tarjan, 1971).

Disease complexes

Few reports exist of interactions between *R. similis* and other rhizosphere organisms

(Feder and Feldmesser, 1961). Feldmesser *et al.* (1959) obtained indirect evidence that secondary fungal invaders play a key role in the disease complex when they treated infected seedlings with the fungicide captan which increased nematode population levels as well as root and top weights of plants. Root lesions are quickly infected by fungi and other rhizosphere inhabitants (Feder *et al.*, 1956; DuCharme, 1968). *R. similis* population levels declined in the presence of mycorrhizal fungi, probably due to enhanced phosphorus uptake because the effect was also obtained on plants growing with supplemental phosphorus (Smith and Kaplan, 1988). Similarly, citrus plant tolerance to *R. similis* appears to be enhanced by mycorrhizal infection when soils are deficient in phosphorus (O'Bannon and Tomerlin, 1971; O'Bannon and Nemec, 1979).

Biotypes

All burrowing nematode-resistant rootstocks support low numbers of *R. similis*, and populations of *R. similis* have broken resistance in Milam lemon, Ridge Pineapple, Albritton sweet orange and Kuharski Carrizo citrange rootstocks (Kaplan and O'Bannon, 1985; Kaplan, 1994b). The pathogenicity of these biotypes was established in pot studies; however, the incidence of resistance-breaking populations on resistant varieties in the field has not been investigated adequately.

Economic importance and damage threshold levels

R. similis and a lesion nematode, *Pratylenchus coffeae*, appear to be the most virulent nematode parasites of citrus worldwide (O'Bannon *et al.*, 1976). However, since *R. similis* distribution on citrus is restricted to Florida, the nematode's economic impact is slight on the world market. In 1972, it was estimated that *R. similis* caused 0.1–0.2% yield losses in the world citrus industry (Cohn, 1972). In infested orchards, the losses were estimated to be of the order of 40–70% for oranges and slightly higher for grapefruit (DuCharme, 1968). The damage by spreading decline within orchards has been mitigated in recent years by improved management practices described below. Unfortunately, the discontinuation of programmes to prevent migration of burrowing nematode from infested to uninfested orchards has increased the rate of spread of this pest.

Management

Management of spreading decline currently focuses on restricting the spread of the nematode through planting-stock certification, sanitation, proper orchard management, use of resistant rootstocks and use of nematicides. Previous practices in the USA emphasized chemical management of the nematode through state-directed efforts known as the 'push and treat' and 'buffer' programmes. Both programmes relied heavily on intensive sampling to ascertain the limits of infested areas accurately. In the push and treat programme, infested trees and a margin of uninfested trees were destroyed, the soil was treated with high rates of various halogenated hydrocarbon soil fumigants and, prior to replanting on resistant rootstocks, the soil was maintained under bare fallow for at least 6 months (Poucher *et al.*, 1967). Buffers are corridors of land 5–18 m wide created between infested and non-infested locations, in which no plants are permitted to grow. Citrus roots within the buffer zones even at great depth were killed by frequent chemical treatment at high rates (Suit and Brooks, 1957; Poucher *et al.*, 1967). The programmes were expensive (as much as US$20,000/ha in 1977) and illustrate the damage caused by this disease. Nevertheless, these programmes limited the spread of the nematode by more than 90% (O'Bannon, 1977). In 1983, both programmes were discontinued due to the discovery that the nematicides being used were contaminating local drinking water wells. Subsequent efforts to maintain barriers using methyl bromide and mechanical root pruning proved too costly (Duncan *et al.*, 1990).

Based on the potential threat of spreading decline to citrus on Florida's ridge, avoiding infestation by *R. similis* should be a high management priority. Planting stock must be certified as pest free. Nurseries are regularly sampled and inspected to remain certified. Commercial movement of soil within and into citrus-producing areas requires certification that the site of origin is pest free. A cost–benefit analysis of the value of the certification programme in reducing potential losses to burrowing nematode estimated a 14:1 return on investment resulting in increased yield worth US$40 million/year (Lehman, 1995). Equipment used in infested orchards should be reserved for that purpose when possible or disinfested between operations (Esser, 1984).

In Florida, with the exception of the ridge area, citrus is commonly grown in shallow soils that limit root development to the surface soil horizons. The fact that *R. similis* damages primarily the deeper (below 45 cm) portion of the citrus root system provides the opportunity to manage spreading decline with cultural or management practices designed to support a healthy, shallow root system. Infested orchards in which sound practices are employed have remained economically viable (Tarjan and O'Bannon, 1977), and may out-produce annual state production averages (Bryan, 1966). Practices which have been suggested include: use of herbicides and mowing rather than cultivation for weed management to avoid cutting surface roots (Tarjan and Simmons, 1966); frequent use of supplemental irrigation to provide sufficient water to the surface root system (Bryan, 1966, 1969); and use of an optimum fertility schedule, preferably through frequent fertigation to maintain nutrients in the shallow rhizosphere.

Three rootstocks are recommended for use against spreading decline, Milam lemon, Ridge Pineapple sweet orange and Kuharski Carrizo citrange. The occurrence of resistance-breaking populations of the burrowing nematode indicates a need for rootstocks with additional resistance genes (Kaplan and O'Bannon, 1985).

Systemic nematicides such as oxamyl are used by some growers to reduce *R. similis* in deeper roots and have been demonstrated to increase yield (O'Bannon and Tomerlin, 1977; O'Bannon and Tarjan, 1979).

Diagnosis and sampling

In Florida, root samples are commonly processed to ascertain whether *R. similis* is present in an orchard because the nematode is highly endoparasitic. Laboratories traditionally obtain samples to depths of 120 cm to obtain roots most likely to contain high populations of the nematode (Poucher *et al.*, 1967). The procedure requires expensive, mechanized equipment and it has since been demonstrated that processing a larger amount of roots near the soil surface (that are acquired easily and inexpensively with a shovel) can more accurately detect nematode-infected trees than processing the smaller amount of roots obtained deeper in the soil (Duncan *et al.*, 1994c). Visual stratification of orchards based on tree decline symptoms is important in sampling for *R. similis*. Random sampling is inappropriate because determination of population levels is generally not the goal of sampling for burrowing nematodes but rather delimiting an area of infestation. Intensive sampling of suspicious trees increases the chance of detecting the nematode, whose population level can be quite low during some periods.

Pratylenchus

Three species of lesion nematodes, *Pratylenchus coffeae*, *P. brachyurus* and *P. vulnus*, have been demonstrated to damage citrus. *P. coffeae* is easily the most pathogenic (Plate 12C). It is widespread, having been reported on citrus in the USA (O'Bannon *et al.*, 1972), India (Siddiqi, 1964), Japan (Yokoo and Ikegemi, 1966), Oman (Mani *et al.*, 1997) South Africa (Milne, 1982) and Taiwan (Huang and Chang, 1976). Variation among *P. coffeae* populations is receiving increased attention (Golden *et al.*, 1992; Duncan *et al.*, 1998, 1999). A lesion nematode, thought to be *P. coffeae*, was detected recently on citrus in Sao Paulo State, Brazil and found to infest about 1% of the nurseries and orchards (Campos *et al.*, 2002). The nematode in Brazil was renamed *P. jaehni* (Inserra *et al.*, 2001) and it appears to be very similar to lesion nematodes from coffee in Sao Paulo (Duncan *et al.*, 1999), although the host ranges differ (Silva and Inomoto, 2002). *P. jaehni* is associated with unthrifty citrus trees; however, its virulence on citrus and economic importance remain to be characterized. Putative *P. coffeae* associated with native vegetation in Florida, which threatened the nematode-free certification of some citrus nurseries, were found to be genetically distinct from *P. coffeae*, incapable of reproducing on citrus, and some populations are probably undescribed species (Inserra *et al.*, 1996, 1998; Duncan *et al.*, 1999).

In the USA, damage by *P. coffeae* has been observed in Florida, where the nematode has been detected in only a few groves (O'Bannon and Tarjan, 1985). In South Africa, the nematode has not been associated with economic problems (Milne, 1982) as it has in other regions where it is found. Infection occurs primarily in the feeder roots where all motile stages of the nematode penetrate cortical tissue both inter- and intracellularly. If penetration of the root tip occurs, the meristem is destroyed and lateral roots are often initiated. The nematodes can be found in vascular tissues only when localized populations are unusually high. Cortical invasion results in extensive cavities, but vascular tissues remain intact until invaded by secondary organisms.

P. coffeae appears to be obligatorily amphimictic, with males feeding in the roots and comprising 30–40% of the population (Radewald *et al.*, 1971b; Inserra *et al.*, 2001). Reproduction of *P. coffeae* is highest when soil temperatures are relatively high (26–30°C). At these temperatures, populations complete the life cycle in less than 1 month and may reach levels as high as 10,000 nematodes/g of root (O'Bannon and Tomerlin, 1969b; Radewald *et al.*, 1971a). The nematode can survive in roots in soil for at least 4 months (Radewald *et al.*, 1971a).

In pot studies, *P. coffeae* reduced root weights by as much as half and plant growth by 38% (Siddiqi, 1964; O'Bannon and Tomerlin, 1969b; Radewald *et al.*, 1971a). In the field, damage by *P. coffeae* can be severe. Growth reduction of young trees during 4 years in the field ranged from 49 to 80% depending on the rate of growth of the nematode on different rootstocks. Again, depending on the rootstock, numbers of fruits during the first bearing years ranged from threefold to 20-fold differences between infected and non-infected trees (O'Bannon and Tomerlin, 1973). Soil types ranging from sands to sandy loams did not affect the pathogenicity of *P. coffeae* to rough lemon roots (O'Bannon *et al.*, 1976). Reported migration of the nematode through soil was relatively slow, of the order 1 m/year (Tarjan, 1971; O'Bannon and Tomerlin, 1973; O'Bannon, 1980), although the rate of spread of decline symptoms in groves is greater. The limited distribution of *P. coffeae* in Florida citrus is partly due to a rootstock certification programme and may also be due to competition with the more widespread *T. semipenetrans*. In a survey within a grove, the two species appeared to be mutually exclusive although exclusion of one species by the other was not observed in experiments (Kaplan and Timmer, 1982). No commercial rootstocks resistant to the nematode are available, although some selections of a Microcitrus hybrid and perhaps of *Poncirus trifoliata* appear to have some resistance (O'Bannon and Esser, 1975).

P. brachyurus has a biology similar to that of *P. coffeae*. Although well distributed worldwide, *P. brachyurus* varies in its distribution in citrus. In Florida, the nematode was present in 90% of groves sampled (Tarjan and O'Bannon, 1969) while it has not been reported from citrus groves in South Africa, even though it is widespread in that country (Milne, 1982). It is a proven pathogen of seedlings in greenhouse trials (Brooks and Perry, 1967; Tarjan and O'Bannon, 1969; Radewald *et al.*, 1971a; Tomerlin and O'Bannon, 1974; Frederick and Tarjan, 1975), and on young trees in the field (O'Bannon *et al.*, 1974). It is generally not considered to be a problem on mature citrus, although it was suggested that other sources of plant stress such as severe drought may exacerbate damage by this species to mature trees (O'Bannon *et al.*, 1974). When populations of *P. brachyurus* in mature Valencia orange trees on rough lemon rootstock were controlled with aldicarb, trees suffered less frost damage during a severe winter and subsequent yields were increased (Wheaton *et al.*, 1985; Childers *et al.*, 1987). It is unclear, however, what other factors may have been affected by the systemic pesticide.

Like *P. coffeae*, *P. brachyurus* reproduces best at temperatures above 25°C and can affect seedling growth in coarse and medium texture soils. Movement of *P. brachyurus* through soil is not as rapid as that of *P. coffeae* (O'Bannon, 1980) and citrus is not as good a host for this nematode; populations in roots frequently are one-tenth of those of *P. coffeae* (Radewald *et al.*, 1971a).

To date, *P. vulnus* has been found associated with citrus in Italy (Inserra and Vovlas, 1974) and California (Siddiqui *et al.*, 1973), and was shown to be capable of causing severe damage to nursery seedlings (Inserra and Vovlas, 1977). As with other species of *Pratylenchus*, the nematode is pathogenic in a range of soils from sand to sandy clay loam. Biology, population growth rates and root damage are similar to those described for *P. coffeae*. Since the nematode does not appear to be widespread in citrus orchards in Italy, certification of nursery stock to be free of the pathogen has been suggested.

Belonolaimus longicaudatus

Belonolaimus longicaudatus can damage citrus by greatly reducing the fibrous root abundance of trees. Sting nematodes occur in fewer than 10% of Florida citrus orchards (Esser *et al.*, 1993), but their incidence in regions with sandy soil was estimated to be as high as 64% (Duncan *et al.*, 1996). Sting nematodes are widely distributed on a number of cultivated and non-cultivated host plants in the south-eastern USA. They are intimately associated with the citrus root system, and can be spread on infested planting stock, even when the roots are devoid of soil (Kaplan, 1985). In nurseries, relatively low populations (40 nematodes/dm^3 of soil) can cause aboveground symptoms of stunted, chlorotic plants (Kaplan, 1985). The nematode is ectoparasitic, feeding on root tips of citrus. Root systems of infested trees appear very coarse due to a reduction in the number of lateral roots and swollen fibrous roots (Plate 12D). Fibrous roots also have swellings at or near terminals as well as multiple apices. The epidermis may slough easily due to secondary infection. Histological examination has shown several meristematic zones at root tips, with tissue disorganization that includes hyperplastic tissue, cavities and extensive vascular formation. Cell disruption at the cavity borders results in cytoplasm leakage into these spaces and suggests them to be the possible site of feeding (Standifer and Perry, 1960; Kaplan, 1985).

Sting nematodes are associated with severe stunting of trees on all known rootstocks in the field (Standifer and Perry, 1960; Esser and Simpson, 1984; Kaplan, 1985; Duncan *et al.*, 1996), and cause similar symptoms in pot experiments (Standifer and Perry, 1960; Abu-Gharbieh and Perry, 1970). The economic importance of sting nematodes may be increasing due to changing cultural practices that favour maintaining a cover crop in the row middle. Twice as many orchards in which row middles were mowed were positive for sting nematode compared with orchards in which middles were cultivated for weed control. Newly planted orchards often contain patches of stunted trees (Plate 12E and F). Tree condition and yield in these orchards are inversely related to population density of sting nematodes. Growth of the stunted trees usually remains poor for several years, after which they resume normal growth. Soil water potential beneath heavily infested young trees with few roots is consistently higher than beneath lightly infested trees with dense roots and high transpiration rates. Thus, when young trees are planted in locations with high numbers of sting nematodes, roots are continually damaged until the trees manage to develop a root system dense enough to cause nematodes to move deeper into the soil due to periodic moisture deficit in surface soil (Duncan *et al.*, 1996).

Pre-plant soil fumigation and post-plant nematicide treatments have alleviated symptoms of sting nematode parasitism (Bistline *et al.*, 1967; D.T. Kaplan, USA, 1989, personal communication). Hot water treatment for 5 min at 49°C was sufficient to kill *B. longicaudatus* and has been suggested as an eradication method for bare root seedlings (Kaplan, 1985).

Meloidogyne

Root knot nematodes (*Meloidogyne* spp.) capable of attacking citrus are very limited in distribution. These nematodes are endoparasites, causing root galls. Although there have been several reports of the common species of root knot nematodes (*M. incognita*, *M. javanica* and *M. arenaria*) developing or reproducing on citrus (Minz, 1956; Den Ouden, 1965; Whitehead, 1968; Scotto la Massèse, 1969; Gill, 1971; De Brito *et al.*, 2000), they appear to be problems in only a few localized regions in China and the Far East. An apparently pathogenic species of root knot nematode was reported from Taiwan and New Delhi where it caused elongated galls on citrus roots. The nematode was given the common name 'Asiatic pyroid citrus nematode' and was found to be able to complete its life cycle on several citrus and other plant species including maize and sweet potato. Control measures suggested at the time focused on the use of a number of trap crops as cover crops since *Crotalaria* sp., strawberry, groundnut (peanut) and soybean were found to be non-hosts even though the nematode invades the roots (Chitwood and Toung, 1960). *Meloidogyne fujianensis* (Pan, 1985) and *M. oteifae* (Pan, 1984) have been reported from China on *C. reticulata*, with the former species parasitizing up to 60% of citrus trees surveyed.

A more common situation in which root knot nematodes may cause problems in citrus was reported by Van Gundy *et al.* (1959) who found that *M. incognita*, *M. javanica* and *M. arenaria* infected roots of Troyer citrange and sour orange causing small galls, but without reproducing. Galls on plants in the field were associated with unthrifty plant growth, but were found to be due to infection by populations that were supported on weed hosts. This work was later supported by that of Inserra *et al.* (1978) who observed extensive root damage due to invasion of citrus roots by *M. javanica* even though no reproduction occurred, and in Israel (Orion and Cohn, 1975) where potted citrus responded to a specialized *M. javanica* race with hypersensitivity and failure of giant cell formation. Nevertheless, the threat posed to citrus production by races of the nematode capable of reproducing on citrus was sufficient to warrant an eradication effort in California of a population of *M. javanica* found to be supported by a dooryard citrus tree (Gill, 1971).

Xiphinema

A large number of nematode species of the genus *Xiphinema* have been reported from the citrus rhizosphere (Baines *et al.*, 1978). These nematodes are all ectoparasitic. Very little research has been done regarding the pathogenicity of these nematodes to citrus even though high populations of some species have been associated consistently with citrus in California, South Africa and Sudan (Yassin, 1974; Cohn, 1976; Baines *et al.*, 1978; Milne, 1982). Most species of *Xiphinema* predominate in lighter textured soils (Cohn, 1969). In South Africa, control of *X. brevicollum* with DBCP did not result in marked tree quality improvement (Milne, 1982). In Sudan, high populations of *X. brevicollum* were associated with declining grapefruit trees. Subsequent pot studies resulted in similar root symptoms of stubby, swollen roots, and root abundance was greatly reduced by the nematode (Yassin,

1974). Similarly, high populations of *X. vulgare* are associated with declining citrus trees in Florida and caused necrosis and severe reduction of the root systems of seedlings in pots (Leone *et al.*, 1997). The nematode was shown to reproduce on citrus, but required 274 days at 24°C to complete its life cycle (Coiro *et al.*, 2002). *X. brevicolle* and *X. index* reduced sour orange seedling size by nearly half in pot studies in Israel (Cohn and Orion, 1970). Feeder root abundance on infested plants is severely reduced. Damage is primarily to epidermal and outer cortical cells which become necrotic and give a typically dark appearance to damaged roots (Cohn, 1970; Cohn and Orion, 1970; Baines *et al.*, 1978).

Trichodorus and *Paratrichodorus*

Low levels of *Trichodorus* and *Paratrichodorus* spp. are often encountered in soil samples from citrus (Baines *et al.*, 1959; Malo, 1961; Colbran, 1965). Population levels may increase above the normal levels in recently fumigated soil (Perry, 1953; Standifer and Perry, 1960). *P. lobatus* has also been found in high numbers in citrus nurseries in Australia where it is widespread in nurseries and orchards (Stirling, 1976). *P. porosus*, *P. lobatus* and *P. minor* have been reported to reduce root elongation and cause stubby root symptoms without evidence of necrosis on citrus in pot studies (Standifer and Perry, 1960; Stirling, 1976; Baines *et al.*, 1978). Despite decreasing feeder root weight in a pot study, *P. lobatus* did not affect taproot or seedling weights, nor were population levels in a nursery correlated with tree size (Stirling, 1976). However, nursery trees infested with the nematode at levels of 1500/500 cm^3 of soil had reduced root systems, poor leaf colour and tended to wilt during the day. Only one other report, based on the response of young trees to soil fumigation, implicates stubby root nematodes as possible pathogens of consequence in the field (Meagher, 1969).

Many dorylaimid nematode species are vectors of plant viruses. Despite a number of attempts, no nematode transmission of citrus viruses has yet been demonstrated.

Hemicycliophora and *Caloosia*

A number of species of *Hemicycliophora* have been identified from the citrus rhizosphere. *H. arenaria* is a species native to plants in the desert valleys of southern California that causes damage in citrus nurseries (McElroy *et al.*, 1966). The nematode was closely studied (Van Gundy, 1959) and quarantined to prevent its spread to other areas of that state. It appears to have a wide host range (ten of 19 hosts tested) although the rutaceous host status is variable. *Citrus limon*, *C. aurantifolia*, *C. reticulata* and *Severinia buxifolia* are susceptible, while *Poncirus trifoliata*, *C. aurantium*, *C. paradisi* and *C. sinensis* are resistant (Van Gundy and Rackham, 1961). The nematode feeds in large numbers at root tips whose roots typically develop around galls arising from hyperplasia. Seedling growth in pot studies was reduced by 35%. *Caloosia nudata* causes similar symptoms on citrus in Australia (Colbran, 1963). *H. arenaria* can be eradicated from root systems with hot water dips (10 min 46°C); pre-plant soil fumigation is very effective and a number of rootstocks resistant to the nematode are available (Van Gundy and McElroy, 1969).

References

Abu-Gharbieh, W.I. and Perry, V.G. (1970) Host differences among Florida populations of *Belonolaimus longicaudatus* Rau. *Journal of Nematology* 2, 209–216.

Al Hinai, M.S. and Mani, A. (1998) Seasonal population changes and management of *Tylenchulus semipenetrans* using organic amendments and fenamiphos. *Nematologia Mediterranea* 26, 179–184.

Anonymous (2002) *Citrus Fruit Fresh and Processed Annual Statistics 2002*. Food and Agriculture Organization of the United Nations, Rome, Italy.

Baghel, P.P.S. and Bhatti, D.S. (1982) Vertical and horizontal distribution of phytonematodes associated with citrus. *Indian Journal of Nematology* 12, 339–344.
Baines, R.C. (1964) Controlling citrus nematode with DBCP increases yields. *California Citrograph* 49, 222–233.
Baines, R.C. (1974) The effect of soil type on movement and infection rate of larvae of *Tylenchulus semipenetrans. Journal of Nematology* 6, 60–62.
Baines, R.C., Foote, F.J. and Martin, J.P. (1956) Fumigate soil before replanting citrus for control of the citrus nematode. *Citrus Leaves* 36, 6–8, 24, 27.
Baines, R.C., Van Gundy, S.D. and Sher, S.A. (1959) Citrus and avocado nematodes. *California Agriculture* 13, 16–18.
Baines, R.C., Martin, J.P., DeWolfe, T.A., Boswell, S.B. and Garber, M.J. (1962) Effect of high doses of D-D on soil organisms and the growth and yield of lemon trees. *Phytopathology* 52, 723.
Baines, R.C., Klotz, L.J., DeWolfe, T.A., Small, R.H. and Turner, G.O. (1966) Nematocidal and fungicidal properties of some soil fumigants. *Phytopathology* 56, 691–698.
Baines, R.C., DeWolfe, T.A., Klotz, L.J., Bitters, W.P., Small, R.H. and Garber, M.J. (1969a) Susceptibility of six *Poncirus trifoliata* selections and Troyer citrange to a biotype of the citrus nematode and growth response on fumigated soil. *Phytopathology* 59, 1016–1017.
Baines, R.C., Miyakawa, T., Cameron, J.W. and Small, R.H. (1969b) Biotypes of the citrus nematode. *Proceedings of the 1st International Citrus Symposium*, Riverside, California 2, 955–956.
Baines, R.C., Van Gundy, S.D. and DuCharme, E.P. (1978) Nematodes attacking citrus. In: Reuther, W., Calavan, E.C. and Carman, G.E. (eds) *The Citrus Industry*, Volume IV. University of California, Division of Agricultural Science, pp. 321–345.
Bello, A., Navas, A. and Belart, C. (1986) Nematodes of citrus-groves in the Spanish Levante Ecological study focused to their control. In: Cavalloro, R. and Di Martino, E. (eds) Proceedings of the Expert's Meeting, 26–29 March 1985, Acireale, *Integrated Pest Control in Citrus Groves.* A.A. Balkema Publishing Co., Rotterdam, pp. 217–226.
Bistline, F.W., Collier, B.L. and Dieter, C.E. (1967) Tree and yield response to control of a nematode complex including *Belonolaimus longicaudatus* in replanted citrus. *Nematologica* 13, 137–138.
Bridge, J., Machon, D. and Djatmiadi, D. (1990) A new nematode problem of citrus caused by *Radopholus* sp. in Java. *Nematologica* 36, 336.
Brooks, T.L. and Perry, V.G. (1962) Apparent parthenogenetic reproduction of the burrowing nematode *Radopholus similis. Soil Crop Science Society of Florida Proceedings* 22, 160–162.
Brooks, T.L. and Perry, V.G. (1967) Pathogenicity of *Pratylenchus brachyurus* to citrus. *Plant Disease Reporter* 51, 569–573.
Bryan, O.C. (1966) Soil moisture – the key factor in the production of nematode-infested groves. *Citrus and Vegetable Magazine* 29, 39–40.
Bryan, O.C. (1969) Living with the burrowing nematode. *Citrus and Vegetable Magazine* 33, 29–38.
Caldwell, M.M., Richards, J.H. and Beyschlag, W. (1991) Hydraulic lift: ecological implications of water efflux from roots. In: Atkinson, D. (ed.) *Plant Root Growth: an Ecological Perspective.* Blackwell Scientific, Oxford, pp. 423–436.
Childers, C.C., Duncan, L.W., Wheaton, T.A. and Timmer, L.W. (1987) Arthropod and nematode control with Aldicarb on Florida citrus. *Journal of Economic Entomology* 80, 1064–1071.
Chitwood, B.G. and Toung, M.C. (1960) Host–parasite interactions of the Asiatic pyroid citrus nema. *Plant Disease Reporter* 44, 848–854.
Cobb, N.A. (1913) Notes on *Mononchus* and *Tylenchulus. Journal of the Washington Academy of Science* 3, 287–288.
Cobb, N.A. (1914) Citrus-root nematode. *Journal of Agricultural Research* 2, 217–230.
Coelho, Y.D.S., Paguio, O.D.L.R., Zem, A.C. and Filho, H.P.S. (1983) Citrus decline and nematode population on citrus. *Fitopatologia Brasileira* 8, 367–370.
Cohn, E. (1964) Penetration of the citrus nematode in relation to root development. *Nematologica* 10, 594–600.
Cohn, E. (1965a) On the feeding and histopathology of the citrus nematode. *Nematologica* 11, 47–54.
Cohn, E. (1965b) The development of the citrus nematode on some of its hosts. *Nematologica* 11, 593–600.
Cohn, E. (1966) Observations on the survival of free-living stages of the citrus nematode. *Nematologica* 12, 321–327.
Cohn, E. (1969) The occurrence and distribution of species of *Xiphinema* and *Longidorus* in Israel. *Nematologica* 15, 179–192.

Cohn, E. (1970) Observations on the feeding and symptomatology of *Xiphinema* and *Longidorus* on selected host roots. *Journal of Nematology* 2, 167–173.

Cohn, E. (1972) Nematode diseases of citrus. In: Webster, J.M. (ed.) *Economic Nematology*. Academic Press, London, pp. 215–244.

Cohn, E. (1976) *Report of Investigations on Nematodes of Citrus and Subtropical Fruit Crops in South Africa.* Citrus and Subtropical Fruit Research Institute, Nelspruit.

Cohn, E. and Orion, D. (1970) The pathological effect of representative *Xiphinema* and *Longidorus* species on selected host plants. *Nematologica* 16, 423–428.

Cohn, E., Minz, G. and Monselise, S.P. (1965) The distribution, ecology and pathogenicity of the citrus nematode in Israel. *Israel Journal of Agricultural Research* 15, 187–200.

Cohn, E., Feder, W.A. and Mordechai, M. (1968) The growth response of citrus to nematicide treatments. *Israel Journal of Agricultural Research* 18, 19–24.

Cohn, E., Hough, A. and Mulder, N. (1976) Elimination of nematodes and other plant pathogens from irrigation water by filtration techniques. *13th International Nematology Symposium*, Dublin, Ireland, 5–11 September, pp. 16–17.

Coiro, M.I., Lamberti, F., Sasanelli, N., Duncan, L.W. and Agostinelli, A. (2002) Reproduction and longevity of *Xiphinema vulgare* (Nematoda). *Nematologia Mediterranea* 30, 91–95.

Colbran, R.C. (1963) Studies of plant and soil nematodes. 6. Two new species from citrus orchards. *Queensland Journal of Agricultural Science* 20, 469–474.

Colbran, R.C. (1965) Studies of plant and soil nematodes. 9. *Trichodorus lobatus* n.sp. (Nematoda: Trichodoridae), a stubby-root nematode associated with citrus and peach trees. *Queensland Journal of Agricultural and Animal Sciences* 22, 273–276.

Cronjé, C., Le Roux, H.F., Truter, M., Van Heerden, I. and Phillips, H. (2002) Long-term effect of preplant soil solarisation on growth of replant citrus trees in South Africa. *African Plant Protection* 8, 41–49.

Das, T.K. and Mukhopadhyaya, M.C. (1985) Influence of citrus species on the morphological variations of *Tylenchulus semipenetrans*. *Indian Journal of Nematology* 15, 114–116.

Davide, R.G. (1971) *Survey of the Distribution of Different Plant Parasitic Nematodes Associated with the Citrus Decline in the Philippines.* A report of NSDB project No. 2203, University of Philippines, College of Agriculture, Laguna.

Davide, R.G. and Dela Rose, A.G. (1976) Host–parasite relationships and control of the citrus nematode in the Philippines. II. Evaluation of nematode damage on citrus. University of Philippines, College of Agriculture, Laguna, *NSDB Technical Journal* 1, 8–15.

Davis, R.M. (1984) Distribution of *Tylenchulus semipenetrans* in a Texas grapefruit orchard. *Journal of Nematology* 16, 313–317.

Davis, R.M. (1985) Citrus nematode control in Texas. *Citrograph* July, pp. 212–213.

Davis, R.M. and Wilhite, H.S. (1985) Control of *Tylenchulus semipenetrans* on citrus with fenamiphos and oxamyl. *Plant Disease* 69, 974–976.

Davis, R.M., Heald, C.M. and Timmer, L.W. (1982) Chemical control of the citrus nematode on grapefruit. *Journal of the Rio Grande Valley Horticultural Society* 35, 59–63.

de Brito, J.A., de Lourdes Mendes, M. and Rodrigues, R. (2000) Rea Hao de Porta-enxerto de *Citrus* spp. a *Meloidogyne incognita* RaHas 2 e 4 e *M. javanica*. *Nematologia Brasileira* 24, 253–256.

de Campos, A.S., dos Santos, J.M. and Duncan, L.W. (2002) Nematodes of citrus in open nurseries and orchards in Sao Paulo State, Brazil. *Nematology* 4, 263–264.

Den Ouden, H. (1965) An infestation on citrus in Surinam caused by *Meloidogyne exigua*. *Surinamese Landbouw* 13, 34.

DuCharme, E.P. (1955) Sub-soil drainage as a factor in the spread of the burrowing nematode. *Proceedings of the Florida State Horticultural Society* 68, 29–31.

DuCharme, E.P. (1959) Morphogenesis and histopathology of lesions induced on citrus roots by *Radopholus similis*. *Phytopathology* 49, 388–395.

DuCharme, E.P. (1967) Annual population periodicity of *Radopholus similis* in Florida citrus groves. *Plant Disease Reporter* 51, 1031–1034.

DuCharme, E.P. (1968) Burrowing nematode decline of citrus, a review. In: Smart, G.C. and Perry, V.G. (eds) *Tropical Nematology*. University of Florida Press, Gainesville, Florida, pp. 20–37.

DuCharme, E.P. (1969) Temperature in relation to *Radopholus similis* (Nematoda) spreading decline of citrus. *Proceedings of the 1st International Citrus Symposium* 2, 979–983.

DuCharme, E.P. and Birchfield, W. (1956) Physiologic races of the burrowing nematode. *Phytopathology* 46, 615–616.

DuCharme, E.P. and Price, W.C. (1966) Dynamics of multiplication of *Radopholus similis*. *Nematologica* 12, 113–121.

DuCharme, E.P. and Suit, R.F. (1967) Population fluctuations of burrowing nematodes in Florida citrus groves. *Proceedings of the Florida State Horticultural Society* 80, 63–67.

Duncan, L.W. (1986) The spatial distribution of citrus feeder roots and of the citrus nematode, *Tylenchulus semipenetrans*. *Revue de Nématologie* 9, 233–240.

Duncan, L.W. (1989) Effect of Fenamiphos placement on *Tylenchulus semipenetrans* and yield in a Florida citrus orchard. *Supplement to the Journal of Nematology* 21 (4S), 703–706.

Duncan, L.W. (1999) Nematode diseases of citrus. In: Timmer, L.W. and Duncan, L.W. (eds) *Citrus Health Management*. APS Press, St Paul, Minnesota, pp. 136–148.

Duncan, L.W. and Eissenstat, D.M. (1993) Responses of *Tylenchulus semipenetrans* to citrus fruit removal: implications for carbohydrate competition. *Journal of Nematology* 25, 7–14.

Duncan, L.W. and El-Morshedy, M.M. (1996) Population changes of *Tylenchulus semipenetrans* under localized versus uniform drought in the citrus root zone. *Journal of Nematology* 28, 360–368.

Duncan, L.W., Ferguson, J.J., Dunn, R.A. and Noling, J.W. (1989) Application of Taylor's Power Law to sample statistics of *Tylenchulus semipenetrans* in Florida citrus. *Supplement to the Journal of Nematology* 21 (4S), 707–711.

Duncan, L.W., Kaplan, D.T. and Noling, J.W. (1990) Maintaining barriers to the spread of *Radopholus citrophilus* in Florida citrus orchards. *Nematropica* 20, 71–88.

Duncan, L.W., Graham, J.H. and Timmer, L.W. (1993) Seasonal patterns associated with *Tylenchulus semipenetrans* and *Phytophthora parasitica* in the citrus rhizosphere. *Phytopathology* 83, 573–581.

Duncan, L.W., El-Morshedy, M.M. and McSorley, R. (1994a) Sampling citrus fibrous roots and *Tylenchulus semipenetrans*. *Journal of Nematology* 26, 442–451.

Duncan, L.W., Inserra, R.N., O'Bannon, J.H. and El-Morshedy, M.M. (1994b) Reproduction of a Florida population of *Tylenchulus semipenetrans* on resistant citrus rootstocks. *Plant Disease* 78, 1067–1071.

Duncan, L.W., Toole, J.D., Inserra, R.N., Castle, W.S. and O'Bannon, J.H. (1994c) Comparing two sampling methods to detect *Radopholus citrophilus* in Florida citrus. *Proceedings of the Soil Crop Science Society of Florida* 53, 42–45.

Duncan, L.W., Mashela, P., Ferguson, J., Graham, J., Abou-Setta, M.M. and El-Morshedy, M.M. (1995) Estimating crop loss in orchards with patches of mature citrus trees infected by *Tylenchulus semipenetrans*. *Nematropica* 25, 43–51.

Duncan, L.W., Noling, J.W., Inserra, R.N. and Dunn, D. (1996) Spatial patterns of *Belonolaimus* spp. among and within citrus orchards on Florida's central ridge. *Journal of Nematology* 28, 352–359.

Duncan, L.W., Inserra, R.N. and Dunn, D. (1998) Seasonal changes in citrus fibrous root starch concentration and body length of female *Pratylenchus coffeae*. *Nematropica* 28, 263–266.

Duncan, L.W., Inserra, R.N., Thomas, W.K., Dunn, D., Mustika, I., Frisse, L.M., Mendes, M.L., Morris, K. and Kaplan, D.T. (1999) Molecular and morphological analysis of isolates of *Pratylenchus coffeae* and closely related species. *Nematropica* 29, 61–80.

Dunn, D.C., Duncan, L.W. and Romeo, J.T. (1998) Changes in arginine, PAL activity, and nematode behavior in salinity-stressed citrus. *Phytochemistry* 49, 413–417.

Eissenstat, D.M. and Duncan, L.W. (1992) Root growth and carbohydrate responses in bearing citrus trees following partial canopy removal. *Tree Physiology* 10, 245–257.

Elbadri, G.A.A., De Ley, P., Waeyenberge, L., Vierstraete, A., Moens, M. and Vanfleteren, J. (2002) Intraspecific variation in *Radopholus similis* isolates assessed with restriction fragment length polymorphism and DNA sequencing of the internal transcribed spacer region of the ribosomal RNA cistron. *International Journal for Parasitology* 32, 199–205.

El-Borai, F.E., Duncan, L.W. and Graham, J.H. (2002) Infection of citrus roots reduces root infection by *Phytophthora nicotianae*. *Journal of Nematology* 34, 384–389.

El-Borai, F.E., Duncan, L.W., Graham, J.H. and Dickstein, E. (2003) *Tylenchulus semipenetrans* alters the microbial community in the citrus rhizosphere. *Journal of Nematology* 35, 167–177.

Esser, R.P. (1984) How nematodes enter and disperse in Florida nurseries via vehicles. Florida Department of Agriculture and Consumer Services, Division of Plant Industry, *Nematology Circular* No. 109.

Esser, R.P. and Simpson, S.E. (1984) Sting nematode on citrus. Florida Department of Agriculture and Consumer Services, Division of Plant Industry, *Nematology Circular* No. 106.

Esser, R.P., Smith, G.T. and O'Bannon, J.H. (1993) An eleven-year phytoparasitic nematode survey of Florida citrus groves and their environs. *Bulletin 15*, Florida Department of Agriculture and Consumer Services, Gainesville, Florida.

Fallas, G.A., Hahn, M.L., Fargette, M., Burrows, P.R. and Sarah, J.L. (1996) Molecular and biochemical diversity among isolates of *Radopholus* spp. from different areas of the world. *Journal of Nematology* 28, 422–430.

Feder, W.A. and Feldmesser, J. (1956) Root abnormalities caused by burrowing nematode infections. *Phytopathology* 46, 11.

Feder, W.A. and Feldmesser, J. (1961) The spreading decline complex: the separate and combined effects of *Fusarium* spp. and *Radopholus similis* on the growth of Duncan grapefruit seedlings in the greenhouse. *Phytopathology* 51, 724–726.

Feder, W.A., Feldmesser, J. and Walkinshaw, C.H. (1956) Microorganisms isolated from feeder roots of citrus seedlings affected by spreading decline. *Soil Crop Science Society of Florida Proceedings* 16, 127–129.

Feldmesser, J., Rebois, R.V. and Taylor, A.L. (1959) Progress report on growth response of burrowing nematode infected citrus following chemical treatments under greenhouse conditions. *Plant Disease Reporter* 43, 261–263.

Feldmesser, J., Cetas, R.C., Grimm, G.R., Rebois, R.V. and Widden, R. (1960) Movement of *Radopholus similis* into rough lemon feeder roots and in soil, and its relation to *Fusarium* in the roots. *Phytopathology* 50, 635.

Ford, H.W. (1952) The effect of spreading decline on the root distribution of citrus. *Proceedings of the Florida State Horticultural Society* 65, 47–50.

Ford, H.W. (1953) Effect of spreading decline disease on the distribution of feeder roots of orange and grapefruit trees on rough lemon rootstocks. *Journal of the American Society for Horticultural Science* 61, 68–72.

Ford, H.W. (1954a) The influence of rootstock and tree age on root distribution of citrus. *Proceedings of the American Society for Horticultural Science* 63, 137–172.

Ford, H.W. (1954b) Root distribution in relation to the water table. *Proceedings of the Florida State Horticultural Society* 67, 30–33.

Ford, H.W. and Feder, W.A. (1961) Additional citrus rootstock selections that tolerate the burrowing nematode. *Proceedings of the Florida State Horticultural Society* 74, 50–53.

Ford, H.W., Feder, W.A. and Hutchins, P.C. (1960) Citrus varieties, hybrids, species, and relatives evaluated for resistance to the burrowing nematode *Radopholus similis*. *Citrus Experiment Station Mimeo Series 60–13.*

Frederick, J.J. and Tarjan. A.C. (1975) Control of *Pratylenchus brachyurus* in rough lemon seedlings with Dowco 275 (diethylfluropyridyl phosphorothioate). *Nematropica* 5, 10–13.

Galeano, M. (2002) Dinamica di popolazione di *Tylenchulus semipenetrans* e della nematofauna di vita libera nella rizosfera di agrumi in Spagna. *Supplemento Nematologia Mediterranea* 30, 49–53.

Galeano, M., Verdejo-Lucas, S., Sorribas, F.J., Ornat, C., Forner, J.B. and Alcaide, A. (2003) New citrus selections from Cleopatra mandarin × *Poncirus trifoliata* with resistance to *Tylenchulus semipenetrans* Cobb. *Nematology* 5, 227–234.

Garabedian, S., Van Gundy, S.D., Mankau, R. and Radewald, J.D. (1984) Nematodes. In: *Integrated Pest Management for Citrus*. Division of Agriculture and Natural Resources Publications, University of California, Berkeley, California, pp. 129–131.

Gill, H.S. (1971) Occurrence and reproduction of *Meloidogyne javanica* on three species of citrus in California. *Plant Disease Reporter* 55, 607–608.

Golden, M.A., Lopez, C.H.R. and Vilchez, R.H. (1992) Description of *Pratylenchus gutierrezi* n.sp. (Nematoda: Pratylenchidae) from coffee in Costa Rica. *Journal of Nematology* 24, 298–304.

Gottlieb, Y., Cohn, E. and Spiegel-Roy, P. (1986) Biotypes of the citrus nematode (*Tylenchulus semipenetrans* Cobb) in Israel. *Phytoparasitica* 14, 193–198.

Graham, J.H. (1988) Fumigation induced stunting. In: Whiteside, J.O., Garnsey, S.M. and Timmer, L.W. (eds) *Compendium of Citrus Diseases*. APS Press, St Paul, Minnesota, p. 61.

Graham, J.H. and Duncan, L.W. (1997) Suppression of *Phytophthora nicotianae* in citrus roots by the citrus nematode (*Tylenchulus semipenetrans*). *Phytopathology* (Supplement) 87, S35.

Graham, J.H., Timmer, L.W. and Lee, R.F. (1983) Comparison of zinc, water uptake by gravity infusion and syringe injection tests for diagnosis of citrus blight. *Proceedings of the Florida State Horticultural Society* 96, 45–47.

Hahn, M.L., Burrows, P.R., Gnanapragasam, N.C., Bridge, J., Vines, N.J. and Wright, D.J. (1994) Molecular diversity amongst *Radopholus similis* populations from Sri Lanka detected by RAPD analysis. *Fundamental and Applied Nematology* 17, 275–281.

Hamid, G.A., Van Gundy, S.D. and Lovatt, C.J. (1988) Phenologies of the citrus nematode and citrus roots treated with oxamyl. *Proceedings of the International Society of Citriculture* 2, 993–1004.

Hannon, C.I. (1963) Longevity of *Radopholus similis* under field conditions. *Plant Disease Reporter* 47, 812–816.

Hannon, C.I. (1964) Longevity of the citrus-root nematode in Florida. *Soil Crop Science Society of Florida Proceedings* 24, 158–161.

Heald, C.M. and O'Bannon, J.H. (1987) Citrus declines caused by nematodes. V. Slow decline. *Florida Department of Agriculture and Consumer Services, Division of Plant Industry, Nematology Circular* No. 143.

Hearn, C.J. (1986) Production trends around the world. In: Wardowski, W.F., Nagy, S. and Grierson, W. (eds) *Fresh Citrus Fruits*. AVI Publishing Co., Westport, Connecticut, pp. 127–132.

Huang, C.S. and Chang, Y.C. (1976) Pathogenicity of *Pratylenchus coffeae* on sunki orange. *Plant Disease Reporter* 60, 957–960.

Huettel, R.N. and Yaegashi, T. (1988) Morphological differences between *Radopholus citrophilus* and *R. similis*. *Journal of Nematology* 20, 150–157.

Huettel, R.N., Dickson, D.W. and Kaplan, D.T. (1984) *Radopholus citrophilus* n.sp. (Nematoda), a sibling species of *Radopholus similis*. *Proceedings of the Helminthological Society of Washington* 51, 32–35.

Hutchinson, D.J. (1985) Rootstock development, screening and selection for disease tolerance and horticultural characteristics. *Fruit Varieties Journal* 39, 21–25.

Inserra, R.N. and Vovlas, N. (1974) Dannida *Pratylenchus vulnus* su arancio amaro in Puglia. *Nematologia Mediterranea* 2, 183–185.

Inserra, R.N. and Vovlas, N. (1977) Effects of *Pratylenchus vulnus* on the growth of sour orange. *Journal of Nematology* 9, 154–157.

Inserra, R.N., Perotta, G., Vovlas, N. and Catara, A. (1978) Reaction of citrus rootstocks to *Meloidogyne javanica*. *Journal of Nematology* 10, 181–184.

Inserra, R.N., Vovlas, N. and O'Bannon, J.H. (1980) A classification of *Tylenchulus semipenetrans* biotypes. *Journal of Nematology* 12, 283–287.

Inserra, R.N., Vovlas, N., O'Bannon, J.H. and Esser, R.P. (1988) *Tylenchulus graminis* n.sp. and *T. palustris* n.sp. (Tylenchulidae), from native flora of Florida, with notes on *T. semipenetrans* and *T. furcus*. *Journal of Nematology* 20, 266–287.

Inserra, R.N., Duncan, L.W., Vovlas, N. and Loof, P.A.A. (1996) *Pratylenchus loosi* from pasture grasses in Central Florida. *Nematologica* 42, 159–172.

Inserra, R.N., Duncan, L.W., Dunn, D.C., Kaplan, D.T. and Porazinska, D. (1998) *Pratylenchus pseudocoffeae* from Florida and its relationship with *P. gutierrezi* and *P. coffeae*. *Nematologica* 44, 683–712.

Inserra, R.N., Duncan, L.W., Troccoli, A., Dunn, D., Maia Sos Santos, J. and Vovlas, N. (2001) *Pratylenchus jaehni* sp. n. from citrus in Brazil and a redescription of *P. coffeae*. *Nematology* 3, 653–665.

Johnson, A.W. (1998) Degradation of fenamiphos in agricultural production soil. *Supplement to the Journal of Nematology* 30, 40–44.

Kaplan, D.T. (1985) Influence of the sting nematode, *Belonolaimus longicaudatus*, on young citrus trees. *Journal of Nematology* 17, 408–414.

Kaplan, D.T. (1986) Variation in *Radopholus citrophilus* population densities in the citrus rootstock Carrizo citrange. *Journal of Nematology* 18, 31–34.

Kaplan, D.T. (1988) Future considerations for nematode management in citrus. *Proceedings of the International Society of Citriculture* 2, 969–975.

Kaplan, D.T. (1994a) Molecular characterization of the burrowing nematode sibling species, *Radopholus citrophilus* and *R. similis*. In: Lamberti, F., De Giorgi, C. and Bird, D.M. (eds) *Advances in Molecular Plant Nematology*. Plenum Press, New York, pp. 77–83.

Kaplan, D.T. (1994b) An assay to estimate citrus rootstock resistance to burrowing nematodes. *Proceedings of the Florida State Horticulture Society* 107, 85–89.

Kaplan, D.T. and O'Bannon, J.H. (1981) Evaluation and nature of citrus nematode resistance in Swingle cit rumelo. *Proceedings of the Florida State Horticultural Society* 94, 33–36.

Kaplan, D.T. and O'Bannon, J.H. (1985) Occurrence of biotypes in *Radopholus citrophilus*. *Journal of Nematology* 17, 158–162.

Kaplan, D.T. and Opperman, C.H. (1997) Genome similarity implies that citrus-parasitic burrowing nematodes do not represent a unique species. *Journal of Nematology* 29, 430–440.

Kaplan, D.T. and Opperman, C.H. (2000) Reproductive strategies and karyotype of the burrowing nematode, *Radopholus similis*. *Journal of Nematology* 32, 126–133.

Kaplan, D.T. and Timmer, L.W. (1982) Effects of *Pratylenchus coffeae–Tylenchulus semipenetrans* interactions on nematode population dynamics in citrus. *Journal of Nematology* 14, 368–373.

Kaplan, D.T., Vanderspool, M.C. and Opperman, C.H. (1997) Sequence tag site and host range assays demonstrate that *Radopholus similis* and *R. citrophilus* are not reproductively isolated. *Journal of Nematology* 29, 421–429.

Kaplan, D.T., Thomas, W.K., Frisse, L.M., Sarah, J.L., Stanton, J.M., Speijer, P.R., Marin, D.H. and Opperman, C.H. (2000) Phylogenetic analysis of geographically diverse *Radopholus similis* via rDNA sequence reveals a monomorphic motif. *Journal of Nematology* 32, 134–142.

Kirkpatrick, J.C. and Van Gundy, S.D. (1966) Scion and rootstock as factors in the development of citrus nematode populations. *Phytopathology* 56, 438–441.

Labuschagne, N. and Kotze, J.M. (1988) Factors affecting feeder root rot of citrus caused by *Fusarium solani. Proceedings of the International Society of Citriculture*, Tel-Aviv, Israel, March 1988.

Labuschagne, N., Van Der Vegte, F.A. and Kotze, J.M. (1989) Interaction between *Fusarium solani* and *Tylenchulus semipenetrans* on citrus roots. *Phytophylactica* 21, 29–33.

Lehman, P.S. (1995) Role of plant protection organizations in nematode management. *XIX Congress of Brazilian Society of Nematology*. Brazilian Society of Nematology, Rio Quente, Brazil, pp. 137–148.

Leone, A., Miano, V., Lamberti, F., Duncan, L.W., Rich, J.R. and Bleve-Zacheo, T. (1997) Cellular changes induced by *Xiphinema vulgare* in the roots of citrumelo and by *Xiphinema intermedium* in the roots of Bermuda grass. *Nematologia Mediterranea* 25, 199–207.

Le Roux, H.F. (1995) Control of the citrus nematode in South Africa. PhD dissertation, Department of Microbiology and Plant Pathology, University of Pretoria, South Africa.

Le Roux, H.F., Wehner, F.C., Kotzé, J.M. and Grech, N.M. (1991) Combining fosetyl-Al trunk injection or metalaxyl soil drenching with soil application of aldicarb for control of citrus decline. *Plant Disease* 75, 1233–1236.

Le Roux, H.F., Ware, A.B. and Pretorius, M.C. (1998) Comparative efficacy of preplant fumigation and post-plant chemical treatment of replant citrus trees in an orchard infested with *Tylenchulus semipenetrans. Plant Disease* 82, 1323–1327.

Ling, P., Duncan, L.W., Deng, Z., Dunn, D., Hu, X., Huang, S. and Gmitter, F.G. Jr (2000) Inheritance of citrus nematode resistance and its linkage with molecular markers. *Theoretical and Applied Genetics* 100, 1010–1017.

Machmer, J.H. (1958) Effect of soil salinity on nematodes in citrus and papaya plantings. *Journal of the Rio Grande Valley Horticultural Society* 12, 57–60.

Machon, J.E. and Bridge, J. (1996) *Radopholus citri* n.sp. (Tylenchida: Pratylenchidae) and its pathogenicity on citrus. *Fundamental and Applied Nematology* 19, 127–133.

Malo, S.E. (1961) Nematode populations associated with citrus roots in central Florida. *Plant Disease Reporter* 45, 20–23.

Mangat, B.P.S. and Sharma, N.K. (1981) Influence of host nutrition on multiplication and development of citrus nematode. *Indian Phytopathology* 34, 90–91.

Mani, A., Al Hinai, M.S. and Handoo, Z.A. (1997) Occurrence, population density, and distribution of root-lesion nematodes, *Pratylenchus* spp., in the Sultanate of Oman. *Nematropica* 27, 209–219.

Mankau, R. (1968) Effect of nematocides on nematode-trapping fungi associated with the citrus nematode. *Plant Disease Reporter* 52, 851–855.

Marin, D.H., Sutton, T.B. and Barker, K.R. (1998) Dissemination of bananas in Latin America and the Caribbean and its relationship to the occurrence of *Radopholus similis. Plant Disease* 82, 964–974.

Martin, J.P. and Van Gundy, S.D. (1963) Influence of soil phosphorous level on the growth of sweet orange seedlings and the activity of the citrus nematode (*Tylenchulus semipenetrans*). *Soil Science* 96, 128–135.

Mashela, P.W. and Nthangeni, M.E. (2002) Osmolyte allocation in response to *Tylenchulus semipenetrans* infection, stem girdling, and root pruning in citrus. *Journal of Nematology* 34, 273–277.

Mashela, P., Duncan, L.W., Graham, J.H. and McSorley, R. (1992a) Leaching soluble salts increases population densities of *Tylenchulus semipenetrans. Journal of Nematology* 24, 103–109.

Mashela, P., Duncan, L.W. and McSorley, R. (1992b) Salinity reduces resistance to *Tylenchulus semipenetrans* in citrus rootstocks. *Nematropica* 22, 7–12.

McClure, M.A. and Schmitt, M.E. (1996) Control of citrus nematode, *Tylenchulus semipenetrans*, with cadusafos. *Supplement to the Journal of Nematology* 28 (4S), 624–628.

McElroy, F.D., Sher, S.A. and Van Gundy, S.D. (1966) The sheath nematode *Hemicycliophora arenaria*, a native to California soils. *Plant Disease Reporter* 40, 581–583.

McSorley, R. and Parrado, J.L. (1982a) Plans for the collection of nematode soil samples from fruit groves. *Nematropica* 12, 257–267.
McSorley, R. and Parrado, J.L. (1982b) Relationship between two nematode extraction techniques on two Florida soils. *Soil Crop Science Society of Florida Proceedings* 41, 30–36.
Meagher, J.W. (1967) Observations on the transport of nematodes in subsoil drainage and irrigation water. *Australian Journal of Experimental Agricultural Animal Husbandry* 7, 577–579.
Meagher, J.W. (1969) Nematodes as a factor in citrus production in Australia. *Proceedings of the 1st International Citrus Symposium, Riverside, California* 2, 999–1006.
Milne, D.L. (1974) *Citrus Seedbeds: Methyl Bromide and Mycorrhizae.* Citrus and Subtropical Fruit Research Institute, Nelspruit, pp. 9–11.
Milne, D.L. (1982) Nematode pests of citrus. In: Keetch, D.P. and Heyns, J. (eds) *Nematology in Southern Africa.* Republic of South Africa, Department of Agriculture and Fisheries Science Bulletin No. 400, pp. 12–18.
Milne, D.L. and De Villiers, E.A. (1977) Soil application of systemic pesticides for control of thrips and nematodes on citrus. *Citrus and Subtropical Fruit Journal* 518, 9–18.
Milne, D.L. and du Toit, W. (1976) *The Effect of Citrus Nematicides on the Earthworm Population in the Soil.* Citrus and Subtropical Fruit Research Institute, Nelspruit, pp. 13–15.
Milne, D.L. and Willers, P. (1979) Yield and nutritional responses to phenamiphos treatment of citrus infested with citrus nematodes. *Subtropica* 1, 11–14.
Minz, G. (1956) The root-knot nematode, *Meloidogyne* spp. in Israel. *Plant Disease Reporter* 40, 798–801.
Nigh, E.L. Jr (1981a) Relation of citrus nematode to root distribution in flood irrigated citrus. *Journal of Nematology* 13, 451–452.
Nigh, E.L. Jr (1981b) Evaluation of sampling and extraction techniques to determine citrus nematode, *Tylenchulus semipenetrans,* populations. *Journal of Nematology* 13, 451.
O'Bannon, J.H. (1968) The influence of an organic soil amendment on infectivity and reproduction of *Tylenchulus semipenetrans* on two citrus rootstocks. *Phytopathology* 58, 597–601.
O'Bannon, J.H. (1977) Worldwide dissemination of *Radopholus similis* and its importance in crop production. *Journal of Nematology* 9, 16–25.
O'Bannon, J.H. (1980) Migration of *Pratylenchus coffeae* and *P. brachyurus* on citrus and in soil. *Nematropica* 10, 70.
O'Bannon, J.H. and Esser, R.P. (1975) Evaluation of citrus, hybrids, and relatives as hosts of the nematode *Pratylenchus coffeae,* with comments on other hosts. *Nematologia Mediterranea* 3, 113–122.
O'Bannon, J.H. and Ford, H.W. (1976) An evaluation of several *Radopholus similis*-resistant or -tolerant citrus rootstocks. *Plant Disease Reporter* 60, 620–624.
O'Bannon, J.H. and Hutchinson, D.H. (1974) Development of rootstocks resistant to the citrus nematode, *Tylenchulus semipenetrans.* In: Jackson, L.K., Krezdorn, A.H. and Soule, J. (eds) *Proceedings of the 1st International Citrus Short Course.* 24–29 September 1973, Gainesville, Florida, pp. 22–29.
O'Bannon, J.H. and Nemec, S. (1978) Influence of soil pesticides on vesicular-arbuscular mycorrhizae in a citrus soil. *Nematropica* 8, 56–61.
O'Bannon, J.H. and Nemec, S. (1979) The response of *Citrus limon* seedlings to a symbiont, *Glomus etunicatus,* and a pathogen, *Radopholus similis. Journal of Nematology* 11, 270–274.
O'Bannon, J.H. and Tarjan, A.C. (1973) Preplant fumigation for citrus nematode control in Florida. *Journal of Nematology* 5, 88–95.
O'Bannon, J.H. and Tarjan, A.C. (1979) Management of *Radopholus similis* infecting citrus with DBCP or phenamiphos. *Plant Disease Reporter* 63, 456–460.
O'Bannon, J.H. and Tarjan, A.C. (1985) *Citrus Declines Caused by Nematodes in Florida. III. Citrus Slump.* Florida Department of Agriculture and Consumer Services, Division of Plant Industry, Nematology Circular No. 117.
O'Bannon, J.H. and Tomerlin, A.T. (1969a) Movement of *Radopholus similis* on a weed host (*Solanum nigrum*). *Journal of Nematology* 1, 21.
O'Bannon, J.H. and Tomerlin, A.T. (1969b) Population studies on two species of *Pratylenchus* on citrus. *Journal of Nematology* 1, 299–300.
O'Bannon, J.H. and Tomerlin, A.T. (1971) Response of citrus seedlings to *Radopholus similis* in two soils. *Journal of Nematology* 3, 255–259.
O'Bannon, J.H. and Tomerlin, A.T. (1973) Citrus tree decline caused by *Pratylenchus coffeae. Journal of Nematology* 5, 311–316.
O'Bannon, J.H. and Tomerlin, A.T. (1977) Control of the burrowing nematode, *Radopholus similis,* with DBCP and oxamyl. *Plant Disease Reporter* 61, 450–454.

O'Bannon, J.H., Reynolds, H.W. and Leathers, C.R. (1966) Effects of temperature on penetration, development, and reproduction of *Tylenchulus semipenetrans*. *Nematologica* 12, 483–487.

O'Bannon, J.H., Leather, C.R. and Reynolds, H.W. (1967) Interactions of *Tylenchulus semipenetrans* and *Fusarium* species on rough lemon (*Citrus limon*). *Phytopathology* 57, 414–417.

O'Bannon, J.H., Radewald, J.D. and Tomerlin, A.T. (1972) Population fluctuation of three parasitic nematodes in Florida citrus. *Journal of Nematology* 4, 194–199.

O'Bannon, J.H., Tarjan, A.C. and Bistline, F.W. (1974) Control of *Pratylenchus brachyurus* on citrus and tree response to chemical treatment. *Soil Crop Science Society of Florida Proceedings* 33, 65–67.

O'Bannon, J.H., Radewald, J.D., Tomerlin, A.T. and Inserra, R.N. (1976) Comparative influence of *Radopholus similis* and *Pratylenchus coffeae* on citrus. *Journal of Nematology* 8, 58–63.

Orion, D. and Cohn, E. (1975) A resistant response of citrus roots to the root-knot nematode, *Meloidogyne javanica*. *Marcellia* 38, 327–328.

Oteifa, B.A., Shafiee, V.A. and Eissa, F.M. (1965) Efficacy of DBCP flood irrigation in established citrus. *Plant Disease Reporter* 49, 598–599.

Pan, C. (1984) Studies on plant-parasitic nematodes on economically important crops in Fujian. I. Species of root-knot nematodes (*Meloidogyne* species) and their host-plants. *Acta Zoologica Sinica* 30, 159–167.

Pan, C. (1985) Studies on plant-parasitic nematodes on economically important crops in Fujian. III. Description of *Meloidogyne fujianensis* n.sp. (Nematoda: Meloidogynidae) infesting citrus in Najing County. *Acta Zoologica Sinica* 31, 263–268.

Perry, V.G. (1953) Return of the nematodes following fumigation of Florida soils. *Proceedings of the Florida State Horticultural Society* 66, 112–114.

Philis, J. (1969) Control of citrus nematode, *Tylenchulus semipenetrans*, with DBCP in established Cyprus citrus groves. *Plant Disease Reporter* 53, 804–806.

Philis, I. (1989) Yield loss assessment caused by the citrus nematode *Tylenchulus semipenetrans* on Valencia oranges in Cyprus. *Nematologia Mediterranea* 17, 5–6.

Poucher, C., Ford, H.W., Suit, R.F. and DuCharme, E.P. (1967) *Burrowing Nematode in Citrus*. Florida Department of Agriculture, Division of Plant Industry Bulletin No. 7.

Prasad, S.K. and Chawla, M.L. (1966) Observations on the population fluctuations of citrus nematode, *Tylenchulus semipenetrans* Cobb, 1913. *Indian Journal of Entomology* 27, 450–454.

Radewald, J.D., O'Bannon, J.H. and Tomerlin, A.T. (1971a) Temperature effects on reproduction and pathogenicity of *Pratylenchus coffeae* and *P. brachyurus* and survival of *P. coffeae* in roots of *Citrus jambhiri*. *Journal of Nematology* 3, 390–394.

Radewald, J.D., O'Bannon, J.H. and Tomerlin, A.T. (1971b) Anatomical studies of *Citrus jambhiri roots* infected by *Pratylenchus coffeae*. *Journal of Nematology* 3, 409–416.

Reynolds, H.W. and O'Bannon, J.H. (1963a) Factors influencing the citrus replants in Arizona. *Nematologica* 9, 337–340.

Reynolds, H.W. and O'Bannon, J.H. (1963b) Decline of grapefruit trees in relation to citrus nematode populations and tree recovery after chemical treatment. *Phytopathology* 53, 1011–1015.

Salem, A.A.-M. (1980) Observations on the population dynamics of the citrus nematode, *Tylenchulus semipenetrans* in Sharkia Governorate. *Egypt Journal of Phytopathology* 12, 31–34.

Schneider, H. and Baines, R.C. (1964) *Tylenchulus semipenetrans*: parasitism and injury to orange tree roots. *Phytopathology* 54, 1202–1206.

Scotto la Massèse, C. (1969) The principal plant nematodes of crops in the French West Indies. In: Peachey, J.E. (ed.) *Nematodes of Tropical Crops*. Commonwealth Bureau of Helminthology (Great Britain) Technical Communication 40, 168–169.

Scotto la Massèse, C. (1980) Possibilités offertes par les 'courbes isopathologiques' dans l'appréciation de la pathogénie des parasites illustrées par deux exemples en nématologie fruitière. *Revue de Zoologie Agricole* 78, 97–113.

Siddiqi, M.R. (1964) Studies on nematode root-rot of citrus in Uttar Pradesh, India. *Proceedings of the Zoological Society* (Calcutta) 17, 67–75.

Siddiqui, I.A., Sher, S.A. and French, A.M. (1973) *Distribution of Plant Parasitic Nematodes in California*. State of California, Department of Food and Agriculture, Division of Plant Industry.

Silva, R.A. and Inomoto, M.M. (2002) Host-range characterization of two *Pratylenchus coffeae* isolates from Brazil. *Journal of Nematology* 34, 135–139.

Sites, J.W., Reitz, H.J. and Deszyck, E.J. (1951) Some results of irrigation research with Florida citrus. *Proceedings of the Florida State Horticultural Society* 64, 71–79.

Smelt, J.H., Van De Peppel-Groen, A.E., Van Der Pas, L.J.T. and Dijksterhuis, A. (1996) Development and duration of accelerated degradation of nematicides in different soils. *Soil Biology and Biochemistry* 28, 1757–1765.

Smith, G. and Kaplan, D.T. (1988) Phosphorus and burrowing nematode interactions on growth of rough lemon citrus seedlings. *Journal of Nematology* 20, 539–544.

Sorribas, F.J., Verdejo-Lucas, S., Forner, J.B., Alcaide, A., Pons, J. and Ornat, C. (2000) Seasonality of *Tylenchulus semipenetrans* Cobb and *Pasteuria* sp. in citrus orchards in Spain. *Supplement to the Journal of Nematology* 32 (4S), 622–632.

Sorribas, F.J., Verdejo-Lucas, S., Galeano, M., Pastor, J. and Ornat, C. (2003) Effect of 1,3 dichloropropene and rootstocks alone and in combination on *Tylenchulus semipenetrans* and citrus tree growth in a replant management program. *Nematropica* 33, 149–158.

Spiegel-Roy, R., Vardi, A., Elhanati, A., Solel, Z. and Bar-Joseph, M. (1988) Rootstock selection from a Poorman orange × *Poncirus trifoliata* cross. *Proceedings of the International Society of Citriculture* 1, 195–200.

Standifer, M.S. and Perry, V.G. (1960) Some effects of sting and stubby root nematodes on grapefruit roots. *Phytopathology* 50, 152–156.

Stirling, G.R. (1976) *Paratrichodorus lobatus* associated with citrus, peach and apricot trees in South Australia. *Nematologica* 22, 138–144.

Stirling, G.R. and Wachtel, M.F. (1985) Effects of nematicides on citrus nematode *Tylenchulus semipenetrans* and the yield of citrus in South Australia. *Agricultural Record (South Australia)* 12, 6–11.

Suit, R.F. and Brooks, T.L. (1957) Current information relating to barriers for the burrowing nematode. *Proceedings of the Florida State Horticultural Society* 70, 55–57.

Suit, R.F. and DuCharme, E.P. (1953) The burrowing nematode and other parasitic nematodes in relation to spreading decline. *Citrus Leaves* 33, 8–9, 32–33.

Suit, R.F., Hanks, R.W., Tarjan, A.C., Feldman, A.W. and DuCharme, E.P. (1967) Control of spreading decline of citrus. State Project 773. *Florida Agricultural Experiment Station Annual Report*, pp. 228–229.

Swingle, W.T. and Reese, R. (1967) The botany of citrus and its wild relatives. In: Reuther, W., Webber, H.J. and Batchelor, L.D. (eds) *The Citrus Industry*. University of California, Division of Agricultural Science, pp. 190–430.

Tarjan, A.C. (1956) The possibility of mechanical transmission of nematodes in citrus groves. *Proceedings of the Florida State Horticultural Society* 69, 34–37.

Tarjan, A.C. (1961) Longevity of *Radopholus similis* (Cobb) in host-free soil. *Nematologica* 6, 170–175.

Tarjan, A.C. (1971) Migration of three pathogenic citrus nematodes through two Florida soils. *Soil Crop Science Society of Florida Proceedings* 31, 253–255.

Tarjan, A.C. and O'Bannon, J.H. (1969) Observations on meadow nematodes (*Pratylenchus* spp.) and their relation to decline of citrus in Florida. *Plant Disease Reporter* 53, 683–686.

Tarjan, A.C. and O'Bannon, J.H. (1977) Nonpesticidal approaches to nematode control. *Proceedings of the International Society of Citriculture* 3, 848–853.

Tarjan, A.C. and Simmons, P.N. (1966) The effect of interacting cultural practices of citrus trees with spreading decline. *Soil Crop Science Society of Florida Proceedings* 26, 22–31.

Thomason, I.J. and Caswell, E.P. (1987) Principles of nematode control. In: Brown, R.H. and Kerry, B.R. (eds) *Principles and Practice of Nematode Control in Crops*. Academic Press, Australia, pp. 87–130.

Timmer, L.W. (1977) Control of citrus nematode *Tylenchulus semipenetrans* on fine-textured soil with DBCP and oxamyl. *Journal of Nematology* 9, 45–50.

Timmer, L.W. and Davis, R.D. (1982) Estimate of yield loss from the citrus nematode in Texas grapefruit. *Journal of Nematology* 14, 582–585.

Tomerlin, A.T. and O'Bannon, J.H. (1974) Effect of *Radopholus similis* and *Pratylenchus brachyurus* on citrus seedlings in three soils. *Soil Crop Science Society of Florida Proceedings* 33, 95–97.

Toung, M.-C. (1963) A study on seasonal influence in quantitative variation of the citrus nema, *Tylenchulus semipenetrans* Cobb. *National Taiwan University Plant Protection Bulletin* 5, 323–327.

Tsai, B.Y. and Van Gundy, S.D. (1988) Comparison of anhydrobiotic ability of the citrus nematode with other plant parasitic nematodes. *Proceedings of the International Society of Citriculture* 2, 983–992.

Valette, C., Mounport, D., Nicole, M., Sarah, J.L. and Baujard, P. (1998) Scanning electron microscope studies of two African populations of *Radopholus similis* (Nematoda: Pratylenchidae) and proposal of *R. citrophilus* as a junior synonym of *R. similis*. *Fundamental and Applied Nematology* 21, 139–146.

Van Gundy, S.D. (1959) The life history of *Hemicycliophora arenaria* Raski (Nematoda: Criconematidae). *Proceedings of the Helminthology Society of Washington* 26, 67–72.

Van Gundy, S.D. and Kirkpatrick, J.D. (1964) Nature of resistance in certain citrus rootstocks to citrus nematodes. *Phytopathology* 54, 419–427.

Van Gundy, S.D. and Martin, J.P. (1961) Influence of *Tylenchulus semipenetrans* on the growth and chemical composition of sweet orange seedlings in soils of various exchangeable cation ratios. *Phytopathology* 51, 146–151.

Van Gundy, S.D. and Martin, J.P. (1962) Soil texture, pH and moisture effects on the development of citrus nematode (*Tylenchulus semipenetrans*). *Phytopathology* 52, 31.

Van Gundy, S.D. and McElroy, F.D. (1969) Sheath nematode: its biology and control. *Proceedings of the 1st International Citrus Symposium*, Riverside, California 2, pp. 985–989.

Van Gundy, S.D. and Meagher, J.W. (1977) Citrus nematode (*Tylenchulus semipenetrans*) problems worldwide. *1977 International Citrus Congress*, Orlando, Florida.

Van Gundy, S.D. and Rackham, R.L. (1961) Studies on the biology and pathogenicity of *Hemicycliophora arenaria*. *Phytopathology* 51, 393–397.

Van Gundy, S.D. and Tsao, P.H. (1963) Growth reduction of citrus seedlings by *Fusarium solani* as influenced by the citrus nematode and other soil factors. *Phytopathology* 53, 488–489.

Van Gundy, S.D., Thomason, I.J. and Rackham, R.L. (1959) The reaction of three *Citrus* spp. to three *Meloidogyne* spp. *Plant Disease Reporter* 43, 970–971.

Van Gundy, S.D., Stolzy, L.H., Szuszkiewicz, T.E. and Rackham, R.L. (1962) Influence of oxygen supply on survival of plant parasite nematodes in soil. *Phytopathology* 52, 628–632.

Van Gundy, S.D., Martin, J.P. and Taso, P.H. (1964) Some soil factors influencing reproduction of the citrus nematode and growth reduction of sweet orange seedlings. *Phytopathology* 54, 294–299.

Van Gundy, S.D., Bird, A.F. and Wallace, H.R. (1967) Aging and starvation in larvae of *Meloidogyne javanica* and *Tylenchulus semipenetrans*. *Phytopathology* 57, 599–571.

Verdejo-Lucas, S. and Kaplan, D.T. (2002) The citrus nematode: *Tylenchulus semipenetrans*. In: Starr, J.L., Cook, R. and Bridge, J. (eds) *Plant Resistance to Parasitic Nematodes*. CAB International, Wallingford, UK, pp. 207–219.

Verdejo-Lucas, S., Sorribas, F.J., Pons, J., Forner, J.B. and Alcaide, A. (1997) The mediterranean biotypes of *Tylenchulus semipenetrans* in Spanish citrus orchards. *Fundamental and Applied Nematology* 20, 399–404.

Verdejo-Lucas, S., Sorribas, F.J., Forner, J.B., Alcaide, A., Ornat, C. and Galeano, M. (2000) Evaluating resistance to *Tylenchulus semipenetrans* Cobb in hybrid citrus rootstocks in Spain. *Proceedings of the International Society of Citriculture, IX Congress 2000*, Orlando, Florida, 1, pp. 818–822.

Verdejo-Lucas, S., Galeano, M., Sorribas, F.J., Forner, J.B. and Alcaide, A. (2003) Effect on resistance to *Tylenchulus semipenetrans* of hybrid citrus rootstocks subjected to continuous exposure to high population densities of the nematode. *European Journal of Plant Pathology* 109, 427–433.

Vilardebo, A. (1964) Etude sur *Tylenchulus semipenetrans* Cobb au Maroc II. *AL Awamia, Rabat* 11, 31–49.

Vilardebo, A., Sqalli, A. and Devaux, R. (1975) Utilisation du DBCP, du phenamiphos et du prophos contre *Tylenchulus semipenetrans* dans les vergers du Maroc. *Fruits* 30, 313–317.

Walker, G.E. and Morey, B.G. (1999) Effects of chemicals and microbial antagonists on nematodes and fungal pathogens of citrus roots. *Australian Journal of Experimental Agriculture* 39, 629–637.

Wheaton, T.A., Childers, C.C., Timmer, L.W., Duncan, L.W. and Nikdel, S. (1985) Effects of aldicarb on yield, fruit quality, and tree condition on Florida citrus. *Proceedings of the Florida State Horticultural Society* 98, 6–10.

Whitehead, A.G. (1968) Taxonomy of *Meloidogyne* (Nematodea: Heteroderidae) with descriptions of four new species. *Transactions of the Zoological Society of London* 31, 263–401.

Willers, P. (1979) *Influence of Citrus Nematode,* Tylenchulus semipenetrans*, on Navel Yield in Sundays River Valley Orchard.* Citrus and Subtropical Fruit Research Institute, Nelspruit, pp. 9–10.

Yassin, A.M. (1974) A note on *Longidorus* and *Xiphinema* species from the Sudan. *Nematologia Mediterranea* 2, 141–147.

Yokoo, T. (1964) Studies on the citrus nematode (*Tylenchulus semipenetrans* Cobb, 1913) in Japan. *Agricultural Bulletin, Saga University* 20, 71–109.

Yokoo, T. and Ikegemi, Y. (1966) Some observations on growth of the new host plant, snapdragon (*Antirrhinum majus* L.) attacked by root lesion nematode, *Pratylenchus coffeae* and control effect of some nematicides. *Agricultural Bulletin, Saga University* 22, 83–92.

Young, T.W. (1954) An incubation method for collecting migratory endoparasitic nematodes. *Plant Disease Reporter* 38, 794–795.

12 Nematode Parasites of Subtropical and Tropical Fruit Tree Crops*

Fahiem E. El-Borai[1] and Larry W. Duncan[2]
[1]*Plant Protection Department, Faculty of Agriculture, El Zagazig University, Egypt;*
[2]*Citrus Research and Education Center, University of Florida, Lake Alfred, Florida, USA*

This chapter covers tropical and subtropical fruit tree crops, for many of which detailed information concerning nematode damage is relatively scarce. We have included here 11 tree crops which, by virtue of their production value on a world basis or their importance in world trade, may largely be regarded in this context as major crops among the long list of tropical and subtropical fruits which are cultivated worldwide. These include eight fruit, three nut and two vine crops. We also treat here eight additional fruit crops which, by the same measure, may be considered to be of more local significance at the present time, although several of them are attracting increasing attention and hold definite economic potential. We have attempted to emphasize those nematode pests for which some evidence of economic impact exists. A literature review – up to 1992 – of nematodes associated with several tropical and subtropical fruits, is also available (McSorley, 1992). The fruit trees are herein reviewed in alphabetical order of their common names within each section.

Fruit Crops

Avocado

The avocado tree (*Persea americana* Mill.) originates from Central America and its fruit is consumed primarily as a fresh product. The major areas of commercial production today are regions in North, Central and South America (Mexico, Brazil, the USA, the Caribbean Islands, Colombia, Chile and Peru) and some Asian (the Philippines, Indonesia, China and Israel) and African (Ethiopia, Congo, South Africa, Cameroon and Zaire) countries (Ahmed and Barmore, 1980; Knight, 1980; Anonymous, 2002). Total world production in 2002 was reported to be 2,701,439 t, of which 72% was produced in the Americas, 11% in Asia and about 12% in Africa.

Avocado, in comparison with other tree crops, appears to be relatively free of aggressive nematode pests, and it is difficult to determine the economic importance of the identified nematode parasites to avocado production. Nevertheless, Sher (1955) attributed plant damage in California to *Pratylenchus vulnus*, and reduced tree

*A revision of the chapter written by E. Cohn and L.W. Duncan.

 Plant Parasitic Nematodes in Subtropical and Tropical Agriculture, 2nd Edition (eds M. Luc, R.A. Sikora, J. Bridge)

growth was shown to be caused by this nematode, both in greenhouse inoculation experiments and in pre-plant fumigation trials with DD (Sher *et al.*, 1959). *P. vulnus* is more likely to cause problems in avocado orchards planted adjacent to walnut orchards, where infested trees were found to be stunted, had fewer feeder roots and responded dramatically to pre-plant fumigation (Westerdahl, 2003). A similar situation may exist in Spain, where *P. vulnus* was found to reproduce well on horticultural crops such as almond, hazelnut, pistachio, grape, apple, cherry, plum, pear and loquat, as well as on avocado (Pinochet *et al.*, 1992). However, with the exception of pre-plant fumigation, practical nematode control recommendations to growers are unavailable.

Work done in Florida during the mid-1950s implicated *P. brachyurus* and *Radopholus similis* (citrus race) in reduced performance of avocado trees (Young and Ruehle, 1955), and Ducharme and Suit (1953) demonstrated their capacity to create root lesions. Again, however, it appears in retrospect that much of this and other work in Florida (McSorley, 1981) was related to surveys carried out in areas of citrus spreading decline, which at that time was a major economic disaster. No practical conclusions or recommendations regarding these nematode species in commercial avocado orchards have been developed since. Variability in the reaction of avocado varieties to different *R. similis* populations is not well characterized. Milne and Keetch (1976) reported avocado to be a non-host for the banana race of *R. similis* in Natal, South Africa, whereas the nematode was shown to be a pathogen of avocado in pot trials in India (Jasy and Koshy, 1992).

In Israel, populations of *X. brevicollum*, sometimes as high as 500/100 g of soil, are often recovered from around avocado roots, and reduced seedling growth in pots as a result of inoculation with this nematode has been demonstrated (Cohn, 1968). However, post-plant dibromochloropropane (DBCP) treatment in orchards did not consistently improve tree performance.

Interestingly, most of the economically important sedentary plant nematodes do not appear to be serious problems on avocado. *Rotylenchulus reniformis* has been observed on avocado roots in West Africa (Peacock, 1956), where Caveness (1967) found avocado to be a good host, and in Brazil (Sharma, 1978). *Meloidogyne hapla* was detected on avocado roots in New Zealand (Knight, 2001). There is no evidence that *R. reniformis* or *M. hapla* cause economic damage to avocado plants. The genera *Helicotylenchus* and *Meloidogyne* were encountered with some frequency at population densities between 150 and 350 nematodes/100 cm^3 of soil in a survey of avocado nurseries and orchards in Colombia (Saltaren *et al.*, 1999). Although *Meloidogyne incognita* race 2 failed to reproduce or cause galling on potted seedlings, inoculation with *Helicotylenchus dihystera* reduced seedling growth by 20–50% (Saltaren *et al.*, 1999). Similarly, granular nematicides reversed a decline of avocados infested with *H. dihystera*, *Criconemoides xenoplax* and *Xiphinima elongatum* (Willers, 1999), suggesting a need for futher study of the economic importance of ectoparasitic nematodes in avocado.

Possibly, the role of nematodes in damaging avocado roots has been overshadowed by the attention aroused by the severe avocado root disease caused by the soil fungus *Phytophthora cinnamomi* and, as suggested by Milne (1982a), it would be interesting to establish whether plant parasitic nematodes are capable of affecting the severity of this disease or the susceptibility to it of avocado trees.

Fig

The fig, *Ficus carica* L., one of the oldest fruits known to man, originates from the Mediterranean region, and is consumed mainly as a dried fruit (~90%), although some are marketed fresh, and a few are canned or made into preserves (Bolin and King, 1980). Turkey and Egypt produced more than 40% of the estimated 1,081,438 t total annual world production (Anonymous, 2002). Nearly three-quarters of the total is grown in Mediterranean countries, and figs are also grown in Iran, Syria, the USA, Australia and South Africa.

The root knot nematode is probably the most severe nematode problem in fig cultivation, and certainly the best documented. Numerous reports of root knot damage to fig exist from Mediterranean, North and South American countries, as well as from southern Africa, and among the identified species are *Meloidogyne arenaria, M. incognita, M. acrita* and *M. javanica* (McSorley, 1981). The problem is recognized as a major limiting factor in commercial fig production in the USA (Knight, 1980), France (Scotto La Massèse *et al.*, 1984) and Brazil (Ferraz *et al.*, 1982; Campos, 1997). Several measures have been recommended to reduce damage. Preplant fumigation permits better establishment of newly planted trees (Krezdorn and Adriance, 1961; Kodira and Westerdahl, 2002). Nematode populations were considerably reduced in young trees by stem treatments with an experimental paste formulation of phenamiphos (Inserra and O'Bannon, 1974). Partial nematode control and improved rooting on cuttings under nursery conditions were also attained by application of several systemic nematicides (Ferraz *et al.*, 1982).

Work has also been carried out to develop root-knot-resistant rootstocks for fig. Tests in California revealed that while all *F. carica* specimens examined were susceptible to *Meloidogyne*, four other *Ficus* species (*F. racemosa* L., *F. cocculifolia* Baker, *F. gnaphalocarpa* Steud. *ex* Miquel. and *F. palmata* Forsk.) showed a high degree of resistance to unidentified species of root knot nematodes, as well as good graft compatibility with *F. carica* (Krezdorn and Glasgow, 1970). The commercial variety 'LSU Purple' (from 'Hunt' × an unknown California caprifig) is reported to be nematode resistant (O'Rourke, 1992). Fig decline in some parts of Japan appears to result from pathogens in addition to *M. incognita* (Hosomi and Uchiyama, 1998). Twenty rootstocks grown in 'sick' soils exhibited a range of root galling, but growth of the scion 'Masui Dauphine' (used on all rootstocks) was not correlated with gall index. The rootstock 'Zidi' performed well in a number of field trials (Hosomi *et al.*, 2002). In Israel, the fig varieties 'Celeste' and 'Poulette' were considered resistant to the nematode, while the species *F. glomerata* Roxb. was found to exhibit a high degree of tolerance, but showed other unsatisfactory qualities as a rootstock (Gur, 1955). *Heterodera fici* is another nematode pest of fig, which is fairly widely distributed throughout the world, having been reported infesting trees in several Mediterranean countries including France (Scotto La Massèse *et al.*, 1984), Spain (Bello-Perez and Jimenez-Milian, 1963), Italy (Di Vito, 1976) and Turkey (Yuksel, 1981), as well as in California (Sher and Raski, 1956), Brazil (Brancalion *et al.*, 1981) and Soviet Asia (Narbaev and Sidikov, 1985). *H. fici* on fig was detected in one-third of 129 localities sampled throughout Yugoslovia, at densities ranging up to 330 cysts/100 cm^3 of soil (Krnjaic *et al.*, 1997). Leachates from fig roots stimulate egg hatch and emergence of juveniles from cysts (Di Vito and Sasanelli, 1990). The potential pathogenicity of *H. fici* on fig seedlings was demonstrated in pot trials by Di Vito and Inserra (1982) who reported 30% death of plants with an initial nematode population of 8 eggs and juveniles/cm^3 of soil, and 100% plant mortality with an initial nematode density of 64 eggs and juveniles/cm^3 of soil and higher. Thus, while field populations of *H. fici* do not generally appear to attain such damaging levels in orchards, the nematode can be considered a potential threat in fig nurseries, where fig rootstocks are often obtained from seeds. It is also noteworthy that the nematode has caused considerable damage to potted plants of the related *F. elastica* Roxb. (Scotto La Massèse *et al.*, 1984; Narbaev and Sidikov, 1985).

F. carica is the type host of *Xiphinema index* (Thorne and Allen, 1950), and this nematode attains extremely large populations around fig trees in the Mediterranean region. The anatomical changes caused by the nematode on fig roots – in the form of terminal galls and modified cells – as well as the associated biochemical changes have been studied and described in great detail (Poehling *et al.*, 1980; Wyss *et al.*, 1980), as has the feeding behaviour of the nematode on fig roots (Wyss, 1987). Although fig has been shown to be a more favourable host of *X. index* than grapevine (Coiro and Lamberti, 1978; Malan and Meyer, 1999), there does not appear to be as much damage to plant growth. Nevertheless, the nematode is considered of economic importance in California, where trees respond favourably to pre-plant fumigation (Koenning *et al.*, 1994; Kodira and Westedahl, 2003). There is no known virus transmission in fig by this nematode, which is the vector of fanleaf virus disease in grapevine. For this reason, movement of fig trees, especially to grape-producing regions, should require regulatory attention (Hirata *et al.*, 2002).

Other nematode species possibly associated with injury to fig roots are *Paratylenchus hamatus* in California (Thorne and Allen, 1950), *Pratylenchus vulnus*, which has been implicated as a possible pathogen of fig in California (McSorley, 1992; Westerdahl, 2003) and in France (Scotto La Massèse *et al.*, 1984), and *P. coffeae* which recently was found to be widely encountered at relatively high population density in fig in Jiangsu Province, China (Li *et al.*, 1999).

A large number of commercial, ornamental and wild fig species are pollinated during ovipositioning of a variety of wasp species. *Schistonchus caprifici* and several other nematode species complete their life cycles in the haemocoel of many of these wasps and in the fig inflorescences (Giblin-Davis *et al.*, 1995). The large variety of plant-insect species involved in this tritrophic interaction make it an interesting model to study the evolution of parasitism (Herre, 1995; Giblin-Davis *et al.*, 2003). Damage to florets results from nematode parasitism; however, population development in commercial figs is lower than in wild figs, and nematodes are of no known economic importance (Vovlas *et al.*, 1992; Vovlas and Larizza, 1996).

Guava

The common guava (*Psidium guajava* L.) is indigenous to tropical America. It is consumed as fresh fruit and also in processed form as jam, paste, puree, canned shells and juice. It is grown today throughout the tropics and subtropics and is of commercial importance in more than 60 countries (Lazan and Ali, 1998). Accurate statistics on production are not available. In South-east Asia, Thailand appears to be the largest producer, with a production of 100,000 t in 1987 (Kwee and Chong 1990), followed by Indonesia and Java (Verheij and Coronel 1991). Hawaii is the major producer in the USA, with a total production of 46,000 t (Davis, 1988).

The best documented nematode problem affecting guava is that created by the root knot nematode (*Meloidogyne* spp.) which is a recognized limiting factor in commercial guava production in Central American countries. In Cuba, guava production has declined steadily during the past quarter century due to increasing pressure from *M. incognita*, *M. arenaria*, *M. hapla*, *M. javanica* and other species (Cuadra and Quincosa, 1982; Rodriguez *et al.*, 1985; Fernandez Diaz Silveira and Ortega Herrera, 1998). New plantings can become non-productive within 5 years (Shesteperov, 1979). Growers in Cuba use nematicides in older plantings and attempt to establish new orchards in virgin sites. Nurseries increasingly grow containerized seedlings using clean soil. Other areas of the New World reporting root knot nematode problems in guava include Puerto Rico (Ayala, 1969), Mexico (Carillo-Rivera *et al.*, 1990; Avelar *et al.*, 2001), Venezuela (Crozzoli *et al.*, 1991; Casassa *et al.*, 1998; Zoraida Suarez *et al.*, 1999), Brazil (de Moura and de Moura, 1989) and Florida (Ruehle, 1959).

The problem in Cuba was also addressed by screening other *Psidium* species for possible resistant rootstocks and resulted in the commercial use of the rootstock *P. friedrichstalianum* (Berg.) Nied., which evidently shows a high degree of resistance to *Meloidogyne* spp. (Fernandez Diaz-Silveira,

1975). Casassa *et al.* (1998) and Matehus *et al.* (1999) found that *M. incognita* populations were not supported by *P. friedrichstalianum*, and the tolerance limit of the rootstock was 60-fold lower than that of a *P. guajava* cultivar. However, the reaction appears to vary with plant material or nematode population. Gonzales and Sourd (1982) and Villota *et al.* (1997) found *P. friedrichstalianum* to show only moderate tolerance to *Meloidogyne*. Other *Psidium* species – among them *P. cattleianam* Sabine, *P. molle* Bertol., *P. guineercsis* and *P. guayabita* – and cultivars of *P. guajava* were highly susceptible to the nematode (Cuadra and Quincosa, 1982; Babatola and Oyedunmade, 1992; Maranhao *et al.*, 2001).

It is noteworthy that there are fewer reports of major damage to guava by root knot nematodes outside of the Caribbean and America. Although a case of slight root galling by *M. arenaria* was reported by Martin (1959) from central Africa, and management trials for *M. incognita* on guava have been conducted in Malaysia (Tuck, 1998) and Taiwan (Lee *et al.*, 1998), occurrence of root knot nematodes on guava seems to be less common outside of the New World. Sikora (1988) reported heavy galling of guava roots – with associated tree decline – in two isolated regions in Niger, evidently involving a nematode species not found on any vegetable crop in the vicinity (Plate 13A and B). A putative virulent race of *M. arenaria* from West Africa, capable of breaking resistance in various vegetable and field crops, was eventually identified as *M. mayaguensis* (Fargette *et al.*, 1996; Blok *et al.*, 1997). Willers (1997) and Carneiro *et al.* (2001) recently identified *M. mayaguensis* as the cause of severe guava decline in Mpumalanga (Eastern Transvaal), South Africa, and in Pernambuco and Bahia states in Brazil. It is therefore possible that the severe root knot problem in the Americas and the isolated cases in Africa involve specialized and particularly virulent species and races of *Meloidogyne*.

Three other plant parasitic nematodes attacking guava warrant mention: *H. dihystera* was found consistently associated with guava plantations in South Africa and was shown to reduce height and leaf size of guava seedlings in inoculation trials (Wipers and Gretch, 1986). Hamiduzzaman *et al.* (1997) and Khan *et al.* (2001) observed greater damage to guava inoculated with both *H. dihystera* and *Fusarium oxysporum* than with the nematode alone. *Hoplolaimus indicus* was shown in pot experiments to be a pathogen of guava in India (Mahto and Edward, 1979; Nigam *et al.*, 1995), and *Tylenchorhynchus cylindricus*, in numbers of up to 2000 nematodes/100 cm^3 of soil, was found associated with damaged guava trees in Iran (Abivardi, 1973).

Lychee

The lychee (*Litchi chinensis* Sonn.) – also spelled litchee, litchi and, its dried fruit form, 'litchi nut' – is indigenous to southern China and is marketed as fresh, dried and canned fruit. China, India, South-east Asia and South Africa are among the major producer regions. Smaller industries exist in the USA and elsewhere (Menzel and Simpson, 1994). World production of lychees is about 1 Mt, with the bulk of the crop growing in China and India (Partridge, 1997). Over 700,000 t of fresh lychees are consumed annually in Asia and India, and a large proportion is also processed in the form of canned fruits or juice (Waite and Hwang, 2002). Countries such as South Africa, Mauritius, Madagascar, Réunion and Australia export lychees.

Detailed information on economic nematode damage to lychee is available only from South Africa. Milne (1982a) recognized *X. brevicollum* and *Hemicriconemoides mangiferae* as major nematode pests of lychee, causing a severe tree decline syndrome. Typical above-ground symptoms were the presence of many bare twigs and branches, leaf chlorosis, leaf-tip burn, poor flowering and excessive fruit drop, and in some orchards up to 40% of the trees died. Root symptoms were severe stubby root and darkening of the roots, leading eventually to loss of a large proportion of the feeder root mass and consequent interference in the uptake of nutrients and water. *X. brevicollum* feeds more superficially, while *H. mangiferae*, which causes extensive destruction of the cortical tissue, is considered the more severe pathogen. Populations as high as 40,000 *H. mangiferae*/dm^3 of soil and roots and 20,000 *X. brevicollum*/dm^3 of soil and roots were recorded.

Pre-plant soil fumigation with Telone or methyl bromide effectively improved the performance of replants in infested areas. DBCP treatment of established trees induced a favourable growth response and attained good nematode control.

M. javanica infection of lychee roots in orchards – confirmed by inoculations – was encountered, but galls are generally inconspicuous. *Trichodorus* spp. have also been found associated with nursery seedlings.

Mango

Mango (*Mangifera indica* L.), the most important and most widely grown tropical fruit, originates from the Indo-Malaysian region and is today cultivated in most tropical and subtropical countries. It is marketed largely as fresh fruit, but also processed as juice, puree, chutney and pickle (Knight, 1980). Total world production in 2002 was 25,754,509 t (Anonymous, 2002), of which 77% was from Asia, where India produces more than half of the mangos, followed by China, Thailand, Pakistan, the Philippines and Indonesia. African countries such as Nigeria, Egypt, Congo, Madagascar, Sudan and Ethiopia produce about 10% of world production, and numerous countries throughout Central and South America grow about 6% of world supply. Like avocado, mango appears to be relatively free from severe nematode damage, despite the fairly long list of nematode species associated with it (Ghorab *et al.*, 1987; Petit, 1990; Korayem and Koura, 1993; Yin, 1995; Anita and Chaubey, 2003). Probably the most widely distributed nematode associated with mango is *H. mangiferae* (Siddiqi, 1977; McSorley, 1992), which has been shown to be potentially damaging to mango seedlings at a population level of 6 nematodes/cm^3 of soil (Saeed, 1974). The nematode is widespread in mango orchards and on numerous other crops in India, particularly in sandier soils, where population density was strongly and positively related to soil moisture (Ashokkumar *et al.*, 1991). Although the pathogenicity to mango of *H. mangiferae* has been demonstrated (McSorley, 1992), its economic importance in the field is unclear. McSorley *et al.* (1981) reported a wide distribution and strong association between *H. mangiferae* and declining mango in Florida, but the relationship is not always evident (Ashokkumar *et al.*, 1991). The nematode was observed feeding on mango roots together with *X. brevicolle* in South Africa, but chemical treatment of existing trees, while reducing nematode populations, failed to induce a favourable tree response (Milne, 1982b). Mango infested by *H. mangiferae* and *R. reniformis* responded strongly to treatment with systemic nematicides; however, control of the mite *Eriophyes mangiferae* that is involved in a witches broom syndrome may have caused at least part of the response (Noriega *et al.*, 1988).

Economic responses to chemical treatment in mango were reported when using DBCP to control *Hoplolaimus columbus* and *Xiphinema* sp. in Egypt (Shafiee and Osman, 1971), and phenamiphos applications were found effective in controlling *P. brachyurus*, but not *R. reniformis* in Florida (McSorley and Parrado, 1983). *R. reniformis* appears to be the only sedentary nematode that commonly infects mango (McSorley and Parrado, 1983) and, interestingly, soil and root populations on seedlings were reduced effectively by application of the growth regulant ethephon (Badra and Khattab, 1982). Anwar *et al.* (1991) considered *R. reniformis* to be damaging on mango, but McSorley *et al.* (1981) found no relationship between infestation by the nematode and tree decline symptoms. Although detected in soil associated with mango (Maqbool, 1991; McSorely, 1992), there is only one report documenting infection of mango by *M. incognita* (Mani *et al.*, 1995).

Olive

The olive tree, *Olea europaea* L., is apparently a native of Western Asia, and is cultivated primarily in the Mediterranean Basin, largely (~75%) for oil extraction. Total world production of olives increased from 827,300 t in 1985 to 13,976,487 t in 2002 (Anonymous, 2002). Countries bordering the northern Mediterranean (Spain, Italy, Greece and Turkey) grow 75% of the total crop, and smaller industries are important in the Middle East (Jordan, Iraq, Iran and Syria), North Africa (Egypt, Morocco, Algeria and Tunisia) and the Americas (Argentina, the USA, Peru and Mexico).

Olive is a host to more than 70 species of plant parasitic nematodes in 33 genera (Lamberti and Vovlas, 1993). It is the type host for a number of species such as *Rotylenchulus macrosoma*, *Meloidogyne lusitanica*, *M. baetica*, *Helicotylenchus oleae* and *H. neopaxilli*, that are rarely encountered in other crops. Surveys of nematodes encountered in many of the major olive industries (Pena-Santiago, 1990; Nejad *et al.*, 1997; De Abrantes *et al.*, 1998; Kepenekci, 2001; Nico *et al.*, 2002) have detected a number of nematode species that are important pathogens of other crops, and pathogenicity of several species to olive has been demonstrated. Nevertheless, olive is an extremely vigorous plant which thrives in hilly, relatively dry areas where most groves are situated. Under such conditions, nematodes generally occur in small numbers and are apparently of limited economic importance. Also, in old groves, trees are grafted on wild olive rootstocks that may tolerate nematodes to some extent. In newer groves, plants are derived from cuttings of various cultivars that are less tolerant of nematode damage. Use of irrigation in these groves and especially in nurseries increases the impact of nematodes in spite of the propensity of olive to regenerate roots (N. Greco, personal communication; Castillo *et al.*, 1999; Nico *et al.*, 2002).

Two species of root knot, *M. incognita* and *M. javanica*, although occurring only patchily in existing groves (Hashim, 1982), have been shown to reduce seedling growth drastically in inoculation trials and have been identified as a factor to reckon with in olive nurseries (Diab and El-Eraki, 1968; Lamberti and Baines, 1969a; Sasanelli *et al.*, 1997). Indeed, tolerance limits of 0.49–0.61 eggs/cm^3 and minimum yields near 50% were estimated for the effects of *M. javanica* on seedling growth of two olive cultivars (Sasanelli *et al.*, 2002). *M. arenaria* race 2, *M. javanica* and *M. incognita* race 1 caused growth reduction, yellowing and leaf drop in seedlings of two common olive rootstocks (Nico *et al.*, 2003). *M. baetica* is a newly described species from wild olive in Spain that does not develop on several hosts (tomato, chickpea and pea) of more commonly encountered *Meloidogyne* species (Castillo *et al.*, 2003). The economic importance of *M. baetica* is unknown, but it infects common olive rootstocks, and its histopathology and population development on olive are similar to those reported for other root knot species. Lamberti *et al.* (2001) found some evidence that infection by *M. incognita* enhances damage by the wilt-inducing fungus *Verticillium daliae*.

P. vulnus has been implicated by Lamberti (1969) as a factor in olive decline in Italy, and has been demonstrated in inoculation trials as a potential pathogen of olive (Lamberti and Baines, 1969b). Nico *et al.* (2002, 2003) found that *P. vulnus* and *P. penetrans*, frequently encountered in nurseries in Andalusia, reduce seedling growth in pots, suggesting a need to evaluate their effect in the field. Several species of *Helicotylenchus*, particularly *H. dihystera*, *H. digonicus*, *H. erythrinae* and *H. oleae*, have been observed to cause root necrosis (Inserra *et al.*, 1979), and are considered by some workers to be capable of affecting olive tree growth (Graniti, 1955; Diab and El-Eraki, 1968). Species of *Xiphinema* also commonly occur around olive roots, and *X. elongatum* and *X. index* have been shown to affect olive plant growth (Diab and El-Eraki, 1968; Sasanelli *et al.*, 1999).

A number of rather specialized sedentary plant nematodes attack olive. A biotype of the citrus nematode, *Tylenchulus semipenetrans*, infects olive in California and Italy, and although population levels on olive are usually lower than on citrus (Inserra and Vovlas, 1977), unusually high levels of *T. semipenetrans* have been shown to inhibit olive growth (Lamberti *et al.*, 1976; McKenry, 2000). *Trophotylenchulus saltensis* was described from olive roots in Jordan (Hashim, 1983b), and a very specialized cyst nematode, *Heterodera mediterranea*, first recorded from Italy, was shown to be capable of feeding and multiplying on olive roots, in which it forms syncytia and causes disorder of the stelar structure (Vovlas and Inserra, 1983). The first naturally occurring infestation of olive orchards by *H. mediterranea* was detected recently in Spain (Castillo *et al.*, 1999) and, although visual symptoms of pathology were unapparent, the nematode was capable of reducing growth of the cultivar 'Arbequina', but not 'Picual', in pot studies (Castillo and Vovlas, 2002). Two sedentary ectoparasitic nematode species, *Gracilacus peratica* and *Ogma rhombosquamatum*, have been observed to feed on olive roots, and their feeding behaviour has been described in detail (Inserra and Vovlas, 1977; Vovlas and Inserra, 1981); however, there is no evidence of a pathogenic effect. Similarly, three species of *Rotylenchulus* have been studied in detail on olive, namely *R. macrodoratus* (Inserra and Vovlas, 1980), *R. macrosoma* (Cohn and Mordechai, 1988) and *R. reniformis* (Hirschmann *et al.*, 1966), but evidence of actual plant damage is lacking. Wild olive orchards heavily infested by *R. macrosoma* recently were discovered for the first time in Spain (Castillo *et al.*, 2003). Although infected plants did not show above-ground disease symptoms, further study of the potential of *R. macrosoma* to damage olive in nurseries or groves is warranted because the nematode does not yet appear to be widespread.

Measures for practical nematode control in olive have been limited so far to nurseries, where pre-plant fumigation with available nematicides has been recommended for controlling diverse nematode species (Hashim, 1982; McKenry, 2000). Suggestions for bare root dips of seedlings in suspensions of nematicidal chemicals (such as phenamiphos), prior to transfer into groves, have also been offered for reducing root knot nematode infestation (Lamberti and Di Vito, 1972). In California, various soil solarization techniques were as effective as soil fumigation to disinfest olive nursery soils of *T. semipenetrans*, *P. vulnus* and *Criconemoides xenoplax* (Stapleton *et al.*, 1999), and solarization may be an ideal tactic for nursery sanitation in Mediterranean industries. Resistance or tolerance of some olive cultivars to *P. vulnus*, *R. reniformis* and various species of *Meloidogyne* has been reported (Al-Sayeed and Abdel-Hameed, 1991; Pinochet *et al.*, 1992; Robinson *et al.*, 1997; Sasanelli *et al.*, 1997), and improved methods to screen olive explants for nematode resistance have been developed (Sasanelli *et al.*, 2000).

Papaya

The papaya (*Carica papaya* L.) is a native of tropical America and is widely distributed today throughout tropical areas of the world, where it is produced largely for fresh fruit, but is also marketed as a preserve and for juice. Another product of papaya is the enzyme papain, a digestive enzyme which is used as a food tenderizer. Papaya is a very good source of vitamins A and C (Knight, 1980). More than 5,950,722 t of papaya were produced worldwide in 2002, of which 47% were produced in Central and South America (Brazil grew 25% of world production, followed by Mexico, Peru and Cuba), 30% in Asia (mainly India, Indonesia, the Philippines and China) and about 20% in Africa (mainly Nigeria, Ethiopia, Congo, Mozambique and South Africa) (Anonymous, 2002).

Of the several nematodes reported to be associated with papaya, only two genera appear to be economically significant in papaya cultivation. These are the root knot nematode (*Meloidogyne* spp.) and the reniform nematode (*Rotylenchulus* spp.), both of which enjoy a worldwide distribution in papaya plantations. In Hawaii, yield losses to these two species are estimated to be 15–20% (Koenning *et al.*, 1999). It is also noteworthy that *H. dihystera* at densities as high as 200 nematodes/g of root was detected in South Africa, although there are no subsequent reports of pathogenicity (Willers and Neething, 1994).

Heavy root knot infections of papaya, primarily by *M. incognita* and *M. javanica*, have been reported from many countries from all continents (McSorley, 1981). Root galling is often severe; galls can be as large as golf balls (Milne, 1982a). A tolerance limit of 0.16 eggs and juveniles/cm^3 of soil and a minimum yield of 0.77 was estimated for the effect of *M. incognita* race 1 on papaya seedling weight in the greenhouse (Bustillo *et al.*, 2000). Other studies have reported similar levels of seedling damage by *M. incognita* (Ramakrishnan and Rajendran, 1998a), whereas some report much lower minimum yields (Lamberti *et al.*, 1980; Darekar and Mhase, 1986; Singh and Nath, 1997). The discrepancy may be due to differences in experimental conditions, virulence of nematode populations or to inadvertent infestation by other pathogens such as *Fusarium solani* (Khan and Husain, 1991). Root knot nematode causes severe damage in the field (Wolfe and Lynch, 1950), producing root rot, reducing the life expectancy of the plant and drastically decreasing yield levels (Milne, 1982a). Ramakrishnan and Rajendran (1998b) estimated nearly 40% fruit loss in the first crop after planting, with increasing losses likely thereafter due to *M. incognita*. Recommended control measures call for pre-plant soil fumigation, especially in seedbeds, and selection of non-infested planting sites. Post-plant application of systemic nematicides such as carbofuran, aldicarb and fenamiphos effectively reduced root gall formation and increased plant growth and yield (Gupta and Yadav, 1988; Routaray and Das, 1988; Ramakrishnan and Rajendran, 1999). Although less successful than nematicides, organic amendments (particularly neem) and mulches provided some control of root knot nematodes and increased fruit yield (Routaray and Das, 1988; Khan *et al.*, 1997; Ramakrishnan and Rajendran, 1999; Elder *et al.*, 2002; Srivastava, 2002). Most cultivars of *C. papaya* are highly susceptible to *M. incognita* (Babatola, 1985; Iglesias and Perez, 1991). Closely related species such as *Carica quercifolia* and *C. candamarcensis* are also root knot susceptible (McSorley, 1981), although there is a discrepancy regarding *C. cauliflora* (Rosales and Suarez, 1991; Indra and Rajvanshi, 2001). Two studies have found *C. papaya* cv. Pusa 22-3 to be resistant, and several other cultivars are reported to have varying degrees of resistance to *M. incognita* (Reddy *et al.*, 1988; Khan *et al.*, 1995; Ramakrishnan and Rajendran, 1998a,b). Severe *M. incognita* infestations of papaya and tomato were observed wherever they were intercropped (R.A. Sikora, Germany, 2004, personal communication) in fields in North Yemen (Plate 13C).

Reniform nematode infection of papaya, by *R. reniformis*, has also been reported from all continents. *R. parvus* has been identified from Kenya, and unidentified species of *Rotylenchulus* reportedly have been associated with this crop in Thailand and Florida (McSorley, 1981). *R. reniformis* has been implicated in severe plant damage and yield reduction in Puerto Rico (Ayala *et al.*, 1971), and in Trinidad it has been associated with tree death and toppling (Singh and Farrell, 1972). A survey of papaya in five Indian states detected 100% incidence of *R. reniformis* at population densities up to 1025 nematodes/100 g of soil (Ganguly *et al.*, 1997). In Fiji, severe damage by the nematode has been reported in nursery seedlings and young plants (Heinlein, 1982; Vilsoni and Heinlein, 1982), and in Brunei plants have reportedly been killed by a combination of *R. reniformis* and *Phytophthora nicotianae* var. *parasitica* (Anonymous, 1972). In pot experiments, *R. reniformis* at the rate of 620 nematodes/200 cm^3 of soil reduced the growth of papaya seedlings (Karim, 1989). Ramakrishnan and Rajendran (1999) found that *M. incognita* is more pathogenic than *R. reniformis*, and competition between the two mitigated the virulence of *M. incognita*. Despite its widespread occurrence in papaya, there are few reports on management of the nematode in the field. Pre-plant soil fumigation in Hawaii with various chemicals has effectively controlled the nematode and maintained low populations over periods of up to 6 months, with resultant yield increases in 15-month-old plants (Lange, 1960); however, foliar applications of the systemic nematicides phenamiphos and oxamyl in Puerto Rico were not only ineffective in reducing nematode numbers but also showed some phytotoxicity (Ayala *et al.*, 1971). Some cultivars of *C. papaya* appear to be resistant to *R. reniformis* (Patel *et al.*, 1989).

Persimmon

Persimmon belongs to the genus *Diospyros*, of which nearly 190 species are known. Almost all commercial persimmon fruit belongs to the species *D. khaki* L. (hence the common name in Europe, Khaki fruit), although *D. lotus* L. and *D. virginiana* L. are often used as rootstocks. *D. khaki*, known also as the Japanese persimmon, is probably native to China and was introduced early to Japan (Itoo, 1980). It is grown commercially today – largely for fresh, but also dried fruit. The total world production is about 2,328,919 t (Anonymous, 2002), of which approximately 71% is produced by China and 24% equally between Japan and Korea. Smaller, but expanding persimmon industries are being developed in Brazil, Italy, Korea, Israel, Spain, the USA, New Zealand and Australia (George and Nissen, 1990; Anonymous, 1994; Mowat and George, 1994). Little is known about economic nematode damage to persimmon. Although root knot nematode (*Meloidogyne* spp.) and burrowing nematode, *R. similis*, have been reported to parasitize both *D. khaki* and *D. virginiana* (McSorley, 1981; Sethi *et al.*, 1988; Inomoto *et al.*, 1991; Khurramov, 1993), no reports of actual plant damage by these nematodes appear to exist. *Pratylenchus scribneri* was reported to be the most frequently encountered nematode associated with persimmon in Korea (Park *et al.*, 1999).

The only nematode species associated with damage to the crop appears to be the citrus nematode, *T. semipenetrans*, for which persimmon has been reported to be a very susceptible host. Extremely large soil and root populations of *T. semipenetrans* are commonly encountered in unthrifty persimmon orchards in Israel on *D. virginiana* rootstock (Cohn and Minz, 1961) and have also been observed in California on *D. lotus* rootstock (Nesbitt, 1956). A similar observation on *D. lotus* roots was reported in Italy (Di Maio, 1979), where a resultant 20–30% loss in yield was estimated. Persimmon serves as a reservoir host for *T. semipenetrans* in parts of Brazil in which citrus is banned in order to control citrus canker (Inomoto *et al.*, 1991). Citrus nematode has also been reported from persimmon in Chile and New Zealand (Gonzalez, 1988; Knight, 2001).

Although no direct control measures appear to have been tested, it would seem probable that pre- and post-plant chemical applications, as recommended in citrus cultivation, could effectively reduce *T. semipenetrans* populations on persimmon, if such treatments would be considered economically feasible. Other cultural control measures against the nematode in citrus groves could also be relevant to persimmon. No information is as yet available on the level of resistance to the nematode of the various persimmon rootstocks or other *Diospyros* species.

Nut Crops

Cashew

The cashew nut (*Anacardium occidentale* L.) is a native of Brazil, where about 12% of the world crop is produced today. World production in 2002 totalled 1,516,935 t, of which 48% is grown in Asia (where India is the largest producer followed by Vietnam and Indonesia) and 39% in tropical Africa (mainly Nigeria, Tanzania, Côte d'Ivoire, Guinea Bissau and Kenya). El

Salvador and Peru also have small cashew industries (Anonymous, 2002).

Limited information on nematodes attacking cashew exists. High populations of *Criconemoides*, *Xiphinema* and *Scutellonema* have been found around unthrifty trees in Brazil (Lima *et al.*, 1975), and da Ponte (1986) recognized 'xifinematose', caused by *X. index*, as one of the more common diseases of cashew in north-east Brazil, although data on its economic impact are lacking. A recent review of cashew diseases in Brazil concluded that nematodes supported by the plant cause no evident damage (Freire *et al.*, 2002). *Xiphinema ifacolum* suppressed growth of several tree crops in nurseries in Liberia, but not cashew (Lamberti *et al.*, 1992). *R. reniformis*, apparently in its migratory form, was reported from around cashew trees in Costa Rica, but, again, evidence of damage is unclear (Lopez and Azofeifa, 1985; Lopez and Salazar, 1987). *Hemicycliophora attapadii* was described from the cashew rhizosphere in India (Rahaman *et al.*, 1996). It is important to emphasize that cashew has been shown clearly to be immune, or at least highly resistant to different populations of the root knot nematode in West Africa (Netscher, 1981) and in Brazil (da Ponte and Saraiva, 1973).

Macadamia

The macadamia nut (*Macadamia integrifolia* Maiden and Betche and *M. tetraphylla* L. Johnson), native to south-eastern Queensland, Australia, is also called Australian nut and Queensland nut. The total world production of 72,914 t of macadamia nuts in 1997 was 67% greater than in 1992 (Anonymous, 1998). The USA harvested 26,309 t in 1997, mainly grown in Hawaii, and Australia replaced the USA as the world's leading producer that year with 27,500 t (Anonymous, 1998). Despite increasing expansion of macadamia hectarage in key producing regions such as Australia, Guatemala, Kenya and South Africa, virtually no information on nematode damage to this crop is available.

Pistachio

The pistachio tree (*Pistacia vera* L.) is native to western Asia and Asia Minor, where about three-quarters of the 571,150 t world crop was grown in 2002. Iran produced more than half of the world total, followed by Syria, Turkey and a few eastern Mediterranean countries. Since the 1960s, pistachio acreage in California increased rapidly, and by 2002, the USA accounted for just under 24% of the world production (Anonymous, 2002). Pistachio growers often use species of *Pistacia* other than *P. vera* as rootstocks. Some of these, particularly *P. atlantica* Desf. and *P. terebinthus* L., have increased resistance to *Meloidogyne javanica* (Anon., 1975) and possibly to other root knot species (McKenry and Kretsch, 1984), although root galling does occur. McKenry and Kretsch (1984) surveyed pistachio orchards in California for plant parasitic nematodes, and found the common occurrence of *Paratylenchus hamatus*, *Pratylenchus neglectus* (syn. *P. minyus*) and *Xiphinema americanum*. *Meloidogyne* spp. were recovered in a minority of the orchards. They concluded that plant parasitic nematodes did not present a serious problem to pistachio production in California. In Iran, even in areas where *P. vera* is widely used as a rootstock, nematodes are not generally considered of economic importance (Javanshash *et al.*, 2000). However, one report of *M. javanica* from galled roots of declining trees in Iran's Semnan Province recommended the use of nematode-free nursery stock, especially in the region's new plantings (Banihashemi and Kheiri, 1995). Two species of *Pistacia*, *P. lentiscus* and *P. vera*, are natural hosts of *Heterodera mediterranea* in Italy (Vovlas and Inserra, 1983), and *P. vera* roots were reported to be infected and heavily galled by the sedentary nematode *Rotylenchulus macrodoratus* (Vovlas, 1983). *Pistacia vera*, *P. atlantica* and *P. terebinthus* are hosts of *P. vulnus*, but there is no evidence of economic importance (Pinochet *et al.*, 1992).

Vine Crops

Passion fruit, kiwifruit and grape are widely cultivated, fruit-bearing vine crops. Because they are not included in many other nematological reviews, the first two crops are treated here. Excellent reviews of nematodes attacking grape have been written by Brown *et al.* (1993) and Esnard and Zuckerman (1998).

Kiwi

Actinidia deliciosa (A. Chevalier) C.F. Liang *et* A.R. Ferguson, native to China, was known primarily as Chinese gooseberry until 1962 when New Zealand growers began to market the fruit as kiwifruit. Ichang gooseberry, monkey peach and sheep peach are other common names. The fruits are mostly consumed fresh, with smaller markets for the juice, and as flavouring. The plant is a vigorous, woody vine that is long lived, in some cases more than 50 years. It grows and produces fruit best in northern tropical areas. The total world production in 2002 was 1,001,121 t, with 34% produced in Italy, 24% in New Zealand, 15% in Chile and smaller industries in France, Greece, Japan, the USA, Iran and Spain (Anonymous, 2002).

The only widespread nematode damage reported on kiwifruit is caused by *Meloidogyne* spp. In France and Italy, *M. hapla* and *M. arenaria* induce small, discrete root galls whose histopathology is similar to that on other crops. In both countries, root knot infestations were associated with unthrifty plants. *M. hapla* and *M. incognita* are both widespread in kiwi orchards in Chile (Philippi *et al.*, 1996). Although kiwi cv. Hayward seedlings were relatively tolerant of *M. hapla* in pot trials in Chile (Philippi and Budge, 1992), Di Vito *et al.* (1988) found that *M. incognita* race 1 caused serious growth suppression in kiwi cv. Howard, with an estimated tolerance limit of 0.43 eggs and juveniles/cm^3 of soil and minimum yield of 0.45. Contaminated nursery stock has resulted in serious infestation by *Meloidogyne* spp. in kiwi orchards in Greece (Vlachopoulos, 1994). The only other record of damage by species other than root knot nematodes is from a pot experiment in which *Pratylenchus penetrans* decreased kiwi seedling growth (Vrain, 1993). The possibility of interactions between nematodes and major soil-borne pathogens of kiwifruit such as *Agrobacterium tumefaciens* and *Phytophthora cinnamomi* has been suggested (Scotto La Massèse, 1973; Talame, 1976; Mancini *et al.*, 1978).

There are no reports of resistant rootstocks for kiwi. Chemical bare root dips with ethoprop and phenamiphos gave good control of root knot infestations in nursery stock (Dale and van der Mespel, 1972; Grandison, 1983). California kiwi growers are advised to avoid the use of cover crops that are hosts of root knot nematodes and to increase irrigation frequency in infested orchards. Pre-plant application of methyl bromide or dichloropropene, or post-plant treatments with fenamiphos are also recommended (McKenry, 2002). Reports of nematode management in kiwi in Europe and elsewhere have focused on organic farming methods. Mycorrhizal kiwi were shown to be more tolerant of *M. javanica* than non-mycorrhizal seedlings (Verdejo *et al.*, 1990), and several studies report some control of root knot nematodes with various organic amendments, mulches and biological control agents (Cayrol *et al.*, 1991; Gonzalez, 1993; Maccari *et al.*, 1993).

Passionfruit

Two varieties of *Passiflora edulis* Sims are known as passionfruits – purple passionfruit, *P. edulis*, and yellow passionfruit, *P. edulis* f. *flavicarpa*. Other common names for both forms include grenadilla, parcha, chinola, parchita, lilikoi, maracuyá, maracuja, peroba, grenadine and couzou. A woody, shallow-rooted vine, the plant is native to a region from southern Brazil to northern Argentina. In this area, the yellow form is processed for juice and the purple form is consumed fresh. Although purple passionfruit was often preferred initially in other areas of the world, it is more susceptible to some nematodes and to Fusarium wilt, and yields substantially less fruit than the yellow form, so that acceptable selections of both types have been developed. Yellow passionfruit is tropical or near-tropical, and purple passionfruit is subtropical. Plantation life ranges from 3 to 8 years and is strongly affected by management of soil-borne diseases (Morton, 1987). The total world production of passionfruit is 780,000 t, of which 52% is grown in Brazil and 23% in Ecuador. The fruit is also grown commercially in Colombia, Peru, Venezuela, Bolivia, Kenya, Zimbabwe, South Africa, Zambia, Uganda, China, Malaysia, Vietnam, Thailand, Sri Lanka, Indonesia, Mexico, Israel and New Zealand (Frei, 2001).

Although a number of plant parasitic nematodes are reported associated with passionfruit (Boesewinkel, 1977; Loof and Sharma, 1979; Milne, 1982a; Sanchez *et al.*, 1993; Suarez *et al.*, 1993; Knight, 2001), only reniform and root knot nematodes are reported to cause economic damage. Both nematodes can severely limit fruit production and plant longevity. *R. reniformis* was detected in 84% of sites sampled in Fiji (Kirby, 1978), with numbers as high as 36,000 nematodes/200 cm^3 of soil. Yellow passionfruit seedlings growing in naturally infested soil were smaller, had chlorotic leaves and darker roots than plants growing in steamed soil in pot studies. However, no effort was made in this experiment to control the *Phytophthora* species which causes collar rot, the most severe disease of passionfruit. In Brunei, *R. reniformis* is reported to enhance collar rot, and plant life is doubled when infested soil is treated with fenamiphos prior to planting. High populations of the nematode were detected consistently in surveys of experimental field plots (Peregrine and Yunton, 1980). The nematode is also associated with passionfruit in Belize, Colombia and Brazil (Sanchez *et al.*, 1993; Bridge *et al.*, 1996; Sharma *et al.*, 2000). Little is known about cultivar susceptibility. Ten cultivars of *P. edulis* were all hosts of a Brazilian population of *R. reniformis*; however, the nematode did not reduce the growth of any cultivar and stimulated growth of the majority (Sharma *et al.*, 2001). The fine structure of the *R. reniformis*–yellow passionfruit interaction has been described in detail (Suarez *et al.*, 1993).

M. incognita (Reddy *et al.*, 1980), *M. javanica* and *Meloidogyne* sp. (De Villiers and Milne, 1973) appear to vary in pathogenicity to passionfruit. In Kenya, it has been suggested that root knot nematodes are not an economic problem on the crop (Ondieki, 1975), and in Fiji, *M. incognita*, *M. arenaria* and *M. javanica* did not reproduce on yellow passionfruit or affect plant growth in pot studies (Kirby, 1978). Therefore, passionfruit is recommended as a suitable rotation crop in Fiji against root knot nematodes. Significant resistance based on root galling studies was also reported for both yellow and purple passionfruit and *M. incognita* and *M. javanica* in Brazil (Klein *et al.*, 1984; Costa *et al.*, 1997; Sharma *et al.*, 2002). In South Africa, however, *M. javanica* and possibly other species are considered as serious pests on yellow and especially purple passionfruit (Milne, 1982a). It is unclear whether damage is due to initial penetration intolerance of seedling and young plant roots by the nematode or to resistance to parasitism. Methyl bromide fumigation of seedbeds is reported to increase plant growth, and preplant treatment of planting sites resulted in marked yield increase (De Villiers and Milne, 1973). It is suggested that soils be leached after methyl bromide fumigation to avoid phytotoxicity. Use of rootstocks such as *P. caerulea*, which are tolerant to root knot nematodes, has also been suggested (Milne, 1982a; Terblanche *et al.*, 1986). Since the vine is relatively short lived and seedling establishment is of great importance, crop rotations should also be useful for nematode control (Milne, 1982a).

Passionfruit has also been suggested as a good rotation crop in South Africa against *R. similis* which does not infect either *P. edulis* or *P. edulis* f. *flavicarpa* (Milne and Keetch, 1976).

Miscellaneous Fruit Trees

Acerola

The acerola, or West Indian cherry (*Malpighia glabra* L., *Malpighia punicifolia* L. (dwarf) and *Malpighia* spp.), is known in cultivation mainly in the West Indies and tropical Central America, from where it originates, and has more recently been introduced to Hawaii, Brazil, India and Africa (Knight, 1980). It is still very limited in production, but is enjoying increasing interest as a commercial product rich in vitamin C. Puerto Rico and, more recently, Brazil are leading producers, and much of our knowledge on nematodes attacking acerola comes from those countries. Ayala (1969) has reported that the plant can be almost destroyed as a result of root knot nematode (*M. incognita*) infection. When the crop was first introduced from Puerto Rico to north-eastern Brazil, *M. incognita* and *M. arenaria* severely damaged the first commercial plantations (Franco and da Ponte, 1989), and *M. incognita*, *M. arenaria* and *Rhizopus nigricans* (*R. stolonifer*) are considered the major limiting factors to acerola production in Brazil (Holanda *et al.*, 1997). A report has cautioned that *M. incognita* races 1, 2, 3 and 4, *M. javanica*, and *M arenaria* race 2 on acerola nurserystock from north-eastern Brazil are being disseminated to other regions of the country (Costa *et al.*, 1999). Ayala and Ramirez (1964) list *Malpighia* species as hosts of the reniform nematode, *R. reniformis*; however, Ferrraz *et al.* (1989) report that acerola is highly resistant to *R. reniformis*, *R. similis*, *T. semipenetrans* and *M. graminicola*. Root knot nematodes are also recognized as economic pests of acerola in Hawaii (Holtzmann, 1968) and especially in Florida, where preplant soil fumigation was recommended, and a tolerant rootstock, *Malpighia suberosa* L., has been assayed, but found inadequately productive (Ledin, 1963). Phenamiphos treatment was found ineffective in controlling nematodes (McSorley and Parrado, 1982). Several clones of *Malpighia emarginata* DC were resistant to *M. javanica* in Brazil (Gomes *et al.*, 2000).

Breadfruit

Breadfruit and the closely related jackfruit belong to the plant genus *Artocarpus* and are fruit trees of largely local significance throughout the tropics in Africa, Asia, the Pacific islands and South America. Little is known about nematode problems on these plants, but two very important nematodes, the root knot nematode *Meloidogyne* spp. and the reniform nematode, *R. reniformis*, have been reported to attack them (Caveness, 1967; Sharma and Sher, 1973; Razak, 1978; McSorley, 1992). Several species of *Helicotylenchus* have also built up to extremely large populations around breadfruit roots (Caveness, 1967). Coates-Beckford and Pereira (1992) found high densities of *P. coffeae*, along with *M. incognita* and *Helicotylenchus* spp. in the roots and rhizospheres of breadfruit in Jamaica, but tree health appeared to be more related to tree age than to nematode density.

Loquat

The loquat, *Eriobotrya japonica* L., is believed to have originated in China, but has been cultivated in Japan since antiquity. In addition to Japan, which during the 1970s produced between 15,000 and 20,000 t annually, loquats are today produced commercially in many warm climate countries throughout Asia, the Mediterranean region, southern Africa, Australia, and North and South America (Knight, 1980). Despite its considerable, and obviously growing, economic importance, the nematode problems affecting loquat cultivation have not been studied. Perhaps the only potentially pathogenic nematodes known to attack loquat are *Rotylenchulus macrodoratus*, which was found to reproduce and induce histological changes in loquat roots (Inserra and Vovlas, 1980), and *P. vulnus* which reproduced on loquat cv. Nadal (Pinochet *et al.*, 1992).

Mangosteen

A native of Malaysia, the mangosteen (*Garcinia mangostana* L.) is still grown predominantly in South-east Asia, mainly in Thailand, Malaysia, the Philippines and Indonesia, and has also been introduced into Central America. Although not much is known about nematode problems affecting this fruit tree, it is noteworthy that mangosteen has been reported from India as a host of the citrus nematode, *T. semipenetrans* (Chawla *et al.*, 1980).

Pomegranate

The pomegranate (*Punica granatum* L.) originates from Persia, and is cultivated in western and central Asia and in the Mediterranean region; it is also grown commercially in California. The predominant parasitic nematodes affecting pomegranate are the root knot nematodes, *M. incognita*, *M. acrita* and *M. javanica* (McSorley, 1981). In Israel, heavy root galling and visible damage to pomegranate trees in young orchards under irrigation are frequently encountered. In Libya, investigations revealed that out of 12 genera of plant parasitic nematodes commonly present in pomegranate nurseries, *M. incognita* and *M. javanica* were the most widespread. Fenamiphos application gave good control of the root knot nematodes, provided protection to roots for 60 days against nematode invasion and improved fruit yields (Siddiqui and Khan, 1986). Treatment of pomegranate with carbofuran reduced populations of *M. incognita*, *Xiphinema insigne* and *Helicotylencus* spp., and increased yields by one-third (Darekar *et al.*, 1989). Among 23 nematode species found in the rhizosphere of pomegranate in Jordan, Hashim (1983a) reported particularly large populations of *Helicotylenchus pseudorobustus*, *Tylenchorhynchus clarus* and *Longidorus* sp. associated with trees showing severe decline symptoms. However, application of carbofuran did not improve tree performance. Pomegranate has been reported as a host of *H. mangiferae* (Ashokkumar *et al.*, 1991). Despite a range of reactions, no pomegranate cultivars tested have been found to posses strong resistance to *M. incognita* (Verma, 1985; Shelke and Darekar, 2000).

Sapodilla

The sapodilla (*Manilkara zapota* L. Royen) is native to Mexico and Central America, and is today grown largely in tropical America, India and the east Asian tropics. Mexico, the leading producer, supplied an annual crop of 11,217 t in the mid-1970s (Knight, 1980), but its consumption is still limited mainly to the regions where it is cultivated. Some nematode problems of sapodilla were investigated by Saeed (1974), who demonstrated pathogenicity of *H. mangiferae* to sapodilla at a population density of 6 nematodes/cm^3 of soil, and suppressed populations with DBCP treatment for a 10 month period. He also reported population build-up of *Helicotylenchus indicus* and *Pratylenchus* spp. around sapodilla roots. Seasonality of *Hemicriconemoides mangiferae* on the crop in India and Pakistan coincides with rainfall patterns (Saeed and Ghaffar, 1986; Ashokkumar *et al.*, 1991).

Soursop

The soursop, or custard apple (*Annona muricata* L. and other *Annona* species), originated in tropical America and is now distributed in most tropical countries throughout the world. However, international trade in this fruit is very limited. Caveness (1967) found it to be a suitable host for several *Helicotylenchus* species, including *H. cavenessi*. *P. coffeae* has been shown to be the causal agent of 'sudden death' of soursop in Brazil (de Moura *et al.*, 1998), and isolates of *P. coffeae* collected from soursop or from yam were capable of causing the disease (de Moura *et al.*, 1999). Control of *X. ifacolum* and *H. pseudorobustus* with carbofuran did not increase growth of soursop in nurseries in Liberia (Lamberti *et al.*, 1992).

Tamarind

The tamarind (*Tamarindus indica* L.), known particularly for its use as a condiment and as an ingredient of chutneys, probably has an East African origin, but was early introduced to India, where annual production in the early 1960s is said to have averaged 230,000 t (Knight, 1980). It is grown today in most tropical regions throughout the world, and particularly in the Far East. Thailand has become a major producer of tamarind, with a production exceeding 140,000 t (Anonymous, 1998). Of the several nematode species associated with the crop, only *H. mangiferae* has been considered as pathogenic at a population density of 6 nematodes/cm^3 of soil (Saeed, 1974). The tamarind has also been reported as a host of *R. similis* (Sosamma and Koshy, 1977).

References

Abivardi, C. (1973) A stylet nematode, *Tylenchorhynchus cylindricus* Cobb 1913, infesting the common guava, *Psidium guajava* L. in Iran. *Nematologia Mediterranea* 1, 139–140.

Ahmed, E.M. and Barmore, C.R. (1980) Avocado. In: Nagy, S. and Shaw, P.E. (eds) *Tropical and Subtropical Fruits. Composition, Properties and Uses.* AVI Publishing, Inc., Westport, Connecticut, pp. 121–156.

Al-Sayed, A.A. and Abdel-Hameed, S.H. (1991) Resistance and susceptibility of olives to *Meloidogyne incognita* and *Rotylenchulus reniformis. Annals of Agricultural Science Moshtohor* 29, 1221–1226.

Anita and Chaubey, A.K. (2003) Influence of soil temperature and moisture on population dynamics of ectoparasitic nematodes infesting *Mangifera indica. Annals of Plant Protection Sciences* 11, 181–183.

Anonymous (1972) *Annual Report of the Plant Pathologist,* Brunei Department of Agriculture. Bandar Seri Begawan, Brunei.

Anonymous (1975) *CSIRO Division of Horticultural Research Report 1973–75.* Adelaide, Australia.

Anonymous (1994) *Proceedings of the First Meeting of the CIHEAM Cooperative Research Network on Underutilized Fruit Trees.* CIHEAM Instituto Agronomico Mediterraneo de Zaragoza, Zaragoza, Spain.

Anonymous (1998) Macadamia Nuts: Economic and Competitive Conditions Affecting the U.S. Industry. International Trade Commission (ITC) Inv. No. 332–386, USITC Publication, 3129.

Anonymous (FAO) (2002) *Faostat Agriculture Data.* Available at: apps.fao.org/

Anwar, S.A., Gorsi, S., Anwar ul Haq, M., Rehman, T. and Yousuf, P. (1991) Plant parasitic nematodes of some field, vegetable, fruit and ornamental crops. *Journal of Agricultural Research of Lahore* 29, 233–249.

Ashokkumar, P.S., Vadivelu, S., Mehta, U.K., Jeyarajan, R. and Regupathy, A. (1991) Occurrence and seasonal variation of *Hemicriconemoides mangiferae* from tropical and subtropical orchards in Peninsular India. *Nematropica* 21, 167–176.

Avelar, M.J.J., Teliz, O.D. and Zavaleta, M.E. (2001) Pathogens associated to 'guava decline'. *Revista Mexicana de Fitopatologia* 19, 223–229.

Ayala, A. (1969) Nematode problems in Puerto Rican agriculture. In: Abad-Ramos, J. (ed.) *Proceedings of the Symposium on Tropical Nematology.* Agricultural Experiment Station, University of Puerto Rico, Rio Piedras, pp. 135–145.

Ayala, A. and Ramirez, C.T. (1964) Host range, distribution and bibliography of the reniform nematode, *Rotylenchulus reniformis,* with special reference to Puerto Rico. *Journal of Agriculture of the University of Puerto Rico* 48, 140–161.

Ayala, A., Acosta, N. and Adsuar, J.A. (1971) A preliminary report on the response of *Carica papaya* to foliar applications of two systemic nematicides. *Nematropica* 1, 10.

Babatola, J.O. (1985) Effects of the root-knot nematode, *Meloidogyne incognita* on *Carica papaya* seedlings. *Pakistan Journal of Nematology* 3, 87–90.

Babatola, J.O. and Oyedunmade, E.E.A. (1992) Host–parasite relationships of *Psidium guajava* cultivars and *Meloidogyne incognita. Nematologia Mediterranea* 20, 233–235.

Badra, T. and Khattab, M.M. (1982) Chemically-induced resistance to *Rotylenchulus reniformis* by ethephon growth regulant and relevant pathometabolites in mango seedlings. *Nematologia Mediterranea* 10, 49–56.

Banihashemi, Z. and Kheiri, A. (1995) The occurrence of root knot nematode (*Meloidogyne javanica*) on pistachio in Damghan. *Iranian Journal of Plant Pathology* 31, 37–38.

Bello-Perez, A. and Jimenez Millan, F. (1963) *Heterodera fici* Kirjanova, 1954 (Nematoda), nueva especie para la fauna espanola. *Boletin de la Real Sociedad Espanola de Historia Natural. Seccion Biológica* 65, 25–28.

Blok, V.C., Phillips, M.S., McNicol, J.W. and Fargette, M. (1997) Genetic variation in tropical *Meloidogyne* spp. as shown by RAPDs. *Fundamental and Applied Nematology* 20, 127–133.

Boesewinkel, H.J. (1977) New plant disease records in New Zealand: records in the period 1969–76. *New Zealand Journal of Agricultural Research* 20, 583–589.

Bolin, H.R. and King, A.D. Jr (1980) Figs. In: Nagy, S. and Shaw, P.E. (eds) *Tropical and Sub-tropical Fruits. Composition, Properties and Uses.* AVI Publishing, Inc., Westport, Connecticut, pp. 492–505.

Brancalion, A.M., Zanon, J.I. and Zem, A.C. (1981) Ocorrencia do nematoide *Heterodera fici* no Rio Grande do Sul. *Revista de Agricultura, Piracicaba* 56, 4.

Bridge, J., Hunt, D.J. and Hunt, P. (1996) Plant parasitic nematodes of crops in Belize. *Nematropica* 26, 111–119.

Brown, D.J.F., Dalmasso, A. and Trudgill, D.L. (1993) Nematode pests of soft fruits and vines. In: Evans, K., Trudgill, D.L. and Webster, J.M. (eds) *Plant Parasitic Nematodes in Temperate Agriculture.* CAB International, Wallingford, UK, pp. 427–462.

Bustillo, C.Y.Y., Crozzoli, R., Greco, N. and Lamberti, F. (2000) Effect of the root-knot nematode *Meloidogyne incognita* on the growth of papaya (*Carica papaya*) in nurseries. *Nematologia Mediterranea* 28, 163–170.

Campos, V.P. (1997) Nematodes of the fig crop. *Informe Agropecuario Belo Horizonte* 18188, 33–38.

Carillo-Rivera, J., Carillo-Fonseca, C. and Dominguez-Alvarez, J.L. (1990) Nematodes attacking guava trees (*Psidium guajava* L.) and chemical control in the Juchipila Canyon, Zac. Mexico. *Revista Chapingo* 15, 67–68.

Carneiro, R.M.D.G., Wellington, A.M., Maria, R.A.A. and Ana Cristina, M.M.G. (2001) First record of *Meloidogyne mayaguensis* on guava in Brazil. *Nematologia Brasileira* 25, 223–228.

Casassa, A.M., Crozzoli, R., Matheus, J., Bravo, V. and Marin, M. (1998) Effect of the root-knot nematode, *Meloidogyne incognita*, on the growth of guava (*Psidium* spp.) in nurseries. *Nematologia Mediterranea* 26, 237–242.

Castillo, P. and Vovlas, N. (2002) Factors affecting egg hatch of *Heterodera mediterranea* and differential responses of olive cultivars to infestation. *Journal of Nematology* 34, 146–150.

Castillo, P., Vovlas, N., Nico, A.I. and Jimenez Diaz, R.M. (1999) Infection of olive trees by *Heterodera mediterranea* in orchards in southern Spain. *Plant Disease* 83, 8, 710–713.

Castillo, P., Vovlas, N. and Troccoli, A. (2003) The reniform nematode, *Rotylenchulus macrosoma,* infecting olive in southern Spain. *Nematology* 5, 23–29.

Castillo, P., Vovlas, N., Subbotin, S. and Troccoli, A. (2003) A new root knot nematode, *Meloidogyne baetica* n.sp. (Nematoda: Heteroderidae), parasitizing wild Olive in Southern Spain. *Phytopathology* 9, 1093–1102.

Caveness, F.E. (1967) Shadehouse host ranges of some Nigerian nematodes. *Plant Disease Reporter* 51, 33–37.

Cayrol, J.C., Frankowski, J.P., Lanza, R. and Tamonte, M. (1991) Les nematodes en kiwi culture. Essai de lutte biologique avec le champignon nematophage T-350. *Revue Horticole* 313, 54–56.

Chawla, M.L., Samathanam, G.J. and Sharma, S.B. (1980) Occurrence of citrus nematode (*Tylenchulus semipenetrans* Cobb. 1913) on roots of mangosteen (*Garcinia mangostana*). *Indian Journal of Nematology* 10, 240–242.

Coates Beckford, P.L. and Pereira, M.J. (1992) Survey of root inhabiting microorganisms on declining and nondeclining breadfruit *Artocarpus altilis* in Jamaica. *Nematropica* 22, 55–63.

Cohn, E. (1968) *Nematodes as Possible Vectors of Virus Diseases of Citrus and other Fruit Crops.* Second Annual Report to USDA, Project No. A10-CR-72, Bet Dagan, Israel.

Cohn, E. and Minz, G. (1961) Citrus nematode on American persimmon in Israel. *Plant Disease Reporter* 45, 505.

Cohn, E. and Mordécai, M. (1988) Morphology and parasitism of the mature female of *Rotylenchulus macrosomus. Revue de Nématologie* 11, 385–389.

Coiro, M.I. and Lamberti, F. (1978) Reproduction of *Xiphinema index* under different environmental conditions. Proceedings of the 6th Conference on virus and virus diseases of the grapevine. *Monografias Instituto National de Investigaeiones Agrarias* 18, 255–257.

Crozzoli, P.R., Casassa, P.A.M., Rivas, G.D. and Matheus, C.J. (1991) Plant parasitic nematodes associated with Guava plantation in Zulia State Venezuela. *Fitopatologia Venezolana* 4, 2–6.

Cuadra, R. and Quincosa, A. (1982) Comportamiento de differentes especies de *Psidium* como patrones para guayabos resistentes a *Meloidogyne* (Nematoda: Heteroderidae). *Ciencias de la Agricultura* 13, 19–26.

da Costa, D.C., Lima, A.A. and Jesus, R.L.M. (1997) Reacao de especies de maracujazeiro a *Meloidogyne incognita. Nematologia Brasileria* 21, 21–22.

da Costa, D.C., Carneiro, R.M.G., Oliveira, J.R.P., dos Soares Filho, W.S. and de Almeida, F.P. (1999) Identification of populations of *Meloidogyne* spp. in roots of Barbados cherry (*Malpighia punicifolia*). *Nematologia Brasileira* 23, 77–80.

Dale, P.S. and van der Mespel, G.J. (1972) Control of root-knot nematodes on Chinese gooseberry (kiwi fruit), *Actinidia chinensis*, by chemical bare-root dip. *Plant Disease Reporter* 56, 850–851.

da Ponte, J.J. (1986) Mapeamento a importancia das doencas do cajueiro no nordeste do Brash. *Nematologia Brasileira* 10, 59–68.

da Ponte, J.J. and Saraiva, L.M. (1973) Imunidade do cajueiro, *Anacardium occidentale* L., a nematoides do genero *Meloidogyne* Goeldi, 1887. *Ciencia Agronomia, Fortaleza* 3, 35–36.

Darekar, K.S. and Mhase, N.L. (1986) Effect of initial inoculum levels of *Meloidogyne incognita* on growth parameters of papaya seedlings. *International Nematology Network Newsletter* 3, 4–5.

Darekar, K.S., Mhase, N.L. and Shelke, S.S. (1989) Management of nematodes infesting pomegranate. *International Nematology Network Newsletter* 6, 15–17.

Davis, A.R. (1988) Statistics of Hawaiian Agriculture 1987. *Hawaii Agriculture Station Service*, p. 41.

De Abrantes, O.I.M., De-Santos, A.M.S.N., De Almeida, M.T.M. and Vovlas, N. (1998) Phytoparasitic nematodes associated with olives in Portugal. *Revista de Ciencias Agrarias* 21, 279–285.

de Moura, R.M and de Moura, A.M. (1989) Root-knot on guava: a severe disease in Pernambuco State, Brazil. *Nematologia Brasileira* 13, 13–19.

de Moura, R.M., Pedrosa, E.M.R., Lira, R.V., Menezes, M., Freire, F.C.O. and Cardoso, J.R. (1998) The etiology of the sudden death of soursop (*Annona muricata*). *Fitopatologia Brasileira* 23, 173–175.

de Moura, R.M., Guimaraes, L.M.P. and Pedrosa, E.M.R. (1999) Studies in sudden death of the soursop. *Nematologia Brasileira* 23, 62–68.

De Villiers, E.A. and Milne, D.L. (1973) The control of nematodes on passionfruit. *Citrus and Sub-tropical Fruit Journal* 480, 7, 9, 11, 13.

Di Maio, F. (1979) Il nematode *Tylenchulus semipenetrans* Cobb, su radiei di loto. *Informatore Fitopatologico* 29, 13–14.

Di Vito, M. (1976) *Heterodera fici su Ficus carica* in Basilicata. *Nematologia Mediterranea* 4, 107–108.

Di Vito, M. and Inserra, R.N. (1982) Effects of *Heterodera fici* on the growth of commercial fig seedlings in pots. *Journal of Nematology* 14, 416–418.

Di Vito, M. and Sasanelli, N. (1990) The effect of natural and artificial hatching agents on the emegence of juveniles of *Heterodera fici. Nematologia Mediterranea* 18, 55–57.

Di Vito, M., Vovlas, N. and Simeone, A.M. (1988) Effect of root-knot nematode *Meloidogyne incognita* on the growth of kiwi (*Actinidia deliciosa*) in pots. *Advances in Horticultural Science* 2, 109–112.

Diab, K.A. and El-Eraki, S. (1968) Plant parasitic nematodes associated with olive decline in the United Arab Republic. *Plant Disease Reporter* 52, 150–154.

DuCharme, E.P. and Suit, R.F. (1953) Nematodes associated with avocado roots in citrus spreading decline areas. *Plant Disease Reporter* 37, 427–428.

Elder, R.J., Reid, D.J., Macleod, W.N.B. and Gillespie, R.L. (2002) Post-ratoon growth and yield of three hybrid papayas (*Carica papaya* L.) under mulched and bare-ground conditions. *Australian Journal of Experimental Agriculture* 42, 71–81.

Esnard, J. and Zuckerman, B.M. (1998) Small fruits. In: Barker, K.R., Pederson, G.A. and Windham, G.L. (eds) *Plant and Nematode Interactions.* Soil Science Society of America, Inc., American Society of Agronomy, Inc., Crop Science Society of America, Inc., Madison, Wisconsin, pp. 685–725.

Fargette, M., Phillips, M.S., Block, V.C., Waugh, R. and Trudgill, D.L. (1996) An RFLP study of relationships between species, populations, and resistance breaking lines of tropical *Meloidogyne* spp. *Fundamental and Applied Nematology* 19, 193–200.

Fernandez Diaz Silveira, M. (1975) El *Psidium friedrichsthalianum* como patron para guayabo, resistente a los nematodos del genero *Meloidogyne. Revista de Agricultura, Cuba* 3, 80–85.

Fernandez Diaz Silveira, M. and Ortega Herrera, J. (1998) An overview of nematological problems in Cuba. *Nematropica* 28, 151–164.

Ferraz, L.C.C.B., Pereira, F.M. and Valente, J. (1982) Consideraqoes sobre o use de nematicidas em viveiro visando a recuperacao demudas de *Ficus carica* ev. roxo de valinhos infestadas por *Meloidogyne incognita*. In: Lordello, L.G.E. (ed.) *Trabalhos apresentados a VI reunias Brasileira de nematologia, 8–12 feveiro de 1982, Fortaleza. Publicacao No. 6*. Sociedade Brasileira de Nematologia, Piracicaba, SP, Brazil, pp. 13–20.

Ferraz, L.C.C.B., Monteiro, A.R. and Inomoto, A.R. (1989) Hospedabilidade de acerola em relacao a sete especies de fitonematoides. *Nematologia Brasileira* 13, 39–49.

Franco, A. and da Ponte, J.J. (1989) Acerola, *Malpighia glabra* L., a new host of root-knot nematodes. *Nematologia Brasileira* 13, 181–183.

Frei, B. (2001) CEO QUICORNAC S.A. IFU Sydney 2001 World Crop Reports, Passion Fruit. Available at: www.worldtrade.org/afb/products/details_products-81100-q.htm

Freire, F.C.O., Cardoso, J.E., dos Santos, A.A. and Viana, F.M.P. (2002) Diseases of cashew nut plants (*Anacardium occidentale* L.) in Brazil. *Crop Protection* 21, 489–494.

Ganguly, S., Tripathi, M.N. and Ghayyur, U. (1997) The reniform nematode, *Rotylenchulus reniformis* – a threat to papaya cultivation. *Indian Journal of Nematology* 27, 135–136.

George, A.P. and Nissen, R.J. (1990) Persimmon. In: Kose, T.K. and Mitra, S.K. (eds) *Fruits: Tropical and Subtropical*. Naya Prokash, Calcutta, India, pp. 469–487.

Ghorab, A.I., Riad, F.W. and Shohla, G.S. (1987) Seasonal fluctuation of plant parasitic nematodes associated with mango. *Agricultural Research Review* 65, 315–323.

Giblin Davis, R.M., Center, B.J., Nadel, H., Frank, J.H. and Ramirez, W. (1995) Nematodes associated with fig wasps, *Pegoscapus* spp. (Agaonidae), and Syconia of native Floridian figs (*Ficus* spp.). *Journal of Nematology* 27, 1–14.

Giblin Davis, R.M., Davies, R.A., Morris, K. and Thomas, W.K. (2003) Evolution of parasitism in insect-transmitted plant nematodes. *Journal of Nematology* 35, 133–141.

Gomes, J.E., dos Santos, J.M., Perecin, D. and Martins, A.B.G. (2000) Resistance of West Indian cherry (*Malpighia emarginata* DC) clones to *Meloidogyne javanica* under greenhouse conditions. *Nematologia Brasileira* 24, 65–71.

Gonzalez, G. and Sourd, F. (1982) Ensayo de tres especies de *Psidium* y su tolerancia a los nematodos. *Ciencia y Tecnica en la Agricultura, Citricos y Otros Frutales* 5, 13–25.

Gonzalez, H. (1988) Persimmon (*Diospyros kaki* L.), a new host of *Tylenchulus semipenetrans* Cobb, in Chile. *Agricultura Tecnica* 48, 1, 56–57.

Gonzalez, H.R. (1993) Effects of an organic nematicide and the control of parasitic nematodes of kiwi crop. *Revista Fruticola* 14, 67–71.

Grandison, G.S. (1983) Root-knot nematode control on kiwifruit (*Actinidia chinensis*) by chemical bare-root dip. *Plant Disease* 67, 899–900.

Graniti, A. (1955) Un deperimento dell' olivo in Sicilia associato a due specie di nematodi. *Olearia* 9, 114–120.

Gupta, I. and Yadav, B.S. (1988) Pathogenicity of root-knot nematode *Meloidogyne incognita* on papaya. *International Nematology Network Newsletter* 5, 14–15.

Gur, A. (1955) *The Fig*. Departments of Horticulture and Agricultural Education, Ministry of Agriculture, Israel, Bulletin No. 57 (in Hebrew).

Hamiduzzaman, M.M., Meah, M.B. and Ahmad, M.U. (1997) Effect of *Fusarium oxysporum* and nematode interaction on guava wilt. *Bangladesh Journal of Plant Pathology* 13, 1–2, 9–11.

Hashim, Z. (1982) Distribution, pathogenicity and control of nematodes associated with olive. *Revue de Nématologie* 5, 169–181.

Hashim, Z. (1983a) Plant-parasitic nematodes associated with pomegranate (*Punica granatum* L.) in Jordan and an attempt to chemical control. *Nematologia Mediterranea* 11, 199–200.

Hashim, Z. (1983b) Description of *Trophotylenchulus saltensis* n.sp. with a comment on the status of *Trophotylenchulus* Raski, 1957 and a proposal for *Ivotylenchulus* n. gen. (Nematoda: Tylenchida). *Revue de Nématologie* 6, 179–186.

Heinlein, M. (1982) Treatments for the control of plant parasitic nematodes in nursery potting soils. *Fiji Agricultural Journal* 44, 85–88.

Herre, E.A. (1995) Factors affecting the evolution of virulence: nematode parasites of fig wasps as a case study. *Parasitology* 111 (Supplement), S179–S191.

Hirata, K., Suzuki, K., Hisai, J., Matsuda, T. and Saeki, S. (2002) Morphology and morphometrics of juvenile stages of *Xiphinema index* Thorne and Allen (Dorylaimida: Longidoridae) intercepted in import plant quarantine in Japan. *Research Bulletin of the Plant Protection Service Japan* 38, 21–26.

Hirschmann, H., Paschallaki Kourtzi, N. and Triantaphyllou, A.C. (1966) A survey of plant-parasitic nematodes in Greece. *Annales de l'Institut Phytopathologique, Benaki* 7, 144–156.

Holanda, Y.C.A., da Ponte, J.J. and Silveira, F.J. (1997) Disease of the Barbados cherry plant (*Malpighia glabra*) in the State of Ceara, Brazil. *Fitopatologia Brasileira* 22, 453.
Holtzmann, O.V. (1968) Plant–nematode associations previously unreported from Hawaii. *Plant Disease Reporter* 52, 515–518.
Hosomi, A. and Uchiyama, T. (1998) Growth inhibiting factors in sick soil of fig orchards. *Journal of the Japanese Society for Horticultural Science* 67, 44–50.
Hosomi, A., Dan, M. and Kato, A. (2002) Screening of fig varieties for rootstocks resistant to soil sickness. *Journal of the Japanese Society for Horticultural Science* 71, 171–176.
Iglesias, M.P. and Perez, J.A. (1991) Susceptibility of 6 papaya (*Carica papaya*) varieties to *Meloidogyne incognita*. *Proteccion de Plantas* 1, 41–46.
Indra, R. and Rajvanshi, I. (2001) Evaluation of papaya cultivars against root-knot nematode (*Meloidogyne incognita*). *Current Nematology* 12, 1–2, 91–92.
Inomoto, M.M., Monteiro, A.R. and Ferraz, L.C.C.B. (1991) Occurrence of *Tylenchulus semipenetrans* and *Meloidogyne incognita* on persimmon in Brazil. *Nematologia Brasileira* 15, 82–84.
Inserra, R.N. and O'Bannon, J.H. (1974) Systemic activity of phenamiphos for control of *Meloidogyne arenaria* on *Gardenia jasminoides* and *Ficus carica*. *Plant Disease Reporter* 58, 1075–1077.
Inserra, R.N. and Vovlas, N. (1977) Parasitic habits of *Gracilacus peratica* on olive feeder roots. *Nematologia Mediterranea* 5, 345–348.
Inserra, R.N. and Vovlas, N. (1980) The biology of *Rotylenchulus macrodoratus*. *Journal of Nematology* 12, 97–102.
Inserra, R.N., Vovlas, N. and Golden, A.M. (1979) *Helicotylenchus oleae* n.sp. and *H. neopaxilli* n.sp. (Hoplolaimidae), two new spiral nematodes parasitic on olive trees in Italy. *Journal of Nematology* 11, 56–62.
Itoo, S. (1980) Persimmon. In: Nagy, S. and Shaw, P.E. (eds) *Tropical and Sub-tropical Fruits. Composition, Properties and Uses.* AVI Publishing, Inc., Westport, Connecticut, pp. 442–468.
Jasy, T. and Koshy, P.K. (1992) Pathogenicity of the burrowing nematode, *Radopholus similis*, on avocado. *Indian Journal of Nematology* 22, 122–124.
Javanshash, A., Avanzato, D. and Hokmabadi, M.H. (2000) Iran's pistachio industry. *Rivista di Frutticoltura e di Ortofloricoltura* 62, 90–92.
Karim, S.A. (1989) Change in population and effect of reniform nematode on papaya plant. *Mardi Research Journal* 17, 226–233.
Kepenekci, I. (2001) Preliminary list of *Tylenchida* (Nematoda) associated with olive in the Black Sea and the Mediterranean regions of Turkey. *Nematologia Mediterranea* 29, 145–147.
Khan, R.M., Kumar, S. and Reddy, P.P. (2001) Role of plant parasitic nematode(s) and fungi in guava wilt. *Pest Management in Horticultural Ecosystems* 7, 152–161.
Khan, T.A. and Husain, S.I. (1991) Effect of age of papaya seedlings on the development of disease complex caused by *Meloidogyne incognita* and *Fusarium solani*. *Nematologia Mediterranea* 19, 327–329.
Khan, T.A., Khan, S.T. and Saxena, S.K. (1995) Relative response of papaya cultivars to the interaction of *Meloidogyne incognita* and *Fusarium solani*. *Annals of Plant Protection Sciences* 3, 46–51.
Khan, T.A., Khan, S.T., Fazal, M. and Siddiqui, Z.A. (1997) Biological control of *Meloidogyne incognita* and *Fusarium solani* disease complex in papaya using *Paecilomyces lilacinus* and *Trichoderma harzianum*. *International Journal of Nematology* 7, 127–132.
Khurramov, S.K. (1993) Parasitic nematodes of subtropical fruit crops. *Zashchita Rastenii Moskva* 9, 11.
Kirby, M.F. (1978) Reniform and root-knot nematodes on passionfruit in Fiji. *Nematropica* 8, 21–25.
Klein, A.L., Ferraz, L.C.C.B. and DeOliveira, J.C. (1984) Behaviour of different passionfruit plants in relation to the root-knot nematode. *Pesquisa Agropecuaria Brasileira* 19, 207–209.
Knight, K.W.L. (2001) Plant parasitic nematodes associated with six subtropical crops in New Zealand. *New Zealand Journal of Crop and Horticultural Science* 29, 267–275.
Knight, R.J. Jr (1980) Origin and world importance of tropical and sub-tropical fruit crops. In: Nagy, S. and Shaw, P.E. (eds) *Tropical and Sub-tropical Fruits. Composition, Properties and Uses.* AVI Publishing, Inc., Westport, Connecticut, pp. 1–120.
Kodira, U.C. and Westerdahl, B.B. (2003) *UC IPM Pest Management Guidelines: Fig.* UCPMG Publication 3447. Division of Agriculture and Natural Resources, University of California (ANR Publications), Oakland. Available at: www.ipm.ucdavis.edu/PMG/r261200111.html
Koenning, S.R., Overstreet, C., Noling, J.W., Donald, P.A., Becker, J.O. and Fortnum, P.A. (1999) Survey of crop losses in response to phytoparasitic nematodes in the United States for 1994. *Journal of Nematology* 31 (4S), 587–618.

Korayem, A.M. and Koura, F.H.F. (1993) Population studies of plant parasitic nematodes associated with mango in Giza, Egypt. *Afro Asian Journal of Nematology* 3, 148–151.

Krezdorn, A.H. and Adriance, G.W. (1961) *Fig Growing in the South. Agricultural Handbook No. 196.* ARS, USDA, Washington, DC.

Krezdorn, A.H. and Glasgow, S.K. (1970) Propagation of *Ficus carica* on tropical species of *Ficus*. *Proceedings of the Tropical Region, American Society for Horticultural Science* 14, 156–164.

Krnjaic, D., Krnjaic, S. and Bacic, J. (1997) Distribution and population density of fig cyst nematode (*Heterodera fici* Kirjanova) in the region of SR Yugoslavia. *Zastita Bilja* 48, 245–251.

Kwee, L.T. and Chong, K.K. (1990) *Guava in Malaysia: Production, Pests and Diseases.* Tropical Press Ltd, Kuala Lumpur, Malaysia.

Lamberti, F. (1969) Presenza in Italia di un deperimento dell'olivo causato dal nematode *Pratylenchus vulnus*, All. *et* Jens. *Phytopathologia Mediterranea* 8, 232–234.

Lamberti, F. and Baines, R.C. (1969a) Pathogenicity of four species of *Meloidogyne* on three varieties of olive trees. *Journal of Nematology* 1, 111–115.

Lamberti, F. and Baines, R.C. (1969b) Effect of *Pratylenchus vulnus* on the growth of 'Ascolano' and 'Manzanillo' olive trees in a glasshouse. *Plant Disease Reporter* 53, 557–558.

Lamberti, F. and Di Vito, M. (1972) Sanitation of root-knot nematode infected olive-stocks. In: *Acta.r do III Congresso da Uniao Fitopat. Medit.* Oerias, Portugal, 22–28 October 1972, pp. 401–411.

Lamberti, F. and Vovlas, N. (1993) Plant parasitic nematodes associated with olive. *Bulletin OEPP Organisation Europeenne et Mediterraneenne pour la Protection des Plantes* 23, 481–488.

Lamberti, F., Vovlas, N. and Tirro, A. (1976) Infettivita a patogenicita di tre popolazioni italiane di *Tylenchulus semipenetrans su* agrumi ed altri ospiti. *Nematologia Mediterranea* 4, 85–91.

Lamberti, F., Rohini, H.M., Ekanayake, V. and Zacheo, F. (1980) Effect of two *Meloidogyne* species on the growth of pawpaw seedlings. *Indian Journal of Nematology* 10, 225–230.

Lamberti, F., Boiboi, J.B., Ciancio, A., Tuopay, D.K., Jimehez, E.A. and Elia, F. (1992) Plant parasitic nematodes associated with tree crops in Liberia. *Nematologia Mediterranea* 20, 79–85.

Lamberti, F., Ciccarese, F., Sasanelli, N., Ambrico, A., D'Addabbo, T. and Schiavone, D. (2001) Relationships between plant parasitic nematodes and *Verticillium dahliae* on olive. *Nematologia Mediterranea* 29, 1, 3–9.

Lange, A.H. (1960) The effect of fumigation on the papaya replant problem in two Hawaiian soils. *Proceedings of the American Society of Horticultural Science* 75, 305–312.

Lazan, H. and Ali, Z.M. (1998) Guava. In: Shaw, P.E., Chan, H.T. Jr and Nagy, S. (eds) *Tropical and Subtropical Fruits.* AgScience, Inc., Auburndale, Florida, pp. 446–485.

Ledin, R.B. (1963) The Barbados or West Indian cherry. *University of Florida Agricultural Extension Service Circular* 250.

Lee, M.D., Chen, C.H., Tsay, T.T. and Lin, Y.Y. (1998) Survey and control of guava nematode diseases. *Plant Protection Bulletin Taipei* 40, 265–276.

Li, H.M., Xu, J.H., Shen, P.Y., Cheng, H.R., Li, H.M., Xu, J.H., Shen, P.Y. and Cheng, H.R. (1999) Distribution and seasonal dynamic changes of nematode parasites in fig main growing areas in Jiangsu Province. *Journal of Nanjing Agricultural University* 22, 38–41.

Lima, J.A.A., Menezes, M., Karan, M. de Q. and Martins, O.F.G. (1975) Generos de nematoides fitopatogenicos isolados da rizosfera do cajueiro, *Anacardium occidentale* L. *Fitossanidade, Brazil* 1, 32–35.

Loof, P.A.A. and Sharma, R.D. (1979) Plant parasitic nematodes from Bahia State, Brazil: the genus *Xiphinema* Cobb, 1913 (Dorylaimoidea). *Nematologica* 25, 111–127.

Lopez, R. and Azofeifa, J. (1985) Nematodos fitoparasitos asociados con frutales en algunos cantones de la provincia de Alajuela. *Agronomia Costarricense* 9, 193–196.

Lopez, R. and Salazar, L. (1987) Observations on the spatial distribution of plant-parasitic nematodes in fruit trees. *Agronomia Costarricense* 11, 141–147.

Maccari, A. Jr, dos Santos, H.R. and Biasi, L.A. (1993) The study of preference and control of nematode *Meloidogyne arenaria* (Neal, 1889) Chitwood, 1949, in kiwi seedling. *Revista do Setor de Ciencias Agrarias* 12, 1–2, 55–60.

Mahto, Y. and Edward, J.C. (1979) Studies on pathogenicity, host parasite relationship and histopathological changes of some important fruit trees due to predominant phytonematode associated with them (Part 1). *Allahabad Farmer* 50, 403.

Malan, A.P. and Meyer, A.J. (1999) The reproduction and life cycle of a South African population of *Xiphinema index*. *South African Journal for Enology and Viticulture* 20, 57–60.

Mancini, G., Moretti, F. and Catroneo, A. (1978) Histological studies on *Actinidia chinensis* and *Salix alba* infested by *Meloidogyne* spp. *Nematologia Mediterranea* 6, 235–238.

Mani, A. and Al Hinai, M.S. (1995) Mango, a new host of *Meloidogyne incognita*. *Nematologia Mediterranea* 23, 267.

Maqbool, M.A. (1991) *Nematode Pests of Economic Significance Affecting Major Crops of the Countries in Asia and the Pacific Region.* Technical Document Asia and Pacific Plant Protection Commission, No. 140. FAO Regional Office for Asia and the Pacific (RAPA), Bangkok, Thailand.

Maranhao, S.R.V.L., de Moura, R.M., Pedrosa, E.M.R. and de Moura, R.M. (2001) Reaction of guava genotypes in relation to *Meloidogyne incognita* race 1 and *M. mayaguensis*. *Nematologia Brasileira* 25, 191–195.

Martin, G.C. (1959) Plant species attacked by root-knot nematodes (*Meloidogyne* spp.) in the Federation of Rhodesia and Nyasaland. *Nematologica* 4, 122–125.

Matehus, J., Suarez, Z.H., Rosales, L.C., Tong, F., Cassava, A., Bravo, V. and Nava, A. (1999) Histological reaction of *Psidium* spp. selections to *Meloidogyne incognita* in Venezuela. *Nematologia Mediterranea* 27, 247–251.

McKenry, M.V. (2000) *UC IPM Pest Management Guidelines: Olive.* IPM Education and Publications; University of California, Davis. UC ANR Publication 3452. Available at: www.ipm.ucdavis.edu/PMG/r583200111.html

McKenry, M.V. (2002) *Kiwifruit Nematodes. UC IPM Pest Management Guidelines.* UC ANR Publication 3449. Available at: www.ipm.ucdavis.edu/PMG/r430200111.html

McKenry, M.V. and Kretsch, J.O. (1984) Nematodes in pistachio orchards. *California Agriculture* 38, 21.

McSorley, R. (1981) *Plant Parasitic Nematodes Associated with Tropical and Sub-tropical Fruits.* Agricultural Experiment Station Institute of Food and Agricultural Science, University of Florida, Gainesville, Florida, Bulletin 823.

McSorley, R. (1992) Nematological problems in tropical and subtropical fruit tree crops. *Nematropica* 22, 103–116.

McSorley, R. and Parrado, J.L. (1982) Effect of Nemacur on phytoparasitic nematodes of Barbados cherry. *Fungicide and Nematicide Tests, American Phytopathological Society* 39, 90.

McSorley, R. and Parrado, J.L. (1983) Effect of Nemacur on phytoparasitic nematodes of mango, 1982. *Fungicide and Nematicide Tests, American Phytopathological Society* 39, 92–93.

Menzel, C.M. and Simpson, D.R. (1994) Lychee. In: Schaffer, B. and Andersen, P.C. (eds) *Handbook of Environmental Physiology of Fruit Crops.* CRC Press, Boca Raton, Florida, pp. 123–141.

Milne, D.L. (1982a) Nematode pests of litchi. In: Keetch, D.P. and Heyns, J. (eds) *Nematology in Southern Africa Science Bulletin.* Department of Agriculture and Fisheries in Republican South Africa, Bulletin No. 400, pp. 38–41.

Milne, D.L. (1982b) Nematode pests of miscellaneous sub-tropical crops. In: Keetch, D.P. and Heyns, J. (eds) *Nematology in Southern Africa Science Bulletin.* Department of Agriculture Fisheries in Republican South Africa, Bulletin No. 400, pp. 42–46.

Milne, D.L. and Keetch, D.P. (1976) Some observations on the host plant relationships of *Radopholus similis* in Natal. *Nematropica* 6, 13–17.

Morton, J.F. and Curtis, F.D. Jr (eds) (1987) *Fruits of Warm Climates.* Creative Resource System, Inc., Winterville, North Carolina.

Mowat, A.D. and George, P. (1994) Persimmon. In: Shaffer, B. and Andersen, P.C. (eds) *Handbook of Environmental Physiology of Fruit Crops. I: Temperate Crops.* CRC Press, Boca Raton, Florida, pp. 209–232.

Narbaev, Z.N. and Sidikov, D.T. (1985) The possible parasitization of *Heterodera fici* on other crops. *Uzbekskii Biologicheskii Zhurnal* 4, 57–58.

Nejad, S.A.H., Maafi, Z.T. and Barooti, S. (1997) Nematodes associated with olive trees (*Olea europea* L.) in Iran. *Applied Entomology and Phytopathology* 65, 10–12.

Nesbitt, R.B. (1956) New host plants of plant parasitic nematodes in California. *Plant Disease Reporter* 40, 276.

Netscher, C. (1981) Arbres resistants au *Meloidogyne* spp.: utilisation comme brise-vent au Sénégal. *L Agronomie Tropicale Nogent* 36, 175–177.

Nico, A.I., Rapoport, H.F., Jimenez Diaz, R.M. and Castillo, P. (2002) Incidence and population density of plant-parasitic nematodes associated with olive planting stocks at nurseries in southern Spain. *Plant-Disease* 86, 1075–1079.

Nico, A.I., Jimenez Diaz, R.M. and Castillo, P. (2003) Host suitability of the olive cultivars Arbequina and Picual for plant parasitic nematodes. *Journal of Nematology* 35, 29–34.

Nigam, K., Verma, R.S., Verma, A.K. and Sinha, V. (1995) Pathogenicity of *Hoplolaimus* spp. Daday, 1905 to guava (*Psidium gujava*). *Advances in Agricultural Research in India* 3, 158–160.

Noriega, C.D.H., Marban, M.N., Rodriguez, A.J. and De Zarate, L.G.P. (1988) Effect of chemical products on phytonematodes associated with the root and on the mite *Eriophyes mangiferaes* involved in witches broom of the mango cultivar Haden in Iguala Guerrero Mexico. *Revista Mexicana de Fitopatologia* 6, 61–72.

Ondieki, J.J. (1975) Diseases and pests of passionfruit in Kenya. *Acta Horticulturae* 49, 291–293.

O'Rourke, E.N. (1992) LSU Purple, a new fig for the Gulf South. *Louisiana Agriculture* 35, 19–20.

Park, S.D., Zakaullah, K., Kim, J.C., Kim, S.J., Kim, K.J. and Choi, B.S. (1999) Association and density of plant parasitic nematodes in fruit orchards in Gyeongbuk province, Korea. *International Journal of Nematology* 9, 185–190.

Partridge, I. (1997) Tropical fruit becomes major industry. *Rural Research* 174, 9–12.

Patel, H.V., Patel, D.J. and Patel, B.A. (1989) Reaction of papaya varieties to *Rotylenchulus reniformis*. *International Nematology Network Newsletter* 6, 3, 24.

Peacock, F.C. (1956) The reniform nematode in the Gold Coast. *Nematologica* 1, 307–310.

Pena Santiago, R. (1990) Plant parasitic nematodes associated with olive *Olea europaea* L. in the province of Jaen Spain. *Revue de Nématologie* 13, 113–115.

Peregrine, W.T.H. and Yunton, B. (1980) A preliminary note on nematode pests in Brunei. *Tropical Pest Management* 26, 416–419.

Petit, R.P.A. (1990) Survey of plant parasitic nematodes associated with fruit trees of economic importance in Venezuela. *Fitopatologia Venezolana* 3, 2–5.

Philippi, I. and Budge, A. (1992) Effects of *Meloidogyne hapla* on young kiwi plants. *Nematropica* 22, 47–54.

Philippi, I., Latorre, B.A., Perez, G.F. and Castillo, L. (1996) Identification of the root-knot nematodes (*Meloidogyne* spp.) on kiwifruit by isoenzyme analysis in Chile. *Fitopatologia* 31, 96–101.

Pinochet, J., Verdejo, S., Soler, A. and Canals, J. (1992) Host range of a population of *Pratylenchus vulnus* in commercial fruit, nut, citrus and grape rootstocks in Spain. *Journal of Nematology* 24, 693–698.

Poehling, H.M., Wyss, U. and Neuhoff, V. (1980) Microanalysis of free amino acids in the aseptic host–parasite system: *Ficus carica–Xiphinema index* (Nematoda). *Physiological Plant Pathology* 16, 49–61.

Rahaman, P.F., Ahmad, I. and Jairajpuri, M.S. (1996) Three new species of *Hemicycliophora* de Man, 1921 from India. *Nematologica* 42, 24–34.

Ramakrishnan, S. and Rajendran, G. (1998a) Influence of *Meloidogyne incognita* on yield components and physiological functions of papaya. *Nematologia Mediterranea* 26, 225–228.

Ramakrishnan, S. and Rajendran, G. (1998b) Assessment of yield loss due to *Meloidogyne incognita* in papaya under field conditions. *Nematologia Mediterranea* 26, 229–230.

Ramakrishnan, S. and Rajendran, G. (1999) Comparison of different methods of control of *Meloidogyne incognita* in relation to growth and yield of papaya. *Nematologia Mediterranea* 27, 115–118.

Raski, D.J. and Krusberg, L.R. (1984) Nematode parasites of grapes and other small fruits. In: Nickle, W.R. (ed.) *Plant and Insect Nematodes*. Marcel Dekker Inc., New York, pp. 457–506.

Razak, A.R. (1978) Variation in plant response, gall size and form induced by *Meloidogyne* on some Malaysian crops. *Kasetsart Journal* 12, 43–45.

Reddy, P.P., Singly, D.B. and Ravishankar, H.R. (1980) Host records of the root-knot and reniform nematodes. *Haryana Journal of Horticultural Sciences* 9, 134–135.

Reddy, P.P., Iyer, C.P.A. and Subramanyam, M.D. (1988) Evaluation of papaya cultivars and hybrids against *Meloidogyne incognita*. *Indian Journal of Nematology* 18, 381–382.

Robinson, A.F., Inserra, R.N., Caswell Chen, E.P., Vovlas, N. and Troccoli, A. (1997) *Rotylenchulus* species: identification, distribution, host ranges, and crop plant resistance. *Nematropica* 27, 127–180.

Rodrigez, H., Fernandez, E. and Shesteperov, A.A. (1985) Adverse effect of *Meloidogyne* infection on guava (*Psidium guajava* L.) and factors influencing its development. *Byulleten' Vsesoyuznogo Instituta Gel'mintologii im. K.I. Skryabina* 41, 48–56.

Rosales, L.C. and Suarez, H.Z. (2001) Reaction of five *Carica* selections to *Meloidogyne incognita*. *Nematologia Mediterranea* 29, 177–180.

Routaray, B.N. and Das, S.N. (1988) Population management of root-knot nematode *Meloidogyne incognita* on papaya. *International Nematology Network Newsletter* 5, 25–27.

Ruehle, G.D. (1959) Growing guavas in Florida. *Agriculture Extension Service Bulletin* 170, Gainesville, Florida.

Saeed, M. (1974) Studies on some stylet-bearing nematodes associated with sapodilla (*Achraszapota* L.) with special reference to *Hemicriconemoides mangiferae* Siddiqi, 1961. PhD Thesis, University of Karachi.

Saeed, M. and Ghaffar, A. (1986) Seasonal population fluctuations of *Hemicriconemoides mangiferae* and other nematodes associated with sapodilla (*Achras zapota*) in Karachi. *Pakistan Journal of Nematology* 4, 67–74.

Saltaren, G., Luis, F., de Varon Agudelo Francia, H. and de la Torre Fernando, M. (1999) Nematodes associated with crops of avocado (*Persea americana* Mill). *Fitopatologia Colombiana* 22, 68–73.

Sanchez, M.Y., Manyoma, L. and de Agudelo, F.V. (1993) Identification and parasitism of nematodes associated with maracuya *Passiflora edulis* Simse. *Fitopatologia Colombiana* 17, 1–2, 12–20.

Sasanelli, N., Fontanazza, G., Lamberti, F. and D'Addabbo, T. (1997) Reaction of olive cultivars to *Meloidogyne* species. *Nematologia Mediterranea* 25, 183–190.

Sasanelli, N., Coiro, M.I., D'Addabbo, T., Lemos, R.J., Ridolfi, M. and Lamberti, F. (1999) Reaction of an olive cultivar and an olive rootstock to *Xiphinema index*. *Nematologia Mediterranea* 27, 253–256.

Sasanelli, N., D'Addabbo, T., Dell'Orco, P. and Mencuccini, M. (2000) The *in vitro* use of olive explants in screening trials for resistance to the root-knot nematode, *Meloidogyne incognita*. *Nematropica* 301, 101–106.

Sasanelli, N., D'Addabbo, T. and Lemos, R.M. (2002) Influence of *Meloidogyne javanica* on growth of olive cuttings in pots. *Nematropica* 32, 59–63.

Scotto La Massèse, C. (1973) A new European host of *Meloidogyne hapla* and *Rotylenchus robustus*: *Actinidia chinensis*. *Nematologia Mediterranea* 1, 57–59.

Scotto La Massèse, C., Deportes, L., Mercier, S. and Roger, J. (1984) Les principaux ennemis du figuier: les nematodes et maladies. *Phytoma – Défense des Cultures* 354, 39–41.

Sethi, C.L., Gaur, H.S., Kaushal, K.K., Srivastava, A.N. and Khan, E. (1988) Occurrence of root-knot nematodes on fruit plants in association with *Agrobacterium tumefaciens*. *International Nematology Network Newsletter* 5, 12–13.

Shafiee, M.F. and Osman, A. (1971) The use of nematicides and acaricides for the control of nematodes and mites in established mango orchards. *Phytopathologia Mediterranea* 10, 271–273.

Sharma, R.D. (1978) Nematodes associated with avocado (*Persea americana* Mill.) in Cerrado soils. *Societa Brasileira Nematologia Publicaciones No. 3*, pp. 65–70.

Sharma, R.D. and Sher, S.A. (1973) Plant nematodes associated with jackfruit (*Artocarpus heterophyllus* Lam.) in Bahia, Brazil. *Nematropica* 3, 20–23.

Sharma, R.D., Junqueira, N.T.V. and Gomes A.C. (2000) Reaction of passionfruit genotypes to the reniform nematode, *Rotylenchulus reniformis*. *Nematologia Brasileira* 25, 211–215.

Sharma, R.D., Junqueira, N.T.V. and Gomes, A.C. (2001) Pathogenicity and reproduction of *Meloidogyne javanica* on yellow passionfruit hybrid. *Nematologia Brasileira* 25, 247–249.

Sharma, R.D., Junqueira Nilton, N.T.V. and Gomes Antonio, C. (2002) Reaction of passionfruit varieties to the root-knot nematode, *Meloidogyne javanica*. *Nematologia Brasileira* 26, 93–96.

Shelke, S.S. and Darekar, K.S. (2000) Reaction of pomegranate germplasm to root-knot nematode. *Journal of Maharashtra Agricultural Universities* 25, 308–310.

Sher, S.A. (1955) Nematodes attacking avocado. *California Citrograph* 40, 198.

Sher, S.A. and Raski, D.J. (1956) *Heterodera fici* Kirjanova 1954 in California. *Plant Disease Reporter* 40, 700.

Sher, S.A., Foote, F.J. and Boswell, S.B. (1959) A root lesion nematode disease of avocados. *Plant Disease Reporter* 43, 797–800.

Shesteperov, A.A. (1979) Pathogenicity of *Meloidogyne* in Cuba. In: *Gallovye nematody sel'skokhoZyaistvennykh Kul'tur i mery bor'by s nimi*. Materialy Simpoziuma, Dushanbe, USSR, 25–27 September 1979. 'Donish', pp. 65–67.

Siddiqui, M.R. (1977) *Hemicriconemoides mangiferae*. *CIH Description of Plant-parasitic Nematodes*, Set 7 No. 99. Commonwealth Agricultural Bureaux, London.

Siddiqui, Z.A. and Khan, M.W. (1986) A survey of nematodes associated with pomegranate in Lìbya and evaluation of some systemic nematicides for their control. *Pakistan Journal of Nematology* 4, 83–90.

Sikora, R.A. (1988) Presence, distribution and importance of plant parasitic nematodes in irrigated agricultural crops in Niger. *Mededelingen Fakulteit Landbouwetenshappen Rijksuniversiteit Gent* 53, 821–834.

Singh, N.D. and Farrell, K.M. (1972) Occurrence of *Rotylenchulus reniformis* in Trinidad, West Indies. *Plant Disease Reporter* 56, 551.

Singh, U.S. and Nath, R.P. (1997) Pathogenicity of root-knot nematode, *Meloidogyne incognita* on papaya. *Indian Journal of Nematology* 26, 115–116.

Sosamma, V.K. and Koshy, P.K. (1977) Additional hosts of burrowing nematode, *Radopholus similis*, infecting coconut plants in South India. *Plant Disease Reporter* 61, 760–761.

Srivastava, S.S. (2002) Efficacy of organic amendments in the management of root knot nematode (*Meloidogyne incognita*) infecting papaya. *Indian Journal of Nematology* 32, 183–184.
Stapleton, J.J., Ferguson, L., McKenry, M.V., Dougherty, D.S., Stapleton, S.C., Metzidakis, I.T. and Voyiatzis, D.G. (1999) Using solarization to disinfest soil for olive nursery production. *Acta Horticulturae* 474, 589–591.
Suarez, H.Z., Gonzalez, M.S., Rosales, L.C. and Tellechea, V. (1993) Histological alterations induced by *Rotylenchulus reniformis* on *Passiflora edulis* f.sp. *flavicarpa*. *Fitopatologia Venezolana* 6, 11–14.
Suarez, H.Z., Rosales, L.C. and Rondon, A. (1999) Synergistic effect of the fungi *Macrophomina* and *Fusarium* with the root-knot nematodes *Meloidogyne* spp. on decline of guava. *Nematologia Mediterranea* 27, 79–82.
Talame, M. (1976) *Actinidia* attacked by nematodes. *Informatore Agrario* 32, 23281–23282.
Terblanche, J.H., Grech, N., Frean, R., Crabbe, F. and Joubert, A. (1986) Good news for passionfruit industry. *Citrus and Subtropical Fruit Research Institute Information Bulletin No. 164*, pp. 1–2, 4–5.
Thorne, G. and Allen, M.W. (1950) *Paratylenchus hamatus* n.sp. and *Xiphinema index* n.sp., two nematodes associated with fig roots with a note on *Paratylenchus anceps* Cobb. *Proceedings of the Helminthological Society of Washington* 17, 27–35.
Tuck, H.C. (1998) Safe and efficient management systems for plantation pests and diseases. *Planter* 74, 369–385.
Verdejo, S., Calvet, C. and Pinochet, J. (1990) Effect of mycorrhiza on kiwi infected by the nematodes *Meloidogyne hapla* and *M. javanica*. *Buletin de Sanidad Vegetal. Plagas* 16, 619–624.
Verheij, E.W.M. and Coronel, R.E. (1991) *Plant Resources of South East Asia No. 2. Edible Fruits and Nuts.* Prosea Foundation, Wageningen, The Netherlands.
Verma, R.R. (1985) Susceptibility of some pomegranate varieties to root-knot nematode. *Indian Journal of Nematology* 15, 247.
Villota, B.J.V. and de Agudelo, F.V. (1997) Evaluation of guava material (*Psidium guajava* L.) for the damage behaviour of *Meloidogyne incognita*. *Fitopatologia Colombiana* 21, 31–38.
Vilsoni, F. and Heinlein, M. (1982) Influence of initial inoculum levels of the reniform nematode on the growth of mung, pawpaw, pigeonpea and sweet potato. *Fiji Agricultural Journal* 44, 61–66.
Vlachopoulos, E.G. (1994) Plant protection problems caused by phytonematodes in Greece. *Bulletin OEPP* 24, 413–415.
Vovlas, N. (1983) Gall formation on *Pistacia vera* by *Rotylenchulus macrodoratus*. *Journal of Nematology* 15, 148–150.
Vovlas, N. and Inserra, R.N. (1981) Parasitic habits of *Ogma rhombosquamatum* and description of the male. *Journal of Nematology* 13, 87–90.
Vovlas, N. and Inserra, R.N. (1983) Biology of *Heterodera mediterranea*. *Journal of Nematology* 15, 571–576.
Vovlas, N. and Larizza, A. (1996) Relationship of *Schistonchus caprifici* (*Aphelenchoididae*) with fig inflorescences, the fig pollinator *Blastophaga psenes*, and its cleptoparasite *Philotrypesis caricae*. *Fundamental and Applied Nematology* 19, 443–448.
Vovlas, N., Inserra, R.N. and Greco, N. (1992) *Schistonchus caprifici* parasitizing caprifig (*Ficus carica sylvestris*) florets and the relationship with its fig wasp (*Blastophaga psenes*) vector. *Nematologica* 38, 215–226.
Vrain, T.C. (1993) Pathogenicity of *Pratylenchus penetrans* to American ginseng (*Panax quinquefolium*) and to Kiwi (*Actinidia chinensis*). *Canadian Journal of Plant Science* 73, 907–912.
Waite, G.K. and Hwang, J.S. (2002) *Tropical Fruit Pests and Pollinators: Biology, Economic Importance, Natural Enemies and Control.* CAB International, Wallingford, UK.
Westerdahl, B.B. (2003) *UC IPM Pest Management Guidelines: Avocado. UCPMG-Publication 3436.* Division of Agriculture and Natural Resources, University of California (ANR Publications), Oakland, California. Available at: www.ipm.ucdavis.edu/PMG/r261200111.html
Willers, P. (1997) The nematode problem of guava is controlled by the nematicide cadusafos. *Inligtingsbulletin Instituut vir Tropiese en-Subtropiese Gewasse* 293, 10–12.
Willers, P. (1999) Do nematodes play a role in avocados? *Neltropika Bulletin* 306, 7–8.
Willers, P. and Neething, C. (1994) Spiral nematode: a potential papaya pathogen? *Inligtingsbulletin Instituut vir Tropiese en Subtropiese Gewasse, No. 267*, p. 13.
Wipers, P. and Grech, N.M. (1986) Pathogenicity of the spiral nematode *Helicotylenchus dihystera* to guava. *Plant Disease* 70, 352.
Wolfe, H.S. and Lynch, S.J. (1950) Papaya culture in Florida. *Agricultural Extension Service Bulletin* 113, Gainesville, Florida.

Wyss, U. (1987) Video assessment of root cell response to Dorylaimid and Tylenchid nematodes. In: Veetch, J.A. and Dickson, D.W. (eds) *Vistas on Nematology*. Society of Nematologists, Inc., Hyattsville, Maryland, pp. 211–220.

Wyss, U., Lehmann, H. and Jank-Ladwig, R. (1980) Ultrastructure of modified root-tip cells in *Ficus carica*, induced by the ectoparasitic nematode *Xiphinema index*. *Journal of Cell Science* 41, 193–208.

Yin, Y.Q. (1994) The investigation and identification of nematodes on mango in Guangdong. *Guangdong Agricultural Sciences* 2, 39–41.

Young, T.W. and Ruehle, G.D. (1955) The role of burrowing and meadow nematodes in avocado decline. *Plant Disease Reporter* 39, 815–817.

Yuksel, H.S. (1981) *Heterodera fici* Kirjanova 1954 in Aaegean Region. *Journal of Turkish Phytopathology* 10, 45–51.

13 Nematode Parasites of Coconut and other Palms*

Reginald Griffith,[1] Robin M. Giblin-Davis,[2] P.K. Koshy[3] and V.K. Sosamma[3]

[1]Coconut Research, Ministry of Food Production, Marine Exploitation, Central Experiment Station, Centeno, Via Arima PO, Trinidad, West Indies; [2]Fort Lauderdale Research and Education Center, University of Florida/IFAS, 3205 College Avenue, Davie, FL 33314, USA; [3]Division of Nematology, Central Plantation Crops Research Institute, Regional Station, Krishnapuram-690533, Kayangulam, Kerala, India

The botanical order Arecales has but a single family, Arecaceae, also known as Palmae. Palm is the common name for any flowering plant of the family. Although many of the 2800 known species of palms have some particular economic importance to any given local population, only a few are of major economic importance worldwide.

Cocos nucifera L., the coconut palm, which originated in Malaysia, South-east Asia, is widely distributed throughout the tropics.

Elaeis guineensis Jacq., the African oil palm, with its origin in Central Africa, has now been introduced throughout the tropics including Latin America.

Phoenix dactylifera L., date palm, is native to the near East where it has been cultivated for its fruit for nearly 8000 years.

Areca catechu L., arecanut, occurs mainly in the humid regions of Asia and the Malay Islands. It was introduced into India in the pre-Christian era where it is now widely cultivated and used as a masticatory.

Metroxylon spp., the sago palms, provide a starchy food material which is stored in their trunks as they develop to the point of flowering. These palms are hypoxanthic and only mature palms just prior to flowering and death are cut and used for starch production. These palms are also used during and after starch production for palm weevil (*Rhynchophorus bilineatus*) larval culture for human consumption in Papua New Guinea (Giblin-Davis, 2001). Sago palms of this genus are native to the Indonesian archipelago.

The fruits and seeds of eight genera of the world's palms are oil-bearing and can be commercially exploited for oil. Only *Cocos* is entirely of an Old World origin; *Elaeis* has one species (*guineensis*) that is of Old World origin and another (*oleifera*) which belongs to tropical America. The other six genera are considered neotropical. There are many palms that are ornamental and are important in horticulture and landscaping.

*A revision of the chapter by R. Griffith and P.K. Koshy.

Coconut

It is generally accepted that coconut palm originated in South-east Asia and was capable of being transported to the Americas and the West Indies by means of ocean currents, as primarily evidenced by its presence on shores and the water-resistant pericarp of its fruit or coconut. Most cultivated forms, however, have been introduced to the New World by man.

Coconut palm is most adapted to temperatures around 27°C with a diurnal range of about 7°C; it does not thrive at temperatures lower than 20°C and is damaged at temperatures below 15°C. Rainfall requirements are about 2500 mm/year; when less than 1000 mm/year, irrigation is normally necessary. The absence of rain for more than three successive months causes a shedding of young fruit and a reduction in fruit size. The palm does best with about 2000 h of sunlight/year, or an average of 6 h/day. Thus, with few exceptions, cultivation is limited by the 20° parallels of latitude and the 500 m contour line.

The largest producers of coconuts are Indonesia, the Philippines, India and Sri Lanka. Significant quantities are produced, however, from most tropical countries of South and East Asia, East and West Africa and the Pacific. Despite there being some large plantations, coconuts are predominantly a smallholder crop, with the average holding less than 1 ha.

The total world area under coconuts was estimated in 1996 as being 11 Mha, with a total production of over 47 Mt (FAO, 2000). The two biggest producers, Indonesia and the Philippines, have about 3.7 and 3.1 Mha, respectively; India is the third largest producer with nearly 1.8 Mha under cultivation. In all producing countries, coconuts make a significant contribution to the diet in addition to being an important source of export earnings.

Nematodes of Coconut

Many different nematodes have been found in diverse forms of association with the living coconut palm, others have been found associated in different types of symbiosis with insect visitors of the palm, operating and existing in various niches (Govindankutty and Koshy, 1979; Koshy and Banu, 2002), but the major nematode disease affecting the crop is red ring disease caused by *Bursaphelenchus cocophilus*. The only other nematode known to cause severe damage leading to malfunction in the coconut is *Radopholus similis*.

Bursaphelenchus cocophilus

The red ring nematode, *B. cocophilus* (Cobb, 1919) Baujard, 1989, was first described by Cobb (1919) as *Aphelenchus cocophilus* from specimens sent from Grenada. Since then it has been known as *A. (Chitinoaphelenchus) cocophilus* and *Chitinoaphelenchus cocophilus* (Micoletzky, 1922), *Aphelenchoides cocophilus* Goodey (1933) and *Rhadinaphelenchus cocophilus* (Goodey, 1960). Giblin-Davis *et al.* (1989b) presented morphological evidence supporting the similarities between *Rhadinaphelenchus* and *Bursaphelenchus*. Baujard (1989) synonymized the monotypic genus *Rhadinaphelenchus* with *Bursaphelenchus*, creating the new combination, *B. cocophilus*. Hunt (1993) retained the genus *Rhadinaphelenchus* as a monotypic genus on the grounds that: (i) more information was necessary to make a decision about its generic ranking; (ii) it is an economically well known genus with an extensive literature as *Rhadinaphelenchus cocophilus*; (iii) both sexes of the nematode are exceptionally long; (iv) host range is restricted to palms; and (v) the genus *Bursaphelenchus* appears to be an uneasy grouping of organisms. Recent sequence data of D2/D3 expansion segments of the LSU rRNA gene, 18S SSU rRNA, and a partial sequence of COI mitochondrial DNA of 20 species of *Bursaphelenchus*, including *B. cocophilus*, support its inclusion within the *Bursaphelenchus* radiation and not in the monotypic genus of *Rhadinaphelenchus* (Ye *et al.*, 2004d). Thus, the Baujard (1989) designation of *B. cocophilus* is accepted (see Chapter 2 for description).

Brief history of red ring disease

The disease was first reported as occurring in Trinidad by Hart in 1905. The first investigations into its nature were by Stockdale in 1906, who thought that two different diseases were being confused because they both culminated in decay of the bud. One of these was red ring disease, then called root disease, and the other bud rot initiated by *Phytophthora palmivora* Butl. Barrett (1906), however, reported that there were few genuine cases of bud rot among the coconuts in Trinidad and that 95% of the losses were really due to root disease.

Nowell (1919) found that a large number of roots examined from diseased trees in Trinidad contained hundreds of nematodes of the same species. They were also present in the constant red ring found in trees in Grenada and also in the material collected in Trinidad by Rorer (1911). Later, he examined stained sections from many other sources and confirmed Rorer's earlier conclusion that a fungus was not the causal organism, but noted that nematodes of the same species previously observed were constantly present in stems, leaves and roots. The name red ring disease was then used by Nowell (1919) and became established.

Distribution of red ring disease

At present, red ring disease has a restricted distribution in tropical America and has only been reported from the West Indies (Trinidad, Tobago, Grenada and St Vincent) and from Latin America (Venezuela, Guyana, Surinam, French Guyana, Colombia, Ecuador, Peru, Mexico, Brazil, Panama, Nicaragua, Costa Rica, Honduras, Belize and El Salvador). It is also reported that red ring disease occurs in Guatemala, but does not occur in the northern Caribbean islands, Florida, Cuba or other parts of the world (Dean, 1979). There are unconfirmed reports of red ring disease from Barbados, Dominica and Jamaica, but the European and Mediterranean Plant Protection Organization (EPPO) considers the disease to be absent from these countries. Some sources have reported the nematode present in the Bahamas, Dominican Republic and Haiti; however, we consider these questionable assertions. A single diseased palm was discovered in Dominica in 1982 but, after its destruction, there has been no further incidence of the disease in that country. The disease in Barbados claimed to be red ring in 1995 was found to be cedros wilt disease caused by a protozoan flagellate (R. Griffith, unpublished). CAB International stated in their Crop Protection Compendium data sheet that there were unconfirmed reports of the nematode being present in Puerto Rico in 2002; however, CAB International now record it as an 'Absent, unreliable record' from the country. None of the more recent surveys has found the nematode in Puerto Rico.

Symptoms of red ring disease

Young or adolescent coconut palms easily succumb to red ring disease. There is no record of any tree, once affected, having recovered. The disease occurs more commonly in trees 2.5–10 years old, with the greatest incidence in those 4–7 years old. Occasionally, a palm as young as 1.5 years or as old as 20 years or more may be attacked.

The symptoms characteristically described are those for palms of the tall cultivar of coconuts or 'typical' which grow in the West Indian islands. These symptoms differ somewhat in the dwarf variety 'nana' and also some 'Panama talls'. Chlorosis first appears at the tips of the oldest leaves and spreads towards their bases but, occasionally, one of the younger leaves may first be affected. The brown lower leaves may break across the petiole or the lower part of the rachis, or they may become partly dislodged at the base and hang down (Plate 14A). Nuts are shed prematurely either simultaneously with the development of leaf symptoms or slightly before. The crown often topples over about 4–6 weeks after symptoms first appear due to associated severe damage caused internally by the larvae of the palm weevil. However, the trunk remains standing in the field for several months until it decays. At the onset of symptoms, the chlorotic yellow appearance of the leaves around the stem is sometimes indistinguishable from that of trees growing under conditions of poor drainage or during intense drought.

The most characteristic symptoms are the internal lesions. In a cross-section of the stem, they appear as an orange to brick-red coloured ring, 2–4 cm wide, and at a distance of 3–5 cm in from the periphery (Fig. 13.1, Plate 14B). In longitudinal section, the reddened tissue may appear as two united bands joined in the bole forming a U-shape (Fig. 13.2). Lesions at the upper end of the stem in the vicinity of the crown are discrete, appearing first as streaks and then as dots. The meristematic tissue in the bud remains white and apparently healthy. Occasionally, in some older trees, the entire central cylinder of the stem becomes one solid block of red (Plate 14C). There is no putrefaction of the bud associated with red ring disease. In the roots, the normally white soft cortex becomes orange to faint red in colour, and dry and flaky in texture when diseased. In the leaves, a solid core of mottled tissue, dull red to brown in colour, extends from the leaf base for varying distances up to about 75 cm in the petioles.

Fig. 13.1. Characteristic red ring symptoms in a cross-section of coconut stem caused by *Bursaphelenchus cocophilus*. (Photo: R. Giblin-Davis.)

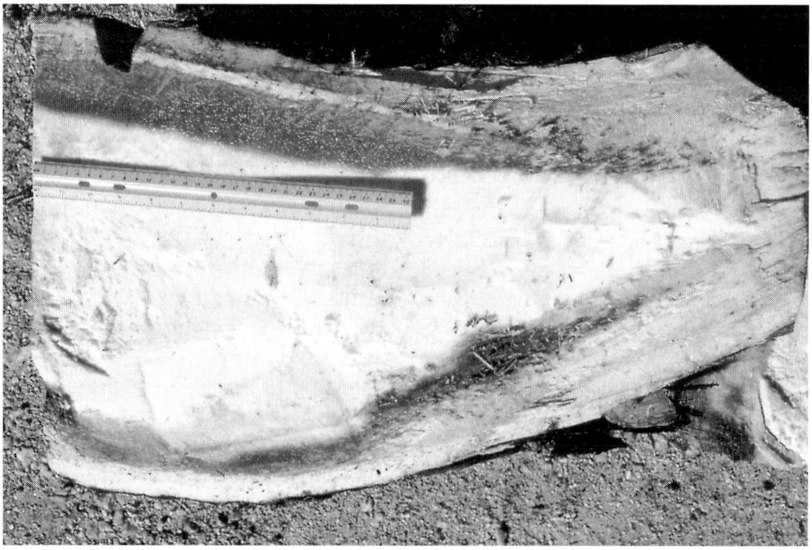

Fig. 13.2. Longitudinal section of coconut stem showing red ring tissues in two united bands joined in the bole forming a U-shape. (Photo: R. Giblin-Davis.)

The disease is not recognizable externally in its very early stages. The roots, stems and leaf petioles are already infested and there is full development of internal symptoms before the first external symptoms become visible. In the dwarf cultivars, the red colour gives way to shades of brown. Thus, instead of a red ring internally, there is a brownish band. The discrete spots are also brownish and the yellow discoloration of the leaves is not often apparent. Generally, the leaves become dried and brown, beginning at the tips of the leaflets and progressing downwards. The yellow dwarf cultivars respond in the same way as the green and the crosses between talls and dwarfs, or between 'Panama tall' and any dwarf. They show a browning instead of a characteristic reddening of the leaves and stem tissue.

Biology of the red ring nematode

The chief vector of the red ring nematode is an insect, the palm weevil (*Rhynchophorus palmarum* L.), and the biology and life cycle of *B. cocophilus* are intimately associated with this and other palm-associated weevils, such as *Dynamis borassi* and *Metamasius hemipterus* (Giblin-Davis, 2001) (Figs 13.3 and 13.4). However, experimentally, it has also been shown that red ring disease can be initiated by the nematodes via the root system.

Studies on the biology of the nematode were initiated by both Cobb and Nowell around 1919. Cobb found that 50% of adult *R. palmarum* and their larvae contained the red ring nematode. As a result, he implicated the palm weevil as being a carrier of the nematode from diseased palms to healthy ones. On the other hand, Nowell's opinion was that the nematode was soil inhabiting.

The general consensus is that *B. cocophilus* does not build up large populations in the soil, as some of the earlier investigators had believed to support their early recommendations of isolated trenches to control the movement of the nematode (Martyn, 1953). It would seem, moreover, that although root infection could be induced artificially, it was not a normal method of initiation of the disease in the field. In the ordinary course of events, the nematodes would not persist in the soil in sufficient numbers to give a reasonable chance of infection. If a persistent source of inoculum was present, e.g. a buried red ring trunk (which remains quite fresh for 2 weeks after burial), or if a high population of *B. cocophilus* (10^4 nematodes/cm^3) is artificially added to the soil in large quantities of water, the nematodes could gain entry through damaged or senescent roots and eventually migrate up into the trunk, producing the usual symptoms.

Transmission of red ring nematode

Larvae of the palm weevil, *R. palmarum*, feed by burrowing through coconut stems and, when this occurs in trees with red ring disease, they can become infested with the nematode. Adult weevils emerging from diseased trees carry the nematode to new sites (Figs 13.3 and 13. 4). Nematodes enter the haemocoel of weevil larvae via the gut tract; in newly emerged adult weevils, the nematodes can be found in the tracheae, gut, body cavity and the region of the ovipositor (Griffith, 1968a).

Survival of the red ring nematode depends on the third stage juvenile. They are sometimes found in tracheal sacs in the insect, from where they can move directly to the ovipositor of the female vector palm weevil (Griffith, 1968a). The percentage of palm weevils associated with red ring nematodes (up to 100%) and the levels of infestation ($>$ 13,000 nematodes/insect) can be quite variable. Evidence has been presented which indicates a connection between weevil size and percentage of weevils carrying the nematodes (Griffith, 1968a). However, other results have shown that variation in levels of infestation apparently are not correlated with other variables, such as weevil size (Giblin-Davis, 1993). The nematodes are putatively injected into the tissues of the coconut tree when the insect deposits its eggs, normally in a leaf axil in the crown of the tree (Griffith, 1968a,b) (Fig. 13.3). The palm

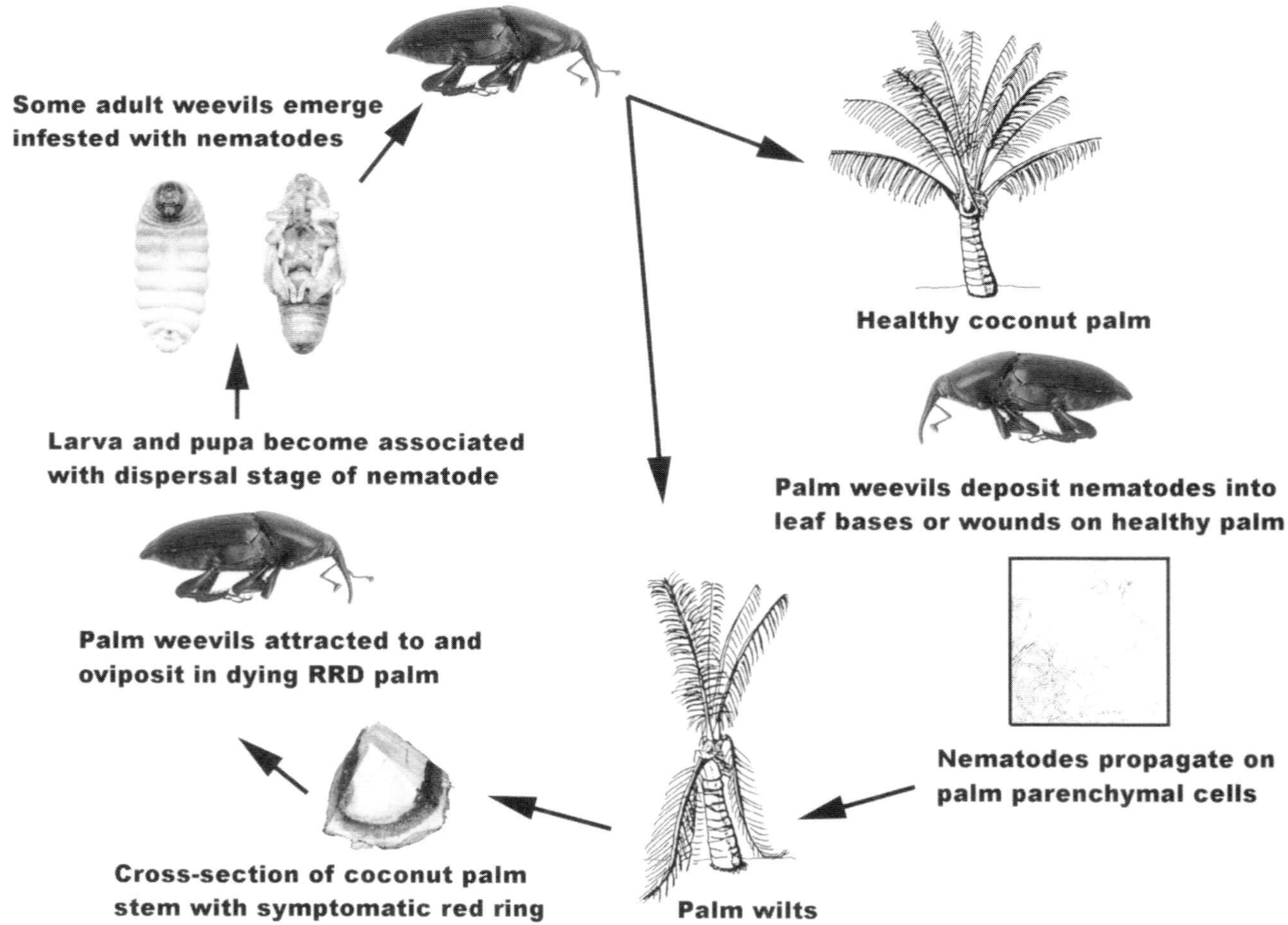

Fig. 13.3. Life cycle of the coconut palm weevil, *Rhyncophorus palmarum*, and transmission of the red ring nematode, *Bursaphelenchus cocophilus*. (R. Giblin-Davis.)

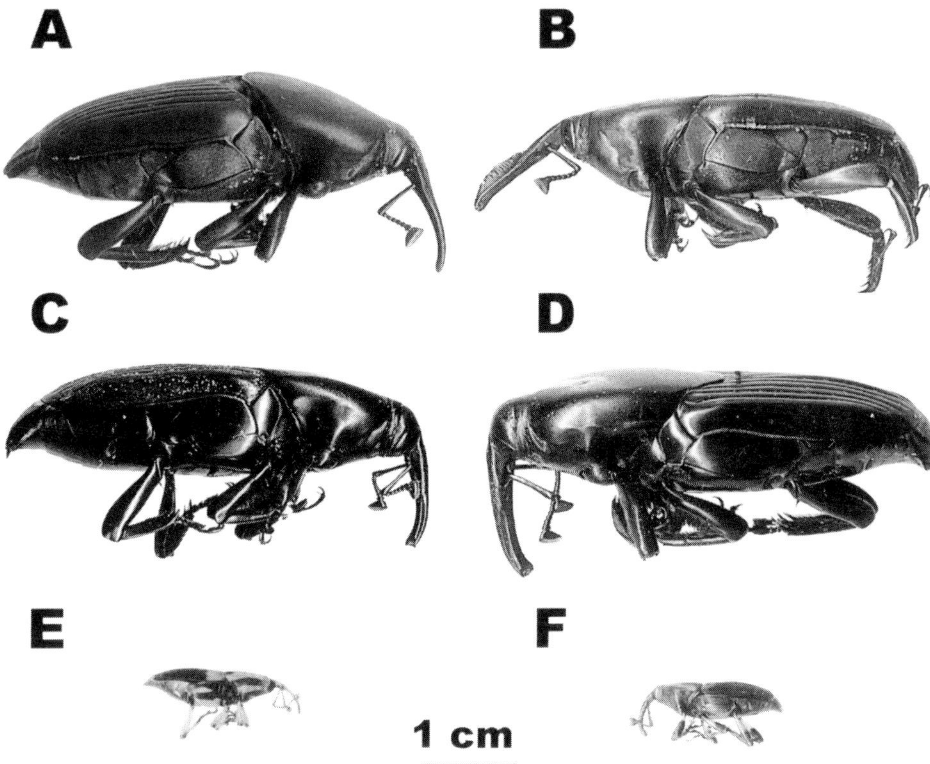

Fig. 13.4. Palm and sugarcane weevils known or suspected to be vectors of the red ring nematode. (A and B) *Rhynchophorus palmarum* female and male, respectively; (C and D) *Dynamis borassi* female and male; (E and F) *Metamasius hemipterus* female and male.

weevil might be considered an obligate transportation host, whereas the coconut palm, in which the nematode multiplies, could be considered as the definitive host.

The palm weevil, *D. borassi* (Fabricius) (Fig. 13.4), can carry close to 2000 red ring nematodes through metamorphosis and is assumed to be a potential vector in coconut where it is sometimes a pest in South America, e.g. Ecuador, Colombia and Brazil, in unopened inflorescences and the crown of coconut (Gerber *et al.*, 1990; Giblin-Davis, 2001).

The weevil, *M. hemipterus* (L.) (Fig. 13.4), has been incriminated as a vector of red ring nematodes in Colombia (Mora *et al.*, 1994) but not in Costa Rica (Bulgarelli *et al.*, 1998) or Trinidad (Hagley, 1963). Trap captures of *M. hemipterus* in African oil palm plantations tend to be up to 35 times higher than *R. palmarum*, but the percentage association and numbers of red ring nematode per weevil were lower with *M. hemipterus* than *R. palmarum* (Mora *et al.*, 1994). *M. hemipterus* may become associated with red ring disease because it attacks the pruned or damaged petioles and frond bases of living coconut, African oil palm, date and other palms (Giblin-Davis, 2001), which are development sites for the red ring nematode (Giblin-Davis *et al.*, 1989a).

Palm and sugarcane weevils in the Dryopthoridae, such as *R. palmarum*, *D. borassi* and *M. hemipterus*, are highly attracted to the volatiles emanating from recently wounded or pruned palms, red-ring-diseased palms, and moist fermenting tissue from palms, sugarcane stalks, various fruits and molasses (Chittenden, 1902; Giblin-Davis, 2001). In addition, males of these weevils produce male- and female-attracting aggregation pheromones identified as methyl-branched secondary alcohols (Giblin-Davis *et al.*, 1996; Giblin-Davis, 2001). The main aggregation pheromones are (4*S*, 2*E*)-6-methyl-2-hepeten-4-ol for *R. palmarum* (Rochat *et al.*, 1991; Oehlschlager *et al.*, 1992b), and (4*S*, 5*S*)-4-methyl-nonan-5-ol for *D. borassi* (Giblin-Davis *et al.*, 1997) and *M. hemipterus* (Rochat *et al.*, 1993; Perez *et al.*, 1997), which are commercially available as synthetic racemic blends at ChemTica Internacional SA (http://www.pheroshop.com/en/home_en.htm or info@pheroshop.com). These pheromones work as synergists with fermenting palm, sugarcane, or pineapple tissue and can be used to create lethal traps that are effective for monitoring or mass trapping efforts. In addition, *R. palmarum*, *D. borassi* and *M. hemipterus* respond to each others' pheromones, increasing the chances that once a weevil finds a stressed tree, other weevils will help to overcome it and use it as a host (Giblin-Davis, 2001). This phenomenon may also increase the chances for spreading red ring disease. In nature, the combination of weevil recruitment through chemical ecology and the killing potential of the red ring nematode may function together as a form of populational mutualism where enhanced reproduction of both partners is the result of the association (Giblin-Davis, 2004).

Soil transmission has been considered as an alternative means of spreading the nematodes. Despite the lack of experimental proof of transmission, some researchers have considered that insect visitors to the decomposing palms, e.g. ants, spiders and many saprophagous and predatory Coleoptera, are vectors of the nematodes to healthy palms. By their normal behaviour, these insects do not encounter healthy trees either directly or indirectly after leaving the decomposing, infected tree.

Biology of B. cocophilus *in coconut tissues*

The nematodes naturally invade only parenchymatous tissue in roots, stems and leaves, and artificially infested nuts. At first, nematodes occur as intercellular parasites in newly invaded tissue, but later they can be found both intercellularly and intracellularly. In many cases, lysigenous cavities are formed in which large numbers of nematodes are present. One gram of such tissue can contain as many as 10,000 nematodes. Nematodes have never been found in xylem vessels nor has there been any evidence of direct damage to the tracheal elements. Despite this, however, many of the vessel elements in the discoloured areas become occluded with tyloses. It has been shown that the uptake of water injected into the stems of trees is much slower in diseased trees than in healthy trees. Thus, one feature of the external symptoms coincides with a pathological condition due to water imbalance in the plant.

The cause of the restriction of nematodes to the narrow band or ring of necrotic tissue in stems has never been explained satisfactorily. Nowell (1923) found no anatomical or physiological factors in trees which might have accounted for it. Martyn (1953) expressed the view that the outer limit of the red zone was determined by the harder tissue at the periphery of the stem and the inner limit was set by aeration and water supply. Nevertheless, occasionally, there is a solid cylinder of discoloured tissue instead of just a band. Nematodes are often found intercellularly in white, apparently healthy tissue for 1 cm on the outside and 2.5 cm on the inside of the red ring tissue. They are less abundant here than in the body of the ring where they are found both intercellularly and intracellularly. It would, therefore, appear that there are other factors which naturally limit the occurrence and activity of the nematode on the outside and inside of the ring. The most outstanding characteristic of all tissue invaded by *B. cocophilus* is the presence of relatively large intercellular spaces. The inadequacy of intercellular space may, therefore, determine the outer limit. The colour of the band appears to be a specific plant chemical reaction to the invasion, and this varies in the talls and dwarf forms of coconut.

Nematodes inoculated into the mesocarp of nuts were found to have a life cycle, from egg to egg, of 9–10 days (Blair, 1964). The red ring nematode can persist in the diseased coconut tissue for about 3 months (Griffith, 1968b). Ashby (1921) found that juveniles were extremely susceptible to desiccation. They died within 6 h of drying and 15 h when provided with small fragments of tissue. The absence of moisture for half an hour only, followed by exposure to a saturated atmosphere for 24 h, resulted in death of the juveniles in nine cases out of ten.

There are other nematodes species besides *B. cocophilus* associated with *R. palmarum*, *D. borass*, and *M. hemipterus* (Griffith, 1968a; Gerber and Giblin-Davis, 1990a,b) that might cause some confusion during dissections. However, all of these additional nematode associates, such as *Rhabditis* spp., have been confirmed to be phoretic saprobionts.

Environmental factors affecting red ring disease

The larvae of the palm weevil often die when they develop in a tree that is attacked by *Phytophthora palmivora* Butl. (bud rot) or *Micrococcus roseus* Ali-Cohen (cedros wilt disease) subsequent to the contracting of red ring disease. Cannibalism in larvae of the palm weevil resulting from overcrowding often affects the number of emerging weevils. It is known that the red strain of *M. roseus* produces disease and septicaemia in affected palm weevils. Some ground lizards also feed on the adult insects.

The heaviest losses due to red ring disease occur at the end of the wet season and in the first 2 or 3 months of the dry season, i.e. between December and March, in Trinidad (Hagley, 1963). The abundance of the disease may be associated with pruning activities or with the activities of other insects that wound the tree first, inducing fermentation to which the palm weevil is attracted for oviposition. The age of the diseased palm is important since the palm weevil rarely becomes infested with nematodes from old trees. Thus, there is never an epidemic in old groves, even if abandoned, despite the fact that older trees can become heavily infested with the palm weevil alone, as seen in Ecuador.

The palm weevil is a pest in its own right and may relate to the environment differently. *R. palmarum* is a pest of the coconut palm, the gru-gru palm and several others. Some of the host palms are wild in the forest and in other uncultivated areas of Latin America, and many represent reservoirs that could become a source of migrant insects. However, epidemics in wild palms have never been reported.

In many Latin American countries, there exist different levels of attack from red ring disease only and palm weevil attack without red ring disease. In Ecuador, where the Creole tall palms seem to have more Panama Tall stock, the palm weevil is a major pest and the adult insects attack healthy trees of any age. In other countries, such intense attack without red ring disease is quite rare, but, in Ecuador, the insect is a pest in a habitat consisting of several other interplanted kinds of food sources for the weevil, such as pineapples, papayas and sugarcane, that are non-hosts for red ring nematode.

The effects of climate on red ring disease incidence are very apparent as one moves from the dry southern Ceara coconut regions to the northerly more humid areas such as Bahia in Brazil. In Ceara, where the dry season extends for 7.5 months, the incidence of red ring is significantly lower than in Rio Grande del Norte, where the dry season is for 5.5 months and less than in Paraiba where the season lasts for 3.5 months. However, in Pernambuco, where the dry season only lasts for 2 months, the incidence is almost as high as in Bahia Sul where there is little or no dry season. The larvae of the palm weevil can develop adequately within the tissues of the coconut trunk; however, the dissemination stage of the adult is affected by the low humidity in the driest regions where one is more likely to find dead adult palm weevils in the field.

Other hosts

Although red ring is primarily a disease of the coconut palm, it has been found in many palms (Table 13.1) including an unidentified species of *Cocos* (probably *syagrus*) in the Botanic Gardens, Grenada (Nowell, 1924), and the date palm, *Phoenix dactylifera*, in the Botanic Gardens, Trinidad. Hagley (1963) found one case of natural infestation of the cabbage palm *Roystonea oleracea*. Disease incidence was reported to be high in the plantation of oil palms, *Elaeis guineensis*, in Venezuela in 1953 (Malaguti, 1953). Nowell (1924) reported successful inoculation of the cabbage and the gru-gru palm, *Acrocomia aculeata*. Latterly, various ornamentals have been artificially inoculated, among them are the Sabal palm, *Sabal palmetto* and the cocorite palm *Mauritia caribea*. The disease has also been found in Brazil on *Attalea cohune*, the Cohune nut.

The palm weevil does not transmit the nematode to any other non-palm host species, e.g. sugarcane, papaya and pineapple. On the other hand, countries such as the USA, which utilize the *Sabal* sp. for decorative purposes in the presence of the American palmetto weevil, *Rhynchophorus cruentatus*, need to ensure proper quarantine measures against both the palm weevil, *R. palmarum*, and the red ring nematode. The date palm, the coconut and the sabal palm have all been reported as hosts of *R. cruentatus*.

Table 13.1. Natural and inoculated host list records for *Bursaphelenchus cocophilus*.

Acrocomia aculeate (gru-gru palm)
Acrocomia intumescens
Attalea cohune (Cohune palm)
Bactris gasipaes
Bactris sp.
Cocos nucifera (coconut)
Cocos sp.
Elaeis guineensis (African oil palm)
Euterpe pacifica (?; = *E. precatoria*, or *Mauritiella pacifica*??)
Jessenia polycarpa
Mauritia flexuosa (Ita palm)
Mauritia caribea (Cocorite)
Mauritia mexicana
Maximiliana maripa
Oenocarpus distichus
Phoenix canariensis (Canary Island date palm)
Phoenix dactylifera (date palm)
Roystonea oleracea (royal or cabbage palm)
Roystonea regia
Sabal palmetto (Sabal palmetto)
Sabal sp.

Epidemiology and general management measures

Red ring disease in new groves generally begins by infection of a 4- to 10-year-old palm by a weevil carrying the nematodes. The most effective management should be implemented during the initial phase of palm weevil and red ring disease infestation to prevent the development of an epiphytotic.

The rate of spread from a primary infector plant depends upon the development of vector palm weevils within the diseased tree. Typically, 3 months after infection, nearby susceptible healthy trees can be infected by a vector(s) emerging from the infector plant (Fig. 13.3). The initially diseased tree remains a source of red ring nematode inoculum for 6–8 weeks after its death as it continues to attract other sugarcane and palm weevils that might become contaminated with red ring nematodes and serve as vectors. Phytosanitary measures of control are critical at this time since disease symptoms are apparent before the vector progeny emerge, and successful intervention can prevent an epiphytotic of red ring disease.

Emerging palm weevils disperse to leaf axils of diseased trees or wounded trees emitting attractive compounds (kairomones such as ethyl acetate and ethanol) (Giblin-Davis *et al.*, 1996; Rochat *et al.*, 2000). In addition, males of *R. palmarum*, *D. borassi* and *M. hemipterus* produce aggregation pheromones that synergize attraction and recruit conspecifics and heterospecifics of both sexes (Giblin-Davis, 2001). These weevils oviposit in newly diseased palms and cause increased insect populations. Control measures relate directly to the abundance of the disease. Since all red-ring-diseased trees are breeding grounds for palm and sugarcane weevils and red ring nematodes, the destruction and removal of these trees and the reduction of their attractiveness is essential to preventing epiphytotics.

Specific management measures for red ring disease in coconut

There are no simple means of controlling red ring disease, and no effective measures are available as yet for control of the nematode in living palms. Control is based on prevention rather than cure by the destruction of infested palm material, and by the trapping and killing of the weevil vectors before they spread the nematodes.

Many trees show yellowing and browning of leaves that may not be due to red ring disease. To prevent unnecessary destruction of trees, a 'core sample' of the trunk should be taken with a 2 cm diameter pipe (see below) to determine the presence of red ring disease and the nematodes before control measures are employed.

INSECTICIDE AND HERBICIDE TREATMENTS. A fundamental principle in the control of the disease is phytosanitary roguing, based primarily on the fact that the diseased palm is the major source of inoculum and the niche for the vector development. After confirmation that a tree has red ring disease (see below), the leaf axils should be sprayed with 0.1% Lannate (Methomyl) solution to kill off the palm weevils living in the crown (Griffith, 1971). Trees should be killed with 100–150 ml (48.3% a.i.) of the herbicide monosodium acid methanearsonate (MSMA) or other herbicide that is injected or placed into the trunk. This usually takes 2–3 weeks. Occasionally, trees injected with MSMA will harbour weevil larvae. Thus, the tree should be cut and sectioned to make sure that weevils are not present. When trees are discovered in advanced stages of the disease or when they are seen in a 'broken neck' condition, they cannot be poisoned with herbicides. Such trees should be cut down and the pieces and remaining stump sprayed thoroughly with an insecticide, such as methomyl, trichlorfon, monocrotophos, carbofuran, carbaryl, imidichloprid or lindane. If the tree is sprayed adequately with an insecticide, all larvae and pupae of the palm weevil that were developing in the diseased tree will be killed. It is recommended that the site be checked every couple of weeks until the palm has decomposed or that the dried out remains of the palm be burned with the aid of kerosene.

MASS TRAPPING OF PALM WEEVILS. Phytosanitation (removal and destruction of red ring-diseased trees) and trapping of weevils using pesticide-treated palm or fruit tissue have been recommended methods of management of red ring disease in coconut for many years (Mariau, 1968; Griffith, 1969; Delgado and Moreno, 1986). Trapping becomes significant in reducing the abundance of palm weevils generally and, dependent upon the density of traps per hectare, catch the smaller percentage of vector weevils which generally visit and infect palms in the near vicinity, one or two trees away, from the source of infection. The identification, synthesis and commercial availability of male-produced aggregation pheromones of palm weevils have significantly improved the efficacy of the older methods of trapping with fermenting tissue alone (Oehlschlager *et al.*, 1993, 2002; Moura *et al.*, 2000). Although more research is needed in coconut, Moura *et al.* (2000) demonstrated in Brazil that by using 54 100-l perimeter traps of pheromone (Rhyncholure; racemic 6-methyl-2-hepten-4-ol; ChemTica International) plus sugarcane around a red ring-diseased coconut plantation (54 ha) for 26 months, the *R. palmarum* capture rate (> 97,000 weevils were captured during this period) remained constant but the red ring disease incidence dropped dramatically. In Tabasco, Mexico using pheromones in guerrero-type coconut tissue traps over several months showed a positive correlation with the number of insects captured and incidence of disease (Perez-Marquez, 1999, personal communication). Future research in the management of red ring disease in coconut palm, which is a very suitable and susceptible host for red ring disease and *R. palmarum*, must examine the efficacy and cost effectiveness of perimeter mass trapping in concert with phytosanitation in different situations (small versus large crop holdings of various aged palms) in rural tropical America.

BIOLOGICAL CONTROL MEASURES. Natural enemies have not been evaluated thoroughly for management of potential palm and sugarcane weevil vectors. Several organisms may hold promise, including entomopathogenic nematodes in the families Steinernematidae and Heterorhabditidae, the prokaryote *Micrococcus roseus*, and tachinid parasites, *Billaea rhynchophorae* and *B. menezesi* (references cited in Giblin-Davis, 2001).

Methods of diagnosis

RECOVERY OF *B. COCOPHILUS* FROM COCONUT TISSUE. The well-established methods for obtaining samples of nematodes from living trees are still used. A stainless steel tube, sharpened at one end, is driven at an angle of 45° at the point selected for sampling. The extracted core is placed in a blender with 50 ml of water and processed for 2 min. The contents of the blender are then poured into a dish and left for 20 min for the nematodes to emerge. The nematodes are then recovered by sieving. The red ring nematodes are often highly mobile in water (swimming and coiling), leading to knots of clumped nematodes or resuspension of nematodes after centrifugation. In coconut and the palmiste palms, the nematodes are most active in the stem tissue except in the very necrotic regions. The core tissue generally shows a red cylinder of necrotic red ring tissue. In the method originally used by Fenwick and Maharaj (1963), diseased coconut stem, petiole or root tissue is chopped into fine pieces about 1 cm in thickness, placed in a large funnel of water, whose stem is closed at one end with a tube and clip, and whose neck has a small plug of cotton acting as a filter. This can be modified by macerating the diseased tissue in a blender to release more nematodes and then screening through a no. 400 USA Standard Testing Sieve (38 μm openings) before backwashing into the funnel. The funnel is allowed to stand overnight before harvest. Schuiling and Van Dinther (1981) offer another modification for extracting red ring nematodes from tissue.

Radopholus similis

The burrowing nematode, *R. similis*, occurs in most tropical and subtropical areas of the world and has been reported from coconut palms in Florida, Jamaica, Sri Lanka and India (Van Weerdt *et al.*, 1959a,b; Ekanayake, 1964; Latta, 1966; Weischer, 1967; Koshy *et al.*, 1975). Koshy (1986) suggested co-evolution of the nematodes along with black pepper and certain cultivars of banana in the western hills of South India. It occurs deep inside the forests on wild black pepper and is widespread on a number of crops such as coconut, arecanut, black pepper, banana, betel vine and ginger in south India.

Symptoms of damage

The burrowing nematode causes non-specific general decline symptoms such as stunting, yellowing, reduction in number and size of leaves and leaflets, delay in flowering, button shedding and reduced yield. *R. similis* infestation produces small, elongate, orange-coloured lesions on tender creamy-white roots. Consequent to nematode parasitization and multiplication, these lesions enlarge and coalesce to cause extensive rotting of the roots (Fig. 13.5). Tender roots of coconut seedlings with heavy infestation become spongy in texture. Surface cracks develop on the semi-hard orange-coloured main roots. Lesions and rotting are confined to the tender portions of the root. Lesions are also not conspicuous on the secondary and tertiary roots since these are narrow and rot quickly on infestation.

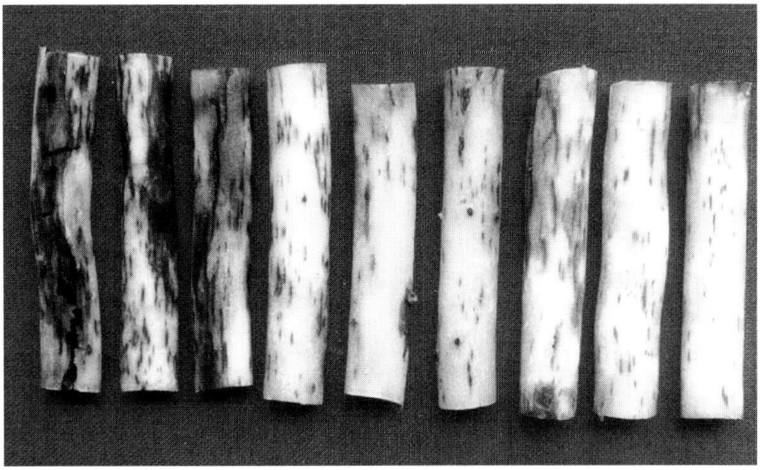

Fig. 13.5. Progressive development of necrotic lesions (from right to left) on coconut roots caused by *Radopholus similis*. (Photo: V.K. Sosamma.)

As many as 4000 nematodes are known to occur in 1 g (2.5 cm length) of main roots. The nematode also attacks the plumule, leaf bases and haustoria of seedlings. The above-ground symptoms being non-specific, the only definite method to identify an infested palm is to look for characteristic lesions on fresh, creamy-white to orange-coloured tender main roots after cleaning and rubbing the epidermis.

R. similis does not enter or penetrate the coconut roots that have developed a hardened or suberized epidermis, but does penetrate the absorbing region behind the root cap covered by very delicate epidermis by lysis of cells. The cavities that form in the outer cortex are always surrounded by deeply stained and heavily suberized cells of irregular shape, whereas those formed in the inner cortex do not have any such deformed darkly stained border cells. The maximum numbers of nematodes and cavities are seen in the outer cortex (Fig. 13.6). Nematodes have not been observed in the stelar region or in the closely packed 4–6 layers of cells outside the endodermis even in heavily infested roots. In the early stage of infection, roots have separate cavities that later merge with each other consequent to feeding and multiplication of nematodes.

Multiple cavities and their coalescence destroy the cortex to a great extent, but the stelar tube remains intact. Eggs and all stages of nematodes with different orientations are seen in the cavities in longitudinal sections (Fig. 13.6) (Koshy and Sosamma, 1982a, 1987; Sosamma and Koshy, 1991, 1998).

Biology and life cycle

The burrowing nematode is a migratory endoparasite and is capable of spending its entire life within roots. Most juvenile stages and adult females including gravid females infest healthy succulent root tips; fourth stage and adult males do not. The nematode takes 25 days at 25–28°C to complete one life cycle (J2 to J2) (Geetha, 1991).

The coconut isolate of *R. similis* from Kerala, India is the 'banana race' as they do not infest *Citrus* spp. or *Poncirus trifoliata* (Koshy and Sosamma, 1977), and has a haploid number (n = 4) of chromosomes (Koshy, 1986; Jasy, 1991). The *R. similis* population from coconut root is easily cultured axenically on carrot discs placed on 1% water agar or 10% tapioca pearl (Koshy and Sosamma, 1980; Banu and Sosamma, 1999). It can also be cultured within the mesocarp of growing tender coconuts without affecting the size or quality of the nuts (Koshy and Sosamma, 1982b).

Fig. 13.6. Longitudinal section of coconut root showing *Radopholus similis* in cavities formed in the roots. (Photo: V.K. Sosamma.)

Survival and means of dissemination

The burrowing nematode survives under field conditions for 6 months in moist soil (27–36°C) and for 1 month in dry soil (29–39°C); it survives for 15 months in moist soil (26–29°C) and for 3 months in dry soil (27–31°C) under glasshouse conditions. The nematode survives in roots of stumps of felled coconut palms for up to 6 months (Sosamma and Koshy, 1986) and as adult females in coconut roots and soil during summer months, causing annual recurrence of infection (Sosamma, 1984).

Coconut seedlings are raised by sowing seed nuts in the interspaces in coconut plantations in Kerala, India. Most of the nurseries in Kerala and Tamil Nadu (south India) are infested by *R. similis* (Sundararaju *et al.*, 1995a,b). One-year-old coconut saplings raised in these infested nurseries harbour large populations of the nematode in roots internal and external to the husk. Such seedlings when distributed for planting help in the dissemination of the nematode over long distances (Koshy and Sosamma, 1978b, 1979).

Environmental factors affecting parasitism

Infested coconut roots yield a maximum number of *R. similis* during October to November and minimum during March to July in India. Factors favourable to nematode multiplication are a mean soil temperature below 25°C and a light rainfall coupled with availability of tender fleshy roots. Nematode populations in roots of individual palms were found to vary considerably during low and high peaks depending upon the age, cultivar and conditions of the palms involved (Koshy and Sosamma, 1978a). The burrowing nematode multiplies well on coconut in loamy sand, followed by riverine alluvium, but least in Kari type soils. However, it causes maximum plant damage in riverine alluvium and the lowest in laterite soil (Sosamma, 1984; Sosamma and Koshy, 1985).

Other hosts

The coconut isolate of *R. similis* has a wide host range including several economically important plants, weeds and trees. Of 115 plant species tested, 48 species belonging to 45 genera in 17 families were recorded as hosts (Koshy and Sosamma, 1975; Sosamma and Koshy, 1977, 1981).

Disease complexes

The fungi *Cylindrocarpon effusum*, *C. lucidum* and *Cylindrocladium clavatum* have been recorded in association with lesions produced by *R. similis* in coconut roots. In pathogenicity studies, the fungus *C. effusum* did not cause any appreciable damage to inoculated seedlings. The fungus, when inoculated simultaneously with the nematode, reduced the rate of multiplication of the nematode and damage to coconut seedlings (Sosamma and Koshy, 1978, 1983; Koshy and Sosamma, 1987; Sosamma, 2000b). *Aphelenchoides aligarhiensis*, *Panagrolaimus rigidus* and *Rhabditis* sp. were isolated from leaf rot disease-affected spindle leaves of coconut in Kerala, India (Nadakkal, 1965; Sosamma, 2000c). Application of Phorate at 2 g a.i./palm to the base of the unopened spear leaf is found helpful in control of the disease. However, the role of nematodes in the disease complex as passive vectors/synergists is yet to be defined (Koshy, 2000; Koshy *et al.*, 2002c).

Economic importance and population damage threshold levels

Surveys of different coconut-growing tracts of Kerala, Karnataka and Tamil Nadu States of India (964,000 ha) revealed the widespread occurrence of *R. similis*. Twenty-four per cent of the root samples yielded *R. similis*, and, of these, 50% yielded one or more *R. similis*/g of root (Koshy *et al.*, 1978; Sosamma, 1984). A 30% increase in yield was recorded by application of *Hydrocarpus* sp. oil cake at 8 kg/palm/year or phorate and aldicarb at 10 g a.i./palm in June–July and October–November to the burrowing nematode-infested coconut palms (Koshy, 1986).

The pathogenicity of *R. similis* on coconut was established by conducting two experiments, the first with a duration of 5 years and the second over a period of 1 year. An initial inoculum level of 62,500 nematodes per seedling caused 4, 22, 76, 18, 25, 40, 48 and 79% reduction with respect to height, girth at collar region, shoot weight, number of leaves, number of leaflets per leaf, leaflet length, lamina length and root weight, respectively, over control plants. The effect of parasitization of the nematode was more pronounced on the root system, especially on the number and mass of feeder roots. The threshold inoculum density required for causing significant reduction of various growth parameters was 100 nematodes in 625 cm^3 or 900 g of soil under field conditions over a period of 5 years.

In the second experiment, an initial inoculum level of 100,000 nematodes caused 40, 55, 20, 65, 20, 48 and 52% reduction with respect to height, shoot weight, number of leaves, leaf area, number of lateral roots, volume and weight of roots, respectively, over control plants over a period of 1 year. Leaf bases and haustoria of seedlings were also infested by nematodes. No appreciable damage was noticed in plants inoculated with the fungus, *C. effusum* alone. The pathogenic threshold level of the axenic *R. similis* population for causing damage to all plant growth parameters was 1000 nematodes per seedling or 10 nematodes/100 cm^3 or 140 g of sandy loam soil under greenhouse conditions. The histopathology of infested roots recorded the presence of nematodes in the cortex in the inter- and intracellular positions (Koshy and Sosamma, 1983, 1987; Sosamma, 1984).

To facilitate normal growth of the plant to flower and exhibit the disease under natural conditions, a detailed pathogenicity trial was initiated in 1.8 m × 1.8 m × 1.2 m field tanks (microplots) over a period of 11 years using axenic inoculum. This experiment, the first of its kind on a perennial crop, clearly brought out the damage potential of burrowing nematodes on coconut. All the uninoculated palms came to flowering during 65–83 months after planting, between leaf axils 31 and 49, whereas four out of the five palms that received an initial inoculum level of 100 nematodes flowered during 67–130 months in the leaf axils from 39 to 56. Two palms each that received an initial inoculum level of 1000 and 10,000 nematodes came to flowering after 108 months and one out of five palms that received an initial inoculum level of 1 lakh (100,000) nematodes also came to flowering after 132 months. None of the palms that received 10 lakh (1,000,000) nematodes came to flowering. The control palms produced a total of 155 inflorescences compared with 67 inflorescences in palms inoculated with 100 nematodes as an initial inoculum level. However, the palms that received an initial inoculum of 1000 nematodes and above did not yield any nuts even 11 years after planting. The control plants produced an average of 125 nuts compared with 37 nuts by palms that were inoculated initially with 100 nematodes. Even 1 nematode in 35,640 cm^3 of soil or 100 nematodes per seedling reduced the yield by 77% (Koshy and Sosamma, 1994, 1996).

Management measures

Management of the burrowing nematode on a perennial palm such as coconut with a massive root system is difficult, especially under the high density multispecies cropping system that exists along the west coast of south India involving susceptible crops such as arecanut, banana, black pepper, betel vine, ginger and turmeric. Unlimited use of nematicides for the control of the burrowing nematode may cause problems of residual toxicity in coconut water and copra (Habeebullah *et al.*, 1983; Sosamma, 1996). Apart from this, it may also lead to residual toxicity in the products of the intercrops. Therefore, control of nematodes by field application of nematicides alone is not a practical proposition.

CULTURAL PRACTICES. The cultural practices existing in Kerala and Karnataka (India) are the application of neem and marotti (*Hydrocarpus*) oil cakes at 2–4 kg/palm/year, farmyard manure at 50 kg/palm/year, and green foliage and tender stem of *Glyricidia maculata* to the basins at 50 kg/palm/year. The growing of green manure crops such as cowpea, *Crotolaria* or *Sesbania* in the basins and interfaces during June to August and ploughing in of the entire crop at flowering help in reducing the burrowing nematode population and enriching the nutritional status of the soil. In addition, the growing of intercrops such as cacao that enrich the soil with sizeable quantities of shed foliage helps in the build-up of beneficial organisms and antagonistic microorganisms that may inhibit nematode multiplication (Koshy *et al.*, 1991a,b, 2002a).

BIOLOGICAL. A significant increase in width and leaf area has been recorded in coconut seedlings that received mycorrhizae alone. An increase in shoot weight, root weight and a decrease in lesion indices occur in seedlings inoculated with mycorrhizae prior to *R. similis*. A mixture of mycorrhizae consisting of multiple endophytes, i.e. *Acaulospora bireticulata*, *Glomus fasciculatum*, *G. macrocarpum*, *G. mosseae*, *G. versiforme*, *Sclerocystis rubiformis* and *Scutellospora nigra*, was found effective in improving the plant growth and reducing *R. similis* infestation of coconut seedlings (Sosamma, 1994).

Minimum growth characters and maximum multiplication of nematodes were recorded in plants that were inoculated with *R. similis* alone. In combined inoculation of mycorrhizae and nematode, maximum growth is recorded in plants inoculated with *A. bireticulata*. The mycorrhizae, *A. bireticulata*, has maximum multiplication on coconut compared with *G. macrocarpum*, *Scutellospora coralloidea* and *S. rubiformis* in nematode-free as well as nematode-inoculated plants. Nematode populations are also low in plants inoculated with *A. bireticulata* (Sosamma, 1994; Sosamma *et al.*, 1998a).

A new isolate of *Pasteuria* parasitizing *R. similis* in Kerala, India has great potential for use in integrated pest management. The infective propagules of *Pasteuria* adhered to the cuticle of adult males, females and juveniles of *R. similis* (Sosamma, 1999, 2000b,d, 2002). Introduction of *Paecilomyces lilacinus*, *Pasteuria penetrans* and mycorrhizae into potting mixture contained in plastic bags in coconut nurseries and again in the planting pit at the time of transplantation of coconut seedlings in the field helps in better establishment of plants and imparts better growth by offering protection against *R. similis* (Koshy, 1998; Koshy *et al.*, 1998a; Sosamma *et al.*, 1998b). *Catenaria vermicola* was also found parasitizing *R. similis* in Kerala (Sosamma, 2000a). Introduction of *P. lilacinus*, *P. penetrans* and vesicular arbuscular mycorrhizae (VAM) into potting mixture contained in poly bags in coconut nurseries and again in planting pits at the time of transplantation of coconut seedlings in the field helps in better establishment of plants and imparts better growth by offering protection against *R. similis* (Koshy, 1998; Sosamma *et al.*, 1998b).

RESISTANCE AND TOLERANCE. All the coconut cultivars (29 exotic, 15 indigenous and 15 hybrids) screened for resistance to *R. similis* in India were found susceptible in varying intensities. The dwarf cultivars Kenthali and Klappawangi recorded the least nematode multiplication and lesion indices. Similar reactions were noticed in hybrids such as Java Giant × Kulasekharam Dwarf Yellow, Kulasekharam Dwarf Yellow × Java Giant, Java Tall × Malayan Yellow Dwarf and San Ramon × Gangabondam (Sosamma *et al.*, 1980, 1988; Sosamma, 1984).

CHEMICAL. Burrowing nematode infestation in coconut nurseries has been detected in India. Increased incidence of *R. similis* can occur when banana is used as a shade crop in coconut nurseries. In these situations, there is possibly a case for treatment of nurseries with nematicides to produce nematode-free seedlings to prevent spread of the nematode into the main field and to uninfested areas.

Past experience has shown that a dip in 1000 ppm dibromochloropropane (DBCP) for 15 min is effective in controlling nematodes in seedlings for *R. similis*-infested coconut nurseries (Koshy and Sosamma, 1978b, 1979). Complete control of *R. similis* can be obtained with soil application of phenamiphos or phorate at 25 kg a.i./ha during September, December and May in infested coconut nurseries (Koshy and Nair, 1979; Koshy *et al.*, 1985).

Summary of management measures

The following measures are suggested towards developing an integrated management schedule for *R. similis* infestation on coconut palms (Koshy, 2002).

- Application of cow dung (50 kg), oil cakes (2–4 kg) and green manuring with *Glyricidia maculata* (50 kg) per palm/year to the basins.
- Growing *Crotolaria juncea*, cowpea or *Sesbania* in the basins and interspaces and incorporating into the soil by ploughing in at flowering stage.
- Application of phorate at 10 g a.i./palm twice yearly (in June–July and in October–November in India).
- Avoid growing bananas as a shade crop in coconut nurseries.
- Use of nematode-free planting material of coconut and other intercrops.
- Use of tolerant or less susceptible cultivars or their hybrids in infested areas.
- Cut and remove all roots external to the husk of seedlings raised in the field before planting.
- Raise coconut seedlings in potting mixture enriched with bioagents such as *P. lilacinus*, *P. penetrans* and mycorrhyzae in plastic bags.
- Introduce bioagents into the planting pits while planting in the main field.
- Apply phorate at 3 g a.i./plant to intercrops such as banana, black pepper and arecanut in June–July and October–November.

Methods of diagnosis

SAMPLING. Soil and root samples for detection of *R. similis* should be collected when maximum populations of the nematode occur (October–November in India). Maximum populations of *R. similis* are found on coconut at a distance of 100 cm from the bole of the palm and at a depth of 50–100 cm. Fifty g of tender, creamy-white to orange-coloured, semi-hard, main roots (~1 cm diameter) showing lesions and rotting should be collected to obtain live populations in large numbers (Koshy *et al.*, 1975).

EXTRACTION. The semi-hard, orange-coloured, main root bits are peeled and sliced longitudinally into 4–8 pieces of 3–5 cm length. These sliced root bits are submerged in water contained in Petri dishes or shallow pans; a temperature of 20–25°C is ideal for extraction of live nematodes from polyphenol-rich coconut roots (Koshy *et al.*, 1975). After every 24 h of incubation, the water needs to be changed; 50% of the population is extracted after 72 h. Most of the nematodes are recovered within 4–7 days.

DETERMINATION OF POPULATIONS AND CROP LOSS. Nematode populations in the tender portions of the main roots can be estimated by staining and blending. Roots may be cut into 2-cm long pieces, sliced longitudinally into eight sections and then stained.

Nematodes for the control of other coconut pests

Entomopathogenic nematodes *Heterorhabditis indica* and *Steinernema* spp. were isolated from soil around coconut in Kerala which were used in the integrated management schedule for the Rhinoceros beetle, *Orycetes rhinoceros* L. and red weevil, *Rhynchophorus ferrugineus* F. (Sosamma and Banu, 1996; Banu *et al.*, 1998; Sosamma, 2000, 2003).

Conclusion and future prospects

The burrowing nematode, *R. similis*, is second in importance to the red ring nematode, *B. cocophilus*, on the basis of its damage potential on coconut. Though the nematode has been reported in association with various coconut diseases (Govindankutty and Koshy, 1979), no detailed investigations have been carried out anywhere else except India. Screening for resistance/tolerance to *R. similis* in coconut cultivars and their hybrids has indicated the availability of possible resistance in some cultivars. Though breeding in coconut is a long-term process, this area could be profitably exploited. Developing an integrated management schedule for the coconut based on subsistence farming systems involving susceptible perennial crops such as arecanut, black pepper, cacao and banana should be the priority area of research.

Oil Palm

The oil palm, *Elaeis guineensis* Jacq., has a natural distribution in West Africa between latitudes 13°N and 12°S from the coast to the Great Lakes. Ecologically, it is found in the transition regions between the rainforest and the savanna. It has also been cultivated extensively in Malaya and Indonesia. Commercial production of oil palm in Central and South America dates back only to the 1960s, though production is expanding in all tropical South America. In the New World, it is a plantation crop with holdings of several hundred to several thousand hectares per unit, whereas in Africa or Asia it can be a large plantation or smallholders crop as with the coconut.

Nematodes of Oil Palm

Generally, the major diseases of the oil palm are found in its area of origin. Curiously, although there are fungal-, bacterial- and suspected viral- or phytoplasma-induced diseases, no records of any economic losses due to nematode damage occur in the Old World. However, *B. cocophilus* causes economic loss in oil palm in Central and South America. Other plant parasitic nematodes have been reported on oil palm in India, Pakistan and other countries (Maqbool, 1991; Salazer *et al.*, 1992; Sundraraju *et al.*, 1995; Sundraraju and Sudha, 1998).

Bursaphelenchus cocophilus

Freeman, in 1925, in Trinidad, appears to have been the first to record the pathogenicity of the red ring nematode on oil palms. Red ring disease caused by *B. cocophilus* has been known from African oil palm from Venezuela since before 1953 (Webster and Gonzales, 1959) from a single plantation of 1000 ha where the disease caused severe losses. Malaguti (1953) demonstrated that African oil palm, which at that time had recently arrived in Latin America from Africa, was invaded by the red ring nematode. Malaguti cites a group of 100 palms showing only 16 doubtful cases in January, but which by August had 22 deaths, nine doubtful or affected cases and only 69 palms remaining healthy. On that Venezuelan estate, about one-third of the total palm population became infected.

Symptoms of red ring disease in oil palms and biology

The coloration in the diseased palm is similar to that of the browning associated with the 'nana' or dwarf cultivar of coconuts, i.e. brownish rather than reddened tissue internally (Plate 14E). Also, the leaves dry out and turn brown instead of the usual yellowing and then browning associated with the tall cultivar of coconuts (Plate 14F). Often, the centre of the crown takes on a dwarfed appearance and the newly opened 'little leaves' become bundled together, the leaflets being twisted, corrugated and adhering to a stiff upright rachis. The developing bunches show necrosis and the inflorescences do not set fruit. The ultimate symptoms of red ring disease in oil palm are similar to those of the coconut palm, but there are some fundamental differences in the progress of the infection that can lead to new and distinct measures for treating the disease in the crop.

Pathogenesis is longer than in the coconut, generally 5–10 months in the more susceptible cultivars. In the coconut, the young 3- to 10-year-old palm is virtually dead within 3 months after infection. In the case of the oil palm, this process can take up to 3–4 years with a palm of the same age group. This is partially because the nematode does not colonize as rapidly in the oil palm tissue as it does in the coconut. Where 5000–10,000 nematodes/g of tissue can be found in the red ring zone of coconut, a similar region in the oil palm often yields less than 500 nematodes/g of tissue. A further difference is that most nematodes are found outside the necrotic zone, even in areas that show no necrosis such as the distal or basal portion of the stem and occasionally in the rachis of the inflorescence (the nematode has not been isolated from the rachis of the coconut panicle). In addition, the red ring disease infection rate is often highest in older oil palms (15 to > 20 years old) (Oehlschlager *et al.*, 2002).

As in the case of the coconut, the most persistent form of the nematode is the third stage juvenile, which can subsist for a long time in the diseased tissue. In the coconut, this juvenile form readily proceeds to the adult in the healthy tissue not showing symptoms. However, in the oil palm, this interval is prolonged for some reason, with the result that colonization of the oil palm is not rapid and pathogenesis is attenuated. A notable feature in accordance with this is that the band of necrotic tissue is usually very narrow and often irregular in shape (Chinchilla, 1988). Eggs appear as usual in the brownish spots that are present in the advancing area of the disease. Such necrotic areas indicate evidence of plant reaction to the cellular damage caused by abundance of the nematode. The nematodes often show no evidence of their presence, and an abundance of nematodes can occur without the plant reacting visibly. Yet, artificial inoculation studies have demonstrated that increased logarithmic strengths of the inoculum correlate inversely with the length of the period for pathogenesis in 8-year-old palms at Centeno, Trinidad.

The palm weevil, *R. palmarum*, as with coconut, is the main vector of *B. cocophilus*. The canopy in an oil palm plantation is always closed, with reduced light intensity and more humidity than in coconuts. This presents ideal conditions for the vector of the nematodes, the palm weevil, which is crepuscular.

In the state of Amazonas, Brazil, the number of weevils infested with nematodes showed high monthly variations and irregular distribution, with higher percentages occurring in November 1988 and September 1991. The relatively low incidence of red ring disease in the area did not suggest any association between the disease and the variations in rainfall measured (Araujo *et al.*, 1998).

Spacial disposition of diseased oil palms

In young 5- to 10-year-old groves, there is a tendency for the diseased palms to be clustered in a 50 m radius that gradually expands. In older groves, however, the diseased trees appear to be distributed at random, giving the impression that the vectors come in from fields that are more susceptible to the simultaneous development of both the nematode and the palm weevil. The major constraint is the poor opportunity for association of the developing weevil larvae in the oil palm with a large number of nematodes. This is a result of the slower rate of colonization of the nematode in the oil palm compared with the coconut palm. Therefore, unlike a coconut estate where most weevils develop in diseased trees which become the main or focal developmental niche (97.3% in Trinidad), the proportion, in oil palm fields, would relate more to the Ecuador situation, with greater percentages of the weevils developing outside the ambit of the nematode. This is markedly so as the incubation period of the pathogen in oil palm is very long and quite variable. The principle of the diseased palm being the main attractive focus for palm weevils would have reduced relevance here; thus, weevil lures would assume a more significant role in disease management in the field. Consequently, weevil trapping in oil palm estates is an important form of control for reduction of the disease. Moreover, location and elimination of sources of infection, other than oil palm, particularly diseased coconut fields, near the affected grove is important. Phytosanitary measures, however, comprise the most utilized method of control in Latin America.

Other hosts for red ring disease in some oil palm estates in Brazil

The wild palm *Oenocarpus distichus* Mart. was found by Schuiling and Van Dinther (1981) to be capable of contracting red ring disease and serving as a host for *R. palmarum*. This is a typical palm of primary and secondary forest of the Amazon estuary. Nematodes are often fewer in number, often less than 100/g of tissue. However, palm weevil larvae found growing in these trees were internally contaminated with red ring nematodes. The weevil *R. palmarum* is reported from 31 plant species, belonging to 12 families, with palms being the main hosts (Sanchez and Cerda, 1993).

Economic importance and damage threshold levels

In Latin America, there is an apparent direct correlation between levels of red ring disease in coconuts and those in oil palms. Countries with high levels of red ring disease in coconut groves also have high levels of red ring in oil palm groves. Generally, in oil palms 8–10 years old, the incidence is around 0.1%, and in palms over 20 years old the incidence is rare. However, in some zones adjoining old coconut establishments, the incidence of disease in oil palms, 11–18 years old, can be as high as 30%. In one parcel of 62 ha of the plantation of Palmeras de la Costa in Colombia, the maximum accumulated disease total for 1987 was 8.3% (Villanueva and Gonzales, 1988).

Little leaf disease of oil palms

The oil palm, as most palms, has a tendency towards producing so-called 'little leaves', the cause of which may be diverse and related to symptoms of other diseases. In Surinam, Van Hoof and Seinhorst (1962) observed that little leaf syndrome was associated with attack by the red ring nematode. Little leaf symptomatic trees can easily be recognized by their erect, short and often deformed leaves with suberized patches especially on the inner side of the leaf stalks (Chinchilla, 1988).

Many *B. cocophilus* have been found on discoloured tissue of young (up to 1.75 m long) folded leaves, still protected from the sun. The nematodes live ectoparasitically in the buds of the palms. In one survey of 50 diseased oil palms cut for investigation, only one did not contain nematodes. *B. cocophilus* was never found on the young leaves of numerous trees that did not suffer from little leaf but were cut for other reasons (Van Hoof and Seinhorst, 1962). Palms exhibiting this symptom can live for many years, but with a reduced leaf emission rate and abortion of inflorescences (Chinchilla, 1988). It has been hypothesized that red ring nematode-induced little leaf is symptomatic of unsuccessful cases of the systemic red ring disease and is more common in African oil palm than coconut where it is rarely observed because coconut is so susceptible and succumbs so easily to red ring disease (Giblin-Davis, 1993).

Management of red ring in oil palm

Generally, control of red ring disease in oil palm is similar to that in coconut by a combination of methods. The destruction of diseased trees is paramount as soon as the symptoms are detectable in order to destroy inoculum.

Injections of systemic nematicides, such as fenamiphos, oxamyl and carbofuran, into little leaf symptomatic palms can help with palm recovery. However, because of the damage to the very young leaves in little leaf palms, the recovery can take between 6 and 8 months (Chinchilla, 1988). This measure is unsuccessful in coconuts because the nematode colonizes the palm tissue too rapidly, but, in oil palm, the slower rate of colonization allows for such a possibility to control the nematode directly. Non-target effects and cost effectiveness must also be considered, and little leaf symptomatic oil palms are usually better removed, providing more light into the canopy and increased productivity to nearby palms (Chinchilla, 1988).

Research by Oehlschlager *et al.* (1992a, 1993, 1995, 2002) in African oil palm plantations in Costa Rica suggests that concerted aggressive phytosanitation and mass trapping with traps baited with sugarcane and synthetic aggregation pheromone (Rhyncholure; racemic 6-methyl-2-hepten-4-ol; ChemTica International) reduce the numbers of *R. palmarum* and change their distribution patterns (from highly aggregated to random) while significantly reducing red ring disease incidence. Initial bimonthly inspections and removal of red-ring-diseased palms did not reduce red ring disease incidence in two large African oil palm plantations in Costa Rica. One year after the initiation of mass trapping at trap densities of about one trap per 5 ha, red ring disease incidence plummeted by more than 80% (Oehlschlager *et al.*, 2002). At mass trapping onset, most *R. palmarum* were captured in 'border' traps of test sites, suggesting removal of potential immigrants into the study area. A combination of perimeter and 'internal' traps appears to be most effective for mass trapping in African oil palm (Oehlschlager *et al.*, 1995). There are many trap designs available for effectively capturing palm and sugarcane weevils (Oehlschlager *et al.*, 1993; Giblin-Davis, 2001). The most important features involve baiting with sugarcane or palm tissue (changed every 2 weeks) and a pheromone release device and making the trap lethal with pesticide treatment of tissue or by using a special trap design with soapy water. The traps must be examined and refreshed at least every 2 weeks with fresh pesticide-treated sugarcane tissue and pheromone (as needed).

Studies to determine whether *Metamasius* sp. is a vector of *B. cocophilus* did not achieve transmission of red ring in oil palms, but the frequency of the nematodes occurring in the insects was significant (Silva, 1991).

Methods of diagnosis

The methods for extraction of *B. cocophilus* from oil palm are similar to those described for the nematode in coconut; however, they are much less accurate because of the lower numbers of nematodes present and the often irregular shape of rings in the oil palm (Chinchilla, 1988). Nematodes also seem to thrive in the petioles.

Date Palm

The date palm, *Phoenix dactylifera* L., is dioecious, and artificial pollination by man has played a significant role in the historical development of the crop. Tissue culture programmes have become important for improving yields. The FAO estimate worldwide production of dates peaked in 1996 at 4,492,000 t. The main contributors were: Iran, which produced 765,000 t; Egypt, 680,000 t; Saudi Arabia, 597,000 t; Iraq, 550,000 t; Pakistan, 533,000 t; Algeria, 361,000 t; and the United Arab Emirates, 240,000 t. Though the palms will grow throughout the tropics, the number of heat units required from the time of blossoming to ripening should be between 4000 and 5500 for various cultivars. Growth of the palm ceases around 10°C. Suitable climatic conditions occur in the dry parts of California where the palm has been grown successfully on a commercial scale. In this introduced environment, the palm has to cope with the new prevailing nematode fauna.

Nematodes of Date Palm

The date palm is affected by numerous pests and diseases wherever it is grown, but nematodes parasitic on date palm, with the exception of root knot nematodes, *Meloidogyne* spp., have not been well studied. However, nematodes have not been found to be a limiting feature in the countries with date as an ancient culture. Root knot nematodes, *Meloidogyne* spp., were found in the Coachella Valley of California on date palms in 1925, where they are now known to be widely distributed in commercial date plantings. Buhrer *et al.* (1933) first reported the occurrence of root knot nematodes on date, and Jensen (1961) found *M. incognita* on roots of date palms in nurseries. Carpenter (1964) reported that root knot nematodes, principally *M. javanica*, can severely damage or kill date palm seedlings.

Young seedlings of 50 date cultivars were susceptible to infection by root knot nematodes; more than 90% of the seedlings were killed prior to emergence when seeds were sown in heavily infested soil. Secondary damage by fungi to roots of field-grown palms infested with the nematodes seemed to be an important factor in the deterioration and death of roots. Minz (1958) reported the occurrence of *M. arenaria*, *M. hapla*, *M. incognita* and *M. javanica* on date palms in Israel. *Meloidogyne* sp. has been reported from Sidi Yaia in Algeria (Lamberti *et al.*, 1975), from the Mauritanian oases of Tayaret and Tejitt (Netscher and Luc, 1974) and from Libya (Fourgani and Edongali, 1989; Edongali, 1996).

The combination of *Thielaviopsis paradoxa* (*Ceratocystis paradox*) with *M. javanica* increased the susceptibility of date palm cutivars to infection by the fungus (Aboud *et al.*, 2002).

Histopathological studies of date palm (*P. dactylifera*) roots infected with *Pratylenchus penetrans*, the root lesion nematode, showed puncture of epidermal cells and disarrangement of cortical cells with large empty, abnormal cavities. Membranous cell walls were wavy and collapsed as the supporting material was destroyed by nematode infection (Khan *et al.*, 2002).

The cellular alteration in *M. incognita*-infected roots of susceptible and resistant date palm cultivars was histologically studied in pot experiments. Giant cell formation was favoured in the susceptible cv. Zaghlool, while in resistant cvs Deglet Noor and Samani, the infected roots reacted to the nematode infection by forming a necrotic area around the invading nematode. In certain cases, malformed small giant cells were observed in association with nematode juvenile stages (Eissa *et al.*, 1998). In Egypt, the largest *M. incognita* populations were found at soil depths of 30–50 and 51–70 cm at 1 and 2 m distances from the trunk of date palm cv. Siwi (Youssef and Eissa, 1994). Thirty-seven species of plant parasitic and free-living nematodes were encountered on date palm in India. Date palm trees infected with *M. incognita* (450 second stage juveniles/250 cm^3 of soil) showed yellowing of leaves and stunted growth (Lal and Mathur, 1986).

A survey of plant parasitic nematodes in the rhizosphere of 30 date palm cultivars in Riyadh, Saudi Arabia, found 18 genera of plant parasitic nematodes in the following descending order of frequency: *Helicotylenchus* (64.9%), *Meloidogyne javanica* (52%), *Hemicriconemoides* (37.8%), *Tylenchorhynchus* (24.3%), *Criconemoides* (16.9%), *Tylenchus* (15.5%), *Aphelenchus* (14.5%), *Hoplolaimus* (10.8%), *Rotylenchulus* (7.4%), *Paratrichodorus* (6.4%), *Pratylenchus* (6.1%), *Trichodorus* (5.7%), *Ditylenchus* and *Longidorus* (3.7%, each), *Zygotylenchus* (2.7%), *Xiphinema* (2%), *Aphelenchoides* (1.7%) and *Paratylenchus* (0.7%). Only *Helicotylenchus* and *Meloidogyne* were found on all the surveyed date palm cultivars (Al-Yahya *et al.*, 2001). A survey in the United Arab Emirates found nematodes associated with diseased date palms (Hashim, 1997). During field surveys of date diseases in coastal regions of Libya, root knot nematodes (*M. incognita* and *M. javanica*), root lesion nematodes (*P. penetrans* and *Pratylenchus* sp.), ring nematode (*Criconemoides* sp.) and others were associated with date palm rhizospheres (Edongali, 1996). A survey carried out in Algeria revealed the occurrence of five species of *Pratylenchus*. The most common species was *P. penetrans*, often associated with date palm (Troccoli *et al.*, 1992). *Criconemoides curvatus* and *Longidorus* sp. nov. were found on date palm in Florida (MacGowan, 1989). In Algeria, Lamberti *et al.* (1975) reported the occurrence of *P. penetrans* on date palm roots in the crescent of oases from Beni Ounif to Biskra, and there are reports of associations with species of *Hemicricononemoides*, *Xiphinema*, *Criconemoides*, *Trichodorus* and *Tylenchus*. *B. cocophilus* is also known to affect the date palm. A specimen in the Botanic Gardens, Trinidad, came down with red ring disease and produced a brownish ring. However, date palms growing in the main production areas prefer a hot dry environment that would limit the activities of the palm weevil, the vector of the red ring nematode, which thrives in areas of high humidity.

Management of date palm nematodes

Nematicides added to cultivated soil were screened for their ability to control soil nematodes. The higher the concentration, the higher the mortality rate. On date palms exhibiting Al-Wijam symptoms, dazomet gave the best control, followed by carbofuran and oxamyl. *Longidorus* spp. were the most susceptible nematodes, followed by *Xiphinema*, but *Meloidogyne* were the most resistant species. However, there was no sign of recovery of treated date palm trees with Al-Wijam disease (Abdulsalam *et al.*, 1996).

Arecanut

Arecanut or betel nut, *Areca catechu* L., occurs in the humid regions of Asia and the Malay islands. It is a masticatory of great antiquity, and betel chewing is a habit of nearly one-third of the world's population. The ripe fruits are sometimes used as an anthelmintic and astringent in Europe.

Nematodes of Arecanut

A number of nematodes have been reported from the rhizosphere of arecanut (Nair, 1964; Weischer, 1967; Pizarro, 1969;

Koshy *et al.*, 1976, 1978, 1981; Reddy, 1978; Sundararaju and Koshy, 1982a; Sundararaju *et al.*, 1984; Dasgupta and Rama, 1987; Rama, 1987; Subramaniyan *et al.*, 1988; McSorley and Dunn, 1989), but only *R. similis* is known to be an important parasite of the palm. A number of other palms have been reported as hosts of *R. similis* (Table 13.2), and it would not be unexpected if nematode problems with some of these other palms became apparent in the years ahead.

Radopholus similis

The burrowing nematode, *R. similis*, was first reported from soil around roots of arecanut palm in Mysore, India by Kumar *et al.* (1971) and later by Koshy *et al.* (1975, 1976) and Sosamma (1998).

Symptoms of damage

The most conspicuous symptoms of *R. similis* infestation are the appearance of lesions and rotting of roots. The nematode produces small, elongate, orange-coloured lesions on the young, succulent, creamy-white to light-orange coloured portion of the main and lateral roots. Subsequently, the adjoining lesions coalesce and cause extensive root rotting. The thick primary roots produced from the bole region of the palm exhibit large, oval sunken, brown to black lesions, 2 mm to 2 cm in length (Plate 14F).

Nematodes occur inter- and intracellularly in the cortex, but do not enter the stelar tissues. Large numbers of nematodes and their eggs are seen in the cavities that develop consequent to nematode feeding in the cortex (Sundararaju, 1984, 2000).

Table 13.2. Palms reported as hosts of the burrowing nematode, *Radopholus similis.*

Archontophoenix cunninghamiana Wendl. and Drude (Seaforthia or Picabeen bungalow palm)
Areca (Actinorhytis) calapparia
Areca catechu L. (Betel-nut palm)
A. langlosiana
A. macrocalyx Beec.
A. normanbyii
A. triandra Roxb.
Chamaedorea cataractarum Martius
C. elegans Martius (parlour palm or Neanthebelia palm)
Cocos nucifera L. (coconut)
Elaeis guineensis Jacq. (African oil palm)
Phoenix canariensis Hort. ex Chabaud (Canary Island date palm)
P. dactylifera L. (date palm)
Rhapis excelsa (Thunb.) Henry (large lady palm)
Roystonea regia (H.B.K.) Cook. (royal palm)
Syagrus romanzoffiana (Cham.) Glassman (queen palm)

Biology and life cycle

The burrowing nematode takes 25–30 days to complete one life cycle (J2–J2) on arecanut seedlings at a temperature range of 21–31°C under glasshouse conditions. Chromosome studies have recorded the presence of a haploid number of chromosomes ($n = 4$) in many isolates of *R. similis* from arecanut roots (Koshy, 1986). The arecanut isolate of *R. similis* belongs to the banana race (Koshy and Sosamma, 1977) and multiplies well on carrot discs maintained on 1% water agar (Koshy and Sosamma, 1980).

The population densities of *R. similis* in arecanut fluctuate; maximum population occurs in roots during October to November and the minimum during March to June in India. Populations are also known to vary between samples, types of roots, palms, groves and soil types during the same period (Koshy and Sosamma, 1978a).

Disease complexes

The fungus *Cylindrocarpon obtusisporum* is found associated with lesions caused by *R. similis* in arecanut roots. The fungus when introduced 3 weeks after nematode inoculation caused more damage to plants compared with inoculations with the nematode alone and it inhibited the rate of multiplication of the nematode (Sundararaju and Koshy, 1984, 1987).

Economic importance and population damage threshold levels

R. similis was recorded from 32% of root samples in the three major arecanut-growing states in south India, with a maximum population of 440 nematodes/g of root. *R. similis* was found in 55, 45, 44, 30 and 11% of root samples from plantations intercropped with banana, black pepper, cardamom, coconut and cacao, respectively, compared with 25% from plantations monocropped to arecanut (Sundararaju, 1984).

The population damage threshold level on arecanut seedling is 100 nematodes/seedling or 1/800 g of laterite soil. The percentage reduction of growth over uninoculated plants at this inoculum level can be 23, 39, 25, 19 and 38% with respect to shoot length, shoot weight, girth at collar region, root length and root weight, respectively, under pot conditions in laterite soil.

Management

RESISTANCE/TOLERANCE. None of the 46 accessions of arecanut germplasm in the CPCRI germplasm collection is immune or highly resistant to *R. similis*. The cultivars Mangala (VTL-3) and Fiji (VTL-26) are highly susceptible, whereas the cultivars Singapore (VTL-17), Solomon Islands-2 (VTL-18c) and Saigon (VTL-27) are less susceptible to *R. similis*; cultivars Indonesia 6 (VTL-11), Mahuva 8 and Andaman-5 (VTL-29e) are tolerant to *R. similis* (Koshy *et al.*, 1979; Sundararaju and Koshy, 1982b). The cultivars Indonesia-6 (VTL-11) and Singapore (VTL-17) are known to yield 15% more nuts over local South Canara cultivar (Anon., 1974). Thus, these cultivars could profitably be recommended for *R. similis*-infested areas. The hybrid VTL-11 × VTL-17 is highly resistant to *R. similis*.

CHEMICAL AND BIOLOGICAL. As arecanut is chewed directly by many consumers, dosage, frequency and time of application of nematicides on arecanut have to be calculated carefully to avoid residues in the nut.

A pot culture experiment carried out under field conditions revealed that fensulfothion and aldicarb at 1 g a.i./seedling applied thrice a year for 3 consecutive years in pots gave control of *R. similis* both in soil and in roots. Increases in plant growth with regard to shoot length, shoot weight, root length, root weight, number of leaves and collar girth with fensulfothion were 46, 168, 33, 173, 25 and 41%, respectively, over control plants after 3 years (Sundararaju and Koshy, 1986a). In a field experiment in India, treatment with fensulfothion at 50 g a.i./palm and aldicarb at 10 g a.i./palm applied during May/June, September/October and December/January for 5 years resulted in control of *R. similis* and a substantial increase in both number and weight of nuts compared with untreated palms (Sundararaju and Koshy, 1986b). However, the nuts were not analysed for their residues, if any, and the cost–benefit ratio has not been determined.

Field experiments were carried out in arecanut monocrop, arecanut + banana and arecanut + banana + pepper to evaluate the efficacy of neem oil cake and phorate singly and in combination for control of *R. similis* in the cropping system. Even though all the treatments were significantly superior over the untreated control, the best treatment in these experiments was 15 g of phorate in combination with 1 kg of neem oil cake, which controlled the *R. similis* population in arecanut and subsidiary crops very well (Sudha and Sundararaju, 1998). A pot trial study to evaluate the combined effect of organic amendments and biocontrol agents, i.e. *P. lilacinus*, *P. penetrans* and arbuscular mycorrhizal fungi (AMF), against *R. similis* infecting arecanut (*A. catechu* var. Mangala) was conducted using sandy loam soil amended with various organic matter. The organic amendments used were neem and marotti oil cakes, leaves and tender shoots of sunnhemp and *Glyricidia*, vermicompost, cow dung and coir pith. Maximum nematode control (95%) was recorded in soil amended with *Glyricidia* leaves and bioagents. Significant reduction in root lesion index and maximum leaf area was recorded in these plants. The percentage increase in height and root growth was maximum in plants grown in coir pith-amended soil with bioagents which was on a par with plants grown in soil enriched with *Glyricidia* leaves and bioagents. A decrease in nematode population was on a par in all treatments receiving organic amendments and bioagents as well as bioagents alone compared with nematode alone. All of the bioagents were re-isolated from all of the treated plants, even after 3 years. Although amendment of soil with organic matter in general was found to increase plant growth and reduce nematode population, the differences were not significant compared with introduction of bioagents in the absence of organic amendments (Koshy *et al.*, 1998c, 2002b).

Summary of management measures

Control of *R. similis* on arecanut is difficult under the high density, multispecies, subsistence farming systems involving perennial crops such as coconut, banana, black pepper, betel vine, cardamom and cacao. Use of nematicides for the control of burrowing nematode on coconut or arecanut may cause problems of residual toxicity. The following control measures are suggested: (i) use of nematode-free planting material of arecanut and other intercrops; (ii) avoiding *R. similis*-susceptible intercrops such as black pepper and banana in infested areas; (iii) use of resistant/tolerant cultivars of arecanut, when available, and other crops in farming systems; (iv) application of 5–10 kg of green manure preferably *Glyricidia* or *Crotolaria*; (v) application of 1 kg of neem oil cake/palm/year; and (vi) application of phorate at 3 g a.i./plant to the root zone of arecanut, banana and black pepper in June–July and October–November in arecanut-based farming systems (Sudha and Sundararaju, 1998; Sundararaju and Sudha, 1998).

Methods of diagnosis

Soil and root samples for detection of *R. similis* should be collected at a distance of 25–75 cm from the bole of the palm at a depth of 25–75 cm when high population densities are present, such as during October/November in India. The method suggested for extraction of *R. similis* from coconut root can also be adopted for arecanut.

Other Palms

Reniform nematode, *Rotylenchulus reniformis*, is a sedentary endoparasitic nematode that sometimes causes concern in field-grown or containerized palms for shipment in the USA. The concern is a regulatory issue rather than one of palm pathology. Most palms, except *Washingtonia robusta* Wendland and *Acoelrrhapphe wrightii* Grisebach and Wendland, are non-hosts for the reniform nematode (Inserra *et al.*, 1994). These hosts, although suitable for nematode reproduction, are devoid of above-ground symptoms, suggesting tolerance. The regulatory problem arises because ornamental palms grow well in southern Florida conditions where reniform nematode is a damaging pest to dicotyledenous field and vegetable crops. Ornamental palms that are contaminated or infested with reniform nematode are subject to quarantine in non-infested areas in Arizona, California and New Mexico that grow valuable and susceptible crops such as cotton. *R. reniformis* has also been found in the rhizosphere of ornamental palms in Egypt (Ismail and Eissa, 1993).

Hot-water treatments were evaluated to disinfect roots and media of potted bamboo or Reed palm, *Chamaedorea seifrizii* Burret, and fishtail palm, *Caryota mitis* Lour. of *R. similis*. A continuous hot-water drenching (50°C for 15 min) of roots and media in pots or hot-water dipping (50°C for 15 min) of bare-rooted plants were successful in eliminating all burrowing nematodes in bamboo palms. Fishtail palms were disinfected of burrowing nematodes after hot-water drenching at 50°C for 13 min. Dipping plants intact in pots in a constant temperature water bath was not effective, as the root temperature remained below the thermal death point for nematodes due to slow heat transfer. No evidence of thermal damage was observed in either palm species drenched with hot water at 50°C for up to 20 min followed by hydro-cooling to ambient temperature. Ambient air-cooling after heat-treatment was detrimental as the residual heat caused both vegetative and root damage in the potted palms. These air-cooled palms suffered reduced growth and required a longer recovery period (Tsang *et al.*, 2003).

General Conclusions and Future Activities on Nematodes of Palms

The foregoing has shown that fatal diseases in palms due to nematodes are unknown except for those palms that are naturally attacked by *B. cocophilus* and its chief insect vector *R. palmarum*. The fact that red ring disease is at present confined to the New World restricts its economic importance to those palms that occur in the area, but others, such as the arecanut palm, are likely to be naturally susceptible even in their areas of origin. Nematodes that have been recorded as pathogenic to palms in their areas of origin are only those that exist in the rhizosphere such as *R. similis* of arecanut and coconut. This problem has not yet been recognized in the New World, but it is very possible that this and other nematode root problems on palms will become apparent in the years ahead. There is a danger of root nematodes, particularly *R. similis* and *R. reniformis*, being introduced through ornamental palms and other hosts.

The major concern of nematologists, plant pathologists and quarantine personnel, therefore, is to ensure against the possibility of red ring disease becoming universal since the likelihood that other species of the palm weevil (e.g. other *Rhynchophorus* spp.) could be vectors to *B. cocophilus* is quite strong. The palms of horticultural value are also susceptible and could in fact increase the likelihood of the disease eventually moving out of Latin America in a palm where symptoms are not so distinct and in which pathogenesis is prolonged. Indeed, there is every probability that symptomless carriers might exist, as in oil palms, that are slowly colonized by the nematode. Another problem occurs with the confusion of similar symptoms in other wilt diseases of palms that can hide the problem of a nematode until it is too late.

The International Bureau of Plant Genetic Resources has been helping in a number of coconut and other palm germplasm collection programmes, and some methods have been developed such as embryo rescue for introduction of clean coconut germplasm for research purposes. Koshy and Kumaran (1997) collected 1342 embryos of 15 accessions from the Indian Ocean islands for the first time. The commercial availability of synthetic male-produced aggregation pheromones for the weevils that can vector red ring nematode allows for monitoring ports of entry for potential vectors.

Generally, as crop plants for small farmers, cordon sanitaires are always necessary for vector-borne pathogens that can have fatal and cumulative effects on the agroecosystems. Thus, control measures for palm diseases must always be inexpensive, effective and readily applicable in all economic circumstances. Essentially, of course, biological control measures and resistant cultivars should always be sought. The stability of the coconut agroecosystem favours management procedures with limited pesticide usage.

References

Abdulsalam, K.S., Abdel-Megeed, M.I., Rezk, M.A., Nageeb, M.A. and Tameem, M. (1996) Efficiency of certain nematicides against nematodes associated with Wijamed date palm trees. *Annals of Agricultural Science Cairo* 41, 475–482.

Aboud, H.M., Said, S.A. and Saleh, H.M. (2002) Role of *Meloidogyne javanica* in predisposing two date palm cultivars to infection by *Thielaviopsis paradoxa*. *Pakistan Journal of Nematology* 20, 49–53.

Al-Yahya, F.A., Ibrahim, A.A.M. and El-Sherbiny, A.A. (2001) Frequency and population density of plant parasitic nematodes associating with date palm cultivars in Riyadh, Saudi Arabia. *Alexandria Journal of Agricultural Research* 46, 127–137.

Anonymous (1974) *Annual Report 1973*. Central Plantation Crops Research Institute, Kasaragod, Kerala, India.

Araujo, J.C.A., Araujo, A.E. and Santos, A.F. (1998) Population dynamics of *Rhynchophorus palmarum* and its association with *Bursaphelenchus cocophilus* on oil palm in the state of Amazonas, Brazil. *Fitopatologia-Brasileira* 23, 23–26.

Ashby, S.P. (1921) Some recent observations on red ring disease. *Agricultural News (Bridgetown)* 20, 350–351.

Banu, J.G. and Sosamma, V.K. (1999) Tapioca pearl as a substitute for agar in culturing *Radopholus similis* on carrot discs. *Indian Journal of Nematology* 29, 98–99.

Banu, J., Sosamma, V.K. and Koshy, P.K. (1998) Natural occurrence of an entomopathogenic nematode, *Heterorhabditis indicus* (Poinar, Karunakar and David) from Kerala, India. *Proceedings of the Third International Symposium of AFRO ASIAN Society of Nematologists (TISAAN) Nematology – Challenges and Opportunities in 21st Century*. Sugarcane Breeding Institute, Coimbatore, 16–19 April 1998, pp. 274–280.

Barrett, O.W. (1906) Notes upon miscellaneous crops. *Proceedings of Agricultural Society of Trinidad* 7, 303.

Baujard, P. (1989) Remarques sur les genres des sous-familles Bursaphelenchina Paramonov, 1964 et Rhadinaphelenchinae Paramonov, 1964 (Nematoda: Aphelenchoididae). *Revue de Nématologie* 12, 323–324.

Blair, G.P. (1964) Red ring disease of the coconut palm. *Journal of Agricultural Society of Trinidad and Tobago* 64, 31–49.

Buhrer, E.M., Cooper, C. and Steiner, G. (1933) A list of plants attacked by the root knot nematode (*Heterodera marioni*). *Plant Disease Reporter* 17, 64–96.

Bulgarelli, J., Chinchilla, C.M. and Oehlschlager, A.C. (1998) The red ring/little leaf syndrome and *Metamasius hemipterus* captures in oil palm in Costa Rica. *ASD Oil Palm Papers* 18, 17–24.

Carpenter, J.B. (1964) Root knot nematode damage to date palm seedlings in relation to germination and stage of development. *Date Growers Institute Report* 41, 10–14.

Chinchilla, C.M., Oehlschlager, A.C. and Gonzalez, L.M. (1992) Use of pheromone-based traps for the management of the American palm weevil, *Rhynchophorus palmarum*, in a commercial oil palm plantation. *Proceedings of the International Oil Palm Congress*. Palm Oil Research Institute of Malaysia, Kuala Lumpur, Malaysia.

Chittenden, F.H. (1902) The palm and palmetto weevils. *US Department of Agriculture Entomology Bulletin* 38, 23–28.

Cobb, N.A. (1919) A newly discovered nematode *Aphelenchus cocophilus* n.sp. connected with a serious disease of the coconut palm. *West Indian Bulletin* 17, 203–210.

Dasgupta, N.K. and Rama, K. (1987) Plant parasitic nematode associated with plantation crops. *Review of Tropical Plant Pathology* 4, 289–304.

Dean, C.G. (1979) *Red Ring Disease of* Cocos nucifera *L. Caused by* Rhadinaphelenchus cocophilus *(Cobb, 1919) Goodey, 1960. An Annotated Bibliography and Review*. Technical Communication No. 47, Commonwealth Institute of Helminthology, St Albans, UK.

Delgado, H.V. and Moreno, F.O. (1986) Evaluacion de atrayentes vegetales y un sistema de trampa para la captura de aldultos de 'gualpa' (*Rhynchophorus palmarum*) insecto-plaga de palma Africana y cocotero. Instituto Nacional de Investigaciones Agropecuarias, Estacion Experimental Santo Domingo, Ecuador, September, *Boletín Técnico* No. 63, 1–10.

Edongali, E.A. (1996) Diseases of date palms (*Phoenix dactylifera* L.) of Libya. *Arab Journal of Plant-Protection* 14, 41–43.

Edongali, E.A. and Dabaj, K.H. (1986) Nematodes associated with date-palms (*Phoenix dactylifera* L.) in some regions of Libya. *Proceedings of the 2nd Symposium of the Date Palm*, Saudi Arabia.

Eissa, M.F.M., El-Sherief, M.A., Abdel-Gawad, M.M., Ismail, A.E. and El-Nagdi, W.M.A. (1998) Histological response of susceptible and resistant date-palm cultivars to *Meloidogyne incognita* infection. *Pakistan Journal of Nematology* 16, 103–109.

Ekanayake, U.B.M. (1964) Annual report of the crop protection officer, Coconut Research Institute for 1963. *Ceylon Coconut Quarterly* 15, 50–57.

FAO (2000) Agriculture: towards 2015/30. Technical Interim Report, April 2000. Available at:

www.fao.org/es/esdat2015/toc-e.htm

Fenwick, D.W. and Maharaj, S. (1963) Recovery of *Rhadinaphelenchus cocophilus* (Cobb, 1919) Goodey, 1960 from coconut tissues. *Journal of Helminthology* 37,11–14.

Fourgani, G.M. and Edongali, E.A. (1989) Speciation of root-knot nematodes (*Meloidogyne* spp.) associated with crop plants in Libya. *International Nematology Network Newsletter* 6 (1), 38–39.

Geetha, S.M. (1991) Studies on the biology, pathogenicity and bio control of different populations of *Radopholus similis*. PhD Thesis, Kerala University, Kerala, India.

Gerber, K. and Giblin-Davis, R.M. (1990a) Association of the red ring nematode, *Rhadinaphelenchus cocophilus*, and other nematode species with *Rhynchophorus palmarum* (Coleoptera: Curculionidae). *Journal of Nematology* 22, 143–149.

Gerber, K. and Giblin-Davis, R.M. (1990b) *Teratorhabditis palmarum* n.sp. (Nemata: Rhabditidae), an associate of *Rhynchophorus palmarum* and *R. cruentatus* (Coleoptera: Curculionidae). *Journal of Nematology* 22, 337–347.

Gerber, K., Giblin-Davis, R.M. and Escobar-Goyes, J. (1990) Association of the red ring nematode, *Rhadinaphelenchus cococophilus*, with weevils from Ecuador and Trinidad. *Nematropica* 20, 39–49.

Giblin-Davis, R.M. (1993) Interactions of nematodes with insects. In: Khan, M.W. (ed.) *Nematode Interactions*. Chapman and Hall, London, pp. 302–344.

Giblin-Davis, R.M. (2001) Borers of palms. In: Howard, F.W., Moore, D., Giblin-Davis, R.M. and Abad, R. (eds) *Insects on Palms*. CAB International, Wallingford, UK, pp. 267–305.

Giblin-Davis, R.M. (2004) Nematodes as mutualists: where and how did these associations evolve? In: Cook, R. and Hunt, D.J. (eds) *Proceedings of the Fourth International Congress of Nematology*, 8–13 June 2002, Tenerife, Spain. *Nematology Monographs and Perspectives* 2 (in press).

Giblin-Davis, R.M., Gerber, K. and Griffith, R. (1989a) *In vivo* and *in vitro* culture of the red ring nematode, *Rhadinaphelenchus cocophilus*. *Nematropica* 19, 135–142.

Giblin-Davis, R.M., Mundo-Ocampo, M., Baldwin, J.G., Gerber, K. and Griffith, R. (1989b) Observations on the morphology of the red ring nematode, *Rhadinaphelenchus cocophilus*. *Revue de Nématologie* 12, 285–292.

Giblin-Davis, R.M., Oehlschlager, A.C., Perez, A., Gries, G., Gries, R., Weissling, T.J., Chinchilla, C.M., Peña, J.E., Hallett, R.H., Pierce, H.D. and Gonzalez, L.M. (1996) Chemical and behavioral ecology of palm weevils (Curculionidae: Rhynchophorinae). *Florida Entomologist* 77, 247–255.

Giblin-Davis, R.M., Gries, R., Gries, G., Peña, E., Pinzon, I., Peña, J.E., Perez, A.L., Pierce, H.D. Jr and Oehlschlager, A.C. (1997) Aggregation pheromone of palm weevil *Dynamis borassi*. *Journal of Chemical Ecology* 23, 2287–2297.

Goodey, J.B. (1960) *Rhadinaphelenchus cocophilus* (Cobb, 1919) n. comb., the nematode associated with 'red-ring' disease of coconut. *Nematologica* 5, 98–102.

Goodey, T. (1933) *Plant Parasitic Nematodes and the Diseases They Cause*. E.P. Dutton and Co. Inc., New York.

Govindankutty, M.P. and Koshy, P.K. (1979) Nematodes associated with the coconut palm (*Cocos nucifera* L.) In: *Technical Bulletin 2 – Nematodes, Fungi, Insects and Mites Associated with the Coconut Palm*. Central Plantation Crops Research Institute, Kasaragod, India, pp. 1–50.

Griffith, R. (1968a) The relationship between the red ring nematode and the palm weevil. *Journal of Agricultural Society of Trinidad and Tobago* 68, 342–356.

Griffith, R. (1968b) The mechanism of transmission of red ring nematode. *Journal of Agricultural Society of Trinidad and Tobago* 68, 437–457.

Griffith, R. (1969) A method of controlling red ring disease in coconuts. *Journal of Agricultural Society of Trinidad and Tobago* 69, 827–845.

Griffith, R. (1971) Red ring disease: the mechanism of spread and recommendations for control. *Nematropica* 1, 2.

Habeebullah, A., Reghupathy, A., Rethinam, P. and Koshy, P.K. (1983) Residues of aldicarb in coconut. *Seminar on Pesticides and Environment*, 4–5 August 1983, Coimbatore, Tamil Nadu, India, 21 (abstract).

Hagley, E. (1963) The role of the palm weevil *Rhynchophorus palmarum* L. as vector of red ring disease in coconut. 1. Result of preliminary observations. *Journal of Economic Entomology* 56, 375–380.

Hart, J.H. (1905) Coconut disease. *Bulletin of Miscellaneous Information, Royal Botanic Gardens, Trinidad* 6, 241–243.

Hashim, A. (1997) Date palm diseases in the United Arab Emirates. *Arab and Near East Plant Protection Newsletter* 25, 29.

Holdeman, Q.L. (1986) *The Burrowing Nematode,* Radopholus similis, sensu lato. Division of Plant Industry. California Department of Food and Agriculture and the Agriculture Commissioners of California.

Hunt, D.J. (1993) *Aphelenchida, Longidoridae and Trichodoridae: Their Systematics and Bionomics.* CAB International, Wallingford, UK.

Inserra, R.N., Dunn, R.A. and Vovlas, N. (1994) Host response of ornamental palms to *Rotylenchulus reniformis. Supplement to the Journal of Nematology* 26, 737–743.

Ismail, A.E. and Eissa, M.F.M. (1993) Plant parasitic nematodes associated with ornamental palms in three botanic gardens in Egypt. *Pakistan Journal of Nematology* 11, 53–59.

Jasy, T. (1991) Host specificity, pathogenicity and morphological variations in the populations of burrowing nematode infesting plantation crops. PhD Thesis, Kerala University, Kerala, India.

Jensen, H.J. (1961) The nematode menace. *American Nursery-man* 114, 7–8, 105–113.

Khan, A., Bilqees, F.M., Samad, M.A., Khatoon, N. and Rizwana, A.G. (2002) Histopathology of date palm roots infected with *Pratylenchus penetrans* root-lesion nematode. *Pakistan Journal of Nematology* 20, 35–39.

Koshy, P.K. (1986) The burrowing nematode *Radopholus similis* (Cobb, 1893) Thorne, 1949. In: Swarup, G. and Dasgupta, D.R. (eds) *Plant Parasitic Nematodes of India – Problems and Progress.* IARI, New Delhi, India, pp. 223–248.

Koshy, P.K. (1998) *Biological Control of Burrowing Nematode on Palms and Black Pepper.* Final Technical Report. USIF Project No. 1 N-AES-779 C.P.C.R.I., Kayangulam, Kerala, India.

Koshy, P.K. (2000) Leaf rot disease of coconut. *Indian Coconut Journal* 31 (2), 4–10.

Koshy P.K. (2002) Dr. G.I. D'Souza memorial lecture on nematode diseases of plantation crops. *Indian Journal of Nematology* 32, 107–117.

Koshy, P.K. and Banu, J.G. (2002) Nematode diversity in plantation crops – the world scenario. In: Jairajpuri, M.S. (ed.) *Nematode Diversity.* Silverline Printers, pp. 390–425.

Koshy, P.K. and Kumaran, P.M. (1997) *Survey and Collection of Coconut Germplasm from Indian Ocean Islands.* Deputation report. Central Plantation Crops Research Institute, Kasaragod, Kerala, India.

Koshy, P.K. and Nair, C.P.R. (1979) Control of *Radopholus similis* (Cobb 1893) Thorne, 1949 in coconut nursery. *Indian Journal of Nematology* 9, 15–19.

Koshy, P.K. and Sosamma, V.K. (1975) Host range of the burrowing nematode, *Radopholus similis* (Cobb, 1893) Thorne, 1949. *Indian Journal of Nematology* 5, 255–257.

Koshy, P.K. and Sosamma, V.K. (1977) Identification of the physiological race of *Radopholus similis* populations infesting coconut, arecanut and banana in Kerala, South India. *Indian Phytopathology* 30, 415–416.

Koshy, P.K. and Sosamma, V.K. (1978a) Studies on the population fluctuations of *Radopholus similis* in coconut and areca roots. *Indian Phytopathology* 31, 180–183.

Koshy, P.K. and Sosamma, V.K. (1978b) Susceptibility of coconut plumule to *Radopholus similis* (Cobb, 1893) Thorne, 1949. *Indian Journal of Nematology* 8, 77–78.

Koshy, P.K. and Sosamma, V.K. (1979) Control of the burrowing nematode *Radopholus similis* on coconut seedlings with DBCP. *Indian Journal of Nematology* 7, 32–33.

Koshy, P.K. and Sosamma, V.K. (1980) Culturing of burrowing nematode, *Radopholus similis* on carrot discs. *Indian Journal of Nematology* 10, 247–249.

Koshy, P.K. and Sosamma, V.K. (1982a) A simple method for inoculation and production of symptoms by *Radopholus similis* on coconut. *Indian Journal of Nematology* 12, 200–203.

Koshy, P.K. and Sosamma, V.K. (1982b) Culturing of *Radopholus similis* within mesocarp of coconut. *Plant Disease* 66, 811.

Koshy, P.K. and Sosamma, V.K. (1983) Pathogenicity of *Radopholus similis* on *Areca catechu. Indian Phytopathology* 36, 469–472.

Koshy, P.K. and Sosamma, V.K. (1987) Pathogenicity of *Radopholus similis* on coconut (*Cocos nucifera* L.) seedlings under green house and field conditions. *Indian Journal of Nematology* 17, 108–118.

Koshy, P.K. and Sosamma, V.K. (1994) Effect of *Radopholus similis* on growth, flowering and yield of coconut. In: *International Symposium on Plantation Crops (PLACROSYM XI).* 30 November–December 1994. NRC for SPICES, Calicut, Kerala, pp. 20–23.

Koshy, P.K. and Sosamma, V.K. (1996) Effect of *Radopholus similis* on growth, flowering and yield of coconut. *Journal of Plantation Crops* 24 (Supplement), 157–165.

Koshy, P.K. and Sosamma, V.K. (2001) Nematode diseases of plantation crops and their management. Paper presented in National Congress on Centenary of Nematology in India. Appraisal and Future Plans, 5–7 December 2001, IARI, New Delhi, pp. 48–49 (abstract).

Koshy, P.K., Sosamma, V.K. and Nair, C.P.R. (1975) Preliminary studies on *Radopholus similis* (Cobb, 1893) Thorne, 1949 infesting coconut and arecanut palms in South India. *Indian Journal of Nematology* 5, 26–35.

Koshy, P.K., Sosamma, V.K. and Sundararaju, P. (1976) Yellow leaf disease: nematological studies. *Arecanut and Spices Bulletin* 8, 37–41.

Koshy, P.K., Sundararaju, P. and Sosamma, V.K. (1978) Occurrence and distribution of *Radopholus similis* (Cobb, 1893) Thorne, 1949 in South India. *Indian Journal of Nematology* 8, 49–58.

Koshy, P.K., Sosamma, V.K. and Sundararaju, P. (1979) Reactions of thirty one *Areca* germplasm collections to the burrowing nematode *Radopholus similis*. *Plant Disease Reporter* 63, 433–435.

Koshy, P.K., Raski, D.J. and Sosamma, V.K. (1981) *Brachydorus swarupi* sp. n. (Nemata Dolichodorinae) from soil about roots of arecanut palms in Kerala state. *Indian Journal of Nematology* 13, 401–404.

Koshy, P.K., Sundararaju, P., Sosamma, V.K. and Ravikumar, K. (1985) Efficacy of four systemic nematicides against *Radopholus similis* in coconut nursery. *Indian Journal of Nematology* 15, 148–151.

Koshy, P.K., Sosamma, V.K. and Sundararaju, P. (1991a) Nematode management in coconut and coconut based cropping system. *Proceedings International Symposium on Coconut Research and Development – II*. Kasaragod, Kerala, India, 26–29 November 1991.

Koshy, P.K., Sosamma, V.K. and Sundararaju, P. (1991b) *Radopholus similis*, the burrowing nematode of coconut. *Journal of Plantation Crops* 19, 139–152.

Koshy, P.K., Sosamma, V.K. and Samuel, R. (1998a) Interactive effect of Vesicular Arbuscular Mycorrhizae (VAM) with *Radopholus similis* on coconut seedlings. *Proceedings of the Third International Symposium of AFRO ASIAN Society of Nematologists (TISAAN) Nematology – Challenges and Opportunities in 21st Century*. Sugarcane Breeding Institute, Coimbatore, April 16–19 1998, pp. 106–110.

Koshy, P.K., Sosamma, V.K. and Samuel, R. (1998b) Standardisation of sampling zone and type of root for arbuscular mycorrhizal fungal estimation on coconut. *PLACROSYM XIII UPASI*. Tea Research Institute, Valapari, 16–18 December, pp. 53–54 (abstract).

Koshy, P.K., Sosamma, V.K., Samuel, R., Lekhakumari, P.R. and Bindu, S.M. (1998c) Effect of biocontrol agents and organic amendments on *Radopholus similis* infecting arecanut. *Proceedings National Symposium on Rational Approaches in Nematode Management for Sustainable Agriculture*, 23–25 November 1998, GAU, Anand, p. 24.

Koshy, P.K., Sosamma, V.K. and Banu, J.G. (2002a) *Milestones of Nematological Research in Kerala*. Seminar on Centenary on Nematological Research in Kerala, 21–22 February 2002, CPCRI, Kayangulam, Kerala, India.

Koshy, P.K., Sosamma, V.K. and Samuel, R. (2002b) Management of *Radopholus similis* infesting arecanut using bio-control agents. *Proceedings National Symposium on Crop Protection and WTO – Indian Perspective*. 22–25 January 2002, CPCRI, Kasaragod, Kerala, India.

Koshy, P.K., Kumar, Y., Jayasree, D., Joseph, U. and Sosamma, V.K. (2002c) Integrated management of leaf rot disease and insect pests on coconut. *Indian Phytopathology* 55, 45–50.

Kumar, A.C., Viswanathan, P.R.K. and D'Souza, G.I. (1971) A study of plant parasitic nematodes of certain commercial crops in coffee tracts of South India. *Indian Coffee* 35, 222–224.

Lal, A. and Mathur, V.K. (1986) Nematodes in the rhizosphere of date palm with description of *Longidoroides bikanerensis* sp.n. (Nematoda: Dorylaimoidea). *Date Palm Journal* 5, 208–214.

Lamberti, F., Greco, N. and Zaouchi, H. (1975) A nematological survey of date palms and other major crops in Algeria. *FAO Plant Protection Bulletin* 23, 156–160.

Latta, R. (1966) Attempts to relate nematodes to lethal yellowing disease of coconut palms in Jamaica. *Tropical Agriculture Trinidad* 43, 59–68.

MacGowan, J.B. (1989) Bureau of nematology: nematode detections of special interest. *Triology Technical Bulletin* 28 (9), 1–3.

Malaguti, G. (1953) 'Pudricion de Cogollo' de la palmera do aceite Africana (*Elaeis guineensis* Jacq.) en Venezuela. *Agronomia Tropical* 3, 13–21.

Maqbool, M.A. (1991) *Nematode Pests of Economic Significance Affecting Major Crops of the Countries in Asia and the Pacific Region*. Technical Document, Asia and Pacific Plant Protection Commission No. 140.

Martyn, E.B. (1953) Red ring disease of coconuts in Trinidad and Tobago. *Tropical Agriculture* 32, 45–53.

Mariau, D. (1968) Methodes de lutte contre le Rhynchophore. *Oleagineux* 23, 443–446.

McSorley, R. and Dunn, R.A. (1989) Effects of root-knot nematodes on *Areca catechu*. *Supplement to the Journal of Nematology* 21, 717–719.

Micoletzky, H. (1922) Die freilebenden erd-Nematoden mit besonderer Berucksichtigung der Steiermark und der Bukowina, zugleich mit einer Revision sämtlicher nicht mariner, freilebender Nematoden in Form von Genus – Beschreibunger und Bestimmungsschlüsseln. *Archiv für Naturgeschichte, Berlin* 87, 1–650.

Minz, G. (1958) Two *Diplodia* species in date palms. *Ktavim* 8, 213–216.

Mora, T.L.S., Calvache, G.H. and Avila, P.M. (1994) The spread of *Rhadinaphelenchus cocophilus* (Cobb) Goodey, causal agent of red ring-little leaf in oil palm. San Carlos de Guaroa (Meta). *Palmas* 15, 15–27.

Moura, J.I.L., Vilela, E.F., Brasil, G.H. and Cangucu, R. (2000) Mass trapping of *Rhynchophorus palmarum* using pheromone in coconut plantation in Brazil. *Proceedings XXI International Congress of Entomology*. Abstract 610. Book 1. Foz do Iguassa, Brazil, p. 154.

Nair, R.B. (1964) Nematodes from arecanut soils. *Arecanut Journal* 15, 87–88.

Netscher, C. and Luc, M. (1974) Nématodes associés aux cultures maraîchéres en Mauritanie. *L'Agronomie Tropicale, Nogent* 29, 697–701.

Nadakkal, A.M. (1965) A note on the occurrence of nematodes on the leaves of coconut palms in Kerala. *Science and Culture* 31, 157–158.

Noriega, C.D.H., Becerra, L.N. and Hernandez, R.F. (1991) Coconut diseases in the Coast of Guerrero, Mexico. *Revista Mexicana de Fitopatologia* 9, 84–93.

Nowell, W. (1919) Red ring or root disease of coconut palms. *West Indian Bulletin* 17, 189–192.

Nowell, W. (1923) *Disease of Crop Plants of the Lesser Antilles*. The West Indian Committee, London.

Nowell, W. (1924) Red ring disease of coconuts. *Proceedings 9th West Indies Conference, Jamaica*, pp. 167–172.

Oehlschlager, A., Chinchilla, C. and Gonzalez, M. (1992a) Management of the American palm weevil *Rhynchophorus palmarum* and the red ring disease in oil palm by pheromone-based trapping. *ASD. Oil Palm Paper* 5, 15–23.

Oehlschlager, A.C., Pierce, H.D. Jr, Morgan, B., Wimalaratne, P.D.C., Slessor, K.N., King, G.G.S., Gries, G., Gries, R., Borden, J.H., Jiron, L.F., Chinchilla, C.M. and Mexzon, R. (1992b) Chirality and field testing rhynchophorol, the aggregation pheromone of the American palm weevil. *Naturwissenschaften* 79, 134–135.

Oehlschlager, A.C., Chinchilla, C.M., Gonzalez, L.M., Jiron, L.F., Mexzon, R. and Morgan, B. (1993) Development of a pheromone-based trap for the American palm weevil, *Rhynchophorus palmarum* (L.). *Journal of Economic Entomology* 86, 1381–1392.

Oehlschlager, A.C., McDonald, R.S., Chinchilla, C.M. and Patschke, S.N. (1995) Influence of a pheromone-based mass trapping system on the distribution of *Rhynchophorus palmarum* (Coleoptera: Curculionidae) in oil palm. *Environmental Entomology* 24, 1005–1012.

Oehlschlager, A.C., Chinchilla, C., Castillo, G., Bulgarelli, J. and Gonzalez, L.M. (2002) Control of red ring disease by mass trapping of *Rhynchophorus palmarum* (Coleoptera: Curculionidae). *Florida Entomologist* 85, 507–513.

Pérez, A.L., Campos-Piedra, Y., Chinchilla, C.M., Oehlschlager, A.C., Gries, G., Gries, R., Castrillo, G., Giblin-Davis, R.M., Peña, J.E., Duncan, R.E., Gonzalez, L.M., Pierce, H.D., McDonald, R. and Andrade, R. (1997) Aggregation pheromones and host kairomones of the West Indian sugarcane weevil, *Metamasius hemipterus sericeus* (Oliv.) (Coleoptera: Curculionidae). *Journal of Chemical Ecology* 22, 357–368.

Pizarro, A.C. (1969) Nematodes associated with roots of coconut and other palms. *Philippine Journal of Plant Industry* 34, 155–158.

Poucher, C., Ford, H.W., Suit, R.F. and Ducharme, E.P. (1967) *Burrowing Nematode in Citrus*. Bulletin No. 7. Division of Plant Industry, Florida Department of Agriculture.

Rama, K. (1987) Ecology of plant parasitic nematodes affecting some major perennial crops in West Bengal. PhD thesis, Visva Bharati.

Raski, D.J., Maggenti, A.R., Koshy, P.K. and Sosamma, V.K. (1980) *Epicharinema keralense* n.gen., n.sp. (Nemata: Tylenchidae) and comments on Atylenchinae and Echphyadophorinae. *Revue de Nématologie* 3, 297–304.

Reddy, D.B. (1978) Insect pests, diseases and nematodes of *Areca catechu* and betel vine (*Piper betle*) recorded in Asia and the Pacific. Technical Document No. 114. Plant Protection Committee for the South East Asia, Asia and Pacific Region. FAO, Bangkok.

Rochat, D., Malosse, C., Lettere, M., Ducrot, P.H., Zagatti, P., Renou, M. and Descoins, C. (1991) Male-produced aggregation pheromone of the American palm weevil, *Rhynchophorus palmarum*: collection, identification, electrophysiological activity and laboratory bioassay. *Journal of Chemical Ecology* 17, 2127–2141.

Rochat, D., Malosse, C., Lettere, M., Ramirez-Lucas, P., Einhorn, J. and Zagatti, P. (1993) Identification of new pheromone-related compounds from volatiles produced by males of four Rhynchophorinae weevils. *Comptes Rendus de l'Academie des Sciences, Paris, Serie II* 316, 1737–1742.

Rochat, D., Meillour, P.N., Estaban-Duran, J.R., Malosse, C., Perthuis, B., Morin, J.P. and Descoins, C. (2000) Identification of pheromone synergists in American palm weevil, *Rhynchophorus palmarum* and attraction of related *Dynamis borassi. Journal of Chemical Ecology* 26, 155–187.

Rorer, J.B. (1911) Diseases of the coconut palm. *Circular 4. Board of Agriculture, Trinidad*, 27–33.

Salazar, S., Mora, O. and Sosa, M. (1992) Diseases of oil palm in the Central West region of Venezuela. *FONAIAP Divulga* 9, 39–40.

Sanchez, P.A. and Cerda, H. (1993) *Bursaphelenchus cocophilus* (Cobb) (Tylenchida: Aphelenchoididae), en Palmeras. *Boletin de Entomologia Venezolana* 8, 1–18.

Schuiling, M. and Van Dinther, J.B.M. (1981) 'Red ring disease' in the Paricatuba oil palm estate, Para, Brazil. A case study. *Zeitschrift für Angewandte Entomologie* 91, 154–169.

Silva, H.M. (1991) *Metamasius* sp. (Coleoptera: Curculionidae) vector of *Rhadinaphelenchus cocophilus* causal agent of the oil palm red ring. *Pesquisa em Andamento Unidade de Execucao de Pesquisa de Ambito Estadual de Belem* No. 17.

Sosamma, V.K. (1984) Studies on the burrowing nematode of coconut. PhD thesis, Kerala University, Trivandrum, Kerala, India.

Sosamma, V.K. (1994) *Interactions Between Vesicular Arbuscular Mycorrhizae (VAM) and Burrowing Nematode on Coconut.* Final Report ICAR AP Cess Fund Scheme 1988–93. CPCRI, Kayangulam, Krishnapuram, Kerala, India.

Sosamma, V.K. (1996) Integrated nematode management – principles and practices. In: *Recent Advances in Integrated Management of Pests and Diseases of Coconut.* CPCRI, Kayangulam, pp. 215–221.

Sosamma, V.K. (1998) Nematode parasites of palms. Paper presented in the group meeting of nematologists working on horticultural crops. 16–18 January 1998, CPCRI, Kayangulam.

Sosamma, V.K. (1999) Occurrence of *Pasteuria* sp. a hyperparasite of *Radopholus similis* and *Heterodera oryzicola* from Kerala. *Annual Meeting and Symposium on Plant Disease Management for Sustainable Agriculture.* 14–16 December 1999, CPCRI, Kayangulam, p. 5 (abstract).

Sosamma, V.K. (2000a) Report of *Catenaria vermicola* from *Radopholus similis* in Kerala. Paper presented in the National Symposium organised by IPS at Directorate of Wheat Research, Agarsain Marg, Karnal, Haryana, p. 153 (abstract).

Sosamma, V.K. (2000b) Nematodes of palms. In: Trivedi, P.C. (ed) *Plant Diseases.* Pointer Publishers, Jaipur, Rajasthan, India, pp. 393–413.

Sosamma, V.K. (2000c) Nematodes associated with spindle leaves of leaf rot affected coconut palms. Paper presented at the National Nematology Symposium on Integrated Nematode Management for Sustainable Agriculture in Changing Agro-ecological and Economic Scenario in the New Millennium, 23–24 November 2000, OUAT, Bhubaneswar.

Sosamma, V.K. (2000d) Host range and specificity of two *Pasteuria* isolates from Kerala, India. Paper presented at the International Conference on Plantation Crops PLACROSYM XIV, 12–16 December 2000, Hyderabad.

Sosamma, V.K. (2002) Nematodes of coconut, arecanut and oil palm. Paper presented at the seminar on Centenary of Nematological Research, 21–22 February 2002, Kerala. CPCRI, Kayangulam, Kerala.

Sosamma, V.K. (2003) *Investigations of Entomophilic Nematodes Associated with Red Palm Weevil,* Rhychophorus ferrugineus *F and Rhinoceros Beetle,* Oryctus rhinoceros *L. on Coconut.* Final Report ICAR Cess Fund Scheme, CPCRI, Kayangulam.

Sosamma, V.K. and Banu, J.G. (1996) Entomophilic nematodes. In: *Recent Advances in Integrated Management of Pests and Diseases of Coconut.* CPCRI, Kayangulam, pp. 230–241.

Sosamma, V.K. and Koshy, P.K. (1977) Additional hosts of the burrowing nematode *Radopholus similis* infesting coconut palm in South India. *Plant Disease Reporter* 61, 760–761.

Sosamma, V.K. and Koshy, P.K. (1978) A note on the association of *Cylindrocarpon effusum* and *C. lucidum* with *Radopholus similis* in coconut. *Indian Phytopathology* 31, 381–382.

Sosamma, V.K. and Koshy, P.K. (1981) Other host records of *Radopholus similis. Indian Journal of Nematology* 11, 73–74.

Sosamma, V.K. and Koshy, P.K. (1983) A note on the occurrence of *Cylindrocladium clavatum* Hodges & May in lesions caused by *Radopholus similis* on coconut roots. *Current Science* 52, 438.

Sosamma, V.K. and Koshy, P.K. (1985) Effects of different types of soils on multiplication of *Radopholus similis* and growth of coconut seedlings. *Indian Journal of Nematology* 15, 217–222.

Sosamma, V.K. and Koshy, P.K. (1986) Survival of *Radopholus similis. National Conference – Plant Parasitic Nematodes of India – Problems and Progress,* 17–20 December 1986. IARI, New Delhi, India, p. 107.

Sosamma, V.K. and Koshy, P.K. (1991) Etiology – nematodes. *Coconut Root (Wilt) Disease Monograph Series No. 3.* CPCRI, Kasaragod, India.

Sosamma, V.K. and Koshy, P.K. (1998) Etiology – nematodes. In: Nampoothiri, K.U.K. and Koshy, P.K. (eds) *Coconut Root (Wilt) Disease,* revised edn. CPCRI, Kasaragod, pp. 21–27.

Sosamma, V.K. and Vasanthi, T. (2001) Effect of bio-control fungi *Trichoderma harzianum* and *T. viride* for the control of *Radopholus similes* on arecanut. Paper presented at the National Symposium on Ecofriendly Approaches for Plant Disease Management, 22–24 January 2001, CAS Botany, University of Madras, Guindy Campus, Chennai, India.

Sosamma, V.K., Koshy, P.K. and Bhaskara Rao, E.V.V. (1980) Susceptibility of coconut cultivars and hybrids to *Radopholus similis* in field. *Indian Journal of Nematology* 10, 250–252.

Sosamma, V.K., Koshy, P.K. and Bhaskara Rao, E.V.V. (1988) Response of coconut cultivars to the burrowing nematode, *Radopholus similis. Indian Journal of Nematology* 18, 136–137.

Sosamma, V.K., Koshy, P.K. and Samuel, R. (1998a) Effects of vesicular arbuscular mycorrhizae, *Glomus mosseae* on nematodes infesting banana under coconut based high density multi species cropping system. *Proceedings of the Third International Symposium of AFRO ASIAN Society of Nematologists (TISAAN) Nematology – Challenges and Opportunities in 21st Century,* Sugarcane Breeding Institute, Coimbatore, India, 16–19 April 1998, pp. 94–97.

Sosamma, V.K., Koshy, P.K., Samuel, R. and Bindu, S.M. (1998b) Interaction of arbuscular mycorrhizal fungi (AMF) isolated from healthy high yielding coconut palms with burrowing nematode, *Radopholus similis* on coconut. *Proceedings National Symposium on Rational Approaches in Nematode Management for Austainable Agriculture,* 23–25 November 1998, GAU, Anand, pp. 34–37.

Sosamma, V.K., Jaijo, M.J. and Rasmi, B. (2000) Occurrence of entomogenous nematodes and an entomopathogenic fungus *Metarhizium flavoviride* associated with Rhinoceros beetle, a pest of coconut. *Proceedings National Nematology Symposium on Integrated Nematode Management for Sustainable Agriculture in Changing Agro-ecoclogical and Economic Scenario in the New Millennium,* 23–24 November 2000, OUAT, Bhubaneswar.

Stockdale, F.A. (1906) Coconut palm disease. *Proceedings of the Agricultural Society of Trinidad and Tobago* 7, 9–31.

Subramaniyan, S., Lakshmanan, P.L. and Vadivelu, S. (1988) Plant parasitic nematodes associated with arecanut palm in Tamil Nadu. *South Indian Horticulture* 36, 89.

Sudha, S. and Sundararaju, P. (1998) Effect of neem oil cake and nematicides for the control of *Radopholus similis* in arecanut based cropping system. *Third International Symposium of Afro Asian Society of Nematologists,* Coimbatore, 16–19 April 1998.

Sundararaju, P. (1984) Studies on the burrowing nematode of arecanut. PhD thesis, Kerala University, Trivandum, Kerala, India.

Sundararaju, P. (2000) Etiology – nematodes. In: Nampoothiri, K.U.K., Ponnamma, K.N. and Chowdappa, P. (eds) *Arecanut Yellow Leaf Disease.* Technical Bulletin No. 39, CPCRI. Kasaragod, Kerala, India, pp. 11–17.

Sundararaju, P. and Koshy, P.K. (1982a) Distribution of phytoparasitic nematodes on arecanut in South India. *Proceedings of the Silver Jubilee Symposium on Arecanut Research and Development,* 12–14 December 1982, Vittal, South Kanara, Karnataka, India, pp. 143–146.

Sundararaju, P. and Koshy, P.K. (1982b) Response of *Areca* collections against *Radopholus similis. Indian Journal of Nematology* 12, 404–405.

Sundararaju, P. and Koshy, P.K. (1984) A note on the association of *Cylindrocarpon obtusisporum* on lesions caused by *Radopholus similis* on arecanut (*Areca catechu* L.) roots. *Indian Phytopathology* 37, 578.

Sundararaju, P. and Koshy, P.K. (1986a) Control of *Radopholus similis* on arecanut seedlings with aldicarb, aldicarb sulfone, carbofuran and fensulfothion. *Indian Journal of Nematology* 16, 4–7.

Sundararaju, P. and Koshy, P.K. (1986b) Effect of different nematicides and neem oil cake in the control of *Radopholus similis* in yellow leaf disease affected arecanut palms. *Indian Journal of Nematology* 16, 44–47.

Sundararaju, P. and Koshy, P.K. (1987) Separate and combined effects of *Radopholus similis* and *Cylindrocarpon obtusisporum* on arecanut seedlings. *Indian Journal of Nematology* 17, 302–305.

Sundararaju, P. and Ratnakaran, K. (2002) Factors influencing the prevalence of the root-lesion nematode,

Pratylenchus coffeae on oil palm (*Elaeis guineensis* Jacq.). *Indian Journal of Nematology* 32, 16–19.

Sundararaju, P. and Sudha, S. (1998) Nematode management in arecanut and arecanut based farming system. *Proceedings of the Third International Symposium of Afro-Asian Society of Nematologists (TISAASN), Nematology: Challenges and Opportunities in 21st Century.* Sugarcane Breeding Institute (ICAR), Coimbatore, India, 16–19 April 1998, pp. 190–192.

Sundararaju, P., Sosamma, V.K. and Koshy, P.K. (1984) Additional host records of root knot nematode, *Meloidogyne incognita. Indian Journal of Nematology* 14, 183–184.

Sundararaju, P., Koshy, P.K., Banu, J.G., Rethnakaran, K. and Sosamma, V.K. (1995a) Plant parasitic nematodes associated with oil palm (*Elaeis guineensis* g.). Abstracts of papers presented in the National Symposium, *Nematode Problems of India – An appraisal of the Nematode Management with Eco-friendly Approaches and Bio-components,* 24–26 March 1995, IARI, New Delhi, India.

Sundararaju, P., Sosamma, V.K., Koshy, P.K., Rethinam, P., Sukumaran, S. and Suja, C.P. (1995b) Occurrence and distribution of *Radopholus similis* in coconut nurseries in South India. Abstracts of papers presented in the National Symposium, *Nematode Problems of India – An Appraisal of the Nematode Management with Eco-friendly Approaches and Bio-components,* 24–26 March 1995, IARI, New Delhi, India.

Troccoli, A., Lamberti, F. and Greco, N. (1992) *Pratylenchus* species occurring in Algeria (Nematoda, Pratylenchidae). *Nematologia Mediterranea* 20, 97–103.

Tsang, M.M.C., Hara, A.H. and Sipes, B. (2003) Hot-water treatments of potted palms to control the burrowing nematode, *Radopholus similis. Crop-Protection* 22, 589–593.

Van Hoof, H.A. and Seinhorst, W. (1962) *Rhadinaphelenchus cocophilus* associated with little leaf of coconut and oil palm. *Netherland Plant Pathology* 68, 251–256.

Van Weerdt, L.G., Birchfield, W.R. and Esser, R.P. (1959a) Observations on some subtropical plant parasitic nematodes in Florida. *Proceedings of the Soil Crop Science Society of Florida* 19, 443–451.

Van Weerdt, L.G., Martinez, A.P. and Esser, R.P. (1959b) Results of a survey designed to determine the etiology of lethal yellowing of *Cocos nucifera. Proceedings of the Florida State Horticultural Society* 72, 421–430.

Villanueva, A. and Gonzalez, A. (1988) Importancia ecónomica del Anillo Rojo en Palmerasde la Costa S.A. y medidas de control practicadas. *Palmeras de la Costa.*

Webster, B.N. and Gonzalez, J. (1959) Investigaciones sobre la enfermedad 'anillo rojo' en cocas y palma de aceite. *9th Convencion de la Asociacion Venezolana el Avancc dc la Ciencia,* Servicio Shell para el Agricultor, Estado Arajua, Venezuela.

Weischer, B. (1967) Report to the Government of India on plant parasitic nematodes. PL 24/52 *Report No TA 2332,* FAO, Rome.

Youssef, M.M.A. and Eissa, M.F.M. (1994) Population dynamics of certain nematodes associated with date-palm in Egypt. *Afro Asian Journal of Nematology* 4, 68–72.

14 Nemato de Parasites of Coffee and Cocoa*

Vicente Paulo Campos[1] and Luc Villain[2]
[1]*Departamento de Fitopatologia, Universidade Federal de Lavras, Caixa Postal 37, 37200-000 Lavras, MG Brazil;* [2]*Cirad-CP, Boulevard de la Lironde, TA 800/PS3, 34398 Montpellier, Cedex 5, France*

Coffee

Coffee is a perennial dicotyledonous shrub or small tree with woody stem, persistent leaves and hermaphrodite flowers that belongs to the genus *Coffea* in the family Rubiaceae. Chevalier (1947) grouped several species of *Coffea* in different sections. The section Eucoffea is the most cultivated species. This section is divided into subsections: the subsection Erythrocoffea includes the species *Coffea arabica*, *C. canephora*, *C. congensis*; Pachycoffea includes *C. liberica* and *C. excelsa*; Mozambicoffea includes *C. racemosa* and *C. salvatrix*; Melonocoffea includes *C. stenophylla*; and Nanocoffea includes *C. montana*, etc. A few species of the section Mascarocoffea such as *C. resinosa* and *C. macrocarpa* have no caffeine alkaloid in the seeds. It is possible that in the future decaffeinated cultivars can be developed from these species.

Phylogenic studies of *Coffea* species based on chloroplast DNA (cpDNA) (Berthou *et al.*, 1983; Cross *et al.*, 1998) and ribosomal DNA (Lashermes *et al.*, 1997) revealed a close relationship between the main branches in the phylogenic trees of this genus and the three main geographical origins of the different species, i.e. Madagascar, western and Central Africa, and East Africa. According to restriction fragment length polymorphism (RFLP) analysis of conservative cpDNA, *C. congensis*, *C. canephora* and *C. eugenioides* seem to be the closest species to *C. arabica*. These studies also confirmed the allotetraploid origin of *C. arabica* and supported the hypothesis suggested by Lashermes *et al.* (1999) that *C. arabica* could come from a hybridization between two *Coffea* species close to *C. eugenioides* and *C. canephora* (Anthony *et al.*, 1999). This allotetraploid origin and reproductive biology, autogamous contrary to allogamous *C. canephora*, could explain the relative low genetic diversity of *C. arabica* (Lashermes *et al.*, 1999).

Seeds of coffee germinate in 3–4 weeks at a temperature of 31–32°C; at 17°C it takes 3 months. The formation of leaves occurs during the whole year, but the ratio of shoot and leaf growth varies with the climatic conditions. Flower formation is induced by photoperiod changes, but dif-

*A revision of part of the chapter by V.P. Campos, P. Sivapalan and N.C. Gnanapragasam.

ferentiation requires short days (< 13–14 h of light). Very high temperature or prolonged drought during the bud dormancy provokes the formation of abnormal or aborted flowers (Anonymous, 1985).

Coffee plants produce fruits containing seeds which, after hulling and washing, are dried, roasted and ground; the powder is used to make the coffee drink. The crop is grown mainly between the Tropics of Cancer and Capricorn. Coffee has been of great relevance to the economy of many tropical countries. Its importance to the total export of countries has decreased in percentage, but the value of coffee exports has increased. Brazil is the major world producer, representing, in 2002, 32.46% of the world production, and South America 68.38% (FAO, 2002).

Current commercial green coffee production depends almost exclusively on two species: *C. arabica* that accounts for 75% of the world coffee exports, and is produced in 60 countries, mostly in South America (Brazil producing more than 40% of the world Arabica production in 2002), Meso America, East and Central Africa; and *C. canephora*, the production of which is mostly concentrated in western Africa and South-east Asia (Vietnam has become the main Robusta producer with almost 35% of the world production) (Anonymous, 1985; FAO, 2002). Other species of minor relevance to world coffee production are *C. racemosa* in Mozambique, *C. stenophylla* in Sierra Leone and the Côte d'Ivoire, *C. excelsa* in the Central African Republic and Vietnam, and *C. liberica* in Guyana, Surinam, Malaysia, the Philippines, São Tomé and Liberia (Krug, 1969).

Arabica coffee, *C. arabica*, originated from the mountain region in the south-west of Ethiopia, southern Sudan and northern Kenya (Thomas, 1942, cited by Anthony *et al.*, 2001; Sylvain, 1959), while Robusta coffee, *C. canephora*, originated from western and Central Africa, from Guinea to Democratic Republic of the Congo (ex Zaire) with two genetic groups, 'Congolese' and 'Guinean' (Berthaud, 1986; Charrier and Eskes, 1997; Anthony *et al.*, 1999). *C. arabica* is an upland species growing best at altitudes of 900–2000 m on the equator with temperatures of 17–25°C and rainfall of 1200–2000 mm. Humid cloudy conditions are preferable. *C. canephora* is not so specific in its requirements, growing from sea level to 1700 m at temperatures of 20–32°C and is better suited to lower altitudes, such as 400 m in Brazil.

Cultivation techniques

Most commercial coffee is planted from seed, and seedlings are raised in nurseries in either beds or bags of plastic or other material. Germination takes 5–10 weeks and seedlings are transplanted to the field when 6–10 months old. Vegetative propagation by cuttings is possible with coffee, but is not the usual practice worldwide. In Brazil, in areas highly infested by *Meloidogyne incognita* and *M. paranaensis* especially in São Paulo and Paraná States, grafting using *C. canephora* cv. Apoatã as rootstock and a commercial cultivar of *C. arabica* as a scion is used in order to have productive plantations. Spacing varies between areas, usually 2–4 m between rows and 0.5–1 m between plants when one seedling is kept per low basin or 'cova', or 2 m apart when there are two plants per cova. New plantations in Brazil have spacing between plants of 0.5 m and of 3.5 m between rows, while in Meso America or Colombia, most of the plantations have a planting density of 1 m × 2 m. Shading is not always necessary for cropping *C. arabica* but it is practised in some areas such as most of the Central American coffee regions located on mountain ranges characterized by a broken topography, frequently volcanic soils and a hot dry season. Shade trees act in these environments as a climatic regulator and contribute to the prevention of soil erosion and even to improve soil characteristics. Shade is used less for *C. canephora*. Other trees or crops, e.g. banana, are used for shading coffee. Mulching is beneficial in non-frosted areas. Pruning is variable and not always done. The most common methods involve cutting the main stem at 0.40 or 1.80 m from the soil or the plagiotropic branches at 0.20 m from the main stem. Trees start bearing after 2.5–3.5 years.

Two characteristics of the physiology of coffee trees are important for cultural practices and should be taken into account for nematode control and nematode-related field studies. One is that coffee berries are produced on second-year wood, i.e. the yield of individual trees greatly depends on potential flowering nodes produced the previous year. Secondly, unusual among woody perennials, that may be due to its deep shade native habitat in the Ethiopian highland forests, is the inability of coffee trees to shed excess fruit in relation to nutritional conditions (Cannell, 1985).

Nematode Parasites of Coffee

Many genera and species of nematodes have been associated with coffee in many countries of the world, including very damaging nematodes causing great losses to the coffee farmers and the local economy of developing countries.

Meloidogyne

Root knot nematodes of the genus *Meloidogyne* are more widely distributed throughout the world in coffee plantations than any other major group of parasitic nematodes (Table 14.1). Furthermore, when their importance is considered on a worldwide basis, they rank high on the list of pests affecting the production of coffee.

Root knot nematode species of coffee can be separated into two categories: (i) the most common, damaging and well-known species on coffee, *M. exigua*, *M. incognita*, *M. coffeicola* and *M. paranaensis*; and (ii) the less widespread species, *M. africana*, *M. decalineata*, *M. megadora*, *M. hapla*, *M. kikuyensis*, *M. inornata*, *M. javanica*, *M oteifae*, *M. arenaria*, *M. thamesi*, *M. arabicida*, *M. konaensis* and *M. mayaguensis*.

Meloidogyne exigua, *M. incognita*, *M. coffeicola* and *M. paranaensis*

Distribution

M. exigua is known to occur in all major coffee-growing countries of South and Central America but is not found outside

Table 14.1. Species of root knot nematodes found on coffee and their distribution.

Meloidogyne species	Country
M. incognita	Brazil, Tanzania, Jamaica, Venezuela, Guatemala, Côte d'Ivoire, India, Costa Rica, El Salvador, Nicaragua, Cuba, USA (Hawaii)
M. exigua	Brazil, Guatemala, Dominican Republic, Nicaragua, Costa Rica, Puerto Rico, Colombia, Peru, El Salvador, Venezuela, Bolivia, Honduras, Panama
M. coffeicola	Brazil
M. javanica	Brazil, Tanzania, Zaire, El Salvador, India, Cuba, São Tomé and Príncipe
M. hapla	Brazil, Tanzania, Zaire, India, Kenya, Congo, Guatemala, El Salvador
M. africana	Kenya, Zaire
M. decalineata	Tanzania, São Tomé and Príncipe
M. kikuyensis	Tanzania
M. arenaria	Jamaica, Cuba, El Salvador
M. megadora	Angola, Uganda
M. inornata	Guatemala
M. oteifae	Zaire
M. thamesi	India
M. paranaensis	Brazil, Guatemala
M. arabicida	Costa Rica
M. konaensis	USA (Hawaii)
M. mayaguensis	Cuba

the Americas (Table 14.1), although a species identified as *M. exigua* was reported from Java in 1931 (Bally and Reydon, 1931). It was the first nematode species found in coffee, when Jobert was invited to study a severe disease of coffee in Rio de Janeiro, Brazil which he showed to be caused by a nematode (Jobert, 1878): the species was described a few years later by Göldi (1889, 1892). In 1929, *M. exigua* was found in São Paulo State (Brazil) (Rahm, 1929). Since then, it has been found in all major coffee-producing states in Brazil (Campos *et al.*, 1985; Campos and Melles, 1987; Campos, 1997), sometimes mixed with other species of *Meloidogyne*. In the 1960s, *M. exigua* was found in coffee plantations in Costa Rica, the Dominican Republic and Venezuela (Salas and Echandi, 1961; Flores and Yépez, 1969; Schieber and Grullon, 1969). In the 1970s, this species was also reported from Guatemala, Peru, El Salvador, Honduras (Pinochet and Ventura, 1980) and Puerto Rico (Sabrego, 1971; Schieber, 1971; Lordello, 1972; Ayala, 1976). However, more recent and large surveys in Central America did not find *M. exigua* in El Salvador and detected this species in just one sample in Guatemala (Hernández, 1997; Sarah, 2003). Moreover, in the report of *M. exigua* as the only *Meloidogyne* species present in Guatemala, Schieber (1966) describes clearly corky roots symptoms without galls, i.e. symptoms totally different from the typical small rounded galls caused by *M. exigua* but similar to symptoms caused by *M. paranaensis* attacks (see below). In the 1980s, *M. exigua* was found in Honduras, Colombia, Nicaragua and Bolivia (Gomez, 1980; Pinochet and Ventura, 1980; Bridge *et al.*, 1982; Vega, 1982) and also Panamá (Marbán-Mendoza, 1995). In Central America, surveys showed that *M. exigua*'s main distribution on coffee seems to extend from southern Costa Rica up to eastern Honduras (Hernández, 1997; Sarah, 2003). This species represents the most widely distributed root knot nematode on coffee in Costa Rica and Nicaragua, where it is present in almost all coffee-producing regions (Flores and López, 1989). In Honduras, *M. exigua* is reported in the border province of El Paraíso, next to Nicaragua (Zelaya-Escoto and Santacreo, 2000).

M. coffeicola was described by Lordello and Zamith (1960) from the coffee plantation of Terra Boa, Paraná State, Brazil. It has not been found outside Brazil. Lordello (1967) found this species attacking coffee in São Paulo State, and in 1983 it was also found in the southern region of Minas Gerais State, Brazil (Guerra Neto *et al.*, 1983), and eradicated thereafter with the elimination of the infested plantation. In 2003, it was found in the Alto Paranaiba region of the same state in Brazil (Castro and Campos, 2004a).

M. incognita was first found attacking coffee in 1960 in Guatemala (Chitwood and Berger, 1960) where its effects were said to be less severe than those of *M. exigua* (Whitehead, 1969b). In 1960, it was reported from the Côte d'Ivoire (Luc and De Guiran, 1960), and then in Tanzania (Whitehead, 1969a), Venezuela (Flores and Yépez, 1969), Jamaica (Hutton *et al.*, 1982) and India (Kumar, 1984), also from Costa Rica (Figueroa, 1988), El Salvador (Pinochet and Guzman, 1987), Panamá and Nicaragua (Marban-Mendoza, 1995), Cuba (Sampedro *et al.*, 1989) and the USA (Hawaii) (Schenck and Schmitt, 1992). In the case of Guatemala and El Salvador, previous *M. incognita* identifications must be considered with caution since more recent surveys did not detect *M. incognita* but revealed the presence of another morphologically closely related species widely distributed in each of these two countries. In Guatemala, it concerns *M. paranaensis* (see below). In El Salvador, it concerns a new species currently under description (R.M.D.G. Carneiro, 2004, personal communication) that shows a perineal pattern similar to the one reported for *M. incognita* but presents a characteristic four-band esterase phenotype (Hernández *et al.*, 1996; Hernández, 1997).

Although *M. incognita* occurs in many coffee-growing areas around the world (Table 14.1), it was in Brazil where its effects on coffee plantations became catastrophic.

M. incognita was first found in 1970 attacking coffee in Pindorama, São Paulo State (Brazil) (Lordello and Mello Filho, 1970). However, this nematode may have been present in coffee in Brazil for some time, as Lordello (1984) pointed out, in many instances the aggressive races of *M. exigua* reported from many locations may actually have been different populations or races of *M. incognita*. In 1971, *M. incognita* was found in Espírito Santo State (Lordello and Hashizume, 1971), in 1972 in Paraná (Lordello Lordello, 1972b), in 1975 in Ceará (Ponte and Castro, 1975), in 1984 in Minas Gerais State (Guerra Neto and D'Antonio, 1984) and in Bahia State (Souza *et al.*, 2000). In 2003, *M. incognita* was found in Rio de Janeiro State only in one plantation of *C. canephora*, but not in *C. arabica* (Barbosa *et al.*, 2003b). Corky root symptoms associated with the presence of *Meloidogyne* sp. were reported in Huatusco region, Veracruz State, Mexico (Teliz-Ortiz *et al.*, 1993). The root knot nematode in question was identified as *M. incognita* by Sanchez (1990); Castillo *et al.* (1995) reported that 60% of the observed specimens had perineal patterns similar to that of *M. incognita*.

M. paranaensis was described by Carneiro *et al.* (1996) from a coffee plantation of Paraná State, Brazil. The species was found attacking coffee in São Paulo in 2001 (Favoreto and Santos, 2001; Kubo *et al.*, 2001), and in 2003 it was found in the South and Alto Paranaiba regions of Minas Gerais State, Brazil (Castro *et al.*, 2003a; Castro and Campos, 2004b). This species is now becoming as catastrophic to coffee plantations as *M. incognita*, especially in the states of Paraná and São Paulo, Brazil. The threat of dissemination of *M. paranaensis* to other coffee-producing states along with the necessity to update the nematode distribution in most coffee-producing states justified money from PNP and D/Café-EMBRAPA of the Brazilian government to finance the largest nematode survey ever done in Brazil. The survey started in 1999 and will be completed by 2005, covering 80% of the coffee-growing area in the states of Paraná, São Paulo, and Minas Gerais and Bahia, Brazil

In coffee plantations of São Paulo and Paraná, Brazil, *M. exigua*, *M. coffeicola*, *M. incognita* and *M. paranaensis* have occurred for many years in separate or mixed populations, with fluctuations in the predominance of each species over the others. Up to four *Meloidogyne* species occur in the same plantation (Otoboni *et al.*, 2003a). In Paraná State, from 1967 to 1970, *M. coffeicola* was found in 16 counties, whereas *M. exigua* was found in only two (Vernalha *et al.*, 1970). Since then, surveys have shown a substantial increase in distribution of *M. incognita* and a decrease of *M. coffeicola* (Lordello *et al.*, 1974; Carneiro and Carneiro, 1982a). It is believed that *M. coffeicola* was eradicated from many plantations during the renewal of damaged coffee after the 1975 great frost. After this period, coffee may have been cultivated in new lands without the nematode being present.

In Paraná State, only *M. incognita* was found (four races) and the so-called IAPAR biotype, today described as a new species, *M. paranaensis*, was found in 545 samples collected in 17 counties (Carneiro *et al.*, 1992). In 2000, from the samples with *Meloidogyne* species, *M. incognita*, *M. exigua*, *M. paranaensis* and *M. javanica* occurred, respectively, at a frequency of 26, 26, 32 and 16% (Survey *et al.*, 2000). From 1999 to 2001, about 210 counties were surveyed in Paraná State, and 657 samples were collected (Krzyzanowski *et al.*, 2001). Thirty-four per cent ot the samples had *Meloidogyne* present; *M. paranaensis* accounted for 44% of the coffee roots infected, followed by 17% of *M. incognita*. In São Paulo State in 1968, *M. exigua* was found in 50 counties and *M. incognita* in only four (Lordello *et al.*, 1968). In 1969, *M. coffeicola* was found in 11 counties (Curi *et al.*, 1969). Since 1970, *M. coffeicola* seems to have disappeared from the coffee plantations of São Paulo according to Lordello (1984), whereas *M. incognita* has become widespread in this state. In Bauru and Marília counties, about 77% of the collected samples contained *M. paranaensis* and *M. incognita* (Kubo *et al.*, 1999). In 2000, *M. incognita*, *M paranaensis*, *M. exigua* and *M. javanica* were identified in São Paulo State (de Oliveira *et al.*, 2000). In 2001, *M. incognita* and *M. paranaensis* was found in 35.5 and 32.9% of the samples collected in Alta Paulista and Nova Alta Paulista regions of São Paulo State, respectively (Oliveira Filho *et al.*, 2001); 30 counties of São Paulo State had an almost even incidence of *M. incognita* with *M. paranaensis* in almost 25% (Favoreto and Santos, 2001). Kubo *et al.* (2001) found *Meloidogyne* species in 56.4% of the 195 samples collected in São Paulo State. From the identified species, *M. exigua* and *M. incognita* were of greatest occurrence, followed by *M. paranaensis* and *M. coffeicola*. *M. incognita* and *M. exigua* were also found predominantly, followed by *M. paranaensis* and *M. javanica*, in 37 counties of São Paulo State (Lordello *et al.*, 2001). Only races 1, 2 and 3 of *M. incognita* were found, with the greatest incidence of race 1. In other surveys, Lordello and Lordello (2001) found two or three times more samples with *M. incognita* and *M. exigua* than with *M. paranaensis* and *M. javanica* in 18 counties of São Paulo State. From 1999 to 2002, *M. exigua*, *M. incognita*, *M. paranaensis* and *M. javanica* have been identified in 20.9, 36.5, 13.0 and 0.9%, respectively, of the infested samples of São Paulo State (Lordello, 2002). *M. paranaensis* has been found in São Paulo and Paraná States in the regions most infested by *M. incognita*, which confirms that there is a mixture of these species which were both identified as *M. incognita* before the description of *M. paranaensis* in 1996 (Carneiro *et al.*, 1996).

Paraná, São Paulo and Minas Gerais States accounted in 2002 for approximately 68.6% of coffee produced in Brazil. *M. exigua* was found to be widespread in the coffee-growing regions of Minas Gerais (Campos *et al.*, 1985; Campos and Lima, 1986; Campos, 1997). It was the only species of *Meloidogyne* found in this state until 1983 when *M. coffeicola* was recorded in Machado (Guerra Neto *et al.*, 1983), and later *M. incognita* was found in the towns of Nova Resende and São Thomas Aquino (Guerra Neto and D'Antonio, 1984). However, *M. incognita* has been restricted to the original sites without any great economic impact on the overall coffee production in Minas Gerais (Campos *et al.*, 1985), and *M. coffeicola* was eradicated from the original site. From 1999 to 2002, 1830 samples were collected in Minas Gerais State, Brazil, and 22% of them had *M. exigua* (Campos, 2002). *M. paranaensis* was found in one county of the South region and in two counties of the Alto Paranaiba region of the same state (Castro *et al.*, 2003a; Castro and Campos, 2004b). *M. coffeicola* was found in two counties of the Alto Paranaiba region of Minas Gerais State (Castro and Campos, 2004a). In another survey done in Minas Gerais State, Brazil, in regions of Triangulo Mineiro and Alto Paranaiba (Pinheiro *et al.*, 2000) and Zona da Mata (Lima, 2002), only *M. exigua* was found. In Minas Gerais State, which produced 51.8% of the total Brazilian coffee in 2002, the predominance of *M. exigua* is highest amongst all coffee-producing-states, but a threat of dissemination of *M. paranaensis* and *M. coffeicola* throughout the state exists. *M. exigua* occurs in Bahia State (Lordello, 1971; Souza *et al.*, 1997), a relatively new coffee-growing region in Brazil, mostly under irrigation. From 1999 to 2002, in a survey carried out in many counties of Bahia State, *M. exigua* and *M. incognita* were found in 57 and 18% of the 316 collected samples, respectively (Souza *et al.*, 2000; Souza, 2002). In Rio de Janeiro State, where *M. exigua* was described by Göldi in 1889, it was found widespread throughout the remaining coffee plantations after the epidemic disaster in the 19th century. In many counties, the infestation was above 50% of the plantations sampled (Barbosa *et al.*, 2003a).

Comparative esterasic and random amplified polymorphic DNA (RAPD) analysis of *Meloidogyne* species, including Guatemalan and Brazilian populations, demonstrated that the most widely distributed *Meloidogyne* species in Guatemala, first studied by Anzueto (1993) and Hernández *et al.* (1996), belong to *M. paranaensis* (Carneiro *et al.*, 2003), though some of the collected Guatemalan populations present a second band on their esterase phenotype compared with the one-band phenotype observed for all Brazilian populations (Carneiro *et al.*, 2003; Sarah, 2003). Corky root symptoms on coffee plants have been described since 1935 in Guatemala by Alvarado (1935).

In Colombia, a root knot nematode population was found to exhibit a perineal pattern similar to that of *M. incognita* but with a response to the North Carolina differential host test similar to that reported for *M. paranaensis* by Carneiro and Almeida (2000) (Vergel-Colon *et al.*, 2000).

The reports of *M. paranaensis* in Guatemala and atypical reports of *M. incognita* in other countries in the Americas suggest that reports on coffee in the Americas of *M. incognita* diagnosed only by perineal patterns should be revised by using esterase phenotypes and differential hosts.

Symptoms of damage

M. exigua causes typical rounded galls (Plate 15A) mostly on new roots formed after the first rains in spring, and continues to produce them into the summer. The galls are initially white to yellowish brown and turn dark brown as the root becomes older. Egg masses are produced in the cortex under the root epidermis. On the Mundo Novo cultivar of *C. arabica* there is no necrosis around the giant cells and there is a tendency for lateral root formation at the region of the gall (Mendes, 1977). Necrotic areas are also to be seen on the galled roots, which may be aggravated by secondary infections, and the section of the root dies. Although many authors have reported that *M. exigua* may not often form galls but instead forms cracks on infected roots (Lordello, 1972), this may be due to a misidentification of the *Meloidogyne* species involved.

Infested seedlings planted in the field show reduced growth and defoliation, and some do not survive the dry season. The management of an infested crop in the field throughout the seedling stage is very difficult. Depending on the soil type, *M. exigua* can cause a serious defoliation of the adult coffee plant, leading to death. In Rio de Janeiro State during the 19th century, *M. exigua* caused the destruction of whole coffee plantations (Göldi, 1889). Young coffee plants in the field seem to suffer more from attack by *M. exigua* than at any other stage.

In Brazil, *M. incognita* causes peeling and cracking of cortical parts of the root tissue in field plants. The cortical cracking results from the hypertrophy of tissues adjacent to the female (Moraes *et al.*, 1973a). Darker dots along the root are observed where the females are located. Egg masses are produced on the root surface. Sometimes, localized swellings on the roots resembling galls are seen on lateral roots. Females feeding in roots kill the surrounding tissues, leading to the death of sections of the root (Fig. 14.1B, Plate 15B) and thus greatly reducing the root system. Young seedlings of coffee grown under the foliage of the infested plants have typical root galls.

The above-ground parts of infested plants in Brazil show foliar chlorosis, leaf fall, general decline, reduced growth and sometimes plants are killed. In São Paulo State, large coffee plantations have been decimated by this nematode, with 5-year-old coffee plantations dying out (Fig. 14.1A). Lordello (1984) has said that *M. incognita* in some areas of São Paulo State is a 'disaster pathogen', becoming the worst enemy of coffee.

In Jamaica, *M. incognita* causes galls on coffee plants, and growth and yield reduction (Hutton *et al.*, 1982).

M. coffeicola causes peeling and cracking of roots but does not produce galls (Plate 15C). The female is easily found in older tissue especially on the taproot. Egg masses are produced on the root surface. Attempts at artificial inoculation of *M. coffeicola* on coffee seedlings have failed. The females lay their eggs outside roots, through cracks that they have induced in the root tissue. The numerous dark spots on infected roots are egg masses of the nematodes. Very few females lay eggs inside the roots. The above-ground part of the infested coffee plant shows yellowing, leaf fall, and there is a general decline of the plant leading to death.

Fig. 14.1. (A) Dying out of 5-year-old coffee plantation infested by *Meloidogyne incognita* (São Paulo State, Brazil). (B) Death of sections of the roots surrounding feeding females. (Photo: V.P. Campos.)

M. paranaensis causes symptoms such as splitting and cracking of the cortical root tissue, especially on the taproot, but it does not produce typical root knot nematode galls on coffee. Egg masses are produced in the root tissues. Necrotic spots occur along the old roots where the females are located. Nematode feeding causes the tissue around the giant cells to die. Symptoms on infected plants include foliar necrosis, leaf drop, general decline, reduced growth and often plant death. In Brazil, large coffee plantations have been severely damaged by this nematode (Carneiro *et al.*, 1996; Gonçalves and Silvarola, 2001). In Guatemala, *M. paranaensis* is responsible for symptoms of damage similar to those in Brazil (Anzueto, 1993).

Among the four most known species of *Meloidogyne* in coffee, only *M. exigua* causes typical galls easy to recognize in the field, but less destruction of the young roots. The other three, *M. incognita*, *M. coffeicola* and *M. paranaensis*, rarely cause typical galls only occasionally light swelling of rootlets. Their typical symptoms are on older roots mostly without side roots resulting from abortion of new rootlets, where females are found (Plates 15C and D), which can lead the researcher to overlook them in the field.

Biology and life cycle

The life cycle of *M. exigua* is very similar to that of the four most common species of the genus *Meloidogyne*. The length of time is longer, taking 32–42 days at 25–30°C to complete the cycle (Lima and Ferraz, 1985). Unlike *M. incognita* and *M. coffeicola*, the egg masses of *M. exigua* are mostly located under the epidermis of coffee roots. *M. exigua* and *M. incognita* have saccate bodies but *M. coffeicola* is more sausage shaped with a long neck, and as much as 1300 μm in length (Fig. 14.2A) (Lordello and Zamith, 1960). The perineal patterns cannot differ-

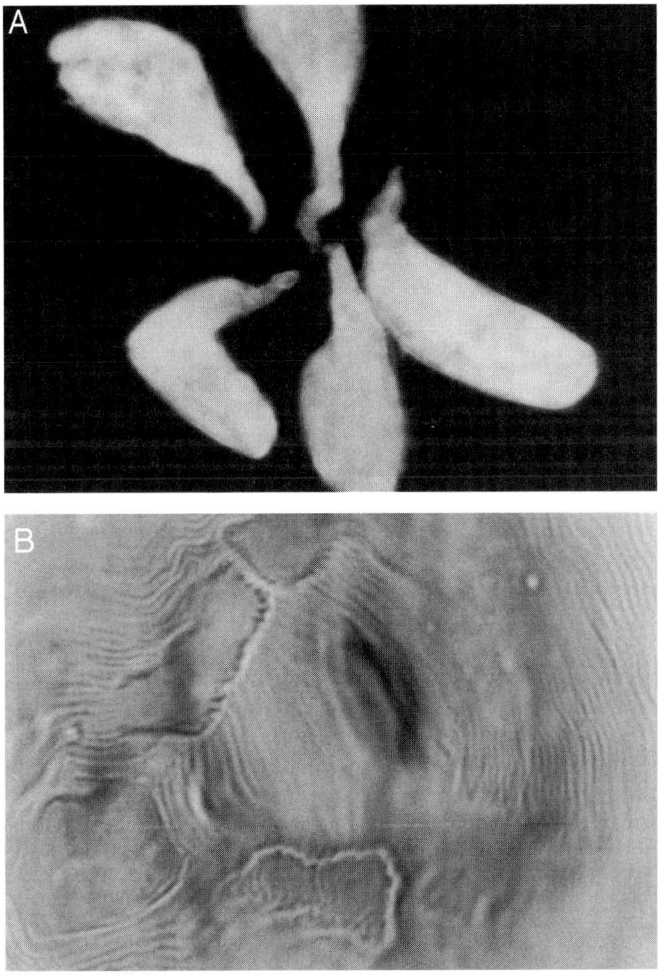

Fig. 14.2. (A) Elongated females of *Meloidogyne coffeicola* – a diagnostic character for species recognition in coffee roots. (B) Perineal pattern of *M. coffeicola* with striae. (Photo: V.P. Campos.)

entiate *M. incognita* from *M. paranaensis* but are distinctly different in all other species (Fig. 14.2B). *M. incognita* is more pathogenic to coffee than *M. exigua* (Moraes and Lordello, 1977), but similar to *M. paranaensis*. Grafted *C. arabica* in *C. canephora* and non-grafted *C. arabica* were cultivated in an *M. coffeicola*-infested area and produced well during the early 9 years; this did not happen in an area infested with *M. incognita* (Carneiro Filho and Yamaguchi, 1995).

Pathotypes, races or biotypes

Most population variations in the pathogenicity of *M. exigua* in coffee reported by many authors in São Paulo State, Brazil, in the past, may be related to misidentification of the nematode (Lordello, 1984). However, Carneiro and Almeida (2000) differentiated two races of *M. exigua* by their efficiency to reproduce on tomato.

In Brazil, four races of *M. incognita* are known to occur in Paraná and São Paulo States (Medina Filho *et al.*, 1981; da Costa *et al.*, 1991; Carneiro *et al.*, 1992). However, in Paraná State, race 2 predominates (Carneiro *et al.*, 1990a), and in São Paulo State race 1 (Monteiro *et al.*, 1995; de Oliveira *et al.*, 2000, 2001a,b; Lordello, 2002). They have been differentiated by the North Carolina differential host test as proposed in Taylor and Sasser (1978). There is no evidence of variations in pathogenicity within *M. coffeicola* populations in the field.

C. arabica cvs Catuaí, Mundo Novo and Bourbon Amarelo, *C. canephora* cvs Robusta, Guarini and Laurenti, and *C. excelsa* are susceptible to *M. incognita* (Moraes *et al.*, 1973b). For many years, researchers in Paraná State, Brazil, called race 5 or biotype IAPAR (Carneiro *et al.*, 1990b, 1992) a more pathogenic pathotype of *M. incognita* in coffee based on differences on the differential host test for races of *M. incognita* (Taylor and Sasser, 1978). In 1996, it was described as a new species, *M. paranaensis* (Carneiro *et al.*, 1996). In Colombia, the so-called race 5 of *M. incognita* was found on coffee (Villalba-Gault *et al.*, 1982). According to the differential host test of Taylor and Sasser (1978) and the root symptomatology, it resembles *M. paranaensis*. Two populations of *M. paranaensis* from soybean and coffee had different rates of reproduction on soybean cultivars, but not on Catuai vermelho coffee (Roese, 2003). The difference in esterase phenotypes between Guatemalan and Brazilian *M. paranaensis* (two versus one band, see above) could be an indication of the existence of different biotypes, but this needs to be confirmed. Differences in pathogenicity on *C. arabica* cv. Caturra (susceptible) and cv. IAPAR 59 (partially resistant) were observed between populations of *M. exigua* originating from different regions of Costa Rica (Alpizar, 2003).

Survival and means of dissemination

Six months after eradication of infested plants, *M. exigua* is not found in the soil (Moraes and Lordello, 1977) and does not survive in soil in the absence of the host for more than 6 months (Alvarenga, 1973). *M. coffeicola* also shows low persistence in the soil (Rebel *et al.*, 1976; Carneiro Filho and Yamaguchi, 1995). *M coffeicola* seems to have a low capacity to infest coffee seedlings and young trees. Thus no one has ever experimented with artificial inoculation on coffee seedlings. However, *M. incognita* causes high infestation on coffee even when infested soil is kept without host plants for 6 months, reducing only to 27% of the initial population (Jaehn and Rebel, 1984).

The method of cultivating coffee in the field by using transplanted seedlings produced in nurseries provides a very efficient dissemination of *Meloidogyne* species on seedling roots, once the nursery is infested. There are many smallholder coffee producers throughout the world, including Brazil, who cannot afford to apply chemicals or any other soil treatment, thus increasing the chance of efficient dissemination of nematodes. In Minas Gerais State, Brazil, since 1996 a law has been enforced by the state government obliging the producers to have a certificate from an official Nematology Laboratory stating the absence of any *Meloidogyne* species in coffee seedling roots. This has decreased the *Meloidogyne* species spreading within the state and avoided the introduction of infested seedlings into non-infested areas.

Environmental factors affecting parasitism

In spite of Whitehead's (1969) statement that coffee is very resistant to *M. incognita*, the rapid distribution and highly destructive nature of this pathogen in Brazil indicates more aggressive pathotypes which are adapted to local environments and to the cultivar of coffee grown. However, when *M. incognita* and *M. paranaensis* originating from coffee roots are cultured on tomato for 2 years consecutively and inoculated back on to coffee, they are no longer pathogenic (Carneiro and Jorge, 2001). Pruning the aerial parts of old coffee plants (> 15 years) is a management strategy used by farmers in Brazil to revitalize them. However, the use of this management in *M. incognita*-, *M. paranaensis*- or *M. exigua*-infested plantations makes the nematode problem worse, due perhaps to the increase of the ratio nematode population × number of viable roots of the stump (Gonçalves and Silvarolla, 2001).

The apparent adaptation and changes of the *M. incognita* population parasitizing coffee in Brazil may have led to a new pathotype, now described as a new species called *M. paranaensis* (Carneiro *et al.*, 1996). Coincidentally, *M. paranaensis* is widespread in Brazil in regions of the states where *M. incognita* also has a wide distribution. However, *M. paranaensis* is not as pathogenic to soybean as it is to coffee. Sandy soil seems to enhance the damage caused by *M. incognita* in Brazil (Jaehn, 1984). Poor management of the coffee crop has increased the damage caused by *M. exigua*. Sandy soil and organic matter depletion seem to enhance the damage caused by *M. paranaensis* in Brazil (Gonçalves, 2000).

Other hosts

In Brazil, rubber tree (*Hevea brasiliensis*) (Santos *et al.*, 1992), *Grevilea robusta* (Santos, 1988), watermelon, onion (Moraes *et al.*, 1972, 1973a), pepper (Lordello, 1964) and the following weeds found in coffee fields have been reported as hosts of *M. exigua*: *Solanum nigrum* (Curi, 1973), *Ipomoea acuminata*, *I. aristolochiaefolia*, *Stachys arvensis*, *Leonorus sibiricus*, *Amaranthus deflexus*, *Galinsoga parviflora*, *Euphorbia heterophylla*, *Taraxacum officinale* (Lima *et al.*, 1985) and *Citrullus vulgaris* (Ponte, 1977). In *I. acuminata*, *S. arvensis* and *L. sibiricus*, the reproduction of *M. exigua* was higher than in *C. arabica* var. Mundo Novo (Lima *et al.*, 1985). In Rio de Janeiro State, where *M. exigua* was first described by Göldi, it was found in the remaining Atlantic forest (Lima *et al.*, 2003). In Colombia, *Commelina diffusa*, *Hydrocotyle* sp., *S. nigrum*, *Inga* sp. and *Cyperus rotundus* are hosts of *M. exigua* (Aragon *et al.*, 1978). Cocoa is a host of *M. exigua* in Bolivia (Bridge *et al.*, 1982). *Miconia* sp., a tree found on a virgin forest in Juntas de Pacuar, Perez Zeledon county, and *Spananthe paniculata* (weed type) are hosts of *M. exigua* in Costa Rica (Lopez and Vilchez, 1991).

M. incognita has a wide host range, infecting many vegetable, grain and fruit crops, weeds and ornamental plants (Ponte, 1977; Nickle, 1984). In Nicaragua, *Desmodium ovalifolium* suppressed *M. incognita* (Herrera and Marban-Mendoza, 1999). However, *M. coffeicola* has been found only on *Eupatorium pauciflorum* and *Psychotria nitidula* (Lordello and Lordello, 1972a; Jaehn *et al.*, 1980), hence Lordello and Zamith (1960) have hypothesized that this species became a pathogen of coffee after the clearing of forests where it was a native species.

In Brazil, soybean (Castro *et al.*, 2003b), *Ilex paraguariensis* (Santiago *et al.*, 2000) *Ageratum conizoides* and *Emilia sonchifolia* (Roese, 2003) are hosts of *M. paranaensis*. In Guatemala, *Impatiens balsamina*, which is a common weed in coffee plantations, is a good host of *M. paranaensis* and has been used successfully for rearing populations of this nematode in pots (L. Villain, unpublished data).

Disease complexes

The fungus *Rhizoctonia solani* inoculated around plants of *C. arabica* or *C. canephora*, after *M. exigua* infestation, caused more root necrosis and defoliation than when both pathogens were inoculated either simultaneously or separately in the greenhouse (Souza, 1977). Isolations from galled roots and histopathological studies 85 and 115 days after inoculations of nematode-infected plants with *R. solani* revealed extensive fungal colonization within the coffee root systems. The fungus *Fusarium oxysporium* f.sp. *coffea* inoculated on to coffee seedling 4 weeks after *M. incognita* increased chlorosis, root necrosis, wilting and stunting. Fungal hyphae were observed in giant cells and xylem vessels (Negson and Acosta, 1989).

Economic importance

Most information on the economic importance of root knot nematodes comes from Brazil where for over 100 years the areas of cultivation with coffee have migrated across the country due to the pressure of nematode damage. In many instances, these nematodes have been the sole cause for convincing the farmer to cease growing coffee. The economic impact of changing to a new crop after nematode infestation is considerable in terms of financial and socioeconomic implications. Investments made on drying machines, an air-drying fruit yard paved with concrete or devices for peeling the coffee berries, etc. are mostly of little use for another crop.

The impact of the incidence of the major species of root knot nematodes on coffee has shifted throughout the years in Brazil. Göldi (1892) reported on the case of the catastrophic disease on coffee in Rio de Janeiro. Since then, the Brazilian farmers have learned to deal with *M. exigua*, but coffee in Rio de Janeiro was replaced by sugarcane and that state is no longer an important coffee producer.

In Colombia, *M. exigua* and *M. javanica* have caused an estimated loss of US$800 million/year on coffee (Barriga, 1976). In Costa Rica, the attacks of *M. exigua* cause general weakening of the trees, with an estimated drop in yields ranging from 10 to 20% (Bertrand *et al.*, 1997).

A traditional coffee location such as the Alta Paulista region of São Paulo State (Brazil) has been changing to other crops including pasture due to the widespread incidence of *M. incognita* (Curi *et al.*, 1977), but grafted coffee on Apoatã cultivar, which is resistant to *M. incognita* and *M. paranaensis* and immune to *M. exigua* (Fazuoli *et al.*, 2002), revived the coffee business in that region.

The outbreak of *M. coffeicola* in Paraná State, Brazil in 1960 which killed many coffee trees (Lordello and Zamith, 1960) had a great economic impact. However, from 1975 until 1990, *M. incognita* spread widely over the best coffee-planting areas in Brazil north of Paraná and west of São Paulo State, causing the destruction of whole plantations and causing farmers to change their crops. In fact, part of this damage was done by *M. paranaensis* described previously as race 5 of *M. incognita*, which became a new species in 1996 (Carneiro *et al.*, 1996). Even though *M. incognita*, *M. paranaensis* and *M. coffeicola* are more pathogenic to coffee than *M. exigua*, *M. exigua* is probably responsible for the greatest losses to coffee production in Brazil because of its widespread occurrence in the most traditional coffee-producing states such as São Paulo and Minas Gerais (Gonçalves and Silvarolla, 2001). The inefficiency of nematicides in reducing the damage done by *M. incognita*, *M. paranaensis* and *M. coffeicola* in coffee plantations in Brazil forced farmers to eradicate plants and to start over again with a new crop with a resistant rootstock, a costly procedure.

In Guatemala, attacks by *M. paranaensis* leads to serious plant mortality on all current *C. arabica* cultivars from the nursery stage. When they concern plants grafted on to common *C. canephora* rootstocks, major damage with plant decay begins when plants start producing, i.e. 3 or 4 years after planting (Villain *et al.*, 1999). Important damage caused by *M.*

paranaensis has also been observed in Guatemala on commercial *C. canephora* plantations, with a progressive decay beginning with chlorosis and production loss. At the time of pruning of these decayed plants, most do not regenerate and die (L. Villain, unpublished data).

A very important source of loss due to root knot nematode is the total destruction of the coffee seedling enforced by law when root knot is found in nurseries. In São Paulo State (Brazil) 3,231,952 seedlings were destroyed from 1976 to 1977 (Gonçalves *et al.*, 1978).

The different types of losses caused by root knot nematodes can be summarized as: (i) yield decreases; (ii) destruction of seedlings in nursery; (iii) unemployment in traditional coffee-producing areas; (iv) decrease of the farmer's income by cultivation of a less profitable crop; (v) losses of investment on equipment or machines specific for this crop; and (vi) increases in the cost of coffee production due to nematicide application and to the use of grafted coffee seedlings. In Brazil, grafted seedlings are four times higher in price than non-grafted. From the research standpoint, yield loss has tended to preoccupy scientists to the exclusion of the other causes of loss.

Control measures

Control of nematodes in a perennial crop is more difficult than in annual or herbaceous crops. The long-term nature of perennial crops makes rotation schemes, which are successfully used with annual crops, impractical. However, from the standpoint of a long-term management, rotation can be useful for a specific nematode when the crop is renewed. With perennial crops, nematodes that survive the control practices have time to recover and build up to destructive levels. Old plants left in the field, weed hosts or surviving roots of excised plants provide a source of nutrient for nematodes and in part negate the effect of control practices.

The controls of coffee root knot nematodes that are used today by many farmers may be considered under four subgroups:

1. Exclusion, including the measures used to keep the parasite from entering the soil in which the host is growing.
2. Application of nematicides, for the elimination or reduction of the parasite level after it has become established in the soil where the host is growing.
3. Grafting on resistant or tolerant cultivars.
4. Other measures under research: breeding coffee for resistance, rotations in areas where old coffee plants have been eradicated, increase of soil organic matter to decrease losses, biological control.

EXCLUSION. In Brazil, the impediment to the movement of infested seedlings into new growing areas was more effective in the past than today. Initially the government financed new coffee plantations by subsidies and imposed the use of new technology and prohibited the planting of coffee: (i) in the area previously planted with coffee or even close to the area; (ii) from seedlings infested with nematodes; and (iii) in regions not recommended for growing this crop. Since 1980, this subsidy has been withdrawn and the government lost their control over planting new coffee plantations. Now the grower has to look independently for information on new technologies from the extension service network, universities, government research companies or other sources. However, the inspection of coffee nurseries in Brazil is still maintained and the law regarding destruction of the infested seedlings is always enforced. In Minas Gerais State, to grow coffee seedlings, producers must have a certificate stating the absence of *Meloidogyne* issued by an official Nematology Laboratory.

The production of seedlings without root knot has relied on using soil in nurseries gathered from areas never previously grown with coffee, especially where pasture is currently grown. Historically, this soil has been sterilized with methyl bromide at the rate of 150 cm^3/m^3 of soil (Moraes *et al.*, 1977b), placed under a plastic cover for 3–4 days and then aerated for 10 days before seeding. Alternative methods of sterilizing soil include the uses of steam and exposure of nursery soil to sun for many weeks during the dry season (Bridge, 1984). The length of time of sun exposure can be reduced by using a sun heat collector (Ghini and Bettiol, 1991); treating infested soils for 2 days with the sun heat collector can reduce populations of *Meloidogyne* spp. in soil to as low as obtained by using methyl bromide (Ghini *et al.*, 1991; Randig *et al.*, 1998). The water source has to be carefully selected, avoiding dams in which runoff water comes from hillsides cultivated with infested coffee plants. Infected seedlings with root knot nematodes should be burned and under no circumstances should they be planted into an area free of damaging nematodes.

The place to establish a new coffee crop has to be very carefully selected, avoiding the recently eradicated old coffee plants, as well as in the proximity of an infested field or on a site at a level below it, where the risk of contamination from runoff water is high. Sometimes a furrow has to be dug to prevent runoff water getting into the infested area. Care has to be taken to wash machines or farm implements used, or that have travelled through infested fields.

NEMATICIDES. Chemicals used today to control nematodes on coffee as on other crops have been mostly restricted to contact or systemic granular products. From the group of fumigants used for controlling nematodes in the past (Anonymous, 1968), methyl bromide has been the most widely used to disinfest nursery soil, but there are now international restrictions on the use of this fumigant.

The systemic insecticides, the organophosphate and organocarbamate chemicals, that have potential for nematode control are rarely phytotoxic at concentrations used for field control. The major disadvantages are that they are water dispersed. Nematicidal activity is usually confined to a shallow root zone or rhizosphere, and is often a result of narcotization and nematode behaviour modification rather than killing. However, disruption of nematode infection and their development and reproduction can temporarily slow or halt increases in nematode numbers. These chemicals give little or no control of fungal or bacterial disease but do provide insecticidal activity depending upon the chemical involved (Van Gundy and McKenry, 1977).

In general, the effective rates of aldicarb, carbofuran, phenamiphos and terbuphos will be in the range of 1.6–6.0 g a.i./plant, in one or two applications during the year. The first application should be at the beginning of the rainy season, followed by the second 3 months later; each time the soil should be wet for the application. A furrow is dug along both sides of the plant row close to the tree where the product is applied and incorporated into the soil, by machine or by hand.

Application of systemic or contact granular nematicides on severely damaged coffee plants, especially those infested by *M. incognita*, has not been effective due to the rapid destruction of large parts of the root system by the nematode (Curi *et al.*, 1977). Poor control also occurs on seedlings infested by *M. incognita* (Jaehn *et al.*, 1984). The yield obtained in *M. incognita*-infested coffee treated with nematicides is far lower than in plants grown in non-infested soil (Gonçalves and Silvarola, 2001). For other *Meloidogyne* species causing similar symptoms to those of *M. paranaensis* and *M. coffeicola*, the use of nematicide as a control measure is not recommended.

For most of the typical gall-forming *Meloidogyne* species, many granular nematicides are effective in decreasing nematode populations a few months after application (Huang *et al.*, 1983). After this time, the populations may increase on treated plants, but the plants have good foliage which seems to be induced by some other action besides controlling the nematodes (Campos and Lima, 1986). New nematicides have been tested for their efficacy against *M. exigua* (Volpato *et al.*, 2001), and some of them have potential to control coffee nematodes (Zem, 1993).

M. exigua-infested coffee treated for 5 consecutive years with nematicides produced 30.9% higher yields than non-treated coffee infested by *M. exigua*. However, nematicide does not eradicate the nematode (Lordello *et al.*, 1990). *Meloidogyne* infection on coffee roots reduces the uptake of fungicides against coffee rust applied via the soil (Otoboni *et al.*, 2001, 2003b).

Granular nematicides when applied in coffee have to be incorporated into the soil under the edge of the foliage toward the stem. Different machines have been developed to do this work. Timing of the application is important since the granular products require water to liberate the active ingredient, therefore application at the beginning of the rainy season (November, in Brazil) is recommended (Campos *et al.*, 1985).

GRAFTING. The widespread distribution and the aggressive parasitism of *M. incognita* in the west of São Paulo has forced researchers in Brazil to seek an efficient control measure other than chemicals. An introduction of *C. canephora* cv. 2258 from the CATIE germplasm collection, Turrialba, Costa Rica, showed high resistance to *M. exigua* and resistance and/or tolerance to several populations of *M. incognita* (Fazuoli, 1986), resistance to race 1, 2 and 3 of *M. incognita* (Gonçalves *et al.* 1996), and to *M. paranaensis* (Fazuoli *et al.*, 2002). The level of resistance of the cv. 2258 was initially 70% but, through selections in the field highly infested with *M. incognita*, this rate was significantly increased. This improved line for rootstock use is resistant to *M. incognita* and *M. paranaensis*, and immune to *M. exigua*, and has been named Apoatã (Fazuoli *et al.*., 2002). *C. arabica* cv. Mundo Novo grafted on to *C. canephora* Apoatã yielded 3.6 times as much as non-grafted plants grown on fields infested with *M. incognita* race 1 (da Costa *et al.*, 1991). In Brazil, in the states highly infested with *M. incognita* and *M. paranaensis* such as São Paulo and Paraná, the planting of grafted coffee is widespread in non-infested areas (Fig. 14.3). In some counties, especially in the west of São Paulo State, the grafted coffee is reviving the coffee business (Campos, 1997). Gonçalves (1995) advised coffee growers not to grow susceptible coffee cultivars in areas infested by *M. incognita* because plants of *C. arabica* will not survive (Fig. 14.4). In short, grafted *C. arabica* on Apoatã rootstock is the only feasible control measure to make possible eco-

Fig. 14.3. Four-year-old grafted *Coffea arabica* on Apoatã rootstocks (*C. canephora*) planted in field naturally infested with *Meloidogyne incognita* (São Paulo State, Brazil). Dead *C. arabica* between the stakes were not grafted. (Photo: V.P. Campos.)

Fig. 14.4. Grafted seedlings of *Coffea arabica* on Apoatã rootstocks. (Photo: V.P. Campos.)

nomic growing of coffee in areas with infested *M. incognita* or *M. paranaensis* in Brazil. However, Apoatã rootstock showed intolerance to *Pratylenchus brachyurus* in greenhouse tests (de Oliveira, 1996). The same *C. canephora* line that was the origin of the Apoatã cultivar in Brazil is also the origin in Central America, as T3561 (2-1) according to the CATIE germplasm collection nomenclature, of a new rootstock cultivar named Nemaya (Anzueto *et al.*, 1996) by crossing with another root knot nematode multiresistant *C. canephora* line, T3751 (1-2) (Fig. 14.5). This rootstock cultivar showed resistance to *M. paranaensis* from Guatemala, at that time identified as *M. incognita*, as well as an undescribed *Meloidogyne* sp. (four-band esterase pheno-

Fig. 14.5. Variability of host status in *Coffea canephora* for a population of *Meloidogyne paranaensis* from Guatemala illustrated by two clone crossing progeny: on the left with one of the parent resistant clones (parent of hybrid rootstock cv. Nemaya) and on the right with two susceptible clones. (Photo: L. Villain.)

type) from El Salvador (Bertrand *et al.*, 2000b). This rootstock cultivar Nemaya also has resistance to *M. exigua* from Costa Rica and *M. incognita* from Nicaragua (Anzueto *et al.*, 1996).

The possibility of using *C. arabica* as a commercial rootstock has been reached with the finding of resistance to *M. paranaensis* in Ethiopian *Coffea arabica* accessions (Anzueto *et al.*, 2001). However, breeding for resistance in *C. canephora*, *C. congensis* and *C. dewevrei* to produce rootstocks resistant to nematodes is of more interest to breeders since these species have abundant root systems, in addition to having a good source of resistance to other pathogen groups (Gonçalves and Silvarola, 2001). However, resistance genes found in wild or semi-wild lines of *C. arabica* from Ethiopia or Yemen could be used in coffee roostock breeding programmes for implementing interspecific hybridizations with resistant diploid *Coffea* spp. lines. For example, Arabusta (*C. canephora* × *C. arabica*) as rootstock germplasm, in addition to the nematode resistance, also has the advantage of a better adaptability than *C. canephora* to the cooler coffee-growing areas and has a very good vigour (Capot, 1972; Berthaud, 1978a,b).

RESISTANCE IN *C. ARABICA*. Resistance to *M. incognita* races, *M. exigua*, *M. paranaensis* and *M. coffeicola* has been found in many Brazilian coffee germplasm lines (Moraes *et al.*, 1973b; Fazuoli and Lordello, 1978; Medina Filho *et al.*, 1981; Fazuoli, 1986; Gonçalves and Ferraz, 1987; Manetti Filho and Carneiro, 1995; Gonçalves *et al.*, 1998b), which makes it possible in the future to obtain better resistant cultivars for either rootstock or direct planting in the field. Several lines of *C. canephora* and *C. congensis* have shown resistance to race 3 of *M. incognita*, and some progeny of *C. canephora*, Sarchimor (derived from crossing Vila Sarchi × Timor Hybrid) and Icatu (advanced line derived from crossing *C. arabica* × *C. canephora*), have shown moderate resistance. Work also has been done on resistance to *M. exigua* (Curi *et al.*, 1970). Different levels of resistance to *M. exigua* have been found on progeny derived from crossing *C. arabica* and *C. canephora* (Ribeiro *et al.*, 2001). Forty-two progeny of Timor hybrid, derived from crossing *C. arabica* × *C. canephora*, were resistant to *M. exigua*. In some of them, eggs were not produced, showing the same behaviour as the parent *C. canephora* and were also resistant to coffee rust (*Hemileia vastatrix*) (Gonçalves *et al.*, 1998b), suggesting the possibility of simultaneous selection for *M. exigua* and *H. vastatrix* resistance (Gonçalves and Pereira, 1998). Among 83 progeny derived from crossing *C. arabica* × *C. canephora*, two of them were immune and homozygous for resistance to *M. exigua*. Four hybrids from crossing Icatu × Sarchimor were resistant. The resistance of *C. canephora* is indeed transferred which does not occur by hybridization with *C. dewevrei* (Silvarola *et al.*, 1998).

Resistance to *M. exigua* has been found in hybrids and progeny from Catuai × Icatu crosses in Honduras (Pineda and Santacreo, 2000; Zelaya-Escoto and Santacreo, 2000). In Costa Rica, 29 wild Ethiopian accessions were evaluated for their resistance to *M. exigua* and all were susceptible, while 14 *C. canephora* accessions showed a high resistance to this nematode (Bertrand *et al.*, 1997, 2001). The same authors observed resistant, susceptible and segregated progeny among Catimor and Sarchimor cultivars, showing that *C. canephora*-introgressed DNA fragments conserved in the lines derived from Timor hybrid were different. These results led to the creation of a *C. arabica* F_1 hybrid by crossing resistant Sarchimor with some Ethiopian lines, combining, for example, resistance genes to *M. exigua* and coffee leaf rust (*H. vastatrix*) with resistance genes to *M. paranaensis* (Bertrand *et al.*, 1999). These F_1 hybrids are currently under selection and field evaluation, but some have already shown satisfactory resistance to leaf rust and some root knot nematodes as well as a high productivity and also a cup quality slightly superior to that of the parents (Montagnon *et al.*, 2002). Resistance to *M. exigua* is controlled by a simply inherited major gene, designated the *Mex-1* locus, in *C. canephora* with, possibly, incomplete dominance (Noir *et al*., 2003). This, the first identified gene of nematode resistance in coffee, represents an important starting point to enhance backcross breeding programmes and thus to perform early marker-assisted selections of resistant seedlings. The *Mex-1* markers may be used in the future for constructing resistant coffee genotypes by holding more than one identified source of nematode resistance. On the other hand, all germplasm of *C. arabica* tested is susceptible to race 3 of *M. incognita* (Gonçalves and Ferraz, 1987). Only tolerance was found in some of the 61 progeny of Icatu coffee tested in the field infested with *M. incognita* race 2 and it is not safe to recommend them for growers (Carneiro, 1995).

Progeny obtained from Icatu vermelho IAC 4160 resulting from crossing *C. arabica* × *C. canephora* were resistant to *M. paranaensis* (Gonçalves *et al*., 1998a). Many *C. arabica* progeny have been tested for resistance to *M. paranaensis* (Mata *et al*., 2000a,b). The best source of resistance to *M. paranaensis* is *C. canephora* and cultivar lines bred from it, such as the Icatu cultivar and lines resulting from crossing the cultivars Catuai × Icatu.

CROP ROTATION AND INTERCROPPING. Moraes *et al*. (1977a) studied rotation with cotton, soybean and maize in *M. exigua*-infested areas and concluded that, after 1 year's rotation with these crops, the grower can return to coffee cultivation. Almeida and Campos (1991a,b) studied rotation with bean, soybean, sorghum and *Panicum maximum* in *M. exigua*-infested areas and also concluded that a 1 year rotation with these crops makes possible a return to cropping susceptible coffee cultivars. However, Carneiro and Carneiro (1982b), who screened 29 crops for rotation in *M. incognita*-infested coffee fields, found that only *Arachis hypogea* and *Ricinus communis* were immune. *Styzolobium deeringianum* and *Crotalaria spectabilis* showed resistance to this nematode. No penetration, galls or egg masses of *M. paranaensis* and *M. incognita* races 1, 2, 3 and 4 were found in *Arachis pintoi* roots by artificial inoculation, which makes it a potential crop for rotation (Santiago *et al*., 2001). In summary, a 1 year rotation in areas infested with *M. exigua* or *M. coffeicola* makes possible the return of planting coffee cultivars susceptible to these species, but this is not possible where land is infested with *M. incognita* or *M. paranaensis*. On the other hand, rotation with non-hosts of *M. incognita* and *M. paranaensis* is a useful tactic to decrease their populations in infested areas before planting coffee grafted on Apoatã, since its resistance is not complete.

Intercropping velvet bean between rows of coffee and incorporation at flowering stage protects coffee from cold wind and improves soil texture, organic matter and nutrients, and thereafter decreases the damage by *M. incognita* and *M. paranaensis* in sandy and soils depleted in organic matter in the west of São Paulo State in Brazil (Fazuoli *et al*., 2002). Replacing organic matter in depleted soils can be a tool to delay coffee eradication due to damage by nematodes.

BIOLOGICAL CONTROL. Biological control is a promising tactic for management of coffee nematodes especially for organic coffee where chemical use for production is prohibited and which has a higher market price. The bacterium *Pasteuria penetrans*, among all nematode antagonistic microorganisms, has the advantage of resistance to heat, drought and pesticides in the field (Campos *et al*., 1998). *P. penetrans* was first found in coffee fields by Baeza-Aragon (1978) in Colombia and by Sharma and Lordello (1992) in Brazil. In Brazil, 21–65% of the second stage juveniles of *M. exigua* in coffee fields were naturally infested by *P. penetrans* throughout the year (Maximiniano *et al*., 2001). In Cuba, isolates of *Verticillium chlamydosporium* isolated from coffee plantations have potential as biological control agents for root knot nematodes of coffee (Hidalgo *et al*., 2000). In Brazil, predators and egg parasitic fungi have been isolated from coffee plantations (Naves and Campos, 1991; Ribeiro and Campos, 1993). The efficacy of *Arthrobotrys conoides*, *A. musiformes*,

Paecillomyces lilacinus and *Verticillium chlamydosporium* in the control of *M. exigua* in coffee was determined by Campos and Campos (1997).

Among the root knot nematodes of coffee in Brazil, *M. incognita* and *M. paranaensis* cause the greatest losses and have become a limiting factor to growing coffee in certain areas due to their greater capacity to destroy the root systems, their ease of dissemination, their high persistence in soil, the inefficiency of chemical control measures and the presence of different biological races in *M. incognita*.

Methods of diagnosis

Diagnosis of the occurrence of *M. exigua* in the field is not difficult because this nematode induces typical rounded galls on roots of infested coffee plants (Plate 15A); this is not the case with *M. incognita*, *M. coffeicola* and *M. paranaensis*. With these latter species, laboratory diagnosis is required to search for *M. incognita*, *M. coffeicola* or *M. paranaensis* in non-galled sections of the root system. *M. coffeicola*, *M. incognita* and *M. paranaensis* are mostly found in older sections of the root, especially the principal root. However, in all cases, the current sampling extraction procedures can be used to recover the second stage juveniles in soil, which helps to identify the disease in combination with the symptomatology on the plant. The perineal pattern is very helpful in the identification of most *Meloidogyne* species but not to separate *M. incognita* from *M. paranaensis*, which is possible by electrophoresis (Carneiro *et al.*, 2000). The elongated form of *M. coffeicola* (Fig. 14.2A) separates it from *M. incognita* and *M. paranaensis* that are all found in older root tissues (Plate 15C). Electrophoresis also allows the detection and identification of non-reported or undescribed root knot nematode species that could be present in the collected coffee root samples, which makes this diagnosis essential for any survey study.

Molecular markers, in this case sequence-characterized amplified region (SCAR) markers originated from transformed RAPD sequences, have been developed for the major *Meloidogyne* species parasitizing coffee in Brazil, i.e. *M. exigua*, *M. incognita* and *M. paranaensis*, which will lead to the production of kits for species identification (Randig *et al.*, 2001, 2002b). Isolation of a species-specific satellite of the *M. exigua* DNA enabled a procedure to identify this species with a single specimen even of juveniles present in the sample to be developed (Randig *et al.*, 2002a). In the case of root knot nematode mixtures, occurring very frequently in field samples from some areas, species detection threshold by SCAR markers is about 1% of the total community population; this is very interesting for *M. exigua* detection because of the species' poor esterase activity (O. Randig and R.M.D.G. Carneiro, unpublished data). Therefore, these DNA diagnostic tools have the advantage of greater accuracy, the possibility to detect almost any stage of the nematode compared with electrophoresis analysis with only females, and an easy and relatively quick procedure without the necessity for a well-equipped molecular laboratory. DNA diagnostic procedures for all other *Meloidogyne* species found in coffee in Brazil are currently being studied, especially the development of SCAR-RAPD markers (Randig *et al.*, 2002a,b). In the future, perhaps all coffee root knot nematodes will be identified by DNA diagnostic methods. Meanwhile, both isoenzyme electrophoresis analysis and molecular markers seem to be necessary and complementary to conduct root knot nematode surveys in coffee-growing areas.

Meloidogyne africana, M. decalineata, M. megadora, M. hapla, M. arenaria, M. kikuyensis, M. inornata, M. javanica, M. oteifae, M. thamesi, M. arabicida, M. konaensis and *M. mayaguensis*

Distribution

Even though relatively few surveys have been done in Africa to provide a good picture of the distribution of nematodes in different countries where coffee is grown, the data available suggest that *M. africana*, *M.*

decalineata, *M. kikuyensis* and *M. megadora* are apparently restricted to relatively few African countries. In Tanzania and Zaire, where more data are available, many species of *Meloidogyne* occur in coffee (Table 14.1), but the other species of this group seem to have restricted ecological requirements limiting their occurrence. *M. decalineata* was the predominant species in Kilimanjaro and the Usambra mountains of northern Tanzania (Swai, 1981). *M. kikuyensis* was also reported from coffee in the region of Kilimanjaro (Swai, 1981). *M. africana* is widespread in Kenya and Zaire (Whitehead, 1959; Lordello, 1972). Bridge (1984) reported the occurrence of *M. decalineata* and other species of *Meloidogyne* in different areas of Tanzania. *M. megadora* is found in Angola and Uganda (Whitehead, 1968a, 1969a).

Meloidogyne has also been found on coffee in Zimbabwe (Way, 1981). *M. hapla* and *M. javanica* are rarely found on coffee in Tanzania, suggesting that there is some resistance in coffee to these species (Whitehead, 1969a,b; Bridge, 1984). *M. hapla* was also detected in a few coffee fields in Guatemala and El Salvador (Hernández, 1997; Villain *et al.*, 2002). *M. arenaria* has been found on coffee in Jamaica (Anonymous, 1963, in Whitehead, 1969b) and more recently in El Salvador and Guatemala (Hernández, 1997; Sarah, 2003). *M. oteifae* occurs in Zaire (Elmiligy, 1968) and *M. inornata* in Guatemala (Schieber and Sosa, 1960). *M. thamesi* has been found in coffee soil in India (Kumar, 1984). *M. arabicida* occurs in Costa Rica with a distribution that seems to be still restricted to a small region around Juan Viñas county (López and Salazar, 1989), *M. konaensis* in the USA (Hawaii) (Eisenback *et al.*, 1995) and *M. mayaguensis* in Cuba (Sampedro *et al.* 1989) as well as in Guatemala and Costa Rica (Sarah, 2003). According to Carneiro (2003), in other hosts, *M. mayaguensis* is widespread in many African countries (Mali, Senegal, South Africa, Côte d'Ivoire and Burkina Faso), as well as in the Americas and Caribbean (Trinidad and Tobago, Martinique, Puerto Rico, Cuba, continental USA, besides Brazil).

Symptoms of damage

M. oteifae forms galls of moderate size on roots of *C. robusta* (Elmiligy, 1968). *M. africana* and *M. decalineata* usually cause small, mainly root tip, galls from 1 to 5 mm in diameter (Fig. 14.6). Affected seedlings are generally stunted, with numerous

Fig. 14.6. Root galls caused by *Meloidogyne decalineata*. (Photo: J. Bridge.)

rootlets behind the affected root tip (Whitehead, 1959). Heavy infestations in mature trees were associated with general unthriftiness, but the nematodes may not have been wholly responsible for this (Whitehead, 1969a,b). *M. africana* attacks *C. arabica* in Kenya, causing poor growth of coffee seedlings (Whitehead, 1959; Anonymous, 1977), and *C. robusta* in Zaire (Whitehead, 1969b). *M. decalineata* causes root galls in *C. canephora* and *C. arabica* in nurseries as well as yellowing of coffee leaves and reduction of plant growth in the field (Lordello and Fazuoli, 1980).

M. hapla causes a slight root galling and swellings in coffee different from other species which occur in Tanzania (Bridge, 1984) (Table 14.1). In Brazil, it causes typical galls with different diameters close to *M. exigua*. Necrosis and induction of lateral roots are also observed close to the nematode galls (Lordello, 1982).

M. arabicida causes numerous swellings evolving into extensive developments of corky tissues on the taproot up to the collar, and on the primary and secondary roots. Considerable cracking of the cortical tissues is also observed (Lopez and Salazar, 1989; Bertrand *et al.*, 2000a). These symptoms are very similar to those observed on coffee trees infected by *M. paranaensis* in Brazil or Guatemala. As with *M. paranaensis*, *M. arabicida* female development leads to peridermal disruption with exterior egg masses (Bertrand *et al.*, 2000a). In the field, infected plants show a progressive decline with leaf chlorosis and leaf fall followed by flowers and fruit fall, leading to plant death within 2–4 years after planting. This species presents a characteristic esterase phenotype (M1F1b) (Hernández, 1997).

M. konaensis causes galls on the roots of infested coffee plants, reduces the proportion of fine roots per root system by 50% and reduces NO_3^- and NH_4^+ uptake by 63 and 54%, respectively (Vaast *et al.*, 1998). It also causes the reduction of shoot and root dry weights of many coffee cultivars (Zhang and Schmitt, 1995a).

M. mayaguensis is the most damaging species among all that occur in Cuba (Rodríguez *et al.*, 1995). In the field, symptoms observed are scattered galls or gall strings on superficial roots, as in the case of *M. exigua* infestations, but also cracking of the cortical tissue on the stem collar.

Biology and life cycle

The development of second stage juveniles of *M. konaensis* to mature females requires 38 and 48 days on coffee at 30 and 26°C, respectively (Zhang and Schmitt, 1995b).

Other hosts

M. arenaria, *M. javanica* and *M. hapla* are found infecting a great number of crops and weeds in many countries of the world (Ponte, 1977; Nickle, 1984). In Africa, *M. africana* is found infecting maize, cowpea, clove, potato, pyrethrum; *M. megadora* in many coffee species; and *M. kikuyensis* in cowpea (Whitehead, 1969a). In Brazil, *M. thamesi* is found infecting cocoa, *Turnera ulmifolia* L., *Spondias lutea*, *Rivina humilis*, *Petiveria hexaglochin* Fisch and Mey and *Leonorus sibiricus* (Ponte, 1977; Lordello, 1984) and *M. inornata* infecting soybean (Ponte, 1977).

In the USA (Hawaii), *Paspalum conjugatum* Berg, *Amaranthus viridis* L. among many cultivated crops are hosts of *M. konaensis* according to Zhang and Schmitt (1994).

In Puerto Rico, aubergine (*Solanum melongena* L.) and tomato (*Lycopersicum esculentum* Mill) are hosts of *M. mayaguensis* (Rammah and Hirschmann, 1988). In Brazil, guava (*Psidium guajava*) is a host of *M. mayaguensis* (Carneiro *et al.*, 2001). In South Africa, commercial crops are hosts of *M. mayaguensis* (Willer, 1997) and it seems to be a polyphagous species (Carneiro, 2003).

Disease complexes

Under controlled inoculation conditions, only the presence of both *M. arabicida* and *F. oxysporum* resulted in corky root symptoms on *C. arabica* cvs Caturra or Catuai. In these controlled conditions, the nematode *M. arabicida* alone only causes gall formation without corky root symptoms. In fields planted with susceptible and resistant cultivars to *M. arabicida*, only susceptible varieties develop the corky root symptoms. Thereafter, predisposition of the plant by *M. arabicida* has a dominant role in this complex aetiology (Bertrand *et al.*, 2000a).

In coffee plants showing corky roots in Mexico, in addition to the *Meloidogyne*, *Pratylenchus* and *Helicotylenchus* present in roots, *Fusarium* and *Trichoderma* were also isolated (Teliz-Ortiz *et al.*, 1993), suggesting a disease complex as shown in Costa Rica.

In the case of *M. arabicida*, *M. paranaensis*, *M. incognita* or *M. coffeicola* parasitism, as observed in different histological studies (Anzueto, 1993; Bertrand *et al.*, 2000a), the development of females close to the surface of the roots with rupture of the cortex, which leads to egg masses emerging out of the root (contrary to *M. exigua* parasitism), may favour the subsequent invasion of secondary pathogens such as *Fusarium*, leading to cracking and corky root symptoms.

Economic importance

Although there is no information available in Tanzania on the actual yield losses caused by nematodes, it is estimated that yield losses of trees severely infested with the African coffee root knot nematodes will be in the region of 20% in optimum conditions, extending to the point of non-productivity (Bridge, 1984). The stress to which trees are subjected because of nematode damage will also cause premature fruit drop, twig dieback and defoliation, nutrient deficiency symptoms and stunted growth. Although *M. arabicida* distribution is restricted to a small area in Costa Rica and has not yet been detected today in other regions or countries, this parasite represents a potential threat for already safe coffee-growing regions. Locally where this nematode is present, its economic impact is so severe that it has led to abandonment of the coffee crop since its appearance in 1978 (Anonymous, 1989).

Management measures

The control measures described for *M. exigua*, *M. incognita*, *M. coffeicola* and *M. paranaensis* are likely to be effective for the control of African root knot nematodes, but application of these measures on a practical basis in African countries is uncertain. However, a test of different *Coffea* species, crosses and selections against root knot nematodes in Tanzania done by Bridge (1984) indicated that some resistance may occur and grafting on to resistant rootstocks could also prove useful in these countries. Reports from the Kenya Coffee Research Station, cited by Whitehead (1968b), suggest resistance to *Meloidogyne* sp. in *C. corrisoi*, *C. conuga* and some lines of *C. congensis* in Angola. Whitehead (1969b) said that coffee is very resistant to both *M. javanica* and *M. kikuyensis*.

Bertrand *et al.* (2002) studied the inheritance of the disease complex known as corky root, composed of the root knot nematode *M. arabicida* and *F. oxysporum*, which causes damage to *C. arabica* in Costa Rica. The resistance to corky root in coffee is heritable. The genetic resistance to *M. arabicida* is an effective strategy against corky root disease complex. By using *C. canephora* rootstocks, it was possible to substantially reduce mortality in the field and reduce by half the number of plants with corky root symptoms. *C. liberica* var. *dewevrei* is resistant to *M. konaensis* (Serracin and Schmitt, 2002). The *C. canephora* rootstocks of Nemaya variety have resistance to *M. arabicida* of Costa Rica (Anzueto *et al.*, 1996). The distribution of *M. arabicida* is still limited to one small region of Costa Rica but, because of its high damaging potential, coffee growers have been warned by Costa Rican authorities to prevent dissemination of this species by transport of infested seedlings.

Pratylenchus

The lesion nematodes, *Pratylenchus* spp., currently known to occur on coffee are *P. coffeae*, *P. brachyurus*, *P. goodeyi*, *P. pratensis*, *P. loosi*, *P. panamaensis* (= *P. gutierrezi*), *P. zeae* and *P. vulnus*.

Distribution

P. coffeae, initially described on coffee in Java by Zimmermann (1898), is still the most widely reported root lesion nematode species in coffee worldwide.

P. coffeae was found in Guatemala (Chitwood and Berger, 1960; Schieber and Sosa, 1960; Schieber, 1966, 1971), El Salvador, where it is considered as the predominant species of nematode on coffee (Abrego and Holdeman, 1961; Whitehead, 1969b; Gutierrez and Jimenez, 1970), Costa Rica (Salas and Echandi, 1961; Tarjan, 1971; Figueroa and Perlaza, 1982), Colombia (Obregon and Rafael, 1936, cited by Sylvain, 1959), Venezuela (Flores and Yépez, 1969) and Brazil, particularly in São Paulo State (Monteiro and Lordello, 1974; Kubo *et al.*, 2002a) but also in Pernambuco State (Moura *et al.*, 2003). *P. coffeae* also occurs in Hawaii (Schenk and Holtzmann, 1990).

In the Caribbean, *P. coffeae* was detected in the Dominican Republic (Schieber and Grullon, 1969), Martinique (Kermarrec and La Massese, 1972), Cuba (Sampedro *et al.*, 1989; Fernandez and Ortega, 1998) and Puerto Rico (Ayala, 1976).

In Asia, besides Java, its typical site (Zimmermann, 1898; Sher and Allen, 1953), where *P. coffeae* became a very damaging and major pest on coffee (Whitehead, 1968b), it was also reported in the Indochina region (Whitehead, 1968a) and India (Palanichamy, 1973).

In Africa, *P. coffeae* is reported in the Democratic Republic of the Congo (Bredo, 1939, cited by Sylvain, 1959) and Tanzania (Bridge, 1984). It also occurs in Madagascar (Whitehead, 1968a). Bridge *et al.* (1997) suggested that this nematode, which has a large pantropic distribution, may have the same geographical origin as banana and plantain, i.e. the Pacific islands and neighbouring Asian countries, from which it may have been spread through plant material transfers.

For a long time, *P. brachyurus* was the only *Pratylenchus* species known to infect coffee in South America (Lordello, 1972). *P. brachyurus* has been found in many regions in Brazil and, to date, it seems to be the most widely distributed root lesion nematode in this country (Lordello and Mello Filho, 1969; Gonçalves *et al.*, 1978; D'Antonio *et al.*, 1980; Campos and Lima, 1986; Campos, 2002; Lima, 2002). In São Paulo State, Brazil, *P. brachyurus* was more widespread than *P. coffeae* (Gonçalves *et al.*, 1978; Kubo *et al.*, 2001, 2002a). In Bauru and Marília counties, about 46% of the collected samples had *P. brachyurus* (Kubo *et al.*, 1999). In Minas Gerais State, Brazil, *P. brachyurus* was found in 20% of the counties sampled (D'Antonio *et al.*, 1980). In the Zona da Mata region of this state, 17% of the collected samples had *P. brachyurus* (Lima, 2002). In the South region of Minas Gerais State, *P. brachyurus*, *P. zeae* and *P. coffeae* were found (Campos, 2002). *P. brachyurus* has also been reported on coffee in the Côte d'Ivoire and Peru (Whitehead, 1968b) and in Hawaii (Schenk and Holtzmann, 1990).

P. pratensis has been reported from one locality in south India by Somasekhar, cited by Whitehead (1968b), and *P. loosi* from Sri Lanka by Hutchinson, cited by Whitehead (1968b). *P. goodeyi* occurs on coffee in Tanzania (Bridge, 1984). *P. vulnus* and *P. zeae* occur on coffee in Brazil (Ferraz, 1980; Monteiro *et al.*, 2001).

Two new species morphologically close to *P. coffeae* were described subsequently on coffee in Panama and Costa Rica, *P. panamaensis* (Siddiqi *et al.*, 1991) and *P. gutierrezi* (Golden *et al.*, 1992), respectively. Neither species has been reported from other sites. Taxonomically, Siddiqi (2000) considers these two species as synonyms. Their pathogenicity on coffee is not known.

Because of the species identification difficulties within this stenomorphic genus (Luc, 1987), many *Pratylenchus* populations have not been identified. For this reason, there are many reports of unidentified species of root lesion nematodes including those in many coffee-growing areas of the world. For example, *Pratylenchus* sp. populations have been reported in many coffee-growing regions of Nicaragua (Sequeira-Bustamente *et al.*, 1979) and Cuba (Sampedro *et al.*, 1989), with severe field damage observations in both countries. In Honduras, a survey in the border region of El Paraiso next to Nicaragua revealed the presence of *Pratylenchus* sp. in 58% of the 860 coffee farms sampled (Padilla and Tronconi, 2002). In Guatemala, surveys revealed that *Pratylenchus* spp. are much more common than *Meloidogyne* spp. and are present in all the coffee-growing regions of this country (Villain *et al.*, 1999; Villain, 2000). *Pratylenchus* spp. are also detected with high frequency in many coffee-growing regions of Costa Rica (Araya, 1994).

P. coffeae and related species complex

Morphological, biological and molecular studies have questioned the taxonomic position of several amphimictic *Pratylenchus* isolates collected on coffee in Central America and Brazil (Villain *et al.*, 1998; Duncan *et al.*, 1999; Hervé, 1997; Villain, 2000; Siciliano-Wilcken *et al.*, 2002a,b).

In Guatemala, three amphimictic and reproductively isolated species have been revealed on coffee, and their identification or description currently is in progress (Villain *et al.*, 1998; Villain, 2000). Five populations from Guatemala, one from El Salvador and one from Costa Rica may belong to the same species because of their interbreeding (between Guatemalan populations) and/or morphological (principally for their divided face pattern observed under a scanning electronic microscope) and genetic similarity (Hervé, 1997; Villain *et al.*, 1998; Duncan *et al.*, 1999; Villain, 2000). This could be a species widely distributed throughout Central America. Moreover, Guatemalan populations belonging to this same species have been observed within a wide range of altitudes, from 450 to 1200 m, and appear to be well adapted to most climatic conditions of the different coffee-growing regions (Villain, 2000). These populations are morphologically different from *P. coffeae sensu lato* and *P. loosi* but similar to *P. pseudocoffeae* (Mizukubo, 1992b), while the population from Costa Rica proved to be genetically distant from the topotypes of *P. gutierrezi* (Villain *et al.*, 1998; Duncan *et al.*, 1999). Two other isolates collected on coffee in Guatemala are morphologically close to *P. coffeae* as described, but do not interbreed. Furthermore, one has a very high degree of pathogenicity on *C. arabica* compared with the other (Villain *et al.*, 2002). Their distribution is still unknown.

In Brazil, some uncertainties on the taxonomic position also persist for certain *Pratylenchus* populations collected on coffee (Duncan *et al.*, 1999; Siciliano-Wilcken *et al.*, 2002a,b) with the recent description of *P. jaehni* (Inserra *et al.*, 2001), a species to which populations collected on coffee are morphologically close.

More globally, different studies on *P. coffeae* and morphologically related species collected on different hosts and different geographic areas led to reconsideration of the taxonomic positions in this amphimictic and morphologically closed species complex (Mizukubo, 1992a,b; Inserra *et al.*, 1998, 2001; Mohotti, 1998; Villain *et al.*, 1998; Duncan *et al.*, 1999; Villain, 2000). Most identifications of these species, including those of *P. coffeae*, frequently are based only on morphological observations under the optic microscope and should therefore be treated with caution. Clarifying the taxonomic status of the *Pratylenchus* populations found on coffee means revising first the species already described with an integrated approach through a complete morphological, biological and molecular characterization and applying the biological concept of species. This includes *P. coffeae* itself and, with the aim of a revision of these species, P. Baujard (1999, personal communication) achieved a survey in Java collecting *Pratylenchus* from many coffee-growing sites including the putative collection site of Zimmermann (1898).

Symptoms of damage

Roots of coffee infected by *P. coffeae* turn yellow then brown, and most lateral roots become rotten. Infected plants look stunted and have a few small chlorotic leaves. The earliest symptoms of infection in the newly transplanted trees are yellowing of leaves, loss of young primary branches and stunting of the shoot. A gradual wilt sets in, followed by death of the whole tree (Whitehead, 1969b). *P. coffeae* reduces NO_3^- uptake rate by 56% and NH^+ uptake rate by 24% in *C. arabica* plants (Vaast *et al.*, 1998).

Severely infected plants may die prematurely. In the field, the symptoms may occur in patches, with reduced yield according to the disease severity. Lesions occur on roots with consequent destruction of the whole root system (Monteiro and Lordello, 1974). In Brazil, *P. coffeae* populations caused root destruction of *C. arabica* seedlings under controlled inoculation conditions (Kubo *et al.*, 2002b). In the north-east of Brazil, a severe infestation of *P. coffeae* on *C. arabica* was reported in fields previously cultivated with yam (*Dioscorea cayenensis*) that were abandoned because of *P. coffeae* attacks (Moura *et al.*, 2003). The authors reported severe root cortex necrosis even in the collar area of the stem and severe decay of trees leading to their death in 70% of the plantation. *P. coffeae* is the most destructive nematode of *C. arabica* in south India (Palanichamy, 1973). In Indonesia, *P. coffeae* is reported as a very destructive nematode to coffee, causing production losses ranging from 29 to 79% on *C. canephora* cv. Robusta plantations due to attack by this nematode (Wiryadiputra, 1990, cited by Toruan-Mathius *et al.*, 1995).

P. brachyurus causes reduced plant and root growth, shedding of leaves and nutritional deficiency (Lordello, 1984).

The influence of infestations of *P. goodeyi*, *P. loosi*, *P. pratensis* and *P. zeae* on coffee growth is not known.

P. brachyurus and *P. zeae* have been reported in soil of *C. arabica* fields previously planted with sugarcane in Hawaii, but populations of both species tended to disappear beyond 3 years after planting, indicating that coffee is a poor host for these root lesion nematodes (Schenk and Schmitt, 1992).

Two of the three unidentified *Pratylenchus* species detected in Guatemala (one morphologically close to *P. pseudocoffeae* and the other one close to *P. coffeae sensu lato*) showed a high reproductive fitness and pathogenicity on *C. arabica* cv. Catuai under controlled inoculation conditions (Villain *et al.*, 1998; Villain, 2000), which tallied with the severity of damage observed in the field (Plate 15E). In a field experiment infested by *Pratylenchus* sp. (*P. pseudocoffeae* morphologically close), a 25–76% tree mortality range was observed 4 years after planting among *C. arabica* cv. Caturra plots without nematode control (Villain *et al.*, 2000). Coffee berry yield of these plots was highly correlated with *Pratylenchus* population densities in coffee tree roots and varied from 6 to 0.5 t/ha for the most infested plots. Moreover, bean size was negatively correlated with nematode population densities. The share of beans retained in 17/64-inch or greater aperture sieves was reduced from 95% for the least infested plots to 65% for the most infested ones (Villain, 2000; Villain *et al.*, 2001a). Thus harm to the coffee bean yield due to *Pratylenchus* attacks is not only quantitative but also qualitative. In contrast, the third unnamed species found in Guatemala (morphologically similar to the *P. coffeae sensu lato*), originating from a northern region of this country, was only very slightly pathogenic on *C. arabica* (Villain, 2000; Villain *et al.*, 2002).

Biology and life cycle

P. coffeae is a bisexual species that reproduces by obligatory amphimixis, so males are frequent, as they are for *P. pratensis*, *P. loosi* and *P. goodeyi*. On the other hand, *P. brachyurus* and *P. zeae* are monosexual species (males absent or rare) that reproduce by mitotic parthenogenesis (Roman and Triantaphyllou, 1969).

For *P. coffeae* and *P. brachyurus*, eggs are laid in root tissues (Roman and Hirschmann, 1969; Loof, 1991). *P. coffeae* eggs hatch in 6–8 days at 28–30°C; first appearance of adults is observed 15 days after hatching at 25–30°C on *Solanum tuberosum* tubers (Siddiqi, 1972). Under these conditions, the average cycle duration is 27 days. For *P. brachyurus*, under optimum temperature conditions (30 or 35°C), one cycle, from adult to adult, takes 4 weeks on maize, while at 10°C the cycle is not completed in 14 weeks (Olowe and Corbett, 1976, cited by Siddiqi, 1976b). *P. zeae* completes its cycle, from egg to maturity, within 35–40 days (Graham, 1951, cited by Siddiqi, 1976a). *P. loosi*, which is more adapted to cooler areas compared with the previously mentioned species, shows a longer life cycle that is completed in 45–48 days, comprising 15–17 days for the eggs to hatch, 15–16 days as juveniles and 15 days as adults before egg laying (Hutchinson and Vythilingam, 1963, cited by Siddiqi, 1977).

Races

Cross-inoculation studies with populations of *P. coffeae* from *C. arabica* in seven different hosts revealed differences in reproduction and pathogenicity, suggesting a physiological specialization in this species (Kumar and Viswanathan, 1972). Differences in aggressiveness among isolates of *P. coffeae* have been reported in Brazil (Silva *et al.*, 2001).

Survival and means of dissemination

When hosts are absent, *P. coffeae* can survive for 8 months in moist soil (Colbran, 1954, cited by Siddiqi, 1972). Different studies on different crops show that soil populations of *P. brachyurus* are strongly influenced by available soil moisture (Siddiqi, 1976b). Means of dissemination of root lesion nematodes are the same as for root knot nematodes, as described above.

Environmental factors affecting parasitism

Sandy soils are more favourable than clay soils for horizontal migrations of *P. brachyurus* and *P. zeae* (Endo, 1967, cited by Siddiqi, 1976a,b).

In Guatemala, population dynamics of *Pratylenchus* sp. (*P. pseudocoffeae* morphologically similar species, see above) were studied during 3 years at two different altitudes (450 and 1200 m). In both sites, two population peaks were observed, one during the dry season and the other at the beginning of the rainy season (Villain, 2000). Population fluctuations were not simply related to rainfall pattern, but were also closely related to the coffee tree phenological rates interacting with climatic factors. The two annual population peaks coincided with root-growing peaks of the coffee trees, while the lowest population levels during the year occurred during coffee berry maturation. This may be linked with some particular characteristics of the physiology of the coffee trees, with the beans working like priority physiological sinks during the maturation, causing an important draining of carbohydrates at the expense of roots and principally secondary roots that stop growing and even die (Cannell and Huxley, 1969; Cannell, 1971). An important fact to mention is that *Pratylenchus* populations show a very large range of increases and decreases in very short times, demonstrating a high parasitic capacity of this nematode in coffee fields.

Soil temperature is an important factor in the development of *Pratylenchus* spp. *P. brachyurus* and *P. zeae* development is faster at 28–35°C than at 15–25°C. *P. coffeae* seems to have its temperature optimum at around 30°C for reproducing, while its parasitic capacity is nearly stopped at 35°C according to observations on *Citrus jambhiri* and *Glycine max* (Radewald *et al.*, 1971; Acosta and Malek, 1978). In contrast, *P. loosi* is less thermophilic and, although it varies according to the host, its temperature optimum seems to be around 20°C (Hutchinson and Vythilingam, 1963, cited by Siddiqi, 1977; Sivapalan and Gnanapragasam, 1975; Gnanapragasam, 1982). Among *Pratylenchus* spp. populations collected in Guatemala, the two populations morphologically close to *P. coffeae* (one highly and the other slightly pathogenic on *C. arabica*) showed the same temperature optimum between 27 and 29°C for reproduction under *in vitro* rearing conditions on carrot discs, while reproduction decreases severely when the temperature falls to 24°C or is raised to 30°C. In contrast, two populations, one from Guatemala and one from El Salvador, that are considered to belong to the same species (*P. pseudocoffeae* morphologically close populations) both showed a temperature optimum between 24 and 27°C (Hervé, 1997; Villain, 2000).

Other hosts

P. brachyurus is found infecting a great number of crops in many countries of the world (Lordello and Mello Filho, 1969; Nickle, 1984). Grasses which commonly occur within coffee plantations in South America such as *Melinis minutiflora* and *Hyparrhenia rufa* are good hosts for this species (Lordello, 1972). *P. coffeae* has a wide host range (Nickle, 1984) as does *P. zeae* (Tenente *et al.*, 2002). *P. coffeae* represents a major pest of other crops such as *Musa* spp. (see Chapter 16), *Citrus* spp. (see Chapter 11), *Dioscorea* spp. (see Chapter 7), *Ipomoea batata* and *Solanum tuberosum* (see Chapter 6). This species is also present on many weeds (Loof, 1991). *P. goodeyi* is an important parasite of banana in many parts of the highlands of East Africa (see Chapter 16). *P. loosi* is an important parasite of tea (*Camellia sinensis*) in many regions of Asia (Whitehead, 1969b; see Chapter 15).

Management measures

Early studies showed the efficacy of oxamyl, phenamiphos and aldicarb for controlling *P. coffeae* in coffee nurseries in El Salvador, and increased yields of coffee were obtained in the second year in plots treated with carbofuran (Abrego, 1974). Good control of *P. coffeae* was also obtained with Nemacur, and it remained effective under field conditions for 90 days after application (Kumar, 1982). On the other hand, Villain *et al.* (2000) observed that terbufos applications (1–2 g/plant) suppressed populations of *Pratylenchus* sp. (*P. pseudocoffeae* morphologically similar population) in coffee roots only during the first 2 years after planting. This resulted in a significant decrease of plant mortality in ungrafted *C. arabica* plots but not in a significant increase in yield.

In a search for resistance genes in *C. arabica*, progeny from Ethiopia (origin of *C. arabica*) and Yemen (first location of *C. arabica* 'domestication'), the two most important geographical centres of *C. arabica* genetic diversity, were highly susceptible to *Pratylenchus* sp. from Guatemala (Anzueto, 1993; Villain *et al.*, 2004).

A variability of resistance to *Pratylenchus* species from Guatemala was observed among different *C. canephora* progeny in accordance with the substantial genetic variability observed within this species (Leroy, 1993). The two reciprocal crossings of the two clone parents of cv. Nemaya (see 'Control measures' above) resulted in the most resistant progeny to *Pratylenchus* sp., which coincided with the CATIE collection introductions that were the most resistant to different *Meloidogyne* spp. from Central America (Bertrand *et al.*, 2000b; see above). Pre- and post-infective resistance factors were observed on *C. canephora* cv. Nemaya (Villain, 2000; Villain *et al.*, 2001b). At the pre-infective stage, early penetration dynamic studies showed that *C. canephora* cv. Nemaya was unattractive for Guatemalan *Pratylenchus* spp. populations compared with the high attractiveness of *C. arabica* cv. Catuai. No histological structure likely to prevent or hinder penetration by nematodes was detected. At the post-infective stage, poor reproduction of *Pratylenchus* spp. on *C. canephora* may be related to the presence of many polyphenols observed in cv. Nemaya roots even without nematodes but never observed in *C. arabica* roots. These components could therefore be involved in constitutive mechanisms of resistance to *Pratylenchus*. In Indonesia, resistance levels of *C. canephora* clones to *P. coffeae* were correlated with their total polyphenol content in roots (Toruan-Mathius *et al.*, 1995). In addition, cambium lignifications close to nematode lesions in the cortical parenchyma were observed on cv. Nemaya roots, suggesting that some inducible resistance mechanisms could also occur (Villain *et al.*, 2001b). If the main *Pratylenchus* resistance factors are linked to the phenolic metabolism, it is to be hoped that this resistance is not very specific (Dalmasso *et al.*, 1992) and would thus provide the plant with an acceptable level of resistance to different *Pratylenchus* species and/or pathotypes, In Brazil, *C. canephora* cv. Conilon is resistant and *C. canephora* cv. Robusta susceptible to *P. coffeae* strain K5 (Tomazini *et al.*, 2003). In India, it was found that *C. canephora* cv. Robusta is more tolerant to *P. coffeae* than *C. arabica* or *C. excelsa* (Anonymous, 1974), hence the use of *C. canephora* cv. Robusta as rootstocks is the most promising means of control (Palanichamy, 1973). In fact, Schieber and Grullon (1969) also suggested the use of *C. canephora* var. Robusta in Guatemala as a source of resistance for rootstocks in grafted plants. Since grafting on to *C. canephora* cv. Robusta at the cotyledon stage was perfected by Reyna (1968), this agronomic practice has become more and more common for controlling nematodes in Guatemala. Grafting on common *C. canephora* rootstock provides efficient control of *Pratylenchus* spp. (Plate 15F), keeping the populations at very low levels even without chemical control (Villain *et al.*, 2000, 2001a). Grafting on *C. canephora* also has the advantage of not affecting either the physical and chemical qualities of the bean or the beverage organoleptic quality of *C. arabica* (Anzueto *et al.*, 2001; Villain *et al.*, 2001a). For the record, grafting was achieved successfully for varieties susceptible to *P. coffeae* much earlier on to Conuga hybrid (*C. congensis* × *C. canephora* cv. Ugandaea) and on to *C. canephora* cv. Robusta in India (De Fluiter, 1947; Pattabhiram, 1949, cited by Siddiqi, 1972).

To prevent serious infestation with these nematodes, the coffee growers should, where possible, disinfest nursery soil and plant seedlings in non-infested field soil. Methyl bromide at rates of 150 cm^3/m^3 of soil has been the most effective means of sterilizing soil, but other methods are available (see 'Management measures' above).

Other nematode parasites of coffee

Among other species of nematodes parasitic to coffee, *Rotylenchulus reniformis* has caused greatest damage to this crop. In the Philippines, *R. reniformis* attacked *C. arabica*, *C. canephora* cv. Robusta and *C. excelsa* with equal severity (Valdez, 1968). In India, it is an important parasite of *C. arabica* (Anonymous, 1966). *R. reniformis* is also reported from coffee seedlings in a commercial nursery in Brazil (Lordello, 1980) and in the field (Campos, 2002), and is also recorded on *Coffea* spp. in the Pacific island countries of New Guinea, Fiji, Tonga and Western Samoa (Bridge, 1988) and in the Côte d'Ivoire (Van Doorsselaere and Samsoen, 1982).

D'Souza and Screenivasan (1965) pointed out that coffee does not grow well in infested fields with an inoculum density of *R. reniformis* greater than 10 nematodes/50 cm^3 of soil. Screening genotypes for resistance has been done. Macedo (1974) found resistance in *C. canephora* cv. Guraini, whereas on cv. Mundo Novo and Catuaí of *C. arabica* a few mature females deposited eggs. No further information on the importance of this nematode and control measures is available.

Whitehead (1968b) commented on the great importance of *Radopholus similis* to coffee in Java reported by Zimmermann (1898). This nematode was considered the most harmful nematode to that country and second only in importance to *P. coffeae*.

Vovlas (1987) reported on the widespread occurrence of *Trophotylenchulus obscurus* as a pest of coffee in São Tomé, West Africa. At feeding, *T. obscurus* introduces its anterior body portion into the peripheral layers of the cortex and the nematode feeds from a single nurse cell, which undergoes senescence and, as a consequence, causes considerable damage to the cortical cells. Dark brown capsules containing eggs, juveniles and males can be observed on the root surface.

Many other parasitic nematode species belonging to the genera *Gracilacus*, *Caloosia*, *Criconemoides*, *Discocriconemella*, *Helicotylenchus*, *Hemicriconemoides Hoplolaimus*, *Longidorus*, *Ogma*, *Paratrichodorus*,

Pratylenchus, *Aorolaimus* (= *Peltamigratus*), *Rotylenchus*, *Scutellonema*, *Trichodorus*, *Tylenchorhynchus*, *Paratylenchus* and *Xiphinema* have been found associated with coffee plants (Luc and de Guiran, 1960; Thorne and Schieber, 1962; Whitehead, 1968b, 1969b; Lordello, 1972; Sharma, 1973; Sharma and Sher, 1973a; Van Doorsselaere and Samsoen, 1981; Bridge *et al.*, 1982; Bridge, 1984; Bridge and Page, 1984; Campos *et al.*, 1987; Vovlas, 1987; Kubo *et al.*, 2001; Campos, 2002). However, information on their pathogenicity, damage, yield loss and possible control measures is lacking.

Conclusions and Future Prospects

Growers must be made aware of the nematode threat to the coffee crop. Certain nematode species, especially those belonging to *Meloidogyne*, which are not considered important today, may become a constraint for coffee production in certain regions in the future. In addition, specific regional coffee ecosystems, poor management and changes in host–parasite relationships may favour the outbreak of a nematode disease in a coffee region. *Meloidogyne* diseases of coffee have been important reasons for the movement of coffee-producing areas in Brazil. Complex disease situations caused by mixtures of many species of *Meloidogyne* do occur in coffee plantations around the world and concomitantly, in some areas, with other very pathogenic species, particularly of the genus *Pratylenchus*. Agricultural scientists need to examine whether coffee nematodes are a problem in their own countries and follow the progress of any nematode disease particularly to avoid the dissemination of the nematode, causing, in consequence, losses which can harm the country's economy.

Improving the awareness of different symptoms of root damage caused by the different parasitic nematodes and a general improvement in laboratory diagnostic services will help in identifying the means of dissemination and the possible unknown damage to coffee.

Identification and better biological characterization, such as pathogenicity, of the inter- and intraspecific biodiversity of plant parasitic nematodes presently found on coffee need to be achieved in most of the coffee-growing countries. This knowledge is of prime importance to improve control measures as part of the development of sustainable coffee crop systems appropriate to the ecological and economic pressures. There is a strong case for the selection of coffee germplasm resistant to nematodes and it is even more true for the future development of fast marker-assisted breeding programmes. Knowledge of coffee parasitic nematode field biodiversity is also a necessity for the development of alternative and complementary control measures such as biological control. Most of the biological agents available for nematode control, such as *Pasteuria penetrans*, fungi or antagonistic plants, show some specificity in their degree of antagonism to plant parasitic nematodes, all the more reason to be aware of the nematode field biodiversity. This knowledge is also important to lay down suitable prophylactic measures restricting distribution areas and the spread of the most pathogenic nematodes. Researchers need to be aware of the complexity of nematode communities in tropical crops (Luc and Reversat, 1985) as is the case in coffee, and the difficulties in identification when using only morphological criteria for many genera such as the two most important in coffee, *Meloidogyne* and *Pratylenchus*. Today, accurate and complementary tools, such as isoenzymatic electrophoresis and molecular analysis, in association with morphological and biological studies, will allow more rapid progress in identification and biosystematics of nematode groups. This is being done mainly with *Meloidogyne*, but progress in taxonomic studies and diagnostic tool development must also be achieved with the other major genus, *Pratylenchus*.

A better future for this crop as far as the nematode diseases are concerned can be attained by the introduction of regulations restricting the planting of infested coffee seedlings in the areas of old, infested coffee plantations. Equally important is the practical mechanism to enforce these regulations.

The use of seedlings for field planting and the perennial nature of the crop increase the risk of severe nematode infestation.

For integrated nematode management in coffee, the selection of coffee germplasm resistant to nematodes is a priority for almost all the most pathogenic nematodes attacking this crop. Grafting commercial cultivars on rootstock resistant or tolerant to damaging nematodes is a very useful control strategy and could be used in other regions with widespread distribution of very destructive nematodes, as has been done in Brazil.

In Brazil, coffee producers are advised to eradicate nematode-infested coffee trees on the spot in the field and replant with *C. canephora* cv. Robusta IAC Apoatã (non-grafted). Susceptible plants that show up should be eliminated from time to time because this species is allogamous (of cross-pollination). The remaining plants will have a good level of resistance to the local nematodes, and seeds of those plants should be harvested and used for rootstock production (Gonçalves and Silvarolla, 2001). The growers can then produce their own seedlings. Care should be taken at the time of seedling planting in the field by: (i) decreasing the spacing between plants in the row, because the rootstocks segregate 10–15% towards susceptibility to the nematode; (ii) using field workers to do the grafting or buying grafted seedling from responsible seedling producers with experience in grafting coffee; and (iii) avoiding planting the seedling too deep into the soil, and eliminating seedlings with crooked roots (Gonçalves and Silvarolla, 2001).

Because of the complexity of nematode communities present in coffee fields, as mentioned above, coffee breeders must take into account the total nematode community present in the field. The use of coffee germplasm that does not present resistance to all the damaging or potentially damaging nematodes could be dangerous, particularly for a perennial crop such as coffee, where there is the possibility of a changing equilibrium between different nematode species due to competition over a relatively long period of time. Competition was observed, for example, in Guatemala between *M. paranaensis* and *Pratylenchus* sp. (Cilas *et al.*, 1993) and in Costa Rica with *M. exigua* and *Pratylenchus* sp. (Bertrand *et al.*, 1998). The latter authors observed that planting cultivars such as Sarchimors, resistant to *M. exigua* but not to *Pratylenchus*, resulted in a large decrease of population densities of *M. exigua* but also in a significant increase of root lesion nematode populations that could be more damaging than *M. exigua*.

With the recent discovery of resistance markers, the possibilities of developing marker-assisted breeding programmes represent important progress for a perennial crop such as coffee because of the long-term need for classical breeding programmes. This should also permit more exhaustive exploration of the wide genetic pool of some *Coffea* species such as *C. canephora* in the search for resistance genes.

However, to ensure a durable management of selected resistance, particularly in a perennial crop, complementary non-chemical control measures must be developed to reduce field inoculum pressures. Biological control, especially with *Pasteuria* sp., could be a promising strategy in the future to control root knot or other nematodes. Crop rotations with non-hosts and/or antagonistic plants could also be an effective way to reduce field nematode inoculum before planting coffee even if it concerns resistant cultivars.

Another important aspect to consider at least in some plant–nematode pathosystems is that some resistance such as that to *Pratylenchus* is of an incomplete nature with a likely oligogenic or polygenic genetic determinism, and so the expression of this type of resistance is probably strongly dependent on environmental factors (Rapilly, 1991). Durable use of these types of resistance must then be planned with appropriate coffee-farming practices, such as appropriate fertilization and suitable soil management, shade in some geographical regions and, more globally, all practices that contribute to conferring an optimum physiological status on the plant.

Cocoa

Cocoa and chocolate are derived from the seeds of *Theobroma cacao*, a small tree indigenous to the forests of Central and South America that belongs to the family Sterculiaceae. The centre of origin is the upper Amazon in South America.

The fruit, which botanically is a berry, usually contains from 20 to 40 seeds, each surrounded by a pulp that is developed from the outer integument of the ovule. The action of yeasts removes the mucilage around the seeds, which facilitates subsequent handling and drying of the beans (Urquhart, 1955).

Within *T. cacao*, genetic diversity can be divided into two main groups and a third one originating from the first two (Anonymous, 2002). (i) The Criollo give warty elongated pods, green or red coloured before ripeness. The white seeds give a fine and aromatic cocoa, but Criollo seeds represent only 1% of the world production. (ii) The Amazonian Forastero comprise the upper and lower Amazon varieties, the most famous variety of which is the Amelonado. The Forastero are widespread in Brazil, western Africa and Asia. The thick husk pods are oval, smooth and green coloured, turning yellow on ripeness. They have dark purple-coloured and flat seeds and constitute most of the common cocoas and about 80% of the world production. (iii) The Trinitario are hybrids originated from the first two groups and are cultivated in all producing countries. They contribute about 20% of the world production. Only 40% of the total world cocoa production comes from improved varieties (Paulin and Eskes, 1995). To ensure cocoa germplasm conservation and utilization, an important world project began in 1998 after acceptance by the Common Fund for Commodities (Eskes *et al.*, 1998). Cocoa is grown in many countries by smallholder farmers of South and Central America, Africa, Asia and Oceania, located mostly between 10° north and south of the equator. The five major world producers are the Côte d'Ivoire (39% of the world production), Ghana (13%), Indonesia and Nigeria (12% each) and Brazil (6%), producing between them 80% of the world total, i.e. 2,809,000 t in 2002 (FAO, 2002).

Cocoa is a lowland crop growing best from sea level to altitudes of 1400 m on the equator with temperatures of 16–34°C and rainfall of 1500–2500 mm. It reacts unfavourably to sudden changes of temperatures or humidity. The main factor limiting the growth of cocoa at the higher altitudes is temperature. The daily variation of temperature should not exceed 9°C (Urquhart, 1955; Braudeau, 1970).

Cultivation techniques

Seed propagation is cheapest. Seed can be planted directly in soil (West Africa), in nursery seedbeds, in baskets or plastic bags. Germination takes 1 or 2 weeks and seedlings are transplanted to the fields when 2–6 months old. Propagation is also possible by cuttings, buddings, grafts and marcots. Spacing varies between areas. Closer spacing is used in Africa such as 2.4 m × 2.4 m, 3 m × 3 m, 3 m × 2–2.5 m, and 4.5 m × 4.5 m. In America and Asia, spacing is predominantly 4 m × 4 m, 3.6 m × 3.6 m and 3 m × 3 m. Shading is commonly used. Thinned natural forest for shading predominates mostly in Africa, while in America, Asia and Oceania the shade trees planted are mostly *Erythrina* spp., *Gliricidia* spp., *Albizzia* spp., *Pithecolobium* spp. and *Leucaena* spp. Managing the shade conditions during the development of the crop is done in some producing countries. Pruning is done to shape or form the young tree, to maintain the subsequent shape or form and to renovate or rehabilitate the tree.

Nematodes of Cocoa

Because of the susceptibility of the germplasm currently grown, world production losses caused by diseases and pests are assessed at 50% (Anonymous, 2002). The most important plant health constraints at world level are currently caused by three fungal diseases: the black pod disease (*Phytophthora* spp.), which is responsible for 30% of losses in the world production with the most virulent species, *P. megakaria*, present in western Africa (Cilas *et al.*, 1998); the witches' broom (*Crinipellis perniciosa*) mainly in South America where it has ruined the cocoa crop of many countries (Pereira, 1998); and the frosty pod disease or moniilia pod rot (*Moniliophthora roreri*), another serious disease in the Americas.

In addition, nematodes such as *Dolichodorus* and *Meloidogyne* species, especially *M. incognita* and to a lesser extent *M. javanica*, have also caused losses in cocoa areas around the world including yield decrease, sudden death of trees and growth retardation of seedlings in nurseries. Many other genera and species of root-feeding nematodes have been found in association with cocoa (Table 14.2) although the pathogenic relationship, for most of them, has not been proved.

Table 14.2. List of endoparasitic and ectoparasitic nematodes associated with cocoa roots.

Allotrichodorus brasiliensis	*Meloidogyne incognita*
Allotrichodorus campanulatus	*Meloidogyne javanica*
Allotrichodorus sharmae	*Meloidogyne thamesi*
Allotrichodorus westindicus	*Neodolichodorus*
Aorolaimus banoae	*Ogma decalineatum*
Aorolaimus holdemani	*Paralongidorus* sp.
Aorolaimus levicaudatus	*Paratrichodorus minor*
Aorolaimus nigeriensis	*Paratylenchus* sp.
Aorolaimus vigiae	*Peltamigratus christiei*
Criconema braziliense	*Peltamigratus holdemani*
Criconema demani	*Peltamigratus macbethi*
Criconemoides ferniae	*Pratylenchus brachyurus*
Criconemoides onoensis	*Pratylenchus coffeae*
Criconemoides paradenoudeni	*Pratylenchus zeae*
Criconemoides paralineolata	*Radopholus similis*
Discocriconemella degrissei	*Rotylenchulus reniformis*
Discocriconemella limitanea	*Rotylenchulus microstriatus*
Dolichodorus minor	*Scutellonema brachyurus*
Eutylenchus africanus	*Scutellonema clathricaudatum*
Helicotylenchus dihystera	*Trichodorus monohystera*
Helicothylenchus erythrinae	*Trophorus imperialis*
Helicotylenchus multicinctus	*Tylenchorhynchus annulatus*
Hemicriconemoides cocophillus	*Tylenchorhynchus queirozi*
Hemicriconemoides mangiferae	*Tylenchus* sp.
Hemicycliophora chilensis	*Xiphidorus minor*
Hemicycliophora loofi	*Xiphinema abeokutae*
Hemicycliophora thienemanni	*Xiphinema americanum*
Heterodera sp.	*Xiphinema bergeri*
Hoplolaimus spp.	*Xiphinema brasiliense*
Longidoroides sp.	*Xiphinema brevicollum*
Longidorus sp.	*Xiphinema ifacolum*
Meloidogyne arenaria	*Xiphinema krugi*
Meloidogyne exigua	*Xiphinema vulgare*

Afolami and Caveness (1983); Badaru *et al.* (1999); Bridge *et al.* (1982); Crozzoli *et al.* (2001); De Waele and Coomans (1993); Freire and Monteiro (1978); Loof and Sharma (1980); Lopez (1994); Lopez *et al.* (1980); Luc and Coomans (1993); Manso *et al.* (1994); Pinochet and Raski (1976); Sharma (1982); Sharma and Loof (1974); Sharma and Sher (1973, 1974a); Sosamma *et al.* (1980a,b); Sudha and Sundararaju (2002); Thorold (1975); Whitehead (1969).

Meloidogyne

Meloidogyne spp. are the most important nematodes of cocoa due to their pathogenicity and wide distribution in cocoa-producing regions.

Distribution

Root knot nematodes have been found in cocoa since 1900 (Ritzema Bos in Sosamma *et al.*, 1980a), and they have been reported from Zaire, São Tomé, Java (Ghesquiére, 1921; Cotterel, 1930; Fluitter and Mulholland, 1941), Ghana, Malawi, Côte d'Ivoire (Edwards, 1955; Luc and de Guiran, 1960; Martin, 1961), Nigeria (Caveness, 1967), Venezuela (Torrealba, 1969), Brazil (Lordello, 1968) and India (Sosamma *et al.*, 1980b).

M. incognita seems to be the most frequently found in cocoa (Luc and de Guiran, 1960; Sharma and Sher, 1974a). It is a common pest in West Africa (Whitehead, 1969) including Nigeria where this species appears as the most economically important nematode on cocoa (Badaru *et al.*, 1999). *M. incognita* is also common in India (Sosamma *et al.*, 1980b; Sudha and Sundararaju, 2002), Malaysia (Razak, 1981) and Venezuela (Crozzoli *et al.*, 2001) and is widespread in cocoa regions of Brazil (Sharma and Sher, 1974a,b; Sharma, 1982). In the cocoa region of Espirito Santo State, Brazil, it is the most frequent nematode in sampled sites (Sharma and Sher, 1974a,b).

However, other species of *Meloidogyne* have also been found on cocoa: *M. exigua* in Bolivia (Bridge *et al.*, 1982), *M. javanica* in Malawi (Corbett, 1961), in Venezuela (Crozzoli *et al.*, 2001) and in Central Africa (Martin, 1961), and *M. arenaria* and *M. thamesi* in Brazil (Sharma, 1979).

Symptoms of damage

In artificially infested seedlings, *M. incognita* causes dieback, stunting, wilting, yellowing of leaves and small leaves. Tiny galls and females with egg masses can be observed on the roots. In Nigeria, seedlings of cv. Amelonado grown in soil inoculated with *M. incognita* show the symptoms from the 16th week, leading to wilting in the 24th week (Afolami, 1981, 1983, 1985). Amazon cultivars also show decay symptoms, but only after the 24th week, and there is no wilting. Sharma and Maia (1976) found that *M. incognita* caused small, rounded and elongated galls with conspicuous egg masses, and stunting in the cv. Catongo. The leaf tips and margins first turn brown and become dried; this spreads to the entire leaves which are eventually shed. The infested plants looked unthrifty, with decreased height, shoot and dry root weights. In Brazil, inoculation of 10,000 *M. incognita* juveniles/plant caused significant reduction in cocoa plant growth within 17 weeks (Anonymous, 1975). Sharma and Maia (1975) found that *M. incognita* was pathogenic to cocoa cv. Catongo, causing growth reductions, small internodes, thin stems, reduction of number and surface area of leaves, reduction of the root system with galls and fewer root hairs. Growth differences were evident 17 weeks after inoculation. Histological studies showed total disorganization of the stele, resulting in serious destruction of the xylem, phloem, pericycle and endodermis. Adult females were found embedded in the cortex, with giant cells around their heads and egg masses deposited on the root surface through ruptures in the cortex. In the field, *M. incognita* produces galls with exposed egg masses on roots, dieback and sudden death of the infested plants. According to Sharma and Sher (1973), when the dieback conditions occur, the trees die down to their roots, which remain alive and send up shoots in the following growing season and also when the dead terminals are pruned off. The syndrome of sudden death disease is permanent wilting, the green leaves suddenly turn yellow and brown, and then dry up to remain hanging. Jimenez-Saenz (1971) and Sharma and Sher (1973) associated the occurrence of sudden death with root knot nematodes.

M. javanica also forms galls on cocoa roots (Martin, 1961). In Malawi, young cocoa trees grew slowly in patches of soil heavily infested with *M. javanica* (Corbett,

1961). Damage symptoms were also observed on cocoa roots infested by *M. exigua* in Bolivia (Bridge *et al.*, 1982).

Nematodes in the nursery can retard the growth of seedlings or may even kill them. The transplantation of nematode-infested seedlings carries nematodes to the plantations where the transplants may die.

Races, means of dissemination, other hosts and economic importance

Among the root knot nematode species found in cocoa, *M. incognita* and *M. arenaria* have host races. Although *M. incognita* has four biological races, no attempt has been made to determine the variation within *M. incognita* populations of infested cocoa fields. Similarly, *M. arenaria*, which is known to have two races in other crops, has not yet been examined for race differentiation in cocoa.

M. javanica, *M. incognita* and *M. arenaria* have wide host ranges (Ponte, 1977; Nickle, 1984), and in many instances the commonly used shade plants, such as banana, may become a source of inoculum in the cocoa plantation (Sosamma *et al.*, 1980a). Corbett (1961) recommended the replacement of banana as a shade for cocoa to reduce the nematode infestation on cocoa in Malawi.

Nursery soil infested with the nematodes will allow the production of infested seedlings which will disseminate nematodes into plantations. Runoff water may also spread nematodes.

Although data on cocoa yield losses caused by nematodes are not yet available, evidence suggests their importance to this crop. Sudden death of cocoa plants in the field has been associated with *Meloidogyne* spp. in many areas of cocoa production, and they could be a limiting factor to productivity and have an economic impact in infested regions.

Other nematode parasites of cocoa

The lesion nematode, *P. brachyurus*, has been widely found in cocoa in Bahia, Brazil (Sharma and Sher, 1973), and also occurs in western Africa (Luc and Guiran, 1960). In Java, *P. coffeae* infects roots of cocoa (Fluitter and Mulholland, 1941). Sudha and Sundararaju (2002) found *P. coffeae* on cocoa in Kerala state, India, and Kumar *et al.* (1971) reported the multiplication of this nematode on cocoa in glasshouse experiments. *P. coffeae* is also reported on cocoa in Indonesia (Siddiqi, 1972), and *P. zeae* occurs in Venezuela (Crozzoli *et al.*, 2001). However, many other root-feeding nematodes have been identified in cocoa (Table 14.2).

Sharma (1971) associated dieback and death of the nursery plants with the presence of *Dolichodorus* sp. (now *D. minor*). The entire root system was reduced, blackened, and showed disintegrated cortex and bead-like gall formation. The galled portion was reddish-brown and hard. In Para State, northern Brazil, *D. minor* was one of the most common nematodes on cocoa; *D. minor* was also reported on cocoa in southeastern Costa Rica (Lopez, 1994).

Helicotylenchus spp. are widespread on cocoa crop in South America and Asia, and *H. pseudorobustus* reproduced on cocoa in Liberia (Lamberti *et al.*, 1992). *H. dihystera* was reported in Bahia State of Brazil as the most widespread species on cocoa, occurring in 70% of the samples (Sharma and Lordello, 1982). Luna (1976) and Campelo and Galli (1980) demonstrated the pathogenicity of this nematode on *T. cacao*. A significant reduction was observed in dry root weight and leaf number 188 days after inoculation of different levels over 20 nematodes/plant; stunting and significant decrease of dry root weight in 20-day-old inoculated seedlings was also observed.

Management

In perennial crops such as cocoa, nematodes that survive the control practices have time to recover and build up again to destructive levels. Hence the most efficient control strategies are: (i) to produce seedlings free of major pathogenic nematodes; and (ii) to cultivate in soils or areas from which the nematodes are absent.

Soil to be used in the nursery can be sterilized by treating with methyl bromide at a rate of 196 cm^3/m^3 of soil (Ferraz, 1979) where the chemical is still registered for use. Using soil collected from areas that are not infested by root knot species and *Pratylenchus* spp. also produces healthy, nematode-free seedlings. Another method is hot air treatment using a hot air sterilizer which raises the temperature to 100°C for 1 h (Sharma, 1975); sun drying or steaming of the soil, as done in coffee nurseries (see above), can also be effective.

The land for cultivating cocoa must be surveyed for important nematodes before transplanting clean seedlings. In the case of established plantations already infested, especially by root knot nematodes, the grower should use nematicides to manage the population level and avoid economic damage.

For agricultural field applications, most fumigant nematicides are no longer used. Where chemicals continue to be used, the emphasis has been on the production and use of contact or systemic nematicides. Past results have shown that application of granular nematicides such as Nemacur 10G, Temik 10G at the rate of 50 mg of commercial product per plant and Terracur 5% at 100 mg/plant in plants infested by *M. incognita* in the glasshouse can reduce the nematode density and increase the numbers of leaves per plant (Sharma and Ferraz, 1977). Tarjan *et al.* (1973) reported an increased yield of 11–96% after field application of Mocap or Terracur at the rate of 34 kg of the commercial product/ha and Nemacur at 22 kg/ha. The products were applied within a cleared area of 1.0 or 1.5 m around each trunk and then incorporated into the upper 2.5 cm of soil. Sosamma *et al.* (1980a) have reported an increase in the number of pods by the application of Dasanit and Nemacur, and an increase of yield by the application of Nemacur, Terracur and Mocap.

Care must be taken with the selection of shade plants, avoiding trees susceptible to root knot or lesion nematodes, for example *Leucaena glauca* and banana (Corbett, 1961; Sosamma *et al.*, 1980a).

Probably the most promising control measure against nematodes, particularly for the most pathogenic species, is to select resistant germplasm as is currently achieved against fungal diseases. Resistance to *M. incognita* has been found in five genotypes of cocoa in Nigeria (Badaru *et al.*, 1999), and further screening of *T. cacao* germplasm for resistance to *M. incognita* has been done in Nigeria (Afomali and Ojo, 1985). In Brazil, of 12 cacao hybrids tested for resistance to *M. incognita*, the hybrid TSH565 × S1C802 appeared to be the most resistant with the smallest gall index and the lowest final population density. The nematode multiplication rate varied from six for this hybrid to 39.6 in SIAL70 × SIAL88 (Anonymous, 1976). In Niger, differences in susceptibility to *M. incognita* were observed among four cultivars in glasshouse inoculation tests (Asare-Nyako and Owusu, 1981).

Methods of diagnosis

Root galling in most cases will be helpful to diagnose the presence of *Meloidogyne* species on cocoa plants. However, the extraction of juveniles of those nematodes from soil can help to confirm their presence and identification. Species-level identification for *Meloidogyne* should be achieved by esterase electrophoresis (see 'Nematode Parasites of Coffee' in this chapter). Other nematodes will be found by sampling soil or roots. For *Pratylenchus* species diagnosis, see also 'Nematode Parasites of Coffee' in this chapter.

Conclusions and Future Prospects

Besides the sudden death of cocoa trees in Bahia State, Brazil, and also some localized occurrence of some damaging nematodes on cocoa plantations in other countries, there is no known wide distribution of nematodes in any specific cocoa region having economic impact. However, the potential pathogenicity of some nematodes, especially *M. incognita*, has already been proved in glasshouse research, and the growers must be aware of this potential threat to their crop.

Emphasis in future research work should be on estimation of yield losses and the distribution of damaging nematodes in specific cocoa regions to obtain a better picture of the economic importance and distribution of these organisms.

The perennial characteristics of the cocoa crop mean that great care should be taken in the preparation of healthy, nematode-free seedlings, and also on the choice of land to be planted. The exclusion approach for preventing damage is cheaper, safer and more efficient.

Many countries have the potential to increase cocoa production, but a profitable crop will require good management of all different agricultural aspects including nematode diseases. Larger markets and a greater competition on a worldwide basis require greater efficiency of production at lower prices, minimizing costs and risks involved. Nematode infestation is a potential constraint by increasing the cost of cocoa production and decreasing yields.

Varietal improvement is a priority for many cocoa-producing countries in order to develop sustainable crop systems, and resistance to diseases such as black pod disease (*Phytophthora* spp.), witches' broom (*C. perniciosa*), monilia pod rot (*M. roreri*) or viral swollen shoot disease are current priorities in most cocoa-breeding programmes, but this is not the case for resistance to nematodes (Nguyen-Ban, 1996; Eskes *et al.*, 1998; Knight, 1998; De Franqueville, 2001). However, intensification of cocoa crop and control of the main diseases by planting resistant germplasm could make the nematode economic impact pass to the foreground. Cocoa breeders should consider including resistance to nematodes in cocoa-breeding programmes in regions where pathogenic nematodes such as *M. incognita* or *P. coffeae* are present.

References

Abrego, L. (1974) Ensayos de selectividade de nematicidas en el combate del *Pratylenchus coffeae* en almaciqueras de café. *Nematropica* 4, 17.

Abrego, L. and Holdeman, Q. (1961) Informe de progresos en el estudio del problema de los nematodos del café en El Salvador. Instituto Salvadoreño de investigaciones del café. *Boletin Informativo, Santa Tecla, El Salvador,* Suppplement 8.

Acosta, N. and Malek, R.B. (1978) Influence of temperature on population development of eight species of *Pratylenchus* on soybean. *Journal of Nematology* 11, 229–232.

Afolami, S.O. (1981) Symptoms of root-knot nematode infection on *Theobroma cacao* L. – a preliminary investigation. In: *Proceedings of the Third Research Planning Conference on Root-knot Nematodes,* Meloidogyne *sp.* 16–20 November, Ibadan, Nigeria, pp. 148–156.

Afolami, S.O. and Caveness, F.E. (1983) The frequency of occurrence and geographical distribution of plant parasitic nematodes associated with *Theobroma cacao* in Nigeria. *Turrialba* 33 (1), 97–100.

Afolami, S.O. and Ojo, A.A. (1985) Screening of *Theobroma cacao* germplasm for resistance against a root-knot nematode, *Meloidogyne incognita,* in Nigeria. In: *Proceedings of the 9th International Cocoa Research Conference, Lome, Togo,* 12–18 February 1984. Cocoa Producers' Alliance, Lagos, Nigeria, pp. 237–242.

Almeida, V.F. and Campos, V.P. (1991a) Reprodutividade de *Meloidogyne exigua* em plantas antagonistas e em culturas de interesse econômico. *Nematologia Brasileira* 15 (1), 24–29.

Almeida, V.F. and Campos, V.P. (1991b) Alternância de culturas e sobrevivência de *Meloidogyne exigua* em áreas de cafezais infestados e erradicados. *Nematologia Brasileira* 15 (1), 30–42.

Alpizar, E. (2003) Evaluación de la Patogenicidad de Poblaciones de *Meloidogyne exigua* en Genotipos de *Coffea arabica* con Diferentes Niveles de Resistencia. MS Thesis, Universidad de Costa Rica, Costa Rica.

Alvarado, J.A. (1935) *Tratado de Caficultura Practica.* Tipografia Nacional, Guatemala.

Alvarenga, A. (1973) Determinação preliminar da longevidade no solo, do nematóide *Meloidogyne exigua.* In: *Congresso Brasileiro de Pesquisas Cafeeiras, IBC – GERCA (EMBRAPA), Poços de Caldas* 2, 45.

Anonymous (1966) *India Coffee Board, 26th Annual Report.* Bangalore, India.

Anonymous (1968) *Principles of Plant and Animal Pest Control*. Vol. 4 – *Control of Plant Parasitic Nematodes*. National Academy of Sciences, Publication No. 1696, Washington, DC.
Anonymous (1974) *India Coffee Board. 33rd Annual Report 1972–1973*. Bangalore, India.
Anonymous (1975) *Technical Report for 1975 of the Cacao Research Centre*. Centro de Pesquisas do Cacau, Itabuna, Brazil.
Anonymous (1976) *Technical Report for 1976 of the Cacao Research Centre*. Centro de Pesquisas do Cacau, Itabuna, Brazil.
Anonymous (1977) *Kenya Coffee Research Foundation. Annual Report 1975–1976*. Ruiru.
Anonymous (1985) *Instituto Brasileiro do Café. Cultura do Café no Brasil. Setor de Progamação Visual e Gráfica (IBC/GERCA)*.
Anonymous (1989) Corchosis del cafeto, una amenaza a la agricultura. Manejo Integrado de Plagas, CATIE. *Boletín Informativo, Turrialba, Costa Rica* 13, 1–3.
Anonymous (2002) Les plantes stimulantes. In: *Mémento de l'Agronome*. Cirad, Montpellier; Gret, Paris; Ministère des Affaires Etrangères, Paris, pp. 1051–1063.
Anthony, F., Astorga, C. and Berthaud, J. (1999) Los recursos genéticos: las bases de una solución genética. In: Bertrand, B. and Rapidel, B. (eds) *Desafios de la Caficultura en Centroamerica*. ICCA-PROMECAFE, San Jose, pp. 369–406.
Anthony, F., Bertrand, B., Quiros, A., Wilches, A., Lashermes, P., Berthaud, J. and Charrier, A. (2001) Genetic diversity of wild coffee (*Coffea arabica* L.) using molecular markers. *Euphytica* 118, 53–65.
Anzueto, F. (1993) Etude de la Résistance du Caféier (*Coffea* sp.) à *Meloidogyne* sp. et *Pratylenchus* sp. PhD Thesis, ENSAR Rennes, France.
Anzueto, F., Bertrand, B., Peña, M., Marban-Mendoza, N., Villain, L. and Ibarra, E.L. (1996) Desarrollo de una variedad porta-injerto resistent a los principales nematodos de America Central. *Simpósio Sobre Caficultura Latino-Americana, San Salvador*, Memoria, Vol. 1.
Anzueto, F., Bertrand, B., Sarah, J.L., Eskes, A.B. and Decazy, B. (2001) Resistance to *Meloidogyne incognita* in Ethiopian *Coffea arabica* accessions. *Euphytica* 118, 1–8.
Aragon, B.C.A., Gomes, M.B. and Caicedo, J.E. (1978) Plantas de la zona cafetera colombiana hospedantes de espécies de *Meloidogyne goeldi*, 1887. *Cenicafé* 29, 35–45.
Araya, M. (1994) Distribución y niveles poblacionales de *Meloidogyne* spp. y *Pratylenchus* spp. en ocho cantones productores de café en Costa Rica. *Agronomia Costarricense* 18, 183–187.
Asare-Nyako, A. and Owusu, G.K. (1981) *Meloidogyne incognita* infection of cocoa seedlings. In: *Proceedings of the Seventh International Cocoa Research Conference*, 4–12 November 1979, Douala, Cameroon. Cocoa Producers' Alliance, Lagosia, Niger, pp. 457–461.
Ayala, A. (1976) State of knowledge of *Meloidogyne* in Puerto Rico. In: *Proceedings of the Regional Conference of the International* Meloidogyne *Project*. Region I, Panamá, pp. 93–95.
Badaru, K., Eran, E.B., Williams, J.A. and Eskes, A. (1999) A review of breeding work on disease resistance of cocoa in Nigeria. In: Bekele, F. and End, M. (eds) *Proceedings of the International Workshop on the Contribution of Disease Resistance to Cocoa Variety Improvement*. International Group for Genetic Improvement of Cocoa, University of Reading, Reading, UK, pp. 167–171.
Baeza-Aragón, C.A. (1978) Parasitismo de *Bacillus penetrans* en *Meloidogyne exigua* estabelecido en *Coffea arabica*. *Cenicafé* 29 (3), 94–97.
Bally, W. and Reydon, G.A. (1931) De tegenwoordige stand van hert Vraagstuk van de Wortelaaltjes in de Koffiecultur. *Archf voor de Koffiercultur, Indonesie* 2, 23–216.
Barbosa, D.H.S.G., Vieira, H.D., Souza, R.M., Silva, C.P., Andrade, W.E.B., Engelhardt, M.A. and Pinto, J.F. (2003a) Levantamento de nematóides de galhas (*Meloidogyne* spp) em áreas cafeeicultoras do estado do Rio de Janeiro. *Congresso Brasileiro de Nematologia, XXIV, Palestras e Resumos*, Petrolina, PE, Brazil, p. 142.
Barbosa, D.H.S.G., Vieira, H.D., Souza, R.M., Silva, C.P., Andrade, W.E.B., Silva, C.P., Pinto, J.F., Engelhardt, M.A. and Araujo, L.C. (2003b) Levantamento de nematóides de galhas (*Meloidogyne* spp) em áreas cafeeiras fluminenses. *Congresso Brasileiro de Pesquisa Cafeeira, XXIX, Resumos*, Araxá, MG, pp. 377–378.
Barriga, R. (1976) Reports by regional investigators: B. South America. In: International *Meloidogyne* Project. *Proceedings of the Research Planning Conference on Root-knot Nematodes* Meloidogyne *spp.* NCSU, USA, pp. 7–9.
Berthaud, J. (1978a) L'hybridation interspécifique entre *Coffea arabica* L. et *Coffea canephora* Pierre. Obtention et comparaison des hybrids triploids, Arabusta et hexaploïdes. Premièe partie. *Café Cacao Thé* 22 (1), 3–12.

Berthaud, J. (1978b) L'hybridation interspécifique entre *Coffea arabica* L. et *Coffea canephora* Pierre. Obtention et comparaison des hybrids triploids, Arabusta et hexaploïdes. Deuxième partie. *Café Cacao Thé* 22 (2), 87–112.

Berthaud, J. (1986) *Les Resources Génétiques pour l'Amélioration des Caféiers Africains Diploïdes.* ORSTOM, Paris, France.

Berthou, F., Mathieu, C. and Vedel, F. (1983) Chloroplast and mitochondrial DNA variation as an indicator of phyloegenetic relationships in the genus *Coffea* L. *Theorical and Applied Genetics* 65, 77–84.

Bertrand, B., Anzueto, F., Peña, M., Anthony, F., Eskes, A.B., Aguilar, G., Bompard, E. and Rafinon, A. (1997) Comportement agronomique et résistance aux principaux déprédateurs des lignées de Sarchimors et Catimors au Costa Rica. *Plante, Recherche, Développement* 4, 312–321.

Bertrand, B., Cilas, C., Herve, G., Anthony, F., Etienne, H. and Villain, L. (1998) Relations entre les populations de deux espèces de nématodes, *Meloidogyne exigua* et *Pratylenchus* sp., dans les racines de *Coffea arabica* au Costa Rica. *Plante, Recherche, Développement* 5, 279–286.

Bertrand, B., Aguilar, G., Santacreo, R. and Anzueto, F. (1999) El mejoramiento genético en América Central. In: Bertrand, B. and Rapidel, B. (eds) *Desafios de la Caficultura en Centroamerica.* ICCA-PROMECAFE, San Jose, CR, pp. 407–456.

Bertrand, B., Nuez, C. and Sarah, J.L. (2000a) Disease complex in coffee involving *Meloidogyne arabicida* and *Fusarium oxysporum. Plant Pathology* 48, 383–388.

Bertrand, B., Duran, M.X.P., Anzueto, F., Cilas, C., Etienne, H., Anthony, F. and Esks, A.B. (2000b) Genetic study of *Coffea canephora* coffee trees resistance to *Meloidogyne incognita* in Guatemala and *Meloidogyne* sp. in El Salvador for selection of rootstock varieties in Central America. *Euphytica* 113 (2), 79–86.

Bertrand, B., Anthony, F. and Lashermes, P. (2001) Breeding for resistance to *Meloidogyne exigua* in *Coffea arabica* by introgression of resistance genes of *Coffea canephora. Plant Pathology* 50, 637–643.

Bertrand, B., Ramirez, G., Topart, P. and Anthony, F. (2002) Resistance of cultivated coffee (*Coffea arabica* and *C. canephora*) trees to corky-root caused by *Meloidogyne arabicida* and *Fusarium oxysporum,* under controlled and field conditions. *Crop Protection* 21, 713–719.

Braudeau, J. (1970) *El Cacao.* Editorial Blume, Barcelona.

Bridge, J. (1984) *Coffee Nematode Survey of Tanzania.* Report on a visit to examine plant parasitic nematodes of coffee in Tanzania, February/March 1984. Commonwealth Institute of Parasitology, UK.

Bridge, J. (1988) Plant parasitic nematode problems in the Pacific islands. *Journal of Nematology* 20, 173–183.

Bridge, J. and Page, S.L.J. (1984) Plant nematode pests of crops in Papua New Guinea. *Journal of Plant Protection in the Tropics* 1, 99–109.

Bridge, J., Page, S.L.J. and Waller, J.M. (1982) *Plant Parasitic Nematodes and Diseases of Crops in the Santa Cruz Department of Bolivia.* UK Overseas Development Administration Report.

Bridge, J., Fogain, R. and Speijer, P. (1997) *Parasites et Ravageurs des* Musa*: Fiche Technique no. 2.* Les nématodes parasites des bananiers. *P. coffeae* (Zimmermann, 1898) Filip. & Schu. Stek., 1941; *P. goodeyi* Sher & Allen, 1953. INIBAP, Parc Scientifique Agropolis II, Montpellier, France.

Campelo, A.M.F. and Galli, F. (1980) Patogenicidade de *Helicotylenchus dihystera* (Cobb) Sher em *Theobroma cacao* L. *Revista Theobroma* 10 (1), 7–14.

Campos, H.D. and Campos, V.P. (1997) Efeito da época e forma de aplicação dos fungos *Arthrobotrys conoides, A. musiformis, Paecilomyces lilacinus* e *Verticillium chlamydosporium* no controle de *Meloidogyne exigua* do cafeeiro. *Fitopatologia Brasileira* 22, 361–365.

Campos, V.P. (1997) Controle de doenças causadas por nematóides. In: Ribeiro do Vale, F.X. and Zambolim, L. (eds) *Controle de Doenças de Plantas,* Volume 1. Editora Universitária, Viçosa, MG, Brazil, pp. 141–170.

Campos, V.P. (2002) *Coffee Nematode Survey in Minas Gerais State, Brazil.* Research report of the grant by PNP&D – café EMBRAPA. Brasilia.

Campos, V.P. and Lima, R.D. (1986) Nematóides parasitas do cafeeiro. In: Rena, A.B., Malavolta, E., Rocha, M. and Yamada, T. (eds) *Cultura do Cafeeiro, Fatores que Afetam a Produtividade.* Associação Brasileira Para Pesquisa da Potassa e do Fosfato, Piracicaba, SP, Brazil, pp. 379–389.

Campos, V.P. and Melles, C.C.A. (1987) Ocorrência e distribuição de espécies de *Meloidogyne* em cafezais dos Campos das Vertentes e do Sul de Minas. *Nematologia Brasileira* 11, 233–241.

Campos, V.P., Lima, R.D. and Almeida, V.F. (1985) Nematóides parasitas do cafeeiro. *Informe Agropecuário. Belo Horizonte* 11, 50–58.

Campos, V.P., Lima, R.D. and Almeida, V.F. (1987) Nematóides parasitas de grandes culturas identificados em localidades de Minas Gerais e São Paulo. *Nematologia Brasileira* 11, 226–232.

Campos, V.P., Souza, J.T. and Souza, R.M. (1998) Controle de fitonematóides por meio de bactérias. In: Luz, W.C. (ed.) *Revisão Anual de Patologia de Plantas*, Volume 6. Passo Fundo, RS, Brazil, pp. 285–327.

Cannell, M.G.R. (1971) Seasonal patterns of growth and developement of Arabica coffee in Kenya. *Kenya Coffee* 36 (422), 68–74.

Cannell, M.G.R. (1985) Physiology of the coffee crop. In: Clifford, M.N. and Willson, K.C. (eds) *Coffee: Botany, Biochemistry and Production of Beans and Beverage*. Croom Helm, New York, pp. 108–134.

Cannell, M.G.R. and Huxley, P.A. (1969) Seasonal differences in the pattern of assimilate movement in branches of *Coffea arabica* L. *Annals of Applied Biology* 64, 345–357.

Capot, J. (1972) L'amélioration du caféier en Côte d'Ivoire. Les hybrides 'Arabusta'. *Café Cacao Thé* 16 (1), 3–18.

Carneiro, R.G. (1995) Reação de progenies de café 'Icatu' a *Meloidogyne incognita* raça 2 em condições de campo. *Nematologia Brasileira* 19 (1–2), 53–59.

Carneiro, R.G. and Carneiro, R.M.D.G. (1982a) Levantamento preliminar dos nematóides do gênero *Meloidogyne* associados a cultura do café no norte do Paraná, no período de 1978 a 1980. *Nematologia Brasileira* 6, 133–139.

Carneiro, R.G. and Carneiro, R.M.D.G. (1982b) Seleção preliminar de plantas para rotação de culturas em áreas infestadas por *Meloidogyne incognita* nos anos de 1979 e 1980. *Nematologia Brasileira* 6, 141–148.

Carneiro, R.G., Antonio, H., Brito, J.A. and Alteia, A.A.K. (1990a) Identificação de espécies e raças de *Meloidogyne* na região noroeste do Estado de São Paulo: resultados preliminares. In: *Congresso Brasileiro de Nematologia. XIV, Resumos*, Londrina, PR, p. 4.

Carneiro, R.G., Brito, J.A., Antonio, H. and Alteia, A.A.K. (1990b) Identificação de novo biótipo de *Meloidogyne* incognita na região noroeste do Estado do Paraná. *Congresso Brasileiro de Nematologia. XIV, Resumos*, Londrina, PR, pp. 1–2.

Carneiro, R.G., Alteia, A.A.K. and Brito, J.A. (1992) Levantamento da ocorrência e da freqüência de espécies e de raças fisiológicas de *Meloidogyne* no Noroeste do Paraná. *Nematologia Brasileira* 16 (1–2), 88–89.

Carneiro, R.M.D.G. (2003) Una visão mundial sobre ocorrência e patogenicidade de *Meloidogyne mayaguensis* em goiabeira e outras culturas. *Congresso Brasileiro de Nematologia, XXIV. Palestras e Resumos*. Petrolina, PE, Brazil, p. 22.

Carneiro, R.M.D.G. and Almeida, M.R.A. (2000) Distribution of *Meloidogyne* sp. on coffee in Brazil: identification, characterization and intraspecific variability. In: *Mejoramiento sostenible de café arabica por los recursos genéticos, asistido por los marcadores moleculares, com enfasis en la resistência a los nemátodos. Turrialba*. Publicacion Especial. CATIE/IRD, Turrialba, pp. 43–48.

Carneiro, R.M.D.G. and Jorge, C.L. (2001) Seletividade fisiológica de populações de *Meloidogyne incognita e Meloidogyne paranaensis* quando multiplicadas durante sucessivas gerações em tomateiros e cafeeiros. *Simpósio de Pesquisa dos Cafés do Brasil, II, Brasilia, Embrapa Café – Resumos*, pp. 82–83.

Carneiro, R.M.D.G., Carneiro, R.G., Abrantes, I.M.O., Santos, M.S.N.A. and Almeida, M.R.A. (1996) *Meloidogyne paranaensis* n.sp. (Nemata: Meloidogynidae), a root knot nematode parasitizing coffee in Brazil. *Journal of Nematology* 28, 177–189.

Carneiro, R.M.D.G., Almeida, M.R.A. and Quénéhervé, P. (2000) Enzyme phenotypes of *Meloidogyne* spp. populations. *Nematology* 2, 645–654.

Carneiro, R.M.D.G., Moreira, W.A., Almeida, M.R.A. and Gomes, A.C.M.M. (2001) Primeiro registro de *Meloidogyne mayaguensis* em goiabeira no Brasil. *Nematologia Brasileira* 25, 223–228.

Carneiro, R.M.D.G., Tigano, M.S., Randig, O., Almeida, M.R.A. and Sarah, J.L. (2004) Identification and genetic diversity of *Meloidogyne* spp. on coffee from Brazil, Central America and Hawaii. *Nematology* 6, 287–298.

Carneiro Filho, F. and Yamaguchi, K. (1995) Comportamento de progenies de *Coffea arabica* enxertados em *Coffea canephora* em área com nematóide *Meloidogyne coffeicola*, Lordello and Zamith, 1960, no Paraná. In: *Congresso Brasileiro de Pesquisas Cafeeiras, XXI, Caxambú. Resumos*, Brasília, pp. 41–42.

Castillo, G.P., Hernández, E.E., Teliz-Ortiz, D., Nieto, D., Obregon, E., Castillo, J. and Ruiz, R. (1995) La corchosis del cafeto en México. In: *XV Simpósio sobre Caficultura Latino-Americana, Memoria, IICA, Tegucigalpa, Honduras. 1992–07–21/24, Xalpa*, Veracruz, México 1.

Castro, J.M.C. and Campos, V.P. (2004a) Ocorrência de *Meloidogyne coffeicola* em cafeeiro na região do Alto Paranaiba em Minas Gerais. *Fitopatologia Brasileira* 29, 227.

Castro, J.M.C. and Campos, V.P. (2004b) Ocorrência de *Meloidogyne paranaensis* em cafeeiro na região do Sul de Minas Gerais. *Summa Phytopathologica* 30, 507.

Castro, J.M.C., Campos, V.P. and Naves, R.L. (2003a) Ocorrência de *Meloidogyne paranaensis* em cafeeiro na região do Alto Paranaiba em Minas Gerais. *Fitopatologia Brasileira* 28, 565.

Castro, J.M.C., Lima, R.D. and Carneiro, R.M.D.G. (2003b) Variabilidade isoenzimática de populações de *Meloidogyne* spp. provenientes de regiões brasileiras produtoras de soja. *Nematologia Brasileira* 27 (1), 1–12.

Caveness, F.E. (1967) *Nematology Studies 1960/1965*. Ministry of Agriculture and Natural Resources. US AID Project, Lagos, Nigeria.

Charrier, A. and Eskes, A. (1997) Les caféiers. In: Charrier, A., Jacquot, M., Hamon, S. and Nicolas, D. (eds) *L'Amélioration des Plantes Tropicales*. CIRAD/ORSTOM, Montpellier, France, pp. 171–196.

Chevalier, A. (1947) Lés caféiers du globe. III. Systématique des caféiers et faux-caféiers, maladies et insectes nusibles, *Encyclopédie Biologique* no. 28. Paris, Paul Lechevalier.

Chitwood, B.G. and Berger, C.A. (1960) Preliminary report on nemic parasites of coffee in Guatemala with suggested and interim control measures. *Plant Disease Reporter* 44, 841–847.

Cilas, C., Villain, L. and Licardie, D. (1993) Etude de la répartition de *Pratylenchus* sp. dans une plantation de caféiers au Guatemala. In: *Quinzième Colloque Scientifique International sur le Café, 1993/06/06–11*, Montpellier, France, pp. 843–847.

Cilas, C., Lanaud, C., Paulin, D., Nyassé, S., N'Goran, J.A., Kébé, B.K., Ducamp, M., Flament, M.K.H., Risterucci, A.M., Pieretti, I., Sounigo, O., Thévenin, J.M. and Despréaux, D. (1998) Resistance pod rot caused by *Phytophthora* spp. The search for resistance components. *Plantations Recherche Développement* 5 441–449.

Corbett, D.C.M. (1961) *Report of Department of Agriculture 1959/1960. Part II.* Nyasaland, pp. 157–158.

Cotterel, G.S. (1930) *Bulletin Department of Agriculture.* Accra Gold Coast.

Cross, J., Combes, M.C., Troulot, P., Anthony, F., Hamon, S., Charrier, A. and Lashermes, P. (1998) Phylogenetic relationships of *Coffea* species: new evidence based on the chloroplast DNA variation analysis. *Molecular Phylogenics and Evolution* 9, 109–117.

Crozzoli, R., Lamberti, F., Greco, N. and Rivas, D. (2001) Plant parasitic nematodes associated with cocoa in Choroni, Cumboto and Cuyagua, Aragua State. *Fitopatologia Venezolana* 14 (1), 5–12.

Curi, S.M. (1973) Novas observações sobre um nematóide do cafeeiro. *O Biológico* 39, 207–209.

Curi, S.M., Monteiro, L.G.E., de Bona, A. and Cintra, A.F. (1969) Levantamento do nematóide do cafeeiro *Meloidogyne coffeicola*, no Estado de São Paulo. *O Biológico* 35, 41–44.

Curi, S.M., Carvalho, A., Moraes, F.P., Monaco, L.C. and de Arruda, H.V. (1970) Novas fontes de resistência genetica de Coffea no controle de nematóide do cafeeiro, *Meloidogyne exigua. O Biológico* 36, 293–295.

Curi, S.M., da Silveira, S.G.P. and Elias, E.G. Jr (1977) Resultados de produção e da proteção do sistema radicular de cafeeiros sob controle químico do nematóide *Meloidogyne incognita* (Kofoid & White, 1969) Chitwood 1949, em condições de campo. *Nematologia Brasileira* 2, 93–99.

da Costa, W.M., Gonçalves, W. and Fazuoli, L.C. (1991) Produção de café Mundo Novo em porta enxertos de *Coffea canephora* em área infestadas por *Meloidogyne incognita* raça 1. *Nematologia Brasileira* 15, 43–50.

Dalmasso, A., Castagnone-Sereno, P. and Abad, P. (1992) Seminar: tolerance and resistance of plants to nematodes – knowledge, needs and prospects. *Nematologica* 38, 466–472.

D'Antonio, A.M., Libeck, P.R., Coelho, A.J.E. and de Paula, V. (1980) Levantamento de nematóides parasitas do cafeeiro que ocorrem no sul de Minas Gerais. In: *Congresso Brasileiro de Pesquisas Cafeeiras, Campos do Jordão. Rio de Janeiro, IBC GERCA*, 440.

De Franqueville, H. (2001) Varietal resistance. In: Mariau, D. (ed.) *Diseases of Tropical Tree Crops*. Science Publishers, Inc., Enfield, New Hampshire, pp. 113–162.

de Moraes, M.V. and Lordello, L.G.E. (1977) Estudo de três populações de nematóides nocivos ao cafeeiro. *Nematologia Brasileira* 2, 249–255.

de Moraes, M.V., Lordello, L.G.E., Piccinin, O.A. and Lordello, R.R.A. (1972) Pesquisas sobre plantas hospedeiras do nematóide do cafeeiro *Meloidogyne exigua* Goeldi, 1887. *Ciência e Cultura* 24, 628–660.

de Moraes, M.V., Lordello, L.G.E., Lordello, R.R.A. and Piccinin, O.A. (1973a) Novas pesquisas sobre plantas hospedeiras do nematóide do cafeeiro, *Meloidogyne exigua* Goeldi, 1887. *Anais da Escola Superior de Agricultura Luiz de Queiroz* 30, 71–75.

de Moraes, M.V., Lordello, L.G.E., Piccinin, O.A. and Lordello, R.R.A. (1973b) Susceptibilidade de cafeeiros ao ataque por um nematóide. *Nematologia Mediterranea* 2, 107–110.

de Moraes, M.V., Lordello, L.G.E., Reis, A.J., Thomaziello, R.A. and Gonçalves, W. (1977a) Ensaio de rotação de culturas para reaproveitamento com cafeeiros de terras infestadas por *Meloidogyne exigua. Nematologia Brasileira* 2, 257–265.

de Moraes, M.V., Thomaziello, R.A., Lordello, L.G.E. and Krinski, J. (1977b) Ensaios de tratamento de solo por produtos químicos para uso em viveiro de café. *Nematologia Brasileira* 2, 231–244.

de Oliveira, C.M.G. (1996) Efeito de Densidades Populacionais de *Pratylenchus brachyurus* (Nemata: Pratylenchidae) no Crescimento de Plântulas de *Coffea arabica* cv. Mundo Novo e *Coffea canephora* cv. Apoatã. MSc Thesis, Escola Superior de Agricultura Luiz de Queiroz, Piracicaba, SP, Brazil.

de Oliveira, C.M.G., Gonçalves, W. and Monteiro, A.R. (2000) Espécies de *Meloidogyne* e raças de *Meloidogyne incognita* em cafezais do Estado de São Paulo. *Fitopatologia Brasileira* 25 (Supplement), 339.

de Oliveira, C.M.G., Gonçalves, W. and Monteiro, A.R. (2001) Espécies de *Meloidogyne* e raças de *Meloidogyne incognita* em cafezais do Estado de São Paulo. *Revista de Agricultura* 76, 155–164.

D'Souza, G.I. and Screenivasan, C.S. (1965) A note on the reniform nematode *Rotylenchulus reniformis*, on arabica coffee in South India. *Indian Coffee* 29, 11–13.

Duncan, L.W., Inserra, R.N., Thomas, W.K., Dunn, D., Mustika, I., Frisse, L.M., Mendes, M.L., Morris, K. and Kaplan, D.T. (1999) Molecular and morphological analysis of isolates of *Pratylenchus coffeae* and closely related species. *Nematropica* 29, 61–80.

Edwards, D. (1955) Further observations in the occurrence of nematodes of the genus *Meloidogyne* in Gold Coast. *Journal of Helminthology* 29, 153–170.

Eisenback, J.D., Bernard, E.C. and Schmitt, D.P. (1995) Description of the kona coffee root-knot nematode, *Meloidogyne konaensis* n.sp. *Journal of Nematology* 26, 363–374.

Elmiligy, I.A. (1968) Three new species of the genus *Meloidogyne*, Goeldi, 1887 (Nematoda: Heteroderidae). *Nematologica* 14, 577–590.

Eskes, B., Engels, J. and Lass, T. (1998) The CFC/ICCO/IPGRI project: a new initiative on cocoa germplasm utilization and conservation. *Plantations Recherche Développement* 5, 412–422.

FAO (2002) Food and Agriculture Organization of the United Nations. World production of green coffee. FAOSTAT Agriculture Data. Available at: apps.fao.org

Favoreto, A.J. and Santos, J.M. (2001) Caracterização bioquímica e morfológica de populações de nematóides de galhas em café (*Coffea arabica* L.) em regiões produtoras do Estado de São Paulo. *Congresso Brasileiro de Nematologia, XXIII*, Marília, São Paulo, p. 93.

Fazuoli, L.C. (1986) Genética e melhoramento do cafeeiro. In: Rena, A.B., Malavolta, E., Rocha, M. and Yamada, T. (eds) *Cultura do Cafeeiro, Fatores que Afetam a Produtividade.* Associação Brasileira para Pesquisa da Potassa e do Fosfato, Piracicaba, SP, Brazil, pp. 87–113.

Fazuoli, L.C. and Lordello, R.R.A. (1978) Fontes de resistência em espécies de cafeeiro a *Meloidogyne exigua. Sociedade Brasileira de Nematologia* 3, 49–52.

Fazuoli, L.C., Medina Filho, H.P., Gonçalves, W., Guerreiro Filho, O. and Silvarolla, M.B. (2002) Melhoramento do cafeeiro: variedades tipo arabica obtidas no Instituto Agronomico de Campinas. In: Zambolim, L. (ed.) *O Ecstado da Arte de Tecnologias na Produção de Café.* Editora UFV, Viçosa, MG, pp. 162–215.

Fernández, M. and Ortega, J. (1998) An overview of nematological problems in Cuba. *Nematropica* 28, 151–163.

Ferraz, E. (1979) *Nematóides do Cacueiro.* IV Curso Internacional de Cacau. Centro de Pesquisa do Cacau. Itabuna, BA, Brazil.

Ferraz, S. (1980) Reconhecimento das espécies de fitonematóides presente nos solos do estado de Minas Gerais. *Experientia* 26 (11), 255–328.

Figueroa, A. (1988) Reconocimiento y análisis de problemas de los namátodos en viveiros de café (*Coffea arabica*). *Memorias XX Congresso ONTA*, Costa Rica, p. 34.

Figueroa, A. and Perlazo, F. (1982) Investigacion sobre *Meloidogyne* en Costa Rica. In: *Proceedings of the Third Research Planning Conference on Root-knot Nematodes,* Meloidogyne *spp. Region I,* 11–15 January, pp. 12–25.

Flores, J.M. and Yépez, T. (1969) *Meloidogyne* in coffee in Venezuela. In: Peachey, J.E. (ed.) *Nematodes of Tropical Crops. Technical Communication no. 40.* Commonwealth Bureau of Helminthology, St Albans, UK, pp. 251–256.

Flores, L. and López, J. (1989) Caracterisación morfológica del nematodo nodulador del cafeto *Meloidogyne exigua* (Nematoda: Heteroderidae). I. Hembras y huevos. *Turrialba* 39, 352–360.

Fluitter, H.J. and Mulholland, J.J. (1941) Gegevens, Verkegen bijihet onderzoek naar de waardplanten van *Tylenchus coffeae. Bergcultures* 15, 1588–1593.

Freire, F.C.O. and Monteiro, A.R. (1978) Nematoides da região Amazonica II: mematóides parasitos e de vida livre associados a pimenta do reino (*Piper nigrum* L.) e ao cacaueiro (*Theobroma cacao* L.). *Acta Amazonica* 8, 561–564.

Ghesquiere, J. (1921) Nouveaux parasites du cacoyer. Maladie vermiculire du cacaoyer *tylenchus (Heterodera) radicicola* [Greef.] et relation aves la maladie du Diplodia. *Bullentin de e Agriculture* 12, 709–718.

Ghini, R. and Bettiol, W. (1991) Coletor solar para desinfestação de substratos. *Summa Phytopathologica* 17, 281–286.

Ghini, R., Inomoto, M.M., Armond, G. and Braga, C.A.S. (1991) Utilização de coletor solar para o controle de *Meloidogyne arenaria. Nematologia Brasileira* 15, 205.

Gnanapragasam, N.C. (1982) Effect of potassium fertilization and of soil temperature on the incidence and pathogenecity of the incidence and pathogenecity of the root lesion nematode, *Pratylenchus loosi* Loof, on tea (*Camelia sinensis* L.). *Tea Quarterly* 51(4), 169–174.

Golden, A.M., Lopez, R. and Vilchez, H. (1992) Description of *Pratylenchus gutierrezi* n.sp. (Nematoda: Pratylenchyda) from coffee in Costa Rica. *Journal of Nematology* 24, 298–304.

Göldi, E.A. (1889) Der kaffeenematode Brasiliens (*Meloidogyne exigua* G.). Zoologische Jahrbücher, abt. *Systematik R* 4, 261–267.

Göldi, E.A. (1892) Relatório sobre a moléstia do cafeeiro na Província do Rio de Janeiro. *Archivos do Museu Nacional Rio de Janeiro* 8 (1987), 7–123.

Gomez, J. (1980) Resultados preliminares sobre a distribuição de fitonematóides en la zona cafetera de los departamentos de Caldas, Risaralda y Quindio en Colombia. *Nematropica* 10, 67.

Gonçalves, W. (1995) Problemas na produção brasileira de café devido a fitonematóides. *Congresso Internacional de Nematologia Tropical, XIX, Resums,* Rio Quente, GO, pp. 216–223.

Gonçalves, W. (2000) Manejo de fitonematóides em cafeeiro no Estado de São Paulo, Brazil. *Congresso Brasileiro de Nematologia, XXII,* Uberlândia, MG, pp. 42–43.

Gonçalves, W. and Ferraz, L.C.B. (1987) Resistência do cafeeiro a nematóides II. Testes de progênies e híbridos para *Meloidogyne incognita* raça 3. *Nematologia Brasileira* 9, 125–142.

Gonçalves, W. and Pereira, A.A. (1998) Resistência de cafeeiro a nematóides IV – reação de cafeeiros derivados de hibrídos Timor a *Meloidogyne exigua. Nematologia Brasileira* 22 (1), 39–50.

Gonçalves, W. and Silvarola, M.B. (2001) Nematóides parasitos do cafeeiro In: Zambolim, L. (ed.) *Tecnologias de Produção de Café com Qualidade.* Editora Universitária, UFV, Viçosa, MG, pp. 199–267.

Gonçalves, W., Thomaziello, R.A., Moraes, M.V., Fernandes, J.A.E., Costa, A.M., Corsi, T., Junqueira, C.A. and Lacerda, L.A.O. (1978) Estimativas de danos ocasionados pelos nematóides do cafeeiro. In: *6° Congresso Brasileiro de Pesquisas Cafeeiras de Ribeirão Preto, São Paulo, Brazil-IBC/GERCA,* pp. 182–186.

Gonçalves, W., Ferraz, L.C.C.B., Lima, M.M.A. and Silvarola, M.B. (1996) Reações de cafeeiros às raças 1, 2 e 3 de *Meloidogyne incognita. Summa Phytopathologica* 22, 172–177.

Gonçalves, W., Fazuoli, L.C., Lima, M.M.A., Silvarola, M.B., Favoreto, A.J., Mota Filho, C. and Guerreiro, G. (1998a) Café IAC 4160 promissora fonte de resistência a *Meloidogyne paranaensis. Congresso Brasileiro de Nematologia, XXI,* Maringá PR, p. 53.

Gonçalves, W., Pereira, A.A. and Mendes, A.N.G. (1998b) Reação de progenies do hibrído Timor a *Meloidogyne exigua. Congresso Brasileiro de Nematologia, XXI,* Maringá, PR, p. 54.

Guerra Neto, E.G. and D'Antonio, A.M. (1984) Nematóides parasitas em lavouras cafeeiras do sul de Minas Gerais. In: *11° Congresso Brasileiro de Pesquisas Cafeeiras, Londrina IBC/GERCA,* p. 171.

Guerra Neto, E.G., D'Antonio, A.M., Almeida, S.R. and Lordello, L.G.E. (1983) Ocorrência do nematóide *Meloidogyne coffeicola* Lordello & Zamith, 1960, em lavouras de café no sul do Estado de Minas Gerais. *Revista de Agricultura* 58, 45–48.

Gutierrez, G. and Jimenez, Q.M.F. (1970) Algumas obrservaciones sobre la infestacion del cafe practicada en Guatemala y El Salvador como medio para el control de nematodes. *Revista Cafetera, Guatemala* 98, 35–47.

Hernández, A., Fargette, M., Molinier, V., Ramenason, H., Decazy, B. and Sarah, J.L. (1996) Enzymatic characterization and reproductive fitness on coffee of root knot nematode populations from Central America. Third International Nematology Congress, 7–12 July 1996, Gosier, Guadeloupe. *Nematropica* 26, 264.

Hernández, F. (1997) Etude de la Variabilité Intra et Interspécifique des Nématodes du Genre *Meloidogyne* Parasites des Caféiers en Amérique Centrale. PhD Thesis, Université des Sciences et Techniques du Languedoc, Montpellier, France.

Herrera, I.C. and Marban-Mendoza, N. (1999) Effect of leguminous cover crops on plant parasitic nematodes associated with coffee in Nicaragua. *Nematropica* 29, 223–232.

Hervé, G. (1997) *Caractérisation Biologique et Moléculaire de Populations de Nématodes Phytoparasites dans les Plantations de Caféiers d'Amérique Centrale.* ENSAR, D.E.A. Rennes, France.
Hidalgo-Diaz, L., Bourne, J.M., Kerry, B.R. and Rodriguez, M.G. (2000) Nematophagous *Verticillium* spp. in soils infested with *Meloidogyne* spp in Cuba: isolation and screening. *International Journal of Pest Management* 46, 277–284.
Huang, S.P., Resende, I.C., de Souza, P.E. and Campos, V.P. (1983) Effect of aldicarb, ethoprop and carbofuran on control of coffee root-knot nematode, *Meloidogyne exigua. Journal of Nematology* 15, 510–514.
Hutton, D.G., Eason-Heath, S.A.E. and Coates-Beckford, P.L. (1982) Control of *Meloidogyne incognita* affecting coffee. In: *Proceedings of the Third Research Planning Conference on Root-knot Nematodes* Meloidogyne *spp., Region I,* 11–15 January, pp. 109–113.
Inserra, R.N., Duncan, L.W., Dunn, D., Kaplan, D.T. and Porazinska, D. (1998) *Pratylenchus pseudocoffeae* from Florida and its relationship with *P. gutierrezi* and *P. coffeae. Nematologica* 44, 683–712.
Inserra, R.N., Duncan, L.W., Troccoli, A., Dunn, D., Maia Dos Santos, J., Kaplan, D.T. and Vovla, N. (2001) *Pratylenchus jaehni* sp. n. from citrus in Brazil and its relationship with *P. coffeae* and *P. loosi. Nematology* 3, 653–665.
Jaehn, A. (1984) Viabilidade do uso de nematicidas em cafezal novo, instalado em solo infestado por *Meloidogyne incognita. Nematologia Brasileira* 8, 275–283.
Jaehn, A. and Rebel, E.K. (1984) Sobrevivência do nematóide de galhas *Meloidogyne incognita* em substrato infestado, para produção de mudas de cafeeiros sadios. *Nematologia Brasileira* 8, 319–324.
Jaehn, A., Rebel, E.K. and Lordello, L.G.E. (1980) A origem do nematóide *Meloidogyne coffeicola. Nematologia Brasileira* 4, 159–161.
Jaehn, A., Rebel, E.K. and Matiello, J.B. (1984) Viabilidade de recuperação de mudas de cafeeiro infestadas por *Meloidogyne incognita* através de nematicidas. *Nematologia Brasileira* 8, 295–300.
Jiménez-Saenz, E. (1971) Relacion entre el ataque de nemátodos y la muerte súbita del cacao (*Theobroma cacao* L.) en Bahia, Brasil. *Turrialba* 19, 225–260.
Jobert, C. (1878) Sur une maladie du caféier observée au Brésil. *Comptes Rendus de la Societé de Biologie, Paris* 87, 941–943.
Kermarrec, A. and La Massese, C.S. (1972) New contributions to the study of the nematode fauna in the French West Indies. *Nematropica* 2 (2), 41–43.
Knight, C. (1998) Sustainable cocoa program. *Plantations Recherche Développement* 5 (6), 387–392.
Koshy, P.K., Geetha, S.M. and Walia, R.K. (1992) *Nematode Pests of Crops.* CBS Publishers and Distributors, Delhi, India, pp. 214–227.
Krug, C.A. (1969) *Estudio Mundial Del Cafe.* FAO, Estudios Agropecuarios no. 76. Organizacion del las Naciones Unidas para la Agricultura y la Alimentacion.
Krzyzanowski, A.A., Figueiredo, R., Santiago, D.C. and Favoreto, L. (2001) Levantamento de espécies e raças de *Meloidogyne* em cafeeiros no Estado do Paraná, *Simpósio de Pesquisa dos Cafés do Brasil,* II, Vitória, ES, p. 81.
Kubo, R.K., Inomoto, M.M., Oliveira, C.M.G. and Antedomênico, S.R. (1999) Ocorrência de fitonematóides em cafezais das regiões de Bauru e Marília. *Fitopatologia Brasileira* 24 (Supplement), 346.
Kubo, R.K., Inomoto, M.M., Oliveira, C.M.G., Antedomênico, S.R. and Monteiro, A.R. (2001) Nematóides associados a cafeeiros do Estado de São Paulo. In: *Congresso Brasileiro de Nematologia XXIII, Resumos,* Marilia, SP, p. 91.
Kubo, R.K., Oliveira, C.M.G., Antedomenico, S.R., Monteiro, A.R. and Inomoto, M.M. (2002a) Distribuição de *Pratylenchus* spp. em cafezais no Estado de São Paulo. *XXXV Congresso Brasileiro de Fitopatologia, Recife, Palestras e Resumos,* p. 230.
Kubo, R., Silva, R., Tomazini, M., Oliveira, C., Mazzafera, P. and Inomoto, M. (2002b) Patogenicidade de *Pratylenchus coffeae* em plântulas de cafeeiro cv. Mundo Novo. *Fitopatologia Brasileira* 28 (1), 41–48.
Kumar, A.C. (1982) Evaluation of Nemacur 5G against the coffee root lesion nematode, *Pratylenchus coffeae. Journal of Coffee Research* 12, 1–7.
Kumar, A.C. (1984) Resistance in coffee to *Meloidogyne* spp. and occurrence of intersexes *In M. thamesi. Nematologica* 30, 108–110.
Kumar, A.C. and Viswanathan, P.R.K. (1972) Studies on physiological races of *Pratylenchus coffeae. Journal of Coffee Research* 2, 10–15.
Kumar, A.C., Viswanathan, P.R.K. and D'Souza, G.I. (1971) A study on plant parasitic nematodes of certain commercial crops in coffee tracts of south India. *Indian Coffee* 35 (6), 222–224.
Lamberti, F., Boiboi, J.B., Ciancio, A., Tuopay, D.K., Jimehez, E.A. and Elia, F. (1992) Plant parasitic nematodes associated with tree crops in Liberia. *Nematologia Mediterranea* 20, 79–85.

Lashermes, P., Combes, M.C., Trouslot, P. and Charrier, A. (1997) Phylogenetic relationships of coffee-tree species (*Coffea* L.) as inferred from ITS sequences of nuclear ribosomal DNA. *Theorical and Applied Genetics* 94, 947–955.

Lashermes, P., Combes, M.C., Robert, J., Trouslot, P., D'Hont, A. and Charrier, A. (1999) Molecular characterization and origin of the *Coffea arabica* L. genome. *Molecular and General Genetics* 261, 259–266.

Leroy, T. (1993) Diversité, Paramètres Génétiques et Amélioration par Sélection Récurrente Réciproque du Caféier. PhD Thesis, Ecole Nationale Supérieure Agronomique de Rennes, Montpellier, France.

Lima, I.M., Souza, R.M., Silva, C.P. and Carneiro, R.M.D.G. (2003) Detecção de *Meloidogyne spp* em áreas nativas de Mata Atlântica do Estado do Rio de Janeiro. *Congresso Brasileiro de Nematologia, XXIV, Paletras e Resumos*, Petrolina, PE, Brazil, p. 156.

Lima, R.D. (2002) *Coffee Nematode Survey in Minas Gerais State, Brazil.* Research Report of the grant by PNP and D-Café EMBRAPA, Brasília.

Lima, R.D. and Ferraz, S. (1985) Biologia de *Meloidogyne exigua* II. Desenvolvimento pós-embriogênico em cafeeiro 'Mundo Novo'. *Revista Ceres* 32 (183), 349–361.

Lima, R.D., Campos, V.P., Huang, S.P. and Melles, C.A.A. (1985) Reprodutividade e parasitismo de *Meloidogyne exigua* em ervas daninhas que ocorrem em cafezais. *Nematologia Brasileira* 9, 63–72.

Loof, P.A.A. (1991) The family Pratylenchidae Thorne, 1949. In: Nickle, W.R. (ed.) *Manual of Agricultural Nematology.* Marcel Dekker, New York, pp. 363–421.

Loof, P.A.A. and Sharma, R.D. (1980) *Discocriconemella* species from Bahia state, Brazil (Nematoda: Criconematidae). *Mededelingen Fakulteit Landbouwwetenschappen Rijksunivérsiteit, Gent* 45, 795–805.

López, R. (1994) Morfología de *Dolichodorus minor* (Nemata: Dolichodoridae) asociado al cacao en el sureste de Costa Rica. *Agronomia Costarricense* 18 (1), 87–92.

López, R. and Salazar, L. (1989) *Meloidogyne arabicida* sp. n. (Nemata: *Heteroderidae*) nativo de Costa Rica. Un nuevo y severo patógeno del cafeto. *Turrialba* 39, 313–323.

López, R. and Vilchez, H. (1991) Two new hosts of the coffee root knot nematode, *Meloidogyne exigua* in Costa Rica. *Agronomica Costarricence* 15 (1–2), 163–166.

Lopez, R., Salazar, F.L. and Asofeifa, C.J. (1980) Observaciones sobre la distribucion espacial de nematodos asociados al cacao en Costa Rica. *Nematropica* 10 (1), 3–4.

Lordello, A.I.L. (2002) *Coffee Nematode survey in São Paulo State, Brazil.* Research Report of the grant by PNP and D-Café EMBRAPA, Brasília.

Lordello, A.I.L. and Lordello, R.R.A. (1991) Raças de *Meloidogyne incognita* identificadas em algumas plantas. *Nematologia Brasileira* 15, 201–202.

Lordello, A.I.L. and Lordello, R.R.A. (2001) Nematóides encontrados em cafezais do Estado de São Paulo. *Nematologia Brasileira* 25, 115.

Lordello, A.I.L., Lordello, R.R.A. and Fazuoli, L.C. (2001) Levantamento de espécies de *Meloidogyne* em cafeeiros no Estado de São Paulo. *Nematologia Brasileira* 25, 116.

Lordello, L.G.E. (1964) Contribuição ao conhecimento dos nematóides que causam galhas em raízes de plantas em São Paulo e Estados vizinhos. *Anais da Escola Superior de Agricultura Luiz de Queiroz* 21, 181–218.

Lordello, L.G.E. (1967) O nematóide coffeicola invade São Paulo. *Revista de Agricultura* 42, 162.

Lordello, L.G.E. (1968) Nematóides associados a uma doença do cacaueiro. *Revista de Agricultura* 43, 154.

Lordello, L.G.E. (1971) Bahia: nematóides que atacam o cafeeiro. *Revista de Agricultura* 46 (4), 175.

Lordello, L.G.E. (1972) Nematode pests of coffee. In: Webster, J.M. (ed.) *Economic Nematology.* Academic Press, London, pp. 268–282.

Lordello, L.G.E. (1980) Estado atual do nematóide reniforme como parasita do cafeeiro. *Revista de Agricultura* 55, 62.

Lordello, L.G.E. (1982) Nova ocorrência do nematóide *Meloidogyne hapla* em cafeeiro. *Revista de Agricultura* 57, 6.

Lordello, L.G.E. (1984) *Nematóides das Plantas Cultivadas*, 8th edn. Livraria Nobel S.A.

Lordello, L.G.E. and Hashizume, H. (1971) Susceptibilidade da variedade Konilon de *Coffea canephora* a um nematóide. *Revista de Agricultura* 46, 157–158.

Lordello, L.G.E. and Lordello, R.R.A. (1972a) Duas plantas hospedeiras novas do nematóide *Meloidogyne coffeicola. Anais da Escola Superior de Agricultura Luiz de Queiroz* 29, 61–62.

Lordello, L.G.E. and Lordello, R.R.A. (1972b) *Meloidogyne incognita* ataca cafeeiro no Paraná. *O Solo* 64, 27.

Lordello, L.G.E. and Mello Filho, A.T. (1969) O capim pangola difunde nematóides. *Revista de Agricultura* 44, 122.

Lordello, L.G.E. and Mello Filho, A.T. (1970) Mais um nematóide ataca o cafeeiro. *Revista de Agricultura* 45, 102.

Lordello, L.G.E. and Zamith, A.P.L. (1960) *Meloidogyne coffeicola* sp. n., a pest of coffee trees in the state of Paraná, Brazil (Nematoda, Heteroderidae). *Revista Brasileira de Biologia* 20, 375–379.

Lordello, L.G.E., Monteiro, A.T. and D'Arce, R.D. (1968) Distribuição geográfica dos nematóides nocivos do cafeeiro. *Revista de Agricultura* 43, 79–82.

Lordello, L.G.E., Carneiro Filho, I., Rebel, E.K., Guidolin, J.A. and Lordello, R.R.A. (1974) Identificação de nematóides em cafezais do Estado do Paraná. *Nematologia Brasileira* 1, 16–24.

Lordello, R.R.A. and Fazuoli, L.C. (1980) *Meloidogyne decalineata* parasita cafeeiro na ilha de São Tomé. *Revista de Agricultura* 55, 238.

Lordello, R.R.A., Lordello, A.I.L., Martins, A.L.M. and Pereira, J.C.V.N.A. (1990) Plantio de cafezal em área infestada por *Meloidogyne exigua*. *Nematologia Brasileira* 14, 18–19.

Luc, M. (1987) A reappraisal of *Tylenchina* (Nemata). 7. The family Pratylenchidae Thorne, 1949. *Revue de Nématologie* 10, 203–218.

Luc, M. and Coomans, A.V. (1993) Description of *Xiphinema abeokutae* n.sp. and redescription of *X. paulistanum* Carvalho, 1965, two rare species (Nemata: Longidoridae). *Fundamental and Applied Nematology* 16, 393–399.

Luc, M. and de Guiran, G. (1960) Les nématodes associés aux plantes de l'ouest Africain. Liste préliminaire. *L'Agronomie Tropicale Nogent* 15, 434–449.

Luc, M. and Reversat, G. (1985) Possibilités et limites des solutions génétiques aux affections provoquées par les nématodes sur les cultures tropicales. In: Création de variétés résistantes aux nématodes des cultures: intérêt, possibilités et limites. 15–16 May 1995, Paris, France. *Comptes Rendus des Séances, Académie d'Agriculture de France* 71, 781–791.

Luna, A.M.F. (1976) Patogenicidade de *Helicotylenchus dihystera* (Cobb 1893) Sher, 1961 em *Theobroma cacao* L. Variedade 'Catongo'. Thesis, Escola Superior de Agricultura 'Luiz de Queiroz', Universidade de Sao Paulo, Brazil.

Macedo, M.C.M. (1974) Susceptibilidade de cafeeiros ao nematóide reniforme. *Solo* 66, 15–16.

Manetti Filho, J. and Carneiro, R.G. (1995) Propagação vegetativa de plantas de *Coffea* sp. para seleção a diferentes raças do nematóide *Meloidogyne incognita*. *Congresso Brasileiro de Pesquisas Cafeeiras, XXI, Resumos*, Caxambu, MG, p. 35.

Manso, E.C., Tenente, R.C.V., Ferraz, L.C.B., Oliveira, R.S. and Mesquita, R. (1994) *Catalogo de Nematóides Fitoparasitos Associados a Diferentes Tipos de Plantas no Brasil.* EMBRAPA-SPI, Brasilia, D.F.

Marban-Mendoza, N. (1995) Nemarodes fitopatogenos de cafe en Centro America e Intentos de su manejo. *XIX Congresso Internacional de Nematologia Tropical. Programa e Anais Rio Quente, Goiás, Brazil*, pp. 224–230.

Martin, G.C. (1961) Plant species attacked by root-knot nematodes (*Meloidogyne* spp.) in the Federation of Rhodesia and Nyasaland. *Nematologica* 6, 130–134.

Mata, J.S., Serra, T., Alteia, M.Z., Petek, M.R., Azevedo, J.A., Fadelli, S. and Colombo, L.A. (2000a) *Cafeeiros arabica* para resistência ao nematóide *Meloidogyne paranaensis. Simpósio de Pesquisa dos Cafés do Brasil, I, Resumos*, Poços de Caldas, MG, pp. 536–539.

Mata, J.S., Sera, T., Azevedo, J.A., Alteia, M.Z., Colombo, L.A., Sanches, R.S., Petek, M.R. and Fadelli, S. (2000b) Seleção de resistência ao nematoide *Meloidogyne paranaensis* EMN95001: IAPARLN 94066 de Catuai × Icatu em área altamente infestada. *Simpósio de Pesquisa dos Cafés do Brazil, I, Resumos*, Poços de Caldas, MG, pp. 515–518.

Maximiniano, C., Campos, V.P., Souza, R.M. and Almeida, A.R. (2001) Flutuação populacional de *Meloidogyne exigua* em cafezal naturalmente infestado por *Pasteuria penetrans. Nematologia Brasileira* 21, 63–69.

Medina Filho, H.P., Fazuoli, L.C. and da Costa, W.M. (1981) Identificação das raças 2,3 e 4 de *Meloidogyne incognita* parasitando cafeeiros. In: *Nono Congresso Brasileiro de Pesquisas Cafeeiras*, São Lourenço, MG, pp. 161–168.

Mendes, B.V. (1977) Observações histopatholófgicas de raízes de cafeeiros parasitadas por *Meloidogyne exígua* Goeldi, 1887. *Nematologica Brasileira* 2, 207–229.

Mizukubo, T. (1992a) Morphological and statistical differentiation of *Pratylenchus coffeae* complex in Japan (Nematoda: Pratylenchida). *Applied Entomology and Zoology* 27, 213–224.

Mizukubo, T. (1992b) *Pratylenchus pseudocoffeae* n.sp. (Nematoda: Pratylenchidae) from composite plants in Japan. *Applied Entomology and Zoology* 27, 437–444.

Mohotti, K.M. (1998) Non-chemical approaches for the management of the root lesion nematode, *Pratylenchus loosi* Loof, 1960 in tea (*Camellia sinensis* (L.) O. Kuntze) with special reference to use of endospore-forming bacterium, *Pasteuria penetrans*. PhD Thesis, University of Reading, UK.

Montagnon, C., Leroy, T., Bertrand, B., Charmetant, P. and Dufour, M. (2002) Recent coffee genetic improvements results. *Plante Recherche Développement* (Special Issue: Research and Coffee Growing), 84–94.

Monteiro, A.R. and Lordello, L.G.E. (1974) Encontro do nematóide *Pratylenchus coffeae* atacando cafeeiro em São Paulo. *Revista de Agricultura* 49, 164.

Monteiro, A.R., Oliveira, C.M.G., Ferraz, L.C.C.B. and Gonçalves, W. (1995) Identificação morfológica de populações de *Meloidogyne* de cafezais paulistas. In: *Congresso Internacional de Nematologia Tropical, Rio Quente, 60, Anais*, p. 82.

Monteiro, A.R., Antedomênico, S.R., Ferraz, L.C.C.B., Inomoto, M.M., Kubo, R.K. and Oliveira, C.M.G. (2001) Primeira ocorrência de *Pratylenchus vulnus* Allen & Jensen, 1951, em cafeeiro. In: *Congresso Brasileiro de Nematologia, XXIII*, Marilia, SP, p. 88.

Moura, R.M., Pedrosa, E.M. and Prado, M.D.C. (2003) Incidência de *Pratylenchus coffeae* causando severa nematose em cafeeiro no nordeste. *Fitopatologia Brasileira* 27, 649.

Naves, R.L. and Campos, V.P. (1991) Ocorrência de fungos predadores de nematóides no Sul de Minas Gerais e estudo da capacidade predatória e crescimento *in vitro* de alguns de seus isolados. *Nematologia Brasileira* 15, 152–162.

Negson, J.A. and Acosta, N. (1989) The *Fusarium oxysporum* f.sp. *coffea–Meloidogyne incognita* complex in 'Bourbon' coffee. *Nematropica* 19 (2), 161–168.

Nguyen-Ban, J. (1996) Cocoa disease and pest control. *Plantations Recherche Développement* 3 (3), 191–195.

Nickle, W.R. (1984) *Plant and Insect Nematodes.* Marcel Dekker, Inc., New York.

Noir, S., Anthony, F., Bertrand, B., Combes, M.C. and Lashermes, P. (2003) Identification of a major gene (Mex-1) from *Coffee canephora* conferring resistance to *Meloidogyne exigua* in *Coffea arabica. Plant Pathology* 52, 97–103.

Oliveira Filho, N.L., de Oliveira, J.C., Otoboni, C.E.M. and Santos, J.M. (2001) Ocorrência de *Meloidogyne paranaensis* nos principais municípios produtores de café na área de abrangência do Escritório de Desenvolvimento Rural de Marilia. In: *Congresso Brasileiro de Nematologia XXIII, Resumo*, Marilia, SP, p. 149.

Otoboni, C.E.M., Lemos, E.G.M., Santos, J.M. and Giroto, F. (2001) Influência do tratamento nematicida em cafeeiros de 8 anos de idade, infectados por *Meloidogyne* sp., sobre a eficácia do controle de ferrugem e bicho mineiro com aplicação via solo de triadimenol and dissulfoton. In: *Congresso Brasileiro de Nematologia, XXIII, Resumo*, Marilia, SP, p. 96.

Otoboni, C.E.M., Santos, J.M. and Otoboni, A.M.M.B. (2003a) Ocorrência de espécies de *Meloidogyne* Goeldi em um mesmo cafezall. *Congresso Brasileiro de Pesquisas Cafeeiras, XXIX, Resumos*, Araxá, MG, pp. 277–278.

Otoboni, C.E.M., Giroto, F. and Santos, J.M. (2003b) Efeito do tratamento químico de solo em cafeeiros sobre o controle de nematóides, ferrugem e bicho mineiro. *Congresso Brasileiro de Pesquisas Cafeeiras, XXIX, Resumos*, Araxá, MG, pp. 278–279.

Padilla, M.R. and Tronconi, N.M. (2002) Diagnóstico nematólogico en la zona cafetalera del departamento de El Paraíso. La corchosis del cafeto en México. In: *VII Seminario Nacional de Generacion y Transferencia de Tecnologia en Caficultura, Memoria, IHCAFE, Tegucigalpa, Honduras*, August 2002, San Pedro Sula, Honduras.

Palanichamy, L. (1973) Nematode problems of coffee in India. *Indian Coffee* 37, 99–100.

Paulin, D. and Eskes, A. (1995) Cocoa: breeding strategies. *Plantations Recherche Développement* 2 (6), 5–18.

Perreira, J.L. (1998) Le développement de la maladie du balai de sorcière du cacoyer: strategies de lutte appliqués. *Plantations Recherche Développement* 5 (6), 435–440.

Pineda, A. and Santacreo, R. (2000) Estabilidad de progenies Sarchimor hibridos, Catuaí × SH 2 SH3 Y retrocruza mientos Catuaí X Icatu. *Simposio Latino Americano de Cafeicultura, XIX, Memorias, San Jose, Costa Rica*, pp. 215–228.

Pinheiro, J.B., Santos, M.A., Santos, C.M. and Lellis, A.M. (2000) Ocorrência de fitonematóides em amostras oriundas de cafezais do triângulo mineiro e Alto Paranaíba. *Simpósio de Pesquisas dos cafés do Brasil, I, Resumos*, Poços de Caldas, MG, pp. 257–259.

Pinochet, J. and Guzman, R. (1987) Nematodos asociados a cultivos agrícolas en El Salvador: su importância y manejo. *Turrialba* 37, 137–146.

Pinochet, J. and Raski, D.J. (1976) *Discocriconemella repleta* n.sp., and the male of *Criconemoides inusitatus* Hoffmann, 1974 (Criconematidae; Nematoda). *Journal of Nematology* 8, 327–335.

Pinochet, J. and Ventura, O. (1980) Nematodes associated with agricultural crops in Honduras. *Turrialba* 30, 43–47.

Ponte, J.J. (1977) *Nematóides das Galhas: Espécies Ocorrentes no Brasil e Seus Hospedeiros*. Coleção Mossoroense, ESAN, RN.

Ponte, J.J. and de Castro, F.E. (1975) Lista adicional de plantas hospedeiras de nematóides das galhas, *Meloidogyne* spp. no Estado do Ceará (Brasil) referente a 1969/74. *Fitossanidade* 1, 29–30.

Radewald, J.D., O'Bannon, J.H. and Tomerlin, T. (1971) Temperature effect on reproduction and pathogenecity of *Pratylenchus coffeae* and *P. brachyurus* and survival of *P. coffeae* in roots of *Citrus jambhiri*. *Journal of Nematology* 3, 390–394.

Rahm, G. (1929) Nematóides parasitas e semiparasitas de diversas plantas cultivadas no Brasil. *Archivos do Instituto Biológico* 2, 67–137.

Rammah, A. and Hirschmann, H. (1988) *Meloidogyne mayaguensis* n.sp. (*Meloidogynidae*), a root knot nematode from Puerto Rico. *Journal of Nematology* 20, 58–69.

Randig, O., Medeiros, C.A. and Sperandio, C.A. (1998) Efeito da desinfestação do solo pelo uso da energia solar sobre nematóides. *Nematologia Brasileira* 22, 1–11.

Randig, O., Leroy, F., Bongiovanni, M. and Carneiro, R.M.D.G. (2001) Caracterização de populações brasileiras de *Meloidogyne* spp. com marcadores moleculares. In: *Congresso Brasileiro de Nematologia XXIII, Resumo*, Marilia, SP, p. 138.

Randig, O., Bongiovanni, M., Carneiro, R.M.D.G., Sarah, J.-L. and Costagnone-Sereno, P. (2002a) A species-specific satellite DNA family in the genome of the coffee root knot nematode *Meloidogyne exigua*: application to molecular diagnostics of the parasite. *Molecular Plant Pathology* 3, 431–437.

Randig, O., Bongiovanni, M., Carneiro, R.M.D.G. and Costagnone-Sereno, P. (2002b) Genetic diversity of root knot nematodes from Brazil and development of SCAR marker specific for the coffee-demaging species. *Genome* 45, 862–870.

Rapilly, F. (1991) *L'Epidémiologie en Pathologie Végétale*. INRA, Paris.

Razak, A.R. (1981) The economic importance and identification of root-knot nematode isolates of Malaysia. In: *Proceedings of the 3rd Research Planning Conference on Root-knot Nematodes,* Meloidogyne *spp., Region VI*, 20–24 July 1981, Jakarta, Indonesia, pp. 31–39.

Rebel, E.K., Gonçalves, J.C. and Lordello, L.G.E. (1976) Considerações sobre o comportamento de *Meloidogyne caffeicola* em mudas, cafezais novos e em recepados. In: *Congresso Brasileiro de Pesquisas Cafeeiras, IV, Caxambu, MG. Resumos IBC-GERCA (RJ)*, pp. 11–12.

Reyna, E.H. (1968) La técnica de injerto hipocotiledonar del cafeto para le control de nematodos. *Café (Lima)* 17 (1), 5–11.

Ribeiro, R.C.F. and Campos, V.P. (1993) Isolamento e identificação e efeito da temperatura no crescimento *'in vitro'* de fungos parasitos de ovos de *Meloidogyne* spp. no Sul de Minas Gerais. *Nematologia Brasileira* 17 (2), 132–138.

Ribeiro, R.C.F., Oliveira, C.H., Pereira, A.A. and Lima, R.D. (2001) Reação de progenies de híbridos interespecíficos de *Coffea arábica* e *Coffea canephora* à *Meloidogyne exígua* (Goeldi, 1887). In: *Congresso Brasileiro de Nematologia, XXIII, Resumos*, Marilia, SP, p. 94.

Rodriguez, M.G., Rodriguez, I. and Sanchez, L. (1995) Espécies del genero *Meloidogyne* que parasitan el cafeto en Cuba. Distribuição geográfica y sintomatologia. *Revista de Proteccion Vegetal* 10 (2), 123–128.

Roese, A.D. (2003) Reação de cultivares de soja (Glycine max L. Merril) e de Espécies de Plantas Daninhas a *Meloidogyne paranaensis*. MSc Thesis, Universidade Federal de Viçosa, Viçosa, MG, Brazil.

Roman, J. and Hirschmann, H. (1969) Morphology and morphometrics of six species of *Pratylenchus*. *Journal of Nematology* 1 (4), 363–386.

Roman, J. and Triantaphyllou, A.C. (1969) Gametogenesis and reproduction of seven species of *Pratylenchus*. *Journal of Nematology* 1 (4), 357–362.

Sabrego, L. (1971) Nematodes, problem in coffee plantations of El Salvador. *Nematropica* 1, 1.

Salas, L.A. and Echandi, E. (1961) Parasitic nematodes in coffee plantations of Costa Rica. *Coffee* 3, 6–9.

Sampedro, J., Perez, J., Fowler, V., Fernandez, E., Gandarila, H., Acosta, O., Lorenzo, E., Basterrechea, M., Garcia, J. and O'Connor, B. (1989) Nematodos parasitos asociados al cultivo del cafeto en Cuba. *Ciencia y Técnica en la Agricultura. Prot. Plant* 12 (4), 59–71.

Sanchez, J.M. (1990) Hongos Asociados a la 'Nigua', Nematodo Agallador (*Meloidogyne incognita*) en Plantas de Café (*Coffea arabica*) Cultivadas con Sombra, en Huatusco, Ver. Méx. Profesional Thesis, U.A.C. Chapingo, México.

Santiago, D.C., Krzyzanowski, A.A. and Homochin, M. (2000) Behavior of *Ilex paraguariensis* st Hilaire, 1822 to *Meloidogyne incognita* and *M. paranaensis* and their influence on development of plantlets. *Brazilian Archives of Biology and Technology* 43 (2), 139–142.

Santiago, D.C., Homochin, M., Krzyzanowski, A.A., Carvalho, S. and Fonseca, I.C.B. (2001) Efeito antagônico de *Arachis pintoi* sobre *Meloidogyne paranaensis* e *Meloidogyne incognita* (Kofoid & White), 1919) Chitwood, 1949, raças 1, 2, 3 e 4. *Anais. XXIII Congresso Brasileiro de Nematologia*, Marília, SP, p. 150.

Santos, B.B. (1988) Nematóides do gênero *Meloidogyne* Goeldi em algumas plantas hospedeiras do Estado do Paraná. *Revista de Agricultura* 63 (1), 37–43.

Santos, J.M., Mattos, C., Barré, L. and Ferraz, S. (1992) *Meloidogyne exigua*, sério patógeno da seringueira Michelin, em Rondonópolis, MT. In: *Congressso Brasileiro de Nematologia, XVI, Resumo*, Lavras, MG, p. 75.

Sarah, J.-L. (2003) Diversity of *Meloidogyne* spp. In: *First Annual Report of Breeding Tools for Durable Resistance to Root-knot Nemartodes (*Meloidogyne *spp.) of Coffee Varieties in Latin America.* INCO-DC/IRD, Montpellier, France, pp. 76–79.

Schenk, S. and Holtzmann, O.V. (1990) Evaluation of potential problems in a changiing agricultural system: nematode control in Hawaiian crops. *Plant Disease* 74, 837–843.

Schenk, S. and Schmitt, P. (1992) Survey of nematodes on coffee in Hawaii. *Supplement to the Journal of Nematology* 24 (4S), 771–775.

Schieber, E. (1966) Nematodos que atacan al café en Guatemala, su distribución, sintomatología y control. *Turrialba* 16 (2), 130–135.

Schieber, E. (1971) The nematode problems of coffee in Guatemala. *Nematologica* 1, 17.

Schieber, E. and Grullon, L. (1969) El problema de nematodos que atacam el café (*Coffea arabica* L.) en la Republica Dominicana. *Turrialba* 19, 513–517.

Schieber, E. and Sosa, O.N. (1960) Nematodes on coffee in Guatemala. *Plant Disease Reporter* 44 (9), 722–723.

Sequeira-Bustamente, F., Schuppener, H., Cuarezma, J. and Zepeda, P. (1979) Nematodos fitoparasitos asociados al cultivo del cafeto (*Coffea arabica* L.) en Nicaragua. *Nematropica* 9 (2), 97.

Serracin, M. and Schmitt, D.P. (2002) *Meloidogyne konaensis* and coffee rootstock interaction at two moisture regimes in four soils. *Nematropica* 32 (1), 65–76.

Sharma, R.D. (1971) Nematode associated with cacao and rubber in Bahia, Brazil. *Revista Theobroma* 1, 43–45.

Sharma, R.D. (1973) Plant parasitic nematodes in the São Francisco Valley, Pernambuco, Brazil. *Nematropica* 3, 51–54.

Sharma, R.D. (1975) Effect of partial soil sterilization on the growth of the cacao (*Theobroma cacao* L.) seeding in the nursery. *Arquivos do Instituto Biológico* 42, 23–30.

Sharma, R.D. (1979) *Informe Técnico-setor de Nematologia 56–60.* CEPEC-CEPLAC, Itabuna, Bahia, Brazil.

Sharma, R.D. (1982) Nematodes associated with cocoa hybrids and clones in Bahia, Brazil. *Nematologia Brasileira* 6, 85–91.

Sharma, R.D. and Ferraz, E.C.A. (1977) Eficácia de nematicidas sistêmicos e não sistêmicos no controle aos nematóides fitoparasitas associados a mudas de cacaueiro. *Nematologia Brasileira* 2, 139–147.

Sharma, R.D. and Loof, P.A.A. (1974) Nematódes of the cocoa region of Bahia, Brazil. III – Plant parasitic and free-living nematodes in the rhizospheres of six different plant species. *Revista Theobroma* 4, 39–43.

Sharma, R.D. and Lordello, L.G.E. (1982) Nematodes associated with cocoa hybrids and clones in Bahia, Brazil. In: *Trabalhos Apresentados a VI Reuniao Brasileira de Nematologia*, 8–12 February 1982, Fortaleza. EMBRAPA/CPAC, Planaltina, DF, Brazil, pp. 85–93.

Sharma, R.D. and Lordello, R.R.A. (1992) Occurrence of *Pasteuria penetrans* in coffee plantations infested by *Meloidogyne exigua* in the State of São Paulo. In: *Congresso Brasileiro de Fitopatologia, XXV, Resumos*, Gramado, p. 183.

Sharma, R.D. and Maia, M. (1975) Pathogenicity of the root-knot nematode, *Meloidogyne incognita*, to cocoa. *Nematropica* 5 (2), 28.

Sharma, R.D. and Maia, M.A.Z. (1976) Pathogenicity of the root-knot nematode *Meloidogyne incognita* on cocoa. *Revista Theobroma* 6, 56–65.

Sharma, R.D. and Sher, S.A. (1973) Nematodes associated with coffee in Bahia, Brazil. *Archivos do Instituto Biológico de São Paulo* 40, 131–135.

Sharma, R.D. and Sher, S.A. (1974a) Nematodes of the cocoa region of Bahia, Brazil. II. Occurrence and distribution of plant parasitic nematodes associated with cocoa (*Theobroma cocoa* L.). *Revista Theobroma* 3, 17–24.

Sharma, R.D. and Sher, S.A. (1974b) Nematóides de região cacaueira do Espírito Santo, Brasil. I. Nematóides associados ao cacaueiro (*Theobroma cacao* L.). *Revista Theobroma* 4, 26–31.

Sher, S.A. and Allen, M.W. (1953) Revision of the genus *Pratylenchus* (Nematoda: Tylenchidae). *University of California Publications in Zoology* 57, 441–470.

Siciliano-Wilcken, S.R., Inomoto, M., Ferraz, L.C. and Oliveira, C.M. (2002a) Morphometry of *Pratylenchus* populations from coffee, banana, ornamental plant and citrus in Brazil. Fourth International Congress of Nematology, Programme and Abstracts, 8–13 June 2002, Tenerife, Canary Islands, Spain. *Nematology* 4 (3), 356.

Siciliano-Wilcken, S.R., Inomoto, M., Ferraz, L.C., Oliveira, C.M. and Mori, E. (2002b) RAPD of *Pratylenchus* populations from coffee, banana, ornamental plant and citrus in Brazil. Fourth International Congress of Nematology, Programme and Abstracts, 8–13 June 2002, Tenerife, Canary Islands, Spain. *Nematology* 4 (3), 163.

Siddiqi, M.R. (1972) *Pratylenchus coffeae*. In: Willmott, S., Gooch, P.S., Siddiqi, M.R. and Franklin, M. (eds) *Descriptions of Plant-parasitic Nematodes*. Set 1, no. 6. Commonwealth Institute of Helminthology, St Albans, UK.

Siddiqi, M.R. (1976a) *Pratylenchus zeae*. In: Willmott, S., Gooch, P.S., Siddiqi, M.R. and Franklin, M. (eds) *Descriptions of Plant-parasitic Nematodes*. Set 1, no. 77. Commonwealth Institute of Helminthology, St Albans, UK.

Siddiqi, M.R. (1976b) *Pratylenchus brachyurus*. In: Willmott, S., Gooch, P.S., Siddiqi, M.R. and Franklin, M. (eds) *Descriptions of Plant-parasitic Nematodes*. Set 1, no. 89. Commonwealth Institute of Helminthology, St Albans, UK.

Siddiqi, M.R. (1977) *Pratylenchus loosi*. In: Willmott, S., Gooch, P.S., Siddiqi, M.R. and Franklin, M. (eds) *Descriptions of Plant-parasitic Nematodes*. Set 1, no. 698. Commonwealth Institute of Helminthology, St Albans, UK.

Siddiqi, M.R. (2000) Tylenchida, *Parasites of Plants and Insects*. CAB International, Wallingford, UK.

Siddiqi, M.R., Dabur, K.R. and Bajaj, H.K. (1991) Descriptions of three new species of *Pratylenchus* Filipjev, 1936 (Nematoda: Pratylenchidae). *Nematologica Mediterranea* 19, 1–7.

Silva, R.A., Inomoto, M.M., Kubo, R.K., Mazzafera, P. and Tomazini, M.D. (2001) Efeito de dois isolados de *Pratylenchus coffeae* na fotossintese e no crescimento de cafeeiros cv. Mundo Novo. In: *Congresso Brasileiro de Nematologia, Resumos*, Marília, SP, p. 87.

Silvarola, M.B., Gonçalves, W. and Marinez, M.A.L. (1998) Resistência do cafeeiro a nematóides V – Reprodução de *Meloidogyne exigua* em cafeeiros derivados da hibridação de *Coffea arabica* com *C. canephora*. *Nematologia Brasileira* 22 (1), 51–59.

Sivapalan, P. and Gnanapragasam, N.C. (1975) The effect of soil temperature and of infestation by *Pratylenchus loosi* on the growth and nutrient status of a susceptible and tolerant variety of young tea (*Camelia sinensis* L.). *Tea Quarterly* 45 (1–2), 29–35.

Sosamma, V.K., Koshy, P.K. and Sundararaju, P. (1980a) Nematodes associated with cacao – a review. *Plant Protection Committee for the South East Asia and Pacific Region, Technical Document* No. 123.

Sosamma, V.K., Koshy, P.K. and Sundararaju, P. (1980b) Plant parasitic nematodes associated with cocoa. *Cocoa Growers Bulletin* 29, 27–30.

Souza, P. (1977) A disease complex of coffee involving *Meloidogyne exigua* and *Rhizoctonia solani*. PhD Thesis, North Carolina State University, Raleigh, North Carolina.

Souza, S.E. (2002) *Coffee Nematode Survery in Bahia State, Brazil* Research Report of the grant by PNP&D-café EMBRAPA, Brasilia.

Souza, S.E., Souza, L.H., Maria, J.S., Lima, E., Castro, J., Santos, F.G., Santos, F.S. and Silva, R.V. (1997) Ocorrência de nematóide das galhas em cafeeiros no município de Barra do Choça, Bahia. *Nematologia Brasileira* 21 (1), 24.

Souza, S.E., Santos, J.M., Matos, R.V., Ramos, J.A., Santos, F.S., Ferraz, R.C.N., Carvalho, G.S. and Oliveira, C.A. (2000) Levantamento preliminar de *Meloidogyne* em cafeeiros do Estado da Bahia Planalto de Vitória da Conquista e Chapada Diamantina. *Simpósio de Pesquisa dos Cafés do Brasil, I, Resumos*, Poços de Caldas, MG, pp. 167–170.

Sudha, S. and Sundararaju, P. (2002) Occurrence and distribution of *Radopholus similis* (Cobb, 1893) Thorne, 1949 and other plant parasitic nematodes in arecanut-based cropping system in Kerala. *Indian Journal of Nematology* 32, 219.
Survey, O.F., Portz, R., Furlanetto, C. and Stangarlin, J.R. (2000) Levantamento de espécies de nematoides do gênero *Meloidogyne* na cultura do café em municipios do Oeste do Paraná. *Fitopatologia Brasileira* 25 (Supplement), 339.
Swai, I.S. (1981) Root knot nematodes, *Meloidogyne* species, in Tanzania. In: *Proceedings of the Third Research Planning Conference on Root-knot Nematodes,* Meloidogyne *spp.* 16–20 November, Ibadan Nigéria, pp. 28–30.
Sylvain, P.G. (1959) The problem of nematodes in coffee productions. *Coffee* 1 (1), 2–13.
Tarjan, A.C. (1971) Some interesting associations of parasitic nematodes with cacao and coffee in Costa Rica. *Nematropica* 1 (1), 5.
Tarjan, A.C., Jimenez, M.F. and Soria, J. (1973) Increasing yields of cacao by application of nematocides. *Turrialba* 23, 138–142.
Taylor, A.L. and Sasser, J.N. (1978) *Biology Identification and Control of Root Knot Nematodes (*Meloidogyne *species). International* Meloidogyne *Project.* North Carolina State University Graphics, Raleigh, North Carolina.
Teliz-Ortiz, D., Castillo-Ponce, G. and Nieto-Angel, D. (1993) La corchosis del cafeto en México. *Revista Mexicana de Fitopatologia* 11 (1), 5–12.
Tenente, R.C.V., Gonzaga, V., Melo, L.A.M.P. and Tenente, M.S.M. (2002) Bibliografía Brasileira de nematóides Vol II. Brasília, DF.: *EMBRAPA Recursos Genéticos e Biotecnologia.*
Thorne, G. and Schieber, E. (1962) American dagger nematode *Xiphinema americanum* on coffee in Guatemala, with suggestions for nematode control in nurseries. *Plant Disease Report* 46 (12), 857.
Thorold, C.A. (1975) *Disease of Cocoa.* Clarendon Press, Oxford.
Tomazini, M.D., Ferraz, L.C.C.B., Silva, R.A., Oliveira, C.M.G., Gonçalves, W. and Inomoto, M.M. (2003) Resistência de genótipos de cafeeiros ao isolado K5 de *Pratylenchus coffee. Congresso Brasileiro de Nematologia, XXIV, Pelestras e Resumos,* Petrolina, PE, Brazil, p. 110.
Torrealba, P.A. (1969) Survey of plant parasitic and free-living nematode genera from Venezuela. In: Peachey, J.E. (ed.) *Nematodes of Tropical Crops.* Commonwealth Agricultural Bureau, Technical Comunication no. 40, St Albans, UK, pp. 257–263.
Toruan-Mathius, N., Pancoro, A., Sudarmadji, D., Mawardi, S. and Hutabarat, T. (1995) Root characteristics and molecular polymorphisms associated with resistance to *Pratylenchus coffeae* in Robusta coffee. *Menara Perkebunan* 63 (2) 43–51.
Urquhart, D.H. (1955) *Cocoa.* Longmans, Green and Co., London.
Vaast, P., Caswell-Chen, E.P. and Zasoski, R.J. (1998) Effects of two endoparasitic nematodes (*Pratylenchus coffeae* and *Meloiodogyne konaensis*) on ammonium and nitrate uptake by Arabica coffee (*Coffea arabica* L.). *Applied Soil Ecology* 10 (1–2), 171–178.
Valdez, R.B. (1968) Stubby roots of coffee seedlings caused by *Rotylenchulus reniformis. Philippines Agriculture* 51, 672–679.
Van Doorsselaere, R. and Samsoen, L. (1981) Some tylenchids from coffee-fields of Ivory Coast, with the descripition of *Hemicriconemaoides snoeecki* sp. (Nematoda: Tylenchida). *Revue de Nématologie* 4, 51–63.
Van Gundy, S.D. and McKenry, M.V. (1977) Action of nematocides. In: Horsfall, J.G. and Cowling, E.B. (eds) *Plant Disease – An Advanced Treatise,* Volume 1. Academic Press, New York, pp. 263–283.
Vega, M.C. (1982) Informe sobre la situacion nematologica en Nicarágua. In: *Proceedings of the Third Research Planning Conference on Root knot Nematodes* Meloidogyne *spp., Region I,* 11–15 January, pp. 6–68.
Vergel-Colon, D.M., Leguizamon-Caycedo, J., Cortina-Guerrero, H. and Torres-Torres, E. (2000) Reconocimiento y frecuencia de *Meloidogyne* spp. en una localidad de la zona cafetera central de Colombia. *Cenicafé* 51 (4), 285–295.
Vernalha, M.M., Zappia, O., Mansur, R. and Rodrigues, S. (1967/1970) Sobre a ocorrência de nematóides do cafeeiro no estado do Paraná. *Arquivos de Biologia e Technologia* 13, 33–38.
Villain, L. (2000) Caractérisation et Bioécologie du Complexe Parasitaire du Genre *Pratylenchus* (Nemata: Pratylenchidae) Présent sur Caféiers (*Coffea* spp.) au Guatemala. PhD Thesis, ENSAR, Rennes, France.
Villain, L., Baujard, P., Molina, A., Pignolet, L. and Sarah, J.L. (1998) Morphological and biological characterization of three *Pratylenchus* spp. populations parasiting coffee trees in Guatemala. In: 24th International Symposium of the European Society of Nematologists, Dundee, UK, 4–9 August 1998. *Nematologica* 44 (5), 600–601.

Villain, L., Anzueto, F., Hernández, A. and Sarah, J.L. (1999) Los nematodos parásitos del cafeto. In: Bertrand, B. and Rapidel, B. (eds) *Desafíos de la caficultura en Centroamérica*. ICCA-PROMECAFE, San Jose, Costa Rica, pp. 327–368.

Villain, L., Molina, A., Sierra, S., Decazy, B. and Sarah, J.L. (2000) Effect of grafting and nematicide treatments on damage by root lesion nematodes (*Pratylenchus* spp.) to *Coffea arabica* L. in Guatemala. *Nematropica* 30 (1), 87–100.

Villain, L., Molina, A., Sierra, S., Decazy, B., Anzueto, F., Guyot, B., Perriot, J.J. and Sarah, J.L. (2001a) Evaluation of grafting and nematicide treatments for the management of a root lesion nematode, *Pratylenchus* sp., in *Coffea arabica* L. plantations in Guatemala. *ASIC, 19ème Colloque Scientifique International sur le Café*. ASIC, Paris, 2001.

Villain, L., Pignolet, L., Michau-Ferrière, N., Figueroa, P., Molina, A. and Sarah, J.L. (2001b) Evidence of resistance factors to *Pratylenchus* spp. on *Coffea canephora* (poster). *XXXIII ONTA Reunion*, 11–15 June 2001, Varadero, Cuba. *Nematropica* 31 (2), 163.

Villain, L., Baujard, P., Anzueto, F., Hernandez, A. and Sarah, J.L. (2002) Integrated protection of coffee plantings in Central America against nematodes. *Plante Recherche Développement* (Special Issue: Research and Coffee Growing), 118–133.

Villain, L., Anzueto, F. and Sarah, J.L. (2004) Resistance to root-lesion nematodes on *Coffea canephora*. In: Cook, R. and Hunt, D.J. (eds) Proceedings of the Fourth International Congress of Nematology, 8–13 June 2002, Tenerife, Spain. *Nematology Monographs and Perspectives* 2 (in press).

Villalba-Gault, D.A., Fernandez-Borrero, O. and Baeza-Aragon, C.A. (1982) Identification de una nueva raza de *Meloidogyne incognita* em *Coffea arabica* variedad Caturra. *Cenicafé, Chinchina* 33 (3), 91–101.

Volpato, A.R., Otoboni, C.E.M., Otoboni, J.A.M., Correa, L.E.A. and Saraiva, R.F. (2001) Eficácia dos produtos Casudafós, Carbofuran e Carbosulfan no controle de *Meloidogyne exígua* e *M. coffeicola* no cafeeiro. In: *Congresso Brasileiro de Nematologia, XXIII, Resumos*, Marilia, SP, p. 92.

Vovlas, N. (1987) Parasitism of *Trophotylenchulus obscurus* on coffee roots. *Revue de Nématologie* 10, 337–342.

Way, J.I. (1981) *Meloidogyne* species in Zimbabwe susceptible crops other than tobacco, taxonomy and resistant crops. In: *Proceedings of the Third Research Planning Conference on Root-knot Nematodes* Meloidogyne *spp. Region IV*, 16–20 November, Ibadan, Nigeria, pp. 13–20.

Whitehead, A.G. (1959) The root-knot nematode of east Africa. *Nematologica* 4, 272–278.

Whitehead, A.G. (1968a) Taxonomy of *Meloidogyne* (Nematodes: Heteroderidae) with descriptions of four new species. *Transactions Zoological Society, London* 31, 263–401.

Whitehead, A.G. (1968b) Nematodea. In: Le Pelley, R.H. (ed.) *Pests of Coffee*. Longmans, Green and Co. Ltd, London, pp. 407–422.

Whitehead, A.G. (1969a) The distribution of root-knot nematodes *Meloidogyne* spp. in tropical Africa. *Nematologica* 15, 315–333.

Whitehead, A.G. (1969b) Nematodes attacking coffee, tea and cocoa and their control. In: Peachey, J.E. (ed.) *Nematodes of Tropical Crops. Technical Communication no. 40*. Commonwealth Bureaux of Helminthology, St Albans, UK, pp. 238–250.

Willer, P. (1997) First record *of Meloidogyne mayaguensis* Rammah and Hirschmann, 1988: Heteroderidae on commercial crops in the Mpumalanga province, South Africa. *Inligtingsbulletin-instituut-vir-Tropiese-en-Subtropiese-Gewasse* 294, 19–20.

Zelaya-Escoto, H.R. and Santacreo, R. (2000) Evaluation of field resistance to *Meloidogyne exigua* in Sarchimor progenis and Catuaí X Icatu back crosses in El Paraiso Honduras. *Simposio Latino Americano de Cafeicultura, XIX Memórias*, San Jose, Costa Rica, pp. 229–242.

Zem, A.C. (1993) Cadusafos: a new highly effective nematicide. *Nematologia Brasileira* 17 (1), 21.

Zhang, F. and Schmitt, D.P. (1994) Host status of 32 plant species to *Meloidogyne konaensis. Supplement to the Journal of Nematology* 26 (48), 744–748.

Zhang, F. and Schmitt, D.P. (1995a) Relationship of *Meloidogyne konaensis* population densities to coffee growth. *Plant Disease* 79, 446–449.

Zhang, F. and Schmitt, D.P. (1995b) Embryogenesis and post infection development of *Meloidogyne konaensis. Journal of Nematology* 27, 103–108.

Zimmermann, A. (1898) De nematoden der koffiewortels. Batavia (=Jakarta, Indonesia) and S'Gravenhage, La Hague, Pays Bas, G. Kolf & Co., coll. *Mededeelingen uit's lands pantentuin* 27.

15 Nematode Parasites of Tea*

Nalini C. Gnanapragasam[1] and Keerthi M. Mohotti[2]
[1]*Cropoptima (Pvt) Ltd, 78/3 Temple Road, Hatton, Sri Lanka;*
[2]*Tea Research Institute of Sri Lanka, St Coombs, Talawakelle, Sri Lanka*

Tea is a beverage crop with two extreme varieties, including the small-leaved China type and the large-leaved Indian or Assam type, both of which belong to the same species, *Camellia sinensis*. Commercial tea populations are polymorphic in origin, derived from *Camellia sinensis* (L) O. Kuntze., *C. assamica* var. *assamica* (Masters) Wight, and *C. assamica* var. *lasiocalyx* (Planch.) Wight, or the hybrids of these different varieties.

Tea is grown presently at latitudes from 27°S (Corrientes, Argentina) to 43°N (Georgia, former USSR), as well as from mean sea level up to an altitude of 2300 m. The tea crop requires well-drained acid soils with a pH range of 4.5–5.5 and reasonably well distributed rainfall, totalling not less than 1000 mm/year.

Cultivation techniques

The population of tea bushes in old tea fields is about 7000/ha, and in many fields the plant population is far below this number due to extensive casualties. The plant population density in the newly planted areas is around 13,000, usually planted along the contour.

When allowed to grow freely, the tea plant could grow to a large tree attaining a height of around 12 m or more. For purposes of commercial exploitation, the plant is kept pruned regularly to be maintained in the form of a bush at a height of around 90 cm.

The unit that is harvested is the tender flush, usually comprising two or three leaves and a bud, and these units are generally harvested at weekly intervals depending on growth rates.

The average yield of tea could range from as low as 500 kg/ha to as high as 6000–7000 kg/ha of made tea/year (which corresponds to ~1600–32,500 kg of green leaf/ha/year). Though broadly similar, the agricultural and manufacturing practices could vary in the different tea-growing areas of the world.

Nematode Species Encountered in Tea

The factors generally limiting nematode reproduction and survival and establishment in a given specific location are known to be very much dependent on the soil environment (Gnanapragasam, 1994a). Due to the wide variability in soil types and climatic conditions under which tea is being cultivated on a commercial scale, the com-

*A revision of part of the chapter by V.P. Campos, P. Sivapalan and N.C. Gnanapragasam.

 Plant Parasitic Nematodes in Subtropical and Tropical Agriculture, 2nd Edition (eds M. Luc, R.A. Sikora, J. Bridge)

plex of nematode populations that attack the tea plant varies very widely and the intensity of attack of the respective species and the degree of the induced pathogenicity could also vary correspondingly. Furthermore, investigations with respect to damage caused by nematodes to the tea crop is limited to only a few countries, whilst the majority of the countries that grow this crop on a commercial scale have not carried out any investigations or surveys on the incidence of these pests.

Several species of plant parasitic nematodes have been encountered in tea soils in the different tea-growing areas of the world. However, no positive evidence of pathogenicity has been established with respect to the majority of these nematodes. The species that are either known or suspected to be pathogenic to tea include the following: *Pratylenchus* spp., *Radopholus similis*, *Meloidogyne* spp., *Hemicriconemoides kanayaensis*, *Rotylenchulus reniformis*, *Helicotylenchus* spp., *Paratylenchus curvitatus*, *Hoplolaimus* sp., *Rotylenchus* sp. and *Xiphinema* sp. (Table 15.1).

Pratylenchus

Species of *Pratylenchus* are known to attack tea growing in almost all parts of the world. Amongst these, *Pratylenchus loosi* is the most serious pest in Sri Lanka (Gadd, 1939; Gadd and Loos, 1946; Loos, 1953a; Sivapalan, 1972). This species of nematode is also recorded as a serious pest of tea in Japan (Kaneko and Ichinohe, 1963; Takagi, 1967, 1969), Iran (Maafi, 1992; Maafi *et al.*, 1999) and Korea (Park *et al.*, 2002).

In Sri Lanka, *P. loosi* is widely distributed amongst tea fields at all altitudes. However, damage to tea is mostly confined to elevations of 900–1800 m, where severe damage and crop loss occur in mature tea, newly planted young fields, as well as in nurseries (Hutchinson and Vythilingam, 1963a; Sivapalan, 1972; Gnanapragasam, 1986a). As a consequence of its distribution and pathogenicity to high elevation tea areas, it is commonly referred to as the 'up-country species of nematode'.

In contrast, in Japan, where tea is cultivated at altitudes of 0–300 m, damage to tea by this species occurs at all locations in view of the fact that this country is located in the cooler temperate zone (Takagi, 1969; Gotoh, 1976).

P. loosi is also known to cause damage to tea in China (Chen Zongmao and Chen Xuefen, 1982; Li, 1985), but, to date, a proper survey has not been carried out and as such the distribution and extent of damage are not well known.

In Darjeeling, India, *P. loosi* was reported for the first time in 1982, but no pathogenicity trials have been carried out (Mukherjea and Dasgupta, 1982).

In Bangladesh, this nematode has been observed to cause symptoms of damage to tea only in nurseries. Nursery soils are, therefore, regularly checked for this species (S.A. Rashid and D.J. Millin, Bangladesh, 1988, personal communication; Huq *et al.*, 1990). Despite such observations, no further attempt has been made to assess the distribution and possible damage in mature tea.

P. loosi was recorded for the first time in Korea in 2000, when it was isolated from the roots and rhizosphere of tea in the districts of Y congam-gun, Jcol Janam-do and Namjcju-gun, Jcju-do, Korea. No pathogenicity trials have yet been reported (Park *et al.*, 2002).

In Iran, *P. loosi* was reported for the first time in 1992 in the tea gardens in Amlash (a region of Guilan Province in North of Iran) near the Caspian sea, which has a subtropical climate (Maafi, 1992). Studies carried out later revealed the presence of this nematode in the entire region of Guilan province ranging from the lower to higher elevations (Nasaj Hosaini, Iran, 2003, personal communication).

Symptoms of damage

Typical symptoms of injury caused by *P. loosi* in both young and mature tea in the field include patches of unthrifty tea (Plate 16A), with the affected plants showing spindly growth with sparse foliage. The leaves are dull, brittle and

Table 15.1. Distribution of nematodes known/suspected to be pathogenic to tea in different geographic regions.

Nematode species	Argentina	Australia	Africa (East)	Africa (South)	Bangladesh	China	India (North-east)	India (South)	Indonesia	Iran	Japan	Kenya	Korea	Malawi	Malaysia	Sri Lanka	Taiwan	Zimbabwe
Helicotylenchus dihystera		+					+				+					+		
Helicotylenchus erythrinae						+	+				+					+	+	
Helicotylenchus sp.			+		+		+	+		+				+		+		+
Hemicriconemoides kanayaensis						+					+						+	
Hoplolaimus sp.			+				+							+		+		
Meloidogyne arenaria						+				+				+		+		+
Meloidogyne brevicauda							+	–								+		
Meloidogyne hapla							+			+								+
Meloidogyne incognita	+	+			+	+	+		+	+	+	+		+		+		+
Meloidogyne javanica		+	+			+	+	+	+	+				+		+		+
Meloidogyne thamesi						+												
Paratylenchus curvitatus							+				+			+		+		
Pratylenchus brachyurus		+					+			+				+				
Pratylenchus loosi					+	+	+			+	+		+			+		
Radopholus similis				+		+			+							+		+
Rotylenchulus reniformis							+							+		+		+
Rotylenchus sp.			+	+			+							+				+
Xiphinema sp.			+	+	+		+				+			+	+	+		+

Fig. 15.1. Stunted tea plant with feeder roots damaged by *Pratylenchus loosi* (left) and uninfested healthy plant (right). (Photo: N.C. Gnanapragasam.)

yellowish in colour. These symptoms are brought about by an altered rate of uptake of essential nutrients by the damaged root system (Fig. 15.1). The heavily infested plants also have a tendency to start the reproductive phase by flowering and setting fruit prematurely. Examination of the roots of such infested plants shows a marked reduction in the growth of feeder roots. The remaining roots appear brown and dried up when compared with the normal healthy roots that are succulent and whitish in colour. Dark brown necrotic patches or lesions of varying size are displayed on peeling the bark of the larger storage roots (Plate 16B). The heavily infested plants either recover very poorly from pruning, remain as unthrifty 'passengers' or fail to recover at all and die (Gadd, 1939; Visser, 1959; Sivapalan, 1967a, 1972; Gnanapragasam, 1986a).

Biology and life cycle

Like other *Pratylenchus* species, *P. loosi* is a migratory endoparasite invading the root cortex of host plants. They move into the soil in search of fresh feeder roots when the parasitized roots are severely damaged or become over-parasitized. Thus it is very common to encounter large populations in the soil in the rhizosphere of infested bushes (at a depth of ~15–25 cm). The nematodes are mostly attracted to the growing parts of the roots where they penetrate and enter near the root tips. According to Seinhorst (1977), it takes the nematode 45–48 days to complete its life cycle, comprising 15–17 days for the eggs to hatch, 15–16 days as juveniles, and 15 days as adults before egg laying. Takagi (1969) and Nasaj Hosaini (Iran, 2003, personal communication) reported the life cycle to be around 40–50 days. Egg laying was found to be delayed in the absence of males (Gadd and Loos, 1941). The optimum temperature range for the highest population build-up and obvious pathogenicity symptoms to occur was found to be at soil temperatures of 18–24°C (Sivapalan and Gnanapragasam, 1975; Nasaj Hosaini, Iran, 2003, personal communication).

Pathotypes (biological races)

Morphological and morphometric studies carried out in Sri Lanka and in Iran on the males and females of *P. loosi* have revealed the possible existence of different pathotypes/strains of *P. loosi* (Pourjam *et al.*, 1997, 1999; Mohotti, 1998; Mohotti *et al.*, 1998, 2002). Although five populations of this species collected from tea soils from geographically different areas including Iran, Japan (Kagoshima prefecture), Japan (Shizuoka prefecture), Sri Lanka, Passara (N.E. monsoonal zone) and Sri Lanka, Talawakelle (S.W. monsoonal zone), respectively, showed morphometrical similarities, observations made under the electron microscope showed distinct variation in the head and tail regions. The intraspecific variability thus observed in the *P. loosi* populations may be attributed to variation in the geographical area, host nutrition and origin of nematodes. The *P. loosi* populations were found to be conspecific with each other and demonstrated a *P. loosi* species complex (Mizukubo, 1998; Mohotti *et al.*, 2002).

Survival and means of dissemination

P. loosi is known to survive in host-free soils in the lesions of the larger old storage roots of tea that are left uncleared, following the uprooting of old tea fields, for as long as 3 years.

One of the most important means of spread of *P. loosi* amongst tea areas is by the dissemination of infested plants to fields from contaminated nurseries. Spread of nematodes could also occur through: (i) movement of infested soil and water – poor soil conservation measures adopted in infested areas, including the use of weeding implements that tend to loosen the soil and inducing erosion and washing down of contaminated soil into areas hitherto uninfested; (ii) uprooting of old tea fields sometimes carried out from the bottom of the slope upwards, thus exposing the newly planted young tea at the bottom to re-infestation from infested old tea still remaining above; and (iii) use of contaminated irrigation water in nurseries (Fig. 15.2) (Gnanapragasam, 1985a, 1989).

Environmental factors affecting pathogenicity

The severity of damage to tea is dependent on the interaction of various factors such as: (i) prevailing climatic conditions; (ii) type of soil in which the tea is growing; (iii) cultural practices; and (iv) age and vigour of the plant (Gnanapragasam, 1988a).

CLIMATIC FACTORS. The distribution of *P. loosi* is determined mainly by soil temperature and soil moisture. The highest population is encountered at altitudes with soil temperatures of 18–24°C. Obvious pathogenicity symptoms are also observed in this temperature regime (Sivapalan and Gnanapragasam, 1975; Gnanapragasam and Manuelpillai, 1984). At temperatures above and below this range, the rate of population build-up is less and, consequently, damage to tea is also reduced (Sivapalan, 1972).

The results of detailed surveys have revealed that the largest population of this species of nematode is encountered in areas with high and well distributed rainfall, and this determines the severity of

Fig. 15.2. Heavy infestation and stunting of nursery plants infested with nematode-contaminated water (B) compared with similar age plants with clean sedimented water (A). (Photo: N.C. Gnanapragasam.)

damage within the same altitude (Hutchinson and Vythilingam, 1963a).

A marked periodic fluctuation in population levels is also observed during the year, and this variation is correlated to the rainfall pattern as well as soil temperature (Sivapalan, 1972) (Fig. 15.3).

TYPE OF SOIL. Nematode damage is known to vary with the type of soil (soil texture) as well as the physical condition of the soil. Damage caused by *P. loosi* was observed to be most severe in clayey ill-drained soils (Sivapalan, 1971).

Under poor soil conditions, the rate of replenishment of roots damaged by nematodes is very much curtailed, resulting in the rapid deterioration of the root system, with the consequent restricted uptake of nutrients, and the plants soon turn out to be mere 'passengers'. Increasing soil acidity has also been observed to aggravate the above condition (Gnanapragasam, 1987a).

INFLUENCE OF CULTURAL PRACTICES. Due to the large genetic variability in seedling tea fields, the pattern of distribution of nematode infestation in such fields is highly clustered. When such old fields are replanted to the genetically uniform high yielding, vegetatively propagated varieties, the spread of infestation could become more uniform, depending on the susceptibility ratings of specific cultivars.

The presence of shade trees and green manure crops amongst tea fields, which form part of the normal cropping pattern, also influences the distribution pattern and the intensity of build-up of this species of nematode (Sivapalan, 1972; Gnanapragasam, 1987b).

Alternative hosts

The presence of other hosts in the vicinity of tea fields also regulates the population levels of *P. loosi*. The presence of crops, such as *Tephrosia vogelii*, *Sesbania cinerascens*, *Cassia elata* and *Acacia* spp., as well as certain weeds, increases the incidence of this nematode species in tea fields (Visser, 1959; Sivapalan, 1972; Gnanapragasam, 1987b; Gnanapragasam *et*

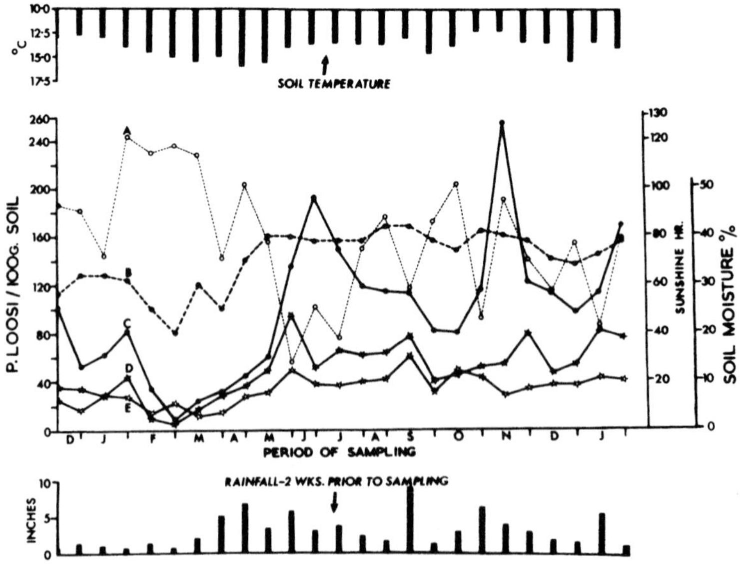

Fig. 15.3. Soil population fluctuation of *Pratylenchus loosi* at varying depths (C = 15 cm, D = 30 cm, E = 45 cm) during different times of the year, as determined by soil temperature, rainfall pattern, sunshine (A) and soil moisture (B).

al., 1989a). Grasses such as Guatemala (*Tripsacum laxum*) and Mana (*Cymbopogon confertiflorus*) are non-hosts and thus do not help in the build-up of this species of nematode. On the other hand, grasses such as *Eragrostis curvula* as well as specific plants such as *Tagetes* spp. (marigold), *Arachis pintoi*, *Tithonia diversifolia* (wild sunflower), *Wedeliya trilobata*, *Vetiveria ziazanoides* (vet-ver), *Adhathoda vasica*, *Ricinis communis*, *Azadirachta indica*, *Madhuca indica*, *Sambucus javanica*, *Plectranthus zeylanicus*, *Indigofera teysamanii*, *Eupatorium inuliformes*, *Calliandra calothyrsus* and *Crotalaria anagyroides* help to reduce the population (Visser and Vythilingam, 1959; Hutchinson, 1962; Kerr, 1963a; Sivapalan, 1972; Gnanapragasam, 1981, 1995, 1997).

Although, *P. loosi* has been reported in coffee (*Coffeae arabica* cv. Catuai) in Quetzeltenco, Guatemala (Anzueto and Sarah, 1992), in Sri Lanka, in areas where coffee is intercropped with tea, no build-up of populations has been encountered so far. It is possible that the species found in Guatemala is a different pathotype to that encountered in Sri Lanka. This needs further confirmation.

P. loosi has also been reported in several other hosts including the roots of *Sorghum vulgare* (Baujard, 1986), groundnut (peanut), millet, cowpea in Senegal, in the Saheline Province of West Africa (Baujard *et al.*, 1990), citrus, *Hibiscus sinensis* and okra in New Delhi (Sethi and Swarup, 1971; Nath *et al.*, 1975), banana (M.R. Siddiqi, 1995, personal communication), cotton, pasture grasses such as bahiagrass (*Paspalum notatum* Fluegge) and maiden cane (*Panicum hemitomon*, J.A. Schultes) (Inserra *et al.*, 1996), apple (*Malus domestica* Kentucky) and grapes (*Vitis vinifera* Griffith) in New South Wales, Australia (Mc Leod *et al.*, 1994), citrus and pear in Japan (Gotoh, 1974) and mango (*Mangifera indica*). However, detailed studies carried out on some of the populations collected from the non-tea hosts has revealed extensive morphological and morphometric variations amongst the studied populations, raising a doubt as to their actual identity as that described as *P. loosi* (Mohotti, 1998). Doubt has also been raised as to the identity of *P. loosi* collected from non-tea hosts (M.R. Siddiqi, 1998, personal communication). Even if some of these species collected from non-tea hosts take on tea, they would not cause any threat to infestation and spread of nematodes in the tea fields as these plants are not normally grown in the vicinity of tea areas.

Disease complexes

Very limited work has been carried out on disease complexes involving nematodes parasitizing tea. The occurrence of a soft root rot disease on mature tea roots, leading to death of affected plants during dry weather, is a disease complex formed by *P. loosi* and a group of three fungi (*Paecilomyces lilacinus* (Thom) Samson, *Paecilomyces* sp. and *Absidia corymbifera* (Cohn) Sacc. and Trotter) (Arulpragasam, 1981). This condition is reported to be brought about by many factors, the primary cause being the predisposition to infestation with *P. loosi* (Arulpragasam and Addaickan, 1983).

A disease complex involving nematode–insect interaction causing yield decline has been observed recently in some of the mid-elevation tea areas of Sri Lanka (200–1000 m). The most serious pests of tea in this region include the insect pest, shot-hole borer (*Euwallaceae fornicatus*), and the plant parasitic nematode *Radopholus similis* and, at the upper limit of this elevation range, *P. loosi* as well. Both the above individual nematode species and the insect pest could by themselves cause severe damage to tea when the population level increases beyond their respective damage threshold level. However, when either of the nematode pests and the insect pest simultaneously attack the tea plant, serious economic damage is brought about at levels well below the respective damage threshold level, and the symptoms of damage also become accentuated (Gnanapragasam, 2002a).

Slow decline of a nematode-tolerant tea cultivar

There have been several instances of slow decline of tea, ultimately leading to death of affected tea bushes, especially following pruning in the high elevation tea areas and in a few of the mid-elevation tea areas of Sri Lanka. In all these instances, the affected tea cultivar is TRI 2025 (a popular high yielding tea cultivar in Sri Lanka, that is known to be only weakly susceptible to *P. loosi*), which has reached the age of about 20 years or more. Although the above-ground symptoms greatly resemble the symptoms brought about by 'soft root rot' disease, the typical pulpy soft appearance of the roots is not present. The typical whitish fungal mycelia in the 'soft root rot'-affected bushes are also not evident amongst the bushes affected by the slow decline of tea. Unlike in the case of 'soft root rot' of tea, where the bushes died after 3–6 months, the death of bushes in this case occurs only after some years. Detailed investigations have revealed that the observed slow decline is brought about by a long-term protracted infestation with *P. loosi*, in high yielding tea fields in which the tea bushes have been subject to other forms of environmental stress (Gnanapragasam, 2002b).

Economic importance and population damage threshold levels

Detailed assessment on crop losses in tea caused by plant parasitic nematodes has been carried out almost entirely in Sri Lanka. Although *P. loosi* has been recovered from several locations, significant damage to tea has been observed mostly at elevations of 900–1800 m. The decline in yield in such areas, though earlier estimated to be in the order of around 225–350 kg of made tea/ha/year (Gadd, 1939; Visser, 1959), could range between 4 and 40% depending on the type of cultivar planted, prevailing climatic conditions, population density of nematode, age and vigour of affected tea bushes, type and condition of soil and pH of soil, etc. (Gnanapragasam, 1988a). The extent of damage is, however, far greater in infested young tea clearings and nurseries, where casualties could range from 60 to 100%, especially when the required sanitary measures are not followed. Of about 55,000–60,000 ha of high elevation tea areas in Sri Lanka, approximately 40–50% are known to suffer obvious damage by this species of nematode. Economic losses caused could be experienced in the remaining high elevation tea areas as well, but such losses have not yet been ascertained, as in most of these areas the observed decline in yield is brought about by more than one factor.

It is difficult to estimate with any precision the population damage threshold of any species of nematode causing an economic loss to a given crop, as this is compounded by an interaction with other environmental factors. In general, a tea plant that is already under stress due to other causes readily succumbs to infestation by even a low population. However, in experiments carried out under controlled conditions in the greenhouse, the damage threshold of *P. loosi* was estimated to be 40 nematodes/100 g of soil, at 24°C, which is the mean temperature of areas between the elevation range of 900 and 1800 m (Gnanapragasam and Manuelpillai, 1984).

Pratylenchus brachyurus

Unlike *P. loosi*, *P. brachyurus* only causes damage to young tea (1- to 3-year-old plants). In north-east India, this species has been detected in the plains of Assam (Basu, 1968). In Sri Lanka, although *P. brachyurus* was detected in the mid-altitude tea areas in the rhizosphere of *Albizia moluccana* trees, the neighbouring tea plants were not infested (Gnanapragasam, 1991a). In Queensland, Australia, where tea was planted relatively recently, this species has been found to attack tea seedlings up to the age of 12 months.

Thereafter, there is no evidence of pathogenicity (P.C. O'Brien, Australia, 1988, personal communication).

A similar observation has also been made in Malawi (Corbett, 1967). The damaged plants are stunted and unthrifty and show characteristic nutrient deficiency symptoms. This nematode attacks mainly the feeder roots and occasionally the taproot as well. During its feeding activity, it moves deep into the root tissue, causing the formation of dark red lesions on the epidermal layer. This species is reported to survive long periods of drought, during which period they remain quiescent (Basu, 1968).

Meloidogyne spp.

Meloidogyne species are the most commonly encountered nematodes in tea in the different tea-growing areas of the world. Most of these species attack only the young nursery plants, whilst the mature tea becomes totally immune, with the plants developing resistance at 12–14 months of age. The only exception is *M. brevicauda*, which is known to attack mature tea very seriously.

Distribution

The first report of root knot nematode infestation in young tea was from south India, where they were found to infest large numbers of tea seedlings (Barber, 1901). In Sri Lanka, large-scale failures in tea nurseries were ascribed to infestations caused by root knot nematodes by Stuart-Light in 1928. Since the 1960s, tea has been propagated by vegetative means, rather than from seeds, and infestation of nursery plants by this species of nematode is seldom encountered in this country. The species that are known to infest young tea in Sri Lanka and north-east India include *M. incognita*, *M. javanica* and *M. arenaria*. On the other hand, *M. hapla* was rarely found to infest tea (Banerjee, 1967; Gnanapragasam, 1985b). Root knot damage was shown to be more abundant at high altitudes than at lower elevations (Basu and Roy, 1976).

In Malawi in 1960, almost all the tea estate samples were infested with *M. javanica* (majority areas), *M. incognita* and *M. arenaria*. As has been reported from other countries, such infested samples were all from tea nurseries (Martin, 1960, 1962). In Zimbabwe, species encountered include *M. incognita*, *M. arenaria* and *M. hapla* (Keetch and Buckley, 1984).

In China, the incidence of root knot nematode damage was found to be about 90% in tea seedlings, and the death rate was estimated at 40% in the seriously affected nurseries. In Yunnan Province, *M. incognita*, *M. javanica* and *M. arenaria* have been reported (Yu Sheng-fu and Xia Bing, 1987). In Zhejian Province, *M. thamesi* has also been found in addition to the other three species (Huan Jin, 1984). In both these provinces, *M. incognita* was found to be more abundant than the other species. In addition to the above species, in China *M. acrita* has also been reported on tea (Chen Zongmao and Chen Xuefen, 1982). *M. incognita* has also been reported from Japan (Takagi, 1967). In Iran, the species of *Meloidogyne* reported to attack tea include *M. incognita*, *M. arenaria*, *M. javanica* and *M. hapla* (Nasaj Hoseini, 2003, personal communication).

Meloidogyne species have been encountered occasionally in tea nurseries in Bangladesh (D.J. Millin and S.A. Rashid, 1988, personal communication). *M. incognita* is the only species of root knot nematode that has been identified from tea roots in nursery beds in Queensland, Australia (P.C. O'Brien, 1988, personal communication). In Kenya, *Meloidogyne* species have been isolated from only one farm amongst the various tea-growing districts (C.O. Othieno, Kenya, 1993, personal communication). *Meloidogyne* spp. have also been reported to damage young tea in Argentina (S.D.P. Kricun, 1988, personal communication).

Symptoms of damage

The species of root knot nematodes that are known to attack only young tea plants form galls on both the taproot and the feeder roots. Some root knot nematode juveniles enter the roots of mature tea bushes but fail to cause giant cells and are apparently unable to complete the moult between the second and third juveniles (Gadd and Loos, 1946). Seedling plants, in which both the taproot and the lateral roots are severely attacked, suffer greater damage than the majority of vegetatively propagated clonal tea plants of similar age, probably because seedling plants possess less than half the root bulk of the clonal plant (Kerr, 1963a).

Although root knot nematodes are root feeders, the collar regions of tea seedlings have been reported to be infested occasionally with *M. incognita* in Assam, India. The females recovered from such infested locations were found to be poorly developed, although they were found to have led to the development of the characteristic galls on such affected stems (Basu, 1976).

Environmental factors affecting pathogenicity of Meloidogyne *spp.*

Since species of *Meloidogyne* have been encountered in almost all tea-growing regions, they seem to be well adapted to different climatic and soil conditions. In China, the optimum soil temperature for pest incidence has been reported to be 20–30°C and in soils with 20% moisture (Rong *et al.*, 1984).

Use of some of the herbicides has been reported to have a suppressing effect on the populations of *Meloidogyne* spp. in tea fields in India (Basu and Gope, 1982; Gope and Borthakur, 1991). Since only short persistent herbicides are used in tea fields, the decline in population is probably due to the eradication of weeds which are good hosts of this species of nematode, rather than direct kill brought about by the herbicides themselves.

Alternative hosts

Species of *Meloidogyne* have the largest number of alternative hosts. However, since they attack only young nursery plants, the presence of alternative hosts in mature tea fields has little influence, other than when soils from such areas are used for nursery plant propagation. On the other hand, rhizospheres of common shade trees in tea fields of Sri Lanka have been found to harbour heavy populations of nematode antagonists (Mohotti, 1998). These shade trees are not hosts of nematodes pathogenic to mature tea, but are susceptible to *Meloidogyne* spp., which in turn are good hosts of many of the nematode anatagonists. Therefore, these shade trees serve as reservoirs to help spread these beneficial agents into the tea fields.

Meloidogyne brevicauda

This species of root knot nematode is the only one that attacks mature tea, and has been so far recorded only in the tea areas of Sri Lanka, north-east India and south India. In Sri Lanka, this species has been recorded in only three plantations, all bordering the same jungle at an altitude of 1500–2000 m (Hutchinson and Vythilingam, 1963b). In south India, it has been recorded in single estates each in the Nilgiris, Wynaad and Karnataka Districts (Venkata Ram, 1963; Mehta and Somasekhar, 1998; Muraleedharan and Selvasundaram, 2001), and in north-east India it has been recorded only in Darjeeling (Mukherjea and Dasgupta, 1982).

Other than Sri Lanka and India, the only country where this species of root knot nematode has been reported is in Apsheron, Azerbaidzhan on saffron (*Crocus sativus*) (Kasimova and Atakishieva, 1980).

Symptoms of damage caused by M. brevicauda

The above-ground symptoms of attack by this species of nematode resemble those brought about by the root lesion nematode. The infested bushes are stunted as a consequence of poor recovery from successive prunes; the leaves are smaller, yellowish

Fig. 15.4. Typical galling of mature tea roots caused by *Meloidogyne brevicauda*. (Photo: N.C. Gnanapragasam.)

and dull in appearance. The roots show the characteristic presence of large galls (Fig. 15.4), many of which display pinhole pits. It is often difficult to isolate living mature females and, when found, they contain only a few eggs (Loos, 1953b).

Biology

The average size of a mature female is about five or six times that of a mature female of *M. incognita* (Fig. 15.5). Despite this massive size, the females are often observed to be empty, with only a few eggs. The mean hatch per egg mass is around ten, whilst in the other common species this is of the order of 200–600 juveniles/egg mass (Gnanapragasam and Manuelpillai, 1981). Males are also extremely rare, and it is possible that eggs develop only following fertilization, which is likely to occur by chance, and those unfertilized fail to produce any eggs. The rest of the life history is very similar to that of the other species of *Meloidogyne*.

Early investigations in Sri Lanka revealed the presence of this species of nematode only in mature seedling tea fields (Loos, 1953b). However, subsequent studies revealed some of the clonal tea also to be susceptible, and signs of infestation and galling became obvious in the tested cultivars only after 10 years from planting in an infested field. The most susceptible cultivar was TRI 2142, which is resistant to the root lesion nematode *P. loosi*. A low level of infestation was also observed in

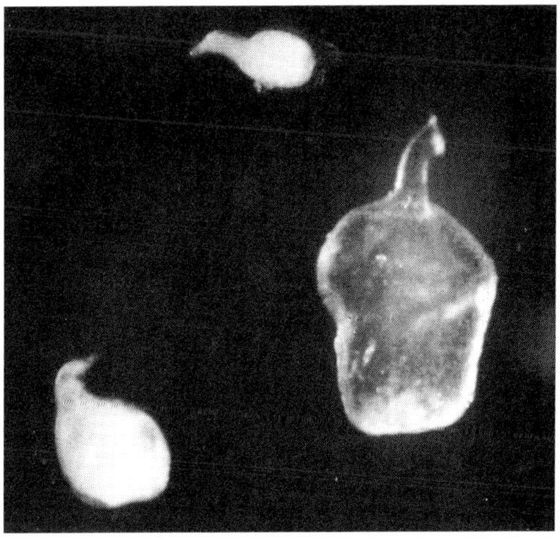

Fig. 15.5. Comparative size of mature female of *Meloidogyne brevicauda* (right) with females of *M. incognita* (left and above). (Photo: N.C. Gnanapragasam.)

cvs K 145, TC9, DT1, TRI 2024 and TRI 2025 (Gnanapragasam *et al.*, 1985). In India, to date, only seedling tea is reported to be infested (Muraleedharan and Selvasundaram, 2001).

Environmental factors affecting parasitism

M. brevicauda needs a cool soil temperature for the build-up of populations. In studies carried out in controlled soil temperatures, successful parasitism of tea plants was observed only at 12°C, whilst no parasitism was found to occur at higher temperatures (Gnanapragasam, 1988a).

Alternative hosts

Despite intensive surveys being carried out for several years in the tea areas for the possible existence of other hosts to *M. brevicauda*, to date none has been found in Sri Lanka or India. Even the weeds checked amongst infested tea fields have been found to be free of this species. The only alternative host reported so far is saffron (*C. sativus*) from Apsheron, Azerbaidzhan (Kasimova and Atakishieva, 1980). However, since saffron is normally not grown in the vicinity of tea fields, it would not pose a threat to spread of infestation.

Economic importance and population damage threshold

No information is yet available on damage threshold. Nevertheless, intensity of damage and associated crop loss seem to be very similar to those caused by *P. loosi*. Taking into consideration the distribution of this nematode and its very limited host range (only tea and saffron), the risk posed by this root knot nematode is small.

Radopholus similis

This species was first reported as a pest of tea in Java, Indonesia (Zimmerman, 1899). Steiner and Buhrer (1933) have also reported tea to be a good host to this species of nematode. The presence of this nematode in tea in Sri Lanka was first reported in 1968, when infestations were observed in young tea fields at an elevation range of 500–1000 m (Sivapalan, 1968). Other surveys have indicated the species to be widely distributed in the tea areas (young and mature) at lower altitudes as well, up to 200 m (Gnanapragasam, 1988a). In the presence of susceptible tea cultivars and under favourable climatic conditions, it is not uncommon to find *R. similis* even at very low altitudes of 50 m (Gnanapragasam, 1990). The species has also been reported from tea in China (Chen Zongmao, 2001, personal communication), Zimbabwe and South Africa (Keetch and Buckley, 1984).

Symptoms of damage

Damage symptoms on tea are very similar to those brought about by *P. loosi*. Parasitized plants are stunted, with pale leaves (Plate 16D), and they go into premature flowering and fruiting, symptoms which are very characteristic of nematode damage to tea (Sivapalan, 1968). The roots of infested plants are sparse and dried up compared with the whitish succulent feeder roots of healthy plants. Although lesions have been observed on the young roots, these are very small compared with those formed by *P. loosi* on tea (Gnanapragasam, 1983).

Biology

As in the case of *P. loosi*, *R. similis* is attracted to the growing part of the tea roots and invades the cortical region, feeding on and destroying the cells. Being an endoparasite, in young tea most of the population is found within the feeder roots. However, when the parasitized roots are severely damaged or when these become over-parasitized, the nematodes move into the soil in search of fresh roots. Therefore, in mature tea fields, large populations could be encountered within roots as well as in the soil in the rhizosphere of infested bushes.

Pathotypes/races

The behaviour of *R. similis* collected from the same hosts in the vicinity of tea areas from different agroecological regions varied. Differential host trials indicated the existence of different biological races of *R. similis* in the tea areas (Gnanapragasam *et al.*, 1991; Gnanapragasam, 1994b). This was confirmed further by molecular analysis (Hahn *et al.*, 1994).

Environmental factors

R. similis appeared to be quite sensitive to cold temperatures and has a poor survival rate in tea at elevations above 1000 m. When both *P. loosi* and *R. similis* are inoculated together on to tea at high elevations, the former takes over rapidly by competitive displacement, with no trace of the latter species within a short period. However, at lower elevations, *R. similis* has been observed in the rhizosphere along with *P. loosi*. In semi-dry areas, *R. similis* also occurs concomitantly with *R. reniformis*. *R. similis* in the tea areas appears to favour uniformly distributed high rainfall. In very wet or dry soil, the population was found to decline (Gnanapragasam, 1993).

Soil type and texture were also found to have significant influence on the reproductive rate and population build-up of this pest. Detailed experiments carried out in a temperature-controlled water bath at 25 ± 1°C revealed a rapid build-up of populations in sandy soil, followed by gravely or loamy soil. There was hardly any build-up in clayey soil. Damage to tea was also found to be significantly more in gravely, sandy and loamy soil (Gnanapragasam, 1990, 1994a).

Means of dissemination and survival

The method of dissemination of *R. similis* in tea is very similar to that of *P. loosi*. However, the survival rate in host-free soil is much shorter for *R. similis*.

Some of the popular cultivars, such as TRI 2025 and TRI 2026, most favoured by the small tea growers and widely planted in the mid- and lower elevations, are particularly susceptible to this species of nematode and have contributed to the spread of this pest. When infested, severe damage is encountered in the nurseries and newly planted young fields causing complete failure in the establishment, as well as in mature tea areas. The use of these tea cultivars is now being discouraged in the areas prone to damage by *R. similis* (Gnanapragasam, 1983, 1995).

Alternative hosts

R. similis is polyphagous, attacking hundreds of plant species. Several weeds and other plants intercropped with tea are suitable hosts to *R. similis*. Amongst these hosts, the most favoured ones are banana (*Musa* spp.), black pepper (*Piper nigrum*) and coconut (*Cocos nucifera*). The other common hosts include *Coffea* spp. (coffee), *Zea mays* (maize), *Saccharum officinarum* (sugarcane), *Pyrus* spp. (pears), *Persea americana* (avocado), *Ananas comosus* (pineapple), *Lycopersicon esculentum* (tomato), *Anthurium andreanum*, *Daucus carota* (carrot), *Areca catechu* (betel nut palm), *Coffea canephora* (Congo coffee tree), *Curcuma longa* (turmeric), *Dioscorea* (yam), *Musa textilis* (manila hemp), *Piper betel* (betel pepper), *Zingiber officinale* (ginger), *Arachis hypogea* (groundnut) and *Solanum nigrum* (nightshade weed) (Gnanapragasam *et al.*, 1991; Gowen *et al.*, Chapter 16; Koshy *et al.*, Chapter 21). Due to the presence of different pathotypes of *R. similis*, populations from some of these hosts were found to not infest tea (Gnanapragasam, 1994b).

Contrary to the situation with *P. loosi*, Guatemala grass (*Tripsacum laxum*), which is often used to recondition old tea fields prior to replanting, was also found to be a host to *R. similis*. *Eragrostis curvula*, marigold (*Tagetes* spp.) and *Vetiveria zizanoides* appear to suppress soil populations of *R. similis* (Gnanapragasam, 1986b, 1987b).

Economic importance and population damage threshold

In the mid- and low elevation tea areas of

Sri Lanka, *R. similis* is becoming as economically important as *P. loosi* is in the high elevation tea areas. Decline amongst several newly planted young tea fields in the mid- and some low elevation tea areas has been associated with moderate to heavy populations of *R. similis*. Since in many tea areas *R. similis* is found associated with *P. loosi*, it is difficult to study the yield loss under field conditions. However, results of pot experiments carried out at 25 ± 1°C revealed severe damage to tea brought about by a low initial population level of 28 nematodes/100 g of soil. In the field, when exposed to additional stress conditions such as drought, poor soil condition and/or attack by other pests, the damage threshold level could be even lower (Gnanapragasam and Herath, 1989).

Hemicriconemoides kanayaensis

H. kanayaensis is one of the important nematode pests of tea in Japan. It was detected originally from the roots of tea seedlings in Kanaya, Shizuoka Prefecture (Nakasono and Ichinohe, 1961) and has been detected in several other tea-planting districts in Japan (Takagi, 1969). The nematode has also been reported in Taiwan (Sivapalan, 1972).

Symptoms of damage

This species of ectoparasitic nematode feeds only on the feeder roots of tea. Continuous feeding by this nematode results in the sloughing off of the root cortex, revealing a brownish discoloured stele (Takagi, 1969). Maximum populations are encountered at a depth of 30 cm (Kaneko and Ichinohe, 1963).

Biology

A single female contains 14–15 eggs. Oviposition studies carried out in the laboratory have shown that this takes place over a period of 15–20 days during the months June/July. The entire life cycle is reported to be completed in 100 days. The ratio of juveniles to adults was found to reach a peak in July (Kaneko and Ichinohe, 1963; Takagi, 1969).

Alternative hosts

Tea is the only reported host of *H. kanayaensis* (Takagi, 1969).

Economic importance

Large numbers of this nematode have been found to result in crop failure in tea (Takagi, 1969). An increase in nitrogenous fertilizer is reported to reduce populations of this species of nematode (Kaneko and Ichinohe, 1963; Takagi, 1969).

Rotylenchulus reniformis

The reniform nematode, *R. reniformis*, was first observed in tea in Indonesia (Java) in 1951, where it was found to be responsible for large-scale casualties in young tea fields (Thorne, 1961). *R. reniformis* was also reported in north-east India; however, the frequency of occurrence was very low in Darjeeling, when compared with other plant parasitic nematodes (Basu and Roy, 1975, 1976). In Sri Lanka, this species was first encountered in a tea nursery in Rakwana in 1960 and subsequently in 19 tea estates at low and mid-elevations below 1200 m (Hutchinson and Vythilingam, 1963b). Although large numbers of nematodes were present in the root zone, no mature females could be detected. Other surveys found the tea-growing areas to be infested in the elevation range 200–900 m. Continuous soil moisture is reported to be essential for the build-up of *R. reniformis*, and reduction in rainfall below 100 mm can bring about a significant reduction in nematode numbers. Under continuous rainfall conditions, juveniles in the range of 200–370/100 g of soil were recovered from infested tea bushes. Low rainfall below 100 mm 1 month prior to sampling can reduce the population to 1–6 nematodes/100 g of soil and sometimes even to non-detectable levels (Gnanapragasam *et al*., 1987a; Gnanapragasam, 1988b).

Symptoms of damage

Only young tea plants were found to be infested with *R. reniformis* in Sri Lanka. The infested plants were stunted, with premature flowering and fruiting. The symptoms of damage were accentuated under poor soil conditions. Examination of the root system revealed that most of the feeder roots were clipped off due to feeding by this nematode. Although a large number of juveniles and immature females were recovered from the root zone, no mature females were found (Gnanapragasam *et al.*, 1987a).

Alternative hosts

R. reniformis has a wide range of hosts, including several common weeds encountered in the tea plantations. Other perennial crops that are sometimes intercropped with tea are good hosts to this species, including pepper (*Piper nigrum*), coffee (*Coffea robusta*) and young clove trees (*Syzgium aromaticum*), as well as grass cover crops, including Guatemala (*Tripsacum laxum*) that is planted in uprooted tea fields for soil reconditioning (Gnanapragasam, 1988b). Species of marigolds (*Tagetes* sp.) have also been reported to be suitable hosts to this nematode in India (Basu and Roy, 1976).

Economic importance

Damage by *R. reniformis* is often found in nursery plants and in newly planted young tea fields, especially on cvs TRI 2025, 2026, 2024. Since these cultivars are no longer recommended for planting in the mid-elevation tea areas of Sri Lanka, the spread of this nematode is now limited.

Other nematodes

Helicotylenchus

Both *Helicotylenchus dihystera* and *H. erythrinae* are commonly encountered in tea soils at all elevations in Sri Lanka (Hutchinson and Vythilingam, 1963b) and in Japan (Takagi, 1969), but no positive evidence of pathogenicity has been reported to tea from these countries. In Queensland, Australia, *H. dihystera* have been reported to affect the growth of young tea seedlings up to 12 months old. No evidence of pathogenicity has been recorded on older plants (P.C. O'Brien, 1988, personal communication). In East Africa, this species of nematode has been reported to be the most common nematode parasite in tea (Hainsworth, 1970). In Darjeeling, India, this species formed the bulk of the nematode fauna in tea soils at all altitudes. Soil samples collected from the rhizophere of weak seedlings had significantly more numbers of nematodes than from that of healthy seedlings; however, no positive evidence of pathogenicity was demonstrated (Basu, 1967).

Paratylenchus curvitatus *(pin nematode)*

The pin nematode, *P. curvitatus*, is also one of the most common and most prevalent plant parasitic nematodes encountered in the rhizophere of tea plants at all elevations in Sri Lanka (Hutchinson and Vythilingam, 1963b), in Japan (Kaneko and Ichinohe, 1963) and in north-east India (Basu, 1967). Although large numbers of this ectoparasitic nematode are encountered in the root zone of both young and mature tea, no positive evidence of pathogenicity has yet been established.

Hoplolaimus, Rotylenchus

In north-east India, these two genera have been found in the zone of weak and stunted seedlings (Basu, 1967). These nematodes have seldom been encountered in tea soils in Sri Lanka and, in locations where they have been found, no correlation has been established between their occurrence and any setback to growth. This species has also been reported in Malawi and East Africa (Hainsworth, 1970).

Xiphinema

Large numbers of this genus have been found in the soils of tea nurseries in north-east India. They have been found to feed at the root tips of feeder roots, resulting in slight swelling of the affected root tips. No further evidence of pathogenicity has been established with respect to this nematode (Basu, 1967). Species of *Xiphinema* have also been reported from tea fields in South Africa (Martin, 1962).

Management of Nematode Parasites in Tea

In most countries, studies on the incidence and pathogenicity of nematodes in tea have been made mostly in nurseries and young tea fields. As such, methods of control have been largely confined to treatment of nursery soils. However, in countries such as Sri Lanka and Japan, where plant parasitic nematodes pose a serious threat to the mature tea crop, various methods have been developed to mitigate their effects on the growth and productivity of tea. In Sri Lanka for the last two decades, an effective integrated management strategy with minimal use of pesticides is being adopted to manage tea nematodes. The different strategies adopted to manage nematodes in the tea-growing areas of the world include: cultural methods, physical methods, resistance and tolerance, chemical control and biological control.

Cultural methods

As long as the tea plant can grow vigorously and produce fresh feeder roots to compensate for those that die prematurely on account of nematode damage, it will be able to withstand parasitism to a significant extent. Therefore, those cultural methods that enhance growth and at the same time curtail nematode soil populations help to sustain productivity at economic levels. Tea fields in which yields have declined, but not to uneconomical levels, benefit most from such practices.

Incorporation of organic matter

Besides helping in the retention of essential soil nutrients and the consequent better nutrient status of the tea plant, large inputs of organic matter, including cattle manure and well-decomposed plant residues, have been reported to suppress the populations level of *P. loosi* (Loos, 1953a; Takagi, 1969).

The incorporation of specific oil cakes, such as margosa seed cake (*Azadirachta indica*), castor oil cake (*Ricinus communis*), mahuva oil cake (*Madhuca indica*), karanj oil cake (*Pongamia glabra*), coconut oil cake, decomposed poultry droppings, decomposed waste tea, as well as plant residues such as freshly harvested 'water hyacinth' (*Eichornia crassipes*), help to curtail damage caused by the root lesion nematode, *P. loosi* (Gnanapragasam, 1987b, 1991a,b, 1994; Mohotti, 1998). Nematode-infested plants grown in organically amended tea soils have enhanced root growth, leaf area and leaf chlorophyll content when compared with control plants treated with nematicides, and there is a decrease in tissue damage (Mohotti *et al.*, 1998, 2000a).

Soil cultivation (forking)

Soils with increasing acidity have a tendency to form a hard pan, and soil compaction impedes the rate of normal replenishment of damaged and dying feeder roots. Tea plants subjected to such conditions suffer most from nematode infestation. Regular forking of such soils helps to break the hard pan and improve soil aeration and the consequent feeder root growth. Tea fields with a hard pan and heavily infested with the root lesion nematode *P. loosi* have recovered remarkably following such treatments (Sivapalan, 1972). An increase in soil aeration brought about by forking is also reported to increase microbial populations in the soil, thereby indirectly helping to reduce populations of nematode (Mohotti, 1998).

Fertilizer application

The provision of balanced fertilizer mixtures influences the physiological status of the plant, which in turn influences the population dynamics of plant parasitic nematodes. An imbalanced supply of potash fertilizer (at lower proportions to increasing levels of nitrogen) was found to enhance the pathogenicity caused by *P. loosi* in tea. The reverse effect was induced by increasing the dosage of potash fertilizer, which also brought about a decline in the population level of this species of nematode (Fig. 15.6) (Gnanapragasam, 1982).

The type of nitrogenous fertilizer applied to tea also influences the population dynamics of *P. loosi* in tea. Application of nitrogen in the form of urea can bring about a significant suppression in the population (Sivapalan, 1980).

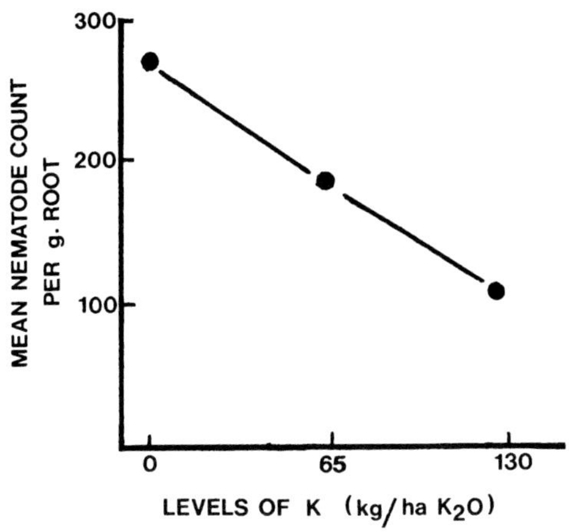

Fig. 15.6. Influence of potash fertilizer on root population of *Pratylenchus loosi* in mature tea. (N.C. Gnanapragasam.)

Cultivation of cover crops

It is customary to plant a grass cover crop for a period of 2 years following uprooting of old tea fields, prior to replanting with young tea. These grass species are meant to improve the physical structure of the soil, improve soil aeration and at the same time add a substantial amount of organic matter that is provided through regular lopping of such grasses. The grass species used for such soil reconditioning include Guatemala grass (*T. laxum*) and Mana (*Cymbopogan confertiflorus*), both of which are non-hosts of *P. loosi* (Visser, 1959; Hutchinson, 1962; Kerr and Vythilingam, 1966). The planting of a non-host also has the added advantage of depriving the nematodes of adequate food and thus helping to bring down the population with time. Since Guatemala grass is a good host of *R. similis*, this grass is not recommended to be planted in areas infested with the nematode (Gnanapragasam, 1995).

Planting of antagonistic crops

The vacant areas amongst nematode-infested old tea fields were sometimes planted to marigolds (*Tagetes erecta* and *T. patula*) to help to reduce nematode populations, prior to infilling such areas with young tea. Since marigold competes for soil moisture and nutrients, this practice is discouraged in young tea fields and such planting is confined to only the older tea areas (Hutchinson, 1964; Hainsworth, 1970). The nematode-suppressing activity of marigold is most effective during its phase of vegetative growth and prior to flowering (Hutchinson, 1961; Sivapalan, 1972).

The planting of *Eragrostis curvula* (which is planted mainly to prevent soil erosion in steep sections and in vacant areas in tea fields) has been found to suppress populations of *P. loosi* (Gnanapragasam, 1981) and *R. similis* (Gnanapragasam, 1986b). This grass has also been reported to suppress populations of *Meloidogyne* sp. in tea fields in Malawi (Anonymous, 1960).

The other trap crops which are recommended to be planted in the vacant areas of tea fields include *Arachis pintoi*, *Tithonia diversifolia* (wild sunflower), *Wedeliya trilobata* and *Vetiveria ziazanodes* (vet-ver) (Gnanapragasam, 1995, 1997).

Plant species such as *Adhathoda vasica*, *Sambucus javanica*, *Indigofera teysamanii*, *Eupatorium inuliformes*, *Calliandra calothyrsus*, *Crotalaria anagyroides* and *Lantana camara* also seem to possess nematicidal properties as these have been found to drastically reduce populations of both *P. loosi* and *R. similis*. These crops do not compete for moisture and nutrients and are suitable to be grown as hedgerow planting in steep areas. Apart from reducing populations of nematodes, these plant species also help to add adequate mulch to help build-up the organic bulk of soil (Gnanapragasam, 1997; Gnanapragasam and Sivapalan, 2001, 2004).

Irrigation

Nursery plants that are irrigated with water collected from ravines that course through infested sections of tea plantations have been found contaminated with nematodes (Gnanapragasam and Jebamalai, 1982). In order to circumvent this danger, in areas prone to nematode infestation, it is a recommended practice to sediment irrigation water in specially built sedimentation tanks for 48 h.

Resting of tea fields

Tea fields are pruned regularly once every 2–4 years, depending on the ambient temperature of the locality. This is a drastic operation, the recovery from which is dependent on the physiological condition of the pruned tea bush. When the tea field is subject to various forms of stress, including nematode parasitism, the affected sections of the tea fields recover poorly, mainly on account of the low carbohydrate reserves. In order to overcome this, such fields are rested prior to pruning for periods ranging from 6 to 8 weeks.

Replanting old tea fields

Tea fields that are uneconomical for further retention are uprooted and replanted to selected tea varieties with specific virtues. If such fields are known to be infested with nematodes, the uprooting of the old tea has to be carried out in such a manner as to ensure the extraction of as many residual roots as is possible. Large root fragments left in the soil harbour nematodes in the periphery of the lesions and such populations are known to remain viable for as long as 2–3 years (Hutchinson, 1960a, 1964). Large-scale failures in newly planted tea areas have been traced to re-infestation from residual populations remaining in old roots (Sivapalan, 1967b). It is, therefore, extremely necessary to ensure a through cleaning up of all residual roots following uprooting.

Re-infestation could also occur rapidly through movement of soil from an infested area higher up on the hill slopes and the crest of the hill. Therefore, when replanting is undertaken in areas prone to nematode infestation, uprooting of tea should commence from the top of the hill downwards, and not vice versa (Gnanapragasam, 1985a).

Physical control

The only physical control method adopted for controlling nematodes in tea is in nurseries. Nursery soil used for propagating young tea plants in India is sometimes heated by spreading the soil on galvanized sheets to temperatures ranging from 60 to 62°C for 5 min (Rao, 1976; Basu, 1978). This method of nursery soil treatment is not practical for large nurseries and is not recommended in Sri Lanka as this could damage the soil tilth.

Resistance and tolerance

Tea cultivars have been assessed to have varying degrees of natural tolerance and resistance to different species of parasitic nematodes in Sri Lanka (Loos, 1953a; Hutchinson, 1960b). Large numbers of tea cultivars have been screened for resistance and tolerance to the root lesion nematode *P. loosi* of tea in Sri Lanka, and several cultivars have been recommended for planting in nematode-infested areas (Kerr and Vythilingam, 1967; Sivapalan, 1967a, 1972; Gnanapagasam, 1986a, 1990, 1991a, 1992, 1995; Anandappa, 1995). Since the early 1980s, several cultivars have also being screened against *R. similis* in Sri Lanka (Gnanapragasam, 1985b) (Table 15.2).

During recent years, the breakdown of resistance has been observed amongst some of the cultivars earlier assessed to be resistant/tolerant, and such breakdown of

Table 15.2. Tea cultivars resistant/tolerant to major tea nematodes.

Nematode species	Tea cultivars		
Pratylenchus loosi	TRI 62/5	TRI 4002	MO 146
	TRI 62/7	TRI 4006	MPA1
	TRI 62/9	TRI 4033	MT 18
	TRI 2025[a]	TRI 4052	N 2
	TRI 2142	TRI 4053	NAY 3
	TRI 3013	TRI 4055	PK 2
	TRI 3014	TRI 4056	W 1/1
	TRI 3016	TRI 4060	WT 26
	TRI 3017	TRI 4066	WY
	TRI 3018	TRI 4070	
	TRI 3019	TRI 4079	
	TRI 3020	B 275	
	TRI 3022	CW 21	
	TRI 3024	CY 9[a]	
	TRI 3047	DG 7	
	TRI 3048	DK 1	
	TRI 3049	DK 16	
	TRI 3055	DN[a]	
	TRI 3059	DT 1[a]	
	TRI 3061	DT 95	
	TRI 3063	DUN 7	
	TRI 3065	DW 12	
	TRI 3069	K 145	
	TRI 3070	KEN 16/3	
	TRI 3072	MO 116	
Radopholus similis	TRI 62/5	TRI 4051	TRI
	TRI 2023	TRI 4052	DT 95
	TRI 2024	TRI 4054	DG 7
	TRI 2027	TRI 4055	DN
	TRI 3030	TRI 4070	N 2
		TRI 4071	
	TRI 3031	TRI 4077	
	TRI 4006	TRI 4078	
	TRI 4024	CH 13	
	TRI 4047	CY 9	

[a]Tolerance was found to break down under specific conditions.

tolerance/resistance was found to have been brought about by the age of plants and adverse environmental factors (Gnanapragasam, 2002b; and unpublished data). Some of the high quality popular cultivars, such as DT 1, earlier assessed to be tolerant to *P. loosi*, are presently found to be very susceptible at the initial stages of growth (N.C. Gnanapragasam, unpublished data). Therefore, continuous monitoring of such chosen cultivars becomes essential to assess such breakdown of tolerance/resistance.

Grafting

In Sri Lanka, some of the high yielding tea cultivars that are renowned for good quality were found to be susceptible to *P. loosi* and/or *R. similis* and therefore not suitable for planting in tea areas. Grafting a nematode-susceptible scion on to a nematode-tolerant/resistant rootstock has been found to induce resistance/tolerance in such combinations, making it possible to use high yielding but nematode-susceptible cultivars in the field (Gnanapragasam, 1992). When choosing such graft combinations, only those varieties which are compatible with each other and suitable for growing in the particular agroclimatic zone are selected.

Chemical control

In a perennial crop that is grown for as long as 50–60 years, chemical control generally proves to be uneconomical, since such costly treatments have to be repeated periodically to sustain populations below economic thresholds. This form of treatment is thus confined to eradicating nematodes from nursery soils and in reducing soil population in the field at the time of planting the young tea.

Chemical control in nurseries

Infested nursery plants are an important source of spread of nematode infestation to fields that may hitherto have been free of infestation. In areas prone to such infestation, it is a routine practice to chemically treat all nursery soils. Nursery soils in the past were treated with fumigants such as methyl bromide, DD soil fumigant, Nemagon 75% E.C. and Dowfume W 85% (Kerr, 1963b; Akbar and Ali, 1965a,b; Sivapalan, 1969; Nara *et al.*, 1973). Since the early 1980s, as an alternative to methyl bromide, Dazomet (Basamid 98% G) was recommended in Sri Lanka (Sivapalan *et al.*, 1980a). Basamid 98% G has also been used effectively to control soil nematodes in Bangladesh tea nurseries (Huq *et al.*, 1990). Soil solarization has also been attempted in Sri Lanka (Vitarana, 2001). However, since large volumes of soil are used in the tea plantations each year for propagating nursery plants, solarization is not practical, and results of such trials have been inconclusive.

Following such chemical treatment in the nursery, the cuttings or seeds are planted after an appropriate interval as specified for the respective chemical. In certain countries such as India and Bangladesh, granular nematicides are added to soils bearing young nursery plants (Rao, 1974, 1976; Basu, 1979; Basu and Gope, 1985; Huq *et al.*, 1990). In Sri Lanka, such treatment of nematode-contaminated nursery plants with granular nematicides is recommended only under very special circumstances; the cultivar involved has to be a tolerant one and infested with only a light population of nematodes and not exhibiting any damage symptoms. Treatment is also limited to instances when these treated plants are to be used only as infillings in fields already having a history of nematode infestation (Gnanapragasam *et al.*, 1987b). In addition to ensuring that tea plants are propagated in nematode-free soil, basic nursery hygiene is also maintained to prevent further contamination (Gnanapragasam, 1989).

Chemical control at planting

Despite soil rehabilitation and minimizing residual populations in the soil, and further confining the replanting to those cultivars that have proven tolerance or resistance to nematode infestation, chemical treatment of planting holes, at planting time, is routinely practised in Sri Lanka. This practice is carried out as an additional measure of insurance against a possible set-back to establishing young plants. The chemicals recommended included Fenamiphos 5% G and Carbofuran 3% G, at the rate of 7 g per planting hole (Sivapalan *et al.*, 1980b). Organophosphate nematicides are also used in Iran to control nematodes in the tea fields (Maafi and Moghadam, 2001).

In Japan, pre-planting nematode control was achieved by fumigating the planting area with ethylene dibromide or DD at 200–300 l/ha at a depth of 20–30 cm (Takagi, 1969).

Chemical control in mature tea

Routine chemical treatment of mature tea is an uneconomical exercise. Nevertheless, each time the tea is pruned, a significant amount of feeder roots decay and, at the time of recovery from pruning, there is a significant growth of new feeder roots that are susceptible to rapid re-infestation, and this has a significant deleterious effect on the rate of recovery from pruning. Therefore, in Sri Lanka, for fields that still have a high yield potential but are subject to moderate to heavy nematode infestation, the current recommendation is to apply a single application of nematicide (fenamiphos 5% G) at 7 g/plant, mixed along with the first application of fertilizer following pruning of such fields (Gnanapragasam, 1987b).

Biological control

Until recently, very little information has been available with regard to control of plant parasitic nematodes in tea by biological agents.

In the past, a sporozoan endoparasite was recorded occasionally from *P. loosi*, but its significance in controlling this pest was not confirmed. The presence of predatory nematodes in tea soils of Sri Lanka was also reported but, since they were not found in large numbers, no attempt was made to investigate their efficiency in controlling plant parasitic nematodes of tea (Gadd and Loos, 1946). The use of fungi in the control of nematodes has been reported by Barua (1983) in northeast India. Kaneko and Ichinohe (1963) have reported a phycomycete fungus to be responsible for as much as 30% reduction in the population of adult females of *H. kanayaensis*.

Although compost and soil amendments have been included in the integrated management programme in Sri Lanka for several years with the intention of helping to increase the natural predators and parasites of nematodes pathogenic to tea, no investigation had been carried out to isolate and identify the predators and parasites involved in such suppression of nematodes.

Several microbial antagonists of plant pathogenic nematodes have been found to be present in Sri Lankan tea soils (Mohotti, 1998; Mohotti *et al.*, 2000b) (Table 15.3). Mohotti (1998) reported the frequency of occurrence of *Pasteuria penetrans* to be relatively low compared with other species in the Sri Lankan soils. The most microbial antagonists encountered are the nematode-trapping fungi (*Arthrobotrys musiformis* Dreshsler, *A. oligospora* Fres., *Arthrobotrys* sp., *Dactylella* sp., *Monacrosporium* sp.) and *Fusarium* sp., *Paecilomyces* sp. and *Trichoderma* sp. *P. penetrans* has also been reported in the tea soils of Iran (Mohotti *et al.*, 1996). The incidence of *P. penetrans* and nematophagous fungi is reported to be high in tea soils when compared with other surveyed agricultural lands of Sri Lanka (Mohotti *et al.*, 1996; Mohotti, 1998). As is to be expected, microbial antagonists of pathogenic nematodes are activated, conserved and enhanced in soils incorporated with organic amendments (Mohotti *et al.*, 2000b).

The host cuticle of *P. loosi*, *R. similis*, *M. brevicauda* and *Pratylenchus* sp. can be encumbered with the endospores of *P. penetrans* (Mohotti, 1998). In *R. similis*, the endospores enter the pseudocoelum. Introducing biological control methods in the integrated management system in tea fields may be effective in controlling plant parasitic nematodes.

Nematode management in organic tea culture/gardens

Recently, there is a high demand in certain countries for high quality organically grown tea, which fetches a significantly higher price than that grown in conventional systems. In organic farming, the plants are grown in an environment free of chemicals. Such an integrated farming system causes the least disruption to the environment, helps to improve soil fertility, helps to enhance microbial activity and can evolve a sustainable farming system. In Sri Lanka, nematode management under such a system is made possible by harnessing eco-friendly alternative methods of control without the use of any nematicides. Some of these recommended practices include: (i) use of non-hosts; (ii) use of resistant/tolerant cultivars of tea; (iii)

Table 15.3. Records of naturally occurring nematode antagonists in tea soils.

Biocontrol group	Organism	Country of report	Reference
Bacteria	*Pasteuria penetrans* group	Iran, Sri Lanka	Barooti (1989); Mohotti (1998)
	Bacillus sp.	India	Pandey *et al.* (2001)
Nematophagous fungi	*Fusarium* sp.	Japan Sri Lanka	Kaneko and Ichinohe (1963); Mohotti (1998)
	Paecilomyces sp.	Japan Sri Lanka	Kaneko and Ichinohe (1963); Mohotti (1998)
	Trichoderma harzianum Rifai	India	CAB International (n.d.)
	Trichoderma koningii	India	Pandey *et al.* (2001)
	Trichoderma sp.	Sri Lanka	Mohotti (1998)
	Trichoderma viride	India	CAB International (n.d.)
	Verticillium sp.	Sri Lanka	CAB International (n.d.); Mohotti (1998)
Nematode-trapping fungi	*Arthrobotrys musiformis* Drechsler	Sri Lanka	Mohotti (1998)
	Arthrobotrys oligospora Fres.	Sri Lanka	Mohotti (1998)
	Arthrobotrys robusta Duddington	Sri Lanka	CAB International (n.d.)
	Arthrobotrys sp.	Sri Lanka	Mohotti (1998)
	Dactylella sp.	Sri Lanka	Mohotti (1998)
	Monacrosporium sp.	Sri Lanka	Mohotti (1998)
Micro arthropods	Tardigrades (Water bears)	Sri Lanka	Hutchinson and Streu (1960)
	Collembolans (Spring tails)	Sri Lanka	Mohotti (2002)
	Mites (Acari)	Sri Lanka	Gadd and Loos (1946); Mohotti (2002)
	Myriapods	Sri Lanka	Mohotti (2002)
Nematodes	*Mononchus* sp., *Diplogaster* sp. and *Dorylaimus* sp.	Sri Lanka	Gadd and Loos (1946);
Miscellaneous	VAM (vesicular arbuscular mycorrhiza)	Sri Lanka, Taiwan	Balasuriya *et al.* (1991); Chang and Young (1992)
	Protozoan	Sri Lanka	Gadd and Loos (1946)

planting of nematode antagonistic plants; (iv) use of botanicals; (v) use of soil amendments; and (vi) use of naturally occurring biological control agents (Mohotti, 1998, 2001; Mohotti *et al.*, 1999; Gnanapragasam and Sivapalan, 2001).

Summary of management methods

Nematode management should commence at the nursery stage itself to ensure that only healthy vigorously growing nematode-free plants are transferred to the field. Nursery plants should be grown in nematode-free soil; proper precaution should be taken to adopt adequate hygienic measures to prevent plants becoming contaminated from adjoining fields; and transportation of nursery plants from one plantation to the other should be discouraged.

Although nematodes cannot be eradicated in a field, it is essential to reduce them below the economic damage threshold to help to avert reduction in crop productivity. Any one of the management strategies by themselves may not be adequate to reduce the population below the economic damage threshold in the field planted to young and mature tea. In Sri Lanka, the management strategy during the past two decades has, therefore, been an integration of the appropriate methods of control most suited for a given environment.

The most useful resources for nematode management in tea fields are:

- Limited use of environmentally friendly chemicals with short soil persistence;
- Planting of nematode-tolerant and -resistant cultivars;
- Proper soil management to maintain soil pH within the range of 4.5–5.0;
- Using potash-enriched fertilizer mixtures;
- Enriching the soil with various organic matter;
- Cultivation of soils by regular forking;
- Use of antagonistic crops and botanicals; and
- Use of biological control agents.

Method of diagnosis

As is the case with the other crops, the above-ground damage symptoms on tea brought about by nematodes are often confused with similar symptoms induced by other factors that tend to restrict root growth. Removal of suspect bushes indicates (if infested) almost complete absence of feeder roots, or if feeder roots are present they would be few and appear dead or dried up. When the bark is peeled lightly, dead brown areas (lesions) can be observed (Plate 16B). Positive diagnosis is made by sampling both soil and roots from affected sections and extracting the nematodes by a modified Baerman funnel technique (Chapter 3). The efficiency of recovery by Baerman funnel techniquo is significantly improved by the addition of small amounts of tea root saponins (1–10 ppm). Further addition beyond 10 ppm suppresses recovery. Storage of soil samples as well as sample size of roots has an effect on the efficiency of recovery (Sivapalan *et al.*, 1979; Gnanapragasam and Sivapalan, 1991).

Soil sampling

Sampling of tea soils in Sri Lanka is carried out routinely in all suspect areas for the three commonly encountered nematodes, *P. loosi*, *R. similis* and the juveniles and immature females of *R. reniformis*. Sampling is usually carried out when the soil is adequately moist at a depth of 15–25 cm and at a distance of 15 cm from the base of the plant. Several samples are collected from a given field, with approximately 25–30 randomly collected samples per 2 ha. If only a section of a field is found to be showing decline symptoms, samples are collected from such specific locations. When collecting the samples, it is necessary to sample the weak as well as the moderately healthy plants, since very weak plants carry only a small population during growth decline. It is essential to include few feeder roots as well as the small root fragments that come with the soil.

Besides recovering the above species of plant parasitic nematodes, several other species of tea nematodes could also be recovered from these samples. However, for a proper sampling of *H. kanayaensis*, sampling should be done deeper at a depth of 30 cm (Kaneko and Ichinohe, 1963; Takagi, 1969).

Root sampling

When a newly planted young tea area is to be sampled for nematode infestation (young tea fields less than 5 years old), it is necessary to collect feeder root samples as well. As many as 25–30 random samples are usually collected per hectare to represent the entire field. If only a section of the field appears to show decline symptoms, collection of samples is confined only to the specific suspect area. A few grams of feeder roots are collected from the rhizosphere of these respective points of sampling and bulked together to form a composite sample.

Detection of Meloidogyne *spp.*

Meloidogyne species can be detected by examining the roots of the suspect tea bushes and checking for the presence of characteristic swellings and/or galls as well as for the presence of females and egg masses clinging on to the outside of the root. Species of *Meloidogyne* are identified by careful dissection and examination of the posterior cuticular pattern ('perineal pattern') in the tail region of the female (Chapter 2). Infestation with *M. brevicauda* can be distinguished easily from the other *Meloidogyne* species by the size of the female, which is significantly larger than the other common species of root knot nematodes encountered in tea fields.

Conclusions and Future Prospects

Although pathogenicity to tea caused by nematodes has been known for several years, the damage caused by this pest is not yet taken seriously by many tea-growing countries.

Therefore, it is necessary to create a greater awareness in all tea-growing countries to recognize this problem, and greater emphasis must be placed on carrying out extensive surveys to identify the presence of different species of pathogenic nematodes and study their interaction with other environmental factors to help quantify crop losses brought about by this pest in different situations. It is only when such visual symptoms of damage are correlated with economic crop loss that the severity of nematode damage becomes more apparent to those who are not aware of the problem.

Molecular and biochemical studies should also be actively pursued to help to identify different races/pathotypes and study their interactions and pathogenicity on different hosts as well as their interactions under different environments. These studies would further strengthen the integrated management programme and help in providing recommendations suited for specific locations.

Recently, some of the tea cultivars assessed to be resistant to pathogenic nematodes have been found to succumb to infestation as a result of the breakdown of resistance/tolerance with age of plants and exposure to adverse environmental factors. Since market requirements for tea pose restrictions with genetically modified organisms, genetic modifications of the tea plant have no future in the management of tea nematodes. Therefore, greater effort should be made to continue to screen cultivars for natural tolerance and susceptibility to this pest under different environmental conditions.

It is also necessary to strengthen studies on biological control further, so that potential formulations of microbial antagonists could be used in nematode management. This will be especially useful in organic farming areas.

Interactions with other scientists in the field of pathology and entomology should also be strengthened to study disease complexes brought about by a combination of pest incidence, such as nematode and insect or nematode and fungi, etc.

Studies should also be carried out to control these pests at the physiological level using pheromones and specific metabolic disruptors.

References

Akbar, K. and Ali, M.A. (1965a) Studies on the control of eelworms in tea nurseries: effect of fumigation with Shell DD on the population of eelworm in the soil. *Tea Journal of Pakistan* 3, 14–18.

Akbar, K. and Ali, M.A. (1965b) Effect of fumigation with Nemagon 75% E.C. and Dowfume W-85 on the populations of eelworms in the nursery soil and health of seedlings. *Tea Journal of Pakistan* 3, 19–25.

Anandappa, T.I. (1995) Botany of tea. In: Sivapalan, P., Gnanapragasam, N.C. and Kathiravetpillai, A. (eds) *Field Guide Book.* Tea Research Institute, Talawakelle, Sri Lanka, pp. 1–9.

Anonymous (1960) Tea Research Station, Tea Association of Nyasaland, Rhodesia. *Quarterly Newsletter* 18.

Anzueto, F. and Sarah, J.L. (1992) A study of the host–parasite relationship of a population of *Pratylenchus loosi* from Guatemala on *Coffea arabica* cv Catuai. *Nematropica* 22, 117.

Arulpragasam, P.V. (1981) *Report of the Tea Research Institute of Sri Lanka – 1981*, pp. 54–58.
Arulpragasam, P.V. and Addaickan, S. (1983) Soft-root-rot. A new disease of tea? *Tea Quarterly Sri Lanka* 52, 52–55.
Balasuriya, A., Arulpragasam, P.V. and Ratnayake, R.M.A. (1991) *Mycorrhiza* in tea. *Tea Bulletin* 11, 3–12.
Banerjee, B. (1967) Eelworm as soil-borne pests of tea. *Proceedings of the 23rd Bienn. Tocklai Conference*. Tea Research Association, Calcutta, India.
Barber, C.A. (1901) A tea eelworm diseases in S. India. *Tropical Agriculturist* 21, 238.
Barooti, S. (1989) Distribution of *Pasteuria penetrans*, a parasite of nematodes Iran. *Iranian Journal of Plant Pathology* 25, 9–10.
Barua, G.C.S. (1983) Fungi in biological control of tea pest and diseases in north east India. *Two and a Bud* 30, 5–7.
Basu, S.D. (1967) Eelworm in tea: some important groups. *Two and a Bud* 14, 84–85.
Basu, S.D. (1968) Eelworms – a progress report on a few more groups found in N.E. Indian tea soils. *Two and a Bud* 15, 70–71.
Basu, S.D. (1976) Root-knot nematode, *Meloidogyne incognita* (Kofoid and White) Chitwood on stem of tea, seedling. *Two and a Bud* 23, 23.
Basu, S.D. (1978) Heat the soil and kill nematodes. *Two and a Bud* 25, 42.
Basu, S.D. (1979) Control of *Meloidogyne incognita* on tea seedlings with some systemic nematicides. *Indian Journal of Nematology* 9, 84.
Basu, S.D. and Gope, B. (1982) Effect of weedicides on nematode populations in tea soils. *Two and a Bud* 29, 62–63.
Basu, S.D. and Gope, B. (1985) Control of root-knot nematode, *Meloidogyne incognita* (Kofoid and White) Chitwood, on tea seedlings with some systemic nematicides. *Two and a Bud* 32, 32–35.
Basu, S.D and Roy, S.K. (1975) A new ectoparasitic nematode in tea soils. *Two and a Bud* 22, 17.
Basu, S.D and Roy, S.K. (1976) Preliminary note on nematode survey in Darjeeling. *Two and a Bud* 23, 58–59.
Baujard, P. (1986) Ecologie des nématodes dans le bassin arachidier du Sénégal. *Revue de Nématologie* 9, 288.
Baujard, P., Mounport, D. and Martiny, B. (1990) Studies on four species of the genus *Pratylenchus* Filip'ev, 1936 (Nemata: Pratylenchidae) under stereoscan microscope. *Revue de Nématologie* 13, 203–210.
CAB International (n.d.) *Herbarium Record Books*. Computerized data of CABI Bioscience (formerly CAB International Institute of Mycology), Egham, UK.
Chang, F. and Young, C. (1992) Effects of VA mycorrhizal fungi and phosphorous-solubilizing bacteria inoculated on growth of tea cuttings in plastic bags. *Taiwan Tea Research Bulletin* 11, 79–89.
Chen, Z. and Chen, X. (1982) Review on tea diseases. *Plant Diseases* 66, 961–965.
Corbett, D.C.M. (1967) Nematodes as plant parasites in Malawi. *PANS* 13, 151–162.
Gadd, C.H. (1939) A destructive root diseases of tea caused by the nematode, *Anguillulina pratensis*. *Tea Quarterly* 12, 131–139.
Gadd, C.H. and Loos, C.A. (1941) Observations on the life history of *Anguillulina pratensis*. *Annals of Applied Biology* 28, 39–51.
Gadd, C.H. and Loos, C.A. (1946) The problem of nematode control in tea plantations. *Tea Quarterly* 18, 3–11.
Gnanapragasam, N.C. (1981) The influence of cultivating *Eragrostis curvula* in nematode infested soil, on the development on the subsequent build-up of populations in replanted tea. *Tea Quarterly* 50, 160–162.
Gnanapragasam, N.C. (1982) Effect of potassium fertilization and soil temperature on the incidence and pathogenicity on the root-lesion nematode, *Pratylenchus loosi* Loof, on tea (*Camellia sinensis* L). *Tea Quarterly* 51, 169–174.
Gnanapragasam, N.C. (1983) Incidence of nematode damage in the mid-country caused by the burrowing nematode, *Radopholus similis*. *Tea Quarterly* 52, 41.
Gnanapragasam, N.C. (1985a) The strategy for replanting eelworm infested areas. *Journal of Tea Science* 54, 46.
Gnanapragasam, N.C. (1985b) *Report of the Tea Research Institute of Sri Lanka – 1985*, pp. 82–96.
Gnanapragasam, N.C. (1986a) Nematode parasites of tea. In: Sivapalan, P., Kulasegaram, S. and Kathiravetpillai, A. (eds) *Hand Book on Tea*. Tea Research Institute of Sri Lanka, Talawakelle, Sri Lanka, pp. 109–123.
Gnanapragasam, N.C. (1986b) *Report of Tea Research Institute of Sri Lanka*, pp. 69–78.

Gnanapragasam, N.C. (1987a) Effect of soil acidity on the population levels of *Pratylenchus loosi*. *International Nematology Network Newsletter* 4, 23–24.

Gnanapragasam, N.C. (1987b) Integrated strategies to manage nematode in tea. In: *Proceedings of the International Conference on Tea Quality and Human Health*, November 1987, Hangzhou, China, pp. 117–183.

Gnanapragasam, N.C. (1988a) Limitation on growth and productivity of tea by plant parasitic nematodes. In: *Proceedings of Regional Tea Scientific Conference*, 19–21 January, 1988, Colombo, Sri Lanka. Tea Research Institute of Sri Lanka, a division of Sri Lanka Tea Board, pp. 123–133.

Gnanapragasam, N.C. (1988b) *Report of the Tea Research Institute of Sri Lanka – 1988*, pp. 54–67.

Gnanapragasam, N.C. (1989) Prevention of dissemination of nematodes pathogenic to tea into hitherto uninfested tea areas. *Tea Bulletin* 9 (1), 20–22.

Gnanapragasam, N.C. (1990) *Report of the Tea Research Institute of Sri Lanka*, pp. 72–73.

Gnanapragasam, N.C. (1991a) *Report of the Tea Research Institute of Sri Lanka – 1991*, pp. 73–81.

Gnanapragasam, N.C. (1991b) Influence of soil amendments in reducing pathogenicity to tea by the root-lesion nematode, *Pratylenchus loosi*. In: *Proceedings of the International Tea Symposium*, Schizuoka, Japan, pp. 684–688.

Gnanapragasam, N.C. (1992) *Report of the Tea Research Institute of Sri Lanka*, pp. 58–67.

Gnanapragasam, N.C. (1993) *Report of the Tea Research Institute of Sri Lanka* – 1993, pp. 68–81.

Gnanapragasam, N.C. (1994a) Effect of environment on population dynamics of plant parasitic nematodes and consequent pathogenicity to tea. *Tea Bulletin* 14, 18–24.

Gnanapragasam, N.C. (1994b) *Report of Tea Research Institute of Sri Lanka*, pp. 52–64.

Gnanapragasam, N.C. (1995) Nematode pests of tea. In: Sivapalan, P., Gnanapragasam, N.C. and Kathiravetpillai, A. (eds) *Field Guide Book*. Tea Research Institute, Talawakelle, Sri Lanka, pp. 83–87.

Gnanapragasam, N.C. (1997) Plant derivatives in managing nematodes attacking tea in Sri Lanka. *International Journal of Nematology* 7, 111–112.

Gnanapragasam, N.C. (2002a) Nematode–insect interaction causing yield decline in tea. *International Journal of Nematology* 12, 117–118.

Gnanapragasam, N.C. (2002b) Slow decline observed in nematode tolerant tea cultivar. *International Journal of Nematology* 12, 232–233.

Gnanapragasam, N.C. and Herath, U.B. (1989) Pathogenicity of the burrowing nematode, *Radopholus similis* to young tea at different initial density. *Sri Lanka Journal of Tea Science* 58, 83–86.

Gnanapragasam, N.C. and Jebamalai, A.S. (1982) Nematode infestation of nursery plant through irrigation water. *Tea Quarterly* 51, 31–33.

Gnanapragasam, N.C. and Manuelpillai, M.E.K. (1981) Relative rates of egg hatch of the root-knot nematode of mature tea, *Meloidogyne brevicauda*, in relation to other species of the nematode. *Tea Quarterly* 50, 160–162.

Gnanapragasam, N.C. and Manuelpillai, M.E.K. (1984) Influence of inoculum level and temperature on the population build-up and pathogenicity of the root-lesion nematode *Pratylenchus loosi* Loof on tea. *Tea Quarterly* 53, 19–22.

Gnanapragasam, N.C. and Sivapalan, P. (1991) Influence of soil types and storage conditions on the recovery of *Pratylenchus loosi* from soil samples. *Afro-Asian Journal of Nematology* 1, 150–153.

Gnanapragasam, N.C. and Sivapalan, P. (2001) Eco-friendly pest management in organic tea farming. *Proceedings of the Fifth IFOAM-Asia Scientific Conference*, 31 October–4 November 2002, Hangzhou, China, pp. 199–201.

Gnanapragasam, N.C. and Sivapalan, P. (2004) Eco-friendly management of tea plantations towards sustainability. In: *Proceedings of the 3rd International Tea Conference in Global Advances in Tea Science*, 20–22 November 2003, Calcutta (in press).

Gnanapragasam, N.C., Sivapalan, P., Jebamalai, A.S. and Dharmasena, W.A.M. (1985) Tolerance/susceptibility of six different tea cultivars to *Meloidogyne brevicauda*, the root-lesion nematode of mature tea. *International Nematology Network* 2, 9–10.

Gnanapragasam, N.C., Ariyaratnam, V. and Navaratne, N. (1987a) *Rotylenchulus reniformis*, the new nematode species causing damage to tea in Sri Lanka. *Tea Bulletin* 7 (2), 34–39.

Gnanapragasam, N.C., Jebamalai, A.S. and Anpalagan, V.T. (1987b) Control of root-lesion nematode, *Pratylenchus loosi* in contaminated tea nursery plants. *Tea Quarterly* 56, 95–97.

Gnanapragasam, N.C., Karunaratne, A.A.C. and Anpalagan, V.T. (1989) Susceptibility of weeds to pathogenic nematodes of tea. *Tea Bulletin* 9, 3–10.

Gnanapragasam, N.C., Prematunga, A.K. and Herath, U.B. (1991) Preliminary survey for alternative hosts of the burrowing nematode, *Radopholus similis*, in the tea areas of Sri Lanka. *Afro-Asian Journal of Nematology* 1, 114–115.

Gope, B. and Borthakur, M. (1991) Long term effects of herbicides on nematode population in tea soil. *Two and a Bud* 38, 1–2, 37–38.

Gotoh, A. (1974) *Pratylenchus* spp. Geographic distribution of *Pratylenchus* spp. (Nematoda:Tylenchida) in Japan. Reprinted from *The Bulletin of the Kyushu Agricultural Experiment Station* Volume 17, pp. 139–224.

Gotoh, A. (1976) A review of plant parasitic nematodes in warm and sub-tropical regions in Japan. *Miscellaneous Bulletin of Kyushu Agricultural Experiment Station* 54, 1–61.

Hahn, M.L., Burrows, P.R., Gnanapragasam, N.C., Bridge, J., Vines, N.J. and Wright, D.J. (1994) Molecular diversity amongst *Radopholus similis* populations from Sri Lanka detected by RAPD analysis. *Fundamental Applied Nematology* 17, 275–281.

Hainsworth, E. (1970) Tea. *Journal of the Tea Boards of E. Africa* 11, 15–17.

Huan Jin (1984) The identification of some species of root-knot nematodes (*Meloidogyne*) on tea seedlings in Zheijiang Province. *Acta Zheijiang Ag. University* 9, 343–350.

Huq, M., Ahmed, M. and Ahmad, A.M. (1990) Control of nematodes in tea nursery. *Tea Journal of Bangladesh* 26, 33–36.

Hutchinson, M.T. (1960a) Nematode samples from estates. *Tea Quarterly* 31, 116–118.

Hutchinson, M.T. (1960b) Resistance and tolerance of tea to nematodes. *Tea Quarterly* 31, 13–18.

Hutchinson, M.T. (1961) Development in the control of the meadow nematodes. *Tea Quarterly* 32, 129–132.

Hutchinson, M.T. (1962) Rehabilitating tea soils – susceptibility of plants now in use to the root-lesion nematode *Pratylenchus loosi*. *Tea Quarterly* 33, 138–140.

Hutchinson, M.T. (1964) Further developments in the control of meadow nematode. *Tea Quarterly* 35, 90–95.

Hutchinson, M.T. and Streu, H.T. (1960) Tardigrades attacking nematodes. *Nematologica* 5, 149.

Hutchinson, M.T. and Vythilingam, M.K. (1963a) The distribution of *Pratylenchus loosi* among tea estates in Ceylon, with particular reference to the altitude. *Tea Quarterly* 34, 68–84.

Hutchinson, M.T. and Vythilingam, M.K. (1963b) Distribution of plant parasitic nematodes in the soils in Ceylon. *Tea Quarterly* 34, 119–126.

Inserra, R.N., Duncan, L.W., Vovlas, N. and Loof, P.A.A. (1996) *Pratylenchus loosi* from pasture grasses in central Florida. *Nematologica* 42, 159–172.

Kaneko, T. and Ichinohe, M. (1963) Notes on the nematode species and their bionomics associated with tea roots in Japan. *Japanese Journal of Applied Entomology and Zoology* 7, 165–174.

Kasimova, G.A. and Atakishieva, Y.Y. (1980) The nematode fauna of *Crocus sativus* in Apsheron. Izvestiya-Akademii-Nauk-Azerbaidzhanskoi-SSR-Azarbajcan-SSR-Elmlar-Akademijasynyn-Harbarlari. *Biologicheskie-Nauki* 1, 94–98.

Keetch, D.P. and Buckley, N.H. (1984) A check list of the plant parasitic nematodes of Southern Africa. *Technical Communication No. 195.* Department of Agricultural Development, South African Government, Republic of South Africa.

Kerr, A. (1963a) *Report of the Tea Research Institute of Ceylon – 1963*, pp. 95–102.

Kerr, A. (1963b) Fumigation of nursery soils. *Tea Quarterly* 34, 150–151.

Kerr, A. and Vythilingam, M.K. (1966) Replanting eelworm infested areas. *Tea Quarterly* 37, 67–72.

Kerr, A. and Vythilingam, M.K. (1967) Resistance of tea cultivars and seedlings to the root-lesion nematode (*Pratylenchus loosi*). *Tea Quarterly* 38, 42–51.

Li, D.Z. (1985) Description of some species of parasitic nematodes of genus *Pratylenchus* on plant roots in Sichuan Province. *Journal of South Western Agricultural College* 2, 49–51.

Loos, C.A. (1953a) Eelworms. *Tea Quarterly* 24, 34–38.

Loos, C.A. (1953b) *Meloidogyne brevicauda* n.sp a cause of root-knot nematode of mature tea in Ceylon. *Proceedings of the Helminthological Society of Washington* 20, 83–91.

Maafi, Z.T. (1992) Extraction of *Pratylenchus loosi* from imported tea plant from Japan. *Iranian Journal of Plant Pathology* 2, 93–94.

Maafi, Z.T. and Moghadam, A.M. (2001) Study on the effect of organophosphate nematicides on the root lesion nematode, *Pratylenchus loosi* in tea plantations. *Indian Journal of Plant Pathology* 37, 29–38.

Martin, G.C. (1960) Plant parasitic nematodes associated with tea production in the southern province of Nyasaland. *Proceedings of the First Inter-African Plant Nematology Conference I.A.P.S.C.*, Publication 86.

Martin, G.C. (1962) Population levels of the nematodes in root and soil around the zone of tea, sugarcane, tobacco and wheat grown in the Federation of Rhodesia and Nyasaland. *Rhodesia Agricultural Journal* 62, 28–35.

Mc Leod, R., Reay, F. and Smyth, J. (1994) *Plant Nematodes of Australia Listed by Plant and by Genus.* NSW Agriculture, Australia, pp. 1–179.

Mehta, U.K. and Somasekhar, N. (1998) Occurrence of root-knot nematode, *Meloidogyne brevicauda* Loos, 1953 in tea plantations of Coonoor, Tamil Nadu, India. *Indian Journal of Nematology* 28, 90–91.

Mizukubo, T. (1998) Morphometric and molecular differentiation of *Pratylenchus loosi* (Nematode: Pratylenchidae) species complex from Japan and Sri Lanka. In: *Proceedings of the 24th International Nematology Symposium of European Society of Nematologists,* 4–9 August 1998, Dundee, UK, p. 72.

Mohotti, K.M. (1998) Non-chemical approaches for the management of the root-lesion nematode, *Pratylenchus loosi,* Loof, 1960 in tea (*Camellia sinensis* L.) O. Kuntze) with special reference to use of endospore forming bacterium, *Pastcuria penetrans.* PhD Thesis, University of Reading, UK.

Mohotti, K.M. (2001) Bio-prospecting of organic farmlands – evidences from Sri Lanka organic tea soils. In: *Proceedings of the Fifth IFOAM-Asia Scientific Conference,* 31 October–4 November 2002, Hangzhou, China, pp. 163–167.

Mohotti, K.M. (2002) Soil biodiversity measurements as potential determinants in organic certification: supportive evidence from Sri Lanka. *Proceedings of the First ARNOA/RDA Conference on 'Asian Organic Agriculture',* 11–15 November 2002, Suweon and Cheonan, Republic of Korea, pp. 75–91.

Mohotti, K.M., Bridge, J. and Gowen, S.R. (1996) Natural occurrence of *Pasteuria penetrans,* a mycelial endospore forming bacterium causing diseases in plant parasitic nematodes in tea fields. In: *Proceedings of the Sixteenth Annual Session of the Institute of Biology, Sri Lanka,* p. 7

Mohotti, K.M., Mohotti, J., Bridge, J. and Gowen, S. (1998) Role of organic matter amendments on recovery, growth and photosynthesis of tea (*Camelia sinensis*) plants infested with the root lesion nematode, *Pratylenchus loosi. Abstracts of Offered Papers in Nematology, Association of Applied Biologists,* 17 December 1998, Linnean Society of London.

Mohotti, K.M., Bridge, J. and Gowen, S.R. (1999) Role of organic farming in managing populations of plant parasitic nematodes and enhancing population densities of their bio-control agents. *Proceedings of the Seminar on Pest Control in the Next Millennium,* June 1999, Colombo, Sri Lanka, pp. 101–102.

Mohotti, K.M., Gowen, S.R. and Bridge, J. (2000a) Histopathology of tea roots infested with *Pratylenchus loosi* due to organic matter treatments. *Abstracts of Papers Presented at the International Meeting on Microbiology of Composting and Other Biodegradation Processes,* 18–20 October 2000, Innsbruck, Austria, p. 40.

Mohotti, K.M., Bridge, J. and Gowen, S.R. (2000b) Development of nematode antagonistic potential in organically amended tea soils. *Abstracts of Papers Presented at the International Meeting on Microbiology of Composting and Other Biodegradation Processes,* 18–20 October 2000, Innsbruck, Austria, p. 16.

Mohotti, K.M., Siddiqi, M.R., Bridge, J. and Gowen, S.R. (2002) Morphological and morphometric variations of *P. loosi sensu lato. Abstracts of Papers Presented at the Fourth International Congress of Nematology,* 8–13 June 2002, Tenerife, Spain. *Nematology* 4, 305.

Mukerjea, B. and Dasgupta, M.K. (1982) Community analyses of plant parasitic nematodes in tea plantation in W. Bengal. *Nematologia Mediterranea* 10, 1–7.

Muraleedharan, N. and Selvasundaram, R. (2001) Nematode parasites of tea. *Planters' Chronicle.* United Planters' Association of Southern India 97, 487–490.

Nakasono, K. and Ichinohe, M. (1961) *Hemicriconemoides kanayaensis* n.sp. associated with tea root in Japan (Nematoda: Criconematidae). *Japanese Journal of Applied Entomology and Zoology* 5, 273–276.

Nara, J., Wargadipura, R., Sukandis, K. and Ruswanda, S. (1973) The efficacy of several nematicides in the tea in the soil media of tea cutting nurseries. *Menara Perkebunan Batania* 4, 297–301.

Nath, R.P., Swarup, G. and Haider, M.G. (1975) Pathogenicity of *Pratylenchus loosi* on okra. *Proceedings of National Academy of Science, India* 45 (A), 167–168.

Pandey, A., Palni, M.S. and Bisht, D. (2001) Dominant fungi in the rhizosphere of established tea bushes and their interaction with the dominant bacteria under *in situ* conditions. *Microbiological Research* 156, 1–6.

Park, B.Y., Choi, D.R., Lee, J.K., Choi, Y.E. and Shin, G.-H. (2002) An unrecorded species of *Pratylenchus loosi* Loof (Tylenchida:Pratylenchidae) from tea in Korea. *Korean Journal of Applied Entomology* 41, 299–303.

Pourjam, E., Kheiri, A. and Geraert, E. (1997) The genus *Pratylenchus* Filp'jev 1936 (Tylenchida): Pratylenchidae from North of Iran. *Mededelingen Faculteit Landbouwwetenschappen University Gent* 62/3a.

Pourjam, E., Waenberge, L., Moens, M. and Geraert, E. (1999) Morphological, morphometrical and molecular study of *Pratylenchus coffeae* and *P. loosi* (Nematoda: Pratylenchidae). Proceedings of 51st International Symposium on Crop Protection Gent, Belgium, 4 May 1999 Part 1. *Mededelingen Faculteit Landbouwkundige en Toegepaste Biologische Wetenschappen Univeriteit Gent* 64, 391–401.

Rao, G.N. (1974) Current pest problems in tea in S. India. *Proceedings of the 19th Scientific Conference, 3–4 September 1973 and Bulletin of the United Planters' Association of South India* 30, 18–31.

Rao, G.N. (1976) Control of nematodes in nursery soils. *Planters' Chronicle. S. India* 7, 257–259.

Rong, W.Z., Juan, J. and Zheng, K.S. (1984) Study of tea seedling disease caused by root-knot nematode, *Meloidogyne* spp. *Acta Phytopathologica Sinica* 14, 225–232.

Seinhorst, J.W. (1977) *Pratylenchus loosi. CIH Descriptions of Plant-parasitic Nematodes. Set 7, No 98.* CIH, St Albans, UK.

Sethi, C.L. and Swarup, G. (1971) Plant parasitic nematodes of North-Western India. III. The genus *Pratylenchus. Indian Phytopathology* 24, 410–412.

Sivapalan, P. (1967a) Nematodes and tea. *Tea Quarterly* 38, 178–185.

Sivapalan, P. (1967b) Nematodes problems in tea. *Tea Quarterly* 38, 260–268.

Sivapalan, P. (1968) Association of *Radopholus similis* with decline in young tea fields. *Plant Disease Reporter* 52, 528.

Sivapalan, P. (1969) Further development in the control of the root lesion nematode, (*Pratylenchus loosi*), in tea nurseries. *Tea Quarterly* 40, 111–114.

Sivapalan, P. (1971) The effects of infestation by *Pratylenchus loosi* and of soil fumigation on the growth of young tea plants in different soil types. *Tea Quarterly* 42, 131–137.

Sivapalan, P. (1972) Nematode pests of tea. In: Webster, J.M. (ed.) *Economic Nematology*. Academic Press, New York, pp. 285–310.

Sivapalan, P. (1980) *Report of the Tea Research Institute of Sri Lanka – 1980*, pp. 83–96.

Sivapalan, P. and Gnanapragasam, N.C. (1975) The effect of soil temperature and infestation by *Pratylenchus loosi* on the growth and nutrient status of a susceptible variety of young tea (*Camellia sinensis* L.). *Tea Quarterly* 45, 29–35.

Sivapalan, P., Gnanapragasam, N.C. and Manivasagar, T. (1979) The influence of tea root saponins on the recovery of *Pratylenchus loosi* from tea feeder roots. *Nematologica* 25, 174–178.

Sivapalan, P., Gnanapragasam, N.C. and Jebamalai, A.S. (1980a) Treatment of nursery soils with Dazomet 98% G (Basamid) as an alternative method of control of the root-lesion nematode, *Pratylenchus loosi* in new plantings. *Tea Quarterly* 49, 134–136.

Sivapalan, P., Gnanapragasam, N.C. and Jebamalai, A.S. (1980b) Chemical control of the root-lesion nematode, *Pratylenchus loosi,* in new tea plantings. *Tea Quarterly* 49, 30–36.

Steiner, G. and Buhrer, E. (1933) The nematode *Tylenchus similis* Cobb as a parasite of tea plant (*Thea sinensis*), its sexual dimorphism, and its nemic associates in the same hosts. *Zeitschrift Parasitenkunde* 5, 412–420.

Stuart-Light, S. (1928) Pests of nurseries: a warning. *Tea Quarterly* 1, 19–22.

Takagi, K. (1967) Control of nematodes in tea. *Shokubutsu Boeki* 2, 429–432.

Takagi, K. (1969) Nematodes confronts tea plantations in Japan. *Japanese Agricultural Research Quarterly* 4, 27–32.

Thorne, G. (1961) *Principles of Nematology*. McGraw Hill Co. Inc., New York.

Venkata Ram, C.S. (1963) *Annual Report of the United Planters' Association of Southern India*. Scientific Department. Appendix, pp. 31–36.

Visser, T. (1959) Observations on the prevalence and control of parasitic eelworms in tea. *Tea Quarterly* 30, 96–107.

Visser, T. and Vythilingam, M.K. (1959) Effects of marigolds and some other crops on the *Pratylenchus loosi* and *Meloidogyne* populations in tea soils. *Tea Quarterly* 30, 30–38.

Vitarana, S.I. (2001) *Report of Entomology Division, Tea Research Institute of Sri Lanka*, pp. 68–81.

Yu, S.-F. and Xia, B. (1987) Preliminary results of identification of causal root-knot nematodes of tea seedlings in Yunnan. *Journal of Yunnan Agricultural University* 2, 35–40.

Zimmermann, A. (1899) Het voorkomen van nematoden in de wortels van sirihen thee I. De aaltjesziekte der sirih (*Piper betel* L) II. *Tylenchus acutocaudatus* Zn in de thee. *Teysmannia* 10, 523–531.

16 Nematode Parasites of Bananas and Plantains*

Simon R. Gowen,[1] Patrick Quénéhervé[2] and Roger Fogain[3]
[1] *School of Agriculture, Policy and Development, University of Reading, Reading RG6 2AT, UK;* [2] *Pôle de Recherche Agronomique de la Martinique (PRAM), Laboratoire de Nématologie Tropicale, IRD, BP 8006, 97259 Fort-de-France, Martinique;* [3] *African Research Centre on Banana and Plantains (CARBAP) PO Box 832, Douala, Cameroon*

Bananas thrive in the lowland tropical regions where rainfall is in excess of 1250 mm/year and there is a mean minimum temperature above 15°C (Stover and Simmonds, 1987). Significant areas of production exist outside these climatic zones such as in the East African highlands, several subtropical countries and in warmer localities beyond the 30° latitudes (Stover and Simmonds, 1987; Robinson, 1996). Bananas originate in South-east Asia and the western Pacific islands where several wild seed-bearing *Musa* spp. still exist in the natural vegetation. There is no firm botanical distinction between the different types of banana, and they are best classified by dividing the many different types into those which are sweet and eaten as a dessert fruit and those which can be eaten only after cooking, or fermented to produce a nutritious type of beer. In many countries, the cooking bananas are known as plantains, but the term is sometimes used ambiguously. All edible bananas are sterile, and are propagated vegetatively. Of the very great number of recognized clones (Stover and Simmonds, 1987; Daniells *et al.*, 2001), some are derived from *Musa acuminata* Colla and others from natural hybridizations of *M. acuminata* and *M. balbisiana* Colla. Currently accepted nomenclature of clones indicates ploidy and genomic origin, with A for *acuminata* and B for *balbisiana*.

Of the most commonly cultivated clones, the dessert bananas that are produced for the international trade (the Cavendish clones) are triploid *Musa* AAA, as is the once popular Gros Michel; Silk, Mysore, Pome and Prata are *Musa* AAB; Sucrier and Pisang Mas, *Musa* AA; and Ney Poovan, *Musa* AB. The plantains that are eaten as a cooked food are *Musa* AAB, but the East African highland clones which are also cooked or brewed are *Musa* AAA. These clones are quite different from Bluggoe and Pisang Awak (*Musa* ABB) which are also cooked, processed or even brewed.

The diversity in diploid, triploid and tetraploid clones has been widened through synthetic hybridizations, some of which are now in cultivation. Besides

*A revision of the chapter by S.R. Gowen and P. Quénéhervé.

 Plant Parasitic Nematodes in Subtropical and Tropical Agriculture, 2nd Edition (eds M. Luc, R.A. Sikora, J. Bridge)

regional field collections, over 1100 accessions of *Musa* are stored in tissue culture by the International Network for the Improvement of Banana and Plantain (INIBAP) at the Katholieke Universiteit, Leuven, Belgium. This collection is held in trust under the auspices of the Food and Agriculture Organization (FAO).

International trade in dessert and cooking bananas amounts to 13 Mt (FAO, 2004) and estimated world production is 100 Mt (FAO, 2004). The principal producing and consuming regions are Asia (27 Mt), South America (20.1 Mt), East Africa (17 Mt), West and Central Africa (11.4 Mt), Central America (6.8 Mt) and the Caribbean (2.7 Mt). The Cavendish varieties that supply the international trade in dessert fruit are all minor variants of one genotype; a convenience for the major export trading companies but a risk from a crop protection point of view. There is no other major fruit or vegetable crop that depends solely on one variety. The entire infrastructure including packaging, refrigeration, shipping and marketing is geared only to the Cavendish varieties. This inhibits the introduction of other varieties that might require different conditions (Loeillet, 2001).

Most bananas are grown for local consumption in mixed cropping systems or as a subsistence crop in gardens. Pure stands of cooking and dessert types usually occur where there is access to urban markets or where the fruit is the major contribution to the diet.

A related crop, abaca (*Musa textilis* Nee), grown for its fibre is of declining importance; more than 80% of the crop (106,000 ha) is grown in the Philippines (FAO, 2004).

The banana root system

Bananas are herbaceous perennials with short underground rhizomes from which grow an adventitious root system of up to 300 first order cord roots. These may grow up to 3 m laterally from the rhizome (corm) and are mostly in the upper 40 cm of soil (Araya *et al.*, 1998). Fewer roots grow vertically or deeper (Summerville, 1939), although rooting density and distribution are influenced by the texture and depth of the topsoil (Irrizary *et al.*, 1981; Weckx, 1982). Second order (lateral) roots develop on the cord roots in the proximal root zones, and short tertiary roots may develop on the secondaries. Diploids and AAA types may have greater numbers of first, second and third order lateral roots as a percentage of the total root length than the AAB dessert and cooking cultivars (Swennen *et al.*, 1986; Draye *et al.*, 1999). This major difference may partially explain the relatively low productivity of many cooking bananas. It is now established that there are genotypic differences in root architecture (Gowen, 1993; Blomme *et al.*, 2000, 2003), a feature that breeders might be able to exploit.

New roots are produced continuously until flowering, which may occur from 7 to 9 months after planting a new crop of the commercial AAA cultivars. The duration of the vegetative phase may be considerably longer if climatic or soil conditions are less favourable and may last more than 1–2 years in the cooler upland regions of East Africa where cooking cultivars are cultivated (INIBAP, 1986). After flowering, the developing inflorescence is sustained by a declining root system in which natural senescence is hastened by the activity of root pathogens. The increasing root growth of the daughter plant (sucker) may be of benefit during this critical phase by providing additional anchorage to the mother plant and also as a supplementary source of nutrients for the maturing fruit (Lavigne, 1987).

Banana propagation techniques

Suckers

The traditional method of propagation is by excising lateral shoots (suckers) that proliferate around parent plants of most banana clones. These shoots are then planted directly into the field. The larger the shoot size, the greater the germination success. Lateral shoots will also develop from the excavated corms of harvested plants; pieces of corm will also serve as new planting material. The disadvantage of this technique is the likelihood that the suckers harbour pests and diseases (Fig. 16.1).

Vitro (tissue culture) plants

The development of the meristem culture (micro-propagation) technique by which tissue generates an abundance of new shoots in sterilized growing media (Israeli *et al.*, 1995) has revolutionized banana propagation. The opportunities for preventing the spread of some pests and diseases is one of several advantages of using plants produced by this system. Most commercial growers have adopted the use of vitro plants whenever replanting becomes necessary. Commercial tissue culture laboratories capable of producing many millions of plants are now established throughout the world. Vitro plants are also being promoted to smallholders in non-exporting countries (Fig. 16.2).

Fig. 16.1. Infected banana suckers.

Fig. 16.2. Production of vitro (tissue culture) banana plants.

Lateral bud proliferation

A low cost alternative to micro-propagation is the technique of promoting shoot development from the lateral buds on the corms of harvested plants. Disease-free corms are removed from the field and planted in boxes of sand or sawdust. After a few months, the lateral shoots begin to develop. Transverse incisions with a clean razor blade across the buds will promote mass production of shoots from these buds. When large enough, the shoots are cut from the corm and allowed to establish as independent plants in clean growing media (Fig. 16.3).

Cropping systems

Bananas may be grown as a permanent crop or on a system of replanting every 3–8 years or longer (Stover and Simmonds, 1987). In many countries, particularly in the Caribbean, Surinam, Côte d'Ivoire, Cameroon and the Pacific islands, bananas and plantains soon become unproductive for reasons related to the soil structure, fertility, drainage and severity of pathogens, so frequent replanting is necessary (Lassoudière 1978; Stover and Simmonds, 1987). Crop longevity is extended if plants are mulched regularly with organic wastes and manures (Wilson *et al.*, 1986), which may explain the long-established banana gardens in many parts of Central and East Africa and elsewhere (Fig. 16.4). The soil conditions for banana cultivation are ideal in the major exporting countries of Latin America and the Philippines and, once established, may remain in production more or less indefinitely.

Cultivation techniques

The intensity of inputs and management for the different farming systems are quite varied and depend on the market or use for which fruit is destined.

Bananas for export

All of the dessert fruit and some cooking bananas grown for the international export trade are managed intensively to ensure high yields of fruit of the correct size, free of skin blemishes and postharvest diseases. Such fruit is usually produced in pure stands at densities maintained at 1700–2000 plants/ha. Routine field operations involve pruning surplus suckers, removal of dead foliage, fruit bunch protection, propping fruiting stems and a regular use of fertilizers, fungicides, nematicides

Fig. 16.3. Production of banana shoots for propagation by low cost lateral bud proliferation.

Fig. 16.4. Banana plantation.

and, when needed, herbicides and insecticides. Irrigation is applied where rainfall is inadequate; a minimum of 100 mm of rain/month is considered ideal.

Non-export bananas

Bananas are a valuable component in mixed farming systems, providing continuity of food, income and employment throughout the year. Fruit can be harvested close to maturity and minor attention is given to fruit size and skin blemishes. Field operations may be done only if necessary to prevent crop loss, although production and fruit quality will be dependent on the extent of sucker pruning, use of fertilizers and crop protection measures.

Bananas as a subsistence crop

There can be few household gardens anywhere in the tropics that do not have one or more clumps of bananas requiring minimal attention other than propping those stems with maturing fruit. Many other crops will thrive alongside bananas benefiting from the shade and the large amount of leaf material that is available for mulching and soil improvement.

Nematodes of Bananas and Plantains

The species of nematodes found to be most detrimental to these crops are those which are involved in the destruction of the primary roots, disrupting the anchorage system and resulting in toppling of the plants. The most widespread and important are *Radopholus similis*, some species of *Pratylenchus* and *Helicotylenchus multicinctus*. As for most tropical crops, nematode parasitism in banana roots is characterized by simultaneous infestations by several species. It is also very common to find some sedentary parasites such as *Meloidogyne* spp. and *Rotylenchulus reniformis* parasitizing the root system. In addition to these five major nematodes parasitic on roots of bananas, there are many other species that have been reported to be associated with *Musa* spp. throughout the world. Until now, none is considered as a serious root pest, although they may be locally important where their densities are very high.

According to the mode of parasitism of the different species, the symptoms will differ from the most severe, such as toppling, to the less obvious, such as prolonging of the vegetative cycle. In situations where toppling is common, crop loss can be extreme because fruit on a fallen plant generally has no value.

Radopholus similis

The disease of banana caused by *R. similis* is known throughout the world by different names, the most common are 'black head toppling disease' and 'toppling disease'. The burrowing nematode, *R. similis*, was first observed by Cobb in necrotic tissue of the roots of *Musa* sp. sent to him in New South Wales from Fiji in July 1891. Since this first record, it has been found widespread in all the tropical and subtropical banana- and plantain-growing regions of the world except Israel, the Canary Islands, the Cape Verde Islands, Cyprus, Crete, Mauritius and Taiwan. It also appears to be absent from some of the important areas of production in the highlands of Eastern Africa. While *R. similis* now occurs in most tropical and subtropical areas of the world, the genus *Radopholus* is indigenous to Australia and New Zealand (Sher, 1968) from where new species have been described recently. Its worldwide distribution is relatively recent (beginning of the 19th century) and is due to the transfer of infected plant material from country to country. The wide distribution of *R. similis* seems often to be correlated with the areas where banana plants of the subgroup Cavendish (AAA) were imported. It is speculated that in Latin America and the Caribbean, *R. similis* was introduced on the cv. Gros Michel and subsequently infested the more susceptible Cavendish cultivars (Marin *et al.*, 1998). The host range of *R. similis* has become wider with exposure to different plant species.

Symptoms of damage

The most obvious symptom of attack of *R. similis* on banana is the toppling over or uprooting of plants (Plate 17A) especially those bearing fruit, but there is a range in gradation in the severity of damage, from the lengthening of the vegetative cycle to the drastic reduction in bunch weight. This reveals two types of damage that can occur in banana plantations; that affecting the anchorage of the plant and, less apparent, the effect on the ability to take up water and nutrients. Macroscopically, several dark red lesions appear on the outer part of the root penetrating throughout the cortex but not in the stele (Plate 17B); adjacent lesions may coalesce and the cortical root tissue atrophies and later turns black. In heavy infestations, the lesion girdles the roots. Nematodes can migrate from infected roots into the corm, causing diffuse black lesions which may then spread around the corm (Loos and Loos, 1960b). Roots emerging become infected as they grow out of the corm. Uprooting occurs commonly in windstorms or if heavy rains loosen the soil. The mechanical stresses on the root system are often increased by the natural angle of leaning which develops as fruit bunches grow. The presence of a number of fungi in nematode-induced lesions probably hastens the destruction of roots and may contribute to toppling disease because fungi colonize the stele which is not penetrated by *R. similis* (Stover, 1972).

Biology and life cycle

R. similis is a migratory endoparasitic species which is able to complete its life cycle within the root cortex. The histopathology of banana roots attacked by *R. similis* was studied by Blake (1961, 1966) and Loos (1962). Penetration occurs mostly near the root tip, but nematodes can invade along the entire length of the root; females and all juvenile stages are infective, although males, which are morphologically degenerate (without a stylet), are probably non-parasitic. After entering the roots of banana, the nematodes occupy an intercellular position in the cortical parenchyma where they feed on the cytoplasm of nearby cells, causing cavities which then coalesce to appear as tunnels.

Invasion of the stele is never observed, even in heavily infected roots. Migration and egg laying are governed by nutritional factors, as females move in search of healthy tissue away from the necrosis. It is within infected tissues that females lay their eggs, with an average of 4–5 eggs/day for 2 weeks. The complete life cycle from egg to egg spans 20–25 days at a temperature range of 24–32°C, the eggs hatch after 8–10 days and the juvenile stages are completed in 10–13 days (Loos, 1962).

Pathotypes/races/biotypes

R. similis has two races, one attacking banana but not citrus, and a 'citrus race' pathogenic to both (DuCharme and Birchfield, 1956). For some years there was controversy concerning the existence of sibling species, but research has not supported this hypothesis (Kaplan and Opperman, 1997; Valette *et al.*, 1998a). Physiological differences in reproductive capabilities and morphological variations of *R. similis* on bananas in Central and South America and elsewhere suggest the existence of different biotypes or isolates on the basis of host preferences and the rate of reproduction (Pinochet, 1979; Tarte *et al.*, 1981; Kaplan and O'Bannon, 1985; Hahn *et al.*, 1996; Marin *et al.*, 1999; Stoffelen *et al.*, 1999).

Survival and means of dissemination

The survival of *R. similis* in soil depends on the effectiveness of the destruction and removal of infected banana roots and corms and weed hosts. Unlike some other species, *R. similis* has no specialized survival strategy outside of its host. Tarjan (1961) and Loos (1961) demonstrated that *R. similis* did not survive in the soil for more than 6 months in the absence of host roots or pieces of live corms. *R. similis* will survive on corms and roots of a previous crop for a long time and, within planting material, it is the major means of re-infestation.

The passive dispersal of the nematode in runoff water and through irrigation systems is potentially serious to growers attempting to rid fields of *R. similis* in areas where infested banana fields are adjacent to new plantings.

Other hosts of Radopholus similis

Most of the banana and plantain cultivars of the edible *Musa* varieties AA, AAA, AB, AAB and ABB are attacked by *R. similis* (Luc and Vilardebó, 1961; Wehunt *et al.*, 1978; Davide and Marasigan, 1985) as well as abaca (Taylor and Loegering, 1953) and other seeded *Musa* species. In the Americas, *R. similis* seems to be confined to *Musa* spp. and to some cultivated plants including ornamentals such as *Anthurium andraeanum* (Bala and Hosein, 1996; Quénéhervé *et al.*, 1997; Sipes and Lichty, 2002). It attacks several crop plants which are important in world commerce and subsistence-type agriculture (Bridge, 1987). O'Bannon (1977) listed more than 250 plants that are susceptible to *R. similis*. *R. similis* is commonly found in Martinique on weeds in banana fields, mainly species of Poaceae, Euphorbiacae and Solanaceae (Quénéhervé *et al.*, 2000b). Its status has been studied extensively from a quarantine point of view (Ayala and Roman, 1963; Edwards and Wehunt, 1971).

Pratylenchus

Several species of *Pratylenchus* root lesion nematodes have been reported attacking *Musa* spp. throughout the world. Among these, only two, *P. coffeae* and *P. goodeyi*, are recognized as damaging pathogens. *P. coffeae* was first observed in roots of plantains in Grenada and described as *Tylenchus musicola* by Cobb in 1919. The demonstration of its pathogenic activity in extensive lesions in the root cortex of abaca was done by Taylor and Loegering (1953) in Costa Rica. *P. goodeyi* was first observed in banana roots in the Canary Islands by de Guiran and Vilardebó (1962) with *P. coffeae* and *P. thornei*. *P. coffeae* is a pan-tropical species. *P. goodeyi* has been observed in every banana-growing area of East Africa (Gichure and Ondieki, 1977; Walker *et al.*,

1984; Bridge, 1988), suggesting that it is indigenous to this area and is usually found at high elevations (Price and Bridge, 1995).

Symptoms of damage

Root lesion nematodes cause symptoms of damage similar to those observed with *R. similis*: stunting of plants, lengthening of the vegetative cycle, reduction in size and number of leaves and in bunch weight, reduction of the productive life of the plantation, and toppling (Plate 17C). Roots heavily infested by *P. coffeae* have extensive black or purple necrosis of epidermal and cortical tissue, often accompanied by secondary rotting and root breakage. Similar necrosis can be observed on the outer parts of the corm (Bridge and Page, 1984). In the Canary Islands, de Guiran and Vilardebó (1962) observed that *P. goodeyi* penetrates the cortical parenchyma of banana roots forming small brownish-red elongated flecks. These feeding areas enlarge and eventually coalesce, so most of the cortical parenchyma is destroyed, impairing root function.

Biology and life cycle

P. coffeae and *P. goodeyi* are migratory endoparasites of the root cortex and banana corm. Nematodes of both sexes and all juvenile stages are invasive. The life cycle is completed within the root. Pinochet (1978) described the histological changes after inoculation of *P. coffeae* on roots of AAB clones. After entering the roots, the nematodes migrate between and within the cells, occupying a position parallel to the stele. They feed on the cytoplasm of neighbouring cells, eventually causing cavities that coalesce. The destruction of the cortical parenchyma of plantain roots by *P. coffeae* is very similar to those effects described by Blake (1961, 1966) for *R. similis* on dessert bananas, except there was no cell enlargement or increase in size of cell nucleus or nucleolus. The life cycle has been discussed in detail on other host plants (Zimmerman, 1898; Gotoh, 1964), and the average life cycle from egg to egg is about 27 days at a temperature range of 25–30°C.

Pathotypes/races/biotypes

There is scarce information on 'biotypes', 'isolates' or 'races' of *P. coffeae*. Wehunt and Edwards (in Stover, 1972) mention the existence of different biotypes or isolates from Honduras and Panama, stated in terms of host preferences related to the infection index on test plants of abaca, plantain and banana. Recent morphological and genomic variation between 32 isolates of *P. coffeae* and closely related species has led to different groupings within the *coffeae* group (Duncan *et al.*, 1999).

Survival and means of dissemination

Root lesion nematodes have also been observed infesting the corm, so dissemination occurs in the same way as described for *R. similis*. Records of the risk of this type of dissemination are reported from the Côte d'Ivoire for *P. coffeae* on dessert bananas and plantains (Adiko, 1988; Fargette and Quénéhervé, 1988) and from East Africa for *P. goodeyi* on highland bananas (Walker *et al.*, 1984; INIBAP, 1986; Bridge, 1988).

Other hosts of Pratylenchus *spp.*

Many other hosts of *Pratylenchus* spp. have been recorded, several of which may be weeds (Fluiter and Mulholland, 1941; Kaplan and MacGowan, 1982; Quénéhervé *et al.*, 1995).

P coffeae is also a major pest of other economic crops including tuber crops (yam and taro, see Chapter 7) and ornamentals (Pinochet and Duarte, 1986; Bala and Hosein, 1996; Quénéhervé *et al.*, 1997).

Helicotylenchus multicinctus

After *R. similis*, the spiral nematode, *H. multicinctus*, is probably the most widespread and abundant nematode on all bananas. *H. multicinctus* and *R. similis* are often encountered together in many dessert banana-growing regions of the world, particularly where bananas are grown under optimal conditions. *H. multicinctus* is often regarded as the main parasitic nematode on bananas where environmental conditions are suboptimal for the crop (and also for *R. similis*) in relation to latitude, temperature and rainfall (McSorley and Parrado, 1986).

Symptoms of damage

The nematodes attack and feed on the outer cells of the root cortex and produce small, characteristic necrotic lesions (Luc and Vilardebó, 1961). Development of root lesions caused by *H. multicinctus* is slow relative to those produced by *R. similis*. Lesions on primary roots are shallow and superficial, like numerous small dashes, reddish-brown to black in colour. However, in heavy infestations, those lesions can coalesce, causing extensive root necrosis in the outer cortex (Plate 17D), and die back; lesions can also be found in the corm (Quénéhervé and Cadet, 1985). The effects of *H. multicinctus* on both banana and plantain can lead to stunting of plants, lengthening of the vegetative cycle, reduction in size of the plant and in bunch weight, and reduction of the productive life of the plantation. Toppling may also occur in situations where there are heavy infestations.

Biology and life cycle

H. multicinctus, unlike most other *Helicotylenchus* species, is regarded as an endoparasitic species which is also able to complete its life cycle within the cortical part of the root where both sexes and all juvenile stages, including eggs, can be found (Zuckerman and Strich-Harari, 1963). The host–parasite relationships of *H. multicinctus* were studied by Blake (1966) who observed that 4 days after inoculation of banana roots, the nematodes were wholly embedded within the cortex, sometimes to a depth of 4–6 cells. Nematodes fed on the cytoplasm of surrounding cells in the root cortex. Infected tissues show various types of cellular damage such as contracted cytoplasm, distorted or ruptured walls and enlarged nucleus but, in contrast to those observed with *R. similis*, histological changes are confined to parenchyma cells close to the epidermis. Damaged cells were often discoloured and became necrotic (Orion *et al.*, 1999).

Pathotypes/races/biotypes

To date, there is no available information on 'biotypes', 'isolates' or 'races' of *H. multicinctus*.

Survival and means of dissemination

Little information exists on the survival of *H. multicinctus* in the absence of a susceptible host. As with *R. similis*, survival occurs on infected corms or on tissue remaining from the previous crop. Infected planting material is also the main means of dissemination.

Other hosts of Helicotylenchus multicinctus

Most of the banana and plantain cultivars of edible *Musa* cultivars of differing ploidy are attacked by *H. multicinctus* (Luc and Vilardebó, 1961; Gowen, 1976; Zem *et al.*, 1981; McSorley and Parrado, 1983). This nematode is also recorded to have a wide host range (Goodey *et al.*, 1965; Stoyanov, 1967), including weeds in banana fields (P. Quénéhervé, Martinique, 2003, personal communication).

Meloidogyne

Root knot nematodes are worldwide in distribution, attacking many economically important crops. On banana, its importance may have been underestimated because of the emphasis on the damage caused by lesion nematodes, by inappropriate sampling and extraction procedures (intended for lesion nematodes) and by the technical problems of apportioning crop loss in mixed infestations. Damage has been noted particularly in greenhouse production systems in North Africa and the Canary Islands (Pinochet *et al.*, 1998). The species most commonly found associated with bananas and plantain are *M. incognita*, *M. arenaria*, *M. javanica* and *M. hapla*. Different species can be observed in the same gall (Pinochet, 1977), and root knot infestations in West Africa (Netscher, 1978; Fargette, 1987), Martinique (Quénéhervé *et al.*, 2000a) and Brazil (Cofcewicz *et al.*, 2001) have been found to be of mixed species. This genus is the second most abundant to be found in banana roots in South Africa (Jones and Milne, 1982) and is the only one in Taiwan (Lin and Tsay, 1985) and in North Yemen (Sikora, 1979) involved in nematode damage to banana plants. It also occurs on abaca in the Philippines (Ocfemia and Calinson, 1928).

Symptoms of damage

The most obvious symptoms are galling on primary and secondary roots (Plate 17E) sometimes causing them to bifurcate and distort. Stunted growth has been attributed to root knot nematodes in India (Sudha and Prabhoo, 1983) and Taiwan (Lin and Tsay, 1985). Sikora (1979) observed higher levels of root rot in plantations in Yemen where *M. incognita* and *Fusarium solani* or *Rhizoctonia* sp. were present concomitantly.

Biology and life cycle

The life cycle, histopathology and aetiology of the disease do not differ significantly on bananas from those reported on other hosts in reviews to which the reader is referred (Bird, 1979; Huang, 1985). In thick, fleshy primary roots, egg masses may not protrude outside the root surface, and multiple cycles can be completed within the same root, depending on the longevity of this root and the severity of necrosis. Pinochet (1977) suggests that, in mixed infestations, the area of influence of this nematode would start between 60 and 90 cm from the rhizome because of the competition with *R. similis* in suppressing or replacing the *Meloidogyne* population. This had also been shown by Luc and Vilardebó (1961) and Quénéhervé (1990).

Survival and means of dissemination

Root knot nematodes have a wide host range, which are usually present in most soils in which bananas are growing. As for other nematodes associated with bananas, survival and dissemination also occur with the planting material on infected roots and corms (Quénéhervé and Cadet, 1985).

Other hosts of Meloidogyne *spp.*

Because of the wide host ranges of root knot nematodes, associations with weeds in banana plantations are more numerous than for other major nematode parasites. Special attention would be needed in maintenance fallows or in selection of cover crops or associate crops in intercropping systems.

Rotylenchulus reniformis

Since the first records of *R. reniformis* on bananas in Puerto Rico by Ayala and Roman (1963), this nematode has now been reported in numerous banana-growing areas. The life cycle and the histopathology and aetiology of the disease do not differ significantly on bananas from those reported on other hosts (Sivakumar and Seshadri, 1974). Juveniles of *R. reniformis* are commonly extracted from the soil and it is generally observed that permanent feeding positions occur mostly on the secondary roots (Ayala, 1962; Edmunds, 1968). As for *Meloidogyne* spp., the effect of this nematode is probably influenced by the presence of other root parasitic nematodes.

Other nematodes

Of the many other species of plant parasitic nematodes found associated with bananas, some are thought to be potentially damaging, but there is no conclusive evidence to show their pest status. Invariably, these nematodes are in mixed communities with species already established as key pests.

Hoplolaimus pararobustus has been found around and within roots and corms of dessert bananas and plantains in different areas of the Côte d'Ivoire (Quénéhervé and Cadet, 1985; Adiko, 1988; Quénéhervé, 1989a,b). Population densities in roots of mature plants have been as high as 200 individuals/g of root (Mateille *et al.*, 1988b). Price (1994a) considers it has minimal damage potential, but this should be critically verified (Bridge, 2000). *Helicotylenchus mucronatus* and *H. microcephalus* have been found to be the cause of root necrosis and stunted growth of bananas at separate sites in Papua New Guinea (Bridge and Page, 1984). *Cephalenchus emarginatus* was found at populations of up to 9000/l of soil taken from around the roots of dessert bananas and plantains in the Côte d'Ivoire (Adiko, 1988; Mateille *et al.*, 1988b; Quénéhervé, 1989a,b). *Heterodera oryzicola*, a pest normally found associated with rice, is found on bananas where these crops grow together. Its pathogenicity has been demonstrated (Charles and Venkitesan, 1993).

Environmental factors affecting parasitism of banana nematodes

On bananas grown under humid, tropical conditions, the major factors affecting nematode populations are abiotic, such as soil type and climate, and biotic, such as plant host status, growth stage, competition with other nematode species and other pests. In subtropical or highland countries, soil temperature is an additional factor influencing parasitism. The parasitism of banana root systems is somewhat different from that of other perennial crops because of the growth habit of the root system in which a succession of fleshy, relatively short-lived roots are produced. Unthriftiness of bananas may result from shallow or poorly drained soils, drought, nutrient deficiency or nutrient imbalance, and symptoms may show on aerial parts of the plant. Such conditions may also cause restriction of root development, and in these situations the presence of nematodes may increase the incidence of toppling as well as exacerbate foliar symptoms. If drainage is poor, high or fluctuating water tables can considerably curtail root growth (Lassoudière and Martin, 1974). Roots in soil saturated for more than 24 h die and rot rapidly. The combination of poor drainage and a nematode problem may result in nematodes and roots being concentrated in the upper layer of soil, resulting in more severe nematode damage.

Influence of soil type

The influence of soil type on nematode community composition has been reviewed by Ferris and Ferris (1974), and Vrain (1986) reviewed the effect of soil moisture content on population dynamics. In general, most information concerning banana nematodes deals with the relationship between soil type and density of nematode species on commercial bananas (Stover and Fielding, 1958; Ayala and Roman, 1963; Varghese and Nair, 1968; Guérout *et al.*, 1976; Davide, 1980; McSorley and Parrado, 1981). In the Côte d'Ivoire, Quénéhervé (1988) showed that, in an organic soil, *H. multicinctus* is predominant in both soil and roots, while on mineral soils *R. similis* predominates. The major differences in nematode community structure occur in the soil. *R. similis* seems less affected by the soil variables because it is strictly an endoparasite. *H. multicinctus* is more frequent in soils characterized by high levels of clay, silt or organic matter and low pH. *H. pararobustus* is more commonly found in coarse volcanic or sandy soils, and *M. incognita* is most abundant in sandy soils.

Influence of climatic factors

Numerous studies have attempted to relate population densities with climatic factors, particularly rainfall; in general, it is assumed that conditions promoting plant root growth will also favour population development. Most extended studies of population dynamics have shown a decline in numbers of *R. similis* during the wet season (Jimenez, 1972; Melin and Vilardebó, 1973; Jaramillo and Figueroa, 1974; McSorley and Parrado, 1981; Hugon *et al.*, 1984; Hunt, in Ambrose, 1984; Quénéhervé, 1989a,b), but the opposite effects have also been reported (Marcelino *et al.*, 1978; Davide and Marasigan, 1985). Similar attempts have been made to correlate population densities of *H. multicinctus* with rainfall, with variable results (Hutton, 1978; McSorley and Parrado, 1981; Badra and Caveness, 1983; Quénéhervé, 1989a,b), but it is a general trend that greater populations can be found in the rainy season. The discrepancies in the relationships between population densities and rainfall may be attributed to difference in soil type, soil temperature, incidence and intensity of rainfall and root growth. Consideration should be given to sampling procedures before initiating such nematode population dynamic studies.

Influence of the root system and physiology of the plant

A relationship has been reported between successive annual peaks in the numbers of *R. similis* in the roots and the active growth of the plant (Jaramillo and Figueroa, 1974), which coincides with the emergence of the banana flower (Melin and Vilardebó, 1973). In Guadeloupe, Hugon *et al.* (1984) observed a relationship between the physiological stage of the banana plant and such climatic factors as temperature and rainfall. Pruning of excess suckers is practised in commercial plantations and this may influence the relative numbers of *R. similis* and *H. multicinctus* in the roots and corms (Mateille *et al.*, 1984). In a study, conducted on both mineral and organic soils in the Côte d'Ivoire, Quénéhervé (1989a,b) has shown differences in the behaviour of the nematodes encountered. *R. similis* acts as the primary root invader, and levels of infestation decrease as the root system ages or decays. Blake (1961) and Loos (1962) showed that migration and egg laying are governed by nutritional factors and that the 'nematodes do not move out of a root so long as they are able to invade healthy tissue'. *R. similis* is able to complete its life cycle in the cortical tissue of the root or the rhizome without a soil phase. After flowering, there is no new root emergence from the main rhizome (Lavigne, 1987), but on the rhizomes of the suckers, prolific root emergence occurs once they have achieved self-reliance (change of the lanceolate leaves to enlarged leaves). In fact, all the factors, endogenous or exogenous, which favour root emergence on banana plants contribute to the build-up of *R. similis* populations (Quénéhervé, 1993a).

Influence of the competition with other parasites

In addition to the various nematodes, other parasites such as fungi and bacteria are present in the roots, and this complex is the cause of root decay. Infestations by nematodes such as *H. multicinctus* may accelerate root decay, thereby restricting the availability of healthy tissue to another endoparasite such as *R. similis*. *H. multicinctus* and *R. similis* often occur together on bananas and plantains in those tropical regions best suited for growth of the crop. Vilardebó and Guérout (1976) noticed that high populations of *H. multicinctus* build up when *R. similis* is locally absent. In the Côte d'Ivoire, it appears that on organic soil, populations of *H. multicinctus* may surpass those of the primary invader *R. similis*. *P. coffeae* has a similar parasitic behaviour to *R. similis* and may compete directly with it. In some parts of the world, this nematode might be the more damaging parasite, such as in Papua New Guinea or like *P. goodeyi* in the Canary Islands (de Guiran and Vilardebó, 1962) or on highland bananas in East Africa (Gichure and Ondieki, 1977; Bridge, 1988; Kashaija *et al.*, 1994). The banana weevil, *Cosmopolites sordidus*, can confuse the diagnosis of a nematode problem because symptoms of damage are similar. With fungi (*Cylindrocarpon* spp., *Fusarium* spp., *Rhizoctonia* spp. and *Cylindrocladium* sp.), the problem becomes even more complex as nematodes and fungi occur within the same cells and infestations result in the same types of discoloration and necrosis (Jones, 2000; Risède and Simoneau, 2004). Often the problem is to define which is the primary or major pathogen. Nematodes create a food base for weak, unspecialized fungal parasites, enabling them to invade the stele and to increase the amount of root necrosis. Differentiation is possible between the deep lesions due to *R. similis*, which are mainly associated with *Fusarium* sp., and the shallow and outer lesions of *H. multicinctus*, which are mainly associated with *Rhizoctonia* sp. (Blake, 1963; Laville, 1964; Stover, 1966; Sikora and Schlosser, 1973; Booth and Stover, 1974; Pinochet and Stover, 1980). Those fungi acting as secondary parasites can increase root breakage and consequently toppling. One of the most devastating fungal diseases affecting commercial bananas (Fusarium wilt or Panama disease) caused by *Fusarium oxysporum* f.sp. *cubense* was formerly observed on the susceptible cv. Gros Michel and forced growers to change to the resistant Cavendish group cultivars between 1950 and 1960. Newhall (1958) and Loos (1959) concluded that the expression of Fusarium wilt on cv. Gros Michel was considerably increased in the presence of *R. similis*, although this was not confirmed from work in the Philippines (Epp, 1987). Three races of *Fusarium* attacking edible banana cultivars have been identified; the latest also infects Cavendish cultivars (Hwang *et al.*, 1984; Stover and Simmonds, 1987; Jones, 2000).

Economic importance

It is uncommon for bananas to be parasitized by monospecific populations, and the relative importance of the different species is not fully understood. In addition to *R. similis*, *H. multicinctus*, *Pratylenchus* spp., *R. reniformis* and *Meloidogyne* spp., populations of other migratory endoparasites, i.e. *H. pararobustus*, or ectoparasites, i.e. *Cephalenchus emarginatus*, may reach high levels. Most evidence of crop loss from field experimentation comes from the use of nematicides which usually decrease populations of all species and can possibly cause other beneficial plant growth effects. The yield responses reported with nematicide applications to dessert and cooking bananas have been up to 275% greater than untreated controls (Tables 2 and 3 of Gowen and Quénéhervé, 1990). The differences in response may be due to several factors, in particular soil type, nematode species and biotype, and climate, and may reflect the losses through uprooting as well as differences in the weights of harvested bunches.

Management measures

The importance of *R. similis* as a widespread cause of banana losses was reported by Leach (1958). Investigations on techniques for its control were made by Vilardebó (1959), Loos and Loos (1960a), Blake (1961) and Luc and Vilardebó (1961). Meanwhile Minz *et al.* (1960) were applying dibromochloropropane (DBCP) for control of *H. multicinctus* in the Jordan valley. Control of the other major endoparasitic genus *Pratylenchus* in the Canary Islands was reported by de Guiran and Vilardebó (1962). Initially, much attention was given to the elimination of nematodes from planting material as it was realized that this was the principal source of infestation by which *R. similis* and other species were distributed through banana-growing regions. The concept of providing nematode-free plant nurseries (Loos and Loos, 1960a) was technically sound, but at that time was never widely successful in practice.

Between 1960 and 1978, the non-phytotoxic fumigant nematicide DBCP was used extensively on commercial bananas, particularly in Central and South America. Treatments were normally applied twice a year usually by hand-held injectors in which the fumigant was injected in 6–8 points at 30–40 cm around individual plants. Less commonly, DBCP was applied through irrigation systems. Hand injection of DBCP was a laborious task requiring constant supervision. Consequently, the granular non-volatile nematicides which are easier to apply began to be used commercially before DBCP was withdrawn from use.

Cultural practices

The opportunity for controlling *R. similis* with cultural techniques is somewhat limited in those areas where bananas are grown continuously, without replanting. In replanted crop systems, control of the *R. similis* populations can be done by total destruction of the previous crop to ensure the elimination of the nematode, followed by a controlled fallow or by rotating with non-host crops.

Fallows may need to last 6 months or longer (Tarjan, 1961; Loos, 1961) and it is essential that all banana roots and suckers are destroyed, which in practice is a difficult task.

Beneficial results were obtained by flooding in Surinam and Côte d'Ivoire (Maas, 1969; Sarah *et al.*, 1983; Mateille *et al.*, 1988a), but this is now an uncommon practice. *R. similis* may be absent from many areas not previously cultivated with bananas. Unwanted introduction of the nematode can be avoided by use of disease-free planting material (Loos and Loos, 1960a), but more reliable is the use of disease-free plants grown by the meristem culture technique. Fallowing is now widely practised where there is available land and *R. similis* is present. In the French Antilles, nematode control in the large commercial banana plantations is currently based on the sanitation of contaminated banana fields, using chemical destruction of existing banana plants and replanting with nematode-free banana plants produced by tissue culture (Chabrier and Quénéhervé, 2003). In Martinique, this fallow/vitro plant regime has extended the field longevity from 3–4 to 6–10 years and already some formerly contaminated banana fields are totally freed from *R. similis*. In Cameroon, when tissue-cultured plants were used after a 1 year fallow, very low populations of *R. similis* were recorded during the two cycles (15–18 months after planting) (Kashaija *et al.*, 1998).

In Taiwan and India, rice may be grown in rotation with bananas, and various rotation combinations have been evaluated in the French Antilles (Ternisien, 1989; Ternisien and Ganry, 1990). Where bananas are grown continuously, i.e. Latin America, or where it would be uneconomic to leave land fallow, crop rotation is uncommon.

Since the work of Loos and contemporaries, most recommendations for banana planting include instructions for the selection and preparation of disease-free suckers. Through tissue culture, such material is now widely available. In commercial cultivations, it is now unlikely that planting material would be taken directly from existing banana fields because of the risk of infestation with nematodes and weevils (*C. sordidus*). In smallholder production systems, the sale and exchange of planting material is common and this contributes to the persistence of the nematode problem. In this case, the recommendation is that if the external tissue of the corm has purple or reddish-brown lesions these, together with root stumps and adhering soil, should be removed with a machete (pared) until only white corm tissue is exposed. The practice of paring suckers should be done away from the field, and severely lesioned corms should be discarded. Similarly, deep lesions and tunnels caused by the weevil larvae should be removed. The paring technique although useful, may never be totally effective in removing all nematode infection.

Organic amendments/mulching/intercropping

Mulching and organic amendments may not have a direct effect on root endoparasitic nematodes. Although there are conflicting opinions on the effects of these treatments on the burrowing and lesion nematodes, it has been shown that mulched nematode-infested plots are likely to produce more than non-infested non-mulched plots, because of the benefit of organic matter on plant growth. Intercropping with leguminous crops, sometimes considered to have repellent action against pests, failed to show such benefit in a 2 year study in Uganda (McIntyre *et al.*, 2001).

Physical treatments – hot water

The immersion of banana suckers in water held at a constant 55°C for periods of 15–25 min has been a commercial practice in Australia and Central and South America (Stover, 1972). The technique is quite difficult to manage because of the critical balance required between a temperature that is lethal to nematodes in the corm tissue and one that causes permanent damage to the plant. This factor can also be important if suckers are not of uniform size. Although widely documented as a control method, hot water treatment has been largely superseded by use of vitro plants.

Resistance and tolerance

There is no widely grown clone of export banana that is known to be resistant to the important nematodes, and genetic improvement in the past has been hindered by the complexity in breeding new banana varieties (Menendez and Shepherd, 1975; Ortiz *et al.*, 1995). Such new varieties have had to have the necessary agronomic and fruit quality attributes to meet the demands of the export trade. New techniques for exploiting genetic resources have been developed in recent years (Persley and De Langhe, 1987; Ganry, 1993; Atkinson *et al.*, 2004), resulting in optimism that breeding objectives and benefits will extend beyond the requirements of the international dessert banana trade. The difficulty for nematologists is that no gene(s) for resistance have yet been recognized which can be a basis for a systematic breeding programme. In addition, nematodes have never been the first priority for breeders, who hitherto have devoted their activities to developing hybrids with resistance to Black Sigatoka (*Mycosphaerella fijiensis*) and Fusarium wilt (Panama disease) (Ganry, 1993). The crisis caused by Panama disease which led to the change of export cultivar from Gros Michel to Cavendish exacerbated the problem of *R. similis* (Leach, 1958; Marin *et al.*, 1998). The problem of *R. similis* then became more acute, and this led to the wide-scale use of soil fumigants and nematicides. Dedicated breeding for resistance to *R. similis* began in Honduras after the discovery of resistance in the diploid cv. Pisang jari buaya (Wehunt *et al.*, 1978), which subsequently was used in hybridization programmes that resulted in an improved hybrid breeding clone SH3142 (*Musa* AA) with resistance to *R. similis* (but not *Pratylenchus*) (Pinochet and Rowe, 1979; Pinochet, 1988). SH3142 has many qualities required by banana breeders and was largely used by the FHIA programme (Rowe and Rosales, 1994; Ortiz *et al.*, 1995). The tetraploid FHIA 1 (*Musa* AAAB) which has SH3142 in its parentage has not been consistently shown to have the same level of resistance in the field (Stanton, 1999). Stanton (1999) also showed that although there was no difference between FHIA 1 and Cavendish cv. Williams in numbers of nematodes per root system, FHIA 1 suffered no reduction in root weight.

Numerous screening studies have been done following different procedures in field trials (Wehunt *et al.*, 1978; Price, 1994b; Binks and Gowen, 1996; Fogain and Gowen, 1997, 1998; Stanton, 1999) and greenhouse experiments (Davide and Marasigan, 1985; Fogain, 1996; Pinochet *et al.*, 1998; Stanton, 1999; Marin *et al.*, 2000; Stoffelen *et al.*, 2000; Van den Berghe *et al.*, 2002; Vianne *et al.*, 2003) with sometimes contradictory results on the resistance status to nematodes of some important accessions. It is often difficult to compare the results from these different screening procedures due to highly variable environmental conditions and biological materials (plants and nematodes). There is evidence to suggest that results of screening studies done on young vitro plants may not be consistent with the results from inoculations done on older plants (Stanton, 1999). Although there are compelling reasons for conducting rapid screening trials on juvenile plants under glasshouse conditions (Speijer and De Waele, 1997), it is clear that the final assessments of nematode susceptibility may have to be made on mature plants in the field as ultimately the characteristics shown under field conditions will be the deciding factor in selection of new varieties. Since the numbers of hybrids produced through conventional banana breeding are relatively small, arguably this is not such an important issue.

Biochemicals that could be associated with resistance to *R. similis* and/or *P. goodeyi* include flavones, catechol, cafeic esters, ferulic acid, lignin, dopamine (Sarah *et al.*, 1997), flavonoids (Valette *et al.*, 1997, 1998b), peroxidase (Mateille, 1994), phenalenone (Binks *et al.*, 1997), phytoanticipin-like compounds (Luis, 1998), condensed tannins, procyanidin and propelargonidin (Collingborn *et al.*, 2000), pre-formed phenolic cells and lignified cell walls (Fogain and Gowen, 1996, 1998).

In summary, there is evidence for resistance to *R. similis* in the Pisang jari buaya group, some other diploid accessions such as Kunnan and some of the *M. acuminata* subspecies *burmanicoides* in India (Sathiamoorthy and Balamohan, 1993) and in the triploid *Musa* AAA Yangambi km 5 which also has resistance to *P. goodeyi* (Fogain and Gowen, 1998). Resistance has also been found to *P. coffeae* in some cultivars belonging to the *burmanicoides* group (P. Quénéhervé, Martinique, 2004, personal communication). Nevertheless, the nematode resistance in these cultivars will be quite difficult to manipulate in breeding improvement programmes. One of several tetraploid AAAA genotypes developed by the Banana Breeding Scheme in Jamaica derived from cv. Highgate, a mutant of Gros Michel, was found to be marginally less susceptible than other clones (Gowen, 1976), and casual observations suggested that tetraploids were less vulnerable to falling over in winds or wet weather. It is possible that the relatively greater width of stems of some tetraploids and perhaps more vigorous root systems confer some tolerance to uprooting (Fig. 16.5). The tetraploids produced by FHIA (Rowe and Rosales, 1994) also show relatively greater vigour than many triploid cultivars but have the advantage of not being as tall as those produced in Jamaica. Field observations in Uganda indicate that they too show tolerance to nematodes (S.R. Gowen, 2004, personal communication).

Fig. 16.5. Strong banana cultivar with wind tolerance.

Chemical

Nematicides have been widely used by growers producing fruit for the international export trade. Nowadays the dependence on a regular use of toxic chemical is less acceptable from regulatory and consumer points of view, and this will eventually lead to a more integrated (and sustainable) system of pest management (Holderness *et al.*, 2000). The need for change is greater in some producing countries such as in the Caribbean than others due to environmental concerns (Ganry, 2001). Nematicide use by producers serving only local markets is less common largely because of the high cost of treatment. Only a few organophosphate, oxime carbamate and carbamate nematicides remain registered for use on export bananas; these are used as either granular or emulsifiable concentrate formulations. The products currently registered are: cadusaphos, fosthiazate, ethoprophos, carbofuran and oxamyl. The method and timing of treatments may vary according to cultural practices (Gowen, 1979), climate (Jaramillo and Figueroa, 1976), crop damage, and knowledge of the nematode population dynamics and plant phenology (Quénéhervé *et al.*, 1991; Quénéhervé, 1993b). When making new plantations using field-collected suckers, which may not be totally nematode free, the best results are achieved if nematicides are applied in the planting hole or mixed with the soil when filling in around the plant which prevents populations increasing to damaging levels (Gowen, 1979). However, in many banana-exporting countries, particularly in Central and South America, the replanting of banana fields is uncommon and nematicide treatments may begin on established crops already supporting high nematode population densities. Under such conditions, the benefits of nematicide use may take several crop cycles to become apparent (Gowen, 1979).

Dosages of 2–3 g a.i./plant are generally used; post-planting applications are made in a 45–100 cm radius around the plant but are not incorporated in the soil.

Established bananas are treated with nematicide every 4–5 months. In mature fields, the granular formulations may be sprinkled in a half circle around the selected follower sucker and not entirely surrounding the mother plant (Fig. 16.6).

Liquid concentrate formulations are available in some countries for use in drip irrigation systems, but this method of application does not have universal approval for reasons of safety, particularly the risks of toxicity for workers handling the products. In the Caribbean, oxamyl 24% L is used with a spot-gun spray applicator directly from disposable containers. A water-based formulation of a 10% concentration L gives similar efficacy (C. Chabrier, Martinique, 2004, personal communication). Repeated use of these compounds has led to a condition known as enhanced degradation in which the active ingredient is rapidly metabolized by soil microflora (Smelt *et al.*, 1987; Suett and Walker, 1988; Anderson and Lafuerza, 1992); this could in some cases be overcome by rotating the use of the different compounds.

Contemporary research in Martinique and Guadeloupe and in West and Central

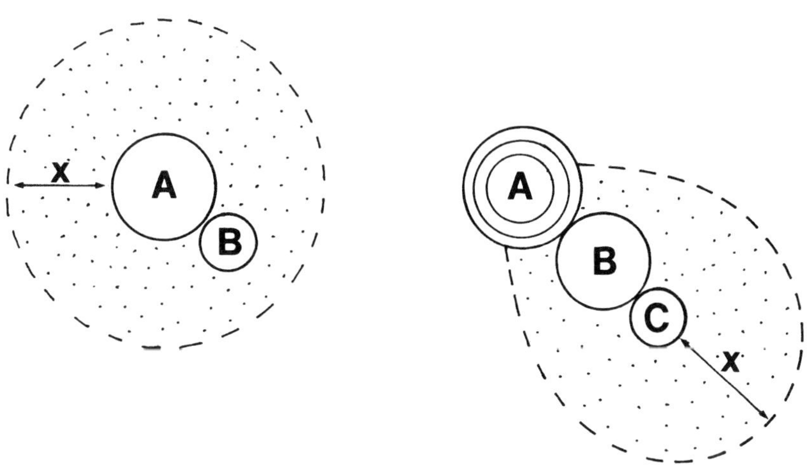

Fig. 16.6. Area of treatment when using granular nematicides on young banana plantation and on ratooning crops. (A) Mother plant. (B) Selected daughter sucker (first ratoon). (C) Selected daughter sucker (to produce second ratoon). (X) Radius of treatment area 35–50 cm.

Africa is directed towards a more rational approach to chemical treatments. This has involved intensive study of nematode populations in banana fields and the decisions on treatments taken only after analysis of the population densities. However, where vitro plants are used, the first applications of nematicides are made only after the first positive record of nematodes, which might be 18–24 months after planting.

There is no conclusive evidence that nematodes have become resistant to nematicides; in banana plantations, the efficiency of soil application is unlikely to be so good as to exert continuous selection pressures on entire populations in roots and soil. The degree of sorption of nematicides in different soil types may influence performance (Hague and Gowen, 1987) and, in light sandy or volcanic ash soils where sorption is low, phytotoxicity might occur. Generally all types are equally effective in sandy or loamy soil, but in peaty soils oxime carbamates may be better than organophosphates (Guérout, 1975; Moss *et al.*, 1975).

In conclusion, we have fewer chemicals available for nematode control and these have to be used with greater attention to the particular needs based on a better understanding of the relationship of nematode density and plant growth. It is unlikely that there will be alternative chemicals developed unless they can be demonstrated to have low toxicity or to be of minor environmental impact. The challenges of finding such chemicals that have specific modes of action on nematode biology or behaviour have not been achieved.

Development of precision application technology in which plants are treated individually at well-defined events such as harvest when the growth of the sucker is stimulated or when nematode infestations are first observed could also be a part of a more integrated pest management system (Quénéhervé *et al.*, 1991).

Biological control

The progress in discovering, characterization and deployment of natural pathogens of migratory endoparasitic nematodes has not been as great as for those that parasitize sedentary endoparasites such as *Meloidogyne* spp. An isolate of *Paecilomyces lilacinus* (Pl 251) originating from the Philippines has been developed commercially, but there are not yet published data on the long-term efficacy under field conditions. Because of the worldwide concerns over the use of nematicides, investigations of potential natural enemies of *R. similis* and *Pratylenchus* spp. have been undertaken. However, none of this has yet been taken to long-term field evaluation. Potential biocontrol agents include *Pseudomonas* spp. (Aalten *et al.*, 1998), Mycorrhizae (Declerck, 1993; Samarao and Martins, 2000; Fogain, 2001; Elsen *et al.*, 2003; Fogain and Njifenjou,

Table 16.1. Methods for maintaining productivity in banana plantations.

A. Established practices for decreasing nematode populations in different banana growing systems.
1. Use of tissue-cultured (vitro) plants
2. Rotation with alternative crops for minimum of 2 years
3. Fallow in the absence of banana 'volunteers' for 10–12 months
4. Selection of disease-free suckers
5. Paring diseased tissue from corms
6. Immersing suckers in hot water
7. Flooding for 8 weeks after having destroyed previous banana crop
8. Applying a nematicide to planting hole and in-fill soil
9. Regular spot applications with nematicides

B. Practices that maintain productivity and vigour.
1. Support plants with bamboo poles or with string guy ropes to prevent plants toppling
2. Regular application of mulches of grass, leaves or organic waste (see Fig. 16.4)
3. Grow cultivars with robust stature and wind tolerance (Fig. 16.5)

2003) and endophytes (Sikora and Schuster, 1998) including *Trichoderma atroviride* and non-pathogenic *F. oxysporum* (Felde *et al.*, 2004; Niere *et al.*, 2004).

Summary of management measures

The different practices used for managing nematodes in bananas are summarized in Table 16.1. In permanent cultivation, the opportunities for control are limited to regular nematicide treatment; however, in subsistence cultivation, the only realistic or economically justifiable techniques for preventing losses from nematodes may be by applying large quantities of mulch to stimulate root growth and by propping fruiting stems. Several of the techniques used for nematode control are also appropriate for controlling the banana borer, which is a widespread pest causing damage to banana corms. The selection of appropriate control techniques will depend largely on the local conditions, availability and reliability of workers and economic considerations. Most control methods depend on the skill and experience of the operators, and may be of little value if the work is not well supervised.

Methods of diagnosis

Sampling

The root systems of bananas are unlike those of short-cycle and other perennial crops, and methods for sampling have to be modified accordingly. Some of the basic principles of sampling are reviewed by Southey (1986) and Quénéhervé and Cadet (1986), and suggested protocols are given by Carlier *et al.* (2002). The growth habit of the banana plant is a clump consisting of a mother plant and a number of lateral (daughter) suckers. The intensity of suckering varies between the different clones, some producing very few (Stover and Simmonds, 1987). A succession of roots develop from the corm of the mother plant and from its suckers until the time of flowering, thereafter the new root growth is only from the daughter suckers. In the field, primary roots may be caused to branch extensively when the dominance of the root apex is disrupted by infection or attack by soil organisms or even unfavourable soil conditions. Samples taken near to the base of the stem of the mother plant will contain roots of different ages and vigour, and consist predominantly of primary roots with relatively smaller quantities of secondary and perhaps no tertiary roots. It is in this region that roots will contain the highest populations of root cortex destroyers, which usually are the 'key pests' (Thomason and Caswell, 1987) against which most control techniques are directed. In an organic soil in the Côte d'Ivoire where *R. similis* and *H. multicinctus* were the principal nematodes, studies of the relative populations in the roots of the different parts of the clump have shown that greater numbers of *R. similis* occur in the roots of the most actively growing suckers. *H. multicinctus* is relatively more numerous in roots of older suckers and harvested plants (Fig. 16.7).

By separating primary roots from the others, Edmunds (1968) showed that by weight the 'secondary' and 'tertiary' roots contained the greater numbers of a mixed population of *R. similis*, *H. multicinctus*, *R. reniformis* and *Meloidogyne* sp. It is possible, however, that the terminology of root types described by Edmunds does not correspond to that described by Swennen *et al.* (1986) who studied root systems of bananas grown hydroponically. Root samples containing large quantities of thin, branching primary roots may therefore contain relatively greater numbers of nematodes than equivalent weights of root consisting of thicker unbranched primaries.

When sampling nematode control experiments in farmers' fields, quantities of roots with adjacent soil are taken from at least ten plants per plot and are bulked to form one composite sample. Samples are normally collected from close to the base of the principal pseudostem at a depth of 5–30 cm where there is an abundance of primary roots and which is within the area over which nemati-

Sucker	Nematodes/g root	
	Rs	*Hm*
A mother plant (harvested)	18	690
B pruned sucker	3	330
C selected daughter sucker (1st ratoon crop unharvested)	39	241
D pruned sucker	276	67
E selected daughter sucker (to produce 2nd ratoon crop)	320	119
F youngest sucker	4	37

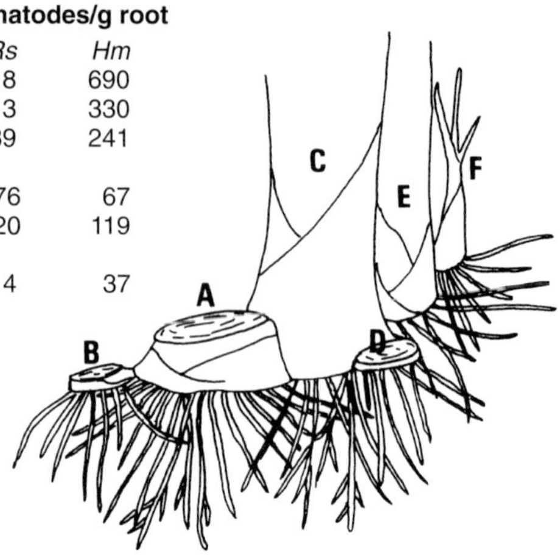

Fig. 16.7. The population levels of *Radopholus similis* (*Rs*) and *Helicotylenchus multicinctus* (*Hm*) in the roots of the different components of a banana clump. From peaty soil, Côte d'Ivoire.

cide treatments are normally applied. Sampling may be done monthly or less frequently, but at a specific stage, e.g. flowering. In more detailed studies of population dynamics of different species over 1 or more years, it may be desirable to analyse separately the roots originating from suckers of different stages of development on single plant clumps and the relative proportions of species along the length of the roots (Quénéhervé, 1990), but this may involve the destructive sampling of entire plants (Quénéhervé and Cadet, 1986). In localities where *R. similis* is known to be the only important root parasite, root sampling may be adequate to represent the population structure as the numbers in soil are relatively low. For other nematodes, particularly *H. multicinctus*, *Pratylenchus* spp., *Meloidogyne* spp. and *R. reniformis*, soil sampling will complement data from root samples.

It is generally accepted that the quality of nematode counts is only as good as the attention given to sampling and extraction. This is particularly true when sampling bananas as it is evident that the task requires careful supervision. In summary, the techniques of sampling bananas and plantains have to be within the capabilities of the available personnel and laboratory facilities. The basic requirements are that sufficient representative plants are sampled (Vilardebó, 1974; Sarah, 1986) and that there is consistency from where the roots (and soil) are taken in relation to position and growth stages of the plant, within samples, and between sampling dates. As a guideline, root sampling might be best done at the time of flowering when the phenology is clearly defined.

Extraction

Samples of banana roots and soil may be collected at locations far from the laboratory. Ideally, processing should be done as quickly as possible and samples should be kept cool and out of direct sunlight during collection and transit. The numbers of *R. similis* and *H. multicinctus* extracted may be affected differentially by the conditions and period of storage prior to processing (Whyte and Gowen, 1974). The techniques used to extract the nematodes of banana may depend on the available laboratory facilities and assistance, and use may be made of non-standard materials purchased locally. This should not prevent or discourage nematologists from adapting a technique which can be used routinely by different operators to give reproducible and equivalent results throughout a period of experimentation. Before initiation of a procedure, it will be necessary to find the optima for sample weight, size of chopped roots, and periods of maceration, incubation, centrifugation or sieving. Banana roots can present some difficulties in extraction if direct maceration and incubation techniques are used. The high levels of phenolic compounds released from chopped or macerated roots can cause depletion in oxygen and thus influence the recovery of nematodes because they may become inactive. This can be partly overcome by adding hydrogen peroxide to the extraction dishes (Gowen and Edmunds, 1973). However, direct recovery techniques by maceration and sieving (Quimi and Villacis, 1977); maceration, sieving and centrifugation (Vilardebó, 1974); and maceration, flocculation–flotation (Escobar and Rodriguez-Kabana, 1980) will be more efficient. The mistifier extraction technique is used in some laboratories for recovering migratory endoparasitic species, and efficiency in recovery improves if the roots are chopped finely. The recovery period may differ for the different species.

Whatever extraction procedure is used, it is important to obtain a representative root sample which should be chopped in 0.5 cm lengths, mixed thoroughly and a 25 g subsample taken for processing. A 24 h period of incubation is sufficient for macerated root samples. Chopped roots should be incubated for 2–4 days and mist extractions may be run for up to 14 days in some laboratories. It is customary to report nematode populations per 100 g of fresh roots, although this quantity is seldom used for extraction.

No specific techniques have been described for extraction or estimation of the sedentary endoparasites *R. reniformis* and *Meloidogyne* spp. in banana roots. Because of the root washing process, the populations of *R. reniformis* can only be estimated from soil samples. With root knot infestations, quantitative data can only be obtained by mist extraction from chopped roots. The many techniques for extracting migratory endoparasites from plant material and the free-living stages in the soil are given by Hooper *et al.* (Chapter 3).

Visual assessments

Where nematologists or laboratory facilities are unavailable, nematode damage is sometimes assessed by recording the incidence of uprooting per hectare per month (Tarte and Pinochet, 1981). This may also be correlated with assessments of necrosis on primary roots and on rhizomes taken from randomly selected plants from a plantation (Stover, 1972; Broadley, 1979; Tarte and Pinochet, 1981; Bridge, 1988; Sikora *et al.*, 1989; Bridge and Gowen, 1993; Speijer and Gold, 1996). Such techniques can be used by those who are familiar with nematode symptoms, but care should be taken not to confuse lesions caused by plant parasitic nematodes with those resulting from other root-infesting pests and pathogens (Fig. 16.8).

Determination of populations and crop loss

Quantification of crop losses attributable to nematodes is difficult because of the close association between species, soil pests and pathogens and with environmental conditions (Ferris, 1981). The nematode parasites of banana can be classified according to the damage caused. The most serious are those that destroy root cortex (*R. similis*, *Pratylenchus* spp. and *H. multicinctus*). Damaged cortex then becomes colonized by fungi which penetrate vascular tissues and hasten the decline in root function. Typically, on an infested plant, all gradations of root damage can be found. The parasitism of *Meloidogyne* spp. and *R. reniformis* may impede the efficiency of roots but does not usually lead to their rapid decomposition. Their location on the thinner roots suggests that damage will affect absorption. Yield losses attributed to these nematodes have not been determined. Many ectoparasitic species probably only browse on the fine secondary and tertiary roots. Despite the large populations recovered from soil, there are no reports of damage causing yield loss.

The damage caused by nematodes in different soil types and the influence of wind exposure can, in terms of uprooting,

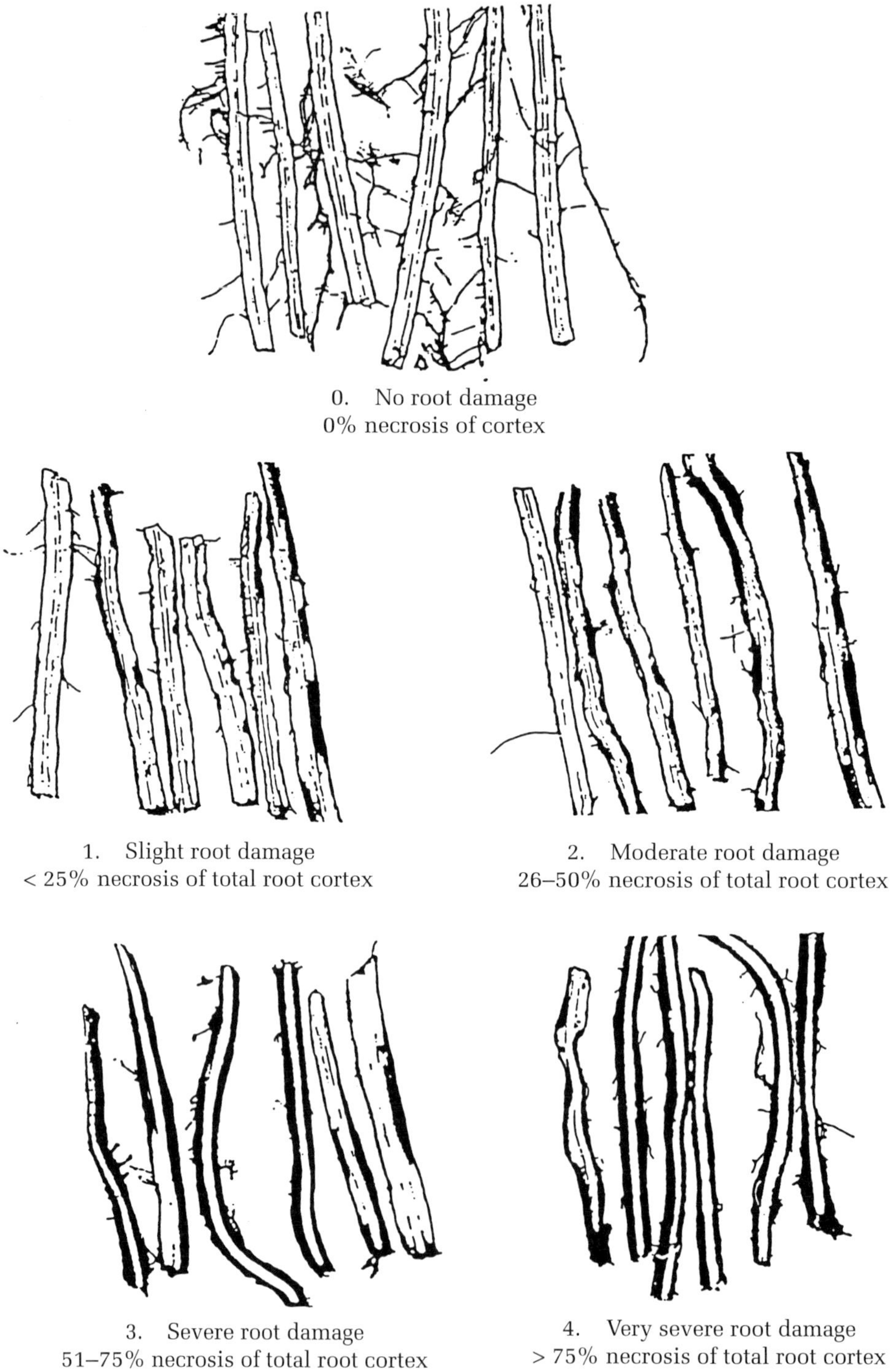

Fig. 16.8. Visual assessment of banana root damage by migratory endoparasitic nematodes using a root index. (Bridge and Gowen, 1993, modified from Broadley, 1973.)

be devastating. The mechanical stresses on the stem and corm of bananas bearing fruit at 2 m or more above the ground are probably considerable. Anchorage may be impaired further by the deliberate removal or suppression of suckers as part of agronomic practice.

Yield loss may be attributed to the smaller size of bunch harvested, but more severe losses occur where banana stems are not propped and the incidence of uprooting is high. Another component of loss is the duration of the vegetative phase, which may be up to 2 months longer in untreated plants over two crop cycles of a replanted banana field infested with *R. similis* and *H. multicinctus* (Gowen, 1975).

There may often, therefore, be direct relationships between nematode populations, root damage and uprooting. In many situations where uprooting occurs, corm necrosis (and consequent root damage) may result from borers (*C. sordidus*). Corm necrosis caused by borers and nematodes can be difficult to distinguish. No universally agreed population damage thresholds have yet been suggested, probably because of the nature of the host plant and of its different parasites in different environments. The nematodes are generally on a continuous reproductive cycle influenced by the vigour of the plant and also by environmental conditions. Similarly, the plant is in a continuous state of aerial growth and root proliferation also mediated by the environment and perhaps foliar and root pathogens. In such situations, it is difficult to introduce concepts of initial inoculum potential linked to crop losses and final population densities as can be shown with some other plant–parasite associations. Nevertheless, in long-term banana experimentation with nematicides, regular sampling can describe population levels that can be compared with crop productivity. From such studies, Guérout (1972) considered that 1000 *R. similis*/100 g of roots was a damage threshold on the AAA cv. Poyo in the Côte d'Ivoire. It might be dangerous to use this value to consider thresholds on other cultivars of banana which may have more or less vigorous root systems. In Latin America, relatively less severe crop losses may be explained by differences in pathogenicity of *R. similis* populations (Pinochet, 1979) or perhaps a different root mycoflora (Felde *et al.*, 2004). It is surprising that in Honduras, Costa Rica and Panama, populations as high as 20,000 per 100 g of roots of AAA cultivars are considered critical (Pinochet, 1987). In the Windward Islands, yield losses can be severe when mixed populations of *R. similis* and *H. multicinctus* exceed 10,000 per 100 g of roots. Despite these differences between regions (and in efficiency of extraction techniques), it is probably not unreasonable to consider root infestations in excess of 2000 per 100 g of roots as a potential cause of crop losses in all commercially grown cultivars. Arguably, any infestation, however small, might be considered as a threat to production over the long term. There is always the likelihood of external influences or events causing crop loss by uprooting. Such losses might be far in excess of those that might be incurred through the general debilitation resulting from the parasitic burden of nematodes feeding in and on the root system.

Conclusions and Future Prospects

Many changes have occurred in the cultivation of bananas and, with increasing interest in the many different types of banana, it may be expected that the areas cultivated for local and regional markets will expand. Since 1961–1965, the combined production of bananas and plantains has increased from 38 to 100 Mt (FAO, 2004). The areas of dessert bananas grown for the international export trade will probably increase marginally, but the spread of some serious diseases is a major threat to production and could destroy the export industry such as has happened in some of the islands of the Pacific (Fullerton, 1987). Export bananas are grown on plantations, but the attention that is necessary for the production and presentation of high quality fruit is closer to that given for horticultural crops. Increasingly, banana plantations will require a well-trained workforce that can adapt to changes in crop management techniques. The wide variability that exists in the many different clones of both dessert and cooking bananas has not been exploited and may show desirable types suited to a broader range of ecological conditions and with useful disease and pest resistance. The International Network for Improvement of Banana and Plantain (INIBAP) coordinates the transfer and evaluation of *Musa* germplasm for disease resistance and genetic improvement. The freer movement of genetic material has been made possible by the development of *in vitro* culture techniques, thus overcoming the fear of further continental and intercontinental movement of some, as yet, uncontrollable pests and diseases. Despite the many years of effort, no new banana has been bred to satisfy the stringent demands of the major banana exporters. International trade is still based on the minor variants of one genotype, *Musa* AAA subgroup Cavendish. Renewed efforts in conventional banana breeding (Shepherd *et al.*, 1987; Bakry *et al.*, 1997) may introduce good agronomic qualities along with pest and disease resistance to cultivars which have a wider acceptance in home or regional markets. However, arguably, there are some uncontentious reasons for the exploitation of the advances in genetic transformation with respect to banana improvement (Tripathi, 2003).

Exploitation of the different sources of resistance to *R. similis* and/or other nematodes particularly *Pratylenchus* spp. should be a major priority. Other plant characters such as root vigour that confers some tolerance to nematodes should also be considered, particularly in programmes for improvement of cooking cultivars. The development of micro-propagation enables the mass production of plants for new commercial plantings. This has considerable advantages over conventional techniques as it ensures that plantations are free (at least initially) from nematode parasites, viruses and borers. The availability of tissue culture plants should enable the critical examination of pathogenicity of the different nematode species (and biotypes) on breeding lines and new cultivars. Results collected so far would suggest that these objectives are quite difficult to achieve. Nematodes will continue to be a major production constraint for most types of banana cropping system. There are no major banana-growing regions in the tropics where *R. similis*, *H. multicinctus* or *Pratylenchus* spp. have not been found. *Meloidogyne* spp. appear to be more damaging in the few special production areas outside the tropics such as Morocco, North Yemen and Cyprus, and in Taiwan and Vietnam. Refinements to nematode management in established plantations have led to a more rational use of nematicides, resulting in lower frequencies of application. Development of precision application technology in which plants are treated individually at well-defined events such as harvest when the growth of the sucker is stimulated or when nematode infestations are first observed could also be a part of a more integrated pest management system. Cost and high mammalian toxicities discourage nematicide use in most growing systems other than for international export. Legislative changes relating to pesticide safety have reduced the number of nematicidal compounds released by the agrochemical companies; many compounds have been withdrawn from the market and this trend will continue. Unfortunately, no new products with novel modes of activity or adequate levels of human or environmental safety have been discovered.

Diversity of cultivars and market opportunities

In the future, there will be a wider choice of varieties available, particularly the new hybrids with resistance to Sigatoka diseases such as produced by the FHIA, IITA and CIRAD breeding programmes. There will be special brands indicating the country or region of origin, and fruit with characteristic sizes, colours and/or flavours that will occupy 'niche markets'. Such varieties and brands are being promoted as organic or 'Fair Trade' and represent an increasing although relatively modest share of the internationally traded fruit.

Nematode control through management but without chemicals will be a primary objective, with efforts to avoid the re-infestation, use of prophylactic measures, physical barriers, and a wider promotion of the use of vitro plants.

Acknowledgements

We thank Barbara Pembroke for drawing the figures, help with the photographs and in the preparation of this chapter. We also extend thanks to M. Amati, C. Bazirake, F. Caveness, H. Chiang, E. Cohn, A. Daudi, R. Hugon, I. Inglis, A. Lassoudière, B. Lubilanji-Tshibamba, Ministry of Agriculture, Mauritius, D. Masaba, E. Mcharo, J. Meyers, F. Ngulu, B. Ngundo, J. Philis, S. Simon and A. Wybou who provided information for the first edition in 1990.

References

Aalten, P.M., Vitour, D., Blanvillain, D., Gowen, S.R. and Sutra, L. (1998) Effect of rhizosphere fluorescent *Pseudomonas* strains on plant-parasitic nematodes *Radopholus similis* and *Meloidogyne* spp. *Letters in Applied Microbiology* 27, 357–361.

Adiko, A. (1988) Plant-parasitic nematodes associated with plantain, *Musa paradisiaca* (AAB), in the Ivory Coast. *Revue de Nematologie* 11, 109–113.

Ambrose, E. (1984) Research and development in banana crop protection (excluding Sigatoka) in the English speaking Caribbean. *Fruits* 39, 234–247.

Anderson, J. and Lafuerza, A. (1992) Microbiological aspects of accelerated pesticide degradation. *Proceeding of International Symposium on Environmental Aspects of Pesticide Microbiology*. Swedish University of Agriculture Science, Uppsala, pp. 184–192.

Araya, M., Vargas, A. and Chaves, A. (1998) Changes in distribution of roots of banana (*Musa* AAA cv Valery) with plant height, distance from the pseudostem and soil depth. *Journal Horticultural Science and Biotechnology* 73, 437–440.

Atkinson, H.J., Grimwood, S., Johnston, K. and Green, J. (2004) Prototype demonstration of transgenic resistance to the nematode *Radopholus similis* conferred on banana by a cystatin. *Transgenic Research* 13, 135–142.

Ayala, A. (1962) Pathogenicity of the reniform nematode on various hosts. *Journal of Agriculture, University of Puerto Rico* 45, 265–299.

Ayala, A. and Roman, J. (1963) Distribution and host range of the burrowing nematode in Puerto Rican soils. *Journal of Agriculture, University of Puerto Rico* 47, 28–37.

Badra, T. and Caveness, F.E. (1983) Effects of dosage sequence on the efficacy of nonfumigant nematicides on plantain yields, and nematode seasonal fluctuations as influenced by rainfall. *Journal of Nematology* 15, 496–502.

Bakry, F., Carreel, F., Caruana, M.L., Côte, F., Jenny, C. and Tezenas du Montcel, H. (1997) Les bananiers. In: Charrier, A., Jacquot, M., Hamon, S. and Nicolas, D. (eds) *L'Amélioration des Plantes Tropicales*. CIRAD-ORSTOM, pp. 109–139.

Bala, G. and Hosein, F. (1996) Plant parasitic nematodes associated with anthuriums and other tropical ornamentals. *Nematropica* 26, 9–14.

Binks, R.H. and Gowen, S.R. (1996) Field evaluation of nematode infestation in *Musa* germplasm at FHIA (La Lima, Honduras). *Infomusa* 5, 15–17.

Binks, R.H., Greenham, J.R., Luis, J.G. and Gowen, S.R. (1997) A phytoalexin from roots of *Musa acuminata* var. Pisang sipulu. *Phytochemistry* 45, 47–49.

Bird, A.F. (1979) Morphology and ultrastructure. In: Lamberti, F. and Taylor, C.E. (eds) *Root-knot Nematodes (*Meloidogyne *species) Systematics, Biology and Control*. Academic Press, London, pp. 59–84.

Blake, C.D. (1961) Root rot of bananas caused by *Radopholus similis* (Cobb) and its control in New South Wales. *Nematologica* 6, 295–310.

Blake, C.D. (1963) Root and corm diseases of bananas. *Agricultural Gazette of New South Wales* 74, 526–531.

Blake, C.D. (1966) The histological changes in banana roots caused by *Radopholus similis* and *Helicotylenchus multicinctus*. *Nematologica* 12, 129–137.

Blomme, G., Swennen, R. and Tenkouano, A. (2000) Assessment of variability in the root system characteristics of banana (*Musa* spp.) according to genome group and ploidy level. *InfoMusa* 9 (2), 4–7.

Blomme, G., Swennen, R. and Tenkouano, A. (2003) Assessment of genotypic variation in the root architecture of *Musa* spp. under field conditions. *InfoMusa* 12 (1), 24–29.

Booth, C. and Stover, R.H. (1974) *Cylindrocarpon musae* sp. nov., commonly associated with burrowing nematode (*Radopholus similis*) lesions on bananas. *Transactions of the British Mycological Society* 63, 503–507.

Bridge, J. (1987) Control strategies in subsistence agriculture. In: Brown, R.H. and Kerry, B.R. (eds) *Principles and Practice of Nematode Control in Crops.* Academic Press, Sydney, Australia, pp. 389–420.

Bridge, J. (1988) Plant nematode pests of banana in East Africa with particular reference to Tanzania. In: *Nematodes and the Borer Weevil in Bananas: Present Status of Rresearch and Outlook. Proceedings of a Workshop,* 7–11 December 1987, Bujumbura, Burundi, pp. 35–39.

Bridge, J. (2000) Nematodes of bananas and plantains in Africa. Research trends and management strategies relating to the small-scale farmer. In: Craenen, K., Ortiz, R., Karamura, E.B. and Vuylsteke, D.R. (eds) *Proceedings of the First International Conference on Banana and Plantain for Africa. Acta Horticulturae* 540, 391–408.

Bridge, J. and Gowen, S.R. (1993) Visual assessment of plant parasitic nematode and weevil damage on bananas and plantain. In: Gold, C.S. and Gemmill, B. (eds) *Biological and Integrated Control of Highland Banana and Plantain Pests and Diseases.* IITA, Cotonou, Benin, pp. 147–154.

Bridge, J. and Page, S.L.J. (1984) Plant nematode pests of crops in Papua New Guinea. *Journal of Plant Protection in the Tropics* 1, 99–109.

Broadley, R.A. (1979) A simple method for estimating banana root rot. *Australian Plant Pathology* 8, 24–25.

Carlier, J., De Waele, D. and Escalant, J.-V. (2002) Global evaluation of *Musa* germplasm for resistance to Fusarium wilt, *Mycosphaerella* leaf spot diseases and nematodes. In: Vézina, A. and Picq, C. (eds) *INIBAP Technical Guidlines 6.* Montpellier, France.

Chabrier, C. and Quénéhervé, P. (2003) Control of the burrowing nematode (*Radopholus similis* Cobb) on banana, impact of the banana field destruction method on the efficiency of the fallowing fallow. *Crop Protection* 22, 121–127.

Charles, J.S.K. and Venkitesan, T.S. (1993) Pathogenicity of *Heterodera oryzicola* (Nemata, Tylenchina) towards banana (*Musa* AAB cv. Nendran). *Fundamental and Applied Nematology* 16, 359–365.

Cofcewicz, E.T., Carneiro, R.M.G., Quénéhervé, P., Augustin, E. and Faria, J.L.C. (2001) Occurrence of *Meloidogyne* spp. in banana producing area in Brazil (abstract). *Nematologia Brasileira* 25, 126.

Collingborn, F.M.B., Gowen, S.R. and Mueller-Harvey, I. (2000) Investigation into the biochemical basis for nematode resistance in roots of three *Musa* cultivars in response to *Radopholus similis* infection. *Journal of Agricultural and Food Chemistry* 48, 5297–5301.

Daniells, J., Jenny, C., Karamura, D. and Tomekpe, K. (2001) *Musalogue, a Catalogue of* Musa *Germplasm. Diversity in the Genus* Musa (compiled by Arnaud, E. and Sharrock, S.). International Network for Improvement of Banana and Plantain (INIBAP), Montpellier, France.

Davide, R.G. (1980) Influence of cultivar, age, soil texture and pH on *Meloidogyne incognita* and *Radopholus similis* on banana. *Plant Disease Reporter* 64, 571–573.

Davide, R.G. and Marasigan, L.Q. (1985) Yield loss assessment and evaluation of resistance of banana cultivars to the nematodes *Radopholus similis* Thome and *Meloidogyne incognita* Chitwood. *Philippine Agriculturalist* 68, 335–349.

Declerck, S. (1993) Prospect for the use of mycorrhized banana plants. *Fruits* 48, 34.

de Guiran, G. and Vilardebó, A. (1962) Le bananier aux Iles Canaries. IV, Les nematodes parasites. *Fruits* 17, 263–277.

Draye, X., Delvaux, B. and Swennen, R. (1999) Distribution of lateral root primordial in root tips of *Musa. Annals of Botany* 84, 393–400.

Ducharme, E.P. and Birchfield, W. (1956) Physiologic races of the burrowing nematode. *Phytopathology* 46, 615–616.

Duncan, L.W., Inserra, R.N., Thomas, W.K., Dunn, D., Mustika, I., Frisse, L.M., Mendes, L.M., Morris, K. and Kaplan, D.T. (1999) Molecular and morphological analysis of isolates of *Pratylenchus coffeae* and closely related species. *Nematropica* 29, 61–80.

Edmunds, J.E. (1968) Nematodes associated with bananas in the Windward Islands. *Tropical Agriculture, Trinidad* 45, 119–124.

Edwards, D.I. and Wehunt, E.J. (1971) Host range of *Radopholus similis* from banana areas of Central America with indications of additional races. *Plant Disease Reporter* 55, 415–418.

Elsen, A., Baimey, H., Swennen, R. and De Waele, D. (2003) Relative mycorrhizal dependency and mycorrhiza–nematode interaction in banana cultivars (*Musa* spp.) differing in nematode susceptibility. *Plant and Soil* 256, 303–313.

Epp, M.D. (1987) Somaclonal variation in bananas, a case study with Fusarium wilt. In: Persley, G.J. and De Langhe, E.A. (eds) *Banana and Plantain Breeding Strategies, Proceedings of an International Workshop,* 13–17 October 1986, Cairns, Australia. ACIAR, Proceedings No. 21, 140–150.

Escobar, J. and Rodriguez-Kabana, R. (1980) Comparacion de un metodo de flotacion con uno de tamizado para la determinacion de *Radopholus similis* en raices de banano. *Nematropica* 10, 86–88.

FAO (2004) FAO Statistical Database. Available at: faostat.fao.org

Fargette, M. (1987) Use of the esterase phenotype in the taxonomy of the genus *Meloidogyne* 2. Esterase phenotypes observed in West Africa populations and their characterization. *Revue de Nematologie* 10, 45–56.

Fargette, M. and Quénéhervé, P. (1988) Populations of nematodes in soils under banana, cv Poyo, in the Ivory Coast. 1. The nematofauna occurring in the banana producing areas. *Revue de Nematologie* 11, 239–244.

Felde, A. zum, Pocasangre, L. and Sikora, R. (2004) The potential use of microbial communities inside suppressive banana plants to increase root health and suppression of the burrowing nematode, *Radopholus similis. Proceeding of the International Symposium – Banana Root Systems, Towards a Better Understanding for its Productive Management,* 3–5 November 2003, Costa Rica. Corbana–INIBAP.

Ferris, H. (1981) Dynamic action thresholds for disease induced by nematodes. *Annual Review of Phytopathology* 19, 427–436.

Ferris, V.R. and Ferris, J.M. (1974) Inter-relationship between nematode and plant communities in agricultural ecosystems. *Agro-Ecosystems* 1, 275–299.

Fluiter, H.J. and Mulholland, J.J. (1941) Gegevens, verkregen bij het onderzoek naar de waardplantenvan *Tylenchus coffeae. Bergcultures* 15, 1588–1593.

Fogain, R. (1996) Screenhouse evaluation of *Musa* for susceptibility to *Radopholus similis.* In: Frison, E.A., Horry, J.P. and De Waele, D. (eds) *Evaluation of Plantains AAB and Diploid AA, AB and BB. Proceeding of the Workshop on New Frontiers in Resistance Breeding for Nematode, Fusarium* and *Sigatoka,* 2–5 October 1995, Kuala Lumpur, Malaysia. INIBAP, Montpellier, France, pp. 79–86.

Fogain, R. (2001) Mycorhization précoce et croissance de deux types de materiel vegetal de plantain. *Cahier d'Etude et de Recherche Francophone en Agriculture* 10, 195–197.

Fogain, R. and Gowen, S.R. (1996) Investigations on possible mechanisms of resistance to nematodes in *Musa. Euphytica* 92, 375–381.

Fogain, R. and Gowen, S.R. (1997) Damage to roots of *Musa* cultivars by *Radopholus similis* with and without protection of nematicides. *Nematropica* 27, 27–32.

Fogain, R. and Gowen, S.R. (1998) Yangambi km 5 (*Musa* AAA, Ibota subgroup), a possible source of resistance to *Radopholus similis* and *Pratylenchus goodeyi. Fundamental and Applied Nematology* 21, 75–80.

Fogain, R. and Njifenjou, S. (2003) Effect of a mycorrhizal *Glomus* sp. on growth of plantain and on the development of *Radopholus similis* under controlled conditions. *African Plant Protection* 9, 27–30.

Fullerton, R.A. (1987) Banana production in selected Pacific Islands. In: Persley, G.J. and De Langhe, E.A. (eds) *Banana and Plantain Breeding Strategies, Proceedings of an International Workshop,* 13–17 October 1986, Cairns, Australia. ACIAR, Proceedings No. 21, pp. 57–62.

Ganry, J. (ed.) (1993) *Breeding Banana and Plantain for Resistance to Diseases and Pests.* CIRAD, Montpellier, France.

Ganry, J. (2001) Maîtrise de la culture du bananier pour une production raisonnée face aux nouveaux défis. *Compte Rendus de l'Académie d'Agriculture de France* 97, 119–127.

Gichure, E. and Ondieki, J.J. (1977) A survey of banana nematodes in Kenya. *Zeitschfrift für Pflanzenkrankheitenund Pflanzenschutz* 84, 724–728.

Goodey, J.B., Franklin, M.T. and Hooper, D.J. (1965) *T. Goodey's the Nematode Parasites of Plants Catalogued Under their Hosts.* Commonwealth Agricultural Bureaux, Slough, UK.

Gotoh, A. (1964) The embryonic and larval development of *Pratylenchus coffeae* (Zimmerman) (Nematoda,Tylenchida). *Japanese Journal of Applied Entomology and Zoology* 8, 26–33.

Gowen, S.R. (1975) Improvement of banana yields with nematicides. *Proceedings 8th British Insecticide and Fungicide Conference*, 1975, Brighton, pp. 121–125.
Gowen, S.R. (1976) Varietal responses and prospects for breeding nematode resistant banana varieties. *Nematropica* 6, 45–49.
Gowen, S.R. (1979) Some considerations of problems associated with the nematode pests of bananas. *Nematropica* 9, 79–91.
Gowen, S.R. (1993) Possible approaches for developing nematode resistance in bananas and plantains. In: Ganry, J. (ed.) *Breeding Banana and Plantain for Resistance to Diseases and Pests.* CIRAD Montpellier, France, pp. 123–128.
Gowen, S.R. and Edmunds, J.E. (1973) An evaluation of some simple extraction techniques and the use of hydrogen peroxide for estimating nematode populations in banana roots. *Plant Disease Reporter* 57, 678–681.
Gowen, S.R. and Quénéhervé, P. (1990) Nematode parasites of bananas, plantains and abaca. In: Luc, M., Sikora, R.A. and Bridge, J. (eds) *Plant Parasitic Nematodes in Tropical and Subtropical Agriculture.* CAB International, Wallingford, UK, pp. 431–460.
Guérout, R. (1972) Relations entre les populations de *Radopholus similis* et la croissance du bananier. *Fruits* 27, 331–337.
Guérout, R. (1975) The soil and its importance on the effect of nematicidal treatments on banana yields. *Nematropica* 5, 22–23.
Guérout, R., Lassoudière, A. and Vilardebó, A. (1976) Efficacite des nematicides sur deux types de sol a caracteristiques particulieres en Cote d'Ivoire. *Fruits* 31, 427–436.
Hague, N.G.M. and Gowen, S.R. (1987) Chemical control of nematodes. In: Brown, R.H. and Kerry, B.R. (eds) *Principles and Practice of Nematode Control in Crops.* Academic Press Sydney, Australia, pp. 131–178.
Hahn, M.L., Sarah, J.-L., Boisseau, M., Vines, N.J., Wright, D.J. and Burrows, P.R. (1996) Reproductive fitness and pathogenicity of selected *Radopholus* populations on two banana cultivars. *Plant Pathology* 45, 223–231.
Holderness, M., Sharrock, S., Frison, E. and Kairo, M. (2000) *Organic Banana 2000, Towards an Organic Banana Initiative in the Caribbean.* INIBAP, Montpellier, France.
Huang, C.S. (1985) Formation, anatomy and physiology of giant cells induced by root-knot nematodes. In: Sasser, J.N. and Carter, C.C. (eds) *An Advanced Treatise on* Meloidogyne. *Volume 1 Biology and Control.* North Carolina State University, Raleigh, North Carolina, pp. 155–164.
Hugon, R., Ganry, J. and Berthe, G. (1984) Dynamique de population du nematode *Radopholus similis* en function du stade de developpement du bananier et du climat. *Fruits* 39, 251–253.
Hutton, D.G. (1978) Influence of rainfall on some plantain nematodes in Jamaica. *Nematropica* 8, 34–39.
Hwang, S.C., Chen, C.L. and Wu, F.L. (1984) An investigation on susceptibility of banana clones to fusarial wilt, freckle and marginal scorch disease in Taiwan. *Plant Protection Bulletin (Taiwan)* 26, 155–161.
INIBAP (1986) *Banana Research in Eastern Africa. Proposal for a Regional Research and Development Network.* International Network for the Improvement of Bananas and Plantains, Montpellier, France.
Irizarry, H., Vicente-Chandler, J. and Silva, S. (1981) Root distribution of plantains growing on five soil types. *Journal of Agriculture, University of Puerto Rico* 65, 29–34.
Israeli, Y., Lahav, E. and Reuveni, O. (1995) *In vitro* culture of bananas. In: Gowen, S. (ed.) *Bananas and Plantains.* Chapman and Hall, London, pp. 147–178.
Jaramillo, R. and Figueroa, A. (1974) Analisis armonico de la densidad de poblacion de *Radopholus similis* (Cobb) Thome en la zona bananera de Guapiles, Costa Rica. *Turrialba* 24, 402–407.
Jaramillo, R. and Figueroa, A. (1976) Relacion entre el balance hidrico y la poblacion de *Radopholus similis* (Cobb) Thome en la zona bananera de Guapiles, Costa Rica. *Turrialba* 26, 187–192.
Jimenez, M.F. (1972) Fluctuaciones anuales de la poblacion de *Radopholus similis* (Cobb) Thome en la zona bananera de Potoci, Costa Rica. *Nematropica* 2, 33–40.
Jones, D.R. (2000) *Diseases of Banana, Abaca and Enset.* CAB International, Wallingford, UK.
Jones, R.K. and Milne, D.L. (1982) Nematode pests of bananas. In: Keetch, D.P. and Heyns, J. (eds) *Nematology in Southern Africa.* Pretoria, Republic of South Africa, pp. 30–37.
Kaplan, D.T. and MacGowan, J.B. (1982) Ability of selected common weeds and ornamentals to host *Pratylenchus coffeae. Nematologica* 12, 165–170.
Kaplan, D.T. and O'Bannon, J.H. (1985) Occurrence of biotypes in *Radopholus citrophilus. Journal of Nematology* 17, 158–162.
Kaplan, D.T. and Opperman, C.H. (1997) Genome similarity implies that citrus-parasitic burrowing nematodes do not represent a unique species. *Journal of Nematology* 29, 430–440.

Kashaija, I.N., Speijer, P.R., Gold, C.S. and Gowen, S.R. (1994) Occurrence, distribution and abundance of plant parasitic nematodes of bananas in Uganda. *African Crop Science Journal* 2, 99–104.

Kashaija, I.N., Fogain, R. and Speijer, P.R. (1998) Habitat management for control of banana nematodes. In: Frison, E.A., Gold, C.S., Karamura, E.B. and Sikora, R.A. (eds) *Mobilizing IPM for Sustainable Banana Production in Africa.* INIBAP, Montpellier, France, pp. 127–136.

Lassoudière, A. (1978) Quelques espects de la croissance et du developpement du bananier 'Poyo' en Cote d'Ivoire. 2. Le systeme radical. *Fruits* 33, 314–338.

Lassoudière, A. and Martin, P. (1974) Problemes de drainage en sols organiques de bananeraie (Agneby, Cote d'Ivoire). *Fruits* 29, 255–266.

Lavigne, C. (1987) Contribution a l'etude du systeme racinaire du bananier. Mise au point de rhizotrons et premiers resultats. *Fruits* 42, 265–271.

Laville, E. (1964) Etude de la mycoflore des racines du bananier 'Poyo'. 1. Etude du systeme radiculaire. *Fruits* 19, 435–449.

Leach, R. (1958) Blackhead toppling disease of bananas. *Nature* 181, 204–205.

Lin, Y.-Y. and Tsay, T.-T. (1985) Studies on banana root-knot nematode disease in the central area of Taiwan. *Journal of the Chinese Society for Horticultural Science* 31, 44–46.

Loeillet, D. (2001) Le commerce de la banane et ses enjeux. *Compte Rendus de l'Académie d'Agriculture de France* 97, 111–118.

Loos, C.A. (1959) Symptom expression of Fusarium wilt disease of the Gros Michel banana in the presence of *Radopholus similis* (Cobb, 1893) Thome, 1949 and *Meloidogyne incognita acrita* Chitwood, 1949. *Proceedings of the Helminthological Society of Washington* 26, 103–111.

Loos, C.A. (1961) Eradictation of the burrowing nematode, *Radopholus similis* from bananas. *Plant Disease Reporter* 45, 457–461.

Loos, C.A. (1962) Studies on the life history and habits of the burrowing nematode, *Radopholus similis*, the cause of blackhead disease of banana. *Proceedings of the Helminthological Society of Washington* 29, 43–52.

Loos, C.A. and Loos, S.B. (1960a) Preparing nematode free banana seed. *Phytopathology* 50, 383–386.

Loos, C.A. and Loos, S.B. (1960b) The blackhead disease of bananas. *Proceedings of the Helminthological Society of Washington* 27, 189–193.

Luc, M. and Vilardebó, A. (1961) Les nematodes associes aux bananiers cultives dans l'ouest Africain. *Fruits* 16, 205–219.

Luis, J.G. (1998) Phenylphenalenone-type phytoalexin and phytoanticipins from susceptible and resistant cultivars of *Musa* species, its potential for engineering resistance to fungi and nematodes into banana. *Acta Horticulturae* 490, 425–430.

Maas, P.W.T. (1969) Two important cases of nematode infestation in Surinam. In: Peachey, J.E. (ed.) *Nematodes of Tropical Crops.* Commonwealth Agricultural Bureaux, Slough, UK, pp. 149–154.

Marcelino, L.L., Viquez, M. and Tarte, R. (1978) Fluctuaciones estacionales de la densidades de poblacion de *Radopholus similis* en raices de banano 'Valery' (*Musa acuminata* AAA) en la zona Pacifica de Panama. *Nematropica* 8, 52–55.

Marin, D.H., Sutton, T.B. and Barker, K.R. (1998) Dissemination of bananas in Latin America and the Caribbean and its relationship to the occurrence of *Radopholus similis. Plant Disease* 82, 964–974.

Marin, D.H., Barker, K.R., Kaplan, D.T., Sutton, T.B. and Opperman, C.H. (1999) Aggressiveness and damage potential of Central American and Caribbean populations of *Radopholus* spp. in banana. *Journal of Nematology* 31, 377–385.

Marin, D.H., Barker, K.R., Kaplan, D.T., Sutton, T.B. and Opperman, C.H. (2000) Development of a standard method for screening for resistance to *Radopholus similis* in bananas. *Plant Disease* 84, 689–693.

Mateille, T. (1994) Biochemical reactions provoked by three phytoparasitic nematodes in roots of *Musa acuminata* (AAA Cavendish group) cvs Poyo and Gros Michel. *Fundamental and Applied Nematology* 17, 283–290.

Mateille, T., Cadet, P. and Quénéhervé, P. (1984) L'influence du recepage du bananier cv. Poyo sur le developpement des populations de *Radopholus similis* et d'*Helicotylenchus multicinctus. Revue de Nematologie* 7, 355–361.

Mateille, T., Foncelle, B. and Ferrer, H. (1988a) Lutte contre les nematodes du bananier par submersion du sol. *Revue de Nematologie* 11, 235–238.

Mateille, T., Quénéhervé, P. and Topart, P. (1988b) Nematicidal treatment of banana AAA cv. Poyo planting material by corm coating. *Revue de Nematologie* 11, 89–92.

Mateille, T., Quénéhervé, P. and Hugon, R. (1994) The development of plant parasitic nematode infestations on micro-propagated banana plants following field control measure in Côte d'Ivoire. *Annals of Applied Biology* 125, 147–159.

McIntyre, B.D., Gold, C.S., Kashaija, I.N., Ssali, H., Night, G. and Bwamiki, D.P. (2001) Effects of legume intercrops on soil-borne pests, biomass, nutrients and soil water in banana. *Biology and Fertility of Soils* 34, 342–348.

McSorley, R. and Parrado, J. (1981) Population fluctuations of plant-parasitic nematodes on bananas in Florida. *Proceedings of the Florida State Horticulture Society* 94, 321–323.

McSorley, R. and Parrado, J.L. (1983) The spiral nematode, *Helicotylenchus multicinctus*, on bananas in Florida and its control. *Proceedings of the Florida State Horticultural Society* 96, 201–207.

McSorley, R. and Parrado, J.L. (1986) *Helicotylenchus multicinctus* on bananas, an international problem. *Nematropica* 16, 73–91.

Melin, P. and Vilardebó, A. (1973) Efficacite de quelques nematicides en bananeraie dans les sols volcaniques de la region du Mungo (Cameroun). *Fruits* 28, 3–17.

Menendez, T. and Shepherd, K. (1975) Breeding new bananas. *World Crops* 27, 104–112.

Minz, G., Ziv, D. and Strich-Harari, D. (1960) Decline of banana plantations caused by spiral nematodes in the Jordan Valley and its control by DBCP. *Ktavim* 10, 147–157.

Moss, S.R., Crump, D. and Whitehead, A.G. (1975) Control of potato cyst nematode *Heterodera rostochiensis* and *H. pallida* in sandy, peaty and silt loam soils by oxime carbamate and organophosphate nematicides. *Annals of Applied Biology* 81, 359–365.

Netscher, C. (1978) Morphological and physiological variability of species of *Meloidogyne* in West Africa and implications for their control. *Mededelingen. Landbouwhogeschool Wageningen* 46, 78–83.

Newhall, A.G. (1958) The incidence of Panama disease of banana in the presence of the root knot and burrowing nematodes (*Meloidogyne* and *Radopholus*). *Plant Disease Reporter* 42, 853–856.

Niere, B., Gold, C.S. and Coyne, D. (2004) Can fungal endophytes control soilborne pests in banana? In: Sikora, R.A., Gowen, S., Hauschild, R. and Kiewnick, S. (eds) *Multitrophic Interactions in Soil and Integrated Control.* IOBC/WPRS Bulletin 27, 203–210.

O'Bannon, J. (1977) Worldwide dissemination of *Radopholus similis* and its importance in crop production. *Journal of Nematology* 9, 16–25.

Ocfemia, G.O. and Calinisan, M.R. (1928) The root knot of abaca or Manila hemp. *Phytopathology* 18, 861–867.

Orion, D., Levy, Y., Israeli, Y. and Fisher, E. (1999) Scanning electron microscope observations on spriral nematode (*Helicotylenchus multicinctus*). *Nematropica* 29, 179–183.

Ortiz, R., Ferris, R.S.B. and Vuylsteke, D.R. (1995) Banana and plantain breeding. In: Gowen, S. (ed.) *Bananas and Plantains.* Chapman and Hall, London, pp. 110–146.

Persley, G.J. and de Langhe, E.A. (1987) *Banana and Plantain Breeding Strategies, Proceedings of an International Workshop,* 13–17 October 1986, Cairns, Australia. ACIAR Proceedings No. 21.

Pinochet, J. (1977) Occurrence and spatial distribution of root-knot nematodes on bananas and plantains in Honduras. *Plant Disease Reporter* 61, 518–520.

Pinochet, J. (1978) Histopathology of the root lesion nematode, *Pratylenchus coffeae* on plantains, *Musa* AAB. *Nematologica* 24, 337–340.

Pinochet, J. (1979) Comparison of four isolates of *Radopholus similis* from Central America on Valery Bananas. *Nematropica* 9, 40–43.

Pinochet, J. (1987) La variabilidad de *Radopholus similis* en banane en las diferentes regiones productoras del mundo. *Memorias VII Reunion ACORBAT,* September 1985, San Jose, Costa Rica. CATIE, Costa Rica, pp. 175–182.

Pinochet, J. (1988) Comments on the difficulty in breeding bananas and plantains for resistance to nematodes. *Revue de Nematologie* 11, 3–5.

Pinochet, J. and Duarte, O. (1986) Additional list of ornamental foliage plants host of the lesion nematode *Pratylenchus coffeae. Nematologica* 16, 11–19.

Pinochet, J. and Rowe, P.R. (1979) Progress in breeding for resistance to *Radopholus similis* in banana. *Nematropica* 9, 76–78.

Pinochet, J. and Stover, R.H. (1980) Fungi lesions caused by burrowing nematodes on bananas and their root and rhizome rotting potential. *Tropical Agriculture, Trinidad* 57, 227–232.

Pinochet, J., Jaizme, M., Fernandez, C., Jaumot, M. and De Waele, D. (1998) Screening bananas for root-knot (*Meloidogyne* spp.) and lesion nematode (*Pratylenchus goodeyi*) resistance for the Canary Islands. *Fundamental and Applied Nematology* 21, 17–23.

Price, N.S. (1994a) Alternate cropping in the management of *Radopholus similis* and *Cosmopolites sordidus*, two important pests of banana and plantain. *International Journal of Pest Management* 40, 237–244.

Price, N.S. (1994b) Field trial evaluation of nematode susceptibility within *Musa*. *Fundamental and Applied Nematology*, 17 391–396.

Price, N.S. and Bridge, J. (1995) *Pratylenchus goodeyi* (Nematoda, Pratylenchidae), a plant-parasitic nematode of the montane highlands of Africa. *Journal of African Zoology* 109, 435–442.

Quénéhervé, P. (1988) Populations of nematodes in soils under banana, cv. 'Poyo', in the Ivory Coast. 2. Influence of soil texture, pH and organic matter on nematode populations. *Revue de Nematologie* 11, 245–251.

Quénéhervé, P. (1989a) Population of nematodes in soils under banana, cv. 'Poyo', in the Ivory Coast. 3. Seasonal dynamics of populations in mineral soil. *Revue de Nematologie* 12, 149–160.

Quénéhervé, P. (1989b) Population of nematodes in soils under banana, cv. 'Poyo', in the Ivory Coast. 4. Seasonal dynamics of populations in organic soil. *Revue de Nematologie* 12, 161–170.

Quénéhervé, P. (1990) Spatial arrangement of nematodes around the banana plant in the Ivory Coast, related comments on the interaction among concomitant phytophagous nematodes. *Acta Oecologica* 11, 875–886.

Quénéhervé, P. (1993a) Banana phenology in relation to phytophagous nematodes. In: Gold, C.S. and Gemmill, B. (eds) *Biological and Integrated Control of Highland Banana and Plantain Pests and Diseases*. IITA, Cotonou, Benin, pp. 218–230.

Quénéhervé, P. (1993b) Nematode management in banana agroecosystems, comments and outlook from the Côte d'Ivoire experience. *Crop Protection* 12, 164–172.

Quénéhervé, P. and Cadet, P. (1985) Localisation des nematodes dans les rhizomes du bananier cv Poyo. *Revue de Nematologie* 8, 3–8.

Quénéhervé, P. and Cadet, P. (1986) Une nouvelle technique d'echantillonnage pour l'etude des nematodes endoparasites du bananier. *Revue de Nematologie* 9, 95–97.

Quénéhervé, P., Mateille, T. and Topart, P. (1991) New approaches to chemical control of nematodes on bananas, field experiments in the Ivory Coast. *Revue de Nématologie* 14, 543–549.

Quénéhervé, P., Drob, F. and Topart, P. (1995) Host status of some weeds to *Meloidogyne* spp., *Helicotylenchus* spp. and *Rotylenchulus reniformis* associated with vegetables cultivated in polytunnels in Martinique. *Nematropica* 25, 149–157.

Quénéhervé, P., Van den Berg, E., Topart, P. and Hostachy, B. (1997) Analyse écologique de la spécificité parasitaire des nematodes phytoparasites associés à quelques plantes ornementales cultivées à la Martinique. *Nematologica* 43, 214–227.

Quénéhervé, P., de Bock, S., Valette, C. and Chabrier, C. (2000a) Enzyme phenotype of *Meloidogyne* spp. associated with bananas in Martinique (abstract). *Nematropica* 30, 145.

Quénéhervé, P., Chabrier, C., Auwerkerken, A. and Marie-Luce, S. (2000b) Weeds as potential reservoirs of *Radopholus similis* in banana fields in Martinique (abstract). *Nematropica* 30, 144.

Quimi, V.H. and Villacis, J. (1977) Estudio comparativo de dos metodos de extraccion del nematodo *Radopholus similis* de las raices de banano. *Nematropica* 7, 44–47.

Risède, J.M. and Simoneau, P. (2004) Pathogenic and genetic diversity of soilborne isolates of *Cylindrocladium* from banana cropping systems. *European Journal of Plant Pathology* 110, 139–154.

Robinson, J.C. (1996) *Bananas and Plantains*. CAB International, Wallingford, UK.

Rowe, P. and Rosales, F. (1994) *Musa* breeding at FHIA. In: Jones, D.R. (ed.) *The Improvement and Testing of* Musa*: a Global Partnership*. INIBAP, Montpellier, France, pp. 117–129.

Samarao, S.S. and Martins, M.A. (2000) Producao de mudas micropropagadas de bananeira (*Musa* sp.) inoculadas com fungos miccorizicos arbusculares. *Revista Brasileira de Fruticultura (Brazil)* 22, 200–204.

Sarah, J.-L. (1986) Repartition spatiale des infestations racinaires de *Radopholus similis* en relation avec la croissance et le developpement du bananier 'Poyo' en Côte d'Ivoire. *Fruits* 41, 427–435.

Sarah, J.-L., Lassoudière, A. and Guérout, R. (1983) La jachere nue et l'immersion du sol, deux methodes interessantes de lutte integree contre *Radopholus similis* (Cobb) dans les bananeraies de sols tourbeux de Cote d'Ivoire. *Fruits* 38, 35–42.

Sarah, J.L., Fogain, R. and Valette, C. (1997) Nematode resistance in bananas, varietal screening and resistance mechanisms. *Fruits* 52, 267–271.

Sathiamoorthy, S. and Balamohan, T.N. (1993) Improvement of banana. In: Chadah, K.L. and Paroek, O.P. (eds) *Advances in Horticulture 1 Fruit Crops*. Malhotra Publishing House, New Delhi, India, pp. 303–335.

Shepherd, K., Dantas, J.L.L. and Alves, E.J. (1987) Banana breeding in Brazil. In: Persley, G.J. and de Langhe, E.A. (eds) *Banana and Plantain Breeding Strategies, Proceedings of an International Workshop,* 13–17 October 1986, Cairns, Australia. ACIAR Proceedings No. 21, pp. 78–83.

Sher, S.A. (1968) Revision of the Genus *Radopholus* Thome, 1949 (Nematoda, Tylenchoidea). *Proceedings of the Helminthological Society of Washington* 35, 219–237.

Sikora, R.A. (1979) Observations on *Meloidogyne* with emphasis on disease complexes, and the effect of host on morphometrics. *Proceedings of the Second Research Planning Conference on Root-knot Nematodes,* Meloidogyne *spp.,* Athens, pp. 93–104.

Sikora, R.A. and Schlosser, E. (1973) Nematodes and fungi associated with root systems of bananas in a state of decline in Lebanon. *Plant Disease Reporter* 57, 615–618.

Sikora, R.A. and Schuster, R.P. (1998) Novel approaches to nematode IPM. In: Frison, E.A., Gold, C.S., Karamura, E.B. and Sikora, R.A. (eds) *Mobilizing IPM for Sustainable Banana Production in Africa.* INIBAP, Montpellier, France, pp. 127–136.

Sipes, B.S. and Lichty, J.S. (2002) *Radopholus similis* damage to *Anthurium andraeanum. Nematropica* 32, 77–81.

Sivakumar, C.V. and Seshadri, A.R. (1974) Histopathology of infection by the reniform nematode *Rotylenchulus reniformis* Linford and Oliveira, 1940 on castor, papaya and tomato. *Indian Journal of Nematology* (1972), 173–181.

Smelt, J.H., Crum, S.J.H., Teunissen, W. and Leistra, M. (1987) Accelerated transformations of aldicarb, oxamyl and ethoprophos after repeated soil treatments. *Crop Protection* 6, 295–303.

Southey, J.F. (1986) Principles of sampling for nematodes. In: Southey, J.F. (ed.) *Laboratory Methods for Work With Plant and Soil Nematodes.* HMSO, London, pp. 1–4.

Speijer, P.R. and De Waele, D. (1997) *Screening of* Musa *Germplasm for Resistance and Tolerance to Nematodes.* INIBAP Technical Guidelines 1. IBPGR, Rome, Italy; INIBAP, Montpellier, France.

Speijer, P.R. and Gold, C.S. (1996) *Musa* root health assessment, a technique for the evaluation of *Musa* germplasm for nematode resistance. In: Frison, E.A., Horry, J.P. and de Waele, D. (eds) *New Frontiers in Resistence Breeding for Nematodes,* Fusarium *and Sigatoka.* INIBAP, Montpellier, France, pp. 62–67.

Stanton, J.M. (1999) Assessment of resistance and tolerance of *in vitro*-propagated banana plants to burrowing nematode, *Radopholus similis. Australian Journal of Experimental Agriculture* 39, 891–895.

Stoffelen, R., Jimenez, M.I., Dierckxsens, C., Tam, V.T.T., Swennen, R. and De Waele, D. (1999) Effect of time and inoculum density on the reproductive fitness of *Pratylenchus coffeae* and *Radopholus similis* populations on carrot disks. *Nematology* 1, 243–250.

Stoffelen, R., Verlinden, R., Xuyen, N.T., Swennen, R. and De Waele, D. (2000) Host plant response of Eumusa and Australimusa bananas (*Musa* spp.) to migratory endoparasitic and root-knot nematodes. *Nematology* 2, 907–916.

Stover, R.H. (1966) Fungi associated with nematodes and non nematode lesions on banana roots. *Canadian Journal of Botany* 44, 1703–1710.

Stover, R.H. (1972) *Banana Plantain and Abaca Diseases.* Commonwealth Mycological Institute, Kew, UK.

Stover, R.H. and Fielding, M.I. (1958) Nematodes associated with root injury of *Musa* sp. in Honduras banana soils. *Plant Disease Reporter* 42, 938–940.

Stover, R.H. and Simmonds, N.W. (1987) *Bananas,* 3rd edn. Longmans, London.

Stoyanov, D. (1967) Especies de nematodos parasitos del platano en Cuba y posibilidades de su control. *Revista Agricultura, Habana* 1, 9–47.

Sudha, S. and Prabhoo, N.R. (1983) *Meloidogyne* (Nematoda, Meloidogynidae) induced root galls of the banana plant *Musa paradisiaca* a study of histopathology. *Proceedings of the Indian Academy of Sciences, Animal Sciences* 92, 467–473.

Suett, D.L. and Walker, A. (1988) Accelerated degradation of soil-applied pesticides – implications for UK horticulture. *Aspects of Applied Biology* 17. Association of Applied Biologists, Warwick, UK, pp. 213–222.

Summerville, W.A.T. (1939) Root distribution of the banana. *Queensland Agricultural Journal* 52, 376–392.

Swennen, R., de Langhe, E., Janssen, J. and Decoene, D. (1986) Study of the root development of some *Musa* cultivars in hydroponics. *Fruits* 41, 515–524.

Tarjan, A.C. (1961) Longevity of *Radopholus similis* (Cobb) in host free soil. *Nematologica* 6, 170–175.

Tarte, R. and Pinochet, J. (1981) *Problemas Nematologicos del Banano. Contribuciones Reciente a su Conocimiento y Combate.* Union Paises Exportadores Banano (UPEB), Panama.

Tarte, R., Pinochet, J., Gabrielli, C. and Ventura, O. (1981) Differences in population increase, host preferences and frequency of morphological variants among isolates of the banana race of *Radopholus similis. Nematropica* 11, 43–52.

Taylor, A.L. and Loegering, W.Q. (1953) Nematodes associated with root lesions in abaca. *Turrialba* 3, 8–13.

Ternisien, E. (1989) Etude des rotations culturales en bananeraie. 2. Impact des cultures de rotation sur la production bananière et l'état sanitaire du sol. *Fruits* 44, 445–454.

Ternisien, E. and Ganry, J. (1990) Rotations culturales en cultures bananière intensive. *Fruits (Spécial Bananes)* 45, 98–102.

Thomason, I.J. and Caswell, E.P. (1987) Principles of nematode control. In: Brown, R.H. and Kerry, B.R. (eds) *Principles and Practice of Nematode Control in Crops*. Academic Press, Sydney, Australia, pp. 87–130.

Tripathi, L. (2003) Genetic engineering for improvement of *Musa* production in Africa. *African Journal of Biotechnology* 2, 503–508.

Valette, C., Nicole, M., Sarah, J.L., Boisseau, M., Boher, B., Fargette, M. and Geiger, J.P. (1997) Ultrastructure and cytochemistry of interactions between banana and the nematode *Radopholus similis*. *Fundamental and Applied Nematology* 20, 67–77.

Valette, C., Mounport, D., Nicole, M., Sarah, J.-L. and Baujard, P. (1998a) Scanning electron microscope study of two African of *Radopholus similis* (Nematoda, Pratylenchidae) and proposal of *R. citrophilus* as a junior synonym of *R. similis*. *Fundamental and Applied Nematology* 21, 137–144.

Valette, C., Andary, C., Geiger, J.P., Sarah, J.L. and Nicole, M. (1998b) Histochemical and cytochemical investigation of phenols in roots of banana infected by the burrowing nematode *Radopholus similis*. *Phytopathology* 88, 1141–1148.

Van den Bergh, I., Nguyet, D.T.M., Tuyet, N.T., Nhi, H.H. and De Waele, D. (2002) Screening of Vietnamese *Musa* germplasm for resistance to root-knot and root lesion nematodes in the glasshouse. *Australasian Plant Pathology* 31, 363–371.

Varghese, K.C. and Nair, M.R.G.K.J. (1968) Studies on the population fluctuations of soil nematodes associated with banana in Kerala State. *Agricultural Research Journal Kerala* 6, 108–112.

Vianne, N., Duran, L.F., Rivera, J.M., Duenas, J., Rowe, P. and De Waele, D. (2003) Responses of banana and plantain cultivars, lines and hybrids to the burrowing nematode *Radopholus similis*. *Nematology* 5, 85–98.

Vilardebó, A. (1959) Note sur la lutte contre les nematodes du bananier en Guinee. *Fruits* 14, 125–126.

Vilardebó, A. (1974) Methode d'essai d'efficacite pratique de nematicides etudies sur *Radopholus similis* Cobb en bananeraie. *Societe Francaise de Phytiatrie and Phytopharmacie* Comm. Essais biol. No. 49, p. 29.

Vilardebó, A. and Guérout, R. (1976) Nematode species in West Africa, Madagascar and Reunion, with some comments on their biology. *Nematropica* 6, 53–54 (abstract).

Vrain, T.C. (1986) Role of soil water in population dynamics of nematodes. In: Leonard, K. and Fry, W. (eds) *Plant Disease Epidemiology*. MacMillan, New York, pp. 101–128.

Walker, P.T., Hebblethwaite, M.J. and Bridge, J. (1984) *Project for Banana Pest Control and Improvement in Tanzania*. EEC Report for the Government of Tanzania. Produced by Tropical Development and Research Institute, London.

Weckx, G. (1982) Invloed van Mulching en Minerale Bemesting op her Wortelstelsel van Plantaan-banaan. Eindwerk, Faculty of Agriculture, Catholic University of Louvain.

Wehunt, E.J., Hutchinson, D.J. and Edwards, D.I. (1978) Reaction of banana cultivars to the burrowing nematode, *Radopholus similis*. *Journal of Nematology* 10, 368–370.

Whyte, E.B. and Gowen, S.R. (1974) Recovery of nematodes from banana roots and soil samples. *Nematropica* 4, 27–31.

Wilson, G.F., Swennen, R. and de Langhe, E. (1986) Effects of mulch and fertilizer on yield and longevity of a medium and giant plantain and a banana cultivar. *Proceedings of Third Conference of the International Association for Research on Plantain and Bananas*, 1985, Abidjan, Ivory Coast, pp. 109–111.

Zem, A.C., Alvez, E.J., Lordello, L.G.E. and Montero, A.R. (1981) Susceptibilidade das bananeiras Prata e Mysore aos nematoides *Radopholus similis* e *Helicotylenchus multicinctus*. *Anais Escola Superior de Agricultura 'Luis de Queroz'* 38, 569–578.

Zimmerman, A.W.P. (1898) Die nematoden der koffiewortels. *Mededelingen's Lands Plantentuin (Buitenzorg)* 27, 1–64.

Zuckerman, B.M. and Strich-Harari, D. (1963) The life stages of *Helicotylenchus multicinctus* (Cobb) in banana roots. *Nematologica* 9, 347–353.

17 Nematode Parasites of Sugarcane*

Patrice Cadet[1] and Vaughan W. Spaull[2]

[1]*Institut de Recherche pour le Développement (IRD), 213 Rue La Fayette, 75480 Paris, Cedex 10, France;* [2]*South African Sugarcane Research Institute, Private Bag X02, Mount Edgecombe 4300, South Africa*

Sugarcane is one of the few crops to provide commercial quantities of food, fibre and fuel. It is grown in more than 80 countries throughout the tropics and subtropics, and in some of these countries it is the principal source of revenue, for example, in the Dominican Republic, Jamaica, Mauritius and Swaziland. The main product of sugarcane is, of course, sugar, the name given to crystals of sucrose. In the 1998/99 season, annual world production of cane sugar exceeded 96,000,000 t, with Brazil and India being the largest producers (Table 17.1).

In many countries, a significant proportion of the sugarcane is used by peasant farmers to produce crude sugar, known as jaggery or panela (Smith, 1978; Sawhney, 1997). More than half of the cane grown in Brazil is used to produce ethyl alcohol (Schmitz *et al.*, 2003). Such alternative uses, together with the area harvested each year, explain the low area to sugar production ratio for some countries.

Sugarcane is a tall, perennial, thick-stemmed grass. Modern cultivars are complex hybrids between *Saccharum officinarum* L. and *S. spontaneum* L. (Butterfield *et al.*, 2001). The centre of origin of these species is probably the New Guinea–east Indonesia area.

Sugarcane plants grow in tufts or stools composed of varying numbers of stalks. At maturity, the stalks are approximately 2–3 m in length and 20–30 mm in diameter. The stalk is composed of a series of nodes each of which carries an axillary bud and a leaf. Carbohydrate is stored in the internodes primarily as sucrose. Modern cultivars of sugarcane normally contain between 11 and 14% sucrose.

Cultivation

Sugarcane is propagated vegetatively by planting setts (stalk cuttings) with two or more nodes. Within a few days, roots develop from primordia around the nodes of the setts. These sett roots support the initial growth of the primary shoots which develop from axillary buds on the setts (Fig. 17.1). Subsequently, tillers arise and these and primary shoots develop shoot roots which soon replace the sett roots. As the shoots grow, they compete for light and space, and a notable proportion die. Those that survive increase in diameter and length. Depending on temperature and available soil moisture, the crop is harvested after

*A revision of the chapter by V.W. Spaull and P. Cadet.

 Plant Parasitic Nematodes in Subtropical and Tropical Agriculture, 2nd Edition (eds M. Luc, R.A. Sikora, J. Bridge)

Table 17.1. Area under sugarcane in the year 2000 and annual sugar production in 1998/99 for the top ten sugar-producing countries in the world.

Country	Area under sugarcane (ha)	Tonnes sugar/ year	Ratio of tonnes sugar/ha
Brazil	5,678,000	19,376,000	3.4
India	4,000,000	16,900,000	4.2
China	1,060,000	8,341,000	7.9
Thailand	927,000	5,478,000	5.9
Australia	508,000	5,150,000	10.1
Mexico	661,000	4,987,000	7.5
Cuba	1,300,000	3,851,000	3.0
Pakistan	1,150,000	3,838,000	3.3
USA	385,800	3,234,000	8.4
South Africa	424,000	2,765,000	6.5
World	19,639,800	96,544,000	4.9

From Licht (2002) and Tew (2003).

approximately 12–24 months, when the sucrose content of the stalk approaches its maximum concentration.

Soon after harvest, new shoots emerge from axillary buds on the stubble and give rise to the ratoon crop. Initially the young shoots are dependent upon the roots of the previous crop (stool roots) but these are replaced by new shoot roots (Fig. 17.1). The crop cycle of sugarcane is normally composed of the plant and, typically, 2–4 ratoon crops. However, the actual number of ratoons harvested before the crop is replaced depends on growing conditions and local cultural practices. There is usually a decline in yield after the first or second ratoon crop. A large proportion of the world's sugarcane is grown under irrigated conditions (Smith, 1978).

Nematodes of Sugarcane

Nematode diversity in sugarcane is greater than in most other cultivated crops, with more than 310 species of 48 genera of endo- and ectoparasitic nematodes having been recorded from its roots and/or rhizosphere. Certain genera are particularly widespread in cane fields, i.e. *Pratylenchus* (with at least 20 species reported from sugarcane worldwide), *Helicotylenchus* (35 spp.) and *Tylenchorhynchus* (36 spp.); several others are common locally, e.g. *Meloidogyne* (seven spp.), *Xiphinema* (52 spp.), *Hoplolaimus* (11 spp.), and *Paratrichodorus* and *Trichodorus* (nine spp.) (Table 17.2).

Sugarcane is normally grown as a continuous monoculture with usually no more than a few months' break between removing the old ratoon crop and replanting the field. Thus conditions tend to favour the development of relatively large populations of selected species. Those most frequently cited as highly pathogenic to sugarcane are *Pratylenchus zeae*, *Meloidogyne incognita* and *M. javanica*.

Pratylenchus

Collectively, species of *Pratylenchus* are the most common plant parasitic nematodes associated with sugarcane (Table 17.2); worldwide, *P. zeae* is the species most frequently encountered.

Symptoms of damage

P. zeae causes conspicuous red, reddish-purple or brown lesions on the roots of cane (Stirling and Blair, 2000). The lesions become necrotic and turn purplish-black, causing the root system to

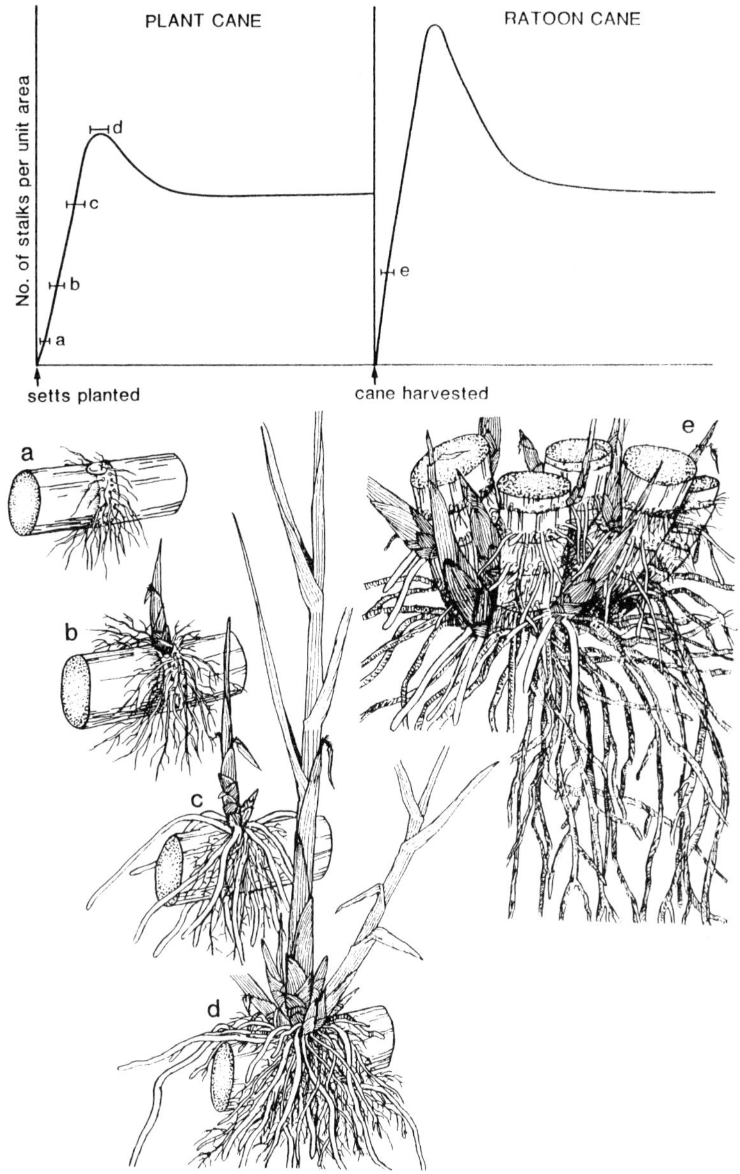

Fig. 17.1. Sequence of events in the early stages of development of plant and ratoon cane. (a) Appearance of sett roots. (b) Emergence of the bud and development of the primary shoot. Establishment of the sett root system. (c) Appearance of the shoot roots on the primary shoot and initiation of tillers. (d) Maximum density of tillers, establishment of the shoot root and disappearance of the sett root system. (e) Stool of ratoon cane showing new shoot arising from lateral buds on the stubble. Shoot roots develop at the base of the new shoots and eventually replace the stool roots (i.e. the shoot roots of the previous crop).

darken in colour. This is associated with a reduction in shoot and root mass and stalk length, as well as a yellowing of the leaves (Valle-Lamboy and Ayala, 1980). Fewer stalks developed on sugarcane growing in microplots infested with *P. zeae* than in uninfested plots (Tarte *et al.*, 1977). *P. zeae* may also adversely affect

Table 17.2. Frequency of occurrence of the more common plant parasitic nematodes associated with sugarcane (%).

	Pratylenchus	*Helicotylenchus*	*Tylenchorhynchus*	*Meloidogyne*	Trichodorids[a]	*Xiphinema*	*Hoplolaimus*	Criconematids[b]	Longidorids[c]	*Paratylenchus*	*Rotylenchulus*	*Hemicycliophora*	*Scutellonema*	Survey
Australia	100	78	79	69	82	37	+[d]	31	3	1	26	2	+	547 fields
Barbados	64	77	16	27	9	+	0	30	+	0	35	0	0	45 fields
Brazil	80	90	+	+	+	+	+	+	0	0	0	+	0	800 samples
Burkina Faso	89	99	74	71	52	74	92	25	0	52	+	+	+	47 fields
Colombia	94	87	42	17	4	0	0	13	31	0	6	0	0	74 fields
Costa Rica	45	88	17	28	22	13	0	35	33	2	0	1	0	146 samples
Cuba	100	100	35	85	25	0	5	25	0	50	25	0	10	20 fields
Egypt	93	43	55	4	0	0	12	55	24	0	45	2	0	550 samples
Fiji	82	79	38	23	7	10	8	28	7	8	21	0	0	390 samples
India	45	89	86	+	+	6	92	+	18	7	+	+	0	150 localities
Côte d'Ivoire	93	97	88	71	27	34	9	48	+	76	+	0	63	49 fields
Japan	62	73	58	47	+	+	+	+	0	6	21	0	+	97 fields
Malaysia	100	50	71	36	14	14	71	36	43	21	+	0	7	14 fields
Mauritius	8	27	6	17	15	61	0	21	28	0	6	10	22	253 samples
Mexico	63	13	37	0	0	8	54	17	0	4	0	0	0	24 fields
Peru	68	93	94	72	+	+	+	+	+	0	0	83	0	10,500 ha
Philippines	99	75	74	12	4	83	45	4	48	2	15	18	11	168 samples
South Africa	96	95	30	71	93	94	8	75	72	9	99	11	97	124 fields
Taiwan	85	71	89	62	76	16	50	25	1	+	0	+	0	17,000 samples
Trinidad	100	91	91	0	0	9	18	0	0	0	0	0	0	11 fields
USA	79	32	93	9	49	0	0	88	0	0	0	0	0	93 fields
Venezuela	87	+	87	+	+	0	0	100	0	0	+	0	0	94 samples
Zimbabwe	85	41	67	41	81	78	19	22	22	0	15	11	59	27 sites
Average	79	72	60	38	31	28	25	36	17	11	17	7	13	Numerical data only

[a]Trichodorids = *Trichodorus* and *Paratrichodorus*.

[b]Criconematids = *Criconemoides* and related genera.

[c]Longidorids = *Longidorus* and *Paralongidorus*

[d]+ = present in survey; 0 = not recorded.

From Lamberti *et al.* (1987), Spaull and Cadet (1990), Blair *et al.* (1999a,b) and Bond *et al.* (2000).

cane quality (Sujatha and Mehta, 1994). Onapitan and Amosu (1982) found that *P. brachyurus* caused damage to the vascular system and destruction of cortical cells, but it did not affect root or shoot mass. In an earlier study, *P. brachyurus* was reported to affect the length and mass of stalks, although no symptoms of damage were evident on the roots (Koike and Roman, 1970).

Environmental factors affecting parasitism and pathogenicity

The number of individuals and frequency of occurrence of species of *Pratylenchus* have been reported to be greater in clay soils than in the light soils in West Africa, Taiwan, Louisiana in the USA and the southern states of India (Hu *et al.*, 1968; Cadet, 1987; Mehta, 1992; Bond *et al.*, 2000), although in Australia and South Africa they were widespread in all soils (Spaull, 1981; Blair *et al.*, 1999a,b). In Martinique, the spatial distribution of *P. zeae* was homogenous in a soil with a normal A horizon but, where this horizon had been removed by levelling, the distribution was more concentrated along the sugarcane rows (Delaville *et al.*, 1996).

P. zeae was more numerous in the sett roots than the shoot roots of plant cane in Burkina Faso and South Africa (Cadet and Spaull, 1985), and in South Africa more individuals were recovered from the shoot roots than from the stool roots of ratoon cane (Spaull and Cadet, 1991). In Australia, densities of *P. zeae* were greater in the roots of plant cane than of ratoon cane, but those in the soil did not differ significantly between crops (Blair *et al.*, 1999a). Penetration of roots by *P. zeae* was greater at higher temperatures (26–28°C), irrespective of soil type (Mehta and Sundararaj, 1990).

Large differences were found in the suitability of different cane cultivars as hosts to *P. zeae*, with many more individuals being recovered from the roots of some cultivars than others (Dinardo-Miranda, 1994; Mehta and Somasekhar, 1998; Blair *et al.*, 1999a).

Disease complexes

The pathogenicity of *P. zeae* to sugarcane is affected by other organisms. Thus, in combination with *Pythium graminicola*, *M. incognita* or both these organisms simultaneously, *P. zeae* had significantly less effect on the mass of cane roots than in their absence (Valle-Lamboy and Ayala, 1980); necrosis of the roots was more than halved when one or both of the other pathogens was also present.

Economic importance

Damage thresholds are not well defined because extraction methods affect nematode counts and environmental factors affect the response to control measures. However, data from Australia suggest that where the number of individuals in the soil was above 100 per 200 g of soil before planting, or above 250 per 200 g of soil or per g of dry roots at mid-season, there may be a significant reduction in cane yield (Stirling and Blair, 2000). Because these levels are exceeded in many fields in Australia, *P. zeae* is considered the primary nematode pathogen of sugarcane in that country (Blair *et al.*, 1999b). *P. zeae* is also of particular importance in Panama (Pinochet, 1987), Burkina Faso and South Africa (Cadet and Spaull, 1985), the USA (Birchfield, 1984) and in the Brazilian state of Pernambuco (de Moura *et al.*, 1999). According to Williams (1963), *P. zeae* was 'one of the most ubiquitous and abundant species associated with sugarcane roots in Mauritius', although, some 20 years later, Lamberti *et al.* (1987) found only *P. brachyurus*, which occurred in just 8% of the sites sampled (Table 17.2).

Meloidogyne

M. incognitia and *M. javanica* have been found in many sugarcane areas, and at least some of the numerous records of unidentified *Meloidogyne* probably refer to one or both of these species. Five other species have been identified from cane: *M. acrita*, *M. arenaria*, *M. hispanica*, *M. kikuyensis* and *M. thamesi*, but none is widespread.

Symptoms of damage

The symptoms of damage are distinct but are usually less easily diagnosed than in many other susceptible crops. Galls formed by *M. incognitia* and *M. javanica* develop on the tips of the sett roots and young shoot roots. They are often small and discrete and not easily detected, except in young plant cane. Williams (1969) illustrated elongated swellings on the tips of sugarcane roots and the proliferation of lateral roots immediately proximal to the gall. In old suberized roots, females may develop at various positions along the root without inducing galling (Martin, 1967). In pot experiments, *M. incognita* and *M. javanica* reduced the top weight and root weight of sugarcane (Valle-Lamboy and Ayala, 1980; Novaretti, 1981). Species of *Meloidogyne* may also reduce the number of tillers developed by sugarcane (Salawu, 1986).

Environmental factors affecting parasitism and pathogenicity

Species of *Meloidogyne* are found more frequently in sandy soils than in finer texture soils (Spaull, 1981; Blair *et al.*, 1999a,b). Greater populations of *M. incognita* and *M. javanica* were recorded in sett roots than shoot roots of plant cane (Cadet and Spaull, 1985).

Pasteuria penetrans was recorded in *Meloidogyne* from a number of sugarcane fields in South Africa (Spaull, 1984). Besides *M. incognita* and *M. javanica*, *P. penetrans* also infected *M. hispanica* but not *M. kikuyensis*. *P. penetrans* was also recorded from *M. incognita* and/or *M. javanica* from sugarcane fields in Mauritius, Louisiana and Papua New Guinea (Williams, 1967; Birchfield, 1984; Bridge, 1986).

Populations of root knot nematode may be influenced by the presence of phytopathogenic fungi. Thus, far fewer *M. javanica* were recorded from the roots of sugarcane infected with the seedling blight fungus, *Curvularia lunata*, than from uninfected plants (Khurana and Singh, 1971). Conversely, the presence of other pathogens favoured colonization of sugarcane roots by *M. incognita*, many more galls being produced in the presence of *P. graminicola* than when the fungus was absent; and when *P. zeae* was also present even more galls were developed, although in both cases the size of the galls was smaller than normal (Valle-Lamboy and Ayala, 1980).

Disease complexes

The effect of *M. javanica* and *C. lunata* on sugarcane was greater when the two organisms were inoculated together than when either was inoculated alone (Khurana and Singh, 1971). A similar interaction was recorded between *M. incognita* and *P. graminicola* on sugarcane seedlings (Apt and Koike, 1962). However, in another study, the combination of *M. incognita* plus *P. graminicola*, *M. incognita* plus *P. zeae* or all three species together had significantly less effect on root mass of sugarcane than when either of the nematodes was acting alone (Valle-Lamboy and Ayala, 1980).

The effect of the combination of *M. incognita* race 1 and ratoon stunting disease (*Leifsonia xyli* subsp. *xyli*) on sugarcane in pots was additive rather than synergistic (Regis and de Moura, 1989).

Economic importance

The same limitations on the use of damage thresholds given for *Pratylenchus* apply to species of *Meloidogyne*. Estimates for *M. javanica* in Australia indicate that where the number of individuals in the soil exceed 100 per 200 g of soil before planting, or exceed 200 per 200 g of soil or per g of dry roots at mid-season, there may be a significant reduction in cane yield (Stirling and Blair, 2000).

Together with *P. zeae*, *M. incognita* and *M. javanica* are probably the most important parasitic nematodes of sugarcane worldwide. Estimates of crop loss due to species of *Meloidogyne* in Mexico, Central and South America, the Caribbean and

South-east Asia ranged from 6 to 9% although these were not supported by experimental data (Sasser, 1979). Cadet and Spaull (2003) compared the yields of two cultivars in each of two field trials situated 800 m apart on similar sandy soil. The nematode fauna was similar in the two sites except that *M. javanica* occurred in one trial. Yield data over a 5 year period indicated that this species was responsible for an annual loss of 30%, equivalent to 15 t cane/ha. In addition, a model derived from the log regression curve of yields from the first to fourth ratoon showed that where *M. javanica* did not occur, the yield of untreated cane remained above a ploughout threshold of 40 t cane/ha for 6 or 8 years instead of for 3 years. Differences were much greater when a nematicide was used (Table 17.3).

Nematode communities

Attention has so far focused on species of *Pratylenchus* and *Meloidogyne*, as they are widespread on sugarcane and generally considered the most damaging plant parasitic nematodes. However, these and other nematodes associated with sugarcane rarely occur alone in the soil but are present in communities comprising a number of species. Surveys from several countries show that the number of genera present in a single soil sample ranges from one to 12 with an average of between 3.2 and 7.9 (Table 17.4). These examples demonstrate the wide diversity of plant parasitic nematodes associated with sugarcane and show that diseases caused by nematodes involve a complex of species.

Free-living nematodes are also part of the nematode community, but they generally occur in lower numbers than plant parasites. Reports from Louisiana, Brazil, Australia, Martinique and South Africa show that free-living nematodes generally represent 30–70% of the total nematode fauna (Showler *et al.*, 1990, 1991; de Moura *et al.*, 1999; Stirling *et al.*, 2001; P. Quénéhervé, Martinique, 2003, personal communication; P. Cadet and V.W. Spaull, unpublished data). Some free-living nematodes are predacious and others are a food source for fungal predators of nematodes, and this may have been one of the reasons why populations of plant parasitic nematodes were suppressed when numbers of free-living nematodes were increased by adding organic matter to soil (Stirling *et al.*, 2003).

Symptoms of damage

The symptoms of nematode damage on the roots of sugarcane are not unlike those observed on other crops. The symptoms listed under the genera in Table 17.5 are observed in pot cultures of single species. However, in field-grown cane, the roots show the combined symptoms of all the nematodes that have been feeding on

Table 17.3. Comparison of the average yield of five nematicide-treated crops and five untreated crops for two cultivars at two sites with similar sandy soil, and the number of annual ratoon crops that can be harvested before reaching a ploughout threshold of 40 t cane/ha.

		Average yield of five crops (t cane/ha)		Years before 40 t cane/ha will be reached	
Site	Cultivar	Untreated	Nematicide	Untreated	Nematicide
LM1	N12	51.2	79.2	8	22
LM2 (*M. javanica*)	N12	39.7	80.9	3	13
LM1	N16	55.4	89.4	6	43
LM2 (*M. javanica*)	N16	37.5	76.4	3	20

Meloidogyne javanica was present at the LM2 site but not at LM1.
From Cadet and Spaull (2003).

Table 17.4. Number of plant parasitic nematodes per individual soil sample from sugarcane fields from various countries.

	Minimum number of genera	Maximum number of genera	Average number of genera
Australia	2	9	5.0
Burkina Faso	3	8	6.4
Chad	3	12	7.9
Côte d'Ivoire	3	9	6.1
Martinique	1	7	4.3
Nigeria	3	10	6.0
Papua New Guinea	1	7	3.2
South Africa	3	9	5.7

Data for: Australia (G.R. Stirling, Queensland, Australia, 2003, personal communication); Burkina Faso (P. Cadet, unpublished data); Chad (G. Reversat, France, 2003, personal communication); Côte d'Ivoire and Martinique (P. Quénéhervé, Martinique, 2003, personal communication); Nigeria (Fademi *et al.*, 1997); Papua New Guinea (Bridge, 1986); and South Africa (V.W. Spaull, unpublished data).

them. Since several species cause similar damage, it is usually not possible to use the symptoms to identify the nematodes responsible. Also, the coloured lesions on the roots associated with the feeding of *P. zeae* and species of *Hoplolaimus* are not unlike the early symptoms produced by root rotting fungi such as *Pythium arrhenomanes* and *Pachymetra chaunorhiza* (Croft and Magarey, 2000; Hoy, 2000). The situation is also complicated by the natural darkening of the epidermis as the root suberizes. In India, chlorosis of the leaves is commonly attributed to nematode damage (Mehta, 1992). In addition, there are some above-ground symptoms that, although not diagnostic, are often associated with the damage caused by nematodes, i.e. the shoots are reduced in number and are stunted (Fig. 17.2), the cane is slow to develop a canopy of leaves and therefore has a more open appearance, and the leaves curl longitudinally and appear spiky. These are also symptoms of drought-stressed cane.

The pathogenicity of many of the plant parasitic nematodes found associated with sugarcane has been demonstrated in pots in the absence of other (reported) organisms (Spaull and Cadet, 1990). The application of this information to the situation in the field is fraught with difficulties as nematodes act alongside other species that can influence their behaviour (Eisenbach, 1993). In addition, attempts to view the nematode community as the functioning unit fail to encompass the myriad of other organisms that reside and interact within the rhizosphere. These include the vast numbers of microorganisms associated with cane roots, some of which are pathogenic to cane (Croft and Magarey, 2000; Hoy, 2000). Also it is as well to note that plant pathogenic bacteria, fungi and nematodes are not the only organisms that cause necrosis and impair the growth of cane roots. Such damage may also result from feeding by the larvae and adults of some Coleoptera, Diptera, Hemiptera, Coccoidea, Collembola, Thysanura and Myriapoda (Wilson, 1969). Also poor root growth may result from an imbalance of soil nutrients, e.g. high levels of aluminium or low levels of phosphorus (Humbert, 1968), or from soil compaction or poor aeration. Only *Meloidogyne* can be diagnosed with confidence because the female may be observed in galls by dissecting the root. In other cases, it may not be prudent to link so-called typical symptoms with one or other species of nematode, as nematodes are only one component of a complex of factors that affect root growth.

Table 17.5. Symptoms of damage caused by various nematode genera on sugarcane in pots and by communities of nematodes in sugarcane fields.

	Pratylenchus[a]	*Meloidogyne*	*Helicotylenchus*	*Tylenchorhynchus*	*Paratrichodorus*	*Xiphinema*	*Hoplolaimus*	Community of nematodes in the field[b]
Reduction in shoot and root mass	+	+	+	+	+	+	+	+
Reduction in number of shoots	+	+						+
Necrosis of cells in root cortex	+		+	+	+	+	+	+
Red/purple/brown/ pink lesions on roots	+		+				+	+
Fewer roots/sparse root system	+		+		+	+	+	+
Stunted roots			+	+	+	+		+
Distorted roots	+		+					+
Galls		+				+		+
Blackening of the roots								+

Not all the symptoms are observed in all situations.
[a]Columns 2–8: summary of data from Spaull and Cadet (1990).
[b]Column 9: unpublished observations on symptoms found in sugarcane fields in: Australia (G.R. Stirling, Queensland, Australia, 2003, personal communication), Brazil (W.R.T. Novaretti, Brazil, 2003, personal communication), Burkina Faso (P. Cadet, unpublished data), Côte d'Ivoire (P. Quénéhervé, Martinique, 2003, personal communication) and South Africa (V.W. Spaull, unpublished data).

Fig. 17.2. Effect of nematodes on the root system of sugarcane collected from a nematicide trial. Plots of the the taller, nematicide-treated cane yielded 80 t cane/ha; plots of untreated cane yielded 21 t cane/ha.

Abiotic soil factors affecting parasitism and pathogenicity

Numerous factors affect the distribution of individual species of plant parasitic nematodes. As a consequence, nematode community composition varies widely from country to country, from one soil type to the next and even over short distances within a field. Communities in sandy soils are more likely to include larger populations of species of *Meloidogyne*, *Hoplolaimus*, *Trichodorus* and/or *Paratrichodorus* than those in the finer textured soils (Spaull, 1981; Mehta, 1992; Blair *et al.*, 1999a,b; Bond *et al.*, 2000). A study conducted in South Africa showed that the distribution of communities containing larger populations of *Meloidogyne* was restricted more by soil type than by climatic or topographic factors (Spaull *et al.*, 2003). In contrast, the distribution of species of *Pratylenchus* and *Helicotylenchus* often appears to be unrelated to soil texture (Spaull, 1981; Blair *et al.*, 1999a), although reports from the USA and India indicate that *Pratylenchus* is more numerous in clay soils (Hall and Irey, 1992; Mehta, 1992). *X. mampara*, one of two common species of *Xiphinema* in South Africa, was found more frequently in clay loams and clays, whereas *X. elongatum* tended to prefer the sandy soils (Spaull and Heyns, 1991). *X. insigne*, the most common species of *Xiphinema* in cane fields in the central and southern Negros Occidental in the Philippines, was as abundant in clay soils as in sandy loams (Estioko and Reyes, 1984). Greater numbers of *Meloidogyne*, *Hemicycliophora*, *Hoplolaimus* and *Paratrichodorus* were recorded in sandy soils compared with soils with high levels of organic matter (Hall and Irey, 1992).

Soil texture seemingly has the greatest influence on, or is the factor most correlated with the pathogenicity of the nematode community. The effect of nematodes on sugarcane is greatest in light textured soils and decreases with increasing clay content (Fig. 17.3). In ratoon cane in Burkina Faso and Côte d'Ivoire, there is no significant response to treatment with a nematicide, irrespective of soil type.

The effect of soil texture on pathogenicity is partly due to the ease of movement of

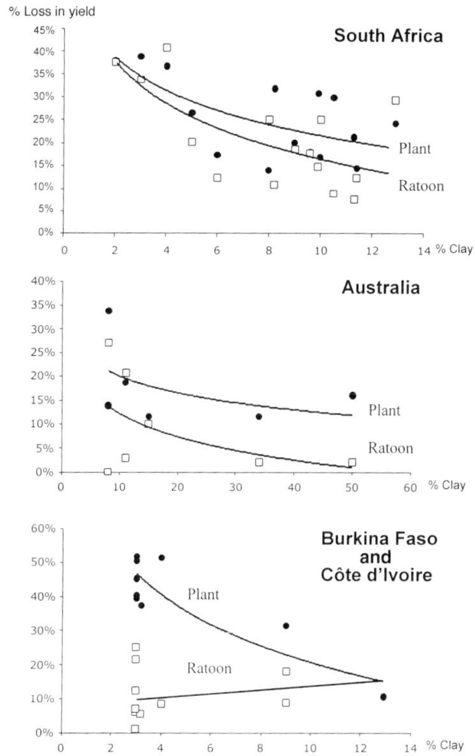

Fig. 17.3. Percentage loss in yield due to nematodes in Australia, South Africa and West Africa according to soil texture shown separately for plant and ratoon crops.

nematodes in sandy soils. In plant cane in West Africa, the invasion of the sett roots by endoparasites was much more rapid in the coarser textured soils. The consequent damage to these roots delayed and disrupted the normal tillering process, with the result that the cane developed fewer stalks (Cadet *et al.*, 1982). In finer textured soil, the endoparasites invaded the sett roots more slowly and caused less damage during the tillering phase. However, the main reason that nematodes have a greater impact in sandy soils is that they have much lower water-holding capacities than heavy textured soils. Plant parasitic nematodes feed on and limit the growth of roots of cane, but the effect of the restricted root system on the uptake of water will be greater in a soil with a low water-holding capacity (Wallace, 1973).

Nematode community structure may be influenced by both altitude and temperature. Thus, in South Africa, communities with larger populations of *P. zeae* and *X. elongatum* tended to occur at altitudes below 300 m where average annual temperatures exceeded 20°C. The reverse was true for communities with larger numbers of *H. dihystera* and a species of *Rotylenchus* (Spaull *et al.*, 2003). In Mauritius, *X. elongatum* was largely confined to altitudes below 250 m where rainfall is less than 2000 mm/year. It was less commonly found in the central, more elevated part of the island, where rainfall was greater and *X. krugi* was widespread (Williams and Luc, 1977; Lamberti *et al.*, 1987). Similarly, *X. americanum s.l.* was not found in sugarcane fields in Hawaii above an altitude of about 230 m (Anonymous, 1961).

Biotic factors affecting parasitism and pathogenicity

As with other crops, the reproductive success of plant parasitic nematodes on sugarcane is affected by a number of biotic factors. The plant itself is the main factor, as there are large differences in the suitability of different cultivars as hosts to certain species (Dinardo-Miranda, 1994; Mehta *et al.*, 1994a). In South Africa, N12 was particularly suitable as a host to *H. dihystera*, and N19 and N27 to *M. javanica* (Rutherford *et al.*, 2002) (Fig. 17.4). However, in Queensland, Australia, cultivars had no effect on the densities of *P. zeae*, except that population densities were lower on CP 51-21 (Blair *et al.*, 1999a).

Stress induced by weeds and viruses may affect the capacity of nematodes to multiply on sugarcane. Thus, Showler *et al.* (1990) found that populations of *Tylenchorhynchus annulatus* on sugarcane were positively correlated with levels of four amino acids produced in response to stresses induced by sugarcane mosaic virus and weeds. Significant correlations were also reported between *P. zeae*, *Criconemoides* spp. and *Helicotylenchus* spp. and a number of free amino acids.

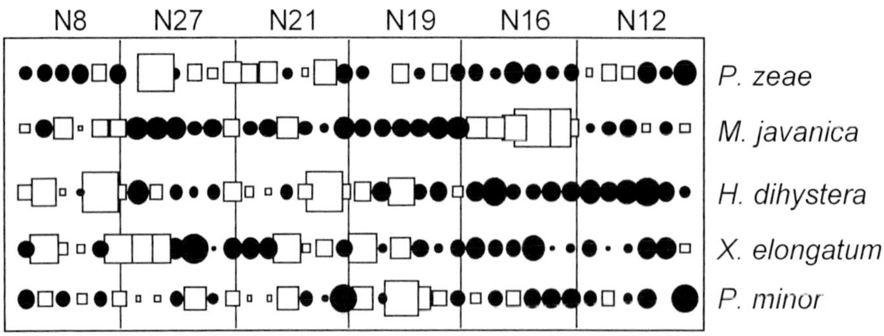

Fig. 17.4. Contrasting population size of the common species of nematodes in six replicate plots of each of six cultivars in a field trial in South Africa. Circles denote population levels above average and squares below average. The size of the symbol is proportional to the absolute value of the component. For each species, differences between replicates per cultivar indicate the natural variation in the field (from Rutherford *et al.*, 2002).

Parasitism of sugarcane by nematodes is also influenced by the crop stage. In most countries, cane is normally cropped over a number of ratoons before the crop is destroyed and the field replanted. During this period, the soil remains largely undisturbed and the balance between the nematode populations within the community may change. In Burkina Faso, over a period of five crops, from planting to the fourth ratoon, the proportion of *Hoplolaimus* in the roots increased from 10% of the endoparasites to about 85%. Associated with this was a decline in the proportions of *Meloidogyne* and, especially, *Pratylenchus* (Cadet, 1985). In the same study, it was found that from the time that the cane was planted to the end of the third ratoon, the numbers of ectoparasites, mainly *Helicotylenchus*, increased fivefold and then declined in the fourth ratoon. The numbers of endoparasites increased in the sett and shoot roots of the plant crop. Thereafter, they increased erratically to the fourth ratoon, with numbers increasing at the start of each ratoon and then declining. Overall, there was an increase from the first to fourth ratoon. Similarly, in the USA, the size of the nematode community was greater in ratoon crops than in the plant crop (Bond *et al.*, 2000). However, in Australia, densities of *P. zeae* were greater in the plant crop than in the first ratoon and there was no consistent effect of crop stage on the other species (Blair *et al.*, 1999a).

Another factor affecting parasitism is the composition of the nematode community. Certain species interfere with each other to the extent that some coexist less frequently than others. In India, Sujatha and Mehta (1993, 1995) found that *P. zeae* occurred less frequently in communities that did not include *Hoplolaimus indicus* and/or *T. annulatus* and was more common when *H. dihystera* was present (Table 17.6).

In sugarcane fields in Côte d'Ivoire, the abundance of *Meloidogyne* was correlated with that of *Paratylenchus* and *Criconemella* and their absence with the presence of *Pratylenchus*. However, numbers of *Meloidogyne* and *Pratylenchus* were not correlated. The latter was more abundant in plant cane and in clay soils, whereas *Meloidogyne* was more abundant in sandier soils and in ratoons (Cadet and Debouzie, 1990).

The amount of damage caused by a nematode may be affected by the occurrence of certain other species in the community. Interactions between *M. incognita* and *P. zeae* have been mentioned previously, while Sujatha and Mehta (1997) observed that concomitant inoculation of *M. javanica* and *P. zeae* caused less crop loss than when each species was cultured in isolation.

The pathogenicity of a nematode community to sugarcane may be reduced when *H. dihystera* is the dominant

Table 17.6. Positive and negative associations between species in sugarcane fields in India.

	P. zeae	*H. dihystera*	*T. annulatus*
Hoplolaimus indicus	–	–	+
Tylenchorhynchus annulatus	–	–	
Helicotylenchus dihystera	+		

From Sujatha and Mehta (1993, 1995).

ectoparasite. Thus, Cadet *et al.* (2002) found that within a field trial with 29 50-m^2 plots, significantly greater yields were recorded from plots that had a higher proportion of *H. dihystera*, relative to the other ectoparasites, and a lower proportion of *M. javanica*, relative to the other endoparasite (Table 17.7). This was true for all the plots, whether or not they had been treated with a nematicide. A similar association of *H. dihystera* with better yielding cane was reported in Burkina Faso by Cadet (1986a). He found that the number of *H. dihystera*, the dominant ectoparasite, was directly proportional to the yield of cane.

Nematode–sugarcane interaction

The roots of sugarcane are normally attacked simultaneously by a number of nematode species, some or all of which may cause serious damage. However, the nematode–plant interaction is complicated by the fact that one root system is replaced by another during the growth of the crop, and this affects nematode population dynamics. To understand the importance of the nematodes and to explain the mechanisms of damage, it is necessary to consider the different components of the nematode community in relation to both the development of the roots and the evolution of those plant parameters that contribute to yield. Yield of sugarcane is a function of the number, length and diameter of the stalks. Root damage by nematodes results in a reduction in the number and length of stalks; occasionally it influences stalk diameter and sucrose content.

Plant cane

Based on studies in Burkina Faso and South Africa, Cadet and Spaull (1985) found that in plant cane the reduction in

Table 17.7. Association between nematode species and sugarcane yield in a field trial in South Africa. Relative proportions (%) calculated separately for ectoparasites and endoparasites (Cadet *et al.*, 2002).

	Low yielding plots (n = 14) 78 t cane/ha				High yielding plots (n = 15) 118 t cane/ha			
	Yield (%)	SE	Abundance/100 cm^3 or g dry weight	SE	Yield (%)	SE	Abundance/100 cm^3 or g dry weight	SE
Roots								
Pratylenchus zeae	14	8	15	7	44	13	15	7
Meloidogyne sp.	86	8	233	92	56	13	96	63
Soil								
Pratylenchus zeae	22	9	9	3	55	14	13	5
Meloidogyne sp.	78	9	86	32	45	14	14	9
Helicotylenchus dihystera	37	3	238	22	55	3	469	67
Xiphinema elongatum	16	2	106	18	10	2	80	13
Paratrichodorus sp.	48	3	316	28	35	4	284	36

the number of stalks takes place primarily during the period of maximum tiller development, i.e. while the cane plant is largely dependent upon the sett root system. A reduction in the length of stalks may also be apparent at this time and, in the presence of certain nematode communities, this increases in magnitude through to harvest. Stalk length may thus be affected by damage to both the sett and the shoot roots.

The results of a number of field trials show that, in Burkina Faso, crop loss in plant cane was due more to a reduction in the number of stalks than to a reduction in the length of stalks, while the reverse was true in South Africa (Fig. 17.5). To explain this difference and to elucidate the roles played by the nematodes in limiting yield of plant cane in the two localities, Cadet and Spaull (1985) related the patterns of change in the nematode populations to the patterns of change in the development of the sugarcane crop. They deduced that:

1. In both Burkina Faso and South Africa, damage to the sett roots by large numbers of *Meloidogyne* and *Pratylenchus* delayed the emergence, and retarded the development of many of the primary shoots, which either produced fewer tillers or were unable to compete successfully with those that developed more rapidly.
2. The suppression of tillering was greater in Burkina Faso than in South Africa because, in the former locality, there was a much greater rate of invasion of the sett roots by endoparasites.
3. *Xiphinema*, and probably *Trichodorus* and *Paratrichodorus*, caused extensive damage to the shoot roots in South Africa which restricted water uptake and thus limited stalk elongation.
4. The dominant ectoparasite in Burkina Faso, *Helicotylenchus dihystera*, had little effect on sugarcane compared with species of *Xiphinema* and trichodorids.
5. Although nematodes caused some damage to the shoot roots in Burkina Faso, this had less effect on water uptake and thus on stalk elongation than in South Africa, because the cane was irrigated.

Ratoon cane

Although plant parasitic nematodes have a marked effect on the plant crop in West Africa, they have little influence on the following ratoon crops (Cadet, 1985). In Brazil and Australia, nematodes have some impact in ratoons, whereas in South Africa, ratoon cane is almost as badly affected by nematodes as is plant cane (Table 17.8). As was done with the plant crop, an attempt was made to understand the relationship between nematodes and ratoon cane by monitoring the nematode populations and the development of the cane in Burkina Faso and South Africa (Spaull and Cadet, 1991). It was deduced that:

1. The notable reduction in length of stalks that occurred in South Africa could be attributed to the considerable damage to the shoot roots caused by *Xiphinema* and probably *Paratrichodorus* species. These ectoparasites were also thought to be responsible for the reduction in the number of stalks in South Africa, since large numbers were present in the soil during the initial period of shoot development. During this short critical period, very few endoparasites were present in the roots.
2. In Burkina Faso, nematodes have relatively little effect on either the number or length of stalks. This was not altogether unexpected since very few endoparasites were recovered from the roots during the entire period of shoot establishment and, although present in large numbers, the dominant ectoparasite *H. dihystera* is considered a weak pathogen of sugarcane.
3. The roots of ratoon cane were not attractive to or suitable for the endoparasitic nematodes, judging from their inactivity during the early stage of growth in both localities. In South Africa, this condition persisted for only 4 weeks, but in Burkina Faso it lasted much longer. It was tentatively suggested that the lack of attraction by the roots was due to the initial inherent, low level of activity of the root system of young ratoon cane. That the activity of the roots in Burkina Faso should have remained at a low level for so long was

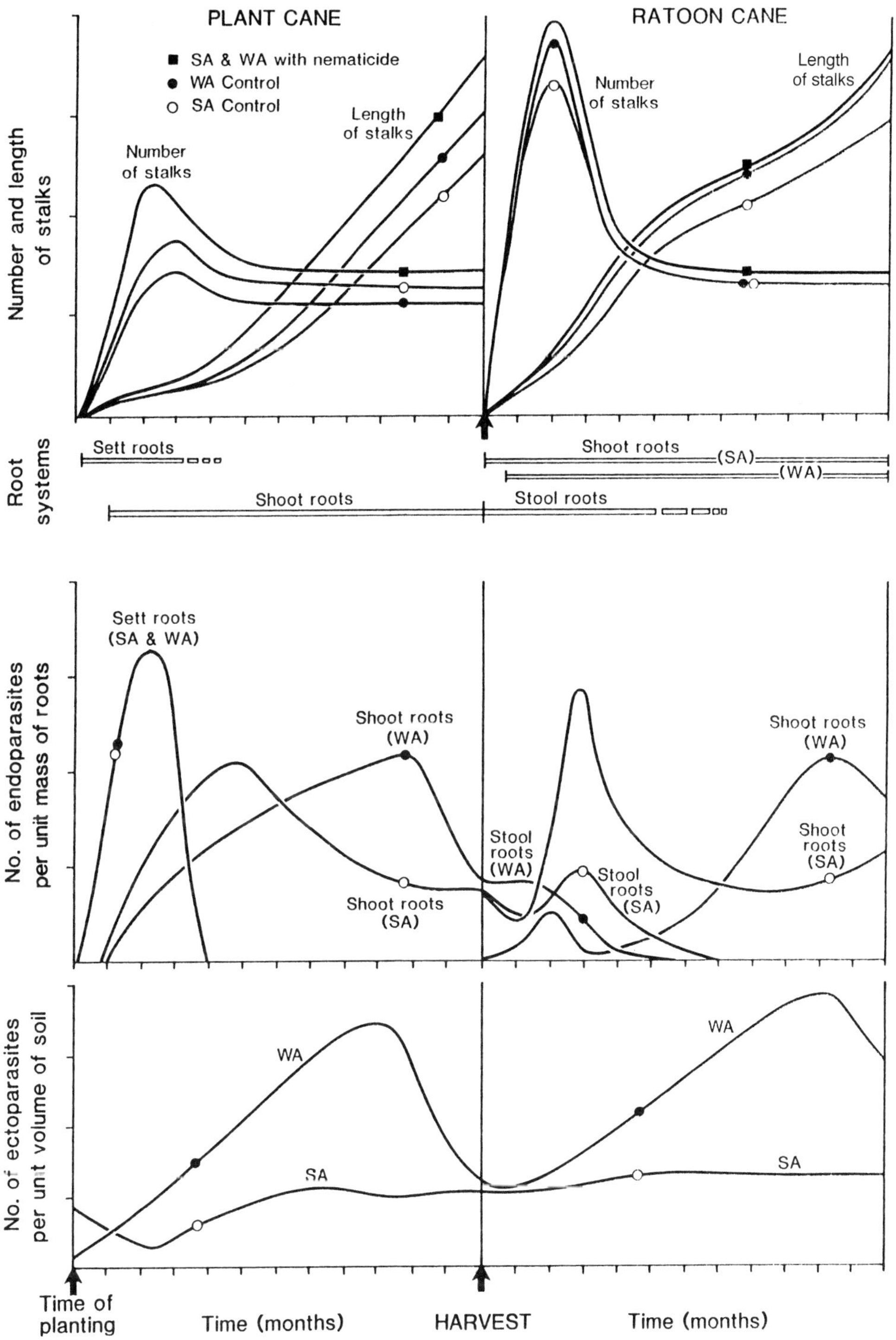

Fig. 17.5. Pictorial representation of the patterns of change in the numbers of nematodes in relation to the patterns of change in the development of sugarcane in South and West Africa.

Table 17.8. Response (%) to treatment with nematicide in four countries.

	Australia	Brazil	Burkina Faso and Côte d'Ivoire	South Africa
Plant crop	23	29	67	46
First ratoon	12	16	9	38
Second ratoon	20	15	0	38
Third ratoon		11	7	69

Data for Australia are from 15 trials in plant cane and two trials in ratoon (Stirling and Blair, 2001); Brazil, one trial (Novaretti, 1982); Burkina Faso and Côte d'Ivoire, 16 trials (P. Cadet, unpublished data); and South Africa, between six and 29 trials (SA Sugarcane Research Institute, unpublished data).

attributed to the height at which the cane is cut in that country (Cadet, 1986a). Whereas in South Africa the stalks are cut at ground level and the shoots and shoot roots are initiated below the ground, in Burkina Faso they are cut approximately 50 mm above ground and most of the new shoots develop from the uppermost buds on the stubble. This takes place beneath the thick blanket of dead leaves (the cane is cut green in Burkina Faso whereas in South Africa the cane leaves are normally burnt at harvest). While the young shoots in Burkina Faso develop rapidly, it is some weeks before the shoot roots reach the ground. During this period, the shoots are reliant upon the large but relatively inactive stool root system (Cadet, 1986b).

To a certain extent, the situation with ratoon cane in Australia is similar to that in West Africa. In Australia during the first few months after planting, soil is moved from the interrow and 'hilled up' on the row. This buries the base of the stalks and facilitates mechanical harvesting. It also means that, after the first harvest, the first ratoon crop develops from a much larger stool than would otherwise be the case. It is assumed that initial shoot dependency on the new root system is reduced by the direct availability of the 'extra' nutrients stored in the stool. This nutritional advantage could explain why the damage is less important in ratoon cane in these two regions. It is possible that a similar situation occurs in Brazil where ratoons also appear less susceptible to nematode damage than plant cane (Table 17.8).

The growth and development of plant and ratoon cane in South and West Africa, and the corresponding fluctuations in the numbers of nematodes in and around the roots are summarized diagrammatically in Fig. 17.5. The direct and indirect consequences of this interaction on the number and length of stalks, two important components of cane yield, are summarized in Fig. 17.6.

Nematode communities and disease complexes

In addition to the interactions with *Pratylenchus* and *Meloidogyne* discussed earlier, broader disease complexes may occur in sugarcane. For example, nematode communities made up of *Pratylenchus*, *Hoplolaimus* and *Tylenchorhynchus* are reported to be associated with species of *Fusarium* and *Acremonium* in the wilt disease complex in India (Mehta, 1992). However, there is no evidence that the widespread and insidious ratoon stunting disease of sugarcane, caused by *Leifsonia xyli* subsp. *xyli*, is exacerbated by nematodes. The combined effect of the disease and a community of nematodes dominated by species of *Helicotylenchus*, *Meloidogyne*, *Paratrichodorus*, *Pratylenchus* and *Xiphinema* was additive rather than synergistic (Spaull and Bailey, 1993).

Control measures

In most countries, sugarcane is cultivated on soils with a relatively high clay or silt content where nematodes have little appar-

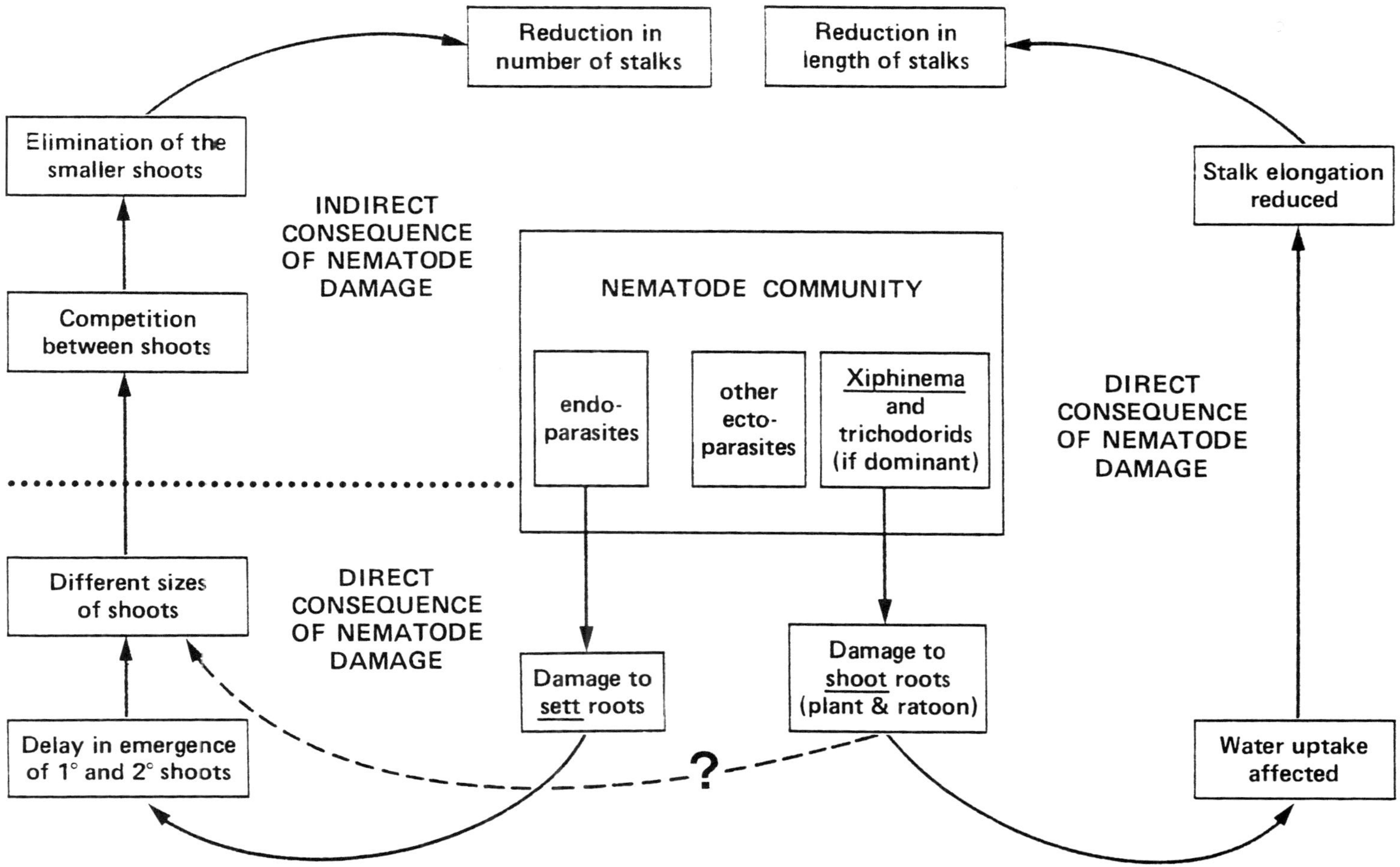

Fig. 17.6. Principal mechanisms of yield loss in sugarcane due to nematodes.

ent effect on growth; sandy soils constitute only a small proportion of the total world area under sugarcane (Rosenfeld, 1956). Thus nematodes have not been considered major pests in all sugarcane-growing areas of the world, although locally they are a serious limiting factor and justify the use of control measures. However, nematode population densities can be as high in clay soils as sandy soils, and there is accumulating evidence that nematodes limit cane yields in such soils (G.R. Stirling, Queensland, Australia, 2003, personal communication; SA Sugarcane Research Institute, unpublished data).

Cultural practices

The problem of growing sugarcane on poor sands in some parts of South Africa was overcome by inverting and mixing the sandy topsoil with a clay subsoil (Anonymous, 1982). Also in South Africa, irrigation considerably improved the yield of cane growing on a poor sand in both nematicide-treated and untreated plots (Donaldson and Turner, 1988). The response to the treatment was smaller than that in plots receiving only rainfall. Time of planting may also influence the effect of nematodes on sugarcane. Thus in Taiwan, judging from the greater response of spring-planted sugarcane to treatment with a nematicide (mean of 33% response in 26 trials) compared with that of cane planted in autumn (mean of 16% response in 31 trials), sugarcane is less tolerant of nematodes when planted in spring (Hu *et al.*, 1968; Hu and Tsai, 1973, 1978, 1982). In Brazil, Novaretti *et al.* (1984) found that whether or not a nematicide was used, the best yields were obtained from cane planted in March (autumn) rather than in December, January, February or April. The second best yield from the control plots and the smallest response to treatment were from cane planted in December.

Fallowing, intercropping and crop rotation

Sugarcane is generally grown as a monoculture and there is normally only a relatively short period between eliminating the previous crop, either chemically, with a herbicide, or physically, with a plough, and planting a new crop. This means that sugarcane-specific pests and pathogens that are present at the end of one crop are simply carried over to the following crop.

If soil is bare fallowed for long periods, population densities of plant parasitic nematodes are reduced and yields increase. However, the partial biological vacuum that is created results in a resurgence in populations of some plant parasitic nematodes, particularly ectoparasites such as *Tylenchorhynchus* and *Paratylenchus* (Stirling *et al.*, 2001). Long periods of bare fallow are not a sustainable option, but a shorter 3–4 month fallow will reduce populations of nematodes, provided soil temperature and moisture conditions are favourable for nematode activity.

Rotating sugarcane with other crops and intercropping is common on the smaller farms in a number of countries including India, Mauritius and Taiwan (Smith, 1978; Parsons, 2003). In Australia, when the sugarcane monoculture was broken with a fallow legume, yield improved by 15–25% (Garside and Bell, 2001). Since the increased yields carried through to subsequent ratoons, the sugar yield forgone by leaving land out of sugarcane for 12 months was more than recovered in the subsequent crops. Grain harvested from the legume also adds to profitability. Nematode control is a contributing factor in the yield response, as legumes such as soybean and groundnut (peanut) reduce populations of several nematode species that attack sugarcane (e.g. *P. zeae* and *T. annulatus*) (Stirling *et al.*, 2001, 2002).

From the perspective of nematode control, the choice of rotation crop will depend on which nematode species is the key pest. Thus legumes that are relatively resistant to most species of *Meloidogyne* (e.g. groundnut or velvet bean) may be the most appropriate options in sandy soils where root knot nematode is the most important pest. In Brazil, for example, a 2 year rotation programme with groundnut and maize proved successful in soils infested with this nematode (de Moura, 1995).

Organic amendments

The addition of crop residues and animal manures to soil invariably improves plant growth and, for this reason, the practice is as old as agriculture itself. The mechanisms involved are complex and involve nutrient inputs, improvements in the cation exchange capacity of the soil, formation and stabilization of soil aggregates, improvements in water infiltration rates and water retention, and suppression of some soil-borne pathogens. Population densities of plant parasitic nematodes are usually reduced by organic amendments, and plants are better able to tolerate attack by nematodes (Stirling, 1991).

In the sugar milling process, vast quantities of crop residues are generated, in particular bagasse, which is primarily cane fibre, and filtercake (millmud), which is the sediment obtained when clarifying the juice expressed from the crushed cane (Qureshi *et al.*, 2001). There are numerous reports of the suppression of nematode populations and an increase in sugarcane yields following the addition of filtercake to the soil (Estioko *et al.*, 1988; Jonathan *et al.*, 1991; Mehta *et al.*, 1994b; Albuquerque *et al.*, 2002). Similar benefits have been reported for other locally available organic materials such as poultry manure, farmyard manure and neem (produced from *Azadirachta indica*) (Salawu, 1992; Mehta and Sundararaj, 1995). The combination of organic amendments and green manure crops has also been effective in reducing numbers of nematodes and increasing yields of sugarcane (Mehta and Sundararaj, 1997; Jonathan *et al.*, 1999). Lower rates of nematicide may be required when used in conjunction with organic amendments (Novaretti, 1992; Salawu, 1992) or greater yields may be achieved by the combination of a nematicide and an organic amendment than either on its own (Novaretti and Nelli, 1985; Cadet *et al.*, 1987a).

The variable nature of organic materials and the complex chemical and biological interactions that occur when they are added to the soil mean that responses to organic amendments are difficult to predict. In Australia, Stirling *et al.* (2003) monitored temporal changes in biological activity and suppressiveness to plant parasitic nematodes in soils amended with sawdust, sugarcane trash, grass hay, legume hay, feedlot manure, poultry manure, chitin and mill mud. Seven months after amendments were incorporated, soils with sawdust, sugarcane trash and grass hay were more suppressive to *M. javanica* than soils amended with nitrogenous materials. Numbers of *P. zeae* in the roots of sugarcane were reduced by 60–90% in some of the amended soils. It was concluded that the quantity, quality and timing of organic inputs influenced the level of nematode control and that amended soils with a fungal dominant biology and high numbers of omnivorous nematodes were most likely to induce suppressiveness.

Resistance

Sugarcane is not attacked by a single nematode species but by a diverse community of plant parasitic nematodes. Breeding for combined resistance, even to the more important components of a community, is therefore likely to be extremely difficult (Luc and Reversat, 1985). Nevertheless, such a combination has been identified in one cultivar in Brazil, SP70-1143, as it is resistant to both *M. javanica* and *P. zeae* and tolerant of *P. brachyurus* (Dinardo-Miranda, 1994; Dinardo-Miranda *et al.*, 1995). This cultivar is widely grown on the sandy soils in Brazil where *M. javanica* is the dominant plant parasitic nematode (G.R. Machado, Brazil, 1989, personal communication). In contrast, the single most widely grown cultivar in Brazil, RB72454 (Tew, 2003), is susceptible to *M. incognita*, *M. javanica* and *P. zeae* (Dinardo-Miranda *et al.*, 1996; Dinardo-Miranda, 1999). Resistance and/or tolerance to species of *Meloidogyne* and *Pratylenchus* have been identified in the cultivar collections of several countries (Suwarno, 1991; Mehta and Somasekhar, 1998; Dinardo-Miranda, 1999). Apart from *Heterodera sacchari* in Nigeria, no attempts have been made to identify resistance to species of other genera (Salawu, 1990).

In India, Sujatha and Mehta (1998) and Kathiresan and Mehta (2002) showed that both resistant and susceptible cultivars respond to pathogenic invasion with qualitative and quantitative changes in peroxidase and acid phosphatase activity in the roots and in the leaves.

Tolerance

While there is only a remote chance of finding cultivars that are resistant to a wide spectrum of plant-feeding nematodes, the selection of tolerant cultivars that grow well in spite of the damage caused by nematodes appears more realistic (Matsuoka, 1980). In fact, the normal selection procedures used by plant breeders tend to select such tolerant cultivars. Cvs N12, N14 and NCo376, which are collectively planted on 60% of the sugar industry in South Africa (Tew, 2003), are tolerant of damage from nematodes (Spaull and Cadet, 2003). Similarly, CP70-321, that occupies 20% of the cane grown in Louisiana and Texas, appears tolerant to several plant parasitic nematodes including species of *Criconemoides*, *Paratrichodorus* and *Tylenchorhynchus* (Koenning *et al.*, 1999). In both Brazil and South Africa, it was calculated that tolerant cultivars reduce the damage caused by nematodes from about 47 to 15% (Matsuoka, 1980; Spaull and Cadet, 2003).

Chemical control

Fumigant and non-fumigant nematicides have been used experimentally on sugarcane in many countries, particularly Australia, Brazil, Burkina Faso, India, Indonesia, Côte d'Ivoire, the Philippines, South Africa and Taiwan. In some instances, responses to treatment were good, especially on sandy soils (Spaull and Cadet, 1990). However, over the past 15 years, almost all the experimental work on chemical control of nematodes has been with non-fumigant nematicides, mostly carbofuran, and most of the reports are from Brazil. Other chemicals investigated include aldicarb, ethoprophos, phenamiphos, terbufos and cadusafos. In many instances, treatment with these nematicides increased yield, especially on sandy soils (Bond *et al.*, 2000; Stirling and Blair, 2001; Cadet and Spaull, 2003). However, due to either their relatively high cost, their non-availability in some countries and the erratic responses that are often obtained, the commercial use of nematicides is restricted to the sandy soils of a few countries, including Australia, Brazil, Burkina Faso and South Africa.

Only a few new chemicals were tested against sugarcane nematodes in the past 10 years. One of them, furfural (2-furfuraldehyde), produced from bagasse, was effective against nematodes under controlled conditions in pots, but not under rainfed conditions in the field (Spaull, 1997). Mehta and Narayanaswamy (1993) showed that the insecticide, phorate reduced nematode populations and increased yield of cane significantly. Similar results were obtained in Brazil (Novaretti and Nelli, 1989) and the USA (Bond *et al.*, 2000).

Time and method of application

PLANT CANE. In Burkina Faso, South Africa and Taiwan, the recommended practice is to apply the nematicide in the furrow at planting (Hu and Tsai, 1973; Moberly and Clowes, 1981; Cadet *et al.*, 1987a). In Australia, however, it is recommended that treatment be applied when the crop is at the 3–5 leaf stage (R.M. Bull and K.J. Chandler, Queensland, Australia, 1988, personal communication). This is based on data from several trials, which indicated that delaying the application of the nematicide until there was slight tillering led to greater yields than those from treatment at planting, or when tillering was well advanced (Bull, 1981). In Burkina Faso, treatment cannot be delayed; when liquid carbofuran is used, it is applied to the soil surface immediately after planting is completed (Cadet *et al.*, 1987b). The results of trials from Australia, Brazil, Burkina Faso, Côte d'Ivoire, South Africa and Taiwan show that treatment with a nematicide at planting may increase not only the yield of the plant crop but also that of the first ratoon (Spaull and Cadet, 1990). This residual or carry-over response is assumed to result from the benefit derived from the more extensive root system of the nematicide-treated plant crop, upon which the following ratoon intially is dependent. This is supported by trial data from South Africa that show that the yield of ratoon crops is significantly correlated with the yield of the plant crop (Cadet and Spaull, 2003).

RATOON CANE. In Burkina Faso, despite causing considerable damage to the plant crop, nematodes have little effect on ratoon cane (Cadet, 1985). In South Africa, however, nematodes are a serious limiting factor in ratoon cane on poor sandy soils, and nematicides are usually reapplied after harvest (Rau and Moberly, 1975). In a series of trials, it was found that for a crop ratooned in the spring, treatment should not be delayed. However, treatment of cane harvested in the drier winter period could be delayed for up to 20 weeks without affecting the response, providing that the nematicide was applied before spring (Spaull and Donaldson, 1983).

Factors affecting response to treatment with nematicides

As discussed previously, the increase in yield of sugarcane following nematicide treatment is generally greater in coarse textured soils (Fig. 17.3). Also it tends to be smaller under dry conditions (Chandler, 1980; Spaull, 1995) and may vary according to whether the plant or ratoon crop is treated. In South Africa, crop loss from nematodes, as measured by response to nematicide treatment, is not affected by crop stage, the damage being similar in plant and ratoon cane. In contrast, in Australia and Brazil, the first ratoons at least are less susceptible to nematodes, and in West Africa they are naturally resistant (Table 17.8).

Where cane growth is limited by inadequate nutrition, the potential response to treatment with a nematicide may not be realized. Thus, in sandy soils in North Queensland, cane did not respond to treatment with aldicarb except where low soil calcium and magnesium levels were corrected (Chandler, 1980). Novaretti *et al.* (1981) found that in Brazilian soils infested with *M. incognita*, the combined application of a mineral fertilizer and carbofuran increased yields over and above the combined response from both treatments alone. There was a direct relationship between increased fertilizer application and increased cane yield when *M. javanica* and other plant parasitic nematodes were controlled with a nematicide (Novaretti *et al.*, 1987), suggesting that nematodes interfered with nutrient uptake by roots.

Effect of nematicides on nematode populations

When sampled at frequent intervals, field trials usually show that nematicides reduce nematode population densities. The level of control depends on the chemical used, soil type, application rate and the method of application. The effects of nematicides usually persist for some weeks to several months, and in some instances they are still detectable at harvest (Chandler, 1980; Cadet, 1985; Novaretti and Nelli, 1985; Bond *et al.*, 2000).

In Burkina Faso, Cadet and Thioulouse (1989) found that over a 5 year period, treatment with nematicides altered the balance between species within the community. At the end of each crop, the endoparasite community was dominated by *Meloidogyne* and *Pratylenchus* in plots where oxamyl, carbofuran or aldicarb had been applied. In contrast, the community in untreated control plots was dominated by populations of *Paratylenchus*, *Hoplolaimus* and *Heterodera* after the second ratoon.

Economics of nematode control with nematicides

Currently, the only option to control nematodes on very poor sandy soils and achieve sustainable economic production of sugarcane is to use a nematicide. The cost of treatment with a nematicide, in Brazil and South Africa for example, is equivalent to about 8 t of cane (W.R.T. Novaretti, Piracicaba, Brazil, 2003, personal communication; V.W. Spaull, unpublished data). In most instances on sandy soils, the response to treatment justifies the cost. However, worldwide concern for the environment, the toxicity of the nematicides, their relatively high cost in relation to the world sugar price and the smaller and more erratic response in better soils have brought into question their continued use. Nevertheless, they still have a vital role to play as a scientific tool for measuring the impact of nematodes on plants and for estimating the level of nematode tolerance in plants. The possible disappearance of nematicides from the market is therefore a matter of concern.

Where nematodes limit the growth of sugarcane, the merits of using a nematicide include benefits other than simply increasing the yield of the treated crop. The residual response of ratoon cane following the treatment of the previous crop has already been mentioned. In regions where ratoon crops are affected by nematodes, the use of nematicides sustains yields over a number of ratoon crops and thus delays the need to replant the cane (Cadet and Spaull, 2003). The improved root system of treated cane increases resistance to drought conditions, may permit the use of smaller quantities of fertilizer (Anonymous, 1984), reduces the cost of weed control due to the more rapid development of a full leaf canopy and provides a thicker and more effective mulch for the following crop because more trash remains after harvest.

Method of diagnosis

Sampling to determine the size and composition of the plant parasitic nematode community must be timed to take into account the dynamics of the root systems of cane. Thus in plant cane, a representative sample of sett roots is required. This can only be taken during the relatively short period after planting when the cane is dependent upon these roots. Samples of shoot roots can be taken at any time during the subsequent growth of the crop. In ratoon cane, the new roots attached to the developing shoots should be distinguished from the old roots of the previous crop, which may persist for several months. Soil samples to a depth of approximately 20 cm are taken close to the row at any time during the growth of the crop. Pre-plant and mid-season threshold levels for species of *Meloidogyne* and *Pratylenchus* have been given by Stirling and Blair (2000). Diagnostic services are available in some countries including Australia and South Africa. In South Africa, nematicide treatment is recommended where symptoms of damage are associated with *Meloidogyne*, *Pratylenchus* or *Xiphinema* in sandy soils.

Determination of crop loss

Based on estimates provided by 65 nematologists from around the world, Sasser and Freckman (1987) reported an annual loss in sugarcane production of 15.3%. This is higher than that of a number of other estimates for individual countries, i.e. Australia, 9% (G.R. Stirling, Queensland, Australia, 2003, personal communication); Peru, 3% (Carbonell, 1978); South Africa, 7.6% (Spaull and Cadet, 2003); USA, 4% (Koenning *et al.*, 1999); and Côte d'Ivoire, 11.0%, but similar to an estimate from Burkina Faso, 14.6% (P. Cadet, unpublished data).

Repeated application of conventional and high rates of nematicides indicates that crop loss estimates in sugarcane are much greater than those derived from treatment with a single (economic) application (Berry *et al.*, 2004) (Fig. 17.7). In addition, there are long-term consequences of the damage caused by nematodes since they not only affect the yield of each crop but also reduce the number of economic ratoons that can be harvested from a single planting (Cadet and Spaull, 2003).

Fig. 17.7. Effect of almost complete eradication of nematodes by repeated application of standard rate of nematicide in a field of sandy soil in KwaZulu Natal, South Africa. The surrounding cane received only a single treatment.

Conclusion and Future Prospects

The world export price of sugar is not much greater now than it was 15 years ago, but production costs have increased substantially. This, together with huge surplus stocks, means that the financial return from growing sugarcane for the world market is much reduced. Sugarcane is one of the most efficient converters of sunlight, water and carbon dioxide into biomass and, unlike fossil fuels, it is a renewable resource. It is already used for a wide range of by-products (Lator, 1986; Wang, 1986; Schmitz *et al.*, 2003) and, with the advent of technology to genetically modify plants, sugarcane has many other possibilities. These include using transgenic cultivars to synthesize not only sucrose but also, for example, certain polyesters (polyhydroxyalkanoates) and protein-based polymers. These would partly replace the plastics produced by the petrochemical industry (Brumbley *et al.*, 2003; Moire *et al.*, 2003). Thus the prospects for sugarcane, as opposed to just sugar, are consequently more promising than the existing world price of sugar would suggest. Increased productivity resulting from nematode control should not, therefore, be neglected.

It is most unlikely that future nematode control recommendations will rely on the highly toxic organophosphate and carbamate nematicides that are used in some countries today. Instead, nematode control will probably be achieved by means of practices that are more target specific. Such practices could include the use of endophytic and rhizospheric bacteria and fungi that are directly antagonistic to nematodes (Kerry, 2000; Kloepper *et al.*, 2003) or nematicidal chemicals derived from such microorganisms (Hallman and Sikora, 1996; Carneiro *et al.*, 1998). Expression of similar nematode-toxic chemicals may be engineered in the plant itself (Opperman and Conkling, 1994). Another possible means of reducing the numbers of nematodes and thus the damage they cause is through the use of genetically modified cultivars that disrupt feeding. For example, the control of sedentary plant parasitic nematodes could be achieved through the nematode-induced expression of a protein that is toxic to the giant cells (Opperman and Conkling, 1994) or, for all types of plant-feeding nematodes, through the engineered inhibition of dietary proteinases (Lilley *et al.*, 1999).

Alternatively, nematode control could be achieved through a low input, integrated fertility management approach in which losses from nematodes are reduced, not completely, but to levels that are both acceptable and sustainable. The key to this approach is biodiversity, which needs to be promoted at three levels: the soil microflora, the plant and the nematode community. The objective will be to restore and sustain 'soil health' and to move away from the practice of an independent treatment for each growth impediment.

1. Soil biological diversity would be enhanced if the crop residues generated from the sugar milling process were returned to the field. Interactions within such an amended soil would reduce plant parasitic nematode pathogenicity, and the nutritional benefit derived from the residues would strengthen the plant's ability to compensate for root damage. Although a better understanding of these processes is still required, encouraging results have already been obtained (Stirling *et al.*, 2003).

2. At plant level, reintroducing plant diversity in the sugarcane monoculture could be achieved with fallows and appropriate rotation or intercrops, which would also enhance biological diversity in the soil. Mitchell *et al.* (2003) showed that decreased plant diversity resulted in an increasing fungal pathogen load across the plant community. Consequently, increased diversity by cropping within-furrow mixtures of cultivars should also be investigated (Wolfe, 2000).

3. The classical approach of selecting for nematode-resistant or tolerant plants in plant breeding programmes has not been adopted for sugarcane anywhere in the world and is unlikely to be. However, sugarcane can grow well in the presence of large numbers of plant-feeding nematodes, which means that the problem can be resolved through nematode community management rather than nematode control. Most such control measures are directed against all the species in the soil (by means of nematicides or organic amendments) or, selectively, against one or a few species (through resistance or crop rotation) and they lead to a reduction in the number of parasites for a varying period of time. However, evidence from Africa shows that reducing the number of parasites may not be necessary to minimize nematode damage. Instead this can be achieved by promoting species that have a mitigating effect on the more pathogenic species within the community (e.g. *H. dihystera*; Cadet *et al.*, 2002). It might be accomplished through the use of transgenic cultivars with genes to, paradoxically, promote the multiplication of a particular phytoparasitic nematode, for example, a mix of proteinase inhibitors that favour *H. dihystera*. Such genes are exactly the opposite of resistance genes. This 'directed tolerance', unlike the use of plant resistance or nematicides, would not create selection pressure or an ecological vacuum, neither of which is desirable. These cultivars would grow in the presence of nematodes, but would not increase the number of pathogenic species. Unlike a resistant or conventional tolerant cultivar, it could be used to advantage everywhere, even when a damaging community is not present. The sustainability of this option is ensured because it is not possible to break down 'anti-resistance'.

Acknowledgements

The authors acknowledge, with thanks, the help of Dr E. Hainzelin (CIRAD), Mrs D. Carslow and Mr S.D. Berry (South Africa), Dr W.R.T. Novaretti (Brazil) and Dr P. Quénéhervé (Martinique) and, in particular, the contribution that Dr G.R. Stirling (Australia) made to this chapter.

References

Albuquerque, P.H.S., Pedrosa, E.M.R. and Moura, R.M. (2002) Host nematode relationships in *Meloidogyne* spp. infested soil amended with stillage and filter cake. *Nematologia Brasileira* 26, 27–34.

Anonymous (1961) *Disease.* Report of the Hawaiian Sugar Planter's Association Experiment Station, pp. 17–23.

Anonymous (1982) Growing sugarcane on Recent Sands. *South African Sugar Journal* 66, 369–370.

Anonymous (1984) Recent sugar pest control research unveils economic benefits. *Sugar y azucar* 79 Supplement, 50–56.

Apt, W.J. and Koike, H. (1962) Pathogenicity of *Helicotylenchus nannus* and its relation with *Pythium graminicola* on sugarcane in Hawaii. *Phytopathology* 52, 798–802.

Berry, S.D., Cadet, P. and Spaull, V.W. (2004) Estimation of nematode damage to sugarcane in sandy soil in KwaZulu-Natal. *African Plant Protection* (in press).

Birchfield, W. (1984) Nematodes parasites of sugarcane. In: Nickel, W.R. (ed.) *Plant Insect Nematodes.* Marcel Dekker, New York, pp. 571–588.

Blair, B.L., Stirling, G.R. and Whittle, P.J.L. (1999a) Distribution of pest nematodes on sugarcane in South Queensland and relationship to soil texture, cultivar, crop age and region. *Australasian Journal of Experimental Agriculture* 39, 43–49.

Blair, B.L., Stirling, G.R., Pattemore, J.A. and Whittle, P.J.L. (1999b) Occurrence of pest nematodes in Burdekin and central Queensland sugarcane fields. *Proceedings of the International Society of Sugar Cane Technologists* 21, 227–233.

Bond, J.P., McGawley, E.C. and Hoy, J.H. (2000) Distribution of plant parasitic nematodes on sugarcane in Louisiana and efficiency of nematicides. *Supplement to the Journal of Nematology* 32, 493–501.

Bridge, J. (1986) *Plant Nematode Survey of Ramu Sugar Limited, Ramu Valley, Papua New Guinea. Report.* CAB International, Institute of Parasitology, St Albans, UK.

Brumbley, S.M., Petrasovits, L.A., Bonaventura, P.A., Shea, M.J.O., Purnell, M.P. and Nielsen, L.K. (2003) Production of polyhydroxyalkanoates in sugarcane. *ISSCT, IVth Molecular Biology Worshop Abstracts,* Montpellier, France.

Bull, R.M. (1981) Studies and observations on nematode control in Bundaberg district. *Proceedings of the Australian Society of Sugar Cane Technologists* 3, 267–274.

Butterfield, M.K., D'Hont, A. and Berding, N. (2001) The sugarcane genome. A synthesis of current understanding, and lessons for breeding and biotechnology. *Proceedings of the South African Sugar Technologists' Association* 75, 1–5.

Cadet, P. (1985) Incidence des nématodes sur les repousses de canne à sucre au Burkina Faso et en Côte d'Ivoire. *Revue de Nématologie* 8, 277–284.

Cadet, P. (1986a) Evolution des nématodes ectoparasites dans la rhizosphère de la canne à sucre au Burkina Faso. *Revue d'Ecologie et de Biologie du Sol* 23, 205–213.

Cadet, P. (1986b) Etude du développement des nématodes endoparasites dans les racines de la canne à sucre au Burkina Faso et en Côte d'Ivoire. *Revue d'Ecologie et de Biologie du Sol* 23, 287–297.

Cadet, P. (1987) Etude comparative des peuplements naturels de nématodes parasites associés à la canne à sucre. *Nematologica* 33, 97–105.

Cadet, P. and Debouzie, D. (1990) Evolution spatio-temporelle d'un peuplement de nématodes parasites de la canne à sucre. *Revue de Nématologie* 13, 77–88.

Cadet, P. and Spaull, V.W. (1985) Studies on the relationship between nematodes and sugarcane in South and West Africa: plant cane. *Revue de Nématologie* 8, 131–142.

Cadet, P. and Spaull, V.W. (2003) Effect of nematodes on sustainability of sugarcane production in South Africa. *Field Crops Research* 83, 91–100.

Cadet, P. and Thioulouse, J. (1989) Traitements nématicides et peuplement de nématodes parasites de la canne à sucre au Burkina Faso. 1. Repousses. *Revue de Nématologie* 12, 35–44.

Cadet, P., Quénéhervé, P. and Merny, G. (1982) Pathogenic action of nematodes on irrigated sugar-cane. *Revue de Nématologie* 5, 205–209.

Cadet, P., Quénéhervé, P. and Hugot, R. (1987a) Incidence agronomique de traitments nématicides sur le rendement des cannes à sucre au Burkina Faso. *Phytoma* 390, 47–49.

Cadet, P., Quénéhervé, P. and Saunier, P. (1987b) Incidence des formulations liquides de nematicides sur le rendement de cannes à sucre en Afrique de l'Ouest. *Phytoma* 392, 56–59.

Cadet, P., Spaull, V.W. and McArthur, D.G. (2002) Role of plant parasitic nematodes and abiotic soil factors in growth heterogeneity of sugarcane on a sandy soil in South Africa. *Plant and Soil* 246, 259–271.

Carbonell, E.E. (1978) The nematodes, pests of sugarcane in Peru. *Proceedings of the International Society of Sugar Cane Technologists* 16, 545–551.

Carneiro, R.M.D.G., de Souza, I.S. and Belarmino, L.C. (1998) Nematicidal activity of *Bacillus* spp strains on juveniles of *Meloidogyne javanica*. *Nematologia Brasileira* 22, 12–21.

Chandler, K.J. (1980) Continued experiments with non-volatile nematicides in North Queensland sugar cane fields. *Proceedings of the Australian Society of Sugar Cane Technologists* 2, 75–82.

Croft, B.J. and Magarey, R.C. (2000) *Pachymetra* root rot. In: Rott, P., Bailey, R.A., Comstock, J.C., Croft, B.J. and Saumtally, A.S. (eds) *A Guide to Sugarcane Diseases*. CIRAD/ISSCT, CIRAD Publications Service, Montpellier, France, pp. 126–130.

Delaville, L., Rossi, J.P. and Quénéhervé, P. (1996) Plant row and soil factors influencing the microspatial patterns of plant parasitic nematodes on sugarcane in Martinique. *Fundamental and Applied Nematology* 19, 321–328.

de Moura, R.M. (1995) Dois anos de rotacao de cultura em campos de cana-de-acucar para controle da meloidoginose. 2. Consideracoes sobre o metodo e reflexos na produtividade agro-industrial da cana-planta. *Fitopatologia Brasileira* 20, 567–600.

de Moura, R.M., Pedrosa, E.M.R., Maranhao, S.R.V.L., Moura, A.M., Macedo, M.E.A. and Silva, E.G. (1999) Nematoides associados a cana-de-acucar no estado de Pernambuco, Brasil. *Nematologia Brasileira* 23, 92–99.

Dinardo-Miranda, L.L. (1994) Hospedabilidade de oito variedades de cana-de-acucar a *Pratylenchus brachyurus* e *P. zeae*. *Nematologia Brasileira* 18, 64–72.

Dinardo-Miranda, L.L. (1999) Reacao de variedades de cana-de-acucar ao parasitismo de *Meloidogyne javanica* e de *M. incognita*. *Nematologia Brasileira* 23, 76–83.

Dinardo-Miranda, L.L., Novaretti, W.R.T., Morelli, J.L. and Nelli, E.J. (1995) Comportamento de variedades de cana-de-acucar em relacao a *Meloidogyne javanica*, em condicoes de campo. *Nematologia Brasileira* 19, 60–66.

Dinardo-Miranda, L.L., Morelli, J.L., de Landell, M.G.A. and de Silva, M.A. (1996) Comportamento de genotipos de cana-de-acucar em relacao a *Pratylenchus zeae*. *Nematologia Brasileira* 20, 52–58.

Donaldson, R.A. and Turner, P.E.T. (1988) A preliminary report of the effect of soil moisture level on responses to Temik and Curaterr. *Proceedings of the South African Sugar Technologists' Association* 62, 164–168.

Eisenbach, J.D. (1993) Interactions between nematodes in cohabitance. In: Khan, M.W. (ed.) *Nematode Interactions*. Chapman and Hall, London, pp. 134–174.

Estioko, R.V. and Reyes, T.T. (1984) Population dynamics of plant-parasitic nematodes associated with sugarcane in Negros Occidental in relation to soil type and weather pattern. *Proceedings of the Philippine Sugar Technologists' Association* 31, 235–252.

Estioko, R.V., Dosayla, R.D. and Jereza, T.S. (1988) Influence of organic materials on population of soil nematodes. *Philsutech Proceedings* 35, 321–327.

Fademi, O.A., Okusanya, B.A. and Agbana, S.B. (1997) Studies on nematodes associated with sugarcane in Nigeria. *Pakistan Journal of Nematology* 15, 15–28.

Garside, A.L. and Bell, M.J. (2001) Fallow legumes in the Australian sugar industry: review of recent research and implications for the sugarcane cropping system. *Proceedings of the Australian Society of Sugar Cane Technologists* 23, 230–235.

Hall, D.G. and Irey, M.S. (1992) Population levels of plant-parasitic nematodes associated with sugarcane in Florida. *American Society of Sugar Cane Technologists* 12, 38–46.

Hallman, J. and Sikora, R.A. (1996) Toxicity of fungal endophyte secondary metabolites to plant parasitic nematodes and soil-borne plant parasitic fungi. *European Journal of Plant Pathology* 102, 155–162.

Hoy, J.W. (2000) *Pythium* root rot. In: Rott, P., Bailey, R.A., Comstock, J.C., Croft, B.J. and Saumtally, A.S. (eds) *A Guide to Sugarcane Diseases*. CIRAD/ISSCT, CIRAD Publications Service, Montpellier, France, pp. 141–146.

Hu, C.H. and Tsai, T.H. (1982) [Control of parasitic nematodes with Vydate, Nemacur and Geofos and its effect on sugarcane yield.] *Report of Taiwan Sugar Research Institute* 95, 9–23.

Hu, C.H. and Tsai, T.H. (1973) [Effects of Terracur P and Thimet on parasitic nematodes and yield increase of sugarcane.] *Report of Taiwan Sugar Research Institute* 60, 61–75.

Hu, C.H. and Tsai, T.H. (1978) [Control of parasitic nematodes with Counter, Furadan and Temik and its effect on sugarcane yield.] *Report of Taiwan Sugar Research Institute* 81, 25–34.

Hu, C.H., Tsai, T.H. and Chu, H.T. (1968) The nematode investigation in sugarcane fields of Taiwan and effects of soil fumigation. *Proceedings of the International Society of Sugar Cane Technologists* 13, 1262–1269.

Humbert, R.P. (1968) *The Growing of Sugarcane.* Elsevier, Amsterdam.

Jonathan, E.I., Krishnamoorthy, S.V., Mandharan, M.L. and Muthukrishnan, K. (1991) Effect of organic amendments on the control of sugarcane nematodes. *Bharatiya Sugar* 16, 37–40.

Jonathan, E.I., Gajendran, G. and Manuel, W.W. (1999) Management of *Helicotylenchus multicinctus* in sugarcane with intercrops and organic amendments. *Nematologia Mediterranea* 27, 221–223.

Kathiresan, T. and Mehta, U.K. (2002) Induction of peroxidase and acid phosphatase activities in resistant and susceptible sugarcane clones infected with *Pratylenchus zeae. International Journal of Nematology* 12, 98–105.

Kerry, B.R. (2000) Rhizosphere interactions and the exploitation of microbial agents for the biological control of plant-parasitic nematodes. *Annual Review of Phytopathology* 38, 423–441.

Khurana, S.M.P. and Singh, S. (1971) Interrelationship of a fungus, *Curvularia lunata* with root-knot nematode, *Meloidogyne javanica* in sugarcane seedling blight. *Annals of the Phytopathological Society of Japan* 37, 313–315.

Kloepper, J.W., Rodriguez, R., Zehnder, G.W., Sikora, E. and Fernandez, C. (1999) Plant root-bacterial interactions in biological control of soilborne diseases and potential extension to systemic and foliar diseases. *Australasian Plant Pathology* 28, 21–26.

Koenning, S.R., Overstreet, C., Noling, J.W., Donald, P.A., Becker, J.O. and Fortnum, B.A. (1999) Survey of crop losses in response to phytoparasitic nematodes in the United States for 1994. *Supplement to the Journal of Nematology* 31, 587–618.

Koike, H. and Roman, J. (1970) Pathogenicity of *Pratylenchus brachyurus* and *Pythium graminicola* to sugarcane. *Phytopathology* 60, 1562–1565.

Lamberti, F., Chinappen, M., Roca, F., Vovlas, N., Di Vito, M. and Bucha, J. (1987) Plant parasitic nematodes found in sugar-cane fields in Mauritius. *FAO Plant Protection Bulletin* 35, 83–91.

Lator, T. (1986) Energy and by-products from sugarcane. *Proceedings of the Inter-American Sugarcane Seminars*, Miami, Florida, September 1986.

Licht, F.O. (2002) World sugar statistics 2002. *Supplement to World Sugar and Sweetener Yearbook 2002.* Agra Europe, London.

Lilley, C.J., Urwin, P.E. and Atkinson, H.J. (1999) Characterization of plant nematode genes: identifying targets for a transgenic defence. *Parasitology* 118, S63–S72.

Luc, M. and Reversat, G. (1985) Possibilités et limites des solutions génétiques aux affections provoquées par les nématodes sur les cultures tropicales. *Comptes Rendus des Séances de l'Académie d'Agriculture de France* 71, 781–791.

Martin, G.C. (1967) Plant parasitic nematodes associated with sugarcane production in Rhodesia. *FAO Plant Protection Bulletin* 15, 45–58.

Matsuoka, S. (1980) Como e quando controlar doencas e nematoides da cana-de-acucar. *Saccharum-stab* 3, 12–22.

Mehta, U.K. (1992) Nematode pests of sugarcane. In: Bhatti, D.S. and Walia, R.K. (eds) *Nematode Pests of Crops.* Vedams Books Ltd, New Delhi, India, pp. 159–176.

Mehta, U.K. and Narayanaswamy, P. (1993) Evaluation of phorate as a nematicide and its persistence pattern in the sugarcane ecosystem. *Pesticide Research Journal* 5, 87–93.

Mehta, U.K. and Somasekhar, N. (1998) Response of sugarcane clones to lesion nematode, *Pratylenchus zeae* parasitism. Nematology, challenges and opportunities in 21st century. *Proceedings of the Third International Symposium of Afro-Asian Society of Nematologists (TISAASN)*, Sugarcane Breeding Institute (ICAR), Coimbatore, India, pp. 140–142.

Mehta, U.K. and Sundararaj, P. (1990) Effect of temperature and soil type on penetration of *Pratylenchus zeae* in sugarcane. *Indian Journal of Nematology* 20, 179–182.

Mehta, U.K. and Sundararaj, P. (1995) Efficacy of some new soil amendments for the control of the lesion nematode in sugarcane. *Nematologia Mediterranea* 23, 321–323.

Mehta, U.K. and Sundararaj, P. (1997) Response of sugarcane cultivars to management practices for controlling *Pratylenchus zeae* at field level. *International Journal of Nematology* 7, 158–161.

Mehta, U.K., Natesan, N. and Sundararaj, P. (1994a) Screening of sugarcane cultivars to *Pratylenchus zeae* for commercial release. *Afro-Asian Journal of Nematology* 4, 109–111.

Mehta, U.K., Sundararaj, P. and Natesan, N. (1994b) Effect of five oil cakes on control of *Pratylenchus zeae* in sugarcane. *Nematologia Mediterranea* 22, 219–220.

Mitchell, C.E., Reich, P.B., Tilman, D. and Groth, J.V. (2003) Effects of elevated CO_2, nitrogen deposition, and decreased species diversity on foliar fungal plant disease. *Global Change Biology* 9, 438–451.

Moberly, P.K. and Clowes, M.St.J. (1981) Trials with nematicides registered for use on sugarcane in South Africa. *Proceedings of the South African Sugar Technologists' Association* 55, 92–98.

Moire, L., Rezzonico, E. and Poirier, Y. (2003) Synthesis of novel biomaterials in plants. *Journal of Plant Physiology* 160, 831–839.

Novaretti, W.R.T. (1981) Efeitos de Diferentes Niveis de Populacoes Inicias de *Meloidogyne javanica* em Duas Veridades de Cana-de-acucar (*Saccharum* spp.) Cultivadas no Estado de Sao Paulo. Dissertation, Mestre em Entomologia, Escola Superior de Agricultura 'Luis de Queiros', Universidade de Sao Paulo.

Novaretti, W.R.T. (1992) Nematoides em cana-de-acucar e seu controle. *Informe Agropecuario (Belo Horizonte)* 16, 37–42.

Novaretti, W.R.T. and Nelli, E.J. (1985) Use of nematicide and filtercake for control of nematodes attacking sugarcane in Sao Paulo state. *Nematologica Brasileira* 9, 175–184.

Novaretti, W.R.T. and Nelli, E.J. (1989) New nematicide tests in sugar cane. *Boletim Tecnico, Copersucar* 47, 19–23.

Novaretti, W.R.T., Coleti, J.T., Lorenzetti, J.M., Gasparini, C.T. and Teodoro Neto, O. (1981) Influencia do ataque de nematoidies na produtividade da cana-de-acucar e seus efeitos na absorcao de nutrients. *Saccharum-stab* 4, 49–55.

Novaretti, W.R.T., Nelli, E.J., Dinardo, L.L. and Carderan, J.O. (1984) Influencia da epoca de plantio da cana-de-acucar no controle quimico de nematoidos. *Nematologia Brasileira* 8, 219–232.

Novaretti, W.R.T., Nelli, E.J. and Gomes, R.S. (1987) Estudos de interacao entre o controle de nematoides e a adubacao da cultura da cana-de acucar. *Nematologia Brasileira* 11, 63–76.

Onapitan, J.A. and Amosu, J.O. (1982) Pathogenicity and histopathology of *Pratylenchus brachyurus* and *Helicotylenchus pseudorobustus* on sugarcane. *Nematropica* 12, 51–60.

Opperman C. and Conkling, M.A. (1994) Nematode-induced plant gene expression and related control strategies. *Fundamental and Applied Nematology* 17, 211–217.

Parsons, M.J. (2003) Successful intercropping of sugarcane. *Proceedings of the South African Sugar Technologists' Association* 77, 77–98.

Pinochet, J. (1987) Management of plant parasitic nematodes in central America: the Panama experience. In: Veetch, J.A. and Dickson, D.W. (eds) *Vistas on Nematology*. Society of Nematologists, Hyattsville, Maryland, pp. 105–113.

Qureshi, M.E., Wegener, M.K. and Mason, F.M. (2001) Economics of sugar mill by-products as a source of nutrients in the Australian sugar industry in Mackay. *Proceedings of the Australian Society of Sugar Cane Technologists* 23, 192–198.

Rau, S. and Moberly, P.K. (1975) Nematicide application to ratoon crops of sugarcane grown in some sandy soils of the Natal sugarbelt. *Proceedings of the South African Sugar Technologists' Association* 49, 171–173.

Regis, E.M.O. and de Moura, R.M. (1989) Efeito conjunto da Meloidoginose e do raquitismo da soqueira em cana-de-acucar. *Nematologia Brasileira* 13, 119–128.

Rosenfeld, A.H. (1956) *Sugarcane Around the World.* University of Chicago Press, Chicago, Illinois.

Rutherford, S., Spaull, V.W., McFarlane, S.A., van Antwerpen, T., Cadet, P. and McFarlane, K. (2002) Use of sugarcane varieties to minimise losses from disease and nematodes. *Proceedings of the South African Sugar Industry Agronomists' Association.* Annual General Meeting. Available at: www.sasa.org.za/sasex/about/agronomy /aapdfs/2002/ rford.pdf

Salawu, E.O. (1986) Reaction of sugarcane varieties to a root-knot nematode, *Meloidogyne incognita. Pakistan Journal of Nematology* 4, 19–26.

Salawu, E.O. (1990) Reaction of selected sugarcane varieties to *Heterodera sacchari* Luc and Merny and related histological changes induced in roots. *Pakistan Journal of Nematology* 6, 9–16.

Salawu, E.O. (1992) Effect of neem leaf extract and ethoprop singly and in combination on *Meloidogyne incognita* and growth of sugarcane. *Pakistan Journal of Nematology* 10, 51–56.

Sasser, J.N. (1979) Economic importance of *Meloidogyne* in tropical countries. In: Lamberti, F. and Taylor, C.E. (eds) *Root-knot Nematodes (*Meloidogyne *species) Systematics, Biology and Control.* Academic Press, London, pp. 359–373.

Sasser, J.N. and Freckman, D.W. (1987) A world perspective on nematology: the role of the society. In: Veetch, J.A. and Dickson, D.W. (eds) *Vistas on Nematology*. Society of Nematology, Hyattsville, Maryland, pp. 7–14.

Sawhney, D.M. (1997) The Indian sugar industry. *Proceedings of the South African Sugar Technologists' Association* 71, xiii–xxiii.
Schmitz, T.G., Schmitz, A. and Seale, J.L. (2003) Brazil's ethanol program. The case of hidden sugar subsidies. *International Sugar Journal* 105, 254–265.
Showler, A.T., Reagan, T.E. and Flynn, J.L. (1991) Augmentation and aldicarb treatment of nematodes in selected sugarcane weed habitats. *Supplement to the Journal of Nematology* 23, 717–723.
Showler, A.T., Reagan, T.E. and Shao, K.P. (1990) Nematode interactions with weeds and sugarcane mosaic virus in Louisiana sugarcane. *Journal of Nematology* 22, 31–38.
Smith, D. (1978) *Cane Sugar World.* Palmer Publications, New York.
Spaull, V.W. (1981) Nematodes associated with sugarcane in South Africa. *Phytophylactica* 13, 175–179.
Spaull, V.W. (1984) Observations on *Bacillus penetrans* infecting *Meloidogyne* in sugarcane fields in South Africa. *Revue de Nématologie* 7, 277–282.
Spaull, V.W. (1995) A revised estimate of crop loss in sugarcane caused by nematodes. *Proccedings of the South African Sugar Technologists' Association* 69, 28–34.
Spaull, V.W. (1997) On the use of furfural to control nematodes in sugarcane. *Proceedings of the South African Sugar Technologists' Association* 71, 96.
Spaull, V.W. and Bailey, R.A. (1993) Combined effect of nematodes and ratoon stunting disease on sugarcane. *Proceedings of the South African Sugar Technologists' Association* 67, 29–33.
Spaull, V.W. and Cadet, P. (1990) Nematodes parasites of sugarcane. In: Luc, M., Sikora, R.A. and Bridge, J. (eds) *Plant Parasitic Nematodes in Subtropical and Tropical Agriculture.* CAB International, Wallingford, UK, pp. 461–491.
Spaull, V.W. and Cadet, P. (1991) Studies on the relationship between nematodes and sugarcane in South and West Africa: ratoon cane. *Revue de Nématologie* 14, 183–186.
Spaull, V.W. and Cadet, P. (2003) Impact of nematodes on sugarcane and the benefit of tolerant varieties. *Proceedings of the South African Sugar Technologists' Association* 77, 230–238.
Spaull, V.W. and Donaldson, R.A. (1983) Relationship between time of nematicide application, numbers of nematodes and response to treatment in ratoon sugarcane. *Proccedings of the South African Sugar Technologists' Association* 57, 123–127.
Spaull, V.W. and Heyns, J. (1991) Distribution of longidorid nematodes in sugarcane fields in South Africa. *Phytophylactica* 23, 167–172.
Spaull, V.W., Cadet, P. and Gers, C. (2003) Factors affecting the distribution of plant feeding nematodes in KwaZulu-Natal. *Proceedings of the South African Sugar Technologists' Association* 77, 220–222.
Stirling, G.R. (1991) *Biological Control of Plant-parasitic Nematodes.* CAB International, Wallingford, UK.
Stirling, G.R. and Blair, B. (2000) Nematodes. In: Rott, P., Bailey, R.A., Comstock, J.C., Croft, B.J. and Saumtally, A.S. (eds) *A Guide to Sugarcane Diseases.* CIRAD/ISSCT, CIRAD Publications Service, Montpellier, France, pp. 299–305.
Stirling, G.R. and Blair, B. (2001) Nematodes are involved in the yield decline syndrome of sugarcane in Australia. *Proceedings of the International Society of Sugar Cane Technologists* 24, 430–433.
Stirling, G.R., Blair, B., Pattemore, J.A., Garside, A.L. and Bell, M.J. (2001) Changes in nematode populations on sugarcane following fallow, fumigation and crop rotation, and implications for the role of nematodes in yield decline. *Australasian Plant Pathology* 30, 323–335.
Stirling, G., Blair, B., Wilson, E. and Stirling, M. (2002) Crop rotation for managing nematode pests and improving soil health in sugarcane cropping systems. *Proceedings of the Australian Society of Sugar Cane Technologists* 24, 129–134.
Stirling, G.R., Wilson, E.J., Stirling, A.M., Pankhurst, C.E., Moody, P.W. and Bell, M.J. (2003) Organic amendments enhance biological suppression of plant-parasitic nematodes in sugarcane soils. *Proceedings of the Australian Society of Sugar Cane Technologists* 25, p.11.
Sujatha, K. and Mehta, U.K. (1993) Association analysis and interrelationship between three ectoparasitic nematodes in sugarcane fields. *Afro-Asian Journal of Nematology* 5, 178–180.
Sujatha, K. and Mehta, U.K. (1994) Influence of the lesion nematode, *Pratylenchus zeae*, on yield and quality characters of two cultivars of sugarcane. *Nematologia Mediterranea* 22, 65–67.
Sujatha, K. and Mehta, U.K. (1995) Patterns of interspecific associations in plant parasitic nematodes of sugarcane ecosystems. *Nematologia Mediterranea* 21, 275–277.
Sujatha, K. and Mehta, U.K. (1997) Relative pathogenicity of *Pratylenchus zeae* and *Meloidogyne javanica* to sugarcane crop. *International Journal of Nematology* 7, 221–224.
Sujatha, K. and Mehta, U.K. (1998) Changes in the activity of peroxidase and polyphenol oxidases in sugarcane roots after infection with *Pratylenchus zeae* and *Meloidogyne javanica. International Journal of Nematology* 8, 102–104.

Suwarno (1991) Uji ketahanan varietas tebu terhadap nematoda bengkak akar *Meloidogyne* spp. *Berita-Pusat Penelitian Perkebunan Gula Indonesia* 5, 12–14.

Tarte, R., Cerrud, D., Rodriguez, I. and Osorio, J.M. (1977) Presencia y parasitismos de *Pratylenchus zeae* en cana de azucar en Panama con indicaciones sobre la susceptibilidad relativa de algunos cultivares. *Turrialba* 27, 259–265.

Tew, T.L. (2003) World sugarcane variety census – year 2000. *Sugar Cane International* March/April, 12–18.

Valle-Lamboy, S. and Ayala, A. (1980) Pathogenicity of *Meloidogyne incognita* and *Pratylenchus zeae*, and their association with *Pythium graminicola* on roots of sugarcane in Puerto Rico. *Journal of Agriculture, Puerto Rico* 64, 338–347.

Wallace, H.R. (1973) *Nematode Ecology and Plant Disease*. Edward Arnold, Oxford.

Wang, L.H. (1986) Recent progress in research and development of by-products utilization at Taiwan Sugar Research Institute. *Taiwan Sugar Research Institute Annual Report* 1985–86, 37–45.

Williams, J.R. (1963) Investigations in Mauritius on the soil inhabiting nematodes associated with sugarcane. *Proceedings of the International Society of Sugar Cane Technologists* 11, 717–722.

Williams, J.R. (1967) Observations on parasitic protozoa in plant-parasitic and free living nematodes. *Nematologica* 13, 336–342.

Williams, J.R. (1969) Nematodes attacking sugarcane. In: Peachey, J.E. (ed.) *Nematodes of Tropical Crops*. Commonwealth Agricultural Bureaux, Farnham Royal, UK, pp. 184–203.

Williams, J.R. and Luc, M. (1977) The species of *Xiphinema* Cobb, 1913 (Nematoda, Longidoridae) in the sugarcane fields of Mauritius. *Occasional Paper Mauritius Sugar Industry Research Institute* 30.

Wilson, G. (1969) Insecticides for the control of soil-inhabiting pests of sugarcane. In: Williams J.R., Metcalfe, J.R., Mungomery, R.W. and Mathes, R. (eds) *Pests of Sugarcane*. Elsevier, Amsterdam, pp. 259–282.

Wolfe, M.S. (2000) Crop strength through diversity. *Nature* 406, 681–682.

18 Nematode Parasites of Tobacco*

Charles S. Johnson,[1] Jennifer Way[2] and Kenneth R. Barker[3]
[1]*Southern Piedmont AREC, Virginia Polytechnic Institute and State University, 2375 Darvills Road, Blackstone, Virginia, USA;* [2]*Tobacco Research Board, PO Box 1909, Harare, Zimbabwe;* [3]*Department of Plant Pathology, North Carolina State University, Raleigh, NC 27695-7616, USA*

Tobacco (*Nicotiana tabacum* L.) is a high value crop that is grown throughout the world for the production of cigarettes, cigars and other products, and may be the most widely grown commercial non-food crop in the world (Shepherd and Barker, 1990). The total world production in 2003 was estimated to be about 6.3 Mt, of which over one-third was produced in China (Anonymous, 2004). The other major tobacco-producing countries include (in order of metric tonnes produced in 2003) Brazil, India, the USA, Zimbabwe, Turkey, Indonesia, Argentina, Greece, Italy and Pakistan (Table 18.1). The sale of cured leaf and manufactured products is a major source of income for many countries, and many governments rely heavily on taxes levied on sales to consumers.

Although the word 'tobacco' usually refers to *N. tabacum*, it may also refer to *N. rustica*, which is grown for similar purposes in some parts of the world (Johnson and Reed, 1994). *N. tabacum* probably originated as a natural hybrid of *N. sylvestris* and *N. tomentosiformis* in Brazil or Central America, and has been under cultivation for many centuries (Ren and Timko, 2001). By the time explorers from Europe came to the Americas, tobacco cultivation was widespread in North, Central and South America, and since then has spread all over the world (Johnson and Reed, 1994). Differences in plant genetics, production practices and environmental characteristics have resulted in separation of the crop into eight classes and 26 types.

Cultivation techniques

Tobacco fields have traditionally been transplanted with seedlings produced in outdoor seedbeds or nurseries, but seedling production is very rapidly switching to hydroponic systems in greenhouses or small outdoor 'float beds' (Reed, 1996; Anonymous, 1998). Although this change eliminates the need for pesticides and/or cultural practices to prevent early nematode parasitism of seedlings, it reduces, but does not eliminate, use of methyl bromide in tobacco transplant production. This material remains necessary to protect tobacco seedlings adequately from diseases caused by fungi such as *Rhizoctonia solani* and *Pythium* species.

*A revision of the chapter by J.A. Shepherd and K.R. Barker.

Table 18.1. Estimated tobacco production in selected countries (after FAOSTAT data, 2004).

	Hectares harvested			Production (t)		
Country	1987	2003	Change (%)	1987	2003	Change (%)
China	1,137,070	1,352,862	18.98	1,967,005	2,487,578	26.47
India	389,200	425,000	9.20	461,800	575,000	24.51
Brazil	294,093	391,508	33.12	395,645	658,251	66.37
USA	237,300	168,440	–29.02	539,260	377,030	–30.08
Indonesia	211,432	156,000	–26.22	112,691	135,000	19.80
Turkey	206,247	193,000	–6.42	184,712	154,000	–16.63
Former USSR	161,000	37,350	–76.80	303,000	69,100	–77.19
Malawi	97,786	122,033	24.80	72,507	69,401	–4.28
Greece	94,520	57,000	–39.70	155,000	121,000	–21.94
Bulgaria	90,468	37,260	–58.81	133,098	58,448	–56.09
Thailand	80,000	41,000	–48.75	67,000	65,000	–2.99
Italy	77,450	36,800	–52.49	162,127	106,250	–34.46
Philippines	68,676	34,585	–49.64	82,779	56,500	–31.75
Former Yugoslavia	61,204	40,224	–34.28	76,000	54,765	–27.94
Zimbabwe	58,242	80,519	38.25	121,320	174,000	43.42
Cuba	55,715	33,942	–39.08	38,790	34,494	–11.08
Argentina	52,658	60,000	13.94	69,765	125,431	79.79
Poland	48,424	10,800	–77.70	113,948	20,000	–82.45
Myanmar	47,826	28,446	–40.52	70,100	48,161	–31.30
Bangladesh	46,330	33,000	–28.77	39,990	40,000	0.03
Japan	42,300	23,410	–44.66	104,400	60,000	–42.53
North Korea	40,000	44,000	10.00	60,000	63,000	5.00
Pakistan	38,996	49,500	26.94	69,249	94,900	37.04
Republic of Korea	35,274	21,000	–40.47	78,039	47,549	–39.07
Romania	34,400	8,800	–74.42	33,100	16,000	–51.66
Mexico	32,913	11,461	–65.18	50,469	21,895	–56.62
Canada	29,540	23,000	–22.14	61,338	60,000	–2.18
South Africa	26,000	14,000	–46.15	37,200	28,400	–23.66
Colombia	21,140	14,700	–30.46	34,870	29,000	–16.83
Spain	19,500	12,430	–36.26	31,900	34,513	8.19
Others	421,245	369,645	–12.25	450,848	491,703	9.06
Total	4,256,949	3,931,715	3.21	6,177,950	6,376,369	3.21

Tobacco is often produced in coarse textured soils with low inherent fertility in order to manage nutrient uptake by the crop more precisely. This characteristic tends to make the crop attractive to farmers, particularly when combined with an often profitable and stable demand. Unfortunately, the warm climates and sandy soils so helpful in managing nitrogen uptake and leaf chemistry (particularly for flue-cured tobacco) also favour reproduction, damage by and survival of plant parasitic nematodes. Air- and fire-cured tobaccos are often grown on heavier soils, but may still suffer economic losses caused by plant parasitic nematodes. In some parts of the world where paddy rice is grown, or where low-lying areas are flooded by tropical rain, tobacco may be planted after the water has receded and be grown without further water. Oriental tobacco, having developed in an area of winter rainfall, is very drought resistant and does not require much extra water during its growth in the field.

Nematodes of Tobacco

Throughout the world, plant parasitic nematodes are found wherever tobacco is grown, but the severity of the problem depends on climate and soil type. A large number of tobacco-producing countries are close to, or within, the inter-tropical zone. The dominant nematodes there are *Meloidogyne* spp., of which the most important are *M. incognita*, *M. javanica* and *M. arenaria*. *M. hapla* and other *Meloidogyne* spp., species of *Pratylenchus*, *Tylenchorhynchus* and *Globodera*, *Ditylenchus dipsaci* and *Aphelenchoides* may cause yield losses in certain restricted areas. Although other nematodes, such as the spiral nematodes (*Helicotylenchus*, *Rotylenchus* and *Scutellonema*), *Rotylenchulus* species, *Tetylenchus* and *Criconemoides* species, have been found in tobacco fields, they are not normally associated with losses. Some species of *Xiphinema*, *Longidorus*, *Trichodorus* and *Paratrichodorus* are reported to transmit viruses to tobacco.

Meloidogyne

Tisdale's report from Florida was one of the first to report the damage that *Meloidogyne* spp. or root knot nematodes can do to tobacco (Shepherd and Barker, 1990). Root knot nematodes were also recognized as serious pests in southern Africa in the late 1920s, and have long been considered important pests in most of the tobacco-growing countries of the tropical and subtropical zone.

A large number of *Meloidogyne* species reproduce on tobacco, but not all are economically important. *M. incognita*, *M. javanica*, *M. arenaria* and *M. hapla* have been most frequently associated with tobacco, with *M. incognita*, *M. javanica* and race 2 of *M. arenaria* considered the most important due to their more widespread distribution, relative reproductive capacity and damage potential (Barker and Lucas, 1984; Johnson, 1998). Juveniles of *M. javanica* can locate and invade tobacco roots more quickly and in larger numbers than *M. arenaria*, which invades roots at a more rapid rate than does *M. incognita* (Johnson, 1998). Significant differences in fecundity were not found among the three species in this research, at least over the first 60 days of a growing season. However, similar relative differences in typical gall size, syncytial shape and structure, and extent of associated root necrosis among the three species may indicate additional differences in the physiological effects of parasitism among *M. arenaria*, *M. javanica* and *M. incognita* (Johnson, 1998).

M. incognita and *M. javanica* are the most widely distributed of the important root knot species (Table 18.2). Their relative importance is largely dependent on the climate, since *M. javanica* has a greater tolerance to drought and high temperature than *M. incognita* (Shepherd and Barker, 1990). *M. arenaria* and *M. hapla* are the next most widely distributed, with *M. hapla* confined to the cooler parts of the world. Root knot nematodes are rare in Canada, with *M. hapla* occurring more than the others. *M. incognita* remains the predominant species in North and South Carolina, although the distribution of other types of root knot nematodes has increased, especially *M. arenaria* (Johnson, 1998). A Florida survey found *M. javanica* in 65% of tobacco fields and *M. incognita* in 33%, but *M. arenaria* was rarely detected (Shepherd and Barker, 1990). *M. arenaria*, *M. javanica* and race 2 of *M. incognita* occur in Cuba (Fernández Díaz-Silveira and Ortega Herrera, 1998). In Brazil, *M. javanica* was identified in 50% of cases and *M. incognita* in 20%, and both together in 25% of samples (Sudo and Espindola, 1987). *Meloidogyne* spp. (*M. incognita* races 1 and 2 and *M. javanica*) are nearly ubiquitous in tobacco fields in Colombia (Barriga-Olivares and Aranda-Ramirez, 2000).

Several surveys of tobacco fields in South Africa and Zimbabwe have shown *M. javanica* to be the dominant problem, although *M. incognita* is also common and important (Shepherd and Barker, 1990). Root knot can be a serious problem on tobacco in areas of Mozambique (Oever *et al.*, 1998). *M. incognita* and *M. javanica* have caused heavy losses to tobacco in Nigeria (Khan, 1990).

Meloidogyne spp. are common in tobacco fields in Italy, but are only a problem in the sandier soils in the northern pro-

Table 18.2. Importance of *Meloidogyne* species in some tobacco-growing countries in 1987.

	Species of *Meloidogyne*			
	arenaria	*hapla*	*incognita*	*javanica*
Africa				
Malagasy			1	2
Malawi				3
Nigeria			2	
South Africa	1	1	3	3
Zimbabwe			1	3
Americas				
Argentina			1	1
Brazil	1		3	3
Canada		1		
Chile			2	2
Colombia			1	1
Cuba			3	
Guatamala			3	
Mexico	2		3	3
Paraguay			3	3
USA	2	1	3	3
Asia and Oceania				
Australia		1	2	2
Bangladesh			2	1
China			2	1
India			3	3
Japan			2	2
Korea			1	
Malaysia			2	2
Pakistan			3	3
Philippines	1		3	2
Thailand	1		2	3
Vietnam			3	3
Europe				
Albania			2	
Bulgaria			2	2
Former Yugoslavia		1	2	2
France	1		2	
Germany			2	2
Greece			2	2
Hungary	1	1	2	2
Italy			3	3
Spain			3	3
Mediterranean countries				
Iraq	1		2	3
Morocco			2	2
Syria			3	3
Turkey			3	3

1, minor importance; 2, moderately or locally important; 3, very important.
Source: Shepherd and Barker (1990).

duction areas of the country (Cimini *et al.*, 1993; S. Miele, personal communication).

M. incognita is the dominant species on tobacco in Sichuan, China, but *M. arenaria*, *M. javanica*, *M. acrita* and *M. hapla* have also been reported as damaging to tobacco (Chen *et al.*, 1991; Jiang and Xing, 1992). *M. incognita*, *M. arenaria*, *M. javanica* and *M. thamesi* are also important problems on tobacco in India (Srivastava, 2001; Hussaini and Krishnamurthy, 2002). In the Philippines, *M. incognita* was reported in 64% of fields and *M. javanica* in 29% (Shepherd and Barker, 1990). Both *M. javanica* and *M. incognita* are found in Sri Lanka, and recent research on resistance to *Meloidogyne* spp. in tobacco cultivars has been reported from Iran (Honarnejad and Shoaei-Deylami, 1997). *M. microcephala*, *M. mayaguensis*, *M. cruciani*, *M. enterolobii*, *M. ethiopica*, *M. paranaensis*, *M. petuniae*, *M. platani*, *M. thamesi*, *M. brasilensis* and *Meloidogyne* sp. are also reported to reproduce on tobacco, but their importance is very restricted (Shepherd and Barker, 1990; Carneiro *et al.*, 1996; Charchar *et al.*, 1999; Charchar and Eisenback, 2002).

Symptoms of damage

The characteristic symptoms of root knot nematode attack are the root galls formed as a reaction to the invasion and feeding by the nematode (Fig. 18.1). These can range from small individual galls to severe distortion and restriction of root development. The size and magnitude of the galls can be a guide to the species involved. Galls induced by *M. hapla* are usually small and affect only a limited portion of the root system. *M. arenaria* causes bead-like galls to form which may involve a large proportion of the root system. Conversely, *M. incognita* and *M. javanica* cause large galls, which may affect 90% or more of the root, with the latter usually causing the more extensive gall formation. Root decay often develops in roots galled by *M. javanica*, *M. incognita* and *M. arenaria* (Fig. 18.2; Plate 19A), whereas decay is usually less severe in roots infested by *M. hapla*.

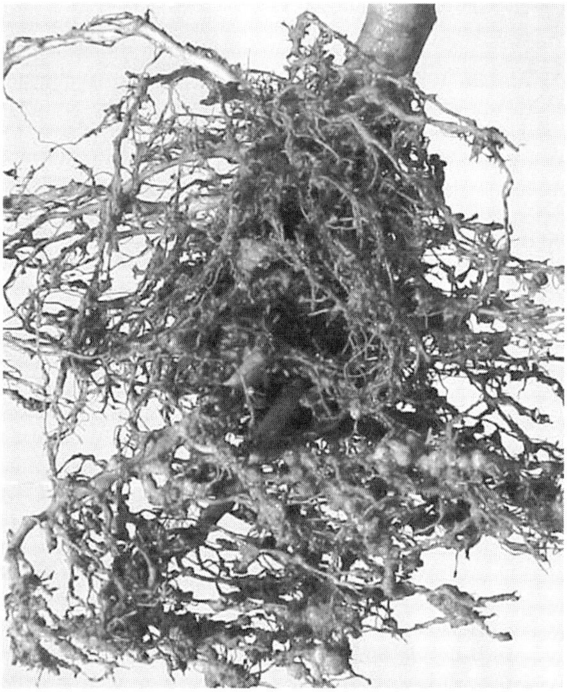

Fig. 18.1. Typical galling caused by *Meloidogyne* species on tobacco roots. (Photo: C. Johnson.)

Fig. 18.2. Galling and root rot of burley tobacco caused by *Meloidogyne* spp. and associated soil microorganisms. Healthier plant on left, severe infection on right. (Photo: C. Johnson.)

The above-ground symptoms of a severe attack are stunted growth, often associated with premature wilting, typically in the afternoon on hot days (Fig. 18.3). These symptoms are often seen in a patchy distribution in the field, unless the infestation is uniformly severe. There may also be signs of nitrogen and potassium deficiency and scorching of the leaf tips and margins. Weeds, which are usually largely controlled by healthy tobacco plants, are able to grow successfully and compete for soil moisture and nutrients. Sucker development is also much suppressed on plants heavily parasitized by nematodes.

Pathotypes

All four pathotypes of *M. incognita* have been recorded on tobacco, but by far the most common is race 1 (Shepherd and Barker, 1990). Race 1 is the most common in North Carolina, although races 2 and 4, which can attack the *M. incognita*-resistant tobaccos, and race 3 are also found. All four pathotypes have been reported in Brazil, and races 1, 2 and 3 in India. In Zimbabwe, races 1 and 3 have been recorded on tobacco, of which the most common is race 3, while in South Africa races 2 and 4 have been identified, with race 4 most commonly found. Race 2 is the more common pathotype recorded when *M. arenaria* is found.

Survival and dissemination

All of the root knot nematodes that attack tobacco have a wide host range and can survive between tobacco crops on many weeds and other crops, especially if tobacco is grown frequently in the same field. The nematodes also spread in soil, remaining on field equipment after cultural operations have been performed, and as eggs and juveniles in irrigation water applied to fields or seedbeds. Although bore-hole or mains water should be clean, surface water from streams or lakes can become contaminated when infested soil is washed into the water source during a heavy rain. Root knot nematodes can also be spread by using improperly prepared compost or dung from animals fed on infected root crops (Shepherd and Barker, 1990).

Disease complexes

Interactions with other microorganisms play an extremely important role in the

Fig. 18.3. Stunting of flue-cured tobacco typical of a root knot nematode 'hot spot'. (Photo: C. Johnson.)

epidemiology and management of *Meloidogyne* spp. on tobacco. Root knot nematode parasitism can exert local and systemic effects on tobacco, changing the structure and function of the root system as well as the physiology of the entire plant, encouraging infection by other pathogens (Johnson, 1998). *M. incognita* does not compete with *Orobanche ramosa* for infection sites on tobacco, so damage from concomitant populations of the two pests is additive (Johnson, 1998). Parasitism by *Meloidogyne* spp., however, increases plant mortality caused by soil-borne fungi such as *Phytophthora parasitica* var. *nicotianae*, causal agent of black shank, one of the most common and damaging diseases of tobacco worldwide (Shepherd and Barker, 1990). Although resistance to root knot can limit black shank development, resistance to black shank does not prevent root knot invasion and subsequent damage (Johnson, 1998). Control of the black shank–root knot disease complex depends upon crop rotation and use of cultivars resistant to both black shank and root knot and/or application of soil pesticides registered for control of both organisms. Increased incidence of Fusarium wilt (*Fusarium oxysporum* var. *nicotianae*) and bacterial wilt (*Ralstonia solanacearum*), often referred to as Granville wilt, are also associated with root knot nematode infection (Shepherd and Barker, 1990). Control of these diseases, where they occur, can be extremely difficult, requiring use of extended crop rotation intervals and high rates of soil fumigants with highly resistant cultivars.

Parasitism by *Meloidogyne* spp. can also enable soil-borne microorganisms that are not normally pathogens of the crop to infect and damage tobacco roots (Shepherd and Barker, 1990). *Rhizoctonia solani* and species of *Curvularia*, *Botrytis*, *Aspergillus*, *Penicillium* and *Trichoderma* were found to cause root necrosis severe enough to result in above-ground symptoms, but only when inoculation was preceded by infection by *M. incognita*. Although increased stem damage by the sore shin fungus *Rhizoctonia solani* has been reported when combined with root knot infection, others indicated no consistent effect of root invasion by *M. javanica* on stem damage by *R. solani* (Shepherd and Barker, 1990).

Root knot parasitism can also exacerbate foliar disease problems. Infection by *M. incognita* predisposes tobacco plants to brown spot caused by *Alternaria alternata*, and the root knot–*Fusarium* disease complex worsens this effect (Barker and Lucas, 1984; Shepherd and Barker, 1990). Infection by *Meloidogyne* spp. may also increase ozone injury (Johnson, 1998). Incidence of tobacco mosaic virus (TMV) may increase in plant populations parasitized by root knot nematodes (Patel and Patel, 1994b). Combined infection by these two pathogens reduces plant growth and yield more than infection by either pathogen alone and alters tobacco leaf chemistry, but these interactions do not appear to be synergistic (Kartono, 1980; Patel and Patel, 1994a,b, 1995). Higher populations of *M. javanica* have been noted in tobacco plants infected with TMV (Goswami and Raychaudhuri, 1973). Tobacco cultivars resistant to races 1 and 3 of *M. incognita* react with a systemic necrosis to infection by the MN strain of potato virus Y (PVY), a possible pleiotropic effect of the root knot resistance gene (Johnson, 1998).

Root knot nematodes occur as populations of a single species or as communities of several species, and must often compete with other plant parasitic nematodes, including other species of *Meloidogyne* (Johnson, 1998). Antagonistic interactions within mixed populations of root knot nematode species may result from competition for feeding sites, further complicated by differences among species in adaptation to environmental conditions (Johnson and Nusbaum, 1970; Hirunsalee *et al.*, 1995b; Ng'ambi *et al.*, 1995). Greenhouse studies indicate that *M. arenaria* out-competes races 1 and 3 of *M. incognita* in parasitizing tobacco roots (Ng'ambi *et al.*, 1995; Johnson, 1998). Although race 1 of *M. arenaria* can damage tobacco, race 2 is the more damaging of the two host races (Hirunsalee *et al.*, 1995a,b). Root knot and lesion nematodes may also compete with each other for penetration sites on tobacco roots (Olthoff *et al.*, 1973). Reproduction of *M. hapla* can be inhibited by concomitant populations of *M. incognita* or *Pratylenchus brachyurus*. Population increases of both *M. incognita* and *P. brachyurus* also may be suppressed when these two species occur together. Similar interactions have been observed among *M. incognita*, *M. javanica* and the reniform nematode *Rotylenchulus reniformis* (S.K. Patel and D.J. Patel, 1991).

Economic importance

Root knot nematodes are common pests of economic importance in tobacco culture, particularly where temperature and soil type favour them (Shepherd and Barker, 1990). They are of limited importance in colder areas such as Canada, where mainly *M. hapla* occurs, or France, where *M. incognita* and *M. arenaria* have been found on tobacco.

Extensive research has been conducted to estimate crop losses quantitatively in tobacco that result from parasitism by *Meloidogyne* species (Johnson, 1998). The relationship between nematode reproduction and yield loss in tobacco is greatest for *M. javanica*, followed by *M. arenaria*, *M. incognita* and *M. hapla*. Losses are generally greatest for tobacco planted in sandy versus clay soils, and when environmental conditions stress the crop (Johnson, 1998). Resistant cultivars suppress nematode reproduction and increase yield, but may still suffer yield loss when initial root knot nematode populations are high, perhaps due to the hypersensitive reaction of resistant roots to attempts by the nematodes to establish feeding sites (Sosa-Moss *et al.*, 1983).

Daulton (1963) stated that field fumigation could increase cured leaf yield in Zimbabwe by 55–1800 kg/ha by controlling *M. javanica*. Currently, yields on small-scale tobacco farms in Zimbabwe (~42% of the crop) average 68% less than yields from commercial-scale farms, largely due to poor nematode control. Recent nematicide tests also indicate the large yield increases that result from controlling *M. javanica* (Table 18.3).

Losses to tobacco from root knot nematodes in the USA have dropped over the past 30 years. Annual loss estimates for North Carolina averaged around 1% from 1970 to 1990, but have been below 0.6% since that time. Root knot nematodes have been estimated to cause yield losses of

Table 18.3. Total final root knot gall ratings and saleable yields from a 2002/03 nematicide test for control of *Meloidogyne javanica* in Zimbabwe.

Nematicide treatment	Rate (ml per plant station)	Root gall rating (0–8)	Yield (kg/ha)
Untreated control	–	7.3	63
Ethylene dibromide, 41%	3	6.8	1582
1,3-D + 35% chloropicrin	5	7.4	976
1,3-D + 35% chloropicrin	6	6.1	1851
1,3-D + 35% chloropicrin	7	5.8	2005
1,3-D	4	6.3	1481

50–60% in some parts of Turkey (Shepherd and Barker, 1990). In Iraq, more than 40% of the tobacco was reported to be infested, with infestation levels going up to 100% in some fields, while in India 25% loss was reported from field infestation and a 50% loss if the infestation started in the seedbed.

Threshold damage levels provide useful guides for managing *M. incognita* with minimal use of pesticides, but nematicide use is commonly recommended whenever *M. arenaria*, *M. javanica* or important root disease complexes are detected (Anonymous, 2000; Imbriani, 2003). A combination of root examination and soil assay results is now recommended in North Carolina and Virginia to assist farmers in determining what measures to employ for nematode control in the current crop (Table 18.4) (Johnson, 2002; Melton and Broadwell, 2002).

Pratylenchus

The migratory endoparasitic root lesion nematodes, *Pratylenchus* species, cause brown root rot of tobacco. Lesion nematodes are less important in the tropical and subtropical regions than the root knot nematodes, but are responsible for significant yield losses in other tobacco-growing areas, such as Canada (Table 18.5). Eleven species (*P. alleni*, *P. brachyurus*, *P. crenatus*, *P. hexincus*, *P. neglectus (minyus)*, *P. penetrans*, *P. pratensis*, *P. scribneri*, *P. thornei*, *P. vulnus* and *P. zeae*) have been reported to occur in tobacco soils from around the world, but the damage potential of *P. alleni*, *P. brachyurus*, *P. scribneri* and *P. zeae* is unclear (Gao *et al.*, 1994; Johnson, 1998).

Above-ground symptoms of *Pratylenchus* attack are very similar to those caused by other tobacco nematodes (Fig. 18.4; Plate

Table 18.4. Suggested action thresholds for races 1 and 3 of *Meloidogyne incognita* in tobacco fields in Virginia.

		Nematodes/500 cm^3 of soil		
Risk of crop loss	% Roots galled	Autumn sample	Spring sample	Control options
Very low	1–10	1–200	1–20	Practise crop rotation and/or plant a resistant cultivar
Low	11–25	201–1000	21–100	Combine crop rotation with a resistant cultivar and/or a nematicide
Moderate	26–50	1001–3000	101–300	Increase crop rotation interval. Use a resistant cultivar and an effective nematicide
High	> 50	> 3000	>300	Maximize intervals between tobacco crops. Use a resistant cultivar and a highly effective nematicide

Table 18.5. Importance of certain plant parasitic nematodes in some tobacco-growing countries.

	Aphelenchoides	*Ditylenchus*	*Globodera*	*Pratylenchus*
Africa				1
Malagasy				2
South Africa				3
Zimbabwe				
Americas				
Brazil	2			2
Canada				2
Chile	1	1		2
Colombia			1	
Mexico				1
Paraguay		1		2
USA			2	1
Asia and Oceania				
Australia				1
China	1	1	2	2
India				1
Korea		1	1	1
Malaysia				2
Pakistan	1	1	2	1
Thailand				1
Vietnam				3
Europe				
Albania		1		
France	2	2	2	
Former Yugoslavia		2	2	1
Germany	1	2		2
Greece			1	
Hungary		1		1
Italy		2	2	2
Mediterranean countries				
Iraq				1
Morocco			2	
Turkey		1		1

1, minor importance; 2, moderately important or locally important; 3, very important.
Adapted from Table 3 of Shepherd and Barker (1990).

19B). Macroscopic root symptoms are also very similar to those of black root rot caused by *Thielaviopsis basicola* (syn. *Chalara elegans*) (Fig. 18.5; Plate 19C). Root lesions caused by *Pratylenchus* spp. first appear as discrete water-soaked areas that attain a yellow colour that darkens to brown over time. Lesions may coalesce to encircle infected roots, causing the cortex to slip off and leave only the vascular cylinder remaining.

The migratory endoparasitic lifestyle of *Pratylenchus* spp. may cause it to interact differently with other soil-borne pathogens of tobacco compared with the sedentary endoparasitic *Meloidogyne* spp. Simultaneous inoculation of black shank-susceptible tobacco with *P. brachyurus* and *Phytophthora parasitica* var. *nicotianae* increased black shank development and severity, but inoculation with *P. brachyurus* or *P. penetrans* prior to *P. parasitica* var. *nicotianae* reduced black shank symptom severity and disease incidence (Inagaki and Powell, 1969; McIntyre and Miller, 1978). Lesion nematode infection did not increase black shank severity on black shank-resistant

Fig. 18.4. Stunting of flue-cured tobacco by *Pratylenchus coffeae* in South Carolina. (Photo: S.A. Lewis.)

cultivars to fungal inoculation. Root wounding due to nematode penetration may have facilitated fungal infection of simultaneously inoculated susceptible cultivars, but prior infection by *P. penetrans* appeared to induce a general, systemic and sustained host response to attack. *P. hexincus* was associated with the development of 'black root rot' (*Thielaviopsis basicola*) in the black turf soils of South Africa when they were wet at planting time (Shepherd and Barker, 1990). Concomitant populations of lesion and root knot nematodes can mutually suppress reproduction, but the specific resistance characteristics of the cultivars involved can significantly change the characteristics of these interactions (Johnson, 1998). Populations of *P. penetrans* can also be suppressed by concomitant populations of *Tylenchorhynchus claytoni* and *Globodera tabacum tabacum* (Johnson, 1998). Although suppression by *G. t. tabacum* is mutual, the antagonistic interaction only reduced population densities of *P. penetrans*.

Although lesion nematodes are rarely considered major tobacco pests (Table 18.5), they can cause significant yield losses when they occur in large numbers (2000/kg of soil) and under appropriate environmental conditions (Olthoff *et al.*, 1973). Losses in gross economic returns were estimated at 11 and 27.5% when initial population densities were 6000 and 18,000/kg of soil, respectively. Up to 70% of the flue-cured tobacco acreage in Ontario, Canada has suffered from brown root rot (Tu *et al.*, 1996). *Pratylenchus* spp. often have a wide host range and, because they can overwinter in plant roots and withstand desiccation, they can remain viable from tobacco crop to tobacco crop.

Globodera

Various tobacco types produced around the world are attacked by members of a species complex of round cyst nematodes. Species within this tobacco cyst nematode (TCN) complex have a narrow host range, only including tobacco and certain members of the Solanaceae. Problems with these nematodes have been reported from Argentina, Mexico and Spain since the 1987 CORESTA survey of tobacco diseases and nematodes (Ambrogioni and D'Erico, 1995; Marché *et al.*, 2001; Espárrago, 2002; C.S. Johnson, unpublished) (Table 18.5). Unfortunately, many reports of these nematodes from around the world do not identify the specific member of the species complex that is involved. The taxonomy of the complex, originally based upon biochemical,

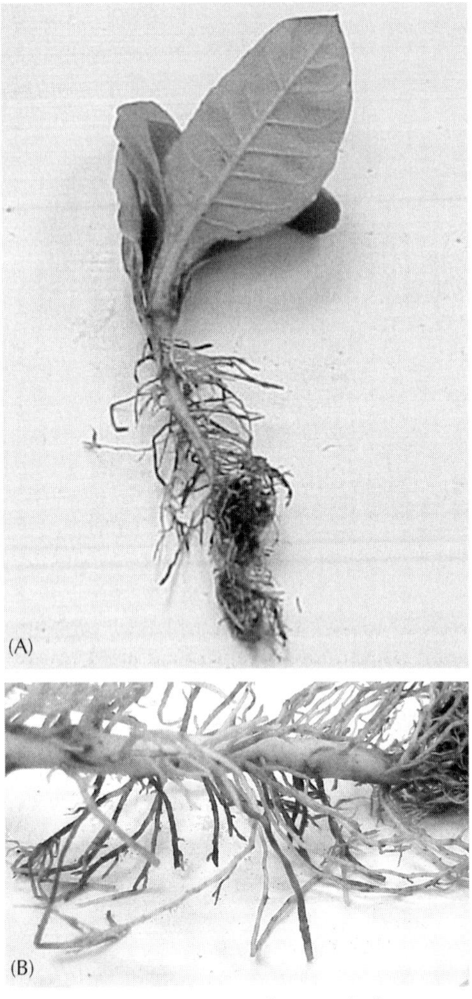

Fig. 18.5. Brown root rot on flue-cured tobacco. (A) Necrotic roots on a young flue-cured tobacco plant; (B) close-up on discrete necrosis of small feeder roots. (Photo: C. Johnson.)

hybridization and morphometric data, has been confirmed more recently using molecular techniques such as restriction fragment length polymorphism (RFLP), random amplified polymorphic DNA (RAPD) and amplified fragment length polymorphism (AFLP) (Thiery and Mugniery, 1996; Thiery *et al.*, 1997; Johnson, 1998; Marché *et al.*, 2001). *G. t. tabacum* has been an important pest of shade-grown tobacco in Connecticut since 1951, and has been identified on tobacco in France (Thiery and Mugniery, 1996; Johnson, 1998). A second subspecies, *G. tabacum solanacearum*, attacks flue-cured tobacco and has spread from Virginia to some counties in North Carolina, and may be present in Mexico (Johnson, 1998; Marché *et al.*, 2001). Crop production practices, soil temperature and antagonistic or competing nematode genera are important factors influencing the spread of this nematode, since *G. t. solanacearum* reproduced similarly in a range of flue-cured tobacco-producing soils from the USA (Rideout *et al.*, 2000a). Another subspecies, *G. tabacum virginiae*, also occurs in Virginia and North Carolina. This member of the TCN species complex does not attack flue-cured cultivars, but reproduces slowly on burley tobacco (Shepherd and Barker, 1990). TCNs have been reported to occur in Morocco, but not from southern Africa (Table 18.5). Limited reproduction by *G. pallida* and *G. rostochiensis* on tobacco has been reported, but this has never been confirmed (Meredith, 1976; Parrott and Miller, 1977).

Above-ground symptoms of TCN parasitism are similar to those associated with severe root knot and lesion nematode infestations (Fig. 18.6). However, TCN-infected plants have small root systems with cysts attached to them, and significant root rot is not observed unless a root disease complex is operating (Fig. 18.7; Plate 19D). TCN cysts are the dried and hardened bodies of adult female nematodes, each of which may contain several hundred eggs. Cysts can range in colour from pearly-white to reddish brown and are barely visible (0.5 mm) to the naked eye. The eggs within these cysts can survive for 11 years or more until stimulated to hatch by temperature and host root exudates, and are difficult to kill with nematicides (Johnson, 1998).

The TCN complex is often associated with increased damage from bacterial wilt, black shank and Fusarium wilt (Elmer *et al.*, 1980; Johnson, 1998) (Fig. 18.8; Plate 19E). TCN increases disease in these interactions via a localized, versus a systemic effect. Although *M. hapla* also increases Fusarium wilt, *G. t. tabacum* increases root infection by *F. oxysporum* to a greater extent, although on wilt-susceptible but not

Fig. 18.6. Stunting of flue-cured tobacco by *Globodera tabacum solanacearum* in Virginia. (Photo: C. Johnson.)

Fig. 18.7. Cysts of *Globodera tabacum solanacearum* on roots of flue-cured tobacco. (A) White females on the roots of a tobacco transplant; (B) brown cysts on tobacco roots. (Photo: C. Johnson.)

resistant tobacco (LaMondia, 1995b). TCNs may also suppress reproduction of mycorrhizal fungi, which are important in the normal development of tobacco (Johnson, 1998). TCNs also interact with concomitant populations of other plant parasitic nematodes, such as *Pratylenchus* spp. (Barker and Lucas, 1984). Although high populations of *P. penetrans* can slow TCN population increase, moderate to high TCN populations can suppress reproduction of *P. penetrans* until the latter are undetectable.

Yield losses of infected tobacco can be very high. Virginia farmers have recorded complete crop failures, but losses generally average 15% (Johnson, 1998). High TCN populations early in the growing season can reduce flue-cured tobacco yield by 25–50%, although tobacco may escape significant losses from moderate populations, especially under favourable growing conditions (Johnson, 1992). Although the relationships between TCN populations and flue-cured tobacco yield may vary considerably across years and cultivars, consistently significant negative correlations have been observed between fresh leaf weight and TCN numbers in soil 6 weeks after transplanting (Wang *et al.*, 1999). Although initial population densities of *G. t. tabacum* below 100 juveniles/cm^3 of soil

Fig. 18.8. Plant mortality in flue-cured tobacco caused by a *Globodera tabacum solanacearum–Fusarium* nematode–disease complex. (Photo: C. Johnson.)

may reduce shade tobacco leaf yield by less than 5%, populations between 500 and 1000 juveniles/cm^3 can reduce yield by 45% (Johnson, 1998). Initial TCN populations below 50 juveniles/cm^3 of soil can reduce shoot weight of shade and broadleaf tobacco by 39 and 14%, respectively (LaMondia, 2002a). Populations above 600 TCN juveniles/cm^3 reduced shoot weight of shade and broadleaf tobacco by 60 and 40%, respectively.

Rotylenchulus reniformis

R. reniformis limits seed germination, nutrient uptake and growth of tobacco seedlings in India, as well as yield and value (Johnson, 1998). This nematode has also been reported on tobacco in Trinidad and North Carolina, but research in North Carolina indicated only moderately increased yield and value arising from nematicide use and suggested that tobacco may not be a good host for *R. reniformis* (Melton and Powell, 1991). Both *M. incognita* and *M. javanica* damage tobacco more than *R. reniformis* and can also suppress reproduction of the reniform nematode (Johnson, 1998).

Ditylenchus

Ditylenchus dipsaci, the stem and bulb nematode, occurs in many countries, but yield loss in tobacco has only been reported from The Netherlands, France, Germany, Switzerland and Serbia (Shepherd and Barker, 1990; Johnson, 1998) (Table 18.5). Nematode isolates extracted from tobacco may not parasitize other crops, such as wheat (*Triticum aestivum* L.), maize (*Zea mays* L.) or sugarbeet (*Beta vulgaris* L. subsp. *vulgaris*) (Johnson, 1998). *Ditylenchus destructor* has also been reported to reproduce on some tobacco genotypes to a limited extent (Johnson, 1998). Invasion of the lower parts of the stem by the nematode causes 'stem break', which is very rarely found in subtropical or tropical countries. The nematode can remain dormant in a cryptobiotic stage for many years and withstand freezing. 'Stem break' is usually associated with cool, damp weather and heavy soils and is only of localized importance, but has been reported to cause losses of up to 54% in parts of north-east France (Shepherd and Barker, 1990).

Other nematodes

A variety of other nematodes have been associated with tobacco in isolated reports or from very restricted production areas. *Aphelenchoides* species have been reported on tobacco in France, Germany, China, Pakistan, Brazil and Chile (Table 18.5). In France, *A. ritzemabosi* has been described as the cause of 'checkered leaf disease' in a localized area near the Atlantic end of the Pyrénées (Shepherd and Barker, 1990). The polygonal leaf blotches bounded by the veins caused by *A. ritzemabosi* are similar to those it causes in chrysanthemums. Various species of *Tylenchorhynchus* have been reported on tobacco in New Zealand, Canada, the USA and India (Shepherd and Barker, 1990; Patel and Patel, 1992, 1999). *Tylenchorhynchus vulgaris* interacts with *Pythium aphanidermatum* to reduce seedling vigour and stand in bidi tobacco plant beds in India (S.B. Patel and H.R. Patel, 1998). Initial populations of 1000 *T. vulgaris*/700 g of soil reduced plant growth and nicotine content (Patel and Patel, 1993). Stunt nematodes have been reported to increase the incidence of Fusarium wilt, but not Granville wilt, and may not damage tobacco directly (Shepherd and Barker, 1990). The spiral nematode is frequently reported from tobacco soil, and *Scutellonema brachyurum* has been reported to reduce growth, but it is considered a very minor pest.

Various nematode species may damage tobacco by vectoring plant viruses (Johnson, 1998). *Paratrichodorus* and *Trichodorus* species vector the 'tobacco rattle' virus in parts of The Netherlands and Germany, and *Xiphinema* and *Longidorus* species are widespread and are also virus vectors. *P. lobatus* is also reported to cause stunting of tobacco in Australia. *X. americanum* is a relatively efficient vector of the tobacco ringspot virus, which is reported in many countries and has localized importance. *L. elongatus* is also reported to damage tobacco in Canada (Shepherd and Barker, 1990).

Management measures

Nematode management measures for tobacco vary widely around the world. Losses to nematode damage are considered slight in some tobacco-growing regions and little attention is paid to nematode control measures. Such regions are often those where tobacco is grown under cool conditions, on heavier soil types, or where the root knot nematode is not widely distributed. However, there are still many countries where nematodes do cause economic losses and little attention is paid to nematode control, especially where itinerant farmers possess insufficient resources to purchase inputs for effective control measures. In other areas, cultural practices such as crop rotation and host resistance are sufficient to limit crop losses to acceptable levels. However, in production regions such as the USA, Australia, and parts of central and southern Africa, where tobacco is an extremely important cash crop and where nematodes, especially root knot, are widely distributed, the entire tobacco-growing cycle can be centred around nematode control. The basic strategy for nematode control for tobacco, in general, is to reduce the initial nematode populations in the soil and/or in transplants and to reduce the subsequent rate of nematode increase.

Cultural control

Cultural practices such as early destruction of tobacco roots after harvest, early and deep ploughing of tobacco fields into high, wide planting ridges before transplanting, early planting, and use of appropriate cover and rotation crops provide the foundation for consistently effective nematode management in tobacco (Shepherd and Barker, 1990; Johnson, 1998). Early destruction of tobacco roots limits nematode reproduction after harvest has been completed, reducing populations in soil awaiting future tobacco crops. Early and deep ploughing of fields exposes nematode populations to adverse temperature and moisture conditions, particularly when pre-transplant cultivation builds elevated beds or ridges into which the crop will be transplanted (Johnson, 1998). Early transplanting may enable tobacco seedlings to begin establishing a functional root system when temperatures in soil are less favourable for nematode hatching and migration. Farmers unable to purchase expensive nematicides or too dependent upon economic returns from tobacco to plant low-value rotation crops can use these methods to reduce, but not eliminate, their losses to nematode parasitism. Cultural nematode management practices also enable farmers who use crop rotation and nematicides to maximize the nematode control benefits from these practices.

Rotating fields away from tobacco reduces soil populations of tobacco parasites, but the effectiveness varies for different nematode species, and nematode management benefits must be balanced with economic and environmental considerations. In general, the longer the time interval between tobacco crops, the better, particularly in regions with longer growing seasons and shorter, milder (frost-free) winters (Barker and Lucas, 1984). Bare fallow reduces nematode populations, but provides no economic return and promotes soil erosion (Shepherd and Barker, 1990; Patel *et al.*, 1994). Weedy fallows can allow nematode populations to increase on alternative hosts, particularly for nematodes with wide host ranges, such as *Meloidogyne* and *Pratylenchus* spp. (Gowda *et al.*, 1995; Johnson, 1998). Although ten accessions of the leguminous shrub *Sesbania sesban* and one accession of *S. macrantha* were poor to moderate hosts of *M. javanica*, a 3 year rotation of 'natural fallow' and maize with tobacco was recommended for management of *M. incognita* and *M. javanica* in Tanzania (Karachi, 1995; Shirima *et al.*, 2000).

Meloidogyne and *Pratylenchus* spp. have wide host ranges, but with significant differences in host range among the species within each genus (Johnson, 1998). In southern Africa, the main nematode pest of tobacco is *M. javanica*, but in many other parts of the world it may be *M. incognita*, *M. arenaria* or one of the *Globodera* species. In cooler tobacco-producing areas, *Globodera* or *Pratylenchus* species or *D. dipsaci* may severely limit yield. The choice of rotation crops is made more difficult when mixtures of root knot nematode species are present, as in the USA, South Africa, Turkey, Brazil, the Philippines, Mexico, Hungary, Iraq, Thailand and Greece.

Small grains and forage grasses such as fescue (*Festuca pratensis*) are recommended commonly to reduce root knot and cyst nematode populations in tobacco fields, although the choice of crops, and even cultivars, to rotate with tobacco should depend on the most important nematode species present and the economic circumstances of the grower (Bertrand, 2002; Fortnum, 2002; Melton and Broadwell, 2002). Care must also be exercised that weed populations that develop between the small grain and tobacco crops do not include nematode hosts such as crabgrass, as such build-up can nullify the beneficial effect of the rotation (Clayton *et al.*, 1944). Pasture grasses protect the soil from erosion better than row crops and, if sown densely enough, will smother weeds which might be nematode hosts. In southern and central Africa, the Ermelo and Umgeni strains of weeping lovegrass (*Eragrostis curvula*), Katambora Rhodes grass (*Chloris gayana*) and Sabi Panic grass (*Panicum maximum*) are recommended, especially when grown for 3 or 4 years before planting tobacco (Shepherd and Barker, 1990). Other grasses, such as some of the *Paspalum* species and *Digitaria decumbens*, are resistant to *M. javanica* and some of the other root knot nematodes, but do not fit well into a tobacco rotation (Shepherd and Barker, 1990). The sunnhemps, *Crotalaria juncea*, *C. spectabilis* and *C. intermedia*, also *C. fulva* and *C. grahamiana* can be used to suppress root knot nematodes (Shepherd and Barker, 1990). The toxins produced by some *Crotalaria* spp. are toxic to livestock and the plants may persist as weeds in subsequent crops (Johnson, 1998). Also, the increased nitrogen status of the soil after a legume is not always desirable for flue-cured tobacco.

Maize is often grown in tobacco rotations, and resistant cultivars can lower populations of *M. javanica* to levels easily controlled by nematicides if grown for 2 years or more (Shepherd and Barker, 1990). However, maize is not generally recommended for root knot control in the USA due to the varying degrees of susceptibility among cultivars to all of the common *Meloidogyne* spp. except *M. hapla* (Bertrand, 2002; Fortnum, 2002; Melton and Broadwell, 2002). Grain sorghum suppressed South Carolina populations of *M. arenaria* race 2 and *M. incognita* race 3, and could be a useful rotation crop for tobacco (Johnson, 1998). In fact, sorghum supported minimum reproduction of *M. arenaria* race 2, race 3 of *M. incognita*, and *M. javanica* (Fortnum *et al.*, 2001b). Rotation with cotton or groundnuts reduced initial populations of *M. javanica* in Zimbabwe (Shepherd and Barker, 1990), but race 2 of *M. arenaria* predominated over *M. incognita* when cotton, maize, sorghum or rye fallow preceded tobacco in South Carolina trials (Fortnum *et al.*, 2001b). Populations of *M. arenaria* race 1 can increase on groundnuts to levels that will cause moderate damage to tobacco (Hirunsalee *et al.*, 1995a). Where vegetable crops highly susceptible to *Meloidogyne* spp. are grown, particularly in peasant agriculture, the damage to subsequent tobacco crops may be severe.

Most *Pratylenchus* species have a wide host range, and this can cause problems in selecting rotation crops. Lesion nematode populations can increase on bluegrass (*Poa* spp.), maize (*Zea mays* L.), rye (*Secale cereale* L.) and many legumes, but barley (*Hordeum vulgare* L.), oat (*Avena sativa* L.) and sweet potato (*Ipomoea batatas* (L.) Lam.) limit reproduction of some *Pratylenchus* species (Johnson, 1998). Rotating tobacco with marigolds (*Tagetes* spp.) reduced populations of *P. penetrans* in Ontario, Canada below the economic threshold for 3 years (Reynolds *et al.*, 2000). Work in both Ontario and Quebec found rotating tobacco with forage pearl millet also to be highly effective in reducing populations of *P. penetrans* (Jagdale *et al.*, 2000; Bélair *et al.*, 2002). The cyst nematodes and *Ditylenchus dipsaci* have very limited host ranges, which facilitate their control by rotation, but their effective survival mechanisms may require the use of non-host crops for a long time. However, use of tomato or resistant tobacco as a trap crop reduced populations of *G. t. tabacum* by 64–84% (LaMondia, 1996b). Planting a trap crop after harvest and prior to seeding a rotation or cover crop was suggested as a practical and effective method to reduce TCN populations.

Physical control

Many of the early attempts to control nematode pests of tobacco, particularly in seedbeds or nurseries, relied on heating the soil either by burning grass and brushwood on the surface or by steaming the soil under a cover. Even though burning was recommended, it was realized that heat penetration was not always enough to kill nematodes at depths below 150 mm (Shepherd and Barker, 1990). However, peasant farmers continue to use burning – or rabbing as it is called in India – as their only method of seedbed control. A split-furrow rabbing method that involved burning husks from tobacco seed, pearl millet or wheat straw and tobacco stalks provided good control of *M. incognita*, *M. javanica* and various weeds in bidi tobacco seedbeds (Patel *et al.*, 1993). Steaming can kill weeds, nematodes, insects and fungal pathogens, but upsets the balance of soil bacteria similar to fumigation under some conditions, possibly leading to increases in soil ammonium and manganese toxicity. Effective penetration is usually about 300 mm but, being slow and expensive, the method is only suitable for seedbeds.

Research results with soil solarization have been variable and very sensitive to environmental factors that are very hard to control, prompting suggestions that this method may not be practical for commercial-scale agriculture (Noling and Becker, 1994). Reports from Cuba, India, Italy and Tanzania, however, indicate that soil solarization can suppress nematode populations in tobacco seedbeds long enough to produce usable transplants (Patel *et al.*, 1995a,b, 2001; Johnson, 1998). Solarization for a minimum of 15 days (and preferably for 40 days or longer) may provide an economic and practical method for small-scale farmers to significantly reduce nematode parasitism on tobacco transplants (Iglesias *et al.*, 1998; Hussaini *et al.*, 2001; Ravindra *et al.*, 2001). Control of *M. incognita*, *M. javanica*, *Rotylenchulus reniformis* and *Tylenchorhynchus vulgaris* from soil solarization (alone or combined with cultural practices such as green manuring and rabbing) has been reported as similar to that from contact nematicides such as cadusafos and phenamiphos (Patel and Patel, 2001; Patel *et al.*, 2001). Nematode populations in soil rebounded more quickly in solarized soil (and in that treated with contact nematicides) than in soil treated with the fumigant dazomet (B.K. Patel and H.R. Patel, 1998, 1999; H.R. Patel and B.N. Patel, 1998). However, solarization may change other properties of soil in tobacco seedbeds as well, and such effects should be accounted for (Patel and Patel, 1997).

Flooding has been associated with a degree of nematode control in places where the tobacco fields are flooded naturally or where tobacco is grown after paddy rice (Shepherd and Barker, 1990). Seventy-five days were needed to reduce the root knot populations by flooding, and some nematodes survived for up to 105 days.

Growers in regions without practical and effective root knot control options for transplant production have been advised to cut off as much of the galled root as possible before planting (Shepherd and Barker, 1990). Although this practice does not provide control, it does reduce nematode parasitism during the critical first few weeks after transplanting, hindering increase of the nematode population while plants are growing in the field (Murthy *et al.*, 1999).

Resistance

Most flue-cured tobacco cultivars planted in the USA are resistant to races 1 and 3 of *M. incognita*, and this resistance has been incorporated into a large number of tobacco cultivars grown throughout the world (Johnson, 1989). Although root exudates of resistant tobacco cultivars may have nematicidal properties, resistance operates by preventing successful establishment of a feeding site rather than by inhibiting penetration (Shukla *et al.*, 1988; Schneider, 1991). Considerable variability has been reported in the reactions of specific root knot-resistant cultivars to various *Meloidogyne* species, and even populations, but resistance to races 1 and 3 of *M. incognita* and host race 1 of *M. arenaria* was found to be conditioned by the same gene (Ng'ambi *et al.*, 1999a). Cultivars possessing this gene have also been found to possess limited resistance to race 2 of *M. arenaria*, slight resistance to *M. javanica*, and to possess more tolerance to attack from *Meloidogyne* species in general (di Vito *et al.*, 1998; Johnson, 1998; Ng'ambi *et al.*, 1999b). Currently, all LK varieties in the Republic of South Africa (RSA) have resistance to *M. incognita* races 1 and 3, and although the RSA is supposed to have only races 2 and 4 of *M. incognita*, this resistance is holding up pretty well (A. Scholtz, personal communication). The gene responsible for resistance to races 1 and 3 of *M. incognita* (*Rk*) was apparently transferred into cultivated tobacco from *N. tomentosa*, and has been mapped to chromosome G (Rufty *et al.*, 1983b; Yi and Rufty, 1988; Yi *et al.*, 1998). Several greenhouse studies suggested that prior infection with *M. arenaria* or *M. hapla* reduced the effectiveness of the *Rk* gene, although prior infection with other species of *Meloidogyne* did not reduce resistance to *M. javanica* in Zimbabwe (Shepherd and Barker, 1990; Johnson, 1998). Reproduction and development of *M. arenaria* and *M. incognita* were mutually suppressed when these species parasitized roots simultaneously, but infection by mixed populations of *Meloidogyne* species did not increase total nematode parasitism of *M. incognita*-resistant cultivars (Ng'ambi *et al.*, 1995). Split-root greenhouse experiments indicated that prior or simultaneous infection of one root portion by *M. arenaria* did not systematically predispose the other root portion to infection by *M. incognita* (Baum *et al.*, 1995b). Resistance due to the *Rk* gene was unaffected by *M. arenaria* race 2 infection across temperatures ranging from 25 to 35°C. Field experiments also indicated that resistance to race 3 of *M. incognita* was not altered by simultaneous inoculation with race 2 of *M. arenaria* (Baum *et al.*, 1995a). Breeding lines resistant to race 3 of *M. incognita* have also been obtained from a *N. repanda* × *N. tabacum* cross, and to races 1 and 4 from accessions of *N. otophora* (Johnson, 1998).

Cultivars with tolerance to *M. javanica* can be effective in fields with low to moderate infections (Shepherd and Barker, 1990; Jack, 1996). Resistance to *M. javanica* has been found in *N. longiflora*, *N. megalosiphon* and *N. repanda*, as well as several other sources (Johnson, 1998). Flue-cured tobacco cultivars with the *Rk* gene and also with resistance to *M. javanica* are now available in Zimbabwe (Kutsaga RK1, Kutsaga RK6, Kutsaga RK8, Kutsaga RK22, Kutsaga RK23, Kutsaga RK26 and Kutsaga RK28). Zimbabwean burley cultivars Banket BRK1, Banket BRK2 and Banket BRK3 also carry resistance to *M. javanica*, but from a different source. Other tobacco cultivars, both burley and flue-cured, are currently being tested. Current breeding programmes in the RSA are also incorporating *M. javanica* resistance into the LK lines. Unfortunately, resistance to *M. incognita* and *M javanica* can break down at extremely high temperatures of 30–35°C, and some strains of *M. javanica* have been reported to parasitize reportedly resistant cultivars (Shepherd and Barker, 1990).

Reduced reproduction by *M. arenaria* has been noted on selections of *N. knightiana*, *N. sanderae* and *N. velutina*, and by both *M. arenaria* and *M. javanica* on *N. glauca*, *N. longiflora*, *N. nudicaulis*, *N. plumbaginifolia* and *N. repanda* (Johnson, 1998). Six tobacco breeding lines were found to be resistant to a North Carolina population of *M. arenaria* race 2 that should be useful in breeding programmes (Ng'ambi *et al.*, 1999b). Resistance to *M. arenaria* and *G. tabacum* is currently being incorporated into cultivated tobacco germplasm, with some promising results against a nematode–*Fusarium* complex (Fig. 18.9; Plate 19F) (G. Esparrago and E.A. Wernsman, personal communication).

Fig. 18.9. Plant mortality from a nematode disease complex in Spain involving *Meloidogyne arenaria*, *Globodera tabacum* and *Fusarium oxysporum*. (Photo: E.A. Wernsman.)

Despite this progress, germplasm resistant to *M. javanica*, *M. arenaria* and *M. mayaguensis* remains limited (Botha, 1996, 2002). The most recent results from the CORESTA Working Group for Nematodes listed seven entries with root gall indices (scale 0–5) below 1 and ten entries with gall indices between 1 and 2 (Table 18.6). Another 30 entries were recorded as having gall indices greater than 2. Unfortunately, a number of the seven more resistant entries are male sterile and cannot be used in breeding programmes.

More and more apparently pleiotropic effects of nematode resistance genes are being reported. The tomato gene *Mi-1.2* confers resistance to some isolates of the potato aphid (*Macrosiphum euphorbiae*) and to the B- and Q-biotypes of the whitefly *Bemisia tabaci*, in addition to *M. incognita*, *M. javanica* and *M. arenaria* (Nombela *et al.*, 2003). Unfortunately, a severe vascular necrosis in response to infection by the M^SN^R strain of potato virus Y (PVY-M^SN^R) appears to be a pleiotropic effect of the *Rk* gene (Rufty *et al.*, 1983a,b). However, resistance to PVY seems to be conditioned by a recessive gene epistatic to the *Rk* gene, and although *N. tomentosa* was the source of the *Rk* gene, accession 58 of that species was also found to be immune to the virus (Rufty *et al.*, 1983b). Resistance to *G. t. solanacearum* from *N. longiflora* is tightly linked to resistance to the wildfire bacterium *Pseudomonas syringae* pv. *tabaci* (Hayes *et al.*, 1997), and the *Ph* gene for resistance to tobacco black shank (*P. parasitica* var. *nicotianae*) has also recently been linked to resistance to TCNs (Johnson and Clarke, 2003). Knowledge of such linkages is crucial to appropriately deploying resistant cultivars to the range of pathogens and pests important in the different tobacco production regions around the world.

Research at the molecular level is increasing our understanding of nematode feeding site establishment in tobacco, and

Table 18.6. 2002 ranking of tobacco entries by the CORESTA Working Group for Nematodes according to galling[a] by *Meloidogyne javanica*.

Gall index ≤ 1		Gall index ≤ 2		Gall index ≤ 3		Gall Index >3	
Entry	Gall index	Entry	Gall index	Entry	Gall index	Entry	Gall index
COLM 54	0.20	T 14	1.20	N20x234	2.05	PVHO2	3.06
Coltab 35	0.32	KRK3	1.26	Domkrag	2.14	MTRA 92	3.12
M 208	0.34	LN 7	1.34	TI 1717	2.18	N8xTL33	3.15
RL2-1-1	0.37	N20x272	1.45	N20x242	2.18	ABL 109	3.24
T 26	0.46	RK8	1.60	OD 694	2.22	Coker 371	3.29
STNCB	0.67	RLC-17	1.66	ODT 73/1/2	2.22	K 326	3.32
NOD 8	0.86	RK6	1.75	K3x272	2.23	KBM 33	3.35
		NODKK3-6	1.88	ODT 62/1/1	2.24	OD 661	3.38
		FLS 89	1.89	WZxRL2-1-1	2.27	KY 907	3.4
		RK1	2.00	T 30	2.28	MTRA 88	3.48
				ODT 4	2.28	MZ 86	3.53
				TL 33	2.46	BM 8410	3.71
				OD 679	2.47	HG	3.74
				OD 490	2.52	B84-1052	3.86
				ODT 73/1/1	2.54		
				NC 95	2.67		
				ABL 34	2.82		
				OD 697	2.90		
				OD 668	2.93		
				TBV 63	3.00		

[a]Gall index scale = 0–5, with 0 = no galling and 5 = 100% root galling.

may enable development of more effective and durable resistance to nematodes, particularly to *Meloidogyne* species (Opperman and Conkling, 1994). This research has shown that the expression of many host genes changes as nematodes attempt to establish their feeding sites (Goddijn *et al.*, 1993). Tobacco plants engineered to constitutively express genes to produce glutamate decarboxylase (GAD) appeared to confer resistance to *M. hapla* (McLean *et al.*, 2003). Identification of host genes expressed only at nematode infection sites (such as *TobRB7*) could be used to target proteins inhibitory or toxic to nematode feeding structures (giant cells) only when and where they would be needed (Opperman *et al.*, 1994).

N. paniculata, *N. glutinosa*, *N. longiflora*, *N. plumbaginifolia*, *N. cordifolia*, *N. miersii*, *N. alata*, *N. repanda* and *N. noctiflora* are resistant to *G. t. solanacearum*, as are several tobacco introductions (Shepherd and Barker, 1990; Herrero *et al.*, 1996; Hayes *et al.*, 1997). Suppressed reproduction by *G. t. solanacearum* has been noted in tobacco cultivars originally developed for resistance to wildfire (*P. syringae* pv. *tabaci*), TMV and tobacco black shank (*P. parasitica* var. *nicotiana*). Wildfire resistance was incorporated into cultivated tobacco from *N. longiflora* and has been linked with resistance to *G. t. solanacearum* (Hayes *et al.*, 1997; LaMondia, 2002b). The TMV resistance in mosaic and *G. t. solanacearum*-resistant flue-cured tobacco cv. 'NC 567' was obtained from *N. glutinosa* (Holmes, 1938). Suppression of *G. t. solanacearum* reproduction has also been linked with the *Ph* gene from *N. plumbaginifolia* for resistance to tobacco black shank (Carlson *et al.*, 1997; Johnson, 2001; Johnson *et al.*, 2002). Resistance against *G. t. solanacearum* has, so far, also been found to be effective against *G. t. tabacum* (LaMondia, 1988, 2002b). Although resistance to *G. t. solanacearum* from *N. longiflora* has been reported to be multigenic (Spasoff *et al.*, 1971; Miller *et al.*, 1972; Crowder *et al.*, 2003), resistance to *G. t. tabacum* from the same sources, and from *N. plumbaginifolia*, appears to be conditioned by a single dominant gene(s) (LaMondia, 1991, 2002b; Crowder *et al.*, 2003). Resistant cultivars are now available (LaMondia, 2000a,b; Johnson, 2002).

Although host root exudates stimulate hatching of TCNs, hatching and penetration are similar for resistant and susceptible cultivars (LaMondia, 1988, 1995a; Wang *et al.*, 1997, 2001). Resistance operates by inhibiting nematode feeding site establishment and possibly subsequent nematode development, and remains effective at 30°C, in contrast to the *Rk* gene for root knot resistance (Wang *et al.*, 2001). Early work associated resistance to *G. t. solanacearum* with severe root necrosis and stunting, but later research found no such association (Wang *et al.*, 1999). Effects of *G. t. solanacearum* on root size were similar on a resistant and a susceptible cultivar, but the increased parasitism in roots of the susceptible cultivar caused greater losses in leaf weight. Crop rotation and nematicides are used with resistant cultivars to reduce initial root damage and increase yields (Johnson *et al.*, 1989; Johnson, 1990; LaMondia, 2002b). To date, nematode biotypes with increased reproductive ability on resistant cultivars have not been detected (Elliott *et al.*, 1986; Rideout *et al.*, 2000b; Syracuse, 2003).

Although high populations of *Pratylenchus* spp. can significantly damage tobacco, it seems to be a relatively poor host for these nematodes. Consequently, research on improved management of brown root rot has focused on crop rotation and nematicides. However, reduced *Pratylenchus* reproduction has been associated with the *Rk* gene for resistance to races 1 and 3 of *M. incognita*, and both partial resistance and tolerance have been noted among tobacco cultivars (Barker and Lucas, 1984).

Although tobacco cultivars have been found to be moderate to good hosts for *Tylenchorynchus vulgaris* in India (Patel and Patel, 1990), tolerance to parasitism was also found (A.S. Patel and H.R. Patel, 1991). *N. longiflora*, *N. glauca* and *N. repanda* are reported to be resistant, while *N. tabacum*, *N. rustica*, *N. nudicales* and

N. plumbaginifolia were rated as tolerant to stunt nematode (S.B. Patel and H.R. Patel, 1995).

Chemical control

Nematicides remain an important tool in many tobacco production areas, despite the widespread use of nematode-resistant cultivars, particularly for root knot control, and the loss of a number of effective products. Some nematicide products are no longer available due to concerns about detrimental effects to the environment, but others have been lost due to problems with continued effectiveness.

Tobacco transplants are now largely produced in greenhouse or 'float-bed' hydroponic systems in many countries, but fumigants continue to be necessary elsewhere for disease, nematode and weed control in tobacco seedbeds. Methyl bromide has been the most widely used seedbed fumigant because of its excellent broad spectrum pest control and ease of use (Shepherd and Barker, 1990), but concerns about atmospheric ozone depletion are restricting, and may eliminate, its availability (Duniway, 2002; Gullino *et al.*, 2003). A number of possible alternative fumigants to methyl bromide are being evaluated, but none has yet emerged as a single replacement pesticide (Table 18.7).

Metam sodium alone or combined with 1,3-dichloropropene (1,3-D) plus chloropicrin provided good disease, nematode and weed control in outdoor seedbeds in the USA (Csinos *et al.*, 1997, 2000). Metam sodium is also being evaluated for seedbed soil disinfestations in Zimbabwe, as well as combinations of metam sodium with ethylene dibromide (EDB) and of 1,3-D with chloropicrin (J.I. Way, 2003, personal communication). Application of cadusafos, carbofuran and fenamiphos reduced parasitism by *M. incognita* and improved plant growth in outdoor seedbeds in India (Swathi *et al.*, 1998; Gowda, 1999), but use of dazomet provided similar or better control of *Meloidogyne* spp. than fenamiphos (Ramakrishnan *et al.*, 1999b). Of course, no disease or weed control would be expected from the non-fumigant nematicides.

Tobacco fields are commonly fumigated for disease and nematode control in tobacco production areas such as the USA and southern Africa. EDB continues to be commonly used in southern Africa, but is not available in the USA, where 1,3-D and chloropicrin are often applied, usually as a mixture of the two compounds within a product (Melton and Broadwell, 2002). Fumigation in the USA often targets nematode disease complexes, particularly bacterial wilt (*Ralstonia solanacearum*), and usually involves row treatments applied 2–3 weeks before transplanting, although some fields are broadcast-

Table 18.7. Possible alternative chemicals to methyl bromide for soil disinfestations (after Duniway, 2002).

	Requiring further development	
Currently available in the USA	MBTOC[a]	Additions[b]
Chloropicrin	Methyl iodide	Other halogenated hydrocarbons
1,3-dichloropropene	Propargyl bromide	Propylene oxide
Methyl isothiocyanate generators:	Ozone	Sulphur dioxide
Metam sodium	Formaldehyde	Peroxyacetic acid
Dazomet	Sodium tetrathiocarbamate	Acrolein (2-propenal)
	Carbon disulphide	Others to be developed
	Anhydrous ammonia	
	Inorganic azides	
	Natural compounds	

[a]Alternatives considered by the 1998 report of the Methyl Bromide Technical Options Committee, United Nations Environmental Programme.
[b]Alternatives added by J.M. Duniway.

fumigated, sometimes as early as the previous autumn (Johnson, 1998; Fortnum and Pullen, 2001). Canadian research on fumigation to control *P. penetrans* has shown that these products influence microbial activity in the soil, particularly that of nitrifying bacteria, early in the growing season (Tu *et al.*, 1995a,b, 1996). These effects generally dissipate by mid-season but, under certain environmental conditions (particularly prolonged cold, wet weather), early inhibition of soil nitrification can increase total alkaloids and decrease reducing sugars in cured leaf, reducing tobacco quality (Shepherd and Barker, 1990). Mixtures of methyl bromide and chloropicrin have been used extensively for nematode control in the USA, but their cost has increased dramatically as the availability of methyl bromide has been restricted, to the extent that these products are no longer economical. Metam sodium is being used successfully on a very limited basis to control *G. t. solanacearum* in sandy loam soils in Virginia, but did not control *Meloidogyne* spp. in heavy clay soils in Zimbabwe (Anonymous, 1979; Johnson and Wilkinson, 2002).

Although non-fumigant nematicides such as aldicarb, fenamiphos, ethoprop and oxamyl do not reduce nematode populations as effectively as fumigants, they have been used extensively in America (Shepherd and Barker, 1990; Lamberti *et al.*, 1993, 2000; Moreno *et al.*, 1994; Fortnum *et al.*, 2001b). A South Carolina study found that increased yield and value from fumigation were greater than any pesticide cost savings associated with use of the non-fumigant nematicide/insecticides (Fortnum *et al.*, 2001a). Residual suppression of nematode parasitism was greater with fenamiphos than aldicarb, and least with carbofuran (Melton *et al.*, 1995). Aldicarb is also used in Zimbabwe, Malawi and South Africa, but is not recommended for use when there is a high root knot nematode population (Shepherd and Barker, 1990). In the USA, the use of aldicarb is restricted to North Carolina and Virginia due to environmental concerns. Oxamyl effectively reduced initial populations of *G. t. tabacum* (LaMondia, 1996a). Use of non-fumigant nematicides has dropped significantly over the past 10 years due to considerations such as increased regulatory restrictions, improved fumigation technology and wider recognition of nematode disease complexes. In addition, annual use of some nematicides seems to have led to reduced effectiveness, probably due to enhanced biodegradation (Davis *et al.*, 1993). Fosthiazate provided good to excellent control of *M. arenaria*, *M. incognita* and *G. t. solanacearum*, but has never been registered in the USA (Johnson, 1995; Pullen and Fortnum, 1999).

Fenamiphos has been recommended in Malawi as a pre-plant nematicide. Fenamiphos and oxamyl have each been used as supplements to fumigation in southern Africa to extend the period of control when a nematode population is high or where there are poor growing conditions early in the season (Johnson, 1998). Non-fumigant nematicides have also been applied as root dips or spot treatments. Where 'stem break' and 'checkered leaf disease' occur in France, suggested seedbed nematicides include aldicarb, dazomet, 1,3-D and metam sodium (Anonymous, 1998). Root knot management in Queensland, Australia usually depends upon crop rotation and early destruction of stalks and roots after harvest, but fenamiphos or cadusafos may be applied when necessary (P. Tonello, personal communication).

Nematicides are rarely used when tobacco is grown on small plots of land by peasant farmers, even when they are recommended. The specific products suggested often vary from one tobacco-producing region to another, although non-fumigant nematicides may be recommended more commonly (Shepherd and Barker, 1990).

Biological control

Research continues to identify practical and effective biological methods for controlling plant parasitic nematodes on tobacco. Research results from Zimbabwe and the USA using *Paecilomyces lilacinus* and *Pasteuria penetrans* to help control *Meloidogyne* spp. on tobacco have not been encouraging, although some suppression of root knot nematodes was observed, and a commercial product (PL plus) is registered in the RSA (Weibelzahl-Fulton *et al.*, 1996; Johnson, 1998; J.I. Way, personal communication). Recent research indicates that tobacco production practices such as topping (removal of the terminal inflorescence) might be useful for increasing the effectiveness of *P. penetrans* against *M. incognita* and *M. javanica* (Rodrigues *et al.*, 2002). Several species of the nematode-trapping fungus *Arthrobotrys* and a strain of *Dactylaria* reduced development of *Meloidogyne mayaguensis* and increased tobacco growth (Duponnois *et al.*, 1997). Incorporation of an endomycorrhizal fungus into the soil of tobacco seed beds has been reported to reduce galling by *M. incognita* and improve transplant growth (Johnson, 1998). A biological nematicide containing dried fermentation solids and solubles from *Myrothecium verrucaria* strain AARC-0255 reduced populations of *M. incognita* and increased plant growth (Melton, 1998; Warrior *et al.*, 1999).

Acibenzolar-S-methyl induces systemic acquired resistance (SAR) in tobacco and is recommended for control of blue mould (*Peronospora tabacina*), tomato spotted wilt virus and wildfire (*P. syringae* pv. *tabaci*), but its potential use for nematode control is largely unexplored. Application of acibenzolar-S-methyl to barley increased barley infection by *Pratylenchus* (Sonnemann *et al.*, 2002) but activated resistance to *Meloidogyne* development in roots of grape (Owen *et al.*, 2002). Harpin is a protein produced by certain bacterial pathogens that induces systemic resistance in some plants, and is registered for use on tobacco in the USA (Jones, 2001). Use of harpin with aldicarb for root knot control did not increase tobacco yield over that from use of aldicarb alone (Melton *et al.*, 2002). However, root dip and soil drench treatments of tomato with plant growth-promoting rhizobacteria (PGPR) induced systemic resistance and reduced penetration of tomato roots by *M. javanica* (Siddiqui and Shaukat, 2002). Combinations of chitosan and PGPR strains promoted growth of tomato, cucumber, pepper and tobacco, and reduced galling by *Meloidogyne* by similar mechanisms (Kloepper *et al.*, 2004). *Rhizobium etli* G12, isolated from a potato rhizosphere soil, induced resistance to *Globodera pallida* (closely related to the *G. tabacum* complex), at least partially by inhibiting nematode penetration (Hasky-Gunther *et al.*, 1998). This same bacterium was found to adversely affect *M. incognita* (Hallmann *et al.*, 2001). Lipopolysaccharides from the cell surface of *R. etli* G12 are responsible for the enhanced resistance to *G. pallida* (Reitz *et al.*, 2000; Reitz and Sikora, 2001). These compounds themselves may prove useful to management of plant parasitic nematodes (Siddiqui and Shaukat, 2003). Mutualistic endophytic fungi have been used to increase resistance in plants to nematodes by adding the antagonistic strains to the transplant production systems (Hallmann *et al.*, 2001). Using this system of targeted application could markedly reduce the costs involved in using biological control.

Root exudates from marigold (*Tagetes* spp.) can reduce populations of *Meloidogyne* and *Pratylenchus* species, but this activity has yet to be utilized in tobacco production (Johnson, 1998). Essential oils from leaves of *N. tabacum* and aqueous extracts from leaves of *Azadirachta indica*, *Melia azedarach* and other plants have shown nematicidal properties against *M. incognita* (Johnson, 1998; Ramakrishnan *et al.*, 1999a).

Plant residues and by-products are also being evaluated as materials to reduce nematode population densities, particularly in tobacco seedbeds managed by small-scale farmers in developing countries. Planting castor prior to seeding beds reduced populations of *Meloidogyne* spp., but was not as effective as fenamiphos (Gowda and Reddy, 1995). Incorporation of sunnhemp (*Crotalaria juncea*) or *Sesbania bispinosa* residue suppressed nematode populations for 70 days after seeding in bidi tobacco seedbeds in India (Patel and Patel, 1999). Although application of oil seed cakes to seedbeds prior to seeding proved ineffective against *M. javanica*, incorporation of neem cakes and soil solarization significantly reduced galling by *M. incognita* and *M. javanica* on bidi tobacco transplants (Krishnamurthy, 1990; Ravindra *et al.*, 2001). Mixtures of chitinous materials with urea and soybean (*Glycine max* (L.) Merr.) meal reduced galling by *M. arenaria* and increased flue-cured tobacco yields in microplot tests, but the effects were less beneficial in larger scale field experiments (Johnson, 1998). Pre-plant incorporation of cottonseed meal-based fertilizer or chitin–urea amendments to shade tobacco fields infested by *G. t. tabacum* also failed to reduce nematode population densities and adversely influenced quality characteristics of the crop (LaMondia, 1994). Organic manures can limit parasitism by *Meloidogyne* spp. and increase tobacco yields, but also have potential for significant phytotoxicity (Johnson, 1998).

Summary of management measures

Consistently reliable nematode control in tobacco requires the use of multiple approaches. Even in areas where nematicide use is routinely necessary, satisfactory control of tobacco nematodes is based upon sound crop rotation plans and, when available, planting resistant cultivars.

Use of healthy transplants is critical to achieving satisfactory tobacco yield and quality. Greenhouse seedling production provides such plants but, when transplants are produced in outdoor seedbeds, these seedbeds should be fumigated to minimize nematode parasitism. Although fumigating seedbeds with 1,3-D or burning brushwood may be cheaper methods to reduce nematode populations than using methyl bromide, other practices may be necessary to also ensure adequate weed control. Tobacco seedbed locations should be rotated, and nematode-resistant crops should be planted in seedbed areas when this is done. Tobacco should only be grown in specific fields for one or two consecutive growing seasons before rotation to other crops that either are non-hosts to the nematode species present or possess resistance. Grass crops are often preferred for rotation with tobacco.

Resistant cultivars are widely available for *M. incognita* races 1 and 3, and have shown at least partial resistance or tolerance to other races and species of *Meloidogyne*. Partial resistance is now also available to *M. javanica* in some areas. Resistance to the TCNs is also available in a number of cultivars now being planted around the world. In many situations, the use of good rotations and a resistant cultivar may be sufficient to limit crop losses to nematodes, but the addition of a nematicide will often help to realize the full yield potential of the tobacco. Nematicides are also thought to provide other benefits, which are often not proven by scientific investigation but are highly valued by tobacco producers. Such benefits include faster early growth of the crop and more uniform crop development. These factors are valued because they tend to reduce the probability of crop losses caused by factors such as weather and other pests or diseases. These factors also lower labour costs by increasing the efficiency of sucker control and harvesting. If the nematode population is high, a fumigant will tend to provide better control than an organophosphate or carbamate nematicide. Multipurpose fumigants are also necessary in fields with a history of nematode disease complexes. Destruction of tobacco roots and stalks as soon after harvest as possible will prevent the nematode populations at the end of the growing season from building up even further to attack the next crop to be planted.

Methods of diagnosis

Selection of appropriate nematode control measures for tobacco depends upon accurate assessment of the nematode population(s) present in fields at transplanting. Bioassays or nematode extraction from soil and/or root samples can be used to detect populations of important nematode parasites of tobacco, and this may be sufficient in areas where a predominant nematode is highly damaging to tobacco, such as *M. javanica* (Shepherd and Barker, 1990). Since bioassays can be quite sensitive compared with nematode assays from soil, bioassays may be the preferred method for monitoring problem nematode populations where *Meloidogyne*, *Globodera* or *Ditylenchus* species are the predominant or only parasitic nematode of concern (Shepherd and Barker, 1990). Bioassays must, however, be performed appropriately and well ahead of time. A minimum of several weeks will be required for any galling or cyst production to become evident. Visual assessment of tobacco roots ploughed out of the soil at the end of the growing season is commonly recommended to US tobacco producers (Johnson, 2002; Melton and Broadwell, 2002) and may, in fact, be a form of bioassay.

In many other situations, satisfactory tobacco yield and quality can be obtained when initial nematode populations are below an estimated economic threshold level. In these cases, reliable quantitative estimates of nematode populations are necessary to optimize nematode management. Nematode assay and advisory services are available in some of the tobacco-growing areas of America, Europe and elsewhere, but are not found in many others. Most tobacco growers in the USA do not assay their fields for nematodes every year (Johnson, 1989). Belief that nematode damage is not a significant risk, routine pesticide use, the high economic value of tobacco and concern about the reliability of nematode assay results have been cited as reasons for this lack of participation. However, periodic nematode assays are often used with field histories to estimate the species composition and relative damage potential of nematode populations in tobacco fields (Fortnum, 2002). Nematode assays from soil are rarely able to differentiate nematodes beyond the genus level, even when such distinctions are important criteria for nematode management decisions. Enzyme phenotypes are now being used to differentiate root knot species in some tobacco-producing areas (Chen *et al.*, 1998). The potential of monoclonal antibodies and molecular DNA-based techniques such as real-time polymerase chain reaction (PCR) currently are being explored for rapidly and reliably identifying the nematode species present in fields (Dong *et al.*, 2001a,b). Some of these techniques are outlined in Chapter 2.

Conclusions and Future Prospects

Significant progress has been made in developing methods for tobacco farmers to reduce the impact of plant parasitic nematodes on tobacco production. Nematodes are considered minor problems in some areas where they were once significant constraints. Nematode resistance is becoming more and more available and, together with cultural practices such as crop rotation and early root and stalk destruction, has enabled growers in many parts of the world to reduce their dependence upon nematicides, particularly in managing populations of *M. incognita* and *G. tabacum*. More effective and durable forms of resistance to plant parasitic nematodes may be available in tobacco cultivars as advances in our understanding at the molecular level enable development of mechanisms to inhibit nematode penetration and development with minimal impact on plant growth.

Nematicide use, however, remains necessary in many situations and areas, particularly southern Africa where *M. javanica* is the predominant nematode and a major problem. For growers in such areas, the dramatic decline in the number of effective nematicides poses a real threat to production. Effective biocontrol agents are not widely available, but research seems to be progressing in terms of identifying potentially effective agents and in determining how these organisms might be used on a practical scale. Nematode management research must continue in order for tobacco producers to reduce potential environmental side effects of tobacco production and to compete in the global market.

Acknowledgements

We would like to acknowledge, with gratitude, the assistance give to us by: K.R. Barker, J. Berenji, P. Bertrand, M. Botha, H. Deeke, T. Fekete, C. Fisher, B.A. Fortnum, I. Mallmann, T.A. Melton, S. Miele, J. Rich, R. Sato, A. Scholtz, R.A. Sikora, P. Tonello, E. van Jaarsveld and E.A. Wernsman.

References

Ambrogioni, L. and D'Erico, F.P. (1995) *Globodera tabacum* (Lownsbery & Lownsbery) Skarbilovich a new plant parasitic nematode in Italy. *Redia* 78, 357–371.

Anonymous (1979) Annual Report and Accounts for the year ended 30th June, 1979. Zimbabwe Rhodesia Tobacco Reasearch Board. *Annual Report and Accounts for the year ended 30th June, 1979.*

Anonymous (1998) Guideline on good plant protection practice – tobacco. *OEPP/EPPO Bulletin* 28, 411–423.

Anonymous (2000) Nematode Guidelines for South Carolina, EC 703. Available at: virtual.clemson.edu/groups/psapublishing/PAGES/PLNTPATH/EC703.pdf

Anonymous (2004) FAOSTAT data. Database for Primary crops (tobacco leaves). Available at: apps.fao.org/faostat/form?collection=Production.Crops.Primary&Domain=Production&servlet=1&hasbulk=0&version=ext&language=EN.

Barker, K.R. and Lucas, G.B. (1984) Nematode parasites of tobacco. *Plant and Insect Nematodes*, 213–242.

Barriga-Olivares, R. and Aranda-Ramirez, A.J. (2000) Important phytoparasitic nematodes in the tobacco zone of Santander and Boyaca, Colombia. *Fitopatologia Colombiana* 24, 10–14.

Baum, T.J., Fortnum, B.A. and Lewis, S.A. (1995a) Characteristics of mixed *Meloidogyne arenaria* and *M. incognita* populations in flue-cured tobacco. *Fundamental and Applied Nematology* 18, 589–594.

Baum, T.J., Fortnum, B.A. and Lewis, S.A. (1995b) Effects of *Meloidogyne arenaria* infection on *M. incognita*-resistance in tobacco. *Fundamental and Applied Nematology* 18, 583–588.

Bélair, G., Fournier, Y., Dauphinais, N. and Dangi, O.P. (2002) Reproduction of *Pratylenchus penetrans* on various rotation crops in Quebec. *Phytoprotection* 83, 111–114.

Bertrand, P. (2002) Tobacco disease control. In: *Georgia Tobacco Grower's Guide for 2003.* University of Georgia.

Botha, M. (1996) Sub-group collaborative study of nematodes. 1996 Report. *Bulletin d'Information-CORESTA* (No. 3/4), 45–46.

Botha, M. (2002) Sub-group collaborative study of nematodes. 2002 Report. *Bulletin d'Information-CORESTA.*

Carlson, S.R., Wolff, M.F., Shew, H.D. and Wernsman, E.A. (1997) Inheritance of resistance to race 0 of *Phytophthora parasitica* var. *nicotianae* from the flue-cured tobacco cultivar Coker 371-Gold. *Plant Disease* 81, 1269–1274.

Carneiro, R.M.D.G., Carneiro, R.G., Abrantes, I.M.d.O., Santos, M.S.N.d.A. and Almeida, M.R.A. (1996) *Meloidogyne paranaensis* n.sp. (Nemata: Meloidogynidae), a root-knot nematode parasitizing coffee in Brazil. *Journal of Nematology* 28, 177–189.

Charchar, J.M. and Eisenback, J.D. (2002) *Meloidogyne brasilensis* n.sp. (Nematoda: Meloidogynidae), a root-knot nematode parasitising tomato cv. Rossol in Brazil. *Nematology* 4, 629–643.

Charchar, J.M., Eisenback, J.D. and Hirschmann, H. (1999) *Meloidogyne petuniae* n.sp. (Nemata: Meloidogynidae), a root-knot nematode parasitic on petunia in Brazil. *Journal of Nematology* 31, 81–91.

Chen, C.M., Chen, P. and Yao, G. (1991) Study on root-knot nematodes on tobacco and their control. *Sichuan Plant Protection* 1/2, 75–78.

Chen, Y.-F., Wu, J.-Y., Hu, X.-Q. and Yu, S.-F. (1998) Using PhastSystem for rapid identification of root-knot nematodes. *Acta Phytopathologica Sinica* 28 (1), 73–77.

Cimini, G., Travaglini, S., de Laurentis, A., Marcucci, A., Gattone, G., Fedele, P., Rainaldi, D., Russo, A. and D'Emilio, A. (1993) Nematological survey of land in the 'Val de Sangro'. *Informatore Agrario* 49, 59–61.

Clayton, E.E., Shaw, K.J., Smith, T.E., Gaines, J.G. and Graham, T.W. (1944) Tobacco disease control by crop rotation. *Phytopathology* 34, 870–883.

Crowder, B.J., Wilkinson, C.A., Johnson, C.S. and Eisenbach, J.D. (2003) Inheritance of resistance to tobacco cyst nematode in flue-cured tobacco. *Crop Science* 43 (4), 1305–1312.

Csinos, A.S., Johnson, W.C., Johnson, A.W., Sumner, D.R., McPherson, R.M. and Gitaitis, R.D. (1997) Alternative fumigants for methyl bromide in tobacco and pepper transplant production. *Crop Protection* 160, 585–594.

Csinos, A.S., Sumner, D.R., Johnson, W.C., Johnson, A.W., McPherson, R.M. and Dowler, C.C. (2000) Methyl bromide alternatives in tobacco, tomato and pepper transplant production. *Crop Protection* 19 (1), 39–49.

Daulton, R.A.C. (1963) Controlling *M. javanica* in Southern Rhodesia. *Rhodesian Agricultural Journal* 60, 150–152.

Davis, R.F., Johnson, A.W. and Wauchope, R.D. (1993) Accelerated degradation of fenamiphos and its metabolites in soil previously treated with fenamiphos. *Journal of Nematology* 25, 679–685.

di Vito, M., Lombardi, D.A., Zaccheo, G. and Catalano, F. (1998) Response of lines and cultivars of tobacco and *Nicotiana* species to Italian populations of *Meloidogyne* species. *Nematologia Mediterranea* 26, 271–274.

Dong, K., Dean, R.A., Fortnum, B.A. and Lewis, S.A. (2001a) Development of PCR primers to identify species of root-knot nematodes: *Meloidogyne arenaria, M. hapla, M. incognita* and *M. javanica. Nematropica* 31, 271–280.

Dong, K., Dean, R.A., Fortnum, B.A. and Lewis, S.A. (2001b) A species-specific DNA probe for the identification of *Meloidogyne hapla. Nematropica* 31, 17–23.

Duniway, J.M. (2002) Status of chemical alternatives to methyl bromide for pre-plant fumigation of soil. *Phytopathology* 92, 1337–1343.

Duponnois, R., Mateille, T. and Bâ, A. (1997) Potential effects of Sahelian nematophagous fungi against *Meloidogyne mayaguensis* on tobacco (*Nicotiana tabacum* L. var. Paraguay × Claro). *Annales du Tabac. Section 2* (No. 29), 61–70.

Elliott, A.P., Phipps, P.M. and Terrill, R. (1986) Effects of continuous cropping of resistant and susceptible cultivars on reproduction potentials of *Heterodera glycines* and *Globodera tabacum solanacearum. Journal of Nematology* 18, 375–379.

Elmer, W.H., Jones, J.B. and Miller, L.I. (1980) Influence of *Pseudomonas solanacearum* and *Globodera tabacum* on growth of tobacco. *Virginia Journal of Science* 31, 78.

Espárrago, G. (2002) First report of the cyst nematode (*Globodera tabacum*) complex on flue-cured tobacco in Spain. *Plant Disease* 86, 1402.

Fernández Díaz-Silveira, M. and Ortega Herrera, J. (1998) An overview of nematological problems in Cuba. *Nematropica* 28, 151–164.

Fortnum, B.A. (2002) Tobacco disease management. In: *South Carolina Tobacco Growers Guide 2003.* Clemson University.

Fortnum, B.A. and Pullen, M.P. (2001) Broadcast fumigation controls root-knot of flue-cured tobacco. American Phytopathological Society. *Abstracts of Presentations, S29, Publication No. P-2001–0205-AMA.*

Fortnum, B.A., Johnson, A.W. and Lewis, S.A. (2001a) Analysis of 1,3-dichloropropene for control of *Meloidogyne* spp. in a tobacco pest management system. *Supplement to the Journal of Nematology* 33, 325–331.

Fortnum, B.A., Lewis, S.A. and Johnson, A.W. (2001b) Crop rotation and nematicides for management of mixed populations of *Meloidogyne* spp. on tobacco. *Supplement to the Journal of Nematology* 33, 318–324.

Gao, X.B., Cheng, H.R. and Fang, Z.D. (1994) Host parasite interactions of *Pratylenchus scribneri* on 10 crops and ability of Nemacur to control nematode populations in corn. *Journal of Nanjing Agricultural University* 17 (1), 43–46.

Goddijn, O.J.M., Lindsey, K., van der Lee, F.M., Klap, J.C. and Sijmons, P.C. (1993) Differential gene expression in nematode-induced feeding structures of transgenic plants harbouring promoter–gusA fusion constructs. *Plant Journal* 4, 863–873.

Goswami, B.K. and Raychaudhuri, S.P. (1973) Host–parasite relationship between tobacco and root-knot nematode, *Meloidogyne javanica* (Treub) Chitwood, influenced by tobacco mosaic virus infection. *Annals of the Phytopathological Society of Japan* 39, 99–102.

Gowda, D.N. (1999) Effective duration of phenamiphos in control of root-knot nematode in tobacco nursery. *Plant Protection Bulletin (Faridabad)* 51 (3/4), 30–31.

Gowda, D.N. and Reddy, V.S. (1995) Effect of castor and phenamiphos in management of root-knot nematode in tobacco nursery. *Indian Journal of Nematology* 25 (2), 221–222.

Gowda, D.N., Kurdikeri, C.B. and Gowda, C.K. (1995) Weeds as hosts of root knot nematode. *Indian Journal of Nematology* 25, 215–216.

Gullino, M.L., Camponogara, A., Gasparrini, G., Rizzo, V., Clini, C. and Garibaldi, A. (2003) Replacing methyl bromide for soil disinfestation – the Italian experience and implications for other countries. *Plant Disease* 87, 1012–1021.

Hallmann, J., Quadt-Hallmann, A., Miller, W.G., Sikora, R.A. and Lindow, S.E. (2001) Endophytic colonization of plants by the biocontrol agent *Rhizobium etli* G12 in relation to *Meloidogyne incognita* infection. *Phytopathology* 91, 415–422.

Hasky-Gunther, K., Hoffmann-Hergarten, S. and Sikora, R.A. (1998) Resistance against the potato cyst nematode *Globodera pallida* systemically induced by the rhizobacteria *Agrobacterium radiobacter* (G12) and *Bacillus sphaericus* (B43). *Fundamental and Applied Nematology* 21, 511–517.

Hayes, A.J., Wilkinson, C.A. and Johnson, C.S. (1997) Evaluation of tobacco accessions for resistance to tobacco cyst nematode and wildfire. *Crop Science* 37, 586–591.

Herrero, S., Rufty, R.C. and Barker, K.R. (1996) Evaluation of tobacco germplasm for resistance to the tobacco cyst nematode, *Globodera tabacum solanacearum. Plant-Disease* 80, 61–65.

Hirunsalee, A., Barker, K.R. and Beute, M.K. (1995a) Effects of peanut–tobacco rotations on population dynamics of *Meloidogyne arenaria* in mixed-race populations. *Journal of Nematology* 27, 178–188.

Hirunsalee, A., Barker, K.R. and Beute, M.K. (1995b) Infection, reproduction potential, and root galling by root-knot nematode species and concomitant populations on peanut and tobacco. *Journal of Nematology* 27, 172–177.

Holmes, F.O. (1938) Inheritance of resistance to tobacco-mosaic disease in tobacco. *Phytopathology* 23, 553–561.

Honarnejad, R. and Shoaei-Deylami, M. (1997) Gene effects and combining ability of some quantitative and qualitative characteristics of tobacco (*Nicotiana tabacum* L.). *Iranian Journal of Agricultural Sciences* 28 (4), 121–145.

Hussaini, S.S. and Krishnamurthy, G.V.G. (2002) Occurrence and distribution of root-knot species of FCV tobacco in Karnataka. *Indian Journal of Nematology* 32 (1), 20–22.

Hussaini, S.S., Bharatha, K.S. and Rao, R.V.V. (2001) Soil solarization – comparison of LDPE cauges and periods for control of root-knot nematode and weeds in FCV tobacco nursery. *Current Nematology* 12, 39–43.

Iglesias, M.P., González, E.F. and Benítez, M.E. (1998) Soil solarization for nematode and weed control on seed beds of tobacco and vegetables. *IDESIA* 15, 35–39.

Imbriani, J.L. (2003) Root-Knot Nematodes on Tobacco. Available at: www.ncagr.com/agronomi/nnote3.htm

Inagaki, H. and Powell, N.T. (1969) Influence of the root-lesion nematode on black shank symptom development in flue-cured tobacco. *Phytopathology* 59, 1350–1355.

Jack, A.M. (1996) The effect of soil fumigation on five flue-cured tobacco cultivars with different levels of resistance to the root-knot nematode (*Meloidogyne javanica*). In: *CORESTA Information Bulletin, Congress Issue, 1996*, p. 180 (abstract).

Jagdale, G.B., Ball-Coelho, B., Potter, J., Brandle, J. and Roy, R.C. (2000) Rotation crop effects on *Pratylenchus penetrans* and subsequent crop yields. *Canadian Journal of Plant Science* 80, 543–549.

Jiang, X.S. and Xing, X.J. (1992) The damage of root-knot nematodes to tobacco and their control in Liangshan. *Xichang Agricultural Science and Technology* 3/4, 23–25.

Johnson, A.W. and Nusbaum, C.J. (1970) Interactions between *Meloidogyne incognita, M. hapla* and *Pratylenchus brachyurus* in tobacco. *Journal of Nematology* 2, 334–340.

Johnson, C.S. (1989) Managing root-knot on tobacco in the southeastern United States. *Supplement to the Journal of Nematology* 21, 604–608.

Johnson, C.S. (1990) Control of *Globodera tabacum solanacearum* by rotating susceptible and resistant flue-cured tobacco cultivars. *Supplement to the Journal of Nematology* 22, 700–706.

Johnson, C.S. (1992) Tobacco cyst nematodes can stimulate yield of flue-cured tobacco. *Phytopathology* 82, 1095 (abstract).

Johnson, C.S. (1995) Use of fosthiazate to control tobacco cyst nematodes in flue-cured tobacco. *Tobacco Science* 39, 69–76.

Johnson, C.S. (1998) Tobacco. In: Barker, K.R., Pederson, G.A. and Windham, G.L. (eds) *Plant and Nematode Interactions, Agronomy Monograph 36.* American Society of Agronomy, Crop Science Society of America, Soil Science Society of America, Madison, Wisconsin.

Johnson, C.S. (2001) Suppressed reproduction of *Globodera tabacum solanacearum* on disease-resistant cultivars of flue-cured tobacco. *Journal of Nematology* 33, 262 (abstract).

Johnson, C.S. (2002) Flue-cured tobacco disease control. In: Reed, T.D., Jones, J.L., Johnson, C.S., Semtner, P.J. and Wilkinson, C.A. (eds) *2004 Flue-cured Tobacco Production Guide.* Virginia Cooperative Extension, Blacksburg, Virginia, pp. 33–49.

Johnson, C.S. and Clarke, C.T. (2003) Host resistance and fumigation to control *Globodera tabacum solanacearum* in 2002. *American Phytopathological Society. Abstracts of Presentations, Publication no. P-2003–0299-AMA* (abstract).

Johnson, C.S. and Reed, T.D. (1994) Tobacco. In: *Encyclopedia of Agricultural Science.* Academic Press, Inc., New York, pp. 323–336.

Johnson, C.S. and Wilkinson, W.B. III (2002) Agri-50 and metam versus standard nematicides to control tobacco cyst nematodes on tobacco in Virginia, 2001. *Fungicide Nematicide Tests* N15.

Johnson, C.S., Komm, D.A. and Jones, J.L. (1989) Control of *Globodera tabacum solanacearum* by alternating host resistance and nematicide. *Journal of Nematology* 21, 16–23.

Johnson, E.S., Wolff, M.F., Wernsman, E.A., Atchley, W.R. and Shew, H.D. (2002) Origin of the black shank resistance gene, *Ph*, in tobacco cultivar Coker 371-Gold. *Plant Disease* 86, 1080–1084.

Jones, J. (2001) Harpin. *Pesticide Outlook* 12 (4), 134–135.

Karachi, M. (1995) *Sesbania* species as potential hosts to root-knot nematode (*Meloidogyne javanica*) in Tanzania. *Agroforestry Systems* 32 (2), 119–125.

Kartono, G. (1980) The influence of a double attack of tobacco mosaic virus and *Meloidogyne incognita* on the growth of tobacco. Pemberitaan Lembaga Penelitian Tanaman Industri (No. 37), 90–99.

Khan, F.A. (1990) Plant parasitic nematodes associated with tobacco (*Nicotiana tabacum*) in Sokoto state of Nigeria. *Pakistan Journal of Nematology* 8, 17–22.

Kloepper, J.W., Reddy, M.S., Rodriguez-Kabana, R., Kenney, D.S., Kokalis-Burelle, N., Martinez-Ochoa, N. and Vavrina, C. (2004) Application for rhizobacteria in transplant production and yield enhancement. *Acta Horticulturae* 631, 219–229.

Krishnamurthy, G.V.G. (1990) Effect of oil seed cakes on the control of root-knot nematode, *Meloidogyne javanica*, in tobacco nursery. In: Chari, M.S. and Ramaprasad, G. (eds) *Botanical Pesticides in Integrated Pest Management: Proceedings of National Symposium*. Indian Society of Tobacco Society, Rajahmundry, India, pp. 434–437.

Lamberti, F., Ekanayake, H.M.R.K., Sasanelli, N. and Zacheo, F. (1993) Control of root-knot nematodes in Sri Lanka. *Nematologia Mediterranea* 21, 227–234.

Lamberti, F., D'Addabbo, T., Greco, P., Sasanelli, N. and Carella, A. (2000) Chemical control of root-knot nematodes. *Acta Horticulturae* (No. 532), 183–187.

LaMondia, J.A. (1988) Tobacco resistance to *Globodera tabacum*. *Annals of Applied Nematology* 2, 77–80 (abstract).

LaMondia, J.A. (1991) Genetics of tobacco resistance to *Globodera tabacum tabacum*. *Plant Disease* 75, 453–454 (abstract).

LaMondia, J.A. (1994) The effects of a chitin–urea soil amendment on *Globodera tabacum tabacum* population changes under shade tobacco. *Tobacco Science* 38, 18–20.

LaMondia, J.A. (1995a) Hatch and reproduction of *Globodera tabacum tabacum* in response to tobacco, tomato, or black nightshade. *Journal of Nematology* 27, 382–386.

LaMondia, J.A. (1995b) Influence of resistant tobacco and tobacco cyst nematodes on root infection and secondary inoculum of *Fusarium oxysporum* f.sp. *nicotianae*. *Plant Disease* 79, 337–340.

LaMondia, J.A. (1996a) Efficacy of band or broadcast Vydate for tobacco cyst nematode control in shade tobacco, 1995. *Fungicide Nematicide Tests* 51, 193.

LaMondia, J.A. (1996b) Trap crops and population management of *Globodera tabacum tabacum*. *Journal of Nematology* 28, 238–243.

LaMondia, J.A. (2000a) Registration of 'Metacomet' tobacco. *Crop Science* 40, 1504–1505.

LaMondia, J.A. (2000b) Registration of 'Poquonock' tobacco. *Crop Science* 40, 1505–1506.

LaMondia, J.A. (2002a) Broadleaf tobacco yield loss in relation to initial *Globodera tabacum tabacum* population density. *Journal of Nematology* 34, 38–42.

LaMondia, J.A. (2002b) Genetics of burley and flue-cured tobacco resistance to *Globodera tabacum tabacum*. *Journal of Nematology* 34, 34–37.

Marché, L., Valette, S., Grenier, E. and Mugniéry, D. (2001) Intra-species DNA polymorphism in the tobacco cyst-nematode complex (*Globodera tabacum*) using AFLP. *Genome* 44, 941–946.

McIntyre, J.L. and Miller, P.M. (1978) Protection of tobacco against *Phytophthora parasitica* var. *nicotianae* by cultivar-nonpathogenic races, cell-free sonicates, and *Pratylenchus penetrans*. *Phytopathology* 68, 235–239.

McLean, M.D., Yevtushenko, D.P., Deschene, A., van Cauwenberghe, O.R., Amina Makhmoudova, Potter, J.W., Bown, A.W. and Shelp, B.J. (2003) Overexpression of glutamate decarboxylase in transgenic tobacco plants confers resistance to the northern root-knot nematode. *Molecular Breeding* 11 (4), 277–285.

Melton, T.A. (1998) Nematicides for control of southern root-knot nematodes, 1996. *Fungicide Nematicide Tests* 53, 280.

Melton, T.A. and Broadwell, A. (2002) Disease management. In: *Flue-cured Tobacco Information 2003*. North Carolina Cooperative Extension Service, Raleigh, North Carolina, pp. 119–143.

Melton, T.A. and Powell, N.T. (1991) Effects of nematicides and cultivars on *Rotylenchulus reniformis* and flue-cured tobacco yield. *Journal of Nematology* 23, 712–716.

Melton, T.A., Barker, K.R., Koenning, S.R. and Powell, N.T. (1995) Temporal efficacy of selected nematicides on *Meloidogyne* species on tobacco. *Journal of Nematology* 27, 263–272.

Melton, T.A., Fisher, T., Imbriani, J., Broadwell, A. and Wilson, J. (2002) Chemical nematicides and fumigants for control of root-knot nematodes, 2001. *Fungicide Nematicide Tests* 57, N18.

Meredith, C.J.A. (1976) Estudio de la reproduccion de *Heterodera rostochiensis* Wollenweber, 1923, en diferentes condiciones ambientales sobre varias solanaceas. Universidad Central de Venezuela.

Miller, L.I., Fox, J.A. and Spasoff, L. (1972) Genetic relationship of resistance to *Heterodera solanacearum* in dark-fired and burley tobacco. *Phytopathology* 62, 778 (abstract).

Moreno, J.J., Verdejo, M.E., Pérez, J.A., Arias, A., Martínez de Velasco, D. and Balvín, M. (1994) Comparison of nematicides in the cultivation of tobacco during three years. *Boletín de Sanidad Vegetal, Plagas* 20, 941–954.

Murthy, K.K., Ramakrishnan, S. and Shenoi, M.M. (1999) Quantification of loss in yield of FCV tobacco due to root-knot infection at seedling stage. *Tobacco Research* 25 (2), 92–94.

Ng'ambi, T.B.S., Rufty, R.C. and Barker, K.R. (1995) Response of root-knot resistant tobacco to concomitant populations of *Meloidogyne* species. *Plant Disease* 79, 1008–1013.

Ng'ambi, T.B.S., Rufty, R.C. and Barker, K.R. (1999a) Genetic analysis of *Meloidogyne arenaria* race 1 resistance in tobacco. *Plant Disease* 83, 810–813.

Ng'ambi, T.B.S., Rufty, R.C., Barker, K.R. and Melton, T.A. (1999b) Identification of sources of resistance to four species of root-knot nematodes in tobacco. *Journal of Nematology* 31, 272–282.

Noling, J.W. and Becker, J.O. (1994) The challenge of research and extension to define and implement alternatives to methyl bromide. *Supplement to the Journal of Nematology* 26, 573–586.

Nombela, G., Williamson, V.M. and Muñiz, M. (2003) The root-knot nematode resistance gene *Mi*-1.2 of tomato is responsible for resistance against the whitefly *Bemisia tabaci*. *Molecular Plant Microbe Interactions* 7, 645–649.

Olthoff, T.H.A., Marks, C.F. and Elliot, J.M. (1973) Relationship between population densities of *Pratylenchus penetrans* and crop losses in flue-cured tobacco in Ontario. *Journal of Nematology* 5, 158–162.

Opperman, C.H. and Conkling, M.A. (1994) Nematode-induced plant gene expression and related control strategies. *Fundamental and Applied Nematology* 17, 211–217.

Opperman, C.H., Taylor, C.G. and Conkling, M.A. (1994) Root-knot nematode-directed expression of a plant root-specific gene. *Science* 263, 221–223.

Owen, K.J., Green, C.D. and Deverall, B.J. (2002) A benzothiadiazole applied to foliage reduced development and egg deposition by *Meloidogyne* spp. in glasshouse-grown grapevine roots. *Australasian Plant Pathology* 31, 47–53.

Parrott, D.M. and Miller, L.I. (1977) NC-95 flue-cured tobacco as a host of *Globodera pallida*. *Virginia Journal of Science* 28, 54 (abstract).

Patel, A.S. and Patel, H.R. (1990) Host range of stunt nematode, *Tylenchorhynchus vulgaris*. *Journal of Mycology and Plant Pathology* 20, 176–178.

Patel, A.S. and Patel, H.R. (1991) Varietal screening of tobacco against stunt nematode. *Tobacco Research* 17, 57–58.

Patel, A.S. and Patel, H.R. (1993) Pathogenicity of stunt nematode on transplants of bidi tobacco. *Tobacco Research* 19 (2), 77–81.

Patel, B.K. and Patel, H.R. (1997) Effect of nematicides, soil solarization, rabbing and green manuring on some physical and chemical properties of soil in nematode infested bidi tobacco nursery. *Tobacco Research* 23 (1/2), 19–26.

Patel, B.K. and Patel, H.R. (1998) Effect of soil solarization, rabbing, nematicides and green manuring on growth and development of bidi tobacco seedlings, root-knot disease, weeds and phytonematodes in nursery. *Indian Journal of Nematology* 28 (1), 15–21.

Patel, B.K. and Patel, H.R. (1999) Effect of physical, cultural and chemical methods of management on population dynamics of phytonematodes in bidi tobacco nursery. *Tobacco Research* 25 (2), 51–60.

Patel, D.J. and Patel, H.R. (1992) Nematode pests of tobacco. In: Bhatti, D.S. and Walia, R.K. (eds) *Nematode Pests of Crops*. CBS Publishers and Distributors, Delhi, India, pp. 177–189.

Patel, D.J., Patel, H.R., Patel, S.K. and Makwana, M.G. (1993) Management of root-knot nematode disease and weeds in bidi tobacco nursery through different rabbing methods using agricultural by-products. *Tobacco Research* 19, 7–11.

Patel, H.R. and Patel, B.N. (1998) Evaluation of dazomet (Basamid G) against rove beetles, damping-off, weeds and nematodes in bidi tobacco nursery. *Journal of Mycology and Plant Pathology* 28 (2), 134–139.

Patel, H.R. and Patel, B.N. (2001) Management of nematodes in bidi tobacco nursery by inimical plants. *Indian Journal of Nematology* 31 (1), 47–51.

Patel, H.R., Lakshminarayana, R. and Makwana, M.G. (1995a) The control of nematodes by soil solarisation for growing tobacco seedlings. *Plasticulture* 106, 2–6.

Patel, H.R., Makwana, M.G. and Patel, B.N. (1995b) The control of nematodes and weeds by soil solarization in tobacco nurseries: effect of the film thickness and of the covering duration. *Plasticulture* 107, 21–27.

Patel, H.R., Patel, B.N., Bhatt, N.A. and Patel, P.M. (2001) Efficacy of nematicides applied at different times in bidi tobacco nursery for management of nematodes. *Indian Journal of Nematology* 31 (1), 75–78.

Patel, K.A. and Patel, B.N. (1994a) Interaction between tobacco mosaic virus and root-knot nematode on growth of bidi tobacco. *Tobacco Research* 20 (2), 127–130.

Patel, K.A. and Patel, B.N. (1994b) Role of root-knot nematode in the development of TMV infection on bidi tobacco. *Tobacco Research* 20 (1), 63–64.

Patel, K.A. and Patel, B.N. (1995) Effect of tobacco mosaic virus and root-knot nematode infection on chemical constituents of bidi tobacco. *Indian Journal of Mycology and Plant Pathology* 25, 228–230.

Patel, P.T., Parmar, M.T. and Patel, B.A. (1994) Build up of plant parasitic nematodes in the bidi tobacco based cropping system. *Tobacco Research* 20 (2), 107–112.

Patel, S.B. and Patel, H.R. (1995) Screening of *Nicotiana* spp. against stunt nematode. *Tobacco Research* 21 (1/2), 86–87.

Patel, S.B. and Patel, H.R. (1998) Interaction of *Tylenchorhynchus vulgaris* and *Pythium aphanidermatum* on bidi tobacco. *Journal of Mycology and Plant Pathology* 28 (2), 177–180.

Patel, S.K. and Patel, D.J. (1991) The interaction between *Meloidogyne incognita, M. javanica* and *Rotylenchulus reniformis. Nematologia Mediterranea* 19, 41–42.

Pullen, M.P. and Fortnum, B.A. (1999) Fosthiazate controls *Meloidogyne arenaria* and *M. incognita* in flue-cured tobacco. *Supplement to the Journal of Nematology* 31, 694–699.

Ramakrishnan, S., Gunasekaran, C.R., Sivagami Vadivelu and Viswanath, S.M. (1999a) Effect of various aqueous leaf extracts on the larval mortality of *Meloidogyne incognita,* infesting tobacco. *Tobacco Research* 25 (1), 37–39.

Ramakrishnan, S., Hussaini, S.S., Viswanath, S.M. and Shenoi, M.M. (1999b) Effect of Basamid G for the control of root-knot nematodes in FCV tobacco nursery. In: *Proceedings of National Symposium on Rational Approaches in Nematode Management for Sustainable Agriculture,* 23–25 November 1998, Anand, India, pp. 121–124.

Ravindra, H., Onkarappa, T., Narayanaswamy, H., Basavaraja, M.K., Ramaswamy, G.R. and Lokamanya, D.S. (2001) Integrated management of root-knot nematode in FCV tobacco nursery. *Tobacco Research* 27 (1), 19–23.

Reed, T.D. (1996) *Float Greenhouse Tobacco – Transplant Production Guide.* Virginia Cooperative Extension, Blacksburg, Virginia, pp. 436–451.

Reitz, M. and Sikora, R.A. (2001) Bacteria-mediated induced systemic resistance in potato against *Globodera pallida. IOBC/wprs Bulletin* 24, 133–138.

Reitz, M., Rudolph, K., Schröder, I., Hoffmann-Hergarten, S., Hallmann, J. and Sikora, R.A. (2000) Lipopolysaccharides of *Rhizobium etli* strain G12 act in potato roots as an inducing agent of systemic resistance to infection by the cyst nematode *Globodera pallida. Applied and Environmental Microbiology* 66, 3515–3518.

Ren, N. and Timko, M.P. (2001) AFLP analysis of genetic polymorphism and evolutionary relationships among cultivated and wild *Nicotiana* species. *Genome* 44, 559–571.

Reynolds, L.B., Potter, J.W. and Ball-Coelho, B.R. (2000) Crop rotation with *Tagetes* sp. is an alternative to chemical fumigation for control of root-lesion nematodes. *Agronomy Journal* 92, 957–966.

Rideout, S.L., Johnson, C.S., Eisenback, J.D. and Reed, T.D. (2000a) Reproduction of *Globodera tabacum solanacearum* in seven flue-cured tobacco producing soils. *Supplement to the Journal of Nematology* 32, 486–492.

Rideout, S.L., Johnson, C.S., Eisenback, J.D. and Wilkinson, C.A. (2000b) Development of selected tobacco cyst nematode isolates on resistant and susceptible cultivars of flue-cured tobacco. *Journal of Nematology* 32, 62–69.

Rodrigues, A.K., de Freitas, L.G. and Ferraz, S. (2002) Effect of cutting the aerial part of *Meloidogyne* spp. host plant and the development of *Pasteuria penetrans. Nematologia Brasileira* 26, 153–160.

Rufty, R.C., Powell, N.T. and Gooding, G.V. Jr (1983a) Relationship between resistance to *Meloidogyne incognita* and a necrotic response to infection by a strain of potato virus Y in tobacco. *Phytopathology* 73, 1418–1423.

Rufty, R.C., Wernsman, E.A. and Powell, N.T. (1983b) A genetic analysis of the association between resistance to *Meloidogyne incognita* and a necrotic response to infection by a strain of potato virus Y in tobacco. *Phytopathology* 73, 1413–1418.

Schneider, S.M. (1991) Penetration of susceptible and resistant tobacco cultivars by *Meloidogyne* juveniles. *Journal of Nematology* 23, 225–228.

Shepherd, J.A. and Barker, K.R. (1990) Nematode parasites of tobacco. In: Luc, M., Sikora, R.A. and Bridge, J. (eds) *Plant Parasitic Nematodes in Subtropical and Tropical Agriculture*. CAB International, Wallingford, UK, pp. 493–517.

Shirima, D.S., Otsyina, R., Mwageni, W.P. and Bridge, J. (2000) Effect of natural and *Sesbania* fallows and crop rotations on the incidence of root-knot nematodes and tobacco production in Tabora, Tanzania. *International Journal of Nematology* 10 (1), 49–54.

Shukla, Y.M., Chakraborty, M.K. and Patel, D.J. (1988) Biological activity and semiochemicals in the root exudates of host plants resistant to root knot nematodes. *Tobacco Research* 14, 51–57.

Siddiqui, I.A. and Shaukat, S.S. (2002) Rhizobacteria-mediated induction of systemic resistance (ISR) in tomato against *Meloidogyne javanica*. *Journal of Phytopathology* 150, 469–473.

Siddiqui, I.A. and Shaukat, S.S. (2003) Suppression of root-knot disease by *Pseudomonas fluorescens* CHA0 in tomato: importance of bacterial secondary metabolite, 2,4-diacetylpholoroglucinol. *Soil Biology and Biochemistry* 35, 1615–1623.

Sonnemann, I., Finkhauser, K. and Wolters, V. (2002) Does induced resistance in plants affect the belowground community? *Applied Soil Ecology* 21, 179–185.

Sosa-Moss, C., Barker, K.R. and Daykin, M.E. (1983) Histopathology of selected cultivars of tobacco infected with *Meloidogyne* species. *Journal of Nematology* 15, 392–397.

Spasoff, L., Fox, J.A. and Miller, L.I. (1971) Multigenic inheritance of resistance to Osborne's cyst nematode. *Journal of Nematology* 3, 329–330 (abstract).

Srivastava, S.S. (2001) Race identification *of Meloidogyne incognita* infecting vegetables. *Indian Journal of Nematology* 31, 91–92.

Sudo, S. and Espindola, E.E. (1987) Survey of root-knot nematode species in tobacco plants grown in the South of Brazil. *CORESTA Meeting: Agronomy and Phytopathology Groups.* CORESTA Information Bulletin, abstract 2086.

Swathi, P., Rao, K.T. and Rao, P.A. (1998) Studies on control of root-knot nematode *Meloidogyne incognita* in tobacco nurseries. *Tobacco Research* 24 (1), 26–30.

Syracuse, A. (2003) Intraspecific variability within *Globodera tabacum solanacearum* and selection for virulence against flue-cured tobacco. MS Thesis, Virginia Polytechnic Institute and State University.

Thiery, M. and Mugniery, D. (1996) Interspecific rDNA restriction fragment length polymorphism in *Globodera* species, parasites of Solanaceous plants. *Fundamental and Applied Nematology* 19, 471–479.

Thiery, M., Fouville, D. and Mugniery, D. (1997) Intra- and interspecific variability in *Globodera*, parasites of solanaceous plants, revealed by random amplified polymorphic DNA (RAPD) and correlation with biological features. *Fundamental and Applied Nematology* 20, 495–504.

Tu, C.M., Marks, C.F., Johnson, P.W. and Elliot, J.M. (1995a) Effects of nematicides on populations of soil microflora and *Pratylenchus penetrans*, soil enzymes and flue-cured tobacco. *Journal of Environmental Science and Health. Part B, Pesticides, Food Contaminants, and Agricultural Wastes* 30, 95–111.

Tu, C.M., Marks, C.F., Johnson, P.W., Gayed, S.K. and Elliot, J.M. (1995b) Effects of pesticides on soil enzymatic activities, *Pratylenchus penetrans* populations, black root rot, and growth of flue-cured tobacco. *Journal of Environmental Science and Health. Part B, Pesticides, Food Contaminants, and Agricultural Wastes* 30, 141–162.

Tu, C.M., Marks, C.F. and Elliot, J.M. (1996) Effects of nematicides on *Pratylenchus penetrans*, soil nitrification, and growth of flue-cured tobacco. *Bulletin of Environmental Contamination and Toxicology* 57, 924–931.

van den Oever, R., van den Berg, E. and Chirruco, J.A. (1998) Plant parasitic nematodes associated with crops grown by smallholders in Mozambique. *Fundamental and Applied Nematology* 21, 645–654.

Wang, J., Johnson, C.S. and Eisenback, J.D. (1997) Enhanced hatching of *Globodera tabacum solanacearum* juveniles by root exudates of flue-cured tobacco. *Journal of Nematology* 29, 484–490 (abstract).

Wang, J., Johnson, C.S., Eisenback, J.D. and Reed, T.D. (1999) Effects of tobacco cyst nematode on growth of flue-cured tobacco. *Journal of Nematology* 31, 326–333.

Wang, J., Johnson, C.S. and Eisenback, J.D. (2001) Effects of host resistance and temperature on development of *Globodera tabacum solanacearum*. *Journal of Nematology* 33 (2/3), 132–136.

Warrior, P., Rehberger, L.A., Beach, M., Grau, P.A., Kirfman, G.W. and Conley, J.M. (1999) Commercial development and introduction of DiTera(R), a new nematicide. *Pesticide Science* 55, 376–379.

Weibelzahl-Fulton, E., Dickson, D.W. and Whitty, E.B. (1996) Suppression of *Meloidogyne incognita* and *M. javanica* by *Pasteuria penetrans* in field soil. *Journal of Nematology* 28 (1), 43–49.
Yi, H.Y. and Rufty, R.C. (1988) RAPD markers elucidate the origin of the root-knot nematode resistance gene (*Rk*) in tobacco. *Tobacco Science* 42, 58–63.
Yi, H.Y., Rufty, R.C., Wernsman, E.A. and Conkling, M.C. (1998) Mapping the root-knot nematode resistance gene (*Rk*) in tobacco with RAPD markers. *Plant Disease* 82, 1319–1322.

19 Nematode Parasites of Pineapple*

Brent S. Sipes,[1] Edward P. Caswell-Chen,[2] Jean-Louis Sarah[3] and Walter J. Apt[1]

[1]*Department of Plant and Environmental Protection Sciences, University of Hawaii, 3190 Maile Way, Honolulu, HI 96822, USA;* [2]*Department of Nematology, University of California, Davis, CA 95616, USA;* [3]*CIRAD/UMR BGPI, TA 41/K, 34398 Montpellier, Cedex 5, France*

The cultivated pineapple *Ananas comosus* L. (Merr.) (Bromeliaceae) is a monocotyledonous perennial herb that probably originated in South America (Collins, 1968). Clones from other groups are often cultivated in small-scale production areas for local consumption. In South America, the fruits of some wild species (e.g. *Bromeiia karatas* 'pinuella') are eaten, whereas others (*Ananas comosus* var. *erectifolius*) are used as fibre crops (Py *et al.*, 1984; Coppens d'Eeckenbrugge and Leal, 2003).

More than 60% of world pineapple production is in Asia. Thailand and the Philippines concentrate on the canned commodity and are the largest producers and exporters in Asia. About 20% of world production is in Mexico, Central and South America, and the Caribbean. Africa produces about 10%, and the largest exporters are the Côte d'Ivoire, South Africa and Kenya. The main producers in the Pacific are Australia (Queensland) and Hawaii (Rohrbach *et al.*, 2003).

Cultivation techniques

Cultivation techniques vary widely. Six major groups of vegetative clones are grown, with 'Cayenne' (typically 'Smooth Cayenne') the most common in commercial production areas of the world (Table 19.1) (Chan *et al.*, 2003). Commercial pineapple plantations recently have begun planting increased hectares of low-acid hybrids to meet changing consumer preferences. These hybrid pineapples have different agronomic characteristics from those of the traditional cultivars, and fertilization and forcing requirements can be quite different compared with what is used with Smooth Cayenne clones. Pest problems may also be different with the hybrids compared with the traditional cultivars, and hybrid behaviour towards plant parasitic nematodes may be different. However, 'Smooth Cayenne' still predominates in large plantations throughout the world and similar cultural practices are generally used. Research on nematode diseases has been conducted primarily in the intensive production systems growing 'Smooth Cayenne'.

*A revision of the chapter by E.P. Caswell, J.-L. Sarah and W.J. Apt.

Table 19.1. The main groups of *Ananas comosus* cultivars grown throughout the world.

	Common name					
	Smooth Cayenne	Singapore Spanish	Queen	Red Spanish	Pérola	Perolera
Synonyms	'Claire' 'Esmeralda' 'Kew' 'Maipuri' 'Saint Michel' 'Sarawak' 'Typhoon'	'Betek' 'Gandul' 'Masmerah' 'Nanas Merah' 'Nangka' 'Red Pine' 'Ruby' 'Singapore Canning'	'Alexandra' 'Buitenzorg' 'Malacca' 'Mauritius' 'Red Ceylon' 'Ripley Queen' 'Victoria'	'Black Spanish' 'Bull Head' 'Cowboy' 'Cuban' 'Cumanesa' 'Havannah' 'Key Largo' 'Native Philippine Red'	'Abacaxi' 'Pernambuco'	'Capachera' 'Lebrija' 'Motilona' 'Tachirense'
Main production zones	Worldwide	South-east Asia	South Africa Australia	Venezuela Carribean basin	Brazil	Colombia Venezuela
Leaves	Broad and short, spineless except near tip	Narrow and long, usually spiny	Narrow and short, very spiny	Narrow and long, spiny or half-spiny	Narrow and long, spiny	Broad and long, spiny
Fruit: size, shape, flesh colour	Large, cylindrical, pale yellow flesh	Small, globular, reddish-yellow flesh	Small, conical–cylindrical, golden-yellow flesh	Medium, barrel, pale yellow flesh	Small to medium, ovoid to conical, white flesh	Large, irregular cylindrical, pale yellow to yellow
Uses	Canning, fresh exports, local consumption	Canning, local consumption	Fresh exports, local consumption	Local consumption	Local consumption	Local consumption

After Chan *et al.* (2003).

Pineapple is cultivated for its 100–200 berry-like fruitlets arranged around a central core continuous with the peduncle (Collins, 1968). Cultivated pineapple is self-sterile and is vegetatively propagated from crowns, slips, suckers or stumps (Dalldorf, 1977; Evans *et al.*, 1988). Crowns are removed at harvest from fruits intended for canning and are commonly used as planting material (seed) for Smooth Cayenne in Hawaii and South Africa. Slips originate axially and are borne on the fruit stalk, becoming visible when the fruit is approximately half developed. The number of slips developing on a plant varies with the clone and the climate (Evans *et al.*, 1988). Slips are used as seed in South Africa. Suckers begin growing at floral differentiation, originating from axillary buds on the stem. They may be removed from the plant after fruit harvest and used as seed (as in the Côte d'Ivoire or South Africa) or left on the plant to produce a ratoon crop as is common in South Africa and Hawaii (Dalldorf, 1977; Anonymous, 1982). Stumps are suckers that have borne a fruit and are used in South Africa for 'Queen' plantings (Dalldorf, 1977).

Pineapple is planted throughout the year in most growing areas. Planting density varies from 15,000 to 120,000 plants/ha in single to triple-row beds, depending on the clone, ecological conditions and production system. Seed is typically planted in two-row beds (rows 40–60 cm apart, beds 120–140 cm centre-to-centre) with densities of 50,000–75,000 plants/ha (Guyot *et al.*, 1974; Lacoeuilhe and Guyot, 1979; Anonymous, 1982; Py *et al.*, 1984; Evans *et al.*, 1988). Beds may be covered with black plastic mulch before planting to retain fumigant and moisture, increase soil temperature and control weeds. Plastic mulch is commonly used in Hawaii. A soil fumigant is usually injected (predominately 1,3-dichloropropene) for nematode control during soil preparation (Côte d'Ivoire) or as the mulch is being laid (Hawaii) (see 'Management measures').

Pineapple is essentially a xerophyte and has stomata and trichomes adapted for reducing water loss, a growth habit allowing collection of rainfall, and a crassulacean acid metabolism. Pineapple has retained epiphytic characteristics such as the ability to absorb water and minerals through the leaves, and a fragile root system (Py *et al.*, 1984). It can be grown successfully in areas with as little as 600 mm annual rainfall. The adventitious root system is not extensive and penetrates the soil to a depth of 5–60 cm and extends 40–80 cm horizontally from the base of the plant (Guérout, 1975); consequently, supplemental irrigation can greatly improve plant growth and yield. Although pineapple can survive poor growing conditions, high levels of nitrogen, potassium and some microelements such as iron are required for profitable yield. Pre-plant fertilizers are placed in the bed during soil preparation, helping to maintain pH in the optimum range of 4.5–5.5, whereas post-plant fertilizers are applied as foliar sprays or through drip irrigation.

Ethylene or other growth regulators are used to force flowering ('forcing') 6–18 months after planting. The time of forcing depends on the climate, the seed and the intended use of the fruit (canning or fresh market) (Anonymous, 1982; Py *et al.*, 1984). Fruits are ready for harvest approximately 5–9 months after forcing. If nematode problems are not severe and soil conditions are adequate, a second crop, the ratoon, can be harvested.

Nematodes of Pineapple

More than 100 species of plant parasitic nematodes have been reported in association with pineapple root systems. The most important species of plant parasitic nematodes in pineapple production are the root knot nematodes, *Meloidogyne javanica* and *M. incognita*, the reniform nematode, *Rotylenchulus reniformis*, and the root lesion nematode, *Pratylenchus brachyurus*.

Other plant parasitic nematodes are associated with pineapple, but most are of limited or unknown pathogenicity. *Helicotylenchus* spp. are commonly found in soil in which pineapple is growing

(Redondo and de Agudelo, 1992; Nath *et al.*, 1997; da Costa *et al.*, 1998; Quesada and Barboza, 1999), and *Helicotylenchus dihystera* has been associated with damage to pineapple in glasshouse studies (Ko and Schmitt, 1993). In South Africa, spiral nematodes (*Helicotylenchus*, *Scutellonema* and *Rotylenchus*) have been reported as problematic (Keetch and Purdon, 1979). *Aorolaimus* sp. has been associated with reddish leaf symptoms in Brazil (da Costa *et al.*, 1998). *Paratylenchus minutus* can be found in exceedingly high numbers in pineapple fields in Hawaii, with population densities more than 5000 per 250 cm^3 of soil (Lindford *et al.*, 1949; B.S. Sipes, unpublished), but has yet to be associated with significant pathology or yield loss. The association of high population densities of a particular plant parasitic nematode with a plant does not prove that the nematode is damaging the plant, but detection of such associations should stimulate research to determine possible damage.

Meloidogyne

The root knot nematode, *M. javanica*, is a severe pathogen of pineapple. It is the most important pineapple nematode in Australia, being widespread in south-east Queensland, and is a significant concern in Mexico, South Africa, Zimbabwe, Thailand, and some areas of the Philippines. *M. javanica* was the main nematode disease problem in Hawaiian pineapple from 1920 until the 1950s when reniform nematode became the primary challenge. *M. javanica* currently is the primary nematode pathogen on only a limited hectarage in Hawaii (Rohrbach and Apt, 1986).

M. incognita has been reported from several pineapple-growing areas, but does not cause serious damage except in some areas of Puerto Rico and Mexico (Ayala *et al.*, 1969; Garcia and Adam, 1972). In the Côte d'Ivoire, *M. incognita* caused damage when some plantations were first established, but its importance there has diminished relative to *P. brachyurus* (Guérout, 1965).

Symptoms of damage

Second stage juveniles infect the primary root tips. Root growth is retarded within 24 h of nematode penetration, and usually a terminal club-shaped gall is produced as the nematode develops (Godfrey and Oliveira, 1932) (Plate 14A). Large galls are not formed, but small, non-terminal fusiform galls may form and cause brooming of the root system (Godfrey, 1936). Second generation juveniles infect lateral roots, causing a reduction of the total root length of the plant, decreased nitrogen absorption and plant growth rate, and reduced yield (Magistad and Oliveira, 1934; Godfrey and Hagan, 1937). Severe infections result in a stunted root system, poor anchorage and plants that are more susceptible to moisture and nutrient stress.

Nematode parasitism should be suspected if symptoms of stress are evident in the foliage despite satisfactory climatic and agronomic conditions. In some cases, careful observation of the roots may permit diagnosis of nematode infection, but nematode sampling is usually required to diagnose the nematode species involved.

Biology and life cycle

Second stage juveniles penetrate roots in the meristematic region of the root tip and become sedentary after 2–3 days (Godfrey and Oliveira, 1932). Development through subsequent moults leads to vermiform adult males and saccate, sedentary females. Reproduction is by mitotic parthenogenesis, and female nematodes produce eggs contained in a gelatinous matrix (see Chapter 2).

Population increase of *Meloidogyne* spp. on pineapple is slow compared with other host plants. Population densities of the nematode remain at pre-plant levels for several months after planting (Fig. 19.1) (Stirling and Nikulin, 1993). After these initially stable population levels, the nematode population enters a linear growth phase and, over the next 6 months, reaches a plateau. The plateau population densities are maintained throughout the remainder of the crop cycle. Significant population decreases do not occur until the pineapple is destroyed and the field fallowed.

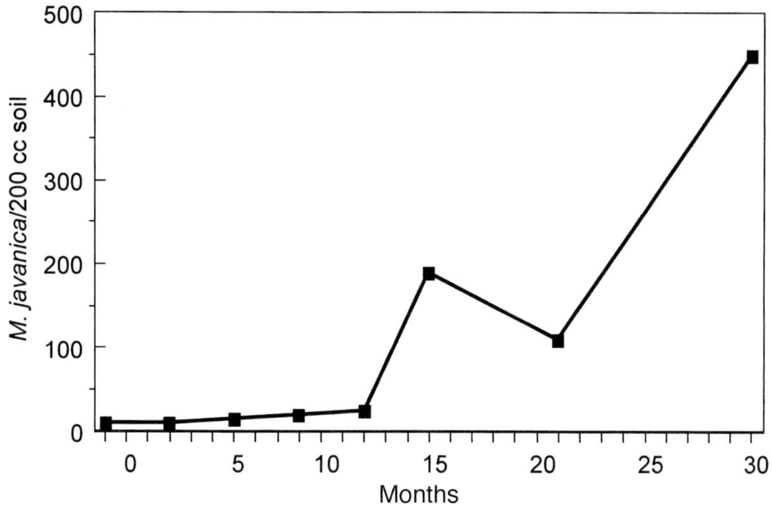

Fig. 19.1. Population increase of *Meloidogyne javanica* on pineapple in Australia.

Pathotypes and races/biotypes

Distinct host responses of *M. javanica* and *M. incognita* towards pineapple cultivars and clones have not been observed. The importance of genetic variation among geographic isolates in root knot nematodes is receiving more attention, leading to recognition that geographic isolates of the same species may not behave in the same way. For example, atypical populations of *M. javanica* that reproduce on cotton, groundnut (peanut) or pepper have been reported (Jepson, 1987). Although *M. javanica* is reported to be a coffee pathogen in many areas of the world, *Coffea arabica* cultivars 'Caturra' and 'Catuai' were non-hosts to a California isolate of *M. javanica* (Araya and Caswell-Chen, 1996). Research to assess the genetic variation that exists among geographic isolates of root knot nematodes is needed to determine the appropriateness of comparing studies from different places.

Survival and dissemination

Eggs in egg masses survive up to approximately 2 h at a relative humidity of 50%, increasing to 8 h at a relative humidity of 90% (Godfrey and Hoshino, 1933). Eggs contained in galled tissue can tolerate 20 days exposure to 90% relative humidity. Exposure to ultraviolet radiation was lethal to eggs, eggs in egg masses, and juvenile stages of the nematode (Godfrey and Hoshino, 1933).

Juveniles of *M. javanica* may survive in desiccated soil without a host for 20–24 weeks, although soil moisture influences survival (Godfrey *et al.*, 1933; Towson and Apt, 1983). The time required to reduce soil populations of *M. javanica* juveniles by 50% in Hawaiian soils was 3, 5, 110, 10 and 3 days at soil moistures of −0.16, −0.30, −1.1, −15 and −92 bars, respectively (Towson and Apt, 1983). *M. javanica* can survive, although at low levels, as long as 2 years in fallow field soil (Godfrey, 1936).

The nematode survives a wide range of temperatures; however, 127 min at 40°C is lethal to juveniles, while 4.5 days at 40°C is lethal to eggs (Hoshino and Godfrey, 1933). Bare pineapple soils in Hawaii may reach 40°C at a depth of 0.6 cm during the summer and, if covered with mulch paper, temperatures greater than 40°C may extend to a depth of 7.5 cm (Hagan, 1933).

The spread of root knot infestation between root systems of adjacent plants is quite slow. Godfrey (1936) observed that up to 7 months were required for an infestation to move 30 cm within a row. The root knot nematodes may be disseminated over long distances in soil adhering to workers' shoes, implements and equipment that is moved from field to field. In South Africa, the nematode is spread by planting infested stumps, so seed material from infested areas is destroyed (Dalldorf, 1977).

Environmental factors affecting parasitism

The minimum temperature for infection by *M. javanica* is approximately 13°C (Godfrey, 1936). *M. javanica* is capable of surviving a wide range of pH levels, and can successfully infect pineapple roots at soil pH of 4.0–8.5, the range of pH at which pineapple is usually grown (Godfrey and Hagan, 1933).

Other hosts

M. javanica has a host range of more than 770 plants, including many economically important crops such as potato, tomato, grape and tobacco.

Disease complexes

Galls of *M. javanica* are subject to secondary invasion by various fungi that cause blackening and drying of the nematode galls, and death of the nematodes within the gall (Godfrey, 1936; Keetch, 1982).

Economic importance and damage threshold

Godfrey (1936), working in Hawaii, suggested that plants could become well established when the population density of root knot nematodes was less than approximately 6 juveniles/cm^3 of soil. He did not directly relate the initial population density to yield, so his estimate cannot actually be considered a damage threshold. Under South African conditions, a single juvenile of *M. javanica* in a root or soil sample is interpreted as a potential problem (Keetch, 1982). In Australia, economically significant crop losses occur in the pineapple ratoon crop when nematode population densities 12 months after planting are greater than 1–5 juveniles/200 cm^3 of soil (Stirling and Kopittke, 2000).

Rotylenchulus reniformis

The reniform nematode, *R. reniformis*, occurs in the tropics, subtropics and warm temperate regions throughout the world. It is the major nematode problem of pineapple in Hawaii and the Philippines (Davide, 1988). Reniform nematode is also important in the Caribbean (e.g. Puerto Rico), in some areas of Thailand, in North Queensland, Australia, and in Oxaca, Mexico. In South Africa, *Rotylenchulus parvus* is more frequently observed but is of no economic importance (Keetch, 1982).

Symptoms of damage

In Hawaii, leaves of infected plants are less erect than those of healthy plants, are reddish in colour and show poor growth. The foliar symptoms are similar to those caused by nutrient or moisture stress. In contrast to the symptoms observed in root knot nematode infections, primary roots of pineapple infected with *R. reniformis* continue to elongate and provide good anchorage for the plant. However, reniform nematode infection inhibits secondary root formation and root systems are poorly developed (Plate 14B). Heavy infestations may result in plant collapse and death. Improper management of reniform populations typically leads to ratoon crop failures in Hawaii.

Biology and life cycle

The reniform nematode has a unique life cycle. Egg hatch is stimulated by root exudates of certain host plants (Kahn, 1985), and second stage juveniles leave the egg and move into the soil. Once in the soil, they undergo three moults without feeding, yielding adult males and 'pre-parasitic' females. Females enter the root system and initiate a feeding site. Females continue to develop, swelling and becoming sedentary (Linford and Oliveira, 1940; Bird, 1984). The mature egg-producing females deposit an average of 60 eggs into a gelatinous matrix (Linford and Oliveira, 1940; Bird, 1984). Although amphimixis appears to be the rule, some populations from Japan are reported as parthenogenetic (Nakasono, 1977, 1983).

Females induce their syncytium in the stele of the root (Robinson *et al.*, 1997). The syncytium is formed from a single endodermal cell that enlarges to incorporate additional cells in the pericycle, vascular parenchyma and sometimes phloem (Rebois *et al.*, 1975; Robinson *et al.*, 1997). Males do not appear to feed at any time.

The population dynamics of the reniform nematode in pineapple are similar to those of *M. javanica* in pineapple (Fig. 19.2). The reniform nematode population density does not increase immediately after pineapples are planted and begin rooting. Nematode populations remain at pre-plant levels for up to 8 months (Sipes and Schmitt, 1994b). After this period of relatively flat population increase, the nematode population enters a linear growth and increases to levels of up to 10,000 nematodes/250 cm^3 within 6 months (Sipes and Schmitt, 1994b). The reniform nematode population remains at these peak levels throughout the crop cycle, showing only slight decreases in population densities at the initiation of pineapple flowering. The delayed population development could be related to endogenous protease inhibitors found in the pineapple roots (Radovich *et al.*, 2004). The protease inhibitor concentration is greater in pineapple infected with reniform nematode than in uninfected plants, suggesting a systemic acquired resistance response to nematode infection by the pineapple (Chinnasri and Sipes, 2004).

Pathotypes and races/biotypes

Distinct races of the reniform nematode are not known, although on the basis of host range and reproductive strategy the existence of races has been suggested (Dasgupta and Seshadri, 1971; Heald, 1978; Nakasono, 1983). There are differences in temperature optima and reproductive behaviour among populations of reniform nematode (Nakasono, 1977, 1983). For example, exposure to low temperatures (15°C) resulted in decreased reproduction in populations from Puerto Rico compared with populations from Louisiana and Texas (Heald and Inserra, 1988).

Survival and dissemination

The reniform nematode tolerates extreme temperatures, and survives extended periods without a host. Reniform nematode populations from Louisiana, Texas and Puerto Rico survived for 6 months without a host at temperatures of –5, –1, 4 and 25°C (Heald and Inserra, 1988). Although the

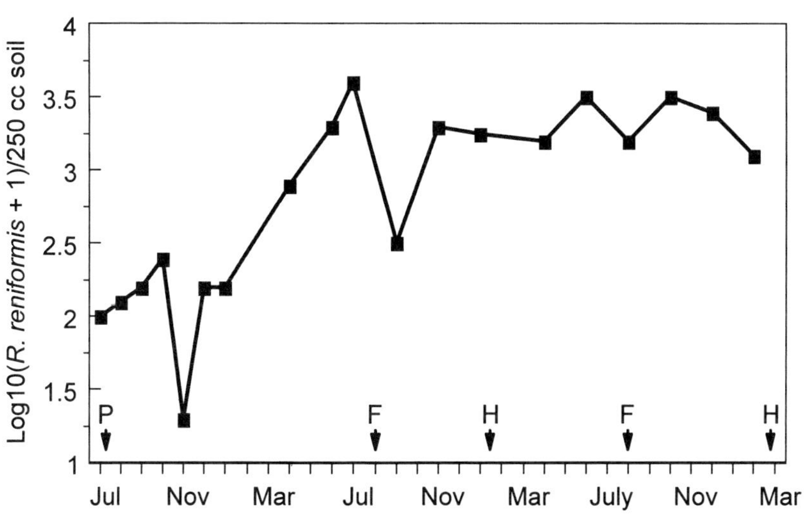

Fig. 19.2. Population increase of *Rotylenchulus reniformis* on pineapple in Hawaii.

reniform nematode is able to survive low soil moisture, soil moistures greater than 7% increase nematode survival at 25°C, but decrease nematode survival at temperatures below freezing (Heald and Inserra, 1988).

Populations of *R. reniformis* can survive for 2 years in fallow soil. Apparently, the nematode survives fallow periods in the egg stage or as anhydrobiotic juvenile stages, depending on soil moisture (Apt, 1976; Tsai and Apt, 1979).

Environmental factors affecting parasitism

The optimum temperature for development is 25–29°C, and reproduction is limited by temperatures above 36°C (Rebois, 1973; Heald and Inserra, 1988). Soil temperatures in pineapple-growing regions are extremely favourable to the development of the reniform nematode.

The reniform nematode did not become a significant agronomic problem in Hawaii until the mid-1950s. The tendency of the pineapple industry to use shorter and shorter fallow periods is thought to have contributed to the increasing problem with reniform nematode (Rohrbach and Apt, 1986). In addition, the pH of pineapple soils steadily decreased from 1930 to 1950 due to the application of ammonium sulphate fertilizers. The pH in some fields in Hawaii was as low as 3.2 by 1950. The optimal pH for reproduction of the reniform nematode in Hawaiian soils is approximately 4.8–5.2 (Rohrbach and Apt, 1986).

Another factor contributing to the increased importance of reniform nematode in Hawaiian pineapple production was soil fumigation. Fumigation with D-D (1,2-dichloropropane, 1,3-dichloropropene mixture), EDB (ethylene dibromide) and DBCP (dibromochloropropane) began in the late 1940s. These soil fumigants undoubtedly suppressed populations of nematode antagonists in the soil (Rohrbach and Apt, 1986). The above-mentioned agricultural practices combined with intensive monoculture appear to have created a soil environment supportive of reniform nematode survival and reproduction. Consequently, in 35–40 years, *R. reniformis* went from an initial limited occurrence to becoming a major limiting factor in Hawaiian pineapple culture.

Other hosts

The reniform nematode has an extensive host range that includes more than 300 plant species (Robinson *et al.*, 1997). Many weed species commonly found in pineapple- and sugarcane-growing areas are hosts (Linford and Yap, 1940; Birchfield and Brister, 1962). Many important crop species, such as soybean, cotton, pigeonpeas and beans, are also hosts.

Economic importance and damage threshold

The reniform nematode is a seriously damaging pathogen of pineapple. In Hawaii, large populations of the nematodes combined with moisture stress can result in complete ratoon failures (Plate 14D) (Rohrbach and Apt, 1986). D-leaf weight, plant height and root biomass did not differ ($P > 0.05$) among a range of reniform nematode populations at 6 or 12 months; however, plant crop fruit yield did differ among the initial population ranges. D-leaf, the youngest mature leave, and plant height are highly correlated to final fruit weight. Pre-plant population densities of *R. reniformis* below 300 nematodes/250 cm^3 of soil damage pineapple but are not the major factor limiting yield (Sipes and Schmitt, 2000). At lower nematode population densities, pineapple yield is limited by soil fertility and inherent soil physical factors. *R. reniformis* becomes the major limiting factor at population densities above 600 nematodes/250 cm^3 of soil (Sipes and Schmitt, 2000).

Pratylenchus

The root lesion nematode, *P. brachyurus*, was described originally from pineapple roots in Hawaii (Godfrey, 1929). It is prevalent and of economic importance throughout the equatorial tropics in the Côte d'Ivoire, Uganda, Hluhluwe in northern

Natal (South Africa) and Brazil (Guérout, 1975; Zem and Reinhardt, 1978; Bafokuzara, 1982; Keetch, 1982; Dinardo-Miranda *et al.*, 1996b). Although present, it is of limited importance in higher latitudes of the subtropics such as the Caribbean, Hawaii, Australia or the Cape Province in South Africa (Guérout, 1975; Keetch, 1982; Rohrbach and Apt, 1986; G.R. Stirling, personal communication).

P. zeae is observed in some pineapple production areas, but there is no information on its pathogenicity to pineapple.

Symptoms of damage

Black lesions caused by *P. brachyurus* develop in the roots at the point of nematode infection. The developing necrosis may extend progressively over the whole surface of the root as the nematodes feed and move through the root. Lesions are surrounded by dead and discoloured epidermal cells and may extend throughout the parenchyma (Godfrey, 1929; Keetch, 1982). In the later stages of infection, the parenchyma is destroyed and the cortex separates from the central cylinder (Guérout, 1975). Secondary roots and root hairs are also destroyed by this nematode, leading to a root system composed of poorly developed primary roots. The damage to parenchyma tissue is not generally visible in the field as pineapple roots are rapidly and heavily suberized.

Infection by *P. brachyurus* decreases plant growth rate, delays leaf emergence and reduces leaf weights 35–40% (Guérout, 1975; Lacoeuilhe and Guérout, 1976; Sarah, 1986). Leaves turn yellow and then red, lose turgidity, and their tips wither (Py *et al.*, 1984). Foliar symptoms result from deficient water and mineral supply to the plant and are especially noticeable if fertilizers are applied as granules to the soil before planting, as fertilizer absorption is suppressed by nematode damage. Foliar application of fertilizer decreases nematode influence on plant growth because leaves absorb nutrients and this compensates for decreased root function (Lacoeuilhe and Guérout, 1976).

Biology and life cycle

P. brachyurus is a migratory endoparasite. Males are rare, and reproduction is by mitotic parthenogenesis (Roman and Triantaphyllou, 1969). The life cycle may be completed within the roots. Thus, large populations can develop quickly and cause the rapid destruction of the cortical parenchyma (Guérout, 1975).

Survival and dissemination

Under laboratory conditions, populations of *P. brachyurus* from the Côte d'Ivoire survive from 20 to 22 months in fallow soil (Feldmesser in Wallace, 1963), as long as viable root fragments are present in the soil (Guérout, 1975). If root fragments are absent from the soil, survival without a host is limited to approximately 7 months. After 35 days at 44°C, only 25–50% of an original South African population survived (Keetch, 1977).

In the Côte d'Ivoire, *P. brachyurus* is sometimes disseminated when infected suckers are used as seed. Generally, the suckers used as seed are uninfested.

Environmental factors affecting parasitism

The optimum temperature for *P. brachyurus* development is 25–30°C (Olowe and Corbett, 1976). This temperature range encompasses the yearly average soil temperatures in the Côte d'Ivoire. Although nematode movement is inhibited by soil temperatures above 40°C (Endo, 1959; Olowe and Corbett, 1976), many Ivorian plantations are located on sandy soils which are very favourable to the movement of *P. brachyurus* when temperatures are adequate.

The soil temperatures in the Côte d'Ivoire are relatively constant and the root lesion nematode responds primarily to changes in soil moisture. If pineapple is planted during the dry season, the nematode populations in the roots will remain at low levels, increasing several weeks after the return of regular rainfalls

(Fig. 19.3A). When planted during the rainy season, nematode population densities in the roots increase rapidly after approximately 3 months (Fig. 19.3B). If soil moisture remains favourable, root population densities remain relatively stable until forcing, and then decline. Approximately 20 mm of rainfall per 10 days is required in the Côte d'Ivoire to maintain high root populations of *P. brachyurus* (Sarah and Hugon, 1991).

Root population densities of *P. brachyurus* increase rapidly in acid soils and very slowly when pH exceeds 5–5.5 (Sarah *et al.*, 1991). Most Ivorian soils are very acid, which may contribute to the prevalence of the nematode in that country. In the Côte d'Ivoire, *P. brachyurus* competitively displaces *Meloidogyne* spp., as the rapid destruction of root tissue by the root lesion nematode seems to prevent the establishment of the root knot nematode (Guérout, 1965).

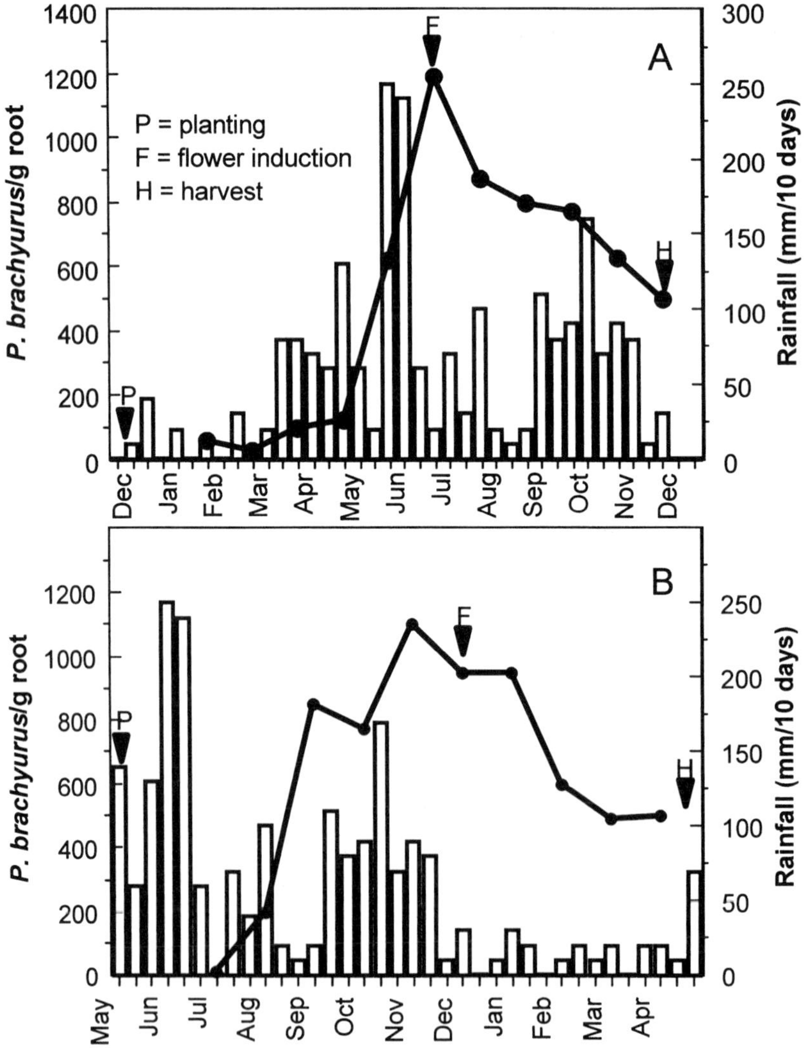

Fig. 19.3. Populations of *Pratylenchus brachyurus* in the roots of *Ananas comosus* cv. Smooth Cayenne related to rainfall in the Côte d'Ivoire. (A) Pineapple planted in December just before the main dry season. (B) Pineapple planted in July at the end of the main rainy season.

Other hosts

The root lesion nematode has a wide host range that includes 100 recorded plant species, many of them grasses found in the natural savannahs of the Côte d'Ivoire (Luc and de Guiran, 1960). Maize and cassava are very good hosts for root lesion nematode, and these plants cannot be used as rotation crops with pineapple in the Côte d'Ivoire (Anonymous, 1987).

Disease complexes

P. brachyurus may infect galls caused by *M. javanica* and cause the rapid breakdown of the gall and death of the root tip (Godfrey, 1929). In the Côte d'Ivoire, Guérout (1975) demonstrated an interaction between *P. brachyurus* and pytheaceous fungi. The fungus–nematode combination results in plant damage greater than that caused by the nematode alone.

Economic importance and damage threshold

In South Africa, inoculation with 200 *P. brachyurus* decreased plant growth by 25% after 10 months. This compares with a decrease of 10% caused by similar inoculation with *M. javanica* (Keetch, 1982). The damage caused by *P. brachyurus* can be severe, with yield losses reaching 30% for the plant crop and 80% for the first ratoon crop in the Côte d'Ivoire (Lacoeuilhe and Guérout, 1976; Sarah, 1986). The damage threshold is partially determined by the planting date because climatic conditions, including soil moisture and temperature, influence nematode population growth rate and the capacity of the plant to tolerate infection. For example, dry conditions combined with *P. brachyurus* infection cause a drastic reduction in sucker development in the Côte d'Ivoire (Sarah, 1987a). The linear relationship between initial population density of *P. brachyurus* and average fruit weight for pineapple planted just before the rainy season in the Côte d'Ivoire (Fig. 19.4) suggests that the damage threshold is very low in that environment (Sarah, 1986).

Management measures

The primary emphasis of nematode management in pineapple is on protection of the young, growing root system. Reduction of nematode inoculum in the soil prior to planting or reduction of nematode population growth rate once plants are established in the field is the goal of management. Preplant control of nematode populations is most important, as damage to the developing roots of the young plant results in poor plant

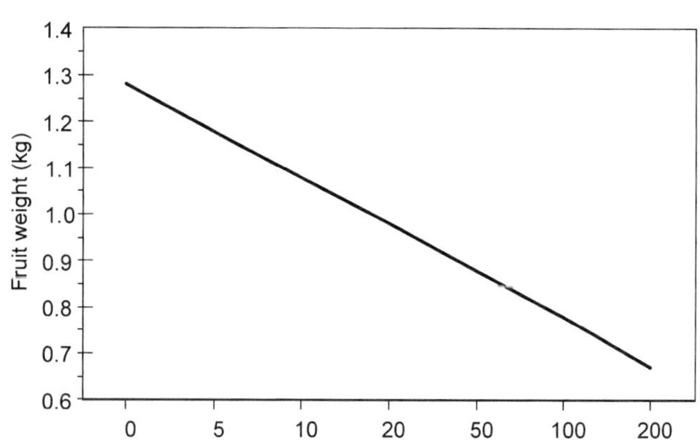

Fig. 19.4. Relationship between pineapple fruit weight and pre-plant soil population densities of *Pratylenchus brachyurus* in the Côte d'Ivoire.

growth (Godfrey, 1936). Pre-plant tactics to suppress nematode inoculum include application of nematicides, rotations with non-host crops, fallowing and soil amendments. Post-plant management options are currently limited to nematicide application.

Cultural practices

Growers usually specialize in pineapple production. Consequently, the crop is grown in long-term monoculture. Some fields in the Côte d'Ivoire have been producing continuous pineapple for 30 years, while fields in Hawaii have produced pineapple for over 80 years.

Pineapple is essentially a perennial plant. After fruiting, the pineapple plant produces a slip that gives rise to the next fruit. This process of producing ratoon stumps can continue indefinitely. Fruit from second ratoons and later tend to be smaller in size than those from the plant crop. Commercial growers decide when to replant based on whether the ratoon fruit has become too small or too sparse. Therefore, pineapple crop cycles can be very long, e.g. 8 years in South Africa. Fields typically are left fallow during the period between pineapple crops (the intercycle). The duration of the intercycle is dictated by economics and pest control considerations. Long crop cycles can be considered to include a long intercycle, while short crop cycles usually have a short intercycle. The success of the intercycle in reducing nematode populations is also influenced by the type of fallow (e.g. clean versus 'natural' fallowing), soil moisture conditions and the host range of the nematode species involved.

Clean fallow

Weed-free fallow can be used to decrease nematode populations, although keeping a field free of weeds is difficult. An additional problem is that pineapple stumps can produce root tissue long after the shoots are destroyed and volunteer pineapple can support nematode reproduction (Ko and Schmitt, 1993; B.S. Sipes and K. Wang, unpublished). Weeds such as nightshade and pigweed (*Amaranthus* spp.) growing during a 1 year fallow period supported high populations of root knot and reniform nematodes in Hawaii (W.J. Apt, unpublished). In the Côte d'Ivoire, *P. brachyurus* was found on 15 common weed species and therefore limits the utility of 'natural' fallowing (Goly and Téhé, 1997). Weed seeds can remain viable for years in a field, and even small seedlings have a root system capable of supporting significant numbers of nematodes. Fields can be kept nearly weed free by application of herbicides, or through periodic cultivation. An added benefit of cultivation is that it brings deeper soil layers to the surface, exposing nematode eggs and juveniles to ultraviolet radiation and desiccation. In addition to soil erosion concerns, a possible problem with deep cultivation is that it brings the deeper weed seed bank to the surface, and may result in increased weed germination. As with all pest management strategies, multiple pests should be considered as appropriate.

Although nematode populations decline during a clean fallow, it is virtually impossible to eradicate nematode populations. Even after fallow periods as long as 2 years, residual inoculum is still present, though difficult to detect (Godfrey, 1936; Guérout, 1975). Additionally, some nematode species have life history strategies that include cryptobiotic capacities, such as dauer stages that allow survival despite environmental extremes, or anhydrobiotic stages that can survive in a quiescent state. For example, *R. reniformis* juveniles can withstand severe dehydration under slow dehydration regimes (Womersley and Ching, 1989). The success of fallowing will depend, to a degree, on the nematode species involved.

The pineapple industry in Hawaii currently uses a 6–12 month clean fallow period between plant cycles. Fallow periods hasten the decline of reniform nematode populations in soil, but moisture plays a role in determining the extent of population decline. *R. reniformis* can survive for as long as 1.5 years in desiccated, fallow soils (Apt, 1976; Tsai and Apt, 1979). In the Côte d'Ivoire, 6 weeks fallow can reduce populations of *P. brachyurus* by half (Guérout, 1975).

Clean fallow can be a problem on large plantations as it is energy intensive and may not be economically justifiable. In addition, erosion, one of the most important problems facing modern agriculture, may be increased considerably by fallow. The absence of a cover crop may reduce soil fertility by slowing the addition of organic matter and decreasing retention of soluble nutrients in the soil. Fallow may decrease the population densities of beneficial microorganisms, such as endomycorrhizae, as has been observed in the Côte d'Ivoire (Sarah, 1987b).

Crop rotation

Because of some of the problems associated with clean fallow, planting non-host cover crops may be desirable. Cover crops may suppress plant parasitic nematode populations, decrease erosion, maintain or enhance soil fertility, and provide a niche for nematode-antagonistic fauna. Some plants produce allelochemicals as root exudates that are actively toxic or inhibitory to nematodes. Other plants produce toxic substances as they break down in the soil and can act as biofumigants. Both monocots and dicots have been evaluated for nematode control in pineapple.

Numerous dicotyledonous plants have been evaluated as potential of intercycle cover crops to reduce nematode numbers in soil. French marigold, *Tagetes patula*, reduced populations of *R. reniformis* (Nakasono, 1973; Ko and Schmitt, 1993), whereas *T. erecta* and *T. polynema* increased populations of the reniform nematode (Wang *et al.*, 2001) as compared with bare fallows. Sunnhemp (*Crotalaria juncea*) has shown the most promise in Hawaii. Sunnhemp is a poor host for *R. reniformis*, has allelopathic effects towards the nematode, enhances antagonistic microorganisms in the soil and adds nitrogen to the soil (Caswell *et al.*, 1991a; Wang *et al.*, 2001, 2002, 2003). In growth chamber experiments, *C. juncea* (PI 207657 and 'Tropic Sun') was resistant to penetration by a California isolate of *M. javanica*, whereas penetration of *Dolichos lablab* and *Sesamum indicum* was significantly lower than penetration of tomato (Araya and Caswell-Chen, 1994b). Greenhouse experiments to assess the reproduction of a California *M. javanica* isolate on *C. juncea* PI 207657 and 'Tropic Sun', *Sesamum indicum*, *D. lablab* and *Elymus glaucus* revealed that individual plants of both *C. juncea* and *S. indicum* supported very limited reproduction of *M. javanica* (Araya and Caswell-Chen, 1994a). The use of such plants that do support limited reproduction might act as a selection pressure on field populations leading to increased virulence in the target nematode. *Brassica napus*, evaluated because of its biofumigation potential, was a poor host to the reniform nematode but an excellent host to *M. javanica* (Wang *et al.*, 2001, 2002). In the Côte d'Ivoire, the legumes *Crotalaria usaramoensis*, *Stylosanthes gracilis* and *Flemingia congesta* reduced populations of *P. brachyurus* after 18 months of growth, increased the nitrogen content of the soil and the subsequent pineapple crop, and increased the fruit weights of the subsequent pineapple crop by 25–30% (Guérout, 1969).

Grasses have also been studied to assess their value as intercycle cover crops. Rhodes grass (*Chloris gayana*) reduced soil populations of *R. reniformis* as well as or better than clean fallow (Caswell *et al.*, 1991a). Rhodes grass is a non-host and is immune to penetration by *R. reniformis*. In glasshouse experiments, root exudates of Rhodes grass applied to tomatoes grown in soil were observed to inhibit reniform nematode hatch and orientation to roots, while French marigold exudates appeared directly toxic to reniform nematode (Caswell *et al.*, 1991a,b). *C. gayana* and *Desmodium unicatum* have been successful as rotation crops to reduce nematode populations (mixed Hoplolaiminae genera and *Meloidogyne* spp.) in the Cape Province (Keetch and Dalldorf, 1980). Pangola grass (*Digitaria decumbens*) has potential as a rotation crop for pineapple as it apparently stimulates eclosion of *M. incognita*, and toxins produced by the roots affect juvenile survival (Ayala *et al.*,

1967; Haroon and Smart, 1983a). Plantings of *D. decumbens* eliminate populations of *M. incognita* after 1 year, and *Criconemella* spp. and *Helicotylenchus* spp. after 18 months. *D. decumbens* is a poor host for *M. javanica* (Haroon and Smart, 1983b), but *P. brachyurus* remained abundant even after 3 years of *D. decumbens* growth (Ayala *et al.*, 1967). Sugarcane is frequently grown in areas where pineapple is produced and is generally considered a non-host for *R. reniformis*. Rotating pineapple with sugarcane may decrease some nematode problems, provided that weed hosts are not present. This strategy was attempted in Hawaii with poor success. Sugarcane is a host for *P. brachyurus* in Hawaii, Brazil and Venezuela. When grown for 6 months, *Panicum maximum* increased pineapple yields better than did 6 months of *Chromolaena odorata* (Asteraceae), even though the latter showed a superior reduction of the nematode population (Anonymous, 1987). This last example demonstrates that the cover crop that gives the best nematode population reduction will not necessarily result in the best yield of the subsequent pineapple crop.

Organic improvements and soil amendments

The addition of organic matter to pineapple soils is beneficial, as the decline of soil organic matter is faster in pineapple soils than under other crops (Py *et al.*, 1984). The addition of organic matter may have direct and indirect effects on nematode populations. For example, adding cassava residues or extracts of neem (*Azadirachta indica*) leaves to soil reduces populations of *P. brachyurus* by 75 and 72%, respectively, in Nigeria (Egunjobi and Larinde, 1975). These are not common amendments to pineapple soils however. Linford (1937) found that adding organic matter to soil increased the activity of nematode-trapping fungi (see 'Biological management', below). Working in Hawaii, H.W. Klemmer and R. Nakano (unpublished) found that incorporating pineapple plant residues into the field (rather than burning them) significantly increased the numbers of nematode antagonists present in the soil. These antagonists reduced reniform nematode populations, but not as effectively as did soil fumigation. Furthermore, the surviving nematode populations rapidly increased during the next crop cycle, with resulting crop damage the equivalent of untreated control plots. Much of the observed beneficial effect of organic matter incorporation is probably due to its stimulatory effect on predators and parasites of nematodes.

Resistance and tolerance

Pineapple has been evaluated for resistance and tolerance to plant parasitic nematodes. Collins and Hagan (1932) assessed the tolerance of several pineapple clones to *M. javanica* by determining the influence of the nematode on root growth. They found that Cayenne was very intolerant of nematode infection, whereas Wild Brazil and an F_1 hybrid from Wild Brazil × Cayenne were much more tolerant, if not immune to damage from nematode infection as measured by shoot weight and root length (Collins and Hagan, 1932; Hagan and Collins, 1935). Collins and Hagan did not assess nematode reproduction in these clones; however, Sipes and Schmitt did (Sipes and Schmitt, 1994a). They found that the same cultivars supported reproduction of *M. javanica* but were tolerant to infection in that plant growth was not affected. *A. comosus* var. *ananasoides* and three other selections were reported as resistant to *M. incognita* in Puerto Rico (Ayala, 1961, 1968; Ayala *et al.*, 1969). Dinardo-Miranda *et al.* (1996a) found that among 13 cultivars evaluated, only 'Huitota' supported significantly lower populations of *M. incognita*. *A. comosus* var. *ananasoides*, 'Venezolana' and two other clonal selections were resistant to Puerto Rican populations of *R. reniformis* (Ayala, 1961, 1968; Ayala *et al.*, 1969). In Hawaii, 18 cultivars were assessed for reniform nematode resistance, including two *A. comosus* var. *ananasoides* lines and two *A. comosus* var. *ananasoides* hybrids, and all supported reniform nematode reproduction (Sipes and Schmitt, 1994a). *A. comosus* var. *ananasoides* is an excellent host for *P. brachyurus* in the Côte d'Ivoire (Py *et al.*, 1984). Different clones, cultivars, species and genera have been tested for resistance to *P. brachyurus* in the Côte d'Ivoire with negative results (Anonymous, 1987). The Queen group of pineapple and *A. comosus* var. *bracteatus* are extremely susceptible to *P. brachyurus*. All of the 14 pineapple cultivars evaluated by Dinardo-Miranda *et al.* (1996b) were good hosts to *P. brachyurus* as well as the 21 pineapple cultivars evaluated by Sarah *et al.* (1997). However cultivars from the Pérola group appeared slightly (although not significantly) less infected in several experiments of the latter study.

Nearly all pineapple cultivars and clones support nematode reproduction. However, the level of reproduction and the tolerance to nematode infection vary widely. Tolerance to reniform and root knot nematodes is manifested in the pineapple as more root growth. Those cultivars and clones that root more vigorously have more tolerance to the nematode. Long-term cultivation of 'Smooth Cayenne' has resulted in an indirect selection for greater tolerance and resistance as compared with other less intensively grown cultivars in Hawaii (Sipes and Schmitt, 1994a).

Nematicides

From the beginning of commercial production in the 1920s, the pineapple industry has adopted a chemical-dependent cropping system. Even today, chemical nematicides remain the primary means of managing plant parasitic nematodes in pineapple, regardless of the nematode species involved. Pre-plant or at-plant soil treatments protect the root system of the young pineapple plant against nematodes that are present. Such treatments can be applied as pre-plant fumigation, at-plant incorporation of granular nematicides, or pre-plant nematicide application via drip irrigation (Rohrbach and Apt, 1986; Apt and Caswell, 1988). An effective nematode management strategy must be based on the crop cycle length and the number of ratoons desired. Research in Hawaii has shown that protecting the root system for a minimum of 6 months is necessary, and 8–12 months of control is preferred, if ratoon crops are to be harvested.

Pre-plant fumigation treatments have changed over the years. At one time, products such as EDB, DBCP and methyl bromide were commonly used in the pineapple industry (Py *et al.*, 1984). Increased environmental concerns and changing government regulations have seen these products removed from the market. Today most pre-plant fumigation involves the application of 1,3-dichloropropene at a rate of 224–336 l/ha in Hawaii and South Africa (Rohrbach and Apt, 1986; Schneider *et al.*, 1995). In Hawaii, minimizing application rates has been achieved with the use of a single chisel and sealing the planting bed with a plastic mulch (Sipes *et al.*, 1993). If the fumigation is successful, it is usually sufficient to protect the plant crop but not subsequent ratoons. In the Côte d'Ivoire, soil incorporation of non-fumigant nematicides is used currently as a pre-plant application. The most commonly used compounds are terbufos, cadusafos and ethoprophos.

Non-fumigant nematicides are typically applied as post-plant treatments, although their usage is also undergoing changes. Generally, post-plant nematicide applications are only necessary in Hawaii if pre-plant fumigation is unsuccessful. In the Côte d'Ivoire, post-plant applications are imperative for a successful plant crop. Post-plant treatments without successful fumigation may not give adequate nematode control (Plates 14E and F). The range and types of non-fumigant nematicides have changed markedly in the past 10 years. In the USA, the Food Quality Protection Act of 1996 has drastically altered the nematicides that are registered for use in pineapple. Almost all carbamate and organophosphate nematicides, such as ethoprophos and fenamiphos, are not registered for use in the USA. Products such as DiTera, derived from the fungus *Myrothecium* sp., and emulsifiable formulations of 1,3-dichloropropene are being evaluated in Hawaii for post-plant nematode control (B.S. Sipes, unpublished).

Research is also underway investigating use of plant systemic acquired resistance as a nematode control tactic.

The systemic properties of some of the non-fumigant nematicides allows for foliar application during any point in the plant growth cycle. Apt is credited with being the first individual to design a pineapple nematode management programme based on the systemic properties of foliar-applied fenamiphos (Zeck, 1971). As early as 1966, W.J. Apt (unpublished) conducted extensive studies with foliar applications of fenamiphos. He obtained control of reniform nematode with foliar applications of fenamiphos at rates of 600–2400 ppm (Zeck, 1971). In the Côte d'Ivoire, foliar application of oxamyl may be as effective as fenamiphos if applied at twice the rate of fenamiphos (Sarah, 1987a). Foliar applications of these organophosphate nematicides currently are not common in Hawaii. Preliminary indications are that applications of 100 ppm of the systemic acquired resistance inducer acibenzolar-s-methyl may hold promise because nematode reproduction is reduced by 50% with a single application (Chinnasri and Sipes, 2002).

Post-plant application of non-fumigant nematicides requires good soil moisture conditions to promote movement in the soil and absorption by the plant, and to ensure that the nematode target is physiologically active (Sarah, 1980).

Application of nematicides through drip irrigation systems has been the focus of research in Hawaii (Apt and Caswell, 1988; Sipes and Schmitt, 1995). The application of the nematicide with water through the drip irrigation system has the advantages of minimizing worker exposure and delivering the product directly to the site of action. Most carbamate and organophosphate products can be applied through drip irrigation (Rohrbach and Apt, 1986; Apt and Caswell, 1988; Sipes and Schmitt, 1995; Sipes, 1996). Post-plant nematicides are most effective when applied to soils having optimal moisture levels. Nematodes are active and the chemical is distributed throughout the rooting zone.

Under some conditions, the non-fumigant nematicides may have phytotoxic side effects, including heart and leaf burns (ethoprophos and fenamiphos), disturbance of growth (isazophos) and flowering (fenamiphos and carbofuran), and decreased sucker production (carbofuran) (Sarah, 1981a,b, 1983, 1987a). The phytotoxicity may result from direct contact with young plant tissues or physiological responses due to the systemic nature of the nematicides. Physiological disturbance caused by carbofuran and other carbamates is well documented in other plants, where the compound inhibits oxidase activity resulting in increased levels of indole acetic acid (Jamet and Piedallu, 1980). Fenamiphos causes the same phenomenon in pineapple (Milne *et al.*, 1977), and this may explain fenamiphos-induced stimulation of growth in the absence of nematodes.

Biological management

Linford, a researcher at the Pineapple Research Institute, was a pioneer in biological control of nematodes. Many nematode-parasitizing fungi have been identified in Hawaiian soils, including *Arthrobotrys oligospora* Fresenius, *Catenaria anguillulae* Sorokin, *Harposporium anguillulae* Lohde and *Stylopage hadra* Drechsler (Linford, 1937). In laboratory and greenhouse experiments, Linford (1937) and Wang *et al.* (2003) examined the potential of incorporating organic matter to stimulate the activity of nematode predators and parasites in the soil. The incorporation of organic matter resulted in increased populations of free-living nematodes that are prey for nematode-parasitizing fungi, resulting in increased fungal populations (Linford, 1937). The addition of chopped pineapple material to soil at a rate of 37–111 kg/m^3 of soil significantly reduced galling caused by root knot nematode as determined by bioassay (Linford, 1937; Linford *et al.*, 1938). Wang *et al.* (2003) found that sunnhemp amendment increased the number of nematode-trapping fungi in the soil. Linford also investigated the potential for using several fungi as manipulable biological control agents (Linford and Yap, 1939). In small pot tests, they observed that addition of *Dactylella ellipsospora* reduced plant injury caused by the root knot nematode, although the results were confounded by the presence of other natural enemies of nematodes in the treatment.

Potential biological control agents must be tested in field soil. Results obtained in greenhouse experiments may differ from those in the field because the activity level of the biological control agent will depend on the biotic and abiotic characteristics of the soil (Linford and Yap, 1939). The majority of Linford's work was completed before the widespread use of soil fumigation, and the above-mentioned caveats are even more important today. Although biological control is a potential component of a nematode management programme in pineapple, it does not currently play a major role in nematode management in any of the world's commercial pineapple cultivation.

Methods of diagnosis

Sampling

Soil samples should be taken before planting to a depth of approximately 30–40 cm with a trowel or soil-sampling tube. Ideally, the soil should be in a condition of good tilth suitable for sampling. A composite soil sample consisting of 30 cores/100 m^2 is adequate for most analyses. If the nematode population density estimates are required to a certain level of accuracy, then pre-treatment samples taken on a quadrant basis can be used to estimate the numbers of samples required for a given degree of accuracy (Barker *et al.*, 1986).

Samples taken from the growing crop are removed from between two plants within the plant row and in the root zone to a depth of approximately 30 cm. Commencing from about 2 months post-plant, samples are taken on a monthly basis in research work. This sampling regime should be followed during the plant crop to allow assessment of nematode population dynamics. A composite sample consists of from 10 cores per 15–20 m of row (as practised in Hawaii) to approximately 12 cores per 30 m of row (as practised in the Côte d'Ivoire). Samples should be placed in plastic bags and protected from temperature extremes until they are processed.

Nematode extraction

The nematode extraction technique used depends on the objectives of the sampling programme, the nematode species present in the soil or the roots, and the stage in the crop cycle. Soil-dwelling root knot nematode juveniles and juvenile and adult stages of the reniform nematode can be recovered by processing known volumes of soil with Baermann funnels, by a combination of Cobb sieving and centrifugation–flotation, or by processing root samples using mist apparatus. Females and associated egg masses can be visualized by staining root segments. Staining females is sometimes inefficient as pineapple roots are heavily suberized and do not clear readily (see Barker *et al.*, 1986). Eggs can be collected from the roots using an NaOCl solution (Barker *et al.*, 1986). The Baermann funnel technique typically yields a lower estimate of reniform nematode population density than the centrifugation–flotation technique. Each techniques allows enumeration of specific root knot or reniform nematode life stages.

Because of the endoparasitic nature of the root lesion nematode, population density estimates are obtained by extracting the life stages of the root lesion nematode from soil and roots using centrifugal flotation with magnesium sulphide (Coolen and d'Herde, 1972; Hendricks *et al.*, 1976). Roots can be macerated or enzymatically digested to release endoparasites for counting (Alvarado and Lopez, 1981; Barker *et al.*, 1986; Araya and Caswell-Chen, 1993).

The inoculum of root lesion nematode prior to planting is sometimes estimated in the Côte d'Ivoire by using a maize bioassay. The bioassay is especially helpful if initial population levels are low, and the bioassay is performed by placing a soil sample into several pots and sowing maize in the pots. The root lesion nematodes are extracted from soil and roots after 5 weeks to allow nematode reproduction, increasing the probability of detecting the nematode.

In some instances, it is desirable to estimate the number of nematodes in the rhizosphere. This can be accomplished by gently shaking the roots to remove adhering soil and then rinsing the remaining soil that is closely associated with the root system into a bucket. This aqueous suspension is passed through a set of nested sieves that subsequently may be subject to centrifugation–flotation.

Determination of populations and crop loss

In the Côte d'Ivoire, studies on nematode damage to pineapple are undertaken in plots of 80–120 plants (two or three beds of 40 plants per double-row bed). Planting distances are 90 cm × 40 cm × 25 cm, yielding a planting density of approximately 61,500 plants/ha for fresh fruits; and 90 cm × 40 cm × 28 cm, yielding a plant density of approximately 55,000 plants/ha for canned fruits. Each treatment should have a minimum of four replications, with five used in general practice. Experimental plots in Hawaii are similar, typically consisting of three or four beds, with the centre bed(s) reserved for yield determination. Each experiment should include appropriate controls; a non-treated control, a standard treatment control (plantation practice) and an irrigated control (if the experiment is irrigated) (Apt and Caswell, 1988).

Observations on plant growth are typically non-destructive, using D-leaf measurements and estimated plant weights. Plants are sometimes uprooted for inspection, or soil profile samples are taken to assess root development and nematode distributions within the soil profile. Soil samples for nematode assessment are taken at random from those beds designated 'non-yield'. Nematode soil samples are taken from the inside edge of these beds in the treated area, while the centre beds are reserved for yield assessment and are not sampled to prevent root system damage. In the Côte d'Ivoire, soil samples are taken on a monthly rotation basis, so that each month soil cores are removed from the soil around plants that have not been sampled previously.

At harvest, fruits are picked, size-classed, and the fruit and crown weights determined per size class. In Hawaii, approximately 100 fruits are harvested per treatment replication, but this depends on the length of the rows in the experiment. In the Côte d'Ivoire, all the fruits of each plot (80–120 fruit) are harvested, and 20 plants per treatment are selected at random for analysis of plant growth, enumeration of fruitlets, size and form of fruits, and fruit analysis (sugar and acidity). The specifics of the analysis are determined by the objectives of the research.

Summary of Nematodes in Pineapple

Plant parasitic nematodes can be devastating to pineapple, reducing total yields and altering fruit size distributions. Nematode control methods have changed dramatically in the last 15 years and are likely to change even more in the next 15 years. Many effective nematicides including several soil fumigants, carbamates and organophosphate nematicides have been removed from the market, and bringing new nematicides to the marketplace is costly. Consequently, alternatives to nematicides are more important than ever. Manipulation of the fallow period with intercycle cover crops and maintenance of soil moisture hold promise for increasing nematode control. Living mulches and nematode antagonists may eventually augment or replace traditional chemical nematicides. Plant resistance and biological (biological agents) control do not seem to be promising approaches for the coming years (short term). Crop management in general and, more particularly, drip irrigation may play an increasingly important role in nematode management by improving the plant's tolerance of nematode damage. However, the most effective means of controlling plant parasitic nematodes on pineapple in intensive production systems remains with chemical nematicides. The use of fumigant nematicides and non-fumigant nematicides, when available, provides very effective nematode management.

References

Alvarado-Soto, M. and Lopez-Chaves, R. (1981) Efficacia de dos metodos y sus modificaciones en la extraccion de nematodos endoparasiticos migratorios de raices de pina y plantano. *Nematropica* 11, 129–136.

Anonymous (1982) Les nematodes. In: *La Culture de l'Ananas d'Exportation en Côte d'Ivoire – Manuel du Planteur.* IRFA, Les Nouvelles Editions Africaines, Abidjan, pp. 21–25.

Anonymous (1987) Complexe sol-racines: les ravageurs. *Fruits, Special Ananas* 42, 669–677.

Apt, W.J. (1976) Survival of reniform nematodes in desiccated soils. *Journal of Nematology* 8, 278.

Apt, W.J. and Caswell, E.P. (1988) Application of nematicides via drip irrigation. *Annals of Applied Nematology (Journal of Nematology 20, Supplement)* 2, 1–13.

Araya, M. and Caswell-Chen, E.P. (1993) Enzymatic digestion of roots for recovery of root-knot nematode developmental stages. *Journal of Nematology* 25, 590–595.

Araya, M. and Caswell-Chen, E.P. (1994a) Host status of *Crotalaria juncea, Sesamum indicum,* and *Dolichos lablab* to *Meloidogyne javanica. Journal of Nematology* 26, 492–497.

Araya, M. and Caswell-Chen, E.P. (1994b) Penetration of *Crotalaria juncea, Sesamum indicum,* and *Dolichos lablab* roots by *Meloidogyne javanica. Journal of Nematology* 26, 238–240.

Araya, M. and Caswell-Chen, E.P. (1996) *Coffea arabica* cvs. Caturra and Catuai nonhosts to a California isolate of *Meloidogyne javanica. Nematropica* 25, 165–171.

Ayala, A. (1961) An analysis of the quantitative and qualitative composition of the nematode populations in pineapple fields in Puerto Rico. *Journal of Agriculture, University of Puerto Rico* 45, 265–299.

Ayala, A. (1968) Nematode problems of pineapple in Puerto Rico. In: Smart, G.C. Jr and Perry, V.G. (eds) *Tropical Nematology.* University of Florida Press, Gainesville, Florida, pp. 49–60.

Ayala, A.A., Roman, J. and Tejera, E.G. (1967) Pangola grass as a rotation crop for pineapple nematode control. *Journal of Agriculture, University of Puerto Rico* 51, 94–96.

Ayala, A., Gonzales-Tejera, E. and Irizarry, H. (1969) Pineapple nematodes and their control. In: Peachey, J.E. (ed.) *Nematodes of Tropical Crops.* CAB International, Wallingford, UK, pp. 210–224.

Bafokuzara, N.D. (1982) Nematodes associated with pineapples in Uganda. *Nematropica* 12, 45–49.

Barker, K.R., Townshend, J.L., Bird, G.W., Thomason, I.J. and Dickson, D.W. (1986) Determining nematode population response to control agents. In: Hickey, K.D. (ed.) *Methods for Evaluating Pesticides for Control of Plant Pathogens.* American Phytopathological Society, St Paul, Minnesota, pp. 283–296.

Birchfield, W. and Brister, L.R. (1962) New hosts and nonhosts of reniform nematode. *Plant Disease Reporter* 46, 683–685.

Bird, A.F. (1984) Growth and moulting in nematodes: moulting and development of the hatched larva of *Rotylenchulus reniformis. Parasitology* 89, 107–119.

Caswell, E.P., DeFrank, J., Apt, W.J. and Tang, C.S. (1991a) Influence of nonhost plants on population decline of *Rotylenchulus reniformis. Journal of Nematology* 23, 91–98.

Caswell, E.P., Tang, C.-S., DeFrank, J. and Apt, W.J. (1991b) The influence of root exudates of *Chloris gayana* and *Tagetes patula* on *Rotylenchulus reniformis. Revue de Nematologie* 14, 581–587.

Chan, Y.K., Coppens d'Eeckenbrugge, G. and Sanewki, G.M. (2003) Breeding and variety improvement. In: Bartholomew, D.P., Paull, R.E. and Rorhbach, K.G. (eds) *The Pineapple: Botany, Production, and Uses.* CAB International Wallingford, UK, pp. 33–55.

Chinnasri, B. and Sipes, B.S. (2004) Effect of a systemic acquired resistance inducer nematodes infecting pineapple. Fourth International Pineapple Symposium. *Acta Horticulturae* (in press).

Collins, J.L. (1968) *The Pineapple.* Leonard Hill, London.

Collins, J.L. and Hagan, H.R. (1932) Nematode resistance of pineapples: varietal resistance of pineapple roots to the nematode *Heterodera radicicola* (Greef) Muller. *Journal of Heredity* 23, 459–511.

Coolen, W.A. and d'Herde, C.J. (1972) *A Method for the Quantitative Extraction of Nematodes from Plant Tissue.* State Agricultural Research Centre, Ghent, Belgium.

Coppens d'Eeckenbrugge, G. and Leal, F. (2003) Morphology, anatomy and taxonomy. In: Bartholomew, D.P., Paull, R.E. and Rorhbach, K.G. (eds) *The Pineapple: Botany, Production, and Uses.* CAB International, Wallingford, UK, pp. 13–32.

da Costa, D.C., dos Santos, J.M. and Sanches, N.F. (1998) Associacao de *Aoroliamus* sp. (Nemata:Hoplolaiminae) com of vermelhao de abacaxizeriro na regiao de Itaberaba, Bahia. *Revista Brasileira de Fruticultura* 20, 119–122.

Dalldorf, E.R. (1977) Selection and preparation of pineapple planting material. In: *Pineapple Cultivation in South Africa.* Citrus and Subtropical Fruit Research Institute, Nelspruit, South Africa, D-3.

Dasgupta, D.R. and Seshadri, A.R. (1971) Races of the reniform nematode, *Rotylenchulus reniformis* Linford & Oliveira, 1940. *Indian Journal of Nematology* 1, 21–24.

Davide, R.G. (1988) Nematode problems affecting agriculture in the Philippines. *Journal of Nematology* 20, 214–218.

Dinardo-Miranda, L.L., Spironello, A. and Martins, A.L.M. (1996a) Adaptação hospedeira de variedades de abacaxi a *Meloidogyne incognita* raça 1. *Bragantia Campinas* 55, 275–278.

Dinardo-Miranda, L.L., Spironello, A. and Martins, A.L.M. (1996b) Reação de variedades de abacaxizeiro a *Pratylenchus brachyurus. Nematologia Brasileira* 20, 1–7.

Egunjobi, O.A. and Larinde, M.A. (1975) Nematodes and maize growth in Nigeria. II. Effects of some amendments on populations of *Pratylenchus brachyurus* and on the growth and production of maize (*Zea mays*) in Nigeria. *Nematologia Mediterranea* 3, 65–73.

Endo, B.Y. (1959) Responses of root-lesion nematodes, *Pratylenchus brachyurus* and *P. zeae,* to various plants and soil types. *Phytopathology* 49, 417–421.

Evans, D.O., Sanford, W.G. and Bartholomew, D.P. (1988) *Pineapple.* University of Hawaii Cooperative Extension Service, Commodity Fact Sheet PIN-3(A).

Garcia-Espinosa, R. and Adam, A.V. (1972) Major diseases of pineapple in Oxaca, Mexico, and their control. *FAG Plant Protection Bulletin* 20, 79–87.

Godfrey, G.H. (1929) A destructive root disease of pineapple and other plants due to *Tylenchus brachyurus* n.sp. *Phytopathology* 19, 611–629.

Godfrey, G.H. (1936) The pineapple root system as affected by the root-knot nematode. *Phytopathology* 26, 408–428.

Godfrey, G.H. and Hagan, H.R. (1933) Influence of soil hydrogen-ion concentrations on infections by *Heterodera radicicola* (Greef) Muller. *Soil Science* 35, 175–184.

Godfrey, G.H. and Hagan, H.R. (1937) Some measurements of detrimental effects of the root-knot nematode on the pineapple plant. *Phytopathology* 27, 515–530.

Godfrey, G.H. and Hoshino, H.M. (1933) Studies on certain environmental relations of the root-knot nematode, *Heterodera radicicola. Phytopathology* 23, 41–62.

Godfrey, G.H. and Oliveira, J.M. (1932) The development of the root-knot nematode in relation to root tissues of pineapple and cowpea. *Phytopathology* 22, 325–348.

Godfrey, G.H., Oliveira, J.M. and Gittel, E.B.H. (1933) The duration of life of the root-knot nematode, *Heterodera radicicola,* in soils subjected to drying. *Soil Science* 35, 185–195.

Goly, P.G. and Téhé, H. (1997) Effets des adventices de l'ananas sur *Pratylencus brachyurus* en Côte d'Ivoire. *Cahiers Agricultures* 6, 199–202.

Guérout, R. (1965) Competition *Pratylenchus brachyurus–Meloidogyne* sp. dans les cutures d'ananas de Côte d'Ivoire. In: *Proceedings of 8th International Nematology Symposium, Antibes.* J. Brill, Leiden, The Netherlands, pp. 64–69.

Guérout, R. (1969) Action des plantes ameliorantes en culture d'ananas. III. Bilan nematologique. *Fruits* 24, 436–443.

Guérout, R. (1975) Nematodes of pineapple: a review. *Pest Articles and News Summaries* 21, 123–140.

Guyot, A., Pinon, A. and Py, C. (1974) L'ananas en Côte d'Ivoire. *Fruits* 29, 85–117.

Hagan, H.R. (1933) Hawaiian pineapple field soil temperatures in relation to the nematode *Heterodera radicicola* (Greef) Muller. *Soil Science* 36, 83–95.

Hagan, H.R. and Collins, J.L. (1935) Studies on varietal resistance of pineapple plants. II. Plant resistance to *Heterodera marioni* (Cornu) Goodey. *Journal of Heredity* 26, 35–46.

Haroon, S. and Smart, G.C. (1983a) Root extracts of pangola digitgrass affect hatch and larval survival of *Meloidogyne incognita. Journal of Nematology* 15, 646–649.

Haroon, S. and Smart, G.C. (1983b) Effects of pangola digitgrass on *Meloidogyne arenaria, M. javanica* and *M. hapla. Journal of Nematology* 15, 649–650.

Heald, C.M. (1978) Effect of the reniform nematode on vegetable yields. *Plant Disease Reporter* 62, 902–904.

Heald, C.M. and Inserra, R.N. (1988) Effect of temperature on infection and survival of *Rotylenchulus reniformis. Journal of Nematology* 20, 356–361.

Hendricks, G.J., Coolen, W.A. and Moernans, R. (1976) Influence de la nature de la densitie des liquides de separation sur les nématodes pendant le processus de centéfugation–flottation. *Nematologia Mediterranea* 4, 31–40.

Hoshino, H.M. and Godfrey, G.H. (1933) Thermal death point of *Heterodera radicicola* in relation to time. *Phytopathology* 23, 260–270.

Jamet, P. and Piedallu, M.A. (1980) Comportement des nématicides organophosphores et carbamates dans le sol et les plantes. Etude bibliographique. *Phytiatrie-Phytopharmacia* 29, 175–199.

Jepson, S.B. (1987) *Identification of Root-knot Nematodes (*Meloidogyne *species).* CAB International, Wallingford, UK.

Kahn, F.A. (1985) Hatching response of *Rotylenchulus reniformis* to root leachates of certain hosts and non-hosts. *Revue de Nematologie* 8, 391–393.

Keetch, D.P. (1977) Nematodes in pineapple. In: *Pineapple Cultivation in South Africa.* Citrus and Subtropical Fruit Research Institute, Nelspruit, South Africa, H-18.

Keetch, D.P. (1982) Nematode pests of pineapple. In: Keetch, D.P. and Heyns, J. (eds) *Nematology in South Africa.* Department of Agriculture and Fisheries, Pretoria, South Africa, pp. 9–29.

Keetch, D.P. and Dalldorf, E.R. (1980) The use of grass and legume rotations for pineapple nematode control in the Eastern Cape. *Citrus and Subtropical Fruit Journal* 557, 10–12.

Keetch, D.P. and Purdon, K.T. (1979) Effect of systemic nematicides on pineapples. *Citrus Subtropical Fruit Research Institute Information Bulletin* 82, 3–4.

Ko, M.P. and Schmitt, D.P. (1993) Pineapple inter-cycle cover crops to reduce plant-parasitic nematode populations. *Acta Horticulturae* 334, 373–382.

Lacoeuilhe, J.J. and Guérout, R. (1976) Action du nematode *Pratylenchus brachyurus* sur la croissance, la nutrition et les rendements de l'ananas. Influence de la localisation de la fumure. *Fruits* 33, 147–156.

Lacoeuilhe, J.J. and Guyot, A. (1979) Les techniques culturales de l'ananas en Côte d'Ivorie. *Fruits* 34, 159–168.

Linford, M.B. (1937) Stimulated activity of natural enemies of nematodes. *Science* 85, 123–124.

Linford, M.B. and Oliveira, J.M. (1940) *Rotylenchulus reniformis,* novo gen., n.sp., a nematode parasite of roots. *Proceedings of the Helminthological Society of Washington* 7, 35–42.

Linford, M.B. and Yap, F. (1939) Root-knot nematode injury restricted by a fungus. *Phytopathology* 29, 596–609.

Linford, M.B. and Yap, F. (1940) Some host plants of the reniform nematode in Hawaii. *Proceedings of the Helminthological Society of Washington* 7, 42–44.

Linford, M.B., Yap, F. and Oliveira, J.M. (1938) Reduction of soil populations of the root-knot nematode during decomposition of organic matter. *Soil Science* 45, 127–141.

Linford, M.B., Oliveira, J.M. and Ishii, M. (1949) *Paratylenchus minutus,* n.sp., a nematode parasitic on roots. *Pacific Science* 3, 111–119.

Luc, M. and de Guiran, G. (1960) Les nématodes associes aux plantes de l'Ouest Africain. *L'Agronomie Tropicale, Nogent* 4, 434–449.

Magistad, O.C. and Oliveira, J.M. (1934) Changes in plant-food intake caused by a population of *Heterodera marioni* (Cornu) Goodey on *Ananas comosus. Phytopathology* 24, 276–283.

Milne, D.L., Van Lelyveld, L.J. and De Villiers, E.A. (1977) The response to peroxidase and indolacetic acid oxidase in pineapple roots to foliar applications of fenamiphos. *Agrochemophysica* 9, 65–70.

Nakasono, K. (1973) Relationship between cultivation of *Tagetes* plants and changes in population of the reniform nematode, *Rotylenchulus reniformis. Japanese Journal of Nematology* 3, 38–41.

Nakasono, K. (1977) Sexual attraction of *Rotylenchulus reniformis* females and some differences in the attractiveness between amphimictic and parthenogenetic populations (Nematoda: Nacobbidae). *Applied Entomology and Zoology* 12, 154–161.

Nakasono, K. (1983) Studies on morphological and physio-ecological variations of the reniform nematode, *Rotylenchulus reniformis* Linford & Oliveira, 1940 with an emphasis on differential distribution of amphimictic and parthenogenetic populations in Japan. *Bulletin of the National Institute of Agricultural Sciences, Japan, Series C* 38, 1–67.

Nath, R.C., Mukherjee, B., Dasgupta, M.K. and Siddiqi, M.R. (1997) Density, diversity and community structure of plant parasitic nematodes in pineapple plantations of Tripura, India. *International Journal of Nematology* 7, 51–61.

Olowe, T. and Corbett, D.C.M. (1976) Aspects of the biology of *Pratylenchus brachyurus* and *P. zeae. Nematologica* 22, 202–211.

Py, C., Lacoeuilhe, J.J. and Teisson, C. (1984) *L'Ananas: sa Culture, ses Produits.* Maisonneuve & Larose, Paris.

Quesada, M. and Barboza, R. (1999) Distribucion espacial de *Helicotylenchus* spp. en suel de una plantacion de piña (*Ananas comosus*) en las zona norte de Costa Rica. *Agronomia Costarricense* 1, 97–103.

Radovich, C., Paull, R. and Sipes, B. (2004) Protease inhibitors and reproduction of *Rotylenchulus reniformis* in pineapple. Fourth International Pineapple Symposium. *Acta Horticulturae* (in press).

Rebois, R.V. (1973) Effect of soil temperature on infectivity and development of *Rotylenchulus reniformis* on resistant and susceptible soybeans, *Glycine max. Journal of Nematology* 5, 10–13.

Rebois, R.V., Madden, P.A. and Eldridge, B.J. (1975) Some ultrastructural changes induced in resistant and susceptible soybean roots following infection by *Rotylenchulus reniformis. Journal of Nematology* 7, 122–139.

Redondo, E.E. and de Agudelo, F.V. (1992) Efecto de los nematodos en el cultivo de la piña (*Ananas comosus*) L. (Merr.). *Fitopatologia Colombiana* 16, 180–192.

Robinson, A.F., Inserra, R.N., Caswell-Chen, E.P., Vovlas, N. and Troccoli, A. (1997) *Rotylenchulus* species: identification, distribution, host ranges, and crop plant resistance. *Nematropica* 27, 127–180.

Rohrbach, K.G. and Apt, W.J. (1986) Nematode and disease problems of pineapple. *Plant Disease* 70, 81–87.

Rohrbach, K.G., Leal, F. and Coppens d'Eeckenbrugge, G. (2003) History, distribution and world production. In: Bartholomew, D.P., Paull, R.E. and Rorhbach, K.G. (eds) *The Pineapple: Botany, Production, and Uses*. CAB International, Wallingford, UK, pp. 1–12.

Roman, J. and Triantaphyllou, A.C. (1969) Gametogenesis and reproduction of seven species of *Pratylenchus. Journal of Nematology* 1, 357–362.

Sarah, J.L. (1980) Utilisation de nématicides endotherapiques dans la lutte contre *Pratylenchus brachyurus* (Godfrey) en culture d'ananas. I. Activite preventive et curative sur les infestations racinaires par applications foliaires. *Fruits* 35, 745–757.

Sarah, J.L. (1981a) Utilisation de nématicides endotherapiques dans la lutte contre *Pratylenchus brachyurus* (Godfrey) en culture d'ananas. II. Effets secondaires d'applications foliaires sur la phase vegetative du cycle de developpement de l'ananas. *Fruits* 36, 275–283.

Sarah, J.L. (1981b) Utilisation de nématicides endotherapiques dans la lutte contre *Pratylenchus brachyurus* (Godfrey) en culture d'ananas. III. Effets secondaires d'applications foliaires sur la response au traitement d'induction florale et sur la floraison. *Fruits* 36, 491–500.

Sarah, J.L. (1983) Utilisation de nématicides endotherapiques dans la lutte contre *Pratylenchus brachyurus* (Godfrey) en culture d'ananas. IV. Effets secondaires d'applications foliaires sur le developpement des plantes apres le traitement d'induction florale et sur la maturation des fruits. *Fruits* 38, 523–540.

Sarah, J.L. (1986) Influence of *Pratylenchus brachyurus* on the growth of development of pineapple in the Ivory Coast. XVIII International Symposium E.S.N., Antibes. *Revue de Nematologie* 9, 308–309.

Sarah, J.L. (1987a) Influence des traitements nematicides sur la production des rejets d'ananas. *Fruits* 42, 71–75.

Sarah, J.L. (1987b) Utilisation d'une jachere travaillee pour lutter contre les nematodes parasites de l'ananas. *Fruits* 42, 357–360.

Sarah, J.L. and Hugon, R. (1991) Dynamique des populations de *Pratylenchus brachyurus* en plantation d'ananas en Côte d'Ivoire. *Fruits,* 46, 241–250.

Sarah, J.L., Osséni, B. and Hugon, R. (1991) Effect of soil pH on development of *Pratylenchus brachyurus* populations in pineapple roots. *Nematropica* 21, 211–216.

Sarah, J.L., Mesnildrey, L., Marguerritte, E. and Boisseau, M. (1997) Une méthode de criblage variétal en laboratoire pour la résistance de l'ananas au nématode *Pratylenchus brachyurus. Acta Horticulturae* 425, 179–186.

Schneider, R.C., Sipes, B.S., Oda, C.H., Green, R.E. and Schmitt, D.P. (1995) Management of 1,3-dichloropropene in pineapple for efficacy and reduced volatile losses. *Crop Protection* 14, 611–618.

Sipes, B.S. (1996) Control of *Rotylenchulus reniformis* in pineapple with fosthiozate. *Fruits* 51, 173–177.

Sipes, B.S. and Schmitt, D.P. (1994a) Evaluation of pineapple, *Ananas comosus,* for host-plant resistance and tolerance to *Rotylenchulus reniformis* and *Meloidogyne javanica. Nematropica* 24, 113–121.

Sipes, B.S. and Schmitt, D.P. (1994b) Population fluctuations of *Rotylenchulus reniformis* and its effects on pineapple yields. *Plant Disease* 78, 895–898.

Sipes, B.S. and Schmitt, D.P. (1995) Evaluation of ethoprop and tetrathiocarbonate for reniform nematode control in pineapple. *Supplement to the Journal of Nematology* 27, 639–644.

Sipes, B.S. and Schmitt, D.P. (2000) *Rotylenchulus reniformis* damage thresholds on pineapple. *Acta Horticulturae* 529, 239–246.

Sipes, B.S., Schmitt, D.P. and Oda, C. (1993) Comparison of single and double chisel injection methods for the control of *Rotylenchulus reniformis* in pineapple. *Supplement to the Journal of Nematology (Annals of Applied Nematology)* 25, 773–777.

Stirling, G.R. and Kopittke, R. (2000) Sampling procedures and damage thresholds for root-knot nematode (*Meloidogyne javanica*) on pineapple. *Australian Journal of Experimental Agriculture* 40, 1003–1010.

Stirling, G.R. and Nikulin, A. (1993) Population dynamics of plant parasitic nematodes in Queensland pineapple fields and the effects of these nematodes on pineapple production. *Australian Journal of Experimental Agriculture* 33, 197–206.

Towson, A.J. and Apt, W.J. (1983) Effect of soil water potential on survival of *Meloidogyne javanica* in fallow soil. *Journal of Nematology* 15, 110–115.

Tsai, B.Y. and Apt, W.J. (1979) Anhydrobiosis of the reniform nematode: survival and coiling. *Journal of Nematology* 11, 316.

Wallace, H.R. (1963) *The Biology of Plant Parasitic Nematodes.* E. Arnold Ltd, London.

Wang, K., Sipes, B.S. and Schmitt, D.P. (2001) Suppression of *Rotylenchulus reniformis* by *Crotalaria juncea, Brassica napus,* and *Tagetes erecta. Nematropica* 31, 235–249.

Wang, K.H., Sipes, B.S. and Schmitt, D.P. (2002) Management of *Rotylenchulus reniformis* in pineapple, *Ananas comosus,* by intercycle cover crops. *Journal of Nematology* 34, 106–114.

Wang, K.H., Sipes, B.S. and Schmitt, D.P. (2003) Enhancement of *Rotylenchulus reniformis* suppressiveness by *Crotalaria juncea* amendment in pineapple soils. *Agriculture, Ecosystems and Environment* 94, 197–203.

Womersley, C. and Ching, C. (1989) Natural dehydration regimes as a prerequisite for the successful induction of anhydrobiosis in the nematode *Rotylenchulus reniformis. Journal of Experimental Biology* 143, 359–372.

Zeck, W.M. (1971) The systemic nematicide potential of Nemacur. *Pfianzenschutz-Nachrichten Bayer* 24, 114–140.

Zem, A.C. and Reinhardt, D.H. (1978) Nematoides associados – cultra do abacaxi no estado da Bahia. In: Mendes, B.V. (ed.) *Resumos dos Trabalhos Cientificos e Conferencias, III Reuniao Brasileira de Nematologia, Sociedade Brasileira de Nematologia et da Escola Superior de Agricultura, Mossoro,* pp. 17–20.

20 Nematode Parasites of Cotton and other Tropical Fibre Crops

James L. Starr,[1] R.G. Carneiro[2] and O. Ruano[3]
[1] *Department of Plant Pathology and Microbiology, Texas A&M University, College Station, TX 77843-2132, USA;* [2] *Recursos Genéticos e Biotecnologia, Brasília, Brazil;* [3] *Instituto Agronômico do Paraná, Londrina, Brazil*

Several fibre crops are important agricultural commodities in the subtropics and tropics, with cotton being the most important one in terms of total production (estimated at nearly 54 Mt for 2002). Although there are four cultivated species of cotton, upland cotton *Gossypium hirsutum* accounts for approximately 90% of the world's production. Other important fibre crops include jute (*Corchorus capsularis* and *C. olitorius*), kenaf (*Hibiscus cannabinus*) and roselle (*Hibiscus sabdariffa*). World production of jute and kenaf was approximately 2.8 Mt and 99,000 t, respectively in 2002. Although multiple nematode species are associated with each of these fibre crops, the root knot nematodes are responsible for the greatest proportion of all reported yield losses due to nematodes.

Cotton

Upland cotton is a relatively drought-tolerant crop by virtue of its long taproot, which may reach depths of greater than 1 m. The importance of the taproot to cotton growth may be a factor in cotton being generally intolerant of the damage caused by parasitic nematodes, especially in climates where soil moisture for crop growth often is limited. Because cotton is grown as a cash crop, it is often grown in a monoculture system that favours the development of a nematode community dominated by one or a few parasitic species (Starr *et al.*, 1993). Monoculture of cotton occurs in both large-scale production systems and in resource-poor production systems in marginal areas of developing countries of Africa and elsewhere. This chapter focuses on nematode species known to suppress yield of *G. hirsutum*. For other reviews of nematodes parasitic on cotton, the reader is referred to Heald and Orr (1984), Bridge (1992) and Starr (1998).

Meloidogyne

Distribution

Of the more than 70 described *Meloidogyne* species, only two are known to be pathogenic to cotton, *M. acronea* and *M. incog-*

*A revision of the chapter by J.L. Starr and S.L.J. Page.

nita (host races 3 and 4). Much of the early literature refers to the subspecies *M. incognita acrita*, which was occasionally given species rank, but *M. incognita acrita* is no longer recognized as a valid taxon and all such reports are now considered to refer to *M. incognita*. Of all of the root knot nematodes, *M. incognita* has the greatest frequency distribution in warm temperate to tropical agroecosystems, accounting for more than 60% of the identified infestations (Sasser and Carter, 1985). Thus, this nematode has been reported from nearly all cotton production regions, especially where soils are coarsely textured (Robinson *et al.*, 1987; Starr *et al.*, 1993). In some cotton production regions of the USA (Starr, 1998) and Brazil (R.G. Carneiro and O. Ruano, unpublished data) with conducive soils, *M. incognita* is present in more than 50% of the cotton fields. In regions of the USA and Brazil where cotton is grown on finely textured soils with higher contents of clay, *M. incognita* is rarely detected. Unfortunately, few other regions have been surveyed in sufficient detail to permit such estimates of frequency distribution. *M. incognita* host race 3 is the most common host race found on cotton (Ruano *et al.*, 1985). In contrast to *M. incognita*, *M. acronea* is known only from the Shire valley in Malawi and other semi-arid regions of southern Africa (Page, 1983). *M. acronea* may be indigenous to this region, which is also a habitat for the wild precursor of some cottons, *G. herbaceum* var. *africaum*.

Symptoms

As with many nematode-incited plant diseases, accurate diagnosis based on foliar symptoms is difficult. The general symptoms of disease include stunting, chlorosis, incipient wilting and a general unthrifty appearance (Fig. 20.1). Silva *et al.* (1997) reported that a common foliar symptom induced by *M. incognita* was a 'speckled' appearance of the interveinal tissues of the leaves (Plate 21A). Root galling of cotton by *M. incognita* is often indistinct (Fig. 20.2, Plate 21B), especially early in a cropping season and with low to moderate levels of infection. Under these conditions, the galls are less than twice the diameter of non-infected roots and are easiest to detect on lateral roots. As the crop nears maturity and the nematode population densities increase, there is an increased frequency of more heavily galled roots and an increase in the size of the galls (Fig. 20.3). The root symptoms induced by *M. acronea* are distinct from those of *M. incognita*. Root galling is very limited in response to *M. acronea*, such that mature females are often exposed on the root surface. Root elongation often ceases following infection by *M. acronea*

Fig. 20.1. Poor stand of cotton due to *Meloidogyne incognita*. (Photo: T.A. Wheeler.)

Fig. 20.2. Moderate root galling of cotton caused by *Meloidogyne incognita*. (Photo: J.L. Starr.)

but there may be a proliferation of lateral roots from the infection site (Fig. 20.4). Roots of cotton infected by *M. acronea* are said to have a 'turned-aside' appearance, which is due to cessation of growth of the taproot accompanied by increased formation of lateral roots (Page, 1983).

When examining cotton roots for symptoms of infection by root knot nematodes, it is important that plants be carefully dug from the soil so as to recover a high proportion of the lateral roots, where root galls are most evident. Pulling plants from the soil to estimate root galling will result in the loss of most of the weak symptomatic lateral roots.

Biology

The biology of *M. incognita* and *M. acronea* is similar to that of other *Meloidogyne* species. *M. acronea* differs from *M. incognita* in that it reproduces almost entirely by amphimixis whereas *M. incognita* is strictly parthenogenetic. *M. incognita* is favoured by warmer soil temperatures (optimum is ~28°C) and does not survive long periods of freezing temperatures. No data are available on optimal temperatures for *M. acronea*, but with its known distribution it is likely to behave similarly with respect to effects of temperature on development and survival.

Population dynamics

In conducive soils with favourable temperatures and adequate moisture, the host status of the crop will govern nematode

Fig. 20.3. Severe root galling of cotton caused by *Meloidogyne incognita*. (Photo: R.G. Smith.)

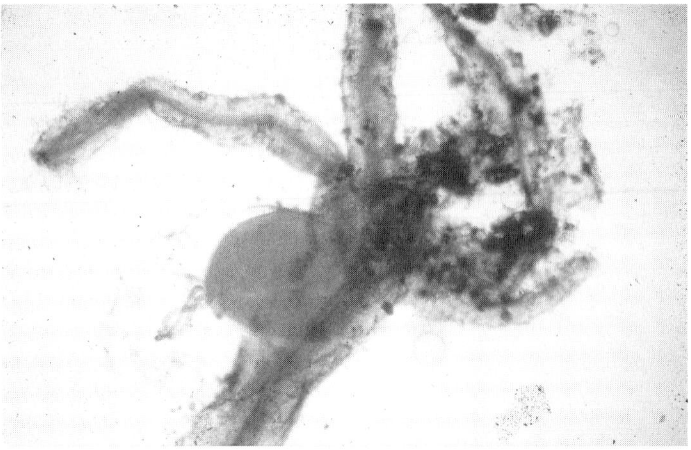

Fig. 20.4. *Meloidogyne acronea* female and proliferation of lateral roots from the feeding site. (Photo: J. Bridge.)

population densities. Except for a few resistant *G. hirsutum* genotypes, cotton is a susceptible host that supports population densities of more than 10^4 eggs and juveniles/500 cm^3 of soil (Veech and Starr, 1986). Population densities may increase several hundred-fold during a cropping season, especially when the initial population densities are less than 10 eggs and juveniles/500 cm^3 of soil. Population densities at crop maturity are inversely related to the initial population densities (Veech and Starr, 1986; Starr *et al.*, 1989). The predominant developmental stage of *M. incognita* populations that can be easily measured during the growing season are eggs (Barker *et al.*, 1987). Because more than 90% of the extractable population may be eggs at this time, it is important when estimating population densities to use soil extraction methods that allow direct egg quantification (Barker *et al.*, 1987) or methods that allow the eggs to hatch (Rodríguez-Kabána and Pope, 1981).

The presence of other nematode species parasitic on cotton can affect the population dynamics of *Meloidogyne* spp. Gay and Bird (1973) reported that *Pratylenchus brachyurus* suppressed population development of *M. incognita*. Bird *et al.* (1974) and Kraus-Schmidt and Lewis (1981) reported that *M. incognita* could not compete with *Hoplolaimus columbus* and that *H. columbus* would replace *M. incognita* as the dominant species in fields infested with both nematodes. Similarly, anecdotal observations suggest that *M. incognita* is a poor competitor in fields also infested with the reniform nematode *Rotylenchulus reniformis*. *M. incognita* populations are also suppressed by fungal pathogens that infect cotton and increase the rates of plant mortality (Starr *et al.*, 1989).

Survival

Most studies of survival of *Meloidogyne* species have focused on winter survival. Winter survival is inversely related to autumn (fall) population densities (Ferris, 1985; Starr and Jeger, 1985) and may be related to reduced partitioning of nutrients from the host into the developing eggs at very high nematode population densities (Starr, 1988). Egg populations decline exponentially after crop harvest during winter months due to the combined effects of hatch and mortality (Starr and Jeger, 1985). Populations of juveniles (J2s) increase initially during the early winter months, before declining in the late winter and early spring. Survival of eggs within the egg mass is enhanced at temperatures less than 20°C in dry soils with low matric potential (–4 bars) due to the inhibition of hatch (Starr, 1993). *M. acronea* also survives the 6–7 month dry season as unhatched eggs within the egg mass or within the body of the dead female that has developed a thickened cuticle. Eggs in these structures are dormant and viable unless the soil becomes very dry (relative humidity of < 97%). The limited distribution of *M. acronea* to the Shire Valley of southern Africa may be related to the alluvial soils of the valley, which have greater moisture-holding capacity than other soils of the region (Page, 1984).

Damage thresholds

Susceptible cotton cultivars are highly intolerant of *M. incognita*, with damage threshold densities in the range of 1–9 individuals/500 cm^3 of soil (Roberts *et al.*, 1985; Starr *et al.*, 1989). Cotton cultivars with resistance to *M. incognita* support lower levels of nematode reproduction but may not have increased tolerance. Koenning *et al.* (2001) and Colyer *et al.* (1997) reported that some cultivars with moderate resistance to *M. incognita* responded to nematicide treatment of infested soil with a significant yield increase. In contrast, Zhou and Starr (2003) reported that, whereas the moderate resistance in the cvs LA 887 and Acala NemX did not affect the damage threshold density, the resistant cultivars were more tolerant than susceptible cultivars. Koenning *et al.* (1996) reported only minor influences of soil type on the damage functions for *M. incognita* on cotton in microplots. Soil type in those experiments had a greater effect on nematode reproduction than on the damage function, with reproduction being greatest in the coarsely textured, sandy soils and least in the more finely textured soils with higher contents of silt and clay.

Disease complexes

M. incognita on cotton is known to be involved in numerous disease complexes, especially with the vascular wilt pathogen *Fusarium oxysporum* f.sp. *vasinfectum* and with several other fungi causing seedling disease (Plate 21C). Roberts *et al.* (1985) reported that the slope of the damage function for *M. incognita* on cotton was more negative in the presence of *F. o. vasinfectum* than in the absence of the wilt pathogen. Starr *et al.* (1989) reported that the interaction between the two pathogens was most evident when the nematode population density exceeded the damage threshold and with intermediate populations of the wilt pathogen. Further, they reported that increased plant mortality early in the growing season was a major consequence of the disease complex. Resistance to Fusarium wilt can be broken by *M. incognita*, and wilt symptoms were more severe, developed more rapidly and with greater frequency when plants were also infected by *M. incognita* (Ruano *et al.*, 1984b). Interactions with seedling pathogens such as *Rhizoctonia solani* and several *Fusarium* and *Pythium* species are well known (Brodie and Cooper, 1964). Interactions with *Thielaviopsis basicola* have also been documented (Walker *et al.*, 1998). In all of these interactions, there is greater incidence of the wilt or seedling disease, with greater yield suppression when cotton is infected with multiple pathogens than when only a single pathogen is present. Few interactions with other pathogens have been reported for *M. acronea*. It has been observed that the cyst-like appearance of mature females of *M. acronea* (Bridge *et al.*, 1976) may be due to the effects of oxidases secreted by *T. basicola*, which cause a tanning reaction of the nematode's cuticle (Page, 1983).

Management measures

CHEMICAL. As a cash crop, management of root knot nematodes in cotton has relied heavily on the use of nematicides. Numerous studies have demonstrated profitable increases in yield in response to nematicide applications (Orr and Robinson, 1984; Lordello and Sabino, 1985). The fumigant 1,3-dichloropropene often provides a greater yield increase and greater suppression of final nematode population densities than does the carbamate non-fumigant, aldicarb (Kinlock and Rich,

1998). Attempts to increase the efficiency of nematicide through the use of variable rates based on initial nematode densities and site-specific treatments have shown only moderate success (Wheeler *et al.*, 1999; Wrather *et al.*, 2002). Variable rate application in a site-specific manner requires intensive sampling to estimate nematode population densities precisely across a field, and such intensive sampling is cost prohibitive with current technology. Additionally, Wheeler *et al.* (2000) reported that one cannot reliably estimate population densities of *M. incognita* in cotton for 3 years based on samples collected in only the first year of the 3 year period; thus fields must be sampled annually.

CROP ROTATION AND SOIL AMENDMENTS. Despite the extensive host range of *M. incognita*, several crop rotation systems that suppress nematode population densities and increase crop yields are known. These include rotations of cotton with groundnut (peanut) (Kirkpatrick and Sasser, 1984), velvet bean (Ferraz, 1964; Silva, 1984) and root knot-resistant cowpea (Duncan and Ferris, 1984). Although some variability in reproduction of *M. incognita* exists among maize (Windham and Williams, 1987) and sorghum (Birchfield, 1983) genotypes, these crops are usually not effective for suppressing the nematode's density when grown in rotation with cotton. In South America, planting cotton in fields previously planted to coffee and infested with *M. incognita* resulted in poor cotton yields (Ruano *et al.*, 1984a). Pearl millet, finger millet, maize, groundnut, guar bean and leucaena bean are poor hosts for *M. acronea* and can be used to suppress nematode densities when grown in rotation with cotton (Page, 1983). Sorghum is a host for *M. acronea* and not a suitable rotation crop. In most crop rotation systems for management of root knot nematodes, the beneficial effects are greater if the non-host crop is grown for at least two seasons before planting susceptible cotton.

Soil amendments with castor bean cakes (Lordello and Sabino, 1985) and with various green manure crops, including some *Tagetes* spp., several *Crotalaria* spp. and velvet beans (*Mucuna pruriens*) (dos Santos and Ruano, 1987), will suppress populations of *Meloidogyne* spp. and other nematodes. Such treatments might be particularly useful in alleviating nematode damage for resource-poor cotton farmers.

RESISTANCE. Although most cotton cultivars grown are susceptible and intolerant of *M. incognita*, numerous sources of resistance have been identified (Shepherd *et al.*, 1988, 1996; Cook *et al.*, 1997; Robinson and Percival, 1997) and a few cultivars (Table 20.1) with useful resistance have been released and are grown commercially. In the cotton production areas of the western USA, the Acala NemX cultivar has been shown to have a competitive yield potential in infested fields and to suppress nematode population densities (Ogallo *et al.*, 1997). Additionally, the use of cotton with resistance to *M. incognita* will reduce yield losses in nematode-susceptible crops

Table 20.1. Cotton cultivars with resistance to the root knot nematode, *Meloidogyne incognita*, and high yield potentials.

Cultivars	References
Acala NemX	Ogallo *et al.* (1999); Zhou and Starr (2003)
Paymaster 1560	Colyer *et al.* (1997); Koeninng *et al.* (2001)
Stoneville LA 887	Koeninng *et al.* (2001); Zhou and Starr (2003)
Stoneville 5599 BR	T.L. Kirkpatrick (personal communication)
IPR 94	Almeida *et al.* (2001); Ruano and Almeida (1999)
IPR 95	Almeida *et al.* (2001); Ruano *et al.* (2001)
IPR 96	Almeida *et al.* (2001); Ruano *et al.* (2001)
IAC 24	Cia *et al.* (2003)

grown in rotation with the resistant cotton (Ogallo *et al.*, 1999). In Brazil, several cotton cultivars not only with resistance to *M. incognita*, but also with resistance to multiple pathogens, have been released (Almeida *et al.*, 2001; Cia *et al.*, 2001).

The *M. incognita*-resistant genotypes Auburn 623 and Clevewilt were reported to be susceptible to *M. acronea* (Page and Bridge, 1994), as were accessions of *G. arboreum*, *G. herbaceum* var. *africanum* and *G. barbadense*. One accession of *G. hirsutum* ('UK 64') was found to have a moderate level of resistance to *M. acronea* (Page and Bridge, 1994).

Rotylenchulus

Distribution and symptoms

Two species of reniform nematodes, *Rotylenchulus parvus* and *R. reniformis*, are confirmed parasites of cotton, but few reports are available concerning the interaction of *R. parvus* and cotton. Both species are distributed widely in the warm temperate to tropical climates of the world. *R. reniformis* is noted especially for being associated with soils with higher silt and clay contents than soils in which root knot nematodes are commonly found (Robinson *et al.*, 1987; Starr *et al.*, 1993). *R. reniformis* is becoming more widespread in the cotton production regions of southern USA (McLean and Lawrence, 2000; Gazaway and McLean, 2003) and in Paraná State in Brazil (W.P. de Almeida, personal communication) and appears to be replacing *M. incognita* as the dominant species in many fields. One unique trait of the reniform nematode is its spatial distribution in infested fields. Tihohod *et al.* (1992) reported that *R. reniformis* has a more uniform distribution in cotton fields than other nematode species. Robinson *et al.* (2000) and Westphal and Smart (2003) have reported that *R. reniformis* is often found relatively deep in the soil profile, in some cases with more that 50% of the population occurring at depths greater than 30 cm.

The symptoms caused by *R. reniformis* are rather nondescript, as is typical for most parasitic nematodes, and may include stunted growth, poorly developed roots and chlorosis. Heavily infected roots may have a 'dirty' appearance, even after rinsing with water, due to the adhesion of soil particles to the egg masses (Fig. 20.5, Plate 21D). Ferraz and

Fig. 20.5. Response of cotton to treatment with nematicide in a field infested with *Rotylenchulus reniformis*.

Monteiro (1995) reported that whereas a 'speckled' appearance may occur in the interveinal tissue, the most visible symptom was a reduction in growth. Due to the relatively uniform distribution of *R. reniformis* in many infested fields, symptomatic plants may not occur in distinct clusters as often occurs with other parasitic nematodes on row crops. Rather, infected and stunted plants may be so uniformly distributed as not to be readily apparent.

Population dynamics and damage thresholds

In general, populations of *R. reniformis* are at a minimum in the late spring and during the first month of a cropping season, and at a maximum as the crop nears maturity. Populations as high as 49,000 individuals/100 g of soil have been detected (Jones *et al.*, 1959). Whereas eggs may comprise more that 90% of the population for *Meloidogyne* species during the mid to late portions of a cropping season for annual crops, for *R. reniformis* eggs seldom comprise more than 50% of the total population (J.L. Starr, unpublished data). *R. reniformis* is noted for its ability to survive periods of drought in an anhydrobiotic state (Apt, 1976; Tsai and Apt, 1979). The distribution of *R. reniformis* to soil depths greater than 45 cm apparently also enhances its survival (Robinson *et al.*, 2003).

Precise damage functions for *R. reniformis* on cotton have not been reported from field tests. A damage threshold of 16 individuals/200 cm^3 of soil has been reported from small pot tests (Sud *et al.*, 1984). Several studies have reported significant increases in cotton yield in response to nematicide application when initial population densities of *R. reniformis* were in the range of 100–250 nematodes/100 cm^3 of soil. Koenning *et al.* (1996) reported that the relationship between initial population densities and seed cotton yield fit a linear model in several soil types, generally with more than 100 individuals/500 cm^3 required to suppress yields by 10%.

Disease complexes

The reniform nematode forms disease complexes with *F. oxysporum* f.sp. *vasinfectum* (Prasad and Padeganur, 1980), *Verticillium dahlia* and with several seedling disease pathogens (Brodie and Cooper, 1964). Sankaralingam and McGawley (1994) reported that the combination of *R. reniformis* and *Rhizoctonia solani* did not affect the severity of seedling disease but did result in lower overall growth and an increase in *R. reniformis* population densities.

Management measures

Management of *R. reniformis* on cotton has relied primarily on nematicides and crop rotation. Numerous studies have reported higher yields of cotton following application of a variety of different nematicides (Kinlock and Rich, 1998; Borges *et al.*, 1999; Oliveira *et al.*, 1999; Seno *et al.*, 1999), including foliar applications of oxamyl (Lawrence and McLean, 2000) (Fig. 20.6). Robinson *et al.* (2002) have reported that, due to the depth of distribution of *R. reniformis* in some soils, yield response to fumigation can be improved by deeper placement of 1,3-dichloropropene. Repeated use of aldicarb has limited its effectiveness in some fields due to the development of a microflora population that can rapidly degrade the material (McLean and Lawrence, 2003). The use of geostatistics to improve the efficiency of nematicide use may eventually lead to improved management systems (Farias *et al.*, 2002).

Several crops can be grown in rotation with cotton to suppress nematode densities and to improve cotton yields, including maize (Westphal and Smart, 2003), reniform-resistant soybean (Davis *et al.*, 2003), sorghum (Thames and Heald, 1974; Birchfield, 1983; Westphal and Smart, 2003), maize intercropped with black velvet bean (Curi, 1980) and wheat (Birchfield, 1983).

No upland cotton cultivars or genotypes with useful levels of resistance to *R. reniformis* are known. The diploid species *G. longicalyx* is highly resistant to *R. reni-*

Fig. 20.6. Egg masses of *Rotylenchulus reniformis* on cotton roots. (Photo: A.F. Robinson.)

formis, but introgression of this resistance into the allotetraploid *G. hirsutum* will be a difficult task (Konaan *et al.*, 2003). Several accessions of the tetraploid *G. barbadense* have moderate levels of resistance to *R. reniformis* (A.F. Robinson, personal communication; Yik and Birchfield, 1984), and crosses between the *G. barbadense* line Tx110 and the root knot-resistant *G. hirsutum* M315 resulted in progeny resistant to both nematodes (Silvey *et al.*, 2003) but lacking the necessary yield potential for commercial release. Cotton accessions with tolerance to parasitism by *R. reniformis* have been identified (Cook *et al.*, 1997; Koenning *et al.*, 2000) and may be useful for limiting yield losses.

Pratylenchus

Pratylenchus brachyurus has been associated with disease of cotton in southern USA and in Brazil. In the USA, there may be differences in the nematode populations across the cotton production area, or possibly a change in host status of more recent cotton cultivars, because reports that provide evidence of pathogenicity on cotton are primarily from the states of Alabama and Georgia, and were prior to 1980 (Graham, 1951; Bird *et al.*, 1971; Hussey and Roncadori, 1978). Starr and Mathieson (1985) were unable to confirm these reports working with a *P. brachyurus* population from Texas. In Brazil, Lordello and Arruda (1957) and Ferraz (1964) reported *P. brachyurus* associated with stunted plants with small stems and poorly developed root systems. Carneiro *et al.* (1990) found that 45% of the cotton fields with sandy soils in Paraná were infested with *P. brachyurus*. In São Paulo State, the disease caused by *P. brachyurus* is known as 'little creeper' because of the mortality of infected plants throughout the season and, whereas control by nematicides is possible, rotations with soybean are not effective because the nematode reproduces well on both crops (Lordello, 1968). *P. brachyurus* also reproduces well on several grain crops, thus most cultivars of maize, sorghum and wheat are not likely to be good rotation crops for suppression of this nematode. Variation in the reaction of cotton cultivars to *P. brachyurus* has been reported (Fuzatto *et al.*, 1997; Goulart *et al.*, 1997).

P. sudanesis was reported as a pathogen of *G. barbadense* but not *G. hirsutum* in Sudan (Yassin, 1974), with potential yield reductions of 56–88% (Yassin, 1980).

Hoplolaimus

Several *Hoplolaimus* species are pathogenic on cotton (Fig. 20.7, Plate 21E). *H. aegypti* is reported from Egypt, *H. columbus* from the USA and Egypt, *H. indicus* from India, and *H. seinhosti* from several

Fig. 20.7. Stunting of cotton plants due to *Hoplolaimus columbus*. (Photo: S.A. Lewis.)

countries in Africa. All species apparently exhibit both ecto- and endoparasitic associations with the host. A damage threshold for 10% yield loss has been reported at 70 *H. columbus*/100 cm^3 of soil (Noe, 1993). Root population densities at 42 days after planting ranging from 200 to 600 nematodes/g of root weight caused yield losses of 2–12% (Mueller and Sullivan, 1988). No resistance has been reported in cotton to *Hoplolaimus* species. Cultural practices such as root destruction immediately after harvest and winter cover crops (Davis *et al.*, 2000) or by alteration of the planting date (Koenning *et al.*, 2003) were not effective in increasing cotton yields in fields infested with *H. columbus*.

Ectoparasitic nematodes

A few species of ectoparasites are pathogenic on cotton, but most of these are limited in distribution and, though of great importance to a given region, are of lesser overall economic importance than root knot and reniform nematodes. The sting nematode, *Belonolaimus longicaudatus*, is an aggressive pathogen of cotton in the sandy soils of south-eastern USA. *B. longicaudatus* will not survive in soils with less than 85% sand content (Robbins and Barker, 1974). Feeding activities of this nematode cause much destruction of cortical tissues, resulting in severely stunted, necrotic root systems and similarly stunted shoot growth. Damage thresholds are low, 1–5 individuals/100 cm^3 of soil (Crow *et al.*, 2000) and the severe damage to cotton by this nematode results in a carrying capacity of only 100 nematodes/100 g of soil (Crow *et al.*, 2000). The sting nematode can be managed by rotation with tobacco, *Crotalaria spectabilis* and *Tagetes minuta* (Good *et al.*, 1965). Tomerlin (1969) reported that various organic soil amendments suppressed *B. longicaudatus*, and a strain of the obligate endoparasitic bacterium *Pasteuria* able to parasitize *B. longicaudatus* has been reported (Giblin-Davis *et al.*, 2001).

Unidentified species of *Longidorus* and *Xiphinema* have been associated with cotton exhibiting poor growth and roots with symptoms to nematode damage in southern Africa (Bridge and Page, 1975).

Kenaf

Kenaf (*Hibiscus cannabinus*) is an important fibre crop in several countries with tropical or subtropical climates. There has been much recent interest in kenaf in the USA as an alternative crop, especially for use in paper production. *M. arenaria*, *M. incognita* and *M. javanica* are recognized pathogens of kenaf in nearly all production regions, causing substantial galling of the roots. Because of the widespread distribution of these species, they represent a potential hazard to kenaf wherever it is grown, especially in sandy soils. Kenaf is also a host for *H. columbus* (Koura *et al.*, 1987) and *H. magnistylus* (Lawrence and McLean, 1990), but yield losses due to these species have not been documented. Kenaf was reported as a non-host for *R. reniformis* in one test (Robinson *et al.*, 1998) and thus suitable for rotation with cotton in fields infested with that nematode, whereas another report lists a 12-fold increase in population densities of *R. reniformis* on kenaf (Lawrence and McLean, 1992).

Several studies have examined the relationship between initial nematode population densities and yield of kenaf. McSorley and Parrado (1986) were able to relate root galling due to *M. incognita* with the growth (height) using the Seinhorst model and observed a damage threshold of eight galls per root system. Di Vito *et al.* (1997) also related the growth of kenaf to *M. incognita* densities using the Seinhorst model and found a damage threshold for shoot weight of 0.18 eggs and J2/cm^3 of soil with a relative minimum yield of 0.1. Zhang and Noe (1996) reported that the growth response of kenaf to infection by *M. arenaria* and *M. incognita* was similar, and Veech (1992) reported that reproduction of the four races of *M. incognita* on kenaf was similar when the Pf/Pi ratios were adjusted for differences in root weight.

In addition to management with nematicides, crop rotation and use of tolerant cultivars can reduce yield losses. Effective rotation crops will vary depending on both the *Meloidogyne* species and the race infesting the field. Thus cotton would be a good rotation crop for fields infested with *M. arenaria*, *M. incognita* races 1 or 2 or *M. javanica*, but would not be suitable if the field is infested with *M. incognita* races 3 or 4. Similarly, groundnut would be suitable for fields infested with *M. incognita*, but not for fields infested with *M. arenaria* race 1 or *M. javanica* race 3. Maize has been reported as a good rotation crop for *M. incognita*-infested fields (Cuadra *et al.*, 1990), but care must be exercised due to the variability in host status among maize inbred lines and hybrids (Windham and Williams, 1987). The availability of cotton, groundnut and soybean cultivars with high levels of resistance to various *Meloidogyne* spp. populations will expand their potential for use as rotation crops.

Several reports have identified sources of resistance to *M. incognita* and *M. javanica* in the kenaf germplasm (Summers *et al.*, 1958; Adeniji, 1970; Adamson *et al.*, 1974; Veech, 1992), but apparently such resistance has not been introgressed into any cultivar. Tolerance has also been identified, and at least one high-yielding cultivar with tolerance has been released (Vawdrey and Stirling, 1992; Lawrence *et al.*, 1994; Cook and Scott, 1995). In fields infested with *M. incognita* and several soil-borne fungal pathogens, the growth of the tolerant SF459 was 55% faster than that of the intolerant standard cultivar, even though both supported high levels of reproduction and had severely galled roots (Cook and Mullin, 1994). In China, an integrated approach to management of root knot nematodes on kenaf has been reported that combines rotation with non-hosts (groundnut, rice, maize and sesame), removal of nematode-infected crop residues, use of nematode-free organic fertilizers and the application of nematicides only in the most severely infested fields (Yu, 1994).

Roselle

Because roselle (*Hibiscus sabdariffa*), also known as sorrel and mesta, is grown in a wide range of tropical environments, it is frequently grown in environments that are conducive to root knot and reniform nematodes. Roselle varies in its susceptibility to *M. incognita*, *M. javanica* and *M. arenaria*. Minton *et al.* (1970) reported that roselle varied in root galling (scale of 1–4) from 3.0 to 3.4 for *M. arenaria*, 1.8 to 2.9 for *M. incognita*, and 1.1 to 1.8 for *M. javanica*. Vawdrey and Stirling (1992) and Adenijii *et al.* (1970) have also reported that several roselle accessions are resistant to *M. incognita* and *M. javanica*. Roselle yields were increased by as much as 2.5 t/ha in fields infested with *M. arenaria* and treated with ethylene dibromide, but no yield response to nematicide treatment was observed in soil infested with *M. javanica* (Minton and Adamson, 1979). Adamson *et al.* (1975) reported that, because of the resistance of roselle, it was an effective rotation crop for management of *M. incognita* and *M. javanica* on kenaf. Roselle is also likely to be effective in the management of *M. incognita* on cotton. In contrast, Heffes *et al.* (1991) reported that the roselle (sorrel) cv. 'Red' was severely galled and supported at least moderate levels of reproduction by a race 1 population of *M. incognita*. Few data are available on the susceptibility of roselle to *R. reniformis*. Heffes *et al.* (1990) reported that Pf/Pi rations for *R. reniformis* were low on the roselle cv. 'White' with moderate reproduction at low Pi levels on cvs 'Red' and 'Pink'.

Jute

Two species of jute, *Corchorus capsularis* and *C. olitorius*, are grown in several tropical regions as a fibre crop; *C. olitorius* is also used as a leafy vegetable crop. Both jute species are hosts to the widespread root knot nematode species *M. arenaria*, *M. incognita* and *M. javanica*, and can be severely galled by these nematodes. *M. hapla* and *M. thamesi* are reported parasitic on jute in China (Lin and Chen, 1992), but jute is reported as resistant to *M. graminicola* (Sperandio and Amaral, 1994). Additionally *H. indicus*, *Helicotylenchus* spp. and *R. reniformis* have been associated with crop damage in India (Mishra *et al.*, 1985). *H. indicus* and *Helicotylenchus* spp. feed endoparasitically in the root cortex, resulting in necrosis and stunting of severely infected root systems.

As is common for root knot nematodes, they are often involved with several soilborne pathogens to cause disease complexes of jute. These pathogens include *Fusarium solani*, *Rhizoctonia bataticola*, *R. solani* and *Ralstonia solanacearum* (*Pseudomonas solanaccarum*), causing a syndrome known as Hoogly wilt (Mandal and Mishra, 2001). In one report, a three-way interaction involving *Macrophomina phaseolina* (*R. bataticola*), *R. solani* and *M. incognita* damage was always more severe when plants were infected by all three pathogens than when infection was by any single pathogen or a combination of two pathogens (Begum *et al.*, 1990).

Several non-traditional approaches have been tested for management of root knot nematodes on jute. Soil amendment with neem cakes at 0.5–1.0 kg/m^2 was effective in control of *Meloidogyne* spp. (Agbakli *et al.*, 1992; Chakraborti, 2001), as was the use of poultry manure combined with rotation with paddy rice (Senapati and Ghosh, 1992). A combination of removal of crop stubble, rotation with rice or wheat, and amendment with poultry manure at 10 t/ha also improves jute yields in soils infested with *M. incognita* or *M. javanica* (Mishra *et al.*, 1987). Growing mustard in rotation with jute can suppress soil populations of *M. incognita* (Khan and Banerjee, 2003). Improvement of soil fertility with addition of nitrogen, potassium and phosphorus has been reported to improve the tolerance of jute to parasitism by root knot nematodes (Balogun and Babatola, 1990). Whereas most cultivated jute varieties appear to be susceptible to a range of *Meloidogyne* spp., some resistance (Laha *et al.*, 1995) or tolerance (Mishra and Chakrabarti, 1987) has been described. The Bast Fiber Crop Research Institute in China has collected hundreds of accessions of jute, kenaf and other fibre crops, along with related plant species. This germplasm collection may provide additional useful sources of resistance (Su, 1993).

References

Adamson, W.C., Stone, E.Q. and Minton, N.A. (1974) Field resistance to the javanese root-knot nematode in kenaf. *Crop Science* 14, 334–335.

Adamson, W.C., Martin, J.A. and Minton, N.A. (1975) Rotation of kenaf and roselle on land infested with root knot nematodes. *Plant Disease Reporter* 59, 130–132.

Adeniji, M.O. (1970) Reaction of kenaf and roselle varieties to the root-knot nematode in Nigeria. *Plant Disease Reporter* 54, 547–549.

Agbakli, R., Maraite, H. and Colin, J.E. (1992) L'utilisation des extraits de neem (*Azadirachta indica*) dans la lutte contre les nématodes a galles (*Meloidogyne* sp.) en culture de legumes feuilles au sud-Benin. *Mededelingen van de Faculeit Landbouwwetenschappen* 57, 913–918.

Almeida, W.P., Pires, J.R., Ruano, O., Turkiewicz, L., Santos, W.J. and Yamaoka, R.S. (2001) IPR 94, IPR 95, e IPR 96: novas cultivares paranaenes de algodoeiro. In: *III Congresso Brasileiro de Algodão, Abstract Books, Volume 2.* Campo Grande, MS 3, pp. 849–852.

Apt, W.J. (1976) Survival of reniform nematodes in desiccated soils. *Journal of Nematology* 8, 28.

Balogun, O.S. and Babatola, J.O. (1990) Pathogenicity of *Meloidogyne incognita* on *Corchorus olitorius. Nematologia Mediterranea* 18, 23–25.

Barker, K.R., Starr, J.L. and Schmitt, D.P. (1987) Usefulness of egg assays in nematode population-density determination. *Journal of Nematology* 19, 130–134.

Begum, H., Sultana, K. and Ahmed, N. (1990) Disease development in jute due to the interaction of fungal pathogens (*Macrophomina phaseolina* and *Rhizoctonia solani*) and root-knot nematode (*Meloidogyne* spp.) and its effect on yield of fiber and seed. *Bangladesh Journal of Jute and Fibre Research* 15, 35–40.

Birchfield, W. (1983) Wheat and grain sorghum varietal reaction to *Meloidogyne incognita* and *Rotylenchulus reniformis. Plant Disease Reporter* 67, 41–42.

Bird, G.W., McCarter, S.M. and Roncadori, R.W. (1971) Role of nematodes and soil-borne fungi in cotton stunt. *Journal of Nematology* 3, 17–22.

Bird, G.W., Brooks, O.L. and Perry, C.E. (1974) Dynamics of concomitant field populations of *Hoplolaimus columbus* and *Meloidogyne incognita. Journal of Nematology* 6, 190–194.

Borges, A., dos Santos, J.C.C. and Scarpellini, J.R. (1999) Controle químico do nematóide reniforme *Rotylenchulus renifomis* com Rugby 200 CS, Marshal 400 SE e Furadan 350 SC na cultura do algodoeiro. *Anais do II Congresso Brasileiro de Algodão,* Ribeirão Preto, pp. 438–440.

Bridge, J. (1992) Nematodes. In: Hillocks, R.J. (ed.) *Cotton Diseases.* CAB International, Wallingford, UK, pp. 331–353.

Bridge, J. and Page, S.L.J. (1975) Plant parasitic nematodes associated with cotton in the lower Shire Valley and other cotton growing regions of Malawi. *United Kingdom Overseas Development Administration Technical Report.*

Bridge, J., Jones, E. and Page, S.L.J. (1976) *Meloidogyne acronea* associated with reduced growth of cotton in Malawi. *Plant Disease Reporter* 60, 5–7.

Brodie, B.B. and Cooper, W.E. (1964) Relation of parasitic nematodes to post-emergence damping off of cotton. *Phytopathology* 54, 1023–1027.

Carneiro, R.G., Antônio, H., Brito, J.A. and Alteia, A.A.K. (1990) Identificação de espécies e raças fisiológocas de *Meloidogyne* na região noroeste do Estado do Paraná: resultados preliminares. *Congresso Brasileiro de Nematologia* 14, 4.

Chakraborti, S. (2001) Integrated management approach for root-knot nematode in jute. *Indian Journal of Nematology* 31, 44–46.

Cia, E., Gridi-Papp, I.L., Chiavegato, E.J., Sabino, N.P., Kondo, J.I., Pizzinato, M.A., Bortoletto, N. and Carvalho, L.H. (2001) Melhoramento do algodoeiro no Estado de São Paulo: obtenção da cultivar IAC 21. *Bragantia* 60, 9–17.

Cia, E., Fuzatto, M.G., Almeida, W.P., Ruano, O., Kondo, J.I., Pizzinatto, M.A., Kasai, F.S. and Foltran, D.E. (2003) Resistência genética a doenças e nematóides em cultivares e linhagensde algodoeiro disponíveis no Brasil. *IV Congresso Brasileiro de Algodão.*

Colyer, P.D., Kirkpatrick, T.L., Caldwell, W.D. and Vernon, P.R. (1997) Influence of nematicide application on the severity of the root-knot nematode–Fusarium wilt complex in cotton. *Plant Disease* 81, 66–70.

Cook, C.G. and Mullin, B.A. (1994) Growth responses of kenaf cultivars in root-knot nematode/soil-borne fungi infested soil. *Crop Science* 34, 1455–1457.

Cook, C.G. and Scott, A.W. Jr (1995) Registration of 'SF459' kenaf. *Crop Science* 35, 1712.

Cook, C.G., Robinson, A.F. and Namken, N.L. (1997) Tolerance to *Rotylenchulus reniformis* and resistance to *Meloidogyne incognita* race 3 in high-yielding breeding lines of upland cotton. *Journal of Nematology* 29, 322–328.

Crow, W.T., Weingartner, D.P., McSorley, R. and Dickson, D.W. (2000) Population dynamics of *Belonolaimus longicaudatus* in a cotton production system. *Journal of Nematology* 32, 210–214.

Cuadra, R., Aguilera, C. and Perez, J.A. (1990) Control de *Meloidogyne incognita* por medio de la rotacion de cultivas y el barbecho. *Ciencias de la Agricultura* 40, 43–49.

Curi, S.M. (1980) Como combater os nematóides. *A Granja* 36, 52–55.

Davis, R.F., Baird, R.E. and McNeill, R.D. (2000) Efficacy of cotton root destruction and winter cover crops for suppression of *Hoplolaimus columbus*. *Supplement to the Journal of Nematology* 32, 550–555.

Davis, R.F., Koenning, S.R., Kemerait, R.C., Cummings, T.D. and Shurley, W.D. (2003) *Rotylenchulus reniformis* management in cotton with crop rotation. *Journal of Nematology* 35, 58–64.

Di Vito, M., Piscioneri, I., Pace, S., Zaccheo, G. and Catalano, F. (1997) Pathogenicity of *Meloidogyne incognita* on kenaf in microplots. *Nematologia Mediterranea* 25, 165–168.

dos Santos, M.A. and Ruano, O. (1987) Reação de plantas usadas como adubos verdes a *Meloidogyne incognita* raçe 3 e a *M. javanica*. *Nematologia Brasileira* 11, 184–197.

Duncan, L.W. and Ferris, H. (1984) Effects of *Meloidogyne incognita* on cotton and cowpeas in rotation. *Proceedings Beltwide Cotton Conferences, Cotton Disease Council* 43, 22–26.

Farias, P.R.S., Sanches-Vila, X., Barbosa, J.C., Viera, S.R., Ferraz, L.C.C.B. and Solis-Delfin, J. (2002) Using geostatistical analysis to evaluate the presence of *Rotylenchulus reniformis* in cotton crops in Brazil: economic implications. *Journal of Nematology* 34, 232–238.

Ferraz, C.A.M. (1964) Nematóides do algodeiro. *Campinas, Secretaria de Agricultura de São Paulo. Boletim Técnio Int. Série* D/5.

Ferraz, L.C.C.B. and Monteiro, A.R.M. (1995) Nematóides. In: Bergamin Filho, A., Kimati, H. and Amorim, L. (eds) *Manual de Fitopatologia*, Volume 1, 3rd edn. Ceres, São Paulo, pp. 168–201.

Ferris, H. (1985) Density dependent nematode seasonal multiplication rates and overwinter survivorship: a critical point model. *Journal of Nematology* 17, 9–100.

Fuzatto, M.G., Cia, E., Chiavegato, E.J., Kondo, J.I. and Pettinelli, A. Jr (1997) Correlação entre produção e sintomas em plantas de algodeiro afetadas por nematóides. *Anais do I Congresso de Algodão*. Campina Grande, PB, pp. 340–342.

Gay, C.M. and Bird, G.W. (1973) Influence of concomitant populations of *Pratylenchus brachyurus* and *Meloidogyne incognita* on root penetration and population dynamics. *Journal of Nematology* 5, 212–217.

Gazaway, W.S. and McLean, K.S. (2003) A survey of plant-parasitic nematodes associated with cotton in Alabama. *Journal of Cotton Science* 7, 1–7.

Giblin-Davis, R.M., Williams, D.S., Wergin, W.P., Dickson, D.W., Hewlett, T.E., Bekal, S. and Becker, J.O. (2001) Ultrastructure and development of *Pasteuria* sp. (S-1 strain), an obligate endoparasite of *Belonolaimus longicaudatus* (Nematoda: Tylechida). *Journal of Nematology* 33, 227–238.

Good, J.M., Minton, N.A. and Jaworski, C.A. (1965) Relative susceptibility of selected cover crops and coastal bermuda grass to plant nematodes. *Phytopathology* 55, 1026–1030.

Goulart, A.M.C., Inomoto, M.M. and Monteiro, A.R. (1997) Hospedabilidade de oito cultivares de algodoeiro a *Pratylenchus brachyurus*. *Nematologia Brasileira* 21, 111–118.

Graham, T.W. (1951) Nematode root rot of tobacco and other plants. *South Carolina Agriculture Experiment Station Bulletin* 390.

Heald, C.M. and Orr, C.C. (1984) Nematode parasites of cotton. In: Nickle, W.R. (ed.) *Plant and Insect Nematodes*. Marcel Decker, New York, pp. 147–166.

Heffes, T.A., Coates-Beckford, P.L. and Hutton, D.G. (1990) Population development and effects of *Rotylenchulus reniformis* on growth of *Amaranthus viridis* and three cultivars of *Hibiscus sabdariffa*. *Nematropica* 20, 95–98.

Heffes, T.A., Coates-Beckford, P.L. and Hutton, D.G. (1991) Effects of *Meloidogyne incognita* on growth and nutrient content of *Amaranthus viridis* and two cultivars of *Hibiscus sabdariffa*. *Nematropica* 21, 7–18.

Hussey, R.S. and Roncadori, R.W. (1978) Interaction of *Pratylenchus brachyurus* and *Gigaspora margarita* on cotton. *Journal of Nematology* 10, 16–20.

Jones, J.E., Newson, L.D. and Finely, E.L. (1959) Effect of the reniform nematode on yield, plant characters, and fiber qualities of cotton. *Agronomy Journal* 51, 353–356.

Khan, M.R. and Banerjee, S. (2003) Dynamics of root-knot nematode, *Meloidogyne incognita*, under a sequence of different crops in West Bengal. *Indian Journal of Nematology* 33, 73–74.

Kinlock, R.A. and Rich, J.R. (1998) Responses of cotton yield and *Meloidogyne incognita* population densities to soil applications of aldicarb and 1,3-dichloropropene in Florida. *Supplement to the Journal of Nematology* 30, 639–642.

Kirkpatrick, T.L. and Sasser, J.N. (1984) Crop rotation and races of *Meloidogyne incognita* on susceptible and resistant cotton. *Journal of Nematology* 16, 323–328.

Koenning, S.R., Walker, S.A. and Barker, K.R. (1996) Impact of soil texture on the reproductive and damage potentials of *Rotylenchulus reniformis* and *Meloidogyne incognita* on cotton. *Journal of Nematology* 28, 527–536.

Koenning, S.R., Barker, K.R. and Bowman, D.T. (2000) Tolerance of selected cotton lines to *Rotylenchulus reniformis. Supplement to the Journal of Nematology* 32, 519–523.

Koenning, S.R., Barker, K.R. and Bowman, D.T. (2001) Resistance as a tactic for management of *Meloidogyne incognita* on cotton in North Carolina. *Journal of Nematology* 33, 126–131.

Koenning, S.R., Edminsten, K.L., Barker, K.R. and Morrison, D.E. (2003) Impact of cotton production systems on management of *Hoplolaimus columbus. Journal of Nematology* 35, 73–77.

Konan, O.N., Ruano, O., Prud'Homme, S., Baudoin, J.P. and Mergeai, G. (2003) Introgression in tetraploid cotton of resistance to the reniform nematode, *Rotylenchulus reniformis* Lind and Oliveira, from the *Gossypium hisutum* × *G. longicalyx* Hutch and Lee × *G. thurberi* Todaro trispecific hybrid. *World Cotton Conference*, 9–13 March 2003, Cape Town, South Africa, p. 8.

Koura, F.H., Abdel-Naga, M. and Abdel-Wani, M. (1987) Effect of population density of *Hoplolaimus columbus* on cotton, flax, and kenaf yield and nematode reproduction. *Egyptian Journal of Phytopathology* 19, 119–126.

Kraus-Schmidt, H. and Lewis, S.A. (1981) Dynamics of concomitant populations of *Hoplolaimus columbus, Scutellonema brachyurum*, and *Meloidogyne incognita* on cotton. *Journal of Nematology* 13, 41–48.

Laha, S.K., Mandal, R.K. and Dasgupta, M.K. (1995) Screening of white jute, *Corchorus capsularis*, germplasm against *Meloidogyne incognita. Nematologia Mediterranea* 23, 51–53.

Lawrence, G.W. and McLean, K.S. (1990) First report of *Hoplolaimus magnistylus* on kenaf in Mississippi. *Plant Disease* 74, 828.

Lawrence, G.W. and McLean, K.S. (1992) Host status and response of kenaf (*Hibiscus cannabinus*) to *Meloidogyne incognita* race 4, *M. javanica, Hoplolaimus magnistylus*, and *Rotylenchulus reniformis. Nematropica* 22, 247–250.

Lawrence, G.W. and McLean, K.S. (2000) Effect of foliar application of oxamyl with aldicarb for the management of *Rotylenchulus reniformis* on cotton. *Supplement to the Journal of Nematology* 32, 542–549.

Lawrence, G.W., Webber, C.L. and McLean, K.S. (1994) Response of four kenaf (*Hibiscus cannabinus*) plant introductions to *Meloidogyne incognita* race 3. *Nematropica* 244, 95–101.

Lin, X.M. and Chen, Q.Q. (1992) Studies on the diseases of root-knot nematodes of crops and a discovery of pathogenic bacteria (*Bacillus penetrans*) of nematodes in Fujian. *Wuyi Science Journal* 9, 261–268 (in Chinese).

Lordello, L.G.E. (1968) *Nematóides das Plantas Cultivadas.* Livraria Nobel.

Lordello, L.G.E. and Arruda, H.V. (1957) Nota prévia acerca da ocorrência de nematóides do gênero *Pratylenchus* en raízes de algodoeiro. *O Solo* 49, 33–34.

Lordello, R.R.A. and Sabino, N.P. (1985) Efeito do controle de nematóides na qualidade da fibra do algodoeiro. *Bragantia* 44, 695–699.

Mandal, R.K. and Mishra, C.D. (2001) Role of different organisms in inducing Hoogly wilt symptoms. *Environment and Ecology* 19, 969–972.

McLean, K.S. and Lawrence, G.W. (1990) First report of *Hoplolaimus magnistylus* on kenaf in Mississippi. *Plant Disease* 74, 828.

McLean, K.S. and Lawrence, G.W. (2000) A survey of plant-parasitic nematodes associated with cotton. *Supplement to the Journal of Nematology* 32, 508–512.

McLean, K.S. and Lawrence, G.W. (2003) Efficacy of aldicarb to *Rotylenchulus reniformis* and biodegradation in cotton field soils. *Journal of Nematology* 35, 65–72.

McSorley, R. and Parrado, J.L. (1986) Relationship between height of kenaf and root galling by *Meloidogyne incognita. Nematropica* 16, 205–211.

Minton, N.A. and Adamson, W.C. (1979) Control of *Meloidogyne javanica* and *M. arenaria* on kenaf and roselle with genetic resistance and nematicides. *Journal of Nematology* 11, 37–41.

Minton, N.A., Adamson, W.C. and White, G.A. (1970) Reaction of kenaf and roselle to three root-knot nematode species. *Phytopathology* 60, 1844–1845.

Mishra, C. and Chakrabarti, N.K. (1987) Screening of jute (*Corchorus capsularis* L.) germplasm against root-knot nematode *Meloidogyne incognita* (Kofoid and White) Chitwood. *Annals of Agricultural Research* 8, 59–63.

Mishra, C., Singh, B. and Chakrabarati, N.K. (1985) Host–parasite relations of some nematode pests with jute (*Corchorus capsularis* and *C. olitorus*). *Annals of Agricultural Research* 6, 162–167.

Mishra, C., Singh, B. and Laha, S.K. (1987) Integrated approach for root-knot nematode management in jute. *Indian Journal of Nematology* 17, 285–287.

Mueller, J.D. and Sullivan, M.J. (1988) Responses of cotton to infection by *Hoplolaimus columbus. Annals of Applied Nematology* 2, 86–89.

Noe, J.P. (1993) Damage functions and population changes of *Hoplolaimus columbus* on cotton and soybean. *Journal of Nematology* 25, 440–445.

Ogallo, J.L., Goodell, P.B., Eckert, J. and Roberts, P.A. (1997) Evaluation of NemX, a new cultivar of cotton with high resistance to *Meloidogyne incognita. Journal of Nematology* 29, 531–537.

Ogallo, J.L., Goodell, P.B., Eckert, J. and Roberts, P.A. (1999) Management of root-knot nematodes with resistant cotton cv. NemX. *Crop Science* 39, 418–421.

Oliveira, C.M.G., Kubo, R.K., Siloto, R.C. and Raga, A. (1999) Eficiência de carbofuran e terbufós sobre nematóides e pragas iniciais na cultura algodoeira. *Revista de Agricultura* 74, 325–244.

Orr, C.C. and Robinson, A.F. (1984) Assessment of cotton losses in Western Texas caused by *Meloidogyne incognita. Plant Disease* 68, 284–285.

Page, S.L.J. (1983) Biological studies of the African cotton root nematode *Meloidogyne acronea*. PhD Thesis, London University.

Page, S.L.J. (1984) Effects of the physical properties of two tropical soils on their permanent wilting point and relative humidity, in relation to survival of *Meloidogyne acronea. Revue de Nématologie* 7, 227–232.

Page, S.L.J. and Bridge, J. (1994) The African root-knot nematode, *Meloidogyne acronea*: its pathogenicity and intra-generic infectivity with *Gossypium. Fundamental and Applied Nematology* 17, 67–73.

Prasad, K.S. and Padeganur, G.M. (1980) Observations on the association of *Rotylenchulus reniformis* with verticillium wilt of cotton. *Indian Journal of Nematology* 10, 91–92.

Robbins, R.T. and Barker, K.R. (1974) The effects of soil type, particle size, temperature and moisture on reproduction of *Belonoliamus longicaudatus. Journal of Nematology* 6, 1–6.

Roberts, P.A., Smith, S.N. and Matthews, W.C. (1985) Quantitative aspects of the interaction of *Meloidogyne incognita* with Fusarium wilt on Acala cotton in field plots. *Proceedings of Beltwide Cotton Production and Research Conferences*, National Cotton Council, Memphis, Tennessee, p. 21.

Robinson, A.F. and Percival, A.E. (1997) Resistance to *Meloidogyne incognita* race 3 and *Rotylenchulus reniformis* in wild accessions of *Gossypium hisutum* and *G. barbadense* from Mexico. *Supplement to the Journal of Nematology* 29, 746–755.

Robinson, A.F., Heald, C.M., Flanagan, S.L., Thames, W.H. and Amador, J. (1987) Geographical distribution of *Rotylenchulus reniformis, Meloidogyne incognita,* and *Tylenchulus semipenetrans* in the lower Rio Grande valley as related to soil type and land use. *Annals of Applied Nematology* 1, 20–25.

Robinson, A.F., Cook, C.G. and Bridges, A.C. (1998) Comparative reproduction of *Meloidogyne incognita* race 3 (cotton root-knot nematode) and *Rotylenchulus reniformis* (reniform nematode) on cotton, kenaf, and sunnhemp. *Proceedings of Beltwide Cotton Conferences,* National Cotton Council, Memphis, Tennessee, pp. 147–148.

Robinson, A.F., Cook, C.G., Kirkpatrick, T.L., McGawley, E.C., Overstreet, C. and Padgett, B. (2000) Vertical distribution of the reniform nematode in the upper 1.5 metres of soil on nine farms in Arkansas, Lousiana, and Texas. *Proceedings of the Beltwide Cotton Conferences* 1, 564.

Robinson, A.F., Cook, C.G., Bradford, J.M., Bridges, A.C. and Bautista, J. (2002) Differences in cotton yield, root growth, and *Rotylenchulus reniformis* following deep soil fumigation. *Proceedings of the Beltwide Cotton Conferences* 3, 564.

Robinson, A.F., Cook, C. and Starr, J. (2003) Depth and texture of soil greatly alter the expression of reniform nematode resistance in key exotic cottons. *Proceedings of the Beltwide Cotton Conferences* 4, 840.

Rodríguez-Kabána, R. and Pope, M.H. (1981) A simple incubation method for the extraction of nematodes from soil. *Nematropica* 11, 175–186.

Ruano, O. and Almeida, W.P. (1999) Sistema para avaliação de resistência múltipla em genótipos de algodoeiro a nematóides e doenças foliares, em casa de vegetação. *Anais do Segundo Congresso Brasileiro de Algodão,* Ribeirão Preto, pp. 488–491.

Ruano, O., Androcioli Filho, A. and Carneiro, R.G. (1984a) Comportamento de algodoeiros em solos de cultivo de cafeeiro infestados com *Meloidogyne incognita*. *Reunião Nacional do Algodão* 3, 169.

Ruano, O., Chaves, G.M., Ferraz, S. and Zambolim, L. (1984b) Reação de algodoeiros a *Meloidogyne incognita, Fusarium oxysporum* f.sp. *vasinfectum* e à associação desses organismos. In: *Renião Nacional do Algodão,* Recife, PE. Resumos. Campina Grande, PB:Embrapa-CNPA, p. 66.

Ruano, O., Chaves, G.M., Ferraz, S. and Zambolim, L. (1985) Distribuição de raças de *Meloidogyne incognita* em áreas algodoeiras nos Estados do Paraná e Goiás. *Fitopatologia Brasileira* 10, 667–670.

Ruano, O., Almeida, W.P. and Muramoto, S.P. (2001) Reação de novos genótipos de algodeiro a nematóides e patógenos foliares em casa-de-vegetação. In: *Anais do III Congresso Brasileiro de Algodão,* Campo Grande, MS, pp. 880–883.

Sankaaralingam, A. and McGawley, E.C. (1994) Interrelationships of *Rotylenchulus reniformis* with *Rhizoctonia solani* on cotton. *Journal of Nematology* 26, 475–485.

Sasser, J.N. and Carter, C.C. (1985) Overview of the International *Meloidogyne* Project, 1975–1984. In: Sasser, J.N. and Carter, C.C. (eds) *An Advanced Treatise on* Meloidogyne, *Volume I Biology and Control.* North Carolina State University Graphics, Raleigh, North Carolina, pp. 19–24.

Senapati, S.K. and Ghosh, S.K. (1992) Integration of different means of root-knot nematodes (*Meloidogyne* spp.) management on jute (*Corchorus capsularis* L.) under natural field conditions. *Crop Research Hissar* 5, 559–564.

Seno, K.C.A., Kondo, E.S. and Lusvarghi Neto, H. (1999) Avaliação de eficácia de diferentes doses do nematicida terbufós no controle do nematóide *Rotylenchulus reniformis* na cultura do algodão (*Gossypium hirsutum*). *Anais do II Congresso Brasileiro de Algodão,* Ribeirão Preto, pp. 435–437.

Shepherd, R.L., McCarty, J.C., Jenkins, J.N. and Parrott, W.L. (1988) Registration of twelve non-photoperiodic lines with root-knot nematode resistant primitive cotton germplasm. *Crop Science* 28, 868–869.

Shepherd, R.L., McCarty, J.C., Jenkins, J.N. and Parrott, W.L. (1996) Registration of nine cotton germplasm lines resistant to root-knot nematodes. *Crop Science* 36, 820.

Silva, N.M. (1984) Efeitos de rotação de culturas de algodão no Estado de São Paulo. In*: Fundação Cargill.* Adubação verde no Brasil, Campinas, pp. 124–128.

Silva, N.M., Fuzatto, M.G., Kondo, J.I., Sabino, J.C., Pettinelli Junior, A. and Gallo, P.B. (1997) A adubação nitrogenada e o sintoma de nematóides no algodoeiro. *Revista Brasileira de Ciência do Solo* 21, 693–697.

Silvey, D.T., Ripple, K., Smith, C.W. and Starr, J.L. (2003) Identification of RFLP loci linked to resistance to *Meloidogyne incognita* and *Rotylenchulus reniformis. Proceedings of the Beltwide Cotton Conferences* 4, 283.

Sperandio, C.A. and Amaral, A.S. (1994) Ocorrencia de *Meloidogyne graminicola* causador da falsa bicheira do arroz irrigado no Rio Grande do Sul. *Lavoura Arrozeira* 47, 18–21.

Starr, J.L. (1988) Relationship between population densities of *Meloidogyne incognita* on cotton and egg viability. *Journal of Nematology* 20, 512–515.

Starr, J.L. (1993) Longevity and recovery of egg masses of *Meloidogyne incognita* during simulated winter survival. *Journal of Nematology* 25, 244–248.

Starr, J.L. (1998) Cotton. In: Pederson, G.A., Windham, G.L. and Barker, K.R. (eds) *Plant–Nematode Interactions.* Agronomy Society of America, Madison, Wisconsin, pp. 359–379.

Starr, J.L. and Jeger, M.J. (1985) Dynamics of overwintering of eggs and juveniles of *Meloidogyne incognita* and *M. arenaria. Journal of Nematology* 17, 252–256.

Starr, J.L. and Matheison, T. (1985) Reproduction of *Pratylenchus brachyurus* on cotton and growth response to infection by the nematode. *45th Cotton Disease Council. Proceedings of Beltwide Cotton Production Research Conferences,* National Cotton Council, Memphis, Tennessee.

Starr, J.L., Jeger, M.J., Martyn, R.D. and Schilling, K. (1989) Effects of *Meloidogyne incognita* and *Fusarium oxysporum* f.sp. *vasinfectum* on plant mortality and yield of cotton. *Phytopathology* 79, 640–646.

Starr, J.L., Heald, C.M., Robinson, A.F., Smith, R.G. and Krausz, J.P. (1993) *Meloidogyne incognita* and *Rotylenchulus reniformis* and associated soil textures from some cotton production areas of Texas. *Supplement to the Journal of Nematology* 25, 895–899.

Su, J.G. (1993) Chinese bast fibre crop genetic resources. *IBPGR Newsletter for Asia, the Pacific, and Oceania* 13, 4–5.

Sud, U.C., Varaprasad, K.J., Seshadri, A.R. and Kher, K.K. (1984) Relationship between initial densities of *Rotylenchulus reniformis* and damage to cotton with fit to Seinhorst curves. *Indian Journal of Nematology* 14, 148–151.

Summers, T.E., Pate, J.B. and Wilson, F.D. (1958) Extent of susceptibility with kenaf, *Hibiscus cannabinus* L., to root-knot nematodes. *Plant Disease Reporter* 42, 591–593.

Thames, W.H. and Heald, C.M. (1974) Chemical and cultural control of *Rotylenchulus reniformis* on cotton. *Plant Disease Reporter* 58, 337–341.

Tihohod, D., Ferraz, L.C.C.B., Barbosa, J.C. and Volpe, C.A. (1992) Manejo de fitonematóides do algodoeiro 2. Distributuição espacial das espécies. *Nematologia Brasileira* 16, 63–73.

Tomerlin, A.H. Jr (1969) The influence of organic amendments on numbers of nematodes and other microorganisms in the soil. PhD Thesis, University of Florida, Gainesville, Florida.

Tsai, B.Y. and Apt, W.J. (1979) Anhydrobiosis of the reniform nematode: survival and coiling. *Journal of Nematology* 11, 316.

Vawdrey, L.L. and Stirling, G.R. (1992) Reaction of kenaf and roselle grown in the Burkin River Irrigation Area to root-knot nematodes. *Australasian Plant Pathology* 21, 8–12.

Veech, J.A. (1992) Reproduction of four races of *Meloidogyne incognita* on *Hibiscus cannabinus. Supplement to the Journal of Nematology* 24, 771–721.

Veech, J.A. and Starr, J.L. (1986) Comparison of development, reproduction and aggressiveness of *Meloidogyne incognita* races 3 and 4 on cotton. *Journal of Nematology* 18, 413–415.

Walker, N.R., Kirkpatrick, T.L. and Rothrock, C.S. (1998) Interaction between *Meloidogyne incognita* and *Thielaviopsis basicola* on cotton (*Gossypium hirsutum*). *Journal of Nematology* 30, 415–422.

Westphal, A. and Smart, J.R. (2003) Depth distribution of *Rotylenchulus reniformis* under different tillage and crop sequence systems. *Phytopathology* 93, 1182–1189.

Wheeler, T.A., Kaufman, H.W., Baugh, B., Kidd, P., Schuster, G. and Siders, K. (1999) Comparison of variable and single-rate application of aldicarb on cotton yield in fields infested with *Meloidogyne incognita. Supplement to the Journal of Nematology* 31, 700–708.

Wheeler, T.A., Baugh, B., Kaufman, H., Schuster, G. and Siders, K. (2000) Variability in time and space of *Meloidogyne incognita* fall population density in cotton fields. *Journal of Nematology* 32, 258–264.

Windham, G.L. and Williams, W.P. (1987) Host suitability of commercial corn hybrids to *Meloidogyne arenaria* and *M. incognita. Annals of Applied Nematology* 1, 13–16.

Wrather, J.A., Stevens, S., Kirkpatrick, T.L. and Kitchen, N.R. (2002) Effects of site-specific application of aldicarb on cotton in a *Meloidogyne incognita*-infested field. *Journal of Nematology* 34, 115–119.

Yassin, A.M. (1974) Role of *Pratylenchus sudanensis* in the syndrome of cotton wilt with reference to its vertical distribution. *Sudan Agricultural Journal* 9, 48–52.

Yassin, A.M. (1980) Studies on the biology and chemical control of root-lesion nematode, *Pratylenchus sudanensis* Loof and Yassin 1970, from the Gezira. *Zeitschrift für angewandte Zoologie* 67, 25–231.

Yik, C.P. and Birchfield, C. (1984) Resistant germplasm in *Gossypium* species and related plants to *Rotylenchulus reniformis. Journal of Nematology* 16, 146–153.

Yu, Y. (1994) The application of integrated control technique of kenaf root-knot nematode. *Guangxi Agriculture Science* 1, 24–25.

Zhang, F. and Noe, J.P. (1996) Damage potential and reproduction of *Meloidogyne incognita* race 3 and *M. arenaria* race 1 on kenaf. *Supplement to the Journal of Nematology* 28, 668–675.

Zhou, E. and Starr, J.L. (2003) A comparison of the damage functions, root galling, and reproduction of *Meloidogyne incognita* on resistant and susceptible cotton cultivars. *Journal of Cotton Science* 7, 224–230.

Zhou, E., Wheeler, T.A. and Starr, J.L. (2000) Root galling and reproduction of *Meloidogyne incognita* populations from Texas on resistant cotton genotypes. *Supplement to the Journal of Nematology* 32, 513–518.

21 Nematode Parasites of Spices, Condiments and Medicinal Plants*

P.K. Koshy,[1] Santhosh J. Eapen[2] and Rakesh Pandey[3]
[1]*Division of Nematology, Central Plantation Crops Research Institute, (Regional Station) Kayangulam, Krishnapuram-690533, Kerala, India;* [2]*Division of Crop Protection, Indian Institute of Spices Research, Calicut-673 012, India; and* [3]*Central Institute of Medicinal and Aromatic Plants, (CIMAP-CSIR), PO CIMAP, Lucknow (UP)-226 015, India*

Spices are strongly flavoured or aromatic substances of plant origin commonly used for seasoning and preserving foodstuffs. They consist of rhizomes, barks, leaves, fruits, seeds and other parts of plants. These plants belong to different families, genera and species. The bulk of the dry matter of their products consists of carbohydrates, volatile oils, fixed oils, proteins, tannins, resins, pigments and mineral elements. These constituents differ in their composition and content in different spices. Most of the spices are crops of the humid tropical regions. India is considered as the home of spices from ancient times and produces a large proportion of all spices. There are innumerable biotic and abiotic problems on spice crops that adversely affect production, including plant parasitic nematodes which can cause considerable damage to some of these crops. Nematode problems of chilli and garlic, which, depending on use, can be considered spices, are not included in this chapter as they are discussed under vegetables. Nematode problems of betel vine (*Piper betle*) and kava (*Piper methysticum*) are also included in this chapter.

Traditional medicines derived from plant sources have gained credibility and have become an important aspect of herbal medicine systems for human health care. The herbal medicine system is widespread in China, India, Japan, Pakistan, Sri Lanka and Thailand. Aroma compounds from botanical sources are increasingly used in cosmeceutical, nutraceutical and the processed food industry due to growing public awareness of the risks involved in the use of synthetic additives. The plant retail for herbs and medicinal plants in the USA is estimated to have a turnover of approximately US$1.6 billion annually. In Europe, about 400,000 t of medicinal plant material is imported from Asia and Africa yearly. The average market value of this plant material is estimated at US$1 billion. Many of the raw materials used in the pharmaceutical industry come from medicinal plants produced on a global scale.

*A revision of the chapter by P.K. Koshy and J. Bridge.

 Plant Parasitic Nematodes in Subtropical and Tropical Agriculture, 2nd Edition (eds M. Luc, R.A. Sikora, J. Bridge)

Black Pepper

Black pepper (*Piper nigrum* L.) is a branching and climbing perennial shrub belonging to the family Piperaceae and is cultivated in the hot and humid parts of the world. India, Indonesia, Vietnam and Brazil, contributing 34, 20, 14 and 10%, respectively, are the major pepper-producing countries in the world today. World production of pepper during 1999 was 219,840 t and covered an area of 466,070 ha (Selvan, 2002). Its origin is considered to be in the hills of south-western India where it is known as the 'king of spices'. It is used in culinary seasonings, as a preservative for meat and other perishable foods, and in medicine. Piperine, the bite factor of pepper, is used to impart a pungent taste to brandy. Pepper oil is used in perfumery. The pepper vine can be propagated either vegetatively or by seed. Raising plants through cuttings is universally adopted. Two pepper vines entwined about a teak wood or concrete post, set in the field, is known as a 'pepper tree'. In India, live trees are used as supports (standards) for climbing pepper.

Nematodes on Black Pepper

Many nematodes have been reported on black pepper (Table 21.1), but the only two known to cause serious damage to the crop are *Radopholus similis* and *Meloidogyne* spp.

Radopholus similis

Association of the burrowing nematode *R. similis* with the yellows disease of pepper was first reported in 1936 and later by Van der Vecht (1950), who made extensive field studies and also demonstrated its pathogenicity under laboratory conditions. The nematode is notorious for being associated with the loss of 22 million pepper vines within 20 years in Bangka Island, Indonesia due to 'yellows disease' (Christie, 1957, 1959). Subsequently, *R. similis* was reported from black pepper from India (D'Souza *et al.*, 1970; Kumar *et al.*, 1971; Venkitesan, 1972; Koshy *et al.*, 1978; Mohandas and Ramana, 1987c; Ramana *et al.*, 1987a; Ramana and Mohandas, 1989), Malaysia, Thailand (Sher *et al.*, 1969; Reddy, 1977) and Sri Lanka (Gnanapragasam *et al.*, 1985). The nematode is also involved in 'slow wilt' disease of black pepper in India, which is almost identical to pepper yellows in Indonesia (Van der Vecht, 1950; Mohandas and Ramana, 1987b) hence, they are dealt with together. Intensive surveys carried out on the role of plant parasitic nematodes in the slow wilt disease complex of black pepper in India showed that high populations of *R. similis* occurred more frequently in slow wilt disease-affected plants than in healthy plants. Discriminate analysis indicated the involvement of *R. similis* in slow wilt disease (Ramana *et al.*, 1987a).

Black pepper was introduced to Indonesia from Kerala, India (Nambiar, 1977) and it is quite likely that the burrowing nematode was also introduced along with the rooted cuttings of black pepper.

Symptoms of damage

The primary symptom of the yellows (slow wilt) disease is the appearance of pale yellow or whitish yellow drooping leaves on the vines. The number of such leaves increases gradually until large numbers of leaves or even the entire foliage becomes yellow (Plate 22A). Yellowing is followed by shedding of leaves, cessation of growth and dieback symptoms (Fig. 21.1). The symptoms are very pronounced when soil moisture is depleted. In the very early stage of the disease in India, the symptoms may disappear with the onset of the south-west monsoon, resulting in an apparently healthy appearance of such plants in the following years because of new leaf growth and shedding of yellowed leaves. This has often given a mistaken impression of the disease being caused by soil moisture stress rather than nematodes. However, within 3–5 years of initiation of yellowing, all the leaves are shed and death of the

Table 21.1. Plant parasitic nematodes associated with spice crops.

Nematode species	Black pepper	Cardamom	Ginger	Turmeric	Fennel	Fenugreek	Coriander	Cumin	Celery	Dill	Vanilla
Aphelenchoides fragariae									•		
A. ritzemabosi									•		
Belonolaimus longicaudatus									•		
Criconema cardamomi		•									
Criconemoides brevistylus						•					
C. onoensis	•										
C. ornatus	•		•	•			•		•		
C. sphaerocephalus				•							
C. xenoplax					•	•					
Discocriconemella limitanea	•										
Ditylenchus destructor									•		
D. dipsaci					•				•		
Dolichodorus sp.	•										
Helicotylenchus abunaami	•		•	•							
H. dihystera	•	•	•	•		•					
H. erythrinae	•		•								
H. indicus						•	•				
H. multicinctus		•		•			•				
H. pseudorobustus	•		•	•							
H. variocaudatus	•										
Hemicriconemoides cocophillus			•	•		•					
H. gaddi	•	•									
H. mangiferae	•										
Hemicycliophora arenaria						•					
Heterodera avenae						•	•				
H. schachtii										•	
Hirschmanniella mucronata							•				
H. oryzae							•				
Hoplolaimus columbus	•			•							
H. indicus	•		•	•		•	•				
H. seinhorsti	•		•								
Longidorus apulus									•	•	
Meloidogyne arenaria		•	•		•	•					

Continued

Table 21.1. *Continued.*

Nematode species	Black pepper	Cardamom	Ginger	Turmeric	Fennel	Fenugreek	Coriander	Cumin	Celery	Dill	Vanilla
M. hapla			•		•		•		•	•	
M. incognita	•	•	•	•	•	•	•	•	•		
M. javanica	•	•	•	•	•	•	•	•	•	•	
M. piperi	•										
Paralongidorus maximus	•										
Paratrichodorus christiei										•	
P. mirzai						•	•				
Pratylenchus brachyurus			•								
P. coffeae	•	•	•	•					•	•	
P. exilis							•				
P. indicus			•								
P. penetrans					•				•		
P. pratensis			•								
P. thornei						•	•				
P. zeae			•			•	•				
Radopholus similis	•	•	•	•							
Rotylenchulus reniformis	•	•	•	•		•	•	•			
Rotylenchus sp.	•	•		•							
Scutellonema siamense	•										
Trichodorus minor		•									
T. primitivus							•		•		
Trophotylenchulus piperis	•										
Tylenchorhynchus mashhoodi	•					•					
T. vulgaris					•						
Tylenchulus semipenetrans	•				•						
Xiphinema americanum			•							•	
X. basiri			•	•			•				
X. brevicollum											•
X. diversicaudatum							•			•	
X. elongatum	•										
X. index			•								
X. insigne			•	•							
X. radicicola	•										

Fig. 21.1. Yellowing and defoliation in black pepper vines affected with yellows or slow decline disease caused by *Radopholus similis*. (Photo: V.K. Sosamma.)

Fig. 21.2. Damage to black pepper cutting (left) caused by *Radopholus similis*. (Photo: V.K. Sosamma.)

vine takes place, and hence the name 'slow wilt' disease. In bearing vines, shedding of spikes (inflorescences) is a major symptom. Large numbers of shed spikes are seen at the base of affected vines. In large plantations, affected patches become conspicuous initially as yellowed plants, and later with large numbers of barren standards that have lost the vines (Plate 22B), or standards supporting dead vines without any leaves. Young and old plants are affected and the replanted vines normally die within 2 years.

The tender thin, white, feeding roots show typical orange to purple coloured lesions. Lesions are not clearly seen on older roots, being brown in colour. The root system exhibits extensive rotting, and the main roots are devoid of fine feeder roots that rot quickly. Extensive necrosis of larger lateral roots develops over time (Fig. 21.2).

Biology and life cycle

The nematode penetrates roots within 24 h of inoculation and the cells around the site of penetration become brown (Venkitesan and Setty, 1977). Nematodes do not enter the stelar portions of the root, but plugging of xylem vessels with a gum-like substance has been reported (Freire and Bridge, 1985a). *R. similis* completes its life cycle within 25 days, in a temperature range of 25–28°C (Geetha, 1991). The black pepper isolate of the nematode is easily cultured on carrot discs at 25°C (Koshy, 1986b). The *R. similis* populations in Indonesia and Kerala (India) have a haploid number (n = 4) of four chromosomes and belong to the 'banana race' (Huettel *et al.*, 1984; Koshy, 1986b; Jasy, 1991; Ramana, 1992).

In India, the maximum nematode population in roots of pepper occurs between September and October and the minimum density between April and May (Ramana, 1986; Mohandas and Ramana, 1988). Low soil temperatures coupled with adequate soil moisture and availability of young tender roots help in the build-up of the population during September–October.

Other hosts

A large number of tree species such as coconut (*Cocos nucifera*), arecanut (*Areca catechu*), jack fruit (*Artocarpus integrifolia*), mango (*Mangifera indica*), gliricidia (*Gliricidia maculata*), dadap (*Erythrina indica*), garuga (*Garuga pinnata*) and Vatta (*Macaranga indica*) are used as live standards. Among these, coconut and arecanut are good hosts of *R. similis*. Crops such as banana, ginger and turmeric that are susceptible to *R. similis* are also intercropped with pepper.

Disease complexes

It has been speculated that yellows disease in Indonesia is caused by a nematode–fungus complex (Hubert, 1957; Bridge, 1978) involving *R. similis*, *Fusarium* spp. and possibly other fungi. There is little direct evidence to support the hypothesis. However, Freire (1982) showed that an Indonesian isolate of *R. similis* predisposed black pepper seedlings to attack by a weakly pathogenic isolate of *Fusarium solani*, causing severe root damage. In addition, Mustika (1992a,b) has clearly demonstrated that *R. similis* alone caused growth reduction and yellow leaves with a stiff droop, but damage was more obvious when *R. similis* acted together with *F. solani*. Studies under simulated field conditions showed that *R. similis* and *Phytophthora capsici* alone or in association resulted in root rotting, leading to slow decline disease (Ramana *et al.*, 1992; Anandaraj *et al.*, 1996a,b).

Economic importance and population damage threshold levels

The slow wilt disease was first reported from the Wynad area in Kerala as early as 1902, and Krishna Menon (1949) reported mortality of up to 10% of the vines due to the disease. Reduction in plant growth has been reported in sterile soil when 55-day-old rooted cuttings of black pepper in pots are inoculated with 2300 nematodes.

The onset of yellows disease in Sumatra, Indonesia is correlated with *R. similis* populations of 2 nematodes/100 g of soil and 25 nematodes/10 g of roots, and *Meloidogyne* spp. populations of 47 nematodes/100 g of soil and 305 nematodes/10 g of roots (Mustika, 1978). Bridge (1978), however, stated that a low population of less than 310 nematodes/10 g of roots may not alone cause the disease. A population level of 250 nematodes/g of roots was constantly recorded with slow-wilt-affected pepper vines in Kerala (Ramana, 1986). In pathogenicity tests, *R. similis* caused significant reduction in the growth and yield of black pepper (Mohandas and Ramana, 1991). Black pepper vines of any age group are susceptible to this nematode (Ramana, 1992). Inoculation with *R. similis* alone reduced growth rate of different cultivars of black pepper (Mustika, 1991).

Management measures

At present, there are no effective control measures for slow wilt or pepper yellows. The price of black pepper is known to fluctuate greatly and, with a fall in prices, the farmer often loses interest in the crop and tends to neglect adoption of even standard agronomic practices. Control methods need to be adopted every year for black pepper, which is a perennial crop, especially under Indian conditions where live standards are used. The perennial multicropping systems involving coconut, arecanut, black pepper, betel vine, banana, ginger, turmeric, etc. that have developed over many years on the west coast of South India are ideal situations where the burrowing nematode multiplies and causes extensive damage to all the susceptible crops. Black pepper, betel vine and banana are crops that succumb to nematode attack early. In later years, the farmers abandon pepper cultivation in arecanut-based farming systems where arecanut is the live standard. Although application of phorate at 3 g a.i./vine twice a year has been found to control *R. similis*, the high density multispecies cropping pattern does not permit use of nematicides, as most of the crops are export oriented and some products are consumed without any processing or cooking, such as banana, betel leaves, etc. This situation is complicated further because arecanut and coconut that are used as live standards are also very good hosts of *R. similis*, which warrants higher dosages and more frequent use of nematicides, especially under irrigated conditions.

CULTURAL. Symptoms of slow wilt and pepper yellows are known to be ameliorated with mulching. Pasril (1976) has recorded an 18% reduction in disease incidence on Bangka Island, Indonesia, after mulching. He also observed a reduction in disease symptoms after application of nematicide with a corresponding increase of yield in the first year of treatment. Addition of chopped leaves of *Glyricidia maculata* (10 g/kg of soil) as green manure reduced populations of *R. similis* and increased plant growth (Jasy and Koshy, 1992).

De Waard (1979) suggested application of fertilizers at a per hectare dose of 400 kg N, 180 kg P, 480 kg K, 425 kg Ca and 112 kg Mg in combination with a mulch for effective control of yellows disease in Bangka, Indonesia. Mustika *et al.* (1984) also reported remission of disease severity when fertilizers were applied to infected vines. Furthermore, foliar yellowing and necrosis of distal ends of laminae of slow wilt-affected vines in Kerala, India were attributed to N and K deficiencies (Wahid *et al.*, 1982).

RESISTANCE AND TOLERANCE. A number of black pepper germplasm accessions, including wild types, were screened against *R. similis* by several workers (Venkitesan and Setty, 1978; Jacob and Kuriyan, 1979a; Koshy and Sundararaju, 1979; Leong, 1986; Paulus *et al.*, 1993). The wild collection Vittal No. 430, *Piper hymenophyllum* and *P. attenuatum*, recorded less than 30% root reduction and a 1.5-fold nematode population increase. The hybrid pepper variety Panniyur-I recorded 91.4% root reduction and a 7.6-fold nematode increase (Venkitesan and Setty, 1978). However, a local cultivar at Peringamala in Kerala, India was not invaded by *R. similis* (Jacob and Kuriyan, 1979b). In Sri Lanka, a black pepper variety, PW 14, was immune to *R. similis* (Gnanapragasam, 1989). No resistance or tolerance was found on screening cultivated and wild germplasm, intercultivar hybrids or open pollinated seedlings, except for *P. colubrinum*, which is now widely used as a rootstock to graft cultivated pepper plants (Ramana *et al.*, 1987b; Ramana, 1992).

CHEMICAL. A number of pesticides have been found effective in reducing *R. similis* populations on black pepper in pot trials as well as in field trials. Aldicarb sulphone at 8 kg a.i./ha was most effective for control of *R. similis* on pepper in pot trials (Venkitesan, 1976; Venkitesan and Setty, 1979). DD, Vapam, Nemagon, Temik, Furadan, Nemacur, Mocap, Hostathione, Dasanit and Dasudin were found to reduce populations of *Meloidogyne* spp. and *R. similis* on *P. nigrum* in greenhouse trials (Mustika and Zainuddin, 1978). Under Indian conditions, aldicarb/carbofuran/phorate at 3 g a.i./vine applied in May/June and again in September/October results in the remission of foliar yellowing and reduction in nematode populations. Among the above three nematicides, phorate is superior (Ramana, 1986; Mohandas and Ramana, 1987a; Lokesh and Gangadharappa, 1995; Sundararaju and Sudha, 1998). The chances of rehabilitating severely affected vines by application of nematicides are low because of heavy damage already caused to the root system and the inability of such plants to put out new roots for quick rejuvenation. Although chemicals have been reported to reduce the nematode population and ameliorate slow wilt symptoms, the cost:benefit ratio has not been calculated.

BIOLOGICAL. There have been few successful attempts to control *R. similis* by using any of the fungal biological control agents, probably due to the migratory endoparasitic nature of this nematode (Geetha, 1991; Ramana, 1994). The mycorrhizal fungus, *Glomus fasciculatum*, suppressed burrowing nematode infestation (Anandaraj *et al.*, 1996c). Recently, rhizobacteria that suppressed *R. similis* infesting black pepper were identified in greenhouse studies (Beena *et al.*, 2003).

Summary of management measures

Integrated methods of nematode management that can be suggested are:

- Planting of nematode-free rooted cuttings raised in nursery mixture sterilized with steam, solar heat or fumigants.
- Uprooting of affected vines and replanting after a period of 9–12 months.
- Use of non-living supports or standards.
- Exclusion of *R. similis*-susceptible trees as standards for trailing black pepper vines, and exclusion of susceptible intercrops such as banana, ginger and turmeric.
- Application of phorate at 3 g a.i./vine with the onset of the monsoon and again after 3 months. The nematicide may be applied after removing the top soil without causing damage to the roots, followed by replacement of the soil. The susceptible intercrops, e.g. banana, may also be treated with nematicides.
- Application of organic amendments, such as 200 g of neem oil cake (*Azadirachta indica*), green foliage (3–5 kg) or farmyard manure (1 kg) per vine.
- Earthing-up after application of nematicides, NPK fertilizers and organic amendments in September/October.

Methods of diagnosis

The presence of nematodes and their association with the disease can be diagnosed by soil sampling at a distance of 25–50 cm from the base of the vine at a depth of 20–30 cm. A soil sample of 200 cm^3 and root sample of 0.5–1.0 g of thin, tender, feeder roots should be taken to obtain maximum nematode population estimates (Koshy, 1986b, 1987a, 1988). Infested roots, showing lesions and rotting, may be split longitudinally and cut to a length of 1–2 cm. When such roots are submerged in water contained in Petri dishes or shallow pans and incubated at 20–25°C, 50% of nematodes are released in 72 h.

Meloidogyne

The root knot nematode, *Meloidogyne* sp., was the first nematode to be recorded on black pepper (Delacroix, 1902) in Cochin, China. In 1906, Butler reported root knot nematodes from black pepper in Wynad, Kerala (India). *M. javanica* and *M. incognita* have been reported from India, Brazil, Sarawak, Borneo, Cochin China, Malaysia, Brunei, Kampuchea, Indonesia, the Philippines, Thailand and Vietnam (Winoto, 1972; Castillo, 1974; Lordello and Silva, 1974; Ichinohe, 1975; Reddy, 1977; Freire and Monteiro, 1978; Kueh and Teo, 1978; Sundararaju *et al.*, 1979a; Ramana and Mohandas, 1983) and *M. arenaria* from Sri Lanka (Lamberti *et al.*, 1983). A new species, *M. piperi*, has been described recently from Kerala, India (Sahoo *et al.*, 2000).

Symptoms of damage

A gradual decline characterized by unthrifty growth and yellowing of leaves are the prominent symptoms. Leaves of vines infested with *Meloidogyne* spp. exhibit dense yellowish discoloration of the interveinal areas, making the leaf veins quite prominent with a deep green colour, whereas leaves of the vines infested with *R. similis* show uniform pale yellow or whitish discoloration and typical drooping (Ramana *et al.*, 1994). Kueh (1990) observed that leaves of root knot nematode-infested vines were held inward and upward and then would drop. *M. incognita* infestation reduced the uptake of nutrients such as P, K, Zn, Mn and Cu (Ferraz *et al.*,

1988). Total chlorophyll content of the leaves was significantly reduced by root knot nematodes, leading to the senescence of leaves (Ferraz and Lordello, 1989). Root systems become heavily galled and the adult females with egg masses are generally enclosed deep within the root tissue (Ramana, 1992; Ramana *et al.*, 1994). In the cv. Panniyur I, the galls are smooth and larger in size compared with the small galls with exposed egg masses, giving a pitted rough appearance to roots of cv. Karimunda.

Other hosts

Among the commercially used standards, *Oroxylum indicum* Vent., *Erythrina lithosperma* Blume, *Ceiba pentandra* (L.) Gaerth. and *Bombax malabaricum* DC. are highly susceptible to root knot nematodes, whereas *Garuga pinnata* Roxb. and *Macaranga indica* Wight are not susceptible. The popular live standards, *Erythrina indica* Lank. and *Gliricidia sepium* (Jacq.) Walp., are less susceptible (Koshy *et al.*, 1977). Large numbers of weeds that are found in pepper gardens have been recorded as hosts of the root knot nematodes (Ramana, 1986).

Disease complexes

Meloidogyne spp. do not significantly enhance the susceptibility of pepper vines to foot rot (Holliday and Mowat, 1963). *M. incognita* and *F. solani* were found associated with black pepper vines in Paraba State, Brazil. Infested plants showed wilting, yellowing of leaves, rotting of stems and roots and cracking of stems; cracked stems 5–10 cm above the soil surface were heavily infected. Joint attack by *R. similis*, *M. incognita* and *Fusarium* sp. caused severe necrosis in the stelar part and resulted in the formation of tyloses that blocked the xylem (Mustika, 1984). Both organisms together were also found to do more harm than either of them alone in other countries (Lopes and Lordello, 1979; Sheela and Venkitesan, 1990; Mustika, 1991, 1992a,b; Zhou and Chi, 1993). Winoto (1972) reported increased susceptibility of *M. incognita*- and *M. javanica*-infested pepper cv. Kuching to *Phytophthora* infection in Malaysia. In India, black pepper plants also showed wilting symptoms quicker when root knot and burrowing nematodes were inoculated first followed by *Phytophthora capsici* (Ramana *et al.*, 1992; Anandaraj *et al.*, 1996a,b). *Rotylenchulus reniformis* was found to inhibit the multiplication of *M. incognita* and the resultant damage on black pepper in autoclaved soil in pots under greenhouse conditions in Brazil (Ferraz and Sharma, 1979). The root gall development and population build-up of *M. incognita* were suppressed in black pepper on inoculation with *R. similis* in succession in sterile soil under pot conditions (Sheela and Venkitesan, 1981).

Economic importance and population damage threshold levels

As much as 91% root knot nematode infestation was reported from Para, Brazil (Ichinohe, 1975) and Kerala, India (Ramana and Mohandas, 1987b; Ramana *et al.*, 1987a). An initial population of ten juveniles per rooted cutting reduces growth by 16%, while a maximum of 50% reduction is observed at an inoculum level of 100,000 over a period of 1 year in sterile soil under potted conditions (Koshy *et al.*, 1979b). *M. incognita* was found highly pathogenic at 100–10,000 juveniles/seedling (Freire and Bridge, 1985c; Mohandas and Ramana, 1991). In Indonesia, yellow symptoms appeared on plants with *Meloidogyne* spp. at population levels of 47 nematodes/100 g of soil and 305 nematodes/10 g of roots (Mustika, 1978).

Management measures

Root knot infestation in black pepper nurseries has been a serious problem in several government nurseries in Kerala, India. Fumigation of nursery potting mixture with methyl bromide is effective in checking the infestation (Koshy, 1974, 1986a; Mohandas and Ramana, 1987a).

CULTURAL. Growing of the non-host cover plant siratro (*Macroptilium atropurpureus*) in the interspace and mulching with Guatemala grass (*Imperata cylindrica*) are recommended to reduce populations of *M. incognita* on black pepper in the Amazonian region (Ichinohe, 1980, 1984). Mulching the basins with *Gliricidia* leaves reduced root knot nematodes in Sri Lanka (Ratnasoma *et al.*, 1991). Application of botanicals such as neem oil cake also can reduce root knot nematodes (Ramana *et al.*, 1992).

RESISTANCE AND TOLERANCE. Among the seven popular cultivars screened, the hybrid cultivar, Panniyur-I was the most susceptible and the cv. Valiakaniakadan was the least susceptible (Koshy and Sundararaju, 1979). The intensity of *M. incognita* damage was less in cultivar Karimunda compared with that of Panniyur-I (Mohandas and Ramana, 1983). Of eight cultivars screened against *M. incognita*, Kalluvalli, Balancotta, Karimunda, Narayakodi and Padapan had fewer galls than Panniyur-I, Cheriyakaniakadan and Kottanadan (Jacob and Kuriyan, 1979a). A total of 101 cultivars, 74 accessions of wild *Piper* sp. and 140 intercultivar hybrids were screened against *M. incognita*, of which one cultivar, CLT-P-812, was found resistant (Ramana and Mohandas, 1986, 1987b; Koshy, 1987b). This cultivar was released as 'Pournami' for cultivation in root knot-infested areas (Ravindran *et al.*, 1992). Some of the wild related species of *Piper* are resistant to root knot nematodes (Ramana, 1992; Paulus *et al.*, 1993).

Infection by nematodes is known to cause biochemical changes in plants (Eapen *et al.*, 1999a). The cv. Cingapura recorded high concentrations of total phenols on inoculation with 6000 *M. incognita* juveniles/pot 95 days after planting, although no resistance was shown (Ferraz *et al.*, 1984). Changes in levels of amino acids, organic acids and sugars in *M. incognita*-inoculated plants compared with uninfected plants were reported by Freire and Bridge (1985b).

CHEMICAL. Most nematicides have been found effective in reducing root knot nematode populations on black pepper, but information on their practical use is limited. Under conditions where a live standard is used, the dosage has to be different depending upon the susceptible/resistant reaction of the standard to the root knot populations. Thus, generalizations on the dosage of nematicides are not possible, and recommendations have to be location specific depending upon the standard, variety of black pepper, rainfall pattern, flowering and harvesting period of black pepper. Green berry yields can be doubled by four applications of carbofuran incorporated into mound soil at 114 g/vine per application in black pepper fields infested with *M. incognita* and *M. javanica* in Malaysia (Kueh and Teo, 1978). Application of Temik 10G at 12.5 g/plant or Furadan 5 G at 50 g/plant twice a year, including at planting around cuttings, reduced populations of *M. incognita* on black pepper in the Amazonian region (Ichinohe, 1980, 1984). Phenamiphos at 1% a.i./vine followed by carbofuran and ethoprophos was effective in controlling nematodes in Malaysia (Leong, 1986) and in Sri Lanka (Ratnasoma *et al.*, 1991).

When aldicarb at 1 g a.i./vine applied twice a year (May/June and October/November) is integrated with fertilizers (N = 100 g, P = 40 g, K = 140 g/vine) in two equal split doses, plus earthing up to 50 cm radius at the base of the vines and mulching the vine base with leaves, there is a reduction in foliar yellowing of 83% and of *M. incognita* juvenile populations by 33–88% (Venkitesan and Jacob, 1985).

BIOLOGICAL. Nematode-free cuttings could be raised by incorporating a biological control agent in the potting mixture. A number of organisms have been tested and found effective in reducing root knot nematodes. Promising among these are *Paecilomyces lilacinus* (Freire and Bridge, 1985d; Ramana, 1994; Sosamma and Koshy, 1997), *Pochonia chlamydosporium* (syn. *Verticillium chlamydosporium*) (Freire and Bridge, 1985d; Sreeja *et al.*, 1996),

Pasteuria penetrans (Ratnasoma *et al.*, 1991; Sosamma and Koshy, 1997), *Bacillus* spp. (Sheela *et al.*, 1993) and *Pseudomonas fluorescens* (Eapen *et al.*, 1997). A number of rhizobacteria that are antagonistic to root knot nematodes have been isolated recently (Beena *et al.*, 2001). Black pepper plants pre-inoculated with arbuscular mycorrhizal fungi such as *Glomus fasciculatum*, *G. etunicatum*, *G. mossae* and *Gigaspora margarita* recorded a significant increase in growth even in the presence of root knot nematodes (Sivaprasad *et al.*, 1990, 1992; Anandaraj *et al.*, 1991).

Other nematodes of black pepper

The other nematodes that have been found associated with black pepper (Table 21.1) in various countries are considered to be of minor economic importance (Timm, 1965; Sher *et al.*, 1969; Castillo, 1974; Sharma and Loof, 1974; Ichinohe, 1975; Reddy, 1977; Bridge, 1978; Sundararaju *et al.*, 1979b; Dasgupta and Rama, 1987; Rama, 1987; Ramana and Mohandas, 1987a). *Trophotylenchulus piperis* has been reported as a widespread parasite of black pepper roots in South India (Mohandas and Ramana, 1982; Mohandas *et al.*, 1985; Ramana and Mohandas, 1987a, 1989; Sundararaju *et al.*, 1997). *T. piperis* completed its life cycle on black pepper roots within 55 days at a room temperature of 24–32°C (Sundararaju *et al.*, 1995). Feeding of this nematode on black pepper roots caused drying and shrinkage of cells in the vicinity of infection (Ramana and Eapen, 1997).

Future prospects

Incorporation of crop rotation systems designed to reduce root knot densities in soil, avoiding susceptible live supports or standards and using resistant cultivars where present, in an integrated nematode management system with minimum or no nematicide application, should be the main thrust of research to increase black pepper yield in areas infested with damaging nematodes.

Cardamom

Cardamom is a fruit (capsule) of the plant *Elettaria cardamomum* Maton, belonging to the family Zingiberaceae. It is a perennial plant having an underground stem (rhizome) with aerial shoots. A mature cardamom plant may measure about 2–4 m in height. Flowers are borne on panicles which emerge directly from the swollen base of the aerial shoot. The fruits are small, trilocular capsules containing 15–20 seeds. Cardamom, known as the 'queen of spices', has its origin in the evergreen rainforests of South India and is basically a shade-loving plant. India and Guatemala are the main producers and exporters of cardamom. Tanzania, Sri Lanka, El Salvador, Vietnam, Laos, Kampuchea and Papua New Guinea are also cardamom growers. The area under cardamom cultivation in India during 1999–2000 was 62,700 ha and the total production was 7800 t (Selvan, 2002). Cardamom is used for flavouring various food preparations, confectionery, beverages, liquors and medicines. Cardamom can be propagated through seedlings as well as suckers. Suckers are better suited for gap filling and multiplication of selected high yielding types.

Nematodes of Cardamom

Nematological investigations on this crop have been undertaken in India, where a number of plant parasitic nematodes have been found (Table 21.1). The most important nematode problem is caused by the root knot nematodes, *Meloidogyne* spp., although the lesion nematode *Pratylenchus coffeae* and the burrowing nematode *R. similis* are also known to cause root rotting (D'Souza *et al.*, 1970; Kumar *et al.*, 1971; Khan and Nanjappa, 1972; Viswanathan *et al.*, 1974; Sundararaju *et al.*, 1979b). Reniform nematode, *R. reniformis*, was also recorded on cardamom (Eapen, 1995a).

Meloidogyne

Widespread occurrence of root knot nematodes *M. incognita* and *M. javanica* has been reported in cardamom nurseries and plantations in India (Kumar *et al.*, 1971; Koshy *et al.*, 1976; Ali and Koshy, 1982a; Ali, 1985, 1986a; Raut and Pande, 1986).

Symptoms of damage

Heavy root knot nematode infestation in mature plants in a plantation causes stunting, reduced tillering, yellowing, premature drying of leaf tips and margins, narrowing of leaf blades, delay in flowering, immature fruit drop and reduction in yield. Unlike several other plant species, galling of roots is not a conspicuous symptom on mature plants. The infested roots, however, exhibit a 'witch's broom' type of excessive branching (Fig. 21.3).

In the primary nurseries, more than 50% of the germinating seeds do not emerge as a consequence of infection of the radicle and plumule by the second stage juveniles of the root knot nematode. The infested seedlings at the two-leaf stage show marginal yellowing and drying of leaves and severe galling of roots. On transplantation to a secondary nursery, they exhibit curling of the unopened leaves. These leaves mostly emerge after the breaking open of the pseudostem. Up to 40% of such seedlings do not establish in the secondary nursery. In secondary nurseries, the infested plants are stunted and yellowed with poor tillering, drying of leaf tips and margins, and heavy galling of root (Ali and Koshy, 1982a; Eapen, 1995b). Young seedlings are more susceptible to root knot nematode attack than mature plants, and galling is more prominent in seedlings (Eapen, 1992). Patches of stunted and weak plants with narrow leaves are a common symptom of nematode infestation in cardamom plantations (Eapen, 1994, 1995b).

Survival and means of dissemination

The heavily shaded, hot, humid atmosphere and continuous availability of soil moisture prevalent in cardamom plantations are congenial conditions for the multiplication of root knot nematodes. Root knot nematode population dynamics in cardamom plantations are influenced by rainfall, soil moisture, soil temperature and

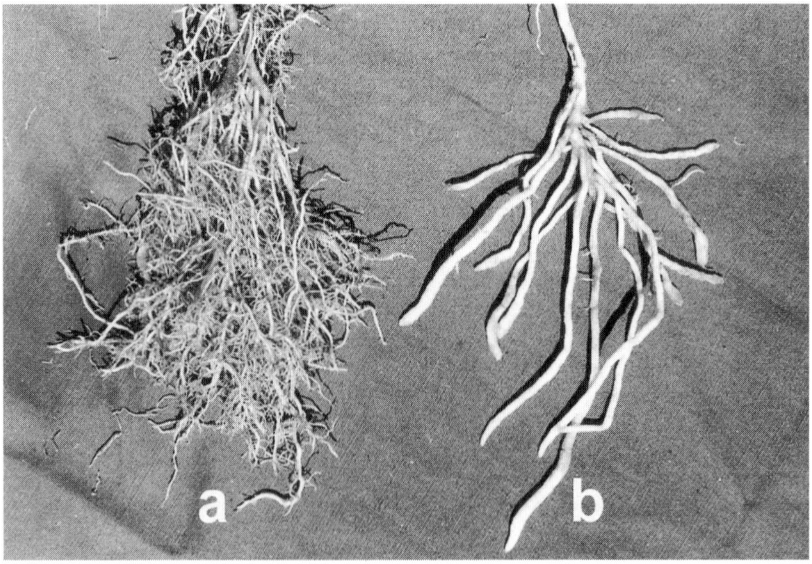

Fig. 21.3. Excessive root growth on cardamom infested with *Meloidogyne* sp. (a) compared with healthy root (b). (Photo: V.K. Sosamma.)

crop phenology. As a result, the root knot nematode population is generally high during the post-monsoon period between November and January (Eapen, 1993). The nematodes are disseminated through infested seedlings and rhizomes used for propagation. Most plantations have their own permanent nursery sites situated in areas having easy access to water sources such as forest streams.

Other hosts

A large number of annual weeds present in the cardamom plantations and the common shade trees, *Erythrina indica* and *E. lithosperma*, are hosts of root knot and help in the build-up of nematode populations (Muniappan, 1993).

Disease complexes

The incidence of rhizome rot and damping-off diseases caused by the fungus *Rhizoctonia solani* increases in the presence of *M. incognita* in the nurseries (Ali, 1986b; Eapen, 1987; Ali and Venugopal, 1992, 1993). The root knot nematode population was found to be 5–10 times higher in virus disease-affected cardamom plants than in healthy plants (Ali, 1989).

Economic importance

A yield loss of 32–47% due to root knot has been reported from the results of a nematicide experiment (Ali, 1985, 1986b). Microplot studies under simulated field conditions showed 46.6% yield loss at an initial inoculum level of 4 nematodes/ 100 cm^3 of soil (Eapen, 1987, 1994).

Management measures

Nematological investigations have helped in creating a general awareness among the planters as well as administrators that the root knot nematode is a major limiting factor. However, planters have not yet adopted recommended control measures. No resistance to root knot nematodes has been found, and the popular cardamom cvs Malabar, Mysore and Vazhuka are all susceptible (Hegde *et al.*, 1993; Eapen, 1995b).

It is advisable to change nursery sites every year, but this is not always practicable in view of the difficulties involved in obtaining suitable sites having facilities for irrigation. Hence, disinfestation of the nursery beds needs to be carried out every year. Disinfestation of nursery beds with fumigant nematicides is effective in controlling root knot infestation in both primary and secondary nurseries (Ali and Koshy, 1982b).

It has been demonstrated that application of aldicarb at 5 kg a.i./ha three times, every 3 months, results in increased growth and vigour of seedlings in both primary and secondary nurseries (Koshy *et al.*, 1979a; Jacob and Chandrasekharan, 1984; Ali, 1986b, 1987). Drenching of nursery beds with fenamiphos also significantly reduced root knot nematodes (Ali, 1986c). Aldicarb, carbofuran and phorate at 5, 10 or 15 kg a.i./ha, respectively, have been applied in primary nurseries of cardamom for control of *M. incognita*. None of the nematicide treatments totally prevented nematode infestation, but there was significant reduction in root knot densities. Aldicarb at the very high level of 15 kg a.i./ha reduced nematode numbers by 90% (Ali, 1987). Application of aldicarb/carbofuran/phorate at 5 and 10 g a.i./plant and neem oil cake at 500 and 1000 g/plant twice a year increases yield of cardamom plants infested with *M. incognita* from 47 to 88%. Maximum yield was obtained from the plants receiving neem oil cake at a rate of 1000 g/plant followed by 500 g/plant (Ali, 1984). However, in another study, application of phorate at 2.5–5.0 g a.i./plant reduced the nematode population and increased the yield by more than 40% (Eapen, 1995b). Since these nematicides do not kill but only inactivate nematodes, repeated use is necessary to ensure good yield.

Cardamom nurseries are ideal for practising biological control. There are reports that *Gigaspora margarita* and *Glomus fasciculatum* reduced *M. incognita* infestation and enhanced growth and vigour of seedlings (Thomas *et al.*, 1989). *P. lilacinus* reduced root knot nematodes by 48.5–57% in pot culture studies and by 19.7% in field studies (Eapen, 1995b; Eapen and Venugopal, 1995). Some native isolates of *Trichoderma harzianum* and other *Trichoderma* spp. are potential antagonists of root knot nematodes. Reduction of root knot nematode infection by this fungus has been clearly shown in laboratory, greenhouse and also in cardamom nurseries (Eapen *et al.*, 2000a,b).

Ginger

Ginger is the rhizome or underground stem of *Zingiber officinale* Rosc., a herbaceous perennial belonging to the family Zingiberaceae. Although the country of origin is not known with certainty, it is presumed to be either India or China. It is grown in many countries of the tropics and subtropics and is used widely in food, beverages, confectionery and medicines. India is the largest producer of dry ginger in the world, contributing about 30% of the world's production. In India, the total area under cultivation during 1999–2000 was 77,610 ha and the total production was 263,170 t (Selvan, 2002). The other ginger-producing countries are Jamaica, Sierra Leone, Nigeria, southern China, Japan, Taiwan and Australia.

Ginger is propagated by seed rhizomes or setts. Seed rhizomes are cut into small pieces of 2.5–5 cm length, weighing 20–25 g each, having one or two good buds. It is grown either as a monocrop or as an intercrop in many farming systems. In India, mulching of ginger beds with green leaves is a traditional practice to enhance the germination of seed rhizomes and conservation of soil moisture. The first mulching is done at the time of planting itself, with green leaves at 10–12 t/ha and repeated with 5 t/ha, 40–90 days after planting, immediately after weeding and application of fertilizers.

Nematodes of Ginger

Plant parasitic nematodes belonging to 17 genera were reported on ginger (Colbran, 1958; Reddy, 1977; Sundararaju *et al.*, 1979b; Rama and Dasgupta, 1985; Kaur, 1987; Ramana and Eapen, 1998); the most important parasites are *Meloidogyne* spp., *R. similis* and *P. coffeae*. In Kerala, *M. incognita* and *R. similis* were the major nematode species found in the rhizosphere of ginger (Mammen, 1973; Charles, 1978; Sheela *et al.*, 1995). *R. reniformis* and *M. incognita* were the dominant plant parasitic nematodes associated with ginger in Orissa (Routaray *et al.*, 1987b). The most prominent nematode pests of ginger in Sikkim (Srivastava *et al.*, 1998) and Himachal Pradesh (Kaur *et al.*, 1989; Khan and Makhnotra, 1998) were *M. incognita* and *P. coffeae*, while in Madhya Pradesh *M. incognita* was the predominant nematode species (Vadhera *et al.*, 1998a). *M. arenaria* was also reported from Himachal Pradesh (Kaur and Sharma, 1988). In west Bengal, *R. reniformis*, *Hoplolaimus indicus* and *P. coffeae* recorded the highest relative density in ginger rhizosphere (Rama and Dasgupta, 1998, 2000).

Meloidogyne

Nagakura (1930) in Japan was the first to report *Meloidogyne* sp. on ginger, and subsequently the species *M. arenaria*, *M. hapla*, *M. incognita* and *M. javanica* have been reported as parasites of ginger in various countries.

Symptoms of damage

The root knot nematodes cause galling and rotting of roots and underground rhizomes. The second stage juveniles of *M. incognita* invade the rhizome through the axils of leaf sheaths in the shoot apex. In fibrous roots, penetration occurs in the area of differentiation and, in fleshy roots, the entire length of root is invaded. In both fleshy and fibrous roots, the nematode develops to maturity in 21 days, but in rhizomes it requires 40 days at 30°C (Cheng and Tu, 1979). Galls are formed on the fibrous roots. Abnormal xylem and hyperplastic parenchyma are observed in all infested tissue except rhizome meristems. Extensive internal lesions are formed in the fleshy roots and rhizomes. Wound cork around the lesions is suberized only in old rhizomes after harvest (Huang, 1966; Shah and Raju, 1977). Infested rhizomes have brown, water-soaked areas in the outer tissues, particularly in the angles between shoots. Nematodes continue to develop after the crop has matured and been harvested, and induce breakdown of the seed rhizomes. Heavily infested plants are stunted, poorly tillered and have chlorotic leaves with marginal necrosis. The affected ginger plants mature, dry faster and die sooner than healthy ones, leaving a poor crop stand at harvest. Infested rhizomes serve as a source of infection and means of dissemination.

Disease complexes

Incidence of rhizome rot of ginger caused by *Pythium aphanidermatum* is reported to be severe when rhizomes are infested with nematodes such as *M. incognita* and *P. coffeae* (Dohroo *et al.*, 1987). However, Doshi and Mathur (1987) could not observe any interaction with these two organisms. Similarly, there was also no interaction between *M. incognita* and *Pythium myriotylu* (Lanjewar and Shukla, 1985). Recent studies have shown that ginger plants inoculated with root knot nematodes developed disease symptoms earlier when inoculated with *P. aphanidermatum* (Ramana *et al.*, 1998). Bacterial wilt of ginger caused by *Ralstonia solanacearum* was also shown to be influenced by *M. incognita* (Samuel and Mathew, 1983); however, there are contradictory reports on the subject (Ramana *et al.*, 1998).

Other hosts

Most of the weeds that are present in ginger-growing areas are known hosts of root knot nematodes.

Economic importance and population damage threshold levels

In Queensland, Australia severe infestation of rhizomes reduces yields by 57% as determined by fumigation (Pegg *et al.*, 1974). Treatment of infested soil with DD before planting nematode-free seed rhizomes has increased yields by 80%. *M. incognita* is widely distributed in ginger fields in India and causes a loss of 46.4% (Charles, 1978). A reduction of 74% rhizome weight has been recorded with an initial inoculum level of 10,000 nematodes/plant over a period of 6 months under potted conditions (Sudha and Sundararaju, 1986).

Both *M. incognita* and *M. hapla* cause significant reduction in shoot length and shoot and root weight following inoculation with root knot nematodes. The economic threshold level of this nematode varied from 2 nematodes/g of soil to 50 larvae/100 ml of soil (Parihar and Yadav, 1986; Sudha and Sundararaju, 1986; Kaur, 1987; Routaray *et al.*, 1987a). At higher initial inoculum levels, *M. incognita* and *M. hapla* cause partial or complete withering of aerial shoots. Typical symptoms of drying and twisting of leaves were observed with *M. arenaria* (Kaur, 1987).

Significant damage is noticeable at 0.5 and 1.25 nematodes/g of soil and above in sterilized soil under potted conditions. The fibrous roots are very much reduced at 2 nematodes/g of soil (Parihar, 1985; Routaray *et al.*, 1987a). Ginger treated with Carbofuran at 1 kg a.i./ha showed an increase of 20% in yield (Makhnotra and Luqman, 1997b). In another study, an avoidable yield loss of 43% was observed at an initial population level of 166 *M. incognita* juveniles/250 g of soil (Sheela *et al.*, 1995).

Management measures

Being an export-oriented crop, the nematodes of ginger have to be managed in an ecofriendly manner. Besides, as ginger is consumed raw, nematicides should be used with extreme care. A careful blend of the following measures may provide adequate management of the nematode problems in this crop.

PRODUCTION OF NEMATODE-FREE PLANTING MATERIAL. Since the seed rhizome generally harbours nematodes, selection of seed rhizomes is very critical for the management of nematodes. Nematode-free planting material should be selected from fields of known history. The control schedule for *M. javanica* involving the use of clean seed and a ginger–taro–fallow rotation has been recommended in Fiji (Haynes *et al.*, 1973). *In vitro* ginger plantlets are used to solve root knot nematode problems in South Africa. Hot water treatment of seed rhizomes at 50–55°C for 10 min was found to reduce the nematode incidence in ginger (Colbran and Davis, 1969; Anonymous, 1971). Disinfestation of rhizomes was also achieved by hot water treatment at 45°C for 3 h (Vadhera *et al.*, 1998a,b).

ORGANIC AMENDMENTS. Mulching or applying well decomposed cattle or poultry manure, compost or neem oil cake reduced nematode build-up (Colbran, 1974; Kaur, 1987; Stirling, 1989; Mohanty *et al.*, 1992; Dohroo *et al.*, 1994; Vadhera *et al.*, 1998b). Growing under sawdust mulch reduced root knot nematode infestation in Australia (Pegg *et al.*, 1974). Pre-plant application of neem cake at 1 t/ha reduced *M. incognita* and increased the yield (Mohanty *et al.*, 1995). Ginger plots mulched with mahaneem leaves at 2.5 kg/m^2 reduced root knot (Das, 1999). Studies in Australia have suggested that root knot on ginger can be controlled by alternating ginger with a green manure crop and applying at least 150 m^3/ha/year of poultry manure (Stirling, 1989; Stirling and Nikulin, 1998). Intercropping bell pepper with ginger significantly reduced both *P. penetrans* and *M. incognita* and improved the yield of ginger (Sharma and Bajaj, 1998). Incorporation of organic materials fortified with biocontrol agents such as *Trichoderma* spp., *P. lilacinus*, *P. chlamydosporia*, etc. is another option to prevent the nematode build-up (Eapen and Ramana, 1996).

HOST RESISTANCE. There are very few reports on resistance in ginger to root knot. In a preliminary evaluation, a few lines of ginger (Accession Nos 36, 59 and 221) were found resistant to *M. incognita* (Eapen *et al.*, 1999b). One of these has been recommended for release as 'IISR Mahima' (Sasikumar *et al.*, 2003).

CHEMICAL CONTROL. Soil fumigation or application of granular pesticides such as fenamiphos or dip treatment with fenamiphos are all recommended for control of nematodes of ginger. The efficacy of several granular nematicides was assessed in Queensland against root knot nematodes (Colbran, 1972; Willers, 1985). Nemacur was found to be the most effective, increasing rhizome yield by up to 15%. Split and late applications at 22.4 kg/ha are more promising than higher doses applied early in the season (Colbran, 1972). A high level of control of root knot nematodes has been obtained with sawdust mulching at a depth of 5–7.5 cm, combined with post-plant application of Nemacur. Application of phenamiphos at 3 kg a.i./ha has resulted in a 70–144% increase in yield of ginger in fields infested with *M. incognita* and *P. coffeae* either singly or in combination (Kaur, 1987).

Dipping ginger rhizomes in fenamiphos at 0.26 and 0.1% a.i. for 30 and 60 min, respectively, controlled root knot nematodes and increased the yield (Willers, 1991). Application of carbofuran at 1 kg a.i./ha 45 days after planting coupled with pre-planting application of neem cake reduced *M. incognita* and increased the yield of ginger (Mohanty *et al.*, 1995).

BIOLOGICAL CONTROL. A large number of bacterial and fungal isolates of biocontrol agents were isolated from ginger fields through random surveys (Ramana *et al.*, 2002). Many of the fungal isolates parasitized root knot nematode egg masses and suppressed their egg hatching. Toxic metabolites of some of them caused mortality of second stage juveniles in addition to direct parasitization. These studies indicated that five biocontrol agents, namely *P. chlamydosporia*, *P. lilacinus*, *Fusarium* sp., *Aspergillus nidulans* and *Scopuloriopsis* sp., reduced root knot nematode populations significantly. Although none of these organisms is registered presently for use on ginger, they are potential tools for nematode management that may become available in the near future.

Radopholus similis

Parasitism of ginger by the burrowing nematode, *R. similis*, was first reported by Hart (1956) in Florida, USA. Later, Butler and Vilsoni (1975) reported heavy infestation of ginger by *R. similis* in Fiji and its further spread through infested seed rhizomes. Occurrence of *R. similis* along with *M. incognita*, *Pratylenchus* sp. and *Helicotylenchus* sp. has also been reported from roots of ginger in India (Charles, 1978; Charles and Kurian, 1982).

Symptoms of damage

Infected plants exhibit stunting, reduced vigour and tillering. The topmost leaves become chlorotic with scorched tips. Affected plants tend to mature and dry out faster than healthy plants. Incipient infections of the rhizomes are evidenced by small, shallow, sunken, water-soaked lesions (Plate 22C) (Vilsoni *et al.*, 1976; Sundararaju *et al.*, 1979a). The nematodes migrate intracellularly through tissues, producing large infection channels or galleries within the rhizomes.

Means of dissemination

R. similis infestation in Fiji of ginger fields appears to have originated through bananas as the areas once used for banana cultivation have been used for growing ginger (Vilsoni *et al.*, 1976). The coconut isolate of *R. similis* in Kerala (India) also reproduces well on ginger (Koshy and Sosamma, 1975, 1977). The perpetuation and dissemination of the nematode is through infested seed rhizomes used for planting.

Economic importance and population damage threshold levels

In Fiji, *R. similis* has been reported from more than 50% of the total area, with a rate of infection ranging from 10 to 50% resulting in yield reductions of about 40%. An initial inoculum level of 10,000 nematodes/plant has been reported to cause 74% reduction in rhizome weight, and an initial inoculum level of ten nematodes per plant reduced shoot weight, root weight and rhizome weight by 43, 56 and 40%, respectively, in a pot experiment (Sundararaju *et al.*, 1979c).

Management measures

Few studies have been done on the control of *R. similis* on ginger, but the measures suggested for control of root knot nematodes, including hot water treatment, could help in reducing the loss.

Pratylenchus coffeae

Several species of *Pratylenchus*, namely *P. brachyurus*, *P. coffeae*, *P. indicus*, *P. pratensis* and *P. zeae*, are reported on ginger (Charles, 1978; Das and Das, 1986; Kaur *et al.*, 1989; Kaur and Sharma, 1990).

Economic importance and symptoms

P. coffeae is reported to cause 'ginger yellows' disease, prevalent in Himachal Pradesh, India (Kaur and Sharma, 1990). The nematode is highly pathogenic to 15-day-old ginger seedlings even at very low initial inoculum levels (Kaur, 1987). Nematode infestation caused yellowing of leaves and dry rot symptoms on rhizomes. Dark, brown necrotic lesions were observed within the infected rhizomes (Kaur and Sharma, 1990).

Turmeric

Turmeric, *Curcuma domestica* Val., is best known as a condiment, although the plant has uses in the social and religious lives of people in South-east Asia, its probable origin. Commercial turmeric is the processed rhizome of *C. domestica*. It is grown mostly in India, and to a small extent in China, Indonesia, Peru and Jamaica. In India, the total area under cultivation during 1999–2000 was 161,300 ha with a production figure of 653,600 t (Selvan, 2002). It is cultivated as either a monocrop or an intercrop in many farming systems.

It is indispensable in the preparation of curry powder, and is an important source of natural yellow dye. It is also used as a colouring additive in the drug, confectionery and food industries. The rhizomes of *C. aromatica* Salisb., a close relative of *C. longa*, are also a source of turmeric.

Nematodes on Turmeric

A number of species of plant parasitic nematodes have been reported in association with turmeric (Nirula and Kumar, 1963; Sundararaju *et al.*, 1979b; Chen *et al.*, 1986; Dasgupta and Rama, 1987; Gunasekharan *et al.*, 1987; Rama, 1987; Routaray *et al.*, 1987b; Bai *et al.*, 1995) of which *Meloidogyne* spp., *R. similis* and *P. coffeae* are of economic importance. *R. reniformis* and *M. incognita* were the most predominant and frequently recorded nematode species in the Chittor and Cuddapah districts of Andhra Pradesh (Mani and Prakash, 1992) and in Bihar (Haider *et al.*, 1995) in India. *R. reniformis* was reported to be more harmful to turmeric than *M. incognita*, and caused a significantly higher reduction in plant growth (Haider *et al.*, 1998a).

Meloidogyne

Two species of root knot nematodes, *M. incognita* and *M. javanica*, have been reported on turmeric, but most investigations have been concerned with *M. incognita*. Turmeric plants infested with *M. incognita* have large root galls (Fig. 21.4), stunted growth, yellowing, marginal and tip drying of leaves, and reduced tillering with galling and rotting of roots. In the field, high densities of *M. incognita* cause yellowing and severe stunting and wilting in large patches. Plants die prematurely, leaving a poor crop stand at harvest. Infested rhizomes tend to lose their bright yellow colour (Mani *et al.*, 1987). Levels of protein, carbohydrate, chlorophyll a and b, and curcumin were lower in plants infested with *M. incognita* (Poornima and Sivagami, 1998a).

The highest nematode multiplication and gall index were seen in peat soils (Poornima and Sivagami, 1998b). The population density of *M. incognita* increased with crop age and decreased with crop senescence (Poornima and Sivagami, 1999).

Economic importance and population damage threshold levels

One hundred juveniles of *M. incognita* caused significant reduction in growth characters of turmeric (Haidar *et al.*, 1998a). Significant reduction in growth and yield of turmeric were noticed in plants inoculated with more than 1000 root knot nematode juveniles/plant (Sudha *et al.*, 1989). When four varieties of turmeric were tested against *M. incognita*, maximum reduction of 18% fresh rhizome weight was observed in Suvarna at 2 juveniles/g of soil (National Research Centre for Spices, 1993). Poornima and Sivagami (1998a) reported that an initial inoculum level of more than 5000 *M. incognita* larvae/plant was highly pathogenic to turmeric. By applying carbofuran at 3 kg a.i./ha, 3 weeks after planting, avoidable yield losses to the extent of 33.61 and 26.30% were observed in turmeric and ginger, respectively (Ray *et al.*, 1995). Avoidable yield loss under field conditions was 45.3% due to *M. incognita* but was

Fig. 21.4. Root galling on turmeric infested with *Meloidogyne* sp. (Photo: V.K. Sosamma.)

only 33.3% in a mixed infestation of *M. incognita* and *R. reniformis* (Bai *et al.*, 1995).

Management measures

RESISTANCE AND TOLERANCE. The cultivars and breeding lines 5379-1-2, 5363-6-3, Kodur, Cheyapuspa 5335-1-7, 5335-27, Ca-17/1, Cli-124/6, Cli-339, Armoor, Duggirala, Guntur-1, Guntur-9, Rajampet, Sugandham and Appalapadu have been reported as resistant to *M. incognita* (Gunasekharan *et al.*, 1987; Mani *et al.*, 1987). The species *C. zedoaria* is more resistant to *M. incognita* than *C. domestica* in China (Chen *et al.*, 1986). In Andhra Pradesh, India, the high yielding varieties such as PCT8, PCT10, Suguna and Sudarshana were free from root knot nematode infestation (Rao *et al.*, 1994). Recently, eight turmeric accessions (Accession Nos 31, 82, 84, 142, 178, 182, 198 and 200) were identified as resistant to root knot nematode (Eapen *et al.*, 1999b).

PHYSICAL. Immersing turmeric rhizomes in hot water at 55°C for 10 min or 45°C for 50 min can kill *M. incognita* inside rhizomes (Chen *et al.*, 1986), and this could be used for establishing nematode-free multiplication plots but is unlikely to be economic for large-scale field use.

CHEMICAL. Application of dibromochloropropane (DBCP; now banned in many countries) at 15 l a.i./ha 15 days prior to planting results in a yield increase of 253–270% compared with a 59–187% increase in yield with application of phenamiphos at 2.5 kg a.i./ha 1 day before planting (Patel *et al.*, 1982). Aldicarb and carbofuran applied at 1 kg a.i./ha increased the yield by 71 and 68%, respectively over control, with a cost:benefit ratio of 1:6 in aldicarb and 1:2 in carbofuran treatments (Gunasekharan *et al.*, 1987). Carbofuran at 4 kg a.i./ha applied in rows to a 4-month-old turmeric crop has resulted in a 81.6% reduction in root knot nematode population as against a 45% increase in untreated plots (Mani *et al.*, 1987). Similarly, application of carbofuran or phorate at 1 kg a.i./ha reduced root knot nematodes (Haidar *et al.*, 1998b).

BIOLOGICAL. The biocontrol agents *Pochonia chlamydosporia* (syn. *Verticillium chlamydosporium*), *Paecilomyces lilacinus*, *Fusarium* sp., *Aspergillus* sp. and *Scopuloriopsis* sp. controlled root knot nematodes in field trials but have not been tested in growers fields (Ramana *et al.*, 2002).

Radopholus similis

Symptoms of damage

Roots of turmeric damaged by *R. similis* become rotted, and most of these decayed roots retain only the epidermis devoid of cortex and stelar portions. The infested plants show a tendency to age and dry faster than healthy plants. Infested rhizomes are of a yolk yellow colour compared with the golden yellow colour of healthy rhizomes and have shallow water-soaked brownish areas on the surface. The scale leaves harbour *R. similis* (Sosamma *et al.*, 1979).

Survival and means of dissemination

The nematodes are disseminated through infested planting material. Populations of *R. similis* from coconut are known to infest turmeric (Koshy and Sosamma, 1975), and the use of turmeric as an intercrop in *R. similis*-infested coconut- and arecanut-based farming systems should be avoided.

Economic importance and population damage threshold levels

Pathogenicity studies show that an initial inoculum level of ten nematodes per plant can cause a reduction of 35% of the rhizome weight after 4 months and a 46% reduction at the end of the season (8 months). With 100,000 nematodes, the extent of reduction in rhizome weight is 65 and 76% after 4 and 8 months, respectively (Sosamma *et al.*, 1979).

Management measures

Control has not been studied under field conditions. However, use of clean, nematode-free rhizomes for planting should be the first step in developing an integrated management system for the burrowing nematode on turmeric.

Pratylenchus coffeae

P. coffeae has been reported to be associated with discoloration (Plate 22D) and rotting of mature rhizomes of 'wild turmeric', *C. aromatica*. In advanced stages of infection, the rhizomes become deep red to dark brown in colour, less turgid and wrinkled with dry rot symptoms. The fingers are more severely affected than the mother rhizomes. Internally, the affected rhizomes show dark brown necrotic lesions (Sarma *et al.*, 1974).

Future prospects

Turmeric has received very little input in terms of nematological research, although *M. incognita*, *M. javanica*, *R. similis* and *P. coffeae* are known to damage the crop. Detailed investigations including surveys, pathogenicity experiments and control through resistant/tolerant cultivars, cultural, chemical and biological methods are warranted.

Other Spices

Although a number of spice crops including tree spices and seed spices are cultivated over large areas in the tropics and subtropics, there is very little information available on the damage and yield loss caused by plant parasitic nematodes on some of these crops. This is not to say that nematode problems do not exist on these crops but only that there has been a lack of nematological investigations. The plant parasitic nematodes that have been reported in association with these crops in surveys and host range studies are given in Table 21.1. Nematodes have been found associated in clove (Ghesquiere, 1921; Goodey *et al.*, 1965; Sharma and Loof, 1974; Bridge, 1978; Sundararaju *et al.*, 1979b), nutmeg (Goffart, 1953; Goodey *et al.*, 1965; Kumar *et al.*, 1971; Sundararaju *et al.*, 1979b; Chawla and Samathanam, 1980), cinnamon (Goffart, 1953; Goodey *et al.*, 1965; Sundararaju *et al.*, 1979b; Chawla and Samathanam, 1980; Dasgupta and Rama, 1987; Rama, 1987), cumin (Swarup *et al.*, 1967; Verma and Prasad, 1969; Shah and Raju, 1977; Shah and Patel, 1979; Patel *et al.*, 1986; Midha and Trivedi, 1991), fennel (Midha and Trivedi, 1991), fenugreek (Chandwani and Reddy, 1967; Krishnamurthy and Elias, 1967; Khan and Khan, 1969, 1973; Mathur *et al.*, 1969; Rashid *et al.*, 1973; Khan, 1975), coriander (Chandwani and Reddy, 1967; Krishnamurthy and Elias, 1967; Sen and Dasgupta, 1977; Das and Sultana, 1979; Midha and Trivedi, 1991) and vanilla (Orton Williams, 1980; Stier, 1984 in Bridge, 1988). All these spices are hosts of *Meloidogyne* spp. The roots of cumin also can be severely galled by *M. incognita* and *M. javanica* (Patel *et al.*, 1986). *Pratylenchus brachyurus* is reported to be a parasite of vanilla in the Pacific island of Tonga, causing reduced growth of vines (Stier, 1984 in Bridge, 1988).

Related Crops

Betel Vine

The betel vine *Piper betle* L. is a perennial, dioecious, semi-woody creeper, probably native of Malaysia. Its leaves are used for chewing, extraction of essential oils such as methyl eugenol and in traditional herbal (ayurvedic) medicines and religious ceremonies. It is grown throughout Asia and also in Africa, the Philippines, Indonesia and the Pacific islands. The area under betel vine cultivation in India is about 30,000 ha with an annual turnover of 7000 million Indian rupees. The yield varies from 7.5 to 22.5 million leaves/ha/year (Shenoy, 1985).

Its cultivation is labour intensive and requires heavy investment. Betel vine is propagated by cuttings of 3–5 nodes from 2-year-old vines. It is trailed on coconut, arecanut or other straight-stemmed plants such as *Sesbania grandiflora* Pers., *Moringa oleifera* Lam and *Erythrina variegata* L. Non-living standards such as bamboo, wooden poles or granite stone supports are also used. The crop is usually heavily manured with farmyard manure, oil cakes, fish manure, sheep manure, etc.

Nematodes on Betel Vine

Numerous plant parasitic nematodes have been reported associated with the betel vine in India and elsewhere (Timm, 1965; Reddy, 1978; Ganguly and Khan, 1983; Sivakumar and Marimuthu, 1984, 1985; Jagdale *et al.*, 1986a,b; Acharya *et al.*, 1988; Ganguly, 1988; Nema, 1997). Nematodes known to cause damage to the crop are *M. incognita*, *R. similis* and *R. reniformis*. Betel vine was also reported as a host for *P. coffeae* (Ganguly and Khan, 1990).

Meloidogyne incognita

M. incognita has been reported to be associated with betel vine decline from all areas in India (Dhande and Sulaiman, 1961; Venkata Rao *et al.*, 1973; Mammen, 1974; Sivakumar and Marimuthu, 1984; Jagdale *et al.*, 1986a).

Symptoms of damage

Infested plants exhibit poor growth, yellowing of leaves, reduced vigour and wilting, with heavy galling and rotting of roots (Jagdale *et al.*, 1986a). Thinly spread foliage with small leaves, yellowing and premature shedding of leaves and stunting were recorded in root knot nematode-infested vines (Acharya and Padhi, 1987a).

Disease complexes

Association of *M. incognita* with severe wilt symptoms of betel vine was reported from India (Mammen, 1974). *M. incognita* predisposed betel vine to root rot caused by *Phytophthora palmivora* (Sivakumar *et al.*, 1987; Marimuthu, 1991; Jonathan *et al.*, 1996) and *P. capsici* (Sitaramaiah and Devi, 1994). Pathogenic association of *M. incognita* with *Sclerotium rolfsii* and *Xanthomonas betlicola* was also reported (Acharya *et al.*, 1987; Sitaramaiah and Devi, 1990). A disease complex involving *M. incognita* and *Colletotrichum* sp. was also reported in betel vine (Ray *et al.*, 1993).

Economic importance and population damage threshold levels

The root knot nematode is damaging to betel vine at an initial inoculum level of 100 juveniles/plant in sterile soil in pots (Jagdale *et al.*, 1985a). The leaf yield of untreated plants showed 38% loss over carbofuran-treated plants (Jonathan *et al.*, 1990). Avoidable yield losses under field conditions in Assam were estimated at 17.95% in terms of number of leaves and 29.06% in terms of fresh weight of leaves (Hazarika *et al.*, 1999b).

Management measures

CULTURAL. A crop rotation of betel vine–rice–banana–rice is helpful in reducing *M. incognita*, *Helicotylenchus* sp. and *Rotylenchulus reniformis* populations on betel vine raised in rice fields (Sivakumar and Marimuthu, 1986a; Sivakumar *et al.*, 1987). Considerable reduction in nematode populations in the soil and number of galls on roots has been reported after application of 50–75 kg of K_2O/ha (Jagdale *et al.*, 1985e; Rabindran *et al.*, 1987). Growing *Tagetes erecta* in the basins of betel vines reduced root knot nematodes (Medhane *et al.*, 1985). Nematode-susceptible standards such as *Sesbania grandiflora* and *S. sesban* should not be used for trailing the vines (Rao *et al.*, 1991). In another study, application of decaffeinated tea waste and mustard oil cake at 1 kg/plant reduced nematode populations and returned significantly higher yields (Hazarika *et al.*, 1999a).

Application of neem oil cake at 1 t/ha and sawdust at 2 t/ha can reduce nematode populations and number of galls and increase the number of leaves harvested significantly (Jagdale *et al.*, 1985b,c; Acharya and Padhi, 1988a). Significant reduction (60%) in the nematode population has been observed in beds amended with chopped and shade-dried leaves of *Calotropis gigantea* at 2.5 t/ha followed by neem oil cake and poultry manure at 44.4 and 40.9%, respectively. Beds amended with *C. gigantea* leaves yielded 14.2 kg of 4840 leaves and with neem oil cake 12.1 kg of 4220 leaves. Soil amendment with sawdust at 2 t/ha + NPK and neem oil cake at 2 t/ha was effective in reducing nematode numbers and increasing yields (Sivakumar and Marimuthu, 1986b; Rana *et al.*, 1991; Murthy and Rao, 1992, 1994). In another study, the highest reduction in nematode population (43%) was obtained with the application of neem seed cake at 0.5 t/ha together with carbofuran at 0.75 kg a.i./ha (Nema, 2001a).

RESISTANCE AND TOLERANCE. The cv. Karpoori is highly susceptible, whereas the cv. Kuljedu had the lowest root knot index and number of egg masses per plant (Jagdale *et al.*, 1985d; Sivakumar *et al.*, 1987). The cvs Kakair, Bangla, Karapaku, Gachipan, Aswani pan and Berhampuri are reported to be tolerant to root knot (Anonymous, 1987). The variety Berhampuri was also reported to be less susceptible to this nematode by other workers (Acharya and Padhi, 1988b). Another cv., Bangla Budagar, was moderately resistant to *M. incognita* (Nema, 2001a).

PHYSICAL. Solarization by mulching the land with 100 gauge black and white polythene before planting for 15 days was found to reduce plant parasitic nematode populations in India (Sivakumar and Marimuthu, 1987; Rao *et al.*, 1996).

BIOLOGICAL. The root knot nematode problem in betel vine was controlled through application of the biocontrol fungus *Paecilomyces lilacinus* (Jonathan *et al.*, 1995; Hazarika, *et al.*, 1998; Nakat *et al.*, 1998; Pathak and Saikia, 1999; Hazarika *et al.*, 2000; Jonathan *et al.*, 2000; Bhatt *et al.*, 2002b). Application of *Trichoderma viride* multiplied on linseed oil cake was also found to be highly effective in reducing the root knot incidence in betel vine (Bhatt *et al.*, 2002a).

CHEMICAL. Field application of aldicarb or carbofuran at 1.5 kg/ha reduced root knot nematode populations (Jagdale *et al.*, 1984). In another experiment, application of aldicarb and carbofuran at 0.75 kg a.i./ha reduced nematode populations by 71 and 55%, respectively, resulting in increased yields. The nematicide, at both levels, degraded to non-detectable levels 41 days after application (Sivakumar *et al.*, 1987). Aldicarb, carbofuran and benfurocarb applied at 1.5, 3.0 or 5.0 kg a.i./ha, respectively, in furrows on either side of the rows can significantly reduce *M. incognita* populations in soil and galling of the roots (Dethe and Pawar, 1987). However, the use of systemic nematicides, i.e. aldicarb and carbofuran, is generally not recommended for betel vine as the leaves are picked continuously and consumed directly without any processing. Because of problems with nematicide residues in leaves (Pattnaik, 1989; Rao *et al.*, 1993; Mahapatra and Awasthi, 1994), root knot nematode infestations on betel vine must be solved by integrated nematode management such as those outlined below:

- Crop rotation wherever possible.
- Use of resistant/tolerant cultivars.
- Use of non-living standards or nematode-resistant live standards for supports.
- Solarization by mulching with 100 gauge clear polythene before planting.
- Application of organic amendments such as neem or *Calotropis* leaves and sawdust at 2 t/ha.
- Supply of nitrogen through neem oil cake at 2 t/ha.

Radopholus similis

The burrowing nematode *R. similis* has been reported to cause 'yellows' or 'slow wilt' disease of betel vine in India. The symptoms produced on betel vine are akin to the symptoms caused by *R. similis* on black pepper vines (Koshy and Sosamma, 1975; Sundararaju and Suja, 1986; Eapen *et al.*, 1987). The integrated management schedules suggested for control of nematodes on black pepper, other than application of nematicides, can be largely adopted with modification to suit local conditions for controlling *R. similis* on betel vine. Inoculation of plants with *Paecilomyces lilacinus* 25 days prior to *R. similis* was effective in reducing plant damage (Sosamma *et al.*, 1994).

Rotylenchulus reniformis

Acharya and Padhi (1987b) and Bhatt *et al.* (2002b) found *R. reniformis* to be pathogenic to betel vine. At inoculum levels of 1000 and 20,000 nematodes/cutting, the reduction in number of leaves was 20 and 60%, respectively. Ganguly (1988) reported *R. reniformis* as the dominant species found associated with five varieties of betel vine in Maharashtra. *R. reniformis* interacted synergistically with *Phytophthora palmivora* to increase vine mortality (Jonathan *et al.*, 1997).

Kava

Kava or Yaqona (*Piper methysticum* Forst.) provides a popular narcotic drink for the peoples of the Pacific islands. The drink is made from the thick roots of this bushy shrub.

Nematodes of Kava

Root knot nematodes, *Meloidogyne* spp., have been found associated with a serious disease of kava, and the nematodes alone can greatly decrease growth of plants in Fiji and Tonga (Stier, 1984 in Bridge, 1988) (Plate 22F). Fliege and Sikora (1981) reported *M. incognita* causing severe root galling of *P. methysticum* in Western Samoa.

Other potentially damaging parasitic nematodes that has been found with kava include *R. reniformis*, *P. coffeae* and *R. similis* (Kirby *et al.*, 1980; Orton Williams, 1980). None of these have as yet been shown to cause economic damage to the crop. Further investigations are necessary to determine the economic importance of nematodes, particularly *Meloidogyne* spp., and their means of control.

Medicinal Plants

Plant parasitic nematodes are associated with all medicinal plants studied to date, and often cause significant damage. However, the magnitude of crop damage has only been established for a few of these nematode–plant interactions (Pandey *et al.*, 2003). Three species of plant parasitic nematodes are considered of economic importance on medicinal plants: the root knot nematodes (*M. incognita* and *M. javanica*), the lesion nematode (*Pratylenchus thornei*) and the stunt nematode (*Tylenchorhynchus vulgaris*). Root knot nematodes are the most important nematode parasites limiting production, with infestations reported on menthol mint, henbanes, basil, opium poppy, aswagandha, sarpgandha, coleus, kinghao, brahmi and musli (Pandey, 1998b, 2003) as well as on jaborandi (*Pilocarpus microphyllos*) (R.A. Sikora, Germany, 2004, personal communication).

Henbanes

Henbanes (*Hyoscyamus muticus*, *H. niger* and *H. albus*) are important tropane alkaloid-bearing plants belonging to the family Solanaceae and one of the chief sources of tropane alkaloids (hyoscine, scopolamine, hyoscyamine, atropine, etc.) obtained from the dried leaves and other plant parts.

Meloidogyne

Although many plant parasitic nematodes have been reported associated with different species of henbane, only root knot nematodes cause serious damage to the crop. Henbanes have been reported to be heavily infested with *M. incognita* and *M. javanica* in India (Pandey, 1990).

Symptoms of damage

Root knot-infested plants of *H. muticus*, *H. niger* and *H. albus* show chlorosis and stunting, and the plants have fewer and smaller leaves and flowers. The roots of infested plants are often severely galled. A pre-plant density of 3–4 juveniles/g of soil caused significant damage to the crop (Haseeb and Pandey, 1989; Pandey, 1990).

Management measures

Crops resistant to root knot should be used in rotation with henbanes to reduce pre-plant nematode densities in the soil. No henbane species screened have proved to be resistant to the nematodes (Pandey, 1998b).

The nematicides carbofuran at 2 kg a.i./ha of soil and monocrotophos at 0.1% in solution have been used to reduce root knot nematode damage to henbane. Monocrotophos was used to soak seeds prior to planting and carbofuran was applied to the soil prior to sowing the crop. The combined treatment effectively reduced root knot infestations (Pandey, 2000a).

When *H. niger* was inoculated with the plant health-promoting rhizobacteria *P. fluorescens* or with one of three species of arbuscular mycorrhizal fungi (*G. aggregatum*, *G. mosseae* or *G. feasiculatum*), *M. incognita* densities were reduced and plant biomass increased. The use of a combination of antagonists proved to be the most effective (Pandey, 1997; Pandey *et al.*, 2000b,c).

In pot tests, essential oils of *Cymbopogon martinii*, *C. wintrianus*, *Ocimum basilicum* and *Mentha arvensis* were effective in reducing *M. incognita* populations and improving the growth of *H. niger*; the oil from *C. martinii* at 2 ml/plant was most effective).

Ashwagandha (*Withania somnifera* L.)

This important medicinal plant is a major source of a number of alkaloids (sominiferine, somnine, withanine, tropine, isopelletierine, cuscohygrine, anaferine, anahygrine, visamine, etc.) and of withanollides, a group of naturally occurring oxygenated ergostane-type steroids. The roots of *W. somnifera* are used locally to treat hiccups, coughing, dropsy, rheumatism and as a sedative. It is also useful for treating inflammatory conditions whereby leaves are used as febrifuge and applied to lesions, painful swellings and sore eyes.

Nematodes of Ashwagandha

During a survey to collect new germplasm, almost all *W. somnifera* plants sampled were found to be galled by *M. incognita* race 2 (R. Pandey, India, 2004, personal communication). Infected plants were chlorotic, stunted, less branched with fewer and smaller leaves. Roots of such plants were severely galled. When the stem touches the soil, it was also found to be infested with the nematode (Fig. 21.5, Plate 22E).

Management measures

Amendments from the neem plant (*Azadirachta indica*), marc from *Artemisia annua* as well as distillates from *Mentha* and *Murrya koengii* were found to reduce *M. incognita* densities on *W. somnifera* (Pandey *et al.*, 2003). The combination of Vermicompost with *Trichoderma harzianum* and *Mentha* distillates with *G. aggregatum* were also found to reduce nematode densities and enhanced the growth of *W. somnifera* significantly. The combined use of these organic amendments with antagonistic microorganisms was considered suitable for nematode management programmes, but still need advanced field testing (Pandey *et al.*, 2003).

Fig. 21.5. Large root galls on *Withania somnifera* infested with *Meloidogyne* sp. (Photo: V.K. Sosamma.)

Fig. 21.6. *Bacopa monnieri* plant infested with *Meloidogyne* sp. (Photo: V.K. Sosamma.)

Brahmi (*Bacopa monnieri*)

B. monnieri L., commonly known as brahmi, is the chief source of baccoside A and B, which are used extensively in formulation of medicines useful against asthma and epilepsy.

Nematodes of Brahmi

Although a number of plant parasitic nematodes are associated with brahmi, only the root knot nematode *M. incognita* causes serious damage to the crop. The nematode causes stunting and leaf chlorosis (Fig. 21.6). In greenhouse trials, a negative correlation between increasing population levels of *M. incognita* and plant growth of *B. monnieri* was demonstrated (Pandey *et al.*, 2003).

Management measures

The amendments and distillates combined with the fungal antagonists discussed above in the section on control in ashwagandha were also successful in reducing root knot densities and enhancing the growth and yield of *B. monnieri* (Pandey *et al.*, 2003).

Chlorophytum borivillianum

C. borivillianum, commonly known as safed musli, is an important medicinal plant belonging to the family Liliaceae. This plant is widely distributed throughout India. The presence of saponins and alkaloids in this plant is of medicinal importance. Progressive farmers in India are cultivating the crop for both the local and international herbal industry. The tuberous root is sold in the market for medicinal use and also saved for planting the next crop. There are several species of *Chlorophytum* grown in India.

Nematodes of *Chlorophytum borivillianum*

Several nematode species are associated with *C. borivillianum*, but the root knot nematode *M. incognita* poses a major threat to successful cultivation of this crop. The nematode parasitizes the fine root system and completes its life cycle within the plant tubers. The nematode, therefore, causes severe tuber loss when present on the crop (Pandey *et al.*, 2003). Plants in infested fields are stunted and have drooping leaves that dry over time.

Management measures

Pandey *et al.* (2003) tested a number of techniques for root knot nematode control in *C. borivillianum* and suggested the following integrated approach to combat the nematode problem:

- Plant healthy tubers that are nematode free;
- Pre-treat the soil before planting with a nematicide; and
- Treat the tubers with a mixture of biological control agents.

Mint (*Mentha* spp.)

Among the different medicinal and aromatic plants, mints are of major pharmaceutical importance due to their many-fold uses. Farmers in the tropics and subtropics can grow mint as a cash crop whenever it fits into a cropping system. The crop generates significant local employment and earns foreign exchange. The main types of mints commercially cultivated in tropical and subtropical countries are: menthol mint (*Mentha arvensis*), peppermint (*Mentha piperita*), spearmint (*Mentha spicata*), scotch spearmint (*Mentha cardiaca*), bergamot mint (*Mentha citrata*) and garden mint (*Mentha viridis*).

Nematodes of Mints

Nematodes have been identified as major pests of several mint species. The important nematodes reducing yield are species of *Meloidogyne*, *Pratylenchus* and *Tylenchorhynchus*. Several other plant parasitic nematodes are associated with these mint species, but are of still unknown economic importance (Pandey, 1999).

Meloidogyne

M. incognita particularly and also *M. javanica* are important parasites on menthol mint wherever it is grown.

Symptoms of damage

Root knot-infested mint plants are stunted and chlorotic, with damage occurring in typical oval patches throughout the field. Root knot-infested suckers or roots bear galls of various sizes (Fig. 21.7) and eggs are clearly visible on the root system under the microscope (Plate 22F).

Fig. 21.7. Suckers of mint, *Mentha arvensis*, infested with *Meloidogyne* sp. (Photo: V.K. Sosamma.)

Biology

The life cycle of *M. incognita* in menthol mint is completed within 28–30 days, with up to four generations developing per season under favourable conditions. Race 2 of *M. incognita* is predominant in the Lucknow area of Uttar Pradesh, India (Pandey *et al.*, 1992).

Survival and dissemination

Because *Meloidogyne* juveniles and eggs survive inside the storage roots and suckers, the nematode is often disseminated in planting material if care is not taken to avoid contamination. Soil adhering to the suckers is also a means of spread. The presence of alternative weed hosts in a field is important in maintaining root knot nematode inoculum between crops.

Environmental factors

Meloidogyne multiplies well in the sandy soils generally used to cultivate menthol mint, therefore, damage caused by root knot nematodes in these regions is often severe (Pandey *et al.*, 1992). Menthol mint is also transplanted in January when temperatures are optimum for nematode infection and development, resulting in 3–4 generations per growing season and high levels of damage.

Economic importance

Meloidogyne species significantly reduce plant growth and oil yield. In addition, *M. incognita* multiplies on all species of *Mentha* as well as on all cultivars (Pandey, 1989). Strong reductions in plant growth as well as in the rate of photosynthesis were found to be directly correlated with initial inoculum densities. *M. incognita* and *M. javanica* caused a 25–30% reduction in oil yield in menthol mint; the quality of the mint oil is also adversely affected by nematode infection (Pandey, 1998a, 2003).

Management of nematodes in menthol mint

Root knot infection was reduced when plants were pre-inoculated with different combinations of arbuscular mycorrhizal fungi (Pandey *et al.*, 1997). The symbionts reduced nematode infection and improved yield.

Successful control of root knot was also achieved with the application of carbofuran at 1.5 kg a.i./ha or with neem cake at 500 kg/ha (Pandey, 2000b, 2003).

Management of *M. incognita* using biological control agents, organic matter and integration of both was studied by Pandey (1995, 1998, 2000). The arbuscular mycorrhizal fungi (*G. aggregatum*, *G. mosseae* and *G. fasiculatum*), the antagonist (*T. harzianum*) and the oil seed cakes from mustard (*Brassica compestris*) and from neem (*A. indica*) along with the nematicide carbofuran were effective in increasing the yield of menthol mint and in reducing root knot densities. Maximum reduction in the nematode population was recorded in the neem cake-treated soil followed by mustard cake, carbofuran and then the biological control agents. Significantly higher levels of yield were obtained in the following order: neem, mustard, *T. harzianum*, *G. aggregatum* and then carbofuran. The use of Vermicompost and different distillation waste products was also found to enhance the growth and yield of different mint species and reduce nematode populations significantly (Pandey *et al.*, 2003). The importance of these control measures to the grower needs further field testing.

Resistance

Germplasms available in the gene bank at the Central Institute of Medicinal and Aromatic Plants (CIMAP) in Lucknow, India were screened for resistance to *M. incognita* (Pandey and Patra, 2001). Moderate to high degrees of resistance were observed on SS-1-4, SS-2-7, SS-15, SS-26, SS-36, *M. piperita* cv. Kukrail, *M. spicata* cv. Neera, *M. spicata* cv. Arka, *M. citrata* cv. Kiran, *M. gracilis* and *M. viridis.*

Non-host crops such as mustard and wheat have been shown to reduce root knot populations and increase yield of menthol mint (Table 21.2). The 'Late Transplanted

Table 21.2. Utilizing crop rotation to increase yield as well as minimizing the root knot nematode populations in menthol mint (Pandey, 2003).

Crop rotation	Net benefit (Rupees)	Root knot index in menthol mint
1 Maize–potato–menthol mint	83,000	+
2 Paddy–potato–menthol mint	80,000	+
3 Paddy–pea–menthol mint	74,000	+++
4 Maize–mustard–menthol mint	76,000	++
5 Pigeonpea–menthol mint	72,000	++
6 Paddy–menthol mint	75,000	++
7 Paddy–wheat–menthol mint	68,000	++

+ = mild , ++ = moderate , +++ = severe infestations.

Mint Technology' developed at CIMAP, which allows farmers to plant these non-host crops, has also greatly benefited crop health and yield. The higher temperatures prevailing during late transplanting (April–July) adversely affect nematode population build-up and infection of the menthol mint crop (Pandey, 2003).

References

Acharya, A. and Padhi, N.N. (1987a) Pathogenic relationship of root knot nematodes, *Meloidogyne incognita* with betelvine (*Piper betle* L.). *Indian Journal of Nematology* 17, 127–128.

Acharya, A. and Padhi, N.N. (1987b) Population threshold of *Rotylenchulus reniformis* in relation to its disease intensity on betelvine. *International Nematology Network Newsletter* 4, 21–23.

Acharya, A. and Padhi, N.N. (1988a) Effect of neem oilcake and sawdust against nematode, *Meloidogyne incognita* on betelvine (*Piper betle* L.). *Indian Journal of Nematology* 18, 105–106.

Acharya, A. and Padhi, N.N. (1988b) Screening of betelvine varieties for resistance against root-knot nematode, *Meloidogyne incognita. Indian Journal of Nematology* 18, 109.

Acharya, A., Dash, S.C. and Padhi, N.N. (1987) Pathogenic association of *Meloidogyne incognita* with *Sclerotium rolfsii* and *Xanthomonas betlicola* on betelvine. *Indian Journal of Nematology* 17, 196–198.

Acharya, A., Padhi, N.N., Swain, P.K. and Dash, S.C. (1988) Occurrence of nematodes on betelvine in Orissa. *Indian Journal of Nematology* 18, 363.

Ali, S.S. (1984) Occurrence of root-knot nematodes in cardamom plantations of Tamil Nadu. In: Bavappa, K.V.A., Iyer, R.D., Das, P.K., Sethuraj, M.R., Ranganathan, V., Nambiar, K.K.N. and Chacko, M.J. (eds) *Proceedings of PLACROSYM-V 1982.* Indian Society for Plantation Crops, Kasaragod, Kerala, India, pp. 615–620.

Ali, S.S. (1985) Preliminary observations on the effect of some systemic nematicides and neem oil cakes in cardamom field infested with root-knot nematodes. In: Sethuraj, M.R. (ed.) *Proceedings of PLACROSYM-VI 1984.* Indian Society for Plantation Crops, Kasaragod, Kerala, India, pp. 215–223.

Ali, S.S. (1986a) Occurrence of root knot nematodes in cardamom plantations of Karnataka. *Indian Journal of Nematology* 16, 269–270.

Ali, S.S. (1986b) Root-knot nematode problem in cardamom and its management. In: *Proceedings of the Second Group Discussions on the Nematological Problems of Plantation Crops.* Central Coffee Research Station, Balehonnur, Karnataka, India, pp. 10–12.

Ali, S.S. (1986c) Evaluation of nemacur against *Meloidogyne incognita* in a cardamom nursery. *Indian Journal of Nematology* 16, 48–50.

Ali, S.S. (1987) Effect of three systemic nematicides against root-knot nematodes in a cardamom nursery. *Nematologia Mediterranea* 15, 155–158.

Ali, S.S. (1989) Influence of 'Katte' mosaic virus of cardamom on the population of *Meloidogyne incognita. Nematologia Mediterranea* 17, 121–122.

Ali, S.S. and Koshy, P.K. (1982a) Occurrence and distribution of root-knot nematodes in cardamom plantations of Kerala. *Nematologia Mediterranea* 10, 107–110.

Ali, S.S. and Koshy, P.K. (1982b) A note on the use of methyl bromide for control of root knot nematodes in cardamom nurseries. *Indian Journal of Nematology* 12, 147–150.

Ali, S.S. and Venugopal, M.N. (1992) Interaction between *Meloidogyne incognita* and *Rhizoctonia solani* in damping off and rhizome rot disease of cardamom seedlings. *Nematologia Mediterranea* 20, 65–66.

Ali, S.S. and Venugopal, M.N. (1993) Prevalence of damping off and rhizome rot disease in nematode infested cardamom nurseries in Karnataka. *Current Nematology* 4, 19–24.

Anandaraj, M., Ramana, K.V. and Sarma, Y.R. (1991) Interaction between vesicular arbuscular mycorrhizal fungi and *Meloidogyne incognita* in black pepper. In: Bagyraj, D.J. and Manjunath, A. (eds) *Mycorrhizal Symbiosis and Plant Growth.* University of Agricultural Sciences, Bangalore, India, pp. 110–112.

Anandaraj, M., Ramana, K.V. and Sarma, Y.R. (1996a) Sequential inoculation of *Phytophthora capsici, Radopholus similis* and *Meloidogyne incognita* in causing slow decline of black pepper. *Indian Phytopathology* 49, 297–299.

Anandaraj, M., Ramana, K.V. and Sarma, Y.R. (1996b) Role of *Phytophthora capsici* in the etiology of slow decline disease of black pepper (*Piper nigrum* L.). *Journal of Plantation Crops* 24 (Supplement), 166–170.

Anandaraj, M., Ramana, K.V. and Sarma, Y.R. (1996c) Suppressive effect of VAM on root damage caused by *Phytophthora capsici, Radopholus similis* and *Meloidogyne incognita* in black pepper. In: Nair, K.S.S., Sharma, J.K. and Varma, R.V. (eds) *Impact of Diseases and Pests in Tropical Forests.* Kerala Forest Research Institute, Peechi, India, pp. 232–238.

Anonymous (1971) *Annual Report 1971.* Department of Agriculture, Fiji.

Anonymous (1987) Improvement of betelvine cultivation in Mahoba. *Progress Report 1986–87.* National Botanical Research Institute, Lucknow, India.

Bai, H., Sheela, M.S. and Jiji, T. (1995) Nemic association and avoidable yield loss in turmeric, *Curcuma longa* L. *Pest Management in Horticultural Ecosystems* 1, 105–110.

Beena, B., Eapen, S.J. and Ramana, K.V. (2001) Antagonistic rhizobacteria of root knot nematodes infesting black pepper (*Piper nigrum* L.). In: *Abstracts – National Congress on Centenary of Nematology in India.* Indian Agricultural Research Institute, New Delhi, India, p. 143.

Beena, B., Eapen, S.J. and Ramana, K.V. (2003) Native rhizobacteria for the biological suppression of *Radopholus similis* infesting black pepper (*Piper nigrum* L.). In: *Abstracts and Short Papers – Sixth International PGPR Workshop,* Calicut, India.

Bhatt, J., Sengupta, S.K. and Chaurasia, R.K. (2002a) Management of *Meloidogyne incognita* by *Trichoderma viride* in betelvine. *Indian Phytopathology* 55, 97–98.

Bhatt, J., Chaurasia, R.K. and Sengupta, S.K. (2002b) Management of *Meloidogyne incognita* by *Paecilomyces lilacinus* and influence of different inoculum levels of *Rotylenchulus reniformis* on betelvine. *Indian Phytopathology* 55, 348–350.

Bridge, J. (1978) *Plant Nematodes Associated with Cloves and Black Pepper in Sumatra and Bangka, Indonesia.* ODM Technical Report on visit to Indonesia 9–19 July 1978. Ministry of Overseas Development, UK.

Bridge, J. (1988) Plant parasitic nematode problems in the Pacific islands. *Journal of Nematology* 20, 173–183.

Butler, E.J. (1906) The wilt disease of pigeon-pea and pepper. *Agricultural Journal of India* 1, 25–36.

Butler, L. and Vilsoni, F. (1975) Potential hosts of burrowing nematode in Fiji. *Fiji Agricultural Journal* 37, 38–39.

Castillo, M.B. (1974) Nematodes associated with stunted black pepper, beans, peas, eggplant and celery in Benguet, Batangas, Laguna and Camarines Sur. *Philippine Agriculturist* 57, 391–402.

Chandwani, G.H. and Reddy, T.S.N. (1967) The host range of root-knot nematode, *Meloidogyne javanica* in tobacco nurseries at Rajahmundry, Andhra Pradesh. *Indian Phytopathology* 20, 383–384.

Charles, J.S. (1978) Studies on the nematode diseases of ginger. MSc Thesis, Kerala Agricultural University, Vellayani, Trivandrum, Kerala, India.

Charles, J.S. and Kuriyan, K.J. (1982) Studies on nematode incidence in ginger. In: Venkataram (ed.) *Proceedings of PLACROSYM-II 1979.* Indian Society for Plantation Crops, Kasaragod, Kerala, India, pp. 50–57.

Chawla, M.L. and Samathanam, C.J. (1980) Two new species of the superfamily Longidoroidea (Dorylaimida: Nematoda) from India. *Indian Journal of Nematology* 10, 205–210.

Chen, C.M., Li, H.Y. and Lii, O.Y. (1986) [The study on root-knot nematodes of common turmeric (*Curcuma domestica* Valet).] *Herald of Agricultural Sciences* 1, 16–22.

Cheng, Y.H. and Tu, C.C. (1979) [Pathogenesis of *Meloidogyne incognita* to edible ginger.] *Journal of Agricultural Research of China* 28, 91–99.

Christie, J.R. (1957) The yellows disease of pepper and spreading decline of citrus. *Plant Disease Reporter* 41, 267–268.

Christie, J.R. (1959) *Plant Nematodes: Their Bionomics and Control.* University of Florida, Gainsville, Florida.

Colbran, R.C. (1958) Studies on plant and soil nematodes. 2. Queensland host records of root-knot nematodes (*Meloidogyne* species). *Queensland Journal of Agricultural Sciences* 15, 101–136.

Colbran, R.C. (1972) Studies on root-knot nematode control in ginger with non-volatile nematicides applied at and after planting. *Queensland Journal of Agriculture and Animal Sciences* 29, 275–280.

Colbran, R.C. (1974) Nematode control in ginger with nematicides, selection of planting material and sawdust mulch. *Queensland Journal of Agriculture and Animal Sciences* 31, 231–235.

Colbran, R.C. and Davis, J.J. (1969) Studies of hot water treatment and soil fumigation for control of root knot in ginger. *Queensland Journal of Agricultural Sciences* 26, 439–445.

Das, N. (1999) Effect of organic mulching on root knot nematode population, rhizome rot incidence and yield of ginger. *Annals of Plant Protection Sciences* 7, 112–114.

Das, S. and Das, S.N. (1986) Host range of *Pratylenchus coffeae. Indian Journal of Nematolology* 16, 180–184.

Das, V.M. and Sultana, S. (1979) Five new species of the genus *Pratylenchus* from vegetable crops of Hyderabad (Andhra Pradesh). *Indian Journal of Nematology* 9, 5–14.

Dasgupta, M.K. and Rama, K. (1987) Plant parasitic nematodes associated with plantation crops. *Review of Tropical Plant Pathology* 4, 289–304.

Delacroix, C. (1902) Une maladie affectant le poivrier (*Piper nigrum*) en Cochinchine. *Agriculture pratique des Pays chauds* 1, 672–680.

Dethe, M.D. and Pawar, V.M. (1987) Control of root-knot nematode on betel vine by granular carbamate pesticides. *International Nematology Network Newsletter* 4, 6–7.

De Waard, P.W.F. (1979) 'Yellows Disease' complex in black pepper on the island of Bangka, Indonesia. *Journal of Plantation Crops* 7, 42–49.

Dhande, G.W. and Sulaiman, M. (1961) Occurrence of root-knot nematodes on betel vine in Maharashtra. *Current Science* 30, 351–362.

Dohroo, N.P., Shyam, K.R. and Bhardwaj, S.S. (1987) Distribution, diagnosis and incidence of rhizome rot complex of ginger in Himachal Pradesh. *Indian Journal of Plant Pathology* 5, 24–25.

Dohroo, N.P., Sharma, O., Sharma, M. and Sarlach, R.S. (1994) Effect of organic amendments of soil on rhizome rot, nematodes and rhizosphere mycoflora of ginger (*Zingiber officinale* Roscoe). *Annals of Biology, Ludhiana* 10, 208–210.

Doshi, A. and Mathur, S. (1987) Symptomatology, interaction and management of rhizome rot of ginger by xenobiotics. *Korean Journal of Plant Protection* 26, 261–265.

D'Souza, G.I., Kumar, A.C., Viswanathan, P.R.K. and Shamanna, H.V. (1970) Relative distribution and prevalence of plant parasitic nematodes in coffee tracts of South Western India. *Indian Coffee* 34, 329–342.

Eapen, S.J. (1987) Plant parasitic nematode problem in small cardamom. In: Mehta, U.K. (ed.) *Proceedings of the Second Group Discussions on the Nematological Problems of Plantation Crops,* Sugarcane Breeding Institute, Coimbatore, Tamil Nadu, India, pp. 43–44.

Eapen, S.J. (1992) Influence of plant age on root knot nematode development in cardamom. *Nematologia Mediterranea* 20, 193–195.

Eapen, S.J. (1993) Seasonal variations of root knot nematode population in a cardamom plantation. *Indian Journal of Nematology* 23, 63–68.

Eapen, S.J. (1994) Pathogenicity of root knot nematode on small cardamom (*Elettaria cardamomum* Maton). *Indian Journal of Nematology* 24, 31–37.

Eapen, S.J. (1995a) A note on the incidence of *Rotylenchulus reniformis* in cardamom plants. *Indian Journal of Nematology* 25, 213.

Eapen, S.J. (1995b) *Investigations on Plant Parasitic Nematodes Associated with Cardamom.* Final Report, Indian Institute of Spices Research, Calicut, Kerala, India

Eapen, S.J. and Ramana, K.V. (1996) Biological control of plant parasitic nematodes of spices. In: Anandaraj, M. and Peter, K.V. (eds) *Biological Control in Spice.* Indian Institute of Spices Research, Calicut, Kerala, India, pp. 20–32.

Eapen, S.J. and Venugopal, M.N. (1995) Field evaluation of *Trichoderma* spp. and *Paecilomyces lilacinus* for control of root knot nematodes and fungal diseases in cardamom nurseries. *Indian Journal of Nematology* 25, 15–16 (abstract).

Eapen, S.J., Geetha, S.M. and Leemoll, M. (1987) A note on the association of *Radopholus similis* with yellow disease of betel vine. *Indian Journal of Nematology* 17, 137–138.

Eapen, S.J., Ramana, K.V. and Sarma, Y.R. (1997) Evaluation of *Pseudomonas fluorescens* isolates for control of *Meloidogyne incognita* in black pepper (*Piper nigrum* L.). In: Edison, S., Ramana, K.V., Sasikumar, B., Nirmal Babu, K. and Eapen, S.J. (eds) *Biotechnology of Spices, Medicinal and Aromatic Plants. Proceedings of the National Seminar on Biotechnology of Spices and Aromatic Plants.* Indian Society for Spices, Calicut, India, pp. 129–133.

Eapen, S.J., Chempakam, B. and Ramana, K.V. (1999a) Role of defense related enzymes in the resistance of black pepper to *Meloidogyne incognita.* In: *Abstracts of Papers – National Seminar on Nematological Research in India,* C.S. Azad University of Agriculture and Technology, Kanpur.

Eapen, S.J., Ramana, K.V., Sasikumar, B. and Johnson, K.G. (1999b) Screening ginger and turmeric germplasm for resistance against root knot nematodes. In: Dhawan, S.C. (ed.) *Proceedings of National Symposium on Rational Approaches in Nematode Management for Sustainable Agriculture,* Nematological Society of India, New Delhi, pp. 142–144.

Eapen, S.J., Beena, B. and Ramana, K.V. (2000a) Preliminary screening of some hyphomycetous fungi for the biological control of root-knot nematodes (*Meloidogyne incognita*) infesting spice crops. In: *Abstracts of Papers – National Nematology Symposium on Integrated Nematode Management,* Orissa University of Agriculture and Technology, Bhubaneswar, Orissa, p. 18.

Eapen, S.J., Venugopal, M.N., Ramana, K.V. and Sarma, Y.R. (2000b) *Trichoderma* spp. for the management of root knot nematode and rhizome rot disease in cardamom nurseries. In: Muraleedharan, N. and Raj Kumar, R. (eds) *Recent Advances in Plantation Crops Research.* Allied Publishers Ltd, Mumbai, India, pp. 382–386.

Ferraz, E.C. de A. and Lordello, L.G.E. (1989) Influencia de *Meloidogyne incognita* (Kofoid and White, 1919) Chitwood, 1949 no teor de clorofila de de pimenteiras-do-reino (*Piper nigrum* L.). *Agrotropica* 1, 57–62.

Ferraz, E.C. de A. and Sharma, R.D. (1979) Interacao e patogenicidade do *Meloidogyne incognita* (Kofoid and White, 1919) Chitwood, 1949 e o *Rotylenchulus reniformis* Linford e Oliveira, 1940 na pimenteira de reino. *Revista Theobroma* 9, 45–53.

Ferraz, E.C. de A., Orchard, J.E. and Lopez, A.S. (1984) Resposta da pimenteria-do-reino a infestacao or *Meloidogyne incognita* un relacao ao teor de jenois totais. *Revista Theobroma* 14, 217–221.

Ferraz, E.C. de A., Lordello, L.G.E. and de Santana, C.J.L. (1988) Absoracao de nutrientes em pimenteiras-do-reino (*Piper nigrum* L.) infestadas por *Meloidogyne incognita* (Kofoid and White, 1919) Chitwood, 1949. *Boletim Tecnico Centro de Pesquisas do Cacau, Brazil,* No. 160, 34 pp.

Fliege, F.H. and Sikora, R.A. (1981) Occurrence and distribution of plant-parasitic nematodes in W. Samoa. *Alafua Agricultural Bulletin* 6, 33–41.

Freire, F.C.O. (1982) Interactions of fungi and nematodes of black pepper (*Piper nigrum* L.). PhD Thesis, Faculty of Science, University of London.

Freire, F.C.O. and Bridge, J. (1985a) Histopathology of black pepper roots infested with *Radopholus similis. Fitopatologia Brasileira* 10, 475–481.

Freire, F.C.O. and Bridge, J. (1985b) Biochemical changes induced in roots and xylem sap of black pepper by *Meloidogyne incognita. Fitopatologia Brasileira* 10, 483–497.

Freire, F.C.O. and Bridge, J. (1985c) Influence of different inoculum levels of *Meloidogyne incognita, Nectria haematococca* f.sp. *piperis* and *Phytophthora palmivora* on black pepper plants. *Fitopatologia Brasileira* 10, 559–575.

Freire, F.C.O. and Bridge, J. (1985d) Parasitism of eggs, females and juveniles of *Meloidogyne incognita* by *Paecilomyces lilacinus* and *Verticillium chlamydosporium. Fitopatologia Brasileira* 10, 577–596.

Freire, F.C.O. and Monteiro, A.R. (1978) Nematodes associaoas a pimenterira-do-reino (*Piper nigrum* L.) na regiao amazonica. III Reuniao de Nematologica. ESAM/SOC. BRAS. NEMAT. *Colecao Mossoroense, Mossoro* 62, 28.

Ganguly, S. (1988) Phytonematodes associated with betelvine and citrus in Nagpur, Maharashtra, India. *Indian Journal of Nematology* 18, 142–144.

Ganguly, S. and Khan, E. (1983) *Trophurus impar* sp.n. and *Scutellonema eclipsi* sp.n. (Nematoda: Tylenchida). *Indian Journal of Nematology* 13, 230–234.

Ganguly, S. and Khan, E. (1990) *Piper betle* – a new host of *Pratylenchus coffeae* in India. *International Nematology Network Newsletter* 7 (4), 21.

Geetha, S.M. (1991) Studies on the biology, pathogenicity and biocontrol of different populations of *Radopholus similis.* PhD Thesis, Kerala University, Trivandrum, Kerala, India.

Ghesquiere, J. (1921) Laboratoire d'entomologie d'Eala (Equateur): rapport de l'entomologiste. *Bulletin Agricole du Congo Belge* 12, 703–732.

Gnanapragasam, N.C. (1989) Varietal response of pepper to infestation by the burrowing nematode, *Radopholus similis. Sri Lanka Journal of Tea Science* 58, 5–8.

Gnanapragasam, N.C., Anpalagan, V.T., Dharmasena, W.A.M., Ariyaratnam, V., Jayasinghe, P.R.R. and Navaratne, N. (1985) *Report of the Tea Research Institute, Sri Lanka.*

Goffart, H. (1953) Beobachtungen an pflanzenschadlichen Nematoden I. *Nachrichten Blatt Pflschutzdienst, Stuttgart* 5, 150–153.

Goodey, J.S., Franklin, M.T. and Hooper, D.J. (1965) *T. Goodey's The Nematode Parasites of Plants Catalogued Under Their Hosts.* CAB, Farnham Royal, UK.

Gunasekharan, C.R., Vadivelu, S. and Jayaraj, S. (1987) Experiments on nematodes of turmeric – a review. *Proceedings of the Third Group Discussions on the Nematological Problems of Plantation Crops,* 29–30 October 1987, Sugarcane Breeding Institute, Coimbatore, Tamil Nadu, India, pp. 45–46.

Haidar, M.G., Jha, R.N. and Nath, R.P. (1995) Studies on the nematodes of spices. 1. Nemic association of turmeric in Bihar and reaction of certain lines to some of the dominating nematodes. *Indian Journal of Nematology* 25, 212–213.

Haidar, M.G., Jha, R.N. and Nath, R.P. (1998a) Studies on the nematodes of spices. Pathogenic effect of root knot (*Meloidogyne incognita*) and reniform nematodes (*Rotylenchulus reniformis*) alone and in combination on turmeric (*Curcuma longa* L.). *Indian Journal of Nematology* 28, 52–55.

Haidar, M.G., Jha, R.N. and Nath, R.P. (1998b) Efficacy of some nematicides and organic soil amendments on nematode population and growth characters of turmeric (*Curcuma longa* L.). *Journal of Research, Birsa Agricultural University* 10, 211–213.

Hart, W.H. (1956) Spreading decline of citrus caused by the burrowing nematode. *Bulletin of the California Department of Agriculture* 45, 263–267.

Haseeb, A. and Pandey, R. (1989) Root-knot disease of henbane, *Hyoscyamus* – a new disease record. *Tropical Pest Management* 35, 212–221.

Haynes, P.H., Partridge, I.J. and Sivan, P. (1973) Ginger production in Fiji. *Fiji Agricultural Journal* 35, 51–56.

Hazarika, K., Dutta, P.K. and Saikia, L. (1998) Field efficacy of *Paecilomyces lilacinus* against root knot nematode in betelvine. *Journal of Phytological Research* 11, 171–173.

Hazarika, K., Dutta, P.K. and Saikia, L. (1999a) Effect of some organic amendments on root-knot nematode of Assamiya pan. *Journal of the Agricultural Science Society of North East India* 12, 207–209.

Hazarika, K., Dutta, P.K. and Saikia, L. (1999b) Yield losses in betelvines due to infestation of *Meloidogyne incognita. Journal of the Agricultural Science Society of North East India* 12, 268–270.

Hazarika, K., Dutta, P.K., Saikia, L. and Deka, S.C. (2000) Biological control of root-knot nematode in betelvine using parasitic fungus, *Paecilomyces lilacinus. Crop Research Hisar* 19, 338–342.

Hegde, L., Krishnappa, K. and Bojappa, K.M. (1993) Reaction of different cultivars of cardamom (*Elettaria cardamomum* Maton) to root knot nematode (*Meloidogyne incognita*). *Indian Journal of Nematology* 23, 127–128.

Holliday, P. and Mowat, W.P. (1963) *Foot Rot of* Piper nigrum *L. (*Phytophthora palmivora*).* Phytopathology Paper No. 5, Commonwealth Mycological Institute, Kew, UK.

Huang, C.S. (1966) Host–parasite relationships of the root-knot nematode in edible ginger. *Phytopathology* 56, 755–759.

Hubert, F.P. (1957) Diseases of some export crops in Indonesia. *Plant Disease Reporter* 41, 55–63.

Huettel, R.N., Dickson, D.W. and Kaplan, D.T. (1984) *Radopholus citrophilus* sp.n. (Nematoda), a sibling species of *Radopholus similis. Proceedings of the Helminthological Society of Washington* 51, 32–35.

Ichinohe, M. (1975) Infestation of black pepper vines by the root-knot nematode, *Meloidogyne incognita,* at Tome-Acu, Para. Brazil. *Japanese Journal of Nematology* 5, 36–39.

Ichinohe, M. (1980) Studies on the root-knot nematode of black-pepper plantation in the Amazon. *Annual Report of the Society of Plant Protection of North Japan* 31, 1–8.

Ichinohe, M. (1984) Integrated control of the root-knot nematode, *Meloidogyne incognita,* on black-pepper plantations in the Amazonian region. *Agriculture, Ecosystems and Environment* 12, 271–283.

Jacob, A. and Chandrasekharan, L.R. (1984) Evaluation of certain granular pesticides for the control of root knot nematodes in cardamom. In: *Proceedings of PLACROSYM-V 1982.* Indian Society for Plantation Crops, Kasaragod, India, pp. 633–637.

Jacob, A. and Kuriyan, J. (1979a) Screening of pepper varieties for resistance against root-knot nematode (*Meloidogyne incognita*). *Agricultural Research Journal of Kerala* 17, 90.

Jacob, A. and Kuriyan, J. (1979b) Survey of nematodes associated with pepper in Kerala. *Agricultural Research Journal of Kerala* 17, 270–271.

Jagdale, G.B., Pawar, A.B. and Darekar, K.S. (1984) Control of root-knot nematode on betel vine with systemic nematicides. *International Nematology Network Newsletter* 1, 4–5.

Jagdale, G.B., Pawar, A.B. and Darekar, K.S. (1985a) Pathogenicity of *Meloidogyne incognita* on betel vine (*Piper betle* L). *Indian Journal of Nematology* 15, 244.

Jagdale, G.B., Pawar, A.B. and Darekar, K.S. (1985b) Studies on control of root-knot nematode on betelvine. *International Nematology Network Newsletter* 2 (1), 10–13.

Jagdale, G.B., Pawar, A.B. and Darekar, K.S. (1985c) Effect of organic amendments on root-knot nematodes infesting betel vine. *International Nematology Network Newsletter* 2, 7–10.

Jagdale, G.B., Pawar, A.B. and Darekar, K.S. (1985d) Screening of some betelvine varieties against the root-knot nematode. *International Nematology Network Newsletter* 2, 22–23.

Jagdale, G.B., Pawar, A.B. and Darekar, K.S. (1985e) Effect of different levels of potash on root-knot nematodes infesting betel vine. *International Nematology Network Newsletter* 2, 8–9.

Jagdale, G.B., Pawar, A.B. and Darekar, K.S. (1986a) Nematodes associated with betelvine in Maharashtra State (India). *International Nematology Network Newsletter* 3, 12–14.

Jagdale, G.B., Pawar, A.B. and Darekar, K.S. (1986b) Seasonal population fluctuation of root-knot nematode on betel vine. *Journal of Maharashtra Agricultural University* 11, 106–107.

Jasy, T. (1991) Host specificity, pathogenicity and morphological variations in the populations of burrowing nematode infesting plantation crops. PhD Thesis, Kerala University, Trivandrum, Kerala, India.

Jasy, T. and Koshy, P.K. (1992) Effect of certain leaf extracts and leaves of *Glyricidia maculata* (H.B. and K.) Steud. as green manure on *Radopholus similis*. *Indian Journal of Nematology* 22, 117–121.

Jonathan, E.I., Nagalakshmi, S. and Padmanabhan, D. (1990) Estimation of yield losses in betelvine due to *Meloidogyne incognita*. *International Nematology Network Newsletter* 7 (4), 26.

Jonathan, E.I., Padmanabhan, D. and Ayyamperumal, A. (1995) Biological control of root-knot nematode on betelvine, *Piper betle*, by *Paecilomyces lilacinus*. *Nematologia Mediterranea* 23, 191–193.

Jonathan, E.I., Sivakumar, M. and Padmanabhan, D. (1996) Interaction of *Meloidogyne incognita* and *Phytophthora palmivora* on betelvine. *Nematologia Mediterranea* 24, 341–343.

Jonathan, E.I., Gajendran, G. and Arulmozhiyan, R. (1997) Interaction of *Rotylenchulus reniformis* and *Phytophthora palmivora* on betelvine. *Nematologia Mediterranea* 25, 191–193.

Jonathan, E.I., Arulmozhiyan, R., Muthusamy, S. and Manuel, W.W. (2000) Field application of *Paecilomyces lilacinus* for the control of *Meloidogyne incognita* on betelvine, *Piper betle*. *Nematologia Mediterranea* 28, 131–133.

Kaur, D.J. and Sharma, N.K. (1988) Occurrence and pathogenicity of *Meloidogyne arenaria* on ginger. *Indian Phytopathology* 41, 467–468.

Kaur, D.J. and Sharma, N.K. (1990) A new report on *Pratylenchus coffeae* – a cause of ginger yellows. *International Nematology Network Newsletter* 7, 15–16.

Kaur, D.J., Sharma, N.K. and Khan, M.L. (1989) Occurrence of *Pratylenchus zeae* Graham 1951 on ginger, *Zingiber officinale* (Rosc.) in Himachal Pradesh. *Indian Journal of Nematology* 19, 68.

Kaur, K.J. (1987) Studies on nematodes associated with ginger (*Zingiber officinale* Rosc.) in Himachal Pradesh. Thesis, Himachal Pradesh University, Shimla, India.

Khan, E. and Nanjappa, C.K. (1972) Four new species of Criconematoidea (Nematoda) from India. *Indian Journal of Nematology* 2, 59–68.

Khan, F.A. (1975) Some new host records of the reniform nematode in India. *Indian Phytopathology* 28, 386–387.

Khan, F.A. and Khan, A.M. (1969) Reaction of certain economically important plants to *Rotylenchulus reniformis*. *Abstracts of All India Nematology Symposium*, 21–22 August 1969, IARI, New Delhi, India, pp. 8–9.

Khan, F.A. and Khan, A.M. (1973) Studies on the reniform nematode, *Rotylenchulus reniformis*. I. Host range and population changes. *Indian Journal of Nematology* 3, 24–30.

Khan, M.L. and Makhnotra, A.K. (1998) Occurrence and distribution of plant nematodes in ginger in Himachal Pradesh. *Indian Journal of Nematology* 28, 87–88.

Kirby, M.F., Kirby, M.E., Siddiqi, M.R. and Loof, P.A.A. (1980) *Fiji Nematode Survey Report: Plant Parasitic Nematode Distributions and Host Associations*. Bulletin No. 68, Ministry of Agriculture and Fisheries, Fiji.

Koshy, P.K. (1974) *Annual Report, 1973*. Central Plantation Crops Research Institute, Kasaragod.

Koshy, P.K. (1986a) *Research on Plant Parasitic Nematodes at the Central Plantation Crops Research Institute*. Kayangulam, Kerala, India, October 1972 to June 1986, Report.

Koshy, P.K. (1986b) The burrowing nematode, *Radopholus similis* (Cobb, 1893) Thorne, 1949. In: Swarup, G. and Dasgupta, D.R. (eds) *Plant Parasitic Nematodes of India – Problems and Progress.* IARI, New Delhi, India, pp. 223–248.

Koshy, P.K. (1987a) Nematological techniques in plantation crops. *Proceedings of the Third Group Discussion on the Nematological Problems of Plantation Crops,* 29–30 October 1987, Sugarcane Breeding Institute, Coimbatore, Tamil Nadu, India, pp. 24–25.

Koshy, P.K. (1987b) Integrated nematode management in plantation crops. *Proceedings of the Third Group Discussion on the Nematological Problems of Plantation Crops,* 29–30 October 1987, Sugarcane Breeding Institute, Coimbatore, Tamil Nadu, India, pp. 49–51.

Koshy, P.K. (1988) Strategies for production of nematode free planting materials in plantation crops. *Workshop on Strategy for Production of Planting Materials in Plantation Crops,* 22–23 January 1988, The Central Plantation Crops Research Institute, Kasaragod, Kerala, India, pp. 37–42.

Koshy, P.K. and Sosamma, V.K. (1975) Host-range of the burrowing nematode *Radopholus similis* (Cobb,1893) Thorne, 1949. *Indian Journal of Nematology* 5, 255–257.

Koshy, P.K. and Sosamma, V.K. (1977) Identification of the physiological race of *Radopholus similis* populations infesting coconut, arecanut and banana in Kerala, South India. *Indian Phytopathology* 30, 415–416.

Koshy, P.K. and Sundararaju, P. (1979) Response of seven black pepper cultivars to *Meloidogyne incognita. Nematologia Mediterranea* 7, 123–135.

Koshy, P.K., Nair, C.P.R., Sosamma, V.K. and Sundararaju, P. (1976) On the incidence of root-knot nematode in cardamom nurseries. *Indian Journal of Nematology* 6, 174–175.

Koshy, P.K., Sosamma, V.K. and Sundararaju, P. (1977) Screening of plants used as pepper standards against root-knot nematode. *Indian Phytopathology* 30, 128–129.

Koshy, P.K., Sundararaju, P. and Sosamma, V.K. (1978) Occurrence and distribution of *Radopholus similis* (Cobb, 1893) Thorne, 1949 in South India. *Indian Journal of Nematology* 8, 49–58.

Koshy, P.K., Sundararaju, P. and Wilson, K.I. (1979a) Efficacy of certain systemic nematicides against *Meloidogyne incognita* in a cardamom nursery. *Nematologia Mediterranea* 7, 195–198.

Koshy, P.K., Premachandran, D., Sosamma, V.K. and Premkumar, T. (1979b) Effect of *Meloidogyne incognita* population on black pepper (*Piper nigrum* L.). *Indian Phytopathology* 32, 221–225.

Krishna Menon, K. (1949) The survey of pollu and root diseases of pepper. *Indian Journal of Agricultural Science* 19, 89–136.

Krishnamurthy, G.V.G. and Elias, N.A. (1967) Host range of *Meloidogyne incognita* causing root-knot on tobacco in Hunsur, Mysore State. *Indian Phytopathology* 20, 374–377.

Kueh, T.K. (1990) Major diseases of black pepper and their management. *The Planter* 66, 59–69.

Kueh, T.K. and Teo, C.H. (1978) Chemical control of root-knot nematodes in *Piper nigrum. The Planter* 54, 237–245.

Kumar, A.C., Viswanalhan, P.R.K. and D'Souza, G.I. (1971) A study of plant parasitic nematodes of certain commercial crops in coffee tracts of South India. *Indian Coffee* 35, 222–224.

Lamberti, F., Rohini, H.M. and Ekanayake, K. (1983) Effect of some plant parasitic nematodes on the growth of black pepper in Sri Lanka. *FAO Plant Protection Bulletin* 31, 163–166.

Lanjewar, R.D. and Shukla, V.N. (1985) Parasitism and interaction between *Pythium myriotylum* and *Meloidogyne incognita* in soft rot complex of ginger. *Indian Journal of Nematology* 15, 170–173.

Leong, C.T.S. (1986) Pepper nematodes. In: *Annual Report for the Year 1984.* Department of Agriculture, Sarawak, Borneo, pp. 74–78.

Lokesh, M.S. and Gangadharappa, P.M. (1995) Management of *Phytophthora* foot rot and nematode diseases in black pepper (*Piper nigrum* L.). *Journal of Spices and Aromatic Crops* 4, 61–63.

Lopes, E.B. and Lordello, L.G.E. (1979) *Meloidogyne incognita, Fusarium solani* f. *piperi,* associados a murcha da pimenteira-do-reino na paraiba. *Revista de Agricultura* 54, 263–268.

Lordello, L.G.E. and Silva, H.M. (1974) Un nematoide nocivo a Pimenta-do-reino. *Revista de Agricultura* 49, 165–166.

Mahapatra, S. and Awasthi, M.D. (1994) Persistence of fenamiphos in betel leaves and soil. *Pesticide Research Journal* 6, 146–148.

Makhnotra, A.K. and Luqman Khan (1997b) Assessment of yield losses in ginger due to *Meloidogyne incognita. Indian Journal of Nematology* 27, 259–260.

Mammen, K.V. (1973) Root gall nematodes as a serious pest of ginger in Kerala. *Current Science* 42, 549.

Mammen, K.V. (1974) On a wilt disease of betel vine in Kerala caused by root-knot nematode. *Agricultural Research Journal of Kerala* 12, 76.

Mani, A. and Prakash, K.S. (1992) Plant parasitic nematodes associated with turmeric in Andhra Pradesh. *Current Nematology* 3, 103–104.

Mani, A., Naidu, P.H. and Madhavachari, S. (1987) Occurrence and control of *Meloidogyne incognita* on turmeric in Andhra Pradesh, India. *International Nematology Network Newsletter* 4, 13–18.

Marimuthu, T. (1991) Fungal nematode wilt complex in betelvine (*Piper betle* L). *Plant Disease Research* 6, 85–88.

Mathur, R.I., Mathur, B.N. and Handa, D.K. (1969) Additions to host records of root-knot nematode. *Nematologica* 15, 160–16l.

Medhane, N.S., Jagdale, G.B., Pawar, A.B. and Darekar, K.S. (1985) Effect of *Tagetes erecta* on root-knot nematodes infecting betelvine. *International Nematology Network Newsletter* 2, 11–12.

Midha, R.L. and Trivedi, P.C. (1991) Estimation of losses caused by *Meloidogyne incognita* on coriander, cumin and fennel. *Current Nematology* 2, 159–162.

Mohandas, C. and Ramana, K.V. (1982) *Trophotylenchulus floridensis* Raski, a new endoparasite of *Piper nigrum* L. from Kerala. *Journal of Plantation Crops* 10, 53–54.

Mohandas, C. and Ramana, K.V. (1983) Effect of different levels of *Meloidogyne incognita* on plant growth of two cultivars of black pepper (*Piper nigrum* L.) Abstracts of papers. *Third Nematology Symposium Organized by the Nematological Society of India*, Himachal Pradesh KrishiViswa Vidyalaya, Solan, Himachal Pradesh, India, pp. 9–10.

Mohandas, C. and Ramana, K.V. (1987a) Slow wilt disease of black pepper and its control. *Indian Cocoa, Arecanut and Spices Journal* 11, 10–11.

Mohandas, C. and Ramana, K.V. (1987b) Pathogenicity, seasonal prevalence and control of plant parasitic nematodes associated with black pepper. *Proceedings of the Third Group Discussions on the Nematological Problems of Plantation Crops*, 29–30 October 1987, Sugarcane Breeding Institute, Coimbatore, Tamil Nadu, India, pp. 44–45.

Mohandas, C. and Ramana, K.V. (1987c) Plant parasitic nematodes associated with black pepper (*Piper nigrum* L.) in Kerala. *Indian Journal of Nematology* 17, 62–66.

Mohandas, C. and Ramana, K.V. (1988) Population behaviour of *Radopholus similis* in roots of black pepper (*Piper nigrum* L.) in Kerala, India. *Indian Journal of Nematology* 18, 18–21.

Mohandas, C. and Ramana, K.V. (1991) Pathogenicity of *Meloidogyne incognita* and *Radopholus similis* on black pepper (*Piper nigrum* L.). *Journal of Plantation Crops* 19, 41–53.

Mohandas, C., Ramana, K.V. and Raski, D.J. (1985) *Trophotylenchulus piperis* n.sp. parasitic on *Piper nigrum* L. in Kerala, India (Nemata: Tylenchulidae). *Revue de Neematologie* 8, 97–102.

Mohanty, K.C., Ray, S., Mohapatra, S.N., Patnaik, P.R. and Ray, P. (1995) Integrated management of root knot nematode in ginger (*Zingiber officinale* Rosc.). *Journal of Spices and Aromatic Crops* 4, 70–73.

Muniappan, R. (1993) Pest and diseases of *Erythrina*: a review. *Journal of Coffee Research* 23, 1–13.

Murthy, M.M. and Rao, K.T. (1992) Control of root knot nematode, *Meloidogyne incognita* on betelvine with certain oil cake amendments and non volatile nematicides. *Indian Journal of Plant Protection* 20, 171–173.

Murthy, M.M. and Rao, K.T. (1994) Oil-cakes and certain non-volatile nematicides in the control of *Meloidogyne incognita* in betelvine. *Indian Journal of Entomology* 56, 36–43.

Mustika, I. (1978) An observation on the relationship between nematode population and yellow disease on black pepper in Bangka. *Pemberitaan Lembaga Penelitian Tanaman Industri, Indonesia* 30, 11–19.

Mustika, I. (1984) Effects of nematodes and fungi on the growth of pepper and yellow disease. *Pemberitaan Lembaga Penelitian Tanaman Industri, Indonesia* 8, 28–37.

Mustika, I. (1991) Response of four black pepper cultivars to infection by *Radopholus similis, Meloidogyne incognita*, and *Fusarium solani*. *Industrial Crops Research Journal* 4, 17–22.

Mustika, I. (1992a) Effects of *Meloidogyne incognita* and *Fusarium solani* on black pepper (*Piper nigrum* L.). *Industrial Crops Research Journal* 4, 7–13.

Mustika, I. (1992b) Interactions of *Radopholus similis* with *Fusarium solani* on black pepper (*Piper nigrum* L.). *Industrial Crops Research Journal* 5, 1–10.

Mustika, I. and Zainuddin, N. (1978) Efficacy tests of some nematicides for the control of nematodes on black pepper. *Pemberitaan Lembaga Penelitian Tanaman Industri, Indonesia* 30, 1–10.

Mustika, I., Sudradjat, D. and Wikanda, A. (1984) Control of pepper yellow disease with fertilizer and pesticides. *Pemberitaan, Penelitian Tanaman Industri, Indonesia* 9, 37–43.

Nagakura, K. (1930) Über den Bau and die Lebengeschichte der *Heterodera radicicola* (Greef) Muller. *Japanese Journal of Zoology* 3, 95–160.

Nakat, R.V., Acharya, A., Rao, D.V.S., Jonathan, E.I., Washnikar, R., Jha, S., Singh, U.S. and Sitaramaiah, K. (1998) Evaluation of *Paecilomyces lilacinus* for the management of root-knot nematode (*Meloidogyne incognita*) on betelvine. In: Reddy, P.P., Kumar, N.K.K. and Verghese, A. (eds) *Advances in IPM for Horticultural Crops*. Proceedings of the First National Symposium on Pest Management in Horticultural Crops: Environmental Implications and Thrusts, Bangalore, India, pp. 308–311.

Nambiar, M.C. (1977) *Proceedings of the National Seminar on Pepper*, 19 December 1977, Central Plantation Crops Research Institute, Kasaragod, Kerala, India, pp. 4–6.

Nema, A.G. (1997) Plant parasitic nematodes associated with betelvine. *Indian Phytopathology* 50, 145–146.

Nema, A.G. (2001a) Reaction of some betelvine (*Piper betle* L.) cultivars to the root knot nematode (*Meloidogyne incognita*). *Advances in Plant Sciences* 14, 605.

Nema, A.G. (2001b) Control of phytonematodes in betelvine (*Piper betle* L.). *Advances in Plant Sciences* 14, 621–622.

Nirula, K.K. and Kumar, R. (1963) Collateral host plant of root-knot nematodes. *Current Science* 32, 221–222.

Orton Williams, K.J. (1980) Plant parasitic nematodes of the Pacific. *UNDP/FAO-SPEC Survey of Agricultural Pests and Diseases in the South Pacific*. Technical Report, Volume 8. Commonwealth Institute of Helminthology, St Albans, UK.

Pandey, R. (1990) Pathogenicity and reproduction of *Meloidogyne incognita* on *Hyoscyamus albus* and rhizospheric association of other phytoparasitic nematodes. *International Nematology Network Newsletter* 7, 34–37.

Pandey, R. (1993) Conceptual models for the management of root-knot and lesion nematodes in important medicinal and aromatic plants. *Afro-Asian Journal of Nematology* 3, 103–106.

Pandey, R. (1995) Application of oil seed cake, pesticides and dry leaf matter on the reproduction potential of *Meloidogyne incognita* and yield of Japanese mint. *Indian Journal of Nematology* 25, 12–13.

Pandey, R. (1997) Ecofriendly approaches for the management of root-knot nematode on black henbane. *Indian Journal of Nematology* 27, 175–178.

Pandey, R. (1998a) Nematode pests of medicinal and aromatic plants and their management in phytonematology – aspects and prospects. In: Trivedi, P.C. (ed.) *Advances in Nematology*. Department of Botany, University of Rajasthan, Jaipur, India, pp. 177–216.

Pandey, R. (1998b) Phytopathological impact of root-knot nematode on some medicinal and aromatic plants. *Journal of Medicinal and Aromatic Crop Sciences* 20, 67–84.

Pandey, R. (1999) Spatial and temporal dynamics of phytonematodes in arable soil planted with Japanese mint. *Journal of Mycology and Plant Pathology* 29, 232–235.

Pandey, R. (2000a) Management of phytonematodes through fungal antagonists. *Journal of Medicinal and Aromatic Plant Sciences* 22, 293–305.

Pandey, R. (2000b) Additive effect of three organic materials and pesticides on the reproduction of *Meloidogyne incognita* and yield of *Mentha arvensis*. *Nematropica* 30, 155–160.

Pandey, R. (2003) Mint nematology – current status and future needs. In: Trivedi, P.C. (ed.) *Advances in Nematology*. Department of Botany, University of Rajasthan, Jaipur, India, pp. 155–166.

Pandey, R. and Patra, N.K. (2001) Screening mint (*Mentha* spp.) accessions against root-knot nematode infection. *Journal of Spices and Aromatic Crops* 10, 55–56.

Pandey, R., Haseeb, A. and Husain, A. (1992) Distribution, pathogenicity and management of *Meloidogyne incognita* on *Mentha arvensis* cv. MAS-1. *Afro-Asian Journal of Nematology* 2, 27–34.

Pandey, R., Singh, H.B. and Gupta, M.L. (1997) Antagonistic impact of vesicular-arbuscular (VAM) on *Meloidogyne incognita* population development in Japanese mint. *International Journal of Tropical Plant Disease* 15, 237–245.

Pandey, R., Singh, H.B., Gupta, M.L., Singh, H.N., Sharma, S. and Kumar, S. (1998) Efficacy of VAM fungi, green manure, organic materials and pesticides in the management of root-knot nematode in menthol mint. In: Dhawan, S.C. and Kaushal, K.K. (eds) *Proceedings of National Symposium on Rational Approaches in Nematode Management for Sustainable Agriculture*. GAU, Anand, India, pp. 47–50.

Pandey, R., Gupta, M.L., Singh, H.B. and Kumar, S. (1999) Interaction potentialities of vesicular-arbuscular mycorrhizal fungi with root knot nematode, *Meloidogyne incognita* on black henbane. *Bioresource Technology* 69, 275–278.

Pandey, R., Kalra, A. and Kumar, S. (2000a) Efficacy of various essential oils in the management of root-knot disease in black henbane. In: *Proceedings of Centennial Conference on Spices and Aromatic Plants*. IISR, Calicut, India, pp. 263–266.

Pandey, R., Singh, H.B. and Kalra, A. (2000b) Microbial management of plant parasitic nematodes. *International Journal of Tropical Plant Disease* 18, 1–23.

Pandey, R., Gupta, M.L., Singh, H.B., Singh, H.N. and Kumar, S. (2000c) Disease management potentiality of VAM fungi to *Meloidogyne incognita* on *Hyoscyamus muticus*. In: *Proceedings of an International Conference on Integrated Plant Disease Management for Sustainable Sgriculture*. IARI, New Delhi, India, pp. 1239–1240.

Pandey, R., Kalra, A., Gupta, M.L. and Sharma, P. (2003) Phytonematodes: major pest of MAPs. In: Mathur, S. (ed.) *Proceedings of First National Interactive Meeting on Medicinal and Aromatic Plants*, CIMAP, Lucknow, India, pp. 188–197.

Parihar, A. (1985) Population level studies of the root-knot nematode, *Meloidogyne incognita* (Kofoid and White, 1919) Chitwood, 1949 on ginger (*Zingiber officinale*, Rosc.). *Indian Journal of Nematology* 15, 292 (abstract).

Parihar, A. and Yadav, B.S. (1986) *Meloidogyne incognita* on ginger (*Zingiber officinale* Rosc.) and its control. *Indian Journal of Mycology and Plant Pathology* 16, 84–86.

Pasril, W. (1976) Studies on yellow disease in black pepper on the island of Bangka. *Pemberitaan Lembaga Penelitian Tanaman Industri, Indonesia* 21, 64–79.

Patel, D.J., Makadia, B.M. and Shah, H.M. (1982) Occurrence of root-knot on turmeric (*Curcuma longa* L.) and its chemical control. *Indian Journal of Nematology* 12, 168–171.

Patel, D.J., Patel, S.T. and Jogani, D.K. (1986) Preliminary screening of cumin varieties against root-knot nematodes. *Indian Journal of Nematology* 16, 130.

Pathak, J.J. and Saikia, B. (1999) Biological control of root-knot nematode of betelvine. *Journal of Interacademicia* 3, 117–120.

Pattnaik, M.R., Senapati, H.K., Pal, A.K. and Acharya, A. (1989) Persistence of carbofuran in betelvine. *Indian Journal of Horticulture* 46, 119–121.

Paulus, A.D., Eng, L., Teo, C.H. and Sim, S.L. (1993) Screening black pepper genotypes and *Piper* spp. for resistance to root knot nematode. *The Black Pepper Industry – Problems and Prospects*. University Pertanian Malaysia Bintulu Campus, Sarawak, Malaysia, pp. 132–139.

Pegg, K.C., Moffett, M.L. and Colbran, R.C. (1974) Disease of ginger in Queensland. *Queensland Agricultural Journal* 100, 611–618.

Poornima, K. and Sivagami, V. (1998a) Pathogenicity of *Meloidogyne incognita* of turmeric (*Curcuma longa* L.). In: *Nematology: Challenges and Opportunities in 21st Century*. Proceedings of the Third International Symposium of Afro-Asian Society of Nematologists (TISAASN), Sugarcane Breeding Institute (ICAR), April 16–19 1998, Coimbatore, India. Afro-Asian Society of Nematologists, Luton, UK, pp. 29–31.

Poornima, K. and Sivagami V. (1998b) Effect of different soil types on *Meloidogyne incognita* in turmeric (*Curcuma longa* L.). *Journal of Plantation Crops* 26, 162–163.

Poornima, K. and Sivagami, V. (1999) Occurrence and seasonal population behaviour of phytonematodes in turmeric (*Curcuma longa* L.). *Pest Management in Horticultural Ecosystems* 5, 42–45.

Rabindran, R., Marimuthu, T., Sivakumar, M. and Natarajan, S. (1987) Role of graded doses of potash on the incidence of *Phytophthora* wilt of betelvine. *Journal of Potassium Research* 3, 155–159.

Rama, K. (1987) Ecology of plant parasitic nematodes affecting some major perennial crops in West Bengal. PhD Thesis, University of Visva-Bharati, India.

Rama, K. and Das Gupta, M.K. (1985) Physical ecology of plant pathogenic nematodes associated with ginger in Darjeeling district, West Bengal. *Indian Journal of Nematology* 15, 284–285 (abstract).

Rama, K. and Das Gupta, M.K. (1998) Population ecology and community structure of plant parasitic nematodes associated with ginger in West Bengal. *Indian Journal of Nematology* 28, 10–14.

Rama, K. and Das Gupta, M.K. (2000) Population ecology and community structure of plant parasitic nematodes associated with ginger in northern West Bengal. *Indian Journal of Nematology* 30, 42–45.

Ramana, K.V. (1986) Slow-wilt disease of black pepper and the role of plant parasitic nematodes in its ctiology. *Proceedings of the Second Group Discussion on Nematological Problems of Plantation Crops*, 24–25 April 1985, Central Coffee Research Station, Balehonnur, Karnataka, India, pp. 17–21.

Ramana, K.V. (1992) *Role of Nematodes in the Incidence of Slow Decline (Slow Wilt Disease) of Black Pepper and Screening Pepper Germplasm Against Nematodes*. Final Report. National Research Centre for Spices, Calicut, Kerala, India.

Ramana, K.V. (1994) Efficacy of *Paecilomyces lilacinus* (Thom.) Samson in suppressing nematode infestations in black pepper (*Piper nigrum* L.). *Journal of Spices and Aromatic Crops* 3, 130–134.

Ramana, K.V. and Eapen, S.J. (1997) *The Parasitic Nematode,* Trophotylenchulus piperis *Mohandas, Ramana and Raski and its Interaction with Black Pepper.* Final Report of the Ad-hoc Project, Indian Institute of Spices Research, Calicut, Kerala, India.

Ramana, K.V. and Eapen, S.J. (1998) Plant parasitic nematodes associated with spices and condiments. In: Trivedi, P.C. (ed.) *Nematode Diseases in Plants.* CBS Publishers and Distributors, New Delhi, pp. 217–251.

Ramana, K.V. and Mohandas, C. (1983) Plant-parasitic nematodes associated with black pepper (*Piper nigrum* L.) in Calicut district, Kerala. *Abstracts of Papers, Third Nematology Symposium, Nematological Society of India.* Himachal Pradesh Krishi Viswa Vidyalaya, Solan, India, p. 9.

Ramana, K.V. and Mohandas, C. (1986) Reaction of black pepper germplasm to root-knot nematode, *Meloidogyne incognita. Indian Journal of Nematology* 16, 138–139.

Ramana, K.V. and Mohandas, C. (1987a) Plant parasitic nematodes associated with black pepper (*Piper nigrum* L.) in Kerala. *Indian Journal of Nematology* 17, 62–66.

Ramana, K.V. and Mohandas, C. (1987b) Nematological problems in black pepper (*Piper nigrum* L.). In: *Proceedings of the Third Group Discussion on Nematological Problems of Plantation Crops,* 29–30 October 1987, Sugarcane Breeding Institute, Coimbatore, Tamil Nadu, India, pp. 43–44.

Ramana, K.V. and Mohandas, C. (1989) Endoparasitic nematodes infecting roots of black pepper (*Piper nigrum*) in two districts of Karnataka, India. *International Nematology Network Newsletter* 6, 33–35.

Ramana, K.V., Mohandas, C. and Balakrishnan, R. (1987a) Role of plant parasitic nematodes in the slow-wilt disease complex of black pepper (*Piper nigrum* L.) in Kerala. *Indian Journal of Nematology* 17, 225–230.

Ramana, K.V., Mohandas, C. and Ravindran, P.N. (1987b) Reaction of black pepper germplasm to the burrowing nematode (*Radopholus similis*). *Journal of Plantation Crops* 15, 65–66.

Ramana, K.V., Sarma, Y.R. and Mohandas, C. (1992) Slow decline disease of black pepper (*Piper nigrum* L.) and role of plant parasitic nematodes and *Phytophthora capsici* in the disease complex. *Journal of Plantation Crops* 20 (Supplement), 65–68.

Ramana, K.V., Mohandas, C. and Eapen, S.J. (1994) *Plant Parasitic Nematodes and Slow Decline Disease of Black Pepper.* Technical Bulletin, National Research Centre for Spices, Calicut, Kerala, India.

Ramana, K.V., Eapen, S.J. and Sarma, Y.R. (1998) Effect of *Meloidogyne incognita, Pythium aphanidermatum* and *Pseudomonas solanacearum* alone and in combinations in ginger. In: Mehta, U.K. (ed.) *Nematology – Challenges and Opportunities in 21st Century.* Afro-Asian Society of Nematologists, Luton, UK, pp. 87–92.

Ramana, K.V., Eapen, S.J. and Beena, B. (2002) *Biological Control of Plant Parasitic Nematodes of Major Spices.* Final Report of ICAR Ad-hoc Project, Indian Institute of Spices Research, Calicut, Kerala, India.

Rana, S., Acharya, A. and Padhi, N.N. (1991) Effect of different sources and levels of nitrogen on the incidence of root-knot of betelvine (*Piper betle* L.). *Indian Journal of Nematology* 21, 128–130.

Rao, B.N., Sultan, M.A. and Reddy, K.N. (1993) Residues of carbofuran in betelvine. *Indian Journal of Plant Protection* 21, 217–219.

Rao, D.V.S., Sitaramaiah, K. and Maiti, S. (1996) Management of plant parasitic nematodes in betelvine gardens through non-chemical methods. *Indian Journal of Nematology* 26, 93–97.

Rao, M.S., Reddy, P.P., Khan, R.M. and Rao, N.N.R. (1991) Role of standards in building of *Meloidogyne incognita* in betelvine. *Current Nematology* 2, 143–144.

Rao, P.S., Krishna, M.R., Srinivas, C., Meenakumari, K. and Rao, A.M. (1994) Short duration, disease-resistant turmerics for northern Telangana. *Indian Horticulture* 39, 55–56.

Rashid, A., Khan, F.A. and Khan, A.M. (1973) Plant parasitic nematodes associated with vegetables, fruits, cereals and other crops in North India. 1. Uttar Pradesh. *Indian Journal of Nematology* 3, 8–23.

Ratnasoma, H.A., Gowen, S.R. and Wijekoon, S. (1991) Evaluation of some potential methods for control of root knot nematodes in black pepper (*Piper nigrum*). *Tropical Agriculturist* 147, 135–142.

Raut, S.P. and Pande, V.S. (1986) A first report of *Meloidogyne incognita* causing root-knot disease of cardamom in Konkan. *Indian Journal of Mycology and Plant Pathology* 16, 167–168.

Ravindran, P.N., Ramana, K.V., Nair, M.K., Babu, K.N. and Mohandas, C. (1992) Pournami – a high yielding black pepper selection tolerant to root-knot nematode (*Meloidogyne incognita*). *Journal of Spices and Aromatic Crops* 1, 136–141.

Ray, P., Dash, S.C. and Acharya, A. (1993) Pathogenic association of *Colletotrichum* sp. with *Meloidogyne incognita* in causation of lethal yellowing complex of betelvine. *Orissa Journal of Agricultural Research* 6, 5–9.

Ray, S., Mohanty, K.C., Mohapatra, S.N., Patnaik, P.R. and Ray, P. (1995) Yield losses in ginger (*Zingiber officinale* Rosc.) and turmeric (*Curcuma longa* L.) due to root knot nematode (*Meloidogyne incognita*). *Journal of Spices and Aromatic Crops* 4, 67–69.

Reddy, D.B. (1977) Pests, diseases and nematodes of major spices and condiments in Asia and the Pacific. *Technical Document No. 114,* Plant Protection Committee for the South East Asia and Pacific Region, FAO, Bangkok.

Reddy, D.B. (1978) Insect pests, diseases and nematodes of areca palm (*Areca catechu* L.) and betel vine (*Piper betle*) recorded in Asia and the Pacific. *Technical Document No. 114.* Plant Protection Committee for the South East Asia and Pacific Region, FAO, Bangkok.

Routaray, B.N., Mohapatra, M.P. and Das, S.N. (1987a) Effects of *Meloidogyne incognita* on ginger, *Zingiber officinale* Rosc. *Indian Journal of Nematology* 17, 327–329.

Routaray, B.N., Sahoo, H. and Das, S.N. (1987b) Nemic associations of ginger and turmeric in Orissa. *Indian Journal of Nematology* 17, 122–123.

Sahoo, N.K., Ganguly, S. and Eapen, S.J. (2000) Description of *Meloidogyne piperi* sp.n. (Nematoda: Meloidogynidae) isolated from the roots of *Piper nigrum* in South India. *Indian Journal of Nematology* 30, 203–209.

Samuel, M. and Mathew, J. (1983) Role and association of root knot nematode *Meloidogyne incognita* in the induction of bacterial wilt of ginger incited by *Pseudomonas solanacearum. Indian Phytopathology* 36, 398–399.

Sarma, Y.R., Nambiar, K.K.N. and Nair, C.P.R. (1974) Brown rot of turmeric. *Journal of Plantation Crops* 2, 33–34.

Sasikumar, B., Saji, K.V., Alice, A., George, J.K., Zachariahk, T.J. and Eapen, S.J. (2003) IISR Mahima and IISR Rejatha – two high yielding and high quality ginger (*Zingiber officinale* Rosc.) cultivars. *Journal of Spices and Aromatic Plants* 12, 34–37.

Selvan, M.T. (2002) *Arecanut and Spices Database.* Directorate of Arecanut and Spices Development, Calicut, India.

Sen, K. and Das Gupta, M.K. (1977) Additional hosts of the root-knot nematode, *Meloidogyne* spp. from India. *Indian Journal of Nematology* 7, 74.

Shah, H.M. and Patel, D.J. (1979) Occurrence of root-knot disease in cumin. *Indian Journal of Nematology* 9, 179–180.

Shah, J.J. and Raju, E.C. (1977) Histopathology of ginger (*Zingiber officinale*) infested by soil nematode *Meloidogyne* sp. *Phyton* 16, 79–84.

Sharma, G.C. and Bajaj, B.K. (1998) Effect of inter-cropping bell-pepper with ginger on plant parasitic nematode populations and crop yields. *Annals of Applied Biology* 133, 199–205.

Sharma, R.D. and Loof, P.A.A. (1974) Nematodes of cocoa region of Bahia, Brazil. IV. Nematodes in the rhizosphere of pepper (*Piper nigrum* L.) and clove (*Eugenia caryophyllata* Thumb). *Revista Theobroma* 4, 26–32.

Sheela, M.S. and Venkitesan, T.S. (1981) Inter-relationships of infectivity between the burrowing and root-knot nematodes in black pepper, *Piper nigrum* L. (Abstr.). *Indian Journal of Nematology* 11, 105.

Sheela, M.S. and Venkitesan, T.S. (1990) Interaction between *Meloidogyne incognita* and the fungus *Fusarium* sp. on black pepper vine (*Piper nigrum* L.). *Indian Journal of Nematology* 20, 184–188.

Sheela, M.S., Venkitesan, T.S. and Mohandas, N. (1993) Status of *Bacillus* spp. as biocontrol agents of root knot nematode (*Meloidogyne incognita*) infesting black pepper (*Piper nigrum* L.). *Journal of Plantation Crops* 21 (Supplement), 218–222.

Sheela, M.S., Bai, H., Jiji, T. and Kuriyan, K.J. (1995) Nematodes associated with ginger rhizosphere and their management in Kerala. *Pest Management in Horticultural Ecosystems* 1, 43–48.

Shenoy, J.R. (1985) Betel vine cultivation. *Indian Cocoa, Arecanut and Spices Journal* 18, 84–85.

Sher, S.A., Chunram, C. and Pholcharoen, S. (1969) Pepper yellows disease and nematodes in Thailand. *FAO Plant Protection Bulletin* 17, 33.

Sitaramaiah, K. and Devi, P. (1990) Influence of root-knot nematode *Meloidogyne incognita* on *Sclerotium* wilt of betelvine. *Indian Journal of Nematology* 20, 230–231.

Sitaramaiah, K. and Devi, P. (1994) Nematodes, soil microorganisms in relation to *Phytophthora* wilt disease in betelvine. *Indian Journal of Nematology* 24, 16–25.

Sivakumar, M. and Marimuthu, T. (1984) Parasitic nematodes associated with betel vine (*Piper betle* L.) in Tamil Nadu. *Madras Agricultural Journal* 71, 108–110.

Sivakumar, M. and Marimuthu, T. (1986a) Population dynamics of phytonematodes associated with betel vine (*Piper betle* L.), banana and paddy with special reference to crop rotation. *Indian Journal of Nematology* 16, 277.

Sivakumar, M. and Marimuthu, T. (1986b) Efficacy of different organic amendments against phytonematodes associated with betel vine (*Piper betle* L.). *Indian Journal of Nematology* 16, 278.

Sivakumar, M. and Marimuthu, T. (1987) Preliminary studies on the effect of solarization on phytonematodes of betel vine. *Indian Journal of Nematology* 17, 58–59.

Sivakumar, M. and Muthukrishnan, T.S. (1985) Record of *Criconemella parvula* (Siddiqi, 1961) De Grisse and Loof 1961 in South India. *Indian Journal of Nematology* 15, 254.

Sivakumar, M., Natarajan, S. and Balasubramaniam, M. (1987) Nematode pests of betel vine (*Piper betle* L.) and their management. In: *Proceedings of the Third Group Discussions on the Nematological Problems of Plantation Crops*, 29–30 October 1987, Sugarcane Breeding Institute, Coimbatore, Tamil Nadu, India, pp. 34–35.

Sivaprasad, P., Jacob, A., Nair, S.K. and George, B. (1990) Influence of VA mycorrhizal colonization on root-knot nematode infestation in *Piper nigrum* L. In: *Trends in Mycorrhizal Research*. Proceedings of the National Conference on Mycorrhiza, 14–16 February 1990, Haryana Agricultural University, Hisar, India, pp. 100–101.

Sivaprasad, P., Jacob, A., Sulochana, K.K., Visalakshi, A. and George, B. (1992) Growth root-knot nematode infestation and phosphorus nutrition in *Piper nigrum* (L.) as influenced by vesicular arbuscular mycorrhizae. In: *Proceedings of the 3rd International Conference on Plant Protection in the Tropics*, 20–23 March 1990, Genting Highlands, Malaysia, Volume 6, pp. 34–37.

Sosamma, V.K. and Koshy, P.K. (1997) Biological control of *Meloidogyne incognita* on black pepper by *Pasteuria penetrans* and *Paecilomyces lilacinus*. *Journal of Plantation Crops* 25, 72–76.

Sosamma, V.K., Sundararaju, P. and Koshy, P.K. (1979) Effect of *Radopholus similis* on turmeric. *Indian Journal of Nematology* 9, 27–31.

Sosamma, V.K., Geetha, S.M. and Koshy, P.K. (1994) Effect of the fungus, *Paecilomyces lilacinus* on the burrowing nematode, *Radopholus similis* infesting betel vine. *Indian Journal of Nematology* 24, 50–53.

Sreeja, T.P., Eapen, S.J. and Ramana, K.V. (1996) Occurrence of *Verticillium chlamydosporium* Goddard in a black pepper (*Piper nigrum* L.) garden in Kerala, India. *Journal of Spices and Aromatic Crops* 5, 143–147.

Srivastava, L.S., Gupta, S.R., Basnet, C.P., Mahato, U.P. and Neopani, B. (1998) Microorganisms associated with ginger in Sikkim. *Journal of Hill Research* 11, 120–122.

Stirling, G.R. (1989) Organic amendments for control of root-knot nematode (*Meloidogyne incognita*) on ginger. *Australasian Plant Pathology* 18, 39–44.

Stirling, G.R. and Nikulin, A. (1998) Crop rotation, organic amendments and nematicides for control of root knot nematodes (*Meloidogyne incognita*) on ginger. *Australasian Plant Pathology* 27, 234–243.

Sudha, S. and Sundararaju, P. (1986) Pathogenicity of *Meloidogyne incognita* on ginger (*Zingiber officinale* Rose.). *Indian Journal of Nematology* 16, 258.

Sudha, S., Koshy, P.K. and Sundararaju, P. (1989) Effect of root-knot nematode, *Meloidogyne incognita* on the growth of turmeric. *Journal of Plantation Crops* 16 (Supplement), 293–295.

Sundararaju, P. and Sudha, S. (1998) Nematode management in arecanut and arecanut based farming system. In: *Nematology: Challenges and Opportunities in 21st Century.* Proceedings of the Third International Symposium of Afro-Asian Society of Nematologists (TISAASN), 16–19 April 1998, Sugarcane Breeding Institute, Coimbatore, India, pp. 190–192.

Sundararaju, P. and Suja, C.P. (1986) Occurrence of *Radopholus similis* on betel vine (*Piper betle* L.) in Karnataka. *Indian Journal of Nematology* 16, 279.

Sundararaju, P., Koshy, P.K. and Sosamma, V.K. (1979a) Plant parasitic nematodes associated with spices. *Journal of Plantation Crops* 7, 15–26.

Sundararaju, P., Koshy, P.K. and Sosamma, V.K. (1979b) Survey of plant parasitic nematodes associated with spices in Kerala and Karnataka. In: *Proceedings of the Second Symposium on Plantation Crops*, Ootacamund, Tamil Nadu, India, pp. 39–44.

Sundararaju, P., Sosamma, V.K. and Koshy, P.K. (1979c) Pathogenicity of *Radopholus similis* on ginger. *Indian Journal of Nematology* 9, 91–94.

Sundararaju, P., Ramana, K.V. and Eapen, S.J. (1995) Development of *Trophotylenchulus piperis* in black pepper roots. *Afro-Asian Journal of Nematology* 5, 166–168.

Sundararaju, P., Ramana, K.V. and Eapen, S.J. (1997) Distribution of the pepper nematode, *Trophotylenchulus piperis* on black pepper (*Piper nigrum* L.) roots. *Indian Journal of Nematology* 27, 116–117.

Swarup, G., Nath, R.P. and Sethi, C.L. (1967) The plant parasitic nematode genus *Rotylenchulus* in India. *Indian Phytopathology* 20, 118–123.

Thomas, G.V., Sundararaju, P., Ali, S.S. and Ghai, S.K. (1989) Individual and interactive effects of VA mycorrhizal fungi and root knot nematodes, *Meloidogyne incognita* on cardamom. *Tropical Agriculture* 66, 21–24.

Timm, R.W. (1965) *A Preliminary Survey of the Plant Parasitic Nematodes of Thailand and the Philippines.* South East Treaty Organization.

Vadhera, I., Tiwari, S.P. and Dave, G.S. (1998a) Plant parasitic nematodes associated with ginger (*Zingiber officinale*) in Madhya Pradesh and denematization of infested rhizome by thermotherapy for management. *Indian Journal of Agricultural Sciences* 68, 367–370.

Vadhera, I., Tiwari, S.P. and Dave, G.S. (1998b) Integrated management of root knot nematode, *Meloidogyne incognita* in ginger. *Indian Phytopathology* 51, 161–163.

Van Der Vecht, J. (1950) Plant parasitic nematodes. In: Karshoven, L.G.E. and Vecht, J.V.D. (eds) [*Diseases of Cultivated Plants in Indonesia Colonies.*] W. van Woeve, The Hague, pp. 16–15.

Venkata Rao, A., Narasimhan, V., Rajagopalan, P., Vidyasekharan, P. and Muthukrishnan, T.S. (1973) Efficacy of nematicides for the control of betel vine wilt. *Indian Journal of Agricultural Science* 43, 1072–1075.

Venkitesan, T.S. (1972) On the occurrence of plant parasitic nematodes associated with different crops in Cannanore district, Kerala. *Agricultural Research Journal of Kerala* 10, 179–180.

Venkitesan, T.S. (1976) Studies on the burrowing nematode, *Radopholus similis* (Cobb, 1893) Thorne, 1949 on pepper (*Piper nigrum* L.) and its role in slow wilt disease. PhD Thesis, Department of Plant Pathology, University of Agricultural Sciences, Bangalore, India.

Venkitesan, T.S. and Jacob, A. (1985) Integrated control of root-knot nematode, infecting pepper vines in Kerala. *Indian Journal of Nematology* 15, 261–262.

Venkitesan, T.S. and Setty, K.G.H. (1977) Pathogenicity of *Radopholus similis* to black pepper (*Piper nigrum*). *Indian Journal of Nematology* 7, 17–26.

Venkitesan, T.S. and Setty, K.G.H. (1978) Reaction of 27 black pepper cultivars and wild forms to the burrowing nematode, *Radopholus similis* (Cobb) Thorne. *Journal of Plantation Crops* 6, 81–83.

Venkitesan, T.S. and Setty, K.G.H. (1979) Control of the burrowing nematode *Radopholus similis* (Cobb) on black pepper. *Pesticides* 13, 40–43.

Verma, S.K. and Prasad, S.K. (1969) The reniform nematode, *Rotylenchulus reniformis* Linford and Oliveira, 1940. II. Studies on control. *Indian Journal of Entomology* 32, 68–72.

Vilsoni, F., McClure, M.A. and Butler, L.D. (1976) Occurrence, host range and histopathology of *Radopholus similis* in ginger (*Zingiber officinale*). *Plant Disease Reporter* 60, 417–420.

Viswanathan, P.R.K., Kumar, A.C. and D'Souza, G.I. (1974) Nematode association of cardamom in South India. *Cardamom News* 6, 19.

Wahid, P.A., Kamalam, N.V. and Venugopal, V.K. (1982) Mineral nutrition of slow wilt affected black pepper (*Piper nigrum* L.). *Journal of Plantation Crops* 10, 21–25.

Willers, P. (1985) Efficient nematode control a step closer to ginger producers. *Information Bulletin, Citrus and Subtropical Fruit Research Institute* No. 160, 4–6.

Willers, P. (1991) Control of root-knot nematode in ginger rhizomes by dip treatment in fenamiphos. *Inlightins Bulletin Navorsingsinstituut vir Sitrus en Subtropiese Vrugte* No. 230, 14–17.

Winoto, S.R. (1972) Effect of *Meloidogyne* species on the growth of *Piper nigrum* L. *Malaysian Agricultural Research* 1, 86–90.

Zhou, K.X. and Chi, P.K. (1993) Identification of the pathogens causing wilt diseases of the common andrographis (*Andrographis paniculata* (Burmf.) Nees.) and black pepper (*Piper nigrum* L.). *Journal of South China Agricultural University* 14, 117–123.

22 Management Practices: an Overview of Integrated Nematode Management Technologies

Richard A. Sikora,[1] John Bridge[2] and James L. Starr[3]
[1]*Soil Ecosystem Phytopathology and Nematology, Institut für Pflanzenkrankheiten, Universität Bonn, Nussallee 9, D-53115 Bonn, Germany;*
[2]*CABI Bioscience, Bakeham Lane, Egham, Surrey TW20 9TY, UK;*
[3]*Department of Plant Pathology & Microbiology, Texas A&M University, College Station, TX 77843-2132, USA*

There has been a drastic change in the direction that management of nematodes has taken since the first version of this book was printed in 1990. Nematode control as used in that first book has expanded from integrated pest management (IPM), which relied heavily on the use of chemical control, to integrated crop management (ICM) which stresses cultural methods of nematode management to bio-management (BM) or bio-system-management (BSM), which favours non-chemical management of nematode populations, to present day integrated nematode management or natural pest management (NPM) strategies (Bridge, 1996; Sikora, 1992). Present day management approaches are more holistic, in that a broad combination of tools is used for nematode management based on farmers' needs rather than scientific ideology. Our thinking has progressed from focusing only on the eradication of nematodes from the soil by chemical means in 'Nematode Control', to the wider view of 'Sustainable Nematode Management' in which some yield loss is accepted. Even though there are obvious differences in all the approaches to nematode control, the ultimate goals remain the same – nematode population reduction and increased yield at cost-effective levels.

The number of management tools used for nematode management has increased drastically in the past 25 years – from a strong reliance on nematicides and straightforward crop rotation – to often highly complex management programmes. This shift in emphasis has occurred because of a number of major changes in nematode–crop interactions:

1. Development of nematode races virulent on a common resistance gene.
2. Shortening of rotations for marketing reasons.
3. Expansion of protected cultivation both under plastic and in soil-free systems.
4. Loss of important non-fumigant nematicides.
5. The ban on the highly effective fumigant methyl bromide.
6. Detection of new and economically important species of nematodes.
7. Spread of important parasites to non-infested regions.

Depending on the crop involved, these factors are often interrelated and have had a major impact on crop yield.

There has also been a marked shift in research on nematode management from an emphasis on nematicides to studies on biological control (Fig. 22.1). This shift is due to many factors, including the absence of new nematicides and increased concerns for environmental quality and human health. The shift has also been due to the removal of a number of highly effective nematicides from the market and the recent banning of the broad-spectrum fumigant methyl bromide. Simultaneously, the development of new nematicides has been slowed down because of: (i) the high level of toxicity required to control a nematode in 3000 t of topsoil; (ii) the high costs of pesticide development and registration; and (iii) the limited market size for these compounds. In addition, for many field crops and in many smallholder situations where nematicides are too costly, alternative management measures are needed.

Increased losses, due to nematode infestations associated with the developments listed above, have generated substantial research to find acceptable alternatives. Old and new alternatives have greatly affected how nematodes are managed and how crops are grown. The development of new resistant varieties in many crops has been promoted. Nematode management based on monitoring threshold levels with standard methods is important. The use of remote sensing followed by treatment with precision farming practices is advancing in importance.

Nematode management is a two-pronged proposition depending on whether your ultimate goal is food or profit and, therefore, is driven by resource availability. Management also varies greatly between the different types of farmers growing crops worldwide: subsistence, resource-limited, conventional small-scale, and modern extensive or high-intensity commercial growers. Flexibility in styling a rotation, therefore, depends on a grower's goals and resources:

- food for survival;
- production for profit;
- availability of management tools;
- resources for procurement; and
- knowledge of control application.

It should not be forgotten that the majority of the world's cultivated land is still farmed by small-scale farmers using traditional methods (Altieri, 1984) and these traditional farming practices can be beneficial in pest management. The occurrence of

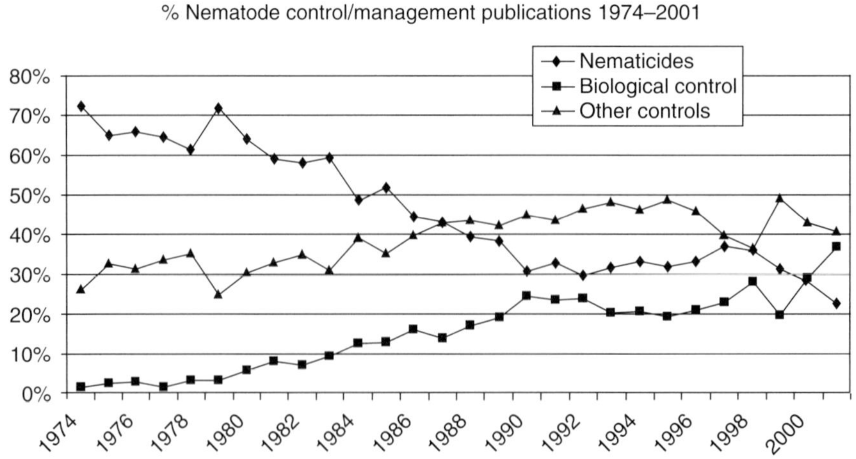

Fig. 22.1. Comparison of the number of publications on research work on nematicides, biological control and general management techniques published between 1974 and 2001.

a serious nematode problem is often one of the first indications that a farming system has become unsustainable (Page and Bridge, 1993). Truly indigenous nematodes generally are not a problem in traditional cropping systems. They become pests normally because of a change in the cropping system, a change in the farming practice, a change in the climate or because they have been introduced. Conversely, the existence of a nematode problem in commercial production is usually well known and the use of any management tool that is cost effective and leads to high yields is acceptable.

This chapter draws heavily on the personal experience of the authors and material presented in the previous chapters in this book. It is not to be seen as a thorough review of the literature. Excellent books have been written and many good reviews published that have been devoted to integrated nematode management (e.g. Brown and Kerry, 1987; Barker *et al.*, 1998; Whitehead, 1998), and should be consulted for guidelines in the structuring of integrated management programmes. Whatever management approach is taken, the ultimate goal is stable or higher yield and, in most cases, increased profit. It is often stated that a farmer is not interested in controlling nematodes but in food production and/or profit depending on the level of production or the production form involved.

The tools used in the management of nematodes can be applied at different times in a cropping cycle or in a rotation sequence. In order to list all the tools known to be effective for nematode control in a logical order, the following headings are used for management approaches:

1. Exclusion, quarantine and diagnosis.
2. Inter-cycle management between susceptible crops.
3. Pre-plant management just prior to planting.
4. At-planting treatments.
5. Plant management.
6. Post-harvest management.
7. Integrated nematode management strategies.

Exclusion, Quarantine and Diagnosis

Exclusion is the most effective and economical means of preventing nematode damage. Preventing the introduction of important pests and/or local spread within a country or region has been effective in the past and needs to be strengthened in the future. The global market for agricultural products with efficient long-distance movement of plant material has supported the spread of important nematode pests around the world and between countries on the same continent. Some recent examples of spread of economically important nematodes are:

- pine wilt nematode *Bursaphelenchus xylophilus* to Europe;
- soybean cyst nematode *Heterodera glycines* to Brazil;
- burrowing nematode *Radopholus similis* to non-infested banana plantations worldwide;
- red ring *Bursaphelenchus cocophilis* to South America; and
- potato cyst nematode *Globodera rostochiensis* to the Philippines.

CAB International publishes up to-date maps showing the worldwide distribution of economically important species. They should be consulted for detailed information on the distribution of species important to quarantine agencies. Distribution maps for species of plant parasitic nematodes not yet widely distributed, but considered by the authors to be of serious importance to agriculture and quarantine agencies, are presented in Figs 22.2–22.9. Distribution maps of important species already having a wide distribution are not presented but should be consulted for more information on other important species.

The reduction in the number of nematologists working in quarantine offices will aggravate this situation in the years to come unless governments react accordingly. Prevention of spread at a local or country level can lead to significant savings in food production, as has been seen in the effectiveness of quarantine of the

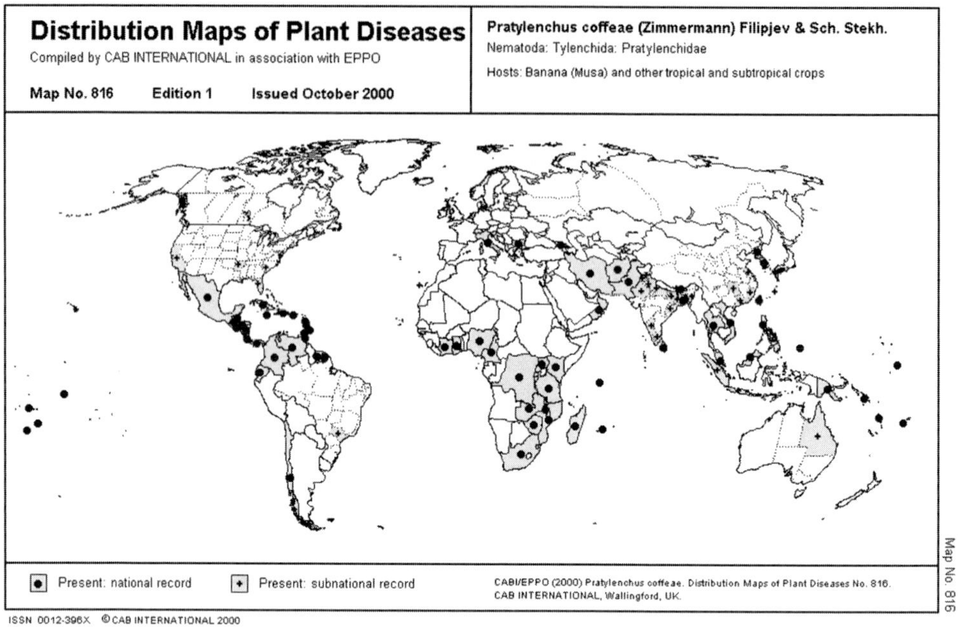

Fig. 22.2. Worldwide distribution of *Pratylenchus coffeae*.

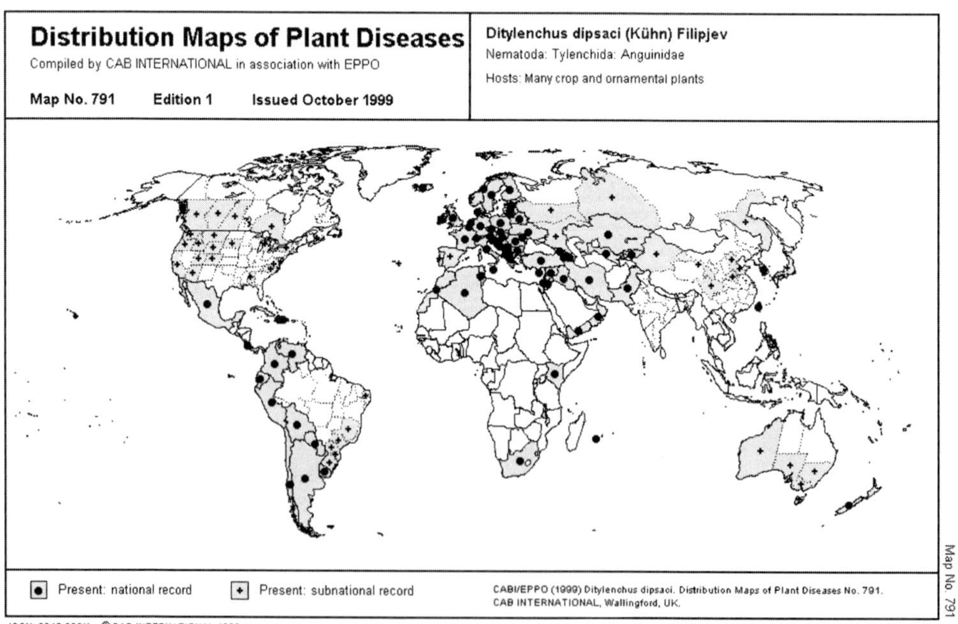

Fig. 22.3. Worldwide distribution of *Ditylenchus dipsaci*.

citrus burrowing nematode, still limited to Florida, the potato cyst nematode that has been isolated in New York state, the extensive measures being made to prevent the spread of the pine wilt nematode *Bursaphelenchus xylophilus* from Portugal

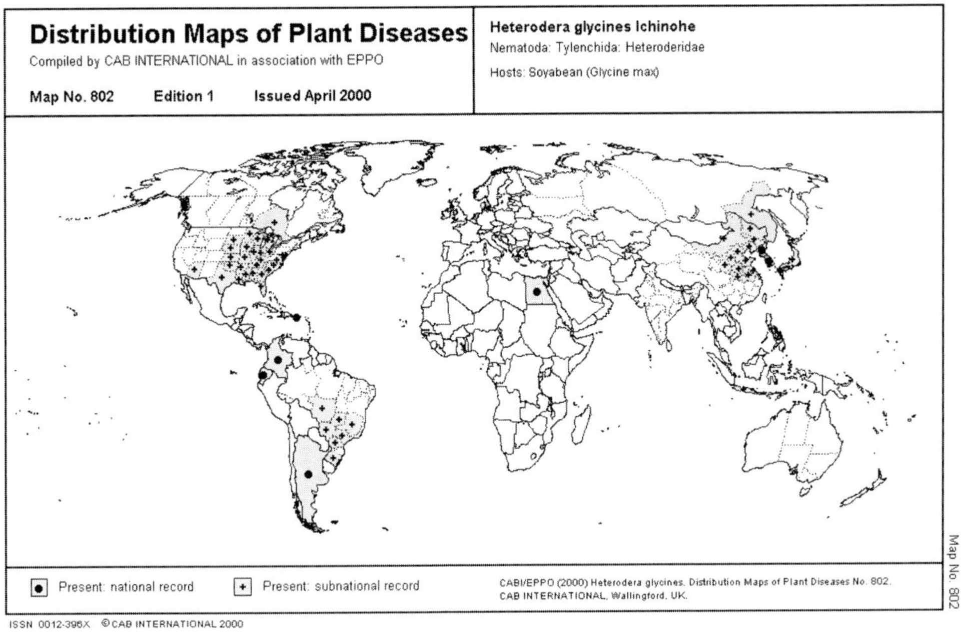

Fig. 22.4. Worldwide distribution of *Heterodera glycines*.

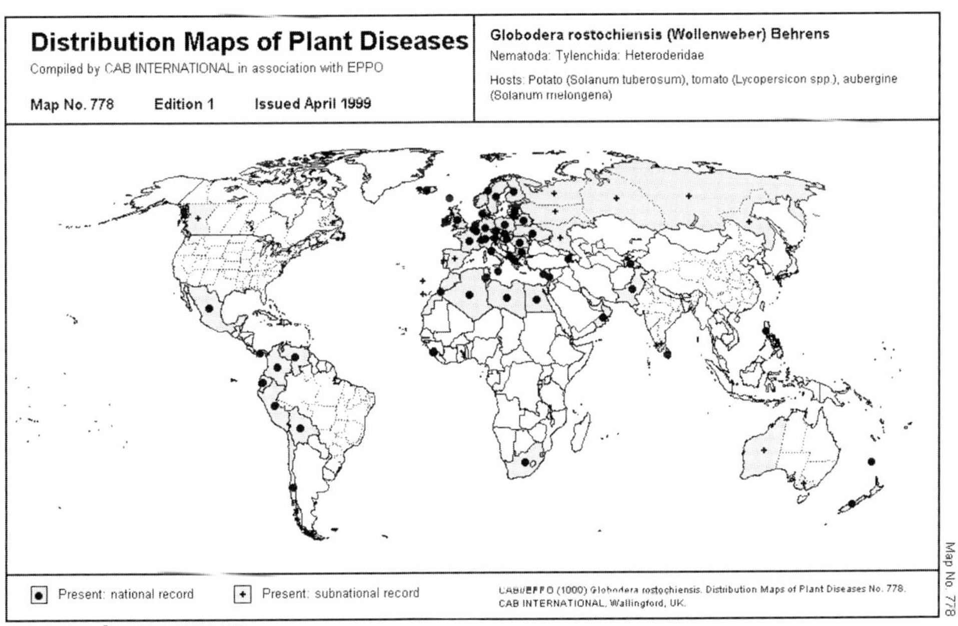

Fig. 22.5. Worldwide distribution of *Globodera rostochiensis*.

to the rest of Europe, and the development of new quarantine laws to limit the spread of *Meloidogyne chitwoodi* within Europe.

Identification of nematodes is critical to quarantine and plant protection in general. The recent description of a new species of

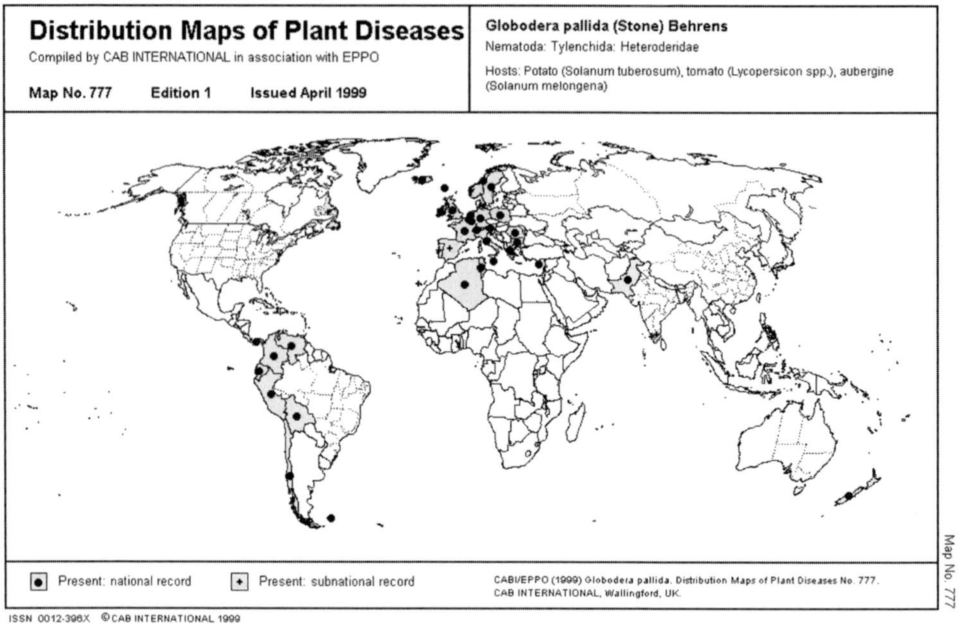

Fig. 22.6. Worldwide distribution of *Globodera pallida*.

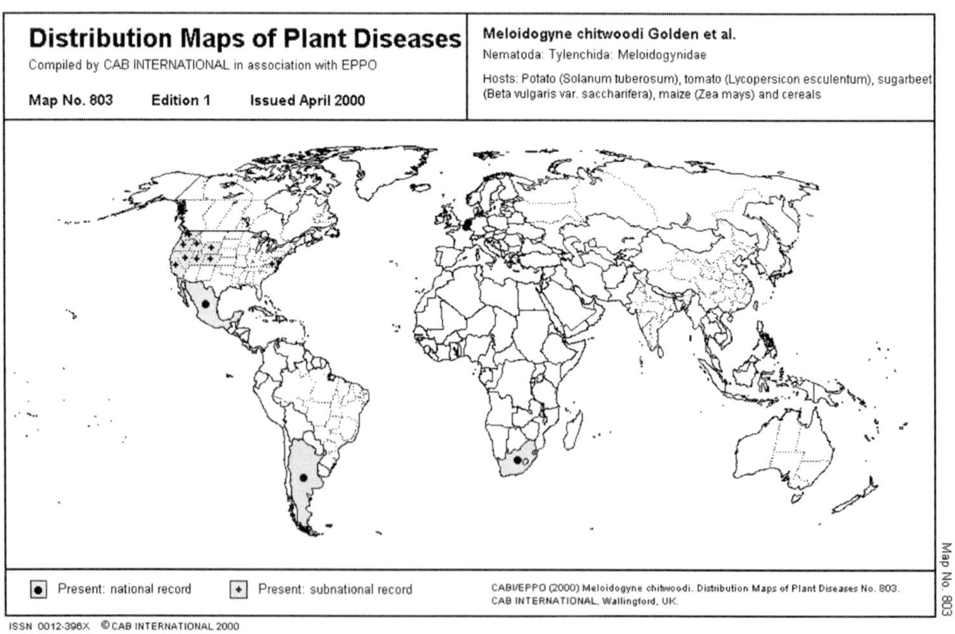

Fig. 22.7. Worldwide distribution of *Meloidogyne chitwoodi*.

root knot nematode, *Meloidogyne floridensis*, which infects all known sources of resistance in tomato and soybean, is of major importance due to the fact that identification could only be accomplished by molecular diagnosis. Up until then it was

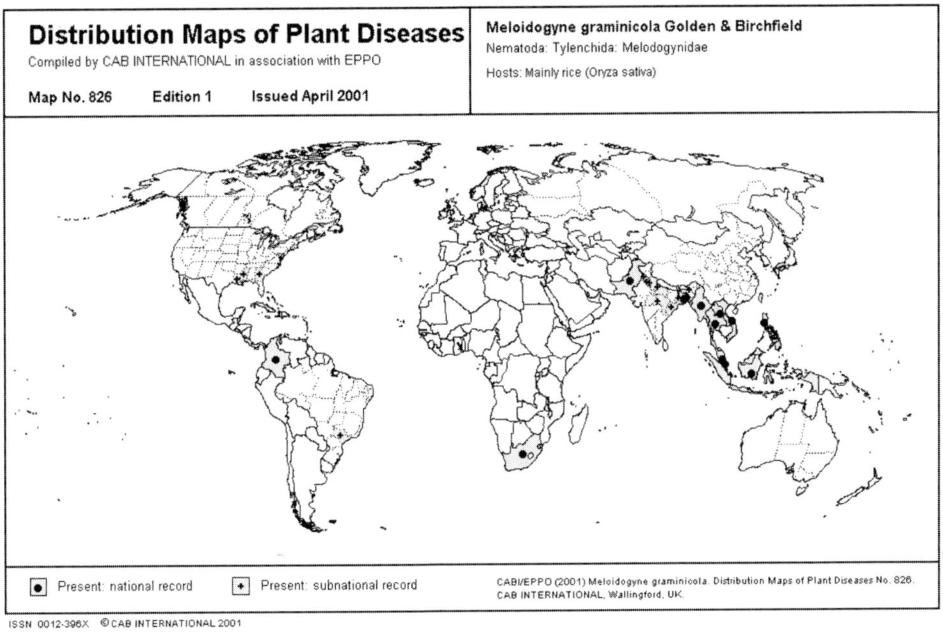

Fig. 22.8. Worldwide distribution of *Meloidogyne graminicola*.

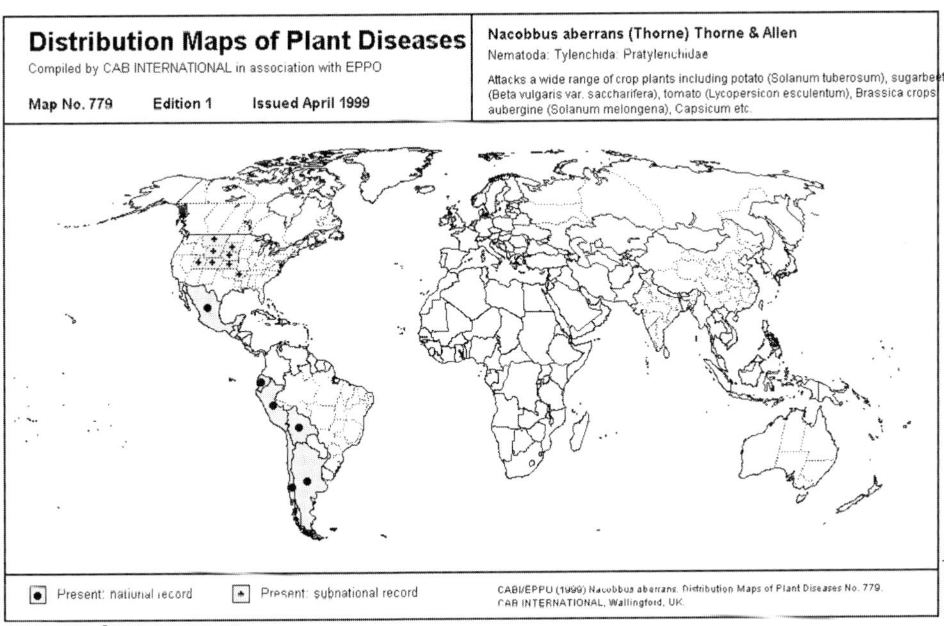

Fig. 22.9. Worldwide distribution of *Nacobbus aberrans*.

considered a race of *M. incognita*. The loss of nematode taxonomists to science will also influence our ability to diagnose problems of this magnitude and will result in crop loss and the need for increased control-measure-related costs over time.

Similar problems with identification exist for *Meloidogyne chitwoodi* and *M. fallax,* and *Globodera rostochiensis* and *G. pallida.*

Inter-cycle Management

The inter-cycle term is used here to describe the time between the multiple crop cycles typical for production in the tropics or subtropics. The term rotation could also be used, but seems to relate more to the temperate regions of the world, where usually one crop is grown on a per-year basis. In tropical and subtropical vegetable production, for instance, as many as six short-cycle crops can be grown in a 12-month period (see Chapter 9). Management tools that can be used in the period between susceptible crops include nematicides, physical means of control and a wide spectrum of crop-based methodologies. In most cases they involve very direct and simple inputs such as fallow, organic amendments or the use of non-host break crops. However, management can also include combinations of approaches that often require information on nematode threshold densities, species composition, crop host spectrum and an economic analysis of the cost to the grower of using multiple inputs.

Physical management tools

Flooding

Nematode densities can drop significantly when soils are flooded for prolonged periods of time. In areas where paddy rice is flooded for prolonged periods of time in the wet season, nematodes are often not a problem in the following dry-season crops (Bridge, 1996). Constant flooding of rice fields for 3 months or more gives acceptable control of root knot nematode for succeeding crops. The degree of root knot damage to processing tomato crops in the Philippines was undetectable in rotations of paddy rice–tomato (R.A. Sikora, unpublished data). Flooding alternated with drying on a 2 to 3 week cycle during the summer has been recommended for vegetables to reduce root knot nematode densities (Noling, 2003) and seems to be more effective than long, continuous flooding cycles. The duration of flooding for effective control may vary with target nematode species. *Meloidogyne graminicola* juveniles are killed after exposure to anaerobic conditions that begin in the soil a few days after flooding (Padgham *et al.*, 2003). The nematode will survive in waterlogged soil, however, for 14 months. *Radopholus similis* can survive in bare soil in the absence of roots for 6 months, and can be controlled efficiently by flooding or planting banana after paddy rice. The duration of flooding for effective control needs to be determined for each nematode species.

Soil tillage

Where the practice is economical, repeated tilling of the soil at regular intervals for 30 days during hot and dry seasons between crops can significantly reduce root knot nematode densities in the upper horizons due to desiccation of eggs and juveniles. One to five deep ploughings was shown to reduce *Heterodera avenae* populations by 9–42% (Mathur *et al.*, 1987). Tillage reduces densities of the target nematode pest as well as secondary pest species, and it also will eliminate alternative weed hosts and volunteer plants from the previous crop. However, one good weed host or one volunteer plant of a susceptible host is often sufficient to maintain a nematode population at threshold densities. Soil tillage and careful mounding-up of the thin top layer of soil into ridges for tobacco beds gave good control of root knot. The upper 5 cm of soil heats to 36°C in the dry season and has only 1% moisture, which leads to total nematode desiccation (Ferris, 1969). Such technology could have application in other crops grown in beds in these types of climates.

Clean fallow

Fallows in plant-free fields are seldom practised due to problems with soil erosion and the simple fact that it is more economical to produce a short-season crop. In addition, clean fallow requires either additional tillage to kill weeds or the use of herbicides, which are both cost factors. Such fallows reduce nematodes more effectively than a field with weed cover due to exclusion of alternative host plants. Clean fallows are most effective in nematode management in the hot, dry summer months between crops. The negative effects on soil conservation will limit the use of clean fallow in many countries.

Organic amendments

Organic amendment is used here to mean organic material incorporated into the soil that comes from external sources such as processing residues or industrial waste products. Organic material added as fresh crop residue and grown in the field in rotation – break, cover, trap, antagonistic or green manure crops – are discussed below. Incorporation into the soil of large amounts of any organic material will reduce nematode densities. Oil cakes, coffee husks (Plate 23A), paper waste, crustacean skeletons, sawdust and chicken manure, amongst others, have been used with some success. Control may be due to any one or more of the following mechanisms:

- toxic and non-toxic compounds present in the organic material;
- toxic metabolites produced during microbial degradation; or
- enhancement of the soil antagonistic potential.

A list of some of the more common organic amendments used for nematode control is given in Table 22.1. Chitin amendments have received much interest in the past as an organic amendment in that they stimulate the antagonistic potential in soil toward nematodes (Culbreath *et al.*, 1985; Rodriguez-Kabana *et al.*, 1987; Spiegel *et al.*, 1987). Organic amendments have also been combined with various biocontrol agents with reports of enhanced levels of control. The use of organic amendments is often limited by availability and, in some cases, by the large quantities needed. In addition to their effects on nematode density, organic amendments also improve soil structure and water-holding capacity, reduce diseases and limit weed growth, all of which ultimately lead to a stronger plant and improved tolerance to nematode attack.

Crop-based management tools

Crop management tools are designed to attain high yield while simultaneously reducing nematode, insect, disease and weed problems, reducing erosion and improving soil fertility. Each production system has different requirements when it

Table 22.1. Important organic soil amendments used for nematode control.

Oil cakes	Agro-industrial wastes	Animal and urban waste	Plant residues
Margosa/neem	Sawdust and tree bark	Chicken manure	Water hyacinth
Mustard	Cellulose waste	Farmyard manure	Seaweed
Groundnut	Sugarcane bagasse	Garden compost	Margosa/neem leaves
Sesame	Sugarcane filtercake	Fish remains	Cabbage leaves
Castor	Rice and coffee husks	Bone meal	Pineapple leaves
Mahuva	Wood ash	Crustacean skeletons	
Cotton seed	Cotton waste	Raw sewage	
Soybean	Cassava peelings	Refuse	
Linseed	Cocoa pods	Urea	
	Tea waste		
	Mycelium waste		
	Potato processing water		
	Sugarbeet processing water		

comes to combating nematode infestations. In addition, the rotation crops used by a grower are planted for different reasons, with the type of rotation crop varying greatly between the tropics and subtropics. Selection is often dependent on the main cash crop in the cropping system. Rotation crops are used to:

- suppress weed growth;
- prevent soil erosion;
- improve soil organic matter levels;
- increase water-holding capacity;
- raise nitrogen concentration directly; or
- control nematodes and other soil-borne pathogens.

Nematode control achieved with crop management is attained by mechanisms including: starvation, trapping, antagonism, stimulation of soil antagonistic potential and/or different degrees of biofumigation. Conversely, in commercial production of many horticultural crops, where fumigation is the backbone of the cropping system and sequential cropping of susceptible vegetable crops is practised, rotation may not be needed.

Weed Fallow

In a normal fallow, weed growth is not usually managed and often leads to extensive biomass production. However, if a few weed species are good hosts for the pest nematode or sufficient volunteer plants of the preceding susceptible crop are present, nematode densities may actually increase during the fallow period. Mulching of the weeds prior to planting of the next crop stimulates the antagonistic potential in the soil and leads to a reduction in inoculum densities. Such fallows are common in the tropics during the rainy periods between major crops. Incorporation and solarization of these weeds has been shown to lead to a significant reduction in root knot in horticultural crops (see Chapter 9).

Non-host crops

Non-host crops are defined here as crops harvested for marketing purposes as opposed to cover crops used for soil conservation, animal grazing or direct nematode control. Rotation with non-host crops is the most important technique used for root knot management worldwide and has been discussed in detail elsewhere (Nusbaum and Ferris, 1973; Barker, 1991; Rodriguez-Kabana, 1992; Noe, 1998). In Table 22.2 a list of primary hosts, host range size and some acceptable rotation crops are given. It should be noted that where multiple nematode parasites are present, a non-host crop used for management of one species may be a good host for the non-target species. Since host susceptibility can also vary amongst populations of a species, testing of a non-host is always warranted before making final recommendations. In the past, many crops considered to be non-hosts of a nematode were found to be moderate hosts.

Rotation with non-hosts can affect the rate of natural attrition and, therefore, the extent of nematode inoculum reduction between susceptible crops. In sugarbeet, barley reduces *Heterodera schachtii* to a greater degree than wheat when used as a non-host crop. Nematode survivability over time in the absence of a host is also very important in designing rotation schemes. Some nematodes can survive long periods in the absence of a host (*Xiphinema*, *Heterodera*, *Globodera*) whereas other nematodes (*Rotylenchulus*, *Meloidogyne*, *Nacobbus*) decrease more rapidly over time. A list of the duration of survival under different conditions has been made by Norton (1978) and is discussed in the various chapters in this book. Survival times for some selected nematode species are given in Table 22.3.

Rotations using moderately resistant or tolerant crops together with highly susceptible vegetable crops have been used for control of root knot. Vegetables considered moderately susceptible or tolerant to root knot are: cabbage, onion, leek, broccoli and amaranthus. Plants considered good host plants of one *Meloidogyne* species in one part of the world are not necessarily hosts to all populations of that species. Because of this large variation in host status within

Table 22.2. A partial list of primary hosts and non-host crops used in rotations for nematode management.

Nematode	Primary hosts	Rotation crops[a]
Meloidogyne incognita	Horticultural crops, cotton, soybean, legumes	Groundnut, T and R cvs, some cereals
M. graminicola	Rice, wheat	Legumes, soybean, jute, sunflower, sweet potato, sesame, okra
M. chitwoodi	Horticultural crops, potato, carrot	Lucerne
M. javanica	Horticultural crops, esp. tomato, legumes	Cotton, groundnut
M. arenaria Race 1	Groundnut	Cotton, maize, sorghum
M. arenaria Race 2	Horticultural crops, soybean	Cotton, groundnut, maize, sorghum
M. hapla	Horticultural crops, esp. carrot, celery, potato	Onion, lettuce, radish
M. artiellia	Chickpea, vegetables	Cotton, potato, oat, maize, lentil, tomato, melon
Heterodera glycines	Soybean	Cereals, most legumes, R cvs
H. schachtii	Sugarbeet, cabbage, rape	Cereals, T and R cvs, horticultural crops
H. cajani	Pigeonpea, chickpea	Cereals
H. avenae	Cereals	Legumes
H. ciceri	Chickpea	Cereals, cotton, horticultural crops
H. oryzae	Rice	Legumes
Globodera rostochiensis	Potato, tomato	Cereals, legumes, R cvs
Hirschmanniella spp.	Rice	Legumes, cereals, cotton, tobacco, sweet potato
Punctodera chalcoensis	Maize, Teosinte	All other grasses and cereals
Scutellonema bradys	Yams	Groundnut, chilli pepper, tobacco, cotton, maize, sorghum
Globodera pallida	Potato, tomato	Cereals, legumes, R cvs
Ditylenchus dipsaci oat race	Onion, legumes, oats	Some cereals, horticultural crops
D. dipsaci giant race	Broadbean	Cereals, horticultural crops
D. angustus	Rice	Jute, legumes
Rotylenchulus reniformis	Cotton, vegetables, pineapple	Sorghum, maize, resistant soybean, sugarcane
Radopholus similis	Banana, ginger, black pepper	Some cereal crops

[a]T, tolerant; R, resistant cultivars.
Sources: Noe, 1998; Anonymous, 2004; CABI Crop Protection Compendium; other chapters in this volume.

Table 22.3. Duration of survival of some plant parasitic nematodes in the absence of a host plant.

Nematode	Survival without a host plant
Anguina tritici	28 years in seed at room temperature
Aphelenchoides besseyi	1–3 years in dry rice seed
Criconemoides xenoplax	2 years in flooded soil
Ditylenchus angustus	4 months in flooded soil
D. dipsaci	Years in dry seed
Globodera rostochiensis, G. pallida	10–15 years in infested soil
Heterodera glycines	84 months in infested soil
Meloidogyne spp.	1–12 months
M. graminicola	5 months in flooded soil
Pratylenchus coffeae	6 months in bare soil
Radopholus similis	6 months in bare soil
Rotylenchulus reniformis	2 years in bare soil; 18 months in dry soil
Xiphinema americanum	49 weeks in soil at 10°C

species of root knot, all crops being considered for rotation must be tested for host status to local populations before rotation schemes are recommended for the field.

Caution must be taken with regard to variation in nematode populations and to the composition of root knot species present in a field. Sometimes the *Meloidogyne* populations are composed of several species that may require different approaches for control. It should be noted that detection of species that make up less than 5% of the population is difficult.

Multiple cropping and mixed cultivars

Multiple cropping is common in subsistence agriculture where food for family consumption is the primary goal. The simultaneous production of many different crops increases the chances of obtaining a crop regardless of environmental calamities such as serious drought or pest and disease occurrence. The multiple cropping systems used do not necessarily lead to a reduction in nematode damage since spacing between susceptible crops is often small (Noe and Sikora, 1990).

The use of alley cropping, on the other hand, could reduce nematode damage in multiple cycles of crops in a year if the survival of the nematode in question (Table 22.3) is limited and the length of the growing season in the two or more crops grown in distinct alleys is sufficiently long. Alley cropping with a cereal in one alley and a susceptible crop in the alternating alley and rotation of the crops in these sections after the first cycle could lead to sufficient reductions of nematode populations. Alley cropping resistant and susceptible vegetable cultivars could also be an alternative approach that could both reduce nematode densities and offset the development of resistance-breaking races. Alley cropping with high-value crops using bare fallow, trap crops or antagonistic fodder crops used for grazing in the alley also needs to be examined. In banana, alley cropping with an alternating fallow seems to have been successful.

The use of precision agricultural technology and remote sensing should allow growers of some crops to plant resistant cultivars in loci of high nematode infestations, e.g. in crops like soybean, wheat, sugarbeet and potato to name a few.

Trap crops

Trap cropping normally targets sedentary nematodes. A good host with quick and extensive root growth is planted for a short duration of time. The crop and planting period must be selected to ensure high nematode penetration and initial development to a non-motile growth stage, usually only a few days after root penetration.

The sedentary juveniles in the root tissue are then killed when the trap crop is terminated by physical destruction or herbicide treatment. Trap cropping, which was originally developed to control cyst nematodes in sugarbeet in the 1800s, has been used for management of nematodes in a number of crops (Table 22.4). Short-season crops are used as trap crops in raised beds in Cuba to control *Meloidogyne* species and rape has been used as a green manure crop to reduce *Heterodera schachtii*. Resistant mustard and oil radish cultivars are also used as a type of trap crop for management of sugarbeet cyst and root knot nematodes in beet crops. Trap crops stimulate hatch and penetration, and reduce nematode density both by trapping and, where present, by resistance mechanisms. They also stimulate the antagonistic potential after biomass incorporation into the soil. Any host that can be planted and then killed by incorporation or herbicide application can be used as a trap crop. However, the following criteria for trap cropping should be present:

- excellent host with extensive root growth to ensure high levels of penetration;
- low cost seed, since yield and/or a green manure effect are not always expected;
- good data on 'day degrees' from penetration until start of nematode egg-laying;
- speedy and complete kill of the root to prevent any reproduction after incorporation; and
- acceptable cost–benefit ratio based on control over a nematicide or resistant cultivar.

It should be noted that if root removal or herbicide killing of the plant is done too close to the start of egg-laying, the females in the surviving root or dying root tissue can lay eggs for a number of days, thereby reducing control efficacy. Therefore, multiple tillage of the soil to promote root death or the use of herbicides that systemically kill root tissue is needed for effective management.

Cover crops

Cover crops are non-hosts that are used mainly to protect the soil from erosion or to suppress weed growth between major crop cycles, or crops used to give some nematode control. They may also be used for animal fodder or as a green manure crop. Cover crops reduce many nematodes just by being non-hosts. However, when incorporated into the soil they can significantly increase the antagonistic potential in the soil. In addition, microbial degradation of organic compounds leads to a form of biofumigation and the production of metabolites that are nematicidal. Major cover crops that have been tested for use are given in Table 22.5.

Antagonistic crops

Plants antagonistic to nematodes are those that are considered to produce antihelminthic compounds with different modes of action (Pandey *et al.*, 2003). The mechanisms responsible for control are often poorly understood and many of the tests made have been conducted *in vitro* with plant extracts. The production and

Table 22.4. Nematode trap crop approaches used in the field for nematode management.

Nematode	Trap crop	In rotation
Heterodera schachtii	*Sinapis alba, Raphanus sativus* ssp. *oleifera*	Sugarbeet and cereals
Meloidogyne incognita	*Solanum nigrum*	Monoculture of African spinach
M. incognita	*Lactuca sativa, Raphanus sativus*	Break crop in protected cultivation
Globodera rostochiensis	*Solanum tuberosum, S. sisymbriifolium*	Potato
G. pallida	*S. sisymbriifolium*	Break crop in potato rotation
Heterodera avenae	*Avena sativa*	Wheat, barley

Table 22.5. Major cover crops used for nematode management.

Nematode	Cover crop
Belonolaimus longicaudatus	*Crotalaria spectabilis*
	Tagetes minuta
Heterodera schachtii	*Fagopyrum esculentum*
	Phacelia tanacetifolia
Hirschmanniella oryzae	*Sesbania rostrata*
	Sphenoclea zeylanica
Meloidogyne spp.	*Aeschynomene americana*
	Chloris gayana
	Crotalaira juncea
	C. spectabilis
	C. intermedia
	Desmodium unicataum
	Digitaria decumbens
	Eragrostis curvula
	Festuca pratensis
	Mucuna pruriens
	M. deeringiana
	Panicum maximum
	Stylosanthes gracilis
M. arenaria	*Paspalum notatum*
M. chitwoodi	*Raphanus sativus*
M. incognita	*Brachiaria plantaginea*
	Cynodon dactylon
	Macroptilium atropurpureum
	Panicum maximum
	Pennisetum purpureum
	Raphanus sativus ssp. *oleifera*
P. brachyurus	*Crotalaria usaramoensis*
	Stylosanthes gracilis
	Flemingia congesta
P. loosi	*Tripsacum laxum*
	Cymbopogon confertiflorus
	Eragrostis curvula
Pratylenchus neglectus	*Raphanus sativus*
Rotylenchulus reniformis	*Chloris gayana*
	Crotalaria juncea
	Tagetes patula

active release of toxic substances while the crop is growing or after incorporation into the soil is usually responsible for control. A large number of plants have been shown to contain nematicidal compounds when extracted from the tissue and tested *in vitro*.

Marigold, sunnhemp, castorbean, partridge pea, asparagus and sesame have been extensively studied for nematode control activity. Sunnhemp is often used as a cover crop and green manure crop and is sometimes considered an antagonistic crop for root knot nematode control. *Crotolaria longirostrata*, for example, when grown as a cover crop and then incorporated into the soil, will reduce root knot galling. Control is probably due to toxins produced during microbial degradation and not by toxic exudates from the plant itself (see chapter 9). In Plate 23C, two types of *Crotolaria* are shown that are used to control root knot in the production of medicinal plants in Brazil. The best studied antagonistic plants are species in the genus *Tagetes* known to produce terthieny and derivatives of bithienyl that are toxic to root knot. Ploeg (1999, 2002) demonstrated that *Tagetes patula*, *T. erecta*, *T. signata* and a *Tagetes* hybrid reduced galling in a subsequent susceptible tomato crop compared to the tomato–tomato rotation.

Biofumigation

This term normally refers to suppression of soil-borne pests and pathogens by biocidal compounds, principally isothiocyanates, released in soil when glucosinolates in cruciferous crop residues are hydrolysed (Kirkegaard *et al.*, 1998). Soil amended with fresh or dried cruciferous residues at 38°C day and 27°C night temperatures reduced *Meloidogyne incognita* galling by 95–100% after 7 days' incubation in controlled environment tests (Stapleton *et al.*, 1998). It should be noted here, however, that many cruciferous plants are good hosts of some important species of *Meloidogyne*.

The term biofumigation is now used more freely whenever volatile substances are produced through microbial degradation of organic amendments that result in significant toxic activity toward a nematode or disease (Anonymous, 1998; Bello, 1998). The release of toxic compounds already present in antagonistic plants used as amendments, e.g. neem, marigold and castor, or the production of toxic compounds due to microbial fermentation of nutrient-rich organic amendments, e.g. velvet bean, sunnhemp or elephant grass, lead to significant levels of nematode control.

Biofumigation under these circumstances is greatest when there is an optimum combination of organic matter, high soil temperature and adequate moisture to promote microbial activity leading to toxin production. In tropical and subtropical production systems, plastic mulch and drip irrigation improve effectiveness of biofumigation. Transporting organic amendments to the field or incorporating cover crops that produce large amounts of biomass into the soil, together with plastic mulch and drip irrigation, should significantly increase the level of control attained.

Biofumigation using fresh marigold as an amendment is used effectively in root knot management in protected cultivation in Morocco (Chapter 9). *Tagetes* is grown in the raised beds prior to the planting of susceptible horticultural crops. The crop is then incorporated into the soil after 2–3 months. The beds are fitted with drip irrigation and covered with plastic mulch. The soil in the bed is then biofumigated under conditions of high temperature and optimum soil moisture (Plate 23D).

Control due to any form of biofumigation is probably the result of multifaceted mechanisms including:

1. Non-host or trap cropping depending on the host status of the plant used.
2. Lethal temperature due to solarization.
3. Nematicidal action of toxic by-products produced during organic matter degradation.
4. Stimulation of antagonists in the soil after biofumigation.

Pre-plant Management

Management tools that are used just prior to sowing or transplanting can have a major impact on plant health in the early stages of plant growth. In many cases such methods are designed to offer protection from infection for 4–5 weeks after germination or transplanting. This length of protection has been shown to be sufficient to ensure good root growth and yield, even if the nematode is not eradicated or killed. Any treatment that reduces nematode densities below the threshold level and keeps the nematode out of the root is of interest at this critical point in the cropping cycle.

Precision and remote sensing

Progress has been made in the use of remote sensing, using infrared and digital thermography, to detect areas in fields (Plate 23E) where plant parasitic nematodes are causing damage (Nicolas *et al.*, 1991; Nutter *et al.*, 2002; Schmitz *et al.*, 2004). This technology coupled with precision farming equipment can increase control efficacy by placing nematicide in the areas of high nematode density (Noe, 1998). The possibility of simultaneously planting mixtures of susceptible and resistant crops, e.g. resistant green manure cultivars in the exact loci where nematodes are above the threshold, would also reduce seed costs and improve management.

Solarization and soil heating

The lethal temperature for control of plant parasitic nematodes is considered to be around 45°C. Heating the soil either with dry or steam heat has been used for many years in protected cultivation to manage root knot nematodes, but the high cost of heating oil has limited its use drastically.

Soil solarization with plastic mulches (Plate 23F), which leads to the development of lethal temperatures in the soil, is being used in some countries for control of root knot and soil-borne diseases (Katan, 1981; Whitehead, 1998). The technique is most effective in regions where high levels of solar energy are available for long periods of time. However, the limited depth to which lethal heat actually penetrates into the soil often restricts control to the upper 5–10 cm layer. Therefore, besides solar energy, root architecture of the crop to be grown and the depth of root knot infestations are important in selecting this approach. Manipulating root growth so that the root system remains shallow and in the upper horizon through breeding or drip irrigation might increase the effectiveness of solarization in the early stages of plant growth. Targeted placement of fertilizer would also affect root architecture.

Solarization will reduce root knot, *Verticillium* wilt and weeds in crops, even though climatic conditions are not considered ideal for soil solarization (Overman and Jones, 1986). Similar results were obtained in Cuba in peri-urban agriculture and in small farm production units using solarization under sub-optimum conditions between July and September (Fernández and Labrada, 1995). Whether the use of solarization under sub-optimal conditions is always effective and economical needs to be ascertained for each situation.

It should be noted that, in many climatic regions and in subsistence agriculture, the costs of using plastic mulches are limiting factors except for eliminating nematodes from soil in seedbeds (Bridge, 1987, 1996).

Fumigant nematicides

Nematicides used in control of root knot nematodes during the pre-planting period are usually fumigants, which are usually liquids and enter the soil water solution from a gas phase. In most cases the fumigants are broad-spectrum contact nematicides effective against adults, juveniles and eggs as well as other pests and diseases plus weeds. There are a number of sources that give excellent reviews on the use of the most common fumigant and non-fumigant nematicides for a broad array of nematodes and crops, which should be consulted for more detail (Johnson, 1985; Hague and Gowen, 1987; Whitehead, 1998; Anonymous, 2004). The most commonly used nematicides are listed in Appendix A at the end of this book.

In some growing areas fumigants are applied under plastic mulch and vegetables are then planted through the mulch into raised beds (Plate 24A). Due to the multiple effects of nematodes, weeds and soil fungi on production in many growing areas, a broad-spectrum fumigant is essential, especially where multiple susceptible crops are grown sequentially. With the loss of methyl bromide, alternative fumigants are being evaluated. When used as directed, fumigants will give excellent nematode control and increase yield significantly. Because registration requirements and efficacy vary with country and crop, no attempt will be made here to list those still being used for the control of root knot nematodes in vegetables. The majority of small farmers, especially those living at the subsistence level, cannot use fumigants because of a lack of capital for equipment, the nematicides or application.

Planting material naturally free of infestation

The production of healthy planting material is of utmost importance in nematode management because nematodes can be found in seeds, tubers, corms or seedlings of many crops. The spread of nematodes can be prevented, or at least reduced, by use of nematode-free seed/planting material and use of nematode-free seedbeds or soils to produce clean seedlings.

If a grower does not have nematode-free areas in his farm, nematode-free planting material can be selected or the nematodes removed from the material before planting. Farmers producing their own seedlings will import fewer nematode problems into their lands than those buying in seedlings which could have become contaminated from infested soils. Planting material that can be guaranteed free of root parasitic nematodes are certain crops propagated vegetatively from stem cuttings, such as sugarcane, sweet potato, cassava and black pepper, that are free of nematodes. The use of tissue culture to produce plantlets, such as banana plantlets, is a particularly effective method of producing nematode-free plants.

Nematodes can produce damage symptoms (surface cracking, surface galls, watery lesions, necrotic spots, blackened roots, galls) in planting material such as bulbs, corms, tubers, seedlings and rootstocks, and farmers recognizing these symptoms as diseased or abnormal generally refrain from using the material for planting (Bridge, 1987, 1996).

Physical removal of tissues infested with nematodes

Examples are found with banana and plantain, yam and taro seed material. The major nematode pests of bananas and plantains (*Radopholus similis*, *Pratylenchus coffeae*, *P. goodeyi*) can be removed from lightly infested planting material by cutting (paring) away roots, soil and purple to black nematode lesions and surrounding tissues from banana corms and suckers used for planting (Plate 24B). In yams (*Dioscorea* spp.), cutting out nematode dry rot lesions caused by *Scutellonema bradys* and *Pratylenchus coffeae* from tubers can be effective in eliminating the nematodes from the seed pieces (Bridge and Page, 1982).

Physical methods of nematode control in planting material

Hot water treatment of planting material can be very effective in controlling nematodes in seeds, bulbs, corms, tubers, rhizomes and rootstocks (Bridge, 1975; Maas, 1987; Whitehead, 1998). Accurate temperature baths and equipment to maintain the correct temperature, which is usually between 44°C to 55°C, are needed. Temperatures and times required for control of some nematodes are given in Table 22.6. The control of *Radopholus similis* in banana corms with hot water baths has been recommended (Plate 23B), but has limited use except by some small growers. A modification for resource-poor growers has been recommended to control migratory endoparasites in banana corms prior to planting in India and East Africa (Prasad and Reddy, 1994; Mbwana *et al.*, 1998) and in groundnuts in Africa (Bridge, 1975). Solarization of tubers has been attempted but is not exact enough to ensure control.

Elimination of nematodes from seedbeds

Infested soils in seedbeds are often the main cause of nematodes being introduced into field soil on infested seedlings. Nematode-free soil for raising seedlings can be obtained from such localities as regularly flooded land. Soil taken from paddy rice production or from river banks is often free of nematodes. The soil should always be examined to make sure it is free of major nematode species, since the soil could be contaminated by runoff water from nearby fields. Soil infested with nematodes can be treated effectively by a range of physical or non-physical techniques (Tables 22.7 and 22.8).

Table 22.6. Heat treatments used to control nematodes in planting material.

Nematode	Treatment
Aphelenchoides besseyi	Rice seed
	Cold soaking 18–25 h, 15 min 51–53°C
	Cold soaking 3 h, 52–57°C
	No soaking, 55–61°C, 10–15 min
Anguina tritici	Wheat seed 4–6 h 54°C 10 min
Ditylenchus dipsaci	Onion bulbs 44–45°C for 3 h
	Garlic cloves 45°C for 20 min
	Shallots sets 44.5°C for 1–2 h
Hirschmanniella miticausa	Taro corms 50°C for 15 min
Meloidogyne spp.	Sweet potato 65 min at 47°C
	Yam 50–51°C 30 min
M. incognita	Sweet potato tubers 65 min at 47°C
M. javanica	Potato tubers 2 h at 46–47.5°C
Pratylenchus coffeae	Yam tubers 46–52°C for 15–20 min
Radopholus similis	Banana corms 55°C for 15–25 min
Scutellonema bradys	Yam tubers 50–55°C for 40 min

Table 22.7. Physical methods used to eliminate nematodes from infested soil.

Physical method of management	Description of method
Steam sterilization	Steam is passed under pressure into the soil under soil surface covers for 30 min for greenhouse high value crops
Application of boiling water	In Bolivia, farmers heat water on wood fires for seedbed treatment, a method also recommended by CIAT (P. Franco, CIAT, 2003, personal communication)
Heat sterilization	A soil sterilizer made from an oil drum and heated by a wood fire can be used to sterilize small amounts of moist soil. When steam rises, a lid added and the fire removed for 1 h
Sun drying and heating	Spreading soil, to a depth of 10 cm, on a soil-free surface exposed to the sun during hot dry season for a minimum of 2 weeks with regular turning will eradicate nematodes
Turning soil to induce nematode desiccation	Nematodes can be killed by the lethal effects of heat from the sun and drying by regularly turning the soil at the end the growing season
Surface burning of plant debris	Heat has to penetrate into the soil to be effective and this requires substantial amounts of slow burning, high temperature output material on the soil surface – wood versus grass

At-planting Management

Date of planting

Planting date is a tool designed to reduce the impact caused by nematode penetration in the early growth stages by taking advantage of nematode inactivity. The fact that the minimum temperature required for *M. incognita* development in the root is significantly lower than the minimum 'activity threshold' of 18°C for *M. incognita* second stage juveniles has been used to alter the date of planting for control of root knot. Changing the normal date of planting to coincide with low soil temperature was considered an important control tactic on carrots (Roberts, 1987). This approach could also be used to limit nematode damage on vegetables in cool upland tropical

Table 22.8. Non-physical methods for soil decontamination.

Non-physical method of management	Description of method
Annual or seasonal rotation of seedbed sites	Rotation of the seedbed areas each season or each year prevents the build-up of soil populations of nematodes
Keeping seedbeds free of weed hosts	Many weeds are hosts for the major nematodes that occur on transplanted crops and their removal from the seedbed is important
Floating seedling tray beds	Production of seedlings in floating trays over nematode-free water in vats will prevent nematode infection
Chemical fumigation	Fumigation with nematicides has been used for many years to eradicate nematodes from infested soil. The recent removal of effective products from the market has affected their use
Sealed container solar heating	Soil is sealed in 5 kg polyethylene bags, which are placed in the sun on a concrete or black plastic surface for at least 2 weeks
Biological enhancement	Biological enhancement of seedlings with beneficial microorganisms antagonistic to nematodes can increase resistance to nematodes

regions. In Zimbabwe, the date of planting of tobacco is regulated to take advantage of cooler periods to reduce root knot infection. This is a technique that could have a major impact in other regions of the world (Shepherd and Barker, 1990).

The early planting of rice at cooler times of the year was effective in reducing *Aphelenchoides besseyi* on rice in the USA, and the early sowing of maize reduced damage caused by the cyst nematode *Punctodera chalcoensis*. Planting date was used to avoid damage by *Globodera rostochiensis* populations that were still in diapause and unable to hatch and penetrate the sequentially planted potato crop in the Philippines (Sikora, 1984). Similar techniques have been developed for other nematodes on wheat and small grains (Johnson and Motsinger, 1990). Delayed planting of cotton also reduced root knot damage and, simultaneously, that of the complex with fungal wilt (Jeffers and Roberts, 1993).

Non-fumigant nematicide treatment

Non-fumigants are granular or liquid formulations that are usually water soluble. Non-fumigant nematicides have either contact or nematistatic and often plant systemic activity against nematodes and insects. In most cases the mechanism of action is associated with suppression of nematode mobility during the period when adequate concentrations are in the soil solution. The non-fumigant nematicides are not effective against the eggs of nematodes and in most cases do not kill the juveniles at the concentrations now being recommended for use. They give the plant a 'head start' by delaying nematode penetration during the highly sensitive seedling or post-transplant stage of plant development. Non-fumigant granular and/or liquid formulations of contact and/or systemic nematicides are suitable for commercial use as well as for use on small farms. The growers however, must be made aware of proper handling and application techniques, as well as time of application, since these materials are highly toxic to humans and the environment when improperly used. Non-fumigant nematicides are often not as effective as fumigants in increasing yields because they do not have broad-spectrum activity and in most cases only inactivate nematodes for short periods of time.

Granular nematicides are either applied broadcast over the soil surface and incorporated into the soil before planting or banded into or over the plant furrow. It is important that users realize human and environmental toxicity can occur and that the presence of residues in the harvested crop is possible if treatment restrictions are not followed.

Liquid formulations allow application by surface and drip irrigation, with the latter of importance to vegetable production. Application through drip irrigation places the material directly in the rhizosphere and can allow treatment during the growing season. It also allows splitting or extending application over specific time intervals to coincide with optimum control. However, many non-fumigants, whilst effective in preventing infections, are not highly effective in suppressing the nematodes' activities once infection has occurred.

Dip treatment or treatment of vegetable transplants in nurseries also has been effective in reducing root knot galling. Efforts are also being made to develop granular formulations that allow seed treatment for nematode management that would greatly reduce the dose needed on a per hectare basis as well as limit environmental impact and crop residues problems. In many short cycle vegetable crops that required protection for 4–5 weeks, this could be an important treatment form.

Bio-enhancement

Biological enhancement of seeds and transplants with arbuscular mycorrhizal fungi (Plate 24C), mutualistic fungal endophytes, plant-health-promoting rhizosphere or mutualistic endophytic bacteria has been shown to increase plant resistance and/or tolerance to nematode infection during plant growth (Sikora and Hoffmann-Hergarten, 1993; Hallmann and Sikora, 1994; Sikora, 1997; Hallmann, 2001).

Tomato and pepper transplant production substrate treated with different formulations of plant-growth-promoting rhizobacteria caused highly significant increases in tomato and pepper growth, vigour and survival in the field, with some formulations reducing numbers of root knot galls on pepper (Kokalis-Burelle *et al.*, 2002). Endophytic bacteria have recently been shown to significantly reduce root knot infection and induce systemic resistance in tomato (Munif *et al.*, 2001).

Enhancement of plants with arbuscular mycorrhizal fungi (Plate 24C), apart from providing plants with nutrients, reduces penetration and development of a number of root knot nematodes in a range of vegetable crops and has effects on burrowing nematode in banana. Mycorrhizal inoculum is now commercially available for this purpose in many countries. Combining mycorrhizal fungi with plant-health-promoting rhizobacteria and mycorrhizal helper bacteria during seedling production and seedling growth has led to increased fungal colonization and root knot control in tomato seedlings (Reimann and Sikora, 2003). Endophytic fungi are prime antagonists for use in biological enhancement of transplants for root knot control in vegetables (Hallmann and Sikora, 1994) and for the treatment of banana tissue culture plants for management of burrowing nematode (Plate 24D) (Niere *et al.*, 1998; Sikora, 2002; Sikora *et al.*, 2003; Zum Felde *et al.*, 2004).

Treatment of fumigated, biofumigated or solarized soil with biologically enhanced transplants would increase overall control, due to the lack of competitive microbial activity in this soil. To be effective, however, biological enhancement requires the existence of commercial biocontrol products, as is the case with mycorrhizal fungi, that can be used by small or large commercial nursery production units that supply bio-enhanced seedlings to growers. In some crops for which large commercial companies produce healthy seedlings for their contract growers, bio-enhancement of planting material could lead to increased yield and reduced pesticide use.

Plant Management

Management of nematode infestations after planting is an important tool for many crops, in particular for perennial crops such as fruit and tree crops and spices. The most important plant management tool is, of course, plant resistance to nematode attack. However, where resistance does not exist other tools have been developed that ensure good yield even in the presence of nematodes in the rhizosphere. There are only a few methodologies that can be used curatively to reduce or inhibit nematode damage once the crop is in the field. However, in some crops they are the mainstay of nematode management strategies, for example, on banana, citrus, fruit and tree crops.

Host resistance

Host resistance, if available in a high-yielding cultivar, should be the foundation upon which other management tactics build. Resistance is typically defined as a plant's ability to inhibit nematode reproduction relative to that on a susceptible genotype (Cook and Evans, 1987; Trudgill, 1991; Roberts, 2002). Thus, resistance is distinct from the effects of nematode parasitism on plant growth and yield. Tolerance and intolerance are most often used to describe a plant's response to parasitism, with a tolerant plant experiencing less yield suppression than an intolerant plant at similar levels of parasitism (Cook and Evans, 1987; Trudgill, 1991; Roberts, 2002). The relationship between resistance and tolerance has not been examined for most resistant genotypes, but in at least a few instances tolerance is inherited independently from resistance (see Trudgill, 1991). Some resistant crop genotypes are known to be relatively intolerant of nematode parasitism (Johnson *et al.*, 1989). The possible combinations of these two distinct traits for any given crop are given in Table 22.9.

Unfortunately, clear distinctions among the four possibilities are not always made, with low to moderate levels of resistance often being referred to as tolerance. Similarly, susceptibility is often equated with intolerance. It is important to consider that both resistance and tolerance are traits that can only be assessed relative to the performance of another genotype of the same species, typically a known susceptible, intolerant genotype. A clear understanding of the differences in these concepts is essential to scientists seeking to advance our understanding of the interaction of plant parasitic nematodes with their hosts and to exploit variation in these relationships to achieve an improvement in crop productivity.

Regardless of the tolerance or intolerance of a resistant crop, in subtropical and tropical environments, where most nematode parasites will complete multiple generations on annual crops, the reduction in total parasitism due to reduced nematode reproduction typically results in increased crop yields. Therefore, intolerant resistant genotypes will likely have the appearance of tolerance. In perennial crops, the long-term effects of reduced nematode reproduction are even greater than in annual crops.

Because resistance typically leads to improved yields in fields infested with nematode population densities that exceed the damage threshold, resistance protects

Table 22.9. Possible combinations of resistance, susceptibility, tolerance and intolerance in a crop genotype with respect to nematode reproduction and plant response to nematode parasitism. (Adapted from Trudgill, 1991.)

	Host growth	
Nematode reproduction	Good	Poor
Good	Tolerant/non-resistant	Intolerant/non-resistant
Poor	Tolerant/resistant	Intolerant/resistant

the genetic yield potential of the crop. This is the most important benefit to be derived from the use of resistance and should be the characteristic that is most appealing to farmers when attempting to convince them to use resistance. However, the benefits to yield potential are also relative. In some cases where the nematode population is less than the damage threshold, a resistant cultivar may have a lower yield potential than that of a high-yielding susceptible cultivar. Thus, resistant cultivars often perform relatively poorly in regional trials conducted to evaluate yield potential of numerous crop cultivars. Such routine yield trials are rarely conducted at sites with nematode populations that exceed the damage threshold. The benefits of resistance can be readily demonstrated in fields with a moderate to severe infestation of the problem nematode species. The best approach for demonstrating the benefits of a resistant cultivar would be to compare the yield of the resistant cultivar that has not received other standard management tactics (e.g. nematicides or crop rotation) to the yield of a susceptible cultivar receiving the standard management tactic in fields heavily infested with the nematode species of concern.

The apparent negative effects of resistance on yield potential are most likely due to linkage drag, whereby genes with negative effects on yield potential are linked to resistance loci. No data are available that show a direct effect of resistance genes on reduced yield potentials. Indeed, as breeding programmes continue to work with resistance, the yield potential of the resistant genotypes usually increases. For example, the first groundnut (peanut) cultivar with resistance to *M. arenaria* was selected from the fifth backcross generation in a breeding programme where resistance was derived from a wild species and introgressed into cultivated groundnut (Simpson and Starr, 2001). Yields of that first release were superior to the best susceptible cultivars in nematode-infested fields, but yields of the resistant cultivar were not competitive in the absence of nematode parasitism (Church *et al.*, 2000). The second released groundnut cultivar with resistance to *M. arenaria* was selected after two additional backcross generations and had yield potentials nearly equal to that of the best susceptible cultivar without any loss of resistance (Fig. 22.10).

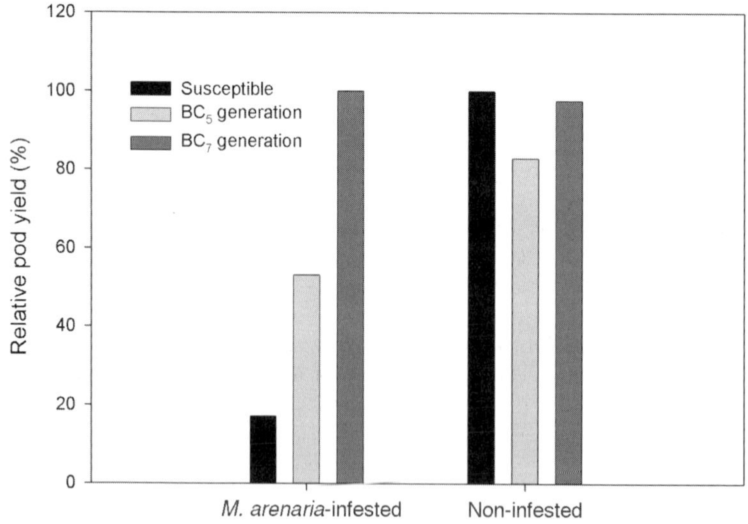

Fig. 22.10. Effect of two additional cycles of backcrosses on yield potential of groundnut with resistance to *Meloidogyne arenaria* introgressed from wild *Arachis* species (from Church *et al.*, 2000).

Efforts to develop soybean cultivars with resistance to *Heterodera glycines* have been in progress for more than 30 years, thus one would expect that resistance would be available in cultivars with the highest yield potentials. In soybean cultivar yield trials conducted at one location in Illinois, five cultivars with resistance to *H. glycines* were among the top 16 cultivars from a total of 45 tested in 2000 (http://web.aces.uiuc.edu/vips/v2CompVar/v2CompVar1.cfm). In 2002, the top five yielding cultivars among 50 tested were all resistant to one or more races of *H. glycines*. Similarly, nearly all modern wheat cultivars contain multiple genes for resistance to fungi and viruses and there is no evidence that these multiple resistance genes have a negative effect on yield. Thus, when there is an apparent yield drag due to resistance, additional breeding effort should enable one to achieve yield potentials equal to those of the best susceptible cultivars.

Because of the differences in the effects of resistance and tolerance on nematode population densities, tolerance and resistance will have different effects on the productivity of cropping systems involving multiple crops with a range of tolerances and levels of resistance. As demonstrated by Ogallo *et al.* (1999), lima beans susceptible to *M. incognita* can be grown successfully following two plantings of a root knot-resistant cotton cultivar, but experienced heavy yield losses when grown following two plantings of a susceptible cultivar. Although a susceptible/tolerant crop will have greater yield than a susceptible/intolerant crop in a nematode-infested field, because of the relatively high level of nematode reproduction on the susceptible/tolerant cultivar, the potential for yield suppression of an intolerant crop following the susceptible/tolerant crop will be similar to that when following a susceptible/intolerant crop. Another possible situation is that crop genotypes with tolerance or low levels of partial resistance may actually result in a greater hazard to a subsequent susceptible crop than when the first crop in the sequence is susceptible and intolerant. When parasitism has little effect on host growth because of tolerance or partial resistance, the nematode densities may be greater than on susceptible/intolerant cultivars that are heavily damaged by the nematodes. Niblack *et al.* (1986) demonstrated this phenomenon in soybean with partial resistance to *M. incognita*, where the nematode population in plots planted to a susceptible cultivar peaked at about 90 days after planting due to severe damage to the plants. In plots planted to a partially resistant cultivar, the nematode population density was still increasing at 120 days after planting.

Resistance, when available, is not a universal solution to nematode management. Because resistance is highly specific, being effective against only a single species or even only one race of a species, it will not control other potential nematode pests in fields with a polyspecific community. This can be a major limitation to the use of resistance, but is not a limitation in cases where the crop only has one major nematode pest species or where a field is infested with only one major pest species. Genetic tolerance may be less specific than resistance and may work against several nematode species, but this hypothesis has not been tested. In crops with partial resistance to one or more nematode species, some yield loss is to be expected at high initial nematode densities, such that resistance must be used in combination with other management tactics to achieve the maximum yield potential. That high levels of resistance are not available is not a reason to avoid resistance; rather, partial resistance may make other management tactics more effective. This is analogous to cases where partial resistance to foliar fungal pathogens, which is of limited value as a sole management tactic, has great value in an integrated programme and permits a reduction in the reliance on fungicides (Maytac and Bailey, 1988). Resistance may lack durability because repeated use of single resistance genes often leads to a shift in the virulence characteristics of the nematode population, such that with time a specific resistance gene is no longer effective. This has been demonstrated with *Globodera* and *Heterodera* species on potato and soybean, respectively (Turner, 1990; Young and Hartwig, 1992), and for root knot nematodes where virulence to the *Mi* gene in tomato in *M. incognita* and *M. javanica* has been identified (Kaloshian *et al.*, 1996; Ornat *et al.*, 2001). However, if the nematode population in a given field or region lacks the appropriate diversity with respect to virulence, then there may not be selection for virulence with repeated use of a given resistance gene. This appears to be the case for the *H1* gene for resistance to *G. rostochiensis* in some regions (Trudgill and Parrott, 1972). Similarly, repeated use of resistance may cause a shift in the species present in a field, with species against which the resistance is not effective becoming dominant. This has been documented for tobacco, where increased use of resistance to *M. incognita* led to an increase in the frequency of *M. javanica* against which the resistance was not effective, and in potato where use of resistance to *G. rostochiensis* led to an increased incidence of *G. pallida* (Trudgill, 1991). Finally, a few resistance genes, most notably the *Mi* gene from tomato, are temperature sensitive, which limits their use in tropical climates.

Resistance is currently available to several nematodes in a relatively limited number of crops (Table 22.10), such that there is a great need for development of resistance to additional nematodes in numerous crops. Although precise data are limited, it appears that available sources of resistance in crops are vastly underutilized; this is true in both highly developed and developing countries. Resistance to cyst nematodes is widely used in potato and soybean in Europe and the USA, respectively. Resistance to *Meloidogyne* species in tomato is widely used commercially in California, but not in many other regions, especially in the tropics. Even though *Mi* is not effective at temperatures above 28°C, it may be used during cooler months in many subtropical and tropical regions. Further, even if *Mi* is effective only during the first several weeks of a growing season before higher temperatures reduce its effectiveness, this period of resistance will be useful when combined with other management tactics. Use of resistance to the cereal cyst nematode *H. avenae* is widespread in Australia and some European countries, and was recently introduced into northern India (J. Nicol, personal communication). Resistance in groundnut gives significant increases in yield over highly susceptible cultivars (Plate 24E). Recently developed resistance in groundnut to *M. arenaria* and *M. javanica* should be useful in Africa, India and South-east Asia, but it is too soon to determine how widely it will be adapted. Resistant rootstocks in perennial crops, such as peach and citrus, have been used successfully for several decades. More recently, the grafting of resistant rootstocks to susceptible scions has been used for management of root knot nematodes on annual crops. This practice is being widely

Table 22.10. A partial list of food crops for which high-yielding cultivars with resistance to one or more nematode species are available.

Crop	Nematode species
Barley	*Heterodera avenae*
Bean, common	*Meloidogyne incognita*
Citrus	*Tylenchulus semipenetrans*
Clover	*Ditylenchus dipsaci*
Cotton	*M. incognita*
Cowpea	*M. incognita*
Groundnut	*M. arenaria, M. javanica*
Lucerne	*D. dipsaci, M. hapla*
Maize	*P. hexincisus*
Peach	*M. incognita*
Potato	*Globodera pallida, G. rostochiensis*
Oat	*D. dipsaci, H. avenae*
Rice	*Aphelenchoides besseyi, D. angustus*
Soybean	*H. glycines, M. arenaria, M. incognita, M. javanica, Rotylenchulus reniformis*
Sweet potato	*M. incognita, R. reniformis*
Tobacco	*Globodera tobacum, M. incognita*
Wheat	*H. avenae, P. neglectus, P. thornei*

used on cucumber, melon, pepper and aubergine in South-east Asia and Morocco.

Unfortunately, in many other cases available resistance is rarely used. Cotton, cowpea (D. Coyne, personal communication) and common bean (A. Marina Torres, personal communication) are examples of resistance to *M. incognita* being developed but apparently rarely used in most tropical countries. Considering the importance of cowpea and common bean as sources of dietary protein and their general susceptibility and intolerance to *M. incognita*, one wonders what are the impediments to greater utilization of resistance. Thus a major challenge appears to be that of getting available resistant cultivars to farmers that need these resources.

Non-fumigant nematicides

Systemic non-fumigant nematicides have been important management tools for control of nematodes ever since their development (Hague and Gowen, 1987; Whitehead, 1998). They are effective in reducing nematode penetration and inhibiting nematode development in the root for a number of weeks after application or during the highly sensitive early growth stage of the plant. The reduction in early root penetration and damage leads to significant root growth and plant resilience to follow-up infection. All nematicides, both past and present are listed at the back of this book (Appendix A) for reference and are discussed in detail in the crop chapters.

Since nematicides are highly toxic and if used improperly can have negative effects on the environment, they should only be used by experienced personnel and following the directions given. It should also be noted that at the doses now used, these nematicides usually only inhibit nematode activity in the soil or in the root for a limited time period. The nematodes recover with time and attack the plant and develop normally, and at the end of the season final population densities are often, but not always, equal to the levels attained without treatment.

These nematicides are often applied to the standing crop to reduce nematode development and damage over time. In banana production, where *Radopholus similis* is a major problem, they may be applied two to three times in one cycle and then annually. In groundnut they are often used to prevent damage to the pods later in the season. However, in most crops they are applied at the time of planting. Application through drip irrigation over an extended period of time during the growth stage has been effective in limiting root knot in vegetable production.

Caution, however, should be taken in that overuse of some non-fumigant nematicides as a management tool has led to microbial breakdown of the compounds and loss of nematicidal efficacy. Therefore, proper application management and rotation of compounds is a requirement for prolonged efficacy.

Grafting

One of the most effective and innovative techniques recently redeveloped for nematode management is the grafting of commercially valuable crop varieties onto nematode- and disease-resistant rootstocks (Plate 24F). Although grafting has been practised since the 1920s in Japan and Korea, it has only recently become highly regarded in protected cultivation for disease, nematode and bacterial wilt control. In Japan, 59% of the cucumber, tomato, aubergine, watermelon and melon grown in protected cultivation are tube-grafted onto rootstocks of various types. Depending on the rootstock, the technique can lead to increased plant vigour and tolerance or resistance to nematodes and diseases. The technique can be used effectively to control root knot, and in many cases, circumvents the long time period needed to breed root knot resistance into all commercially acceptable cultivars. Depending on the price of production it can be very effective in both field and protected cultivation of vegetables. Since nematode pathotypes can develop on these rootstocks, resistance management must be incorporated into the management programme.

Species of *Solanum* have been shown to have a high level of resistance to *M. incognita* and *M. arenaria*, but they are poor hosts for *M. javanica* and have been successfully used as rootstocks. Of seven wild species of *Solanum* tested, three were found to be resistant to *M. incognita*, i.e. *S. sisymbriifolium*, *S. torvum* and *S. toxicarium* (Mian *et al.*, 1995) and also reduce bacterial wilt. Granges and Leger (1996) showed that when susceptible tomatoes were grafted onto rootstocks having resistance to species of *Meloidogyne* and various root pathogens, yield increased 50% and 30% at the beginning and end of harvest when compared to the non-grafted plants, respectively. Grafting could prove to be an alternative management approach in many countries, especially where temperature does not affect the genes controlling nematode resistance.

Improved crop husbandry

Proper fertilization and proper moisture levels help plants to compensate for nematode damage. This is probably true when a plant is attacked by sub-threshold densities as opposed to high densities. Improved fertilizer application, especially nitrogen, has been shown to increase yield in a number of crops infested with nematodes (Brown, 1987). Fertilization, for example, improved the yield of wheat in the presence of *Heterodera avenae* (Gair *et al.*, 1969). Proper plant management after pre-plant nematode control obviously will ensure a stronger root system and thereby reduce the effects of nematode penetration on early stages of plant growth.

Post-harvest Management

Root destruction

Because nematodes can survive and reproduce on the viable root tissue left in the soil after harvest, roots should be eliminated by uprooting and destruction whenever possible. The spread of the nematode to the follow-up crop will be retarded and the overall population density reduced. It has been estimated that when soil temperatures are high, each month that the root system survives causes a tenfold increase in root knot nematode densities. Root knot, for example, can even survive and reproduce in excavated roots and tubers over many weeks in such crops as tomato and pepper, and even in small pieces of sweet potato tubers. Root removal and burning of tobacco roots as well as ploughing the field after harvest to encourage root degradation will reduce the impact of root knot nematodes on the subsequent crop (Shepherd and Barker, 1990).

Time of harvesting

The use of day degrees or the temperature sum needed to complete a life cycle can be used to time the harvest to trap the last life-cycle of a nematode and thereby reduce nematode densities. This has been demonstrated for cyst nematodes that have a long duration life cycle, e.g. the sugarbeet cyst nematode *Heterodera schachtii* and the potato cyst nematode *Globodera rostochiensis*. Using short-maturity groupings of crops such as soybean can also limit population build-up (Koenning *et al.*, 1993). Sikora (1984) suggested using early-maturing potato cultivars to trap *Globodera rostochiensis* in double-cropped potato. The second crop was also used to trap the juveniles still in diapause that emerged late in the growth cycle of the second crop.

Integrated Nematode Management Strategies

Development of nematode integrated management programmes requires analysis of the impact of each individual tool on a nematode population as well as determination of cost–benefit ratios for grower acceptance. In Fig. 22.11 an attempt is made to give an esti-

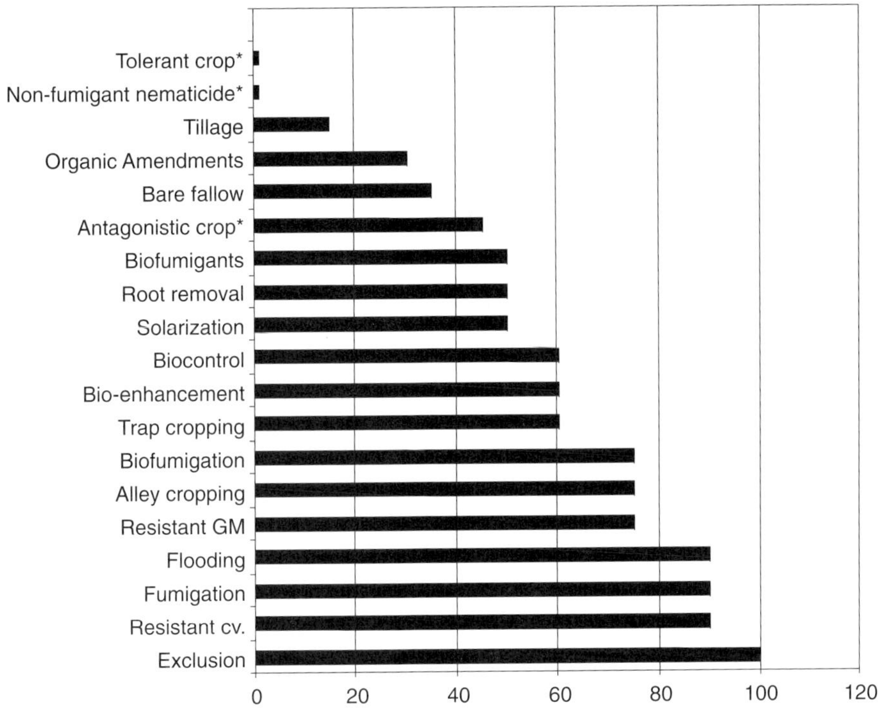

Fig. 22.11. Estimated maximum levels of control of plant parasitic nematodes using practical management methodologies. Asterisk indicates that tolerant cultivars do not reduce nematode levels, non-fumigant nematicides do not kill nematodes.

mate of the maximum impact a management tool can have on a nematode population in the soil after treatment. The estimates of the levels of control have been extracted from chapters dealing with control (Brown and Kerry, 1987; Luc *et al.*, 1990; Evans *et al.*, 1993; Barker *et al.*, 1998; Whitehead, 1998). The estimates are given only as a guideline for the development of new approaches to nematode management. The level of control will vary with environmental factors such as soil type, moisture and temperature, the crop nematode syndrome involved, the crop management programme being use and the proper use of the technology applied. Many factors affect the level of control, for example:

- resistant cultivars will not be effective when races able to break resistance are present;
- non-fumigant nematicides will not work well where microbial breakdown is high;
- fumigants will not be effective where movement of the gas is restricted;
- solarization will be ineffective under low solar energy;
- biofumigation requires large amounts of biomass and high temperatures;
- heat treatment time and temperature must be adequate to kill the nematode and not the plant;
- trap cropping will not work if nematode penetration is limited by poor root growth;
- flooding will only be effective in field crops when long-term flooding is maintained;
- tillage effects will be determined by environmental factors favouring desiccation; and
- non-hosts must be non-hosts.

Every control methodology listed has advantages and disadvantages that have to be understood in the development of a new management programme. In the chapters in this book, integrated control options have been outlined in detail for all crops and nematodes of economic importance in the tropics and subtropics. The chapters, as well as the other references given on the subject, should be consulted when developing a management programme. A list of integrated nematode management approaches that are being used in the field or that have been suggested for use is given in Table 22.11. The list is far from complete, but it shows the broad spectrum of approaches now being used by nematologists.

Nematode management in the future will never again be able to rely on one type of methodology, as it has in the past. Management will require the logical use of effective control methodologies in combinations that are economically acceptable to the grower. The advantages and disadvantages of each methodology have been discussed in this chapter and the chapters in this book.

Table 22.11. Integrated management strategies for nematode management.

Nematode	Main crop	Integrated nematode programme[a]
Heterodera glycines	Soybean	SC–non-host–non-host–RC
H. schachtii	Sugarbeet	SC–winter wheat–winter barley–resistant green manure, e.g. mustard, oil radish
		SC–winter wheat–winter barley–rape trap crop–SC
Globodera rostochiensis	Potato	RC–cereal–cereal–SC
G. pallida		SC–trap crop–RC–cereal–SC–trap crop
		Monoculture–trap crop effect during diapause
		RC–non-host–non-host–TC
Meloidogyne spp.	Vegetables	Fumigation–SC–SC–SC–fumigation
		RC–non-host–SC–non-host
		RC–SC–SC (double-cropping after resistant cultivar)
		Biologicals–root knot TC–susceptible crop
		Soilless culture–water filtration or sterilization
		Fallow–tillage–non-host–SC
		SC–solarization or biofumigation–bioenhancement
		Paddy rice–SC
		Solarization–RC
		Flooding–solarization–SC
		Multiple cropping of short duration SC and trapping
Meloidogyne	Melon	SC–weed fallow–solarization=biofumigation–SC
Radopholus similis	Perennial banana	Nematicide 2–3× per cycle or RC
	One cycle banana	Paddy rice–1 cycle crop from tissue culture
		1 cycle intensive alley-cropping with nematicide or bioenhancement with antagonistic fungal endophytes
	Organic banana	Clean or near clean soil–biofumigation–organic amendment in planting hole–bioenhancement of tissue culture with antagonistic fungal endophytes

[a]SC, susceptible crop; RC, resistant crop; TC, tolerant crop.

References

Altieri, M.A. (1984) Pest management technologies for peasants: a farming system approach. *Crop Protection* 3, 87–94.

Anonymous (1998) *Report of the Methyl Bromide Technical Options Committee – 1998 Assessment of Alternatives to Methyl Bromide.* United Nations Environment Programme, Ozone Secretariat, Nairobi, Kenya, 354 pp.

Anonymous (2004) *Institute of Food and Agricultural Sciences (IFAS) Florida Nematode Control Guide.* Entomology and Nematology Department and Cooperative Extension Service, University of Florida, Gainesville, Florida. Available at: http://edis.ifas.ufl.edu/CV112

Barker, K.R. (1991) Rotation and cropping systems of nematode control: the North Carolina experience–introduction. *Journal of Nematology* 23, 342–343.

Barker, K.R., Pederson, G.A. and Windham, G.L. (eds) (1998) *Plant and Nematode Interactions.* American Society of Agronomy, Inc., Madison, Wisconsin, 771 pp.

Bello, A. (1998) Biofumigation and integrated crop management. In: Bello, A., González, J.A., Arias, M. and Rodríguez-Kábana, R. (eds) *Alternatives to Methyl Bromide for the Southern European Countries.* DG-XI, EU, CSIC, Madrid, pp. 99–126.

Bridge, J. (1975) Hot water treatment to control plant parasitic nematodes of tropical crops. *Meded. Fac. Landbouww. Rijks. Univ. Gent* 40, 249–259.

Bridge, J. (1984) *Coffee Nematode Survey of Tanzania.* Report of a visit February/March 1984. CAB Commonwealth Institute of Parasitology, St Albans, UK, 22 pp.

Bridge, J. (1987) Control strategies in subsistence agriculture. In: Brown, R.H. and Kerry, B.R. (eds) *Principles and Practice of Nematode Control in Crops.* Academic Press, New York, pp. 389–420.

Bridge, J. (1996) Nematode management in sustainable and subsistence agriculture. *Annual Review of Phytopathology* 34, 201–225.

Bridge, J. and Page, S.L.J. (1982) *Plant Nematodes of Papua New Guinea: their Importance as Crop Pests. Report of a Plant Nematode Survey in Papua New Guinea.* CAB Commonwealth Institute of Parasitology, St Albans, UK, 91pp.

Brown, R.H. (1987) Control strategies in low-value crops. In: Brown, R.H. and Kerry, B.R. (eds) (1987) *Principles and Practice of Nematode Control in Crops.* Academic Press, New York, pp. 351–387.

Brown, R.H. and Kerry, B.R. (eds) (1987) *Principles and Practice of Nematode Control in Crops.* Academic Press, New York, 447 pp.

Church, G.T., Simpson, C.E., Burow, M.D., Paterson, A.H. and Starr, J.L. (2000) Use of RFLP markers for identification of individuals homozygous for resistance to *Meloidogyne arenaria* in peanut. *Nematology* 2, 575–580.

Cook, R. and Evans, K. (1987) Resistance and tolerance. In: Brown, R.H. and Kerry, B.R. (eds) *Principles and Practice of Nematode Control in Crops.* Academic Press, New York, pp. 179–231.

Cuadra, R.M., Cruz, X. and Fajardo, J.F. (2000) Cultivos de ciclo corto como plantas trampas para el control del nematodo agallador. *Nematropica* 30, 241–246.

Culbreath, A.K., Rodriguez-Kabana, R. and Morgan-Jones, G. (1985) The use of hemicellulosic waste matter for reduction of the phytotoxic effect of chitin and control of root-knot nematodes. *Nematropica* 15, 49–75.

Daulton, R.A.C. (1955) Progress report on eelworm control experiments. *Rhodesian Tobacco* 11, 21–24.

de Guiran, G. (1970) Le problème *Meloidogyne* sur tabac a Madagascar. Cahiers ORSTOM, *Série Biologie* 11, 187–208.

Di Vito, M., Cianciotta, V. and Zaccheo, G. (1991) The effect of population densities of *Meloidogyne incognita* on yield of susceptible and resistant tomato. *Nematologia Mediterranea* 19, 265–268.

Di Vito, M., Cianciotta, V. and Zaccheo, G. (1992) Yield of susceptible and resistant pepper in microplots infested with *Meloidogyne incognita. Nematropica* 22, 1–6.

Evans, K., Trudgill, D.L. and Webster, J.M. (eds) (1993) *Plant Parasitic Nematodes in Temperate Agriculture.* CAB International, Wallingford, UK, 648 pp.

Fernández, E. and Labrada, R. (1995) Experiencia en el uso de la solarización en Cuba. In: *Memorias del taller 'Solarización del Suelo',* Escuela Agrícola Panamericana 'El Zamorano' Honduras, 18–21 September, pp. 5–6.

Ferris, H. (1969) The vertical distribution of the root-knot nematode (*Meloidogyne javanica*) in Rhodesian sandy soil. *Rhodesian Journal of Agricultural Research* 7, 15–24.

Gair, R., Mathias, P.L. and Harvey, P.N. (1969) Studies of cereal nematode populations and cereal yields under continuous intensive culture. *Annals of Applied Biology* 63, 503–512.

Gardner, J. and Caswell-Chen, E.P. (1993) Penetration, development and reproduction of *Heterodera schachtii* on *Fagopyrum esculentium, Phacelia tanacetifolia, Raphanus sativus, Sinapis alpa* and *Brassica oleracea. Journal of Nematology* 25, 695–702.

Gaur, H.S. and Perry, R.S. (1991) The use of soil solarization for control of plant parasitic nematodes. *Nematological Abstracts* 60, 153–167.

Germani, G., Reversat, G. and Luc, M. (1983) Effect of *Sesbania rostrata* on *Hirschmanniella orytzae* in flooded rice. *Journal of Nematology* 5, 269–271.

Gnanapragasam, N.C. (1981) The influence of cultivating *Eragrostis curvula* in nematode infested soil, on the development and the subsequent build-up of populations in replanted tea. *Tea Quarterly* 50, 160–162.

Granges, A. and Leger, A. (1996) Essai comparatif de plants de tomates greffes a une ou deux tetes et de desinfection du sol a la vapeur en serre. *Revue Suisse de Viticulture, d'Arboriculture et d'Horticulture* 28, 389–392.

Hague, N.G.M. and Gowen, S.R. (1987) Chemical control of nematodes. In: Brown, R.H. and Kerry, B.R. (eds) (1987) *Principles and Practice of Nematode Control.* Academic Press, New York, pp. 131–173.

Hallmann, J. (2001) Plant interactions with endophytic bacteria. In: Jeger, M.J. and Spence, N.J. (eds) *Biotic Interactions in Plant–Pathogen Associations.* CAB International, Wallingford, UK, pp. 87–119.

Hallmann, J. and Sikora, R.A. (1994) Influence of *Fusarium oxysporum*, a mutualistic fungal endophyte, on *Meloidogyne incognita* infection of tomato. *Zeitschrift für Pflanzenkrankheiten und Pflanzenschutz* 101, 475–481.

Hallmann, J., Faupel, A., Krechel, A., Sikora, R.A. and Berg, G. (2004) Endophytic bacteria and biological control of nematodes. In: Sikora, R.A., Gowen, S., Hauschild, R. and Kiewnick, S. (eds) *Multitrophic Interactions in Soil.* Proceeding of the International Organization for Biological Control of Noxious Animals and Plants Workshop, Bad Honnef, Germany 2003. *IOBC-WPRS Bulletin* 27, 83–94.

Hutchinson, M.T. (1962) Rehabilitating tea soils – susceptibility of plants now in use to the root-lesion nematode, *Pratylenchus loosi. Tea Quarterly* 33, 138–142.

Jeffers, D.P. and Roberts, P.A. (1993) Effect of planting date and host genotype on the root-knot nematode–*Fusarium* wilt diseases complex of cotton. *Phytopathology* 83, 645–654.

Johnson, A.W. (1985) The role of nematicides in nematode management. In: Sasser, J.N. and Carter, C.C. (eds) *An Advanced Treatise on* Meloidogyne, Volume I. *Biology and Control.* North Carolina University Graphics, Raleigh, North Carolina, pp. 249–267.

Johnson, A.W. and Motsinger, R.E. (1990) Effects of planting date, small grain crop destruction, fallow, and soil temperature on the management of *Meloidogyne incognita. Journal of Nematology* 22, 348–355.

Johnson, C.S., Komm, D.A. and Jones, J.L. (1989) Control of *Globodera tobacum solanacearum* by alternating resistance and nematicide. *Journal of Nematology* 21, 16–23.

Kaloshian, I., Williamson, V.M.,. Miyao, G., Lawn, D.A. and Westerdahl, B.B. (1996) 'Resistance-breaking' nematodes identified in California tomnatoes. *Califorinia Agriculture* 50(6), 18–19.

Katan, J. (1981) Solar heating (solarization) of soil for control of soilborne pests. *Annual Review of Phytopathology* 19, 211–236.

Kerr, A. and Vythilingam, M.K. (1966) Replanting eelworm infested areas. *Tea Quarterly* 37, 67–72.

Kirkegaard, J.A., Sarwar, M., Matthiessen, J.N., Thomas, G. and Monteiro, A.A. (1998) Assessing the biofumigation potential of crucifers. *Acta Horticulturae* 459, 105–111.

Koch, D.W. and Gray, F.A. (1997) Nematode-resistant oil radish for control of *Heterodera schachtii* in sugar beet–barley rotations. *Journal of Sugar Beet Research* 34, 107–113.

Koenning, S.R., Schmitt, D.P. and Barker, K.R. (1993) Effects of cropping systems on population density of *Heterodera glycines* and soybean yield. *Plant Disease* 77, 780–786.

Kokalis-Burelle, N., Vavrina, C.S., Rosskopf, E.N. and Shelby, R.A. (2002) Field evaluation of plant growth-promoting rhizobacteria amended transplant mixes and soil solarization for tomato and pepper production in Florida. *Plant and Soil* 238, 257–266.

LaMondia, J.A. and Brodie, B.B. (1986) The effect of trap crops and fallow on decline of *Globodera rostochiensis. Annals of Applied Biology* 108, 347–352.

Luc, M., Sikora, R.S. and Bridge, J. (eds) (1990) *Plant Parasitic Nematodes in Subtropical and Tropical Agriculture.* CAB International, Wallingford, UK, 629 pp.

Maas, P.W.T. (1987) Physical methods and quarantine. In: Brown, R.H. and Kerry, B.R. (eds) *Principles and Practice of Nematode Control in Crops.* Academic Press, New York, pp. 265–291.

Mathur, B.N., Handa, D.K. and Swarup, G. (1987) Effect of deep summer ploughings on the cereal cyst nematode, *Heterodera avenae* and yield of wheat in Rajasthan, India. *Indian Journal of Nematology* 17, 292–295.

Maytac, C.A. and Bailey, J.E. (1988) Modification of the peanut leaf spot advisory for use on genotypes with partial resistance. *Phytopathology* 78, 640–644.

Mbwana, A.S.S., Ngode, L., Reddy, K.V.S. and Sikora, R.A. (1998) *A Guide to Growing Bananas in the Eastern African Highlands.* International Centre of Insect Physiology and Ecology, Nairobi, Kenya, 40 pp.

Mian, I.H., Ali, M. and Akhter, R. (1995) Grafting on *Solanum* rootstocks to control root-knot of tomato and bacterial wilt of eggplant. *Bulletin of the Institute of Tropical Agriculture, Kyushu University* 18, 41–47.

Mohandes, C., Rao, Y.S. and Sahu, S.C. (1981) Cultural control of rice root nematodes (*Hirschmanniella* spp.) with *Sphenoclea zeylandica. Proceedings of the Indian Academy of Sciences (Animal Sciences)* 90, 373–376.

Mugniery, D. and Balandras, C. (1984) Examen des possibilités d'eradication du nématode á kystes, *G. pallida* Stone. *Agronomie* 4, 773–778.

Munif, A., Hallmann, J. and Sikora, R.A. (2001) Induced systemic resistance of selected endophytic bacteria against *Meloidogyne incognita* on tomato. *Communications in Applied Biological Sciences* 66, 663–669.

Niblack, T.L., Hussey, R.S. and Boerma, H.R. (1986) Effects of environments, *Meloidogyne incognita* inoculum levels, and *Glycine max* genotype on root-knot nematode–soybean interactions in field microplots. *Journal of Nematology* 18, 338–346.

Nicolas, H., Rivoal, R., Duchesne, J. and Lili, Z. (1991) Detection of *Heterodera avenae* infestations on winter wheat by radiothermometry. *Revue Nematologie* 14, 285–290.

Niere, B.I., Speijer, P.R., Gold, C.S. and Sikora, R.A. (1998) Fungal endophytes from bananas for the biocontrol of *Radopholus similis.* In: *Proceedings of the Banana IPM Meeting, Nelspruit, South* Africa, pp. 313–318.

Noe, J.P. (1998) Crop and nematode-management systems. In: Barker, K.R., Pederson, G.A. and Windham, G.L. (eds) *Plant and Nematode Interactions.* American Society of Agronomy, Inc., Madison, Wisconsin, pp. 159–171.

Noe, J. and Sikora, R.A. (1990) Effects of tropical climates on the distribution and host–parasite relationship of plant parasitic nematodes. In: Luc, M., Sikora, R.A. and Bridge, J. (eds) *Plant Parasitic Nematodes in Subtropical and Tropical Agriculture.* CAB International, Wallingford, UK, pp. 583–597.

Noling, J.W. (2003) *Nematodes and their Management.* Horticultural Sciences Department, UF/IFAS, Florida Extension Service, ENY450 7178. Available at: http://edu.ifas.ufl.edu/CV112p

Norton, D.C. (1978) *Ecology of Plant-parasitic Nematodes.* John Wiley & Sons, New York, 268 pp.

Nusbaum, C.J. and Ferris, H. (1973) The role of cropping systems in nematode population management. *Annual Review of Phytopathology* 11, 423–440.

Nutter, F.W. Jr, Tylka, G.L., Guan, J., Morreira, A.J.D., Marett, C.C., Rosburg, T.R., Basart, J.P. and Chong, C.S. (2002) Use of remote sensing to detect soybean cyst nematode-induced plant stress. *Journal of Nematology* 34, 222–231.

Ogallo, J.L., Goodell, P.B., Eckert, J.W. and Roberts, P.A. (1999) Management of root-knot nematodes with resistant cotton cv. NemX. *Crop Science* 39, 418–421.

Ornat, C., Verdejo-Lucas, S. and Sorribas, F.J. (2001) A population of *Meloidogyne javanica* in Spain virulent to the *Mi* resistance gene in tomato. *Plant Disease* 85, 271–276.

Overman, A.J. and Jones, J.P. (1986) Soil solarization, reaction, and fumigation effects on double-cropped tomato under full-bed mulch. *Proceedings Florida State Horticultural Society* 99, 315–318.

Padgham, J.L., Abawai, G.S. and Duxbury, J.M. (2003) Survival and infectivity of *Meloidogyne graminicola* in flooded and non-flooded soils. *Nematologia Mediterranea* 31, 225–230.

Page, S.L.J. and Bridge, J. (1993) Plant nematodes and sustainability in tropical agriculture. *Experimental Agriculture* 29, 139–154.

Pandey, R., Sikora, R.A., Kalra, A., Singh, H.B. and Pandey, S. (2003) Plants and their products act as major nematode inhibitory agents. In: Trivedi, P.C. (ed.) *Nematode Management in Plants.* Scientific Publishers, Jodhpur, India, pp. 103–131.

Ploeg, A.T. (1999) Greenhouse studies on the effect of marigolds (*Tagetes* spp.) on four *Meloidogyne* species. *Journal of Nematology* 31, 62–69.

Ploeg, A.T. (2002) Effects of selected marigold varieties on root-knot nematodes and tomato and melon yields. *Plant Disease* 86, 505–508.

Prasad, J.S. and Reddy, K.V.S. (1994) Hot water treatment for banana planting material made easier. *Infomusa* 3, 16

Quénéhervé, P., Topart, P. and Martiny, B. (1998) *Mucuna pruriens* and other rotational crops for control of *Meloidogye incognita* and *Rotylenchulus reniformis* in vegetables in polytunnels in Martinique. *Nematropica* 28, 19–30.

Rao, G.N. (1976) Control of nematodes in nursery soils. *The Planter's Chronicle* 71, 257–259.

Rehiayani, A.S. and Hafez, S. (1998) Host status and green manure effect of selected crops on *Meloidogyne chitwoodi* race 2 and *Pratylenchus neglectus. Nematropica* 28, 213–230.

Reimann, S. and Sikora, R.A. (2003) Managing the mycorrhizosphere – an approach to sustainable agriculture after the phase out of methyl bromide. *Communications in Agriculture and Applied Biological Sciences* 68, 129–134.

Roberts, P.A. (1987) The influence of planting date of carrot on *Meloidogyne incognita* reproduction and injury to roots. *Nematologica* 33, 335–342.

Roberts, P.A. (2002) Concepts and consequences of resistance. In: Starr, J.L., Cook, R. and Bridge, J. (eds) *Plant Resistance to Parasitic Nematodes.* CAB International, Wallingford, UK, pp. 23–41.

Rodriguez-Kabana, R. (1992) Cropping systems for the management of phytonematodes. In: Gommers, F.J. and Maas, P.W.T. (eds) *Nematology from Molecule to Ecosystem.* European Society of Nematologists, pp. 219–233.

Rodriguez-Kabana, R., Morgan-Jones, G. and Chet, I. (1987) Biological control of nematodes: soil amendments and microbial antagonists. *Plant and Soil* 100, 237–247.

Rodriguez-Kabana, R., Weaver, C.F., Garcia, R., Robertson, D.G. and Carden, E.L. (1989) Bahiagrass for the management of root-knot and cyst nematodes in soybean. *Nematropica* 19, 185–193.

Schmitz, A., Kiewnick, S., Schlang, J. and Sikora, R.A. (2004) Use of high resolution digital thermography to detect *Heterodera schachtii* infestations in sugar beets. *Communications in Agriculture and Applied Biological Sciences* 69 (in press).

Scholte, K. (2000a) Effect of potato used as a trap crop on potato cyst nematodes and other soil pathogens and on the growth of a subsequent main potato crop. *Annals of Applied Biology* 136, 229–238.

Scholte, K. (2000b) Screening of non-tuber bearing Solanaceae for resistance to and induction of juvenile hatch of potato cyst nematodes and their potential for trap cropping. *Annals of Applied Biology* 136, 239–246.

Seinhorst, J.W. (1965) The relationship between nematode density and damage to plants. *Nematologica* 11, 137–154.

Shepherd, J.A. (1968) A nematode survey of tobacco soils in Rhodesia and Zambia, and the effects of grass–tobacco rotations on nematode populations. *Rhodesian Journal of Agricultural Research* 6, 19–26.

Shepherd, J.A. (1978) TRB Tests new lovegrass strain, and recommends it. *Rhodesian Tobacco Today* 1, 15.

Shepherd, J.A. and Barker, K.R. (1990) Nematode parasites of tobacco. In: Luc, M., Sikora, R.A. and Bridge, J. (eds) *Plant Parasitic Nematodes in Subtropical and Tropical Agriculture.* CAB International, Wallingford, UK, pp. 493–517.

Sikora, R.A. (1984) Importance of diapause, resistant and early maturing potato cultivars, chicken manure and non-host crops for *Globodera rostochiensis* integrated control in the upland tropics. *Communications in Agriculture and Applied Biological Sciences* 49, 613–620.

Sikora, R.A. (1992) Management of the antagonistic potential in agricultural ecosystems for the biological control of plant parasitic nematodes. *Annual Review of Phytopathology* 30, 245–270.

Sikora, R.A. (1997) Biological system management in the rhizosphere: an inside-out/outside-in perspective. *Communications in Agriculture and Applied Biological Sciences* 62, 105–112.

Sikora, R.A. (2002) Strategies for biological system management of nematodes in horticulture crops: fumigate, confuse or ignore them. *Communications in Agricultural and Applied Biological Sciences* 67, 5–18.

Sikora, R.A. and Hoffmann-Hergarten, S. (1993) Biological control of plant parasitic nematodes with plant-health-promoting rhizobacteria. In: Lumsden, R.D. and Vaughn, F.L. (eds) *Pest Management: Biological Based Technologies. Proceedings of Beltsville Symposium XVIII, USDA, Beltsville, Maryland, 2–6 May 1993.* American Chemical Society, Washington, DC, pp. 166–172.

Sikora, R.A., Niere, B. and Kimenju, J. (2003) Endophytic microbial biodiversity and plant nematode management in African agriculture. In: Neuenschwander, P., Borgemeister, C. and Langewald, J. (eds) *Biological Control in IPM Systems in Africa.* CAB International, Wallingford, UK, pp. 179–192.

Simpson, C.E. and Starr, J.L. (2001) Registration of 'COAN' peanut. *Crop Science* 41, 918.

Sivapalan, P. (1967) Nematodes and tea. *Tea Quarterly* 38, 178–185.

Spiegel, Y., Chet, I. and Cohn, E. (1987) Use of chitin for controlling plant parasitic nematodes. II. Mode of action. *Plant and Soil* 98, 337–345.

Stapleton, J.J. and DeVay, J.E. (1986) Soil solarization: a non-chemical approach for management of plant pathogens and pests. *Crop Protection* 5, 190–198.

Stapleton, J., Duncan, R.A. and Johnson, R. (1998) Soil disinfestation with cruciferous amendments and sublethal heating: effects on *Meloidogyne incognita, Sclerotium rolfsii* and *Pythium ultimum.* In: *Proceeding of the 7th International Congress of Plant Pathology,* Edinburgh, UK. *Plant Pathology* 47, 737–742.

Starr, J.L., Morgan, E.R. and Simpson, C.E. (2002) Management of peanut root-knot nematode, *Meloidogyne arenaria,* with host resistance. Online, *Plant Health Progress* doi:10.1094/PHP-2002–1221–01-HM.

Trudgill, D.L. (1991) Resistance and tolerance of plant parasitic nematodes in plants. *Annual Review of Phytopathology* 29, 167–192.

Trudgill, D.L. and Parrot, D.M. (1972) Effects of growing resistant potatoes with the *H1* gene from *Solanum tuberosum* ssp. *andigena* on populations of *Heterodera rostochiensis* pathotype A. *Annals of Applied Biology* 73, 67–75.

Turner, S.J. (1990) The identification and fitness of virulent potato-cyst nematodes (*Globodera pallida*) selected on resistant *Solanum vernei* hybrids for up to eleven generations. *Annals of Applied Biology* 117, 385–397.

Whitehead, A.G. (1998) *Plant Nematode Control.* CAB International, Wallingford, UK, 384 pp.

Young, L.D. and Hartwig, E.E. (1992) Problems and strategies associated with long-term use of nematode resistant cultivars. *Journal of Nematology* 24, 228–233.

Zum Felde, A., Pocasangre, L. and Sikora, R.A. (2004) Use of microbial communities inside suppressive banana plants to increase biocontrol of the burrowing nematode, *Radopholus similis.* In: Rosales, F.E., Pocasangre, L.E., Moens, T. and Vega, L. (eds) *Proceedings of Workshop on Banana Root Systems: towards a better understanding for its productive management,* 3–5 Nov. 2003, San Jose, Costa Rica (in press).

Appendix A – Nematicides*

Richard A. Sikora[1] and Peter Marczok[2]
[1]*Institut für Pflanzenkrankheiten, Universität Bonn, Nussallee 9, D-53115 Bonn, Germany;* [2]*Bayer Crop Science, Research Insecticides, Agriculture Centre, D-40789 Monheim, Germany*

A list is given here that includes most of the chemicals in use in the 2000s for the control of plant parasitic nematodes by nematicidal, nematostatic or nemato-repellent action. Some of these chemicals may also be used to control insects, weeds or other plant pests or diseases. Insecticides, fungicides or herbicides which may also be active against nematodes are not included.

Entries are arranged by common name followed by the preferred chemical name, then other names or codes, which may vary from country to country, and finally by the type of formulation of the chemical.

For information on the usage of a particular nematicide for a crop, refer to the index of this book or to standard reference works such as the *Pesticide Manual* produced by the British Crop Protection Council and CAB International.

Fumigants

Basamid *see* dazomet
1,3-dichloropropene
1,3-D; DCP; Telone, Nematrap, Nematox
Liquid formulation
carbathion *see* metham sodium
chloropicrin
Chloropicrin, Dorochlor
Liquid formulation
dazomet
3,5-dimethyl, 1,3,5-thiadiazine-2-thione
Basamid; Mylone, Gastard
Dust and granular formulation
D-D *see* dichloropropane-dichloropropene
dibromochloropropane* *see* DBCP
DBCP
1,2-dibromo-3-chloropropane
Fumazone, Nemagon
Liquid formulation
dichioropropene-dichloropropane*
DCIP
Nemamort
Liquid and granular formulation
1,2-dichloropropane with 1,3-dichloropropene
D-D
Vidden D
Liquid formulation
Dowfume *see* methyl bromide
Dorochlor *see* Chloropicrin
EDB *see* ethylene dibromide
ethylene dibromide
1,2-dibromethane
EDB; Terrafume
Liquid formulations

*A revision of the Appendix by Peter S. Gooch.

Fumazone *see* DBCP
Gastard *see* Dazomet
metham sodium
monosodium methyldithiocarbamate
carbathion; Vapam, Trimaton
Liquid formulation
methyl bromide*
bromomethane
Dowfume MC
Gas formulation
Methyl iodide
Liquid formulation
methyl isothiocyanate
isothiocyanatomethane
Trapex
Liquid formulation
Mylone *see* dazomet
Nemagon *see* DBCP
Nemamort *see* DCIP
Nematox *see* 1,3-dichloropropene
Nematrap *see* 1,3-dichloropropene
Nitrochloroform, trichloronitromethane
Liquid formulation
Telone 11 *see* 1,3-dichloropropene
Terrafume *see* ethylene dibromide
Trapex *see* methyl isothiocyanate
Trimaton *see* metham sodium
Vapam *see* metham sodium

Non-fumigants

Organophosphates

Acconem *see* Fosthietan
Boltage *see* Pyraclofos
Cadusafos
S,S-di-sec-butyl *O*-ethyl phosphorodithioate
Rugby, Sebufos
Granular and liquid formulations
Counter *see* terbufos
Dasanit *see* fensulfothion
Diamidafos
phenyl *N N*-dimethyl-phosphorodiamidate
Nellite
dichlofenthion
O-(2, 4-dichlorophenyl) *O, O*-diethyl phosophorothioate
Hexanema, Mobilawn
Granular and liquid formulations
ethoprop *see* Ethoprophos
Ethoprophos
O-ethyl S, S′-dipropyl phosphorodithioate
Mocap
Granular and liquid formulations
fenamiphos
ethyl 4-methylthio-*m*-tolyl isopropylphosphoramidate
Nemacur
Granular and liquid formulations
fensulfothion
O, O-diethy1-*O*-4-methylsulfinylphenylphosphorothioate
Terracur P, Dasanit
Granular and liquid formulations
Fostiazate
O-ethyl S-(1-methylpropyl) (2-oxo-3-thiazolidinyl) phosphonothioate
Nematorin, Nemathorin
Granular and liquid formulations
Fosthietan
diethyl 1,3-dithietan-2-ylidenephosphoramidate
Acconem; Nem-a-tak; geofos
geofos *see* Fosthietan
Hexanema *see* dichlofenthion
Isazofos
O-5-chloro-l-isopropyl-1H-1,2, 4-triazol-3-yl *O, O*-diethyl phosophorothioate
Miral
Granular and liquid formulations
Miral *see* isazofos
Mobilawn *see* Dichlofenthion
Mocap *see* Ethoprophos
Nellite *see* Diamidafos
Nem-a-tak *see* Fosthietan
Nemacur *see* fenamiphos
Nemaphos *see* thionazin
Phenamiphos *see* fenamiphos
phorate
O, O-diethyl S-ethylthiomethylphosphorodithioate
Thimet
Granular and liquid formulations
Pyraclofos
O-[1-(4-chlorophenyl)-1H-pyrazole-4-yl]-*O*-ethyl-S-propyl-phosphorothioate
Boltage, Voltage
Granular and liquid formulations

Rugby *see* **Cadusafos**
Sebufos *see* **Cadusafos**
terbufos
S-tert-butylthiomethyl *O, O*-diethylphosphorodithioate
Counter
Granular formulations
Terracur P *see* **fensulfothion**
Thimet *see* **phorate**
thionazin
O, O-diethyl *O*-pyrazin-2-yl phosphorothioate
Nemaphos; Zinophos
Granular and liquid formulations
Voltage *see* **Pyraclofos**
Zinophos *see* **thionazin**

Carbamates

aldicarb
2-methyl-2-(methylthio) propionaldehyde *O*-(methy1carbamoyl)-oxime
Temik
Granular formulations
Aldoxycarb
2-methyl-2-methylpropionaldehyde *O*-methylcarbamoyloxime
Standak
Flowable formulation
carbofuran
2,3-dihydro-2,2-dimethylbenzofuran-7-yl methylcarbamate
Curaterr; Furadan, Yaltox
Granular and flowable formulations
cloethocarb
2-(2-chloro-l-methoxyethoxy)phenyl methylcarbamate
Lance
Granular formulation
Curaterr *see* **carbofuran**
Furadan *see* **carbofuran**
Lance *see* **cloethocarb**
oxamyl
S-methyl *N′, N′*-dimethyl-*N*-[(methylcarbamoyl)oxy]-1-thio-oxamimidate
Vydate
Granular and liquid formulations
Standak *see* **Aldoxycarb**
Temik *see* **aldicarb**
Vydate *see* **oxamyl**
Yaltox *see* **carbofuran**

*The manufacture and/or use of these compounds has been either banned or is being considered for removal from the market. They may no longer be available, but in some cases may be obtainable locally in some countries under other brand names.

Off-patent compounds may be available under brand names not listed in this index. The omission of other product names or formulations does not imply that they might not be suitable as nematicides.

Nematicides should only be used with strict adherence to the safety precautions recommended by the manufacturer. Many nematicides are toxic to human beings and livestock and should always be treated with respect. This list is presented as a general guide and not a complete list of all products available in the past or present.

Appendix B – Plant Parasitic Nematode Genera and Species Cited

Michel Luc[1] and David J. Hunt[2]
[1]*6 rue Boutard, 92200 Neuilly-sur-Seine, France;* [2]*CABI Bioscience, Bakeham Lane, Egham, Surrey TW20 9TY, UK*

All genera and species of plant parasitic nematodes cited in the book are listed alphabetically below. They are followed by their 'authorities', i.e. the name(s) of the author(s) of the original description, in some cases followed by the name(s) of the author(s) who have published the more recent valid taxonomic name, i.e. by placing the species in another genus. In such cases, the original authorities are placed in parentheses. Both authorities are followed by the year of publication of their respective works.

The most common synonyms are also listed alphabetically as 'cf.', and referred to after '=' below the valid name.

For each genus, the group to which it pertains is indicated as follows:

Tyl. = Tylenchina
A. = Aphelenchina
L. = Longidoridae (Dorylaimina)
P. = Panagrolaimidae (Rhabditida)
T. = Tylencholaimidae (Dorylaimina)
Tri. = Trichodoridae (Diphtherophorina)

Achlysiella Hunt, Bridge & Machon, 1989 [Tyl., Pratylenchidae]
Achlysiella williamsi (Siddiqi, 1964) Hunt, Bridge & Machon, 1989
= *Radopholus williamsi* Siddiqi, 1964
Afenestrata Baldwin & Bell, 1985 [Tyl., Heteroderidae]
Allotrichodorus Rodriguez-M, Sher & Siddiqi, 1978 [Tri.]
Allotrichodorus brasiliense Rashid, De Waele & Coomans, 1986
Allotrichodorus campanulatus Rodriguez-M, Sher & Siddiqi, 1978
Allotrichodorus sharmae Rashid, De Waele & Coomans, 1986
Allotrichodorus westindicus
cf. *Ecuadorus westindicus*
Amplimerlinius Siddiqi, 1976 [Tyl., Belonolaimidae]
Anguina Scopoli, 1777 [Tyl., Anguinidae]
Anguina agrostis (Steinbuch, 1799) Filipjev, 1936
Anguina tritici (Steinbuch, 1799) Chitwood, 1935
Aorolaimus Sher, 1963 [Tyl., Hoplolaimidae]
= *Peltamigratus* Sher, 1964
Aorolaimus banoae (Rashid, Geraert & Sharma, 1987) Baujard, Castillo, Doucet, Martiny, Mounport & N'Diaye, 1991
= *Peltamigratus banoae* Rashid, Geraert & Sharma, 1987
Aorolaimus holdemani (Sher, 1964) Fortuner, 1987
= *Peltamigratus holdemani* Sher, 1964
Aorolaimus levicaudatus (Bittencourt & Huang, 1986) Baujard, Castillo, Doucet, Martiny, Mounport & N'Diaye, 1991
= *Peltamigratus levicaudatus* Bittencourt & Huang, 1986

Aorolaimus luci (Sher, 1964) Fortuner, 1987
= *Peltamigratus luci* Sher, 1964
Aorolaimus nigeriensis (Sher, 1964) Fortuner, 1987
= *Peltamigratus nigeriensis* Sher, 1964
Aorolaimus vigiae (Rashid, Geraert & Sharma, 1987) Baujard, Castillo, Doucet, Martiny, Mounport & N'Diaye, 1991
= *Peltamigratus vigiae* Rashid, Geraert & Sharma, 1987
Aphasmatylenchus Sher, 1965 [Tyl., Hoplolaimidae]
Aphasmatylenchus liberiensis Baujard, Vovlas, Mounport & Martiny, 1998
Aphasmatylenchus nigeriensis Sher, 1965
Aphasmatylenchus straturatus Germani, 1970
Aphelenchoides Fischer, 1894 [A., Aphelenchoididae]
Aphelenchoides aligarhiensis Siddiqi, Husain & Khan, 1967
Aphelenchoides arachidis Bos, 1977
Aphelenchoides besseyi Christie, 1942
= *Aphelenchoides oryzae* Yokoo, 1948
Aphelenchoides bicaudatus (Imamura, 1931) Filipjev & Schuurmans Stekhoven, 1941
Aphelenchoides fragariae (Ritzema-Bos, 1890) Christie, 1932
Aphelenchoides ritzemabosi (Schwartz, 1911) Steiner & Buhrer, 1932
Aphelenchus Bastian, 1865 [A., Aphelenchidae]
Aphelenchus avenae Bastian, 1865
Atalodera Wouts & Sher, 1971 [Tyl., Heteroderidae]
= *Thecavermiculatus* Robbins, 1978
Atalodera andina (Golden, Franco, Jatala & Astogaza, 1983) de Souza & Huang, 1994
= *Thecavermiculatus andinus* Golden, Franco, Jatala & Astogaza, 1983
Basirolaimus
cf. *Hoplolaimus*
Belonolaimus Steiner, 1949 [Tyl., Belonolaimidae]
= *Ibipora* Monteiro & Lordello, 1977
Belonolaimus euthychilus Rau, 1963
Belonolaimus gracilis Steiner, 1949
Belonolaimus longicaudatus Rau, 1958
Belonolaimus maritimus Rau, 1963
Belonolaimus nortoni Rau, 1963
Bursaphelenchus Fuchs, 1937 [A., Aphelenchoididae]
= *Rhadinaphelenchus* Goodey, 1960
Bursaphelenchus cocophilus (Cobb, 1919) Baujard, 1989
= *Rhadinaphelenchus cocophilus* (Cobb, 1919) Goodey, 1960
Bursaphelenchus mucronatus Mamiya & Enda, 1979
Bursaphelenchus xylophilus (Steiner & Buhrer, 1934) Nickle, 1970
Cacopaurus Thorne, 1943 [Tyl., Tylenchulidae]
Cactodera Krall & Krall, 1978 [Tyl., Heteroderidae]
Cactodera amaranthi (Stoyanov, 1972) Krall & Krall, 1978
Caloosia Siddiqi & Goodey, 1964 [Tyl., Criconematidae]
Caloosia exilis Mathur, Khan, Nand & Prasad, 1969
Caloosia heterocephala
cf. *Caloosia paxi*
Caloosia nudata (Colbran, 1963) Brzeski, 1974
= *Hemicycliophora nudata* Colbran, 1963
Caloosia paradoxa (Luc, 1958) Brzeski, 1974
= *Hemicycliophora paradoxa* Luc, 1958
Caloosia paxi Mathur, Khan, Nand & Prasad, 1969
= *Caloosia heterocephala* Rao & Mohandas, 1976
Cephalenchus Goodey, 1962 [Tyl., Tylenchidae]
Cephalenchus emarginatus (Cobb, 1893) Geraert, 1968
Cephalenchus hexalineatus (Geraert, 1962) Geraert & Goodey, 1964
Criconema Hofmänner & Menzel, 1914 [Tyl., Criconematidae]
Criconema braziliense (Raski & Pinochet, 1976) Raski & Luc, 1985
= *Mesocriconema braziliense* Raski & Pinochet, 1976
Criconema cardamomi (Khan & Nanjappa, 1972) Raski & Luc, 1985
Criconema coorgi (Khan & Nanjappa, 1972) Raski & Luc, 1985
Criconema corbetti (De Grisse, 1967) Raski & Luc, 1985
= *Nothocriconema corbetti* De Grisse, 1967

Criconema crassianulatum (de Guiran, 1963) Raski & Luc, 1985
Criconema demani Micoletzky, 1925
Criconema jaejuense (Choi & Geraert, 1975) Raski & Luc, 1985
= *Nothocriconema jaejuense* Choi & Geraert, 1975
Criconemella
cf. *Criconemoides*
Criconemoides Taylor, 1936 [Tyl., Criconematidae]
= *Macroposthonia* de Man, 1880 (gen. dub.)
= *Criconemella* De Grisse & Loof, 1965
= *Mesocriconema* Andrássy, 1965
Criconemoides annulatus Cobb in Taylor, 1936 nec *Macroposthonia annulata* de Man, 1880 [sp. inq.]
Criconemoides axestis Fassuliotis & Williamson, 1959
= *Criconemella axestis* (Fassuliotis & Williamson, 1959) Luc & Raski, 1981
Criconemoides brevistylus Singh & Khera, 1976
Criconemoides curvatus Raski, 1952
= *Criconemella curvata* (Raski, 1952) Luc & Raski, 1981
Criconemoides denoudeni Heyns, 1962
Criconemoides dherdei (De Grisse, 1967) Luc, 1970
Criconemoides ferniae Luc, 1959
= *Criconemella ferniae* (Luc, 1959) Raski & Luc, 1981
Criconemoides incisus Raski & Golden, 1956
= *Criconemella incisa* (Raski & Golden, 1956) Luc & Raski, 1961
= *Macroposthonia incisa* (Raski & Golden, 1956) De Grisse, 1967
Criconemoides informis (Micoletzky, 1922) Taylor, 1936
= *Hoplolaimus informis* Micoletzky, 1922
= *Criconemella informis* (Micoletzky, 1922) Ebsary, 1991
= *Macroposthonia informis* (Micoletzky, 1922) De Grisse & Loof, 1965
Criconemoides obtusicaudatus Heyns, 1962
Criconemoides onoensis Luc, 1959
= *Criconemella onoensis* (Luc, 1959) Luc & Raski, 1981
= *Macroposthonia onoensis* (Luc, 1959) De Grisse & Loof, 1965
Criconemoides ornatus Raski, 1958
= *Criconemella ornata* (Raski, 1958) Luc & Raski, 1981
= *Macroposthonia ornata* (Raski, 1958) De Grisse & Loof, 1965
Criconemoides palustris Luc, 1970
= *Criconemella palustris* (Luc, 1970) Raski & Luc, 1981
Criconemoides paradenoudeni (Rashid, Geraert & Sharma, 1987) n. comb.
= *Criconemella paradenoudeni* Rashid, Geraert & Sharma, 1987
= *Macroposthonia paradenoudeni* (Rashid, Geraert & Sharma, 1987) Siddiqi, 2000
Criconemoides paragoodeyi (Choi & Geraert, 1975) Loof & De Grisse, 1989
= *Criconemella paragoodeyi* Choi & Geraert, 1975
Criconemoides paralineolatus (Rashid, Geraert & Sharma, 1987) n. comb.
= *Criconemella paralineolata* Rashid, Geraert & Sharma, 1987
= *Macroposthonia paralineolata* (Rashid, Geraert & Sharma, 1987) Siddiqi, 2000
Criconemoides pseudohercyniensis De Grisse & Koen, 1964
= *Criconemella pseudohercyniensis* (De Grisse & Koen, 1964) Raski & Luc, 1981
Criconemoides rusticus (Micoletzky, 1915) Taylor, 1936
= *Criconemella rustica* (Micoletzky, 1915) Luc & Raski, 1981
Criconemoides sphaerocephala Taylor, 1936
= *Criconemella sphaerocephala* (Taylor, 1936) Luc & Raski, 1981
= *Macroposthonia sphaerocephala* (Taylor, 1936) De Grisse & Loof, 1965
Criconemoides tescorum de Guiran, 1963
= *Macroposthonia tescorum* (de Guiran, 1963) De Grisse & Loof, 1965
Criconemoides xenoplax Raski, 1952
= *Criconemella xenoplax* (Raski, 1952) Luc & Raski, 1981
= *Macroposthonia xenoplax* (Raski, 1952) De Grisse & Loof, 1965

Crossonema
cf. *Ogma*
Discocriconemella De Grisse & Loof, 1965 [Tyl., Criconematidae]
Discocriconemella degrissei Loof & Sharma, 1980
Discocriconemella elettariae Sharma & Edward, 1985
Discocriconemella limitanea (Luc, 1959) De Grisse & Loof, 1965
Ditylenchus Filipjev, 1936 [Tyl., Anguinidae]
Ditylenchus africanus Wendt, Swart, Vrain & Webster, 1995
Ditylenchus allii
cf. *Ditylenchus dipsaci*
Ditylenchus angustus (Butler, 1913) Filipjev, 1936
Ditylenchus destructor Thorne, 1945
Ditylenchus dipsaci (Kühn, 1857) Filipjev, 1936
= *Ditylenchus allii* (Beijerinck, 1883) Tarjan, 1960
= *Ditylenchus fragariae* Kirjanova, 1951
Ditylenchus fragariae
cf. *Ditylenchus dipsaci*
Ditylenchus humuli Skarbilovich, 1872
Ditylenchus myceliophagus Goodey, 1958
Ditylenchus procerus (Bally & Reydon, 1931) Filipjev, 1936
Dolichodorus Cobb, 1914 [Tyl., Dolichodoridae]
Dolichodorus heterocephalus Cobb, 1914
Dolichodorus minor Loof & Sharma, 1975
Ecuadorus Siddiqi, 2002 [Tri.]
Ecuadorus westindicus (Rodriguez-M, Sher & Siddiqi, 1978) Siddiqi, 2002
= *Allotrichodorus westindicus* (Rodriguez-M, Sher & Siddiqi, 1978) Rashid, De Waele & Coomans, 1986
= *Nanidorus westindicus* Rodriguez-M, Sher & Siddiqi, 1978
Eutylenchus Cobb, 1913 [Tyl., Atylenchidae]
Eutylenchus africanus Sher, Corbett & Colbran, 1966
Globodera Skarbilovich, 1959 [Tyl., Heteroderidae]
Globodera pallida Stone, 1973
= *Heterodera pallida* Stone, 1973
Globodera rostochiensis (Wollenweber, 1923) Skarbilovich, 1959
Globodera tabacum tabacum (Lownsbery & Lownsbery, 1954) Skarbilovich, 1959
= *Globodera tabacum* (Lownsbery & Lownsbery, 1954) Skarbilovich, 1959
Globodera tabacum solanacearum (Miller & Gray, 1972) Behrens, 1975
= *Globodera solanacearum* (Miller & Gray, 1972) Behrens, 1975
Globodera virginiae (Miller & Gray, 1968) Stone, 1973
= *Globodera tabacum virginiae* (Miller & Gray, 1968) Stone, 1973
Gracilacus Raski, 1962 [Tyl., Tylenchulidae]
Gracilacus peratica Raski, 1962
Halenchus Cobb *in* Cobb, 1933 [Tyl., Anguinidae]
Helicotylenchus Steiner, 1945 [Tyl., Hoplolaimidae]
= *Rotylenchoides* Whitehead, 1958
Helicotylenchus abunaamai Siddiqi, 1972
Helicotylenchus affinis (Luc, 1960) Fortuner, 1984
= *Rotylenchoides affinis* Luc, 1960
Helicotylenchus astriatus Khan & Nanjappa, 1972
Helicotylenchus brevis (Whitehead, 1958) Fortuner, 1984
= *Rotylenchoides brevis* Whitehead, 1958
Helicotylenchus cavenessi Sher, 1966
Helicotylenchus crenacauda Sher, 1966
Helicotylenchus digitiformis Ivanova, 1967
Helicotylenchus digonicus Perry in Perry, Darling & Thorne, 1959
Helicotylenchus dihystera (Cobb, 1893) Sher, 1961
Helicotylenchus egyptiensis Tarjan, 1964
Helicotylenchus erythrinae (Zimmermann, 1904) Golden, 1956
Helicotylenchus indicus Siddiqi, 1963
Helicotylenchus intermedius (Luc, 1960) Siddiqi & Husain, 1964
= *Rotylenchoides intermedius* Luc, 1960
Helicotylenchus microcephalus Sher, 1966
Helicotylenchus mucronatus Siddiqi, 1963
Helicotylenchus multicinctus (Cobb, 1893) Golden, 1956
Helicotylenchus neopaxilli Inserra, Vovlas & Golden, 1975
Helicotylenchus oleae Inserra, Vovlas & Golden, 1979

Helicotylenchus paracanalis Sauer & Winoto, 1975
Helicotylenchus pseudorobustus (Steiner, 1914) Golden, 1956
Helicotylenchus serenus Siddiqi, 1963
Helicotylenchus sharafati Mulk & Jairajpuri, 1975
Helicotylenchus variocaudatus (Luc, 1960) Fortuner, 1984
= *Rotylenchoides variocaudatus* Luc, 1960
Hemicriconemoides Chitwood & Birchfield, 1957 [Tyl., Criconematidae]
Hemicriconemoides chitwoodi Esser, 1960
Hemicriconemoides cocophillus (Loos, 1949) Chitwood & Birchfield, 1957
Hemicriconemoides gaddi (Loos, 1949) Chitwood & Birchfield, 1957
Hemicriconemoides kanayaensis Nakasono & Ichinohe, 1961
Hemicriconemoides mangiferae Siddiqi, 1961
Hemicriconemoides mehdii Suryawanshi, 1971
Hemicriconemoides snoecki Van Doorsselaere & Samsoen, 1982
Hemicycliophora de Man, 1921 [Tyl., Criconematidae]
Hemicycliophora arenaria Raski, 1958
Hemicycliophora argiensis Khan & Nanjappa, 1972
Hemicycliophora attapadii Rahaman, Ahmad & Jairajpuri, 1996
Hemicycliophora chathami Yeates, 1978
Hemicycliophore chilensis Brzeski, 1974
Hemicycliophore loofi Maas, 1970
Hemicycliophora nudata
cf. *Caloosia nudata*
Hemicycliophora parvana Tarjan, 1952
Hemicycliophora penetrans Thorne, 1955
Hemicycliophora poranga Monteiro & Lordello, 1978
Hemicycliophora similis Thorne, 1955
Hemicycliophora thienemanni (Schneider, 1925) Loos, 1948
Hemicycliophora typica de Man, 1921
Hemicycliophora utkali Ray & Das, 1981
Heterodera Schmidt, 1871 [Tyl., Heteroderidae]
Heterodera aucklandica Wouts & Sturhan, 1995
Heterodera australis Subbotin, Sturhan, Rumpenhorst & Moens, 2002
Heterodera avenae Wollenweber, 1924
Heterodera bifenestrata Cooper, 1956
Heterodera cajani Koshy, 1967
= *Heterodera vigni* Edward & Misra, 1968
Heterodera ciceri Vovlas, Greco & di Vito, 1985
Heterodera cruciferae Franklin, 1945
Heterodera delvii Jairajpuri, Khan, Setty & Govindu, 1979
Heterodera elachista Ohshima, 1974
Heterodera fici Kirjanova, 1954
Heterodera filipjevi (Madzhidov, 1981) Steiner & Stelter, 1984
Heterodera gambiensis Merny & Netscher, 1976
Heterodera glycines Ichinohe, 1952
Heterodera goettingiana Liebscher, 1892
Heterodera graminis Stynes, 1971
Heterodera hordecalis Anderson, 1975
Heterodera latipons Franklin, 1969
Heterodera lespedezae Golden & Cobb, 1963
Heterodera marioni
cf. *Meloidogyne marioni*
Heterodera mediterranea Vovlas, Inserra & Stone, 1981
Heterodera mani Mathews, 1971
Heterodera mothi Khan & Husain, 1965
Heterodera oryzae Luc & Berdon Brizuela, 1961
Heterodera oryzicola Rao & Jayaprakash, 1978
Heterodera pakistanensis Maqbool & Shahina, 1986
Heterodera pallida
cf. *Globodera pallida*
Heterodera punctata
cf. *Punctodera punctata*
Heterodera sacchari Luc & Merny, 1963
Heterodera schachtii A. Schmidt, 1871
Heterodera skohensis Kaushal, Sharma & Singh, 2000
Heterodera sorghi Jain, Sethi, Swarup & Srivastava, 1982
Heterodera swarupi Sharma, Siddiqi, Rahaman & Ansari, 1999
Heterodera trifolii Goffart, 1932
Heterodera vigni
cf. *Heterodera cajani*

Heterodera zeae Koshy, Swarup & Sethi, 1971
Hirschmanniella Luc & Goodey, 1964 [Tyl., Pratylenchidae]
Hirschmanniella asteromucronata Rasjivin, Fernandez, Ortega & Quincoa, 1981
Hirschmanniella belli Sher, 1968
Hirschmanniella caudacrena
  cf. *Hirschmanniella mexicana*
Hirschmanniella diversa Sher, 1968
Hirschmanniella dubia
  cf. *Hirschmanniella magna*
Hirschmanniella furcata Razjivin, Fernandez, Ortega & Quincosa, 1981
Hirschmanniella gracilis (de Man, 1880) Luc & Goodey, 1964
Hirschmanniella imamuri Sher, 1968
Hirschmanniella indica
  cf. *Hirschmanniella mucronata*
Hirschmanniella kaverii
  cf. *Hirschmanniella mucronata*
Hirschmanniella magna Siddiqi, 1966
  = *Hirschmanniella dubia* Khan, 1972
Hirschmanniella mangalorensis
  cf. *Hirschmanniella mucronata*
Hirschmanniella marina Sher, 1968
Hirschmanniella mexicana (Chitwood, 1961) Sher, 1968
  = *Hirschmanniella caudacrena* Sher, 1968
Hirschmanniella microtyla Sher, 1968
Hirschmanniella miticausa Bridge, Mortimer & Jackson, 1984
Hirschmanniella mucronata (Das, 1960) Luc & Goodey, 1964
  = *Hirschmanniella indica* Ahmad, 1974
  = *Hirschmanniella kaverii* Sivakumar & Khan, 1982
  = *Hirschmanniella mangalorensis* Mathur & Prasad, 1971
Hirschmanniella nana
  cf. *Hirschmanniella oryzae*
Hirschmanniella nghetinhiensis Eroshenko & Chau *in* Eroshenko, Tyau, Tkhan & Kan, 1985
Hirschmanniella obesa Razjivin, Fernandez, Ortega & Quincosa, 1981
Hirschmanniella ornata Eroshenko & Chau *in* Eroshenko, Tyau, Tkhan & Kan, 1985
Hirschmanniella oryzae (van Breda de Haan, 1902) Luc & Goodey, 1964
  = *Hirschmanniella nana* Siddiqi, 1966
Hirschmanniella shamimi Ahmad, 1972
Hirschmanniella spinicaudata (Schuurmans Stekhoven, 1944) Luc & Goodey, 1964
Hirschmanniella thornei Sher, 1968
Hirschmanniella truncata Razjivin, Fernandez, Ortega & Quincosa, 1981
Hoplolaimus von Daday, 1905 [Tyl., Hoplolaimidae]
  = *Basirolaimus* Shamsi, 1979
Hoplolaimus aegypti Shaflee & Koura, 1970
  = *Basirolaimus aegypti* (Shaflee & Koura, 1970) Shamsi, 1979
Hoplolaimus clarissimus Fortuner, 1973
  = *Basirolaimus clarissimus* (Fortuner, 1973) Shamsi, 1979
Hoplolaimus columbus Sher, 1963
Hoplolaimus dimorphicus Mulk & Jairajpuri, 1976
  = *Basirolaimus dimorphicus* (Mulk & Jairajpuri, 1976) Shamsi, 1979
Hoplolaimus dubius Chaturvedi & Khera, 1979
  = *Basirolaimus dubius* (Chatuverdi & Khera, 1979) Siddiqi, 1986
Hoplolaimus galeatus (Cobb, 1913) Filipjev & Schuurmans Stekhoven, 1941
Hoplolaimus indicus Sher, 1963
  = *Basirolaimus indicus* (Sher, 1963) Shamsi, 1979
Hoplolaimus magnistylus Robbins, 1982
Hoplolaimus pararobustus (Schuurmans Stekhoven & Teunissen, 1938) Sher, 1963
Hoplolaimus seinhorsti Luc, 1958
  = *Basirolaimus seinhorsti* (Luc, 1958) Shamsi, 1979
Hypsoperine
  cf. *Meloidogyne*
Ibipora
  cf. *Belonolaimus*
Longidoroides Khan, Chawla & Saha, 1978 [L.]
Longidorus (Micoletzky, 1922) Thorne & Swanger, 1936 [L.]
Longidorus africanus Merny, 1966
Longidorus apulus Lamberti & Bleve-Zacheo, 1977
Longidorus elongatus (de Man, 1876) Thorne & Swanger, 1936
Longidorus fursti Heyns, Coomans, Hutsebaut & Swart, 1987

Longidorus israelensis Peneva, Orion, Shlevin, Bar-Eyal & Brown, 1998
Longidorus laevicapitatus Williams, 1959
Longidorus leptocephalus Hooper, 1961
Longidorus maximus
cf. *Paralongidorus maximus*
Longidorus pisi Edward, Misra & Singh, 1964
= *Longidorus siddiqii* Aboul-Eid, 1970
Longidorus siddiqii
cf. *Longidorus pisi*
Longidorus vineacola Sturhan & Weischer, 1964
Macroposthonia
cf. *Criconemoides*
Meloidogyne Goeldi, 1892 [Tyl., Meloidogynidae]
= *Hypsoperine* Sledge & Golden, 1964
Meloidogyne acrita
cf. *Meloidogyne incognita*
Meloidogyne acronea Coetzee, 1956
Meloidogyne africana Whitehead, 1960
Meloidogyne arabidicida López & Salazar, 1989
Meloidogyne arenaria (Neal, 1889) Chitwood, 1949
Meloidogyne arenaria thamesi
cf. *Meloidogyne thamesi*
Meloidogyne artiellia Franklin, 1961
Meloidogyne baetica Castillo, Vovlas, Subbotin & Troccoli, 2003
Meloidogyne brasiliensis Charchar & Eisenback, 2002
Meloidogyne brevicauda Loos, 1953
Meloidogyne chitwoodi Golden, O'Bannon, Santo & Finley, 1980
Meloidogyne coffeicola Lordello & Zamith, 1960
Meloidogyne cruciani Garcia-Martinez, Taylor & Smart, 1982
Meloidogyne decalineata Whitehead, 1958
Meloidogyne enterolobii Yang & Eisenback, 1983
Meloidogyne ethiopica Whitehead, 1968
Meloidogyne exigua Goeldi, 1892
Meloidogyne fallax Karssen, 1996
Meloidogyne floridensis Handoo, Nyczepir, Esmenjaud, van der Beek, Castagnone-Sereno, Carta, Skantar & Higgins, 2004
Meloidogyne fujianensis Pan, 1985
Meloidogyne grahami
cf. *Meloidogyne incognita*
Meloidogyne graminicola Golden & Birchfield, 1965
Meloidogyne graminis (Sledge & Golden, 1964) Whitehead, 1968
Meloidogyne hainanensis Liao JinLing & Feng ZhiXin, 1995
Meloidogyne hapla Chitwood, 1949
Meloidogyne haplanaria Eisenback, Bernard, Starr, Lee & Tomaszewski, 2003
Meloidogyne hispanica Hirschmann, 1986
Meloidogyne incognita (Kofoid & White, 1919) Chitwood, 1949
= *Meloidogyne incognita acrita* Chitwood, 1949
= *Meloidogyne acrita* Chitwood, 1949
= *Meloidogyne grahami* Golden & Slana, 1978
Meloidogyne inornata Lordello, 1956
Meloidogyne javanica (Treub, 1885) Chitwood, 1949
Meloidogyne kikuyensis De Grisse, 1961
Meloidogyne konaensis Eisenback, Bernard & Schmitt, 1995
Meloidogyne lini Yang, Hu & Zhu, 1988
Meloidogyne lusitanica Abrantes & Santos, 1991
Meloidogyne marioni (Cornu, 1879) Chitwood & Oteifa, 1952 (sp. inq.)
= *Heterodera marioni* (Cornu, 1879) Marcinowski, 1909
Meloidogyne mayaguensis Rammah & Hirschmann, 1988
Meloidogyne megadora Whitehead, 1958
Meloidogyne microcephala Cliff & Hirschmann, 1984
Meloidogyne microtyla Mulvey, Townshend & Porter, 1975
Meloidogyne naasi Franklin, 1965
Meloidogyne oryzae Maas, Sanders & Dede, 1978
Meloidogyne oteifae Elmiligy, 1968
Meloidogyne ottersoni (Thorne, 1969) Franklin, 1971
Meloidogyne paranaensis Carneiro, Carneiro, Abrantes, Santos & Almeida, 1996
Meloidogyne petuniae Charchar, Eisenback & Hirschmann, 1999
Meloidogyne platani Hirschmann, 1982
Meloidogyne salasi Lopez-Chaves, 1985
Meloidogyne spartinae (Rau & Fassuliotis, 1965) Whitehead, 1968

Meloidogyne thamesi Chitwood in Chitwood, Specht & Havis, 1952
= *Meloidogyne arenaria thamesi* Chitwood *in* Chitwood, Specht & Havis, 1952
Meloidogyne triticoryzae Gaur, Saha & Khan, 1993
Merlinius Siddiqi, 1970 [Tyl., Belonolaimidae]
Merlinius brevidens (Allen, 1955) Siddiqi, 1970
Merlinius cylindricus (Ivanova, 1962) Siddiqi, 1970
Mesocriconema
cf. *Criconemoides*
Monotrichodorus
cf. *Trichodorus*
Nacobbus Thorne & Allen, 1944 [Tyl., Pratylenchidae]
Nacobbus aberrans (Thorne, 1935) Thorne & Allen, 1944
Nacobbus bolivianus Lordello, Zamith & Boock, 1961
Nacobbus dorsalis Thorne & Allen, 1944
Neodolichodorus Andrássy, 1976 [Tyl. Dolichodoridae]
Ogma Southern, 1914 [Tyl., Criconematidae]
= *Crossonema* Khan, Chawla & Saha, 1976
Ogma decalineatum (Chitwood, 1957) Andrássy, 1979
Ogma rhombosquamatum (Mehta & Raski, 1971) Andrássy, 1979
Ogma taylatum (Khan, Chawla & Saha, 1976) Siddiqi, 1986
= *Crossonema taylatum* Khan, Chawla & Saha, 1976
Panagrolaimus Fuchs, 1930 [P.]
Panagrolaimus rigidus (Schneider, 1866) Thorne, 1937
Paralongidorus Siddiqi, Hooper & Khan, 1963 [L.]
= *Siddiqia* Khan, Chawla & Saha, 1968
Paralongidorus australis Stirling & McCulloch, 1985
Paralongidorus citri (Siddiqi, 1959) Siddiqi, Hooper & Khan, 1963
Paralongidorus lutensis Hunt & Rahman, 1991
Paralongidorus maximus (Bütschli, 1874) Siddiqi, 1964
= *Longidorus maximus* (Bütschli, 1874) Thorne & Swanger, 1936
Paralongidorus natalensis (Jacobs & Heyns, 1982) Luc & Doucet, 1984
Paralongidorus oryzae Verma, 1973
Paralongidorus zenobiae Hunt & Rahman, 1991
Paratrichodorus Siddiqi, 1974 [Tri.]
Paratrichodorus allius (Jensen, 1963) Siddiqi, 1974
= *Trichodorus allius* Jensen, 1963
Paratrichodorus anemones (Loof, 1965) Siddiqi, 1974
= *Trichodorus anemones* Loof, 1965
Paratrichodorus christiei
cf. *Paratrichodorus minor*
Paratrichodorus lobatus (Colbran, 1965) Siddiqi, 1974
= *Trichodorus lobatus* Colbran, 1965
Paratrichodorus minor (Colbran, 1956) Siddiqi, 1974
= *Trichodorus minor* Colbran, 1956
= *Paratrichodorus christiei* (Allen, 1957) Siddiqi, 1974
= *Trichodorus christiei* Allen, 1957
Paratrichodorus mirzai (Siddiqi, 1960) Siddiqi, 1974
= *Trichodorus mirzai* Siddiqi, 1960
Paratrichodorus pachydermus (Seinhorst, 1954) Siddiqi, 1974
= *Trichodorus pachydermus* Seinhorst, 1954
Paratrichodorus porosus (Allen, 1957) Siddiqi, 1974
= *Trichodorus porosus* Allen, 1957
Paratrichodorus teres (Hooper, 1962) Siddiqi, 1974
= *Trichodorus teres* Hooper, 1962
Paratrophurus Arias, 1970 [Tyl., Belonolaimidae]
Paratrophurus acristylus Siddiqi & Siddiqui, 1983
Paratylenchus Micoletzky, 1922 [Tyl., Tylenchulidae]
Paratylenchus besoekianus Bally & Reydon, 1931
Paratylenchus curvitatus van der Linde, 1938 (*sp. inq.*)
Paratylenchus hamatus Thorne & Allen, 1950
Paratylenchus minutus Linford *in* Linford, Oliveira & Ishii, 1949
Peltamigratus
cf. *Aorolaimus*

Pratylenchoides Winslow, 1958 [Tyl., Pratylenchidae]
Pratylenchoides leiocauda Sher, 1970
Pratylenchus Filipjev, 1936 [Tyl., Pratylenchidae]
Pratylenchus alleni Ferris, 1981
Pratylenchus andinus Lordello, Zamith & Boock, 1961
Pratylenchus barkati Das & Sultana, 1979
Pratylenchus brachyurus (Godfrey, 1929) Filipjev & Schuurmans Stekhoven, 1941
Pratylenchus coffeae (Zimmermann, 1898) Filipjev & Schuurmans Stekhoven, 1941
Pratylenchus crenatus Loof, 1960
Pratylenchus dasi Fortuner, 1985
Pratylenchus delattrei Luc, 1958
Pratylenchus exilis Das & Sultana, 1979
Pratylenchus fallax Seinhorst, 1968
Pratylenchus flakkensis Seinhorst, 1968
Pratylenchus goodeyi Sher & Allen, 1953
Pratylenchus gutierrezi
cf. *Pratylenchus panamaensis*
Pratylenchus hexincisus Taylor & Jenkins, 1957
Pratylenchus indicus Das, 1960 *(sp. inq.)*
Pratylenchus jaehni Inserra, Duncan, Troccoli, Dunn, Maia dos Santos, Kaplan & Vovlas, 2001
Pratylenchus loosi Loof, 1960
Pratylenchus mediterraneus Corbett, 1983
Pratylenchus minyus
cf. *Pratylenchus neglectus*
Pratylenchus neglectus (Rensch, 1924) Filipjev & Schuurmans Stekhoven, 1941
= *Pratylenchus minyus* Sher & Allen, 1953
Pratylenchus panamaensis Siddiqi, Dabur & Bajaj, 1991
= *Pratylenchus gutierrezi* Golden, López Ch. & Vilchez R., 1992
Pratylenchus penetrans (Cobb, 1917) Filipjev & Schuurmans Stekhoven, 1941
Pratylenchus pinguicaudatus Corbett, 1969
Pratylenchus pratensis (de Man, 1880) Filipjev, 1936
Pratylenchus pseudocoffeae Mizukobo, 1992
Pratylenchus pseudopratensis Seinhorst, 1968
= *Pratylenchus sefaensis* Fortuner, 1974
Pratylenchus scribneri Steiner, 1943
Pratylenchus sefaensis
cf. *Pratylenchus pseudopratensis*
Pratylenchus singhi Das & Sultana, 1979
Pratylenchus sudanensis Loof & Yassin, 1971
Pratylenchus thornei Sher & Allen, 1953
Pratylenchus vulnus Allen & Jensen, 1951
Pratylenchus zeae Graham, 1951
Punctodera Mulvey & Stone, 1976 [Tyl., Heteroderidae]
Punctodera chalcoensis Stone, Sosa-Moss & Mulvey, 1976
Punctodera punctata (Thorne, 1928) Mulvey & Stone, 1976
= *Heterodera punctata* Thorne, 1928
Quinisulcius Siddiqi, 1971 [Tyl., Belonolaimidae]
Radopholus Thorne, 1949 [Tyl., Pratylenchidae]
Radopholus bridgei Siddiqi & Hahn, 1995
Radopholus citri Machon & Bridge, 1996
Radopholus duriophilus Nguyen, Subbotin, Madani, Trinh & Moens, 2003
Radopholus inaequalis Sauer, 1958
Radopholus musicola Stanton, Mundo-Ocampo, Baldwin & Kaplan, 2001
Radopholus rotundisemenus Sher, 1968
Radopholus similis Cobb, 1913
= *Radopholus similis similis* Cobb, 1913
= *Radopholus similis citrophilus* Huettel, Dickson & Kaplan, 1984
Radopholus citrophilus
cf. *Radopholus similis*
Radopholus vangundyi Sher, 1968
Radopholus williamsi
cf. *Achlysiella williamsi*
Rhadinaphelenchus
cf. *Bursaphelenchus*
Rotylenchoides
cf. *Helicotylenchus*
Rotylenchulus Linford & Oliveira, 1940 [Tyl., Hoplolaimidae]
Rotylenchulus borealis Loof & Oostenbrink, 1962
Rotylenchulus macrodoratus Dasgupta, Raski & Sher, 1968
Rotylenchulus macrosoma Dasgupta, Raski & Sher, 1968
Rotylenchulus parvus (Williams, 1960) Sher, 1961
Rotylenchulus reniformis Linford & Oliveira, 1940

Rotylenchus Filipjev, 1936 [Tyl., Hoplolaimidae]
Rotylenchus buxophilus Golden, 1956
Rotylenchus caudaphasmidius Sher, 1965
Rotylenchus microstriatus Siddiqi & Corbett, 1983
Scutellonema Andrássy, 1958 [Tyl., Hoplolaimidae]
Scutellonema aberrans
  cf. *Scutellonema clathricaudatum*
Scutellonema africanum Smit, 1971
Scutellonema brachyurus (Steiner, 1938) Andrássy, 1958
Scutellonema bradys (Steiner & LeHew, 1933) Andrássy, 1958
  = *Scutellonema blaberum* (Steiner, 1937) Andrássy, 1958
Scutellonema cavenessi Sher, 1964
Scutellonema clathricaudatum Whitehead, 1960
  = *Scutellonema aberrans* (Whitehead, 1960) Sher 1961
Scutellonema magniphasma Sher, 1965
Scutellonema siamense Timm, 1965
Scutellonema unum Sher, 1964
Senegalonema Germani, Luc & Baldwin, 1984 [Tyl., Hoplolaimidae]
Siddiqia
  cf. *Paralongidorus*
Thecavermiculatus
  cf. *Atalodera*
Trichodorus Cobb, 1913 [Tri.]
  = *Monotrichodorus* Andrássy, 1976
Trichodorus borneoensis Hooper, 1962
Trichodorus monohystera Allen, 1957
  = *Monotrichodorus monohystera* (Allen, 1957) Andrássy, 1976
Trichodorus porosus
  cf. *Paratrichodorus porosus*
Trichodorus primitivus (de Man, 1880) Micoletzky, 1922
Trichodorus similis Seinhorst, 1963
Trichodorus viruliferus Hooper, 1963
Trophotylenchulus Raski, 1957 [Tyl., Tylenchulidae]
Trophotylenchulus obscurus (Colbran, 1961) Cohn & Kaplan, 1983
Trophotylenchulus piperis Mohandas, Ravana & Raski, 1985
Trophotylenchulus saltensis Hashim, 1984
Trophurus Loof, 1956 [Tyl., Belonolaimidae]
Trophurus imperialis Loof, 1956
Tylencholaimus de Man, 1876 [T.]
Tylencholaimus asymmetricus Khan & Ahmad, 1994
Tylenchorhynchus Cobb, 1913 [Tyl., Belonolaimidae]
Tylenchorhynchus acutus Allen, 1955
Tylenchorhynchus annulatus (Cassidy, 1930) Golden, 1971
  = *Tylenchorhynchus martini* Fielding, 1956
Tylenchorhynchus cicerus Kakar, Khan & Siddiqi, 1995
Tylenchorhynchus brassicae Siddiqi, 1961
Tylenchorhynchus brevilineatus Williams, 1960
  = *Tylenchorhynchus indicus* Siddiqi, 1961
Tylenchorhynchus capitatus Allen, 1955
Tylenchorhynchus cicerus Kakar, Khan & Siddiqi, 1995
Tylenchorhynchus clarus Allen, 1955
Tylenchorhynchus clavicaudatus Seinhorst, 1963
Tylenchorhynchus claytoni Steiner, 1937
Tylenchorhynchus crassicaudatus Williams, 1960
Tylenchorhynchus cylindricus Cobb, 1913
Tylenchorhynchus elegans Siddiqi, 1961
Tylenchorhynchus iarius Saha, Gaur & Lal, 1998
Tylenchorhynchus indicus
  cf. *Tylenchorhynchus brevilineatus*
Tylenchorhynchus karnalensis Saha, Singh, Lal & Kaushal, 2002
Tylenchorhynchus martini
  cf. *Tylenchorhynchus annulatus*
Tylenchorhynchus mashhoodi Siddiqi & Basir, 1959
Tylenchorhynchus nudus Allen, 1955
Tylenchorhynchus obtusus (Siddiqi, 1978) Fortuner & Luc, 1987
Tylenchorhynchus oryzae Kaul & Waliullah, 1995
Tylenchorhynchus queirozi Monteiro & Lordello, 1976
Tylenchorhynchus swarupi Singh & Khera, 1978
Tylenchorhynchus vulgaris Upadhyay, Swarup & Sethi, 1972
Tylenchulus Cobb, 1913 [Tyl., Tylenchulidae]

Tylenchulus graminis Inserra, Vovlas, O'Bannon & Esser, 1988
Tylenchulus palustris Inserra, Vovlas, O'Bannon & Esser, 1988
Tylenchulus semipenetrans Cobb, 1913
Tylenchus Bastian, 1865 [Tyl., Tylenchidae]
Xiphidorus Monteiro, 1976 [L.]
Xiphidorus minor Rashid, Coomans & Sharma, 1986
Xiphinema Cobb, 1913 [L.]
Xiphinema abeokutae Luc & Coomans, 1993
Xiphinema americanum sensu lato Cobb, 1913
Xiphinema attorodorum Luc, 1961
Xiphinema basilgoodeyi Coomans, 1964
Xiphinema basiri Siddiqi, 1959
Xiphinema bergeri Luc, 1973
Xiphinema brevicollum Lordello & da Costa, 1961
  = *Xiphinema diffusum* Lamberti & Bleve-Zacheo, 1979
  = *Xiphinema parvum* Lamberti, Cicancio, Agostinelli & Coiro, 1992
  = *Xiphinema pseudoguirani* Lamberti, Cicancio, Agostinelli & Coiro, 1992
Xiphinema cavenessi Luc, 1973
Xiphinema denoudeni
  cf. *Xiphinema krugi*
Xiphinema diffusum
  cf. *Xiphinema brevicollum*
Xiphinema diversicaudatum (Micoletzky, 1927) Thorne, 1939
Xiphinema ebriense Luc, 1958
Xiphinema elongatum Schuurmans Stekhoven & Teunissen, 1938
Xiphinema guirani Luc & Williams, 1978
Xiphinema heynsi Siddiqi, 1979
Xiphinema ifacolum Luc, 1961
Xiphinema imitator Heyns, 1965
Xiphinema index Thorne & Allen, 1950
Xiphinema insigne Loos, 1949
Xiphinema krugi Lordello, 1955
  = *Xiphinema denoudeni* Loof & Maas, 1972
Xiphinema longicaudatum Luc, 1961
Xiphinema mammatum Siddiqi, 1979
Xiphinema mampara Heyns, 1979
Xiphinema mediterraneum
  cf. *Xiphinema pachtaicum*
Xiphinema neobasiri Siddiqi, 1979
Xiphinema nigeriense Luc, 1961
Xiphinema orbum Siddiqi, 1964
Xiphinema oryzae Bos & Loof, 1985
Xiphinema pachtaicum (Tulaganov, 1938) Kirjanova, 1951
  = *Xiphinema mediterraneum* Martelli & Lamberti, 1967
Xiphinema paritaliae Loof & Sharma, 1979
Xiphinema parvum
  cf. *Xiphinema brevicollum*
Xiphinema pseudoguirani
  cf. *Xiphinema brevicollum*
Xiphinema radicicola Goodey, 1936
Xiphinema savanicola Luc & Southey, 1981
Xiphinema seredouense Luc, 1975
Xiphinema setariae Luc, 1958
  = *Xiphinema vulgare* Tarjan, 1964
Xiphinema vulgare
  cf. *Xiphinema setariae*
Zygotylenchus Siddiqi, 1963 [Tyl., Pratylenchidae]
Zygotylenchus guevarai (Tobar Jiménez, 1963) Braun & Loof, 1968

Index

The page numbers in *italics* refer to figures and tables.